Complete Solutions Manual

for

Gustafson and Frisk's

Intermediate Algebra

Eighth Edition

Michael G. Welden
Rock Valley College

Australia • Brazil • Canada • Mexico • Singapore • Spain • United Kingdom • United States

Printed in the United States of America

1 2 3 4 5 6 7 11 10 09 08 07

Printer: Thomson/West

ISBN-13: 978-0-495-11795-7
ISBN-10: 0-495-11795-1

Thomson Higher Education
10 Davis Drive
Belmont, CA 94002-3098
USA

Preface

This manual contains detailed solutions to all of the exercises of the text *Intermediate Algebra*, eighth edition, by R. David Gustafson and Peter D. Frisk.

Many of the exercises in the text may be solved using more than one method, but it is not feasible to list all possible solutions in this manual. Also, some of the exercises may have been solved in this manual using a method that differs slightly from that presented in the text. There are a few exercises in the text whose solutions may vary from person to person. Some of these solutions may not have been included in this manual. For the solution to an exercise like this, the notation "answers may vary" has been included.

If you are a student using this manual, please remember that only reading a solution does not teach you how to solve a problem. To repeat a commonly used phrase, mathematics is not a spectator sport. You MUST make an honest attempt to solve each exercise in the text without using this manual first. This manual should be viewed more or less as a last resort. Above all, DO NOT simply copy the solution from this manual onto your own paper. Doing so will not help you learn how to do the exercise, nor will it help you to do better on quizzes or tests.

I would like to thank the members of the mathematics faculty at Rock Valley College and Laura Localio of Brooks/Cole Publishing Company for their help and support. This solutions manual was prepared using EXP 5.1.

This book is dedicated to my parents, Ed and Carol Welden, for their constant encouragement and support throughout my life.

May your study of this material be successful and rewarding.

Michael G. Welden

Contents

Exercise 1.1 (page 11)

1. $\frac{6}{8} = \frac{\cancel{2} \cdot 3}{\cancel{2} \cdot 4} = \frac{3}{4}$

2. $\frac{15}{20} = \frac{\cancel{5} \cdot 3}{\cancel{5} \cdot 4} = \frac{3}{4}$

3. $\frac{32}{40} = \frac{\cancel{8} \cdot 4}{\cancel{8} \cdot 5} = \frac{4}{5}$

4. $\frac{56}{72} = \frac{\cancel{8} \cdot 7}{\cancel{8} \cdot 9} = \frac{7}{9}$

5. $\frac{1}{4} \cdot \frac{3}{5} = \frac{1 \cdot 3}{4 \cdot 5} = \frac{3}{20}$

6. $\frac{3}{5} \cdot \frac{20}{27} = \frac{3 \cdot 20}{5 \cdot 27} = \frac{\cancel{3} \cdot \cancel{5} \cdot 4}{\cancel{5} \cdot \cancel{3} \cdot 9} = \frac{4}{9}$

7. $\frac{2}{3} \div \frac{3}{7} = \frac{2}{3} \cdot \frac{7}{3} = \frac{2 \cdot 7}{3 \cdot 3} = \frac{14}{9}$

8. $\frac{3}{5} \div \frac{9}{15} = \frac{3}{5} \cdot \frac{15}{9} = \frac{\cancel{3} \cdot \cancel{3} \cdot \cancel{5}}{\cancel{5} \cdot \cancel{3} \cdot \cancel{3}} = 1$

9. $\frac{5}{9} + \frac{4}{9} = \frac{5+4}{9} = \frac{9}{9} = 1$

10. $\frac{16}{7} - \frac{2}{7} = \frac{16-2}{7} = \frac{14}{7} = 2$

11. $\frac{2}{3} + \frac{4}{5} = \frac{2 \cdot 5}{3 \cdot 5} + \frac{4 \cdot 3}{5 \cdot 3} = \frac{10}{15} + \frac{12}{15} = \frac{22}{15}$

12. $\frac{7}{9} - \frac{2}{5} = \frac{7 \cdot 5}{9 \cdot 5} - \frac{2 \cdot 9}{5 \cdot 9} = \frac{35}{45} - \frac{18}{45} = \frac{17}{45}$

13. set

14. natural

15. even

16. odd

17. natural; 1; itself

18. composite; 1; prime

19. 0

20. 0

21. rational

22. irrational

23. $<$

24. $\geq$

25. $\approx$

26. $-x$

27. natural: $1, 2, 9$

28. whole: $0, 1, 2, 9$

29. integer: $-3, 0, 1, 2, 9$

30. rational number: $-3, 0, \frac{2}{3}, 1, 2, 9$

31. irrational number: $\sqrt{3}$

32. real number: $-3, 0, \frac{2}{3}, 1, \sqrt{3}, 2, 9$

33. even natural number: 2

34. odd integer: $-3, 1, 9$

35. prime number: 2

36. composite number: 9

37. odd composite number: 9

38. even prime number: 2

39.

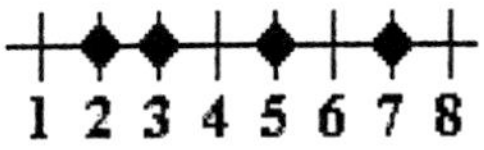

1 2 3 4 5 6 7 8

40.

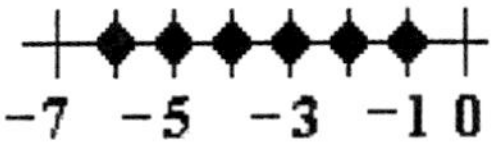

−7 −5 −3 −1 0

41.

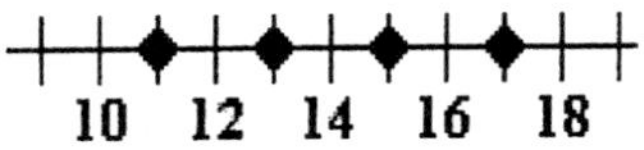

10 12 14 16 18

42.

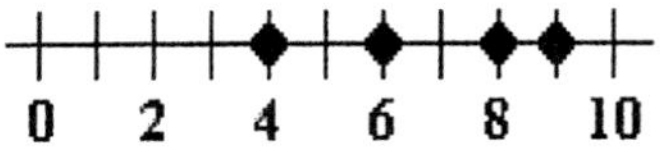

0 2 4 6 8 10

43. $\frac{7}{8} = 7 \div 8 = 0.875$; terminating

44. $\frac{7}{3} = 7 \div 3 = 2.\overline{3}$; repeating

45. $-\frac{11}{15} = -11 \div 15 = -0.7\overline{3}$; repeating

46. $-\frac{19}{16} = -1.1875$; terminating

47. $5 < 9$

48. $9 > 0$

49. $-5 > -10$

50. $-3 < 10$

51. $-7 < 7$

52. $0 > -5$

53. $6 > -6$

54. $-6 < -2$

55. $19 > 12 \Rightarrow 12 < 19$

56. $-3 \geq -5 \Rightarrow -5 \leq -3$

57. $-6 \leq -5 \Rightarrow -5 \geq -6$

58. $-10 < 13 \Rightarrow 13 > -10$

59. $5 \geq -3 \Rightarrow -3 \leq 5$

60. $0 \leq 12 \Rightarrow 12 \geq 0$

61. $-10 < 0 \Rightarrow 0 > -10$

62. $-4 > -8 \Rightarrow -8 < -4$

63. $\{x|x > 3\} \Rightarrow$ 3

64. $\{x|x < 0\} \Rightarrow$ 0

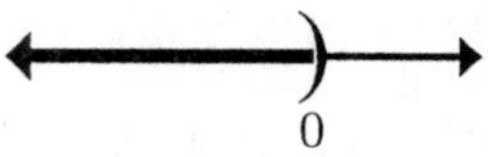

65. $\{x|x \leq 7\} \Rightarrow$ 7

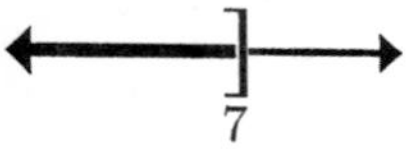

66. $\{x|x \geq -2\} \Rightarrow$  −2

67. $[-5, \infty) \Rightarrow$ −5

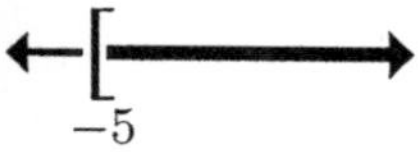

68. $(-\infty, 9] \Rightarrow$ 9

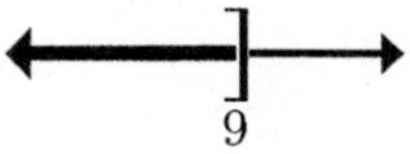

69. $\{x|2 < x < 5\} \Rightarrow$ 2 5

70. $[0, 5) \Rightarrow$ 0 5

71. $[-6, 9] \Rightarrow$ −6 9

72. $\{x|-1 < x \leq 3\} \Rightarrow$ −1 3

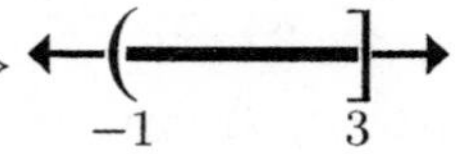

73. $\{x|x < -3 \text{ or } x > 3\}$

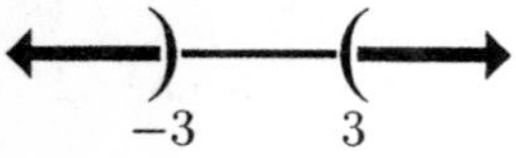

−3 3

74. $(-\infty, -4] \cup (2, \infty)$

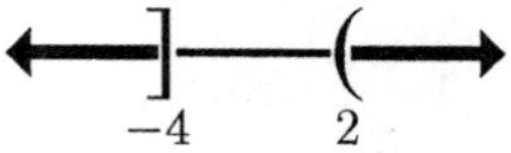

−4 2

75. $(-\infty, -6] \cup [5, \infty)$

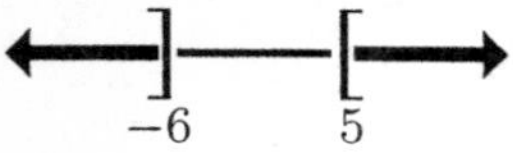

−6 5

76. $\{x|x < -2 \text{ or } x \geq 3\}$

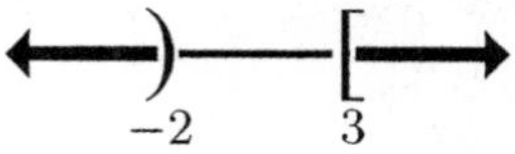

−2 3

77. $|20| = 20$

78. $|-20| = 20$

79. $-|-6| = -(+6) = -6$

80. $-|8| = -(+8) = -8$

81. $|-5| + |-2| = 5 + 2 = 7$

82. $|12| + |-4| = 12 + 4 = 16$

83. $|-5| \cdot |4| = 5 \cdot 4 = 20$

84. $|-6| \cdot |-3| = 6 \cdot 3 = 18$

85. $|x| = 3 \Rightarrow x = 3$ or $x = -3$

86. $|x| = 7 \Rightarrow x = 7$ or $x = -7$

87. If $x = |x|$, then $x \geq 0$.

88. If $x + |x| = 0$, then $|x| = -x$ and $x \leq 0$.

89.

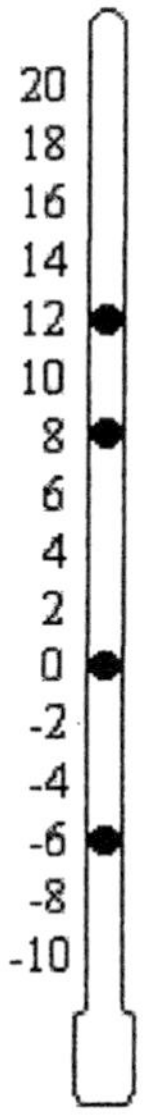

90. **a.** Mayapán fell.
b. Maya culture began.

91. **Answers may vary.**

92. **Answers may vary.**

93. **Answers may vary.**

94. **Answers may vary.**

95. If $|x| < 50$, then $x \in \{-49, -48, -47, ..., -1, 0, 1, ..., 47, 48, 49\}$. There are 99 integers.

96. If $20 < |x| < 40$ with x odd, then $x \in \{-39, -37, ..., -21, 21, 23, ..., 39\}$. There are 20 integers.

97. **Answers may vary.**

98. **a.** Not always true
b. Always true
c. Always true

Exercise 1.2 (page 24)

1. $\{x|x > 4\} \Rightarrow$ 

2. $(-\infty, -5] \Rightarrow$ 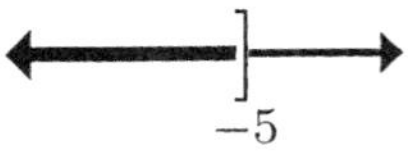

3. $(2, 10] \Rightarrow$ 

4. $\{x|-4 \leq x \leq 4\} \Rightarrow$ 

SECTION 1.2

5. Cost of gasoline $= 32(2.29) = 73.28$; Cost of oil $= 3(1.35) = 4.05$
Tax on oil $= 0.05(4.05) = 0.2025 \approx 0.20$
Total cost $= 73.28 + 4.05 + 0.20 = 77.53 \Rightarrow$ The total cost is \$77.53.

6. The amount over 29,700 $= 57760 - 29700 = 28060$.
Tax $= 4090 + 0.25(28060) = 4090 + 7015 = 11105$.
Her tax bill is \$11,105.

7. absolute; common

8. subtract

9. change; add

10. positive

11. negative

12. 0

13. mean; median; mode

14. $C = \pi D = 2\pi r$

15. $(a \cdot b) \cdot c = a \cdot (b \cdot c)$

16. $a + b = b + a$

17. $a(b + c) = ab + ac$

18. 0

19. 1

20. a

21. $-3 + (-5) = -(3 + 5) = -8$

22. $2 + (+8) = +(2 + 8) = 10$

23. $-7 + 2 = -(7 - 2) = -5$

24. $3 + (-5) = -(5 - 3) = -2$

25. $-3 - 4 = -3 + (-4) = -7$

26. $-11 - (-17) = -11 + (+17) = 6$

27. $-33 - (-33) = -33 + (+33) = 0$

28. $14 - (-13) = 14 + (+13) = 27$

29. $-2(6) = -12$

30. $3(-5) = -15$

31. $-3(-7) = 21$

32. $-2(-5) = 10$

33. $\frac{-8}{4} = -2$

34. $\frac{25}{-5} = -5$

35. $\frac{-16}{-4} = 4$

36. $\frac{-5}{-25} = \frac{1}{5}$

37. $\frac{1}{2} + \left(-\frac{1}{3}\right) = \frac{3}{6} + \left(-\frac{2}{6}\right) = \frac{1}{6}$

38. $-\frac{3}{4} + \left(-\frac{1}{5}\right) = -\frac{15}{20} + \left(-\frac{4}{20}\right) = -\frac{19}{20}$

39. $\frac{1}{2} - \left(-\frac{3}{5}\right) = \frac{5}{10} + \frac{6}{10} = \frac{11}{10}$

40. $\frac{1}{26} - \frac{11}{13} = \frac{1}{26} - \frac{22}{26} = -\frac{21}{26}$

41. $\frac{1}{3} - \frac{1}{2} = \frac{2}{6} - \frac{3}{6} = -\frac{1}{6}$

42. $\frac{7}{8} - \left(-\frac{3}{4}\right) = \frac{7}{8} + \frac{6}{8} = \frac{13}{8}$

43. $\left(-\frac{3}{5}\right)\left(\frac{10}{7}\right) = -\frac{30}{35} = -\frac{6}{7}$

44. $\left(-\frac{6}{7}\right)\left(-\frac{5}{12}\right) = \frac{30}{84} = \frac{5}{14}$

45. $\frac{3}{4} \div \left(-\frac{3}{8}\right) = \frac{3}{4}\left(-\frac{8}{3}\right) = -\frac{24}{12} = -2$

46. $-\frac{3}{5} \div \frac{7}{10} = -\frac{3}{5} \cdot \frac{10}{7} = -\frac{30}{35} = -\frac{6}{7}$

47. $-\frac{16}{5} \div \left(-\frac{10}{3}\right) = -\frac{16}{5}\left(-\frac{3}{10}\right) = \frac{48}{50} = \frac{24}{25}$

48. $-\frac{5}{24} \div \frac{10}{3} = -\frac{5}{24} \cdot \frac{3}{10} = -\frac{15}{240} = -\frac{1}{16}$

49. $3 + 4 \cdot 5 = 3 + 20 = 23$

50. $5 \cdot 3 - 6 \cdot 4 = 15 - 24 = -9$

51. $3 - 2 - 1 = 1 - 1 = 0$

52. $5 - 3 - 1 = 2 - 1 = 1$

53. $3 - (2 - 1) = 3 - 1 = 2$

54. $5 - (3 - 1) = 5 - 2 = 3$

55. $2 - 3 \cdot 5 = 2 - 15 = -13$

56. $6 + 4 \cdot 7 = 6 + 28 = 34$

57. $8 \div 4 \div 2 = 2 \div 2 = 1$

58. $100 \div 10 \div 5 = 10 \div 5 = 2$

59. $8 \div (4 \div 2) = 8 \div 2 = 4$

60. $100 \div (10 \div 5) = 100 \div 2 = 50$

61. $2 + 6 \div 3 - 5 = 2 + 2 - 5 = 4 - 5 = -1$

62. $6 - 8 \div 4 - 2 = 6 - 2 - 2 = 4 - 2 = 2$

63. $(2 + 6) \div (3 - 5) = 8 \div (-2) = -4$

64. $(6 - 8) \div (4 - 2) = -2 \div 2 = -1$

65. $\frac{3(8+4)}{2 \cdot 3 - 9} = \frac{3(12)}{6 - 9} = \frac{36}{-3} = -12$

66. $\frac{5(4-1)}{3 \cdot 2 + 5 \cdot 3} = \frac{5(3)}{6 + 15} = \frac{15}{21} = \frac{5}{7}$

67. $\frac{100(2-4)}{1{,}000 \div 10 \div 10} = \frac{100(-2)}{100 \div 10} = \frac{-200}{10} = -20$

68. $\frac{8(3) - 4(6)}{5(3) + 3(-7)} = \frac{24 - 24}{15 + (-21)} = \frac{0}{-6} = 0$

69. $\text{mean} = \frac{7+5+9+10+8+6+6+7+9+12+9}{11} = \frac{88}{11} = 8$

70. In order: $5, 6, 6, 7, 7, 8, 9, 9, 9, 10, 12$. Since there is an odd number of measurements, the median is the middle measurement. The median is 8.

71. In order: $5, 6, 6, 7, 7, 8, 9, 9, 9, 10, 12$. The mode is 9.

72. In order: $8, 8, 10, 12, 12, 12, 14, 16, 16, 23, 23, 26$. Since there is an even number of measurements, the median is computed using the middle two measurements: $\frac{12+14}{2} = \frac{26}{2} = 13$.

73. In order: $8, 8, 10, 12, 12, 12, 14, 16, 16, 23, 23, 26$. The mode is 12.

74. $\text{mean} = \frac{8+12+23+12+10+16+26+12+14+8+16+23}{12} = \frac{180}{12} = 15$

75. $ab + cd = (3)(-2) + (-1)(2)$
$= -6 + (-2) = -8$

76. $ad + bc = (3)(2) + (-2)(-1) = 6 + 2 = 8$

77. $a(b + c) = 3[-2 + (-1)] = 3(-3) = -9$

78. $d(b + a) = 2(-2 + 3) = 2(1) = 2$

79. $\frac{ad + c}{cd + b} = \frac{3(2) + (-1)}{-1(2) + (-2)} = \frac{6 + (-1)}{-2 + (-2)}$
$= \frac{5}{-4} = -\frac{5}{4}$

80. $\frac{ab + d}{bd + a} = \frac{3(-2) + 2}{(-2)(2) + 3} = \frac{-6 + 2}{-4 + 3} = \frac{-4}{-1} = 4$

81. $\frac{ac - bd}{cd - ad} = \frac{3(-1) - (-2)(2)}{(-1)(2) - 3(2)} = \frac{-3 - (-4)}{-2 - 6}$
$= \frac{1}{-8} = -\frac{1}{8}$

82. $\frac{bc - ad}{bd + ac} = \frac{(-2)(-1) - 3(2)}{(-2)(2) + 3(-1)} = \frac{2 - 6}{-4 + (-3)}$
$= \frac{-4}{-7} = \frac{4}{7}$

83. $C = \frac{N(N-1)}{2} = \frac{200(200-1)}{2} = \frac{200(199)}{2} = \frac{\not{2} \cdot 100(199)}{\not{2}} = 19{,}900$ comparisons

84. $C = \frac{N(N-1)}{2} = \frac{10000(10000-1)}{2} = \frac{10000(9999)}{2} = \frac{99{,}990{,}000}{2} = 49{,}995{,}000$ comparisons

85. $P = a + b + c = (23.5 + 37.2 + 39.7)$ ft $= 100.4$ ft

86. $P = a + b + c + d = (43.27 + 47.37 + 50.21 + 52.93)$ cm $= 193.78$ cm

87. commutative property of addition

88. associative property of multiplication

89. distributive property

90. commutative property of multiplication

91. additive identity property

92. distributive property

93. multiplicative inverse property

94. commutative property of addition

95. associative property of addition

96. multiplicative identity property

97. commutative property of multiplication

98. additive inverse property

Problems 99-102 are to be solved using a calculator. The keystrokes needed to solve each problem using a TI-83 graphing calculator appear in each solution. There may be other solutions. Keystrokes for other calculators may be slightly different.

99. [(] [3] [7] [.] [9] [+] [2] [5] [.] [2] [)] [+] [1] [4] [.] [3] [ENTER] {77.4}
[3] [7] [.] [9] [+] [(] [2] [5] [.] [2] [+] [1] [4] [.] [3] [)] [ENTER] {77.4}
associative property of addition

SECTION 1.2

100. [7] [.] [1] [×] [(] [3] [.] [9] [+] [8] [.] [8] [)] [ENTER] {90.17}
[7] [.] [1] [×] [3] [.] [9] [+] [7] [.] [1] [×] [8] [.] [8] [ENTER] {90.17}
distributive property

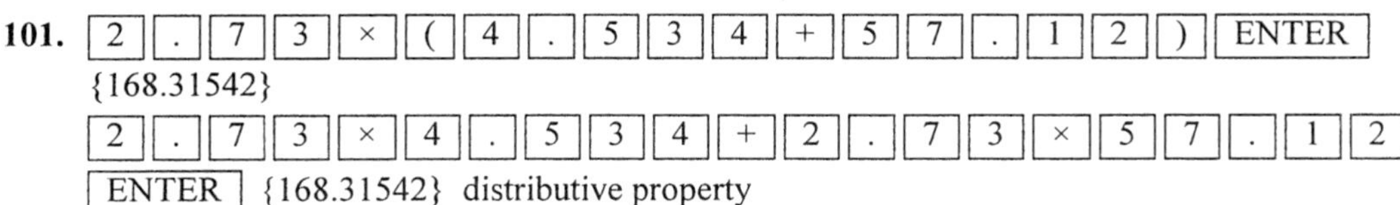

101. [2] [.] [7] [3] [×] [(] [4] [.] [5] [3] [4] [+] [5] [7] [.] [1] [2] [)] [ENTER]
{168.31542}
[2] [.] [7] [3] [×] [4] [.] [5] [3] [4] [+] [2] [.] [7] [3] [×] [5] [7] [.] [1] [2]
[ENTER] {168.31542} distributive property

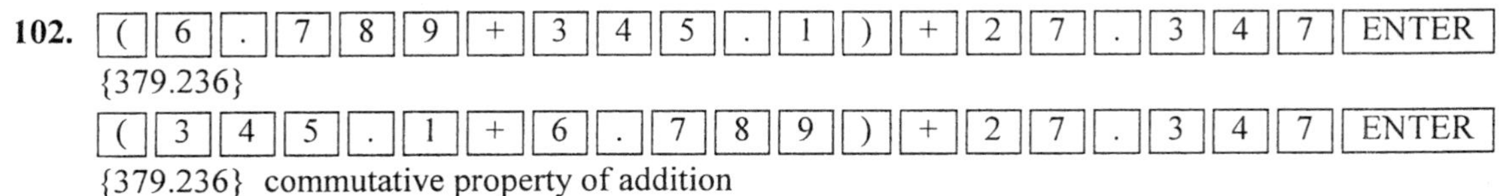

102. [(] [6] [.] [7] [8] [9] [+] [3] [4] [5] [.] [1] [)] [+] [2] [7] [.] [3] [4] [7] [ENTER]
{379.236}
[(] [3] [4] [5] [.] [1] [+] [6] [.] [7] [8] [9] [)] [+] [2] [7] [.] [3] [4] [7] [ENTER]
{379.236} commutative property of addition

103. $(+22.25) + (+39.75) = +62$
He earned \$62.

104. $(-13.5) + (-11.5) = -25$
-25 pounds

105. $(+17) + (-13) = +4$
The temperature rose 4°.

106. $38 + 19 = 57$
The flag has traveled 57 feet.

107. $-3(-4) = +12$
It was 12° warmer 3 hours ago.

108. $15(-30) = -450$
He lost \$450.

109. $-(5 \cdot 60)(+23) = -6900$
There were 6900 fewer gallons of water.

110. $-(2 \cdot 60)(-12) = +1440$
There were 1440 more gallons of water.

The keystrokes needed to solve problems 111-120 using a TI-83 graphing calculator appear in each solution. There may be other solutions. Keystrokes for other calculators may be slightly different.

111. [(–)] [2] [3] [0] [0] [+] [1] [7] [5] [0] [+] [1] [8] [7] [5] [ENTER] {1325.}
The army gained 1325 meters.

112. [9] [4] [–] [8] [×] [8] [ENTER] {30.} He had \$30 before starting work.

113. [4] [3] [7] [.] [3] [7] [+] [1] [2] [5] [.] [1] [8] [+] [1] [3] [7] [.] [2] [6] [+] [1]
[4] [5] [.] [5] [6] [–] [1] [1] [7] [.] [1] [1] [–] [1] [8] [3] [.] [4] [9] [–] [1] [2]
[2] [.] [8] [9] [ENTER] {421.88} Her ending balance is \$421.88.

114. [3] [7] [+] [2] [4] [–] [1] [1] [+] [1] [7] [–] [2] [1] [ENTER] {46.}
It gained 46 points.

115. [(] [1] [5] [2] [5] [+] [7] [8] [5] [+] [1] [6] [2] [8] [+] [1] [2] [1] [4] [+] [9]
[1] [7] [+] [1] [1] [9] [7] [)] [÷] [6] [ENTER] {1211.} The mean of daily sales is \$1211.

116. (2 . 5 + 1 0 5 . 1 + 7 4 . 9 + 1 3 7 . 4 + 5 2 . 6) ÷ 5 ENTER {74.5} The mean length is 74.5 centimicrons.

117. (7 5 + 8 2 + 8 7 + 8 0 + 7 6) ÷ 5 ENTER {80.} His mean score is 80.

118. (2 9 8 + 2 8 7 + 3 1 0 + 3 0 2 + 3 0 3) ÷ 5 ENTER {300.} The mean weight of the offensive line is 300 pounds.

119. (1 0 0 0 0 0 + 4 × 1 0 0 0 0) ÷ 5 ENTER {28000.} The ad is misleading. Although the mean wage of all workers including the businessman is \$28,000 as stated, a better measure of the average would be the median wage, which is \$10,000.

120. Assume the student receives a grade of 100% on the last test. Compute his mean.
(7 8 + 8 5 + 8 8 + 9 6 + 1 0 0) ÷ 5 ENTER {89.4} He cannot receive an average of 90% (and get an A).

121. $P = 4s = 4(7.5)\text{ cm} = 30\text{ cm}$

122. $C = \pi D = \pi(25)\text{ m} = 25\pi\text{ m} \approx 25(3.1416)\text{ m} = 78.54\text{ m}$

123. Answers may vary.

124. Answers may vary.

125. The mean increases by 7 as well. This property is always true.

126. The mean is multiplied by 7 as well. This property is always true.

127. The median is often the most appropriate average to use when one or two extremely high or low values occur. The following are situations when computing the median might be appropriate:
- finding an average salary of all employees when there are a few highly paid employees
- finding an average household income for a city
- finding an average score of all students on a test when there is one extremely high score

128. The mode is often the most appropriate average to use when it is not practical to consider or list all of the values, or if the most popular value is desired. The following are situations when computing the mode might be appropriate:
- finding the average female shoe size in the United States
- finding the flavor of ice cream the average person purchases
- finding the vacation destination of the average person

Exercise 1.3 (page 36)

1. $a + b + c = 4 + (-2) + 5 = 2 + 5 = 7$

2. $a - 2b - c = 4 - 2(-2) - 5$
$= 4 + 4 - 5 = 3$

SECTION 1.3

3. $\dfrac{ab+2c}{a+b} = \dfrac{4(-2)+2(5)}{4+(-2)} = \dfrac{-8+10}{2} = \dfrac{2}{2} = 1$

4. $\dfrac{ac-bc}{6ab+b} = \dfrac{4(5)-(-2)(5)}{6(4)(-2)+(-2)} = \dfrac{20-(-10)}{-48+(-2)} = \dfrac{30}{-50} = -\dfrac{3}{5}$

5. base; exponent

6. factor

7. x^{m+n}

8. x^{mn}

9. $x^n y^n$

10. $\dfrac{x^n}{y^n}$

11. 1

12. $\dfrac{1}{a}$

13. x^{m-n}

14. $\left(\dfrac{5}{4}\right)^3$

15. $A = s^2$

16. $A = lw$

17. $A = \frac{1}{2}bh$

18. $A = \frac{1}{2}h(b_1 + b_2)$

19. $A = \pi r^2$

20. $V = s^3$

21. $V = lwh$

22. $V = \frac{4}{3}\pi r^3$

23. $V = Bh$

24. $V = \frac{1}{3}Bh$

25. $V = \frac{1}{3}Bh$

26. base

27. base $= 5$, exponent $= 3$

28. base $= 7$, exponent $= 2$

29. base $= x$, exponent $= 5$

30. base $= -t$, exponent $= 4$

31. base $= b$, exponent $= 6$

32. base $= 3xy$, exponent $= 5$

33. base $= -mn^2$, exponent $= 3$

34. base $= -p^2q$, exponent $= 2$

35. $3^2 = 3 \cdot 3 = 9$

36. $3^4 = 3 \cdot 3 \cdot 3 \cdot 3 = 81$

37. $-3^2 = -1 \cdot 3^2 = -1 \cdot 3 \cdot 3 = -9$

38. $-3^4 = -1 \cdot 3^4 = -1 \cdot 3 \cdot 3 \cdot 3 \cdot 3 = -81$

39. $(-3)^2 = (-3)(-3) = 9$

40. $(-3)^3 = (-3)(-3)(-3) = -27$

41. $5^{-2} = \dfrac{1}{5^2} = \dfrac{1}{25}$

42. $5^{-4} = \dfrac{1}{5^4} = \dfrac{1}{625}$

43. $-5^{-2} = -1 \cdot 5^{-2} = -\dfrac{1}{5^2} = -\dfrac{1}{25}$

44. $-5^{-4} = -1 \cdot 5^{-4} = -\dfrac{1}{5^4} = -\dfrac{1}{625}$

45. $(-5)^{-2} = \dfrac{1}{(-5)^2} = \dfrac{1}{25}$

46. $(-5)^{-4} = \dfrac{1}{(-5)^4} = \dfrac{1}{625}$

47. $8^0 = 1$

48. $-9^0 = -1 \cdot 9^0 = -1 \cdot 1 = -1$

49. $(-8)^0 = 1$

50. $(-9)^0 = 1$

51. $(-2x)^5 = (-2)^5 x^5 = -32x^5$

52. $(-3a)^3 = (-3)^3 a^3 = -27a^3$

SECTION 1.3

53. $(-2x)^6 = (-2)^6x^6 = 64x^6$

54. $(-3y)^5 = (-3)^5y^5 = -243y^5$

55. $x^2x^3 = x^{2+3} = x^5$

56. $y^3y^4 = y^{3+4} = y^7$

57. $k^0k^7 = 1 \cdot k^7 = k^7$

58. $x^8x^{11} = x^{8+11} = x^{19}$

59. $x^2x^3x^5 = x^{2+3+5} = x^{10}$

60. $y^3y^7y^2 = y^{3+7+2} = y^{12}$

61. $p^9pp^0 = p^9p^1 \cdot 1 = p^{9+1} = p^{10}$

62. $z^7z^0z = z^7 \cdot 1 \cdot z^1 = z^{7+1} = z^8$

63. $aba^3b^4 = a^{1+3}b^{1+4} = a^4b^5$

64. $x^2y^3x^3y^2 = x^{2+3}y^{3+2} = x^5y^5$

65. $(-x)^2y^4x^3 = (-1)^2x^2y^4x^3 = x^{2+3}y^4 = x^5y^4$

66. $-x^2y^7y^3x^{-2} = -x^{2+(-2)}y^{7+3} = -x^0y^{10} = -y^{10}$

67. $(x^4)^7 = x^{4 \cdot 7} = x^{28}$

68. $(y^7)^5 = y^{7 \cdot 5} = y^{35}$

69. $(b^{-8})^9 = b^{-8 \cdot 9} = b^{-72} = \dfrac{1}{b^{72}}$

70. $(z^{12})^2 = z^{12 \cdot 2} = z^{24}$

71. $(x^3y^2)^4 = (x^3)^4(y^2)^4 = x^{3 \cdot 4}y^{2 \cdot 4} = x^{12}y^8$

72. $(x^2y^5)^2 = (x^2)^2(y^5)^2 = x^{2 \cdot 2}y^{5 \cdot 2} = x^4y^{10}$

73. $(r^{-3}s)^3 = (r^{-3})^3s^3 = r^{-3 \cdot 3} = r^{-9}s^3 = \dfrac{s^3}{r^9}$

74. $(m^5n^2)^{-3} = (m^5)^{-3}(n^2)^{-3} = m^{5(-3)}n^{2(-3)} = m^{-15}n^{-6} = \dfrac{1}{m^{15}n^6}$

75. $(a^2a^3)^4 = (a^{2+3})^4 = (a^5)^4 = a^{5 \cdot 4} = a^{20}$

76. $(bb^2b^3)^4 = (b^{1+2+3})^4 = (b^6)^4 = b^{6 \cdot 4} = b^{24}$

77. $(y^3y^{-1})^4 = (y^{3+(-1)})^4 = (y^2)^4 = y^{2 \cdot 4} = y^8$

78. $(z^{-2}z^5)^3 = (z^{-2+5})^3 = (z^3)^3 = z^{3 \cdot 3} = z^9$

79. $(x^2)^3(x^3)^2 = x^{2 \cdot 3}x^{3 \cdot 2} = x^6x^6 = x^{6+6} = x^{12}$

80. $(a^{-2})^4(a^3)^2 = a^{-2 \cdot 4}a^{3 \cdot 2} = a^{-8}a^6 = a^{-8+6} = a^{-2} = \dfrac{1}{a^2}$

81. $(-d^2)^3(d^{-3})^3 = (-1)^3(d^2)^3d^{-3 \cdot 3} = -1 \cdot d^{2 \cdot 3}d^{-9} = -1 \cdot d^6d^{-9} = -1 \cdot d^{6+(-9)} = -d^{-3} = -\dfrac{1}{d^3}$

82. $(c^3)^2(c^4)^{-2} = c^{3 \cdot 2}c^{4(-2)} = c^6c^{-8} = c^{6+(-8)} = c^{-2} = \dfrac{1}{c^2}$

83. $(3x^3y^4)^3 = 3^3(x^3)^3(y^4)^3 = 27x^{3 \cdot 3}y^{4 \cdot 3} = 27x^9y^{12}$

84. $\left(\frac{1}{2}a^2b^5\right)^4 = \left(\frac{1}{2}\right)^4(a^2)^4(b^5)^4 = \frac{1}{16}a^{2 \cdot 4}b^{5 \cdot 4} = \frac{1}{16}a^8b^{20}$

85. $\left(-\frac{1}{3}mn^2\right)^6 = \left(-\frac{1}{3}\right)^6 m^6 \left(n^2\right)^6 = \frac{1}{729} m^6 n^{2 \cdot 6} = \frac{1}{729} m^6 n^{12}$

86. $\left(-3p^2q^3\right)^5 = (-3)^5 \left(p^2\right)^5 \left(q^3\right)^5 = -243 p^{2 \cdot 5} q^{3 \cdot 5} = -243 p^{10} q^{15}$

87. $\left(\dfrac{a^3}{b^2}\right)^5 = \dfrac{(a^3)^5}{(b^2)^5} = \dfrac{a^{15}}{b^{10}}$

88. $\left(\dfrac{a^2}{b^3}\right)^4 = \dfrac{(a^2)^4}{(b^3)^4} = \dfrac{a^8}{b^{12}}$

89. $\left(\dfrac{a^{-3}}{b^{-2}}\right)^{-2} = \dfrac{(a^{-3})^{-2}}{(b^{-2})^{-2}} = \dfrac{a^6}{b^4}$

90. $\left(\dfrac{k^{-3}}{k^{-4}}\right)^{-1} = \left(k^{-3-(-4)}\right)^{-1} = \left(k^1\right)^{-1}$

$= k^{-1} = \dfrac{1}{k}$

91. $\dfrac{a^8}{a^3} = a^{8-3} = a^5$

92. $\dfrac{c^7}{c^2} = c^{7-2} = c^5$

93. $\dfrac{c^{12}c^5}{c^{10}} = \dfrac{c^{17}}{c^{10}} = c^{17-10} = c^7$

94. $\dfrac{a^{33}}{a^2a^3} = \dfrac{a^{33}}{a^5} = a^{33-5} = a^{28}$

95. $\dfrac{m^9m^{-2}}{\left(m^2\right)^3} = \dfrac{m^7}{m^6} = m^{7-6} = m^1 = m$

96. $\dfrac{a^{10}a^{-3}}{a^5a^{-2}} = \dfrac{a^7}{a^3} = a^{7-3} = a^4$

97. $\dfrac{1}{a^{-4}} = a^4$

98. $\dfrac{3}{b^{-5}} = 3b^5$

99. $\dfrac{3m^5m^{-7}}{m^2m^{-5}} = \dfrac{3m^{-2}}{m^{-3}} = 3m^{-2-(-3)} = 3m$

100. $\dfrac{\left(2a^{-2}\right)^3}{a^3a^{-4}} = \dfrac{8a^{-6}}{a^{-1}} = 8a^{-6-(-1)} = 8a^{-5} = \dfrac{8}{a^5}$

101. $\left(\dfrac{4a^{-2}b}{3ab^{-3}}\right)^3 = \left(\dfrac{4bb^3}{3aa^2}\right)^3 = \left(\dfrac{4b^4}{3a^3}\right)^3 = \dfrac{4^3\left(b^4\right)^3}{3^3\left(a^3\right)^3} = \dfrac{64b^{12}}{27a^9}$

102. $\left(\dfrac{2ab^{-3}}{3a^{-2}b^2}\right)^2 = \left(\dfrac{2aa^2}{3b^2b^3}\right)^2 = \left(\dfrac{2a^3}{3b^5}\right)^2 = \dfrac{2^2\left(a^3\right)^2}{3^2\left(b^5\right)^2} = \dfrac{4a^6}{9b^{10}}$

103. $\left(\dfrac{3a^{-2}b^2}{17a^2b^3}\right)^0 = 1$

104. $\dfrac{a^0+b^0}{2(a+b)^0} = \dfrac{1+1}{2 \cdot 1} = \dfrac{2}{2} = 1$

105. $\left(\dfrac{-2a^4b}{a^{-3}b^2}\right)^{-3} = \left(\dfrac{a^{-3}b^2}{-2a^4b}\right)^3 = \dfrac{\left(a^{-3}\right)^3\left(b^2\right)^3}{(-2)^3\left(a^4\right)^3b^3} = \dfrac{a^{-9}b^6}{-8a^{12}b^3} = \dfrac{b^3}{-8a^{21}} = -\dfrac{b^3}{8a^{21}}$

106. $\left(\dfrac{-3x^4y^2}{-9x^5y^{-2}}\right)^{-2} = \left(\dfrac{-9x^5y^{-2}}{-3x^4y^2}\right)^2 = \left(\dfrac{3x}{y^4}\right)^2 = \dfrac{3^2x^2}{\left(y^4\right)^2} = \dfrac{9x^2}{y^8}$

107. $\left(\dfrac{2a^3b^2}{3a^{-3}b^2}\right)^{-3} = \left(\dfrac{3a^{-3}b^2}{2a^3b^2}\right)^3 = \left(\dfrac{3}{2a^6}\right)^3 = \dfrac{3^3}{2^3\left(a^6\right)^3} = \dfrac{27}{8a^{18}}$

108. $\left(\dfrac{3x^5y^2}{6x^5y^{-2}}\right)^{-4} = \left(\dfrac{6x^5y^{-2}}{3x^5y^2}\right)^4 = \left(\dfrac{2}{y^4}\right)^4 = \dfrac{2^4}{(y^4)^4} = \dfrac{16}{y^{16}}$

109. $\dfrac{(3x^2)^{-2}}{x^3x^{-4}x^0} = \dfrac{3^{-2}(x^2)^{-2}}{x^{-1}} = \dfrac{3^{-2}x^{-4}}{x^{-1}} = 3^{-2}x^{-4-(-1)} = 3^{-2}x^{-3} = \dfrac{1}{3^2x^3} = \dfrac{1}{9x^3}$

110. $\dfrac{y^{-3}y^{-4}y^0}{(2y^{-2})^3} = \dfrac{y^{-7}}{8y^{-6}} = \dfrac{1}{8}y^{-7-(-6)} = \dfrac{1}{8}y^{-1} = \dfrac{1}{8y}$

111. $\dfrac{a^na^3}{a^4} = \dfrac{a^{n+3}}{a^4} = a^{n+3-4} = a^{n-1}$

112. $\dfrac{b^9b^7}{b^n} = \dfrac{b^{16}}{b^n} = b^{16-n}$

113. $\left(\dfrac{b^n}{b^3}\right)^3 = \dfrac{(b^n)^3}{(b^3)^3} = \dfrac{b^{3n}}{b^9} = b^{3n-9}$

114. $\left(\dfrac{a^2}{a^n}\right)^4 = \dfrac{(a^2)^4}{(a^n)^4} = \dfrac{a^8}{a^{4n}} = a^{8-4n}$

115. $\dfrac{a^{-n}a^2}{a^3} = \dfrac{a^{-n+2}}{a^3} = a^{-n+2-3} = a^{-n-1}$

$\left(\text{an equivalent form is } \dfrac{1}{a^{n+1}}\right)$

116. $\dfrac{a^na^{-2}}{a^4} = \dfrac{a^{n-2}}{a^4} = a^{n-2-4} = a^{n-6}$

117. $\dfrac{a^{-n}a^{-2}}{a^{-4}} = \dfrac{a^{-n-2}}{a^{-4}} = a^{-n-2-(-4)} = a^{2-n}$

118. $\dfrac{a^n}{a^{-3}a^5} = \dfrac{a^n}{a^2} = a^{n-2}$

Problems 119-130 are to be solved using a calculator. The keystrokes needed to solve each problem using a TI-83 graphing calculator appear in each solution. There may be other solutions. Keystrokes for other calculators may be slightly different.

119. [1] [.] [2] [3] [^] [6] [ENTER]
{3.462825992}

120. [.] [0] [5] [3] [7] [^] [4] [ENTER]
{0.000008315}

121. [(–)] [6] [.] [2] [5] [^] [3] [ENTER]
{–244.140625}

122. [(] [(–)] [2] [5] [.] [1] [)] [^] [5]
[ENTER] {–9962506.263}

123. [3] [.] [6] [8] [^] [0] [ENTER] {1.}

124. [2] [.] [1] [^] [4] [×] [2] [.] [1] [^] [3] [ENTER] {180.1088541}
[2] [.] [1] [^] [7] [ENTER] {180.1088541}

125. [7] [.] [2] [x^2] [×] [2] [.] [7] [x^2] [ENTER] {377.9136}
[(] [7] [.] [2] [×] [2] [.] [7] [)] [x^2] [ENTER] {377.9136}

126. [3] [.] [7] [x^2] [+] [4] [.] [8] [x^2] [ENTER] {36.73}
[(] [3] [.] [7] [+] [4] [.] [8] [)] [x^2] [ENTER] {72.25}

127. [3] [.] [2] [x^2] [×] [3] [.] [2] [^] [(–)] [2] [ENTER] {1}

128. [(] [5] [.] [9] [^] [3] [)] [x^2] [ENTER] {42180.533641}
[5] [.] [9] [^] [6] [ENTER] {42180.533641}

129. [7] [.] [2] [3] [^] [(–)] [3] [ENTER] {0.002645971171398}
[1] [÷] [(] [7] [.] [2] [3] [^] [3] [)] [ENTER] {0.002645971171398}

130. [(] [5] [.] [4] [÷] [2] [.] [7] [)] [^] [(–)] [4] [ENTER] {0.0625}
[(] [2] [.] [7] [÷] [5] [.] [4] [)] [^] [4] [ENTER] {0.0625}

131. $x^2y^3 = (-2)^2(3)^3 = 4(27) = 108$

132. $x^3y^2 = (-2)^3(3)^2 = -8(9) = -72$

133. $\dfrac{x^{-3}}{y^3} = \dfrac{1}{x^3y^3} = \dfrac{1}{(-2)^3(3)^3} = \dfrac{1}{-8(27)} = -\frac{1}{216}$

134. $\dfrac{x^2}{y^{-3}} = \dfrac{x^2y^3}{1} = x^2y^3 = (-2)^2(3)^2 = 4(27) = 108$

135. $(xy^2)^{-2} = x^{-2}y^{-4} = \dfrac{1}{x^2y^4} = \dfrac{1}{(-2)^2(3)^4} = \dfrac{1}{4(81)} = \dfrac{1}{324}$

136. $-y^3x^{-2} = -\dfrac{y^3}{x^2} = -\dfrac{3^3}{(-2)^2} = -\dfrac{27}{4}$

137. $(-yx^{-1})^3 = (-1)^3y^3(x^{-1})^3 = -y^3x^{-3} = -\dfrac{y^3}{x^3} = -\dfrac{3^3}{(-2)^3} = -\dfrac{27}{-8} = \dfrac{27}{8}$

138. $(-y)^3x^{-2} = (-1)^3y^3x^{-2} = -\dfrac{y^3}{x^2} = -\dfrac{3^3}{(-2)^2} = -\dfrac{27}{4}$

139. $A = lw = (3\,\text{m})(5\,\text{m}) = 15\,\text{m}^2$

140. $A = s^2 = (6\,\text{in.})^2 = 36\,\text{in.}^2$

141. $A = \pi r^2 = \pi(6\,\text{cm})^2 = \pi(36\,\text{cm}^2) \approx 113\,\text{cm}^2$

142. $A = \dfrac{1}{2}bh = \dfrac{1}{2}(12\,\text{in.})(8\,\text{in.}) = (6\,\text{in.})(8\,\text{in.}) = 48\,\text{in.}^2$

143. $A = \frac{1}{2}h(b_1 + b_2) = \frac{1}{2}(5\,\text{cm})(6\,\text{cm} + 12\,\text{cm}) = \frac{1}{2}(5\,\text{cm})(18\,\text{cm}) = \frac{1}{2}(90\,\text{cm}^2) = 45\,\text{cm}^2$

144. $A = lw + \frac{1}{2}\pi r^2 = (10\,\text{cm})(6\,\text{cm}) + \frac{1}{2}\pi(5\,\text{cm})^2 = (60 + \frac{25}{2}\pi)\,\text{cm}^2 \approx 99\,\text{cm}^2$

145. $A = s^2 + \frac{1}{2}bh = (15\,\text{cm})^2 + \frac{1}{2}(15\,\text{cm})(10\,\text{cm}) = (225 + 75)\,\text{cm}^2 = 300\,\text{cm}^2$

146. $A = \frac{1}{2}bh + \frac{1}{2}\pi r^2 = \frac{1}{2}(4\,\text{cm})(8\,\text{cm}) + \frac{1}{2}\pi(4\,\text{cm})^2 = (16 + 8\pi)\,\text{cm}^2 \approx 41\,\text{cm}^2$

147. $V = s^3 = (7\,\text{m})^3 = 343\,\text{m}^3$

148. $V = \frac{4}{3}\pi r^3 = \frac{4}{3}\pi(20\,\text{cm})^3 \approx 33{,}510\,\text{cm}^3$

149. $V = lwh = (6\,\text{ft})(6\,\text{ft})(10\,\text{ft}) = 360\,\text{ft}^3$

150. $V = \frac{1}{3}Bh = \frac{1}{3}s^2h = \frac{1}{3}(8\,\text{cm})^2(6\,\text{cm})$
$= 128\,\text{cm}^3$

151. $V = \frac{1}{3}Bh = \frac{1}{3}\pi r^2h = \frac{1}{3}\pi(4\,\text{ft})^2(10\,\text{ft}) = \frac{1}{3}(160\,\text{ft}^3)\pi \approx 168\,\text{ft}^3$

152. $V = Bh = \pi r^2h = \pi(3\,\text{m})^2(11\,\text{m}) = 99\pi\,\text{m}^3 \approx 311\,\text{m}^3$

153. $V = Bh + \frac{1}{2}\cdot\frac{4}{3}\pi r^3 = \pi r^2h + \frac{2}{3}\pi r^3 = \pi(6\,\text{m})^2(20\,\text{m}) + \frac{2}{3}\pi(6\,\text{m})^3 \approx 2714\,\text{m}^3$

154. $V = \frac{1}{3}Bh + \frac{1}{3}Bh = \frac{1}{3}\pi r^2h + \frac{1}{3}\pi r^2h = \frac{1}{3}\pi(5\,\text{in.})^2(8\,\text{in.}) + \frac{1}{3}\pi(5\,\text{in.})^2(8\,\text{in.}) \approx 419\,\text{in.}^3$

Problems 155-156 are to be solved using a calculator. The keystrokes needed to solve each problem using a TI-83 graphing calculator appear in each solution. There may be other solutions. Keystrokes for other calculators may be slightly different.

155. [5] [0] [0] [0] [(] [1] [+] [.] [1] [1] [)] [^] [5] [0] [ENTER]
{922824.1337} The balance will be $922,824.13.

156. [1] [0] [0] [0] [0] [0] [0] [(] [1] [+] [.] [0] [9] [)] [^] [(−)] [5] [0] [ENTER]
{13448.5389} The deposit should be $13,448.54.

157-160. Answers may vary.

161. $2^{-1} + 3^{-1} - 4^{-1} = \frac{1}{2} + \frac{1}{3} - \frac{1}{4} = \frac{6}{12} + \frac{4}{12} - \frac{3}{12} = \frac{7}{12}$

162. $(3^{-1} + 4^{-1})^{-2} = \left(\frac{1}{3} + \frac{1}{4}\right)^{-2} = \left(\frac{7}{12}\right)^{-2} = \left(\frac{12}{7}\right)^2 = \frac{144}{49}$

163. Let $x = 2$, $m = 3$ and $n = 4$: $x^m + x^n = 2^3 + 2^4 = 24$ while $x^{m+n} = 2^{3+4} = 2^7 = 128$.

164. Let $x = 2$, $y = 3$ and $m = 4$: $x^m + y^m = 2^4 + 3^4 = 97$ while $(x+y)^m = (2+3)^4 = 5^4 = 625$.

Exercise 1.4 (page 45)

1. $\frac{3}{4} = 3 \div 4 = 0.75$

2. $\frac{4}{5} = 4 \div 5 = 0.8$

3. $\frac{13}{9} = 13 \div 9 = 1.\overline{4}$

4. $\frac{14}{11} = 14 \div 11 = 1.\overline{27}$

5. $3^2 + 4^3 + 2^4 = 9 + 64 + 16 = 89$

6. $\dfrac{5ab - 4ac - 2}{3bc + abc} = \dfrac{5(-2)(-3) - 4(-2)(4) - 2}{3(-3)(4) + (-2)(-3)(4)} = \dfrac{30 - (-32) - 2}{-36 + 24} = \dfrac{60}{-12} = -5$

7. 10^n

8. four

9. left

10. $>$

SECTION 1.4

11. $3900 = 3.9 \times 10^3$

12. $1700 = 1.7 \times 10^3$

13. $0.0078 = 7.8 \times 10^{-3}$

14. $0.068 = 6.8 \times 10^{-2}$

15. $-45{,}000 = -4.5 \times 10^4$

16. $-547{,}000 = -5.47 \times 10^5$

17. $-0.00021 = -2.1 \times 10^{-4}$

18. $-0.00078 = -7.8 \times 10^{-4}$

19. $17{,}600{,}000 = 1.76 \times 10^7$

20. $89{,}800{,}000 = 8.98 \times 10^7$

21. $0.0000096 = 9.6 \times 10^{-6}$

22. $0.000046 = 4.6 \times 10^{-5}$

23. $$\begin{aligned} 323 \times 10^5 &= 3.23 \times 10^2 \times 10^5 \\ &= 3.23 \times 10^7 \end{aligned}$$

24. $$\begin{aligned} 689 \times 10^9 &= 6.89 \times 10^2 \times 10^9 \\ &= 6.89 \times 10^{11} \end{aligned}$$

25. $$\begin{aligned} 6000 \times 10^{-7} &= 6.0 \times 10^3 \times 10^{-7} \\ &= 6.0 \times 10^{-4} \end{aligned}$$

26. $$\begin{aligned} 765 \times 10^{-5} &= 7.65 \times 10^2 \times 10^{-5} \\ &= 7.65 \times 10^{-3} \end{aligned}$$

27. $$\begin{aligned} 0.0527 \times 10^5 &= 5.27 \times 10^{-2} \times 10^5 \\ &= 5.27 \times 10^3 \end{aligned}$$

28. $$\begin{aligned} 0.0298 \times 10^3 &= 2.98 \times 10^{-2} \times 10^3 \\ &= 2.98 \times 10^1 \end{aligned}$$

29. $$\begin{aligned} 0.0317 \times 10^{-2} &= 3.17 \times 10^{-2} \times 10^{-2} \\ &= 3.17 \times 10^{-4} \end{aligned}$$

30. $$\begin{aligned} 0.0012 \times 10^{-3} &= 1.2 \times 10^{-3} \times 10^{-3} \\ &= 1.2 \times 10^{-6} \end{aligned}$$

31. $2.7 \times 10^2 = 270$

32. $7.2 \times 10^3 = 7200$

33. $3.23 \times 10^{-3} = 0.00323$

34. $6.48 \times 10^{-2} = 0.0648$

35. $7.96 \times 10^5 = 796{,}000$

36. $9.67 \times 10^6 = 9{,}670{,}000$

37. $3.7 \times 10^{-4} = 0.00037$

38. $4.12 \times 10^{-5} = 0.0000412$

39. $5.23 \times 10^0 = 5.23$

40. $8.67 \times 10^0 = 8.67$

41. $23.65 \times 10^6 = 23{,}650{,}000$

42. $75.6 \times 10^{-5} = 0.000756$

43. $$\frac{(4000)(30{,}000)}{0.0006} = \frac{(4 \times 10^3)(3 \times 10^4)}{6 \times 10^{-4}} = \frac{(4)(3)}{6} \cdot \frac{10^3 10^4}{10^{-4}} = 2 \times 10^{11}$$

44. $$\frac{(0.0006)(0.00007)}{21{,}000} = \frac{(6 \times 10^{-4})(7 \times 10^{-5})}{2.1 \times 10^4} = \frac{(6)(7)}{2.1} \cdot \frac{10^{-4} 10^{-5}}{10^4} = 20 \times 10^{-13} = 2 \times 10^{-12}$$

45. $$\begin{aligned} \frac{(640{,}000)(2{,}700{,}000)}{120{,}000} = \frac{(6.4 \times 10^5)(2.7 \times 10^6)}{1.2 \times 10^5} = \frac{(6.4)(2.7)}{1.2} \cdot \frac{10^5 10^6}{10^5} &= 14.4 \times 10^6 \\ &= 1.44 \times 10^7 \end{aligned}$$

46. $\dfrac{(0.0000013)(0.000090)}{0.00039} = \dfrac{(1.3 \times 10^{-6})(9 \times 10^{-5})}{3.9 \times 10^{-4}} = \dfrac{(1.3)(9)}{3.9} \cdot \dfrac{10^{-6}10^{-5}}{10^{-4}} = 3 \times 10^{-7}$

47. $\dfrac{(0.006)(0.008)}{0.0012} = \dfrac{(6 \times 10^{-3})(8 \times 10^{-3})}{1.2 \times 10^{-3}} = \dfrac{(6)(8)}{1.2} \cdot \dfrac{10^{-3}10^{-3}}{10^{-3}} = 40 \times 10^{-3} = 0.04$

48. $\dfrac{(600)(80{,}000)}{120{,}000} = \dfrac{(6 \times 10^{2})(8 \times 10^{4})}{1.2 \times 10^{5}} = \dfrac{(6)(8)}{1.2} \cdot \dfrac{10^{2}10^{4}}{10^{5}} = 40 \times 10^{1} = 400$

49. $\dfrac{(220{,}000)(0.000009)}{0.00033} = \dfrac{(2.2 \times 10^{5})(9 \times 10^{-6})}{3.3 \times 10^{-4}} = \dfrac{(2.2)(9)}{3.3} \cdot \dfrac{10^{5}10^{-6}}{10^{-4}} = 6 \times 10^{3} = 6000$

50. $\dfrac{(0.00024)(96{,}000{,}000)}{640{,}000{,}000} = \dfrac{(2.4 \times 10^{-4})(9.6 \times 10^{7})}{6.4 \times 10^{8}} = \dfrac{(2.4)(9.6)}{6.4} \cdot \dfrac{10^{-4}10^{7}}{10^{8}} = 3.6 \times 10^{-5}$

$= 0.000036$

51. $\dfrac{(320{,}000)^2(0.0009)}{12{,}000^2} = \dfrac{(3.2 \times 10^{5})^2(9 \times 10^{-4})}{(1.2 \times 10^{4})^2} = \dfrac{[(3.2)^2 \times 10^{10}](9 \times 10^{-4})}{(1.2)^2 \times 10^{8}}$

$= \dfrac{(3.2)^2(9)}{(1.2)^2} \cdot \dfrac{10^{10}10^{-4}}{10^{8}}$

$= 64 \times 10^{-2} = 0.64$

52. $\dfrac{(0.000012)^2(49{,}000)^2}{0.021} = \dfrac{(1.2 \times 10^{-5})^2(4.9 \times 10^{4})^2}{2.1 \times 10^{-2}} = \dfrac{[(1.2)^2 \times 10^{-10}][(4.9)^2 \times 10^{8}]}{2.1 \times 10^{-2}}$

$= \dfrac{(1.2)^2(4.9)^2}{2.1} \cdot \dfrac{10^{-10}10^{8}}{10^{-2}}$

$= 16.464 \times 10^{0} = 16.464$

53. $23{,}437^3 \approx 1.2874 \times 10^{13}$

54. $0.00034^4 \approx 1.3 \times 10^{-14}$

55. $(63{,}480)(893{,}322) \approx 5.671 \times 10^{10}$

56. $(0.0000413)(0.0000049)^2 \approx 9.9 \times 10^{-16}$

57. $\dfrac{(69.4)^8(73.1)^2}{(0.0043)^3} \approx 3.6 \times 10^{25}$

58. $\dfrac{(0.0031)^4(0.0012)^5}{(0.0456)^{-7}} \approx 9.4 \times 10^{-35}$

59. gamma ray, x-ray, visible light, infrared, radio wave

60. $93{,}000{,}000 \text{ mi} = \dfrac{93000000 \text{ mi}}{1} \cdot \dfrac{5280 \text{ ft}}{1 \text{ mi}} = (9.3 \times 10^{7})(5.28 \times 10^{3}) \text{ ft} = 49.104 \times 10^{10} \text{ ft}$

$= 4.910 \times 10^{11} \text{ ft}$

61. $3.31 \times 10^4 \text{ cm/sec} = \dfrac{3.31 \times 10^4 \text{ cm}}{1 \text{ sec}} \cdot \dfrac{60 \text{ sec}}{1 \text{ min}} \cdot \dfrac{60 \text{ min}}{1 \text{ hr}} = \dfrac{(3.31 \times 10^4)(6 \times 10^1)(6 \times 10^1) \text{ cm}}{1 \text{ hr}}$

$$\begin{aligned} &= (3.31)(6)(6) \cdot (10^4 10^1 10^1) \text{ cm/hour} \\ &= 119.16 \times 10^6 \text{ cm/hour} \\ &= 1.1916 \times 10^8 \text{ cm/hour} \\ &\approx 1.19 \times 10^8 \text{ cm/hour} \end{aligned}$$

62. $V = lwh = (3000 \text{ mm})(7000 \text{ mm})(4000 \text{ mm}) = (3 \times 10^3)(7 \times 10^3)(4 \times 10^3) \text{ mm}^3$

$$\begin{aligned} &= (3)(7)(4) \cdot (10^3)(10^3)(10^3) \text{ mm}^3 \\ &= 84 \times 10^9 \text{ mm}^3 \\ &= 8.4 \times 10^{10} \text{ mm}^3 \approx 8 \times 10^{10} \text{ mm}^3 \end{aligned}$$

63. Mass of one million protons $=$ Mass of one proton $\cdot$ 1,000,000 $= 1.67248 \times 10^{-24} \cdot 1 \times 10^6 \text{ g}$

$$= 1.67248 \times 10^{-18} \text{ g}$$

64. $30{,}000{,}000{,}000 \text{ cm/sec} = \dfrac{3 \times 10^{10} \text{ cm}}{1 \text{ sec}} \cdot \dfrac{1 \text{ mile}}{160{,}000 \text{ cm}} \cdot \dfrac{60 \text{ sec}}{1 \text{ min}} \cdot \dfrac{60 \text{ min}}{1 \text{ hour}}$

$$\begin{aligned} &= \frac{(3 \times 10^{10})(6 \times 10^1)(6 \times 10^1) \text{ mile}}{1.6 \times 10^5 \text{ hour}} \\ &= \frac{(3)(6)(6)}{1.6} \cdot \frac{10^{10} 10^1 10^1}{10^5} \text{ miles/hour} \\ &= 67.5 \times 10^7 \text{ miles/hour} \\ &= 6.75 \times 10^8 \text{ miles/hour} \approx 7 \times 10^8 \text{ miles per hour} \end{aligned}$$

65. $235{,}000 \text{ miles} = \dfrac{2.35 \times 10^5 \text{ mile}}{1} \cdot \dfrac{5280 \text{ ft}}{1 \text{ mile}} \cdot \dfrac{12 \text{ in}}{1 \text{ ft}} = \dfrac{(2.35 \times 10^5)(5.28 \times 10^3)(1.2 \times 10^1) \text{ in}}{1}$

$$\begin{aligned} &= (2.35)(5.28)(1.2) \cdot 10^5 10^3 10^1 \text{ in.} \\ &= 14.8896 \times 10^9 \text{ in.} \\ &= 1.48896 \times 10^{10} \text{ in.} \approx 1.49 \times 10^{10} \text{ in.} \end{aligned}$$

66. $149{,}700{,}000 \text{ km} = \dfrac{1.497 \times 10^8 \text{ km}}{1} \cdot \dfrac{0.6214 \text{ mile}}{1 \text{ km}} = (1.497 \times 10^8)(6.214 \times 10^{-1}) \text{ miles}$

$$\begin{aligned} &= (1.497)(6.214) \cdot 10^8 10^{-1} \text{ miles} \\ &= 9.302358 \times 10^7 \text{ miles} \approx 9.302 \times 10^7 \text{ miles} \end{aligned}$$

67. $95{,}000 \text{ km} = \dfrac{9.5 \times 10^4 \text{ km}}{1} \cdot \dfrac{0.6214 \text{ mile}}{1 \text{ km}} = (9.5 \times 10^4)(6.214 \times 10^{-1}) \text{ miles}$

$$\begin{aligned} &= (9.5)(6.214) \cdot 10^4 10^{-1} \text{ miles} \\ &= 59.033 \times 10^3 \text{ miles} \approx 5.9 \times 10^4 \text{ miles} \end{aligned}$$

68. $$378{,}196\text{ km} = \frac{3.78196\times 10^5\text{ km}}{1}\cdot\frac{1\text{ mile}}{1.61\text{ km}}\cdot\frac{5280\text{ ft}}{1\text{ mile}}\cdot\frac{12\text{ in}}{1\text{ ft}}$$
$$= \frac{(3.78196\times 10^5)(5.28\times 10^3)(1.2\times 10^1)}{1.61\times 10^0}\text{ in.}$$
$$= \frac{(3.78196)(5.28)(1.2)}{1.61}\cdot\frac{10^5 10^3 10^1}{10^0}\text{ in.}$$
$$= 14.88353947826\times 10^9\text{ in.} \approx 1.489\times 10^{10}\text{ in.}$$

69. $$1\text{ in.} = \frac{1\text{ in.}}{1}\cdot\frac{25.4\text{ mm}}{1\text{ in.}}\cdot\frac{1\text{ angstrom}}{0.0000001\text{ mm}} = \frac{2.54\times 10^1}{1\times 10^{-7}} = 2.54\times 10^8 \approx 3\times 10^8\text{ angstroms}$$

70. $$0.6\text{ AU} = \frac{6\times 10^{-1}\text{ AU}}{1}\cdot\frac{9.3\times 10^7\text{ mi}}{1\text{ AU}} = (6)(9.3)\cdot 10^{-1}10^7\text{ mi} = 55.8\times 10^6\text{ mi} \approx 6\times 10^7\text{ mi}$$
$$18\text{ AU} = \frac{1.8\times 10^1\text{ AU}}{1}\cdot\frac{9.3\times 10^7\text{ mi}}{1\text{ AU}} = (1.8)(9.3)\cdot 10^1 10^7\text{ mi} = 16.74\times 10^8\text{ mi} \approx 1.7\times 10^9\text{ mi}$$

71. $$\frac{3{,}574{,}000{,}000}{18{,}000}\text{ hr} = \frac{3.574\times 10^9}{1.8\times 10^4}\text{ hr} = \frac{3.574}{1.8}\cdot\frac{10^9}{10^4}\text{ hr} = 1.98555555556\times 10^5\text{ hr}$$
$$1.98555555556\times 10^5\text{ hr} = \frac{1.98555555556\times 10^5\text{ hr}}{1}\cdot\frac{1\text{ day}}{24\text{ hr}}\cdot\frac{1\text{ yr}}{365\text{ day}}$$
$$= \frac{(1.98555555556\times 10^5)}{(2.4\times 10^1)(3.65\times 10^2)}\text{ years}$$
$$= \frac{1.98555555556}{(2.4)(3.65)}\cdot\frac{10^5}{10^1 10^2}\text{ years}$$
$$= 0.2266615930999\times 10^2\text{ years} \approx 23\text{ years}$$

72. $$300{,}000{,}000\text{ m/sec} = \frac{3\times 10^8\text{ m}}{1\text{ sec}}\cdot\frac{60\text{ sec}}{1\text{ min}}\cdot\frac{60\text{ min}}{1\text{ hr}}\cdot\frac{24\text{ hr}}{1\text{ day}}\cdot\frac{365\text{ day}}{1\text{ yr}}$$
$$= (3\times 10^8)(6\times 10^1)(6\times 10^1)(2.4\times 10^1)(3.65\times 10^2)\text{ m/yr}$$
$$= (3)(6)(6)(2.4)(3.65)\cdot 10^8 10^1 10^1 10^1 10^2\text{ m/yr}$$
$$= 946.08\times 10^{13}\text{ m/yr} \approx 9\times 10^{15}\text{ m/yr}$$

73. $$186{,}000\text{ mi/sec} = \frac{1.86\times 10^5\text{ mi}}{1\text{ sec}}\cdot\frac{60\text{ sec}}{1\text{ min}}\cdot\frac{60\text{ min}}{1\text{ hr}}\cdot\frac{24\text{ hr}}{1\text{ day}}\cdot\frac{365\text{ day}}{1\text{ yr}}$$
$$= (1.86\times 10^5)(6\times 10^1)(6\times 10^1)(2.4\times 10^1)(3.65\times 10^2)\text{ mi/yr}$$
$$= (1.86)(6)(6)(2.4)(3.65)\cdot 10^5 10^1 10^1 10^1 10^2\text{ mi/yr} = 586.5696\times 10^{10}\text{ mi/yr}$$
$$1.3\text{ p} = \frac{1.3\times 10^0\text{ p}}{1}\cdot\frac{3.26\text{ ly}}{1\text{ p}} = 1.3(3.26\cdot 586.5696\times 10^{10})\text{ mi} \approx 2.5\times 10^{13}\text{ miles}$$

74. Each time the comet is close to the sun, it loses

$\frac{10^7 \text{ g}}{1 \text{ sec}} \cdot \frac{60 \text{ sec}}{1 \text{ min}} \cdot \frac{60 \text{ min}}{1 \text{ hr}} \cdot \frac{24 \text{ hr}}{1 \text{ day}} \cdot \frac{10 \text{ days}}{1} = 8.64 \times 10^{12}$ grams.

Then, the number of times the comet can get close to the sun before it all evaporates is

$\frac{10^{16} \text{ grams}}{8.64 \times 10^{12} \text{ grams}} = \frac{1}{8.64} \cdot \frac{10^{16}}{10^{12}} = 0.1157407407407 \times 10^4 \approx 1157$ times

Since each time is 50 years apart, the expected life is $1157 \cdot 50 = 57{,}850$, or about 60,000 years.

75. **Answers may vary.**

76. **Answers may vary.**

77. Answer will depend on calculator.

78. Answer will depend on calculator.

Exercise 1.5 (page 55)

1. $(-4)^3 = (-4)(-4)(-4) = -64$

2. $-3^3 = -1 \cdot 3^3 = -1 \cdot 27 = -27$

3. $\left(\frac{x+y}{x-y}\right)^0 = 1$

4. $(x^2x^3)^4 = (x^5)^4 = x^{20}$

5. $\left(\frac{x^2x^5}{x^3}\right)^2 = \left(\frac{x^7}{x^3}\right)^2 = (x^4)^2 = x^8$

6. $\left(\frac{x^4y^3}{x^5y}\right)^3 = \left(\frac{y^2}{x}\right)^3 = \frac{y^6}{x^3}$

7. $(2x)^{-3} = \frac{1}{(2x)^3} = \frac{1}{8x^3}$

8. $\left(\frac{x^2}{y^5}\right)^{-4} = \left(\frac{y^5}{x^2}\right)^4 = \frac{y^{20}}{x^8}$

9. equation

10. satisfies (or solves)

11. equivalent

12. c; c

13. c; $\frac{b}{c}$

14. term

15. Like

16. coefficients; variables

17. identity

18. no

19.
$$\begin{aligned} 3x + 2 &= 17 \\ 3(5) + 2 &\stackrel{?}{=} 17 \\ 15 + 2 &\stackrel{?}{=} 17 \\ 17 &= 17 \end{aligned}$$
5 is a solution.

20.
$$\begin{aligned} 7x - 2 &= 33 \\ 7(5) - 2 &\stackrel{?}{=} 33 \\ 35 - 2 &\stackrel{?}{=} 33 \\ 33 &= 33 \end{aligned}$$
5 is a solution.

21.
$$\begin{aligned} \frac{3}{5}x - 5 &= -2 \\ \frac{3}{5}(5) - 5 &\stackrel{?}{=} -2 \\ 3 - 5 &\stackrel{?}{=} -2 \\ -2 &= -2 \end{aligned}$$
5 is a solution.

22.
$$\begin{aligned} \frac{2}{5}x + 12 &= 8 \\ \frac{2}{5}(5) + 12 &\stackrel{?}{=} 8 \\ 2 + 12 &\stackrel{?}{=} 8 \\ 14 &\neq 8 \end{aligned}$$
5 is not a solution.

23.
$$\begin{aligned} x + 6 &= 8 \\ x + 6 - 6 &= 8 - 6 \\ x &= 2 \end{aligned}$$

24.
$$\begin{aligned} y - 7 &= 3 \\ y - 7 + 7 &= 3 + 7 \\ y &= 10 \end{aligned}$$

25.
$$\begin{aligned} a - 5 &= 20 \\ a - 5 + 5 &= 20 + 5 \\ a &= 25 \end{aligned}$$

26.
$$\begin{aligned} b + 4 &= 18 \\ b + 4 - 4 &= 18 - 4 \\ b &= 14 \end{aligned}$$

27.
$$\begin{aligned} 2u &= 6 \\ \frac{2u}{2} &= \frac{6}{2} \\ u &= 3 \end{aligned}$$

28.
$$\begin{aligned} 3v &= 12 \\ \frac{3v}{3} &= \frac{12}{3} \\ v &= 4 \end{aligned}$$

29.
$$\begin{aligned} \frac{x}{4} &= 7 \\ 4 \cdot \frac{x}{4} &= 4 \cdot 7 \\ x &= 28 \end{aligned}$$

30.
$$\begin{aligned} \frac{x}{6} &= 8 \\ 6 \cdot \frac{x}{6} &= 6 \cdot 8 \\ x &= 48 \end{aligned}$$

31.
$$\begin{aligned} 3x + 1 &= 3 \\ 3x + 1 - 1 &= 3 - 1 \\ 3x &= 2 \\ \frac{3x}{3} &= \frac{2}{3} \\ x &= \tfrac{2}{3} \end{aligned}$$

32.
$$\begin{aligned} 8x - 2 &= 13 \\ 8x - 2 + 2 &= 13 + 2 \\ 8x &= 15 \\ \frac{8x}{8} &= \frac{15}{8} \\ x &= \tfrac{15}{8} \end{aligned}$$

33.
$$\begin{aligned} 2x + 1 &= 13 \\ 2x + 1 - 1 &= 13 - 1 \\ 2x &= 12 \\ \frac{2x}{2} &= \frac{12}{2} \\ x &= 6 \end{aligned}$$

34.
$$\begin{aligned} 2x - 4 &= 16 \\ 2x - 4 + 4 &= 16 + 4 \\ 2x &= 20 \\ \frac{2x}{2} &= \frac{20}{2} \\ x &= 10 \end{aligned}$$

35.
$$\begin{aligned} 3(x - 4) &= -36 \\ 3x - 12 &= -36 \\ 3x - 12 + 12 &= -36 + 12 \\ 3x &= -24 \\ \frac{3x}{3} &= \frac{-24}{3} \\ x &= -8 \end{aligned}$$

36.
$$\begin{aligned} 4(x + 6) &= 84 \\ 4x + 24 &= 84 \\ 4x + 24 - 24 &= 84 - 24 \\ 4x &= 60 \\ \frac{4x}{4} &= \frac{60}{4} \\ x &= 15 \end{aligned}$$

37.
$$\begin{aligned} 3(r - 4) &= -4 \\ 3r - 12 &= -4 \\ 3r - 12 + 12 &= -4 + 12 \\ 3r &= 8 \\ \frac{3r}{3} &= \frac{8}{3} \\ r &= \tfrac{8}{3} \end{aligned}$$

38.
$$\begin{aligned} 4(s - 5) &= -3 \\ 4s - 20 &= -3 \\ 4s - 20 + 20 &= -3 + 20 \\ 4s &= 17 \\ \frac{4s}{4} &= \frac{17}{4} \\ s &= \tfrac{17}{4} \end{aligned}$$

39. like terms; $2x + 6x = 8x$

40. not like terms

41. not like terms

42. like terms; $-3t^2 + 12t^2 = 9t^2$

43. like terms; $3x^2 + (-5x^2) = -2x^2$

44. not like terms

45. not like terms

46. like terms; $-4x + (-5x) = -9x$

47.
$$\begin{aligned} 3a - 22 &= -2a - 7 \\ 3a + 2a - 22 &= -2a + 2a - 7 \\ 5a - 22 &= -7 \\ 5a - 22 + 22 &= -7 + 22 \\ 5a &= 15 \\ \frac{5a}{5} &= \frac{15}{5} \\ a &= 3 \end{aligned}$$

48.
$$\begin{aligned} a + 18 &= 6a - 3 \\ a - 6a + 18 &= 6a - 6a - 3 \\ -5a + 18 &= -3 \\ -5a + 18 - 18 &= -3 - 18 \\ -5a &= -21 \\ \frac{-5a}{-5} &= \frac{-21}{-5} \\ a &= \tfrac{21}{5} \end{aligned}$$

49.
$$\begin{aligned} 2(2x+1) &= 15+3x \\ 4x+2 &= 15+3x \\ 4x-3x+2 &= 15+3x-3x \\ x+2 &= 15 \\ x+2-2 &= 15-2 \\ x &= 13 \end{aligned}$$

50.
$$\begin{aligned} -2(x+5) &= 30-x \\ -2x-10 &= 30-x \\ -2x+x-10 &= 30-x+x \\ -x-10 &= 30 \\ -x-10+10 &= 30+10 \\ -x &= 40 \\ x &= -40 \end{aligned}$$

51.
$$\begin{aligned} 3(y-4)-6 &= y \\ 3y-12-6 &= y \\ 3y-18 &= y \\ 3y-y-18 &= y-y \\ 2y-18 &= 0 \\ 2y-18+18 &= 0+18 \\ 2y &= 18 \\ \frac{2y}{2} &= \frac{18}{2} \\ y &= 9 \end{aligned}$$

52.
$$\begin{aligned} 2x+(2x-3) &= 5 \\ 2x+2x-3 &= 5 \\ 4x-3 &= 5 \\ 4x-3+3 &= 5+3 \\ 4x &= 8 \\ \frac{4x}{4} &= \frac{8}{4} \\ x &= 2 \end{aligned}$$

53.
$$\begin{aligned} 5(5-a) &= 37-2a \\ 25-5a &= 37-2a \\ 25-5a+2a &= 37-2a+2a \\ 25-3a &= 37 \\ 25-25-3a &= 37-25 \\ -3a &= 12 \\ \frac{-3a}{-3} &= \frac{12}{-3} \\ a &= -4 \end{aligned}$$

54.
$$\begin{aligned} 4a+17 &= 7(a+2) \\ 4a+17 &= 7a+14 \\ 4a-7a+17 &= 7a-7a+14 \\ -3a+17 &= 14 \\ -3a+17-17 &= 14-17 \\ -3a &= -3 \\ \frac{-3a}{-3} &= \frac{-3}{-3} \\ a &= 1 \end{aligned}$$

55.
$$\begin{aligned} 4(y+1) &= -2(4-y) \\ 4y+4 &= -8+2y \\ 4y-2y+4 &= -8+2y-2y \\ 2y+4 &= -8 \\ 2y+4-4 &= -8-4 \\ 2y &= -12 \\ \frac{2y}{2} &= \frac{-12}{2} \\ y &= -6 \end{aligned}$$

56.
$$\begin{aligned} 5(r+4) &= -2(r-3) \\ 5r+20 &= -2r+6 \\ 5r+2r+20 &= -2r+2r+6 \\ 7r+20 &= 6 \\ 7r+20-20 &= 6-20 \\ 7r &= -14 \\ \frac{7r}{7} &= \frac{-14}{7} \\ r &= -2 \end{aligned}$$

57.
$$\begin{aligned}2(a-5)-(3a+1)&=0\\2a-10-3a-1&=0\\-a-11&=0\\-a-11+11&=0+11\\-a&=11\\a&=-11\end{aligned}$$

58.
$$\begin{aligned}8(3a-5)-4(2a+3)&=12\\24a-40-8a-12&=12\\16a-52&=12\\16a-52+52&=12+52\\16a&=64\\\frac{16a}{16}&=\frac{64}{16}\\a&=4\end{aligned}$$

59.
$$\begin{aligned}3(y-5)+10&=2(y+4)\\3y-15+10&=2y+8\\3y-5&=2y+8\\3y-2y-5&=2y-2y+8\\y-5&=8\\y-5+5&=8+5\\y&=13\end{aligned}$$

60.
$$\begin{aligned}2(5x+2)&=3(3x-2)\\10x+4&=9x-6\\10x-9x+4&=9x-9x-6\\x+4&=-6\\x+4-4&=-6-4\\x&=-10\end{aligned}$$

61.
$$\begin{aligned}9(x+2)&=-6(4-x)+18\\9x+18&=-24+6x+18\\9x+18&=6x-6\\9x-6x+18&=6x-6x-6\\3x+18&=-6\\3x+18-18&=-6-18\\3x&=-24\\\frac{3x}{3}&=\frac{-24}{3}\\x&=-8\end{aligned}$$

62.
$$\begin{aligned}3(x+2)-2&=-(5+x)+x\\3x+6-2&=-5-x+x\\3x+4&=-5\\3x+4-4&=-5-4\\3x&=-9\\\frac{3x}{3}&=\frac{-9}{3}\\x&=-3\end{aligned}$$

63.
$$\begin{aligned}-4p-2(3p+5)&=-6p+2(p+2)\\-4p-6p-10&=-6p+2p+4\\-10p-10&=-4p+4\\-10p+4p-10&=-4p+4p+4\\-6p-10&=4\\-6p-10+10&=4+10\\-6p&=14\\\frac{-6p}{-6}&=\frac{14}{-6}\\p&=-\tfrac{7}{3}\end{aligned}$$

64.
$$\begin{aligned}2q-3(q-5)&=5(q+2)-7\\2q-3q+15&=5q+10-7\\-q+15&=5q+3\\-q-5q+15&=5q-5q+3\\-6q+15&=3\\-6q+15-15&=3-15\\-6q&=-12\\\frac{-6q}{-6}&=\frac{-12}{-6}\\q&=2\end{aligned}$$

65.
$$\begin{aligned} 4+4(n+2) &= 3n-2(n-5) \\ 4+4n+8 &= 3n-2n+10 \\ 4n+12 &= n+10 \\ 4n-n+12 &= n-n+10 \\ 3n+12 &= 10 \\ 3n+12-12 &= 10-12 \\ 3n &= -2 \\ n &= -\tfrac{2}{3} \end{aligned}$$

66.
$$\begin{aligned} 4x-2(3x+2) &= 2(x+3) \\ 4x-6x-4 &= 2x+6 \\ -2x-4 &= 2x+6 \\ -2x-2x-4 &= 2x-2x+6 \\ -4x-4 &= 6 \\ -4x-4+4 &= 6+4 \\ -4x &= 10 \\ \frac{-4x}{-4} &= \frac{10}{-4} \\ x &= -\tfrac{5}{2} \end{aligned}$$

67.
$$\begin{aligned} \frac{1}{2}x-4 &= -1+2x \\ 2\left(\frac{1}{2}x-4\right) &= 2(-1+2x) \\ 2\left(\frac{1}{2}x\right)+2(-4) &= 2(-1)+2(2x) \\ x-8 &= -2+4x \\ x-4x-8 &= -2+4x-4x \\ -3x-8 &= -2 \\ -3x-8+8 &= -2+8 \\ -3x &= 6 \\ \frac{-3x}{-3} &= \frac{6}{-3} \\ x &= -2 \end{aligned}$$

68.
$$\begin{aligned} 2x+3 &= \frac{2}{3}x-1 \\ 3(2x+3) &= 3\left(\frac{2}{3}x-1\right) \\ 3(2x)+3(3) &= 3\left(\frac{2}{3}x\right)-3(1) \\ 6x+9 &= 2x-3 \\ 6x-2x+9 &= 2x-2x-3 \\ 4x+9 &= -3 \\ 4x+9-9 &= -3-9 \\ 4x &= -12 \\ \frac{4x}{4} &= \frac{-12}{4} \\ x &= -3 \end{aligned}$$

69.
$$\begin{aligned} \frac{x}{2}-\frac{x}{3} &= 4 \\ 6\left(\frac{x}{2}-\frac{x}{3}\right) &= 6(4) \\ 6\left(\frac{x}{2}\right)-6\left(\frac{x}{3}\right) &= 24 \\ 3x-2x &= 24 \\ x &= 24 \end{aligned}$$

70.
$$\begin{aligned} \frac{x}{2}+\frac{x}{3} &= 10 \\ 6\left(\frac{x}{2}+\frac{x}{3}\right) &= 6(10) \\ 6\left(\frac{x}{2}\right)+6\left(\frac{x}{3}\right) &= 60 \\ 3x+2x &= 60 \\ 5x &= 60 \\ \frac{5x}{5} &= \frac{60}{5} \\ x &= 12 \end{aligned}$$

71.

$$\begin{aligned} \frac{x}{6}+1 &= \frac{x}{3} \\ 6\left(\frac{x}{6}+1\right) &= 6\left(\frac{x}{3}\right) \\ 6\left(\frac{x}{6}\right)+6(1) &= 2x \\ x+6 &= 2x \\ x-2x+6 &= 2x-2x \\ -x+6 &= 0 \\ -x+6-6 &= 0-6 \\ -x &= -6 \\ x &= 6 \end{aligned}$$

72.

$$\begin{aligned} \frac{3}{2}(y+4) &= \frac{20-y}{2} \\ 2\left[\frac{3}{2}(y+4)\right] &= 2\left(\frac{20-y}{2}\right) \\ 3(y+4) &= 20-y \\ 3y+12 &= 20-y \\ 3y+y+12 &= 20-y+y \\ 4y+12 &= 20 \\ 4y+12-12 &= 20-12 \\ 4y &= 8 \\ \frac{4y}{4} &= \frac{8}{4} \\ y &= 2 \end{aligned}$$

73.

$$\begin{aligned} 5-\frac{x+2}{3} &= 7-x \\ 3\left(5-\frac{x+2}{3}\right) &= 3(7-x) \\ 3(5)-3\left(\frac{x+2}{3}\right) &= 3(7)+3(-x) \\ 15-(x+2) &= 21-3x \\ 15-x-2 &= 21-3x \\ 13-x &= 21-3x \\ 13-x+3x &= 21-3x+3x \\ 13+2x &= 21 \\ 13-13+2x &= 21-13 \\ 2x &= 8 \\ \frac{2x}{2} &= \frac{8}{2} \\ x &= 4 \end{aligned}$$

74.

$$\begin{aligned} 3x-\frac{2(x+3)}{3} &= 16-\frac{x+2}{2} \\ 6\left[3x-\frac{2(x+3)}{3}\right] &= 6\left(16-\frac{x+2}{2}\right) \\ 6(3x)-6\left(\frac{2x+6}{3}\right) &= 6(16)-6\left(\frac{x+2}{2}\right) \\ 18x-2(2x+6) &= 96-3(x+2) \\ 18x-4x-12 &= 96-3x-6 \\ 14x-12 &= 90-3x \\ 14x+3x-12 &= 90-3x+3x \\ 17x-12 &= 90 \\ 17x-12+12 &= 90+12 \\ 17x &= 102 \\ \frac{17x}{17} &= \frac{102}{17} \\ x &= 6 \end{aligned}$$

75.

$$\begin{aligned} \frac{4x-2}{2} &= \frac{3x+6}{3} \\ 6\left(\frac{4x-2}{2}\right) &= 6\left(\frac{3x+6}{3}\right) \\ 3(4x-2) &= 2(3x+6) \\ 12x-6 &= 6x+12 \\ 12x-6x-6 &= 6x-6x+12 \\ 6x-6 &= 12 \\ 6x-6+6 &= 12+6 \\ 6x &= 18 \\ \frac{6x}{6} &= \frac{18}{6} \\ x &= 3 \end{aligned}$$

76.

$$\begin{aligned} \frac{t+4}{2} &= \frac{2t-3}{3} \\ 6\left(\frac{t+4}{2}\right) &= 6\left(\frac{2t-3}{3}\right) \\ 3(t+4) &= 2(2t-3) \\ 3t+12 &= 4t-6 \\ 3t-4t+12 &= 4t-4t-6 \\ -t+12 &= -6 \\ -t+12-12 &= -6-12 \\ -t &= -18 \\ t &= 18 \end{aligned}$$

77.

$$\begin{aligned} \frac{a+1}{3}+\frac{a-1}{5} &= \frac{2}{15} \\ 15\left(\frac{a+1}{3}+\frac{a-1}{5}\right) &= 15\left(\frac{2}{15}\right) \\ 15\left(\frac{a+1}{3}\right)+15\left(\frac{a-1}{5}\right) &= 2 \\ 5(a+1)+3(a-1) &= 2 \\ 5a+5+3a-3 &= 2 \\ 8a+2 &= 2 \\ 8a+2-2 &= 2-2 \\ 8a &= 0 \\ \frac{8a}{8} &= \frac{0}{8} \\ a &= 0 \end{aligned}$$

78.

$$\begin{aligned} \frac{2z+3}{3}+\frac{3z-4}{6} &= \frac{z-2}{2} \\ 6\left(\frac{2z+3}{3}+\frac{3z-4}{6}\right) &= 6\left(\frac{z-2}{2}\right) \\ 6\left(\frac{2z+3}{3}\right)+6\left(\frac{3z-4}{6}\right) &= 3(z-2) \\ 2(2z+3)+3z-4 &= 3z-6 \\ 4z+6+3z-4 &= 3z-6 \\ 7z+2 &= 3z-6 \\ 7z-3z+2 &= 3z-3z-6 \\ 4z+2 &= -6 \\ 4z+2-2 &= -6-2 \\ 4z &= -8 \\ \frac{4z}{4} &= \frac{-8}{4} \\ z &= -2 \end{aligned}$$

79.

$$\begin{aligned} \frac{5a}{2}-12 &= \frac{a}{3}+1 \\ 6\left(\frac{5a}{2}-12\right) &= 6\left(\frac{a}{3}+1\right) \\ 6\left(\frac{5a}{2}\right)-6(12) &= 6\left(\frac{a}{3}\right)+6(1) \\ 3(5a)-72 &= 2a+6 \\ 15a-72 &= 2a+6 \\ 15a-2a-72 &= 2a-2a+6 \\ 13a-72 &= 6 \\ 13a-72+72 &= 6+72 \\ 13a &= 78 \\ \frac{13a}{13} &= \frac{78}{13} \\ a &= 6 \end{aligned}$$

80.

$$\begin{aligned} \frac{5a}{6}-\frac{5}{2} &= -\frac{1}{2}-\frac{a}{6} \\ 6\left(\frac{5a}{6}-\frac{5}{2}\right) &= 6\left(-\frac{1}{2}-\frac{a}{6}\right) \\ 6\left(\frac{5a}{6}\right)+6\left(-\frac{5}{2}\right) &= 6\left(-\frac{1}{2}\right)+6\left(-\frac{a}{6}\right) \\ 5a+3(-5) &= -3-a \\ 5a-15 &= -3-a \\ 5a+a-15 &= -3-a+a \\ 6a-15 &= -3 \\ 6a-15+15 &= -3+15 \\ 6a &= 12 \\ \frac{6a}{6} &= \frac{12}{6} \\ a &= 2 \end{aligned}$$

81.

$$\begin{aligned} 9.8-15z &= -15.7 \\ 9.8-9.8-15z &= -15.7-9.8 \\ -15z &= -25.5 \\ \frac{-15z}{-15} &= \frac{-25.5}{-15} \\ z &= 1.7 \end{aligned}$$

82.

$$\begin{aligned} 0.05a+0.25 &= 0.77 \\ 0.05a+0.25-0.25 &= 0.77-0.25 \\ 0.05a &= 0.52 \\ \frac{0.05a}{0.05} &= \frac{0.52}{0.05} \\ a &= 10.4 \end{aligned}$$

83.
$$\begin{aligned} 0.45 &= 16.95 - 0.25(75 - 3a) \\ 0.45 &= 16.95 - 18.75 + 0.75a \\ 0.45 &= -1.8 + 0.75a \\ 0.45 + 1.8 &= -1.8 + 1.8 + 0.75a \\ 2.25 &= 0.75a \\ \frac{2.25}{0.75} &= \frac{0.75a}{0.75} \\ 3 &= a \end{aligned}$$

84.
$$\begin{aligned} 3.2 + x &= 0.25(x + 32) \\ 3.2 + x &= 0.25x + 8 \\ 3.2 + x - 0.25x &= 0.25x - 0.25x + 8 \\ 3.2 + 0.75x &= 8 \\ 3.2 - 3.2 + 0.75x &= 8 - 3.2 \\ 0.75x &= 4.8 \\ \frac{0.75x}{0.75} &= \frac{4.8}{0.75} \\ x &= 6.4 \end{aligned}$$

85.
$$\begin{aligned} 0.09x + 0.14(10000 - x) &= 1275 \\ 0.09x + 1400 - 0.14x &= 1275 \\ -0.05x + 1400 &= 1275 \\ -0.05x + 1400 - 1400 &= 1275 - 1400 \\ -0.05x &= -125 \\ \frac{-0.05x}{-0.05} &= \frac{-125}{-0.05} \\ x &= 2500 \end{aligned}$$

86.
$$\begin{aligned} 0.04(20) + 0.01l &= 0.02(20 + l) \\ 0.8 + 0.01l &= 0.4 + 0.02l \\ 0.8 + 0.01l - 0.01l &= 0.4 + 0.02l - 0.01l \\ 0.8 &= 0.4 + 0.01l \\ 0.8 - 0.4 &= 0.4 - 0.4 + 0.01l \\ 0.4 &= 0.01l \\ \frac{0.4}{0.01} &= \frac{0.01l}{0.01} \\ 40 &= l \end{aligned}$$

87.
$$\begin{aligned} 4(2 - 3t) + 6t &= -6t + 8 \\ 8 - 12t + 6t &= -6t + 8 \\ 8 - 6t &= -6t + 8 \\ -6t + 8 &= -6t + 8 \end{aligned}$$
Identity

88.
$$\begin{aligned} 2x - 6 &= -2x + 4(x - 2) \\ 2x - 6 &= -2x + 4x - 8 \\ 2x - 6 &= 2x - 8 \\ 2x - 2x - 6 &= 2x - 2x - 8 \\ -6 &= -8 \end{aligned}$$
Contradiction

89.
$$\begin{aligned} \frac{a+1}{4} + \frac{2a-3}{4} &= \frac{a}{2} - 2 \\ 4\left(\frac{a+1}{4} + \frac{2a-3}{4}\right) &= 4\left(\frac{a}{2} - 2\right) \\ 4\left(\frac{a+1}{4}\right) + 4\left(\frac{2a-3}{4}\right) &= 4\left(\frac{a}{2}\right) + 4(-2) \\ a + 1 + 2a - 3 &= 2a - 8 \\ 3a - 2 &= 2a - 8 \\ 3a - 2a - 2 &= 2a - 2a - 8 \\ a - 2 &= -8 \\ a - 2 + 2 &= -8 + 2 \\ a &= -6 \end{aligned}$$

90.
$$\begin{aligned} \frac{y-8}{5} + 2 &= \frac{2}{5} - \frac{y}{3} \\ 15\left(\frac{y-8}{5} + 2\right) &= 15\left(\frac{2}{5} - \frac{y}{3}\right) \\ 15\left(\frac{y-8}{5}\right) + 15(2) &= 15\left(\frac{2}{5}\right) + 15\left(-\frac{y}{3}\right) \\ 3(y - 8) + 30 &= 3(2) - 5y \\ 3y - 24 + 30 &= 6 - 5y \\ 3y + 6 &= 6 - 5y \\ 3y + 5y + 6 - 6 &= 6 - 6 - 5y + 5y \\ 8y + 0 &= 0 \\ 8y &= 0 \\ \frac{8y}{8} &= \frac{0}{8} \\ y &= 0 \end{aligned}$$

91.
$$\begin{aligned}3(x-4)+6&=-2(x+4)+5x\\3x-12+6&=-2x-8+5x\\3x-6&=3x-8\\3x-3x-6&=3x-3x-8\\-6&=-8\end{aligned}$$
Contradiction

92.
$$\begin{aligned}2(x-3)&=\frac{3}{2}(x-4)+\frac{x}{2}\\2\cdot 2(x-3)&=2\left[\frac{3}{2}(x-4)+\frac{x}{2}\right]\\4(x-3)&=2\left[\frac{3}{2}(x-4)\right]+2\left(\frac{x}{2}\right)\\4x-12&=3(x-4)+x\\4x-12&=3x-12+x\\4x-12&=4x-12\end{aligned}$$
Identity

93.
$$\begin{aligned}y(y+2)+1&=y^2+2y+1\\y^2+2y+1&=y^2+2y+1\end{aligned}$$
Identity

94.
$$\begin{aligned}x(x-3)&=x^2-2x+1-(5+x)\\x^2-3x&=x^2-2x+1-5-x\\x^2-3x&=x^2-3x-4\\x^2-x^2-3x&=x^2-x^2-3x-4\\-3x&=-3x-4\\-3x+3x&=-3x+3x-4\\0&=-4\end{aligned}$$
Contradiction

95.
$$A=lw$$
$$\frac{A}{l}=\frac{lw}{l}$$
$$\frac{A}{l}=w,\text{ or } w=\frac{A}{l}$$

96.
$$p=4s$$
$$\frac{p}{4}=\frac{4s}{4}$$
$$\frac{p}{4}=s,\text{ or } s=\frac{p}{4}$$

97.
$$\begin{aligned}V&=\frac{1}{3}Bh\\3V&=3\cdot\frac{1}{3}Bh\\3V&=Bh\\\frac{3V}{h}&=\frac{Bh}{h}\\\frac{3V}{h}&=B,\text{ or } B=\frac{3V}{h}\end{aligned}$$

98.
$$\begin{aligned}b&=\frac{2A}{h}\\h\cdot b&=h\cdot\frac{2A}{h}\\bh&=2A\\\frac{bh}{2}&=\frac{2A}{2}\\\tfrac{1}{2}bh&=A,\text{ or } A=\tfrac{1}{2}bh\end{aligned}$$

99.
$$I=prt$$
$$\frac{I}{pr}=\frac{prt}{pr}$$
$$\frac{I}{pr}=t,\text{ or } t=\frac{I}{pr}$$

100.
$$I=prt$$
$$\frac{I}{pt}=\frac{prt}{pt}$$
$$\frac{I}{pt}=r,\text{ or } r=\frac{I}{pt}$$

101.
$$\begin{aligned} p &= 2l + 2w \\ p - 2l &= 2l - 2l + 2w \\ p - 2l &= 2w \\ \frac{p-2l}{2} &= \frac{2w}{2} \\ \frac{p-2l}{2} &= w, \text{ or } w = \frac{p-2l}{2} \end{aligned}$$

102.
$$\begin{aligned} p &= 2l + 2w \\ p - 2w &= 2l + 2w - 2w \\ p - 2w &= 2l \\ \frac{p-2w}{2} &= \frac{2l}{2} \\ \frac{p-2w}{2} &= l, \text{ or } l = \frac{p-2w}{2} \end{aligned}$$

103.
$$\begin{aligned} A &= \frac{1}{2}h(B+b) \\ 2A &= 2 \cdot \frac{1}{2}h(B+b) \\ 2A &= h(B+b) \\ \frac{2A}{h} &= \frac{h(B+b)}{h} \\ \frac{2A}{h} &= B + b \\ \frac{2A}{h} - b &= B + b - b \\ \frac{2A}{h} - b &= B, \text{ or } B = \frac{2A}{h} - b \end{aligned}$$

104.
$$\begin{aligned} A &= \frac{1}{2}h(B+b) \\ 2A &= 2 \cdot \frac{1}{2}h(B+b) \\ 2A &= h(B+b) \\ \frac{2A}{h} &= \frac{h(B+b)}{h} \\ \frac{2A}{h} &= B + b \\ \frac{2A}{h} - B &= B - B + b \\ \frac{2A}{h} - B &= b, \text{ or } b = \frac{2A}{h} - B \end{aligned}$$

105.
$$\begin{aligned} y &= mx + b \\ y - b &= mx + b - b \\ y - b &= mx \\ \frac{y-b}{m} &= \frac{mx}{m} \\ \frac{y-b}{m} &= x, \text{ or } x = \frac{y-b}{m} \end{aligned}$$

106.
$$\begin{aligned} y &= mx + b \\ y - b &= mx + b - b \\ y - b &= mx \\ \frac{y-b}{x} &= \frac{mx}{x} \\ \frac{y-b}{x} &= m, \text{ or } m = \frac{y-b}{x} \end{aligned}$$

107.
$$\begin{aligned} l &= a + (n-1)d \\ l &= a + nd - d \\ l - a + d &= a - a + nd - d + d \\ l - a + d &= nd \\ \frac{l-a+d}{d} &= \frac{nd}{d} \\ \frac{l-a+d}{d} &= n, \text{ or } n = \frac{l-a+d}{d} \end{aligned}$$

108.
$$\begin{aligned} l &= a + (n-1)d \\ l - a &= a - a + (n-1)d \\ l - a &= (n-1)d \\ \frac{l-a}{n-1} &= \frac{(n-1)d}{n-1} \\ \frac{l-a}{n-1} &= d, \text{ or } d = \frac{l-a}{n-1} \end{aligned}$$

109.

$$\begin{aligned} S &= \frac{a - lr}{1 - r} \\ (1-r)S &= (1-r) \cdot \frac{a - lr}{1 - r} \\ S - rS &= a - lr \\ S - Sr - a &= a - a - lr \\ S - Sr - a &= -lr \\ \frac{S - Sr - a}{-r} &= \frac{-lr}{-r} \\ \frac{-S + Sr + a}{r} &= l, \text{ or } l = \frac{a - S + Sr}{r} \end{aligned}$$

110.

$$\begin{aligned} C &= \frac{5}{9}(F - 32) \\ \frac{9}{5}C &= \frac{9}{5} \cdot \frac{5}{9}(F - 32) \\ \frac{9}{5}C &= F - 32 \\ \frac{9}{5}C + 32 &= F - 32 + 32 \\ \frac{9}{5}C + 32 &= F, \text{ or } F = \frac{9}{5}C + 32 \end{aligned}$$

111.

$$\begin{aligned} S &= \frac{n(a + l)}{2} \\ 2S &= 2 \cdot \frac{n(a + l)}{2} \\ 2S &= n(a + l) \\ 2S &= na + nl \\ 2S - na &= na - na + nl \\ 2S - na &= nl \\ \frac{2S - na}{n} &= \frac{nl}{n} \\ \frac{2S - na}{n} &= l, \text{ or } l = \frac{2S - na}{n} \end{aligned}$$

112.

$$\begin{aligned} S &= \frac{n(a + l)}{2} \\ 2S &= 2 \cdot \frac{n(a + l)}{2} \\ 2S &= n(a + l) \\ \frac{2S}{a + l} &= \frac{n(a + l)}{a + l} \\ \frac{2S}{a + l} &= n, \text{ or } n = \frac{2S}{a + l} \end{aligned}$$

113.

$$\begin{aligned} F &= \frac{GmM}{d^2} \\ d^2F &= d^2 \cdot \frac{GmM}{d^2} \\ Fd^2 &= GmM \\ \frac{Fd^2}{GM} &= \frac{GmM}{GM} \\ \frac{Fd^2}{GM} &= m, \text{ or } m = \frac{Fd^2}{GM} \end{aligned}$$

114.

$$\begin{aligned} G &= U - TS + pV \\ G - U - pV &= U - U - TS + pV - pV \\ G - U - pV &= -TS \\ \frac{G - U - pV}{-T} &= \frac{-TS}{-T} \\ \frac{-G + U + pV}{T} &= S, \text{ or } S = \frac{U + pV - G}{T} \end{aligned}$$

115.

$$\begin{aligned} F &= \frac{9}{5}C + 32 \\ F - 32 &= \frac{9}{5}C + 32 - 32 \\ F - 32 &= \frac{9}{5}C \\ \frac{5}{9}(F - 32) &= \frac{5}{9} \cdot \frac{9}{5}C \\ \frac{5}{9}(F - 32) &= C, \text{ or } C = \frac{5}{9}(F - 32) \end{aligned}$$

$F = 32°$: $C = \frac{5}{9}(32 - 32) = \frac{5}{9}(0) = 0°$

$F = 70°$: $C = \frac{5}{9}(70 - 32) = \frac{5}{9}(38) \approx 21.1°$

$F = 212°$: $C = \frac{5}{9}(212 - 32) = \frac{5}{9}(180) = 100°$

116.

$$A = p + prt$$
$$A - p = p - p + prt$$
$$A - p = prt$$
$$\frac{A-p}{pr} = \frac{prt}{pr}$$
$$\frac{A-p}{pr} = t, \text{ or } t = \frac{A-p}{pr}$$

Let $A = 2000$, and substitute the values for r:

$$r = 0.05\text{: } t = \frac{2000 - 1000}{1000(0.05)} = \frac{1000}{50} = 20 \text{ years}$$
$$r = 0.07\text{: } t = \frac{2000 - 1000}{1000(0.07)} = \frac{1000}{70} \approx 14.29 \text{ years}$$
$$r = 0.10\text{: } t = \frac{2000 - 1000}{1000(0.10)} = \frac{1000}{1000} = 10 \text{ years}$$

117.

$$C = 0.07n + 6.50$$
$$C - 6.50 = 0.07n + 6.50 - 6.50$$
$$C - 6.50 = 0.07n$$
$$\frac{C - 6.50}{0.07} = \frac{0.07n}{0.07}$$
$$\frac{C - 6.50}{0.07} = n, \text{ or } n = \frac{C - 6.50}{0.07}$$

$$C = 49.97\text{: } n = \frac{49.97 - 6.50}{0.07} = \frac{43.47}{0.07} = 621 \text{ kwh}$$
$$C = 76.50\text{: } n = \frac{76.50 - 6.50}{0.07} = \frac{70}{0.07} = 1000 \text{ kwh}$$
$$C = 125\text{: } n = \frac{125 - 6.50}{0.07} = \frac{118.50}{0.07} \approx 1692.9 \text{ kwh}$$

118.

$$n = \frac{5000C - 17{,}500}{6}$$
$$6n = 6 \cdot \frac{5000C - 17{,}500}{6}$$
$$6n = 5000C - 17{,}500$$
$$6n + 17{,}500 = 5000C - 17{,}500 + 17{,}500$$
$$6n + 17{,}500 = 5000C$$
$$\frac{6n + 17{,}500}{5000} = \frac{5000C}{5000}$$
$$\frac{6n + 17{,}500}{5000} = C, \text{ or } C = \frac{6n + 17{,}500}{5000}$$

$$n = 500\text{: } C = \frac{6(500) + 17{,}500}{5000} = \$4.10$$
$$n = 1200\text{: } C = \frac{6(1200) + 17{,}500}{5000} = \$4.94$$
$$n = 2500\text{: } C = \frac{6(2500) + 17{,}500}{5000} = \$6.50$$

119.

$$E = IR$$
$$\frac{E}{I} = \frac{IR}{I}$$
$$\frac{E}{I} = R, \text{ or } R = \frac{E}{I}$$
$$R = \frac{56}{7} = 8 \text{ ohms}$$

120.

$$A = P(1 + rt)$$
$$\frac{A}{1+rt} = \frac{P(1+rt)}{1+rt}$$
$$\frac{A}{1+rt} = P, \text{ or } P = \frac{A}{1+rt}$$
$$P = \frac{6693.75}{1 + (0.055)(5)} = \$5250$$

121.
$$a = 180\left(1 - \frac{2}{n}\right)$$
$$a = 180 - \frac{360}{n}$$
$$a - 180 = 180 - 180 - \frac{360}{n}$$
$$a - 180 = -\frac{360}{n}$$
$$n(a-180) = n\left(-\frac{360}{n}\right)$$
$$n(a-180) = -360$$
$$\frac{n(a-180)}{a-180} = \frac{-360}{a-180}$$
$$n = \frac{360}{180-a}$$
$$n = \frac{360}{180-a} = \frac{360}{180-135} = 8 \text{ sides}$$

122.
$$P = \frac{E^2}{R}$$
$$RP = R \cdot \frac{E^2}{R}$$
$$RP = E^2$$
$$\frac{RP}{P} = \frac{E^2}{P}$$
$$R = \frac{E^2}{P}$$
$$R = \frac{60^2}{4.8} = 750 \text{ ohms}$$

123. Answers may vary.

124. Answers may vary.

125. The distributive property was not used correctly.

126. Only full terms may be subtracted from both sides of the equation.

Exercise 1.6 (page 65)

1. $\left(\frac{3x^{-3}}{4x^2}\right)^{-4} = \left(\frac{4x^2}{3x^{-3}}\right)^4 = \left(\frac{4x^5}{3}\right)^4 = \frac{256x^{20}}{81}$

2. $\left(\frac{r^{-3}s^2}{r^2r^3s^{-4}}\right)^{-5} = \left(\frac{r^2r^3s^{-4}}{r^{-3}s^2}\right)^5 = \left(\frac{r^5s^{-4}}{r^{-3}s^2}\right)^5 = \left(\frac{r^8}{s^6}\right)^5 = \frac{r^{40}}{s^{30}}$

3. $\frac{a^m a^3}{a^2} = \frac{a^{m+3}}{a^2} = a^{m+3-2} = a^{m+1}$

4. $\left(\frac{b^n}{b^3}\right)^3 = \left(b^{n-3}\right)^3 = b^{3n-9}$

5.	$5x + 4$	**6.**	$0.06x$	**7.**	$40x$	**8.**	$90°$
9.	$180°$	**10.**	acute	**11.**	complementary	**12.**	$180°$
13.	right	**14.**	isosceles	**15.**	vertex	**16.**	equal

17. Let $x =$ length of 1st piece.
Then $2x$, $4x$ and $8x =$ the other lengths.

$$\boxed{\text{Sum of lengths}} = \boxed{\text{total length}}$$
$$x + 2x + 4x + 8x = 60$$
$$15x = 60$$
$$x = 4$$

The longest piece has a length of $8x$, or $8(4) = 32$ feet.

18. Let $x =$ length of 1st piece. Then $x + 3$, $x + 6$ and $x + 9 =$ the other lengths.

$$\boxed{\text{Sum of lengths}} = \boxed{\text{total length}}$$
$$x + x + 3 + x + 6 + x + 9 = 186$$
$$4x + 18 = 186$$
$$4x = 168$$
$$x = 42$$

The pieces have lengths of 42 feet, 45 feet, 48 feet and 51 feet.

19. Let $x =$ the length of the shorter piece.
Then $2x + 1 =$ the length of the other piece.

$$\boxed{\text{shorter length}} + \boxed{\text{other length}} = \boxed{\text{total length}}$$
$$x + 2x + 1 = 22$$
$$3x + 1 = 22$$
$$3x = 21$$
$$x = 7$$

The pieces have lengths of 7 feet and 15 feet.

20. Let $x =$ the length of the shorter piece.
Then $3x + 2 =$ the length of the other piece.

$$\boxed{\text{shorter length}} + \boxed{\text{other length}} = \boxed{\text{total length}}$$
$$x + 3x + 2 = 30$$
$$4x + 2 = 30$$
$$4x = 28$$
$$x = 7$$

The pieces have lengths of 7 feet and 23 feet.

21. Let $x =$ the cost of the VCR.
Then $x + 55 =$ the cost of the TV.

$$\boxed{\text{VCR cost}} + \boxed{\text{TV cost}} = \boxed{\text{total cost.}}$$
$$x + x + 55 = 655$$
$$2x + 55 = 655$$
$$2x = 600$$
$$x = 300$$

The TV costs $x + 55 = 300 + 55 = \$355$.

22. Let $w =$ the cost of the woods.
Then $w + 40 =$ the cost of the irons.

$$\boxed{\text{woods cost}} + \boxed{\text{irons cost}} = \boxed{\text{total cost}}$$
$$w + w + 40 = 590$$
$$2w + 40 = 590$$
$$2w = 550$$
$$w = 275$$

The irons cost $w + 40 = 275 + 40 = \$315$.

23. Percent markdown $= \dfrac{\text{amount of markdown}}{\text{regular price}} = \dfrac{726 - 580.80}{726} = \dfrac{145.20}{726} = 0.20 = 20\%$.

24. Percent markdown $= \dfrac{\text{amount of markdown}}{\text{regular price}} = \dfrac{983 - 737.25}{983} = \dfrac{245.75}{983} = 0.25 = 25\%$.

25. Percent markup $= \dfrac{\text{amount of markup}}{\text{regular price}} = \dfrac{40 - 12}{12} = \dfrac{28}{12} \approx 2.33 = 233\%$.

26. Percent markup $= \dfrac{\text{amount of markup}}{\text{regular price}} = \dfrac{30 - 18}{18} = \dfrac{12}{18} \approx 0.67 = 67\%$.

27. Let $B = \#$ of shares in Big Bank. Then $500 - B = \#$ of shares in Safe Savings.

$$\boxed{\text{Big Bank shares}} + \boxed{\text{Safe Savings shares}} = \boxed{\text{Total value}}$$

$$\begin{aligned} 115B + 97(500 - B) &= 53900 \\ 115B + 48500 - 97B &= 53900 \\ 18B &= 5400 \\ B &= 300 \end{aligned}$$

The student owns 300 shares of Big Bank and 200 shares of Safe Savings.

28. Let $S = \#$ of stock shares. Then $12000 - S = \#$ of bond shares.

$$\boxed{\text{Stock value}} + \boxed{\text{Bond value}} = \boxed{\text{Total value}}$$

$$\begin{aligned} 12S + 15(12000 - S) &= 165000 \\ 12S + 180000 - 15S &= 165000 \\ -3S &= -15000 \\ S &= 5000 \end{aligned}$$

The fund owns 5000 shares of stock fund and 7000 shares of bond fund.

29. Let $B = \#$ of scientific model sold. Then $85 - B = \#$ of graphing model sold.

$$\boxed{\text{Value of scientific}} + \boxed{\text{Value of graphing}} = \boxed{\text{Total value}}$$

$$\begin{aligned} 15B + 67(85 - B) &= 3875 \\ 15B + 5695 - 67B &= 3875 \\ -52B &= -1820 \\ B &= 35 \end{aligned}$$

The store sold 35 scientific calculators and 50 graphing calculators.

30. Let $m =$ the bags of mixture sold. Then $19 - m =$ bags of pure bluegrass sold.

$$\boxed{\text{Value of mixture}} + \boxed{\text{Value of bluegrass}} = \boxed{\text{Total value}}$$

$$\begin{aligned} 245m + 347(19 - m) &= 5369 \\ 245m + 6593 - 347m &= 5369 \\ -102m &= -1224 \\ m &= 12 \end{aligned}$$

The company sold 12 bags of the mixture and 7 bags of pure bluegrass.

31. Let $r =$ the number of roses he buys.

$$\boxed{\text{Cost of roses}} + \boxed{\text{Delivery charge}} = \boxed{\text{Total cost}}$$

$$\begin{aligned} 1.25r + 5 &= 21.25 \\ 1.25r &= 16.25 \\ r &= 13 \end{aligned}$$

The man can buy 13 roses.

32. Let $m =$ the number of miles he drives.

$$\boxed{\text{Mileage cost}} + \boxed{\text{Daily cost}} = \boxed{\text{Total cost}}$$

$$\begin{aligned} 0.39m + 49.95 &= 147.45 \\ 0.39m &= 97.50 \\ m &= 250 \end{aligned}$$

The man can drive 250 miles.

33. Let $m =$ the number of miles he drives.

$$\boxed{\text{Mileage cost}} + \boxed{\text{Daily cost}} = \boxed{\text{Total cost}}$$

$$\begin{aligned} 0.30m + 2 \cdot 12 &= 42 \\ 0.30m &= 18 \\ m &= 60 \end{aligned}$$

$$\boxed{\text{Mileage cost}} + \boxed{\text{Daily cost}} = \boxed{\text{Total cost}}$$

$$\begin{aligned} 0.30m + 2 \cdot 12 &= 60 \\ 0.30m &= 36 \\ m &= 120 \end{aligned}$$

He can drive 60 miles for \$42 and 120 miles for \$60.

34. Let $d =$ the number of deliveries she makes.

$$\boxed{\text{Delivery pay}} + \boxed{\text{Daily pay}} = \boxed{\text{Total pay}}$$

$$\begin{aligned} 0.60d + 5 &= 17 \\ 0.60d &= 12 \\ d &= 20 \end{aligned}$$

$$\boxed{\text{Delivery pay}} + \boxed{\text{Daily pay}} = \boxed{\text{Total pay}}$$

$$\begin{aligned} 0.60d + 5 &= 23 \\ 0.60d &= 18 \\ d &= 30 \end{aligned}$$

She must make 10 additional deliveries per day (from 20 to 30) to make \$23 per day.

35. Let $w =$ the width. Then $2w =$ the length.

$$\boxed{\text{Perimeter of garden}} = 72$$
$$w + 2w + w + 2w = 72$$
$$6w = 72$$
$$w = 12$$

The dimension are 12 m by 24 m.

36. Let $w =$ the width. Then $3w =$ the length.

$$\boxed{\text{Perimeter of garden}} = 96$$
$$w + 3w + w + 3w = 96$$
$$8w = 96$$
$$w = 12$$

The dimension are 12 m by 36 m.

37. Let $w =$ the width. Then $2w =$ the length.

$$\boxed{\text{Total amount of fence}} = 624$$
$$w + 2w + w = 624$$
$$4w = 624$$
$$w = 156$$

The dimension are 156 feet by 312 feet.

38.

$$\boxed{\text{Total amount of fence}} = 150$$
$$x + x + x + x + x + 5 + x + 5 + x = 150$$
$$7x + 10 = 150$$
$$7x = 140$$
$$x = 20$$

The dimensions are 20 feet by 45 feet.

39.

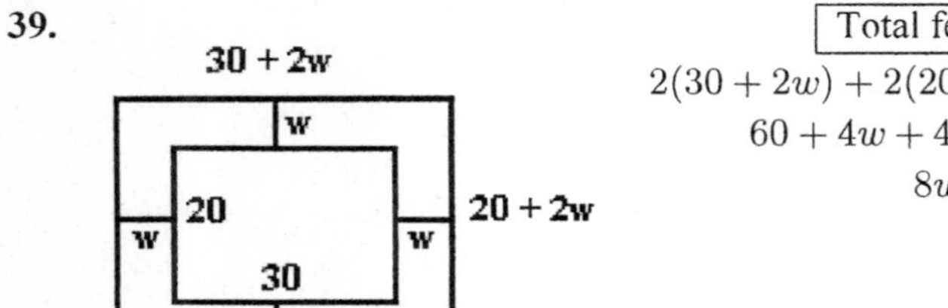

$$\boxed{\text{Total fencing}} = 180$$
$$2(30 + 2w) + 2(20 + 2w) = 180$$
$$60 + 4w + 40 + 4w = 180$$
$$8w + 100 = 180$$
$$8w = 80$$
$$w = 10$$

The walkway will be 10 feet wide.

40.

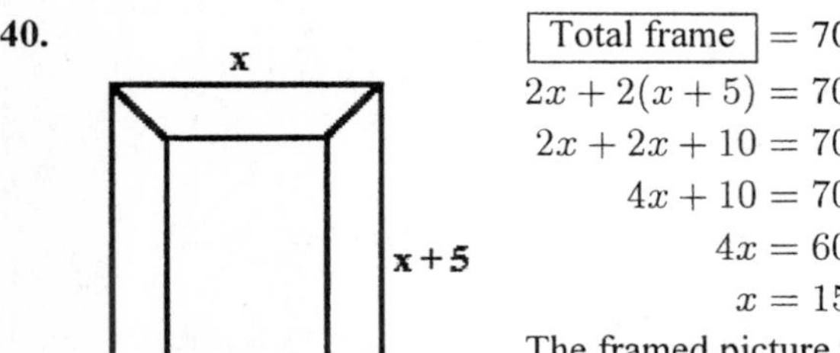

$$\boxed{\text{Total frame}} = 70$$
$$2x + 2(x + 5) = 70$$
$$2x + 2x + 10 = 70$$
$$4x + 10 = 70$$
$$4x = 60$$
$$x = 15$$

The framed picture will be $x = 15$ inches wide.

41. Let $x =$ the measure of the smaller angle. Then $x + 35 =$ the other measure.

$$\boxed{\begin{array}{c}\text{The sum of}\\ \text{the angles}\end{array}} = 180°$$
$$x + x + 35 = 180$$
$$2x + 35 = 180$$
$$2x = 145$$
$$x = 72.5$$

The smaller angle has a measure of 72.5°.

42.

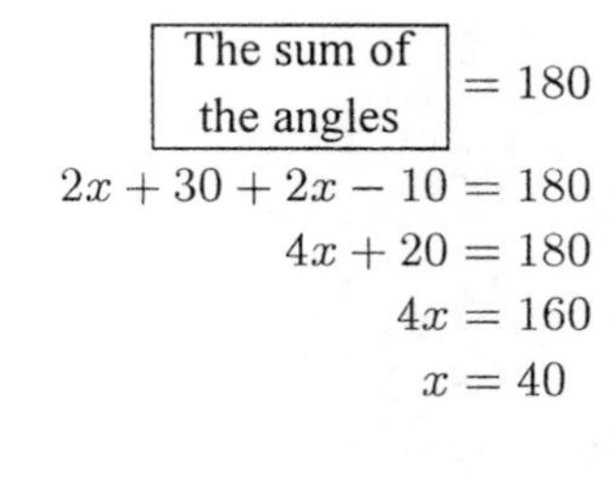

$$\boxed{\begin{array}{c}\text{The sum of}\\ \text{the angles}\end{array}} = 180$$
$$2x + 30 + 2x - 10 = 180$$
$$4x + 20 = 180$$
$$4x = 160$$
$$x = 40$$

43. Let $x =$ the measure of the smaller angle. Then $x + 22 =$ the other measure.

$$\boxed{\text{sum of angles}} = 90°$$
$$x + x + 22 = 90$$
$$2x + 22 = 90$$
$$2x = 68$$
$$x = 34$$

The larger angle has a measure of 56°.

44. The sum of all three angles in any triangle must be 180°. Since a right triangle has a right angle with a measure of 90°, the other two angles must have measures which add up to the other 90° in the triangle, and are then complementary.

$$\boxed{\text{sum of acute angles}} = 90$$
$$2x + 15 + x = 90$$
$$3x = 75$$
$$x = 25 \quad \angle A \text{ has a measure of } 25°.$$

45.

$$\boxed{\text{The sum of the measures of } \angle 1 \text{ and } \angle 2} = 180°$$
$$x + 50 + 2x - 20 = 180°$$
$$3x + 30 = 180°$$
$$3x = 150°$$
$$x = 50°$$

46. Let $x =$ the measure of $\angle 3$.

The measure of $\angle 2 = 2x - 20 = 80°$

$$\boxed{\text{The sum of the measures of } \angle 2 \text{ and } \angle 3} = 180°$$
$$80 + x = 180°$$
$$x = m(\angle 3) = 100°$$

47. Let $x =$ the measure of each angle.

$$\boxed{\text{Sum of measures of all angles}} = 180°$$
$$3x = 180°$$
$$x = 60°$$

Each angle has a measure of 60°.

48.

$$\boxed{\text{Measure of } \angle 1} = \boxed{\text{Measure of } \angle 3}$$
$$3x + 10 = 5x - 10$$
$$-2x = -20$$
$$x = 10$$

49.

$$\boxed{\text{Measure of } \angle 2} = \boxed{\text{Measure of } \angle 4}$$
$$6x + 20 = 8x - 20$$
$$-2x = -40$$
$$x = 20$$
$$\text{m}(\angle 2) = 6x + 20 = 6(20) + 20 = 140°$$
$$\text{m}(\angle 1) = 180 - 140 = 40°$$

50. Let $x =$ the measure of each angle.

$$\boxed{\text{Sum of measures of all angles}} = 360°$$
$$2x + 240 = 360°$$
$$2x = 120°$$
$$x = 60°$$

51. Let $x =$ the original height. Then $3x =$ the new height.

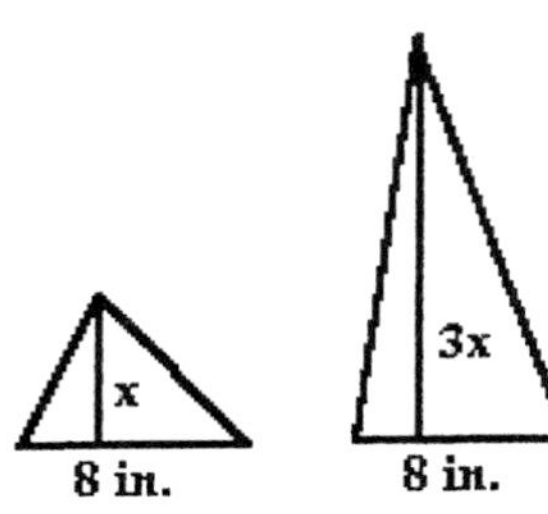

$$\boxed{\text{Old Area}} + 96 = \boxed{\text{New Area}}$$
$$\frac{1}{2}(8)(x) + 96 = \frac{1}{2}(8)(3x)$$
$$4x + 96 = 12x$$
$$96 = 8x$$
$$12 = x$$

The original height is 12 inches.

52. $\boxed{\text{Area of rectangle}} = \frac{1}{2} \cdot \boxed{\text{Area of trapezoid (total)}}$

$$(12)(7) = \tfrac{1}{2} \cdot \tfrac{1}{2}(7)([(15 + w) + 15]$$
$$84 = \tfrac{1}{4} \cdot 7(30 + w)$$
$$336 = 210 + 7w$$
$$126 = 7w$$
$$18 = w \Rightarrow w = 18 \text{ inches}$$

53.

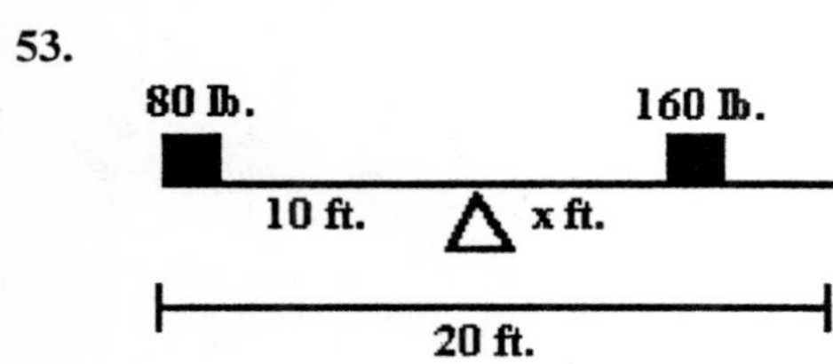

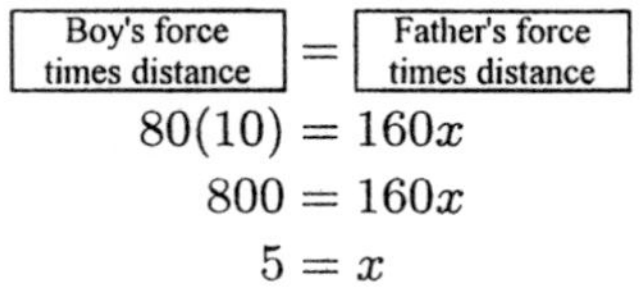

$\boxed{\text{Boy's force times distance}} = \boxed{\text{Father's force times distance}}$

$$80(10) = 160x$$
$$800 = 160x$$
$$5 = x$$

The father should sit 5 feet from the fulcrum.

54.

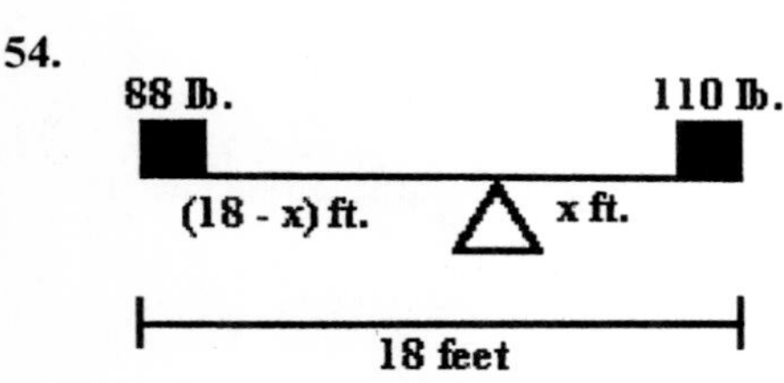

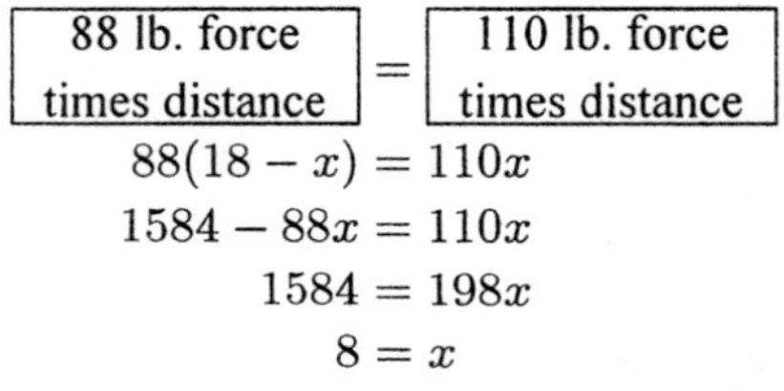

$\boxed{\text{88 lb. force times distance}} = \boxed{\text{110 lb. force times distance}}$

$$88(18 - x) = 110x$$
$$1584 - 88x = 110x$$
$$1584 = 198x$$
$$8 = x$$

The fulcrum should be 8 feet from the greater force.

55.

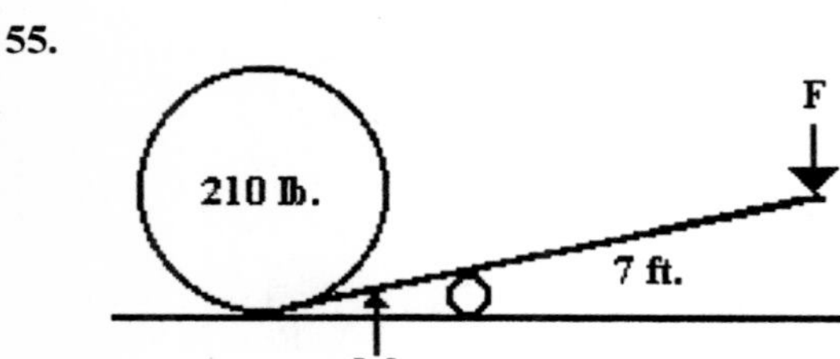

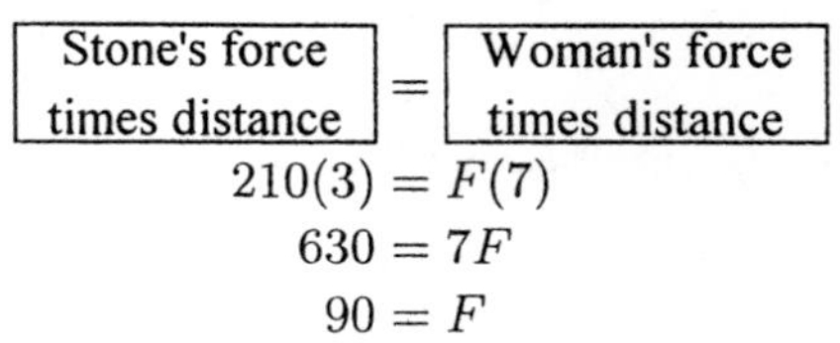

$\boxed{\text{Stone's force times distance}} = \boxed{\text{Woman's force times distance}}$

$$210(3) = F(7)$$
$$630 = 7F$$
$$90 = F$$

The woman needs to exert 90 pounds of force.

56.

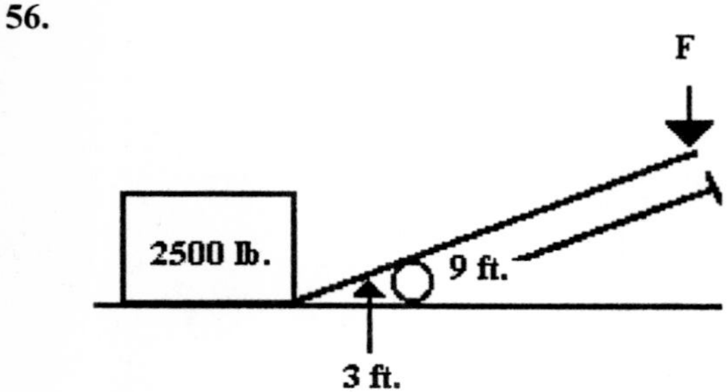

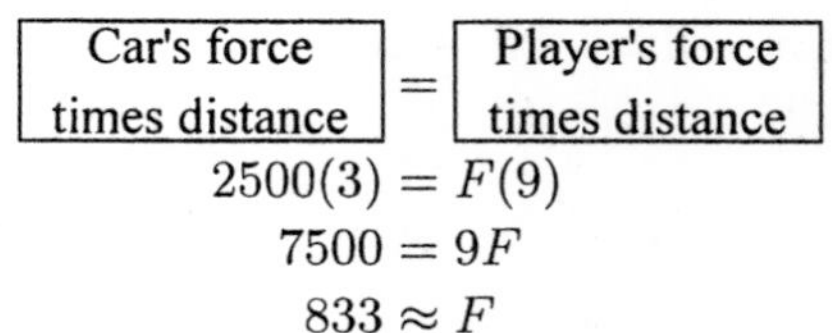

$\boxed{\text{Car's force times distance}} = \boxed{\text{Player's force times distance}}$

$$2500(3) = F(9)$$
$$7500 = 9F$$
$$833 \approx F$$

The player needs to exert 833 pounds of force, but he weighs only 350 pounds. He cannot do it.

57. $\boxed{\text{Sum of forces times distances on left}} = \boxed{\text{Sum of forces times distances on right}}$

$$\begin{aligned} 100(12) + 70(8) &= 40(x) + 200(8) \\ 1200 + 560 &= 40x + 1600 \\ 160 &= 40x \\ 4 &= x \quad \text{The distance should be 4 feet.} \end{aligned}$$

58. Let $x =$ Kim's weight.

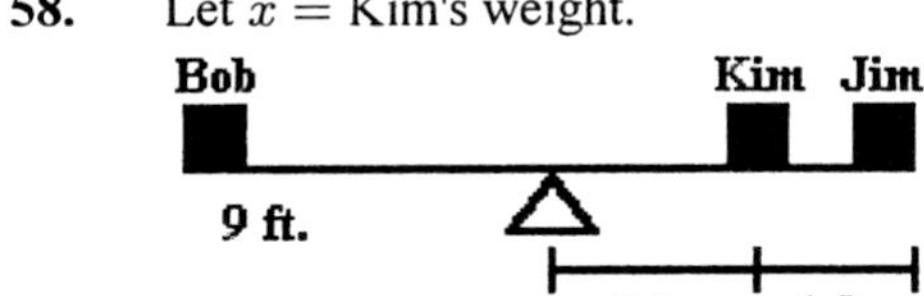

$\boxed{\begin{array}{c}\text{Bob's force}\\ \times\text{ distance}\end{array}} = \boxed{\begin{array}{c}\text{Kim's force}\\ \times\text{ distance}\end{array}} + \boxed{\begin{array}{c}\text{Jim's force}\\ \times\text{ distance}\end{array}}$

$$\begin{aligned} 200(9) &= x(5) + 160(9) \\ 1800 &= 5x + 1440 \\ 360 &= 5x \\ 72 &= x \quad \text{Kim weighs 72 pounds.} \end{aligned}$$

59.
$$\begin{aligned} F &= \frac{5}{9}(F - 32) \\ 9F &= 5(F - 32) \\ 9F &= 5F - 160 \\ 4F &= -160 \\ F &= -40 \end{aligned}$$
The temperatures are the same at $-40°$.

60. $\boxed{\begin{array}{c}\text{Area of}\\ \text{Left}\end{array}} = \boxed{\begin{array}{c}\text{Area of}\\ \text{Right}\end{array}}$

$$\begin{aligned} 11w &= 8(w + 3) \\ 11w &= 8w + 24 \\ 3w &= 24 \\ w &= 8 \end{aligned}$$
The widths are 8 feet and 11 feet.

61. **Answers may vary.**

62. **Answers may vary.**

63.
$$\begin{aligned} 40(6) + 10x &= 50(4) \\ 240 + 10x &= 200 \\ 10x &= -40 \\ x &= -4 \end{aligned}$$

64. This means that the 10 pound force must be located 4 feet to the **right** of the fulcrum.

Exercise 1.7 (page 74)

1.
$$\begin{aligned} 9x - 3 &= 6x \\ 9x - 6x - 3 &= 6x - 6x \\ 3x - 3 + 3 &= 0 + 3 \\ 3x &= 3 \\ x &= 1 \end{aligned}$$

2.
$$\begin{aligned} 7a + 2 &= 12 - 4(a - 3) \\ 7a + 2 &= 12 - 4a + 12 \\ 7a + 2 &= -4a + 24 \\ 11a &= 22 \\ a &= 2 \end{aligned}$$

3.
$$\begin{aligned} \frac{8(y - 5)}{3} &= 2(y - 4) \\ 3 \cdot \frac{8(y - 5)}{3} &= 3 \cdot 2(y - 4) \\ 8(y - 5) &= 6(y - 4) \\ 8y - 40 &= 6y - 24 \\ 2y &= 16 \\ y &= 8 \end{aligned}$$

4.
$$\begin{aligned} \frac{t - 1}{3} &= \frac{t + 2}{6} + 2 \\ 6 \cdot \frac{t - 1}{3} &= 6 \cdot \frac{t + 2}{6} + 6 \cdot 2 \\ 2(t - 1) &= t + 2 + 12 \\ 2t - 2 &= t + 14 \\ t &= 16 \end{aligned}$$

5. principal; rate; time

6. $d = rt$; distance; rate; time

7. value; price; number

8. butterfat; butterfat; butterfat

9. Let $x =$ amount invested at 4%. Then $12{,}000 - x =$ amount invested at 6%.

$$\boxed{\text{Interest at 4\%}} + \boxed{\text{Interest at 6\%}} = \boxed{\text{Total interest}}$$
$$\begin{aligned} 0.04x + 0.06(12000 - x) &= 680 \\ 0.04x + 720 - 0.06x &= 680 \\ -0.02x &= -40 \\ x &= \frac{-40}{-0.02} \\ x &= 2000 \end{aligned}$$

She invested \$2000 in the money market account (4%) and \$10,000 in the CD (6%).

10. Let $x =$ amount invested at 7%. Then $14{,}000 - x =$ amount invested at 10%.

$$\boxed{\text{Interest at 7\%}} + \boxed{\text{Interest at 10\%}} = \boxed{\text{Total interest}}$$
$$\begin{aligned} 0.07x + 0.10(14000 - x) &= 1280 \\ 0.07x + 1400 - 0.10x &= 1280 \\ -0.03x &= -120 \\ x &= \frac{-120}{-0.03} \\ x &= 4000 \end{aligned}$$

He invested \$4000 at 7% interest and \$10,000 at 10% interest.

11. Let $r =$ the needed interest rate for the remainder of the money.

$$\boxed{\text{Interest at 7\%}} + \boxed{\text{Interest at } r\%} = \boxed{\text{Total interest}}$$
$$\begin{aligned} 0.07(6000) + r(10000) &= 1500 \\ 420 + 10000r &= 1500 \\ 10000r &= 1080 \\ r &= \frac{1080}{10000} \\ r &= 0.1080 \end{aligned}$$

She needs to earn 10.8% interest on the remainder of the money.

12. Let $x =$ amount invested at 7%. Then $2x =$ amount invested at 10%.

$$\boxed{\text{Interest at 7\%}} + \boxed{\text{Interest at 10\%}} = \boxed{\text{Total interest}}$$
$$\begin{aligned} 0.07x + 0.10(2x) &= 4050 \\ 0.07x + 0.2x &= 4050 \\ 0.27x &= 4050 \\ x &= \frac{4050}{0.27} \\ x &= 15000 \end{aligned}$$

He invested \$15,000 at 7% and \$30,000 at 10%. The total inheritance was \$45,000.

13. Let $x =$ the amount she has to invest. She needs $x + 3000$ to earn 11% interest.

$$\boxed{\text{Interest at 11\%}} = 2 \cdot \boxed{\text{Interest at 7.5\%}}$$
$$\begin{aligned} 0.11(x + 3000) &= 2(0.075)(x) \\ 0.11x + 330 &= 0.15x \\ -0.04x &= -330 \\ x &= \frac{-330}{-0.04} \\ x &= 8250 \end{aligned}$$

She has \$8,250 on hand to invest.

14. Let $r =$ the needed interest rate for the remainder of the money.

$$\boxed{\text{Interest at 8\%}} + \boxed{\text{Interest at } r\%} = \boxed{\text{Total interest}}$$
$$\begin{aligned} 0.08(10000) + r(30000) &= 3500 \\ 800 + 30000r &= 3500 \\ 30000r &= 2700 \\ r &= \frac{2700}{30000} \\ r &= 0.09 \end{aligned}$$

He needs to earn 9% interest on the remainder of the money.

15. Let $s =$ the # of student tickets sold. Then $200 - s =$ the # of adult tickets sold.

$\boxed{\text{Student } \$} + \boxed{\text{Adult } \$} = \boxed{\text{Total } \$}$

$$2s + 4(200 - s) = 750$$
$$2s + 800 - 4s = 750$$
$$-2s + 800 = 750$$
$$-2s = -50$$
$$s = \frac{-50}{-2}$$
$$r = 25$$

25 student tickets were sold.

16. Let $a =$ the # of adult tickets sold. Then $140 - a =$ the # of student tickets sold.

$\boxed{\text{Adult } \$} + \boxed{\text{Student } \$} = \boxed{\text{Total } \$}$

$$2.5a + 1.5(140 - a) = 290$$
$$2.5a + 210 - 1.5a = 290$$
$$a + 210 = 290$$
$$a = 80$$

80 adult tickets were sold.

17. Let $t =$ time for the cars to meet.

	Rate	Time	Dist.
First car	50	t	$50t$
Second car	48	t	$48t$

$\boxed{\text{1st dist.}} + \boxed{\text{2nd dist.}} = \boxed{\text{Total dist.}}$

$$50t + 48t = 343$$
$$98t = 343$$
$$t = \tfrac{343}{98}$$
$$t = \tfrac{7}{2} = 3\tfrac{1}{2}$$

It will take the cars $3\frac{1}{2}$ hours to meet.

18. Let $t =$ time to lose contact.

	Rate	Time	Dist.
First team	20	t	$20t$
Second team	25	t	$25t$

$\boxed{\text{1st dist.}} + \boxed{\text{2nd dist.}} = \boxed{\text{Distance apart}}$

$$20t + 25t = 90$$
$$45t = 90$$
$$t = \tfrac{90}{45} = 2$$

They will lose contact after 2 hours.

19. Let $t =$ time car travels.
Then $t + 1 =$ time cyclist travels.

	Rate	Time	Dist.
Cyclist	18	$t+1$	$18(t+1)$
Car	45	t	$45t$

$\boxed{\text{Cyclist distance}} = \boxed{\text{Car distance}}$

$$18(t + 1) = 45t$$
$$18t + 18 = 45t$$
$$-27t = -18$$
$$t = \tfrac{-18}{-27}$$
$$t = \tfrac{2}{3}$$

It will take $\frac{2}{3}$ of an hour to pass the cyclist.

20. Let $t =$ time for the cars to travel.

	Rate	Time	Dist.
North car	50	t	$50t$
South car	60	t	$60t$

$\boxed{\text{North dist.}} + \boxed{\text{South dist.}} = \boxed{\text{Dist. between cars}}$

$$50t + 60t = 165$$
$$110t = 165$$
$$t = \tfrac{165}{110}$$
$$t = \tfrac{3}{2} = 1\tfrac{1}{2}$$

It will take the cars $1\frac{1}{2}$ hours to be 165 miles apart.

21. Let t = time for them to run.

	Rate	Time	Dist.
1st Runner	12	t	$12t$
2nd Runner	10	t	$10t$

[1st dist.] − [2nd dist.] = [Dist. between them]

$$12t - 10t = 0.25$$
$$2t = \tfrac{1}{4}$$
$$t = \tfrac{1}{8}$$

It will take $\frac{1}{8}$ hour to be $\frac{1}{4}$ mile apart.

22. Let t = time 1st truck travels.
Then $t - 2$ = time 2nd truck travels.

	Rate	Time	Dist.
1st truck	50	t	$50t$
2nd truck	56	$t-2$	$56(t-2)$

[1st truck distance] + [2nd truck distance] = [Distance between trucks]

$$50t + 56(t-2) = 683$$
$$50t + 56t - 112 = 683$$
$$106t = 795$$
$$t = \tfrac{795}{106} = \tfrac{15}{2} = 7\tfrac{1}{2}$$

It will take $7\frac{1}{2}$ hours to be 683 miles apart.

23. Let t = time to return.

	Rate	Time	Dist.
Upstream	8	3	24
Downstream	16	t	$16t$

[Downstream distance] = [Upstream distance]

$$16t = 24$$
$$t = \tfrac{24}{16} = \tfrac{3}{2} = 1\tfrac{1}{2}$$

It will take the rider $1\frac{1}{2}$ hours to return.

24. Let t = her time walking north.

	Rate	Time	Dist.
North	3	t	$3t$
Return	4	$3.5-t$	$4(3.5-t)$

[North distance] = [Return distance]

$$3t = 4(3.5-t)$$
$$3t = 14 - 4t$$
$$7t = 14$$
$$t = 2$$

[Total distance] $= 3t + 4(3.5-t)$
$= 3(2) + 4(3.5-2)$
$= 12$

She walked a total of 12 miles.

25. Let t = the time at the slower rate.

	Rate	Time	Dist.
Slower	45	t	$45t$
Faster	55	$8-t$	$55(8-t)$

[Slower dist.] + [Faster dist.] = [Total dist.]

$$45t + 55(8-t) = 400$$
$$45t + 440 - 55t = 400$$
$$-10t = -40$$
$$t = \frac{-40}{-10} = 4$$

He traveled 4 hours at 45 mph, and 4 hours at 55 mph.

26. Let r = the speed of the current.

	Rate	Time	Dist.
Downstream	$18+r$	4	$4(18+r)$
Upstream	$18-r$	5	$5(18-r)$

[Downstream distance] = [Upstream distance]

$$4(18+r) = 5(18-r)$$
$$72 + 4r = 90 - 5r$$
$$9r = 18$$
$$r = \frac{18}{9} = 2$$

The current has a speed of 2 mph.

27. Let $x =$ pounds of cheaper candy. Then $30 - x =$ pounds of other candy.

	Price per pound	# pounds	Value
Cheaper	2.95	x	$2.95x$
Other	3.10	$30 - x$	$3.10(30 - x)$
Mixture	3.00	30	$3.00(30)$

Value of cheaper + Value of other = Value of mixture

$$2.95x + 3.10(30 - x) = 3.00(30)$$
$$2.95x + 93 - 3.1x = 90$$
$$-0.15x = -3$$
$$x = \tfrac{-3}{-0.15} = 20 \Rightarrow 30 - x = 30 - 20 = 10$$

The owner should use 20 pounds of the cheaper candy and 10 pounds of the other candy.

28. Let $x =$ price per pound of more expensive candy.

	Price per pound	# pounds	Value
Cheaper	3.80	32	$3.80(32) = 121.6$
Other	x	12	$12x$
Mixture	3.89	44	$3.89(44) = 171.16$

Value of cheaper + Value of other = Value of mixture

$$121.6 + 12x = 171.16$$
$$12x = 49.56$$
$$x = \tfrac{49.56}{12} = 4.13$$

The more expensive candy costs $4.13 per pound.

29. Let $x =$ ounces of water added (0% alcohol).

Alcohol at start + Alcohol added = Alcohol at end

$$0.15(20) + 0(x) = 0.10(20 + x)$$
$$3 + 0 = 2 + 0.10x$$
$$1 = 0.10x$$
$$x = \frac{1}{0.10} = 10$$

10 ounces of water should be added.

30. Let $x =$ gallons of water boiled away.

Salt at start − Salt boiled away = Salt at end

$$0.02(300) - 0(x) = 0.03(300 - x)$$
$$6 - 0 = 9 - 0.03x$$
$$-3 = -0.03x$$
$$x = \frac{-3}{0.03} = 100$$

100 gallons of water should be boiled away.

31. Let $x =$ gallons of cream used.
Then $20 - x =$ gallons of 2% milk used.

Butterfat in cream + Butterfat in 2% milk = Butterfat in 4% milk

$$0.22x + 0.02(20 - x) = 0.04(20)$$
$$0.22x + 0.4 - 0.02x = 0.8$$
$$0.2x = 0.4$$
$$x = \frac{0.4}{0.2} = 2$$

2 gallons of cream should be used.

32. Let $x =$ grams of acid added (100% acid).

Acid in 65% sol. + Acid added = Acid in 75% sol.

$$0.65(60) + 1.00x = 0.75(60 + x)$$
$$39 + x = 45 + 0.75x$$
$$0.25x = 6$$
$$x = \frac{6}{0.25} = 24$$

24 grams of acid should be added.

33. Let $x =$ the number of additional problems the student gets correct.

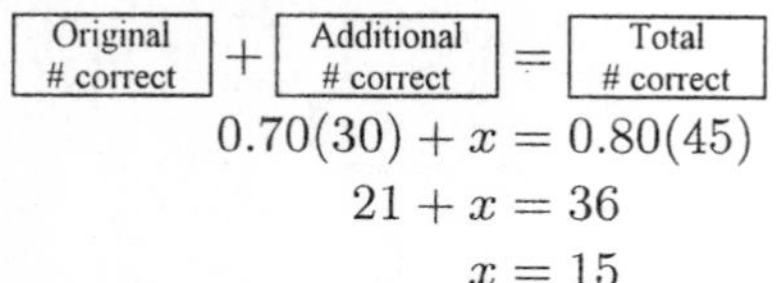

$$0.70(30) + x = 0.80(45)$$
$$21 + x = 36$$
$$x = 15$$

The student must get all 15 of the additional problems correct to receive a grade of 80%.

34. Let $x =$ the number of additional problems the student gets correct.

[Original # correct] + [Additional # correct] = [Total # correct]

$$0.60(20) + x = 0.70(40)$$
$$12 + x = 28$$
$$x = 16$$

The student must get 16 of the additional problems correct to receive a grade of 70%.

35. Let $x =$ number of points earned on the final.

[Original points] + [Points on final] = [Points at end]

$$375 + x = 0.90(450)$$
$$375 + x = 405$$
$$x = 30$$

She must earn 30 points on the final to receive a grade of 90% (A).

36. Let $x =$ number of points earned on the final.

[Original points] + [Points on final] = [Points at end]

$$435 + x = 0.80(600)$$
$$435 + x = 480$$
$$x = 45$$

He must earn 45 points on the final to receive a grade of 80% (B).

37. Let $w =$ the wholesale price of the textbook.

[Wholesale price] + [Markup] = [Retail price]

$$w + 0.30w = 65$$
$$1.3w = 65$$
$$w = \frac{65}{1.3} = 50$$

The bookstore pays \$50 for the book.

38. Let $w =$ the wholesale price of the textbook.

[Wholesale price] + [Markup] = [Retail price]

$$w + 0.40w = 39.20$$
$$1.4w = 39.20$$
$$w = \frac{39.20}{1.4} = 28$$

The bookstore pays \$28 for the book.

39. Let $x =$ space in front of first partition. Then $x + 3$ and $x + 6$ are the other spaces.

[Lengths of spaces] + [Lengths of partitions] = [Total length]

$$x + x + 3 + x + 6 + 2(0.5) = 28$$
$$3x + 10 = 28$$
$$3x = 18$$
$$x = 6$$

The first space should be 6 inches.

40. Let $x =$ first space from top. Then $x + 6$, $x + 12$, $x + 18$ and $x + 24$ are the others.

[Lengths of spaces] + [Lengths of partitions] = [Total length]

$$x + x + 6 + x + 12 + x + 18 + x + 24 + 4(0.75) = 8(12)$$
$$5x + 63 = 96$$
$$5x = 33$$
$$x = \frac{33}{5} = 6\frac{3}{5}$$

The bottom shelf should be $x + 24 = 6\frac{3}{5} + 24 = 30\frac{3}{5}$ inches from the floor.

41. **Answers may vary.**

42. **Answers may vary.**

43. We need to know how far or how long he drove while at the other rate in order to solve the problem.

44. This problem can be solved like **#16** in the exercises from this section. However, the solution to the equation is not a whole number. Since you cannot have a fractional number of tickets sold, there is no solution to the problem. Another way to see this is to notice that the ticket prices are even numbers, yet the total amount sold is reported to be an odd number ($245). You cannot add together even numbers (the ticket prices) to get an odd total (the total dollar amount sold).

Chapter 1 Summary (page 78)

1. whole numbers: $0, 1, 2, 4$

2. natural numbers: $1, 2, 4$

3. rational numbers: $-4, -\frac{2}{3}, 0, 1, 2, 4$

4. integers: $-4, 0, 1, 2, 4$

5. irrational numbers: π

6. real numbers: $-4, -\frac{2}{3}, 0, 1, 2, \pi, 4$

7. negative numbers: $-4, -\frac{2}{3}$

8. positive numbers: $1, 2, \pi, 4$

9. prime numbers: 2

10. composite numbers: 4

11. even integers: $-4, 0, 2, 4$

12. odd integers: 1

13.

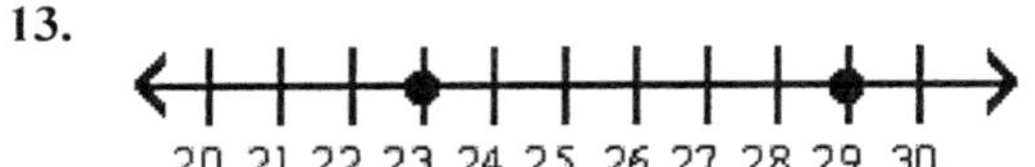

14.

15. $\{x|x \geq -4\} \Rightarrow$ -4

16. $\{x|-2 < x \leq 6\} \Rightarrow$

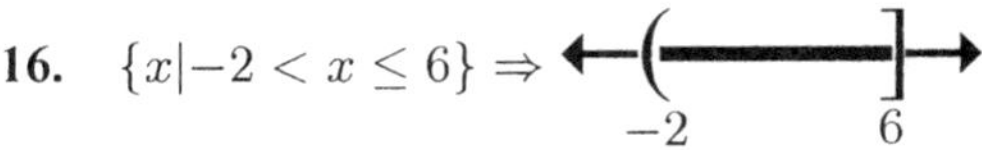

17. $(-2, 3) \Rightarrow$ -2 3

18. $[2, 6] \Rightarrow$ 2 6

19. $\{x|x > 2\} \Rightarrow$ 2

20. $(-\infty, -1) \Rightarrow$ -1

21. $(-\infty, 0] \cup (2, \infty) \Rightarrow$ 0 2

22. $|0| = 0$

23. $|-1| = 1$

24. $|8| = 8$

25. $-|8| = -(8) = -8$

26. $3 + (+5) = +8$

27. $-6 + (-3) = -(6 + 3) = -9$

28. $-15 + (-13) = -(15 + 13) = -28$

29. $25 + 32 = +(25 + 32) = +57$

30. $-2+5=+(5-2)=3$

31. $3+(-12)=-(12-3)=-9$

32. $8+(-3)=+(8-3)=+5$

33. $7+(-9)=-(9-7)=-2$

34. $-25+12=-(25-12)=-13$

35. $-30+35=+(35-30)=+5$

36. $-3-10=-3+(-10)=-13$

37. $-8-(-3)=-8+3=-5$

38. $27-(-12)=27+12=39$

39. $38-(-15)=38+15=53$

40. $(+5)(+7)=+35$

41. $(-6)(-7)=+42$

42. $\frac{-16}{-4}=4$

43. $\frac{-25}{-5}=5$

44. $4(-3)=-12$

45. $-3(8)=-24$

46. $\frac{-8}{2}=-4$

47. $\frac{8}{-4}=-2$

48. $-4(3-6)=-4(-3)=12$

49. $3[8-(-1)]=3(9)=27$

50. $-[4-2(6-4)]=-[4-2(2)]=-[4-4]=-[0]=0$

51. $3[-5+3(2-7)]=3[-5+3(-5)]=3[-5+(-15)]=3[-20]=-60$

52. $\frac{3-8}{10-5}=\frac{-5}{5}=-1$

53. $\frac{-32-8}{6-16}=\frac{-40}{-10}=4$

54. $\text{mean}=\frac{12+13+14+14+15+15+15+17+19+20}{10}=\frac{154}{10}=15.4$

55. $\text{median}=\frac{15+15}{2}=\frac{30}{2}=15$

56. $\text{mode}=15$

57. Yes. The mean, median and mode **could** be the same value for a group of numbers.

58. $\frac{3a-2b}{cd}=\frac{3(5)-2(-2)}{(-3)(2)}=\frac{15-(-4)}{-6}=\frac{19}{-6}=-\frac{19}{6}$

59. $\frac{3b+2d}{ac}=\frac{3(-2)+2(2)}{5(-3)}=\frac{-6+4}{-15}=\frac{-2}{-15}=\frac{2}{15}$

60. $\frac{ab+cd}{c(b-d)}=\frac{5(-2)+(-3)(2)}{-3(-2-2)}=\frac{-10+(-6)}{-3(-4)}=\frac{-16}{12}=-\frac{4}{3}$

61. $\frac{ac-bd}{a(d+c)}=\frac{5(-3)-(-2)(2)}{5[2+(-3)]}=\frac{-15-(-4)}{5(-1)}=\frac{-11}{-5}=\frac{11}{5}$

62. distributive property

63. commutative property of addition

64. associative property of addition

65. additive identity property

66. additive inverse property

67. commutative property of multiplication

68. associative property of multiplication

69. multiplicative identity property

70. multiplicative inverse property

71. double negative rule

72. $3^6 = 3 \cdot 3 \cdot 3 \cdot 3 \cdot 3 \cdot 3 = 729$

73. $-2^6 = -1 \cdot 2^6 = -1 \cdot 2 \cdot 2 \cdot 2 \cdot 2 \cdot 2 \cdot 2 = -1 \cdot 64 = -64$

74. $(-4)^3 = (-4)(-4)(-4) = -64$

75. $-5^{-4} = -1 \cdot 5^{-4} = -1 \cdot \frac{1}{5^4} = -\frac{1}{625}$

76. $(3x^4)(-2x^2) = 3(-2)x^4x^2 = -6x^{4+2} = -6x^6$

77. $(-x^5)(3x^3) = -1(3)x^5x^3 = -3x^{5+3} = -3x^8$

78. $x^{-4}x^3 = x^{-4+3} = x^{-1} = \frac{1}{x}$

79. $x^{-10}x^{12} = x^{-10+12} = x^2$

80. $(3x^2)^3 = 3^3(x^2)^3 = 27x^6$

81. $(4x^4)^4 = 4^4(x^4)^4 = 256x^{16}$

82. $(-2x^2)^5 = (-2)^5(x^2)^5 = -32x^{10}$

83. $-(-3x^3)^5 = -(-3)^5(x^3)^5 = -(-243)x^{15} = 243x^{15}$

84. $(x^2)^{-5} = x^{-10} = \frac{1}{x^{10}}$

85. $(x^{-4})^{-5} = x^{20}$

86. $(3x^{-3})^{-2} = 3^{-2}x^6 = \frac{x^6}{3^2} = \frac{x^6}{9}$

87. $(2x^{-4})^4 = 2^4x^{-16} = \frac{16}{x^{16}}$

88. $\frac{x^6}{x^4} = x^{6-4} = x^2$

89. $\frac{x^{12}}{x^7} = x^{12-7} = x^5$

90. $\frac{a^7}{a^{12}} = a^{7-12} = a^{-5} = \frac{1}{a^5}$

91. $\frac{a^4}{a^7} = a^{4-7} = a^{-3} = \frac{1}{a^3}$

92. $\frac{y^{-3}}{y^4} = y^{-3-4} = y^{-7} = \frac{1}{y^7}$

93. $\frac{y^5}{y^{-4}} = y^{5-(-4)} = y^9$

94. $\frac{x^{-5}}{x^{-4}} = x^{-5-(-4)} = x^{-1} = \frac{1}{x}$

95. $\frac{x^{-6}}{x^{-9}} = x^{-6-(-9)} = x^3$

96. $(3x^2y^3)^2 = 3^2(x^2)^2(y^3)^2 = 9x^4y^6$

97. $(-3a^3b^2)^{-4} = \frac{1}{(-3a^3b^2)^4} = \frac{1}{(-3)^4(a^3)^4(b^2)^4} = \frac{1}{81a^{12}b^8}$

98. $\left(\frac{3x^2}{4y^3}\right)^{-3} = \left(\frac{4y^3}{3x^2}\right)^3 = \frac{(4y^3)^3}{(3x^2)^3} = \frac{4^3(y^3)^3}{3^3(x^2)^3} = \frac{64y^9}{27x^6}$

99. $\left(\frac{4y^{-2}}{5y^{-3}}\right)^3 = \frac{4^3(y^{-2})^3}{5^3(y^{-3})^3} = \frac{64y^{-6}}{125y^{-9}} = \frac{64y^3}{125}$

100. $19{,}300{,}000{,}000 = 1.93 \times 10^{10}$

101. $0.0000000273 = 2.73 \times 10^{-8}$

102. $7.2 \times 10^7 = 72{,}000{,}000$

103. $8.3 \times 10^{-9} = 0.0000000083$

104. $\frac{270{,}000{,}000 \text{ persons}}{1} \cdot \frac{1640 \text{ gallons}}{1 \text{ person}} = \left(2.7 \times 10^8\right)\left(1.64 \times 10^3\right) \text{ gallons} = 4.428 \times 10^{11} \text{ gallons}$

105.
$$\begin{aligned} 5x + 12 &= 37 \\ 5x + 12 - 12 &= 37 - 12 \\ 5x &= 25 \\ \frac{5x}{5} &= \frac{25}{5} \\ x &= 5 \end{aligned}$$

106.
$$\begin{aligned} -3x - 7 &= 20 \\ -3x - 7 + 7 &= 20 + 7 \\ -3x &= 27 \\ \frac{-3x}{-3} &= \frac{27}{-3} \\ x &= -9 \end{aligned}$$

107.
$$\begin{aligned} 4(y - 1) &= 28 \\ 4y - 4 &= 28 \\ 4y - 4 + 4 &= 28 + 4 \\ 4y &= 32 \\ \frac{4y}{4} &= \frac{32}{4} \\ y &= 8 \end{aligned}$$

108.
$$\begin{aligned} 3(x + 7) &= 42 \\ 3x + 21 &= 42 \\ 3x + 21 - 21 &= 42 - 21 \\ 3x &= 21 \\ \frac{3x}{3} &= \frac{21}{3} \\ x &= 7 \end{aligned}$$

109.
$$\begin{aligned} 13(x - 9) - 2 &= 7x - 5 \\ 13x - 117 - 2 &= 7x - 5 \\ 13x - 119 &= 7x - 5 \\ 13x - 7x - 119 &= 7x - 7x - 5 \\ 6x - 119 + 119 &= -5 + 119 \\ 6x &= 114 \\ \frac{6x}{6} &= \frac{114}{6} \\ x &= 19 \end{aligned}$$

110.
$$\begin{aligned} \frac{8(x - 5)}{3} &= 2(x - 4) \\ 3 \cdot \frac{8(x - 5)}{3} &= 3 \cdot 2(x - 4) \\ 8(x - 5) &= 6(x - 4) \\ 8x - 40 &= 6x - 24 \\ 8x - 6x - 40 &= 6x - 6x - 24 \\ 2x - 40 + 40 &= -24 + 40 \\ 2x &= 16 \\ \frac{2x}{2} &= \frac{16}{2} \\ x &= 8 \end{aligned}$$

111.
$$\begin{aligned}
\frac{3y}{4} - 13 &= -\frac{y}{3} \\
12\left(\frac{3y}{4} - 13\right) &= 12\left(-\frac{y}{3}\right) \\
12\left(\frac{3y}{4}\right) - 12(13) &= -4y \\
3(3y) - 156 &= -4y \\
9y - 156 &= -4y \\
9y + 4y - 156 &= -4y + 4y \\
13y - 156 + 156 &= 0 + 156 \\
13y &= 156 \\
\frac{13y}{13} &= \frac{156}{13} \\
y &= 12
\end{aligned}$$

112.
$$\begin{aligned}
\frac{2y}{5} + 5 &= \frac{14y}{10} \\
10\left(\frac{2y}{5} + 5\right) &= 10\left(\frac{14y}{10}\right) \\
10\left(\frac{2y}{5}\right) + 10(5) &= 14y \\
2(2y) + 50 &= 14y \\
4y + 50 &= 14y \\
4y - 14y + 50 &= 14y - 14y \\
-10y + 50 - 50 &= 0 - 50 \\
-10y &= -50 \\
\frac{-10y}{-10} &= \frac{-50}{-10} \\
y &= 5
\end{aligned}$$

113.
$$\begin{aligned}
0.07a &= 0.10 - 0.04(a-3) \\
0.07a &= 0.10 - 0.04a + 0.12 \\
0.07a &= 0.22 - 0.04a \\
0.07a + 0.04a &= 0.22 \\
0.11a &= 0.22 \\
\frac{0.11a}{0.11} &= \frac{0.22}{0.11} \\
a &= 2
\end{aligned}$$

114.
$$\begin{aligned}
0.12x + 0.06(50000 - x) &= 4080 \\
0.12x + 3000 - 0.06x &= 4080 \\
0.06x + 3000 &= 4080 \\
0.06x + 3000 - 3000 &= 4080 - 3000 \\
0.06x &= 1080 \\
\frac{0.06x}{0.06} &= \frac{1080}{0.06} \\
x &= 18000
\end{aligned}$$

115.
$$\begin{aligned}
V &= \frac{4}{3}\pi r^3 \\
3V &= 3\left(\frac{4}{3}\pi r^3\right) \\
3V &= 4\pi r^3 \\
\frac{3V}{4\pi} &= \frac{4\pi r^3}{4\pi} \\
\frac{3V}{4\pi} &= r^3, \text{ or } r^3 = \frac{3V}{4\pi}
\end{aligned}$$

116.
$$\begin{aligned}
V &= \frac{1}{3}\pi r^2 h \\
3V &= 3\left(\frac{1}{3}\pi r^2 h\right) \\
3V &= \pi r^2 h \\
\frac{3V}{\pi r^2} &= \frac{\pi r^2 h}{\pi r^2} \\
\frac{3V}{\pi r^2} &= h, \text{ or } h = \frac{3V}{\pi r^2}
\end{aligned}$$

117.

$$v = \frac{1}{6}ab(x+y)$$
$$6v = 6 \cdot \frac{1}{6}ab(x+y)$$
$$6v = ab(x+y)$$
$$\frac{6v}{ab} = \frac{ab(x+y)}{ab}$$
$$\frac{6v}{ab} = x+y$$
$$\frac{6v}{ab} - y = x + y - y$$
$$\frac{6v}{ab} - y = x, \text{ or } x = \frac{6v}{ab} - y$$

118.

$$V = \pi h^2\left(r - \frac{h}{3}\right)$$
$$\frac{V}{\pi h^2} = \frac{\pi h^2(r - \frac{h}{3})}{\pi h^2}$$
$$\frac{V}{\pi h^2} = r - \frac{h}{3}$$
$$\frac{V}{\pi h^2} + \frac{h}{3} = r - \frac{h}{3} + \frac{h}{3}$$
$$\frac{V}{\pi h^2} + \frac{h}{3} = r, \text{ or } r = \frac{V}{\pi h^2} + \frac{h}{3}$$

119. Let x = the length of the first piece.
Then $3x$ = the length of the other piece.

[Sum of lengths] = [Total length]

$$x + 3x = 20$$
$$4x = 20$$
$$x = 5$$

He should cut the board 5 feet from one end.

120. Let w = the width of the rectangle. Then $w + 4$ = the length of the rectangle.

[Perimeter] = 28

$$w + w + w + 4 + w + 4 = 28$$
$$4w + 8 = 28$$
$$4w = 20$$
$$w = 5$$

The dimensions are 5 meters by 9 meters, for an area of 45 m^2.

121.

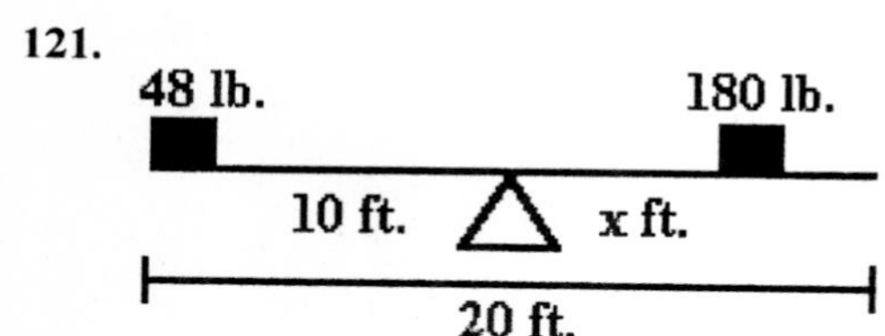

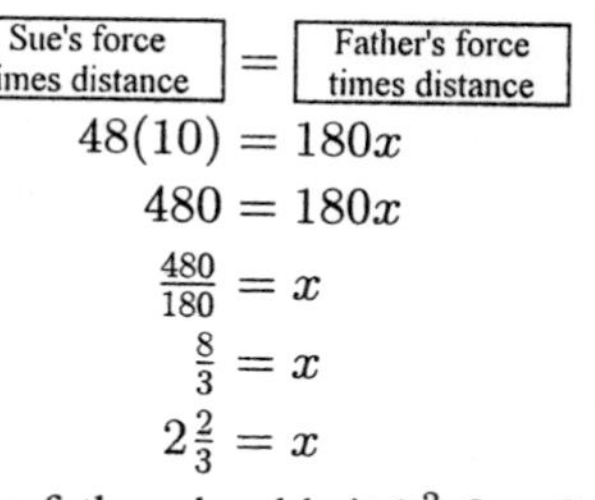

[Sue's force times distance] = [Father's force times distance]

$$48(10) = 180x$$
$$480 = 180x$$
$$\frac{480}{180} = x$$
$$\frac{8}{3} = x$$
$$2\tfrac{2}{3} = x$$

The father should sit $2\frac{2}{3}$ feet from the fulcrum.

122. Let x = the amount she invests at 10%.
Then $25{,}000 - x$ = amount invested at 9%.

[Interest at 10%] + [Interest at 9%] = [Total interest]

$$0.10x + 0.09(25000 - x) = 2430$$
$$0.10x + 2250 - 0.09x = 2430$$
$$0.01x = 180$$
$$x = 18000$$

She invested \$18,000 at 10% interest and \$7000 at 9% interest.

123. Let x = liters of water added.

[Alcohol at start] + [Alcohol added] = [Alcohol at end]

$$0.12(20) + 0(x) = 0.08(20 + x)$$
$$2.4 = 1.6 + 0.08x$$
$$0.8 = 0.08x$$
$$10 = x$$

10 liters of water should be added.

124. Let $t =$ time for them to travel.

	Rate	Time	Dist.
Car	55	t	$55t$
Motorcycle	40	t	$40t$

Car distance − Motorcycle distance = Distance between them

$$55t - 40t = 5$$
$$15t = 5$$
$$t = \frac{5}{15} = \frac{1}{3}$$

It will take $\frac{1}{3}$ of an hour for the vehicles to be 5 miles apart.

Chapter 1 Test (page 82)

1. 1, 2 and 5 are natural numbers.

2. $\sqrt{7}$ is an irrational number.

3.

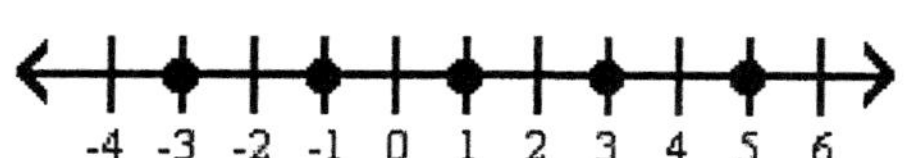

4.

1 2 3 4 5 6 7 8 9 10 11 12

5. $\{x|x > 4\} \Rightarrow$ 4

6. $[-3, \infty) \Rightarrow$ -3

7. $\{x|-2 \le x < 4\} \Rightarrow$ -2 4

8. $(-\infty, -1] \cup [2, \infty) \Rightarrow$ -1 2

9. $-|8| = -8$

10. $|-5| = 5$

11. $7 + (-5) = 2$

12. $-5(-4) = 20$

13. $\frac{12}{-3} = -4$

14. $-4 - \frac{-15}{3} = -4 - (-5) = -4 + 5 = 1$

15. mean $= \frac{-2 + 0 + 2 + (-2) + 3 + (-1) + (-1) + 1 + 1 + 2}{10} = \frac{3}{10} = 0.3$

16. $-2, -2, -1, -1, 0, 1, 1, 2, 2, 3$: median $= \frac{0 + 1}{2} = \frac{1}{2} = 0.5$

17. $ab = (2)(-3) = -6$

18. $a + bc = 2 + (-3)(4) = 2 + (-12) = -10$

19. $ab - bc = 2(-3) - (-3)(4) = -6 - (-12)$
$= -6 + 12 = 6$

20. $\frac{-3b + a}{ac - b} = \frac{-3(-3) + 2}{2(4) - (-3)} = \frac{9 + 2}{8 + 3} = \frac{11}{11}$
$= 1$

21. commutative property of addition

22. distributive property

23. $x^3 x^5 = x^{3+5} = x^8$

24. $(2x^2y^3)^3 = 2^3(x^2)^3(y^3)^3 = 8x^6y^9$

25. $\left(m^{-4}\right)^2 = m^{-8} = \dfrac{1}{m^8}$

26. $\left(\dfrac{m^2n^3}{m^4n^{-2}}\right)^{-2} = \left(\dfrac{m^4n^{-2}}{m^2n^3}\right)^2 = \left(\dfrac{m^2}{n^5}\right)^2 = \dfrac{m^4}{n^{10}}$

27. $4{,}700{,}000 = 4.7 \times 10^6$

28. $0.00000023 = 2.3 \times 10^{-7}$

29. $6.53 \times 10^5 = 653{,}000$

30. $24.5 \times 10^{-3} = 0.0245$

31.
$$\begin{aligned}
9(x+4)+4 &= 4(x-5)\\
9x+36+4 &= 4x-20\\
9x+40 &= 4x-20\\
9x-4x+40 &= 4x-4x-20\\
5x+40 &= -20\\
5x+40-40 &= -20-40\\
5x &= -60\\
\tfrac{5x}{5} &= \tfrac{-60}{5}\\
x &= -12
\end{aligned}$$

32.
$$\begin{aligned}
\frac{2y+3}{3}+\frac{3y-4}{6} &= \frac{y-2}{2}\\
6\left(\frac{2y+3}{3}+\frac{3y-4}{6}\right) &= 6\left(\frac{y-2}{2}\right)\\
6\left(\frac{2y+3}{3}\right)+6\left(\frac{3y-4}{6}\right) &= 3(y-2)\\
2(2y+3)+3y-4 &= 3y-6\\
4y+6+3y-4 &= 3y-6\\
7y+2 &= 3y-6\\
7y-3y+2 &= 3y-3y-6\\
4y+2 &= -6\\
4y+2-2 &= -6-2\\
4y &= -8\\
\frac{4y}{4} &= \frac{-8}{4}\\
y &= -2
\end{aligned}$$

33.
$$\begin{aligned}
\frac{y-1}{5}+2 &= \frac{2y-3}{3}\\
15\left(\frac{y-1}{5}+2\right) &= 15\left(\frac{2y-3}{3}\right)\\
15\left(\frac{y-1}{5}\right)+15(2) &= 5(2y-3)\\
3(y-1)+30 &= 10y-15\\
3y-3+30 &= 10y-15\\
3y+27 &= 10y-15\\
3y-10y+27 &= 10y-10y-15\\
-7y+27-27 &= -15-27\\
-7y &= -42\\
\tfrac{-7y}{-7} &= \tfrac{-42}{-7}\\
y &= 6
\end{aligned}$$

34.
$$\begin{aligned}
400+1.5x &= 500+1.25x\\
400+1.5x-1.25x &= 500+1.25x-1.25x\\
400+0.25x &= 500\\
400-400+0.25x &= 500-400\\
0.25x &= 100\\
\frac{0.25x}{0.25} &= \frac{100}{0.25}\\
x &= 400
\end{aligned}$$

35.
$$\begin{aligned} p &= L + \frac{s}{f}i \\ p - L &= L - L + \frac{s}{f}i \\ p - L &= \frac{s}{f}i \\ f(p-L) &= f \cdot \frac{s}{f}i \\ f(p-L) &= si \\ \frac{f(p-L)}{s} &= \frac{si}{s} \\ \frac{f(p-L)}{s} &= i, \text{ or } i = \frac{f(p-L)}{s} \end{aligned}$$

36. Let $w =$ the width of the rectangle. Then $w + 5 =$ the length of the rectangle.

$$\begin{aligned} \boxed{\text{Perimeter}} &= 26 \\ w + w + w + 5 + w + 5 &= 26 \\ 4w + 10 &= 26 \\ 4w &= 16 \\ w &= 4 \end{aligned}$$

The dimensions are 4 cm by 9 cm, for an area of 36 cm^2.

37. Let $x =$ the amount invested at 9%. Then $10000 - x =$ the amount at 8%.

$$\boxed{\text{Interest at 9\%}} + \boxed{\text{Interest at 8\%}} = \boxed{\text{Total interest}}$$

$$\begin{aligned} 0.09x + 0.08(10000 - x) &= 860 \\ 0.09x + 800 - 0.08x &= 860 \\ 0.01x &= 60 \\ x &= 6000 \end{aligned}$$

He has \$6000 invested at 9% interest, and **\$4000 invested at 8% interest.**

38. Let $x =$ liters of water added.

$$\boxed{\text{Salt at start}} + \boxed{\text{Salt added}} = \boxed{\text{Salt at end}}$$

$$\begin{aligned} 0.05(20) + 0(x) &= 0.01(20 + x) \\ 1 + 0 &= 0.2 + 0.01x \\ 0.8 &= 0.01x \\ 80 &= x \end{aligned}$$

80 liters of water should be added.

Exercise 2.1 (page 97)

1. $-3 - 3(-5) = -3 + 15 = 12$

2. $(-5)^2 + (-5) = 25 + (-5) = 20$

3. $\frac{-3 + 5(2)}{9 + 5} = \frac{-3 + 10}{14} = \frac{7}{14} = \frac{1}{2}$

4. $|-1 - 9| = |-10| = 10$

5.
$$\begin{aligned} -4x + 7 &= -21 \\ -4x + 7 - 7 &= -21 - 7 \\ -4x &= -28 \\ \frac{-4x}{-4} &= \frac{-28}{-4} \\ x &= 7 \end{aligned}$$

6.
$$\begin{aligned} P &= 2l + 2w \\ P - 2l &= 2l - 2l + 2w \\ P - 2l &= 2w \\ \frac{P-2l}{2} &= \frac{2w}{2} \\ \frac{P-2l}{2} &= w, \text{ or } w = \frac{P-2l}{2} \end{aligned}$$

7. ordered pair

8. y-

9. origin

10. quadrants

11. rectangular coordinate

12. graphing or plotting

13. the y-intercept

14. x-axis

15. vertical

16. horizontal

17. sub 1

18. $\left(\frac{a+c}{2}, \frac{b+d}{2}\right)$

19-26.

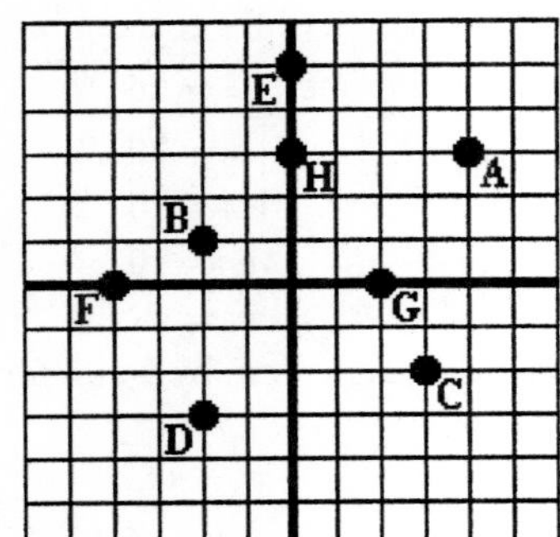

27. $(2,4)$

28. $(-5,5)$

29. $(-2,-1)$

30. $(4,-4)$

31. $(4,0)$

32. $(-5,-3)$

33. $(0,0)$

34. $(0,-4)$

35. $y=-x+4$
$y=-(-1)+4=1+4=5$
$y=-(0)+4=0+4=4$
$y=-2+4=2$

x	y
-1	5
0	4
2	2

36. $y=x-2$
$y=-2-2=-4$
$y=0-2=-2$
$y=4-2=2$

x	y
-2	-4
0	-2
4	2

37. $y=2x-3$
$y=2(-1)-3=-2-3=-5$
$y=2(0)-3=0-3=-3$
$y=2(3)-3=6-3=3$

x	y
-1	-5
0	-3
3	3

38. $y=-\frac{1}{2}x+\frac{5}{2}$
$y=-\frac{1}{2}(-3)+\frac{5}{2}=\frac{3}{2}+\frac{5}{2}=\frac{8}{2}=4$
$y=-\frac{1}{2}(-1)+\frac{5}{2}=\frac{1}{2}+\frac{5}{2}=\frac{6}{2}=3$
$y=-\frac{1}{2}(3)+\frac{5}{2}=-\frac{3}{2}+\frac{5}{2}=\frac{2}{2}=1$

x	y
-3	4
-1	3
3	1

39. From Exercise #35:

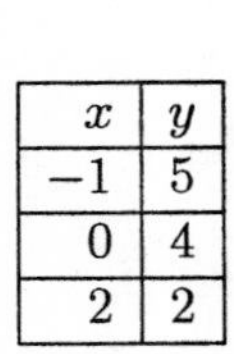

x	y
-1	5
0	4
2	2

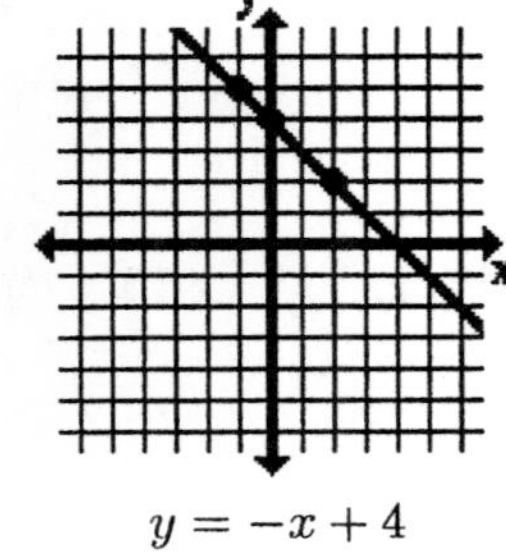

$y=-x+4$

40. From Exercise #36:

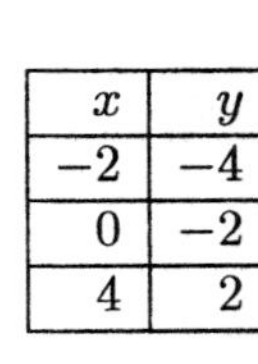

x	y
-2	-4
0	-2
4	2

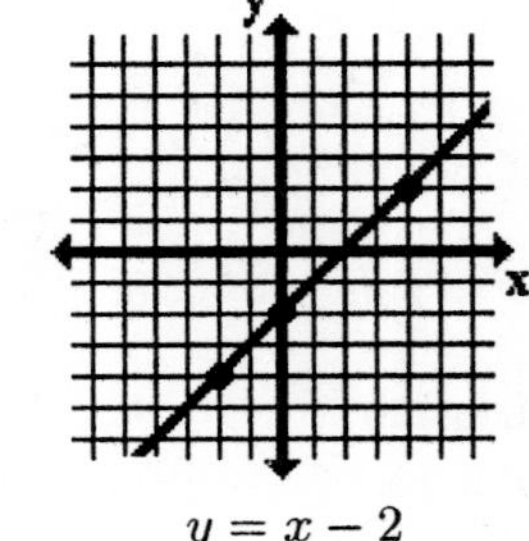

$y=x-2$

41. From Exercise #37:

x	y
-1	-5
0	-3
3	3

$y = 2x - 3$

42. From Exercise #38:

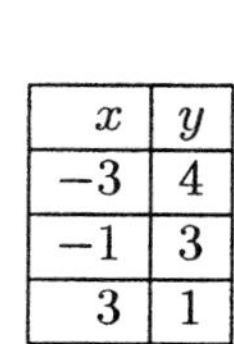

x	y
-3	4
-1	3
3	1

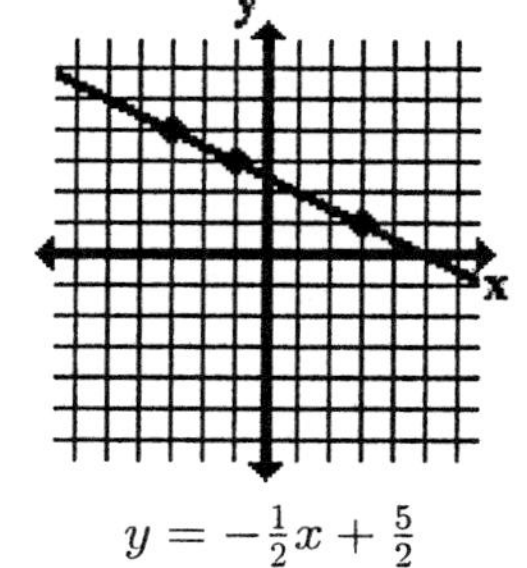

$y = -\frac{1}{2}x + \frac{5}{2}$

43.

$$\begin{aligned} 3x + 4y &= 12 \\ 3x + 4(0) &= 12 \\ 3x &= 12 \\ x &= 4 \end{aligned}$$

x-intercept: $(4, 0)$

$$\begin{aligned} 3x + 4y &= 12 \\ 3(0) + 4y &= 12 \\ 4y &= 12 \\ y &= 3 \end{aligned}$$

y-intercept: $(0, 3)$

$3x + 4y = 12$

44.

$$\begin{aligned} 4x - 3y &= 12 \\ 4x - 3(0) &= 12 \\ 4x &= 12 \\ x &= 3 \end{aligned}$$

x-intercept: $(3, 0)$

$$\begin{aligned} 4x - 3y &= 12 \\ 4(0) - 3y &= 12 \\ -3y &= 12 \\ y &= -4 \end{aligned}$$

y-intercept: $(0, -4)$

$4x - 3y = 12$

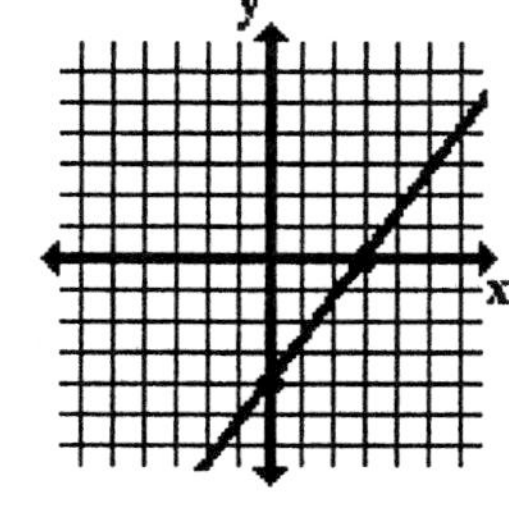

45.

$$\begin{aligned} y &= -3x + 2 \\ 0 &= -3x + 2 \\ 3x &= 2 \\ x &= \tfrac{2}{3} \end{aligned}$$

x-intercept: $\left(\frac{2}{3}, 0\right)$

$$\begin{aligned} y &= -3x + 2 \\ y &= -3(0) + 2 \\ y &= 2 \end{aligned}$$

y-intercept: $(0, 2)$

$y = -3x + 2$

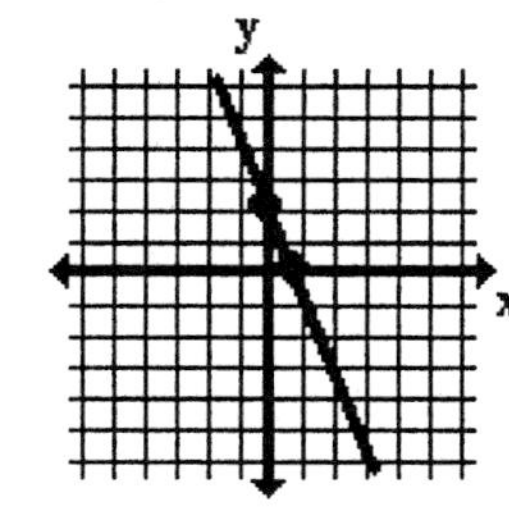

46.

$y = 2x + 3$
$0 = 2x + 3$
$-2x = 3$
$x = \frac{3}{-2} = -\frac{3}{2}$
x-intercept: $\left(-\frac{3}{2}, 0\right)$

$y = 2x + 3$
$y = 2(0) + 3$
$y = 3$
y-intercept: $(0, 3)$

$y = 2x + 3$

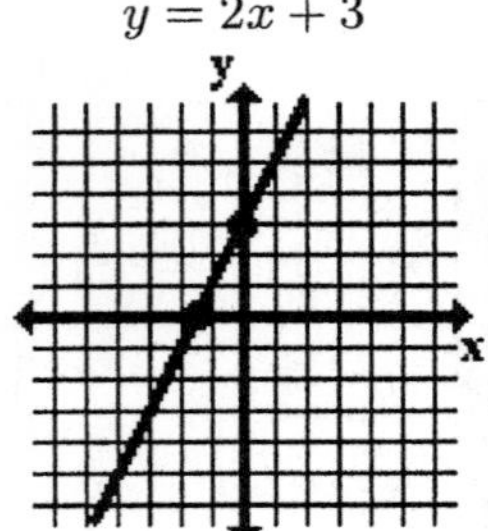

47.

$y = \frac{3}{2}x$
$0 = \frac{3}{2}x$
$2(0) = 2 \cdot \frac{3}{2}x$
$0 = 3x$
$x = \frac{0}{3} = 0$
x-intercept: $(0, 0)$

Pick a value for x such as $x = 4$.
$y = \frac{3}{2}x$
$y = \frac{3}{2}(4)$
$y = 6$
point: $(4, 6)$

$y = \frac{3}{2}x$

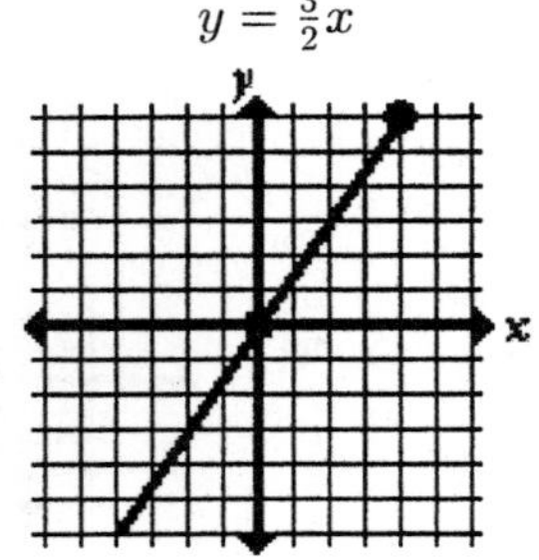

48.

$y = -\frac{2}{3}x$
$0 = -\frac{2}{3}x$
$3(0) = 3\left(-\frac{2}{3}x\right)$
$0 = -2x$
$x = \frac{0}{-2} = 0$
x-intercept: $(0, 0)$

Pick a value for x such as $x = 6$.
$y = -\frac{2}{3}x$
$y = -\frac{2}{3}(6)$
$y = -4$
point: $(6, -4)$

$y = -\frac{2}{3}x$

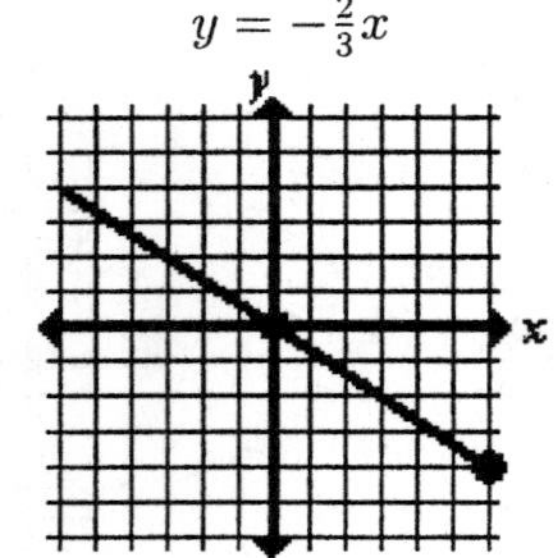

49.

$3y = 6x - 9$
$3(0) = 6x - 9$
$0 = 6x - 9$
$-6x = -9$
$x = \frac{-9}{-6} = \frac{3}{2}$
x-intercept: $\left(\frac{3}{2}, 0\right)$

$3y = 6x - 9$
$3y = 6(0) - 9$
$3y = 0 - 9$
$3y = -9$
$y = \frac{-9}{3} = -3$
y-intercept: $(0, -3)$

$3y = 6x - 9$

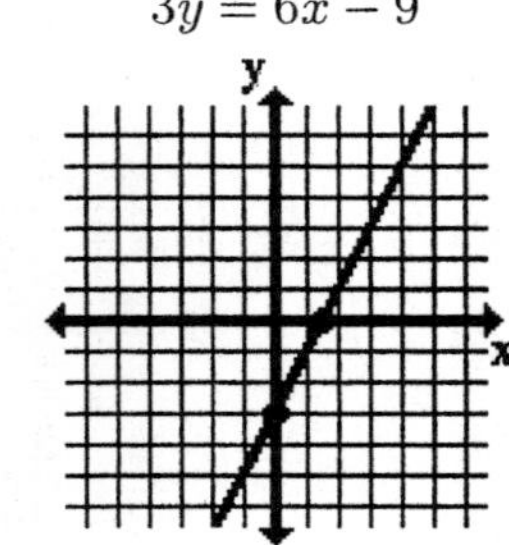

50.

$$\begin{aligned} 2x &= 4y - 10 \\ 2x &= 4(0) - 10 \\ 2x &= -10 \\ x &= -5 \end{aligned}$$

x-intercept: $(-5, 0)$

$$\begin{aligned} 2x &= 4y - 10 \\ 2(0) &= 4y - 10 \\ 0 &= 4y - 10 \\ -4y &= -10 \\ y &= \tfrac{-10}{-4} = \tfrac{5}{2} \end{aligned}$$

y-intercept: $\left(0, \frac{5}{2}\right)$

$2x = 4y - 10$

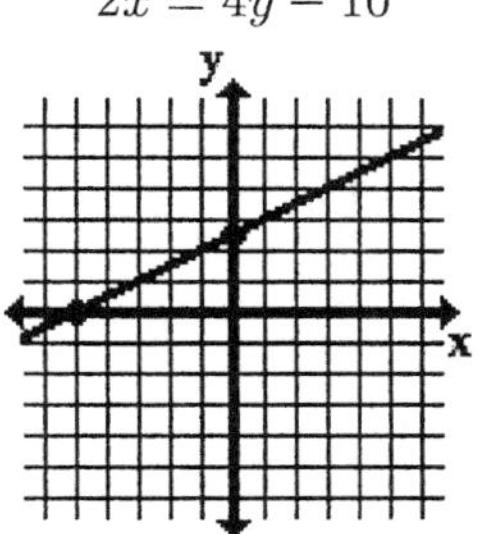

51.

$$\begin{aligned} 3x + 4y - 8 &= 0 \\ 3x + 4(0) - 8 &= 0 \\ 3x - 8 &= 0 \\ 3x &= 8 \\ x &= \tfrac{8}{3} \end{aligned}$$

x-intercept: $\left(\frac{8}{3}, 0\right)$

$$\begin{aligned} 3x + 4y - 8 &= 0 \\ 3(0) + 4y - 8 &= 0 \\ 4y - 8 &= 0 \\ 4y &= 8 \\ y &= 2 \end{aligned}$$

y-intercept: $(0, 2)$

$3x + 4y - 8 = 0$

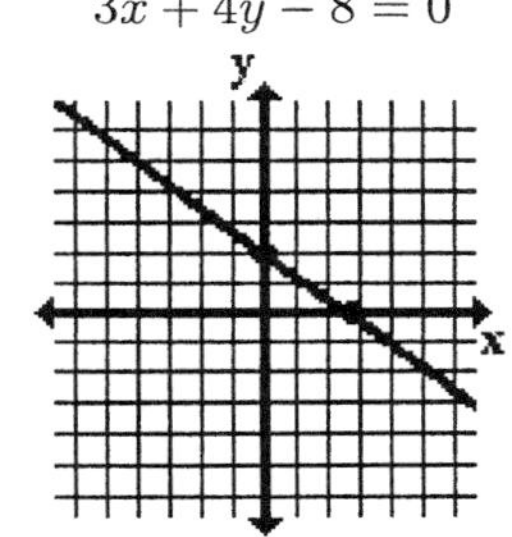

52.

$$\begin{aligned} -2y - 3x + 9 &= 0 \\ -2(0) - 3x + 9 &= 0 \\ -3x + 9 &= 0 \\ -3x &= -9 \\ x &= 3 \end{aligned}$$

x-intercept: $(3, 0)$

$$\begin{aligned} -2y - 3x + 9 &= 0 \\ -2y - 3(0) + 9 &= 0 \\ -2y + 9 &= 0 \\ -2y &= -9 \\ y &= \tfrac{-9}{-2} = \tfrac{9}{2} \end{aligned}$$

y-intercept: $\left(0, \frac{9}{2}\right)$

$-2y - 3x + 9 = 0$

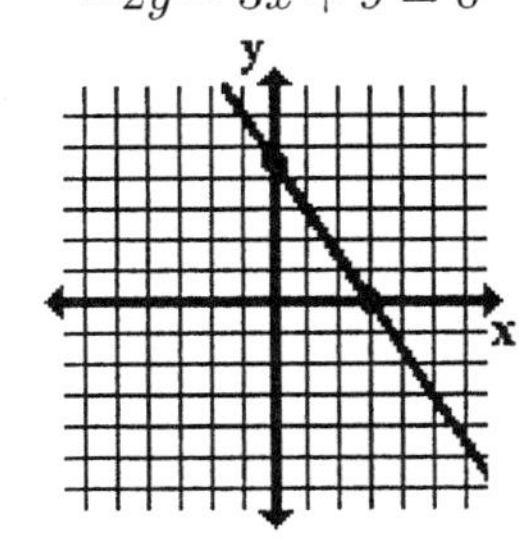

53. $x = 3$

vertical line with x-coordinate of 3

$x = 3$

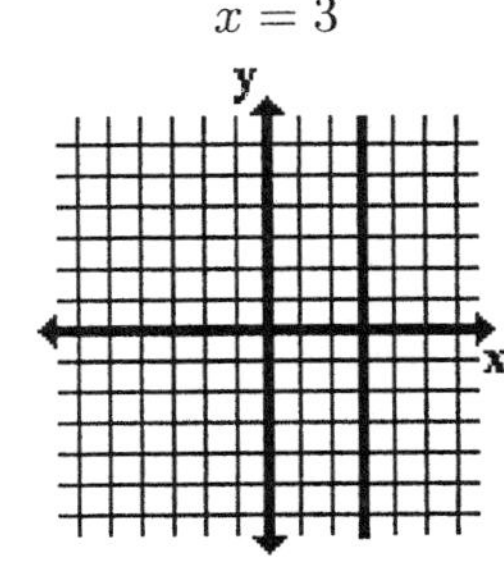

54. $y = -4$

horizontal line with y-coordinate of -4

$y = -4$

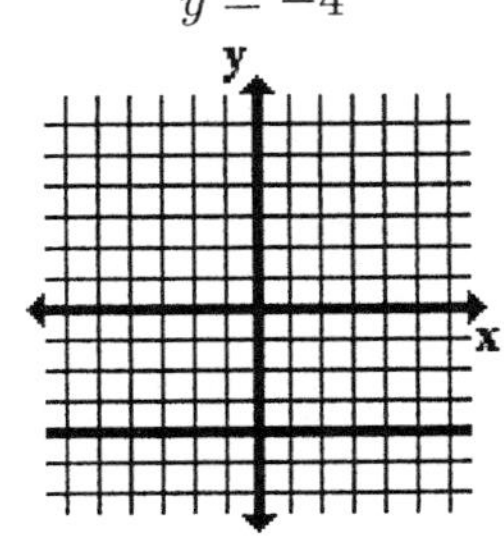

55. $-3y+2=5$
$-3y=3$
$y=-1$
horizontal line with y-coordinate of -1

$y=-1$

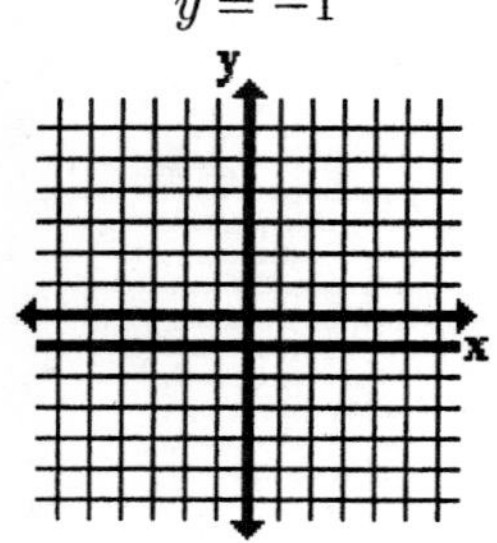

56. $-2x+3=11$
$-2x=8$
$x=-4$
vertical line with x-coordinate of -4

$x=-4$

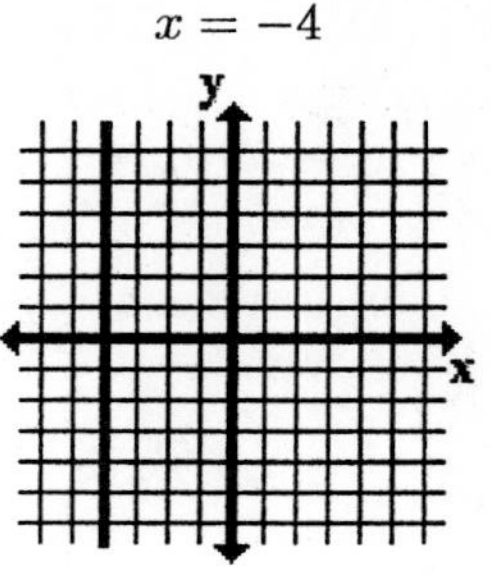

57. $x=\frac{x_1+x_2}{2}=\frac{0+6}{2}=\frac{6}{2}=3$
$y=\frac{y_1+y_2}{2}=\frac{0+8}{2}=\frac{8}{2}=4$
midpoint: $(3,4)$

58. $x=\frac{x_1+x_2}{2}=\frac{10+0}{2}=\frac{10}{2}=5$
$y=\frac{y_1+y_2}{2}=\frac{12+0}{2}=\frac{12}{2}=6$
midpoint: $(5,6)$

59. $x=\frac{x_1+x_2}{2}=\frac{6+12}{2}=\frac{18}{2}=9$
$y=\frac{y_1+y_2}{2}=\frac{8+16}{2}=\frac{24}{2}=12$
midpoint: $(9,12)$

60. $x=\frac{x_1+x_2}{2}=\frac{10+2}{2}=\frac{12}{2}=6$
$y=\frac{y_1+y_2}{2}=\frac{4+(-2)}{2}=\frac{2}{2}=1$
midpoint: $(6,1)$

61. $x=\frac{x_1+x_2}{2}=\frac{2+5}{2}=\frac{7}{2}$
$y=\frac{y_1+y_2}{2}=\frac{4+8}{2}=\frac{12}{2}=6$
midpoint: $\left(\frac{7}{2},6\right)$

62. $x=\frac{x_1+x_2}{2}=\frac{5+8}{2}=\frac{13}{2}$
$y=\frac{y_1+y_2}{2}=\frac{9+13}{2}=\frac{22}{2}=11$
midpoint: $\left(\frac{13}{2},11\right)$

63. $x=\frac{x_1+x_2}{2}=\frac{-2+3}{2}=\frac{1}{2}$
$y=\frac{y_1+y_2}{2}=\frac{-8+4}{2}=\frac{-4}{2}=-2$
midpoint: $\left(\frac{1}{2},-2\right)$

64. $x=\frac{x_1+x_2}{2}=\frac{-5+7}{2}=\frac{2}{2}=1$
$y=\frac{y_1+y_2}{2}=\frac{-2+3}{2}=\frac{1}{2}$
midpoint: $\left(1,\frac{1}{2}\right)$

65. $x=\frac{x_1+x_2}{2}=\frac{-3+(-5)}{2}=\frac{-8}{2}=-4$
$y=\frac{y_1+y_2}{2}=\frac{5+(-5)}{2}=\frac{0}{2}=0$
midpoint: $(-4,0)$

66. $x=\frac{x_1+x_2}{2}=\frac{2+4}{2}=\frac{6}{2}=3$
$y=\frac{y_1+y_2}{2}=\frac{-3+(-8)}{2}=\frac{-11}{2}=-\frac{11}{2}$
midpoint: $\left(3,-\frac{11}{2}\right)$

67.

$x=\frac{x_1+x_2}{2}$	$y=\frac{y_1+y_2}{2}$
$-2=\frac{-8+x_2}{2}$	$3=\frac{5+y_2}{2}$
$-4=-8+x_2$	$6=5+y_2$
$4=x_2$	$1=y_2$

Q has coordinates $(4,1)$.

68.

$x=\frac{x_1+x_2}{2}$	$y=\frac{y_1+y_2}{2}$
$6=\frac{x_1+(-5)}{2}$	$-5=\frac{y_1+(-8)}{2}$
$12=x_1-5$	$-10=y_1-8$
$17=x_1$	$-2=y_1$

P has coordinates $(17,-2)$.

69.

X	Y1
-2.5	-4.75
-2	-3.5
-1.5	-2.25
-1	-1
-.5	.25
0	1.5
.5	2.75

X=-2.5

When $x = -1$, $y = -1$.

70.

X	Y1
-2.5	-4.7
-2	-4.4
-1.5	-4.1
-1	-3.8
-.5	-3.5
0	-3.2
.5	-2.9

X=-2.5

When $x = -1$, $y = -3.8$.

71.
$$\begin{aligned} 3.2x - 1.5y &= 2.7 \\ -1.5y &= -3.2x + 2.7 \\ y &= \frac{-3.2x + 2.7}{-1.5} \end{aligned}$$

X	Y1
-2.5	-7.133
-2	-6.067
-1.5	-5
-1	-3.933
-.5	-2.867
0	-1.8
.5	-.7333

X=-2.5

When $x = -1$, $y = -3.933$.

72.
$$\begin{aligned} -1.7x + 3.7y &= -2.8 \\ 3.7y &= 1.7x - 2.8 \\ y &= \frac{1.7x - 2.8}{3.7} \end{aligned}$$

X	Y1
-2.5	-1.905
-2	-1.676
-1.5	-1.446
-1	-1.216
-.5	-.9865
0	-.7568
.5	-.527

X=-2.5

When $x = -1$, $y = -1.216$.

73. $y = 3.7x - 4.5$; x-intercept: $(1.22, 0)$

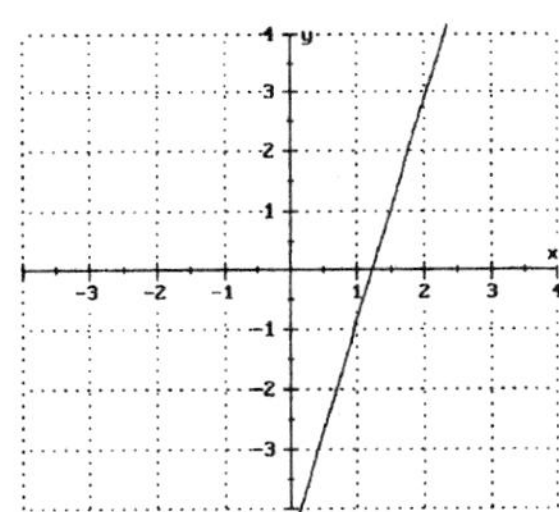

74. $y = \frac{3}{5}x + \frac{5}{4}$; x-intercept: $(-2.08, 0)$

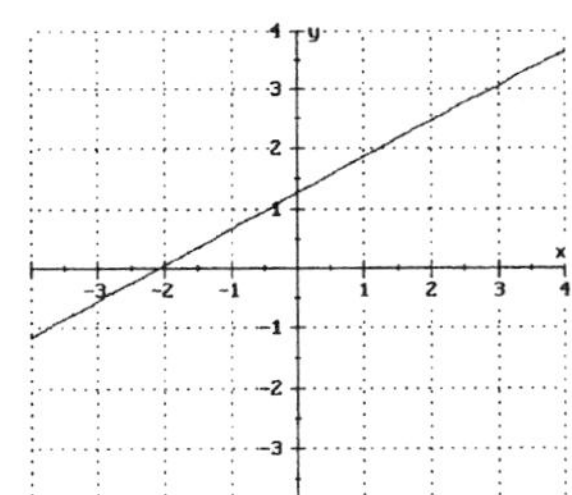

75. $1.5x - 3y = 7$; x-intercept: $(4.67, 0)$

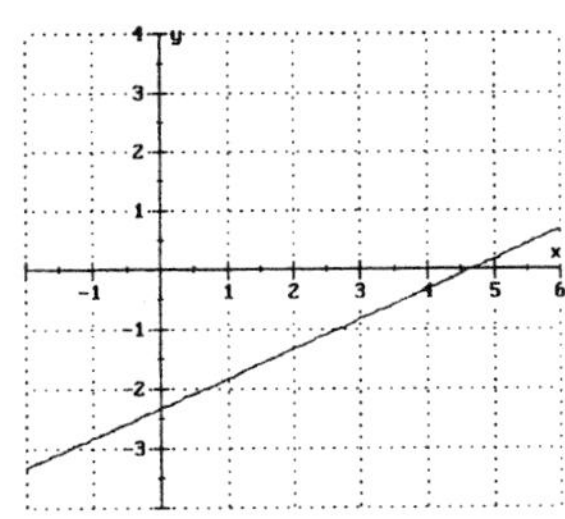

76. $0.3x + y = 7.5$; x-intercept: $(25.00, 0)$

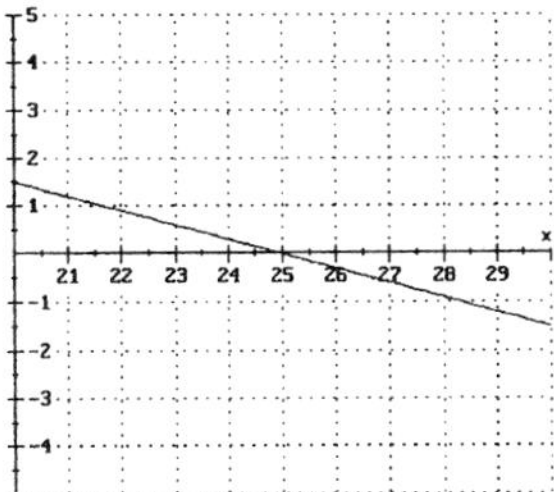

77.

x	2	4	5	6
y	12	24	30	36

Graph the ordered pairs and connect with a line:

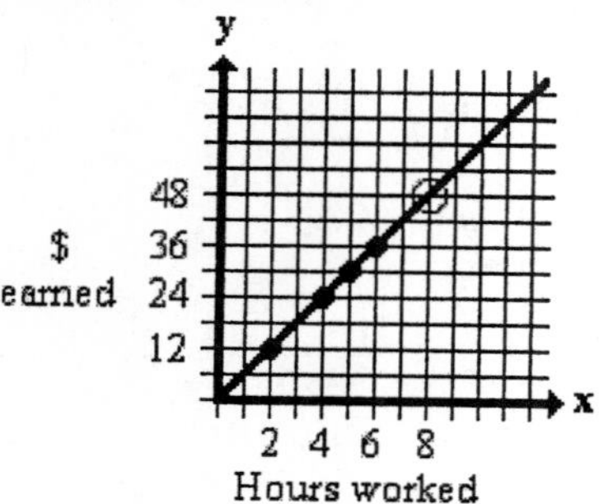

The point $(8, 48)$ is on the graph. The student will earn \$48 for 8 hours of work.

78.

x	2	4	5	6
y	30	60	75	90

Graph the ordered pairs and connect with a line:

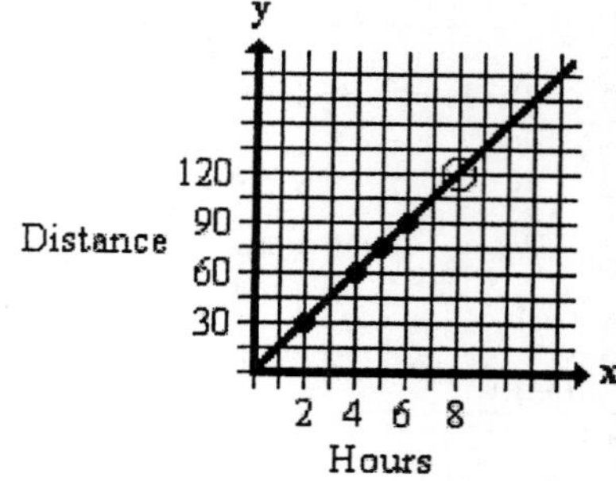

The point $(8,120)$ is on the graph. The biker can go 120 miles in 8 hours.

79.

x	0	1	3
y	15,000	12,000	6000

Graph the ordered pairs and connect with a line:

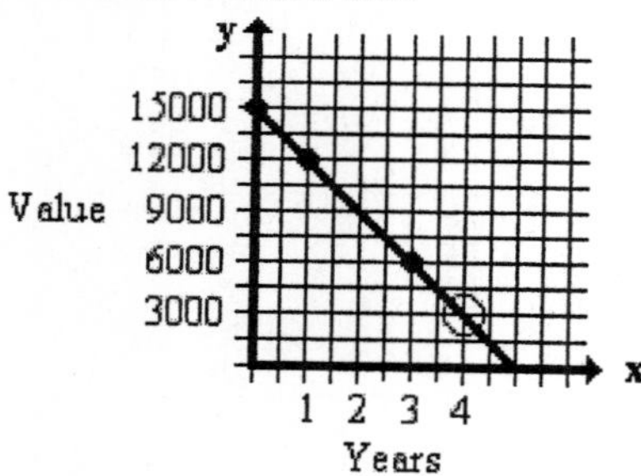

The point $(4, 3000)$ is on the graph. The car will be worth \$3000 in 4 years.

80.

x	0	1	4
y	1000	1050	1200

Graph the ordered pairs and connect with a line:

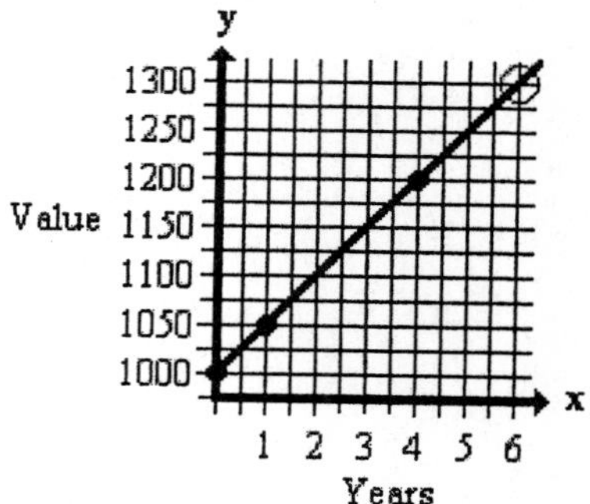

The point $(6, 1300)$ is on the graph. The account will be worth \$1300 in 6 years.

81. To find the value of the house after 5 years, let $x = 5$:

$$\begin{aligned} y &= 7500x + 125{,}000 \\ &= 7500(5) + 125{,}000 \\ &= 37{,}500 + 125{,}000 \\ &= 162{,}500 \end{aligned}$$

It will be worth \$162,500 after 5 years.

82. To find when the car will be worthless, let $y = 0$:

$$\begin{aligned} y &= -1360x + 17{,}000 \\ 0 &= -1360x + 17{,}000 \\ 1360x &= 17{,}000 \\ x &= \frac{17{,}000}{1360} = 12.5 \end{aligned}$$

The car will be worthless after 12.5 years.

83. To find the number of TVs sold, let $p = 150$:

$$p = -\frac{1}{10}q + 170$$
$$150 = -\frac{1}{10}q + 170$$
$$\frac{1}{10}q = 20$$
$$q = 200$$

200 TVs will be sold at a price of \$150.

84. To find the number of TVs produced, let $p = 150$:

$$p = \frac{1}{10}q + 130$$
$$150 = \frac{1}{10}q + 130$$
$$20 = \frac{1}{10}q$$
$$200 = q$$

200 TVs will be produced at a price of \$150.

85. The shuttle is launched from the point $(3, 0)$ along the line $x = 3$. Let $x = 3$ in the equation: $y = 2x + 6 = 2(3) + 6 = 6 + 6 = 12 \Rightarrow$ The shuttle will stay in range for 12 miles.

86. $y = 5x + 2$

x	y
1	7
2	12
3	17

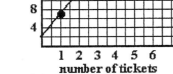

cost (\$)

number of tickets

It will cost \$22 to buy 4 tickets.

87. $y = 0.25x + 5$

x	y
4	6
8	7
12	8
16	9

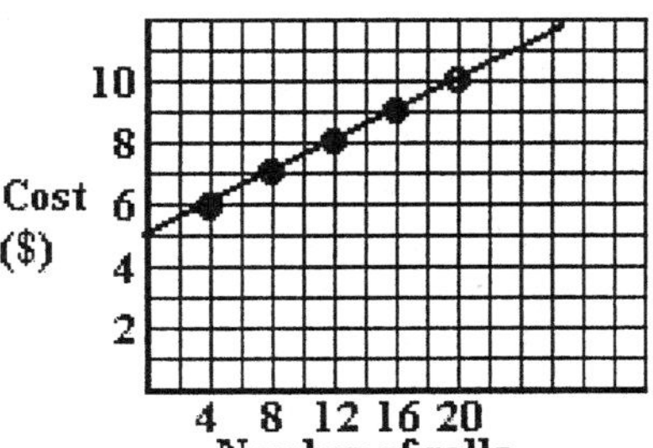

It will cost \$10 to make 20 calls.

88. To find the expenditure needed, let $n = 350$:

$$n = 430 - 0.005d$$
$$350 = 430 - 0.005d$$
$$-80 = -0.005d$$
$$\frac{-80}{-0.005} = \frac{-0.005d}{-0.005}$$

$16,000 = d \Rightarrow$ An expenditure of \$16,000 would reduce the number of incidents to 350.

89. **a.** The s-intercept occurs when $t = 0$. This will indicate the number of swimmers (in millions) when $t = 0$, or during 1990. Since $s = 65.5$ when $t = 0$, there were 65.5 million swimmers in 1990.

b. Find s when $t = 12$.

$$s = -0.9t + 65.5$$
$$= -0.9(12) + 65.5$$
$$= -10.8 + 65.5 = 54.7$$

There were 54.7 million swimmers in 2002.

90. **a.** The a-intercept occurs when $t = 0$. This will indicate the acres of farmland when $t = 0$, or during 1990. Since $a = 983{,}000{,}000$ when $t = 0$, there were 983,000,000 acres of farmland in 1990.

b. Find a when $t = 8$.
$$\begin{aligned} a &= -3{,}700{,}000t + 983{,}000{,}000 \\ &= -3{,}700{,}000(8) + 983{,}000{,}000 \\ &= -29{,}600{,}000 + 983{,}000{,}000 \\ &= 953{,}400{,}000 \end{aligned}$$
There were 953,400,000 acres of farmland in 1998.

91. **Answers may vary.**

92. **Answers may vary.**

93. If the line $y = ax + b$ passes through only quadrants I and II, then $a = 0$ and $b > 0$.

94. First, find the midpoint of $\overline{PQ}$ (call it M): $\left(\frac{a+c}{2}, \frac{b+d}{2}\right)$. Next, find the midpoint of the segment between P and M: $\left(\frac{a+\frac{a+c}{2}}{2}, \frac{b+\frac{b+d}{2}}{2}\right) = \left(\frac{3a+c}{4}, \frac{3b+d}{4}\right)$. Finally, find the midpoint of the segment between M and Q: $\left(\frac{\frac{a+c}{2}+c}{2}, \frac{\frac{b+d}{2}+d}{2}\right) = \left(\frac{a+3c}{4}, \frac{b+3d}{4}\right)$.

Exercise 2.2 (page 109)

1. $(x^3y^2)^3 = x^9y^6$

2. $\left(\frac{x^5}{x^3}\right)^3 = (x^2)^3 = x^6$

3. $(x^{-3}y^2)^{-4} = x^{12}y^{-8} = \frac{x^{12}}{y^8}$

4. $\left(\frac{x^{-6}}{y^3}\right)^{-4} = \left(\frac{y^3}{x^{-6}}\right)^4 = (x^6y^3)^4 = x^{24}y^{12}$

5. $\left(\frac{3x^2y^3}{8}\right)^0 = 1$

6. $\left(\frac{x^3x^{-7}y^{-6}}{x^4y^{-3}y^{-2}}\right)^{-2} = \left(\frac{x^{-4}y^{-6}}{x^4y^{-5}}\right)^{-2} = \left(\frac{x^4y^{-5}}{x^{-4}y^{-6}}\right)^2 = (x^8y^1)^2 = x^{16}y^2$

7. y; x

8. change

9. $\frac{y_2 - y_1}{x_2 - x_1}$

10. rise

11. run

12. horizontal

13. vertical

14. positive

15. Parallel

16. perpendicular; reciprocals

17. $m = \frac{\Delta y}{\Delta x} = \frac{5-(-3)}{2-(-2)} = \frac{8}{4} = 2$

18. $m = \frac{\Delta y}{\Delta x} = \frac{-3-4}{2-(-3)} = \frac{-7}{5} = -\frac{7}{5}$

19. $m = \dfrac{\Delta y}{\Delta x} = \dfrac{9-0}{3-0} = \dfrac{9}{3} = 3$

20. $m = \dfrac{\Delta y}{\Delta x} = \dfrac{0-6}{0-9} = \dfrac{-6}{-9} = \dfrac{2}{3}$

21. $m = \dfrac{\Delta y}{\Delta x} = \dfrac{1-8}{6-(-1)} = \dfrac{-7}{7} = -1$

22. $m = \dfrac{\Delta y}{\Delta x} = \dfrac{8-(-8)}{3-(-5)} = \dfrac{16}{8} = 2$

23. $m = \dfrac{\Delta y}{\Delta x} = \dfrac{2-(-1)}{-6-3} = \dfrac{3}{-9} = -\dfrac{1}{3}$

24. $m = \dfrac{\Delta y}{\Delta x} = \dfrac{0-(-8)}{-5-0} = \dfrac{8}{-5} = -\dfrac{8}{5}$

25. $m = \dfrac{\Delta y}{\Delta x} = \dfrac{5-5}{-9-7} = \dfrac{0}{-16} = 0$

26. $m = \dfrac{\Delta y}{\Delta x} = \dfrac{-8-(-8)}{3-2} = \dfrac{0}{1} = 0$

27. $m = \dfrac{\Delta y}{\Delta x} = \dfrac{-2-(-5)}{-7-(-7)} = \dfrac{3}{0}$: undefined

28. $m = \dfrac{\Delta y}{\Delta x} = \dfrac{14-(-5)}{3-3} = \dfrac{19}{0}$: undefined

29. $m = \dfrac{\Delta y}{\Delta x} = \dfrac{2.5-3.7}{3.7-2.5} = \dfrac{-1.2}{1.2} = -1$

30. $m = \dfrac{\Delta y}{\Delta x} = \dfrac{-1.7-(-2.3)}{2.3-1.7} = \dfrac{0.6}{0.6} = 1$

31. Find two points on the line:

Let $x = 0$:	Let $y = 0$:
$3x + 2y = 12$	$3x + 2y = 12$
$3(0) + 2y = 12$	$3x + 2(0) = 12$
$2y = 12$	$3x = 12$
$y = 6$	$x = 4$
$(0,6)$	$(4,0)$

$$m = \frac{\Delta y}{\Delta x} = \frac{6-0}{0-4} = \frac{6}{-4} = -\frac{3}{2}$$

32. Find two points on the line:

Let $x = 0$:	Let $y = 0$:
$2x - y = 6$	$2x - y = 6$
$2(0) - y = 6$	$2x - (0) = 6$
$-y = 6$	$2x = 6$
$y = -6$	$x = 3$
$(0,-6)$	$(3,0)$

$$m = \frac{\Delta y}{\Delta x} = \frac{-6-0}{0-3} = \frac{-6}{-3} = 2$$

33. Find two points on the line:

Let $x = 0$:	Let $y = 0$:
$3x = 4y - 2$	$3x = 4y - 2$
$3(0) = 4y - 2$	$3x = 4(0) - 2$
$0 = 4y - 2$	$3x = -2$
$2 = 4y$	$x = -\frac{2}{3}$
$\frac{1}{2} = y$	$(-\frac{2}{3}, 0)$
$(0, \frac{1}{2})$	

$$m = \frac{\Delta y}{\Delta x} = \frac{\frac{1}{2}-0}{0-(-\frac{2}{3})} = \frac{\frac{1}{2}}{\frac{2}{3}} = \frac{3}{4}$$

34. Find two points on the line:

Let $x = 0$:	Let $x = 1$:
$x = y$	$x = y$
$0 = y$	$1 = y$
$(0,0)$	$(1,1)$

$$m = \frac{\Delta y}{\Delta x} = \frac{1-0}{1-0} = \frac{1}{1} = 1$$

35. Find two points on the line:

Let $x = 0$: $y = \frac{x-4}{2}$; $y = \frac{0-4}{2}$; $y = \frac{-4}{2}$; $y = -2$; $(0, -2)$

Let $y = 0$: $y = \frac{x-4}{2}$; $0 = \frac{x-4}{2}$; $0 = x - 4$; $4 = x$; $(4, 0)$

$$m = \frac{\Delta y}{\Delta x} = \frac{-2-0}{0-4} = \frac{-2}{-4} = \frac{1}{2}$$

36. Find two points on the line:

Let $x = 0$: $x = \frac{3-y}{4}$; $0 = \frac{3-y}{4}$; $0 = 3 - y$; $y = 3$; $(0, 3)$

Let $y = 0$: $x = \frac{3-y}{4}$; $x = \frac{3-0}{4}$; $x = \frac{3}{4}$; $\left(\frac{3}{4}, 0\right)$

$$m = \frac{\Delta y}{\Delta x} = \frac{3-0}{0-\frac{3}{4}} = \frac{3}{-\frac{3}{4}} = -4$$

37. $4y = 3(y+2)$

$4y = 3y + 6$

$y = 6$

horizontal line $\Rightarrow m = 0$

38. Simplify the equation:

$$x + y = \frac{2-3y}{3}$$
$$3(x+y) = 3 \cdot \frac{2-3y}{3}$$
$$3x + 3y = 2 - 3y$$
$$3x + 6y = 2$$

Find two points on the line:

Let $x = 0$: $3x + 6y = 2$; $3(0) + 6y = 2$; $6y = 2$; $y = \frac{2}{6} = \frac{1}{3}$; $\left(0, \frac{1}{3}\right)$

Let $y = 0$: $3x + 6y = 2$; $3x + 6(0) = 2$; $3x = 2$; $x = \frac{2}{3}$; $\left(\frac{2}{3}, 0\right)$

$$m = \frac{\Delta y}{\Delta x} = \frac{\frac{1}{3}-0}{0-\frac{2}{3}} = \frac{\frac{1}{3}}{-\frac{2}{3}} = -\frac{1}{2}$$

39. negative

40. 0

41. positive

42. positive

43. undefined

44. negative

45. $m_1 \neq m_2 \Rightarrow$ not parallel

$m_1 \cdot m_2 = 3\left(-\frac{1}{3}\right) = -1 \Rightarrow$ perpendicular

46. $m_1 \neq m_2 \Rightarrow$ not parallel

$m_1 \cdot m_2 = \frac{1}{4}(4) = 1 \Rightarrow$ not perpendicular

47. $m_1 \neq m_2 \Rightarrow$ not parallel

$m_1 \cdot m_2 = 4(0.25) = 1 \Rightarrow$ not perpendicular

48. $m_1 = m_2 \Rightarrow$ parallel

49. $m_1 = m_2 \Rightarrow$ parallel

50. $m_1 \neq m_2 \Rightarrow$ not parallel

$m_1 \cdot m_2 = \frac{3.2}{-9.1} \cdot \frac{-9.1}{3.2} = 1$

$\Rightarrow$ not perpendicular

51. $m_{\overline{PQ}} = \dfrac{\Delta y}{\Delta x} = \dfrac{2-4}{4-3} = \dfrac{-2}{1} = -2$
same slope $\Rightarrow$ parallel

52. $m_{\overline{PQ}} = \dfrac{\Delta y}{\Delta x} = \dfrac{5-4}{8-6} = \dfrac{1}{2}$
opposite reciprocal slope $\Rightarrow$ perpendicular

53. $m_{\overline{PQ}} = \dfrac{\Delta y}{\Delta x} = \dfrac{5-1}{6-(-2)} = \dfrac{4}{8} = \dfrac{1}{2}$
opposite reciprocal slope $\Rightarrow$ perpendicular

54. $m_{\overline{PQ}} = \dfrac{\Delta y}{\Delta x} = \dfrac{-5-4}{-3-3} = \dfrac{-9}{-6} = \dfrac{3}{2}$
neither parallel nor perpendicular

55. $m_{\overline{PQ}} = \dfrac{\Delta y}{\Delta x} = \dfrac{6-4}{6-5} = \dfrac{2}{1} = 2$
neither parallel nor perpendicular

56. $m_{\overline{PQ}} = \dfrac{\Delta y}{\Delta x} = \dfrac{-9-3}{4-(-2)} = \dfrac{-12}{6} = -2$
same slope $\Rightarrow$ parallel

57. $m_{\overline{PQ}} = \dfrac{\Delta y}{\Delta x} = \dfrac{8-4}{4-(-2)} = \dfrac{4}{6} = \dfrac{2}{3}$
$m_{\overline{PR}} = \dfrac{\Delta y}{\Delta x} = \dfrac{12-4}{8-(-2)} = \dfrac{8}{10} = \dfrac{4}{5}$
Since they do not have the same slope, the three points are not on the same line.

58. $m_{\overline{PQ}} = \dfrac{\Delta y}{\Delta x} = \dfrac{6-10}{0-6} = \dfrac{-4}{-6} = \dfrac{2}{3}$
$m_{\overline{PR}} = \dfrac{\Delta y}{\Delta x} = \dfrac{8-10}{3-6} = \dfrac{-2}{-3} = \dfrac{2}{3}$
Since they have the same slope and a common point, the three points are on the same line.

59. $m_{\overline{PQ}} = \dfrac{\Delta y}{\Delta x} = \dfrac{0-10}{-6-(-4)} = \dfrac{-10}{-2} = 5$
$m_{\overline{PR}} = \dfrac{\Delta y}{\Delta x} = \dfrac{5-10}{-1-(-4)} = \dfrac{-5}{3} = -\dfrac{5}{3}$
Since they do not have the same slope, the three points are not on the same line.

60. $m_{\overline{PQ}} = \dfrac{\Delta y}{\Delta x} = \dfrac{-10-(-13)}{-8-(-10)} = \dfrac{3}{2}$
$m_{\overline{PR}} = \dfrac{\Delta y}{\Delta x} = \dfrac{-16-(-13)}{-12-(-10)} = \dfrac{-3}{-2} = \dfrac{3}{2}$
Since they have the same slope and a common point, the three points are on the same line.

61. $m_{\overline{PQ}} = \dfrac{\Delta y}{\Delta x} = \dfrac{8-4}{0-(-2)} = \dfrac{4}{2} = 2$
$m_{\overline{PR}} = \dfrac{\Delta y}{\Delta x} = \dfrac{12-4}{2-(-2)} = \dfrac{8}{4} = 2$
Since they have the same slope and a common point, the three points are on the same line.

62. $m_{\overline{PQ}} = \dfrac{\Delta y}{\Delta x} = \dfrac{-12-(-4)}{0-8} = \dfrac{-8}{-8} = 1$
$m_{\overline{PR}} = \dfrac{\Delta y}{\Delta x} = \dfrac{-20-(-4)}{8-8} = \dfrac{-16}{0} \Rightarrow$ und.
Since they do not have the same slope, the three points are not on the same line.

63. On the x-axis, all y-coordinates are 0. The equation is $y = 0$, and the slope is 0.

64. On the y-axis, all x-coordinates are 0. The equation is $x = 0$, and the slope is undefined.

65. Call the points $A(-3,4)$, $B(4,1)$ and $C(-1,-1)$. Compute these slopes:

$$m_{\overline{AB}} = \frac{\Delta y}{\Delta x} = \frac{1-4}{4-(-3)} = \frac{-3}{7} = -\frac{3}{7}$$

$$m_{\overline{AC}} = \frac{\Delta y}{\Delta x} = \frac{-1-4}{-1-(-3)} = \frac{-5}{2} = -\frac{5}{2}$$

$$m_{\overline{BC}} = \frac{\Delta y}{\Delta x} = \frac{-1-1}{-1-4} = \frac{-2}{-5} = \frac{2}{5}$$

Since $\overline{AC}$ and $\overline{BC}$ are perpendicular, it is a right triangle.

66. Call the points $A(0,0)$, $B(12,0)$ and $C(13,12)$. Compute these slopes:

$$m_{\overline{AB}} = \frac{\Delta y}{\Delta x} = \frac{0-0}{12-0} = \frac{0}{12} = 0$$
$$m_{\overline{AC}} = \frac{\Delta y}{\Delta x} = \frac{12-0}{13-0} = \frac{12}{13}$$
$$m_{\overline{BC}} = \frac{\Delta y}{\Delta x} = \frac{12-0}{13-12} = \frac{12}{1} = 12$$

Since none of the sides are perpendicular, it is not a right triangle.

67. Call the points $A(a,0)$, $B(0,a)$, $C(-a,0)$ and $D(0,-a)$. Compute these slopes:

$$m_{\overline{AB}} = \frac{\Delta y}{\Delta x} = \frac{a-0}{0-a} = \frac{a}{-a} = -1$$
$$m_{\overline{BC}} = \frac{\Delta y}{\Delta x} = \frac{0-a}{-a-0} = \frac{-a}{-a} = 1$$
$$m_{\overline{CD}} = \frac{\Delta y}{\Delta x} = \frac{-a-0}{0-(-a)} = \frac{-a}{a} = -1$$
$$m_{\overline{DA}} = \frac{\Delta y}{\Delta x} = \frac{0-(-a)}{a-0} = \frac{a}{a} = 1$$

Thus, $\overline{AB} \perp \overline{BC}$, $\overline{BC} \perp \overline{CD}$, $\overline{CD} \perp \overline{DA}$, and $\overline{DA} \perp \overline{AB}$.

68. Call the points $A(2b,a)$, $B(b,b)$ and $C(a,0)$. Compute these slopes:

$$m_{\overline{AB}} = \frac{\Delta y}{\Delta x} = \frac{b-a}{b-2b} = \frac{b-a}{-b} = -\frac{b-a}{b}$$
$$m_{\overline{AC}} = \frac{\Delta y}{\Delta x} = \frac{0-a}{a-2b} = \frac{-a}{a-2b} = -\frac{a}{a-2b}$$
$$m_{\overline{BC}} = \frac{\Delta y}{\Delta x} = \frac{b-0}{b-a} = \frac{b}{b-a}$$

Since $\overline{AB}$ and $\overline{BC}$ are perpendicular, it is a right triangle.

69. Call the points $A(0,0)$, $B(0,a)$, $C(b,c)$ and $D(b,a+c)$. Compute these slopes:

$$m_{\overline{AB}} = \frac{\Delta y}{\Delta x} = \frac{0-a}{0-0} = \frac{-a}{0} \Rightarrow \text{undefined}$$
$$m_{\overline{BD}} = \frac{\Delta y}{\Delta x} = \frac{a-(a+c)}{0-b} = \frac{a-a-c}{-b} = \frac{-c}{-b} = \frac{c}{b}$$
$$m_{\overline{DC}} = \frac{\Delta y}{\Delta x} = \frac{c-(a+c)}{b-b} = \frac{c-a-c}{0} = \frac{-a}{0} \Rightarrow \text{undefined}$$
$$m_{\overline{CA}} = \frac{\Delta y}{\Delta x} = \frac{c-0}{b-0} = \frac{c}{b}$$

Thus, $\overline{AB} \parallel \overline{DC}$ and $\overline{BD} \parallel \overline{CA}$ and the figure is a parallelogram.

70. Call the points $A(0,0)$, $B(0,b)$, $C(8,b+2)$ and $D(12,3)$. Compute these slopes:

$m_{\overline{AB}} = \frac{\Delta y}{\Delta x} = \frac{0-b}{0-0} = \frac{-b}{0} \Rightarrow$ undefined

$m_{\overline{BC}} = \frac{\Delta y}{\Delta x} = \frac{b-(b+2)}{0-8} = \frac{b-b-2}{-8} = \frac{-2}{-8} = \frac{1}{4}$

$m_{\overline{CD}} = \frac{\Delta y}{\Delta x} = \frac{b+2-3}{8-12} = \frac{b-1}{-4} = -\frac{b-1}{4}$

$m_{\overline{DA}} = \frac{\Delta y}{\Delta x} = \frac{0-3}{0-12} = \frac{-3}{-12} = \frac{1}{4}$

Thus, $\overline{BC} \parallel \overline{DA}$ and the figure is a trapezoid.

71. slope $= \frac{\text{rise}}{\text{run}} = \frac{32\text{ ft}}{1\text{ mi}} = \frac{32\text{ ft}}{5280\text{ ft}} = \frac{1}{165}$

72. slope $= \frac{\text{rise}}{\text{run}} = \frac{3\text{ ft}}{12\text{ ft}} = \frac{1}{4}$

73. slope $= \frac{\text{rise}}{\text{run}} = \frac{18\text{ ft}}{5\text{ ft}} = \frac{18}{5}$

74. $m = \frac{\Delta y}{\Delta x} = \frac{2}{50} = 0.04$; $m = \frac{\Delta y}{\Delta x} = \frac{5}{50} = 0.1$; $m = \frac{\Delta y}{\Delta x} = \frac{8}{50} = 0.16$

75. $m = \frac{\Delta y}{\Delta x} = \frac{1.4}{50} = \frac{14}{500} = \frac{7}{250}$

$\frac{7}{250}$ degree per year

76. $m = \frac{\Delta y}{\Delta x} = \frac{0.7}{50} = \frac{7}{500}$

$\frac{7}{500}$ degree per year

77. Let x represent the number of years, and let y represent the enrollment. Then we know two points on the line: $(1,8)$ and $(5,20)$. Find the slope:

$m = \frac{\Delta y}{\Delta x} = \frac{20-8}{5-1} = \frac{12}{4} = 3$ students per year.

78. **a.** $m = \frac{\Delta y}{\Delta x} = \frac{3}{16}$ **b.** $m = \frac{\Delta y}{\Delta x} = \frac{1.5}{12} = \frac{1}{8}$ (each part)

c. Design #1 has just one level, but the slope is steep. Design #2 has slopes which are less steep, but there are two levels.

79. Let x represent the number of years, and let y represent the sales. Then we know two points on the line: $(1,85000)$ and $(3,125000)$. Find the slope:

$m = \frac{\Delta y}{\Delta x} = \frac{125000-85000}{3-1} = \frac{40000}{2} =$ \$20,000 per year.

80. Let x represent the number of years, and let y represent the price. Then we know two points on the line: $(-10,5700)$ and $(-2,1499)$. Find the slope:

$m = \frac{\Delta y}{\Delta x} = \frac{5700-1499}{-10-(-2)} = \frac{4201}{-8} \approx -\$525.13 \Rightarrow$ The cost is decreasing about \$525.13 per year.

81. **Answers may vary.**

82. **Answers may vary.**

83. Find two points on the line:

Let $x = 0$:
$$\begin{aligned} Ax + By &= C \\ A(0) + By &= C \\ By &= C \\ y &= \tfrac{C}{B} \end{aligned}$$
$\left(0, \frac{C}{B}\right)$ is on the line.

Let $y = 0$:
$$\begin{aligned} Ax + By &= C \\ Ax + B(0) &= C \\ Ax &= C \\ x &= \tfrac{C}{A} \end{aligned}$$
$\left(\frac{C}{A}, 0\right)$ is on the line.

$$m = \frac{\Delta y}{\Delta x} = \frac{\frac{C}{B} - 0}{0 - \frac{C}{A}} = \frac{\frac{C}{B}}{-\frac{C}{A}} = -\frac{A}{B}$$

84. Find two points on the line:

Let $x = 0$:
$$\begin{aligned} y &= mx + b \\ y &= m(0) + b \\ y &= b \end{aligned}$$
$(0, b)$ is on the line.

Let $y = 0$:
$$\begin{aligned} y &= mx + b \\ 0 &= mx + b \\ -b &= mx \\ \tfrac{-b}{m} &= x \end{aligned}$$
$\left(-\frac{b}{m}, 0\right)$ is on the line.

$$m = \frac{\Delta y}{\Delta x} = \frac{b - 0}{0 - \left(-\frac{b}{m}\right)} = \frac{b}{\frac{b}{m}} = m$$

85. If the three points are on a line, then all slopes must be equal:

$$m = \frac{\Delta y}{\Delta x} = \frac{7 - 10}{5 - 7} = \frac{-3}{-2} = \frac{3}{2} \qquad m = \frac{\Delta y}{\Delta x} = \frac{a - 7}{3 - 5} = \frac{a - 7}{-2}$$

$$\begin{aligned} \frac{a - 7}{-2} &= \frac{3}{2} \\ -2 \cdot \frac{a - 7}{-2} &= -2 \cdot \frac{3}{2} \\ a - 7 &= -3 \\ a &= 4 \end{aligned}$$

86. $m_{\overline{AB}} = \dfrac{\Delta y}{\Delta x} = \dfrac{3 - 7}{1 - (-2)} = \dfrac{-4}{3} = -\dfrac{4}{3}$. Then $m_{\overline{CD}}$ must equal $\dfrac{3}{4}$.

$$m_{\overline{CD}} = \frac{\Delta y}{\Delta x} = \frac{b - (-1)}{4 - 8} = \frac{b + 1}{-4} = \frac{3}{4}$$

$$\begin{aligned} -4 \cdot \frac{b + 1}{-4} &= -4 \cdot \frac{3}{4} \\ b + 1 &= -3 \\ b &= -4 \end{aligned}$$

Exercise 2.3 (page 122)

1.
$$\begin{aligned} 3(x + 2) + x &= 5x \\ 3x + 6 + x &= 5x \\ 4x + 6 &= 5x \\ 6 &= x \end{aligned}$$

2.
$$\begin{aligned} 12b + 6(3 - b) &= b + 3 \\ 12b + 18 - 6b &= b + 3 \\ 6b + 18 &= b + 3 \\ 5b + 18 &= 3 \\ 5b &= -15 \\ b &= -3 \end{aligned}$$

3.
$$\begin{aligned}\frac{5(2-x)}{3}-1&=x+5\\ \frac{10-5x}{3}&=x+6\\ 3\cdot\frac{10-5x}{3}&=3(x+6)\\ 10-5x&=3x+18\\ 10&=8x+18\\ -8&=8x\\ -1&=x\end{aligned}$$

4.
$$\begin{aligned}\frac{r-1}{3}&=\frac{r+2}{6}+2\\ 6\cdot\frac{r-1}{3}&=6\left(\frac{r+2}{6}+2\right)\\ 2(r-1)&=6\cdot\frac{r+2}{6}+6(2)\\ 2r-2&=r+2+12\\ 2r-2&=r+14\\ r-2&=14\\ r&=16\end{aligned}$$

5. We want the alloy to contain 25% gold. Let $x =$ the amount of copper added to the alloy.

[Gold at start] + [Gold added] = [Gold at end]

$$\begin{aligned}20+0&=0.25(60+x)\\ 20&=15+0.25x\\ 5&=0.25x\\ 20&=x\end{aligned}$$

20 oz of copper should be added to the alloy.

6. Let $x =$ the lb of cheaper coffee used. Then $80-x =$ the lb of the other coffee.

[Value of \$3.25 coffee] + [Value of \$3.85 coffee] = [Value of mixture]

$$\begin{aligned}3.25x+3.85(80-x)&=272\\ 3.25x+308-3.85x&=272\\ -0.60x+308&=272\\ -0.60x&=-36\\ x&=60\end{aligned}$$

60 lb of the \$3.25 coffee should be used.

7. $y-y_1=m(x-x_1)$

8. $y=mx+b$

9. $Ax+By=C$

10. same

11. perpendicular

12. depreciation

13.
$$\begin{aligned}y-y_1&=m(x-x_1)\\ y-7&=5(x-0)\\ y-7&=5x\\ -5x+y&=7\\ 5x-y&=-7\end{aligned}$$

14.
$$\begin{aligned}y-y_1&=m(x-x_1)\\ y-(-2)&=-8(x-0)\\ y+2&=-8x\\ 8x+y&=-2\end{aligned}$$

15.
$$\begin{aligned}y-y_1&=m(x-x_1)\\ y-0&=-3(x-2)\\ y&=-3x+6\\ 3x+y&=6\end{aligned}$$

16.
$$\begin{aligned}y-y_1&=m(x-x_1)\\ y-0&=4(x-(-5))\\ y&=4(x+5)\\ y&=4x+20\\ -4x+y&=20\\ 4x-y&=-20\end{aligned}$$

17. Notice that the line goes through $P(2,5)$ and the point $(-1,3)$. Find the slope:

$$m = \frac{\Delta y}{\Delta x} = \frac{5-3}{2-(-1)} = \frac{2}{3}$$

Use point-slope form to find the equation:

$$\begin{aligned} y - y_1 &= m(x - x_1) \\ y - 5 &= \frac{2}{3}(x-2) \\ 3(y-5) &= 3 \cdot \frac{2}{3}(x-2) \\ 3y - 15 &= 2x - 4 \\ -2x + 3y &= 11 \\ 2x - 3y &= -11 \end{aligned}$$

18. Notice that the line goes through $P(-3,2)$ and the point $(0,0)$. Find the slope:

$$m = \frac{\Delta y}{\Delta x} = \frac{2-0}{-3-0} = \frac{2}{-3} = -\frac{2}{3}$$

Use point-slope form to find the equation:

$$\begin{aligned} y - y_1 &= m(x - x_1) \\ y - 0 &= -\frac{2}{3}(x-0) \\ y &= -\frac{2}{3}x \\ 3y &= -2x \\ 2x + 3y &= 0 \end{aligned}$$

19. Find the slope of the line:

$$m = \frac{\Delta y}{\Delta x} = \frac{4-0}{4-0} = \frac{4}{4} = 1$$

Use point-slope form to find the equation:

$$\begin{aligned} y - y_1 &= m(x - x_1) \\ y - 0 &= 1(x-0) \\ y &= x \end{aligned}$$

20. Find the slope of the line:

$$m = \frac{\Delta y}{\Delta x} = \frac{-5-0}{-5-0} = \frac{-5}{-5} = 1$$

Use point-slope form to find the equation:

$$\begin{aligned} y - y_1 &= m(x - x_1) \\ y - 0 &= 1(x-0) \\ y &= x \end{aligned}$$

21. Find the slope of the line:

$$m = \frac{\Delta y}{\Delta x} = \frac{4-(-3)}{3-0} = \frac{7}{3}$$

Use point-slope form to find the equation:

$$\begin{aligned} y - y_1 &= m(x - x_1) \\ y - 4 &= \tfrac{7}{3}(x-3) \\ y - 4 &= \tfrac{7}{3}x - 7 \\ y &= \tfrac{7}{3}x - 3 \end{aligned}$$

22. Find the slope of the line:

$$m = \frac{\Delta y}{\Delta x} = \frac{0-(-8)}{4-6} = \frac{8}{-2} = -4$$

Use point-slope form to find the equation:

$$\begin{aligned} y - y_1 &= m(x - x_1) \\ y - 0 &= -4(x-4) \\ y &= -4x + 16 \end{aligned}$$

23. Find the slope of the line:

$$m = \frac{\Delta y}{\Delta x} = \frac{4-(-5)}{-2-3} = \frac{9}{-5} = -\frac{9}{5}$$

Use point-slope form to find the equation:

$$\begin{aligned} y - y_1 &= m(x - x_1) \\ y - 4 &= -\tfrac{9}{5}(x-(-2)) \\ y - 4 &= -\tfrac{9}{5}(x+2) \\ y - 4 &= -\tfrac{9}{5}x - \tfrac{18}{5} \\ y &= -\tfrac{9}{5}x - \tfrac{18}{5} + 4 \\ y &= -\tfrac{9}{5}x + \tfrac{2}{5} \end{aligned}$$

24. Find the slope of the line:

$$m = \frac{\Delta y}{\Delta x} = \frac{3-(-5)}{2-(-3)} = \frac{8}{5}$$

Use point-slope form to find the equation:

$$\begin{aligned} y - y_1 &= m(x - x_1) \\ y - 3 &= \tfrac{8}{5}(x-2) \\ y - 3 &= \tfrac{8}{5}x - \tfrac{16}{5} \\ y &= \tfrac{8}{5}x - \tfrac{16}{5} + 3 \\ y &= \tfrac{8}{5}x - \tfrac{1}{5} \end{aligned}$$

25. $y = mx + b$
$y = 3x + 17$

26. $y = mx + b$
$y = -2x + 11$

27.
$$\begin{aligned} y &= mx + b \\ 5 &= -7(7) + b \\ 5 &= -49 + b \\ 54 &= b \\ y &= -7x + 54 \end{aligned}$$

28.
$$\begin{aligned} y &= mx + b \\ -5 &= 3(-2) + b \\ -5 &= -6 + b \\ 1 &= b \\ y &= 3x + 1 \end{aligned}$$

29.
$$\begin{aligned} y &= mx + b \\ -4 &= 0(2) + b \\ -4 &= b \\ y &= 0x + (-4) \\ y &= -4 \end{aligned}$$

30.
$$\begin{aligned} y &= mx + b \\ 0 &= -7(0) + b \\ 0 &= 0 + b \\ 0 &= b \\ y &= -7x + 0 \\ y &= -7x \end{aligned}$$

31. Find the slope of the line:
$$m = \frac{\Delta y}{\Delta x} = \frac{8 - 10}{6 - 2} = \frac{-2}{4} = -\frac{1}{2}$$
$$\begin{aligned} y &= mx + b \\ 8 &= -\frac{1}{2}(6) + b \\ 8 &= -3 + b \\ 11 &= b \\ y &= -\frac{1}{2}x + 11 \end{aligned}$$

32. Find the slope of the line:
$$m = \frac{\Delta y}{\Delta x} = \frac{5 - (-6)}{-4 - 2} = \frac{11}{-6} = -\frac{11}{6}$$
$$\begin{aligned} y &= mx + b \\ 5 &= -\frac{11}{6}(-4) + b \\ 5 &= \frac{22}{3} + b \\ -\frac{7}{3} &= b \\ y &= -\frac{11}{6}x - \frac{7}{3} \end{aligned}$$

33. $y + 1 = x$
$y = x - 1$
$m = 1; b = -1 \Rightarrow (0, -1)$

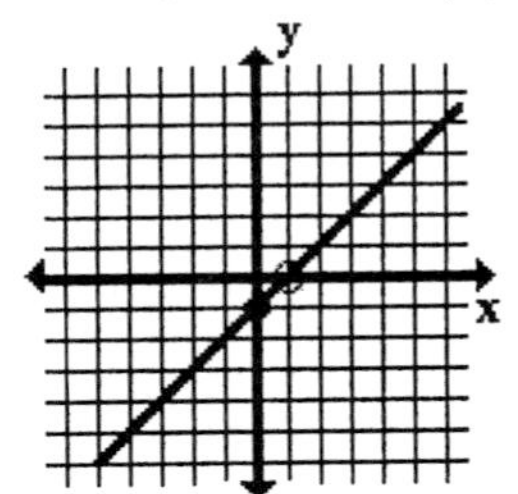

34. $x + y = 2$
$y = -x + 2$
$m = -1; b = 2 \Rightarrow (0, 2)$

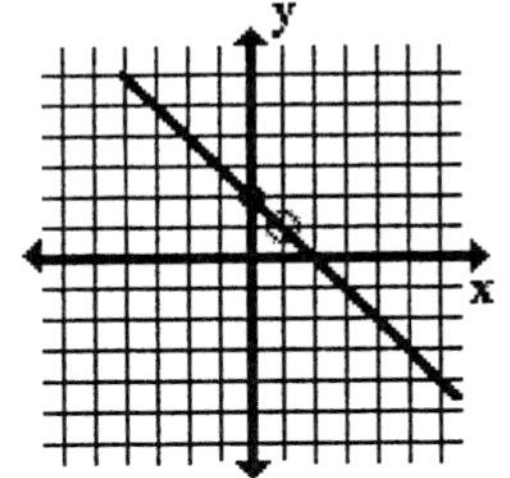

35.
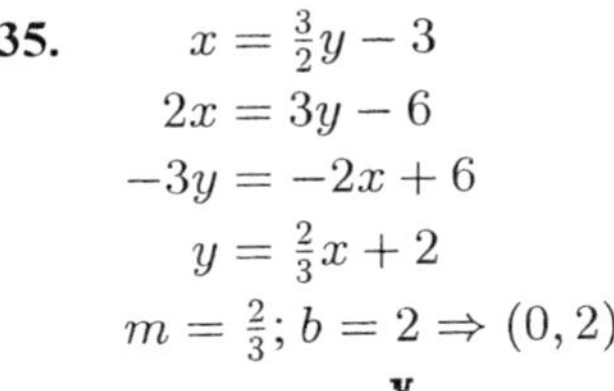
$$\begin{aligned} x &= \tfrac{3}{2}y - 3 \\ 2x &= 3y - 6 \\ -3y &= -2x + 6 \\ y &= \tfrac{2}{3}x + 2 \end{aligned}$$
$m = \frac{2}{3}; b = 2 \Rightarrow (0, 2)$

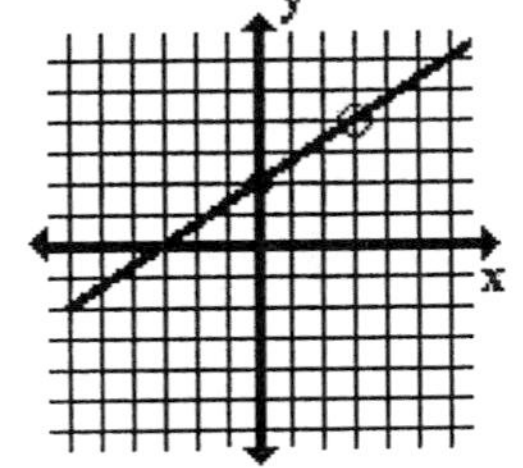

36. $x = -\frac{4}{5}y + 2$
$5x = -4y + 10$
$4y = -5x + 10$
$y = -\frac{5}{4}x + \frac{5}{2}$
$m = -\frac{5}{4};\ b = \frac{5}{2} \Rightarrow \left(0, \frac{5}{2}\right)$

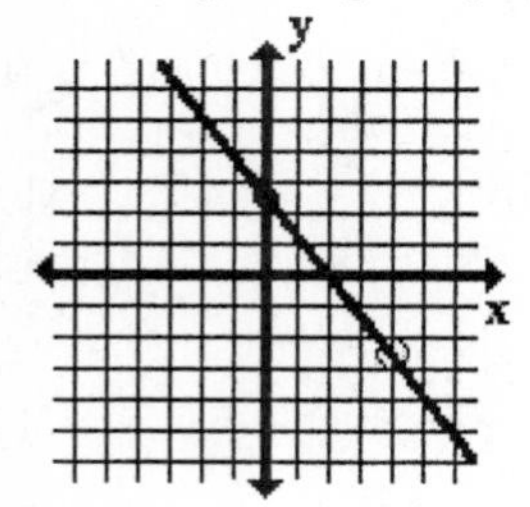

37. $3(y-4) = -2(x-3)$
$3y - 12 = -2x + 6$
$3y = -2x + 18$
$y = -\dfrac{2}{3}x + 6$
$m = -\frac{2}{3};\ b = 6 \Rightarrow (0, 6)$

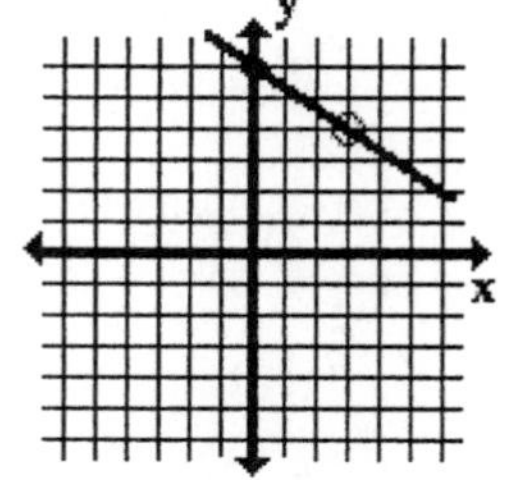

38. $-4(2x+3) = 3(3y+8)$
$-8x - 12 = 9y + 24$
$-9y = 8x + 36$
$y = -\frac{8}{9}x - 4$
$m = -\frac{8}{9};\ b = -4$
$(0, -4)$

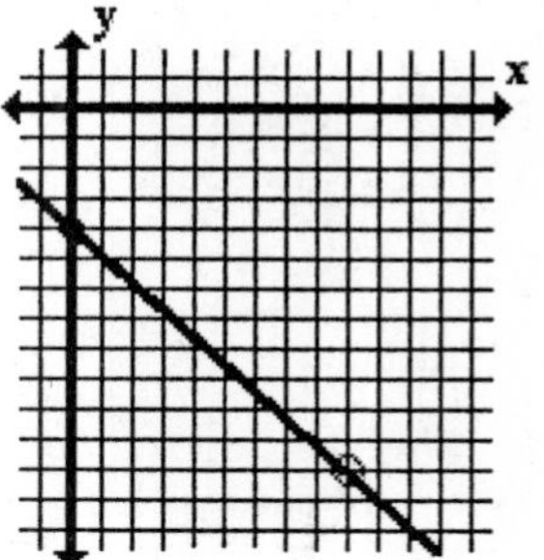

39. $3x - 2y = 8$
$-2y = -3x + 8$
$y = \frac{3}{2}x - 4$
$m = \frac{3}{2};\ b = -4 \Rightarrow (0, -4)$

40. $-2x + 4y = 12$
$4y = 2x + 12$
$y = \frac{1}{2}x + 3$
$m = \frac{1}{2};\ b = 3 \Rightarrow (0, 3)$

41. $-2(x + 3y) = 5$
$-2x - 6y = 5$
$-6y = 2x + 5$
$y = -\frac{1}{3}x - \frac{5}{6}$
$m = -\frac{1}{3};\ b = -\frac{5}{6} \Rightarrow \left(0, -\frac{5}{6}\right)$

42. $5(2x - 3y) = 4$
$10x - 15y = 4$
$-15y = -10x + 4$
$y = \frac{2}{3}x - \frac{4}{15}$
$m = \frac{2}{3};\ b = -\frac{4}{15} \Rightarrow \left(0, -\frac{4}{15}\right)$

43. $x = \dfrac{2y-4}{7}$
$7x = 2y - 4$
$-2y = -7x - 4$
$y = \frac{7}{2}x + 2$
$m = \frac{7}{2};\ b = 2 \Rightarrow (0, 2)$

44. $3x + 4 = -\dfrac{2(y-3)}{5}$
$15x + 20 = -2(y-3)$
$15x + 20 = -2y + 6$
$2y + 20 = -15x + 6$
$2y = -15x - 14$
$y = -\frac{15}{2}x - 7$
$m = -\frac{15}{2};\ b = -7 \Rightarrow (0, -7)$

45. $y = 3x + 4 \qquad y = 3x - 7$
$m = 3 \qquad m = 3$
parallel

46. $y = 4x - 13 \qquad y = \frac{1}{4}x + 13$
$m = 4 \qquad m = \frac{1}{4}$
neither

47. $x + y = 2$ $\quad y = x + 5$

$y = -x + 2 \quad m = 1$

$m = -1$

perpendicular

48. $x = y + 2 \quad y = x + 3$

$-y = -x + 2 \quad m = 1$

$y = x - 2$

$m = 1$

parallel

49. $y = 3x + 7 \quad 2y = 6x - 9$

$m = 3 \quad y = 3x - \dfrac{9}{2}$

$m = 3$

parallel

50. $2x + 3y = 9 \quad 3x - 2y = 5$

$3y = -2x + 9 \quad -2y = -3x + 5$

$y = -\frac{2}{3}x + 3 \quad y = \frac{3}{2}x - \frac{5}{2}$

$m = -\frac{2}{3} \quad m = \frac{3}{2}$

perpendicular

51. $x = 3y + 4 \quad y = -3x + 7$

$-3y = -x + 4 \quad m = -3$

$y = \frac{1}{3}x - \frac{4}{3}$

$m = \frac{1}{3}$

perpendicular

52. $3x + 6y = 1 \quad y = \frac{1}{2}x$

$6y = -3x + 1 \quad m = \frac{1}{2}$

$y = -\frac{1}{2}x + \frac{1}{6}$

$m = -\frac{1}{2}$

neither

53. $y = 3 \quad x = 4$

horizontal line vertical line

perpendicular

54. $y = -3 \quad y = -7$

horizontal line horizontal line

parallel

55. $x = \dfrac{y-2}{3} \quad 3(y-3) + x = 0$

$3x = y - 2 \quad 3y - 9 + x = 0$

$-y = -3x - 2 \quad 3y = -x + 9$

$y = 3x + 2 \quad y = -\frac{1}{3}x + 3$

$m = 3 \quad m = -\frac{1}{3}$

perpendicular

56. $2y = 8 \quad 3(2 + x) = 2(x + 2)$

$y = 4 \quad 6 + 3x = 2x + 4$

horizontal line $\quad x = -2$

vertical line

perpendicular

57. Find the slope of the given line:

$y = 4x - 7 \Rightarrow m = 4$

Use the parallel slope.

$y - y_1 = m(x - x_1)$

$y - 0 = 4(x - 0)$

$y = 4x$

58. Find the slope of the given line:

$x = -3y - 12$

$3y = -x - 12$

$y = -\frac{1}{3}x - 4 \Rightarrow m = -\frac{1}{3}$

Use the parallel slope.

$y - y_1 = m(x - x_1)$

$y - 0 = -\frac{1}{3}(x - 0)$

$y = -\frac{1}{3}x$

59. Find the slope of the given line:

$$4x - y = 7$$
$$-y = -4x + 7$$
$$y = 4x - 7 \Rightarrow m = 4$$

Use the parallel slope.

$$y - y_1 = m(x - x_1)$$
$$y - 5 = 4(x - 2)$$
$$y - 5 = 4x - 8$$
$$y = 4x - 3$$

60. Find the slope of the given line:

$$y + 3x = -12$$
$$y = -3x - 12 \Rightarrow m = -3$$

Use the parallel slope.

$$y - y_1 = m(x - x_1)$$
$$y - 3 = -3(x - (-6))$$
$$y - 3 = -3(x + 6)$$
$$y - 3 = -3x - 18$$
$$y = -3x - 15$$

61. Find the slope of the given line:

$$x = \frac{5}{4}y - 2$$
$$4x = 5y - 8$$
$$-5y = -4x - 8$$
$$y = \tfrac{4}{5}x + \tfrac{8}{5} \Rightarrow m = \tfrac{4}{5}$$

Use the parallel slope.

$$y - y_1 = m(x - x_1)$$
$$y - (-2) = \tfrac{4}{5}(x - 4)$$
$$y + 2 = \tfrac{4}{5}x - \tfrac{16}{5}$$
$$y = \tfrac{4}{5}x - \tfrac{16}{5} - 2$$
$$y = \tfrac{4}{5}x - \tfrac{26}{5}$$

62. Find the slope of the given line:

$$x = -\tfrac{3}{4}y + 5$$
$$4x = -3y + 20$$
$$3y = -4x + 20$$
$$y = -\tfrac{4}{3}x + \tfrac{20}{3} \Rightarrow m = -\tfrac{4}{3}$$

Use the parallel slope.

$$y - y_1 = m(x - x_1)$$
$$y - (-5) = -\tfrac{4}{3}(x - 1)$$
$$y + 5 = -\tfrac{4}{3}x + \tfrac{4}{3}$$
$$y = -\tfrac{4}{3}x + \tfrac{4}{3} - 5$$
$$y = -\tfrac{4}{3}x - \tfrac{11}{3}$$

63. Find the slope of the given line:

$$y = 4x - 7 \Rightarrow m = 4$$

Use the perpendicular slope.

$$y - y_1 = m(x - x_1)$$
$$y - 0 = -\tfrac{1}{4}(x - 0)$$
$$y = -\tfrac{1}{4}x$$

64. Find the slope of the given line:

$$x = -3y - 12$$
$$3y = -x - 12$$
$$y = -\tfrac{1}{3}x - 4 \Rightarrow m = -\tfrac{1}{3}$$

Use the perpendicular slope.

$$y - y_1 = m(x - x_1)$$
$$y - 0 = 3(x - 0)$$
$$y = 3x$$

65. Find the slope of the given line:

$$4x - y = 7$$
$$-y = -4x + 7$$
$$y = 4x - 7 \Rightarrow m = 4$$

Use the perpendicular slope.

$$y - y_1 = m(x - x_1)$$
$$y - 5 = -\tfrac{1}{4}(x - 2)$$
$$y - 5 = -\tfrac{1}{4}x + \tfrac{1}{2}$$
$$y = -\tfrac{1}{4}x + \tfrac{1}{2} + 5$$
$$y = -\tfrac{1}{4}x + \tfrac{11}{2}$$

66. Find the slope of the given line:

$$y + 3x = -12$$
$$y = -3x + 12 \Rightarrow m = -3$$

Use the perpendicular slope.

$$y - y_1 = m(x - x_1)$$
$$y - 3 = \tfrac{1}{3}(x - (-6))$$
$$y - 3 = \tfrac{1}{3}(x + 6)$$
$$y - 3 = \tfrac{1}{3}x + 2$$
$$y = \tfrac{1}{3}x + 5$$

67. Find the slope of the given line:

$$x = \tfrac{5}{4}y - 2$$
$$4x = 5y - 8$$
$$-5y = -4x - 8$$
$$y = \tfrac{4}{5}x + \tfrac{8}{5} \Rightarrow m = \tfrac{4}{5}$$

Use the perpendicular slope.

$$y - y_1 = m(x - x_1)$$
$$y - (-2) = -\tfrac{5}{4}(x - 4)$$
$$y + 2 = -\tfrac{5}{4}x + 5$$
$$y = -\tfrac{5}{4}x + 3$$

68. Find the slope of the given line:

$$x = -\tfrac{3}{4}y + 5$$
$$4x = -3y + 20$$
$$3y = -4x + 20$$
$$y = -\tfrac{4}{3}x + \tfrac{20}{3} \Rightarrow m = -\tfrac{4}{3}$$

Use the perpendicular slope.

$$y - y_1 = m(x - x_1)$$
$$y - (-5) = \tfrac{3}{4}(x - 1)$$
$$y + 5 = \tfrac{3}{4}x - \tfrac{3}{4}$$
$$y = \tfrac{3}{4}x - \tfrac{3}{4} - 5$$
$$y = \tfrac{3}{4}x - \tfrac{23}{4}$$

69. Find the slope of the 1st line. $m = -\frac{A}{B} = -\frac{4}{5}$ Find the slope of the 2nd line. $m = -\frac{A}{B} = -\frac{5}{-4} = \frac{5}{4}$ perpendicular

70. Find the slope of the 1st line. $m = -\frac{A}{B} = -\frac{9}{-12} = \frac{3}{4}$ Find the slope of the 2nd line. $m = -\frac{A}{B} = -\frac{3}{-4} = \frac{3}{4}$ parallel

71. Find the slope of the 1st line. $m = -\frac{A}{B} = -\frac{2}{3}$ Find the slope of the 2nd line. $m = -\frac{A}{B} = -\frac{6}{9} = -\frac{2}{3}$ parallel

72. Find the slope of the 1st line. $m = -\frac{A}{B} = -\frac{5}{6}$ Find the slope of the 2nd line. $m = -\frac{A}{B} = -\frac{6}{5}$ neither

73. The line $y = 3$ is horizontal, so any perpendicular line will be vertical.
Find the midpoint of the described segment:

$$x = \frac{x_1 + x_2}{2} = \frac{2 + (-6)}{2} = \frac{-4}{2} = -2; \quad y = \frac{y_1 + y_2}{2} = \frac{4 + 10}{2} = \frac{14}{2} = 7$$

The vertical line through the point $(-2, 7)$ is $x = -2$.

74. The line $y = -8$ is horizontal, so any parallel line will be horizontal.
Find the midpoint of the described segment:

$$x = \frac{x_1 + x_2}{2} = \frac{-4 + (-2)}{2} = \frac{-6}{2} = -3; \quad y = \frac{y_1 + y_2}{2} = \frac{2 + 8}{2} = \frac{10}{2} = 5$$

The horizontal line through the point $(-3, 5)$ is $y = 5$.

75. The line $x = 3$ is vertical, so any parallel line will be vertical.
Find the midpoint of the described segment:

$$x = \frac{x_1 + x_2}{2} = \frac{2 + 8}{2} = \frac{10}{2} = 5; \quad y = \frac{y_1 + y_2}{2} = \frac{-4 + 12}{2} = \frac{8}{2} = 4$$

The vertical line through the point $(5, 4)$ is $x = 5$.

76. The line $x = 3$ is vertical, so any perpendicular line will be horizontal.
Find the midpoint of the described segment:

$$x = \frac{x_1 + x_2}{2} = \frac{-2+4}{2} = \frac{2}{2} = 1; \; y = \frac{y_1 + y_2}{2} = \frac{2+(-8)}{2} = \frac{-6}{2} = -3$$

The horizontal line through $(1, -3)$ is $y = -3$.

77.

$$\begin{aligned} Ax + By &= C \\ By &= -Ax + C \\ \frac{By}{B} &= \frac{-Ax + C}{B} \\ y &= -\frac{A}{B}x + \frac{C}{B} \end{aligned}$$

$$m = -\frac{A}{B}; \; b = \frac{C}{B}$$

78. To find the x-intercept, set $y = 0$:

$$\begin{aligned} Ax + By &= C \\ Ax + B(0) &= C \\ Ax &= C \\ x &= \frac{C}{A} \end{aligned}$$

The x-intercept is $\left(\frac{C}{A}, 0\right)$.

79. Let x represent the number of years since the truck was purchased, and let y represent the value of the truck. Then we know two points on the depreciation line: $(0, 19984)$ and $(8, 1600)$. Find the slope and then use point-slope form to find the depreciation equation.

$$m = \frac{\Delta y}{\Delta x} = \frac{19984 - 1600}{0 - 8} = \frac{18384}{-8} = -2298$$

$$\begin{aligned} y - y_1 &= m(x - x_1) \\ y - 19984 &= -2298(x - 0) \\ y &= -2298x + 19984 \end{aligned}$$

80. Let x represent the number of years since the computer was purchased, and let y represent the value of the computer. Then we know two points on the depreciation line: $(0, 2350)$ and $(5, 200)$. Find the slope and then use point-slope form to find the depreciation equation.

$$m = \frac{\Delta y}{\Delta x} = \frac{2350 - 200}{0 - 5} = \frac{2150}{-5} = -430$$

$$\begin{aligned} y - y_1 &= m(x - x_1) \\ y - 2350 &= -430(x - 0) \\ y &= -430x + 2350 \end{aligned}$$

81. Let x represent the time in years after 1987, and let y represent the value of the painting. Then we know two points on the appreciation line: (0, 36225000) and (20, 72450000). Find the slope and then use point-slope form to find the appreciation equation.

$$m = \frac{\Delta y}{\Delta x} = \frac{72450000 - 36225000}{20 - 0} = \frac{36225000}{20} = 1811250$$

$$\begin{aligned} y - y_1 &= m(x - x_1) \\ y - 36225000 &= 1811250(x - 0) \\ y &= 1{,}811{,}250x + 36{,}225{,}000 \end{aligned}$$

82. Let x represent the number of years since the house was built, and let y represent the value of the house. Since the house is expected to appreciate \$4000 per year, the slope of the appreciation line is $m = 4000$. We also know one point on the appreciation line: (2, 122000). Use point-slope form to find the equation of the appreciation line.

$$\begin{aligned} y - y_1 &= m(x - x_1) \\ y - 122000 &= 4000(x - 2) \\ y - 122000 &= 4000x - 8000 \\ y &= 4{,}000x + 114{,}000 \end{aligned}$$

83. Let x represent the number of years since the painting was purchased, and let y represent the value of the painting. Then we know two points on the appreciation line: $(0, 250000)$ and $(5, 500000)$. Find the slope and then use point-slope form to find the appreciation equation.

$$m = \frac{\Delta y}{\Delta x} = \frac{500000 - 250000}{5 - 0} = \frac{250000}{5} = 50000$$

$$\begin{aligned} y - y_1 &= m(x - x_1) \\ y - 250000 &= 50000(x - 0) \\ y &= 50{,}000x + 250{,}000 \end{aligned}$$

84. Let x represent the number of years since the house was purchased, and let y represent the value of the house. Then we know two points on the appreciation line: $(0, 142000)$ and $(8, 284000)$. Find the slope and then use point-slope form to find the appreciation equation.

$$m = \frac{\Delta y}{\Delta x} = \frac{284000 - 142000}{8 - 0} = \frac{142000}{8} = 17750$$

$$\begin{aligned} y - y_1 &= m(x - x_1) \\ y - 142000 &= 17750(x - 0) \\ y &= 17{,}750x + 142{,}000 \end{aligned}$$

85. Let x represent the number of years since the TV was purchased, and let y represent the value of the TV. Then we know two points on the depreciation line: $(0, 1750)$ and $(3, 800)$. Find the slope and then use point-slope form to find the depreciation equation.

$$m = \frac{\Delta y}{\Delta x} = \frac{1750 - 800}{0 - 3} = \frac{950}{-3} = -\frac{950}{3}$$

$$\begin{aligned} y - y_1 &= m(x - x_1) \\ y - 1750 &= -\tfrac{950}{3}(x - 0) \\ y &= -\tfrac{950}{3}x + 1750 \end{aligned}$$

86. Let x represent the number of years since the lawn mower was purchased, and let y represent its value. Then we know two points on the depreciation line: $(0, 450)$ and $(10, 0)$. Find the slope and then use point-slope form to find the depreciation equation.

$$m = \frac{\Delta y}{\Delta x} = \frac{450 - 0}{0 - 10} = \frac{450}{-10} = -45$$

$$\begin{aligned} y - y_1 &= m(x - x_1) \\ y - 450 &= -45(x - 0) \\ y &= -45x + 450 \end{aligned}$$

Now substitute $x = 6.5$ to find the value after 6.5 years:

$y = -45(6.5) + 450 = -292.50 + 450 =$ \$157.50

87. Let x represent the number of years since the copy machine was purchased, and let y represent its value. Since the value decreases \$180 per year, the slope of the depreciation line is -180. Since the copier is worth \$1750 new, the y-intercept is 1750. Thus the depreciation equation can be found:

$$\begin{aligned} y &= mx + b \\ y &= -180x + 1750 \end{aligned}$$

Now let $x = 7$: $y = -180x + 1750 = -180(7) + 1750$

$= -1260 + 1750 =$ \$490

88. Let x represent the number of years since the machine was purchased, and let y represent its value. Then we know two points on the depreciation line: $(0, 47600)$ and $(15, 500)$. The rate of depreciation is the slope of the line:

$m = \frac{\Delta y}{\Delta x} = \frac{47600 - 500}{0 - 15} = \frac{47100}{-15} = -3140$ The rate of depreciation is \$3,140 per year.

89. Let x represent the number of years since the house was purchased, and let y represent its value. Since the value increases \$4000 per year, the slope of the depreciation line is 4000. We also know a point on the line: $(2, 122000)$. Use point-slope form to find the appreciation equation.

$$\begin{aligned} y - y_1 &= m(x - x_1) \\ y - 122000 &= 4000(x - 2) \\ y - 122000 &= 4000x - 8000 \\ y &= 4000x + 114000 \end{aligned}$$

Substitute $x = 10$ and find its value:

$$\begin{aligned} y = 4000x + 114000 &= 4000(10) + 114000 \\ &= 40000 + 114000 = \$154{,}000 \end{aligned}$$

90. The relationship between hours and the total cost is linear. Let x represent the number of hours for a service call, and let y represent the total cost. Then we know two points on the line: $(2, 143)$ and $(5, 320)$. The slope of the line will be the hourly rate.

$m = \frac{\Delta y}{\Delta x} = \frac{320 - 143}{5 - 2} = \frac{177}{3} = 59$ The hourly rate is \$59.

91. The relationship between number of copies and the total cost is linear. Let x represent the number of copies (in hundreds), and let y represent the total cost. The slope of the line will be the charge per one hundred copies, or 15. We also know a point on the line: $(3, 75)$. Use point-slope form to find the equation of the line:

$$\begin{aligned} y - y_1 &= m(x - x_1) \\ y - 75 &= 15(x - 3) \\ y - 75 &= 15x - 45 \\ y &= 15x + 30 \end{aligned}$$

Let $x = 10$ and find the cost:

$y = 15x + 30 = 15(10) + 30 = 150 + 30 = \$180$

92. Let x represent the population, and let y represent the number of burglaries. We know two points on the graph of the line: $(77000, 575)$ and $(87000, 675)$. Find the slope, and use point-slope form to find the equation of the line.

$m = \frac{\Delta y}{\Delta x} = \frac{675 - 575}{87000 - 77000} = \frac{100}{10000} = \frac{1}{100}$

$$\begin{aligned} y - y_1 &= m(x - x_1) \\ y - 675 &= \tfrac{1}{100}(x - 87000) \\ y - 675 &= \tfrac{1}{100}x - 870 \\ y &= \tfrac{1}{10}x - 195 \end{aligned}$$

Let $x = 110000$ and find the number of burglaries:

$$\begin{aligned} y = \tfrac{1}{100}x - 195 = \tfrac{1}{100}(110000) - 195 &= 1100 - 195 \\ &= 905 \text{ burglaries} \end{aligned}$$

93-100. Answers may vary.

101. To pass through II and IV, the slope must be negative. To pass through II and not III, the y-intercept must be positive. Thus $a < 0$ and $b > 0$.

102. To pass through I and IV only, the line must be vertical (and the slope undefined). Also, when solved for x, the equation should equal a positive number. Thus $B = 0$, and A and C have the same sign.

Exercise 2.4 (page 133)

1.
$$\begin{aligned} \frac{y+2}{2} &= 4(y+2) \\ 2 \cdot \frac{y+2}{2} &= 2 \cdot 4(y+2) \\ y+2 &= 8(y+2) \\ y+2 &= 8y+16 \\ -7y &= 14 \\ y &= -2 \end{aligned}$$

2.
$$\begin{aligned} \frac{3z-1}{6} - \frac{3z+4}{3} &= \frac{z+3}{2} \\ 6\left(\frac{3z-1}{6} - \frac{3z+4}{3}\right) &= 6 \cdot \frac{z+3}{2} \\ 3z-1-2(3z+4) &= 3(z+3) \\ 3z-1-6z-8 &= 3z+9 \\ -3z-9 &= 3z+9 \\ -6z &= 18 \\ z &= -3 \end{aligned}$$

3.
$$\begin{aligned} \frac{2a}{3} + \frac{1}{2} &= \frac{6a-1}{6} \\ 6\left(\frac{2a}{3} + \frac{1}{2}\right) &= 6 \cdot \frac{6a-1}{6} \\ 2(2a)+3 &= 6a-1 \\ 4a+3 &= 6a-1 \\ -2a &= -4 \\ a &= 2 \end{aligned}$$

4.
$$\begin{aligned} \frac{2x+3}{5} - \frac{3x-1}{3} &= \frac{x-1}{15} \\ 15\left(\frac{2x+3}{5} - \frac{3x-1}{3}\right) &= 15 \cdot \frac{x-1}{15} \\ 3(2x+3)-5(3x-1) &= x-1 \\ 6x+9-15x+5 &= x-1 \\ -10x &= -15 \\ x &= \tfrac{-15}{-10} = \tfrac{3}{2} \end{aligned}$$

5. input **6.** y **7.** x **8.** y

9. function; input; output **10.** domain **11.** range **12.** y

13. 0 **14.** cannot **15.** $mx+b$ **16.** slope; y-intercept

17. $y = 2x+3$ **is a function**, since each value of x corresponds to exactly one value of y.

18. $y = -1$ **is a function**, since each value of x corresponds to exactly one value of y.

19. $y = 2x^2$ **is a function**, since each value of x corresponds to exactly one value of y.

20. $y^2 = x+1$ **is not a function**, since $x = 3$ corresponds to both $y = 2$ and $y = -2$.

21. $y = 3+7x^2$ **is a function**, since each value of x corresponds to exactly one value of y.

22. $y^2 = 3-2x$ **is not a function**, since $x = -3$ corresponds to both $y = 3$ and $y = -3$.

23. $x = |y|$ **is not a function**, since $x = 2$ corresponds to both $y = 2$ and $y = -2$.

24. $y = |x|$ **is a function**, since each value of x corresponds to exactly one value of y.

25. $f(x) = 3x$
$f(3) = 3(3) = 9$
$f(-1) = 3(-1) = -3$

26. $f(x) = -4x$
$f(3) = -4(3) = -12$
$f(-1) = -4(-1) = 4$

27. $f(x) = 2x - 3$
$f(3) = 2(3) - 3 = 3$
$f(-1) = 2(-1) - 3 = -5$

28. $f(x) = 3x - 5$
$f(3) = 3(3) - 5 = 4$
$f(-1) = 3(-1) - 5 = -8$

29. $f(x) = 7 + 5x$
$f(3) = 7 + 5(3) = 22$
$f(-1) = 7 + 5(-1) = 2$

30. $f(x) = 3 + 3x$
$f(3) = 3 + 3(3) = 12$
$f(-1) = 3 + 3(-1) = 0$

31. $f(x) = 9 - 2x$
$f(3) = 9 - 2(3) = 3$
$f(-1) = 9 - 2(-1) = 11$

32. $f(x) = 12 + 3x$
$f(3) = 12 + 3(3) = 21$
$f(-1) = 12 + 3(-1) = 9$

33. $f(x) = x^2$
$f(2) = 2^2 = 4$
$f(3) = 3^2 = 9$

34. $f(x) = x^2 - 2$
$$\begin{aligned} f(2) &= 2^2 - 2 = 4 - 2 \\ &= 2 \end{aligned}$$
$$\begin{aligned} f(3) &= 3^2 - 2 = 9 - 2 \\ &= 7 \end{aligned}$$

35. $f(x) = x^3 - 1$
$$\begin{aligned} f(2) &= 2^3 - 1 = 8 - 1 \\ &= 7 \end{aligned}$$
$$\begin{aligned} f(3) &= 3^3 - 1 = 27 - 1 \\ &= 26 \end{aligned}$$

36. $f(x) = x^3$
$f(2) = 2^3 = 8$
$f(3) = 3^3 = 27$

37. $f(x) = (x + 1)^2$
$f(2) = (2 + 1)^2 = 3^2 = 9$
$f(3) = (3 + 1)^2 = 4^2 = 16$

38. $f(x) = (x - 3)^2$
$$\begin{aligned} f(2) &= (2 - 3)^2 = (-1)^2 \\ &= 1 \end{aligned}$$
$f(3) = (3 - 3)^2 = 0^2 = 0$

39. $f(x) = 2x^2 - x$
$$\begin{aligned} f(2) &= 2(2)^2 - 2 \\ &= 2(4) - 2 \\ &= 8 - 2 = 6 \end{aligned}$$
$$\begin{aligned} f(3) &= 2(3)^2 - 3 \\ &= 2(9) - 3 \\ &= 18 - 3 = 15 \end{aligned}$$

40. $f(x) = 5x^2 + 2x$
$$\begin{aligned} f(2) &= 5(2)^2 + 2(2) \\ &= 5(4) + 4 \\ &= 20 + 4 = 24 \end{aligned}$$
$$\begin{aligned} f(3) &= 5(3)^2 + 2(3) \\ &= 5(9) + 6 \\ &= 45 + 6 = 51 \end{aligned}$$

41. $f(x) = |x| + 2$
$$\begin{aligned} f(2) &= |2| + 2 \\ &= 2 + 2 = 4 \end{aligned}$$
$$\begin{aligned} f(-2) &= |-2| + 2 \\ &= 2 + 2 = 4 \end{aligned}$$

42. $f(x) = |x| - 5$
$$\begin{aligned} f(2) &= |2| - 5 \\ &= 2 - 5 = -3 \end{aligned}$$
$$\begin{aligned} f(-2) &= |-2| - 5 \\ &= 2 - 5 = -3 \end{aligned}$$

43. $f(x) = x^2 - 2$
$$\begin{aligned} f(2) &= 2^2 - 2 \\ &= 4 - 2 = 2 \end{aligned}$$
$$\begin{aligned} f(-2) &= (-2)^2 - 2 \\ &= 4 - 2 = 2 \end{aligned}$$

44. $f(x) = x^2 + 3$
$$\begin{aligned} f(2) &= 2^2 + 3 \\ &= 4 + 3 = 7 \end{aligned}$$
$$\begin{aligned} f(-2) &= (-2)^2 + 3 \\ &= 4 + 3 = 7 \end{aligned}$$

45. $f(x) = \dfrac{1}{x + 3}$
$f(2) = \dfrac{1}{2 + 3} = \dfrac{1}{5}$
$f(-2) = \dfrac{1}{-2 + 3} = \dfrac{1}{1} = 1$

46. $f(x) = \dfrac{3}{x-4}$

$f(2) = \dfrac{3}{2-4} = -\dfrac{3}{2}$

$f(-2) = \dfrac{3}{-2-4} = -\dfrac{3}{6} = -\dfrac{1}{2}$

47. $f(x) = \dfrac{x}{x-3}$

$f(2) = \dfrac{2}{2-3} = \dfrac{2}{-1} = -2$

$f(-2) = \dfrac{-2}{-2-3} = \dfrac{-2}{-5} = \dfrac{2}{5}$

48. $f(x) = \dfrac{x}{x^2+2}$

$f(2) = \dfrac{2}{2^2+2} = \dfrac{2}{4+2} = \dfrac{2}{6} = \dfrac{1}{3}$

$f(-2) = \dfrac{-2}{(-2)^2+2} = \dfrac{-2}{4+2} = \dfrac{-2}{6} = -\dfrac{1}{3}$

49. $g(x) = 2x$
$g(w) = 2w$
$g(w+1) = 2(w+1) = 2w+2$

50. $g(x) = -3x$
$g(w) = -3w$
$g(w+1) = -3(w+1) = -3w-3$

51. $g(x) = 3x-5$
$g(w) = 3w-5$
$g(w+1) = 3(w+1)-5 = 3w+3-5 = 3w-2$

52. $g(x) = 2x-7$
$g(w) = 2w-7$
$g(w+1) = 2(w+1)-7 = 2w+2-7 = 2w-5$

53. $f(3)+f(2) = [2(3)+1]+[2(2)+1] = 6+1+4+1 = 12$

54. $f(1)-f(-1) = [2(1)+1]-[2(-1)+1] = 2+1-[-2+1] = 3-(-1) = 4$

55. $f(b)-f(a) = [2b+1]-[2a+1] = 2b+1-2a-1 = 2b-2a$

56. $f(b)+f(a) = [2b+1]+[2a+1] = 2b+2a+2$

57. $f(b)-1 = 2b+1-1 = 2b$

58. $f(b)-f(1) = [2b+1]-[2(1)+1] = [2b+1]-[2+1] = 2b+1-3 = 2b-2$

59. $f(0)+f(-\frac{1}{2}) = [2(0)+1]+\left[2(-\frac{1}{2})+1\right] = 0+1-1+1 = 1$

60. $f(a)+f(2a) = [2(a)+1]+[2(2a)+1] = 2a+1+4a+1 = 6a+2$

61. domain $= \{-2, 4, 6\}$; range $= \{3, 5, 7\}$

62. domain $= \{0, 1, 3\}$; range $= \{2, 4\}$

63. To ensure that the denominator is not 0, x cannot equal 4. The domain is $(-\infty, 4) \cup (4, \infty)$. Because the numerator will never be 0, the range is $(-\infty, 0) \cup (0, \infty)$.

64. To ensure that the denominator is not 0, x cannot equal -1. The domain is $(-\infty, -1) \cup (-1, \infty)$. Because the numerator will never be 0, the range is $(-\infty, 0) \cup (0, \infty)$.

65. not a function

66. The domain is the set of all x-coordinates on the graph. The domain is $(-\infty, \infty)$.
The range is the set of all y-coordinates on the graph. The range is $[-1, \infty)$.
Since each vertical line passes through the graph at most once, it is the graph of a function.

67. The domain is the set of all x-coordinates on the graph. The domain is $(-\infty, \infty)$.
The range is the set of all y-coordinates on the graph. The range is $(-\infty, \infty)$.
Since each vertical line passes through the graph at most once, it is the graph of a function.

68. The domain is the set of all x-coordinates on the graph. The domain is $(-\infty, \infty)$.
The range is the set of all y-coordinates on the graph. The range is $(-\infty, 2]$.
Since each vertical line passes through the graph at most once, it is the graph of a function.

69. $f(x) = 2x - 1$
$\mathrm{D} = (-\infty, \infty)$
$\mathrm{R} = (-\infty, \infty)$

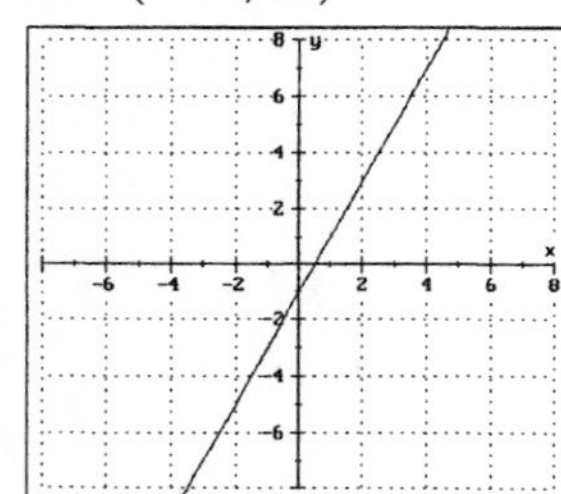

70. $f(x) = -x + 2$
$\mathrm{D} = (-\infty, \infty)$
$\mathrm{R} = (-\infty, \infty)$

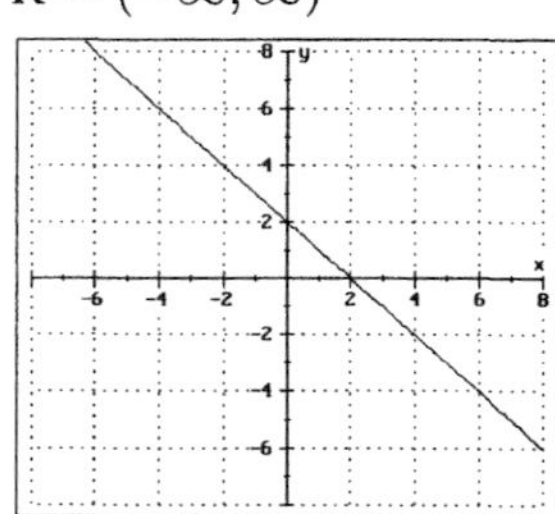

71. $2x - 3y = 6$
$\mathrm{D} = (-\infty, \infty)$
$\mathrm{R} = (-\infty, \infty)$

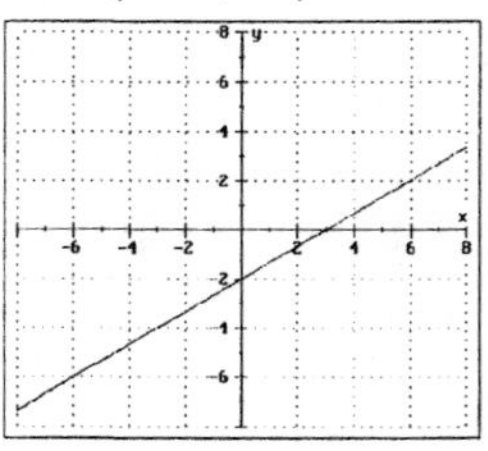

72. $3x + 2y = -6$
$\mathrm{D} = (-\infty, \infty)$
$\mathrm{R} = (-\infty, \infty)$

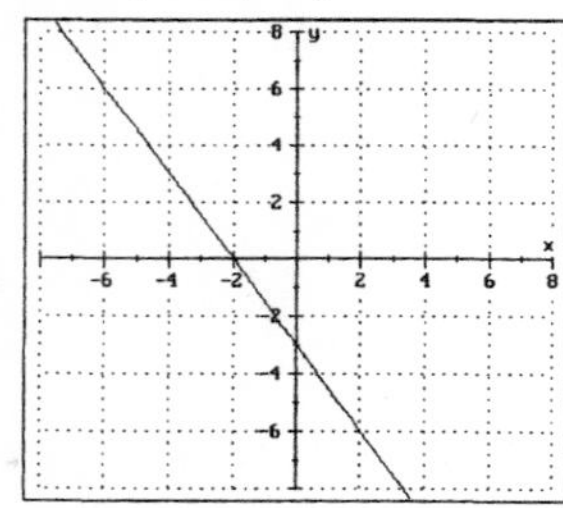

73. $y = 3x^2 + 2 \Rightarrow$ This is not a linear function because the exponent on x is 2.

74. $y = \dfrac{x-3}{2} \Rightarrow y = \dfrac{1}{2}x - \dfrac{3}{2} \Rightarrow$ This is a linear function $\left(m = \dfrac{1}{2}, b = -\dfrac{3}{2}\right)$.

75. $x = 3y - 4 \Rightarrow y = \dfrac{1}{3}x + \dfrac{4}{3} \Rightarrow$ This is a linear function $\left(m = \dfrac{1}{3}, b = \dfrac{4}{3}\right)$.

76. $x = \dfrac{8}{y} \Rightarrow y = \dfrac{8}{x} \Rightarrow y = 8x^{-1} \Rightarrow$ This is not a linear function because the exponent on x is -1.

77. Let $t = 3$ and find s.

$s = f(t) = -16t^2 + 256t$
$= -16(3)^2 + 256(3)$
$= -16(9) + 768$
$= -144 + 768$
$= 624$

The bullet will have a height of 624 ft.

78. Let $t = 20$ and find s.

$s = f(t) = -16t^2 + 512t + 64$
$= -16(20)^2 + 512(20) + 64$
$= -16(400) + 10240 + 64$
$= -6400 + 10240 + 64$
$= 3904$

The shell will have a height of 3904 ft.

79. Let $t = 1.5$ and find h.

$h = -16t^2 + 32t$
$h = -16(1.5)^2 + 32(1.5)$
$h = -16(2.25) + 48$
$h = -36 + 48 = 12$

The dolphin will be 12 feet above the water.

80. Let $m = 3$ and $v = 6$ and find E.

$E = \frac{1}{2}mv^2$
$E = \frac{1}{2}(3)(6)^2$
$E = \frac{1}{2}(3)(36)$
$E = \frac{1}{2}(108) = 54$

The kinetic energy is 54 joules.

81. Let $C = 25$ and find F.

$F(C) = \frac{9}{5}C + 32$
$= \frac{9}{5}(25) + 32$
$= 45 + 32$
$= 77$

The temperature is $77°F$.

82. Let $F = 14$ and find C.

$C(F) = \frac{5}{9}F - \frac{160}{9}$
$= \frac{5}{9}(14) - \frac{160}{9}$
$= \frac{70}{9} - \frac{160}{9} = -\frac{90}{9} = -10$

The temperature is $-10°\ C$.

83. Set $R(x)$ equal to $C(x)$ and solve for x.

$R(x) = C(x)$
$120x = 57.50x + 12000$
$62.50x = 12000$
$x = \frac{12000}{62.50}$
$x = 192$

The company must sell 192 players.

84. Let $x =$ the number of tires sold.
Set $R(x)$ equal to $C(x)$ and solve for x.

$R(x) = C(x)$
$130x = 93.50x + 15512.50$
$36.50x = 15512.50$
$x = \frac{15512.50}{36.50} = 425$

The company must sell 425 tires.

85. **a.** $I(h) = 0.10h + 50$

b. $I(115) = 0.10(115) + 50$
$= 11.5 + 50 = 61.5$

He makes $61.50.

86. **a.** $C(s) = 102s + 14{,}000$

b. $C(1{,}800) = 102(1{,}800) + 14{,}000$
$= 183{,}600 + 14{,}000$
$= 197{,}600$

The house will cost $197,600.

87. **Answers may vary.**

88. **Answers may vary.**

89. $f(x) + g(x) = 2x + 1 + x^2 = x^2 + 2x + 1 = g(x) + f(x)$

90. $f(x) - g(x) = 2x + 1 - x^2$; $g(x) - f(x) = x^2 - (2x + 1) = x^2 - 2x - 1$. They are not equal.

Exercise 2.5 (page 145)

1. 41, 43, 47 **2.** $(a+b)+c = a+(b+c)$ **3.** $a \cdot b = b \cdot a$

4. 0 **5.** 1 **6.** $\frac{3}{5}$

7. squaring **8.** cubing **9.** absolute value **10.** vertical

11. horizontal **12.** 5; up **13.** 2; down **14.** 5; to the right

15. 4; to the left **16.** both; x-coordinate

17. $f(x) = x^2 - 3$
Shift $y = x^2$ down 3.

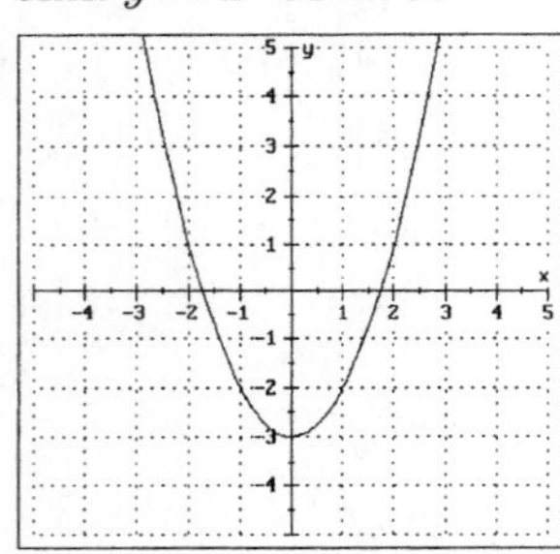

18. $f(x) = x^2 + 2$
Shift $y = x^2$ up 2.

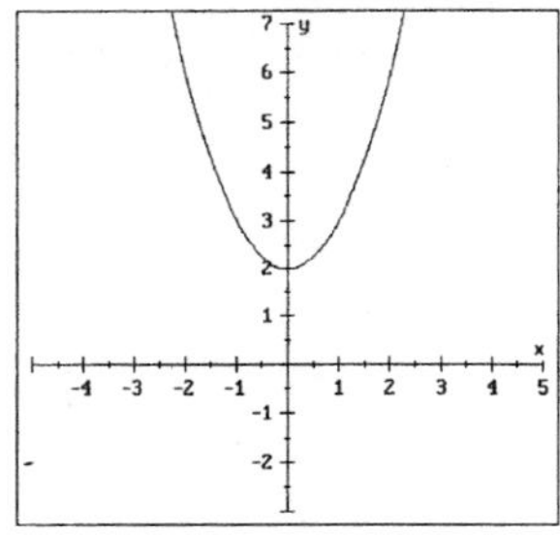

19. $f(x) = (x-1)^3$
Shift $y = x^3$ right 1.

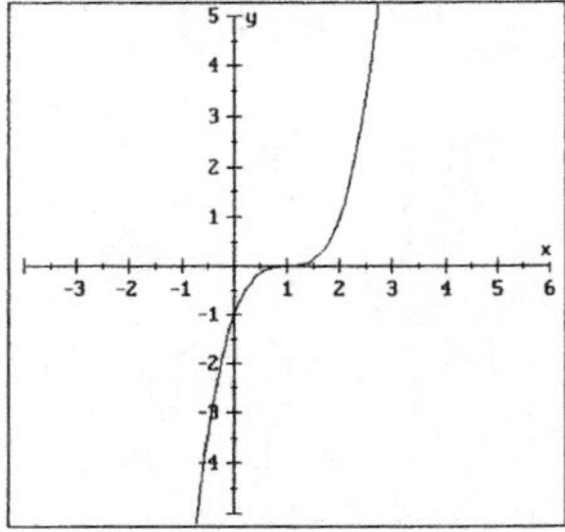

20. $f(x) = (x+1)^3$
Shift $y = x^3$ left 1.

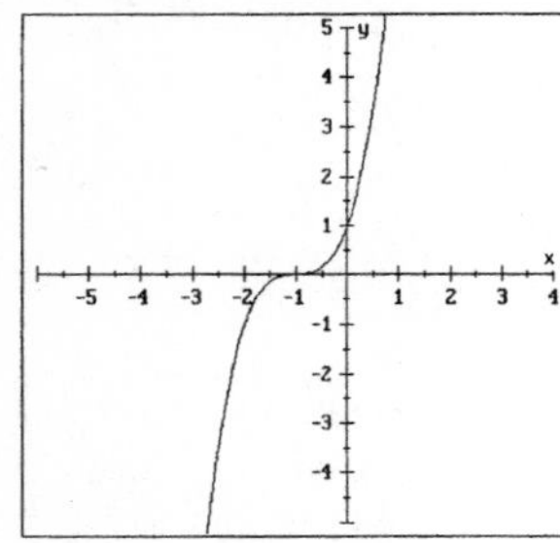

21. $f(x) = |x| - 2$
Shift $y = |x|$ down 2.

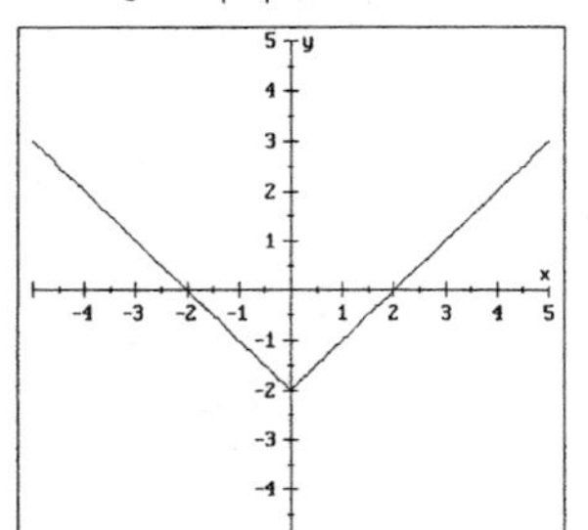

22. $f(x) = |x| + 1$
Shift $y = |x|$ up 1.

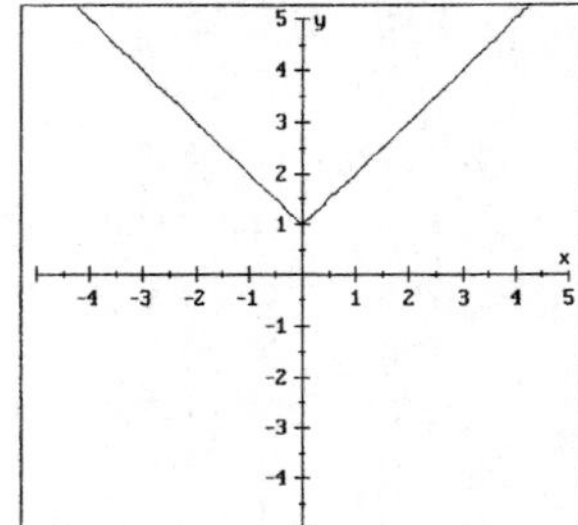

23. $f(x) = |x - 1|$
Shift $y = |x|$ right 1.

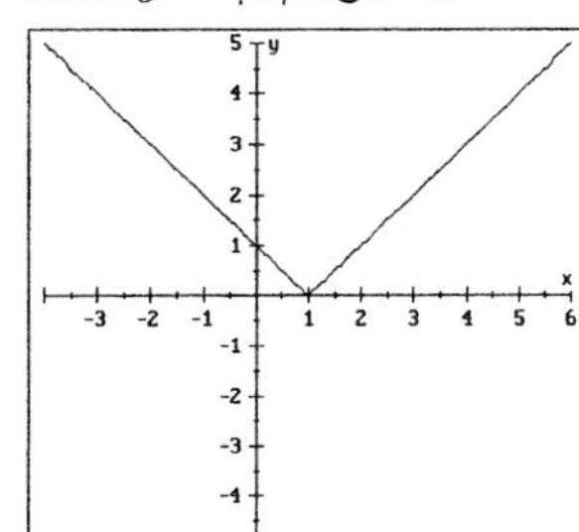

24. $f(x) = |x + 2|$
Shift $y = |x|$ left 2.

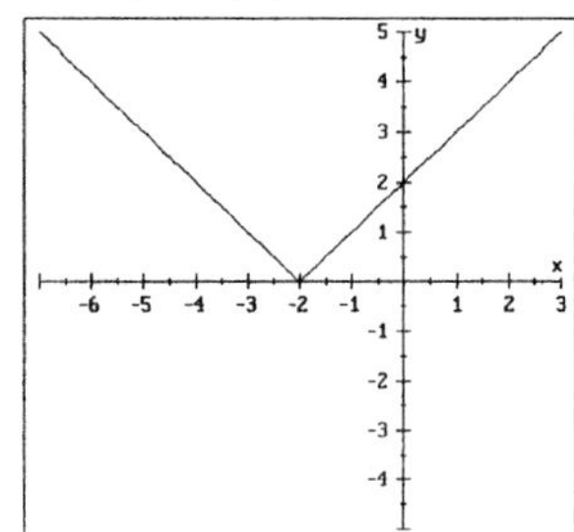

25. $f(x) = x^2 + 8$
x: $[-4, 4]$, y: $[-4, 4]$

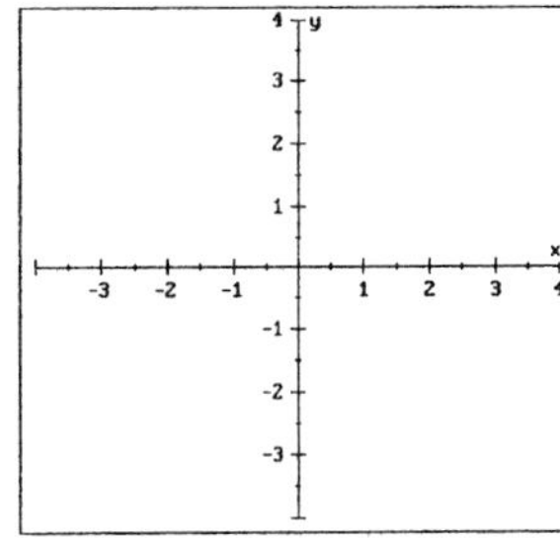

x: $[-7, 7]$, y: $[-2, 12]$

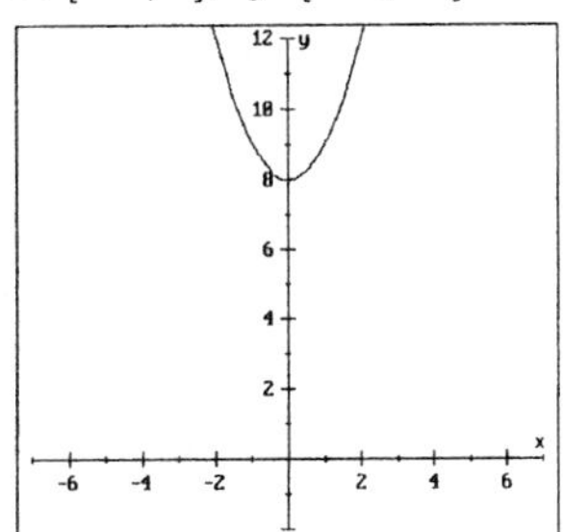

26. $f(x) = x^3 - 8$
x: $[-4, 4]$, y: $[-4, 4]$

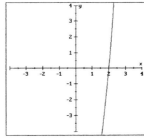

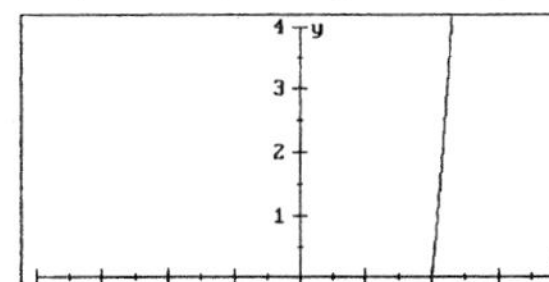

x: $[-7, 7]$, y: $[-12, 2]$

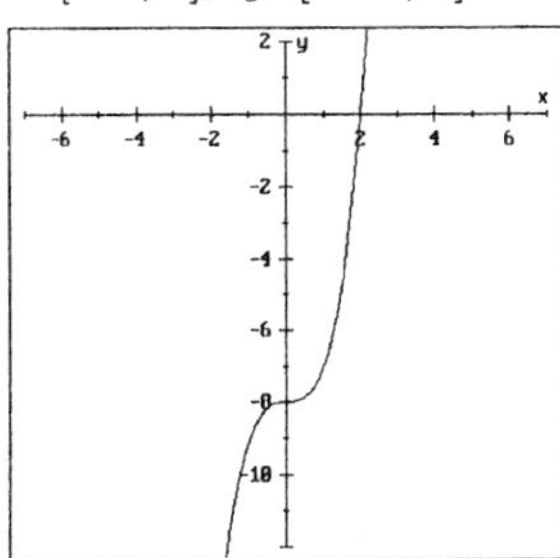

27. $f(x) = |x + 5|$
x: $[-4, 4]$, y: $[-4, 4]$

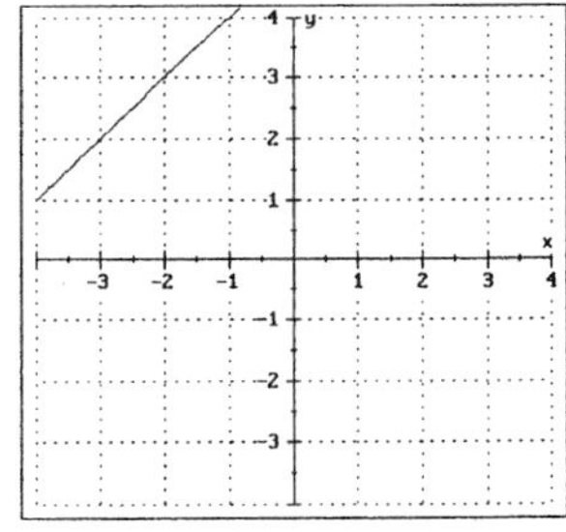

x: $[-12, 2]$, y: $[-4, 10]$

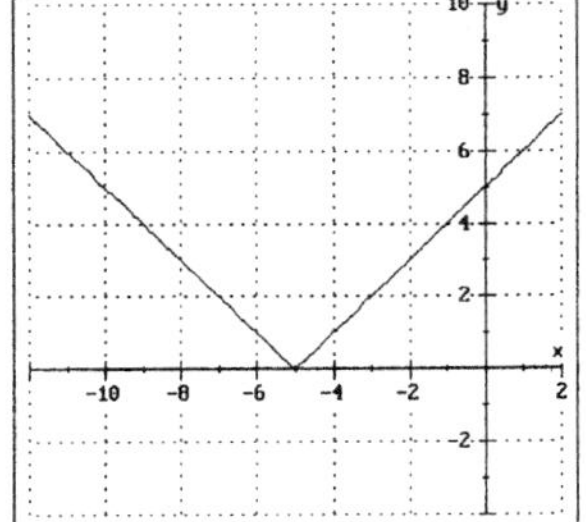

28. $f(x) = |x - 5|$
x: $[-4, 4]$, y: $[-4, 4]$

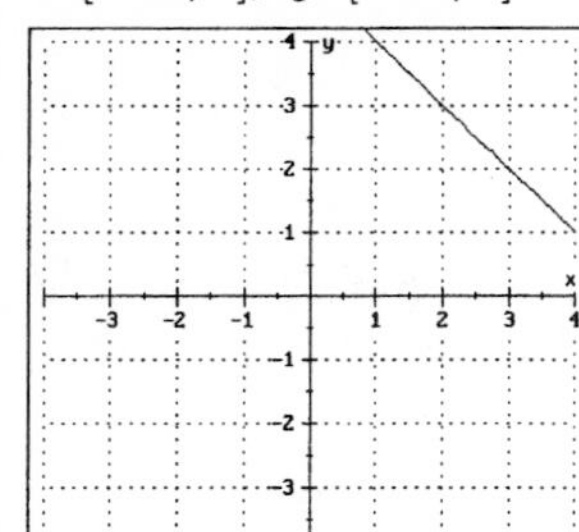

$f(x) = |x - 5|$
x: $[-2, 12]$, y: $[-4, 10]$

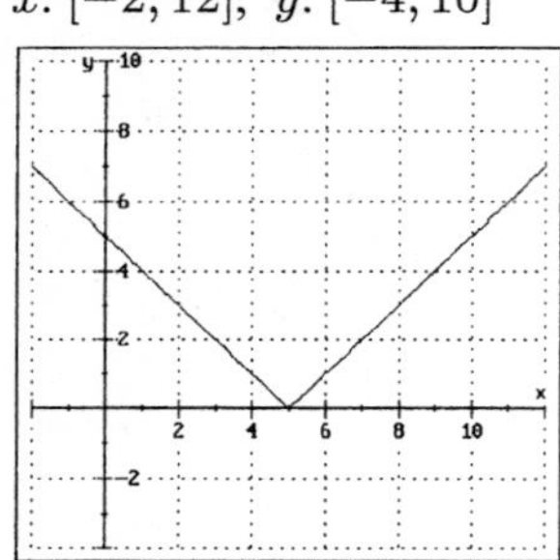

29. $f(x) = (x - 6)^2$
x: $[-4, 4]$, y: $[-4, 4]$

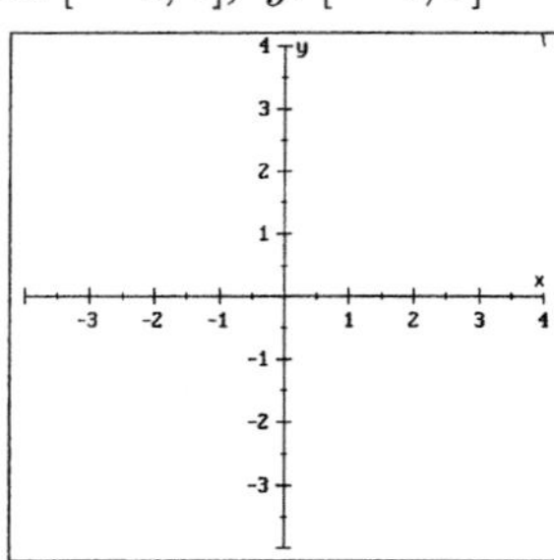

x: $[-2, 12]$, y: $[-2, 12]$

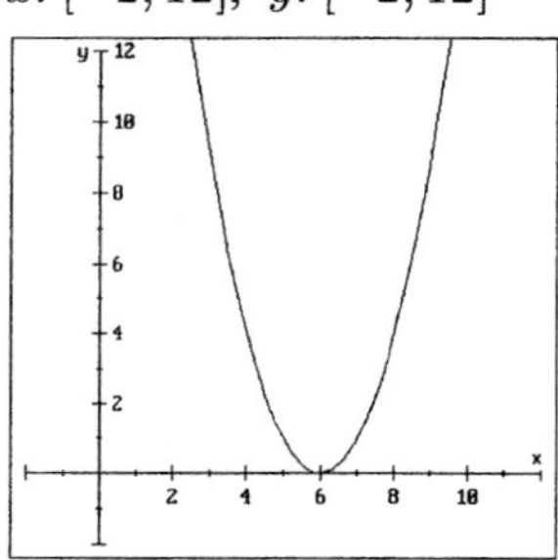

30. $f(x) = (x + 9)^2$
x: $[-4, 4]$, y: $[-4, 4]$

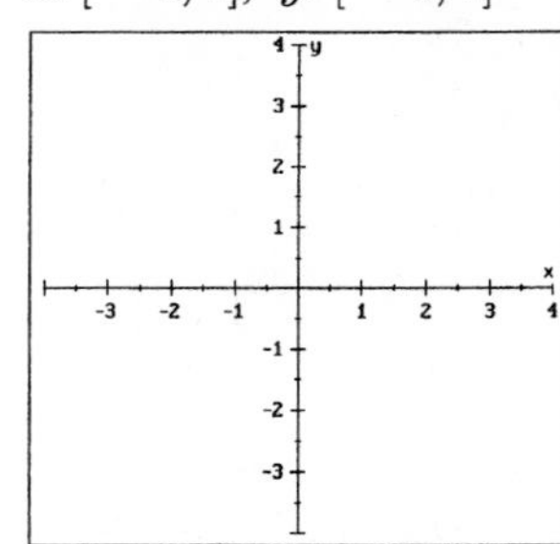

x: $[-13, 1]$, y: $[-3, 11]$

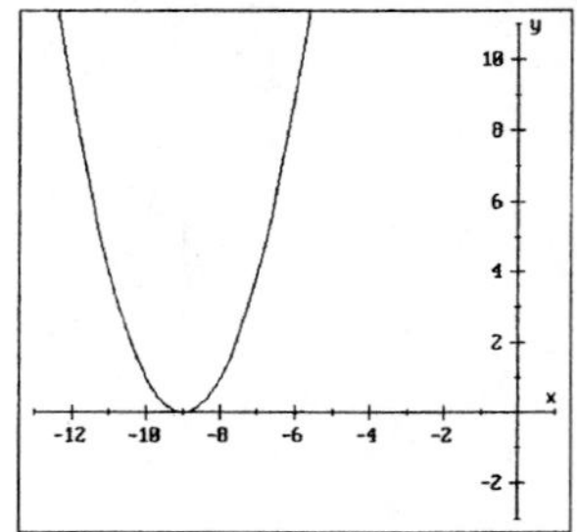

31. $f(x) = x^3 + 8$
x: $[-4, 4]$, y: $[-4, 4]$

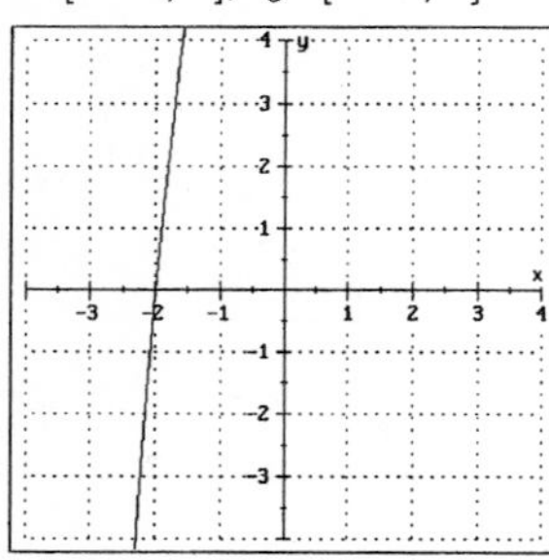

x: $[-10, 10]$, y: $[-4, 16]$

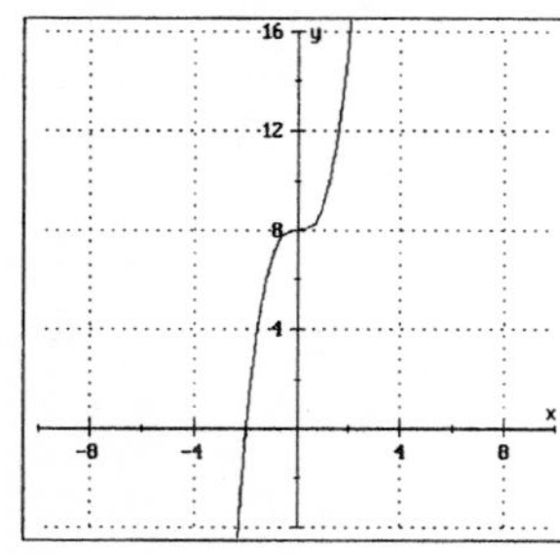

32. $f(x) = x^3 - 12$
x: $[-4, 4]$, y: $[-4, 4]$

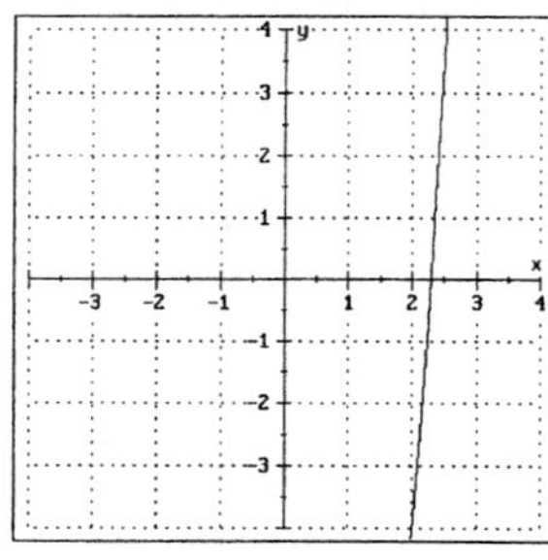

x: $[-10, 10]$, y: $[-16, 4]$

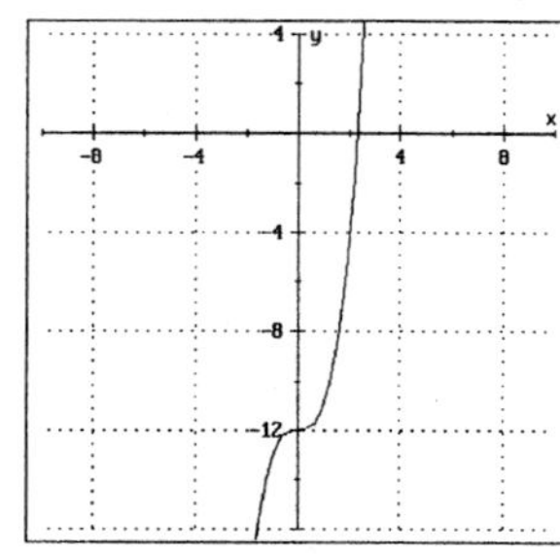

33. $f(x) = x^2 - 5$
Shift $f(x) = x^2$ D 5.

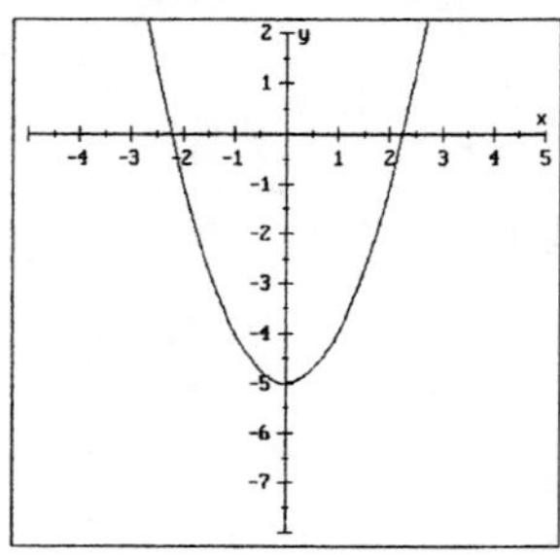

34. $f(x) = x^3 + 4$

Shift $f(x) = x^3$ U 4.

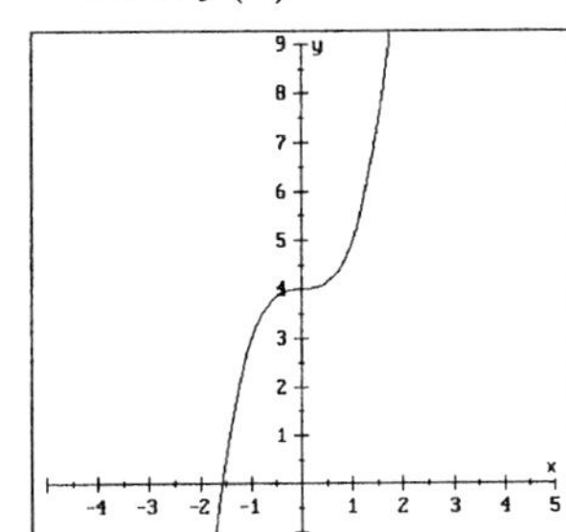

35. $f(x) = (x-1)^3$

Shift $f(x) = x^3$ R 1.

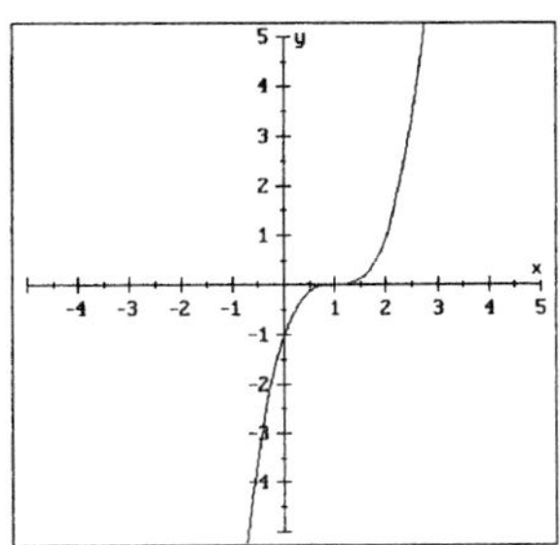

36. $f(x) = (x+4)^2$

Shift $f(x) = x^2$ L 4.

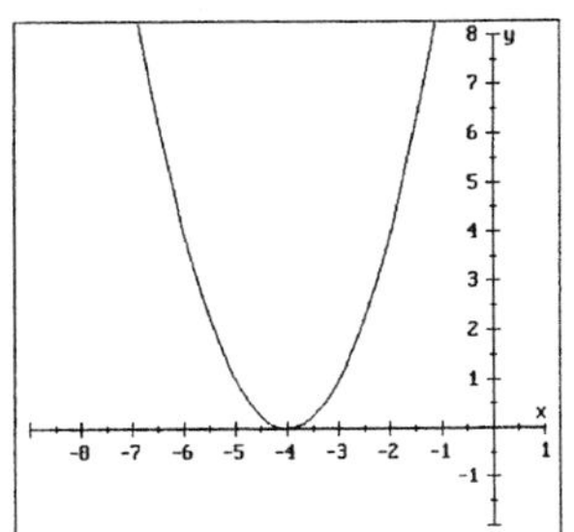

37. $f(x) = |x-2| - 1$

Shift $f(x) = |x|$ D 1, R 2.

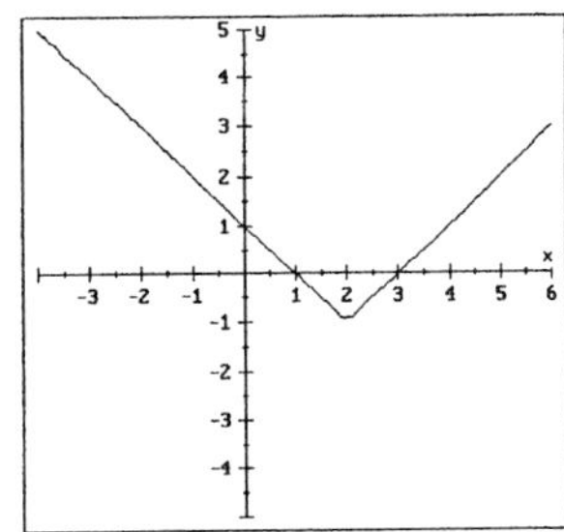

38. $f(x) = (x+2)^2 - 1$

Shift $f(x) = x^2$ D 1, L 2.

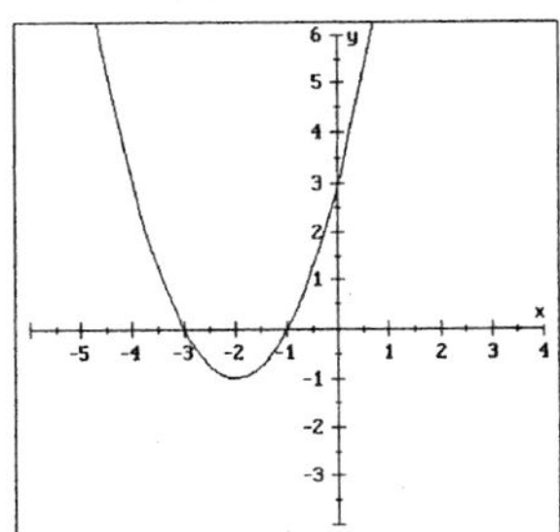

39. $f(x) = (x+1)^3 - 2$

Shift $f(x) = x^3$ D 2, L 1.

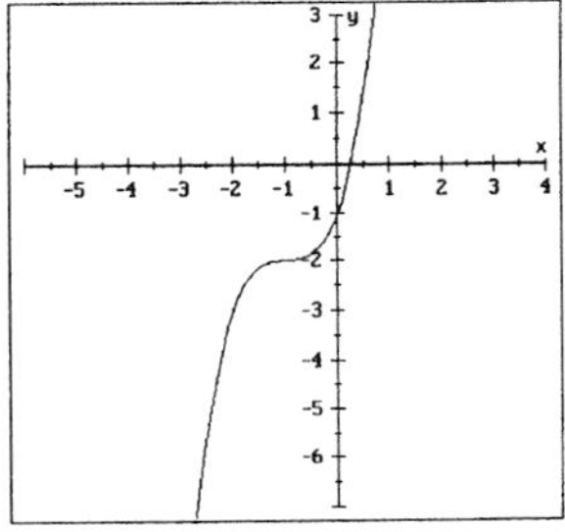

40. $f(x) = |x+4| + 3$

Shift $f(x) = |x|$ U 3, L 4.

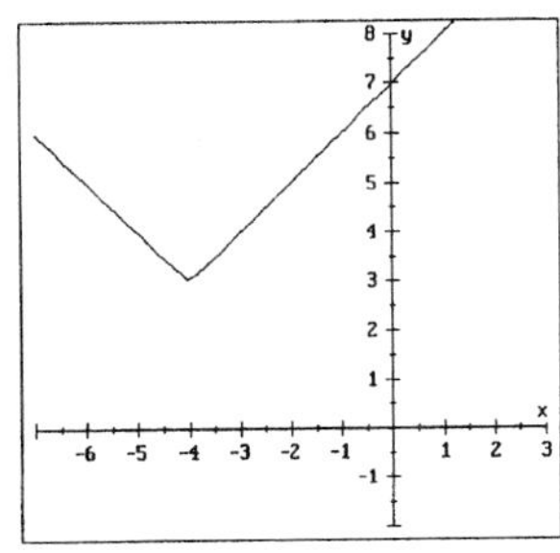

41. $f(x) = -|x| + 1$

Shift $f(x) = -|x|$ U 1.

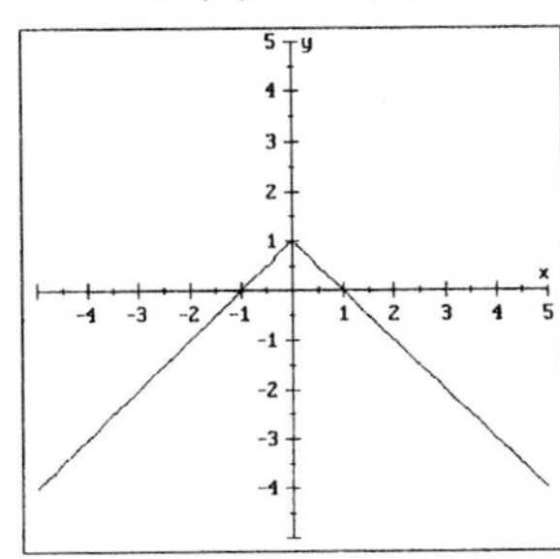

42. $f(x) = -x^3 - 2$

Shift $f(x) = -x^3$ D 2.

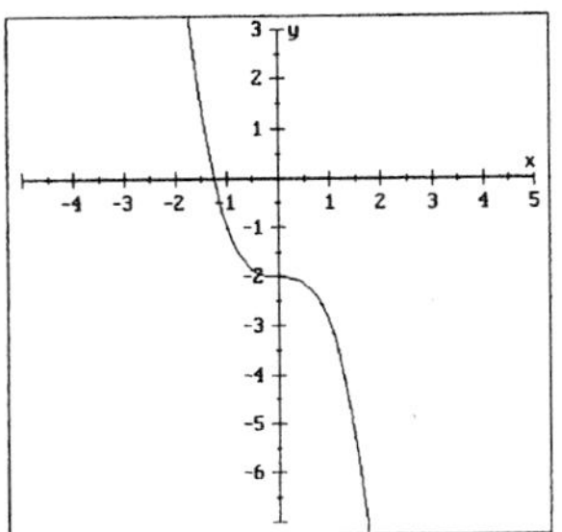

43. $f(x) = -(x-1)^2$
Shift $f(x) = -x^2$ R 1.

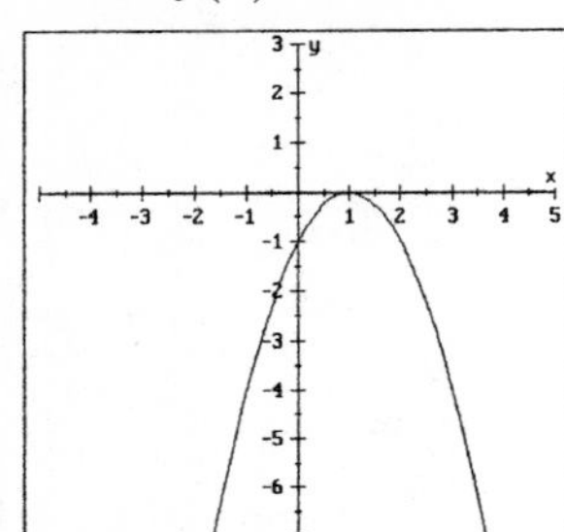

44. $f(x) = -(x+2)^2$
Shift $f(x) = -x^2$ L 2.

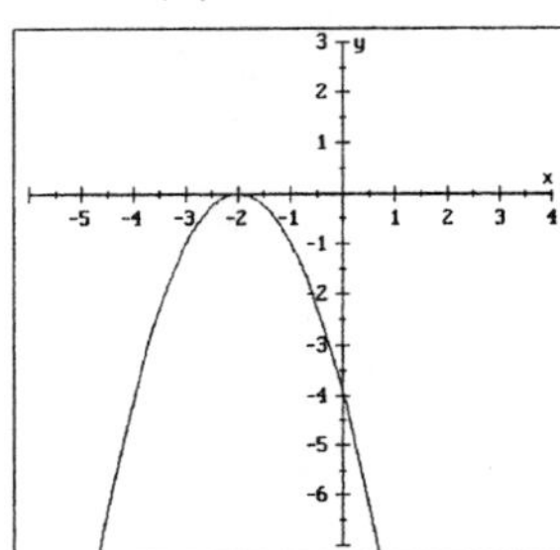

45. Graph $y = 3x + 6$ and $y = 0$, and find the x-coordinates of any points of intersection.

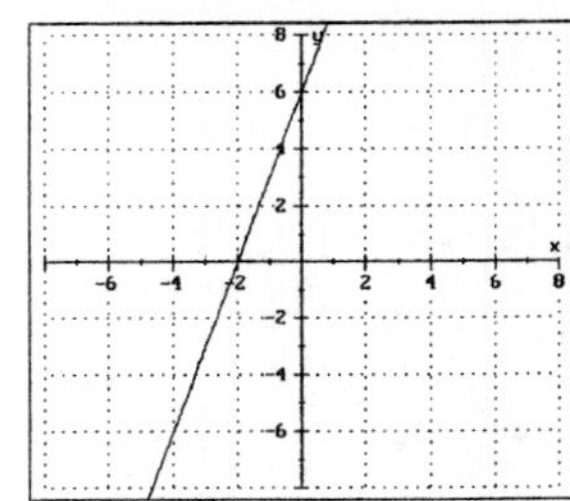

solution: $x = -2$

46. Graph $y = 7x - 21$ and $y = 0$, and find the x-coordinates of any points of intersection.

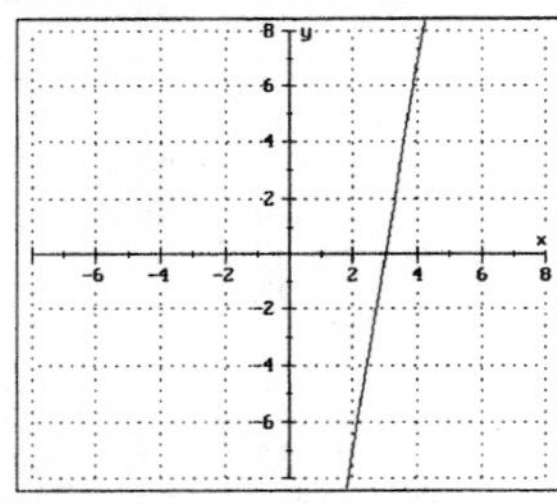

solution: $x = 3$

47. Graph $y = 4(x-1)$ and $y = 3x$, and find the x-coordinates of any points of intersection.

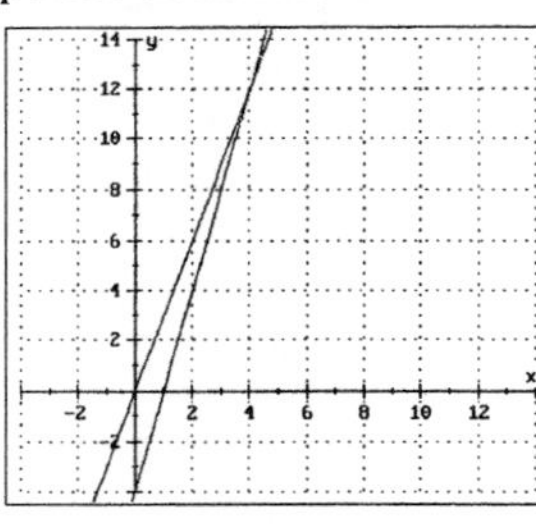

solution: $x = 4$

48. Graph $y = 4(x-3) - x$ and $y = x - 6$, and find the x-coordinates of any points of intersection.

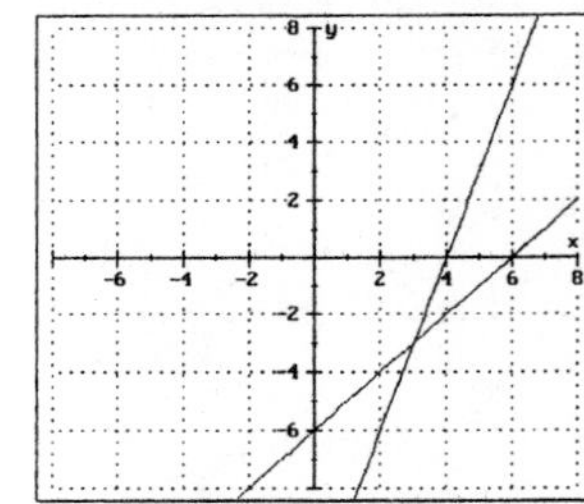

solution: $x = 3$

49. Graph $y = 11x + 6(3-x)$ and $y = 3$, and find the x-coordinates of any points of intersection.

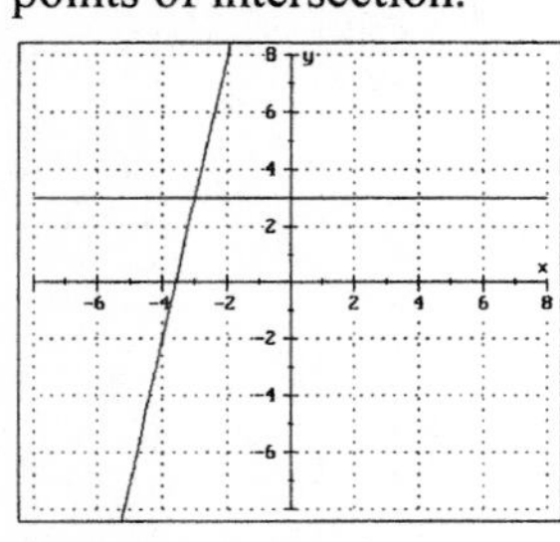

solution: $x = -3$

50. Graph $y = 2(x+2)$ and $y = 2(1-x) + 10$, and find the x-coordinates of any points of intersection.

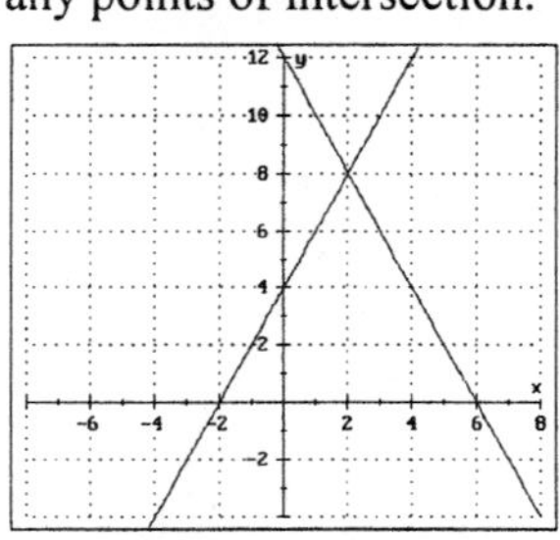

solution: $x = 2$

51-60. Answers may vary.

Chapter 2 Summary (page 148)

1.

x	y
-9	$\boxed{-10}$
$\boxed{-6}$	-8
-3	$\boxed{-6}$
$\boxed{0}$	-4
3	$\boxed{-2}$
$\boxed{6}$	0
$\boxed{9}$	2

$$\begin{aligned} 2x - 3y &= 12 \\ 2(-9) - 3y &= 12 \\ -18 - 3y &= 12 \\ -3y &= 30 \\ y &= -10 \end{aligned} \qquad \begin{aligned} 2x - 3y &= 12 \\ 2x - 3(-8) &= 12 \\ 2x + 24 &= 12 \\ 2x &= -12 \\ x &= -6 \end{aligned} \qquad \begin{aligned} 2x - 3y &= 12 \\ 2(-3) - 3y &= 12 \\ -6 - 3y &= 12 \\ -3y &= 18 \\ y &= -6 \end{aligned}$$

$$\begin{aligned} 2x - 3y &= 12 \\ 2x - 3(-4) &= 12 \\ 2x + 12 &= 12 \\ 2x &= 0 \\ x &= 0 \end{aligned} \qquad \begin{aligned} 2x - 3y &= 12 \\ 2(3) - 3y &= 12 \\ 6 - 3y &= 12 \\ -3y &= 6 \\ y &= -2 \end{aligned} \qquad \begin{aligned} 2x - 3y &= 12 \\ 2x - 3(0) &= 12 \\ 2x &= 12 \\ x &= 6 \end{aligned} \qquad \begin{aligned} 2x - 3y &= 12 \\ 2x - 3(2) &= 12 \\ 2x - 6 &= 12 \\ 2x &= 18 \\ x &= 9 \end{aligned}$$

2.

$$\begin{aligned} x + y &= 4 \\ 0 + y &= 4 \\ y &= 4 \end{aligned} \qquad \begin{aligned} x + y &= 4 \\ x + 0 &= 4 \\ x &= 4 \end{aligned}$$

$(0, 4)$ $(4, 0)$

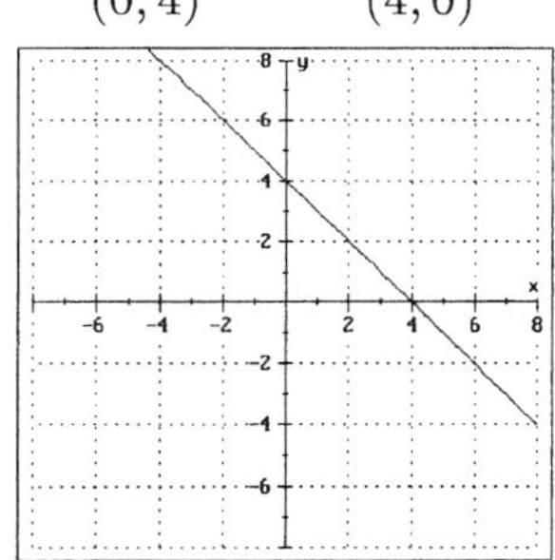

3.

$$\begin{aligned} 2x - y &= 8 \\ 2(0) - y &= 8 \\ -y &= 8 \\ y &= -8 \end{aligned} \qquad \begin{aligned} 2x - y &= 8 \\ 2x - 0 &= 8 \\ 2x &= 8 \\ x &= 4 \end{aligned}$$

$(0, -8)$ $(4, 0)$

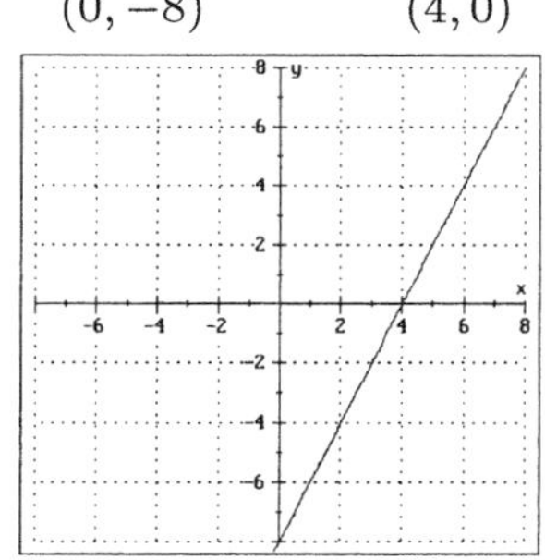

4.

$$\begin{aligned} y &= 3x + 4 \\ y &= 3(0) + 4 \\ y &= 0 + 4 \\ y &= 4 \end{aligned} \qquad \begin{aligned} y &= 3x + 4 \\ y &= 3(1) + 4 \\ y &= 3 + 4 \\ y &= 7 \end{aligned}$$

$(0, 4)$ $(1, 7)$

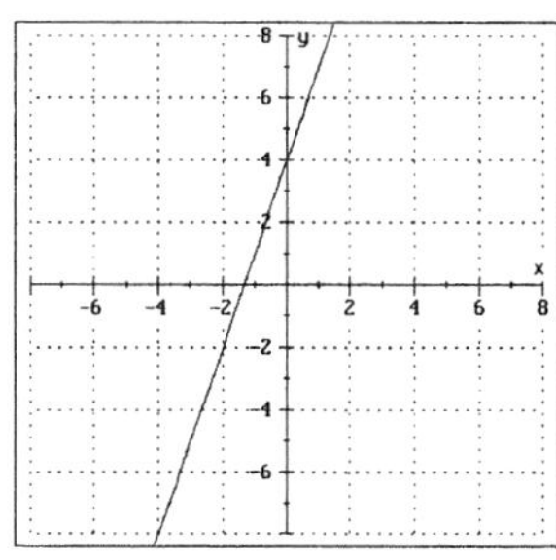

5.

$$\begin{aligned} x &= 4 - 2y \\ 0 &= 4 - 2y \\ 2y &= 4 \\ y &= 2 \\ &(0, 2) \end{aligned} \qquad \begin{aligned} x &= 4 - 2y \\ x &= 4 - 2(0) \\ x &= 4 \\ &(4, 0) \end{aligned}$$

6. $y = 4$ (horizontal)

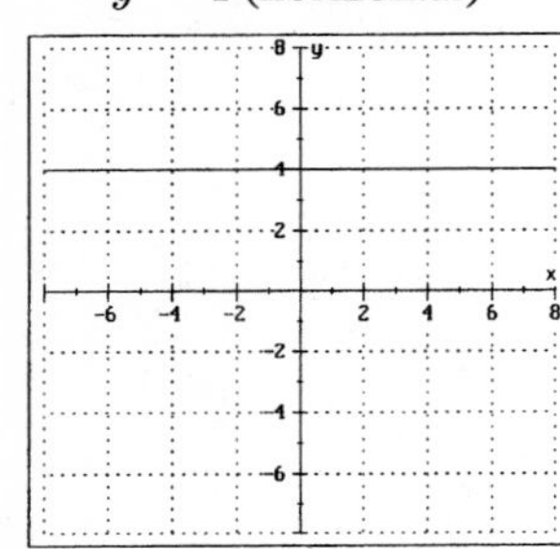

7. $x = -2$ (vertical)

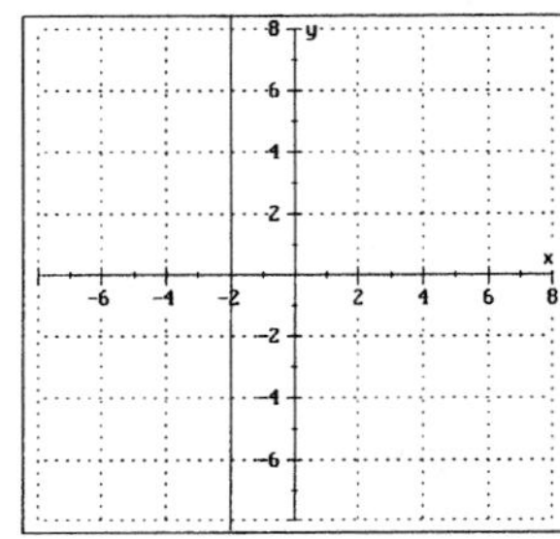

8.

$$\begin{aligned} 2(x + 3) &= x + 2 \\ 2x + 6 &= x + 2 \\ x &= -4 \text{ (vertical)} \end{aligned}$$

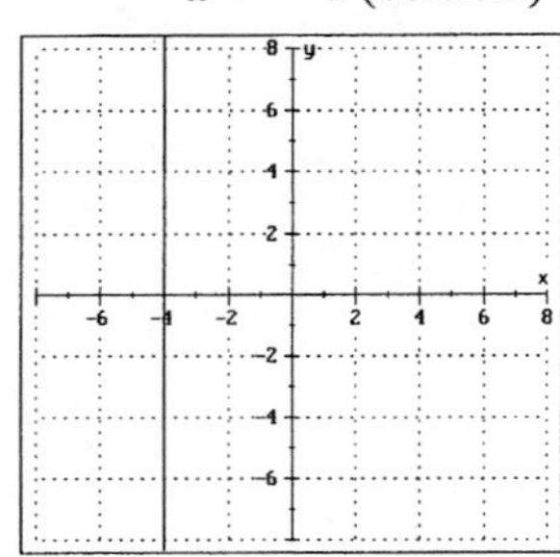

9.

$$\begin{aligned} 3y &= 2(y - 1) \\ 3y &= 2y - 2 \\ y &= -2 \text{ (horiozntal)} \end{aligned}$$

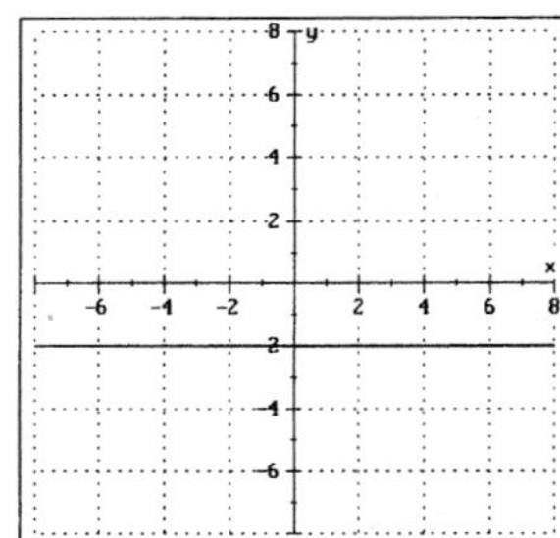

10. $x = \frac{x_1 + x_2}{2} = \frac{-3 + 6}{2} = \frac{3}{2}$; $y = \frac{y_1 + y_2}{2} = \frac{5 + 11}{2} = \frac{16}{2} = 8$; midpoint: $\left(\frac{3}{2}, 8\right)$

11. $m = \frac{\Delta y}{\Delta x} = \frac{8 - 5}{5 - 2} = \frac{3}{3} = 1$

12. $m = \frac{\Delta y}{\Delta x} = \frac{12 - (-2)}{6 - (-3)} = \frac{14}{9}$

13. $m = \frac{\Delta y}{\Delta x} = \frac{-6 - 4}{-5 - (-3)} = \frac{-10}{-2} = 5$

14. $m = \frac{\Delta y}{\Delta x} = \frac{-9 - (-4)}{-6 - 5} = \frac{-5}{-11} = \frac{5}{11}$

15. $m = \dfrac{\Delta y}{\Delta x} = \dfrac{4-4}{8-(-2)} = \dfrac{0}{10} = 0$

16. $m = \dfrac{\Delta y}{\Delta x} = \dfrac{8-(-4)}{-5-(-5)} = \dfrac{12}{0}$

undefined slope

17.
$$\begin{aligned} 2x - 3y &= 18 \\ -3y &= -2x + 18 \\ \frac{-3y}{-3} &= \frac{-2x}{-3} + \frac{18}{-3} \\ y &= \tfrac{2}{3}x - 6 \Rightarrow m = \tfrac{2}{3} \end{aligned}$$

18.
$$\begin{aligned} 2x + y &= 8 \\ y &= -2x + 8 \Rightarrow m = -2 \end{aligned}$$

19.
$$\begin{aligned} -2(x-3) &= 10 \\ -2x + 6 &= 10 \\ -2x &= 4 \\ x &= -2 \end{aligned}$$
The slope is undefined (vertical).

20.
$$\begin{aligned} 3y + 1 &= 7 \\ 3y &= 6 \\ y &= 2 \end{aligned}$$
$m = 0$ (horizontal)

21. $m_1 \neq m_2 \Rightarrow$ not parallel
$m_1 \cdot m_2 = 4\left(-\tfrac{1}{4}\right) = -1 \Rightarrow$ perpendicular

22. $m_1 = m_2 \Rightarrow$ parallel

23. $m_1 \neq m_2 \Rightarrow$ not parallel
$m_1 \cdot m_2 = 0.5\left(-\tfrac{1}{2}\right) = -0.25$
$\Rightarrow$ not perpendicular

24. $m_1 \neq m_2 \Rightarrow$ not parallel
$m_1 \cdot m_2 = 5(-0.2) = -1$
$\Rightarrow$ perpendicular

25. Let $y =$ sales and let $x =$ year of business.
$$m = \frac{\Delta y}{\Delta x} = \frac{y_2 - y_1}{x_1 - x_1} = \frac{130{,}000 - 65{,}000}{4-1} = \frac{65{,}000}{3} \approx \$21{,}666.67 \text{ per year}$$

26.
$$\begin{aligned} y - y_1 &= m(x - x_1) \\ y - 5 &= 3(x+8) \\ y - 5 &= 3x + 24 \\ -3x + y &= 29 \\ 3x - y &= -29 \end{aligned}$$

27. $m = \dfrac{\Delta y}{\Delta x} = \dfrac{4-(-9)}{-2-6} = \dfrac{13}{-8} = -\dfrac{13}{8}$
$$\begin{aligned} y - y_1 &= m(x - x_1) \\ y - 4 &= -\frac{13}{8}(x+2) \\ 8(y-4) &= 8\left(-\frac{13}{8}\right)(x+2) \\ 8y - 32 &= -13(x+2) \\ 8y - 32 &= -13x - 26 \\ 13x + 8y &= 6 \end{aligned}$$

28. Find the slope of the given line:

$$3x - 2y = 7$$
$$-2y = -3x + 7$$
$$y = \tfrac{3}{2}x - \tfrac{7}{2} \Rightarrow m = \tfrac{3}{2}$$

Use the parallel slope:

$$y - y_1 = m(x - x_1)$$
$$y + 5 = \tfrac{3}{2}(x + 3)$$
$$2(y + 5) = 2 \cdot \tfrac{3}{2}(x + 3)$$
$$2y + 10 = 3(x + 3)$$
$$2y + 10 = 3x + 9$$
$$-3x + 2y = -1 \Rightarrow 3x - 2y = 1$$

29. Find the slope of the given line:

$$3x - 2y = 7$$
$$-2y = -3x + 7$$
$$y = \tfrac{3}{2}x - \tfrac{7}{2} \Rightarrow m = \tfrac{3}{2}$$

Use the perpendicular slope:

$$y - y_1 = m(x - x_1)$$
$$y + 5 = -\tfrac{2}{3}(x + 3)$$
$$3(y + 5) = 3(-\tfrac{2}{3})(x + 3)$$
$$3y + 15 = -2(x + 3)$$
$$3y + 15 = -2x - 6$$
$$2x + 3y = -21$$

30. Let x represent the number of years since the copy machine was purchased, and let y represent its value. We know two points on the line: $(0, 8700)$ and $(5, 100)$.

Find the slope:

$$m = \frac{\Delta y}{\Delta x} = \frac{8700 - 100}{0 - 5} = \frac{8600}{-5} = -1720$$

Find the equation of the line:

$$y - y_1 = m(x - x_1)$$
$$y - 8700 = -1720(x - 0)$$
$$y = -1720x + 8700$$

31. $y = 6x - 4$ **is a function**, since each value of x corresponds to exactly one value of y.

32. $y = 4 - x$ **is a function**, since each value of x corresponds to exactly one value of y.

33. $y^2 = x$ **is not a function**, since $x = 9$ corresponds to both $y = 3$ and $y = -3$.

34. $|y| = x^2$ **is not a function**, since $x = 2$ corresponds to both $y = 4$ and $y = -4$.

35. $f(x) = 3x + 2$
$f(-3) = 3(-3) + 2 = -9 + 2 = -7$

36. $g(x) = x^2 - 4$
$g(8) = 8^2 - 4 = 64 - 4 = 60$

37. $g(x) = x^2 - 4$
$g(-2) = (-2)^2 - 4 = 4 - 4 = 0$

38. $f(x) = 3x + 2$
$f(5) = 3(5) + 2 = 15 + 2 = 17$

39. Since any value can be substituted for x, the domain is $(-\infty, \infty)$. The range is also $(-\infty, \infty)$.

40. Since any value can be substituted for x, the domain is $(-\infty, \infty)$. The range is also $(-\infty, \infty)$.

41. Since any value can be substituted for x, the domain is $(-\infty, \infty)$. Since $x^2 \geq 0$, $x^2 + 1$ must be greater than or equal to 1. Thus the range is $[1, \infty)$.

42. Since $x = 2$ makes the denominator equal to 0, the domain is $(-\infty, 2) \cup (2, \infty)$. Since the numerator will never equal 0, the fraction will never equal 0. The range is $(-\infty, 0) \cup (0, \infty)$.

43. Since $x = 3$ makes the denominator equal to 0, the domain is $(-\infty, 3) \cup (3, \infty)$. Since the numerator will never equal 0, the fraction will never equal 0. The range is $(-\infty, 0) \cup (0, \infty)$.

44. x can be any real number, so the domain is $(-\infty, \infty)$. y always has a value of 7, so the range is $\{7\}$.

45. Since each vertical line passes through the graph at most once, it is a function.

46. Since a vertical line can pass through the graph more than once, it is not a function.

47. Since a vertical line can pass through the graph more than once, it is not a function.

48. Since each vertical line passes through the graph at most once, it is a function.

49. $f(x) = x^2 - 3$
Shift $y = x^2$ down 3.

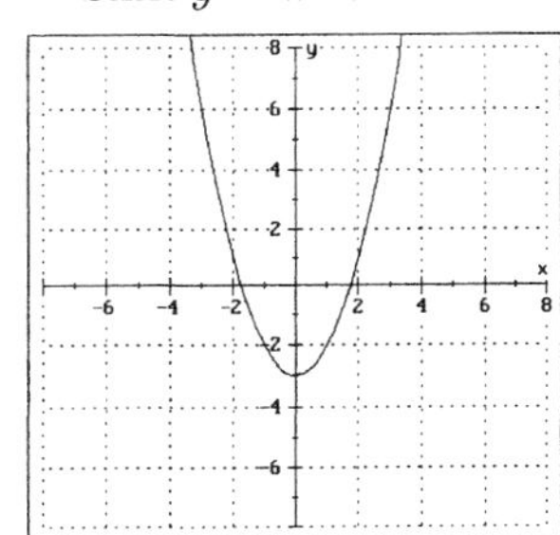

50. $f(x) = |x| - 4$
Shift $y = |x|$ down 4.

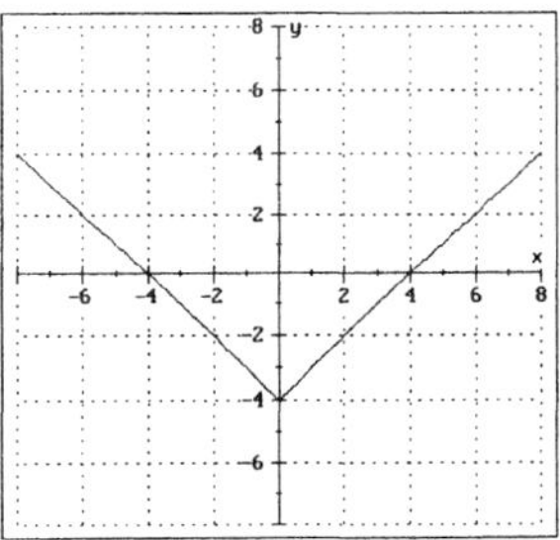

51. $f(x) = (x - 2)^3$
Shift $y = x^3$ right 2.

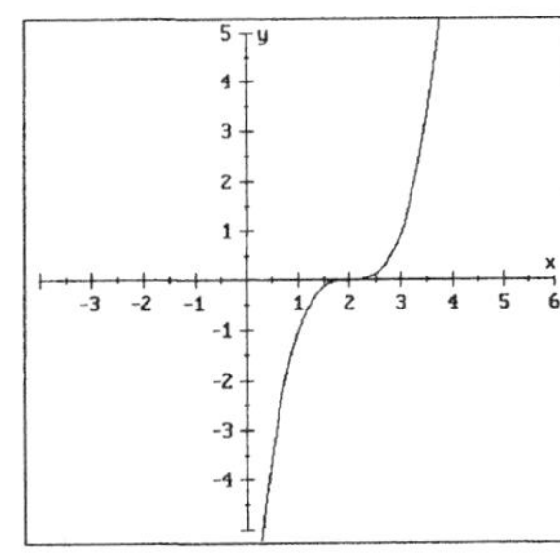

52. $f(x) = (x + 4)^2 - 3$
Shift $y = x^2$ down 3, left 4.

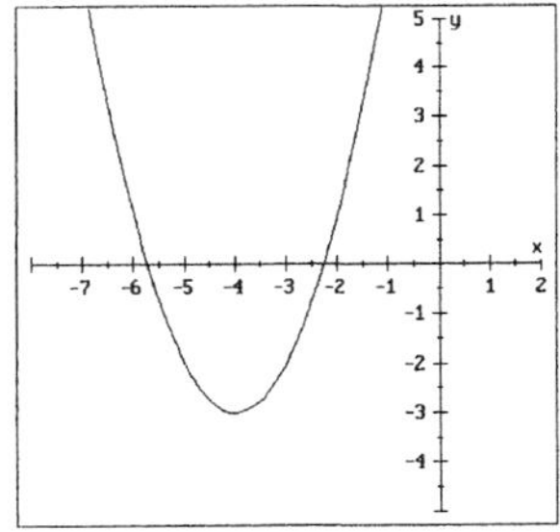

53-56. Compare your graphs to the graphs in numbers **49-52.**

57. $y = 3x + 2 \Rightarrow$ linear $(m = 3, b = 2)$

58. $y = \dfrac{x+5}{4} \Rightarrow y = \dfrac{1}{4}x + \dfrac{5}{4} \Rightarrow$ linear $\left(m = \frac{1}{4}, b = \frac{5}{4}\right)$

59. $4x - 3y = 12 \Rightarrow -3y = -4x + 12 \Rightarrow y = \dfrac{4}{3}x - 4 \Rightarrow$ linear $\left(m = \dfrac{4}{3}, b = -4\right)$

60. $y = x^2 - 25 \Rightarrow$ not linear (The exponent on x is 2.)

61. $f(x) = -|x-3|$

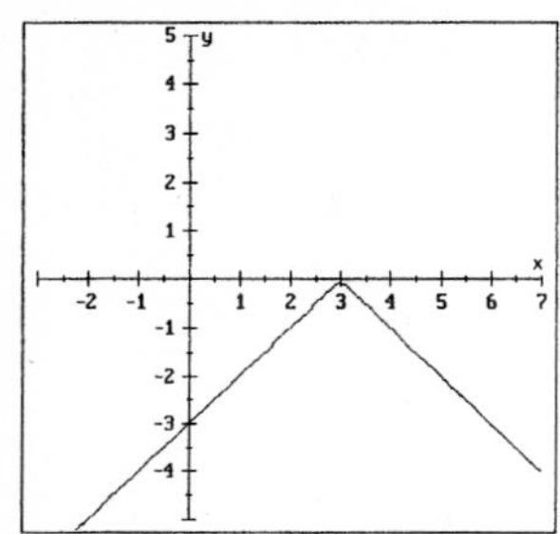

Chapter 2 Test (page 152)

1.

$$\begin{aligned} 2x - 5y &= 10 \\ 2(0) - 5y &= 10 \\ -5y &= 10 \\ y &= -2 \\ (0, -2)& \end{aligned} \qquad \begin{aligned} 2x - 5y &= 10 \\ 2x - 5(0) &= 10 \\ 2x &= 10 \\ x &= 5 \\ (5, 0)& \end{aligned}$$

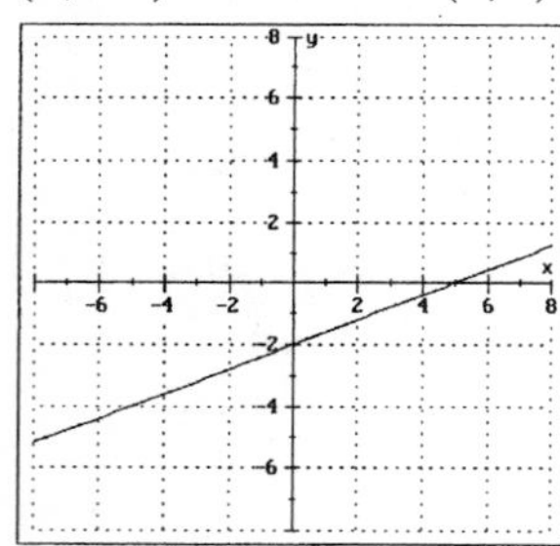

2.

$$\begin{aligned} y &= \frac{x-3}{5} \\ 0 &= \frac{x-3}{5} \\ 5(0) &= 5 \cdot \frac{x-3}{5} \\ 0 &= x - 3 \\ 3 &= x \end{aligned} \qquad \begin{aligned} y &= \frac{x-3}{5} \\ y &= \frac{0-3}{5} \\ y &= -\frac{3}{5} \end{aligned}$$

x-intercept: $(3, 0)$ $\quad$ y-intercept: $\left(0, -\frac{3}{5}\right)$

3. $x = \dfrac{x_1 + x_2}{2} = \dfrac{-3+4}{2} = \dfrac{1}{2}$; $y = \dfrac{y_1 + y_2}{2} = \dfrac{3 + (-2)}{2} = \dfrac{1}{2}$; midpoint: $\left(\frac{1}{2}, \frac{1}{2}\right)$

4. $m = \dfrac{\Delta y}{\Delta x} = \dfrac{4-8}{-2-6} = \dfrac{-4}{-8} = \dfrac{1}{2}$

5.

$$\begin{aligned} 2x - 3y &= 8 \\ -3y &= -2x + 8 \\ y &= \frac{2}{3}x - \frac{8}{3} \Rightarrow m = \frac{2}{3} \end{aligned}$$

6. The graph of $x = 12$ is a vertical line, so the slope is undefined.

7. The graph of $y = 12$ is a horizontal line, so the slope is 0.

8.
$$\begin{aligned} y-y_1 &= m(x-x_1) \\ y+5 &= \frac{2}{3}(x-4) \\ y+5 &= \frac{2}{3}x-\frac{8}{3} \\ y &= \frac{2}{3}x-\frac{8}{3}-5 \\ y &= \frac{2}{3}x-\frac{23}{3} \end{aligned}$$

9.
$$\begin{aligned} m &= \frac{\Delta y}{\Delta x} = \frac{6-(-10)}{-2-(-4)} = \frac{16}{2} = 8 \\ y-y_1 &= m(x-x_1) \\ y-6 &= 8(x+2) \\ y-6 &= 8x+16 \\ -8x+y &= 22 \\ 8x-y &= -22 \end{aligned}$$

10.
$$\begin{aligned} -2(x-3) &= 3(2y+5) \\ -2x+6 &= 6y+15 \\ 6y+15 &= -2x+6 \\ 6y &= -2x-9 \\ y &= \tfrac{-2}{6}x-\tfrac{9}{6} \\ y &= -\tfrac{1}{3}x-\tfrac{3}{2} \end{aligned}$$
$m=-\frac{1}{3},\ b=-\frac{3}{2} \Rightarrow \left(0,-\frac{3}{2}\right)$

11.
$$\begin{aligned} 4x-y &= 12 \\ -y &= -4x+12 \\ y &= 4x-12 \Rightarrow m=4 \end{aligned}$$
$y=\frac{1}{4}x+3 \Rightarrow m=\frac{1}{4}$
neither parallel nor perpendicular

12. $y=-\frac{2}{3}x+4 \Rightarrow m=-\frac{2}{3}$
$2y=3x-3$
$y=\frac{3}{2}x-\frac{3}{2} \Rightarrow m=\frac{3}{2}$
perpendicular

13. $y=\frac{3}{2}x-7 \Rightarrow m=\frac{3}{2}$
Use the parallel slope:
$$\begin{aligned} y-y_1 &= m(x-x_1) \\ y-0 &= \tfrac{3}{2}(x-0) \\ y &= \tfrac{3}{2}x \end{aligned}$$

14. $y=-\frac{2}{3}x-7 \Rightarrow m=-\frac{2}{3}$
Use the perpendicular slope:
$$\begin{aligned} y-y_1 &= m(x-x_1) \\ y-6 &= \tfrac{3}{2}(x+3) \\ y-6 &= \tfrac{3}{2}x+\tfrac{9}{2} \\ y &= \tfrac{3}{2}x+\tfrac{21}{2} \end{aligned}$$

15. $|y|=x$ **is not a function**, since $x=2$ corresponds to both $y=2$ and $y=-2$.

16. Domain: $(-\infty,\infty)$; Range: $[0,\infty)$

17. Domain: $(-\infty,\infty)$; Range:$(-\infty,\infty)$

18. $f(x)=3x+1$
$f(3)=3(3)+1=9+1=10$

19. $g(x)=x^2-2$
$g(0)=0^2-2=0-2=-2$

20. $f(x)=3x+1$
$f(a)=3a+1$

21. $g(x)=x^2-2$
$g(-x)=(-x)^2-2=x^2-2$

22. function

23. not a function

24. $f(x) = x^2 - 1$
Shift $y = x^2$ down 1.

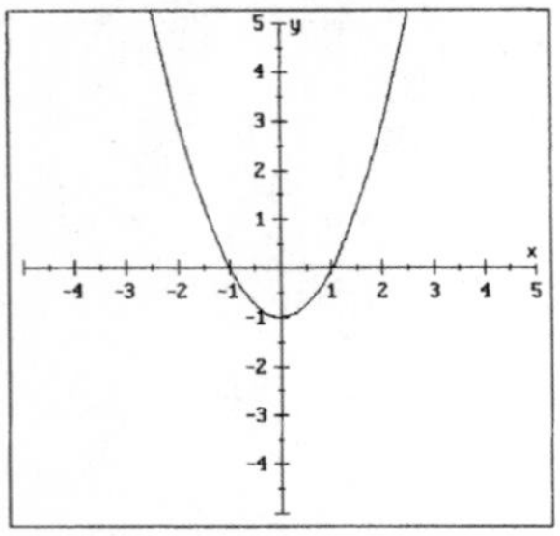

25. $f(x) = -|x + 2|$
Shift $y = -|x|$ left 2.

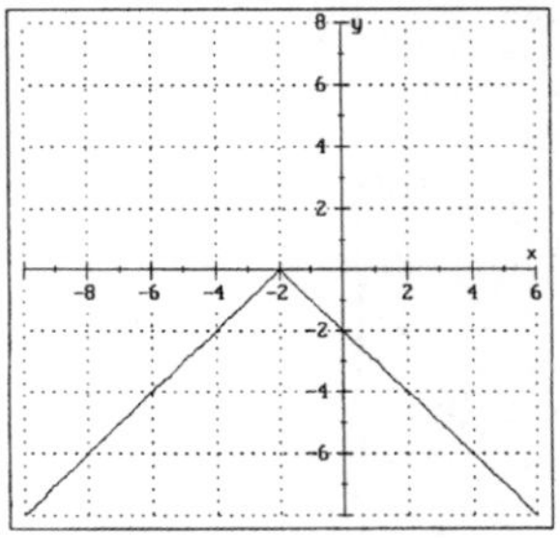

Cumulative Review Exercises (page 153)

1. natural numbers: $1, 2, 6, 7$

2. whole numbers: $0, 1, 2, 6, 7$

3. rational numbers: $-2, 0, 1, 2, \frac{13}{12}, 6, 7$

4. irrational numbers: $\sqrt{5}, \pi$

5. negative numbers: -2

6. real numbers: $-2, 0, 1, 2, \frac{13}{12}, 6, 7, \sqrt{5}, \pi$

7. prime numbers: $2, 7$

8. composite numbers: 6

9. even numbers: $-2, 0, 2, 6$

10. odd numbers: $1, 7$

11. $\{x | -2 < x \le 5\} \Rightarrow$ 

12. $[-5, 0) \cup [3, 6] \Rightarrow$ 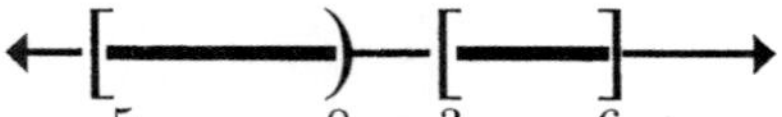

13. $-|5| + |-3| = -5 + 3 = -2$

14. $\dfrac{|-5| + |-3|}{-|4|} = \dfrac{5+3}{-4} = \dfrac{8}{-4} = -2$

15. $2 + 4 \cdot 5 = 2 + 20 = 22$

16. $\dfrac{8-4}{2-4} = \dfrac{4}{-2} = -2$

17. $20 \div (-10 \div 2) = 20 \div (-5) = -4$

18. $\dfrac{6 + 3(6+4)}{2(3-9)} = \dfrac{6 + 3(10)}{2(-6)} = \dfrac{6+30}{-12}$
$= \dfrac{36}{-12} = -3$

19. $-x - 2y = -2 - 2(-3) = -2 + 6 = 4$

20. $\dfrac{x^2 - y^2}{2x + y} = \dfrac{2^2 - (-3)^2}{2(2) + (-3)} = \dfrac{4-9}{4-3} = \dfrac{-5}{1}$
$= -5$

CUMULATIVE REVIEW EXERCISES

21. associative property of addition

22. distributive property

23. commutative property of addition

24. associative property of multiplication

25. $(x^2y^3)^4 = (x^2)^4(y^3)^4 = x^8y^{12}$

26. $\dfrac{c^4c^8}{(c^5)^2} = \dfrac{c^{12}}{c^{10}} = c^2$

27. $\left(-\dfrac{a^3b^{-2}}{ab}\right)^{-1} = \left(-\dfrac{a^2}{b^3}\right)^{-1} = -\dfrac{b^3}{a^2}$

28. $\left(\dfrac{-3a^3b^{-2}}{6a^{-2}b^3}\right)^0 = 1$

29. $0.00000497 = 4.97 \times 10^{-6}$

30. $9.32 \times 10^8 = 932{,}000{,}000$

31.
$$\begin{aligned} 2x - 5 &= 11 \\ 2x &= 16 \\ x &= 8 \end{aligned}$$

32.
$$\begin{aligned} \frac{2x-6}{3} &= x + 7 \\ 3 \cdot \frac{2x-6}{3} &= 3(x+7) \\ 2x - 6 &= 3x + 21 \\ -x &= 27 \\ x &= -27 \end{aligned}$$

33.
$$\begin{aligned} 4(y-3) + 4 &= -3(y+5) \\ 4y - 12 + 4 &= -3y - 15 \\ 4y - 8 &= -3y - 15 \\ 7y &= -7 \\ y &= -1 \end{aligned}$$

34.
$$\begin{aligned} 2x - \frac{3(x-2)}{2} &= 7 - \frac{x-3}{3} \\ 6(2x) - 6 \cdot \frac{3x-6}{2} &= 6(7) - 6 \cdot \frac{x-3}{3} \\ 12x - 3(3x-6) &= 42 - 2(x-3) \\ 12x - 9x + 18 &= 42 - 2x + 6 \\ 3x + 18 &= 48 - 2x \\ 5x &= 30 \\ x &= 6 \end{aligned}$$

35.
$$\begin{aligned} S &= \frac{n(a+l)}{2} \\ 2S &= 2 \cdot \frac{n(a+l)}{2} \\ 2S &= n(a+l) \\ \frac{2S}{n} &= \frac{n(a+l)}{n} \\ \frac{2S}{n} &= a + l \\ \frac{2S}{n} - l &= a, \text{ or } a = \frac{2S}{n} - l \end{aligned}$$

36.
$$\begin{aligned} A &= \frac{1}{2}h(b_1 + b_2) \\ 2A &= 2 \cdot \frac{1}{2}h(b_1 + b_2) \\ 2A &= h(b_1 + b_2) \\ \frac{2A}{b_1 + b_2} &= \frac{h(b_1 + b_2)}{b_1 + b_2} \\ \frac{2A}{b_1 + b_2} &= h, \text{ or } h = \frac{2A}{b_1 + b_2} \end{aligned}$$

CUMULATIVE REVIEW EXERCISES

37. Let x represent the first even integer. Then $x+2$ and $x+4$ represent the others.

$$\boxed{\text{1st}}+\boxed{\text{2nd}}+\boxed{\text{3rd}}=\boxed{\text{Sum of the integers}}$$

$$\begin{aligned} x+x+2+x+4&=90 \\ 3x+6&=90 \\ 3x&=84 \\ x&=28 \end{aligned}$$

The integers are 28, 30 and 32.

38. Let w represent the width of the rectangle. Then $3w$ represents the length.

$$2\cdot\boxed{\text{Length}}+2\cdot\boxed{\text{Width}}=\boxed{\text{Perimeter}}$$

$$\begin{aligned} 2(3w)+2w&=112 \\ 6w+2w&=112 \\ 8w&=112 \\ w&=14 \end{aligned}$$

The dimensions are 14 cm by 42 cm.

39. $2x-3y=6$

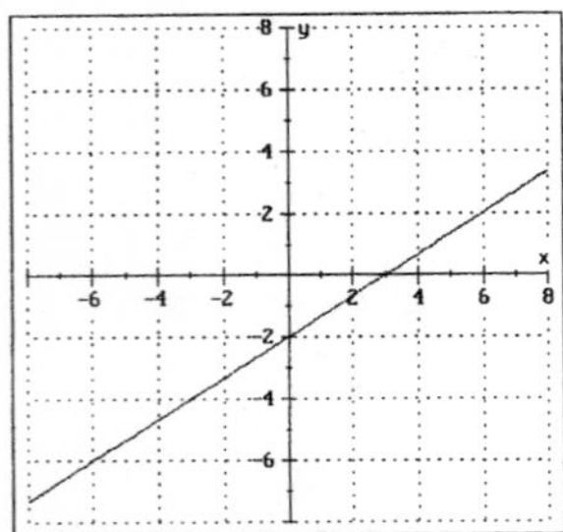

The equation defines a function.

40. $m=\dfrac{\Delta y}{\Delta x}=\dfrac{5-(-9)}{-2-8}=\dfrac{14}{-10}=-\dfrac{7}{5}$

41. Find the slope and use slope-intercept form:

$$m=\frac{\Delta y}{\Delta x}=\frac{5-(-9)}{-2-8}=\frac{14}{-10}=-\frac{7}{5}$$

$$\begin{aligned} y-y_1&=m(x-x_1) \\ y-5&=-\frac{7}{5}(x+2) \\ y-5&=-\frac{7}{5}x-\frac{14}{5} \\ y&=-\frac{7}{5}x+\frac{11}{5} \end{aligned}$$

42. Find the slope of the given line, and use that slope to find the equation of the desired line:

$$\begin{aligned} 3x+y&=8 \\ y&=-3x+8\Rightarrow m=-3 \\ y-y_1&=m(x-x_1) \\ y-3&=-3(x+2) \\ y-3&=-3x-6 \\ y&=-3x-3 \end{aligned}$$

43.

$$\begin{aligned} f(x)&=3x^2+2 \\ f(-1)&=3(-1)^2+2 \\ f(-1)&=3(1)+2=3+2=5 \end{aligned}$$

44.

$$\begin{aligned} g(x)&=2x-1 \\ g(0)&=2(0)-1=0-1=-1 \end{aligned}$$

45.

$$\begin{aligned} g(x)&=2x-1 \\ g(t)&=2t-1 \end{aligned}$$

46.

$$\begin{aligned} f(x)&=3x^2+2 \\ f(-r)&=3(-r)^2+2=3r^2+2 \end{aligned}$$

47. $y = -x^2 + 1$

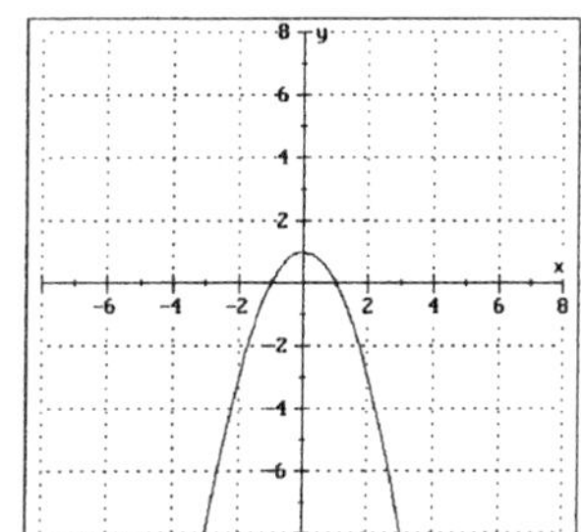

The equation describes a function.
domain: $(-\infty, \infty)$; range: $(-\infty, 1]$

48. $y = \left|\frac{1}{2}x - 3\right|$

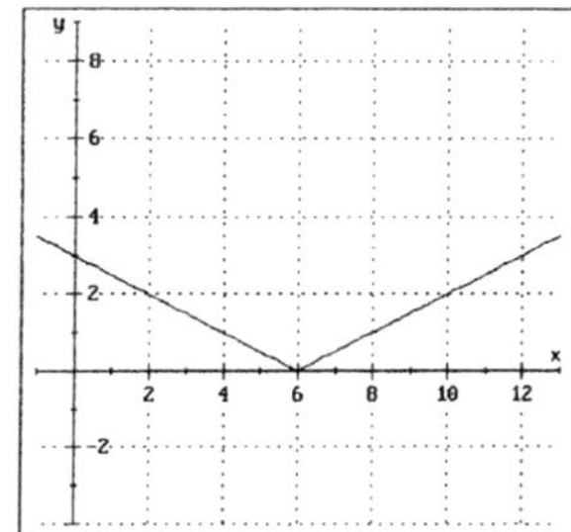

The equation describes a function.
domain: $(-\infty, \infty)$; range: $[0, \infty)$

Exercise 3.1 (page 162)

1. $93{,}000{,}000 = 9.3 \times 10^7$

2. $0.0000000236 = 2.36 \times 10^{-8}$

3. $$\begin{aligned} 345 \times 10^2 &= 3.45 \times 10^2 \times 10^2 \\ &= 3.45 \times 10^4 \end{aligned}$$

4. $$\begin{aligned} 752 \times 10^{-5} &= 7.52 \times 10^2 \times 10^{-5} \\ &= 7.52 \times 10^{-3} \end{aligned}$$

5. system

6. consistent

7. inconsistent

8. independent

9. dependent

10. equivalent

11. $$\begin{aligned} y &= 2x \\ 2 &\stackrel{?}{=} 2(1) \\ 2 &= 2 \end{aligned} \qquad \begin{aligned} y &= \frac{1}{2}x + \frac{3}{2} \\ 2 &\stackrel{?}{=} \frac{1}{2}(1) + \frac{3}{2} \\ 2 &\stackrel{?}{=} \frac{1}{2} + \frac{3}{2} \\ 2 &= \frac{4}{2} \end{aligned}$$

$(1, 2)$ is a solution.

12. $$\begin{aligned} y &= 3x + 5 \\ 2 &\stackrel{?}{=} 3(-1) + 5 \\ 2 &\stackrel{?}{=} -3 + 5 \\ 2 &= 2 \end{aligned} \qquad \begin{aligned} y &= x + 4 \\ 2 &\stackrel{?}{=} -1 + 4 \\ 2 &\neq 3 \end{aligned}$$

$(-1, 2)$ is not a solution.

13. $$\begin{aligned} y &= \frac{1}{2}x - 2 \\ -3 &\stackrel{?}{=} \frac{1}{2}(2) - 2 \\ -3 &\stackrel{?}{=} 1 - 2 \\ -3 &\neq -1 \end{aligned}$$

$(2, -3)$ is not a solution.

14. $$\begin{aligned} 4x - y &= -19 \\ 4(-4) - 3 &\stackrel{?}{=} -19 \\ -16 - 3 &\stackrel{?}{=} -19 \\ -19 &= -19 \end{aligned} \qquad \begin{aligned} 3x + 2y &= -6 \\ 3(-4) + 2(3) &\stackrel{?}{=} -6 \\ -12 + 6 &\stackrel{?}{=} -6 \\ -6 &= -6 \end{aligned}$$

$(-4, 3)$ is a solution.

15. $\begin{cases} x+y=6 \\ x-y=2 \end{cases}$

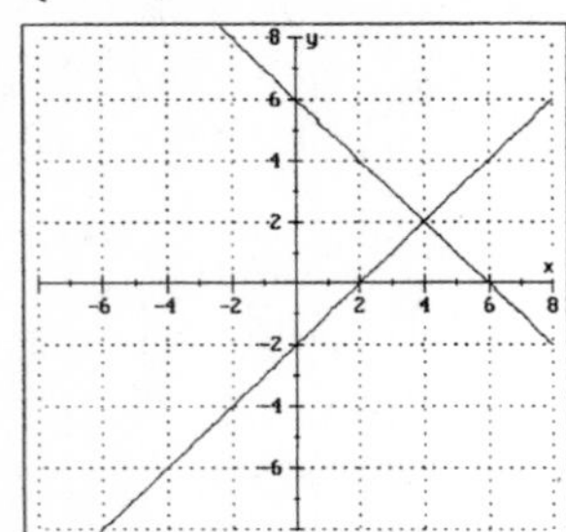

$(4,2)$ is the solution.

16. $\begin{cases} x-y=4 \\ 2x+y=5 \end{cases}$

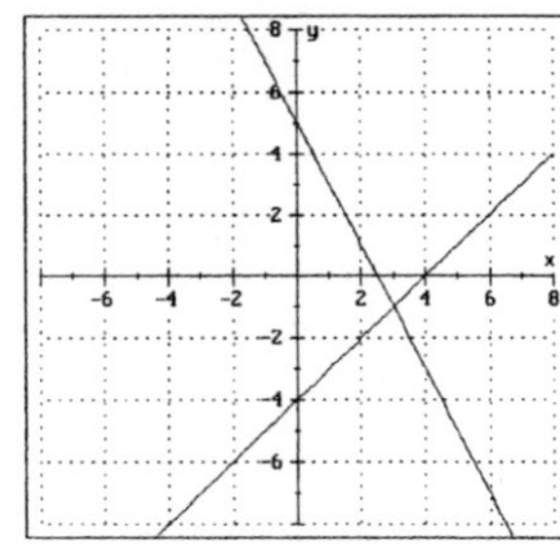

$(3,-1)$ is the solution.

17. $\begin{cases} 2x+y=1 \\ x-2y=-7 \end{cases}$

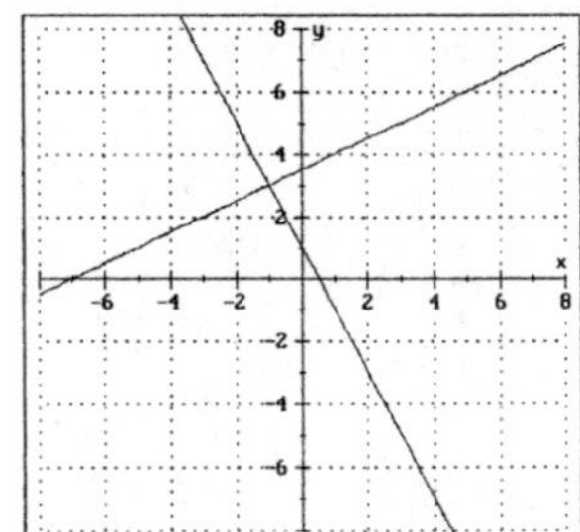

$(-1,3)$ is the solution.

18. $\begin{cases} 3x-y=-3 \\ 2x+y=-7 \end{cases}$

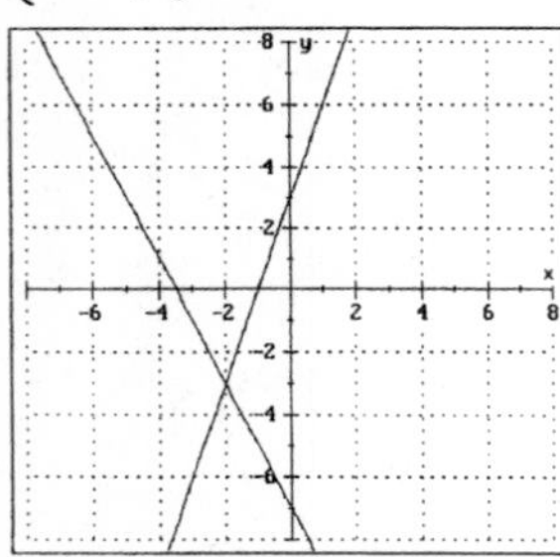

$(-2,-3)$ is the solution.

19. $\begin{cases} 2x+3y=0 \\ 2x+y=4 \end{cases}$

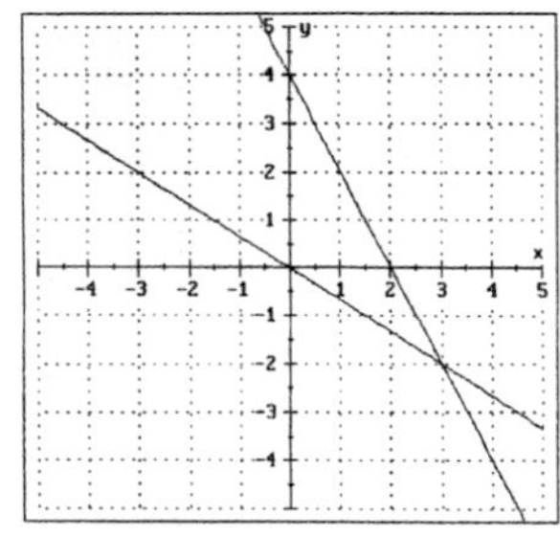

$(3,-2)$ is the solution.

20. $\begin{cases} 3x-2y=0 \\ 2x+3y=0 \end{cases}$

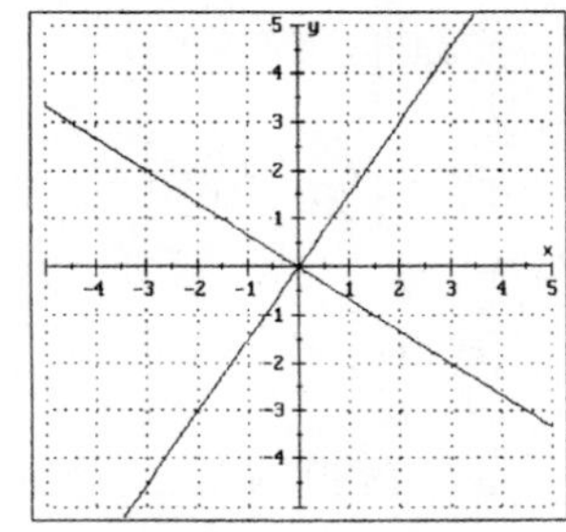

$(0,0)$ is the solution.

21. $\begin{cases} x=13-4y \\ 3x=4+2y \end{cases}$

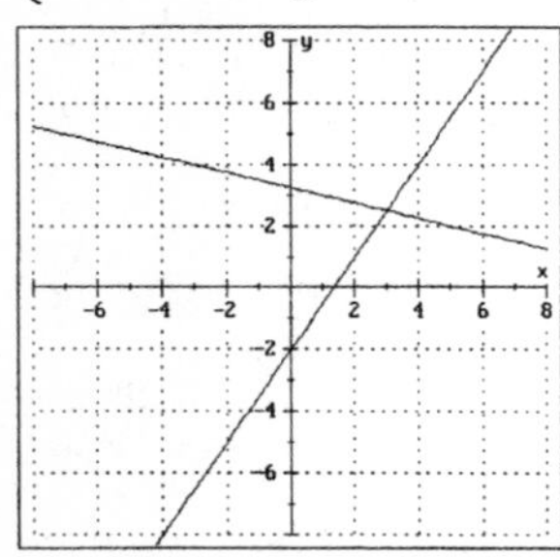

$\left(3,\frac{5}{2}\right)$ is the solution.

22. $\begin{cases} 3x=7-2y \\ 2x=2+4y \end{cases}$

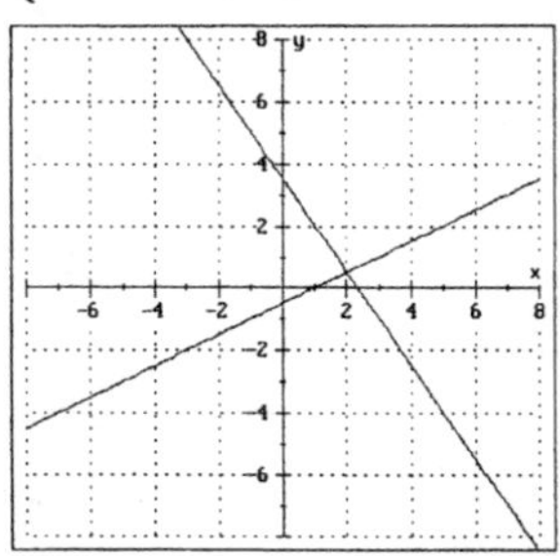

$\left(2,\frac{1}{2}\right)$ is the solution.

23. $\begin{cases} x=3-2y \\ 2x+4y=6 \end{cases}$

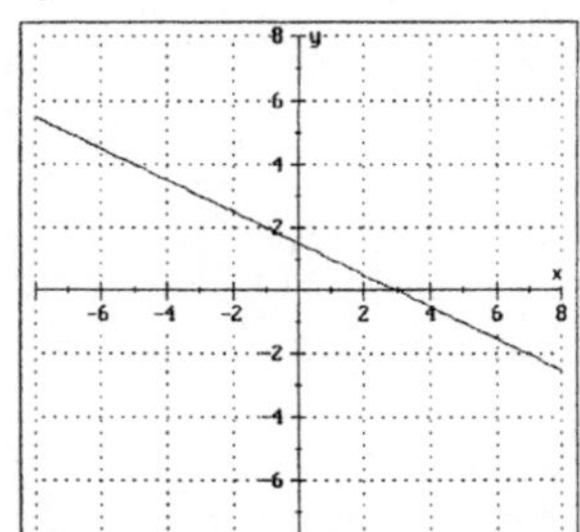

dependent system

24. $\begin{cases} 3x = 5 - 2y \\ 3x + 2y = 7 \end{cases}$

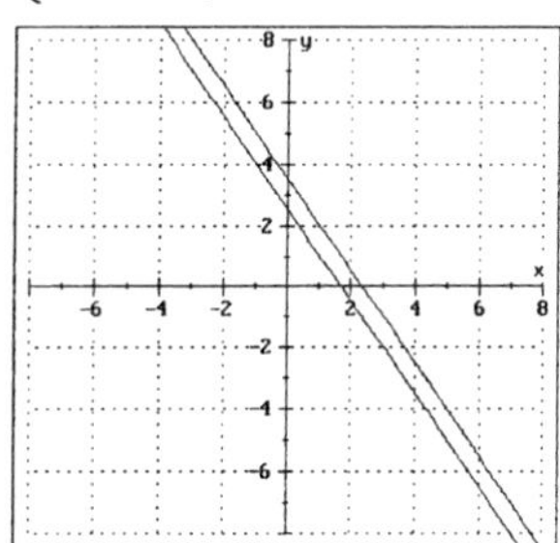

inconsistent system

25. 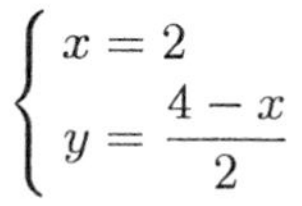

$\begin{cases} x = 2 \\ y = \dfrac{4 - x}{2} \end{cases}$

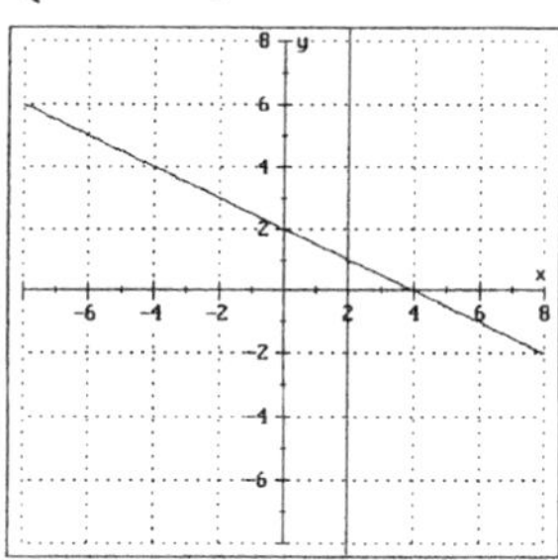

$(2, 1)$ is the solution.

26. $\begin{cases} y = -2 \\ x = \dfrac{4 + 3y}{2} \end{cases}$

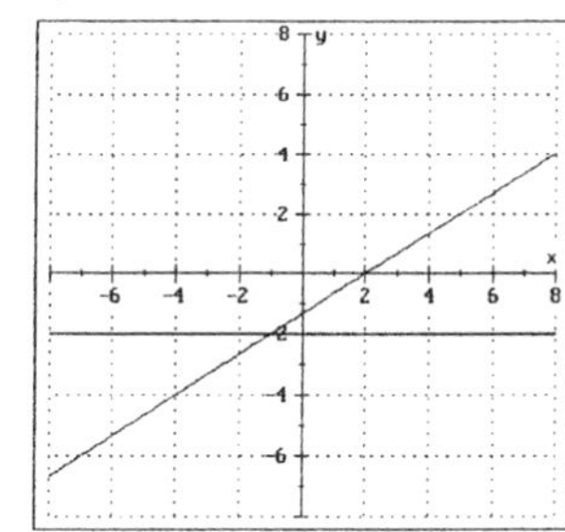

$(-1, -2)$ is the solution.

27. $\begin{cases} y = 3 \\ x = 2 \end{cases}$

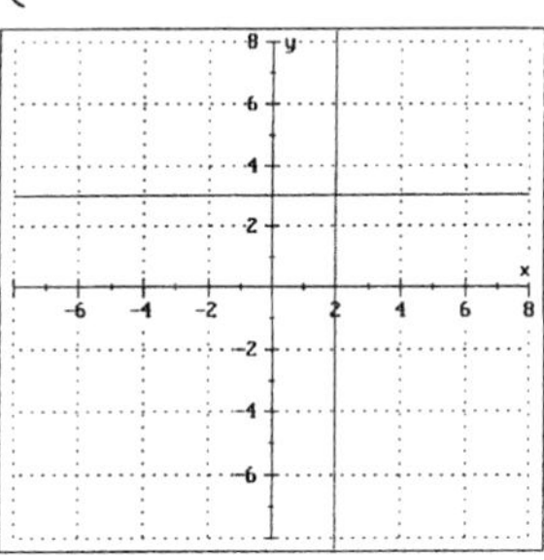

$(2, 3)$ is the solution.

28. 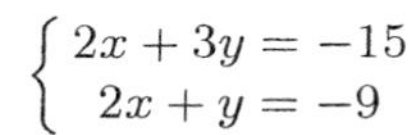

$\begin{cases} 2x + 3y = -15 \\ 2x + y = -9 \end{cases}$

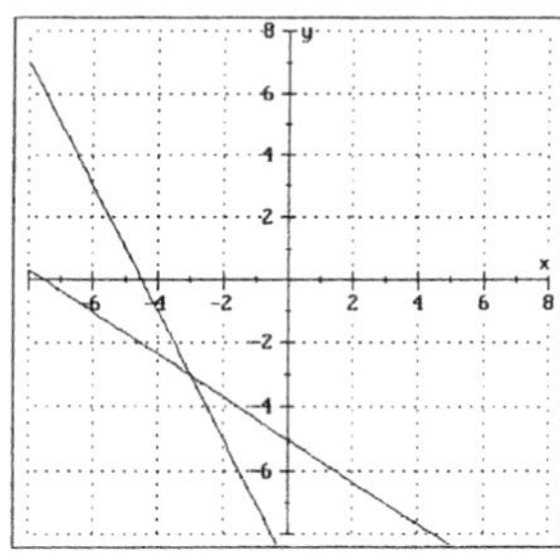

$(-3, -3)$ is the solution.

29. $\begin{cases} x = \dfrac{11 - 2y}{3} \\ y = \dfrac{11 - 6x}{4} \end{cases}$

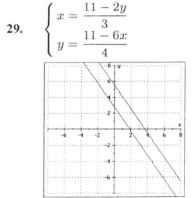

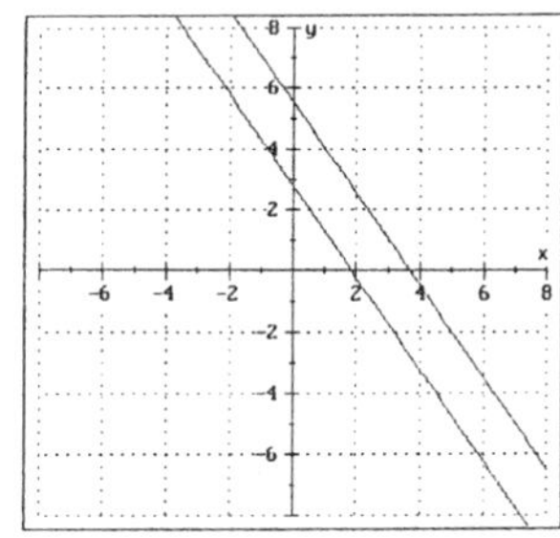

inconsistent system

30. $\begin{cases} x = \dfrac{1 - 3y}{4} \\ y = \dfrac{12 + 3x}{2} \end{cases}$

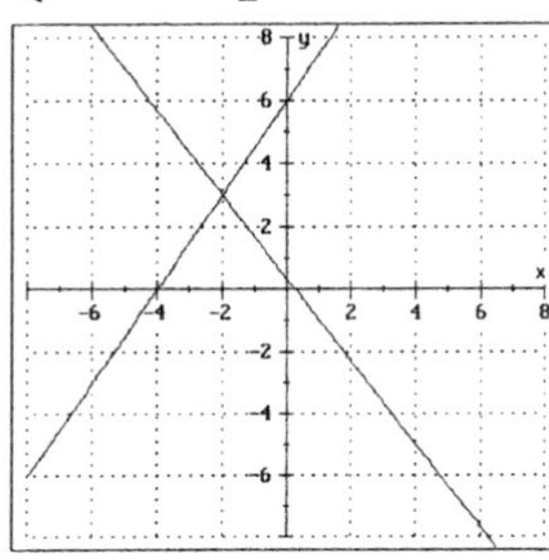

$(-2, 3)$ is the solution.

31. $\begin{cases} \dfrac{5}{2}x + y = \dfrac{1}{2} \\ 2x - \dfrac{3}{2}y = 5 \end{cases}$

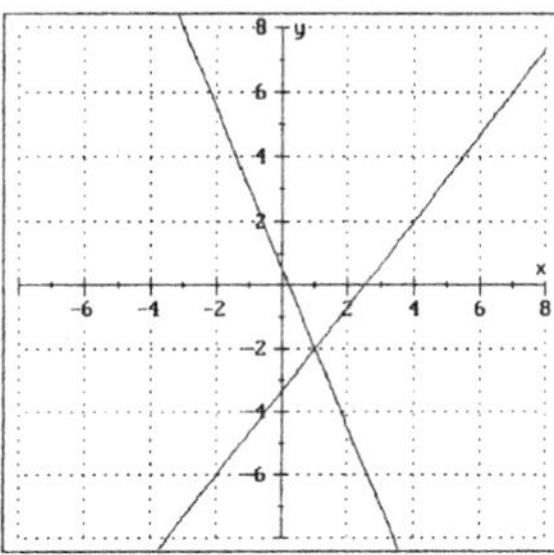

$(1, -2)$ is the solution.

32. $\begin{cases} \dfrac{5}{2}x + 3y = 6 \\ y = \dfrac{24 - 10x}{12} \end{cases}$

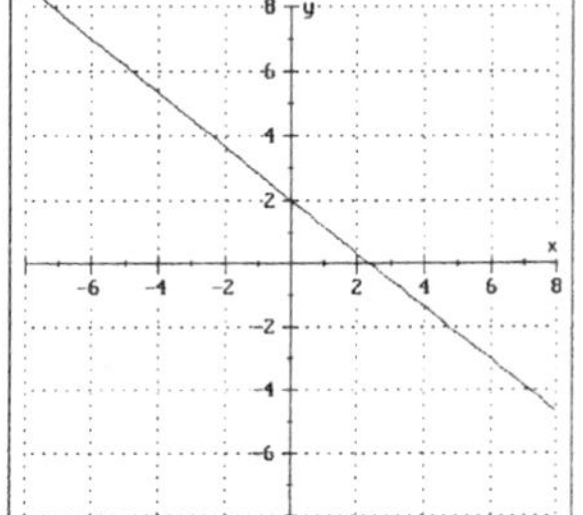

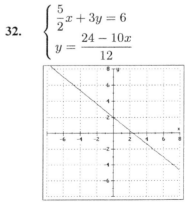

dependent system

33. 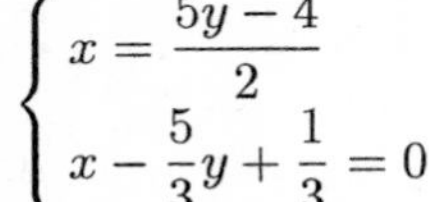
$$\begin{cases} x = \dfrac{5y-4}{2} \\ x - \dfrac{5}{3}y + \dfrac{1}{3} = 0 \end{cases}$$

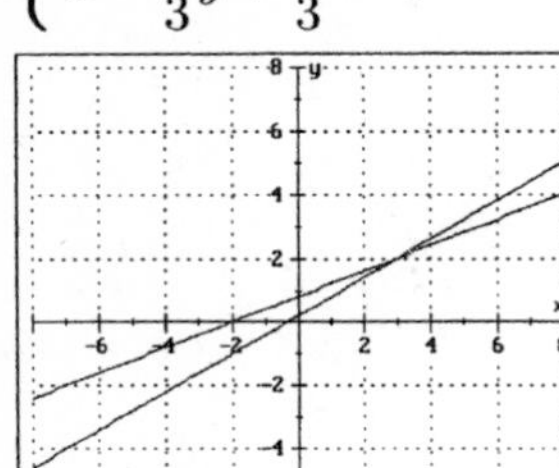

$(3, 2)$ is the solution.

34. $\begin{cases} 2x = 5y - 11 \\ 3x = 2y \end{cases}$

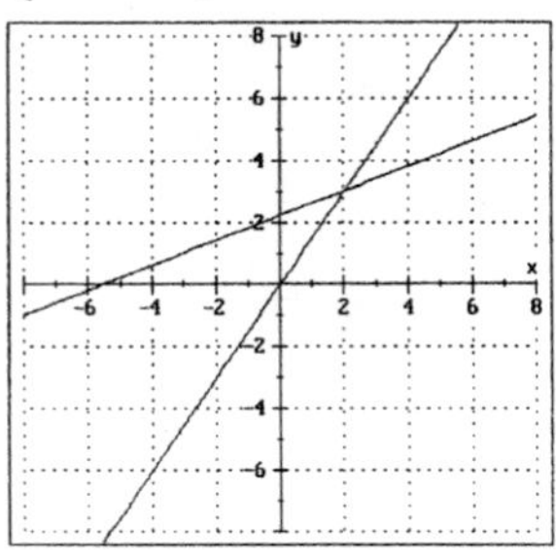

$(2, 3)$ is the solution.

35. $\begin{cases} x = -\dfrac{3}{2}y \\ x = \dfrac{3}{2}y - 2 \end{cases}$

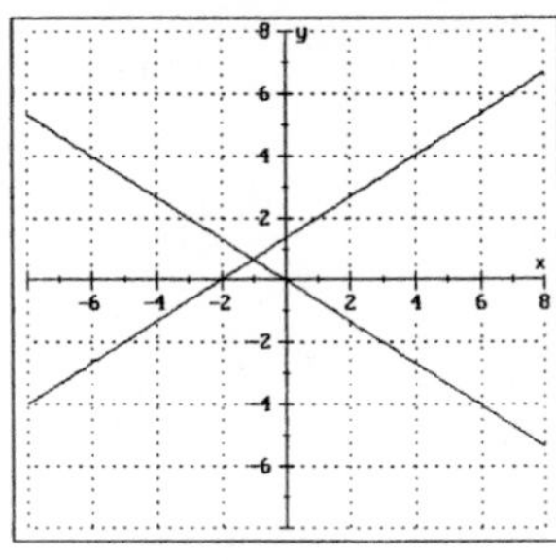

$\left(-1, \frac{2}{3}\right)$ is the solution.

36. $\begin{cases} x = \dfrac{3y-1}{4} \\ y = \dfrac{4-8x}{3} \end{cases}$

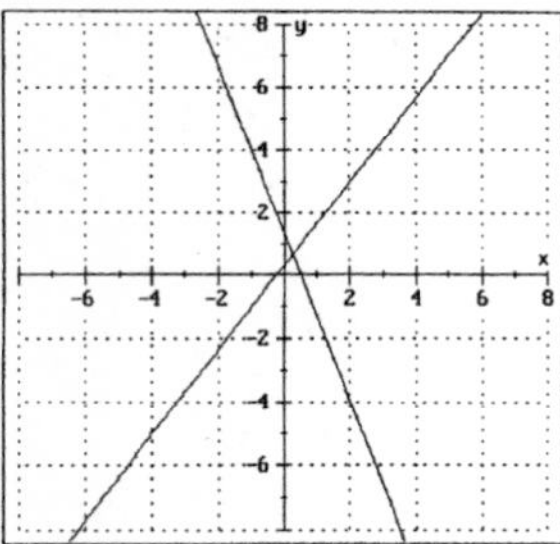

$\left(\frac{1}{4}, \frac{2}{3}\right)$ is the solution.

37. 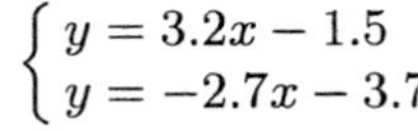
$$\begin{cases} y = 3.2x - 1.5 \\ y = -2.7x - 3.7 \end{cases}$$

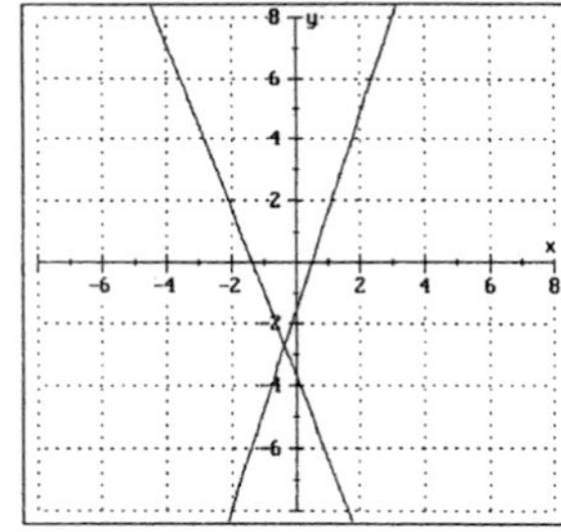

solution: $(-0.37, -2.69)$

38. $\begin{cases} y = -0.45x + 5 \\ y = 5.55x - 13.7 \end{cases}$

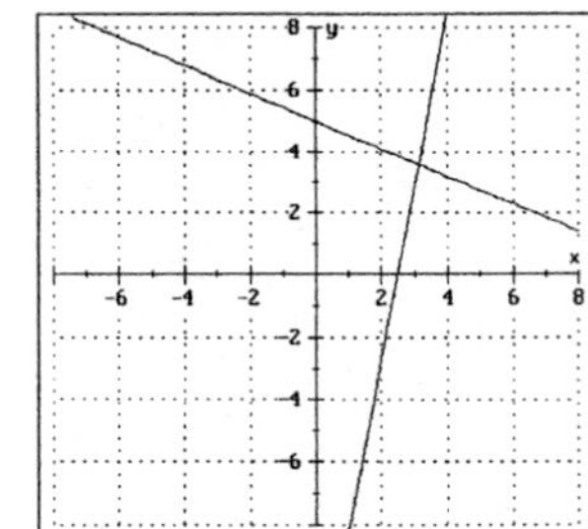

solution: $(3.12, 3.60)$

39. $\begin{cases} 1.7x + 2.3y = 3.2 \\ y = 0.25x + 8.95 \end{cases}$

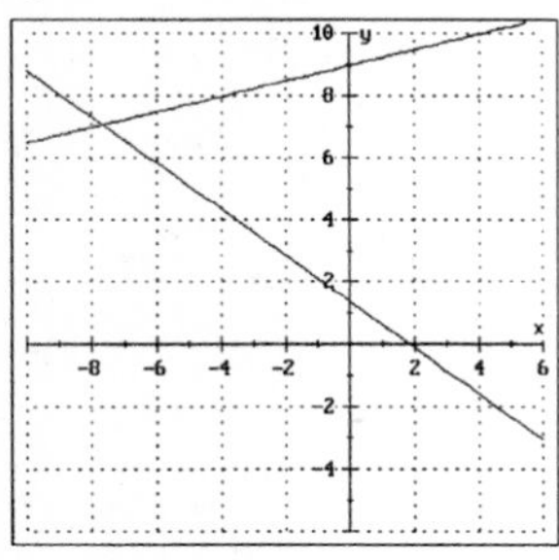

solution: $(-7.64, 7.04)$

40. $\begin{cases} 2.75x = 12.9y - 3.79 \\ 7.1x - y = 35.76 \end{cases}$

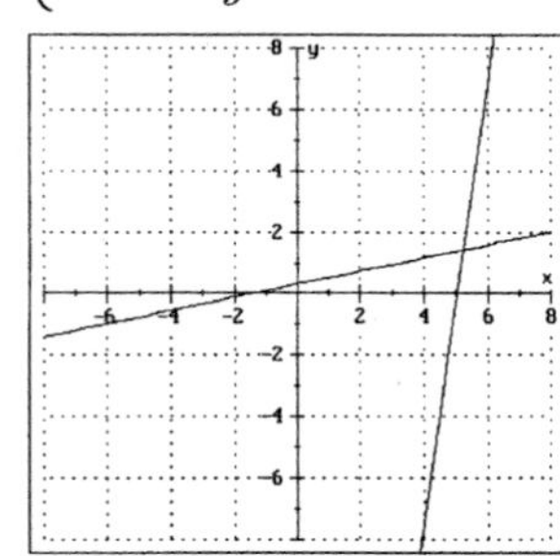

solution: $(5.24, 1.41)$

41. Graph $y = \frac{2}{3}x - 3$ and $y = -15$ and find the x-coordinate of any point of intersection.

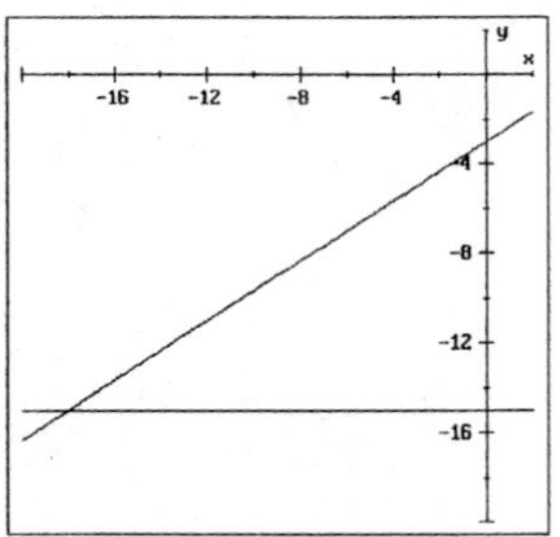

solution: $b = -18$

42. Graph $y = -7(x - 2)$ and $y = 8$ and find the x-coordinate of any point of intersection.

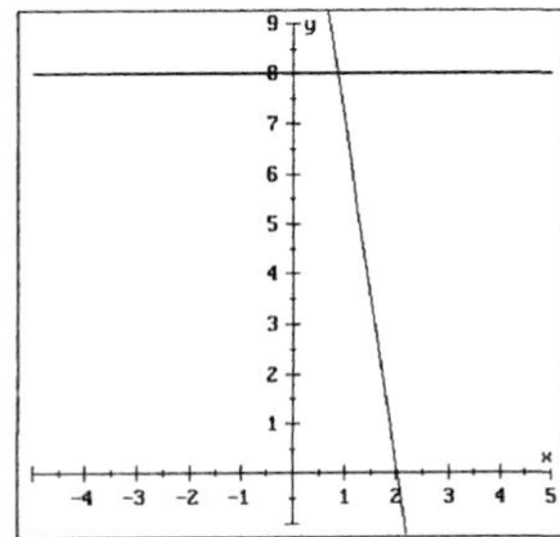

solution: $a \approx 0.857$

43. Graph $y = 2(2x + 1)$ and $y = 3x + 15$ and find the x-coordinate of any point of intersection.

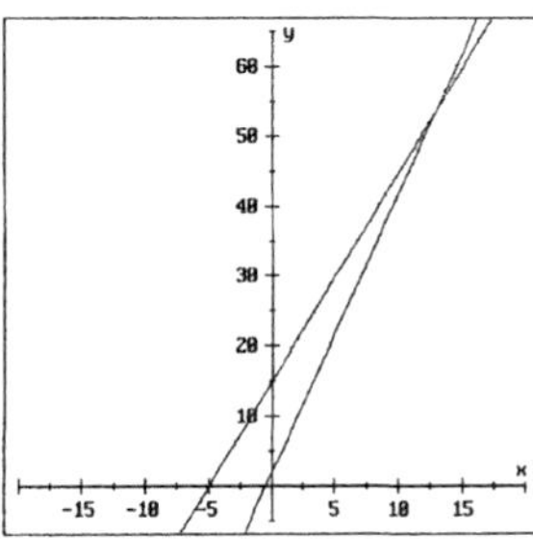

solution: $x = 13$

44. Graph $y = 2(x - 5)$ and $y = 3x + 1$ and find the x-coordinate of any point of intersection.

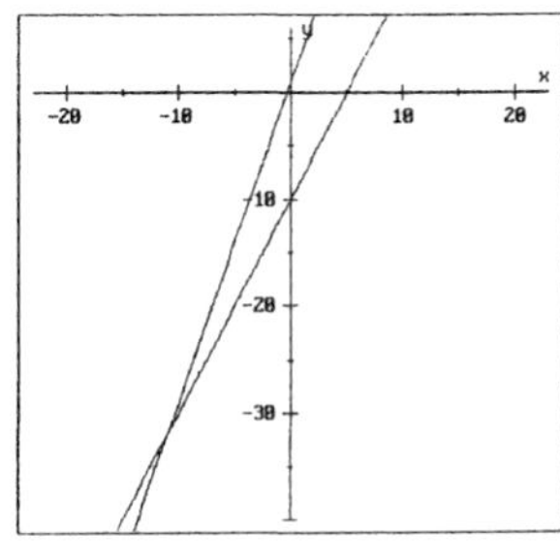

solution: $a = -11$

45. **a.** The point $(15, 2.0)$ is on the graph of the cost function, so it costs \$2 million to manufacture 15,000 cameras.

b. The point $(20, 3.0)$ is on the graph of the revenue function, so there is a revenue of \$3 million when 20,000 cameras are sold.

c. The graphs of the cost function and the revenue function meet at the point $(10, 1.5)$, so the revenue and cost functions are equal for 10,000 cameras.

46. **a.** The graphs meet at the approximate point $('01, 63)$.

b. The point of intersection indicates that in 2001, the number of take-out meals and the number of on-premise meals per person per year were equal, at 63 each per person per year.

47. $\begin{cases} 2x + 3y = 6 \\ 2x - 3y = 9 \end{cases}$

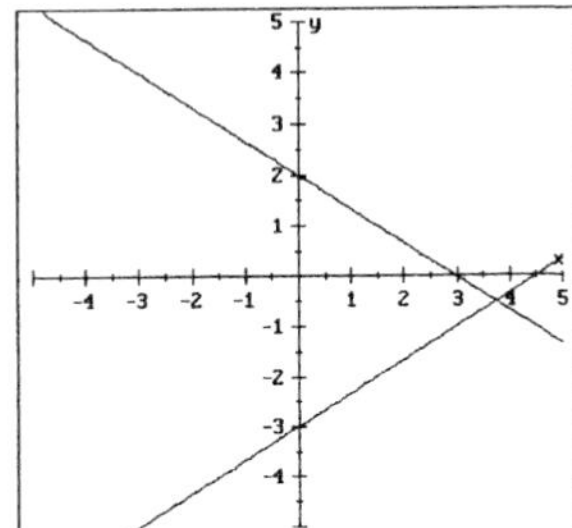

a. There is a possibility of a collision.

b. The danger point is at $(3.75, -0.5)$.

c. The collision is not certain, since the ships could be there at different times.

48. $\begin{cases} y = \frac{2}{5}x - 2 \\ x = \frac{5y+7}{2} \end{cases}$

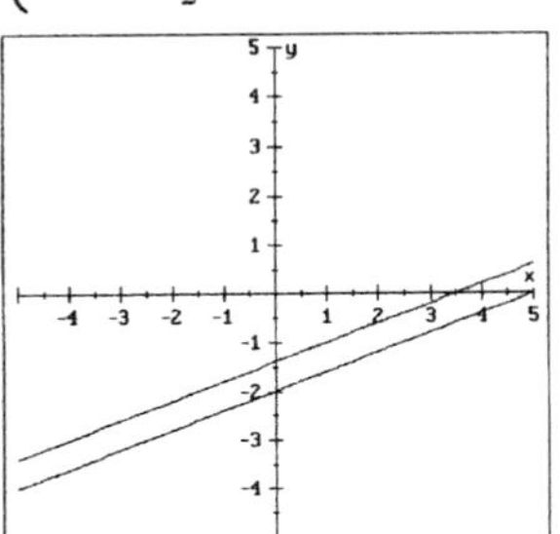

There is no chance of collision. (The lines are parallel.)

49. Let $x =$ the number of hours spent and let $y =$ the total cost.
$$\begin{cases} y = 50x \\ y = 40x + 30 \end{cases}$$
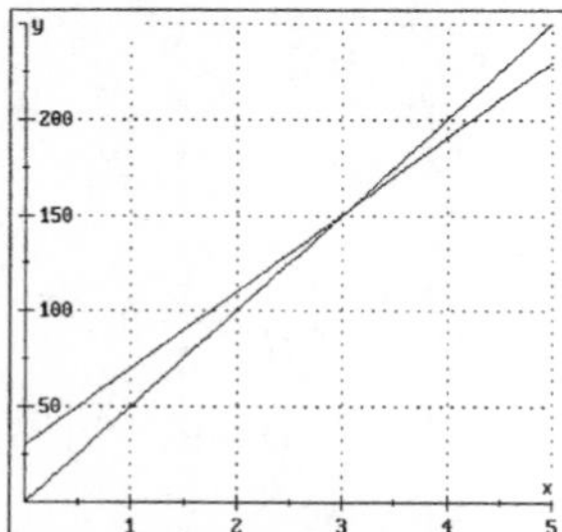
The lines meet at $(3, 150)$. Thus, the repair takes 3 hours.

50. Let $x =$ the number of trees and let $y =$ the number of bushes.
$$\begin{cases} x + y = 25 \\ 100x + 50y = 1500 \end{cases}$$
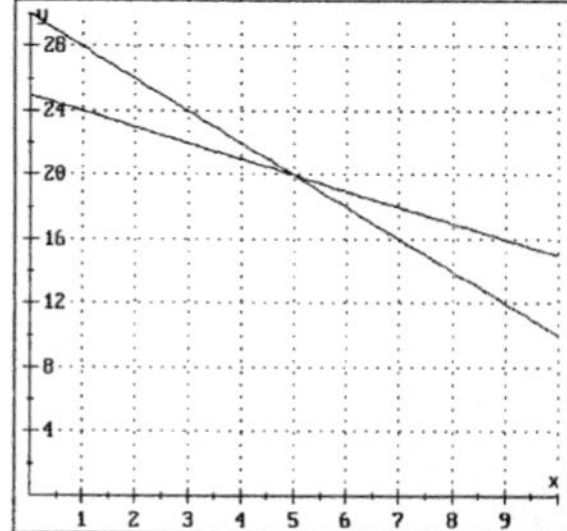
The lines meet at $(5, 20)$. Thus, he planted 5 trees and 20 bushes.

51. $$\begin{cases} y = 200x + 400 \\ y = 280x \end{cases}$$
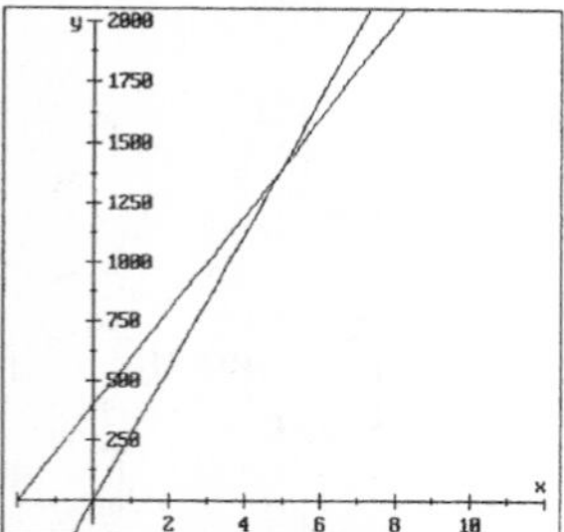
The lines meet at $(5, 1400)$.

52. The college needs to offer more than 5 sections.

53. **Answers may vary.**

54. **Answers may vary.**

55. One possible answer:
$$\begin{cases} x + y = -3 \\ x - y = -7 \end{cases}$$

56. One possible answer:
$$\begin{cases} y = 3x + 17 \\ 2y = 6x + 34 \end{cases}$$

Exercise 3.2 (page 176)

1. $$(a^2a^3)^2(a^4a^2)^2 = (a^5)^2(a^6)^2 = a^{10}a^{12} = a^{22}$$

2. $$\left(\frac{a^2b^3c^4d}{ab^2c^3d^4}\right)^{-3} = \left(\frac{ab^2c^3d^4}{a^2b^3c^4d}\right)^3 = \left(\frac{d^3}{abc}\right)^3 = \frac{d^9}{a^3b^3c^3}$$

SECTION 3.2

3. $\left(\frac{-3x^3y^4}{x^{-5}y^3}\right)^{-4} = \left(\frac{x^{-5}y^3}{-3x^3y^4}\right)^4 = \left(\frac{1}{-3x^8y}\right)^4 = \frac{1}{81x^{32}y^4}$

4. $\frac{3t^0 - 4t^0 + 5}{5t^0 + 2t^0} = \frac{3(1) - 4(1) + 5}{5(1) + 2(1)} = \frac{3 - 4 + 5}{5 + 2} = \frac{4}{7}$

5. setup; unit

6. break

7. parallelogram

8. Opposite

9. Opposite

10. Consecutive

11. $\begin{cases} (1) & y = x \\ (2) & x + y = 4 \end{cases}$

Substitute $y = x$ from (1) into (2):

$x + \boldsymbol{y} = 4$

$x + \boldsymbol{x} = 4$

$2x = 4$

$x = 2$

Substitute this and solve for y:

$y = x = 2$

Solution: $(2, 2)$

12. $\begin{cases} (1) & y = x + 2 \\ (2) & x + 2y = 16 \end{cases}$

Substitute $y = x + 2$ from (1) into (2):

$x + 2\boldsymbol{y} = 16$

$x + 2\boldsymbol{(x + 2)} = 16$

$x + 2x + 4 = 16$

$3x = 12$

$x = 4$

Substitute this and solve for y:

$y = x + 2 = 4 + 2 = 6$

Solution: $(4, 6)$

13. $\begin{cases} (1) & x - y = 2 \\ (2) & 2x + y = 13 \end{cases}$

Substitute $x = y + 2$ from (1) into (2):

$2\boldsymbol{x} + y = 13$

$2\boldsymbol{(y + 2)} + y = 13$

$2y + 4 + y = 13$

$3y = 9$

$y = 3$

Substitute this and solve for x:

$x = y + 2 = 3 + 2 = 5$

Solution: $(5, 3)$

14. $\begin{cases} (1) & x - y = -4 \\ (2) & 3x - 2y = -5 \end{cases}$

Substitute $x = y - 4$ from (1) into (2):

$3\boldsymbol{x} - 2y = -5$

$3\boldsymbol{(y - 4)} - 2y = -5$

$3y - 12 - 2y = -5$

$y = 7$

Substitute this into and solve for x:

$x = y - 4 = 7 - 4 = 3$

Solution: $(3, 7)$

15. $\begin{cases} (1) & x + 2y = 6 \\ (2) & 3x - y = -10 \end{cases}$

Substitute $x = -2y + 6$ from (1) into (2):

$$\begin{aligned} 3\boldsymbol{x} - y &= -10 \\ 3(\mathbf{-2y+6}) - y &= -10 \\ -6y + 18 - y &= -10 \\ -7y &= -28 \\ y &= 4 \end{aligned}$$

Substitute this and solve for x:

$x = -2y + 6 = -2(4) + 6 = -2$

Solution: $(-2, 4)$

16. $\begin{cases} (1) & 2x - y = -21 \\ (2) & 4x + 5y = 7 \end{cases}$

Substitute $y = 2x + 21$ from (1) into (2):

$$\begin{aligned} 4x + 5\boldsymbol{y} &= 7 \\ 4x + 5(\mathbf{2x+21}) &= 7 \\ 4x + 10x + 105 &= 7 \\ 14x &= -98 \\ x &= -7 \end{aligned}$$

Substitute this and solve for y:

$y = 2x + 21 = 2(-7) + 21 = 7$

Solution: $(-7, 7)$

17. $\begin{cases} (1) & 3x = 2y - 4 \\ (2) & 6x - 4y = -4 \end{cases}$

Substitute $x = \dfrac{2y-4}{3}$ from (1) into (2):

$$\begin{aligned} 6\boldsymbol{x} - 4y &= -4 \\ 6\left(\mathbf{\frac{2y-4}{3}}\right) - 4y &= -4 \\ 2(2y-4) - 4y &= -4 \\ 4y - 8 - 4y &= -4 \\ -8 &= -4 \end{aligned}$$

Impossible $\Rightarrow$ no solution

18. $\begin{cases} (1) & 8x = 4y + 10 \\ (2) & 4x - 2y = 5 \end{cases}$

Substitute $x = \dfrac{4y+10}{8}$ from (1) into (2):

$$\begin{aligned} 4\boldsymbol{x} - 2y &= 5 \\ 4\left(\mathbf{\frac{4y+10}{8}}\right) - 2y &= 5 \\ \frac{4y+10}{2} - 2y &= 5 \\ 2\left(\frac{4y+10}{2}\right) - 2(2y) &= 2(5) \\ 4y + 10 - 4y &= 10 \\ 10 &= 10 \end{aligned}$$

Always true $\Rightarrow$ dependent equations

19. $\begin{cases} (1) & 3x - 4y = 9 \\ (2) & x + 2y = 8 \end{cases}$

Substitute $x = -2y + 8$ from (2) into (1):

$$\begin{aligned} 3\boldsymbol{x} - 4y &= 9 \\ 3(\mathbf{-2y+8}) - 4y &= 9 \\ -6y + 24 - 4y &= 9 \\ -10y &= -15 \\ y &= \tfrac{3}{2} \end{aligned}$$

Substitute this and solve for x:

$x = -2y + 8 = -2\left(\dfrac{3}{2}\right) + 8 = 5$

Solution: $\left(5, \frac{3}{2}\right)$

20. $\begin{cases} (1) & 3x - 2y = -10 \\ (2) & 6x + 5y = 25 \end{cases}$

Substitute $x = \dfrac{2y-10}{3}$ from (1) into (2):

$$\begin{aligned} 6\boldsymbol{x} + 5y &= 25 \\ 6\left(\mathbf{\frac{2y-10}{3}}\right) + 5y &= 25 \\ 2(2y-10) + 5y &= 25 \\ 4y - 20 + 5y &= 25 \\ 9y &= 45 \\ y &= 5 \end{aligned}$$

Substitute this and solve for x :

$x = \dfrac{2y-10}{3} = \dfrac{2(5)-10}{3} = \dfrac{0}{3} = 0$

Solution: $(0, 5)$

21. $\begin{cases} (1) & 2x + 2y = -1 \\ (2) & 3x + 4y = 0 \end{cases}$

Substitute $y = \dfrac{-2x-1}{2}$ from (1) into (2):

$$\begin{aligned} 3x + 4\boldsymbol{y} &= 0 \\ 3x + 4\left(\frac{\boldsymbol{-2x-1}}{\boldsymbol{2}}\right) &= 0 \\ 3x + 2(-2x-1) &= 0 \\ 3x - 4x - 2 &= 0 \\ -x &= 2 \\ x &= -2 \end{aligned}$$

Substitute this and solve for y:

$$y = \frac{-2x-1}{2} = \frac{-2(-2)-1}{2} = \frac{3}{2}$$

Solution: $\left(-2, \frac{3}{2}\right)$

22. $\begin{cases} (1) & 5x + 3y = -7 \\ (2) & 3x - 3y = 7 \end{cases}$

Substitute $y = \dfrac{-5x-7}{3}$ from (1) into (2):

$$\begin{aligned} 3x - 3\boldsymbol{y} &= 7 \\ 3x - 3\left(\frac{\boldsymbol{-5x-7}}{\boldsymbol{3}}\right) &= 7 \\ 3x - (-5x-7) &= 7 \\ 3x + 5x + 7 &= 7 \\ 8x &= 0 \\ x &= 0 \end{aligned}$$

Substitute this and solve for y:

$$y = \frac{-5x-7}{3} = \frac{-5(0)-7}{3} = -\frac{7}{3}$$

Solution: $\left(0, -\frac{7}{3}\right)$

23.

$$\begin{array}{rl} x - y = & 3 \\ x + y = & 7 \\ \hline 2x \quad = & 10 \\ x \quad = & 5 \end{array}$$

Substitute and solve for y:

$$\begin{aligned} x + y &= 7 \\ 5 + y &= 7 \\ y &= 2 \end{aligned}$$

The solution is $(5, 2)$.

24.

$$\begin{array}{rl} x + y = & 1 \\ x - y = & 7 \\ \hline 2x \quad = & 8 \\ x \quad = & 4 \end{array}$$

Substitute and solve for y:

$$\begin{aligned} x + y &= 1 \\ 4 + y &= 1 \\ y &= -3 \end{aligned}$$

The solution is $(4, -3)$.

25.

$$\begin{array}{rl} 2x + y = & -10 \\ 2x - y = & -6 \\ \hline 4x \quad = & -16 \\ x \quad = & -4 \end{array}$$

Substitute and solve for y:

$$\begin{aligned} 2x + y &= -10 \\ 2(-4) + y &= -10 \\ -8 + y &= -10 \\ y &= -2 \end{aligned}$$

The solution is $(-4, -2)$.

26.

$$\begin{array}{rl} x + 2y = & -9 \\ x - 2y = & -1 \\ \hline 2x \quad = & -10 \\ x \quad = & -5 \end{array}$$

Substitute and solve for y:

$$\begin{aligned} x + 2y &= -9 \\ -5 + 2y &= -9 \\ 2y &= -4 \\ y &= -2 \end{aligned}$$

The solution is $(-5, -2)$.

27.

$$\begin{array}{ll} 2x + 3y = 8 \Rightarrow \times(2) & 4x + 6y = 16 \\ 3x - 2y = -1 \Rightarrow \times(3) & 9x - 6y = -3 \\ \hline & 13x \quad = 13 \\ & x \quad = 1 \end{array}$$

$$\begin{aligned} 2x + 3y &= 8 \\ 2(1) + 3y &= 8 \\ 3y &= 6 \\ y &= 2 \end{aligned}$$

Solution: $\boxed{(1, 2)}$

SECTION 3.2

28. $$\begin{array}{lll} 5x-2y=19 \Rightarrow \times(2) & 10x-4y=38 & 3x+4y=1 \\ \underline{3x+4y=\ \ 1} & \underline{\ \ 3x+4y=\ \ 1} & 3(3)+4y=1 \\ & 13x \quad = 39 & 4y=-8 \\ & x \quad =\ \ 3 & y=-2 \end{array}$$

Solution: $\boxed{(3,-2)}$

29. $$\begin{array}{lll} 4x+9y=\ \ 8 & 4x+\ \ 9y=\ \ 8 & 2x-6y=-3 \\ \underline{2x-6y=-3} \Rightarrow \times(-2) & \underline{-4x+12y=\ \ 6} & 2x-6\left(\frac{2}{3}\right)=-3 \\ & 21y=14 & 2x-4=-3 \\ & y=\frac{14}{21}=\frac{2}{3} & 2x=1 \\ & & x=\frac{1}{2} \end{array}$$

Solution: $\boxed{\left(\frac{1}{2},\frac{2}{3}\right)}$

30. $$\begin{array}{lll} 4x+6y=5 \Rightarrow \times(-2) & -8x-12y=-10 & 4x+6y=5 \\ \underline{8x-9y=3} & \underline{\ \ 8x-\ \ 9y=\ \ \ 3} & 4x+6\left(\frac{1}{3}\right)=5 \\ & -21y=\ \ -7 & 4x+2=5 \\ & y=\ \frac{-7}{-21}=\frac{1}{3} & 4x=3 \\ & & x=\frac{3}{4} \end{array}$$

Solution: $\boxed{\left(\frac{3}{4},\frac{1}{3}\right)}$

31. $$\begin{array}{lll} 8x-4y=16 \Rightarrow & 8x-4y=16 \Rightarrow & 8x-4y=\ \ 16 \\ \underline{2x-\ \ 4=\ \ y} \Rightarrow & \underline{2x-\ \ y=\ \ 4} \Rightarrow \times(-4) & \underline{-8x+4y=-16} \\ & & 0=\ \ \ 0 \Rightarrow \boxed{\text{Dependent equations}} \end{array}$$

32. $$\begin{array}{lll} 2y-3x=-13 \Rightarrow & -3x+2y=-13 & 3x-4y=17 \\ \underline{3x-17=\ \ 4y} \Rightarrow & \underline{\ \ 3x-4y=\ \ 17} & 3x-4(-2)=17 \\ & -2y=\ \ \ 4 & 3x+8=17 \\ & y=\ \ -2 & 3x=9 \\ & & x=3 \end{array}$$

Solution: $\boxed{(3,-2)}$

33. $$\begin{array}{llll} x=\frac{3}{2}y+5 \Rightarrow \times(2) & 2x=3y+10 \Rightarrow & 2x-3y=10 \Rightarrow & 2x-3y=\ \ 10 \\ \underline{2x\ -\ 3y=8} \Rightarrow & \underline{2x-3y=\ \ 8} \Rightarrow & \underline{2x-3y=\ \ 8} \Rightarrow \times(-1) & \underline{-2x+3y=-8} \\ & & & 0\neq\ \ 2 \Rightarrow \boxed{\text{No solution}} \end{array}$$

34. $$\begin{array}{llll} x=\frac{2}{3}y \quad \Rightarrow \times 3 & 3x=2y \quad \Rightarrow & 3x-2y=0 \Rightarrow & 3x-2y=\ \ 0 \\ \underline{y=\ 4x+5} \Rightarrow & \underline{-4x+\ \ y=5} \Rightarrow & \underline{-4x+\ \ y=5} \Rightarrow \times 2 & \underline{-8x+2y=\ 10} \\ & & & -5x \quad =\ 10 \\ & & & x \quad =-2 \end{array}$$

$y=4x+5=4(-2)+5=-8+5=-3$ Solution: $\boxed{(-2,-3)}$

35. $$\begin{array}{lll} \frac{x}{2}+\frac{y}{2}=\ \ 6 \Rightarrow \times 2 & x+y=\ 12 & x+y=12 \\ \underline{\frac{x}{2}-\frac{y}{2}=-2} \Rightarrow \times 2 & \underline{x-y=-4} & 4+y=12 \\ & 2x \quad =\ \ 8 & y=8 \\ & x \quad =\ \ 4 & \end{array}$$

Solution: $\boxed{(4,8)}$

36. $\frac{x}{2} - \frac{y}{3} = -4 \Rightarrow \times 6 \quad 3x - 2y = -24$

$\frac{x}{2} + \frac{y}{9} = 0 \Rightarrow \times 18 \quad 9x + 2y = 0$

$$\begin{aligned} 12x &= -24 \\ x &= -2 \end{aligned}$$

$$\begin{aligned} 9x + 2y &= 0 \\ 9(-2) + 2y &= 0 \\ -18 + 2y &= 0 \\ 2y &= 18 \\ y &= 9 \end{aligned}$$

Solution: $\boxed{(-2, 9)}$

37. $\frac{3}{4}x + \frac{2}{3}y = 7 \Rightarrow \times 12 \quad 9x + 8y = 84 \Rightarrow \times 2 \quad 18x + 16y = 168$

$\frac{3}{5}x - \frac{1}{2}y = 18 \Rightarrow \times 10 \quad 6x - 5y = 180 \Rightarrow \times(-3) \quad -18x + 15y = -540$

$$\begin{aligned} 31y &= -372 \\ y &= -12 \end{aligned}$$

$$\begin{aligned} 6x - 5y &= 180 \\ 6x - 5(-12) &= 180 \\ 6x + 60 &= 180 \\ 6x &= 120 \\ x &= 20 \end{aligned}$$

Solution: $\boxed{(20, -12)}$

38. $\frac{2}{3}x - \frac{1}{4}y = -8 \Rightarrow \times 12 \quad 8x - 3y = -96 \Rightarrow \quad 8x - 3y = -96$

$\frac{1}{2}x - \frac{3}{8}y = -9 \Rightarrow \times 16 \quad 8x - 6y = -144 \Rightarrow \times(-1) \quad -8x + 6y = 144$

$$\begin{aligned} 3y &= 48 \\ y &= 16 \end{aligned}$$

$$\begin{aligned} 8x - 3y &= -96 \\ 8x - 3(16) &= -96 \\ 8x - 48 &= -96 \\ 8x &= -48 \\ x &= -6 \end{aligned}$$

Solution: $\boxed{(-6, 16)}$

39. $\frac{3x}{2} - \frac{2y}{3} = 0 \Rightarrow \times 6 \quad 9x - 4y = 0 \Rightarrow \quad 9x - 4y = 0$

$\frac{3x}{4} + \frac{4y}{3} = \frac{5}{2} \Rightarrow \times 12 \quad 9x + 16y = 30 \Rightarrow \times(-1) \quad -9x - 16y = -30$

$$\begin{aligned} -20y &= -30 \\ y &= \tfrac{-30}{-20} \\ y &= \tfrac{3}{2} \end{aligned}$$

$$\begin{aligned} 9x - 4y &= 0 \\ 9x - 4\left(\tfrac{3}{2}\right) &= 0 \\ 9x - 6 &= 0 \\ 9x &= 6 \\ x &= \tfrac{6}{9} = \tfrac{2}{3} \end{aligned}$$

Solution: $\boxed{\left(\tfrac{2}{3}, \tfrac{3}{2}\right)}$

40. $\frac{3x}{5} + \frac{5y}{3} = 2 \Rightarrow \times 15 \quad 9x + 25y = 30$

$\frac{6x}{5} - \frac{5y}{3} = 1 \Rightarrow \times 15 \quad 18x - 25y = 15$

$$\begin{aligned} 27x &= 45 \\ x &= \tfrac{45}{27} = \tfrac{5}{3} \end{aligned}$$

$$\begin{aligned} 9x + 25y &= 30 \\ 9\left(\tfrac{5}{3}\right) + 25y &= 30 \\ 15 + 25y &= 30 \\ 25y &= 15 \\ y &= \tfrac{15}{25} = \tfrac{3}{5} \end{aligned}$$

Solution: $\boxed{\left(\tfrac{5}{3}, \tfrac{3}{5}\right)}$

41.

$$\begin{array}{l} \frac{2}{5}x - \frac{1}{6}y = \frac{7}{10} \Rightarrow \times 30 \quad 12x - 5y = 21 \Rightarrow \times 3 \qquad 36x - 15y = 63 \\ \frac{3}{4}x - \frac{2}{3}y = \frac{19}{8} \Rightarrow \times 24 \quad 18x - 16y = 57 \Rightarrow \times(-2) \quad -36x + 32y = -114 \\ \hline \qquad\qquad\qquad\qquad\qquad\qquad 17y = -51 \\ \qquad\qquad\qquad\qquad\qquad\qquad y = -3 \end{array}$$

$$\begin{aligned} 12x - 5y &= 21 \\ 12x - 5(-3) &= 21 \\ 12x + 15 &= 21 \\ 12x &= 6 \\ x &= \tfrac{6}{12} = \tfrac{1}{2} \Rightarrow \text{Solution: } \boxed{\left(\tfrac{1}{2}, -3\right)} \end{aligned}$$

42.

$$\begin{array}{l} \frac{5}{6}x + \frac{2}{3}y = \frac{7}{6} \Rightarrow \times 6 \quad 5x + 4y = 7 \Rightarrow \times 7 \quad 35x + 28y = 49 \\ \frac{10}{7}x - \frac{4}{9}y = \frac{17}{21} \Rightarrow \times 63 \quad 90x - 28y = 51 \Rightarrow \qquad 90x - 28y = 51 \\ \hline \qquad\qquad\qquad\qquad\qquad\qquad 125x = 100 \\ \qquad\qquad\qquad\qquad\qquad\qquad x = \tfrac{100}{125} = \tfrac{4}{5} \end{array}$$

$$\begin{aligned} 5x + 4y &= 7 \\ 5 \cdot \frac{4}{5} + 4y &= 7 \\ 4 + 4y &= 7 \\ 4y &= 3 \\ y &= \tfrac{3}{4} \Rightarrow \text{Solution: } \boxed{\left(\tfrac{4}{5}, \tfrac{3}{4}\right)} \end{aligned}$$

43. Let $x = 0.3333\overline{3}$. Then $10x = 3.3333\overline{3}$.

$$\begin{aligned} 10x &= 3.3333\overline{3} \\ x &= 0.3333\overline{3} \\ \hline 9x &= 3 \\ \frac{9x}{9} &= \frac{3}{9} \\ x &= \tfrac{1}{3} \end{aligned}$$

44. Let $x = 0.2929\overline{29}$. Then $100x = 29.2929\overline{29}$.

$$\begin{aligned} 100x &= 29.2929\overline{29} \\ x &= 0.2929\overline{29} \\ \hline 99x &= 29 \\ \frac{99x}{99} &= \frac{29}{99} \\ x &= \tfrac{29}{99} \end{aligned}$$

45. Let $x = -0.34898\overline{989}$.

Then $100x = -34.89898\overline{989}$.

$$\begin{aligned} 100x &= -34.89898\overline{989} \\ x &= \ \ -0.34898\overline{989} \\ \hline 99x &= -34.55 \\ \frac{99x}{99} &= -\frac{34.55}{99} \\ x &= -\frac{34.55(100)}{99(100)} \\ x &= -\frac{3455}{9900} = -\frac{691}{1980} \end{aligned}$$

The fraction is $-\frac{691}{1980}$.

46. Let $x = -2.34747\overline{47}$.

Then $100x = -234.74747\overline{47}$

$$\begin{aligned} 100x &= -234.74747\overline{47} \\ x &= \ \ -2.34747\overline{47} \\ \hline 99x &= -232.4 \\ \frac{99x}{99} &= -\frac{232.4}{99} \\ x &= -\frac{232.4(10)}{99(10)} \\ x &= -\frac{2324}{990} = -\frac{1162}{495} \end{aligned}$$

The fraction is $-\frac{1162}{495}$.

For problems #47-50, begin each problem by letting $m = \frac{1}{x}$ and $n = \frac{1}{y}$. Solve for m and n, and then solve for x and y.

47.
$$\begin{aligned} \frac{1}{x} + \frac{1}{y} = \frac{5}{6} &\Rightarrow m + n = \frac{5}{6} \\ \frac{1}{x} - \frac{1}{y} = \frac{1}{6} &\Rightarrow m - n = \frac{1}{6} \\ \hline 2m &= \frac{6}{6} \\ 2m &= 1 \\ m &= \frac{1}{2} \end{aligned}$$

Solve for n:
$$\begin{aligned} m + n &= \frac{5}{6} \\ \frac{1}{2} + n &= \frac{5}{6} \\ n &= \frac{5}{6} - \frac{1}{2} \\ n &= \frac{1}{3} \end{aligned}$$

Solve for x:
$$\begin{aligned} m &= \frac{1}{x} \\ \frac{1}{2} &= \frac{1}{x} \\ 2 &= x \end{aligned}$$

Solve for y:
$$\begin{aligned} n &= \frac{1}{y} \\ \frac{1}{3} &= \frac{1}{y} \\ 3 &= y \end{aligned}$$

Solution: $\boxed{(2,3)}$

48.
$$\begin{aligned} \frac{1}{x} + \frac{1}{y} = \frac{9}{20} &\Rightarrow m + n = \frac{9}{20} \\ \frac{1}{x} - \frac{1}{y} = \frac{1}{20} &\Rightarrow m - n = \frac{1}{20} \\ \hline 2m &= \frac{10}{20} \\ 2m &= \frac{1}{2} \\ m &= \frac{1}{4} \end{aligned}$$

Solve for n:
$$\begin{aligned} m + n &= \frac{9}{20} \\ \frac{1}{4} + n &= \frac{9}{20} \\ n &= \frac{9}{20} - \frac{1}{4} \\ n &= \frac{4}{20} = \frac{1}{5} \end{aligned}$$

Solve for x:
$$\begin{aligned} m &= \frac{1}{x} \\ \frac{1}{4} &= \frac{1}{x} \\ 4 &= x \end{aligned}$$

Solve for y:
$$\begin{aligned} n &= \frac{1}{y} \\ \frac{1}{5} &= \frac{1}{y} \\ 5 &= y \end{aligned}$$

Solution: $\boxed{(4,5)}$

49.

$$\begin{array}{llll} \frac{1}{x}+\frac{2}{y}=-1 \Rightarrow & m+2n=-1 \Rightarrow & & m+2n= \ -1 \\ \frac{2}{x}-\frac{1}{y}=-7 \Rightarrow & 2m \ -n=-7 \Rightarrow & \times 2 & 4m-2n=-14 \\ & & & 5m \qquad = -15 \\ & & & m \qquad = \ -3 \end{array}$$

Solve for n:

$$\begin{aligned} m+2n&=-1\\ -3+2n&=-1\\ 2n&=2\\ n&=1 \end{aligned}$$

Solve for x:

$$m=\frac{1}{x} \qquad -3=\frac{1}{x} \qquad -\frac{1}{3}=x$$

Solve for y:

$$n=\frac{1}{y} \qquad 1=\frac{1}{y} \qquad 1=y$$

Solution: $\boxed{(-\frac{1}{3},1)}$

50.

$$\begin{array}{llll} \frac{3}{x}-\frac{2}{y}=-30 \Rightarrow & 3m-2n=-30 \Rightarrow & \times(-3) & -9m+6n= \ \ 90 \\ \frac{2}{x}-\frac{3}{y}=-30 \Rightarrow & 2m-3n=-30 \Rightarrow & \times 2 & 4m-6n=-60 \\ & & & -5m \qquad = \ \ 30 \\ & & & m \qquad = \ -6 \end{array}$$

Solve for n:

$$\begin{aligned} 3m-2n&=-30\\ 3(-6)-2n&=-30\\ -18-2n&=-30\\ -2n&=-12\\ n&=6 \end{aligned}$$

Solve for x:

$$m=\frac{1}{x} \qquad -6=\frac{1}{x} \qquad -\frac{1}{6}=x$$

Solve for y:

$$n=\frac{1}{y} \qquad 6=\frac{1}{y} \qquad \frac{1}{6}=y$$

Solution: $\boxed{(-\frac{1}{6},\frac{1}{6})}$

51. Let $x=$ the cost of the pair of shoes and $y=$ the cost of the sweater.

(1) $x+y=98$

(2) $y=x+16$

$$\begin{aligned} x+\boldsymbol{y}&=98\\ x+\boldsymbol{x+16}&=98\\ 2x&=82\\ x&=41 \end{aligned} \qquad \begin{aligned} y&=x+16\\ y&=41+16\\ y&=57 \end{aligned}$$

The sweater cost $57.

52. Let $x=$ the cost of a reel and $y=$ the cost of a rod.

$$\begin{array}{lrl} 2x+5y=270 \Rightarrow \times(-2) & -4x- \ 10y &= -540 \\ 4x+2y=220 \Rightarrow & 4x+ \ \ 2y &= \ \ 220 \\ & -8y &= -320 \\ & y &= \ \ \ 40 \end{array} \qquad \begin{aligned} 2x+5y&=270\\ 2x+5(40)&=270\\ 2x+200&=270\\ 2x&=70\\ x&=35 \end{aligned}$$

A reel costs $35, while a rod costs $40.

53. (1) $R_1+R_2=1375$

(2) $R_1=R_2+125$

$$\begin{aligned} \boldsymbol{R_1}+R_2&=1375\\ \boldsymbol{R_2+125}+R_2&=1375\\ 2R_2&=1250\\ R_2&=625 \end{aligned} \qquad \begin{aligned} R_1&=R_2+125\\ R_1&=625+125\\ R_1&=750 \end{aligned}$$

The resistances are $R_1=750$ ohms and $R_2=625$ ohms.

54. Let $x =$ the amount stored in the first compartment and $y =$ the amount stored in the second.

$$\begin{array}{llll} (1)\ x+y=950 & \boldsymbol{x}+y=950 & x=y+150 & \text{The compartments} \\ (2)\ \quad x=y+150 & \boldsymbol{y+150}+y=950 & x=400+150 & \text{hold 400 pounds and} \\ & 2y=800 & x=550 & \text{550 pounds.} \\ & y=400 & & \end{array}$$

55. Let $l =$ the length of the field and $w =$ the width of the field.

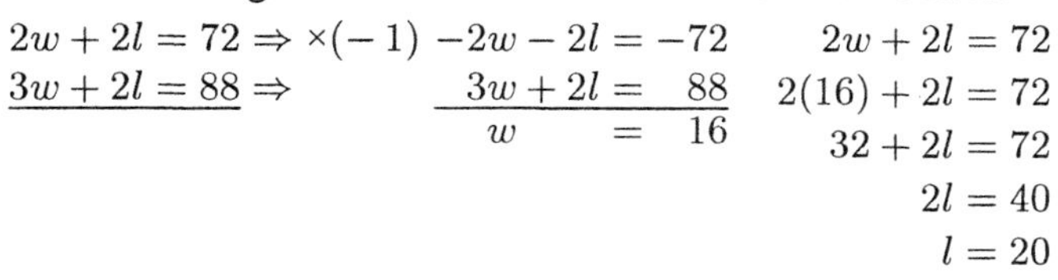

$$\begin{array}{lrr} 2w+2l=72 \Rightarrow \times(-1) & -2w-2l=-72 & 2w+2l=72 \\ \underline{3w+2l=88} \Rightarrow & 3w+2l=\ \ 88 & 2(16)+2l=72 \\ & w\ \ =\ \ 16 & 32+2l=72 \\ & & 2l=40 \\ & & l=20 \end{array}$$

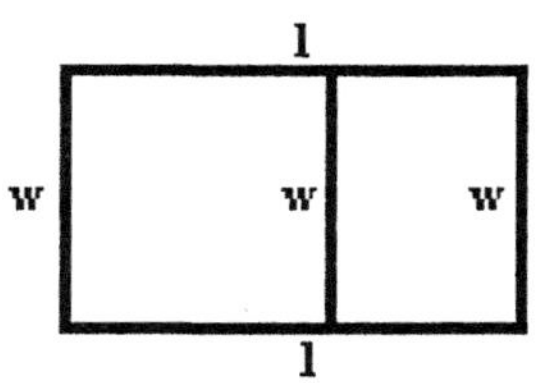

The dimensions of the field are 20 meters by 16 meters.

56. Let $x =$ the measure of one of the acute angles and $y =$ the measure of the other acute angle.

$$\begin{array}{llll} (1)\ x+y=90 & \boldsymbol{x}+y=90 & x=2y+15 & \text{The } \textbf{difference} \\ (2)\ \quad x=2y+15 & \boldsymbol{2y+15}+y=90 & x=2(25)+15 & \text{between the angles} \\ & 3y=75 & x=50+15 & \text{is } 65-25=40^\circ. \\ & y=25 & x=65 & \end{array}$$

57. Let $x =$ the amount invested at 10% and $y =$ the amount invested at 12%.

$$\begin{array}{lll} x+\quad y=8000 \Rightarrow & x+\quad y=\ 8000 \Rightarrow \times(-10) & -10x-10y=-80000 \\ \underline{0.10x+0.12y=\ 900} \Rightarrow \times 100 & \underline{10x+12y=90000} \Rightarrow & \underline{10x+12y=\ \ 90000} \\ & & 2y=\ \ 10000 \\ & & y=\ \ 5000 \end{array}$$

$x+y=8000$

$x+5000=8000$

$x=3000 \Rightarrow$ \$3000 was invested at 10%, while \$5000 was invested at 12%.

58. Let $x =$ the amount invested at 6% and $y =$ the amount invested at 7.5%.

$$\begin{array}{lll} x+\quad y=12000 \Rightarrow & x+\quad y=\ 12000 \Rightarrow \times(-60) & -60x-60y=-720000 \\ \underline{0.06x+0.075y=\ 810} \Rightarrow \times 1000 & \underline{60x+75y=810000} \Rightarrow & \underline{60x+75y=\ \ 810000} \\ & & 15y=\ \ 90000 \\ & & y=\ \ 6000 \end{array}$$

$x+y=12000$

$x+6000=12000$

$x=6000 \Rightarrow$ \$6000 was invested at 6%, while \$6000 was invested at 7.5%.

59. Let $x =$ the # of ounces of the 8% solution and $y =$ the # of ounces of the 15% solution.

$$\begin{array}{lll} x + y = 100 \Rightarrow & x + y = 100 \Rightarrow \times(-8) & -8x - 8y = -800 \\ \underline{0.08x + 0.15y = 12.2} \Rightarrow \times 100 & \underline{8x + 15y = 1220} \Rightarrow & \underline{8x + 15y = 1220} \\ & & 7y = 420 \\ & & y = 60 \end{array}$$

$$x + y = 100$$
$$x + 60 = 100$$
$x = 40 \Rightarrow$ 40 oz of the 8% and 60 oz of the 15% solution should be used.

60. Let $x =$ the # of pounds of the \$2 candy and $y =$ the # of pounds of the \$4 candy.

$$\begin{array}{llll} x + y = 60 \Rightarrow \times(-2) & -2x - 2y = -120 & \text{Solve for } x: & \text{30 pounds of each kind of candy} \\ \underline{2x + 4y = 180} \Rightarrow & \underline{2x + 4y = 180} & x + y = 60 & \text{should be used.} \\ & 2y = 60 & x + 30 = 60 & \\ & y = 30 & x = 30 & \end{array}$$

61. Let $c =$ the speed of the car and $p =$ the speed of the plane. Remember: distance $=$ rate $\cdot$ time, so time $= \dfrac{\text{distance}}{\text{rate}}$. Form one equation from the fact that the car travels 50 miles in the same time that the plane travels 180 miles:

Car time $=$ Plane time $\Rightarrow \dfrac{50}{c} = \dfrac{180}{p} \Rightarrow 50p = 180c \Rightarrow 50p - 180c = 0$.

Form a second equation from the relationship given between the rates: $p = c + 143$.

$$\begin{array}{lll} (1) \quad 50p - 180c = 0 & 50\boldsymbol{p} - 180c = 0 & \text{The car's speed is 55 mph.} \\ (2) \quad p = c + 143 & 50\boldsymbol{(c + 143)} - 180c = 0 & \\ & 50c + 7150 - 180c = 0 & \\ & -130c = -7150 & \\ & c = 55 & \end{array}$$

62. Let $c =$ the distance the car has traveled and $t =$ the distance the truck has traveled.

$$\begin{array}{lll} (1) \quad c + t = 350 & \boldsymbol{c} + t = 350 & c = t + 70 \\ (2) \quad c = t + 70 & \boldsymbol{t + 70} + t = 350 & c = 140 + 70 \\ & 2t = 280 & c = 210 \Rightarrow \text{The car has traveled 210 miles.} \\ & t = 140 & \end{array}$$

63. Let $r =$ the number of racing bikes and $m =$ the number of mountain bikes.

$$\begin{array}{lll} 60r + 90m = 15900 \Rightarrow \times(-7) & -420r - 630m = -111300 & 60r + 90m = 15900 \\ \underline{55r + 70m = 13075} \Rightarrow \times 9 & \underline{495r + 630m = 117675} & 60(85) + 90m = 15900 \\ & 75r = 6375 & 5100 + 90m = 15900 \\ & r = 85 & 90m = 10800 \\ & & m = 120 \end{array}$$

85 racing bikes and 120 mountain bikes can be built.

64. Let $A =$ the # of grams of Mix A and $B =$ the # of grams of Mix B.

$$\begin{array}{lll} 0.12A + 0.15B = 15 \Rightarrow \times 100 & 12A + 15B = 1500 \Rightarrow & 12A + 15B = 1500 \\ 0.09A + 0.05B = 7.5 \Rightarrow \times 100 & 9A + 5B = 750 \Rightarrow \times(-3) & -27A - 15B = -2250 \\ \hline & & -15A = -750 \\ & & A = 50 \end{array}$$

$$\begin{aligned} 12A + 15B &= 1500 \\ 12(50) + 15B &= 1500 \\ 600 + 15B &= 1500 \\ 15B &= 900 \\ B &= 60 \Rightarrow \text{50 grams of mix A and 60 grams of mix B should be used.} \end{aligned}$$

65. Let $x =$ the number of plates produced at the break-even point. Let $C_1 =$ the cost of the first machine: $C_1 = 300 + 2x$. Let $C_2 =$ the cost of the second machine: $C_2 = 500 + x$. To find the break point, find x such that $C_1 = C_2$:

$$\begin{aligned} C_1 &= C_2 \\ 300 + 2x &= 500 + x \\ x &= 200 \Rightarrow \text{The break point is 200 plates.} \end{aligned}$$

66. Let $x =$ the number of books produced at the break-even point. Let $C_1 =$ the cost of the first press: $C_1 = 210 + 5.98x$. Let $C_2 =$ the cost of the second press: $C_2 = 350 + 5.95x$. To find the break point, find x such that $C_1 = C_2$:

$$\begin{aligned} C_1 &= C_2 \\ 210 + 5.98x &= 350 + 5.95x \\ 0.03x &= 140 \\ x &= 4666\tfrac{2}{3} \Rightarrow \Rightarrow \text{The break point is } 4666\tfrac{2}{3} \text{ books.} \end{aligned}$$

67. Let $x =$ the number of computers sold at the break-even point. Let $C =$ the costs of the store: $C = 8925 + 850x$. Let $R =$ the revenue of the store: $R = 1275x$. To find the break-even point, find x such that $C = R$:

$$\begin{aligned} C &= R \\ 8925 + 850x &= 1275x \\ 8925 &= 425x \\ 21 &= x \Rightarrow \text{The break-even point is 21 computers.} \end{aligned}$$

68. Let $x =$ the number of permanents sold at the break-even point. Let $C =$ the costs of the shop: $C = 2101.20 + 23.60x$. Let $R =$ the revenue of the shop: $R = 44x$. To find the break-even point, find x such that $C = R$:

$$\begin{aligned} C &= R \\ 2101.20 + 23.60x &= 44x \\ 2101.20 &= 20.40x \\ 103 &= x \Rightarrow \text{The break-even point is 103 permanents.} \end{aligned}$$

69. Let $x =$ the number of pieces of software sold at the break-even point. Let $C =$ the costs of the business: $C = 18375 + 5.45x$. Let $R =$ the revenue of the business: $R = 29.95x$. To find the break-even point, find x such that $C = R$:

$$\begin{aligned} C &= R \\ 18375 + 5.45x &= 29.95x \\ 18375 &= 24.50x \\ 750 &= x \Rightarrow \text{The break-even point is 750 pieces of software.} \end{aligned}$$

70. Let $x =$ the number of sets of CDs sold at the break-even point. Let $C =$ the costs of the company: $C = 3 \cdot 35000 + 18.95x = 105000 + 18.95x$. Let $R =$ the revenue of the business: $R = 3 \cdot 15x = 45x$. To find the break-even point, find x such that $C = R$:

$$\begin{aligned} C &= R \\ 105000 + 18.95x &= 45x \\ 105000 &= 26.05x \\ 4030.71 &= x \Rightarrow \text{The break-even point is 4030.71 sets of CDs.} \end{aligned}$$

The investors must sell 4031 sets of CDs in order to make a profit.

71. Let $x =$ the number of gallons of paint A sold at the break-even point. Let $C =$ the costs of the company: $C = 32500 + 13x$. Let $R =$ the revenue of the business: $R = 18x$. To find the break-even point, find x such that $C = R$:

$$\begin{aligned} C &= R \\ 32500 + 13x &= 18x \\ 32500 &= 5x \\ 6500 &= x \Rightarrow \text{The break-even point is 6500 gallons of paint A.} \end{aligned}$$

72. Let $x =$ the number of gallons of paint B sold at the break-even point. Let $C =$ the costs of the company: $C = 80600 + 5x$. Let $R =$ the revenue of the business: $R = 18x$. To find the break-even point, find x such that $C = R$:

$$\begin{aligned} C &= R \\ 80600 + 5x &= 18x \\ 80600 &= 13x \\ 6200 &= x \Rightarrow \text{The break-even point is 6200 gallons of paint B.} \end{aligned}$$

73. Calculate the profit made using each process (profit = revenue − cost)
A: revenue − cost $= 18(6000) - (32500 + 13 \cdot 6000) = 108000 - 110500 = -2500$
B: revenue − cost $= 18(6000) - (80600 + 5 \cdot 6000) = 108000 - 110600 = -2600$
Since the loss is less with process A, process A should be used.

74. Calculate the profit made using each process (profit = revenue − cost)
A: revenue − cost $= 18(7000) - (32500 + 13 \cdot 7000) = 126000 - 123500 = 2500$
B: revenue − cost $= 18(7000) - (80600 + 5 \cdot 7000) = 126000 - 115600 = 10400$
Since the profit is more with process B, process B should be used.

75. Let $x =$ the number of pumps using process A sold at the break-even point. Let $C =$ the costs of the company: $C = 12390 + 29x$. Let $R =$ the revenue of the business: $R = 50x$. To find the break-even point, find x such that $C = R$:

$$\begin{aligned} C &= R \\ 12390 + 29x &= 50x \\ 12390 &= 21x \\ 590 &= x \Rightarrow \text{The break-even point is 590 pumps using process A.} \end{aligned}$$

76. Let $x =$ the number of pumps using process B sold at the break-even point. Let $C =$ the costs of the company: $C = 20460 + 17x$. Let $R =$ the revenue of the business: $R = 50x$. To find the break-even point, find x such that $C = R$:

$$\begin{aligned} C &= R \\ 20460 + 17x &= 50x \\ 20460 &= 33x \\ 620 &= x \Rightarrow \text{The break-even point is 620 pumps using process B.} \end{aligned}$$

77. Calculate the profit made using each process (profit = revenue – cost)
A: revenue – cost $= 550(50) - (12390 + 550 \cdot 29) = 27500 - 28340 = -840$
B: revenue – cost $= 550(50) - (20460 + 550 \cdot 17) = 27500 - 29810 = -2310$
Since the loss is less with process A, process A should be used.

78. Calculate the profit made using each process (profit = revenue – cost)
A: revenue – cost $= 600(50) - (12390 + 600 \cdot 29) = 30000 - 28340 = 210$
B: revenue – cost $= 600(50) - (20460 + 600 \cdot 17) = 30000 - 30660 = -660$
Since the profit is more with process A, process A should be used.

79. Calculate the profit made using each process (profit = revenue – cost)
A: revenue – cost $= 650(50) - (12390 + 650 \cdot 29) = 32500 - 31240 = 1260$
B: revenue – cost $= 650(50) - (20460 + 650 \cdot 17) = 32500 - 31510 = 990$
Since the profit is more with process A, process A should be used.

80. Let $x =$ the number of pumps sold when Process A and Process B have equal profits.
Let $P_A =$ the profit using Process A. $P_A = 50x - (12390 + 29x) = 21x - 12390$.
Let $P_B =$ the profit using Process B. $P_B = 50x - (20460 + 17x) = 33x - 20460$.
Set $P_A = P_B$ and solve for x:

$$\begin{aligned} P_A &= P_B \\ 21x - 12390 &= 33x - 20460 \\ 8070 &= 12x \\ 672.5 &= x \end{aligned}$$

Since Process A is better than Process B for x-values less than 672.5 (see Exercises 77-79), Process B will be better than Process A for x-values greater than 672.5, starting with 673 pumps sold.

SECTION 3.2

81. Let $x =$ the measure of the first angle and $y =$ the measure of the second angle.

$$\begin{aligned} x + y &= 180 \\ x - y &= 110 \\ \hline 2x &= 290 \\ x &= 145 \end{aligned} \qquad \begin{aligned} x + y &= 180 \\ 145 + y &= 180 \\ y &= 35 \end{aligned}$$

The angles have measures of 35° and 145°.

82. Let $x =$ the measure of the first angle and $y =$ the measure of the second angle.

$$\begin{aligned} (1)\quad x + y &= 90 \\ (2)\quad x &= y + 16 \end{aligned} \qquad \begin{aligned} \boldsymbol{x} + y &= 90 \\ \boldsymbol{y + 16} + y &= 90 \\ 2y &= 74 \\ y &= 37 \end{aligned} \qquad \begin{aligned} x &= y + 16 \\ x &= 37 + 16 \\ x &= 53 \end{aligned}$$

The angles have measures of 37° and 53°.

83. $\angle A$ and $\angle B$ are supplementary $\Rightarrow 2x + y + 3x = 180 \Rightarrow 5x + y = 180$.
$\angle A$ and $\angle D$ are supplementary $\Rightarrow 2x + y + y = 180 \Rightarrow 2x + 2y = 180$.

$$\begin{aligned} 5x + \ \ y &= 180 \Rightarrow \times(-2) \\ 2x + 2y &= 180 \Rightarrow \end{aligned} \qquad \begin{aligned} -10x - 2y &= -360 \\ 2x + 2y &= \ \ 180 \\ \hline -8x &= -180 \\ x &= \ \ 22.5 \end{aligned}$$

Solve for y:

$$\begin{aligned} 2x + 2y &= 180 \\ 2(22.5) + 2y &= 180 \\ 45 + 2y &= 180 \\ 2y &= 135 \\ y &= 67.5 \end{aligned}$$

84. $\angle CAB$ and $\angle ACD$ are congruent $\Rightarrow x - 2y = 50$.
$\angle DAC$ and $\angle ACB$ are congruent $\Rightarrow x + y = 80$.

$$\begin{aligned} x - 2y &= 50 \Rightarrow \\ x + \ \ y &= 80 \Rightarrow \times 2 \end{aligned} \qquad \begin{aligned} x - 2y &= \ \ 50 \\ 2x + 2y &= 160 \\ \hline 3x &= 210 \\ x &= \ \ 70 \end{aligned}$$

Solve for y:

$$\begin{aligned} x + y &= 80 \\ 70 + y &= 80 \\ y &= 10 \end{aligned}$$

85. Let $x =$ the measure of the angle of the range of motion and $y =$ the measure of the second angle.

$$\begin{aligned} (1)\quad x + y &= 90 \\ (2)\quad x &= 4y \end{aligned} \qquad \begin{aligned} \boldsymbol{x} + y &= 90 \\ \boldsymbol{4y} + y &= 90 \\ 5y &= 90 \\ y &= 18 \end{aligned} \qquad \begin{aligned} x &= 4y \\ x &= 4(18) \\ x &= 72 \end{aligned}$$

The range of motion is 72°.

86. Let $x =$ the measure of the first angle and $y =$ the measure of the second angle.

$$\begin{aligned} (1)\quad x + y &= 180 \\ (2)\quad x &= 2y - 15 \end{aligned} \qquad \begin{aligned} \boldsymbol{x} + y &= 180 \\ \boldsymbol{2y - 15} + y &= 180 \\ 3y &= 195 \\ y &= 65 \end{aligned} \qquad \begin{aligned} x &= 2y - 15 \\ x &= 2(65) - 15 \\ x &= 130 - 15 \\ x &= 115 \end{aligned}$$

The angles have measures of 65° and 115°.

87. Set $X_L = X_C$:

$$\begin{aligned} X_L &= X_C \\ 2\pi f L &= \frac{1}{2\pi f C} \\ 2\pi f L \cdot 2\pi f C &= \frac{1}{2\pi f C} \cdot 2\pi f C \\ 4\pi^2 f^2 LC &= 1 \\ \frac{4\pi^2 f^2 LC}{4\pi^2 LC} &= \frac{1}{4\pi^2 LC} \\ f^2 &= \frac{1}{4\pi^2 LC} \end{aligned}$$

88. Let $x =$ sales to make the checks equal.

$$\begin{aligned} 0.07x &= 150 + 0.02x \\ 0.05x &= 150 \\ x &= 3000 \end{aligned}$$

\$3000 in sales will result in equal checks.

89-92. Answers may vary.

Exercise 3.3 (page 187)

1. $m = \frac{\Delta y}{\Delta x} = \frac{-4-5}{-2-3} = \frac{-9}{-5} = \frac{9}{5}$

2.
$$\begin{aligned} y - y_1 &= m(x - x_1) \\ y - 5 &= \tfrac{9}{5}(x-3) \\ 5(y-5) &= 9(x-3) \\ 5y - 25 &= 9x - 27 \\ -9x + 5y &= -2 \Rightarrow 9x - 5y = 2 \end{aligned}$$

3. $f(0) = 2(0)^2 + 1 = 2(0) + 1 = 0 + 1 = 1$

4. $f(-2) = 2(-2)^2 + 1 = 2(4) + 1 = 8 + 1 = 9$

5. $f(s) = 2s^2 + 1$

6. $f(2t) = 2(2t)^2 + 1 = 2(4t^2) + 1 = 8t^2 + 1$

7. plane

8. dependent; infinitely

9. infinitely

10. no

11.
$$\begin{aligned} x - y + z &= 2 \\ 2 - 1 + 1 &\stackrel{?}{=} 2 \\ 2 &= 2 \end{aligned}
\qquad
\begin{aligned} 2x + y - z &= 4 \\ 2(2) + 1 - 1 &\stackrel{?}{=} 4 \\ 4 + 1 - 1 &\stackrel{?}{=} 4 \\ 4 &= 4 \end{aligned}
\qquad
\begin{aligned} 2x - 3y + z &= 2 \\ 2(2) - 3(1) + 1 &\stackrel{?}{=} 2 \\ 4 - 3 + 1 &\stackrel{?}{=} 2 \\ 2 &= 2 \end{aligned}$$

$(2, 1, 1)$ is a solution to the system.

12.
$$\begin{aligned} 2x + 2y + 3z &= -1 \\ 2(-3) + 2(2) + 3(-1) &\stackrel{?}{=} -1 \\ -6 + 4 - 3 &\stackrel{?}{=} -1 \\ -5 &\neq -1 \end{aligned}$$

$(-3, 2, -1)$ is not a solution.

13.

$$\begin{array}{ll} (1) & x+y+z=4 \\ (2) & 2x+y-z=1 \\ (3) & 2x-3y+z=1 \end{array} \quad \begin{array}{ll} (1) & x+y+z=4 \\ (2) & 2x+y-z=1 \\ \hline (4) & 3x+2y\quad=5 \end{array} \quad \begin{array}{ll} (2) & 2x+y-z=1 \\ (3) & 2x-3y+z=1 \\ \hline (5) & 4x-2y\quad=2 \end{array}$$

$$\begin{array}{ll} (4) & 3x+2y=5 \\ (5) & 4x-2y=2 \\ \hline & 7x\quad=7 \\ & x\quad=1 \end{array} \quad \begin{array}{r} 3x+2y=5 \\ 3(1)+2y=5 \\ 3+2y=5 \\ 2y=2 \\ y=1 \end{array} \quad \begin{array}{r} x+y+z=4 \\ 1+1+z=4 \\ 2+z=4 \\ z=2 \end{array}$$

$\boxed{\text{The solution is } (1,1,2).}$

14.

$$\begin{array}{ll} (1) & x+y+z=4 \\ (2) & x-y+z=2 \\ (3) & x-y-z=0 \end{array} \quad \begin{array}{ll} (1) & x+y+z=4 \\ (2) & x-y+z=2 \\ \hline (4) & 2x\quad+2z=6 \end{array} \quad \begin{array}{ll} (1) & x+y+z=4 \\ (3) & x-y-z=0 \\ \hline (5) & 2x\quad=4 \\ & x\quad=2 \end{array}$$

$$\begin{array}{lr} (4) & 2x+2z=6 \\ & 2(2)+2z=6 \\ & 4+2z=6 \\ & 2z=2 \\ & z=1 \end{array} \quad \begin{array}{r} x+y+z=4 \\ 2+y+1=4 \\ 3+y=4 \\ y=1 \end{array}$$

$\boxed{\text{The solution is } (2,1,1).}$

15.

$$\begin{array}{ll} (1) & 2x+2y+3z=10 \\ (2) & 3x+y-z=0 \\ (3) & x+y+2z=6 \end{array} \quad \begin{array}{ll} (1) & 2x+2y+3z=10 \\ 3\cdot(2) & 9x+3y-3z=0 \\ \hline (4) & 11x+5y\quad=10 \end{array} \quad \begin{array}{ll} (3) & x+y+2z=6 \\ 2\cdot(2) & 6x+2y-2z=0 \\ \hline (5) & 7x+3y\quad=6 \end{array}$$

$$\begin{array}{l} 11x+5y=10 \Rightarrow \times 3 \\ 7x+3y=6 \Rightarrow \times(-5) \\ \hline \end{array} \quad \begin{array}{r} 33x+15y=30 \\ -35x-15y=-30 \\ \hline -2x\quad=0 \\ x\quad=0 \end{array} \quad \begin{array}{r} 11x+5y=10 \\ 11(0)+5y=10 \\ 0+5y=10 \\ 5y=10 \\ y=2 \end{array} \quad \begin{array}{r} x+y+2z=6 \\ 0+2+2z=6 \\ 2+2z=6 \\ 2z=4 \\ z=2 \end{array}$$

Solution: $\boxed{(0,2,2)}$

16.

$$\begin{array}{ll} (1) & x-y+z=4 \\ (2) & x+2y-z=-1 \\ (3) & x+y-3z=-2 \end{array} \quad \begin{array}{ll} (1) & x-y+z=4 \\ (2) & x+2y-z=-1 \\ \hline (4) & 2x+y\quad=3 \end{array} \quad \begin{array}{ll} (3) & x+y-3z=-2 \\ 3\cdot(1) & 3x-3y+3z=12 \\ \hline (5) & 4x-2y\quad=10 \end{array}$$

$$\begin{array}{l} 2x+y=3 \Rightarrow \times 2 \\ 4x-2y=10 \Rightarrow \\ \hline \end{array} \quad \begin{array}{r} 4x+2y=6 \\ 4x-2y=10 \\ \hline 8x\quad=16 \\ x\quad=2 \end{array} \quad \begin{array}{r} 2x+y=3 \\ 2(2)+y=3 \\ 4+y=3 \\ y=-1 \end{array} \quad \begin{array}{r} x-y+z=4 \\ 2-(-1)+z=4 \\ 3+z=4 \\ z=1 \end{array}$$

Solution: $\boxed{(2,-1,1)}$

17.

$$\begin{array}{ll} (1) & a+b+2c=7 \\ (2) & a+2b+c=8 \\ (3) & 2a+b+c=9 \end{array} \qquad \begin{array}{rrcr} (1) & a+\ \ b+2c &=& 7 \\ -2\cdot(2) & -2a-4b-2c &=& -16 \\ \hline (4) & -a-3b \quad &=& -9 \end{array} \qquad \begin{array}{rrcr} (1) & a+\ \ b+2c &=& 7 \\ -2\cdot(3) & -4a-2b-2c &=& -18 \\ \hline (5) & -3a-\ \ b \quad &=& -11 \end{array}$$

$$\begin{array}{l} -a-3b=\ \ -9 \Rightarrow \times(-3) \\ \underline{-3a-\ \ b=-11 \Rightarrow} \end{array} \quad \begin{array}{rcr} 3a+9b &=& 27 \\ -3a-\ \ b &=& -11 \\ \hline 8b &=& 16 \\ b &=& 2 \end{array} \quad \begin{array}{rcl} -a-3b &=& -9 \\ -a-3(2) &=& -9 \\ -a-6 &=& -9 \\ -a &=& -3 \\ a &=& 3 \end{array} \quad \begin{array}{rcl} 2a+b+c &=& 9 \\ 2(3)+2+c &=& 9 \\ 6+2+c &=& 9 \\ 8+c &=& 9 \\ c &=& 1 \end{array}$$

Solution: $\boxed{(3,2,1)}$

18.

$$\begin{array}{ll} (1) & 2a+3b+c=2 \\ (2) & 4a+6b+2c=5 \\ (3) & a-2b+c=3 \end{array} \qquad \begin{array}{rrcr} (2) & 4a+6b+2c &=& 5 \\ -2\cdot(1) & -4a-6b-2c &=& -4 \\ \hline (4) & 0 &=& 1 \end{array}$$

Since equation (4) is always false, there is no solution. The system is inconsistent.

19.

$$\begin{array}{ll} (1) & 2x+y-z=1 \\ (2) & x+2y+2z=2 \\ (3) & 4x+5y+3z=3 \end{array} \qquad \begin{array}{rrcr} (2) & x+2y+2z &=& 2 \\ 2\cdot(1) & 4x+2y-2z &=& 2 \\ \hline (4) & 5x+4y \quad &=& 4 \end{array} \qquad \begin{array}{rrcr} (3) & 4x+5y+3z &=& 3 \\ 3\cdot(1) & 6x+3y-3z &=& 3 \\ \hline (5) & 10x+8y \quad &=& 6 \end{array}$$

$$\begin{array}{l} 5x+4y=4 \Rightarrow \times(-2) \\ \underline{10x+8y=6 \Rightarrow} \end{array} \quad \begin{array}{rcr} -10x-8y &=& -8 \\ 10x+8y &=& 6 \\ \hline 0 &=& -2 \end{array}$$

Since this equation is always false, there is no solution. The system is inconsistent.

20.

$$\begin{array}{ll} (1) & 3x-y-2z=12 \\ (2) & x+y+6z=8 \\ (3) & 2x-2y-z=11 \end{array} \qquad \begin{array}{rrcr} (1) & 3x-y-2z &=& 12 \\ (2) & x+y+6z &=& 8 \\ \hline (4) & 4x \quad +4z &=& 20 \end{array} \qquad \begin{array}{rrcr} 2\cdot(2) & 2x+2y+12z &=& 16 \\ (3) & 2x-2y-\ \ z &=& 11 \\ \hline (5) & 4x \quad +11z &=& 27 \end{array}$$

$$\begin{array}{l} 4x+\ \ 4z=20 \Rightarrow \times(-1) \\ \underline{4x+11z=27 \Rightarrow} \end{array} \quad \begin{array}{rcr} -4x-\ \ 4z &=& -20 \\ 4x+11z &=& 27 \\ \hline 7z &=& 7 \\ z &=& 1 \end{array} \quad \begin{array}{rcl} 4x+4z &=& 20 \\ 4x+4(1) &=& 20 \\ 4x &=& 16 \\ x &=& 4 \end{array} \quad \begin{array}{rcl} x+y+6z &=& 8 \\ 4+y+6(1) &=& 8 \\ y+10 &=& 8 \\ y &=& -2 \end{array}$$

Solution: $\boxed{(4,-2,1)}$

21.

$$\begin{array}{ll} (1) & 4x+3z=4 \\ (2) & 2y-6z=-1 \\ (3) & 8x+4y+3z=9 \end{array} \qquad \begin{array}{rrcr} (2) & 2y-6z &=& -1 \\ 2\cdot(1) & 8x \quad +6z &=& 8 \\ \hline (4) & 8x+2y \quad &=& 7 \end{array} \qquad \begin{array}{rrcr} (2) & 2y-6z &=& -1 \\ 2\cdot(3) & 16x+\ \ 8y+6z &=& 18 \\ \hline (5) & 16x+10y \quad &=& 17 \end{array}$$

$$\begin{array}{l} 8x+\ \ 2y=\ \ 7 \Rightarrow \times(-2) \\ \underline{16x+10y=17 \Rightarrow} \end{array} \quad \begin{array}{rcr} -16x-\ \ 4y &=& -14 \\ 16x+10y &=& 17 \\ \hline 6y &=& 3 \\ y &=& \frac{1}{2} \end{array} \quad \begin{array}{rcl} 8x+2y &=& 7 \\ 8x+2\left(\frac{1}{2}\right) &=& 7 \\ 8x+1 &=& 7 \\ 8x &=& 6 \\ x &=& \frac{3}{4} \end{array} \quad \begin{array}{rcl} 4x+3z &=& 4 \\ 4\left(\frac{3}{4}\right)+3z &=& 4 \\ 3+3z &=& 4 \\ 3z &=& 1 \\ z &=& \frac{1}{3} \end{array}$$

Solution: $\boxed{\left(\frac{3}{4},\frac{1}{2},\frac{1}{3}\right)}$

22.

$$\begin{array}{ll} (1) & 2x+3y+2z=1 \\ (2) & 2x-3y+2z=-1 \\ (3) & 4x+3y-2z=4 \end{array} \quad \begin{array}{ll} (1) & 2x+3y+2z=\ \ 1 \\ (2) & \underline{2x-3y+2z=-1} \\ (4) & 4x \qquad\ \ +4z=\ \ 0 \end{array} \quad \begin{array}{ll} (2) & 2x-3y+2z=-1 \\ (3) & \underline{4x+3y-2z=\ \ 4} \\ (5) & 6x \qquad\qquad\ =\ 3 \\ & x \qquad\qquad\ \ = \frac{3}{6}=\frac{1}{2} \end{array}$$

$$\begin{aligned} 4x+4z&=0 \\ 4\left(\tfrac{1}{2}\right)+4z&=0 \\ 2+4z&=0 \\ 4z&=-2 \\ z&=-\tfrac{2}{4}=-\tfrac{1}{2} \end{aligned} \qquad \begin{aligned} 2x+3y+2z&=1 \\ 2\left(\tfrac{1}{2}\right)+3y+2\left(-\tfrac{1}{2}\right)&=1 \\ 1+3y-1&=1 \\ 3y&=1 \\ y&=\tfrac{1}{3} \end{aligned}$$

Solution: $\boxed{\left(\frac{1}{2},\frac{1}{3},-\frac{1}{2}\right)}$

23.

$$\begin{array}{ll} (1) & 2x+3y+4z=6 \\ (2) & 2x-3y-4z=-4 \\ (3) & 4x+6y+8z=12 \end{array} \quad \begin{array}{ll} (3) & 4x+6y+8z=\ \ 12 \\ -2\cdot(1) & \underline{-4x-6y-8z=-12} \\ (4) & \qquad\qquad\quad 0=\ \ 0 \end{array}$$

Since equation (4) is always true, the equations are dependent, and there are infinitely many solutions.

24.

$$\begin{array}{ll} (1) & x-3y+4z=2 \\ (2) & 2x+y+2z=3 \\ (3) & 4x-5y+10z=7 \end{array} \quad \begin{array}{ll} (1) & x-3y+\ 4z=\ 2 \\ 3\cdot(2) & \underline{6x+3y+\ 6z=\ 9} \\ (4) & 7x \qquad +10z=11 \end{array} \quad \begin{array}{ll} (3) & 4x-5y+10z=\ \ 7 \\ 5\cdot(2) & \underline{10x+5y+10z=15} \\ (5) & 14x \qquad\ +20z=22 \end{array}$$

$$\begin{array}{l} 7x+10z=11 \Rightarrow \times(-2) \\ \underline{14x+20z=22 \Rightarrow} \end{array} \quad \begin{array}{r} -14x-20z=-22 \\ \underline{14x+20z=\ \ 22} \\ 0=\ \ 0 \end{array}$$

Since this equation is always true, the equations are dependent, and there are infinitely many solutions.

25.

$$\begin{array}{lll} x+\frac{1}{3}y+z=13 & \Rightarrow\times 3 & (1)\ 3x+y+3z=39 \\ \frac{1}{2}x-y+\frac{1}{3}z=-2 & \Rightarrow\times 6 & (2)\ 3x-6y+2z=-12 \\ x+\frac{1}{2}y-\frac{1}{3}z=2 & \Rightarrow\times 6 & (3)\ 6x+3y-2z=12 \end{array}$$

$$\begin{array}{ll} (2) & 3x-6y+\ 2z=-12 \\ 6\cdot(1) & \underline{18x+6y+18z=\ 234} \\ (4) & 21x \qquad +20z=\ 222 \end{array} \quad \begin{array}{ll} (2) & 3x-6y+2z=-12 \\ 2\cdot(3) & \underline{12x+6y-4z=\ \ 24} \\ (5) & 15x \qquad -2z=\ \ 12 \end{array}$$

$$\begin{array}{l} 21x+20z=222 \Rightarrow \\ \underline{15x-\ \ 2z=\ \ 12} \Rightarrow \times 10 \end{array} \quad \begin{array}{r} 21x+20z=222 \\ \underline{150x-20z=120} \\ 171x \qquad =342 \\ x \qquad =\ \ 2 \end{array} \quad \begin{aligned} 15x-2z&=12 \\ 15(2)-2z&=12 \\ 30-2z&=12 \\ -2z&=-18 \\ z&=9 \end{aligned} \quad \begin{aligned} 3x+y+3z&=39 \\ 3(2)+y+3(9)&=39 \\ 6+y+27&=39 \\ y&=6 \end{aligned}$$

Solution: $\boxed{(2,6,9)}$

26.

$$\begin{array}{lll} x-\frac{1}{5}y-z=9 & \Rightarrow\times 5 & (1)\ 5x-y-5z=45 \\ \frac{1}{4}x+\frac{1}{5}y-\frac{1}{2}z=5 & \Rightarrow\times 20 & (2)\ 5x+4y-10z=100 \\ 2x+y+\frac{1}{6}z=12 & \Rightarrow\times 6 & (3)\ 12x+6y+z=72 \end{array}$$

continued on next page...

26. continued...

$$\begin{array}{rl} (2) & 5x+4y-10z=100 \\ 4\cdot(1) & 20x-4y-20z=180 \\ \hline (4) & 25x \qquad -30z=280 \end{array} \qquad \begin{array}{rl} (3) & 12x+6y+ \quad z= \ 72 \\ 6\cdot(1) & 30x-6y-30z=270 \\ \hline (5) & 42x \qquad -29z=342 \end{array}$$

$$\begin{array}{l} 25x-30z=280 \Rightarrow \times 29 \\ 42x-29z=342 \Rightarrow \times(-30) \end{array} \quad \begin{array}{rcr} 725x-870z &=& 8120 \\ -1260x+870z &=& -\ 10260 \\ \hline -535x &=& -2140 \\ x &=& 4 \end{array} \quad \begin{array}{r} 25x-30z=280 \\ 25(4)-30z=280 \\ 100-30z=280 \\ -30z=180 \\ z=-6 \end{array}$$

$$\begin{array}{r} 5x-y-5z=45 \\ 5(4)-y-5(-6)=45 \\ 20-y+30=45 \\ y=5 \end{array}$$

Solution: $\boxed{(4,5,-6)}$

27. Let $x =$ the first integer, $y =$ the second integer and $z =$ the third integer.

$$\begin{array}{rcl} x+y+z=18 & \Rightarrow & (1)\ x+y+z=18 \\ z=4y & \Rightarrow & (2)\ -4y+z=0 \\ y=x+6 & \Rightarrow & (3)\ -x+y=6 \end{array} \qquad \begin{array}{rr} (1) & x+\ y+z=18 \\ (3) & -x+\ y \qquad = \ 6 \\ \hline (4) & 2y+z=24 \end{array}$$

$$\begin{array}{l} 2y+z=24 \Rightarrow \times 2 \\ -4y+z= \ 0 \Rightarrow \end{array} \quad \begin{array}{r} 4y+2z=48 \\ -4y+ \ z= \ 0 \\ \hline 3z=48 \\ z=16 \end{array} \quad \begin{array}{r} 2y+z=24 \\ 2y+16=24 \\ 2y=8 \\ y=4 \end{array} \quad \begin{array}{r} x+y+z=18 \\ x+4+16=18 \\ x=-2 \end{array}$$

The integers are $-2, 4$ and 16.

28. Let $x =$ the first integer, $y =$ the second integer and $z =$ the third integer.

$$\begin{array}{rl} (1) & x+y+z=48 \\ (2) & 2x+y+z=60 \\ (3) & x+2y+z=63 \end{array} \quad \begin{array}{rrcr} (1) & x+y+z &=& 48 \\ -1\cdot(2) & -2x-y-z &=& -60 \\ \hline (4) & -x &=& -12 \\ & x &=& 12 \end{array} \quad \begin{array}{rrcr} (1) & x\ +y+z &=& 48 \\ -1\cdot(3) & -x-2y-z &=& -63 \\ \hline (5) & -y &=& -15 \\ & y &=& 15 \end{array}$$

$$\begin{array}{r} x+y+z=48 \\ 12+15+z=48 \\ z=21 \end{array}$$

The integers are $12, 15$ and 21.

29. Let A, B and C represent the measures of the three angles.

$$\begin{array}{rcl} A+B+C=180 & \Rightarrow & (1)\ A+B+C=180 \\ A=B+C-100 & \Rightarrow & (2)\ A-B-C=-100 \\ C=2B-40 & \Rightarrow & (3)\ -2B+C=-40 \end{array} \qquad \begin{array}{rrcr} (1) & A+B+C &=& 180 \\ (2) & A-B-C &=& -100 \\ \hline (4) & 2A &=& 80 \\ & A &=& 40 \end{array}$$

$$\begin{array}{rr} & A+B+C=180 \\ & 40+B+C=180 \\ (5) & B+C=140 \end{array} \quad \begin{array}{rrcr} (-1)\cdot 3 & 2B-C &=& 40 \\ (5) & B+C &=& 140 \\ \hline & 3B &=& 180 \\ & B &=& 60 \end{array} \quad \begin{array}{r} B+C=140 \\ 60+C=140 \\ C=80 \end{array}$$

The angles have measures of $40°, 60°$ and $80°$.

30. Let $A =$ the measure of $\angle A$, let $B =$ the measure of B and let $C =$ the measure of $\angle C$.

$$\begin{array}{rcl} A+B+C+40=360 & \Rightarrow & (1)\ A+B+C=320 \\ A=B & \Rightarrow & (2)\ A-B=0 \\ C=A+20 & \Rightarrow & (3)\ -A+C=20 \end{array}$$

$$\begin{array}{lrcl} (1) & A+B+C & = & 320 \\ (2) & A-B & = & 0 \\ \hline (4) & 2A\quad +C & = & 320 \end{array}$$

$$\begin{array}{lrcl} 2\cdot(3) & -2A+2C & = & 40 \\ (4) & 2A+\ C & = & 320 \\ \hline & 3C & = & 360 \\ & C & = & 120 \end{array} \qquad \begin{aligned} 2A+C&=320\\ 2A+120&=320\\ 2A&=200\\ A&=100 \end{aligned} \qquad \begin{aligned} A-B&=0\\ 100-B&=0\\ 100&=B \end{aligned}$$

The angles have measures of 100°, 100°, 120° and 40°.

31. Let $A =$ the units of food A, $B =$ the units of food B and $C =$ the units of food C.

$$\begin{array}{lrl} (1) & A+2B+2C=11 & \text{(fat)} \\ (2) & A+B+C=6 & \text{(carbohydrate)} \\ (3) & 2A+B+2C=10 & \text{(protein)} \end{array}$$

$$\begin{array}{lrcl} (1) & A+2B+2C & = & 11 \\ -2\cdot(2) & -2A-2B-2C & = & -12 \\ \hline (4) & -A & = & -1 \\ & A & = & 1 \end{array}$$

$$\begin{array}{lrcl} (1) & A+2B+2C & = & 11 \\ -2\cdot(3) & -4A-2B-4C & = & -20 \\ \hline (5) & -3A\quad -2C & = & -9 \end{array} \qquad \begin{aligned} -3A-2C&=-9\\ -3(1)-2C&=-9\\ -3-2C&=-9\\ -2C&=-6\\ C&=3 \end{aligned} \qquad \begin{aligned} A+B+C&=6\\ 1+B+3&=6\\ B+4&=6\\ B&=2 \end{aligned}$$

1 unit of food A, 2 units of food B and 3 units of food C should be used.

32. Let $A =$ the units of food A, $B =$ the units of food B and $C =$ the units of food C.

$$\begin{array}{lrl} (1) & 2A+3B+C=14 & \text{(fat)} \\ (2) & A+2B+C=9 & \text{(carbohydrate)} \\ (3) & 2A+B+2C=9 & \text{(protein)} \end{array}$$

$$\begin{array}{lrcl} (1) & 2A+3B+\ C & = & 14 \\ -2\cdot(2) & -2A-4B-2C & = & -18 \\ \hline (4) & -\ B-\ C & = & -4 \end{array}$$

$$\begin{array}{lrcl} (3) & 2A+\ B+2C & = & 9 \\ -2\cdot(2) & -2A-4B-2C & = & -18 \\ \hline (5) & -3B & = & -9 \\ & B & = & 3 \end{array} \qquad \begin{aligned} -B-C&=-4\\ -3-C&=-4\\ -C&=-1\\ C&=1 \end{aligned} \qquad \begin{aligned} A+2B+C&=9\\ A+2(3)+1&=9\\ A+7&=9\\ A&=2 \end{aligned}$$

2 units of food A, 3 units of food B and 1 unit of food C should be used.

33. Let $x =$ the number of \$5 statues, $y =$ the number of \$4 statues and $z =$ the number of \$3 statues.

$$\begin{array}{lrl} (1) & x+y+z=180 & \text{(total number made)} \\ (2) & 5x+4y+3z=650 & \text{(total cost)} \\ (3) & 20x+12y+9z=2100 & \text{(total revenue)} \end{array}$$

$$\begin{array}{lrcl} -3\cdot(1) & -3x-3y-3z & = & -540 \\ (2) & 5x+4y+3z & = & 650 \\ \hline (4) & 2x+\ y & = & 110 \end{array}$$

$$\begin{array}{lrcl} -9\cdot(1) & -9x-\ 9y-9z & = & -1620 \\ (3) & 20x+12y+9z & = & 2100 \\ \hline (5) & 11x+\ 3y & = & 480 \end{array} \qquad \begin{array}{l} 2x+\ y=110 \Rightarrow \times(-3) \\ \underline{11x+3y=480} \Rightarrow \end{array} \qquad \begin{array}{rcl} -6x-3y & = & -330 \\ 11x+3y & = & 480 \\ \hline 5x & = & 150 \\ x & = & 30 \end{array}$$

$$\begin{aligned} 2x+y&=110\\ 2(30)+y&=110\\ 60+y&=110\\ y&=50 \end{aligned} \qquad \begin{aligned} x+y+z&=180\\ 30+50+z&=180\\ z&=100 \end{aligned}$$

30 of the \$5, 50 of the \$4 and 100 of the \$3 statues should be made.

34. Let x = the number of \$4 balls, y = the number of \$3 balls and z = the number of \$2 balls.

$$\begin{array}{lll} (1) & x+y+z=1125 & \text{(total number made)} \\ (2) & 4x+3y+2z=2425 & \text{(total cost)} \\ (3) & 12x+9y+8z=9275 & \text{(total profit)} \end{array}$$

$$\begin{array}{lrcl} -2\cdot(1) & -2x-2y-2z &=& -2250 \\ (2) & 4x+3y+2z &=& 2425 \\ \hline (4) & 2x+y &=& 175 \end{array}$$

$$\begin{array}{lrcl} -8\cdot(1) & -8x-8y-8z &=& -9000 \\ (3) & 12x+9y+8z &=& 9275 \\ \hline (5) & 4x+y &=& 275 \end{array}$$

$$\begin{array}{l} 2x+y=175 \Rightarrow \times(-1) \\ \underline{4x+y=275 \Rightarrow} \end{array} \quad \begin{array}{rcl} -2x-y &=& -175 \\ 4x+y &=& 275 \\ \hline 2x &=& 100 \\ x &=& 50 \end{array}$$

$$\begin{array}{rcl} 2x+y &=& 175 \\ 2(50)+y &=& 175 \\ 100+y &=& 175 \\ y &=& 75 \end{array} \qquad \begin{array}{rcl} x+y+z &=& 1125 \\ 50+75+z &=& 1125 \\ 125+z &=& 1125 \\ z &=& 1000 \end{array}$$

50 of the \$4, 75 of the \$3 and 1000 of the \$2 balls should be made.

35. Let x = the number of \$5 tickets, y = the number of \$3 tickets and z = the number of \$2 tickets.

$$\begin{array}{lll} (1) & x+y+z=750 & \text{(total sold)} \\ (2) & x=2z \Rightarrow x-2z=0 & \text{(twice as many)} \\ (3) & 5x+3y+2z=2625 & \text{(total revenue)} \end{array}$$

$$\begin{array}{lrcl} 2\cdot(1) & 2x+2y+2z &=& 1500 \\ (2) & x \qquad -2z &=& 0 \\ \hline (4) & 3x+2y &=& 1500 \end{array}$$

$$\begin{array}{lrcl} (2) & x \qquad -2z &=& 0 \\ (3) & 5x+3y+2z &=& 2625 \\ \hline (5) & 6x+3y &=& 2625 \end{array}$$

$$\begin{array}{l} 3x+2y=1500 \Rightarrow \times(-2) \\ \underline{6x+3y=2625 \Rightarrow} \end{array} \quad \begin{array}{rcl} -6x-4y &=& -3000 \\ 6x+3y &=& 2625 \\ \hline -y &=& -375 \\ y &=& 375 \end{array}$$

$$\begin{array}{rcl} 3x+2y &=& 1500 \\ 3x+2(375) &=& 1500 \\ 3x+750 &=& 1500 \\ 3x &=& 750 \\ x &=& 250 \end{array} \qquad \begin{array}{rcl} x+y+z &=& 750 \\ 250+375+z &=& 750 \\ 625+z &=& 750 \\ z &=& 125 \end{array}$$

250 of the \$5, 375 of the \$3 and 125 of the \$2 tickets were sold.

36. Let x = the pounds of peanuts, y = the pounds of cashews and z = the pounds of Brazil nuts.

$$\begin{array}{lll} (1) & x+y+z=50 & \text{(lbs)} \\ (2) & 3x+9y+9z=300 \quad (50\cdot 6) & \text{(cost)} \\ (3) & y=x-15 \Rightarrow -x+y=-15 & \text{(cashews/peanuts)} \end{array}$$

$$\begin{array}{lrcl} -3\cdot(1) & -3x-3y-3z &=& -150 \\ (2) & 3x+9y+9z &=& 300 \\ \hline (4) & 6y+6z &=& 150 \end{array} \qquad \begin{array}{lrcl} (1) & x+y+z &=& 50 \\ (3) & -x+y &=& -15 \\ \hline (5) & 2y+z &=& 35 \end{array}$$

$$\begin{array}{l} 6y+6z=150 \Rightarrow \\ \underline{2y+z=35 \Rightarrow \times(-3)} \end{array} \quad \begin{array}{rcl} 6y+6z &=& 150 \\ -6y-3z &=& -105 \\ \hline 3z &=& 45 \\ z &=& 15 \end{array}$$

Solve for y:

$$\begin{array}{rcl} 2y+z &=& 35 \\ 2y+15 &=& 35 \\ 2y &=& 20 \\ y &=& 10 \end{array} \qquad \begin{array}{rcl} x+y+z &=& 50 \\ x+10+15 &=& 50 \\ x+25 &=& 50 \\ x &=& 25 \end{array}$$

25 lbs of peanuts, 10 lbs of cashews and 15 lbs of Brazil nuts should be used.

37. Let x = the number of totem poles, y = the number of bears and z = the number of deer.

$$\begin{array}{lll} (1) & 2x+2y+z=14 & \text{(carving)} \\ (2) & x+2y+2z=15 & \text{(sanding)} \\ (3) & 3x+2y+2z=21 & \text{(painting)} \end{array} \qquad \begin{array}{lrl} -2\cdot(1) & -4x-4y-2z= & -28 \\ (2) & x+2y+2z= & 15 \\ \hline (4) & -3x-2y\quad = & -13 \end{array}$$

$$\begin{array}{lrl} -2\cdot(1) & -4x-4y-2z= & -28 \\ (3) & 3x+2y+2z= & 21 \\ \hline (5) & -x-2y\quad = & -7 \end{array} \qquad \begin{array}{r} -3x-2y=-13 \Rightarrow \times(-1) \\ -x-2y=\ -7 \Rightarrow \end{array} \qquad \begin{array}{rl} 3x+2y= & 13 \\ -x-2y= & -7 \\ \hline 2x\quad = & 6 \\ x\quad = & 3 \end{array}$$

$$\begin{array}{r} 3x+2y=13 \\ 3(3)+2y=13 \\ 9+2y=13 \\ 2y=4 \\ y=2 \end{array} \qquad \begin{array}{r} 2x+2y+z=14 \\ 2(3)+2(2)+z=14 \\ 6+4+z=14 \\ z=4 \end{array}$$

3 totem poles, 2 bears and 4 deer should be made.

38. Let x = the number of coats, y = the number of shirts and z = the number of slacks.

$$\begin{array}{lll} (1) & 20x+15y+10z=6900 & \text{(cutting)} \\ (2) & 60x+30y+24z=16800 & \text{(sewing)} \\ (3) & 5x+12y+6z=3900 & \text{(packaging)} \end{array} \qquad \begin{array}{lrl} -3(1) & -60x-45y-30z= & -20700 \\ (2) & 60x+30y+24z= & 16800 \\ \hline (4) & -15y-6z= & -3900 \end{array}$$

$$\begin{array}{lrl} -4(3) & -20x-48y-24z= & -15600 \\ (1) & 20x+15y+10z= & 6900 \\ \hline (5) & -33y-14z= & -8700 \end{array} \qquad \begin{array}{r} -15y-6z=-3900 \Rightarrow \times(-7) \\ -33y-14z=-8700 \Rightarrow \times 3 \end{array} \qquad \begin{array}{rl} 105y+42z= & 27300 \\ -99y-42z= & 26100 \\ \hline 6y\quad = & 1200 \\ y\quad = & 200 \end{array}$$

Solve for z:

$$\begin{array}{r} -15y-6z=-3900 \\ -15(200)-6z=-3900 \\ -3000-6z=-3900 \\ -6z=-900 \\ z=150 \end{array} \qquad \begin{array}{r} 5x+12y+6z=3900 \\ 5x+12(200)+6(150)=3900 \\ 5x+2400+900=3900 \\ 5x+3300=3900 \\ 5x=600 \\ x=120 \end{array}$$

120 coats, 200 shirts and 150 slacks should be made.

39. Let x = the % of nitrogen, y = the % of oxygen and z = the % of other gases.

$$\begin{array}{lll} (1) & x+y+z=100 & \\ (2) & x=3(y+z)+12 \Rightarrow & x-3y-3z=12 \\ (3) & z=y-20 \Rightarrow & -y+z=-20 \end{array} \qquad \begin{array}{lrl} (1) & x+y+z= & 100 \\ (3) & -y+z= & -20 \\ \hline (4) & x\quad +2z= & 80 \end{array}$$

$$\begin{array}{lrl} 3\cdot(1) & 3x+3y+3z= & 300 \\ (2) & x-3y-3z= & 12 \\ \hline (5) & 4x\quad = & 312 \\ & x\quad = & 78 \end{array} \qquad \begin{array}{r} x+2z=80 \\ 78+2z=80 \\ 2z=2 \\ z=1 \end{array} \qquad \begin{array}{r} x+y+z=100 \\ 78+y+1=100 \\ y+79=100 \\ y=21 \end{array}$$

The composition is 78% nitrogen, 21% oxygen, and 1% other gases.

40. Let x = passes from Steve Young, y = passes from Joe Montana and z = passes from Rich Gannon.

$$\begin{array}{llll} (1) & x = y + 30 \Rightarrow \quad x - y = 30 & (3) & x + y + z = 156 \\ (2) & y = z + 39 \Rightarrow \quad y - z = 39 & (1) & x - y \qquad = 30 \\ (3) & x + y + z = 156 & (4) & 2x \qquad + z = 186 \end{array}$$

$$\begin{array}{rl} (3) & x + y + z = 156 \\ -1 \cdot (2) & -y + z = -39 \\ (5) & x \quad + 2z = 117 \end{array} \qquad \begin{array}{l} 2x + z = 186 \Rightarrow \times(-2) \\ x + 2z = 117 \Rightarrow \end{array} \qquad \begin{array}{rl} -4x - 2z &= -372 \\ x + 2z &= 117 \\ -3x &= -255 \\ x &= 85 \end{array}$$

$$\begin{array}{rl} 2x + z &= 186 \\ 2(85) + z &= 186 \\ 170 + z &= 186 \\ z &= 16 \end{array} \qquad \begin{array}{l} y = z + 39 \\ y = 16 + 39 \\ y = 55 \end{array}$$

He caught 85 passes from Young, 55 passes from Montana, and 16 passes from Gannon.

41. Substitute the coordinates of each point for x and y in the equation $y = ax^2 + bx + c$.

$$\begin{array}{lll} y = ax^2 + bx + c & y = ax^2 + bx + c & y = ax^2 + bx + c \\ 0 = a(0)^2 + b(0) + c & -4 = a(2)^2 + b(2) + c & 0 = a(4)^2 + b(4) + c \\ 0 = c & -4 = 4a + 2b + c & 0 = 16a + 4b + c \end{array}$$

Solve the system of equations formed from the three equations:

$$\begin{array}{rl} (1) & c = 0 \\ (2) & 4a + 2b + c = -4 \\ (3) & 16a + 4b + c = 0 \end{array} \qquad \begin{array}{l} 4a + 2b = -4 \Rightarrow \times(-2) \\ 16a + 4b = 0 \Rightarrow \end{array} \qquad \begin{array}{rl} -8a - 4b &= 8 \\ 16a + 4b &= 0 \\ 8a &= 8 \\ a &= 1 \end{array} \qquad \begin{array}{rl} 16a + 4b &= 0 \\ 16(1) + 4b &= 0 \\ 16 + 4b &= 0 \\ 4b &= -16 \\ b &= -4 \end{array}$$

The equation is $y = x^2 - 4x$.

42. Substitute the coordinates of each point for x and y in the equation $y = ax^2 + bx + c$.

$$\begin{array}{lll} y = ax^2 + bx + c & y = ax^2 + bx + c & y = ax^2 + bx + c \\ 3 = a(-1)^2 + b(-1) + c & 1 = a(1)^2 + b(1) + c & 7 = a(3)^2 + b(3) + c \\ 3 = a - b + c & 1 = a + b + c & 7 = 9a + 3b + c \end{array}$$

Solve the system of equations formed from the three equations:

$$\begin{array}{rl} (1) & a - b + c = 3 \\ (2) & a + b + c = 1 \\ (3) & 9a + 3b + c = 7 \end{array} \qquad \begin{array}{rl} (1) & a - b + c = 3 \\ (2) & a + b + c = 1 \\ (4) & 2a \quad + 2c = 4 \end{array} \qquad \begin{array}{rl} 3 \cdot (1) & 3a - 3b + 3c = 9 \\ (3) & 9a + 3b + c = 7 \\ (5) & 12a \quad + 4c = 16 \end{array}$$

$$\begin{array}{l} 2a + 2c = 4 \Rightarrow \times(-2) \\ 12a + 4c = 16 \Rightarrow \end{array} \qquad \begin{array}{rl} -4a - 4c &= -8 \\ 12a + 4c &= 16 \\ 8a &= 8 \\ a &= 1 \end{array} \qquad \begin{array}{rl} 2a + 2c &= 4 \\ 2(1) + 2c &= 4 \\ 2c &= 2 \\ c &= 1 \end{array} \qquad \begin{array}{rl} a + b + c &= 1 \\ 1 + b + 1 &= 1 \\ 2 + b &= 1 \\ b &= -1 \end{array}$$

The equation is $y = x^2 - x + 1$.

43. Substitute the coordinates of each point for x and y in the equation $x^2+y^2+cx+dy+e=0$.

$$\begin{aligned} x^2+y^2+cx+dy+e&=0 \\ (1)^2+(3)^2+c(1)+d(3)+e&=0 \\ 1+9+c+3d+e&=0 \\ c+3d+e&=-10 \end{aligned} \qquad \begin{aligned} x^2+y^2+cx+dy+e&=0 \\ (3)^2+(1)^2+c(3)+d(1)+e&=0 \\ 9+1+3c+d+e&=0 \\ 3c+d+e&=-10 \end{aligned}$$

$$\begin{aligned} x^2+y^2+cx+dy+e&=0 \\ (1)^2+(-1)^2+c(1)+d(-1)+e&=0 \\ 1+1+c-d+e&=0 \\ c-d+e&=-2 \end{aligned} \qquad \begin{aligned} (1)\quad c+3d+e&=-10 \\ (2)\quad 3c+d+e&=-10 \\ (3)\quad c-d+e&=-2 \end{aligned}$$

$$\begin{array}{rrcr} (1) & c+3d+e & = & -10 \\ -1\cdot(2) & -3c-\ d-e & = & 10 \\ \hline (4) & -2c+2d & = & 0 \end{array} \qquad \begin{array}{rrcr} (1) & c+3d+e & = & -10 \\ -1\cdot(3) & -c+\ d-e & = & 2 \\ \hline (5) & 4d & = & -8 \\ & d & = & -2 \end{array} \qquad \begin{aligned} -2c+2d&=0 \\ -2c+2(-2)&=0 \\ -2c-4&=0 \\ -2c&=4 \\ c&=-2 \end{aligned}$$

$$\begin{aligned} c+3d+e&=-10 \\ -2+3(-2)+e&=-10 \\ -2-6+e&=-10 \\ -8+e&=-10 \\ e&=-2 \end{aligned}$$

The equation is $x^2+y^2-2x-2y-2=0$.

44. Substitute the coordinates of each point for x and y in the equation $x^2+y^2+cx+dy+e=0$.

$$\begin{aligned} x^2+y^2+cx+dy+e&=0 \\ (0)^2+(0)^2+c(0)+d(0)+e&=0 \\ e&=0 \end{aligned} \qquad \begin{aligned} x^2+y^2+cx+dy+e&=0 \\ (3)^2+(3)^2+c(3)+d(3)+e&=0 \\ 9+9+3c+3d+e&=0 \\ 3c+3d+e&=-18 \end{aligned}$$

$$\begin{aligned} x^2+y^2+cx+dy+e&=0 \\ (6)^2+(0)^2+c(6)+d(0)+e&=0 \\ 36+6c+e&=0 \\ 6c+e&=-36 \end{aligned} \qquad \begin{aligned} (1)\quad e&=0 \\ (2)\quad 3c+3d+e&=-18 \\ (3)\quad 6c+e&=-36 \end{aligned} \qquad \begin{aligned} 6c+e&=-36 \\ 6c+0&=-36 \\ c&=-6 \end{aligned}$$

$$\begin{aligned} 3c+3d+e&=-18 \\ 3(-6)+3d+0&=-18 \\ -18+3d&=-18 \\ 3d&=0 \\ d&=0 \end{aligned}$$

The equation is $x^2+y^2-6x=0$.

45. **Answers may vary.**

46. **Answers may vary.**

47.

$$\begin{array}{ll} (1) & x+y+z+w=3 \\ (2) & x-y-z-w=-1 \\ (3) & x+y-z-w=1 \\ (4) & x+y-z+w=3 \end{array}$$

$$\begin{array}{rl} (1) & x+y+z+w=3 \\ (2) & x-y-z-w=-1 \\ \hline & 2x = 2 \\ & x = 1 \end{array} \qquad \begin{array}{rl} (1) & x+y+z+w=3 \\ (3) & x+y-z-w=1 \\ \hline & 2x+2y = 4 \end{array}$$

$$\begin{array}{rl} (1) & x+y+z+w=3 \\ (4) & x+y-z+w=3 \\ \hline & 2x+2y+2w=6 \end{array} \qquad \begin{array}{r} 2x+2y=4 \\ 2(1)+2y=4 \\ 2y=2 \\ y=1 \end{array} \qquad \begin{array}{r} 2x+2y+2w=6 \\ 2(1)+2(1)+2w=6 \\ 4+2w=6 \\ 2w=2 \\ w=1 \end{array} \qquad \begin{array}{r} x+y+z+w=3 \\ 1+1+z+1=3 \\ 3+z=3 \\ z=0 \end{array}$$

The solution is $(1,1,0,1)$.

48.

$$\begin{array}{ll} (1) & 2x+y+z+w=3 \\ (2) & x-2y-z+w=-3 \\ (3) & x-y-2z-w=-3 \\ (4) & x+y-z+2w=4 \end{array}$$

Add (1) and (2):

$$\begin{array}{r} 2x+y+z+w=3 \\ x-2y-z+w=-3 \\ \hline 3x-y+2w=0 \end{array}$$

Add $2\cdot(1)$ and (3):

$$\begin{array}{r} 4x+2y+2z+2w=6 \\ x-y-2z-w=-3 \\ \hline 5x+y+w=3 \end{array}$$

Add (1) and (4):

$$\begin{array}{r} 2x+y+z+w=3 \\ x+y-z+2w=4 \\ \hline 3x+2y+3w=7 \end{array}$$

$$\begin{array}{ll} (5) & 3x-y+2w=0 \\ (6) & 5x+y+w=3 \\ (7) & 3x+2y+3w=7 \end{array} \qquad \begin{array}{ll} (5) & 3x-y+2w=0 \\ (6) & 5x+y+w=3 \\ \hline (8) & 8x+3w=3 \end{array}$$

$$\begin{array}{ll} 2\cdot(5) & 6x-2y+4w=0 \\ (7) & 3x+2y+3w=7 \\ \hline (9) & 9x+7w=7 \end{array} \qquad \begin{array}{ll} 7\cdot(8) & 56x+21w=21 \\ -3\cdot(9) & -27x-21w=-21 \\ \hline (4) & 28x=0 \\ & x=0 \end{array} \qquad \begin{array}{r} 8x+3w=3 \\ 8(0)+3w=3 \\ 3w=3 \\ w=1 \end{array}$$

$$\begin{array}{r} 5x+y+w=3 \\ 5(0)+y+1=3 \\ y=2 \end{array} \qquad \begin{array}{r} 2x+y+z+w=3 \\ 2(0)+2+z+1=3 \\ z=0 \end{array}$$

The solution is $(0,2,0,1)$.

Exercise 3.4 (page 197)

Note: The notation $3R_1 + R_3 \Rightarrow R_2$ means to multiply Row #1 of the previous matrix, add that result to Row #3 of the previous matrix, and write the final result in Row #2 of the current matrix.

1. $93{,}000{,}000 = 9.3 \times 10^7$

2. $0.00045 = 4.5 \times 10^{-4}$

3. $63 \times 10^3 = 6.3 \times 10^1 \times 10^3 = 6.3 \times 10^4$

4. $0.33 \times 10^3 = 3.3 \times 10^{-1} \times 10^3 = 3.3 \times 10^2$

5. matrix

6. elements

7. 3; columns

8. square

9. augmented

10. triangular

11. type 1

12. multiplying

13. nonzero

14. augmented; back

15. $\begin{bmatrix} 2 & 1 & 1 \\ 5 & 4 & 1 \end{bmatrix} \Rightarrow \overset{R_2 + (-R_1) \Rightarrow R_2}{\begin{bmatrix} 2 & 1 & 1 \\ 3 & 3 & \boxed{0} \end{bmatrix}}$

16. $\begin{bmatrix} -1 & 3 & 2 \\ 1 & -2 & 3 \end{bmatrix} \Rightarrow \overset{R_1 + R_2 \Rightarrow R_2}{\begin{bmatrix} -1 & 3 & 2 \\ \boxed{0} & 1 & 5 \end{bmatrix}}$

17. $\begin{bmatrix} 3 & -2 & 1 \\ -1 & 2 & 4 \end{bmatrix} \Rightarrow \overset{2R_2 \Rightarrow R_2}{\begin{bmatrix} 3 & -2 & 1 \\ -2 & 4 & \boxed{8} \end{bmatrix}}$

18. $\begin{bmatrix} 2 & 1 & -3 \\ 2 & 6 & 1 \end{bmatrix} \Rightarrow \overset{3R_1 \Rightarrow R_1}{\begin{bmatrix} 6 & 3 & \boxed{-9} \\ 2 & 6 & 1 \end{bmatrix}}$

19. $\left[\begin{array}{cc|c} 1 & 1 & 2 \\ 1 & -1 & 0 \end{array}\right] \Rightarrow \overset{R_1 + (-R_2) \Rightarrow R_2}{\left[\begin{array}{cc|c} 1 & 1 & 2 \\ 0 & 2 & 2 \end{array}\right]} \Rightarrow \overset{\frac{1}{2}R_2 \Rightarrow R_2}{\left[\begin{array}{cc|c} 1 & 1 & 2 \\ 0 & 1 & 1 \end{array}\right]}$

From R_2, $y = 1$. From R_1:

$$x + y = 2$$
$$x + 1 = 2 \Rightarrow x = 1$$

The solution is $(1, 1)$.

20. $\left[\begin{array}{cc|c} 1 & 1 & 3 \\ 1 & -1 & -1 \end{array}\right] \Rightarrow \overset{R_1 + (-R_2) \Rightarrow R_2}{\left[\begin{array}{cc|c} 1 & 1 & 3 \\ 0 & 2 & 4 \end{array}\right]} \Rightarrow \overset{\frac{1}{2}R_2 \Rightarrow R_2}{\left[\begin{array}{cc|c} 1 & 1 & 3 \\ 0 & 1 & 2 \end{array}\right]}$

From R_2, $y = 2$. From R_1:

$$x + y = 3$$
$$x + 2 = 3 \Rightarrow x = 1$$

The solution is $(1, 2)$.

21. $\left[\begin{array}{cc|c} 1 & 2 & -4 \\ 2 & 1 & 1 \end{array}\right] \Rightarrow \overset{-2R_1 + R_2 \Rightarrow R_2}{\left[\begin{array}{cc|c} 1 & 2 & -4 \\ 0 & -3 & 9 \end{array}\right]} \Rightarrow \overset{-\frac{1}{3}R_2 \Rightarrow R_2}{\left[\begin{array}{cc|c} 1 & 2 & -4 \\ 0 & 1 & -3 \end{array}\right]}$

From R_2, $y = -3$. From R_1:

$$x + 2y = -4$$
$$x + 2(-3) = -4$$
$$x - 6 = -4 \Rightarrow x = 2$$

The solution is $(2, -3)$.

22. $\left[\begin{array}{cc|c} 2 & -3 & 16 \\ -4 & 1 & -22 \end{array}\right] \Rightarrow \overset{2R_1 + R_2 \Rightarrow R_2}{\left[\begin{array}{cc|c} 2 & -3 & 16 \\ 0 & -5 & 10 \end{array}\right]} \Rightarrow \overset{-\frac{1}{5}R_2 \Rightarrow R_2}{\left[\begin{array}{cc|c} 2 & -3 & 16 \\ 0 & 1 & -2 \end{array}\right]}$

From R_2, $y = -2$. From R_1:

$$2x - 3y = 16$$
$$2x - 3(-2) = 16$$
$$2x + 6 = 16$$
$$2x = 10 \Rightarrow x = 5$$

The solution is $(5, -2)$.

23. $\left[\begin{array}{cc|c} 3 & 4 & -12 \\ 9 & -2 & 6 \end{array}\right] \overset{-3R_1+R_2 \Rightarrow R_2}{\Rightarrow} \left[\begin{array}{cc|c} 3 & 4 & -12 \\ 0 & -14 & 42 \end{array}\right] \overset{-\frac{1}{14}R_2 \Rightarrow R_2}{\Rightarrow} \left[\begin{array}{cc|c} 3 & 4 & -12 \\ 0 & 1 & -3 \end{array}\right]$

From R_2, $y = -3$. From R_1:

$$\begin{aligned} 3x + 4y &= -12 \\ 3x + 4(-3) &= -12 \\ 3x - 12 &= -12 \\ 3x &= 0 \Rightarrow x = 0 \end{aligned}$$

The solution is $(0, -3)$.

24. $\left[\begin{array}{cc|c} 5 & -4 & 10 \\ 1 & -7 & 2 \end{array}\right] \overset{R_1+(-5R_2) \Rightarrow R_2}{\Rightarrow} \left[\begin{array}{cc|c} 5 & -4 & 10 \\ 0 & 31 & 0 \end{array}\right] \overset{\frac{1}{31}R_2 \Rightarrow R_2}{\Rightarrow} \left[\begin{array}{cc|c} 5 & -4 & 10 \\ 0 & 1 & 0 \end{array}\right]$

From R_2, $y = 0$. From R_1:

$$\begin{aligned} 5x - 4y &= 10 \\ 5x - 4(0) &= 10 \\ 5x &= 10 \Rightarrow x = 2 \end{aligned}$$

The solution is $(2, 0)$.

25. $\left[\begin{array}{cc|c} 5 & -2 & 4 \\ 2 & -4 & -8 \end{array}\right] \overset{\frac{1}{2}R_2 \Rightarrow R_2}{\Rightarrow} \left[\begin{array}{cc|c} 5 & -2 & 4 \\ 1 & -2 & -4 \end{array}\right] \overset{R_1+(-5R_2) \Rightarrow R_2}{\Rightarrow} \left[\begin{array}{cc|c} 5 & -2 & 4 \\ 0 & 8 & 24 \end{array}\right] \overset{\frac{1}{8}R_2 \Rightarrow R_2}{\Rightarrow} \left[\begin{array}{cc|c} 5 & -2 & 4 \\ 0 & 1 & 3 \end{array}\right]$

From R_2, $y = 3$. From R_1:

$$\begin{aligned} 5x - 2y &= 4 \\ 5x - 2(3) &= 4 \\ 5x &= 10 \Rightarrow x = 2 \end{aligned}$$

The solution is $(2, 3)$.

26. $\left[\begin{array}{cc|c} 2 & -1 & -1 \\ 1 & -2 & 1 \end{array}\right] \overset{R_1+(-2R_2) \Rightarrow R_2}{\Rightarrow} \left[\begin{array}{cc|c} 2 & -1 & -1 \\ 0 & 3 & -3 \end{array}\right] \overset{\frac{1}{3}R_2 \Rightarrow R_2}{\Rightarrow} \left[\begin{array}{cc|c} 2 & -1 & -1 \\ 0 & 1 & -1 \end{array}\right]$

From R_2, $y = -1$. From R_1:

$$\begin{aligned} 2x - y &= -1 \\ 2x - (-1) &= -1 \\ 2x &= -2 \Rightarrow x = -1 \end{aligned}$$

The solution is $(-1, -1)$.

27. $\begin{cases} 5a = 24 + 2b \\ 5b = 3a + 16 \end{cases} \Rightarrow \begin{cases} 5a - 2b = 24 \\ -3a + 5b = 16 \end{cases}$

$\left[\begin{array}{cc|c} 5 & -2 & 24 \\ -3 & 5 & 16 \end{array}\right] \overset{2R_2+R_1 \Rightarrow R_1}{\Rightarrow} \left[\begin{array}{cc|c} -1 & 8 & 56 \\ -3 & 5 & 16 \end{array}\right] \overset{-R_1 \Rightarrow R_1}{\Rightarrow} \left[\begin{array}{cc|c} 1 & -8 & -56 \\ -3 & 5 & 16 \end{array}\right] \overset{3R_1+R_2 \Rightarrow R_2}{\Rightarrow} \left[\begin{array}{cc|c} 1 & -8 & -56 \\ 0 & -19 & -152 \end{array}\right]$

$\overset{-\frac{1}{19}R_2 \Rightarrow R_2}{\Rightarrow} \left[\begin{array}{cc|c} 1 & -8 & -56 \\ 0 & 1 & 8 \end{array}\right]$

From R_2, $b = 8$. From R_1:

$$\begin{aligned} a - 8b &= -56 \\ a - 8(8) &= -56 \\ a - 64 &= -56 \Rightarrow a = 8 \end{aligned}$$

The solution is $(8, 8)$.

28. $\begin{cases} 3m = 2n + 16 \\ 2m = -5n - 2 \end{cases} \Rightarrow \begin{cases} 3m - 2n = 16 \\ 2m + 5n = -2 \end{cases}$

$$\left[\begin{array}{cc|c} 3 & -2 & 16 \\ 2 & 5 & -2 \end{array}\right] \Rightarrow \overset{-R_2 + R_1 \Rightarrow R_1}{\left[\begin{array}{cc|c} 1 & -7 & 18 \\ 2 & 5 & -2 \end{array}\right]} \Rightarrow \overset{-2R_1 + R_2 \Rightarrow R_2}{\left[\begin{array}{cc|c} 1 & -7 & 18 \\ 0 & 19 & -38 \end{array}\right]} \Rightarrow \overset{\frac{1}{19}R_2 \Rightarrow R_2}{\left[\begin{array}{cc|c} 1 & -7 & 18 \\ 0 & 1 & -2 \end{array}\right]}$$

From R_2, $n = -2$. From R_1:

$$m - 7n = 18$$
$$m - 7(-2) = 18$$
$$m + 14 = 18 \Rightarrow m = 4$$

The solution is $(4, -2)$.

29.

$$\left[\begin{array}{ccc|c} 1 & 1 & 1 & 6 \\ 1 & 2 & 1 & 8 \\ 1 & 1 & 2 & 9 \end{array}\right] \Rightarrow \overset{\substack{-R_1 + R_2 \Rightarrow R_2 \\ -R_1 + R_3 \Rightarrow R_3}}{\left[\begin{array}{ccc|c} 1 & 1 & 1 & 6 \\ 0 & 1 & 0 & 2 \\ 0 & 0 & 1 & 3 \end{array}\right]}$$

From R_3, $z = 3$. From R_2, $y = 2$. From R_1:

$$x + y + z = 6$$
$$x + 2 + 3 = 6$$
$$x + 5 = 6$$
$$x = 1$$

The solution is $(1, 2, 3)$.

30.

$$\left[\begin{array}{ccc|c} 1 & -1 & 1 & 2 \\ 1 & 2 & -1 & 6 \\ 2 & -1 & -1 & 3 \end{array}\right] \Rightarrow \overset{\substack{-R_1 + R_2 \Rightarrow R_2 \\ -2R_1 + R_3 \Rightarrow R_3}}{\left[\begin{array}{ccc|c} 1 & -1 & 1 & 2 \\ 0 & 3 & -2 & 4 \\ 0 & 1 & -3 & -1 \end{array}\right]} \Rightarrow \overset{-3R_3 + R_2 \Rightarrow R_3}{\left[\begin{array}{ccc|c} 1 & -1 & 1 & 2 \\ 0 & 3 & -2 & 4 \\ 0 & 0 & 7 & 7 \end{array}\right]} \Rightarrow \overset{\frac{1}{7}R_3 \Rightarrow R_3}{\left[\begin{array}{ccc|c} 1 & -1 & 1 & 2 \\ 0 & 3 & -2 & 4 \\ 0 & 0 & 1 & 1 \end{array}\right]}$$

From R_3, $z = 1$. From R_2:

$$3y - 2z = 4$$
$$3y - 2(1) = 4$$
$$3y - 2 = 4$$
$$3y = 6$$
$$y = 2$$

From R_1:

$$x - y + z = 2$$
$$x - 2 + 1 = 2$$
$$x - 1 = 2$$
$$x = 3$$

The solution is $(3, 2, 1)$.

31. $$\left[\begin{array}{ccc|c} 2 & 1 & 3 & 3 \\ -2 & -1 & 1 & 5 \\ 4 & -2 & 2 & 2 \end{array}\right] \overset{\substack{R_1 + R_2 \Rightarrow R_2 \\ -2R_1 + R_3 \Rightarrow R_3}}{\Rightarrow} \left[\begin{array}{ccc|c} 2 & 1 & 3 & 3 \\ 0 & 0 & 4 & 8 \\ 0 & -4 & -4 & -4 \end{array}\right] \overset{\substack{-\frac{1}{4}R_3 \Rightarrow R_2 \\ \frac{1}{4}R_2 \Rightarrow R_3}}{\Rightarrow} \left[\begin{array}{ccc|c} 2 & 1 & 3 & 3 \\ 0 & 1 & 1 & 1 \\ 0 & 0 & 1 & 2 \end{array}\right]$$

From R_3, $z = 2$. From R_2: From R_1: The solution is $(-1, -1, 2)$.

$$\begin{aligned} y + z &= 1 \\ y + 2 &= 1 \\ y &= -1 \end{aligned} \qquad \begin{aligned} 2x + y + 3z &= 3 \\ 2x + (-1) + 3(2) &= 3 \\ 2x - 1 + 6 &= 3 \\ 2x + 5 &= 3 \\ 2x &= -2 \\ x &= -1 \end{aligned}$$

32. $$\left[\begin{array}{ccc|c} 3 & 2 & 1 & 8 \\ 6 & -1 & 2 & 16 \\ -9 & 1 & -1 & -20 \end{array}\right] \overset{\substack{-2R_1 + R_2 \Rightarrow R_2 \\ 3R_1 + R_3 \Rightarrow R_3}}{\Rightarrow} \left[\begin{array}{ccc|c} 3 & 2 & 1 & 8 \\ 0 & -5 & 0 & 0 \\ 0 & 7 & 2 & 4 \end{array}\right] \overset{\substack{-\frac{1}{5}R_2 \Rightarrow R_2 \\ \frac{7}{5}R_2 + R_3 \Rightarrow R_3}}{\Rightarrow} \left[\begin{array}{ccc|c} 3 & 2 & 1 & 8 \\ 0 & 1 & 0 & 0 \\ 0 & 0 & 2 & 4 \end{array}\right] \overset{\frac{1}{2}R_3 \Rightarrow R_3}{\Rightarrow} \left[\begin{array}{ccc|c} 3 & 2 & 1 & 8 \\ 0 & 1 & 0 & 0 \\ 0 & 0 & 1 & 2 \end{array}\right]$$

From R_3, $z = 2$. From R_2, $y = 0$. From R_1: The solution is $(2, 0, 2)$.

$$\begin{aligned} 3x + 2y + z &= 8 \\ 3x + 2(0) + 2 &= 8 \\ 3x + 2 &= 8 \\ 3x &= 6 \\ x &= 2 \end{aligned}$$

33. $$\left[\begin{array}{ccc|c} 3 & 1 & -3 & 5 \\ 1 & -2 & 4 & 10 \\ 1 & 1 & 1 & 13 \end{array}\right] \overset{\substack{-3R_2 + R_1 \Rightarrow R_2 \\ -3R_3 + R_1 \Rightarrow R_3}}{\Rightarrow} \left[\begin{array}{ccc|c} 3 & 1 & -3 & 5 \\ 0 & 7 & -15 & -25 \\ 0 & -2 & -6 & -34 \end{array}\right] \overset{\frac{2}{7}R_2 + R_3 \Rightarrow R_3}{\Rightarrow} \left[\begin{array}{ccc|c} 3 & 1 & -3 & 5 \\ 0 & 7 & -15 & -25 \\ 0 & 0 & -\frac{72}{7} & -\frac{288}{7} \end{array}\right]$$

From R_3: From R_2: From R_1:

$$\begin{aligned} -\frac{72}{7}c &= -\frac{288}{7} \\ -\frac{7}{72}\left(-\frac{72}{7}c\right) &= -\frac{7}{72}\left(-\frac{288}{7}\right) \\ c &= 4 \end{aligned} \qquad \begin{aligned} 7b - 15c &= -25 \\ 7b - 15(4) &= -25 \\ 7b &= 35 \\ b &= 5 \end{aligned} \qquad \begin{aligned} 3a + b - 3c &= 5 \\ 3a + 5 - 3(4) &= 5 \\ 3a - 7 &= 5 \\ 3a &= 12 \\ a &= 4 \end{aligned}$$

The solution is (4, 5, 4).

34.
$$\left[\begin{array}{ccc|c} 2 & 1 & -3 & -1 \\ 3 & -2 & -1 & -5 \\ 1 & -3 & -2 & -12 \end{array}\right] \Rightarrow \begin{array}{l} R_3 \Rightarrow R_1 \\ -2R_3 + R_1 \Rightarrow R_2 \\ -3R_3 + R_2 \Rightarrow R_3 \end{array} \left[\begin{array}{ccc|c} 1 & -3 & -2 & -12 \\ 0 & 7 & 1 & 23 \\ 0 & 7 & 5 & 31 \end{array}\right] \Rightarrow \begin{array}{l} -R_2 + R_3 \Rightarrow R_3 \end{array} \left[\begin{array}{ccc|c} 1 & -3 & -2 & -12 \\ 0 & 7 & 1 & 23 \\ 0 & 0 & 4 & 8 \end{array}\right]$$

From R_3: $4c = 8$, $c = 2$

From R_2: $7b + c = 23$, $7b + 2 = 23$, $7b = 21$, $b = 3$

From R_1: $a - 3b - 2c = -12$, $a - 3(3) - 2(2) = -12$, $a - 13 = -12$, $a = 1$

The solution is (1, 3, 2).

35.
$$\left[\begin{array}{ccc|c} 3 & -2 & 4 & 4 \\ 1 & 1 & 1 & 3 \\ 6 & -2 & -3 & 10 \end{array}\right] \Rightarrow \begin{array}{l} -3R_2 + R_1 \Rightarrow R_2 \\ -2R_1 + R_3 \Rightarrow R_3 \end{array} \left[\begin{array}{ccc|c} 3 & -2 & 4 & 4 \\ 0 & -5 & 1 & -5 \\ 0 & 2 & -11 & 2 \end{array}\right] \Rightarrow \begin{array}{l} \frac{2}{5}R_2 + R_3 \Rightarrow R_3 \end{array} \left[\begin{array}{ccc|c} 3 & -2 & 4 & 4 \\ 0 & -5 & 1 & -5 \\ 0 & 0 & -\frac{53}{5} & 0 \end{array}\right]$$

From R_3, $z = 0$.

From R_2: $-5y + z = -5$, $-5y + 0 = -5$, $-5y = -5$, $y = 1$

From R_1: $3x - 2y + 4z = 4$, $3x - 2(1) + 4(0) = 4$, $3x - 2 = 4$, $3x = 6$, $x = 2$

The solution is $(2, 1, 0)$.

36.
$$\left[\begin{array}{ccc|c} 2 & 3 & -1 & -8 \\ 1 & -1 & -1 & -2 \\ -4 & 3 & 1 & 6 \end{array}\right] \Rightarrow \begin{array}{l} -2R_2 + R_1 \Rightarrow R_2 \\ 2R_1 + R_3 \Rightarrow R_3 \end{array} \left[\begin{array}{ccc|c} 2 & 3 & -1 & -8 \\ 0 & 5 & 1 & -4 \\ 0 & 9 & -1 & -10 \end{array}\right] \Rightarrow \begin{array}{l} -\frac{9}{5}R_2 + R_3 \Rightarrow R_3 \end{array} \left[\begin{array}{ccc|c} 2 & 3 & -1 & -8 \\ 0 & 5 & 1 & -4 \\ 0 & 0 & -\frac{14}{5} & -\frac{14}{5} \end{array}\right] \Rightarrow \begin{array}{l} -\frac{5}{14}R_3 \Rightarrow R_3 \end{array} \left[\begin{array}{ccc|c} 2 & 3 & -1 & -8 \\ 0 & 5 & 1 & -4 \\ 0 & 0 & 1 & 1 \end{array}\right]$$

From R_3, $z = 1$.

From R_2: $5y + z = -4$, $5y + 1 = -4$, $5y = -5$, $y = -1$

From R_1: $2x + 3y - z = -8$, $2x + 3(-1) - 1 = -8$, $2x - 3 - 1 = -8$, $2x = -4$, $x = -2$

The solution is $(-2, -1, 1)$.

37. $\begin{bmatrix} 1 & 1 & | & 3 \\ 3 & -1 & | & 1 \\ 2 & 1 & | & 4 \end{bmatrix} \overset{\substack{-3R_1+R_2 \Rightarrow R_2 \\ -2R_1+R_3 \Rightarrow R_3}}{\Rightarrow} \begin{bmatrix} 1 & 1 & | & 3 \\ 0 & -4 & | & -8 \\ 0 & -1 & | & -2 \end{bmatrix} \overset{\substack{-\frac{1}{4}R_2 \Rightarrow R_2 \\ -4R_3+R_2 \Rightarrow R_3}}{\Rightarrow} \begin{bmatrix} 1 & 1 & | & 3 \\ 0 & 1 & | & 2 \\ 0 & 0 & | & 0 \end{bmatrix}$

From R_2, $y = 2$. From R_1: The solution is $(1, 2)$.

$$\begin{aligned} x + y &= 3 \\ x + 2 &= 3 \\ x &= 1 \end{aligned}$$

38. $\begin{bmatrix} 1 & -1 & | & -5 \\ 2 & 3 & | & 5 \\ 1 & 1 & | & 1 \end{bmatrix} \overset{\substack{-2R_1+R_2 \Rightarrow R_2 \\ -R_1+R_3 \Rightarrow R_3}}{\Rightarrow} \begin{bmatrix} 1 & -1 & | & -5 \\ 0 & 5 & | & 15 \\ 0 & 2 & | & 6 \end{bmatrix} \overset{\substack{\frac{1}{5}R_2 \Rightarrow R_2 \\ -\frac{2}{5}R_2+R_3 \Rightarrow R_3}}{\Rightarrow} \begin{bmatrix} 1 & -1 & | & -5 \\ 0 & 1 & | & 3 \\ 0 & 0 & | & 0 \end{bmatrix}$

From R_2, $y = 3$. From R_1: The solution is $(-2, 3)$.

$$\begin{aligned} x - y &= -5 \\ x - 3 &= -5 \\ x &= -2 \end{aligned}$$

39. $\begin{bmatrix} 2 & -1 & | & 4 \\ 1 & 3 & | & 2 \\ -1 & -4 & | & -2 \end{bmatrix} \overset{\substack{-2R_2+R_1 \Rightarrow R_2 \\ 2R_3+R_1 \Rightarrow R_3}}{\Rightarrow} \begin{bmatrix} 2 & -1 & | & 4 \\ 0 & -7 & | & 0 \\ 0 & -9 & | & 0 \end{bmatrix} \overset{\substack{-\frac{1}{7}R_2 \Rightarrow R_2 \\ -\frac{7}{9}R_2+R_3 \Rightarrow R_3}}{\Rightarrow} \begin{bmatrix} 2 & -1 & | & 4 \\ 0 & 1 & | & 0 \\ 0 & 0 & | & 0 \end{bmatrix}$

From R_2, $y = 0$. From R_1: The solution is $(2, 0)$.

$$\begin{aligned} 2x - y &= 4 \\ 2x - 0 &= 4 \\ 2x &= 4 \\ x &= 2 \end{aligned}$$

40. $\begin{bmatrix} 3 & -2 & | & 5 \\ 1 & 2 & | & 7 \\ -3 & -1 & | & -11 \end{bmatrix} \overset{\substack{-3R_2+R_1 \Rightarrow R_2 \\ R_1+R_3 \Rightarrow R_3}}{\Rightarrow} \begin{bmatrix} 3 & -2 & | & 5 \\ 0 & -8 & | & -16 \\ 0 & -3 & | & -6 \end{bmatrix} \overset{\substack{-\frac{1}{8}R_2 \Rightarrow R_2 \\ -\frac{3}{8}R_2+R_3 \Rightarrow R_3}}{\Rightarrow} \begin{bmatrix} 3 & -2 & | & 5 \\ 0 & 1 & | & 2 \\ 0 & 0 & | & 0 \end{bmatrix}$

From R_2, $y = 2$. From R_1: The solution is $(3, 2)$.

$$\begin{aligned} 3x - 2y &= 5 \\ 3x - 2(2) &= 5 \\ 3x - 4 &= 5 \\ 3x &= 9 \Rightarrow x = 3 \end{aligned}$$

41.
$$\begin{bmatrix} 2 & 1 & | & 7 \\ 1 & -1 & | & 2 \\ -1 & 3 & | & -2 \end{bmatrix} \xRightarrow[2R_3 + R_1 \Rightarrow R_3]{-2R_2 + R_1 \Rightarrow R_2} \begin{bmatrix} 2 & 1 & | & 7 \\ 0 & 3 & | & 3 \\ 0 & 7 & | & 3 \end{bmatrix} \xRightarrow[-\frac{7}{3}R_2 + R_3 \Rightarrow R_3]{\frac{1}{3}R_2 \Rightarrow R_2} \begin{bmatrix} 2 & 1 & | & 7 \\ 0 & 1 & | & 1 \\ 0 & 0 & | & -4 \end{bmatrix}$$

R_3: $0x + 0y = -4$, or $0 = -4$, which is an impossible equation. NO SOLUTION

42.
$$\begin{bmatrix} 3 & -1 & | & 2 \\ -6 & 3 & | & 0 \\ -1 & 2 & | & -4 \end{bmatrix} \xRightarrow[3R_3 + R_1 \Rightarrow R_3]{2R_1 + R_2 \Rightarrow R_2} \begin{bmatrix} 3 & -1 & | & 2 \\ 0 & 1 & | & 4 \\ 0 & 5 & | & -10 \end{bmatrix} \xRightarrow{-5R_2 + R_3 \Rightarrow R_3} \begin{bmatrix} 3 & -1 & | & 2 \\ 0 & 1 & | & 4 \\ 0 & 0 & | & -30 \end{bmatrix}$$

R_3: $0x + 0y = -30$, or $0 = -30$, which is an impossible equation. NO SOLUTION

43.
$$\begin{bmatrix} 1 & 3 & | & 7 \\ 1 & 1 & | & 3 \\ 3 & 1 & | & 5 \end{bmatrix} \xRightarrow[-3R_1 + R_3 \Rightarrow R_3]{-R_2 + R_1 \Rightarrow R_2} \begin{bmatrix} 1 & 3 & | & 7 \\ 0 & 2 & | & 4 \\ 0 & -8 & | & -16 \end{bmatrix} \xRightarrow[4R_2 + R_3 \Rightarrow R_3]{\frac{1}{2}R_2 \Rightarrow R_2} \begin{bmatrix} 1 & 3 & | & 7 \\ 0 & 1 & | & 2 \\ 0 & 0 & | & 0 \end{bmatrix}$$

From R_2, $y = 2$. From R_1:

$$x + 3y = 7$$
$$x + 3(2) = 7$$
$$x + 6 = 7 \Rightarrow x = 1$$

The solution is $(1, 2)$.

44.
$$\begin{bmatrix} 1 & 1 & | & 3 \\ 1 & -2 & | & -3 \\ 1 & -1 & | & 1 \end{bmatrix} \xRightarrow[-R_3 + R_1 \Rightarrow R_3]{-R_2 + R_1 \Rightarrow R_2} \begin{bmatrix} 1 & 1 & | & 3 \\ 0 & 3 & | & 6 \\ 0 & 2 & | & 2 \end{bmatrix} \xRightarrow[-\frac{2}{3}R_2 + R_3 \Rightarrow R_3]{\frac{1}{3}R_2 \Rightarrow R_2} \begin{bmatrix} 1 & 1 & | & 3 \\ 0 & 1 & | & 2 \\ 0 & 0 & | & -2 \end{bmatrix}$$

R_3: $0x + 0y = -2$, or $0 = -2$, which is an impossible equation. NO SOLUTION

45.
$$\begin{bmatrix} 1 & 2 & 3 & | & -2 \\ -1 & -1 & -2 & | & 4 \end{bmatrix} \xRightarrow{R_1 + R_2 \Rightarrow R_2} \begin{bmatrix} 1 & 2 & 3 & | & -2 \\ 0 & 1 & 1 & | & 2 \end{bmatrix}$$

From R_2:

$$y + z = 2$$
$$y = 2 - z$$

From R_1:

$$x + 2y + 3z = -2$$
$$x + 2(2 - z) + 3z = -2$$
$$x + 4 - 2z + 3z = -2$$
$$x + z = -6$$
$$x = -6 - z$$

The solution is $(-6 - z, 2 - z, z)$.

46. $\left[\begin{array}{ccc|c} 2 & -4 & 3 & 6 \\ -4 & 6 & 4 & -6 \end{array}\right] \overset{2R_1+R_2 \Rightarrow R_2}{\Rightarrow} \left[\begin{array}{ccc|c} 2 & -4 & 3 & 6 \\ 0 & -2 & 10 & 6 \end{array}\right] \overset{-\frac{1}{2}R_2 \Rightarrow R_2}{\Rightarrow} \left[\begin{array}{ccc|c} 2 & -4 & 3 & 6 \\ 0 & 1 & -5 & -3 \end{array}\right]$

From R_2:

$$y - 5z = -3$$
$$y = 5z - 3$$

From R_1:

$$2x - 4y + 3z = 6$$
$$2x - 4(5z-3) + 3z = 6$$
$$2x - 20z + 12 + 3z = 6$$
$$2x = 17z - 6$$
$$x = \frac{17z-6}{2}$$

The solution is $\left(\frac{17}{2}z - 3, 5z - 3, z\right)$.

47. $\left[\begin{array}{ccc|c} 1 & -1 & 0 & 1 \\ 0 & 1 & 1 & 1 \\ 1 & 0 & 1 & 2 \end{array}\right] \overset{-R_1+R_3 \Rightarrow R_3}{\Rightarrow} \left[\begin{array}{ccc|c} 1 & -1 & 0 & 1 \\ 0 & 1 & 1 & 1 \\ 0 & 1 & 1 & 1 \end{array}\right] \overset{-R_2+R_3 \Rightarrow R_3}{\Rightarrow} \left[\begin{array}{ccc|c} 1 & -1 & 0 & 1 \\ 0 & 1 & 1 & 1 \\ 0 & 0 & 0 & 0 \end{array}\right]$

From R_2:

$$y + z = 1$$
$$y = 1 - z$$

From R_1:

$$x - y = 1$$
$$x - (1 - z) = 1$$
$$x - 1 + z = 1$$
$$x + z = 2$$
$$x = 2 - z$$

The solution is $(2 - z, 1 - z, z)$.

48. $\left[\begin{array}{ccc|c} 1 & 0 & 1 & 1 \\ 1 & 1 & 0 & 2 \\ 2 & 1 & 1 & 3 \end{array}\right] \overset{\substack{-R_1+R_2 \Rightarrow R_2 \\ -2R_1+R_3 \Rightarrow R_3}}{\Rightarrow} \left[\begin{array}{ccc|c} 1 & 0 & 1 & 1 \\ 0 & 1 & -1 & 1 \\ 0 & 1 & -1 & 1 \end{array}\right] \overset{R_2+(-R_3) \Rightarrow R_3}{\Rightarrow} \left[\begin{array}{ccc|c} 1 & 0 & 1 & 1 \\ 0 & 1 & -1 & 1 \\ 0 & 0 & 0 & 0 \end{array}\right]$

From R_2:

$$y - z = 1$$
$$y = z + 1$$

From R_1:

$$x + z = 1$$
$$x = 1 - z$$

The solution is $(1 - z, z + 1, z)$.

49. Let $x =$ the measure of the first angle and $y =$ the measure of the second angle. Form and solve this system of equations: $\begin{cases} x + y = 90 \\ y = x + 46 \end{cases} \Rightarrow \begin{cases} x + y = 90 \\ -x + y = 46 \end{cases}$

$\left[\begin{array}{cc|c} 1 & 1 & 90 \\ -1 & 1 & 46 \end{array}\right] \overset{R_1+R_2 \Rightarrow R_2}{\Rightarrow} \left[\begin{array}{cc|c} 1 & 1 & 90 \\ 0 & 2 & 136 \end{array}\right] \overset{\frac{1}{2}R_2 \Rightarrow R_2}{\Rightarrow} \left[\begin{array}{cc|c} 1 & 1 & 90 \\ 0 & 1 & 68 \end{array}\right]$

From R_2, $y = 68$.

From R_1:

$$x + y = 90$$
$$x + 68 = 90$$
$$x = 22$$

The angles have measures of 22° and 68°.

50. Let $x =$ the measure of the first angle and $y =$ the measure of the second angle. Form and solve this system of equations: $\begin{cases} x+y=180 \\ y=x+28 \end{cases} \Rightarrow \begin{cases} x+y=180 \\ -x+y=28 \end{cases}$

$$\left[\begin{array}{cc|c} 1 & 1 & 180 \\ -1 & 1 & 28 \end{array}\right] \overset{R_1+R_2 \Rightarrow R_2}{\Rightarrow} \left[\begin{array}{cc|c} 1 & 1 & 180 \\ 0 & 2 & 208 \end{array}\right] \overset{\frac{1}{2}R_2 \Rightarrow R_2}{\Rightarrow} \left[\begin{array}{cc|c} 1 & 1 & 180 \\ 0 & 1 & 104 \end{array}\right]$$

From R_2, $y = 104$.

From R_1:

$$\begin{aligned} x+y &= 180 \\ x+104 &= 180 \\ x &= 76 \end{aligned}$$

The angles have measures of 76° and 104°.

51. Let A, B and C represent the measures of the three angles.

$$\begin{cases} A+B+C=180 \\ B=A+25 \\ C=2A-5 \end{cases} \Rightarrow \begin{cases} A+B+C=180 \\ -A+B=25 \\ -2A+C=-5 \end{cases}$$

$$\left[\begin{array}{ccc|c} 1 & 1 & 1 & 180 \\ -1 & 1 & 0 & 25 \\ -2 & 0 & 1 & -5 \end{array}\right] \overset{\substack{R_1+R_2 \Rightarrow R_2 \\ 2R_1+R_3 \Rightarrow R_3}}{\Rightarrow} \left[\begin{array}{ccc|c} 1 & 1 & 1 & 180 \\ 0 & 2 & 1 & 205 \\ 0 & 2 & 3 & 355 \end{array}\right] \overset{-R_2+R_3 \Rightarrow R_3}{\Rightarrow} \left[\begin{array}{ccc|c} 1 & 1 & 1 & 180 \\ 0 & 2 & 1 & 205 \\ 0 & 0 & 2 & 150 \end{array}\right] \overset{\frac{1}{2}R_3 \Rightarrow R_3}{\Rightarrow} \left[\begin{array}{ccc|c} 1 & 1 & 1 & 180 \\ 0 & 2 & 1 & 205 \\ 0 & 0 & 1 & 75 \end{array}\right]$$

From R_3, $C = 75$.

From R_2:

$$\begin{aligned} 2B+C &= 205 \\ 2B+75 &= 205 \\ 2B &= 130 \\ B &= 65 \end{aligned}$$

From R_1:

$$\begin{aligned} A+B+C &= 180 \\ A+65+75 &= 180 \\ A+140 &= 180 \\ A &= 40 \end{aligned}$$

The angles have measures of 40°, 65° and 75°.

52. Let A, B and C represent the measures of the three angles.

$$\begin{cases} A+B+C=180 \\ A=B-10 \\ B=C-10 \end{cases} \Rightarrow \begin{cases} A+B+C=180 \\ A-B=-10 \\ B-C=-10 \end{cases}$$

$$\left[\begin{array}{ccc|c} 1 & 1 & 1 & 180 \\ 1 & -1 & 0 & -10 \\ 0 & 1 & -1 & -10 \end{array}\right] \overset{-R_2+R_1 \Rightarrow R_2}{\Rightarrow} \left[\begin{array}{ccc|c} 1 & 1 & 1 & 180 \\ 0 & 2 & 1 & 190 \\ 0 & 1 & -1 & -10 \end{array}\right] \overset{-2R_3+R_2 \Rightarrow R_3}{\Rightarrow} \left[\begin{array}{ccc|c} 1 & 1 & 1 & 180 \\ 0 & 2 & 1 & 190 \\ 0 & 0 & 3 & 210 \end{array}\right] \overset{\frac{1}{3}R_3 \Rightarrow R_3}{\Rightarrow} \left[\begin{array}{ccc|c} 1 & 1 & 1 & 180 \\ 0 & 2 & 1 & 190 \\ 0 & 0 & 1 & 70 \end{array}\right]$$

From R_3, $C = 70$.

From R_2:

$$\begin{aligned} 2B+C &= 190 \\ 2B+70 &= 190 \\ 2B &= 120 \\ B &= 60 \end{aligned}$$

From R_1:

$$\begin{aligned} A+B+C &= 180 \\ A+60+70 &= 180 \\ A+130 &= 180 \\ A &= 50 \end{aligned}$$

The angles have measures of 50°, 60° and 70°.

53. Plug the coordinates of the points into the general equation to form and solve a system of equations.

$y = ax^2 + bx + c$ $\quad$ $1 = a(0)^2 + b(0) + c$ $\quad$ $1 = c$

$y = ax^2 + bx + c$ $\quad$ $2 = a(1)^2 + b(1) + c$ $\quad$ $2 = a + b + c$

$y = ax^2 + bx + c$ $\quad$ $4 = a(-1)^2 + b(-1) + c$ $\quad$ $4 = a - b + c$

$$\begin{cases} a+b+c=2 \\ a-b+c=4 \\ c=1 \end{cases} \Rightarrow \left[\begin{array}{ccc|c} 1 & 1 & 1 & 2 \\ 1 & -1 & 1 & 4 \\ 0 & 0 & 1 & 1 \end{array}\right] \overset{-R_2+R_1 \Rightarrow R_2}{\Rightarrow} \left[\begin{array}{ccc|c} 1 & 1 & 1 & 2 \\ 0 & 2 & 0 & -2 \\ 0 & 0 & 1 & 1 \end{array}\right] \overset{\frac{1}{2}R_2 \Rightarrow R_2}{\Rightarrow} \left[\begin{array}{ccc|c} 1 & 1 & 1 & 2 \\ 0 & 1 & 0 & -1 \\ 0 & 0 & 1 & 1 \end{array}\right]$$

From R_3, $c = 1$.

From R_2, $b = -1$. From R_1:

$$\begin{aligned} a+b+c &= 2 \\ a+(-1)+1 &= 2 \\ a &= 2 \end{aligned}$$

The equation is $y = 2x^2 - x + 1$.

54. Plug the coordinates of the points into the general equation to form and solve a system of equations.

$y = ax^2 + bx + c$ $\quad$ $1 = a(0)^2 + b(0) + c$ $\quad$ $1 = c$

$y = ax^2 + bx + c$ $\quad$ $1 = a(1)^2 + b(1) + c$ $\quad$ $1 = a + b + c$

$y = ax^2 + bx + c$ $\quad$ $-1 = a(-1)^2 + b(-1) + c$ $\quad$ $-1 = a - b + c$

$$\begin{cases} a+b+c=1 \\ a-b+c=-1 \\ c=1 \end{cases} \Rightarrow \left[\begin{array}{ccc|c} 1 & 1 & 1 & 1 \\ 1 & -1 & 1 & -1 \\ 0 & 0 & 1 & 1 \end{array}\right] \overset{-R_2+R_1 \Rightarrow R_2}{\Rightarrow} \left[\begin{array}{ccc|c} 1 & 1 & 1 & 1 \\ 0 & 2 & 0 & 2 \\ 0 & 0 & 1 & 1 \end{array}\right] \overset{\frac{1}{2}R_2 \Rightarrow R_2}{\Rightarrow} \left[\begin{array}{ccc|c} 1 & 1 & 1 & 1 \\ 0 & 1 & 0 & 1 \\ 0 & 0 & 1 & 1 \end{array}\right]$$

From R_3, $c = 1$.

From R_2, $b = 1$. From R_1:

$$\begin{aligned} a+b+c &= 1 \\ a+1+1 &= 1 \\ a &= -1 \end{aligned}$$

The equation is $y = -x^2 + x + 1$.

55. Let $x =$ the measure of the first angle and $y =$ the measure of the second angle. Form and solve this system of equations: $\begin{cases} x+y=180 \\ y=x+28 \end{cases} \Rightarrow \begin{cases} x+y=180 \\ -x+y=28 \end{cases}$

$$\left[\begin{array}{cc|c} 1 & 1 & 180 \\ -1 & 1 & 28 \end{array}\right] \overset{R_1+R_2 \Rightarrow R_2}{\Rightarrow} \left[\begin{array}{cc|c} 1 & 1 & 180 \\ 0 & 2 & 208 \end{array}\right] \overset{\frac{1}{2}R_2 \Rightarrow R_2}{\Rightarrow} \left[\begin{array}{cc|c} 1 & 1 & 180 \\ 0 & 1 & 104 \end{array}\right]$$

From R_2, $y = 104$.

From R_1:

$$\begin{aligned} x+y &= 180 \\ x+104 &= 180 \\ x &= 76 \end{aligned}$$

The angles have measures of 76° and 104°.

56. Let $x =$ the number of dogs and $y =$ the number of cats. Form and solve this system of equations:

$$\begin{cases} x + y = 135 \\ y = x + 15 \end{cases} \Rightarrow \begin{cases} x + y = 135 \\ -x + y = 15 \end{cases}$$

$$\left[\begin{array}{cc|c} 1 & 1 & 135 \\ -1 & 1 & 15 \end{array}\right] \overset{R_1 + R_2 \Rightarrow R_2}{\Rightarrow} \left[\begin{array}{cc|c} 1 & 1 & 135 \\ 0 & 2 & 150 \end{array}\right] \overset{\frac{1}{2}R_2 \Rightarrow R_2}{\Rightarrow} \left[\begin{array}{cc|c} 1 & 1 & 135 \\ 0 & 1 & 75 \end{array}\right]$$

From R_2, $y = 75$. From R_1:

$$x + y = 135$$
$$x + 75 = 135$$
$$x = 60$$

There are 60 million dogs and 75 million cats.

57. Let $x =$ the number of nickels, $y =$ the number of dimes, and $z =$ the number of quarters.

$$\begin{cases} x + y + z = 64 \\ 5x + 10y + 25z = 600 \\ 10x + 5y + 25z = 500 \end{cases}$$

$$\left[\begin{array}{ccc|c} 1 & 1 & 1 & 64 \\ 5 & 10 & 25 & 600 \\ 10 & 5 & 25 & 500 \end{array}\right] \overset{\substack{-5R_1 + R_2 \Rightarrow R_2 \\ -10R_1 + R_3 \Rightarrow R_3}}{\Rightarrow} \left[\begin{array}{ccc|c} 1 & 1 & 1 & 64 \\ 0 & 5 & 20 & 280 \\ 0 & -5 & 15 & -140 \end{array}\right] \overset{R_2 + R_3 \Rightarrow R_3}{\Rightarrow} \left[\begin{array}{ccc|c} 1 & 1 & 1 & 64 \\ 0 & 5 & 20 & 280 \\ 0 & 0 & 35 & 140 \end{array}\right] \overset{\substack{\frac{1}{5}R_2 \Rightarrow R_2 \\ \frac{1}{35}R_3 \Rightarrow R_3}}{\Rightarrow} \left[\begin{array}{ccc|c} 1 & 1 & 1 & 64 \\ 0 & 1 & 4 & 56 \\ 0 & 0 & 1 & 4 \end{array}\right]$$

From R_3, $z = 4$.

From R_2, $y + 4z = 56$

$$y + 4(4) = 56$$
$$y + 16 = 56$$
$$y = 40$$

From R_3, $x + y + z = 64$

$$x + 40 + 4 = 64$$
$$x + 44 = 64$$
$$x = 20$$

There are 20 nickels, 40 dimes, and 4 quarters.

58. Let $x =$ the number of founder's circle seats, $y =$ the number of box seats, and $z =$ the number of promenade seats.

$$\begin{cases} x + y + z = 800 \\ 30x + 20y + 10z = 13000 \\ 40x + 30y + 25z = 23000 \end{cases}$$

$$\left[\begin{array}{ccc|c} 1 & 1 & 1 & 800 \\ 30 & 20 & 10 & 13000 \\ 40 & 30 & 25 & 23000 \end{array}\right] \overset{\substack{-30R_1 + R_2 \Rightarrow R_2 \\ -40R_1 + R_3 \Rightarrow R_3}}{\Rightarrow} \left[\begin{array}{ccc|c} 1 & 1 & 1 & 800 \\ 0 & -10 & -20 & -11000 \\ 0 & -10 & -15 & -9000 \end{array}\right] \overset{-R_2 + R_3 \Rightarrow R_3}{\Rightarrow} \left[\begin{array}{ccc|c} 1 & 1 & 1 & 800 \\ 0 & -10 & -20 & -11000 \\ 0 & 0 & 5 & 2000 \end{array}\right]$$

From R_3, $5z = 2000$

$$z = 400$$

From R_2, $-10y - 20z = -11000$

$$-10y - 20(400) = -11000$$
$$-10y - 8000 = -11000$$
$$-10y = -3000$$
$$y = 300$$

From R_3, $x + y + z = 800$

$$x + 300 + 400 = 800$$
$$x + 700 = 800$$
$$x = 100$$

There are 100 founder's circle, 300 box, and 400 promenade seats.

59. **Answers may vary.**

60. **Answers may vary.**

61. The last equation represents the equation $0x + 0y + 0z = k$, or $0 = k$. If $k = 0$, then the system can be solved. However, if $k \neq 0$, the system will have no solution.

62. Consider the system $\begin{cases} x + y + z = 3 \\ -x - y - z = 4 \end{cases}$. The system has 2 equations and 3 variables, so there are fewer equations than variables. When the two equations are added together, the resulting equation is $0x + 0y + 0z = 7$, or $0 = 7$, which is an impossible equation. Thus, the system has no solution.

Exercise 3.5 (page 207)

1.
$$\begin{aligned} 3(x+2) - (2-x) &= x - 5 \\ 3x + 6 - 2 + x &= x - 5 \\ 4x + 4 &= x - 5 \\ 3x &= -9 \\ x &= -3 \end{aligned}$$

2.
$$\begin{aligned} \frac{3}{7}x &= 2(x+11) \\ 7 \cdot \frac{3}{7}x &= 7 \cdot 2(x+11) \\ 3x &= 14(x+11) \\ 3x &= 14x + 154 \\ -11x &= 154 \\ x &= -14 \end{aligned}$$

3.
$$\begin{aligned} \frac{5}{3}(5x+6) - 10 &= 0 \\ 3 \cdot \frac{5}{3}(5x+6) - 3 \cdot 10 &= 3 \cdot 0 \\ 5(5x+6) - 30 &= 0 \\ 25x + 30 - 30 &= 0 \\ 25x &= 0 \\ x &= 0 \end{aligned}$$

4.
$$\begin{aligned} 5 - 3(2x-1) &= 2(4+3x) - 24 \\ 5 - 6x + 3 &= 8 + 6x - 24 \\ 8 - 6x &= 6x - 16 \\ -12x &= -24 \\ x &= 2 \end{aligned}$$

5. number

6. $ad - bc$

7. $\begin{vmatrix} a_2 & c_2 \\ a_3 & c_3 \end{vmatrix}$

8. row; column

9. $\begin{vmatrix} 3 & 4 \\ 2 & -3 \end{vmatrix}$

10. dependent; inconsistent

11.
$$\begin{aligned} \begin{vmatrix} 2 & 3 \\ -2 & 1 \end{vmatrix} &= 2(1) - 3(-2) \\ &= 2 + 6 = 8 \end{aligned}$$

12.
$$\begin{aligned} \begin{vmatrix} 3 & -2 \\ -2 & 4 \end{vmatrix} &= 3(4) - (-2)(-2) \\ &= 12 - 4 = 8 \end{aligned}$$

13.
$$\begin{aligned} \begin{vmatrix} -1 & 2 \\ 3 & -4 \end{vmatrix} &= -1(-4) - 2(3) \\ &= 4 - 6 = -2 \end{aligned}$$

14.
$$\begin{aligned} \begin{vmatrix} -1 & -2 \\ -3 & -4 \end{vmatrix} &= -1(-4) - (-2)(-3) \\ &= 4 - 6 = -2 \end{aligned}$$

15.
$$\begin{vmatrix} x & y \\ y & x \end{vmatrix} = x(x) - y(y) = x^2 - y^2$$

16. $\begin{vmatrix} x+y & y-x \\ x & y \end{vmatrix} = (x+y)y - (y-x)x = xy + y^2 - (yx - x^2) = xy + y^2 - xy + x^2 = y^2 + x^2$

17. $\begin{vmatrix} 1 & 0 & 1 \\ 0 & 1 & 0 \\ 1 & 1 & 1 \end{vmatrix} = 1\begin{vmatrix} 1 & 0 \\ 1 & 1 \end{vmatrix} - 0\begin{vmatrix} 0 & 0 \\ 1 & 1 \end{vmatrix} + 1\begin{vmatrix} 0 & 1 \\ 1 & 1 \end{vmatrix} = 1(1) - 0(0) + 1(-1) = 1 - 0 - 1 = 0$

18. $\begin{vmatrix} 1 & 2 & 0 \\ 0 & 1 & 2 \\ 0 & 0 & 1 \end{vmatrix} = 1\begin{vmatrix} 1 & 2 \\ 0 & 1 \end{vmatrix} - 2\begin{vmatrix} 0 & 2 \\ 0 & 1 \end{vmatrix} + 0\begin{vmatrix} 0 & 1 \\ 0 & 0 \end{vmatrix} = 1(1) - 2(0) + 0(0) = 1 - 0 + 0 = 1$

19. $\begin{vmatrix} -1 & 2 & 1 \\ 2 & 1 & -3 \\ 1 & 1 & 1 \end{vmatrix} = -1\begin{vmatrix} 1 & -3 \\ 1 & 1 \end{vmatrix} - 2\begin{vmatrix} 2 & -3 \\ 1 & 1 \end{vmatrix} + 1\begin{vmatrix} 2 & 1 \\ 1 & 1 \end{vmatrix} = -1(4) - 2(5) + 1(1) = -4 - 10 + 1$

$= -13$

20. $\begin{vmatrix} 1 & 2 & 3 \\ 1 & 2 & 3 \\ 1 & 2 & 3 \end{vmatrix} = 1\begin{vmatrix} 2 & 3 \\ 2 & 3 \end{vmatrix} - 2\begin{vmatrix} 1 & 3 \\ 1 & 3 \end{vmatrix} + 3\begin{vmatrix} 1 & 2 \\ 1 & 2 \end{vmatrix} = 1(0) - 2(0) + 1(0) = 0 - 0 + 0 = 0$

21. $\begin{vmatrix} 1 & -2 & 3 \\ -2 & 1 & 1 \\ -3 & -2 & 1 \end{vmatrix} = 1\begin{vmatrix} 1 & 1 \\ -2 & 1 \end{vmatrix} - (-2)\begin{vmatrix} -2 & 1 \\ -3 & 1 \end{vmatrix} + 3\begin{vmatrix} -2 & 1 \\ -3 & -2 \end{vmatrix} = 1(3) + 2(1) + 3(7) = 3 + 2 + 21$

$= 26$

22. $\begin{vmatrix} 1 & 1 & 2 \\ 2 & 1 & -2 \\ 3 & 1 & 3 \end{vmatrix} = 1\begin{vmatrix} 1 & -2 \\ 1 & 3 \end{vmatrix} - 1\begin{vmatrix} 2 & -2 \\ 3 & 3 \end{vmatrix} + 2\begin{vmatrix} 2 & 1 \\ 3 & 1 \end{vmatrix} = 1(5) - 1(12) + 2(-1) = 5 - 12 - 2 = -9$

23. $\begin{vmatrix} 1 & 2 & 3 \\ 4 & 5 & 6 \\ 7 & 8 & 9 \end{vmatrix} = 1\begin{vmatrix} 5 & 6 \\ 8 & 9 \end{vmatrix} - 2\begin{vmatrix} 4 & 6 \\ 7 & 9 \end{vmatrix} + 3\begin{vmatrix} 4 & 5 \\ 7 & 8 \end{vmatrix} = 1(-3) - 2(-6) + 3(-3) = -3 + 12 - 9 = 0$

24. $\begin{vmatrix} 1 & 4 & 7 \\ 2 & 5 & 8 \\ 3 & 6 & 9 \end{vmatrix} = 1\begin{vmatrix} 5 & 8 \\ 6 & 9 \end{vmatrix} - 4\begin{vmatrix} 2 & 8 \\ 3 & 9 \end{vmatrix} + 7\begin{vmatrix} 2 & 5 \\ 3 & 6 \end{vmatrix} = 1(-3) - 4(-6) + 7(-3) = -3 + 24 - 21 = 0$

25. $\begin{vmatrix} a & 2a & -a \\ 2 & -1 & 3 \\ 1 & 2 & -3 \end{vmatrix} = a\begin{vmatrix} -1 & 3 \\ 2 & -3 \end{vmatrix} - 2a\begin{vmatrix} 2 & 3 \\ 1 & -3 \end{vmatrix} + (-a)\begin{vmatrix} 2 & -1 \\ 1 & 2 \end{vmatrix} = a(-3) - 2a(-9) - a(5)$

$= -3a + 18a - 5a = 10a$

26. $\begin{vmatrix} 1 & 2b & -3 \\ 2 & -b & 2 \\ 1 & 3b & 1 \end{vmatrix} = 1\begin{vmatrix} -b & 2 \\ 3b & 1 \end{vmatrix} - 2b\begin{vmatrix} 2 & 2 \\ 1 & 1 \end{vmatrix} + (-3)\begin{vmatrix} 2 & -b \\ 1 & 3b \end{vmatrix} = 1(-7b) - 2b(0) - 3(7b)$

$= -7b - 0 - 21b = -28b$

27. $\begin{vmatrix} 1 & a & b \\ 1 & 2a & 2b \\ 1 & 3a & 3b \end{vmatrix} = 1\begin{vmatrix} 2a & 2b \\ 3a & 3b \end{vmatrix} - a\begin{vmatrix} 1 & 2b \\ 1 & 3b \end{vmatrix} + b\begin{vmatrix} 1 & 2a \\ 1 & 3a \end{vmatrix} = 1(0) - a(b) + b(a) = 0 - ab + ab = 0$

28. $\begin{vmatrix} a & b & c \\ 0 & b & c \\ 0 & 0 & c \end{vmatrix} = a\begin{vmatrix} b & c \\ 0 & c \end{vmatrix} - b\begin{vmatrix} 0 & c \\ 0 & c \end{vmatrix} + c\begin{vmatrix} 0 & b \\ 0 & 0 \end{vmatrix} = a(bc) - b(0) + c(0) = abc$

29. $x = \dfrac{\begin{vmatrix} 6 & 1 \\ 2 & -1 \end{vmatrix}}{\begin{vmatrix} 1 & 1 \\ 1 & -1 \end{vmatrix}} = \dfrac{-6-2}{-1-1} = \dfrac{-8}{-2} = 4$; $y = \dfrac{\begin{vmatrix} 1 & 6 \\ 1 & 2 \end{vmatrix}}{\begin{vmatrix} 1 & 1 \\ 1 & -1 \end{vmatrix}} = \dfrac{2-6}{-2} = \dfrac{-4}{-2} = 2$; solution: $(4, 2)$

30. $x = \dfrac{\begin{vmatrix} 4 & -1 \\ 5 & 1 \end{vmatrix}}{\begin{vmatrix} 1 & -1 \\ 2 & 1 \end{vmatrix}} = \dfrac{4-(-5)}{1-(-2)} = \dfrac{9}{3} = 3$; $y = \dfrac{\begin{vmatrix} 1 & 4 \\ 2 & 5 \end{vmatrix}}{\begin{vmatrix} 1 & -1 \\ 2 & 1 \end{vmatrix}} = \dfrac{5-8}{3} = \dfrac{-3}{3} = -1$; solution: $(3, -1)$

31. $x = \dfrac{\begin{vmatrix} 1 & 1 \\ -7 & -2 \end{vmatrix}}{\begin{vmatrix} 2 & 1 \\ 1 & -2 \end{vmatrix}} = \dfrac{-2-(-7)}{-4-1} = \dfrac{5}{-5} = -1$; $y = \dfrac{\begin{vmatrix} 2 & 1 \\ 1 & -7 \end{vmatrix}}{\begin{vmatrix} 2 & 1 \\ 1 & -2 \end{vmatrix}} = \dfrac{-14-1}{-5} = \dfrac{-15}{-5} = 3$

solution: $(-1, 3)$

32. $x = \dfrac{\begin{vmatrix} -3 & -1 \\ -7 & 1 \end{vmatrix}}{\begin{vmatrix} 3 & -1 \\ 2 & 1 \end{vmatrix}} = \dfrac{-3-7}{3-(-2)} = \dfrac{-10}{5} = -2$; $y = \dfrac{\begin{vmatrix} 3 & -3 \\ 2 & -7 \end{vmatrix}}{\begin{vmatrix} 3 & -1 \\ 2 & 1 \end{vmatrix}} = \dfrac{-21-(-6)}{5} = \dfrac{-15}{5} = -3$

solution: $(-2, -3)$

33. $x = \dfrac{\begin{vmatrix} 0 & 3 \\ -4 & -6 \end{vmatrix}}{\begin{vmatrix} 2 & 3 \\ 4 & -6 \end{vmatrix}} = \dfrac{0-(-12)}{-12-12} = \dfrac{12}{-24} = -\dfrac{1}{2}$; $y = \dfrac{\begin{vmatrix} 2 & 0 \\ 4 & -4 \end{vmatrix}}{\begin{vmatrix} 2 & 3 \\ 4 & -6 \end{vmatrix}} = \dfrac{-8-0}{-24} = \dfrac{-8}{-24} = \dfrac{1}{3}$

solution: $\left(-\frac{1}{2}, \frac{1}{3}\right)$

34. $x = \dfrac{\begin{vmatrix} -1 & -3 \\ 4 & 3 \end{vmatrix}}{\begin{vmatrix} 4 & -3 \\ 8 & 3 \end{vmatrix}} = \dfrac{-3-(-12)}{12-(-24)} = \dfrac{9}{36} = \dfrac{1}{4}$; $y = \dfrac{\begin{vmatrix} 4 & -1 \\ 8 & 4 \end{vmatrix}}{\begin{vmatrix} 4 & -3 \\ 8 & 3 \end{vmatrix}} = \dfrac{16-(-8)}{36} = \dfrac{24}{36} = \dfrac{2}{3}$

solution: $\left(\frac{1}{4}, \frac{2}{3}\right)$

35. $\begin{cases} y = \dfrac{-2x+1}{3} \\ 3x - 2y = 8 \end{cases} \Rightarrow \begin{cases} 3y = -2x+1 \\ 3x-2y = 8 \end{cases} \Rightarrow \begin{cases} 2x+3y = 1 \\ 3x-2y = 8 \end{cases}$

$x = \dfrac{\begin{vmatrix} 1 & 3 \\ 8 & -2 \end{vmatrix}}{\begin{vmatrix} 2 & 3 \\ 3 & -2 \end{vmatrix}} = \dfrac{-2-24}{-4-9} = \dfrac{-26}{-13} = 2$; $y = \dfrac{\begin{vmatrix} 2 & 1 \\ 3 & 8 \end{vmatrix}}{\begin{vmatrix} 2 & 3 \\ 3 & -2 \end{vmatrix}} = \dfrac{16-3}{-13} = \dfrac{13}{-13} = -1$; solution: $(2, -1)$

36. $\begin{cases} 2x+3y = -1 \\ x = \dfrac{y-9}{4} \end{cases} \Rightarrow \begin{cases} 2x+3y = -1 \\ 4x = y-9 \end{cases} \Rightarrow \begin{cases} 2x+3y = -1 \\ 4x-y = -9 \end{cases}$

$x = \dfrac{\begin{vmatrix} -1 & 3 \\ -9 & -1 \end{vmatrix}}{\begin{vmatrix} 2 & 3 \\ 4 & -1 \end{vmatrix}} = \dfrac{1-(-27)}{-2-12} = \dfrac{28}{-14} = -2$; $y = \dfrac{\begin{vmatrix} 2 & -1 \\ 4 & -9 \end{vmatrix}}{\begin{vmatrix} 2 & 3 \\ 4 & -1 \end{vmatrix}} = \dfrac{-18-(-4)}{-14} = \dfrac{-14}{-14} = 1$

solution: $(-2, 1)$

37. $\begin{cases} y = \dfrac{11-3x}{2} \\ x = \dfrac{11-4y}{6} \end{cases} \Rightarrow \begin{cases} 2y = 11-3x \\ 6x = 11-4y \end{cases} \Rightarrow \begin{cases} 3x+2y = 11 \\ 6x+4y = 11 \end{cases}$

$x = \dfrac{\begin{vmatrix} 11 & 2 \\ 11 & 4 \end{vmatrix}}{\begin{vmatrix} 3 & 2 \\ 6 & 4 \end{vmatrix}} = \dfrac{44-22}{12-12} = \dfrac{22}{0} \Rightarrow$ denominator $= 0$, numerator $\neq 0 \Rightarrow$ no solution

38. $\begin{cases} x = \dfrac{12-6y}{5} \\ y = \dfrac{24-10x}{12} \end{cases} \Rightarrow \begin{cases} 5x = 12-6y \\ 12y = 24-10x \end{cases} \Rightarrow \begin{cases} 5x+6y = 12 \\ 10x+12y = 24 \end{cases}$

$x = \dfrac{\begin{vmatrix} 12 & 6 \\ 24 & 12 \end{vmatrix}}{\begin{vmatrix} 5 & 6 \\ 10 & 12 \end{vmatrix}} = \dfrac{144-144}{60-60} = \dfrac{0}{0} \Rightarrow$ denominator $= 0$, numerator $= 0 \Rightarrow$ dependent equations

39. $\begin{cases} x = \dfrac{5y-4}{2} \\ y = \dfrac{3x-1}{5} \end{cases} \Rightarrow \begin{cases} 2x = 5y-4 \\ 5y = 3x-1 \end{cases} \Rightarrow \begin{cases} 2x-5y=-4 \\ -3x+5y=-1 \end{cases}$

$$x = \frac{\begin{vmatrix} -4 & -5 \\ -1 & 5 \end{vmatrix}}{\begin{vmatrix} 2 & -5 \\ -3 & 5 \end{vmatrix}} = \frac{-20-5}{10-15} = \frac{-25}{-5} = 5;\ y = \frac{\begin{vmatrix} 2 & -4 \\ -3 & -1 \end{vmatrix}}{\begin{vmatrix} 2 & -5 \\ -3 & 5 \end{vmatrix}} = \frac{-2-12}{-5} = \frac{-14}{-5} = \frac{14}{5}$$

solution: $(5, \frac{14}{5})$

40. $\begin{cases} y = \dfrac{1-5x}{2} \\ x = \dfrac{3y+10}{4} \end{cases} \Rightarrow \begin{cases} 2y = 1-5x \\ 4x = 3y+10 \end{cases} \Rightarrow \begin{cases} 5x+2y=1 \\ 4x-3y=10 \end{cases}$

$$x = \frac{\begin{vmatrix} 1 & 2 \\ 10 & -3 \end{vmatrix}}{\begin{vmatrix} 5 & 2 \\ 4 & -3 \end{vmatrix}} = \frac{-3-20}{-15-8} = \frac{-23}{-23} = 1;\ y = \frac{\begin{vmatrix} 5 & 1 \\ 4 & 10 \end{vmatrix}}{\begin{vmatrix} 5 & 2 \\ 4 & -3 \end{vmatrix}} = \frac{50-4}{-23} = \frac{46}{-23} = -2$$

solution: $(1, -2)$

Note: In the following problems, D stands for the denominator determinant, while N_x, N_y and N_z stand for the numerator determinants for x, y and z, respectively.

41. $$D = \begin{vmatrix} 1 & 1 & 1 \\ 1 & 1 & -1 \\ 1 & -1 & 1 \end{vmatrix} = 1\begin{vmatrix} 1 & -1 \\ -1 & 1 \end{vmatrix} - 1\begin{vmatrix} 1 & -1 \\ 1 & 1 \end{vmatrix} + 1\begin{vmatrix} 1 & 1 \\ 1 & -1 \end{vmatrix} = 1(0) - 1(2) + 1(-2) = -4$$

$$N_x = \begin{vmatrix} 4 & 1 & 1 \\ 0 & 1 & -1 \\ 2 & -1 & 1 \end{vmatrix} = 4\begin{vmatrix} 1 & -1 \\ -1 & 1 \end{vmatrix} - 1\begin{vmatrix} 0 & -1 \\ 2 & 1 \end{vmatrix} + 1\begin{vmatrix} 0 & 1 \\ 2 & -1 \end{vmatrix} = 4(0) - 1(2) + 1(-2) = -4$$

$$N_y = \begin{vmatrix} 1 & 4 & 1 \\ 1 & 0 & -1 \\ 1 & 2 & 1 \end{vmatrix} = 1\begin{vmatrix} 0 & -1 \\ 2 & 1 \end{vmatrix} - 4\begin{vmatrix} 1 & -1 \\ 1 & 1 \end{vmatrix} + 1\begin{vmatrix} 1 & 0 \\ 1 & 2 \end{vmatrix} = 1(2) - 4(2) + 1(2) = -4$$

$$N_z = \begin{vmatrix} 1 & 1 & 4 \\ 1 & 1 & 0 \\ 1 & -1 & 2 \end{vmatrix} = 1\begin{vmatrix} 1 & 0 \\ -1 & 2 \end{vmatrix} - 1\begin{vmatrix} 1 & 0 \\ 1 & 2 \end{vmatrix} + 4\begin{vmatrix} 1 & 1 \\ 1 & -1 \end{vmatrix} = 1(2) - 1(2) + 4(-2) = -8$$

$$x = \frac{N_x}{D} = \frac{-4}{-4} = 1;\ y = \frac{N_y}{D} = \frac{-4}{-4} = 1;\ z = \frac{N_z}{D} = \frac{-8}{-4} = 2 \Rightarrow \text{solution: } (1, 1, 2)$$

42. $$D = \begin{vmatrix} 1 & 1 & 1 \\ 1 & -1 & 1 \\ 1 & -1 & -1 \end{vmatrix} = 1\begin{vmatrix} -1 & 1 \\ -1 & -1 \end{vmatrix} - 1\begin{vmatrix} 1 & 1 \\ 1 & -1 \end{vmatrix} + 1\begin{vmatrix} 1 & -1 \\ 1 & -1 \end{vmatrix} = 1(2) - 1(-2) + 1(0) = 4$$

$$N_x = \begin{vmatrix} 4 & 1 & 1 \\ 2 & -1 & 1 \\ 0 & -1 & -1 \end{vmatrix} = 4\begin{vmatrix} -1 & 1 \\ -1 & -1 \end{vmatrix} - 1\begin{vmatrix} 2 & 1 \\ 0 & -1 \end{vmatrix} + 1\begin{vmatrix} 2 & -1 \\ 0 & -1 \end{vmatrix} = 4(2) - 1(-2) + 1(-2) = 8$$

continued on next page...

42. continued

$$N_y = \begin{vmatrix} 1 & 4 & 1 \\ 1 & 2 & 1 \\ 1 & 0 & -1 \end{vmatrix} = 1\begin{vmatrix} 2 & 1 \\ 0 & -1 \end{vmatrix} - 4\begin{vmatrix} 1 & 1 \\ 1 & -1 \end{vmatrix} + 1\begin{vmatrix} 1 & 2 \\ 1 & 0 \end{vmatrix} = 1(-2) - 4(-2) + 1(-2) = 4$$

$$N_z = \begin{vmatrix} 1 & 1 & 4 \\ 1 & -1 & 2 \\ 1 & -1 & 0 \end{vmatrix} = 1\begin{vmatrix} -1 & 2 \\ -1 & 0 \end{vmatrix} - 1\begin{vmatrix} 1 & 2 \\ 1 & 0 \end{vmatrix} + 4\begin{vmatrix} 1 & -1 \\ 1 & -1 \end{vmatrix} = 1(2) - 1(-2) + 4(0) = 4$$

$$x = \frac{N_x}{D} = \frac{8}{4} = 2;\ y = \frac{N_y}{D} = \frac{4}{4} = 1;\ z = \frac{N_z}{D} = \frac{4}{4} = 1 \Rightarrow \text{solution: } (2, 1, 1)$$

43. $$D = \begin{vmatrix} 1 & 1 & 2 \\ 1 & 2 & 1 \\ 2 & 1 & 1 \end{vmatrix} = 1\begin{vmatrix} 2 & 1 \\ 1 & 1 \end{vmatrix} - 1\begin{vmatrix} 1 & 1 \\ 2 & 1 \end{vmatrix} + 2\begin{vmatrix} 1 & 2 \\ 2 & 1 \end{vmatrix} = 1(1) - 1(-1) + 2(-3) = -4$$

$$N_x = \begin{vmatrix} 7 & 1 & 2 \\ 8 & 2 & 1 \\ 9 & 1 & 1 \end{vmatrix} = 7\begin{vmatrix} 2 & 1 \\ 1 & 1 \end{vmatrix} - 1\begin{vmatrix} 8 & 1 \\ 9 & 1 \end{vmatrix} + 2\begin{vmatrix} 8 & 2 \\ 9 & 1 \end{vmatrix} = 7(1) - 1(-1) + 2(-10) = -12$$

$$N_y = \begin{vmatrix} 1 & 7 & 2 \\ 1 & 8 & 1 \\ 2 & 9 & 1 \end{vmatrix} = 1\begin{vmatrix} 8 & 1 \\ 9 & 1 \end{vmatrix} - 7\begin{vmatrix} 1 & 1 \\ 2 & 1 \end{vmatrix} + 2\begin{vmatrix} 1 & 8 \\ 2 & 9 \end{vmatrix} = 1(-1) - 7(-1) + 2(-7) = -8$$

$$N_z = \begin{vmatrix} 1 & 1 & 7 \\ 1 & 2 & 8 \\ 2 & 1 & 9 \end{vmatrix} = 1\begin{vmatrix} 2 & 8 \\ 1 & 9 \end{vmatrix} - 1\begin{vmatrix} 1 & 8 \\ 2 & 9 \end{vmatrix} + 7\begin{vmatrix} 1 & 2 \\ 2 & 1 \end{vmatrix} = 1(10) - 1(-7) + 7(-3) = -4$$

$$x = \frac{N_x}{D} = \frac{-12}{-4} = 3;\ y = \frac{N_y}{D} = \frac{-8}{-4} = 2;\ z = \frac{N_z}{D} = \frac{-4}{-4} = 1 \Rightarrow \text{solution: } (3, 2, 1)$$

44. $$D = \begin{vmatrix} 1 & 2 & 2 \\ 2 & 1 & 2 \\ 2 & 2 & 1 \end{vmatrix} = 1\begin{vmatrix} 1 & 2 \\ 2 & 1 \end{vmatrix} - 2\begin{vmatrix} 2 & 2 \\ 2 & 1 \end{vmatrix} + 2\begin{vmatrix} 2 & 1 \\ 2 & 2 \end{vmatrix} = 1(-3) - 2(-2) + 2(2) = 5$$

$$N_x = \begin{vmatrix} 10 & 2 & 2 \\ 9 & 1 & 2 \\ 1 & 2 & 1 \end{vmatrix} = 10\begin{vmatrix} 1 & 2 \\ 2 & 1 \end{vmatrix} - 2\begin{vmatrix} 9 & 2 \\ 1 & 1 \end{vmatrix} + 2\begin{vmatrix} 9 & 1 \\ 1 & 2 \end{vmatrix} = 10(-3) - 2(7) + 2(17) = -10$$

$$N_y = \begin{vmatrix} 1 & 10 & 2 \\ 2 & 9 & 2 \\ 2 & 1 & 1 \end{vmatrix} = 1\begin{vmatrix} 9 & 2 \\ 1 & 1 \end{vmatrix} - 10\begin{vmatrix} 2 & 2 \\ 2 & 1 \end{vmatrix} + 2\begin{vmatrix} 2 & 9 \\ 2 & 1 \end{vmatrix} = 1(7) - 10(-2) + 2(-16) = -5$$

$$N_z = \begin{vmatrix} 1 & 2 & 10 \\ 2 & 1 & 9 \\ 2 & 2 & 1 \end{vmatrix} = 1\begin{vmatrix} 1 & 9 \\ 2 & 1 \end{vmatrix} - 2\begin{vmatrix} 2 & 9 \\ 2 & 1 \end{vmatrix} + 10\begin{vmatrix} 2 & 1 \\ 2 & 2 \end{vmatrix} = 1(-17) - 2(-16) + 10(2) = 35$$

$$x = \frac{N_x}{D} = \frac{-10}{5} = -2;\ y = \frac{N_y}{D} = \frac{-5}{5} = -1;\ z = \frac{N_z}{D} = \frac{35}{5} = 7 \Rightarrow \text{solution: } (-2, -1, 7)$$

45. $D = \begin{vmatrix} 2 & 1 & -1 \\ 1 & 2 & 2 \\ 4 & 5 & 3 \end{vmatrix} = 2\begin{vmatrix} 2 & 2 \\ 5 & 3 \end{vmatrix} - 1\begin{vmatrix} 1 & 2 \\ 4 & 3 \end{vmatrix} + (-1)\begin{vmatrix} 1 & 2 \\ 4 & 5 \end{vmatrix} = 2(-4) - 1(-5) - 1(-3) = 0$

$N_x = \begin{vmatrix} 1 & 1 & -1 \\ 2 & 2 & 2 \\ 3 & 5 & 3 \end{vmatrix} = 1\begin{vmatrix} 2 & 2 \\ 5 & 3 \end{vmatrix} - 1\begin{vmatrix} 2 & 2 \\ 3 & 3 \end{vmatrix} + (-1)\begin{vmatrix} 2 & 2 \\ 3 & 5 \end{vmatrix} = 1(-4) - 1(0) - 1(4) = -8$

$x = \dfrac{N_x}{D} = \dfrac{-8}{0} \Rightarrow$ denominator $= 0$, numerator $\neq 0 \Rightarrow$ no solution.

46. $D = \begin{vmatrix} 4 & 0 & 3 \\ 0 & 2 & -6 \\ 8 & 4 & 3 \end{vmatrix} = 4\begin{vmatrix} 2 & -6 \\ 4 & 3 \end{vmatrix} - 0\begin{vmatrix} 0 & -6 \\ 8 & 3 \end{vmatrix} + 3\begin{vmatrix} 0 & 2 \\ 8 & 4 \end{vmatrix} = 4(30) - 0(48) + 3(-16) = 72$

$N_x = \begin{vmatrix} 4 & 0 & 3 \\ -1 & 2 & -6 \\ 9 & 4 & 3 \end{vmatrix} = 4\begin{vmatrix} 2 & -6 \\ 4 & 3 \end{vmatrix} - 0\begin{vmatrix} -1 & -6 \\ 9 & 3 \end{vmatrix} + 3\begin{vmatrix} -1 & 2 \\ 9 & 4 \end{vmatrix} = 4(30) - 0(51) + 3(-22) = 54$

$N_y = \begin{vmatrix} 4 & 4 & 3 \\ 0 & -1 & -6 \\ 8 & 9 & 3 \end{vmatrix} = 4\begin{vmatrix} -1 & -6 \\ 9 & 3 \end{vmatrix} - 4\begin{vmatrix} 0 & -6 \\ 8 & 3 \end{vmatrix} + 3\begin{vmatrix} 0 & -1 \\ 8 & 9 \end{vmatrix} = 4(51) - 4(48) + 3(8) = 36$

$N_z = \begin{vmatrix} 4 & 0 & 4 \\ 0 & 2 & -1 \\ 8 & 4 & 9 \end{vmatrix} = 4\begin{vmatrix} 2 & -1 \\ 4 & 9 \end{vmatrix} - 0\begin{vmatrix} 0 & -1 \\ 8 & 9 \end{vmatrix} + 4\begin{vmatrix} 0 & 2 \\ 8 & 4 \end{vmatrix} = 4(22) - 0(8) + 4(-16) = 24$

$x = \dfrac{N_x}{D} = \dfrac{54}{72} = \dfrac{3}{4};\ y = \dfrac{N_y}{D} = \dfrac{36}{72} = \dfrac{1}{2};\ z = \dfrac{N_z}{D} = \dfrac{24}{72} = \dfrac{1}{3} \Rightarrow$ solution: $\left(\dfrac{3}{4}, \dfrac{1}{2}, \dfrac{1}{3}\right)$

47. $D = \begin{vmatrix} 2 & 1 & 1 \\ 1 & -2 & 3 \\ 1 & 1 & -4 \end{vmatrix} = 2\begin{vmatrix} -2 & 3 \\ 1 & -4 \end{vmatrix} - 1\begin{vmatrix} 1 & 3 \\ 1 & -4 \end{vmatrix} + 1\begin{vmatrix} 1 & -2 \\ 1 & 1 \end{vmatrix} = 2(5) - 1(-7) + 1(3) = 20$

$N_x = \begin{vmatrix} 5 & 1 & 1 \\ 10 & -2 & 3 \\ -3 & 1 & -4 \end{vmatrix} = 5\begin{vmatrix} -2 & 3 \\ 1 & -4 \end{vmatrix} - 1\begin{vmatrix} 10 & 3 \\ -3 & -4 \end{vmatrix} + 1\begin{vmatrix} 10 & -2 \\ -3 & 1 \end{vmatrix}$

$= 5(5) - 1(-31) + 1(4) = 60$

$N_y = \begin{vmatrix} 2 & 5 & 1 \\ 1 & 10 & 3 \\ 1 & -3 & -4 \end{vmatrix} = 2\begin{vmatrix} 10 & 3 \\ -3 & -4 \end{vmatrix} - 5\begin{vmatrix} 1 & 3 \\ 1 & -4 \end{vmatrix} + 1\begin{vmatrix} 1 & 10 \\ 1 & -3 \end{vmatrix}$

$= 2(-31) - 5(-7) + 1(-13) = -40$

$N_z = \begin{vmatrix} 2 & 1 & 5 \\ 1 & -2 & 10 \\ 1 & 1 & -3 \end{vmatrix} = 2\begin{vmatrix} -2 & 10 \\ 1 & -3 \end{vmatrix} - 1\begin{vmatrix} 1 & 10 \\ 1 & -3 \end{vmatrix} + 5\begin{vmatrix} 1 & -2 \\ 1 & 1 \end{vmatrix} = 2(-4) - 1(-13) + 5(3) = 20$

$x = \dfrac{N_x}{D} = \dfrac{60}{20} = 3;\ y = \dfrac{N_y}{D} = \dfrac{-40}{20} = -2;\ z = \dfrac{N_z}{D} = \dfrac{20}{20} = 1 \Rightarrow$ solution: $(3, -2, 1)$

48. $D = \begin{vmatrix} 3 & 2 & -1 \\ 2 & -1 & 7 \\ 2 & 2 & -3 \end{vmatrix} = 3\begin{vmatrix} -1 & 7 \\ 2 & -3 \end{vmatrix} - 2\begin{vmatrix} 2 & 7 \\ 2 & -3 \end{vmatrix} + (-1)\begin{vmatrix} 2 & -1 \\ 2 & 2 \end{vmatrix} = 3(-11) - 2(-20) - 1(6)$

$= 1$

$N_x = \begin{vmatrix} -8 & 2 & -1 \\ 10 & -1 & 7 \\ -10 & 2 & -3 \end{vmatrix} = -8\begin{vmatrix} -1 & 7 \\ 2 & -3 \end{vmatrix} - 2\begin{vmatrix} 10 & 7 \\ -10 & -3 \end{vmatrix} + (-1)\begin{vmatrix} 10 & -1 \\ -10 & 2 \end{vmatrix}$

$= -8(-11) - 2(40) - 1(10) = -2$

$N_y = \begin{vmatrix} 3 & -8 & -1 \\ 2 & 10 & 7 \\ 2 & -10 & -3 \end{vmatrix} = 3\begin{vmatrix} 10 & 7 \\ -10 & -3 \end{vmatrix} - (-8)\begin{vmatrix} 2 & 7 \\ 2 & -3 \end{vmatrix} + (-1)\begin{vmatrix} 2 & 10 \\ 2 & -10 \end{vmatrix}$

$= 3(40) + 8(-20) - 1(-40) = 0$

$N_z = \begin{vmatrix} 3 & 2 & -8 \\ 2 & -1 & 10 \\ 2 & 2 & -10 \end{vmatrix} = 3\begin{vmatrix} -1 & 10 \\ 2 & -10 \end{vmatrix} - 2\begin{vmatrix} 2 & 10 \\ 2 & -10 \end{vmatrix} + (-8)\begin{vmatrix} 2 & -1 \\ 2 & 2 \end{vmatrix}$

$= 3(-10) - 2(-40) - 8(6) = 2$

$x = \frac{N_x}{D} = \frac{-2}{1} = -2$; $y = \frac{N_y}{D} = \frac{0}{1} = 0$; $z = \frac{N_z}{D} = \frac{2}{1} = 2 \Rightarrow$ solution: $(-2, 0, 2)$

49. $D = \begin{vmatrix} 2 & 3 & 4 \\ 2 & -3 & -4 \\ 4 & 6 & 8 \end{vmatrix} = 2\begin{vmatrix} -3 & -4 \\ 6 & 8 \end{vmatrix} - 3\begin{vmatrix} 2 & -4 \\ 4 & 8 \end{vmatrix} + 4\begin{vmatrix} 2 & -3 \\ 4 & 6 \end{vmatrix} = 2(0) - 3(32) + 4(24) = 0$

$N_x = \begin{vmatrix} 6 & 3 & 4 \\ -4 & -3 & -4 \\ 12 & 6 & 8 \end{vmatrix} = 6\begin{vmatrix} -3 & -4 \\ 6 & 8 \end{vmatrix} - 3\begin{vmatrix} -4 & -4 \\ 12 & 8 \end{vmatrix} + 4\begin{vmatrix} -4 & -3 \\ 12 & 6 \end{vmatrix} = 6(0) - 3(16) + 4(12) = 0$

$x = \frac{N_x}{D} = \frac{0}{0} \Rightarrow$ denominator $= 0$, numerator $= 0 \Rightarrow$ dependent equations.

50. $\begin{cases} x - 3y + 4z - 2 = 0 \Rightarrow & x - 3y + 4z = 2 \\ 2x + y + 2z - 3 = 0 \Rightarrow & 2x + y + 2z = 3 \\ 4x - 5y + 10z - 7 = 0 \Rightarrow & 4x - 5y + 10z = 7 \end{cases}$

$D = \begin{vmatrix} 1 & -3 & 4 \\ 2 & 1 & 2 \\ 4 & -5 & 10 \end{vmatrix} = 1\begin{vmatrix} 1 & 2 \\ -5 & 10 \end{vmatrix} - (-3)\begin{vmatrix} 2 & 2 \\ 4 & 10 \end{vmatrix} + 4\begin{vmatrix} 2 & 1 \\ 4 & -5 \end{vmatrix} = 1(20) + 3(12) + 4(-14) = 0$

$N_x = \begin{vmatrix} 2 & -3 & 4 \\ 3 & 1 & 2 \\ 7 & -5 & 10 \end{vmatrix} = 2\begin{vmatrix} 1 & 2 \\ -5 & 10 \end{vmatrix} - (-3)\begin{vmatrix} 3 & 2 \\ 7 & 10 \end{vmatrix} + 4\begin{vmatrix} 3 & 1 \\ 7 & -5 \end{vmatrix} = 2(20) + 3(16) + 4(-22) = 0$

$x = \frac{N_x}{D} = \frac{0}{0} \Rightarrow$ denominator $= 0$, numerator $= 0 \Rightarrow$ dependent equations.

51. $\begin{cases} x+y=1 \Rightarrow & x+y=1 \\ \frac{1}{2}y+z=\frac{5}{2} \Rightarrow & y+2z=5 \\ x-z=-3 \Rightarrow & x-z=-3 \end{cases}$

$$D=\begin{vmatrix} 1 & 1 & 0 \\ 0 & 1 & 2 \\ 1 & 0 & -1 \end{vmatrix}=1\begin{vmatrix} 1 & 2 \\ 0 & -1 \end{vmatrix}-1\begin{vmatrix} 0 & 2 \\ 1 & -1 \end{vmatrix}+0\begin{vmatrix} 0 & 1 \\ 1 & 0 \end{vmatrix}=1(-1)-1(-2)+0(-1)=1$$

$$N_x=\begin{vmatrix} 1 & 1 & 0 \\ 5 & 1 & 2 \\ -3 & 0 & -1 \end{vmatrix}=1\begin{vmatrix} 1 & 2 \\ 0 & -1 \end{vmatrix}-1\begin{vmatrix} 5 & 2 \\ -3 & -1 \end{vmatrix}+0\begin{vmatrix} 5 & 1 \\ -3 & 0 \end{vmatrix}=1(-1)-1(1)+0(3)=-2$$

$$N_y=\begin{vmatrix} 1 & 1 & 0 \\ 0 & 5 & 2 \\ 1 & -3 & -1 \end{vmatrix}=1\begin{vmatrix} 5 & 2 \\ -3 & -1 \end{vmatrix}-1\begin{vmatrix} 0 & 2 \\ 1 & -1 \end{vmatrix}+0\begin{vmatrix} 0 & 5 \\ 1 & -3 \end{vmatrix}=1(1)-1(-2)+0(-5)=3$$

$$N_z=\begin{vmatrix} 1 & 1 & 1 \\ 0 & 1 & 5 \\ 1 & 0 & -3 \end{vmatrix}=1\begin{vmatrix} 1 & 5 \\ 0 & -3 \end{vmatrix}-1\begin{vmatrix} 0 & 5 \\ 1 & -3 \end{vmatrix}+1\begin{vmatrix} 0 & 1 \\ 1 & 0 \end{vmatrix}=1(-3)-1(-5)+1(-1)=1$$

$x=\frac{N_x}{D}=\frac{-2}{1}=-2$; $y=\frac{N_y}{D}=\frac{3}{1}=3$; $z=\frac{N_z}{D}=\frac{1}{1}=1 \Rightarrow$ solution: $(-2,3,1)$

52. $\begin{cases} 3x+4y+14z=7 \Rightarrow & 3x+4y+14z=7 \\ -\frac{1}{2}x-y+2z=\frac{3}{2} \Rightarrow & -x-2y+4z=3 \\ x+\frac{3}{2}y+\frac{5}{2}z=1 \Rightarrow & 2x+3y+5z=2 \end{cases}$

$$D=\begin{vmatrix} 3 & 4 & 14 \\ -1 & -2 & 4 \\ 2 & 3 & 5 \end{vmatrix}=3\begin{vmatrix} -2 & 4 \\ 3 & 5 \end{vmatrix}-4\begin{vmatrix} -1 & 4 \\ 2 & 5 \end{vmatrix}+14\begin{vmatrix} -1 & -2 \\ 2 & 3 \end{vmatrix}=3(-22)-4(-13)+14(1)$$
$$=0$$

$$N_x=\begin{vmatrix} 7 & 4 & 14 \\ 3 & -2 & 4 \\ 2 & 3 & 5 \end{vmatrix}=7\begin{vmatrix} -2 & 4 \\ 3 & 5 \end{vmatrix}-4\begin{vmatrix} 3 & 4 \\ 2 & 5 \end{vmatrix}+14\begin{vmatrix} 3 & -2 \\ 2 & 3 \end{vmatrix}=7(-22)-4(7)+14(13)=0$$

$x=\frac{N_x}{D}=\frac{0}{0} \Rightarrow$ denominator $=0$, numerator $=0 \Rightarrow$ dependent equations.

53. $\begin{cases} 2x-y+4z+2=0 \Rightarrow & 2x-y+4z=-2 \\ 5x+8y+7z=-8 \Rightarrow & 5x+8y+7z=-8 \\ x+3y+z+3=0 \Rightarrow & x+3y+z=-3 \end{cases}$

$$D=\begin{vmatrix} 2 & -1 & 4 \\ 5 & 8 & 7 \\ 1 & 3 & 1 \end{vmatrix}=2\begin{vmatrix} 8 & 7 \\ 3 & 1 \end{vmatrix}-(-1)\begin{vmatrix} 5 & 7 \\ 1 & 1 \end{vmatrix}+4\begin{vmatrix} 5 & 8 \\ 1 & 3 \end{vmatrix}=2(-13)+1(-2)+4(7)=0$$

$$N_x=\begin{vmatrix} -2 & -1 & 4 \\ -8 & 8 & 7 \\ -3 & 3 & 1 \end{vmatrix}=-2\begin{vmatrix} 8 & 7 \\ 3 & 1 \end{vmatrix}-(-1)\begin{vmatrix} -8 & 7 \\ -3 & 1 \end{vmatrix}+4\begin{vmatrix} -8 & 8 \\ -3 & 3 \end{vmatrix}=-2(-13)+1(13)+4(0)$$
$$=39$$

$x=\frac{N_x}{D}=\frac{39}{0} \Rightarrow$ denominator $=0$, numerator $\neq 0 \Rightarrow$ no solution.

54. $\begin{cases} \frac{1}{2}x+y+z+\frac{3}{2}=0 \Rightarrow & x+2y+2z=-3 \\ x+\frac{1}{2}y+z-\frac{1}{2}=0 \Rightarrow & 2x+y+2z=1 \\ x+y+\frac{1}{2}z+\frac{1}{2}=0 \Rightarrow & 2x+2y+z=-1 \end{cases}$

$$D=\begin{vmatrix} 1 & 2 & 2 \\ 2 & 1 & 2 \\ 2 & 2 & 1 \end{vmatrix}=1\begin{vmatrix} 1 & 2 \\ 2 & 1 \end{vmatrix}-2\begin{vmatrix} 2 & 2 \\ 2 & 1 \end{vmatrix}+2\begin{vmatrix} 2 & 1 \\ 2 & 2 \end{vmatrix}=1(-3)-2(-2)+2(2)=5$$

$$N_x=\begin{vmatrix} -3 & 2 & 2 \\ 1 & 1 & 2 \\ -1 & 2 & 1 \end{vmatrix}=-3\begin{vmatrix} 1 & 2 \\ 2 & 1 \end{vmatrix}-2\begin{vmatrix} 1 & 2 \\ -1 & 1 \end{vmatrix}+2\begin{vmatrix} 1 & 1 \\ -1 & 2 \end{vmatrix}=-3(-3)-2(3)+2(3)=9$$

$$N_y=\begin{vmatrix} 1 & -3 & 2 \\ 2 & 1 & 2 \\ 2 & -1 & 1 \end{vmatrix}=1\begin{vmatrix} 1 & 2 \\ -1 & 1 \end{vmatrix}-(-3)\begin{vmatrix} 2 & 2 \\ 2 & 1 \end{vmatrix}+2\begin{vmatrix} 2 & 1 \\ 2 & -1 \end{vmatrix}=1(3)+3(-2)+2(-4)=-11$$

$$N_z=\begin{vmatrix} 1 & 2 & -3 \\ 2 & 1 & 1 \\ 2 & 2 & -1 \end{vmatrix}=1\begin{vmatrix} 1 & 1 \\ 2 & -1 \end{vmatrix}-2\begin{vmatrix} 2 & 1 \\ 2 & -1 \end{vmatrix}+(-3)\begin{vmatrix} 2 & 1 \\ 2 & 2 \end{vmatrix}=1(-3)-2(-4)-3(2)=-1$$

$$x=\frac{N_x}{D}=\frac{9}{5};\ y=\frac{N_y}{D}=\frac{-11}{5};\ z=\frac{N_z}{D}=\frac{-1}{5} \Rightarrow \text{solution: } \left(\tfrac{9}{5},-\tfrac{11}{5},-\tfrac{1}{5}\right)$$

55. $\begin{vmatrix} x & 1 \\ 3 & 2 \end{vmatrix}=1$

$2x-3=1$

$2x=4$

$x=2$

56. $\begin{vmatrix} x & -x \\ 2 & -3 \end{vmatrix}=-5$

$-3x-(-2x)=-5$

$-x=-5$

$x=5$

57. $\begin{vmatrix} x & -2 \\ 3 & 1 \end{vmatrix}=\begin{vmatrix} 4 & 2 \\ x & 3 \end{vmatrix}$

$x-(-6)=12-2x$

$x+6=12-2x$

$3x=6$

$x=2$

58. $\begin{vmatrix} x & 3 \\ x & 2 \end{vmatrix}=\begin{vmatrix} 3 & 2 \\ 1 & 1 \end{vmatrix}$

$2x-3x=3-2$

$-x=1$

$x=-1$

59. $\begin{cases} 2x+y=180 \\ y=x+30 \end{cases} \Rightarrow \begin{cases} 2x+y=180 \\ -x+y=30 \end{cases}$

$$x=\frac{\begin{vmatrix} 180 & 1 \\ 30 & 1 \end{vmatrix}}{\begin{vmatrix} 2 & 1 \\ -1 & 1 \end{vmatrix}}=\frac{180-30}{2-(-1)}=\frac{150}{3}=50;\ y=\frac{\begin{vmatrix} 2 & 180 \\ -1 & 30 \end{vmatrix}}{\begin{vmatrix} 2 & 1 \\ -1 & 1 \end{vmatrix}}=\frac{60-(-180)}{3}=\frac{240}{3}=80$$

60. Let x = number of cheaper phones and let y = number of more expensive phones.
$$\begin{cases} x+y=360 \\ 67x+100y=29400 \end{cases}$$

$$x = \frac{\begin{vmatrix} 360 & 1 \\ 29400 & 100 \end{vmatrix}}{\begin{vmatrix} 1 & 1 \\ 67 & 100 \end{vmatrix}} = \frac{36000-29400}{100-67} = \frac{6600}{33} = 200$$

The warehouse had 200 of the \$67 phones and 160 of the \$100 phones.

$$y = \frac{\begin{vmatrix} 1 & 360 \\ 67 & 29400 \end{vmatrix}}{\begin{vmatrix} 1 & 1 \\ 67 & 100 \end{vmatrix}} = \frac{29400-24120}{33} = \frac{5280}{33} = 160$$

61. Let x = the amount invested in HiTech, y = the amount invested in SaveTel and z = the amount invested in HiGas. Form and solve the following system of equations:
$$\begin{cases} x+y+z=20000 \Rightarrow & x+y+z=20000 \\ 0.10x+0.05y+0.06z=0.066(20000) \Rightarrow & 10x+5y+6x=132000 \\ y+z=3x \Rightarrow & -3x+y+z=0 \end{cases}$$

$$D = \begin{vmatrix} 1 & 1 & 1 \\ 10 & 5 & 6 \\ -3 & 1 & 1 \end{vmatrix} = 1\begin{vmatrix} 5 & 6 \\ 1 & 1 \end{vmatrix} - 1\begin{vmatrix} 10 & 6 \\ -3 & 1 \end{vmatrix} + 1\begin{vmatrix} 10 & 5 \\ -3 & 1 \end{vmatrix} = 1(-1)-1(28)+1(25) = -4$$

$$N_x = \begin{vmatrix} 20000 & 1 & 1 \\ 132000 & 5 & 6 \\ 0 & 1 & 1 \end{vmatrix} = 20000\begin{vmatrix} 5 & 6 \\ 1 & 1 \end{vmatrix} - 1\begin{vmatrix} 132000 & 6 \\ 0 & 1 \end{vmatrix} + 1\begin{vmatrix} 132000 & 5 \\ 0 & 1 \end{vmatrix}$$
$$= 20000(-1)-1(132000)+1(132000) = -20000$$

$$N_y = \begin{vmatrix} 1 & 20000 & 1 \\ 10 & 132000 & 6 \\ -3 & 0 & 1 \end{vmatrix} = 1\begin{vmatrix} 132000 & 6 \\ 0 & 1 \end{vmatrix} - 20000\begin{vmatrix} 10 & 6 \\ -3 & 1 \end{vmatrix} + 1\begin{vmatrix} 10 & 132000 \\ -3 & 0 \end{vmatrix}$$
$$= 1(132000)-20000(28)+1(396000) = -32000$$

$$N_z = \begin{vmatrix} 1 & 1 & 20000 \\ 10 & 5 & 132000 \\ -3 & 1 & 0 \end{vmatrix} = 1\begin{vmatrix} 5 & 132000 \\ 1 & 0 \end{vmatrix} - 1\begin{vmatrix} 10 & 132000 \\ -3 & 0 \end{vmatrix} + 20000\begin{vmatrix} 10 & 5 \\ -3 & 1 \end{vmatrix}$$
$$= 1(-132000)-1(396000)+20000(25) = -28000$$

$$x = \frac{N_x}{D} = \frac{-20000}{-4} = 5000;\ y = \frac{N_y}{D} = \frac{-32000}{-4} = 8000;\ z = \frac{N_z}{D} = \frac{-28000}{-4} = 7000$$

He should invest \$5000 in HiTech, \$8000 in SaveTel and \$7000 in HiGas.

62. Let x = the 12-month amount, y = the 24-month amount and z = the 36-month amount. Form and solve the following system of equations:
$$\begin{cases} x+y+z=30000 \Rightarrow & x+y+z=30000 \\ 0.06x+0.07y+0.08z=\frac{7\frac{1}{3}}{100}(30000) \Rightarrow & 6x+7y+8z=220000 \\ z=5x \Rightarrow & -5x+z=0 \end{cases}$$

continued on next page...

62. continued

$$D=\begin{vmatrix} 1 & 1 & 1 \\ 6 & 7 & 8 \\ -5 & 0 & 1 \end{vmatrix} = 1\begin{vmatrix} 7 & 8 \\ 0 & 1 \end{vmatrix} - 1\begin{vmatrix} 6 & 8 \\ -5 & 1 \end{vmatrix} + 1\begin{vmatrix} 6 & 7 \\ -5 & 0 \end{vmatrix} = 1(7)-1(46)+1(35)=-4$$

$$N_x=\begin{vmatrix} 30000 & 1 & 1 \\ 220000 & 7 & 8 \\ 0 & 0 & 1 \end{vmatrix} = 30000\begin{vmatrix} 7 & 8 \\ 0 & 1 \end{vmatrix} - 1\begin{vmatrix} 220000 & 8 \\ 0 & 1 \end{vmatrix} + 1\begin{vmatrix} 220000 & 7 \\ 0 & 0 \end{vmatrix}$$

$$= 30000(7) - 1(220000) + 1(0) = -10000$$

$$N_y=\begin{vmatrix} 1 & 30000 & 1 \\ 6 & 220000 & 8 \\ -5 & 0 & 1 \end{vmatrix} = 1\begin{vmatrix} 220000 & 8 \\ 0 & 1 \end{vmatrix} - 30000\begin{vmatrix} 6 & 8 \\ -5 & 1 \end{vmatrix} + 1\begin{vmatrix} 6 & 220000 \\ -5 & 0 \end{vmatrix}$$

$$= 1(220000) - 30000(46) + 1(1100000) = -60000$$

$$N_z=\begin{vmatrix} 1 & 1 & 30000 \\ 6 & 7 & 220000 \\ -5 & 0 & 0 \end{vmatrix} = 1\begin{vmatrix} 7 & 220000 \\ 0 & 0 \end{vmatrix} - 1\begin{vmatrix} 6 & 220000 \\ -5 & 0 \end{vmatrix} + 30000\begin{vmatrix} 6 & 7 \\ -5 & 0 \end{vmatrix}$$

$$= 1(0) - 1(1100000) + 30000(35) = -50000$$

$$x=\frac{N_x}{D}=\frac{-10000}{-4}=2500;\ y=\frac{N_y}{D}=\frac{-60000}{-4}=15000;\ z=\frac{N_z}{D}=\frac{-50000}{-4}=12500$$

She should invest \$2500 for 12 months, \$15,000 for 24 months and \$12,500 for 36 months.

63. $\begin{vmatrix} 2 & -3 & 4 \\ -1 & 2 & 4 \\ 3 & -3 & 1 \end{vmatrix} = -23$

64. $\begin{vmatrix} -3 & 2 & -5 \\ 3 & -2 & 6 \\ 1 & -3 & 4 \end{vmatrix} = -7$

65. $\begin{vmatrix} 2 & 1 & -3 \\ -2 & 2 & 4 \\ 1 & -2 & 2 \end{vmatrix} = 26$

66. $\begin{vmatrix} 4 & 2 & -3 \\ 2 & -5 & 6 \\ 2 & 5 & -2 \end{vmatrix} = -108$

67. **Answers may vary.**

68. **Answers may vary.**

69.

$$\begin{vmatrix} x & y & 1 \\ -2 & 3 & 1 \\ 3 & 5 & 1 \end{vmatrix} = 0$$

$$x\begin{vmatrix} 3 & 1 \\ 5 & 1 \end{vmatrix} - y\begin{vmatrix} -2 & 1 \\ 3 & 1 \end{vmatrix} + 1\begin{vmatrix} -2 & 3 \\ 3 & 5 \end{vmatrix} = 0$$

$$x(3-5) - y(-2-3) + 1(-10-9) = 0$$

$$-2x + 5y - 19 = 0$$

$$2x - 5y = -19$$

Verify that both points satisfy this equation.

70. $\frac{1}{2}\begin{vmatrix} 0 & 0 & 1 \\ 3 & 0 & 1 \\ 0 & 4 & 1 \end{vmatrix} = \frac{1}{2}\left(0\begin{vmatrix} 0 & 1 \\ 4 & 1 \end{vmatrix} - 0\begin{vmatrix} 3 & 1 \\ 0 & 1 \end{vmatrix} + 1\begin{vmatrix} 3 & 0 \\ 0 & 4 \end{vmatrix}\right) = \frac{1}{2}(0-0+12) = \frac{1}{2}(12) = 6$

Since the triangle is a right triangle (graph the vertices), the area $= \frac{1}{2}bh = \frac{1}{2}(3)(4) = \frac{1}{2}(12) = 6$

71. $\begin{vmatrix} 1 & 0 & 2 & 1 \\ 2 & 1 & 1 & 3 \\ 1 & 1 & 1 & 1 \\ 2 & 1 & 1 & 1 \end{vmatrix} = 1\begin{vmatrix} 1 & 1 & 3 \\ 1 & 1 & 1 \\ 1 & 1 & 1 \end{vmatrix} - 0\begin{vmatrix} 2 & 1 & 3 \\ 1 & 1 & 1 \\ 2 & 1 & 1 \end{vmatrix} + 2\begin{vmatrix} 2 & 1 & 3 \\ 1 & 1 & 1 \\ 2 & 1 & 1 \end{vmatrix} - 1\begin{vmatrix} 2 & 1 & 1 \\ 1 & 1 & 1 \\ 2 & 1 & 1 \end{vmatrix}$

$= 1(0) - 0(???) + 2(-2) - 1(0) = -4$

72. $\begin{vmatrix} 1 & 2 & -1 & 1 \\ -2 & 1 & 3 & -1 \\ 0 & 1 & 1 & 2 \\ 2 & 0 & 3 & 1 \end{vmatrix} = 1\begin{vmatrix} 1 & 3 & -1 \\ 1 & 1 & 2 \\ 0 & 3 & 1 \end{vmatrix} - 2\begin{vmatrix} -2 & 3 & -1 \\ 0 & 1 & 2 \\ 2 & 3 & 1 \end{vmatrix} + (-1)\begin{vmatrix} -2 & 1 & -1 \\ 0 & 1 & 2 \\ 2 & 0 & 1 \end{vmatrix} - 1\begin{vmatrix} -2 & 1 & 3 \\ 0 & 1 & 1 \\ 2 & 0 & 3 \end{vmatrix}$

$= 1(-11) - 2(24) - 1(4) - 1(-10) = -53$

Chapter 3 Summary (page 211)

1. $\begin{cases} 2x + y = 11 \\ -x + 2y = 7 \end{cases}$

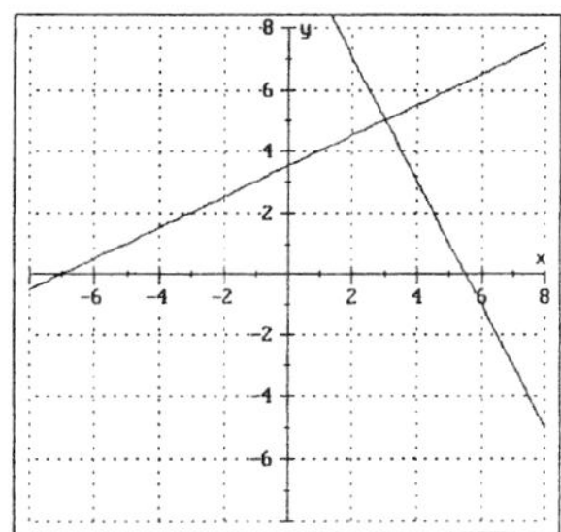

$(3, 5)$ is the solution.

2. $\begin{cases} 3x + 2y = 0 \\ 2x - 3y = -13 \end{cases}$

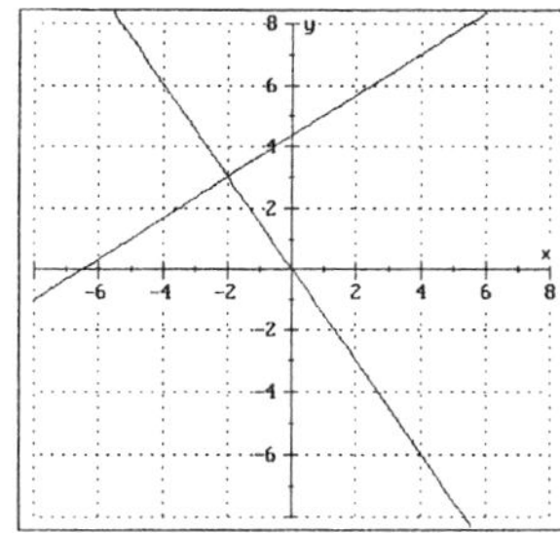

$(-2, 3)$ is the solution.

3. $\begin{cases} \frac{1}{2}x + \frac{1}{3}y = 2 \\ y = 6 - \frac{3}{2}x \end{cases}$

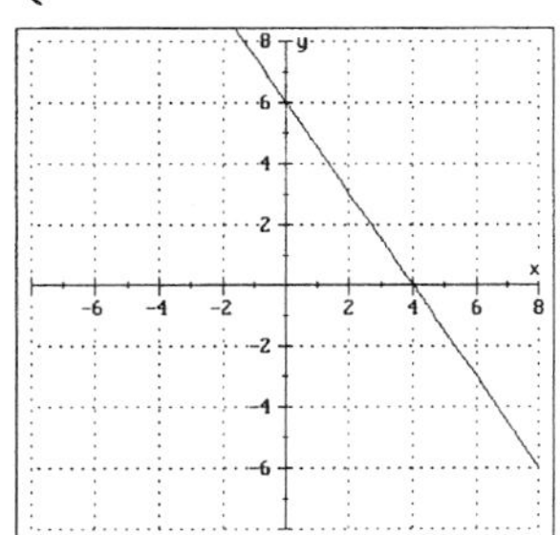

dependent equations

4. $\begin{cases} \frac{1}{3}x - \frac{1}{2}y = 1 \\ 6x - 9y = 2 \end{cases}$

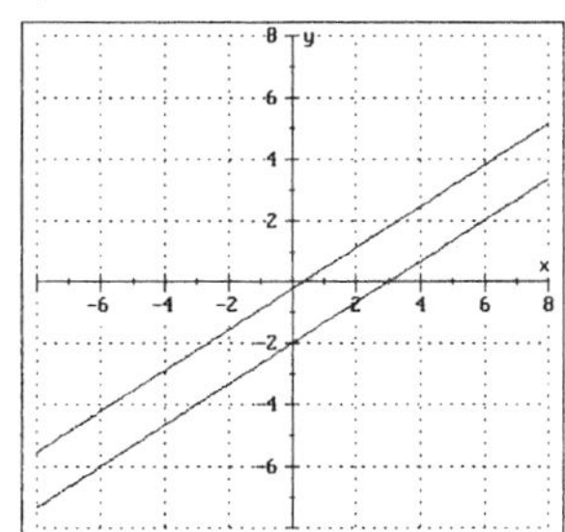

inconsistent system

5. $\begin{cases} (1) & y = x + 4 \\ (2) & 2x + 3y = 7 \end{cases}$

Substitute $y = x + 4$ from (1) into (2):

$$\begin{aligned} 2x + 3\mathbf{y} &= 7 \\ 2x + 3(\mathbf{x+4}) &= 7 \\ 2x + 3x + 12 &= 7 \\ 5x &= -5 \\ x &= -1 \end{aligned}$$

Substitute this and solve for y:

$y = x + 4 = -1 + 4 = 3$

Solution: $(-1, 3)$

6. $\begin{cases} (1) & y = 2x + 5 \\ (2) & 3x - 5y = -4 \end{cases}$

Substitute $y = 2x + 5$ from (1) into (2):

$$\begin{aligned} 3x - 5\mathbf{y} &= -4 \\ 3x - 5(\mathbf{2x+5}) &= -4 \\ 3x - 10x - 25 &= -4 \\ -7x &= 21 \\ x &= -3 \end{aligned}$$

Substitute this and solve for y:

$y = 2x + 5 = 2(-3) + 5 = -1$

Solution: $(-3, -1)$

7. $\begin{cases} (1) & x + 2y = 11 \\ (2) & 2x - y = 2 \end{cases}$

Substitute $x = -2y + 11$ from (1) into (2):

$$\begin{aligned} 2\mathbf{x} - y &= 2 \\ 2(\mathbf{-2y+11}) - y &= 2 \\ -4y + 22 - y &= 2 \\ -5y &= -20 \\ y &= 4 \end{aligned}$$

Substitute this and solve for x:

$x = -2y + 11 = -2(4) + 11 = 3$

Solution: $(3, 4)$

8. $\begin{cases} (1) & 2x + 3y = -2 \\ (2) & 3x + 5y = -2 \end{cases}$

Substitute $x = \frac{-3y-2}{2}$ from (1) into (2):

$$\begin{aligned} 3\mathbf{x} + 5y &= -2 \\ 3 \cdot \mathbf{\tfrac{-3y-2}{2}} + 5y &= -2 \\ 3(-3y - 2) + 10y &= -4 \\ -9y - 6 + 10y &= -4 \\ y &= 2 \end{aligned}$$

Substitute this and solve for y:

$x = \frac{-3y-2}{2} = \frac{-3(2)-2}{2} = \frac{-8}{2} = -4$

Solution: $(-4, 2)$

9.
$$\begin{array}{lcr} x + y = -2 \Rightarrow \times(-2) & & -2x - 2y = 4 \\ \underline{2x + 3y = -3} \Rightarrow & & \underline{2x + 3y = -3} \\ & & y = 1 \end{array}$$

Substitute and solve for x:

$x + y = -2$

$x + 1 = -2$

$x = -3$ Solution: $\boxed{(-3, 1)}$

10.
$$\begin{array}{lr} 3x + 2y = 1 \Rightarrow \times 3 & 9x + 6y = 3 \\ \underline{2x - 3y = 5} \Rightarrow \times 2 & \underline{4x - 6y = 10} \\ & 13x \quad = 13 \\ & x \quad = 1 \end{array}$$

Substitute and solve for y:

$$\begin{aligned} 3x + 2y &= 1 \\ 3(1) + 2y &= 1 \\ 3 + 2y &= 1 \\ 2y &= -2 \\ y &= -1 \end{aligned}$$

Solution: $\boxed{(1, -1)}$

11.
$$\begin{array}{lr} x + \frac{1}{2}y = 7 \Rightarrow \times 2 & 2x + y = 14 \Rightarrow \\ \underline{-2x = 3y - 6} \Rightarrow & \underline{-2x - 3y = -6} \Rightarrow \\ & -2y = 8 \\ & y = -4 \end{array}$$

Solve for x:

$$\begin{aligned} 2x + y &= 14 \\ 2x + (-4) &= 14 \\ 2x &= 18 \\ x &= 9 \end{aligned}$$

$\boxed{\text{Solution: } (9, -4)}$

12. $y = \dfrac{x-3}{2} \Rightarrow \times 2 \quad 2y = x - 3 \Rightarrow -x + 2y = -3$

$x = \dfrac{2y+7}{2} \Rightarrow \times 2 \quad 2x = 2y + 7 \Rightarrow 2x - 2y = 7$

$x = 4$

Solve for y:

$y = \dfrac{x-3}{2}$

$y = \dfrac{4-3}{2} = \dfrac{1}{2}$ Solution: $\boxed{(4, \frac{1}{2})}$

13.

$$\begin{array}{ll} (1) & x + y + z = 6 \\ (2) & x - y - z = -4 \\ (3) & -x + y - z = -2 \end{array} \qquad \begin{array}{ll} (1) & x + y + z = 6 \\ (2) & x - y - z = -4 \\ \hline (4) & 2x = 2 \\ & x = 1 \end{array} \qquad \begin{array}{ll} (1) & x + y + z = 6 \\ (3) & -x + y - z = -2 \\ \hline (5) & 2y = 4 \\ & y = 2 \end{array}$$

$$\begin{aligned} x + y + z &= 6 \\ 1 + 2 + z &= 6 \\ 3 + z &= 6 \\ z &= 3 \end{aligned}$$

Solution: $\boxed{(1, 2, 3)}$

14.

$$\begin{array}{ll} (1) & 2x + 3y + z = -5 \\ (2) & -x + 2y - z = -6 \\ (3) & 3x + y + 2z = 4 \end{array} \qquad \begin{array}{ll} (1) & 2x + 3y + z = -5 \\ (2) & -x + 2y - z = -6 \\ \hline (4) & x + 5y = -11 \end{array} \qquad \begin{array}{ll} (3) & 3x + y + 2z = 4 \\ 2 \cdot (2) & -2x + 4y - 2z = -12 \\ \hline (5) & x + 5y = -8 \end{array}$$

$$\begin{array}{ll} x + 5y = -11 \Rightarrow \times(-1) & -x - 5y = 11 \\ x + 5y = -8 \Rightarrow & x + 5y = -8 \\ \hline & 0 = 3 \end{array}$$

Since this is an impossible equation, there is no solution. It is an inconsistent system.

15.

$$\left[\begin{array}{cc|c} 1 & 2 & 4 \\ 2 & -1 & 3 \end{array}\right] \xRightarrow{-2R_1 + R_2 \Rightarrow R_2} \left[\begin{array}{cc|c} 1 & 2 & 4 \\ 0 & -5 & -5 \end{array}\right] \xRightarrow{-\frac{1}{5}R_2 \Rightarrow R_2} \left[\begin{array}{cc|c} 1 & 2 & 4 \\ 0 & 1 & 1 \end{array}\right]$$

From R_2, $y = 1$. From R_1:

$$\begin{aligned} x + 2y &= 4 \\ x + 2(1) &= 4 \\ x &= 2 \end{aligned}$$

Solution: $\boxed{(2, 1)}$

16.

$$\left[\begin{array}{ccc|c} 1 & 1 & 1 & 6 \\ 2 & -1 & 1 & 1 \\ 4 & 1 & -1 & 5 \end{array}\right] \xRightarrow[-4R_1 + R_3 \Rightarrow R_3]{-2R_1 + R_2 \Rightarrow R_2} \left[\begin{array}{ccc|c} 1 & 1 & 1 & 6 \\ 0 & -3 & -1 & -11 \\ 0 & -3 & -5 & -19 \end{array}\right] \xRightarrow[-R_3 + R_2 \Rightarrow R_3]{-R_2 \Rightarrow R_2} \left[\begin{array}{ccc|c} 1 & 1 & 1 & 6 \\ 0 & 3 & 1 & 11 \\ 0 & 0 & 4 & 8 \end{array}\right] \xRightarrow{\frac{1}{4}R_3 \Rightarrow R_3} \left[\begin{array}{ccc|c} 1 & 1 & 1 & 6 \\ 0 & 3 & 1 & 11 \\ 0 & 0 & 1 & 2 \end{array}\right]$$

From R_3, $z = 2$. From R_2:

$$\begin{aligned} 3y + z &= 11 \\ 3y + 2 &= 11 \\ 3y &= 9 \\ y &= 3 \end{aligned}$$

From R_1:

$$\begin{aligned} x + y + z &= 6 \\ x + 3 + 2 &= 6 \\ x &= 1 \end{aligned}$$

Solution: $\boxed{(1, 3, 2)}$

17. $$\left[\begin{array}{cc|c} 1 & 1 & 3 \\ 1 & -2 & -3 \\ 2 & 1 & 4 \end{array}\right] \overset{\substack{-R_1+R_2 \Rightarrow R_2 \\ -2R_1+R_3 \Rightarrow R_3}}{\Rightarrow} \left[\begin{array}{cc|c} 1 & 1 & 3 \\ 0 & -3 & -6 \\ 0 & -1 & -2 \end{array}\right] \overset{\substack{-\frac{1}{3}R_2 \Rightarrow R_2 \\ -3R_3+R_2 \Rightarrow R_3}}{\Rightarrow} \left[\begin{array}{cc|c} 1 & 1 & 3 \\ 0 & 1 & 2 \\ 0 & 0 & 0 \end{array}\right]$$

From R_2, $y = 2$. From R_1:
$$x + y = 3$$
$$x + 2 = 3$$
$$x = 1$$

Solution: $\boxed{(1, 2)}$

18. $$\left[\begin{array}{ccc|c} 1 & 2 & 1 & 2 \\ 2 & 5 & 4 & 5 \end{array}\right] \overset{-2R_1+R_2 \Rightarrow R_2}{\Rightarrow} \left[\begin{array}{ccc|c} 1 & 2 & 1 & 2 \\ 0 & 1 & 2 & 1 \end{array}\right]$$

From R_2:
$$y + 2z = 1$$
$$y = 1 - 2z$$

From R_1:
$$x + 2y + z = 2$$
$$x + 2(1 - 2z) + z = 2$$
$$x + 2 - 4z + z = 2$$
$$x - 3z = 0$$
$$x = 3z$$

Solution: $\boxed{(3z, 1 - 2z, z)}$

19. $\begin{vmatrix} 2 & 3 \\ -4 & 3 \end{vmatrix} = 2(3) - 3(-4) = 6 + 12 = 18$

20. $\begin{vmatrix} -3 & -4 \\ 5 & -6 \end{vmatrix} = -3(-6) - (-4)(5) = 18 - (-20) = 18 + 20 = 38$

21. $\begin{vmatrix} -1 & 2 & -1 \\ 2 & -1 & 3 \\ 1 & -2 & 2 \end{vmatrix} = -1\begin{vmatrix} -1 & 3 \\ -2 & 2 \end{vmatrix} - 2\begin{vmatrix} 2 & 3 \\ 1 & 2 \end{vmatrix} + (-1)\begin{vmatrix} 2 & -1 \\ 1 & -2 \end{vmatrix} = -1(4) - 2(1) - 1(-3) = -3$

22. $\begin{vmatrix} 3 & -2 & 2 \\ 1 & -2 & -2 \\ 2 & 1 & -1 \end{vmatrix} = 3\begin{vmatrix} -2 & -2 \\ 1 & -1 \end{vmatrix} - (-2)\begin{vmatrix} 1 & -2 \\ 2 & -1 \end{vmatrix} + 2\begin{vmatrix} 1 & -2 \\ 2 & 1 \end{vmatrix} = 3(4) + 2(3) + 2(5) = 28$

23. $x = \dfrac{\begin{vmatrix} 10 & 4 \\ 1 & -3 \end{vmatrix}}{\begin{vmatrix} 3 & 4 \\ 2 & -3 \end{vmatrix}} = \dfrac{-30 - 4}{-9 - 8} = \dfrac{-34}{-17} = 2; y = \dfrac{\begin{vmatrix} 3 & 10 \\ 2 & 1 \end{vmatrix}}{\begin{vmatrix} 3 & 4 \\ 2 & -3 \end{vmatrix}} = \dfrac{3 - 20}{-17} = \dfrac{-17}{-17} = 1$

Solution: $\boxed{(2, 1)}$

24. $x = \dfrac{\begin{vmatrix} -17 & -5 \\ 3 & 2 \end{vmatrix}}{\begin{vmatrix} 2 & -5 \\ 3 & 2 \end{vmatrix}} = \dfrac{-34+15}{4+15} = \dfrac{-19}{19} = -1$; $y = \dfrac{\begin{vmatrix} 2 & -17 \\ 3 & 3 \end{vmatrix}}{\begin{vmatrix} 2 & -5 \\ 3 & 2 \end{vmatrix}} = \dfrac{6+51}{19} = \dfrac{57}{19} = 3$

Solution: $\boxed{(-1,3)}$

25. $D = \begin{vmatrix} 1 & 2 & 1 \\ 2 & 1 & 1 \\ 1 & 1 & 2 \end{vmatrix} = 1\begin{vmatrix} 1 & 1 \\ 1 & 2 \end{vmatrix} - 2\begin{vmatrix} 2 & 1 \\ 1 & 2 \end{vmatrix} + 1\begin{vmatrix} 2 & 1 \\ 1 & 1 \end{vmatrix} = 1(1) - 2(3) + 1(1) = -4$

$N_x = \begin{vmatrix} 0 & 2 & 1 \\ 3 & 1 & 1 \\ 5 & 1 & 2 \end{vmatrix} = 0\begin{vmatrix} 1 & 1 \\ 1 & 2 \end{vmatrix} - 2\begin{vmatrix} 3 & 1 \\ 5 & 2 \end{vmatrix} + 1\begin{vmatrix} 3 & 1 \\ 5 & 1 \end{vmatrix} = 0(1) - 2(1) + 1(-2) = -4$

$N_y = \begin{vmatrix} 1 & 0 & 1 \\ 2 & 3 & 1 \\ 1 & 5 & 2 \end{vmatrix} = 1\begin{vmatrix} 3 & 1 \\ 5 & 2 \end{vmatrix} - 0\begin{vmatrix} 2 & 1 \\ 1 & 2 \end{vmatrix} + 1\begin{vmatrix} 2 & 3 \\ 1 & 5 \end{vmatrix} = 1(1) - 0(3) + 1(7) = 8$

$N_z = \begin{vmatrix} 1 & 2 & 0 \\ 2 & 1 & 3 \\ 1 & 1 & 5 \end{vmatrix} = 1\begin{vmatrix} 1 & 3 \\ 1 & 5 \end{vmatrix} - 2\begin{vmatrix} 2 & 3 \\ 1 & 5 \end{vmatrix} + 0\begin{vmatrix} 2 & 1 \\ 1 & 1 \end{vmatrix} = 1(2) - 2(7) + 0(1) = -12$

$x = \dfrac{N_x}{D} = \dfrac{-4}{-4} = 1$; $y = \dfrac{N_y}{D} = \dfrac{8}{-4} = -2$; $z = \dfrac{N_z}{D} = \dfrac{-12}{-4} = 3 \Rightarrow$ solution: $(1,-2,3)$

26. $D = \begin{vmatrix} 2 & 3 & 1 \\ 1 & 3 & 2 \\ 1 & -1 & -1 \end{vmatrix} = 2\begin{vmatrix} 3 & 2 \\ -1 & -1 \end{vmatrix} - 3\begin{vmatrix} 1 & 2 \\ 1 & -1 \end{vmatrix} + 1\begin{vmatrix} 1 & 3 \\ 1 & -1 \end{vmatrix} = 2(-1) - 3(-3) + 1(-4)$

$= 3$

$N_x = \begin{vmatrix} 2 & 3 & 1 \\ 7 & 3 & 2 \\ -7 & -1 & -1 \end{vmatrix} = 2\begin{vmatrix} 3 & 2 \\ -1 & -1 \end{vmatrix} - 3\begin{vmatrix} 7 & 2 \\ -7 & -1 \end{vmatrix} + 1\begin{vmatrix} 7 & 3 \\ -7 & -1 \end{vmatrix}$

$= 2(-1) - 3(7) + 1(14) = -9$

$N_y = \begin{vmatrix} 2 & 2 & 1 \\ 1 & 7 & 2 \\ 1 & -7 & -1 \end{vmatrix} = 2\begin{vmatrix} 7 & 2 \\ -7 & -1 \end{vmatrix} - 2\begin{vmatrix} 1 & 2 \\ 1 & -1 \end{vmatrix} + 1\begin{vmatrix} 1 & 7 \\ 1 & -7 \end{vmatrix}$

$= 2(7) - 2(-3) + 1(-14) = 6$

$N_z = \begin{vmatrix} 2 & 3 & 2 \\ 1 & 3 & 7 \\ 1 & -1 & -7 \end{vmatrix} = 2\begin{vmatrix} 3 & 7 \\ -1 & -7 \end{vmatrix} - 3\begin{vmatrix} 1 & 7 \\ 1 & -7 \end{vmatrix} + 2\begin{vmatrix} 1 & 3 \\ 1 & -1 \end{vmatrix}$

$= 2(-14) - 3(-14) + 2(-4) = 6$

$x = \dfrac{N_x}{D} = \dfrac{-9}{3} = -3$; $y = \dfrac{N_y}{D} = \dfrac{6}{3} = 2$; $z = \dfrac{N_z}{D} = \dfrac{6}{3} = 2 \Rightarrow$ solution: $(-3,2,2)$

Chapter 3 Test (page 213)

1. $\begin{cases} 2x + y = 5 \\ y = 2x - 3 \end{cases}$

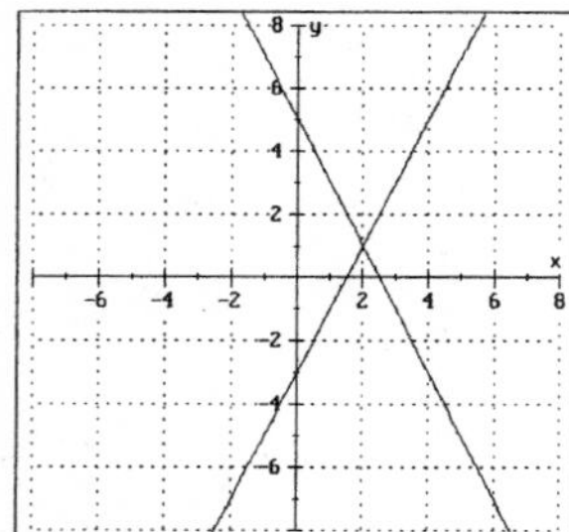

$(2, 1)$ is the solution.

2. $\begin{cases} (1) & 2x - 4y = 14 \\ (2) & x = -2y + 7 \end{cases}$

Substitute $x = -2y + 7$ from (2) into (1):

$$\begin{aligned} 2x - 4y &= 14 \\ 2(\mathbf{-2y + 7}) - 4y &= 14 \\ -4y + 14 - 4y &= 14 \\ -8y &= 0 \\ y &= 0 \end{aligned}$$

Substitute this and solve for x:

$$\begin{aligned} x &= -2y + 7 \\ x &= -2(0) + 7 = 7 \end{aligned}$$

Solution: $(7, 0)$

3.
$$\begin{aligned} 2x + 3y &= -5 \Rightarrow \times 2 \\ 3x - 2y &= \ \ 12 \Rightarrow \times 3 \end{aligned} \qquad \begin{aligned} 4x + 6y &= -10 \\ 9x - 6y &= 36 \\ \hline 13x &= 26 \\ x &= 2 \end{aligned} \qquad \begin{aligned} 2x + 3y &= -5 \\ 2(2) + 3y &= -5 \\ 4 + 3y &= -5 \\ 3y &= -9 \\ y &= -3 \end{aligned}$$

Solution: $\boxed{(2, -3)}$

4.
$$\begin{aligned} \frac{x}{2} - \frac{y}{4} &= -4 \Rightarrow \times 4 \\ x + y &= -2 \Rightarrow \end{aligned} \qquad \begin{aligned} 2x - y &= -16 \\ x + y &= -2 \\ \hline 3x &= -18 \\ x &= -6 \end{aligned} \qquad \begin{aligned} x + y &= -2 \\ -6 + y &= -2 \\ y &= 4 \end{aligned}$$

Solution: $\boxed{(-6, 4)}$

5.
$$\begin{aligned} 3(x + y) &= x - 3 \Rightarrow \\ -y &= \frac{2x + 3}{3} \Rightarrow \end{aligned} \qquad \begin{aligned} 2x + 3y &= -3 \\ 2x + 3y &= -3 \end{aligned} \qquad \begin{aligned} 2x + 3y &= -3 \Rightarrow \\ 2x + 3y &= -3 \Rightarrow \times(-1) \end{aligned} \qquad \begin{aligned} 2x + 3y &= -3 \\ -2x - 3y &= 3 \\ \hline 0 &= 0 \end{aligned}$$

The equation is an identity, so the system has infinitely many solutions. $\Rightarrow$ dependent equations

6. See #5. Since the system has at least one solution, it is a consistent system.

7. $\begin{bmatrix} 1 & 2 & -1 \\ 2 & -2 & 3 \end{bmatrix} \Rightarrow \overset{-3R_1 + R_2 \Rightarrow R_2}{\begin{bmatrix} 1 & 2 & -1 \\ -1 & -8 & \boxed{6} \end{bmatrix}}$

8. $\begin{bmatrix} -1 & 3 & 6 \\ 3 & -2 & 4 \end{bmatrix} \Rightarrow \overset{-2R_1 + R_2 \Rightarrow R_2}{\begin{bmatrix} -1 & 3 & 6 \\ 5 & -8 & \boxed{-8} \end{bmatrix}}$

9. $\left[\begin{array}{ccc|c} 1 & 1 & 1 & 4 \\ 1 & 1 & -1 & 6 \\ 2 & -3 & 1 & -1 \end{array}\right]$

10. $\begin{bmatrix} 1 & 1 & 1 \\ 1 & 1 & -1 \\ 2 & -3 & 1 \end{bmatrix}$

11. $$\left[\begin{array}{cc|c}1 & 1 & 4\\ 2 & -1 & 2\end{array}\right] \overset{-2R_1+R_2 \Rightarrow R_2}{\Rightarrow} \left[\begin{array}{cc|c}1 & 1 & 4\\ 0 & -3 & -6\end{array}\right] \overset{-\frac{1}{3}R_2 \Rightarrow R_2}{\Rightarrow} \left[\begin{array}{cc|c}1 & 1 & 4\\ 0 & 1 & 2\end{array}\right]$$

From R_2, $y = 2$. From R_1:

$$x + y = 4$$
$$x + 2 = 4$$
$$x = 2$$

Solution: $\boxed{(2,2)}$

12. $$\left[\begin{array}{cc|c}1 & 1 & 2\\ 1 & -1 & -4\\ 2 & 1 & 1\end{array}\right] \overset{\substack{-R_2+R_1 \Rightarrow R_2\\ -2R_1+R_3 \Rightarrow R_3}}{\Rightarrow} \left[\begin{array}{cc|c}1 & 1 & 2\\ 0 & 2 & 6\\ 0 & -1 & -3\end{array}\right] \overset{\substack{\frac{1}{2}R_2 \Rightarrow R_2\\ 2R_3+R_2 \Rightarrow R_3}}{\Rightarrow} \left[\begin{array}{cc|c}1 & 1 & 2\\ 0 & 1 & 3\\ 0 & 0 & 0\end{array}\right]$$

From R_2, $y = 3$. From R_1:

$$x + y = 2$$
$$x + 3 = 2$$
$$x = -1$$

Solution: $\boxed{(-1,3)}$

13. $$\begin{vmatrix}2 & -3\\ 4 & 5\end{vmatrix} = 2(5) - (-3)(4) = 10 - (-12) = 22$$

14. $$\begin{vmatrix}-3 & -4\\ -2 & 3\end{vmatrix} = -3(3) - (-4)(-2) = -9 - 8 = -17$$

15. $$\begin{vmatrix}1 & 2 & 0\\ 2 & 0 & 3\\ 1 & -2 & 2\end{vmatrix} = 1\begin{vmatrix}0 & 3\\ -2 & 2\end{vmatrix} - 2\begin{vmatrix}2 & 3\\ 1 & 2\end{vmatrix} + 0\begin{vmatrix}2 & 0\\ 1 & -2\end{vmatrix} = 1(6) - 2(1) + 0(-4) = 4$$

16. $$\begin{vmatrix}2 & -1 & 1\\ 3 & 1 & 0\\ 0 & 1 & 2\end{vmatrix} = 2\begin{vmatrix}1 & 0\\ 1 & 2\end{vmatrix} - (-1)\begin{vmatrix}3 & 0\\ 0 & 2\end{vmatrix} + 1\begin{vmatrix}3 & 1\\ 0 & 1\end{vmatrix} = 2(2) + 1(6) + 1(3) = 13$$

17. $$\begin{vmatrix}-6 & -1\\ -6 & 1\end{vmatrix}$$

18. $$\begin{vmatrix}1 & -1\\ 3 & 1\end{vmatrix}$$

19. $$x = \frac{\begin{vmatrix}-6 & -1\\ -6 & 1\end{vmatrix}}{\begin{vmatrix}1 & -1\\ 3 & 1\end{vmatrix}} = \frac{-6-6}{1-(-3)} = \frac{-12}{4} = -3$$

20. $$y = \frac{\begin{vmatrix}1 & -6\\ 3 & -6\end{vmatrix}}{\begin{vmatrix}1 & -1\\ 3 & 1\end{vmatrix}} = \frac{-6-(-18)}{1-(-3)} = \frac{12}{4} = 3$$

21. $D = \begin{vmatrix} 1 & 1 & 1 \\ 1 & 1 & -1 \\ 2 & -3 & 1 \end{vmatrix} = 1\begin{vmatrix} 1 & -1 \\ -3 & 1 \end{vmatrix} - 1\begin{vmatrix} 1 & -1 \\ 2 & 1 \end{vmatrix} + 1\begin{vmatrix} 1 & 1 \\ 2 & -3 \end{vmatrix} = 1(-2) - 1(3) + 1(-5) = -10$

$N_x = \begin{vmatrix} 4 & 1 & 1 \\ 6 & 1 & -1 \\ -1 & -3 & 1 \end{vmatrix} = 4\begin{vmatrix} 1 & -1 \\ -3 & 1 \end{vmatrix} - 1\begin{vmatrix} 6 & -1 \\ -1 & 1 \end{vmatrix} + 1\begin{vmatrix} 6 & 1 \\ -1 & -3 \end{vmatrix}$

$= 4(-2) - 1(5) + 1(-17) = -30$

$x = \dfrac{N_x}{D} = \dfrac{-30}{-10} = 3$

22. See #21. $D = -10$.

$N_z = \begin{vmatrix} 1 & 1 & 4 \\ 1 & 1 & 6 \\ 2 & -3 & -1 \end{vmatrix} = 1\begin{vmatrix} 1 & 6 \\ -3 & -1 \end{vmatrix} - 1\begin{vmatrix} 1 & 6 \\ 2 & -1 \end{vmatrix} + 4\begin{vmatrix} 1 & 1 \\ 2 & -3 \end{vmatrix} = 1(17) - 1(-13) + 4(-5) = 10$

$z = \dfrac{N_z}{D} = \dfrac{10}{-10} = -1$

Exercise 4.1 (page 223)

1. $\left(\dfrac{t^3t^5t^{-6}}{t^2t^{-4}}\right)^{-3} = \left(\dfrac{t^2t^{-4}}{t^3t^5t^{-6}}\right)^3 = \left(\dfrac{t^{-2}}{t^2}\right)^3 = \left(\dfrac{1}{t^4}\right)^3 = \dfrac{1}{t^{12}}$

2. $\left(\dfrac{a^{-2}b^3a^5b^{-2}}{a^6b^{-5}}\right)^{-4} = \left(\dfrac{a^6b^{-5}}{a^{-2}b^3a^5b^{-2}}\right)^4 = \left(\dfrac{a^6b^{-5}}{a^3b}\right)^4 = \left(\dfrac{a^3}{b^6}\right)^4 = \dfrac{a^{12}}{b^{24}}$

3. Let $x =$ the number of pies made.

$\boxed{\text{Expenses}} = \boxed{\text{Income}}$

$1200 + 3.40x = 5.95x$

$1200 = 2.55x$

$470.59 = x$

He must sell at least 471 pies.

4. Let $x =$ the amount at 7%. Then $15000 - x$ is the amount at 8%.

$0.07x(2) + 0.08(15000 - x)(2) = 2200$

$0.14x + 2400 - 0.16x = 2200$

$-0.02x = -200$

$x = 10000$

She invested \$10,000 at 7% interest.

5. $\neq$

6. $>$

7. $<$

8. $\leq$

9. $\geq$

10. $a = b; a > b$

11. $a < c$

12. positive

13. reversed

14. linear

15. $c < x; x < d$

16. one

17. open

18. closed

19.
$$x + 4 < 5$$
$$x + 4 - 4 < 5 - 4$$
$$x < 1$$
solution set: $\{x|x < 1\}$

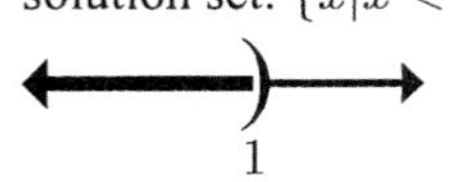

20.
$$x - 5 > 2$$
$$x - 5 + 5 > 2 + 5$$
$$x > 7$$
solution set: $\{x|x > 7\}$

7

21.
$$x + 2 \geq -3$$
$$x + 2 - 2 \geq -3 - 2$$
$$x \geq -5$$
solution set: $\{x|x \geq -5\}$

−5

22.
$$x - 3 \leq 5$$
$$x - 3 + 3 \leq 5 + 3$$
$$x \leq 8$$
solution set: $\{x|x \leq 8\}$

8

23.
$$2x + 3 < 9$$
$$2x + 3 - 3 < 9 - 3$$
$$2x < 6$$
$$\frac{2x}{2} < \frac{6}{2}$$
$$x < 3$$
solution set: $\{x|x < 3\}$

3

24.
$$5x - 1 \geq 19$$
$$5x - 1 + 1 \geq 19 + 1$$
$$5x \geq 20$$
$$\frac{5x}{5} \geq \frac{20}{5}$$
$$x \geq 4$$
solution set: $\{x|x \geq 4\}$

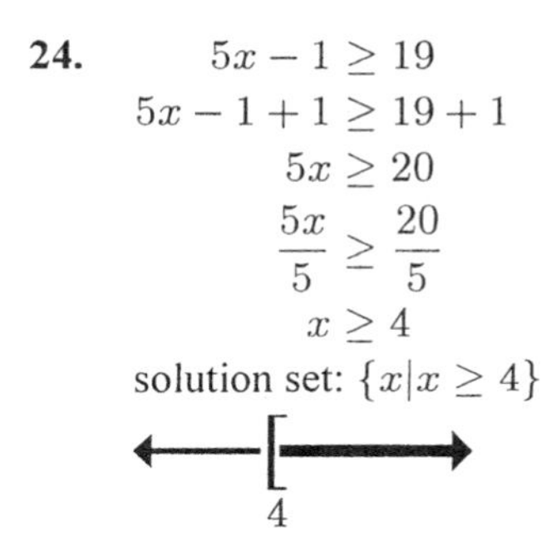

25.
$$-3x - 1 \leq 5$$
$$-3x - 1 + 1 \leq 5 + 1$$
$$-3x \leq 6$$
$$\frac{-3x}{-3} \geq \frac{6}{-3}$$
$$x \geq -2$$
solution set: $\{x|x \geq -2\}$

−2

26.
$$-2x + 6 \geq 16$$
$$-2x + 6 - 6 \geq 16 - 6$$
$$-2x \geq 10$$
$$\frac{-2x}{-2} \leq \frac{10}{-2}$$
$$x \leq -5$$
solution set: $\{x|x \leq -5\}$

−5

27.
$$5x - 3 > 7$$
$$5x - 3 + 3 > 7 + 3$$
$$5x > 10$$
$$\frac{5x}{5} > \frac{10}{5}$$
$$x > 2$$
solution set: $\{x|x > 2\}$

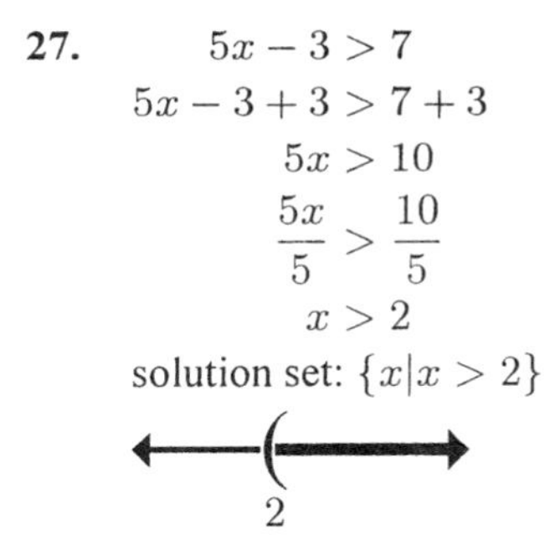

28.
$$7x - 9 < 5$$
$$7x - 9 + 9 < 5 + 9$$
$$7x < 14$$
$$\frac{7x}{7} < \frac{14}{7}$$
$$x < 2$$
solution : $\{x|x < 2\}$
set

2

29.
$$-4 < 2x < 8$$
$$\frac{-4}{2} < \frac{2x}{2} < \frac{8}{2}$$
$$-2 < x < 4$$
solution : $\{x|-2 < x < 4\}$
set

−2 4

30.
$$-3 \leq 3x < 12$$
$$\frac{-3}{3} \leq \frac{3x}{3} < \frac{12}{3}$$
$$-1 \leq x < 4$$
solution : $\{x|-1 \leq x < 4\}$
set

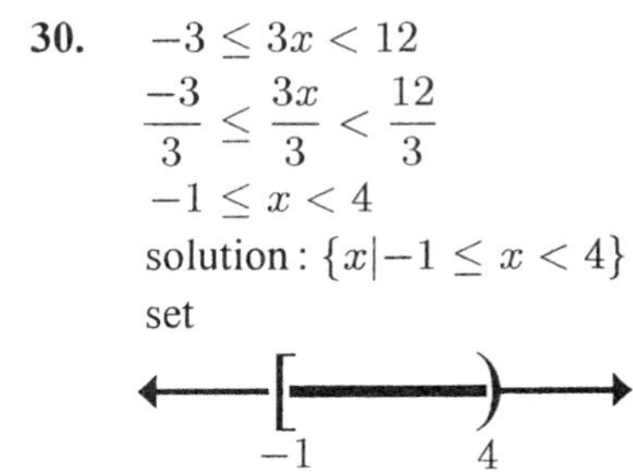

31.

$$\begin{aligned} 8x + 30 &> -2x \\ 8x + 2x + 30 &> -2x + 2x \\ 10x + 30 - 30 &> 0 - 30 \\ 10x &> -30 \\ \frac{10x}{10} &> \frac{-30}{10} \\ x &> -3 \end{aligned}$$

solution set: $(-3, \infty)$

32.

$$\begin{aligned} 5x - 24 &\le 6 \\ 5x - 24 + 24 &\le 6 + 24 \\ 5x &\le 30 \\ \frac{5x}{5} &\le \frac{30}{5} \\ x &\le 6 \end{aligned}$$

solution set: $(-\infty, 6]$

33.

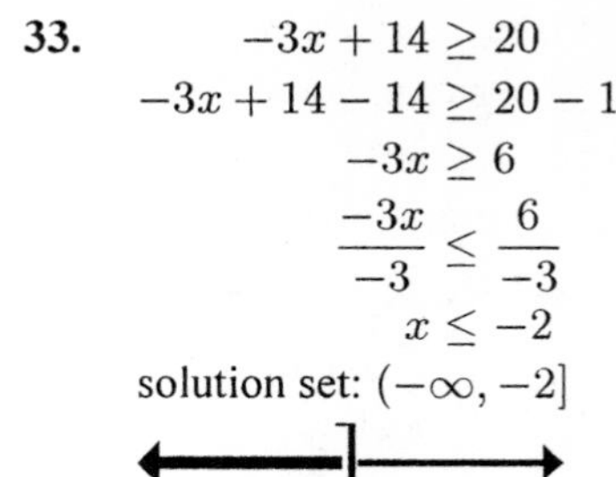

$$\begin{aligned} -3x + 14 &\ge 20 \\ -3x + 14 - 14 &\ge 20 - 14 \\ -3x &\ge 6 \\ \frac{-3x}{-3} &\le \frac{6}{-3} \\ x &\le -2 \end{aligned}$$

solution set: $(-\infty, -2]$

34.

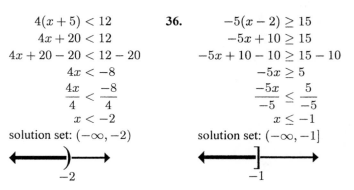

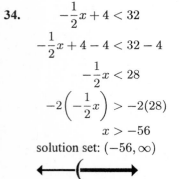

$$\begin{aligned} -\frac{1}{2}x + 4 &< 32 \\ -\frac{1}{2}x + 4 - 4 &< 32 - 4 \\ -\frac{1}{2}x &< 28 \\ -2\left(-\frac{1}{2}x\right) &> -2(28) \\ x &> -56 \end{aligned}$$

solution set: $(-56, \infty)$

35.

$$\begin{aligned} 4(x + 5) &< 12 \\ 4x + 20 &< 12 \\ 4x + 20 - 20 &< 12 - 20 \\ 4x &< -8 \\ \frac{4x}{4} &< \frac{-8}{4} \\ x &< -2 \end{aligned}$$

solution set: $(-\infty, -2)$

36.

$$\begin{aligned} -5(x - 2) &\ge 15 \\ -5x + 10 &\ge 15 \\ -5x + 10 - 10 &\ge 15 - 10 \\ -5x &\ge 5 \\ \frac{-5x}{-5} &\le \frac{5}{-5} \\ x &\le -1 \end{aligned}$$

solution set: $(-\infty, -1]$

37.

$$\begin{aligned} 3(z - 2) &\le 2(z + 7) \\ 3z - 6 &\le 2z + 14 \\ 3z - 6 + 6 &\le 2z + 14 + 6 \\ 3x - 2z &\le 2z - 2z + 20 \\ z &\le 20 \end{aligned}$$

solution set: $(-\infty, 20]$

38.

$$\begin{aligned} 5(3 + z) &> -3(z + 3) \\ 15 + 5z &> -3z - 9 \\ 15 - 15 + 5z &> -3z - 9 - 15 \\ 5z + 3z &> -3z + 3z - 24 \\ 8z &> -24 \\ \frac{8z}{8} &> \frac{-24}{8} \\ z &> -3 \end{aligned}$$

solution set: $(-3, \infty)$

39.
$$
\begin{aligned}
-11(2-b) &< 4(2b+2)\\
-22+11b &< 8b+8\\
-22+22+11b &< 8b+8+22\\
11b-8b &< 8b-8b+30\\
3b &< 30\\
\frac{3b}{3} &< \frac{30}{3}\\
b &< 10
\end{aligned}
$$
solution set: $(-\infty, 10)$

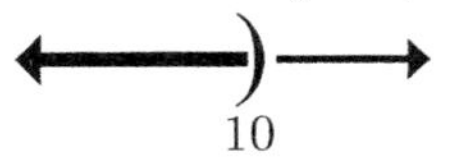

40.
$$
\begin{aligned}
-9(h-3)+2h &\le 8(4-h)\\
-9h+27+2h &\le 32-8h\\
-7h+27 &\le 32-8h\\
-7h+27-27 &\le 32-27-8h\\
-7h+8h &\le 5-8h+8h\\
h &\le 5
\end{aligned}
$$
solution set: $(-\infty, 5]$

5

41.
$$
\begin{aligned}
\frac{1}{2}y+2 &\ge \frac{1}{3}y-4\\
6\left(\frac{1}{2}y+2\right) &\ge 6\left(\frac{1}{3}y-4\right)\\
3y+12 &\ge 2y-24\\
3y+12-12 &\ge 2y-24-12\\
3y-2y &\ge 2y-2y-36\\
y &\ge -36
\end{aligned}
$$
solution set: $[-36, \infty)$

-36

42.
$$
\begin{aligned}
\frac{1}{4}x-\frac{1}{3} &\le x+2\\
12\left(\frac{1}{4}x-\frac{1}{3}\right) &\le 12(x+2)\\
3x-4 &\le 12x+24\\
3x-4+4 &\le 12x+24+4\\
3x-12x &\le 12x-12x+28\\
-9x &\le 28\\
\frac{-9x}{-9} &\ge \frac{28}{-9}\\
x &\ge -\tfrac{28}{9}
\end{aligned}
$$
solution set: $[-\tfrac{28}{9}, \infty)$

$-\frac{28}{9}$

43.

$$
\begin{aligned}
\frac{2}{3}x + \frac{3}{2}(x-5) &\le x \\
6\left[\frac{2}{3}x + \frac{3}{2}(x-5)\right] &\le 6x \\
4x + 9(x-5) &\le 6x \\
4x + 9x - 45 &\le 6x \\
13x - 45 + 45 &\le 6x + 45 \\
13x - 6x &\le 6x - 6x + 45 \\
7x &\le 45 \\
\frac{7x}{7} &\le \frac{45}{7} \\
x &\le \tfrac{45}{7}
\end{aligned}
$$

solution set: $\left(-\infty, \tfrac{45}{7}\right]$

$\frac{45}{7}$

44.

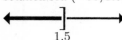

$$
\begin{aligned}
\frac{5}{9}(x+3) - \frac{4}{3}(x-3) &\ge x - 1 \\
9\left[\frac{5}{9}(x+3) - \frac{4}{3}(x-3)\right] &\ge 9(x-1) \\
5(x+3) - 12(x-3) &\ge 9x - 9 \\
5x + 15 - 12x + 36 &\ge 9x - 9 \\
-7x + 51 &\ge 9x - 9 \\
-7x + 51 - 51 &\ge 9x - 9 - 51 \\
-7x - 9x &\ge 9x - 9x - 60 \\
-16x &\ge -60 \\
\frac{-16x}{-16} &\le \frac{-60}{-16} \\
x &\le \tfrac{15}{4}
\end{aligned}
$$

solution set: $\left(-\infty, \tfrac{15}{4}\right]$

$\frac{15}{4}$

45.

$$
\begin{aligned}
0.4x + 0.4 &\le 0.1x + 0.85 \\
0.4x - 0.1x + 0.4 &\le 0.1x - 0.1x + 0.85 \\
0.3x + 0.4 - 0.4 &\le 0.85 - 0.4 \\
0.3x &\le 0.45 \\
\frac{0.3x}{0.3} &\le \frac{0.45}{0.3} \\
x &\le 1.5
\end{aligned}
$$

solution set: $(-\infty, 1.5]$

1.5

46.

$$
\begin{aligned}
0.05 - 0.5x &\ge -0.7 - 0.8x \\
0.05 - 0.5x + 0.8x &\ge -0.7 - 0.8x + 0.8x \\
0.05 - 0.05 + 0.3x &\ge -0.7 - 0.05 \\
0.3x &\ge -0.75 \\
\frac{0.3x}{0.3} &\ge \frac{-0.75}{0.3} \\
x &\ge -2.5
\end{aligned}
$$

solution set: $[-2.5, \infty)$

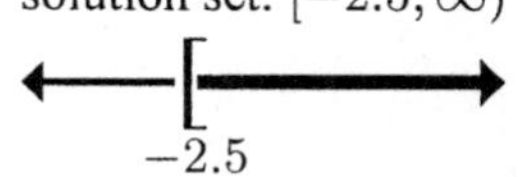

47.

$$
\begin{aligned}
-2 < -b + 3 &< 5 \\
-2 - 3 < -b + 3 - 3 &< 5 - 3 \\
-5 < -b &< 2 \\
\frac{-5}{-1} > \frac{-b}{-1} &> \frac{2}{-1} \\
5 > b > -2, \text{ or } -2 < b &< 5
\end{aligned}
$$

solution set: $(-2, 5)$

−2 5

48.

$$
\begin{aligned}
2 < -t - 2 &< 9 \\
2 + 2 < -t - 2 + 2 &< 9 + 2 \\
4 < -t &< 11 \\
\frac{4}{-1} > \frac{-t}{-1} &> \frac{11}{-1} \\
-4 > t > -11, \text{ or } -11 < t &< -4
\end{aligned}
$$

solution set: $(-11, -4)$

−11 −4

49.

$$15 > 2x - 7 > 9$$
$$15 + 7 > 2x - 7 + 7 > 9 + 7$$
$$22 > 2x > 16$$
$$\frac{22}{2} > \frac{2x}{2} > \frac{16}{2}$$
$$11 > x > 8, \text{ or } 8 < x < 11$$

solution set: $(8, 11)$

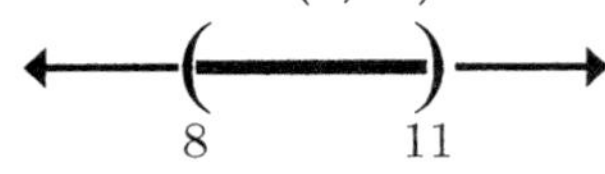

50.

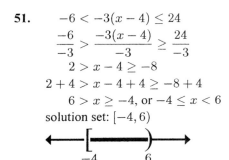

$$25 > 3x - 2 > 7$$
$$25 + 2 > 3x - 2 + 2 > 7 + 2$$
$$27 > 3x > 9$$
$$\frac{27}{3} > \frac{3x}{3} > \frac{9}{3}$$
$$9 > x > 3, \text{ or } 3 < x < 9$$

solution set: $(3, 9)$

3 9

51.

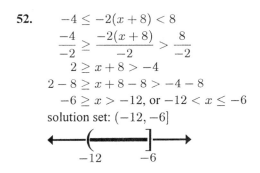

$$-6 < -3(x - 4) \leq 24$$
$$\frac{-6}{-3} > \frac{-3(x-4)}{-3} \geq \frac{24}{-3}$$
$$2 > x - 4 \geq -8$$
$$2 + 4 > x - 4 + 4 \geq -8 + 4$$
$$6 > x \geq -4, \text{ or } -4 \leq x < 6$$

solution set: $[-4, 6)$

−4 6

52.

$$-4 \leq -2(x + 8) < 8$$
$$\frac{-4}{-2} \geq \frac{-2(x+8)}{-2} > \frac{8}{-2}$$
$$2 \geq x + 8 > -4$$
$$2 - 8 \geq x + 8 - 8 > -4 - 8$$
$$-6 \geq x > -12, \text{ or } -12 < x \leq -6$$

solution set: $(-12, -6]$

−12 −6

53. $0 \geq \frac{1}{2}x - 4 > 6$

This inequality indicates that $0 \geq 6$ (by the transitive property). Since this is not possible, there is no solution to the inequality.

54.

$$-6 \leq \frac{1}{3}a + 1 < 0$$
$$3(-6) \leq 3\left(\frac{1}{3}a + 1\right) < 3(0)$$
$$-18 \leq a + 3 < 0$$
$$-18 - 3 \leq a + 3 - 3 < 0 - 3$$
$$-21 \leq a < -3$$

solution set: $[-21, -3)$

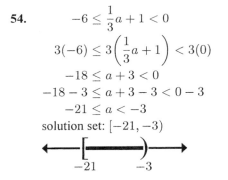

−21 −3

55.

$$\begin{aligned}
0 &\le \frac{4-x}{3} \le 2 \\
3(0) &\le 3\left(\frac{4-x}{3}\right) \le 3(2) \\
0 &\le 4-x \le 6 \\
0-4 &\le 4-4-x \le 6-4 \\
-4 &\le -x \le 2 \\
\frac{-4}{-1} &\ge \frac{-x}{-1} \ge \frac{2}{-1} \\
4 &\ge x \ge -2, \text{ or } -2 \le x \le 4
\end{aligned}$$

solution set: $[-2, 4]$

56.

$$\begin{aligned}
-2 &\le \frac{5-3x}{2} \le 2 \\
2(-2) &\le 2\left(\frac{5-3x}{2}\right) \le 2(2) \\
-4 &\le 5-3x \le 4 \\
-4-5 &\le 5-5-3x \le 4-5 \\
-9 &\le -3x \le -1 \\
\frac{-9}{-3} &\ge \frac{-3x}{-3} \ge \frac{-1}{-3} \\
3 &\ge x \ge \tfrac{1}{3}, \text{ or } \tfrac{1}{3} \le x \le 3
\end{aligned}$$

solution set: $\left[\frac{1}{3}, 3\right]$

57.

$$x+3 < 3x-1 < 2x+2$$

$$\begin{aligned}
x+3 &< 3x-1 \\
x+3-3 &< 3x-1-3 \\
x-3x &< 3x-3x-4 \\
-2x &< -4 \\
\frac{-2x}{-2} &> \frac{-4}{-2} \\
x &> 2
\end{aligned}
\qquad \text{and} \qquad
\begin{aligned}
3x-1 &< 2x+2 \\
3x-1+1 &< 2x+2+1 \\
3x-2x &< 2x-2x+3 \\
x &< 3
\end{aligned}$$

If $x > 2$ **and** $x < 3$, then $2 < x < 3$. The solution set is $(2, 3)$.

58.

$$x-1 \le 2x+4 \le 3x-1$$

$$\begin{aligned}
x-1 &\le 2x+4 \\
x-1+1 &\le 2x+4+1 \\
x-2x &\le 2x-2x+5 \\
-x &\le 5 \\
\frac{-x}{-1} &\ge \frac{5}{-1} \\
x &\ge -5
\end{aligned}
\qquad \text{and} \qquad
\begin{aligned}
2x+4 &\le 3x-1 \\
2x+4-4 &\le 3x-1-4 \\
2x-3x &\le 3x-3x-5 \\
-x &\le -5 \\
\frac{-x}{-1} &\ge \frac{-5}{-1} \\
x &\ge 5
\end{aligned}$$

If $x \ge -5$ **and** $x \ge 5$, then $x \ge 5$. The solution set is $[5, \infty)$.

59.

$$4x \geq -x+5 \geq 3x-4$$

$$\begin{aligned} 4x &\geq -x+5 \\ 4x+x &\geq -x+x+5 \\ 5x &\geq 5 \\ \frac{5x}{5} &\geq \frac{5}{5} \\ x &\geq 1 \end{aligned} \qquad \text{and} \qquad \begin{aligned} -x+5 &\geq 3x-4 \\ -x+5-5 &\geq 3x-4-5 \\ -x-3x &\geq 3x-3x-9 \\ -4x &\geq -9 \\ \frac{-4x}{-4} &\leq \frac{-9}{-4} \\ x &\leq \tfrac{9}{4} \end{aligned}$$

If $x \geq 1$ **and** $x \leq \frac{9}{4}$, then $1 \leq x \leq \frac{9}{4}$. The solution set is $\left[1, \frac{9}{4}\right]$.

60.

$$x+2 < -\frac{1}{3}x < \frac{1}{2}x$$

$$\begin{aligned} x+2 &< -\frac{1}{3}x \\ 3(x+2) &< 3\left(-\frac{1}{3}x\right) \\ 3x+6 &< -x \\ 3x+6-6 &< -x-6 \\ 3x+x &< -x+x-6 \\ 4x &< -6 \\ x &< -\tfrac{3}{2} \end{aligned} \qquad \text{and} \qquad \begin{aligned} -\frac{1}{3}x &< \frac{1}{2}x \\ 6\left(-\frac{1}{3}x\right) &< 6\left(\frac{1}{2}x\right) \\ -2x &< 3x \\ -2x-3x &< 3x-3x \\ -5x &< 0 \\ \frac{-5x}{-5} &> \frac{0}{-5} \\ x &> 0 \end{aligned}$$

Since it is impossible to have $x < -\frac{3}{2}$ **and** $x > 0$, there is no solution.

61.

$$5(x+1) \leq 4(x+3) < 3(x-1)$$

$$\begin{aligned} 5(x+1) &\leq 4(x+3) \\ 5x+5-5 &\leq 4x+12-5 \\ 5x-4x &\leq 4x-4x+7 \\ x &\leq 7 \end{aligned} \qquad \text{and} \qquad \begin{aligned} 4(x+3) &< 3(x-1) \\ 4x+12-12 &< 3x-3-12 \\ 4x-3x &< 3x-3x-15 \\ x &< -15 \end{aligned}$$

If $x \leq 7$ **and** $x < -15$, then $x < -15$. The solution set is $(-\infty, -15)$.

-15

62.
$$-5(2+x) < 4x+1 < 3x$$

$$\begin{aligned} -5(2+x) &< 4x+1 \\ -10+10-5x &< 4x+1+10 \\ -5x-4x &< 4x-4x+11 \\ -9x &< 11 \\ \frac{-9x}{-9} &> \frac{11}{-9} \\ x &> -\tfrac{11}{9} \end{aligned} \quad \text{and} \quad \begin{aligned} 4x+1 &< 3x \\ 4x+1-1 &< 3x-1 \\ 4x-3x &< 3x-3x-1 \\ x &< -1 \end{aligned}$$

If $x > -\frac{11}{9}$ **and** $x < -1$, then $-\frac{11}{9} < x < -1$. The solution set is $\left(-\frac{11}{9}, -1\right)$.

$-\frac{11}{9}$ $\quad -1$

63.
$$\begin{aligned} 3x+2 &< 8 \\ 3x+2-2 &< 8-2 \\ 3x &< 6 \\ x &< 2 \end{aligned} \quad \text{or} \quad \begin{aligned} 2x-3 &> 11 \\ 2x-3+3 &> 11+3 \\ 2x &> 14 \\ x &> 7 \end{aligned}$$

If $x < 2$ **or** $x > 7$, then the solution set is $(-\infty, 2) \cup (7, \infty)$.

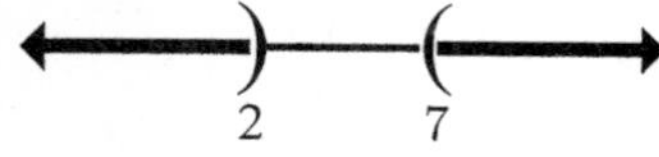

64.
$$\begin{aligned} 3x+4 &< -2 \\ 3x+4-4 &< -2-4 \\ 3x &< -6 \\ x &< -2 \end{aligned} \quad \text{or} \quad \begin{aligned} 3x+4 &> 10 \\ 3x+4-4 &> 10-4 \\ 3x &> 6 \\ x &> 2 \end{aligned}$$

If $x < -2$ **or** $x > 2$, then the solution set is $(-\infty, -2) \cup (2, \infty)$.

-2 $\quad 2$

65.
$$\begin{aligned} -4(x+2) &\geq 12 \\ -4x-8+8 &\geq 12+8 \\ -4x &\geq 20 \\ \frac{-4x}{-4} &\leq \frac{20}{-4} \\ x &\leq -5 \end{aligned} \quad \text{or} \quad \begin{aligned} 3x+8 &< 11 \\ 3x+8-8 &< 11-8 \\ 3x &< 3 \\ x &< 1 \end{aligned}$$

If $x \leq -5$ **or** $x < 1$, then $x < 1$. The solution set is $(-\infty, 1)$.

66.

$$5(x-2) \geq 0 \qquad \text{and} \qquad -3x < 9$$
$$5x - 10 + 10 \geq 0 + 10 \qquad \frac{-3x}{-3} > \frac{9}{-3}$$
$$5x \geq 10 \qquad x > -3$$
$$x \geq 2$$

If $x \geq 2$ **and** $x > -3$, then $x \geq 2$. The solution set is $[2, \infty)$.

67.

$$x < -3 \quad \text{and} \quad x > 3$$

It is impossible for x to be both less than -3 and greater than 3. There is no solution.

68.

$$x < 3 \quad \text{or} \quad x > -3$$

Every real number satisfies at least one of these conditions. The solution set is $(-\infty, \infty)$.

69. Let $x =$ the number of additional hours the person rents the rototiller.

$$15.50 + 7.95x < 50$$
$$7.95x < 34.50$$
$$x < 4.34$$

The person may rent the rototiller for 4 additional hours, or for 5 total hours.

70. Let $x =$ the number of additional hours the truck is rented. Solve this inequality:

$$29.95 + 8.95x < 110$$
$$8.95x < 80.05$$
$$x < 8.94$$

The person may rent the truck for 8 additional hours, or for 9 total hours.

71. Let $x =$ the number of children.

$$205 + 175 + 90x \leq 750$$
$$90x \leq 370$$
$$x \leq 4.11$$

At most 4 children can ride along.

72. Let $x =$ the number of boxes.

$$165 + 80x \leq 900$$
$$80x \leq 735$$
$$x \leq 9.1875$$

At most 9 boxes can be carried.

73. Let $x =$ the amount invested at 9%.

$$0.08(10000) + 0.09x > 1250$$
$$800 + 0.09x > 1250$$
$$0.09x > 450$$
$$x > 5000$$

She must invest more than \$5000.

74. Let $x =$ the amount invested at 8.75%.

$$0.055(8900) + 0.0875x > 1500$$
$$489.50 + 0.0875x > 1500$$
$$0.0875x > 1010.50$$
$$x > 11548.571$$

He must invest more than \$11,548.57.

75. Let $x =$ the # of compact discs bought.

$$175 + 8.50x \leq 330$$
$$8.50x \leq 155$$
$$x \leq 18.24$$

He can buy at most 18 discs.

76. Let $x =$ the # of DVDs bought.

$$1695.95 + 19.95x \leq 2000$$
$$19.95x \leq 304.05$$
$$x \leq 15.24$$

She can buy at most 15 DVDs.

77. Let $x =$ the student's score on 4th exam.

$$\frac{70 + 77 + 85 + x}{4} \geq 80$$
$$232 + x \geq 320$$
$$x \geq 88$$

The student needs a score of at least 88.

78. Let $x =$ the student's score on 5th exam.

$$\frac{70 + 79 + 85 + 88 + x}{5} > 80$$
$$322 + x > 400$$
$$x > 78$$

She needs a score greater than 78.

79. Let $x =$ the # of hours he works at the library. Then $20 - x =$ the # of hours he works construction.

$$5x + 9(20 - x) > 125$$
$$5x + 180 - 9x > 125$$
$$-4x > -55$$
$$x < 13.75$$

He can work up to 13 full hours at the library.

80. Let $x =$ the # of hours for the backhoe, and $40 - x =$ the # of hours for the bulldozer.

$$300x + 500(40 - x) \geq 18500$$
$$300x + 20000 - 500x \geq 18500$$
$$-200x \geq -1500$$
$$x \leq 7.5$$

The backhoe can be used at most 7 hrs/wk.

81. Let $x =$ the total bill. Under Plan 1, the employee will pay \$100, plus 30% of the rest (the amount over \$100). This amount could be represented by $100 + 0.30(x - 100)$. Similarly, under Plan 2 the employee would pay $200 + 0.20(x - 200)$.

$$\boxed{\text{Amount paid in Plan 1}} > \boxed{\text{Amount paid in Plan 2}}$$
$$100 + 0.30(x - 100) > 200 + 0.20(x - 200)$$
$$100 + 0.30x - 30 > 200 + 0.20x - 40$$
$$70 + 0.30x > 160 + 0.20x$$
$$0.10x > 90$$
$$x > 900$$

If the total bill is over \$900, then Plan 2 is better than Plan 1.

82. Let $x =$ the total bill. Under Plan 1, the employee will pay \$200, plus 30% of the rest (the amount over \$200). This amount could be represented by $200 + 0.30(x - 200)$. Similarly, under Plan 2 the employee would pay $400 + 0.20(x - 400)$.

$$\boxed{\text{Amount paid in Plan 1}} > \boxed{\text{Amount paid in Plan 2}}$$
$$200 + 0.30(x - 200) > 400 + 0.20(x - 400)$$
$$200 + 0.30x - 60 > 400 + 0.20x - 80$$
$$140 + 0.30x > 320 + 0.20x$$
$$0.10x > 180$$
$$x > 1800$$

If the total bill is over \$1800, then Plan 2 is better than Plan 1.

83. $2x + 3 < 5$; Graph $y = 2x + 3$ and $y = 5$ and find the x-coordinates when the first graph is below the second:

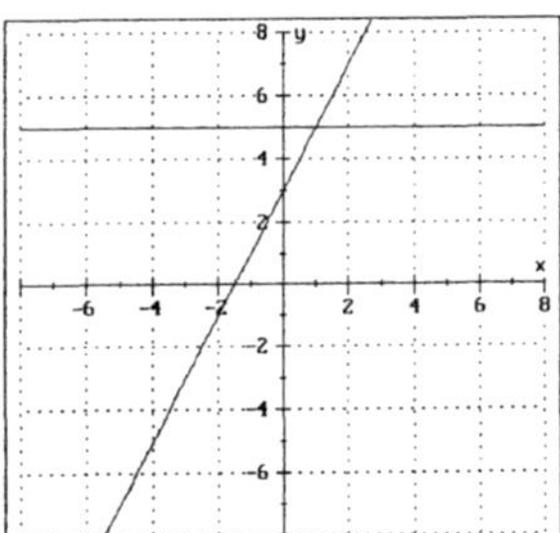

Solution: $x < 1$

84. $3x - 2 > 4$; Graph $y = 3x - 2$ and $y = 4$ and find the x-coordinates when the first graph is above the second:

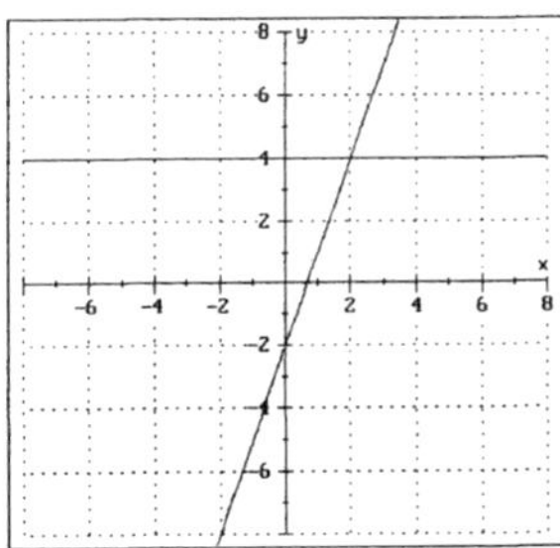

Solution: $x > 2$

85. $5x + 2 \geq -18$; Graph $y = 5x + 2$ and $y = -18$ and find the x-coordinates when the first graph is on or above the second:

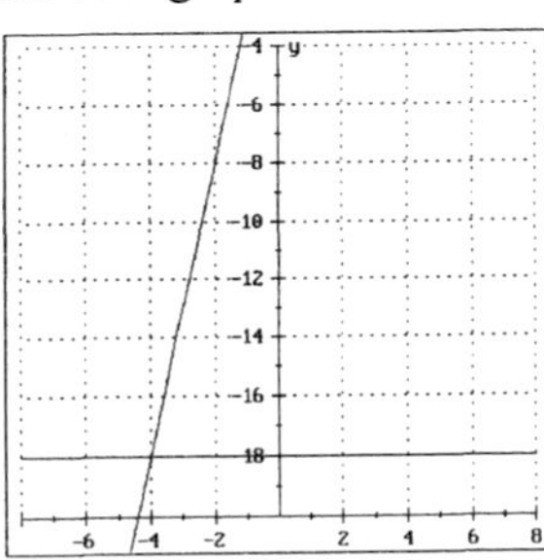

Solution: $x \geq -4$

86. $3x - 4 \leq 20$; Graph $y = 3x - 4$ and $y = 20$ and find the x-coordinates when the first graph is on or below the second:

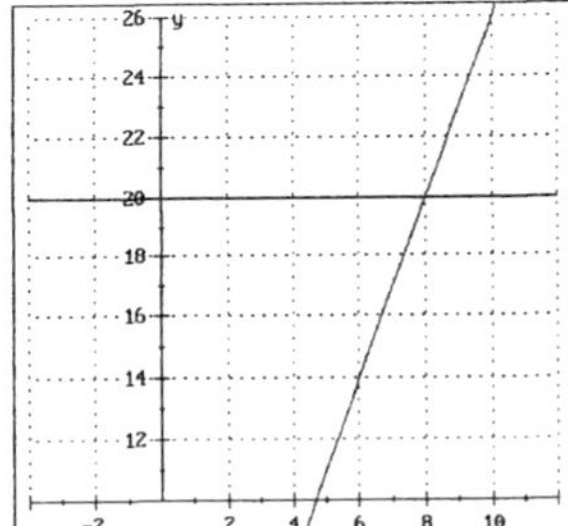

Solution: $x \leq 8$

87. We have values of $\sigma = 120$ and $E = 20$. Substitute into the formula:

$$\frac{3.84\sigma^2}{N} < E^2$$
$$\frac{3.84(120)^2}{N} < 20^2$$
$$\frac{55296}{N} < 400$$
$$55296 < 400N$$
$$138.24 < N$$

The sample size must be at least 139.

88. We have values of $\sigma = 120$ and $E = 10$. Substitute into the formula:

$$\frac{3.84\sigma^2}{N} < E^2$$
$$\frac{3.84(120)^2}{N} < 10^2$$
$$\frac{55296}{N} < 100$$
$$55296 < 100N$$
$$552.96 < N$$

The sample size must be at least 553.

89. **Answers may vary.**

90. **Answers may vary.**

91. Let $x = 2$.
Then $x > -3$, but $x^2 = 2^2 = 4 < 9$.

92. If $x > 2$, then it must be true that $x^2 > 4$.

93. Transitive: $=$, $\leq$ and $\ngeq$

94. The solution is not correct because when both sides are multiplied by $\mathbf{3x}$, and the order of the inequality is not changed, it is being assumed that $x > 0$. However, $x < 0$ would also be a solution of the inequality. The solution is $(-\infty, 0) \cup (3, \infty)$.

Exercise 4.2 (page 237)

1.
$$\begin{aligned} 3(2a-1) &= 2a \\ 6a - 3 &= 2a \\ 4a &= 3 \\ a &= \frac{3}{4} \end{aligned}$$

2.
$$\begin{aligned} \frac{t}{6} - \frac{t}{3} &= -1 \\ 6\left(\frac{t}{6} - \frac{t}{3}\right) &= 6(-1) \\ t - 2t &= -6 \\ -t &= -6 \\ t &= 6 \end{aligned}$$

3.
$$\begin{aligned} \frac{5x}{2} - 1 &= \frac{x}{3} + 12 \\ 6\left(\frac{5x}{2} - 1\right) &= 6\left(\frac{x}{3} + 12\right) \\ 15x - 6 &= 2x + 72 \\ 13x &= 78 \\ x &= 6 \end{aligned}$$

4.
$$\begin{aligned} 4b - \frac{b+9}{2} &= \frac{b+2}{5} - \frac{8}{5} \\ 10\left(4b - \frac{b+9}{2}\right) &= 10\left(\frac{b+2}{5} - \frac{8}{5}\right) \\ 40b - 5(b+9) &= 2(b+2) - 2(8) \\ 40b - 5b - 45 &= 2b + 4 - 16 \\ 33b &= 33 \\ b &= 1 \end{aligned}$$

5.
$$\begin{aligned} A &= p + prt \\ A - p &= prt \\ \frac{A-p}{pr} &= t, \text{ or } t = \frac{A-p}{pr} \end{aligned}$$

6.
$$\begin{aligned} P &= 2w + 2l \\ P - 2w &= 2l \\ \frac{P-2w}{2} &= l, \text{ or } l = \frac{P-2w}{2} \end{aligned}$$

7. x

8. $-x$

9. 0

10. right; down

11. reflected

12. $x = k$ or $x = -k$

13. $a = b$ or $a = -b$

14. $-k < x < k$

15. $x \leq -k$ or $x \geq k$

16. no

17. $|8| = 8$

18. $|-18| = 18$

19. $-|2| = -2$

20. $-|-20| = -20$

21. $-|-30| = -30$

22. $-|25| = -25$

23. $|\pi - 4| = -(\pi - 4) = 4 - \pi$
($\pi - 4$ is less than zero.)

24. $|2\pi - 4| = 2\pi - 4$
($2\pi - 4$ is greater than zero.)

25. $f(x) = |x| - 2$; Shift $f(x) = |x|$ down 2.

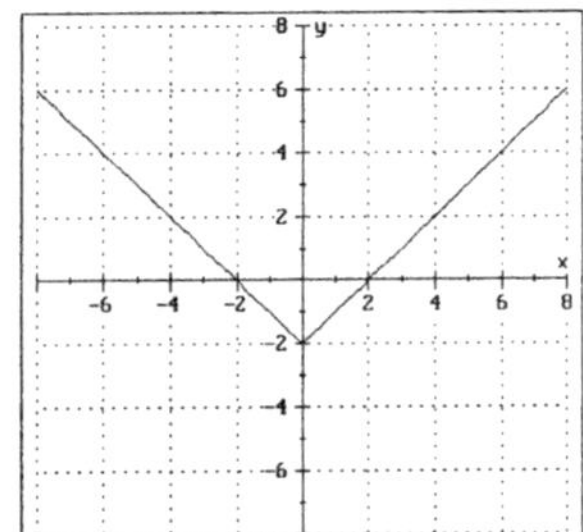

26. $f(x) = -|x| + 1$; Reflect $f(x) = |x|$ about the x-axis and shift up 1.

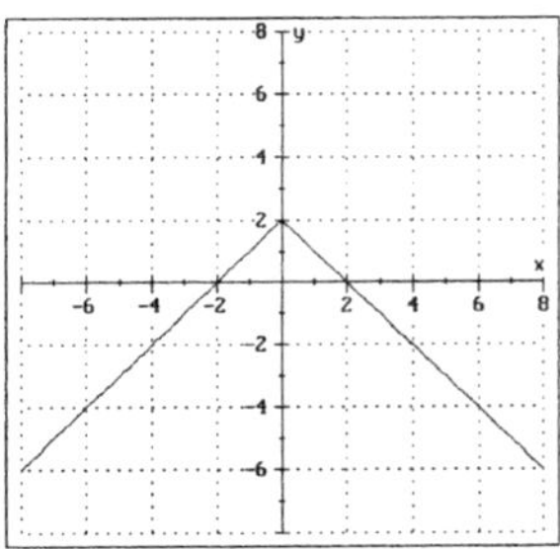

27. $f(x) = -|x + 4|$; Reflect $f(x) = |x|$ about the x-axis and shift left 4.

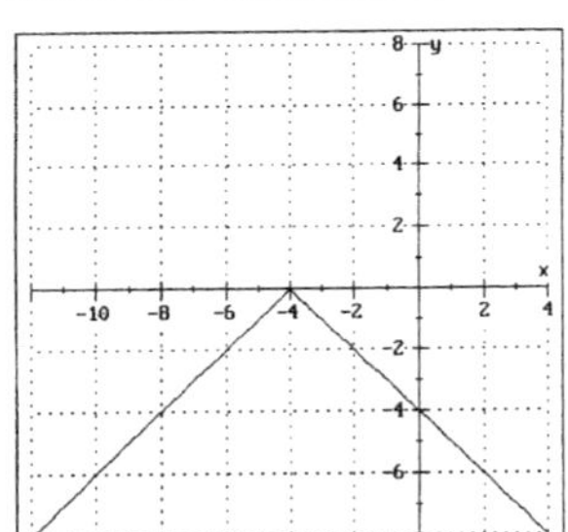

28. $f(x) = |x - 1| + 2$; Shift $f(x) = |x|$ right 1 and up 2.

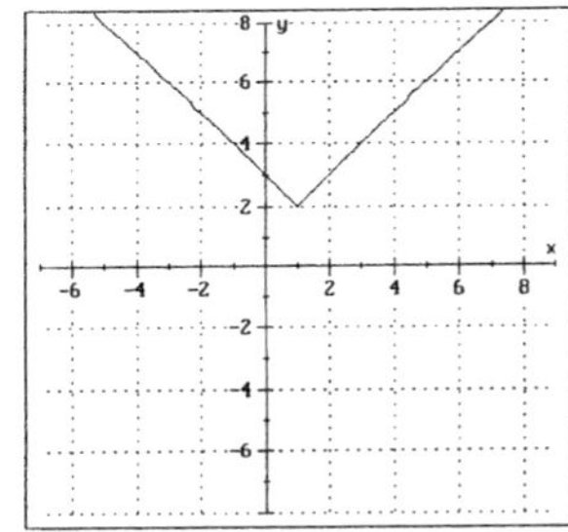

29.
$$|x| = 4$$
$$x = 4 \quad \textbf{or} \quad x = -4$$

30.
$$|x| = 9$$
$$x = 9 \quad \textbf{or} \quad x = -9$$

31.
$$|x - 3| = 6$$
$$x - 3 = 6 \quad \textbf{or} \quad x - 3 = -6$$
$$x = 9 \qquad\qquad x = -3$$

32.
$$|x + 4| = 8$$
$$x + 4 = 8 \quad \textbf{or} \quad x + 4 = -8$$
$$x = 4 \qquad\qquad x = -12$$

33.
$$|2x - 3| = 5$$
$$2x - 3 = 5 \quad \textbf{or} \quad 2x - 3 = -5$$
$$2x = 8 \qquad\qquad 2x = -2$$
$$x = 4 \qquad\qquad x = -1$$

34.
$$|4x - 4| = 20$$
$$4x - 4 = 20 \quad \textbf{or} \quad 4x - 4 = -20$$
$$4x = 24 \qquad\qquad 4x = -16$$
$$x = 6 \qquad\qquad x = -4$$

35.
$$|3x + 2| = 16$$
$$3x + 2 = 16 \quad \textbf{or} \quad 3x + 2 = -16$$
$$3x = 14 \qquad\qquad 3x = -18$$
$$x = \tfrac{14}{3} \qquad\qquad x = -6$$

36.
$$|5x - 3| = 22$$
$$5x - 3 = 22 \quad \textbf{or} \quad 5x - 3 = -22$$
$$5x = 25 \qquad\qquad 5x = -19$$
$$x = 5 \qquad\qquad x = -\tfrac{19}{5}$$

37. $\left|\frac{7}{2}x+3\right|=-5$

Since an absolute value cannot be negative, there is no solution.

38. $|2x+10|=0$

$$\begin{array}{lcl} 2x+10=0 & \textbf{or} & 2x+10=-0 \\ 2x=-10 & & 2x=-10 \\ x=-5 & & x=-5 \end{array}$$

39. $\left|\frac{x}{2}-1\right|=3$

$$\begin{array}{lcl} \frac{x}{2}-1=3 & \textbf{or} & \frac{x}{2}-1=-3 \\ \frac{x}{2}=4 & & \frac{x}{2}=-2 \\ x=8 & & x=-4 \end{array}$$

40. $\left|\frac{4x-64}{4}\right|=32$

$$\begin{array}{lcl} \frac{4x-64}{4}=32 & \textbf{or} & \frac{4x-64}{4}=-32 \\ 4x-64=128 & & 4x-64=-128 \\ 4x=192 & & 4x=-64 \\ x=48 & & x=-16 \end{array}$$

41. $|3-4x|=5$

$$\begin{array}{lcl} 3-4x=5 & \textbf{or} & 3-4x=-5 \\ -4x=2 & & -4x=-8 \\ x=-\frac{1}{2} & & x=2 \end{array}$$

42. $|8-5x|=18$

$$\begin{array}{lcl} 8-5x=18 & \textbf{or} & 8-5x=-18 \\ -5x=10 & & -5x=-26 \\ x=-2 & & x=\frac{26}{5} \end{array}$$

43. $|3x+24|=0$

$$\begin{array}{lcl} 3x+24=0 & \textbf{or} & 3x+24=-0 \\ 3x=-24 & & 3x=-24 \\ x=-8 & & x=-8 \end{array}$$

44. $|x-21|=-8$

Since an absolute value cannot be negative, there is no solution.

45. $\left|\frac{3x+48}{3}\right|=12$

$$\begin{array}{lcl} \frac{3x+48}{3}=12 & \textbf{or} & \frac{3x+48}{3}=-12 \\ 3x+48=36 & & 3x+48=-36 \\ 3x=-12 & & 3x=-84 \\ x=-4 & & x=-28 \end{array}$$

46. $\left|\frac{x}{2}+2\right|=4$

$$\begin{array}{lcl} \frac{x}{2}+2=4 & \textbf{or} & \frac{x}{2}+2=-4 \\ \frac{x}{2}=2 & & \frac{x}{2}=-6 \\ x=4 & & x=-12 \end{array}$$

47. $|2x+1|-3=12$

$|2x+1|=15$

$$\begin{array}{lcl} 2x+1=15 & \textbf{or} & 2x+1=-15 \\ 2x=14 & & 2x=-16 \\ x=7 & & x=-8 \end{array}$$

48. $|3x-2|+1=11$

$|3x-2|=10$

$$\begin{array}{lcl} 3x-2=10 & \textbf{or} & 3x-2=-10 \\ 3x=12 & & 3x=-8 \\ x=4 & & x=-\frac{8}{3} \end{array}$$

49. $|x+3|+7=10$

$|x+3|=3$

$$\begin{array}{lcl} x+3=3 & \textbf{or} & x+3=-3 \\ x=0 & & x=-6 \end{array}$$

50. $|2-x|+3=5$

$|2-x|=2$

$$\begin{array}{lcl} 2-x=2 & \textbf{or} & 2-x=-2 \\ -x=0 & & -x=-4 \\ x=0 & & x=4 \end{array}$$

51.

$$\begin{aligned}\left|\frac{3}{5}x-4\right|-2&=-2\\ \left|\frac{3}{5}x-4\right|&=0\end{aligned}$$

$$\frac{3}{5}x-4=0 \quad \textbf{or} \quad \frac{3}{5}x-4=-0$$

$$\frac{3}{5}x=4 \qquad\qquad \frac{3}{5}x=4$$

$$x=\tfrac{20}{3} \qquad\qquad x=\tfrac{20}{3}$$

52.

$$\begin{aligned}\left|\frac{3}{4}x+2\right|+4&=4\\ \left|\frac{3}{4}x+2\right|&=0\end{aligned}$$

$$\frac{3}{4}x+2=0 \quad \textbf{or} \quad \frac{3}{4}x+2=-0$$

$$\frac{3}{4}x=-2 \qquad\qquad \frac{3}{4}x=-2$$

$$x=-\tfrac{8}{3} \qquad\qquad x=-\tfrac{8}{3}$$

53.

$$|2x+1|=|3x+3|$$

$$2x+1=3x+3 \quad \textbf{or} \quad 2x+1=-(3x+3)$$

$$-x=2 \qquad\qquad 2x+1=-3x-3$$

$$x=-2 \qquad\qquad 5x=-4$$

$$x=-\tfrac{4}{5}$$

54.

$$|5x-7|=|4x+1|$$

$$5x-7=4x+1 \quad \textbf{or} \quad 5x-7=-(4x+1)$$

$$x=8 \qquad\qquad 5x-7=-4x-1$$

$$9x=6$$

$$x=\tfrac{2}{3}$$

55.

$$|3x-1|=|x+5|$$

$$3x-1=x+5 \quad \textbf{or} \quad 3x-1=-(x+5)$$

$$2x=6 \qquad\qquad 3x-1=-x-5$$

$$x=3 \qquad\qquad 4x=-4$$

$$x=-1$$

56.

$$|3x+1|=|x-5|$$

$$3x+1=x-5 \quad \textbf{or} \quad 3x+1=-(x-5)$$

$$2x=-6 \qquad\qquad 3x+1=-x+5$$

$$x=-3 \qquad\qquad 4x=4$$

$$x=1$$

57.

$$|2-x|=|3x+2|$$

$$2-x=3x+2 \quad \textbf{or} \quad 2-x=-(3x+2)$$

$$-4x=0 \qquad\qquad 2-x=-3x-2$$

$$x=0 \qquad\qquad 2x=-4$$

$$x=-2$$

58.

$$|4x+3|=|9-2x|$$

$$4x+3=9-2x \quad \textbf{or} \quad 4x+3=-(9-2x)$$

$$6x=6 \qquad\qquad 4x+3=-9+2x$$

$$x=1 \qquad\qquad 2x=-12$$

$$x=-6$$

59.
$$\left|\frac{x}{2}+2\right| = \left|\frac{x}{2}-2\right|$$
$$\frac{x}{2}+2 = \frac{x}{2}-2 \quad \textbf{or} \quad \frac{x}{2}+2 = -\left(\frac{x}{2}-2\right)$$
$$0 = -4 \qquad\qquad \frac{x}{2}+2 = -\frac{x}{2}+2$$
(no solution from this part) $\qquad x = 0$

60.
$$|7x+12| = |x-6|$$
$$7x+12 = x-6 \quad \textbf{or} \quad 7x+12 = -(x-6)$$
$$6x = -18 \qquad 7x+12 = -x+6$$
$$x = -3 \qquad 8x = -6$$
$$x = -\tfrac{3}{4}$$

61.
$$\left|x+\tfrac{1}{3}\right| = |x-3|$$
$$x+\tfrac{1}{3} = x-3 \quad \textbf{or} \quad x+\tfrac{1}{3} = -(x-3)$$
$$3x+1 = 3x-9 \qquad x+\tfrac{1}{3} = -x+3$$
$$0 = -10 \qquad 3x+1 = -3x+9$$
(no solution from this part) $\qquad 6x = 8$
$$x = \tfrac{4}{3}$$

62.
$$\left|x-\tfrac{1}{4}\right| = |x+4|$$
$$x-\tfrac{1}{4} = x+4 \quad \textbf{or} \quad x-\tfrac{1}{4} = -(x+4)$$
$$4x-1 = 4x+16 \qquad x-\tfrac{1}{4} = -x-4$$
$$0 = 17 \qquad 4x-1 = -4x-16$$
(no solution from this part) $\qquad 8x = -15$
$$x = -\tfrac{15}{8}$$

63. $|3x+7| = -|8x-2|$
Since an absolute value cannot be negative, there is no solution.

64. $-|17x+13| = |3x-14|$
Since an absolute value cannot be negative, there is no solution.

65. $|2x| < 8$
$-8 < 2x < 8$
$\frac{-8}{2} < \frac{2x}{2} < \frac{8}{2}$
$-4 < x < 4$
solution set: $(-4, 4)$

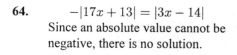

66. $|3x| < 27$
$-27 < 3x < 27$
$\frac{-27}{3} < \frac{3x}{3} < \frac{27}{3}$
$-9 < x < 9$
solution set: $(-9, 9)$

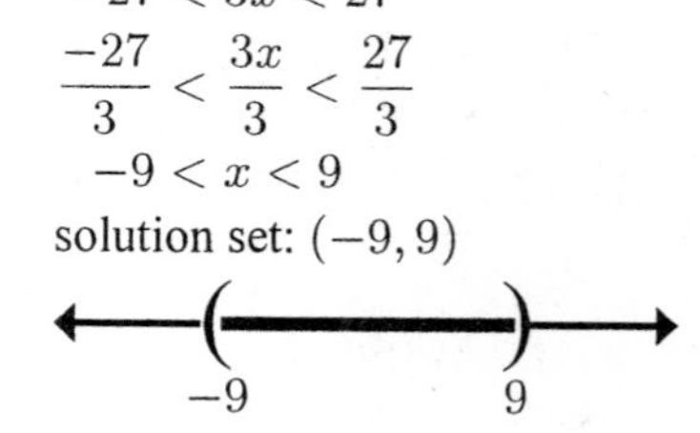

67. $|x+9| \leq 12$

$$-12 \leq x+9 \leq 12$$
$$-12-9 \leq x+9-9 \leq 12-9$$
$$-21 \leq x \leq 3$$

solution set: $[-21, 3]$

68. $|x-8| \leq 12$

$$-12 \leq x-8 \leq 12$$
$$-12+8 \leq x-8+8 \leq 12+8$$
$$-4 \leq x \leq 20$$

solution set: $[-4, 20]$

69. $|3x+2| \leq -3$; Since an absolute value can never be negative, this inequality has no solution.

70. $|3x-2| < 10$

$$-10 < 3x-2 < 10$$
$$-10+2 < 3x-2+2 < 10+2$$
$$-8 < 3x < 12$$
$$\frac{-8}{3} < \frac{3x}{3} < \frac{12}{3}$$
$$-\frac{8}{3} < x < 4$$

solution set: $\left(-\frac{8}{3}, 4\right)$

71. $|4x-1| \leq 7$

$$-7 \leq 4x-1 \leq 7$$
$$-7+1 \leq 4x-1+1 \leq 7+1$$
$$-6 \leq 4x \leq 8$$
$$\frac{-6}{4} \leq \frac{4x}{4} \leq \frac{8}{4}$$
$$-\frac{3}{2} \leq x \leq 2$$

solution set: $\left[-\frac{3}{2}, 2\right]$

72. $|5x-12| < -5$; Since an absolute value can never be negative, this inequality has no solution.

73. $|3-2x| < 7$

$$-7 < 3-2x < 7$$
$$-7-3 < 3-3-2x < 7-3$$
$$-10 < -2x < 4$$
$$\frac{-10}{-2} > \frac{-2x}{-2} > \frac{4}{-2}$$
$$5 > x > -2, \text{ or } -2 < x < 5$$

solution set: $(-2, 5)$

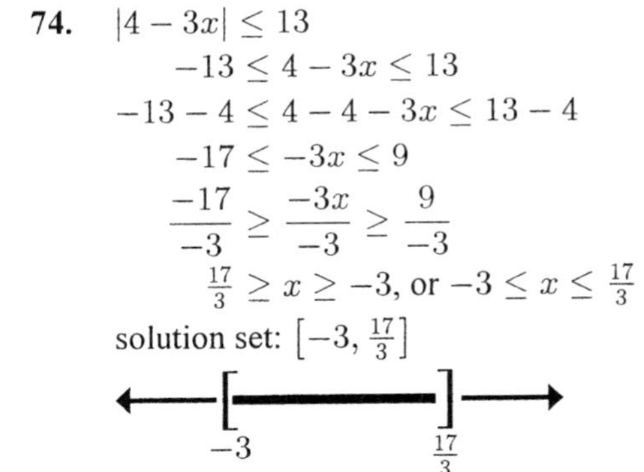

74. $|4-3x| \leq 13$

$$-13 \leq 4-3x \leq 13$$
$$-13-4 \leq 4-4-3x \leq 13-4$$
$$-17 \leq -3x \leq 9$$
$$\frac{-17}{-3} \geq \frac{-3x}{-3} \geq \frac{9}{-3}$$
$$\frac{17}{3} \geq x \geq -3, \text{ or } -3 \leq x \leq \frac{17}{3}$$

solution set: $\left[-3, \frac{17}{3}\right]$

75.

$$|5x|+2 > 7$$
$$|5x| > 5$$

$5x < -5$ **or** $5x > 5$

$x < -1$ $\quad$ $x > 1$

solution set: $(-\infty, -1) \cup (1, \infty)$

76.

$$|7x|-3 > 4$$
$$|7x| > 7$$

$7x < -7$ **or** $7x > 7$

$x < -1$ $\quad$ $x > 1$

solution set: $(-\infty, -1) \cup (1, \infty)$

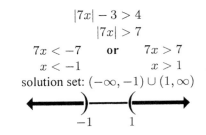

77.

$$|x - 12| > 24$$
$$x - 12 < -24 \quad \textbf{or} \quad x - 12 > 24$$
$$x < -12 \qquad\qquad x > 36$$

solution set: $(-\infty, -12) \cup (36, \infty)$

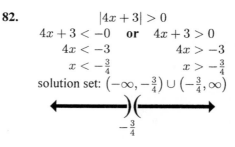

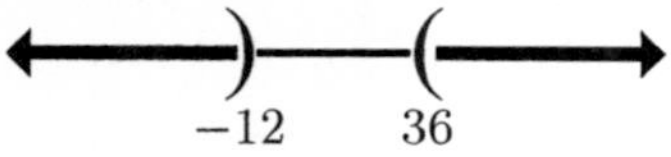

−12 36

78.

$$|x + 5| \geq 7$$
$$x + 5 \leq -7 \quad \textbf{or} \quad x + 5 \geq 7$$
$$x \leq -12 \qquad\qquad x \geq 2$$

solution set: $(-\infty, -12] \cup [2, \infty)$

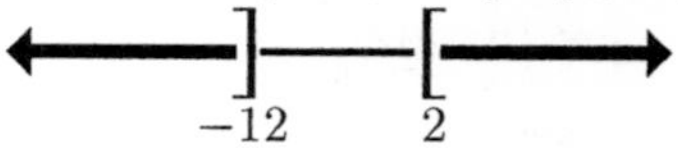

−12 2

79.

$$|3x + 2| > 14$$
$$3x + 2 < -14 \quad \textbf{or} \quad 3x + 2 > 14$$
$$3x < -16 \qquad\qquad 3x > 12$$
$$x < -\tfrac{16}{3} \qquad\qquad x > 4$$

solution set: $\left(-\infty, -\frac{16}{3}\right) \cup (4, \infty)$

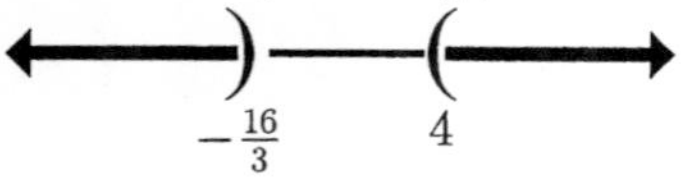

$-\frac{16}{3}$ 4

80.

$$|2x - 5| > 25$$
$$2x - 5 < -25 \quad \textbf{or} \quad 2x - 5 > 25$$
$$2x < -20 \qquad\qquad 2x > 30$$
$$x < -10 \qquad\qquad x > 15$$

solution set: $(-\infty, -10) \cup (15, \infty)$

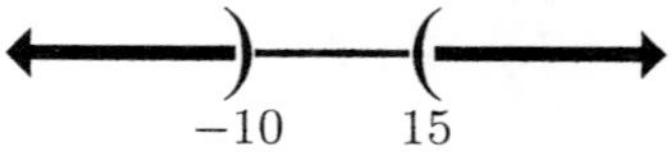

−10 15

81.

$$|4x + 3| > -5$$

Since an absolute value is always at least 0, this inequality is true for all real numbers.

solution set: $(-\infty, \infty)$

0

82.

$$|4x + 3| > 0$$
$$4x + 3 < -0 \quad \textbf{or} \quad 4x + 3 > 0$$
$$4x < -3 \qquad\qquad 4x > -3$$
$$x < -\tfrac{3}{4} \qquad\qquad x > -\tfrac{3}{4}$$

solution set: $\left(-\infty, -\frac{3}{4}\right) \cup \left(-\frac{3}{4}, \infty\right)$

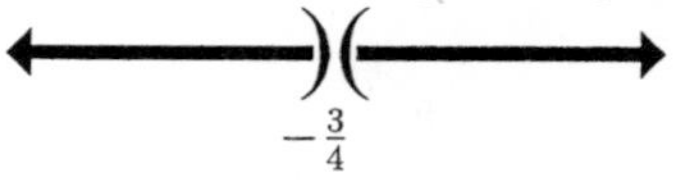

$-\frac{3}{4}$

83.

$$|2 - 3x| \geq 8$$
$$2 - 3x \leq -8 \quad \textbf{or} \quad 2 - 3x \geq 8$$
$$-3x \leq -10 \qquad\qquad -3x \geq 6$$
$$x \geq \tfrac{10}{3} \qquad\qquad x \leq -2$$

solution set: $(-\infty, -2] \cup \left[\frac{10}{3}, \infty\right)$

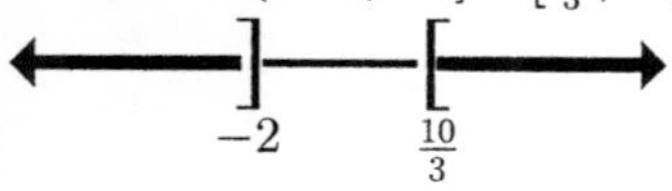

−2 $\frac{10}{3}$

84.

$$|-1 - 2x| > 5$$
$$-1 - 2x < -5 \quad \textbf{or} \quad -1 - 2x > 5$$
$$-2x < -4 \qquad\qquad -2x > 6$$
$$x > 2 \qquad\qquad x < -3$$

solution set: $(-\infty, -3) \cup (2, \infty)$

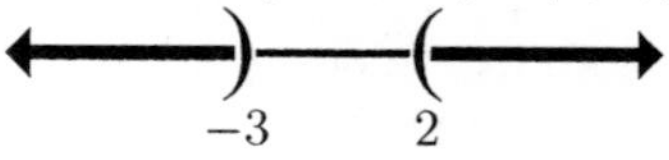

−3 2

85.

$$-|2x - 3| < -7$$
$$|2x - 3| > 7$$
$$2x - 3 < -7 \quad \textbf{or} \quad 2x - 3 > 7$$
$$2x < -4 \qquad\qquad 2x > 10$$
$$x < -2 \qquad\qquad x > 5$$

solution set: $(-\infty, -2) \cup (5, \infty)$

−2 5

86.

$$-|3x + 1| < -8$$
$$|3x + 1| > 8$$
$$3x + 1 < -8 \quad \textbf{or} \quad 3x + 1 > 8$$
$$3x < -9 \qquad\qquad 3x > 7$$
$$x < -3 \qquad\qquad x > \tfrac{7}{3}$$

solution set: $(-\infty, -3) \cup \left(\frac{7}{3}, \infty\right)$

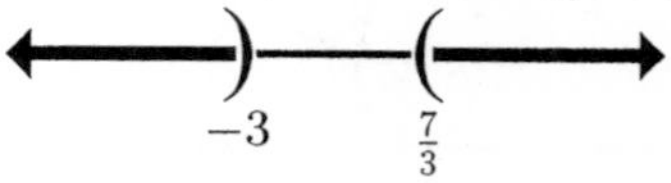

−3 $\frac{7}{3}$

87.

$$|8x - 3| > 0$$

$$8x - 3 < -0 \quad \textbf{or} \quad 8x - 3 > 0$$

$$8x < 3 \qquad\qquad 8x > 3$$

$$x < \tfrac{3}{8} \qquad\qquad x > \tfrac{3}{8}$$

solution set: $\left(-\infty, \frac{3}{8}\right) \cup \left(\frac{3}{8}, \infty\right)$

$\frac{3}{8}$

88.

$$|7x + 2| > -8$$

Since an absolute value is always at least 0, this inequality is true for all real numbers.

solution set: $(-\infty, \infty)$

0

89.

$$\left|\frac{x-2}{3}\right| \le 4$$

$$-4 \le \frac{x-2}{3} \le 4$$

$$-12 \le x - 2 \le 12$$

$$-10 \le x \le 14$$

solution set: $[-10, 14]$

-10 $\quad$ 14

90.

$$\left|\frac{x-2}{3}\right| > 4$$

$$\frac{x-2}{3} < -4 \quad \textbf{or} \quad \frac{x-2}{3} > 4$$

$$x - 2 < -12 \qquad\qquad x - 2 > 12$$

$$x < -10 \qquad\qquad x > 14$$

solution set: $(-\infty, -10) \cup (14, \infty)$

-10 $\quad$ 14

91.

$$|3x + 1| + 2 < 6$$

$$|3x + 1| < 4$$

$$-4 < 3x + 1 < 4$$

$$-5 < 3x < 3$$

$$-\tfrac{5}{3} < x < 1$$

solution set: $\left(-\frac{5}{3}, 1\right)$

$-\frac{5}{3}$ $\quad$ 1

92.

$$|3x - 2| + 2 \ge 0$$

$$|3x - 2| \ge -2$$

Since an absolute value is always at least 0, this inequality is true for all real numbers.

solution set: $(-\infty, \infty)$

0

93.

$$3|2x + 5| \ge 9$$

$$|2x + 5| \ge 3$$

$$2x + 5 \le -3 \quad \textbf{or} \quad 2x + 5 \ge 3$$

$$2x \le -8 \qquad\qquad 2x \ge -2$$

$$x \le -4 \qquad\qquad x \ge -1$$

solution set: $(-\infty, -4] \cup [-1, \infty)$

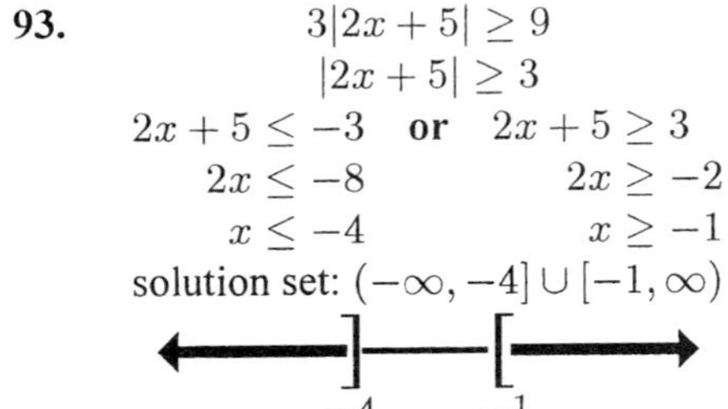

94.

$$-2|3x - 4| < 16$$

$$|3x - 4| > -8$$

Since an absolute value is always at least 0, this inequality is true for all real numbers.

solution set: $(-\infty, \infty)$

0

95. $|5x-1|+4 \le 0$

$|5x-1| \le -4$

Since an absolute value can never be negative, this inequality has no solution.

96. $-|5x-1|+2<0$

$-|5x-1|<-2$

$|5x-1|>2$

$5x-1<-2$ **or** $5x-1>2$

$5x<-1$ $\quad$ $5x>3$

$x<-\frac{1}{5}$ $\quad$ $x>\frac{3}{5}$

solution set: $\left(-\infty,-\frac{1}{5}\right)\cup\left(\frac{3}{5},\infty\right)$

$-\frac{1}{5}$ $\quad$ $\frac{3}{5}$

97. $\left|\frac{1}{3}x+7\right|+5>6$

$\left|\frac{1}{3}x+7\right|>1$

$\frac{1}{3}x+7<-1$ **or** $\frac{1}{3}x+7>1$

$x+21<-3$ $\quad$ $x+21>3$

$x<-24$ $\quad$ $x>-18$

solution set: $(-\infty,-24)\cup(-18,\infty)$

-24 $\quad$ -18

98. $\left|\frac{1}{2}x-3\right|-4<2$

$\left|\frac{1}{2}x-3\right|<6$

$-6<\frac{1}{2}x-3<6$

$-3<\frac{1}{2}x<9$

$-6<x<18$

solution set: $(-6,18)$

-6 $\quad$ 18

99. $\left|\frac{1}{5}x-5\right|+4>4$

$\left|\frac{1}{5}x-5\right|>0$

$\frac{1}{5}x-5<-0$ **or** $\frac{1}{5}x-5>0$

$\frac{1}{5}x<5$ $\quad$ $\frac{1}{5}x>5$

$x<25$ $\quad$ $x>25$

solution set: $(-\infty,25)\cup(25,\infty)$

25

100. $\left|\frac{1}{6}x+6\right|+2<2$

$\left|\frac{1}{6}x+6\right|<0$

Since an absolute value can never be negative, this inequality has no solution.

101. $\left|\frac{1}{7}x+1\right|\le 0$

Since an absolute value can never be less than zero, the only solution for this inequality is when the absolute value is equal to 0.

$\left|\frac{1}{7}x+1\right|=0$

$\frac{1}{7}x+1=0$ **or** $\frac{1}{7}x+1=-0$

$\frac{1}{7}x=-1$ $\quad$ $\frac{1}{7}x=-1$

$x=-7$ $\quad$ $x=-7$

solution set: $[-7,-7]$

-7

102. $|2x+1|+2\le 2$

$|2x+1|\le 0$

Since an absolute value can never be less than zero, the only solution for this inequality is when the absolute value is equal to 0.

$|2x+1|=0$

$2x+1=0$ **or** $2x+1=-0$

$2x=-1$ $\quad$ $2x=-1$

$x=-\frac{1}{2}$ $\quad$ $x=-\frac{1}{2}$

solution set: $\left[-\frac{1}{2},-\frac{1}{2}\right]$

$-\frac{1}{2}$

103. $\left|\frac{x-5}{10}\right| \le 0$

Since an absolute value can never be less than zero, the only solution for this inequality is when the absolute value is equal to 0.

$\left|\frac{x-5}{10}\right| = 0$

$\frac{x-5}{10} = 0$ **or** $\frac{x-5}{10} = -0$

$x - 5 = 0$ $\quad$ $x - 5 = 0$

$x = 5$ $\quad$ $x = 5$

solution set: $[5, 5]$

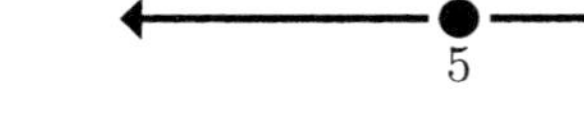

104. $\left|\frac{3}{5}x - 2\right| + 3 \le 3$

$\left|\frac{3}{5}x - 2\right| \le 0$

Since an absolute value can never be less than zero, the only solution for this inequality is when the absolute value is equal to 0.

$\left|\frac{3}{5}x - 2\right| = 0$

$\frac{3}{5}x - 2 = 0$ **or** $\frac{3}{5}x - 2 = -0$

$\frac{3}{5}x = 2$ $\quad$ $\frac{3}{5}x = 2$

$x = \frac{10}{3}$ $\quad$ $x = \frac{10}{3}$

solution set: $[\frac{10}{3}, \frac{10}{3}]$

$\frac{10}{3}$

105. $-4 < x < 4 \Rightarrow |x| < 4$

106. $x < -4$ or $x > 4 \Rightarrow |x| > 4$

107. $x + 3 < -6$ or $x + 3 > 6 \Rightarrow |x + 3| > 6$

108. $-5 \le x - 3 \le 5 \Rightarrow |x - 3| \le 5$

109. $|d - 5| \le 1$

$-1 \le d - 5 \le 1$

$-1 + 5 \le d - 5 + 5 \le 1 + 5$

$4 \text{ ft} \le d \le 6 \text{ ft}$

110. $|a - 30000| \le 1500$

$-1500 \le a - 30000 \le 1500$

$-1500 + 30000 \le a - 30000 + 30000 \le 1500 + 30000$

$28500 \text{ ft} \le a \le 31500 \text{ ft}$

111. $|t - 78^\circ| \le 8^\circ$

$-8^\circ \le t - 78^\circ \le 8^\circ$

$-8^\circ + 78^\circ \le t - 78^\circ + 78^\circ \le 8^\circ + 78^\circ$

$70^\circ \le t \le 86^\circ$

112. $|t - 40^\circ| < 80^\circ$

$-80^\circ < t - 40^\circ < 80^\circ$

$-80^\circ + 40^\circ < t - 40^\circ + 40^\circ < 80^\circ + 40^\circ$

$-40^\circ < t < 120^\circ$

113. $0.6^\circ - 0.5^\circ \le c \le 0.6^\circ + 0.5^\circ$

$-0.5^\circ \le c - 0.6^\circ \le 0.5^\circ$

$|c - 0.6^\circ| \le 0.5^\circ$

114. $0.25 - 0.015 \le x \le 0.25 + 0.015$

$-0.015 \le x - 0.25 \le 0.015$

$|x - 0.25| \le 0.015$

115-118. Answers may vary.

119. $|x| + k = 0$

$|x| = -k$

If this equation has exactly two solutions, then $-k > 0$, or $k < 0$.

120. $|x| + k = 0$

$|x| = -k$

If this equation has exactly one solution, then $-k = 0$, or $k = 0$.

121. $|x| + |y| > |x + y|$ when x and y have different signs.

122. $|x| + |y| = |x + y|$ when x and y have the same sign.

Exercise 4.3 (page 246)

1.
$$\begin{aligned} x + y &= 4 \\ x - y &= 2 \\ \hline 2x \quad &= 6 \\ x \quad &= 3 \end{aligned}$$
Substitute and solve for y:
$$\begin{aligned} x + y &= 4 \\ 3 + y &= 4 \\ y &= 1 \end{aligned}$$
The solution is $(3, 1)$.

2.
$$\begin{aligned} 2x - \ y &= -4 \Rightarrow \times 2 \\ x + 2y &= \ \ 3 \end{aligned} \qquad \begin{aligned} 4x - 2y &= -8 \\ x + 2y &= \ \ 3 \\ \hline 5x \quad &= -5 \\ x \quad &= -1 \end{aligned}$$
Substitute and solve for y:
$$\begin{aligned} x + 2y &= 3 \\ -1 + 2y &= 3 \\ 2y &= 4 \\ y &= 2 \end{aligned}$$
The solution is $(-1, 2)$.

3.
$$\begin{aligned} 3x + \ y &= \ \ 3 \Rightarrow \times 3 \\ 2x - 3y &= 13 \end{aligned} \qquad \begin{aligned} 9x + 3y &= \ \ 9 \\ 2x - 3y &= 13 \\ \hline 11x \quad &= 22 \\ x \quad &= \ \ 2 \end{aligned}$$
Substitute and solve for y:
$$\begin{aligned} 3x + y &= 3 \\ 3(2) + y &= 3 \\ 6 + y &= 3 \\ y &= -3 \end{aligned}$$
The solution is $(2, -3)$.

4.
$$\begin{aligned} 2x - 5y &= \ \ 8 \Rightarrow \times 2 \\ 5x + 2y &= -9 \Rightarrow \times 5 \end{aligned} \qquad \begin{aligned} 4x - 10y &= \ \ 16 \\ 25x + 10y &= -45 \\ \hline 29x \quad &= -29 \\ x \quad &= \ \ -1 \end{aligned}$$
Substitute and solve for y:
$$\begin{aligned} 5x + 2y &= -9 \\ 5(-1) + 2y &= -9 \\ -5 + 2y &= -9 \\ 2y &= -4 \\ y &= -2 \end{aligned}$$
The solution is $(-1, -2)$.

5. linear

6. half-planes

7. edge

8. $y = \frac{1}{2}x - 2$

9. $y > x + 1$

10. $y < 2x - 1$

11. $y \geq x$

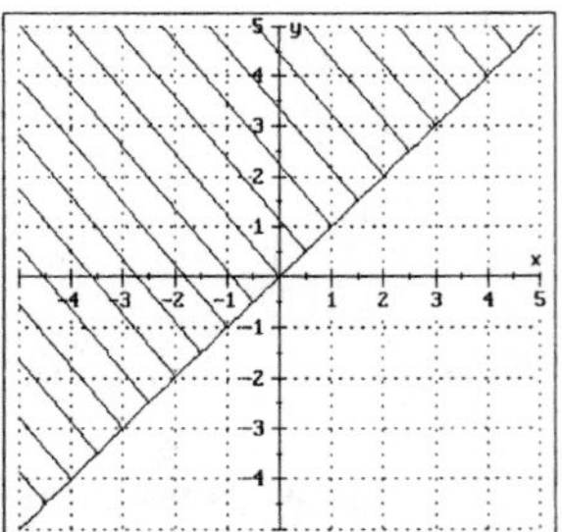

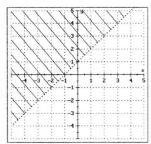

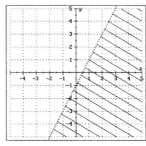

SECTION 4.3

12. $y \le 2x$

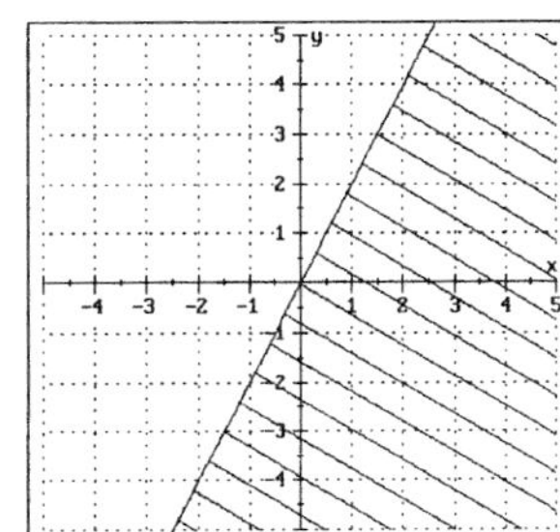

13. $2x + y \le 6$

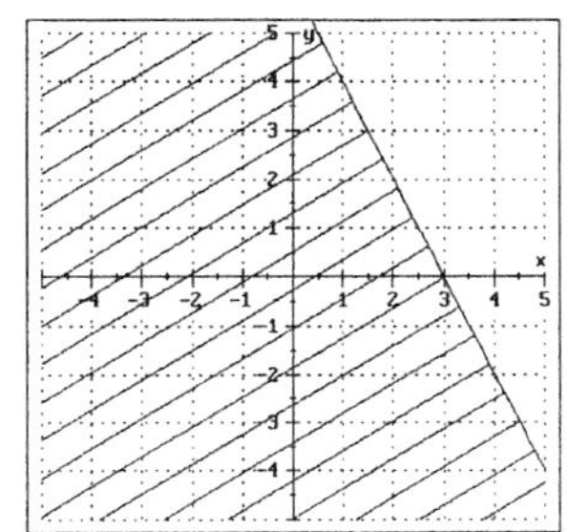

14. $x - 2y \ge 4$

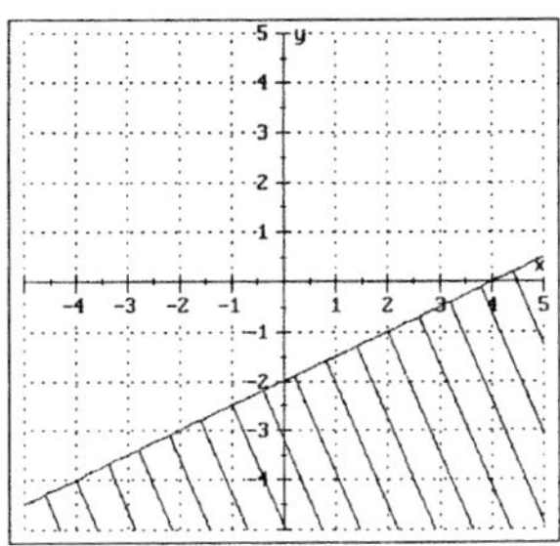

15. $3x \ge -y + 3$

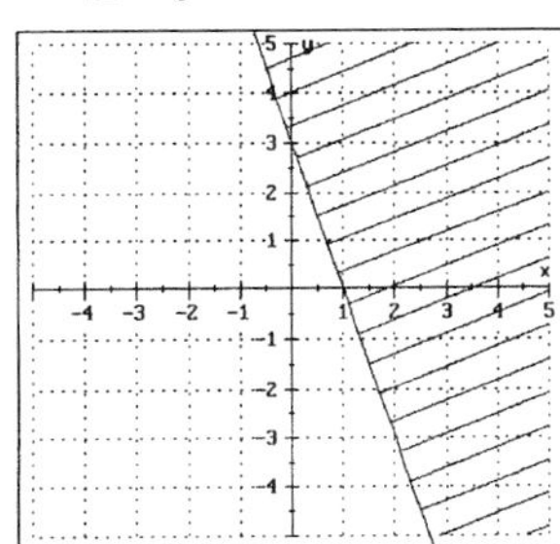

16. $2x \le -3y - 12$

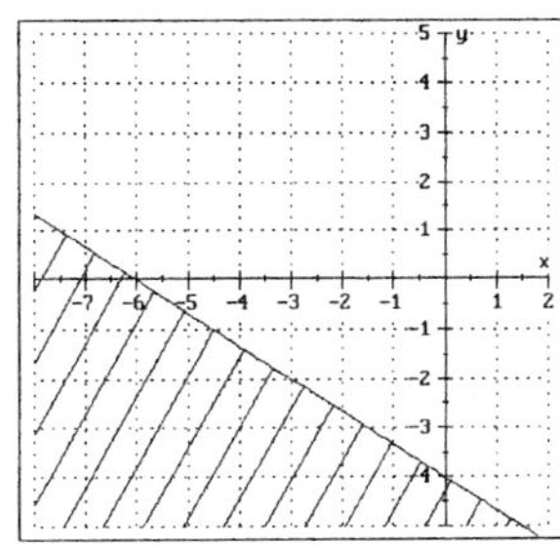

17. $y \ge 1 - \frac{3}{2}x$

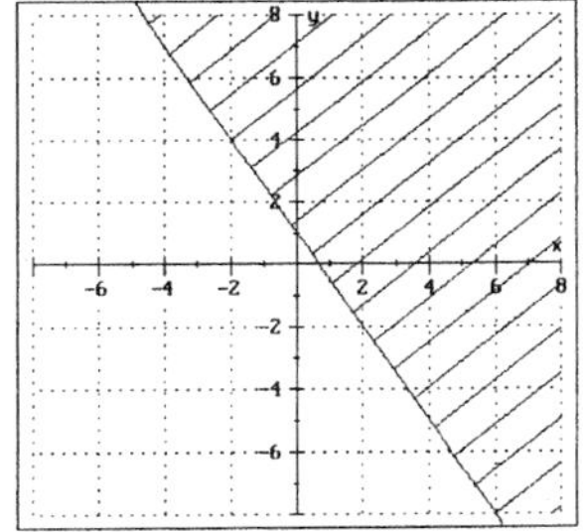

18. $y < \frac{1}{3}x - 1$

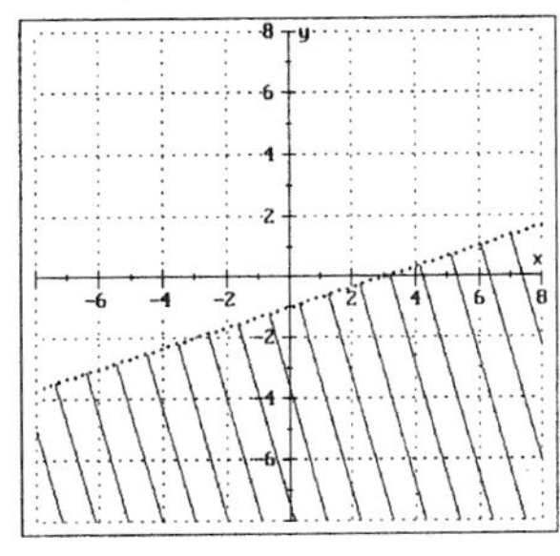

19. $0.5x + 0.5y \le 2$

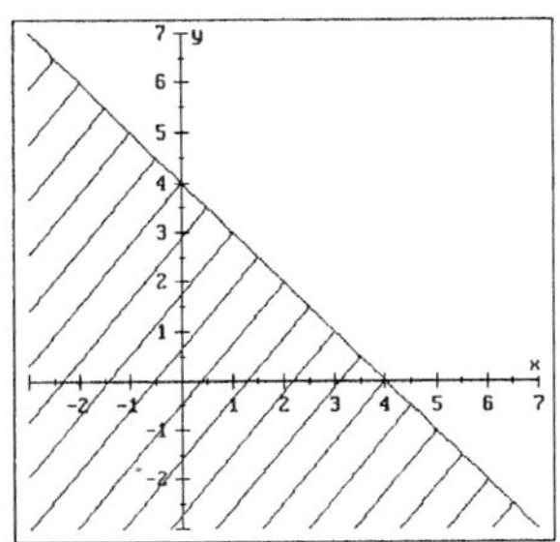

20. $0.5x + y > 1.5 + x$

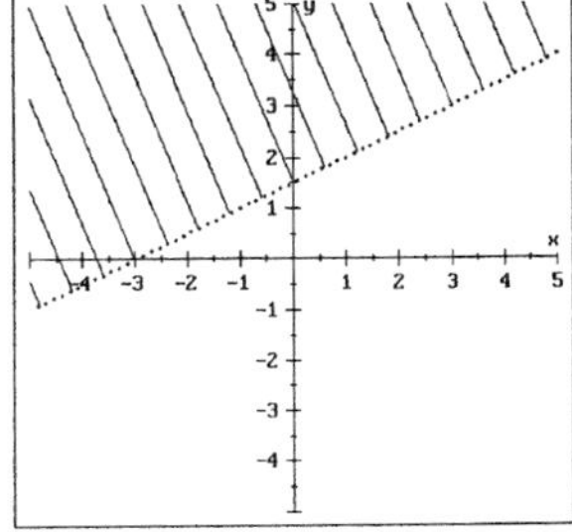

21. $x < 4$

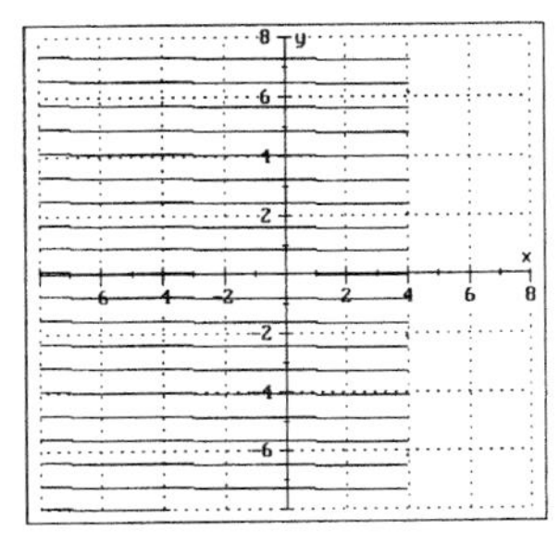

22. $y \ge -2$

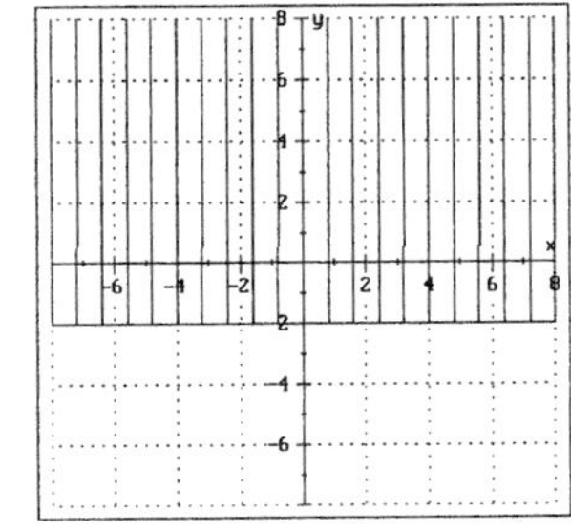

23. $-2 \le x < 0$

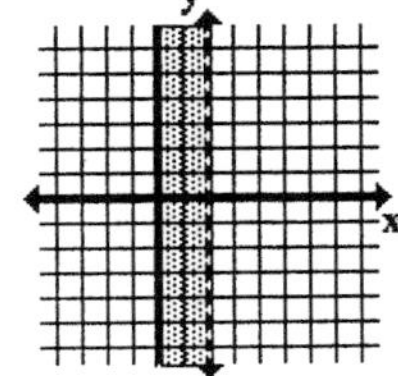

24. $-3 < y \le -1$

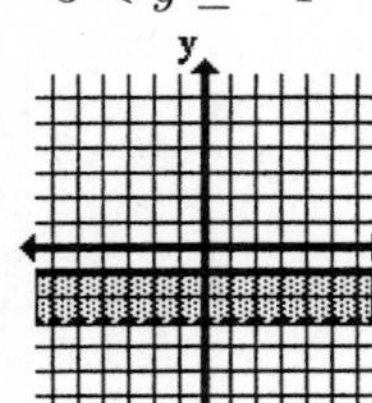

25. $y < -2$ or $y > 3$

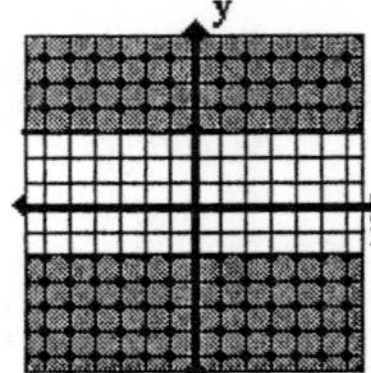

26. $-x \le 1$ or $x \ge 2$
$x \ge -1$ or $x \ge 2$

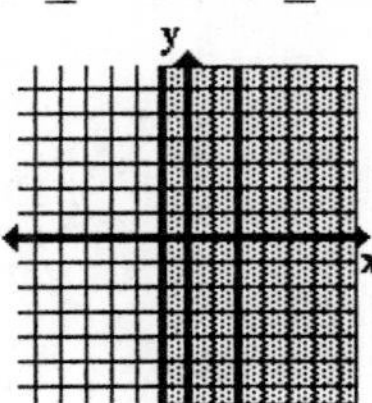

27. The boundary line goes through $(2, 0)$ and $(0, 3)$. Use point-slope form to find the equation:

$$m = \frac{\Delta y}{\Delta x} = \frac{3-0}{0-2} = \frac{3}{-2} = -\frac{3}{2}$$
$$y - y_1 = m(x - x_1)$$
$$y - 0 = -\frac{3}{2}(x - 2)$$
$$y = -\frac{3}{2}x + 3$$
$$3x + 2y = 6$$

Pick a point which is shaded, and substitute the coordinates for x and y:

$$3x + 2y \;\square\; 6$$
$$3(2) + 2(1) \;\square\; 6$$
$$5 + 2 > 6$$

Since the line is dotted, do not include the line: $\boxed{3x + 2y > 6}$

28. The boundary line goes through $(-1, 0)$ and $(0, 2)$. Use point-slope form to find the equation:

$$m = \frac{\Delta y}{\Delta x} = \frac{2-0}{0-(-1)} = \frac{2}{1} = 2$$
$$y - y_1 = m(x - x_1)$$
$$y - 2 = 2(x - 0)$$
$$y - 2 = 2x$$
$$2x - y = -2$$

Pick a point which is shaded, and substitute the coordinates for x and y:

$$2x - y \;\square\; -2$$
$$2(0) - 3 \;\square\; -2$$
$$-3 < -2$$

Since the line is not dotted, include the line: $\boxed{2x - y \le -2}$

29. The boundary line is $x = 3$. Since points with x-coordinates less than 3 are shaded, and since the line is not dotted, the inequality is $\boxed{x \le 3}$.

30. The boundary line is $y = -2$. Since points with y-coordinates greater than -2 are shaded, and since the line is dotted, the inequality is $\boxed{y > -2}$.

31. The boundary line goes through $(0, 0)$ and $(1, 1)$. Use point-slope form to find the equation:

$$m = \frac{\Delta y}{\Delta x} = \frac{1-0}{1-0} = \frac{1}{1} = 1$$
$$y - y_1 = m(x - x_1)$$
$$y - 0 = 1(x - 0)$$
$$y = x$$

Pick a point which is shaded, and substitute the coordinates for x and y:

$$y \;\square\; x$$
$$0 < 1$$

Since the line is not dotted, include the line: $\boxed{y \le x}$

32. The boundary line goes through $(0, 0)$ and $(1, -1)$. Use point-slope form to find the equation:

$$m = \frac{\Delta y}{\Delta x} = \frac{-1-0}{1-0} = \frac{-1}{1} = -1$$

$$y - y_1 = m(x - x_1)$$
$$y - 0 = -1(x - 0)$$
$$y = -x$$

Pick a point which is shaded, and substitute the coordinates for x and y:

$y \square -x$

$1 > 0$

Since the line is dotted, do not include the line:

$\boxed{y > -x}$

33. The points between the lines $x = -2$ and $x = 3$ are shaded. Both lines are not dotted, so they should be included.

$\boxed{-2 \leq x \leq 3}$

34. The points between the lines $y = -2$ and $y = 3$ are shaded. Both lines are dotted, so they should not be included.

$\boxed{-2 < y < 3}$

35. The points above the line $y = -1$ (dotted) and below the line $y = -3$ (not dotted) are shaded. $\boxed{y > -1 \text{ or } y \leq -3}$

36. The points to the left of the line $x = 1$ and to the right of the line $x = 3$ (neither is dotted) are shaded. $\boxed{x \leq 1 \text{ or } x \geq 3}$

37. $y < 0.27x - 1$

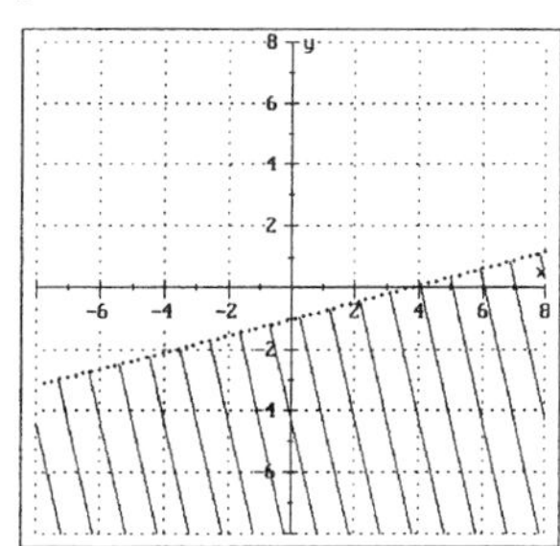

38. $y > -3.5x + 2.7$

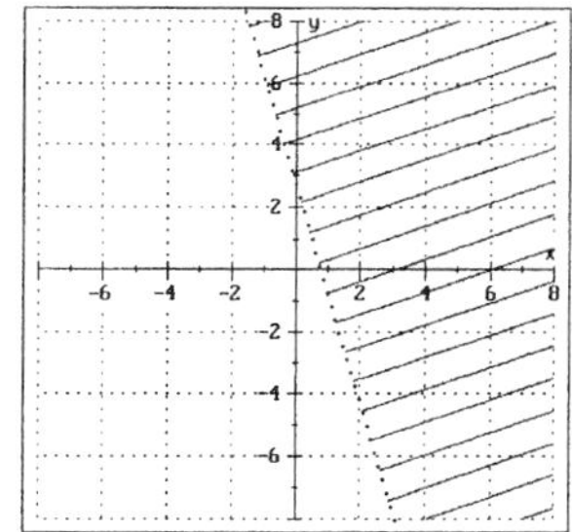

39. $y \geq -2.37x + 1.5$

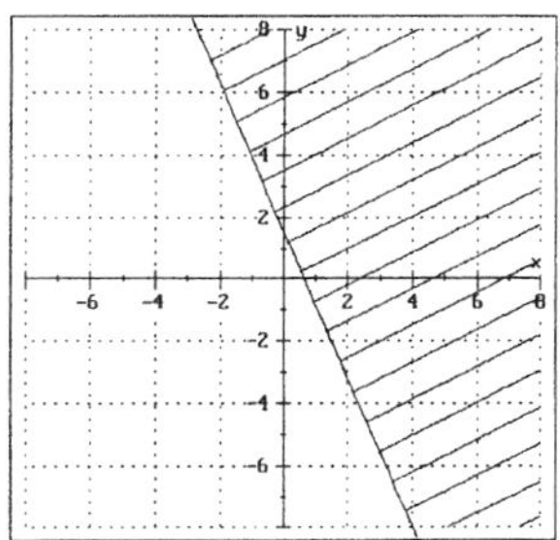

40. $y \leq 3.37x - 1.7$

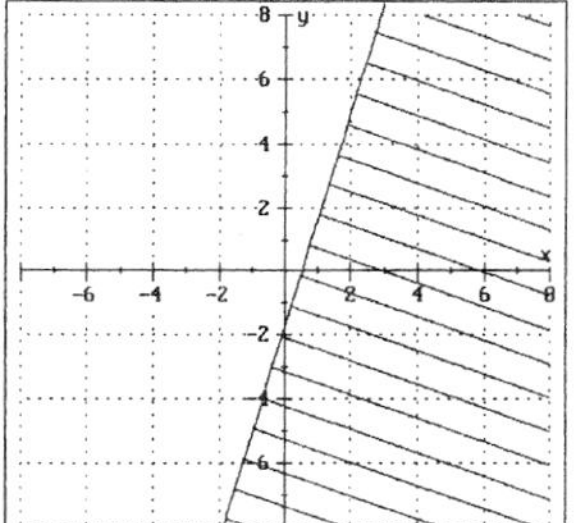

41. Let $x =$ the number of simple returns completed, and let $y =$ the number of complicated returns completed. The inequality is $x + 3y \leq 9$. Some ordered pairs are: $(1, 1)$, $(2, 1)$, $(2, 2)$

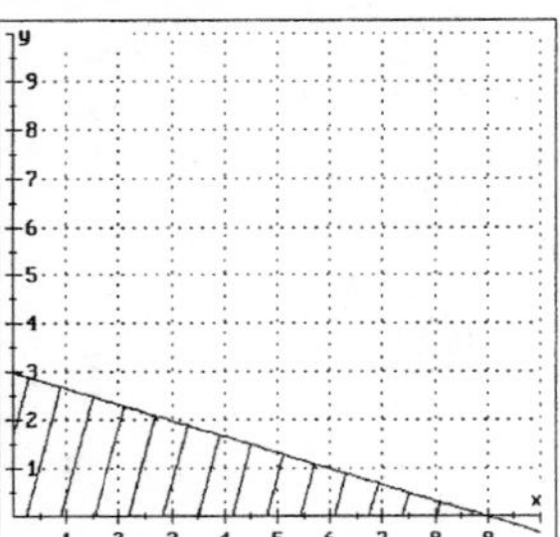

42. Let $x =$ the number of maple trees sold, and let $y =$ the number of pine trees sold. The inequality is $100x + 125y > 2000$. Some ordered pairs are: $(0, 17)$, $(5, 20)$, $(15, 10)$

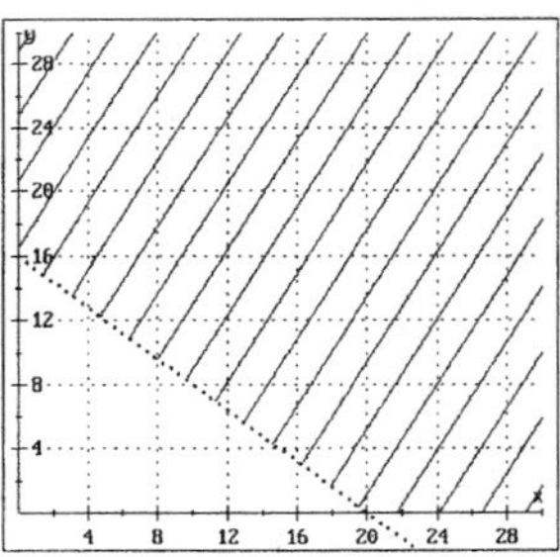

43. Let $x =$ the number of hours she uses the first, and let $y =$ the number of hours she uses the second. The inequality is $6x + 7y \leq 42$. Some ordered pairs are: $(2, 2)$, $(3, 3)$, $(5, 1)$

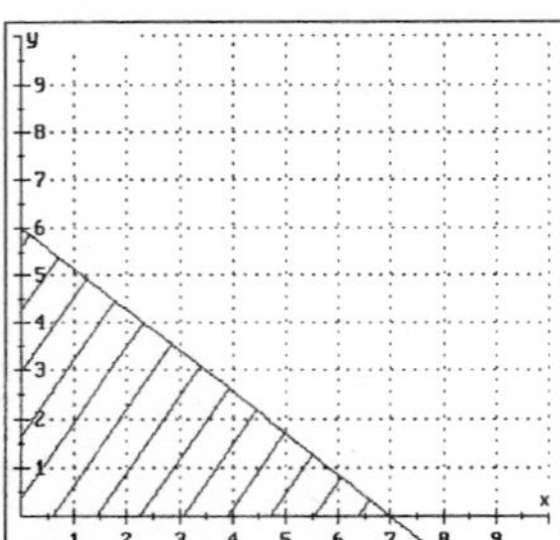

44. Let $x =$ the number of rods made, and let $y =$ the number of reels made. The inequality is $10x + 15y \geq 1200$. Some ordered pairs are: $(40, 80)$, $(80, 80)$, $(120, 40)$

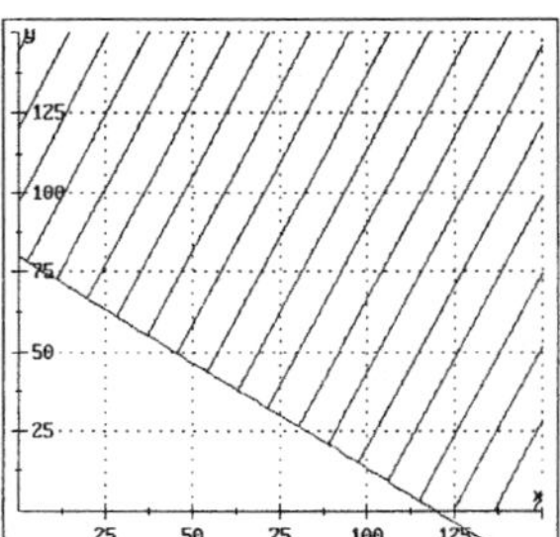

45. Let $x =$ the number of shares of Traffico, and let $y =$ the number of shares of Cleanco. The inequality is $50x + 60y \leq 6000$. Some ordered pairs are: $(40, 20)$, $(60, 40)$, $(80, 20)$

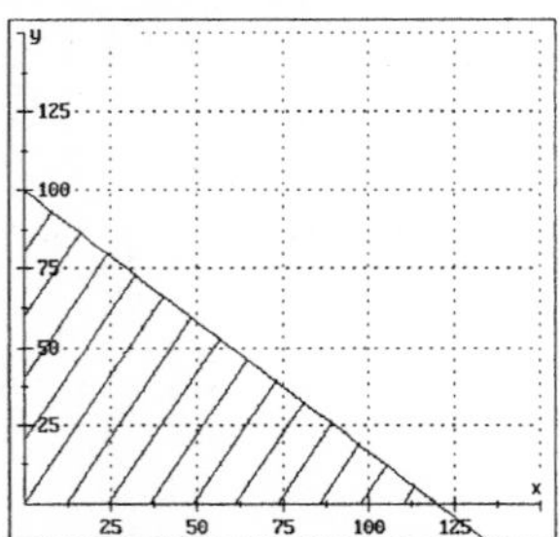

46. Let $x =$ the number of reserved tickets, and let $y =$ the number of general tickets. The inequality is $6x + 4y \geq 10200$. Some ordered pairs are: $(1800, 0)$, $(1000, 1500)$, $(2000, 2000)$

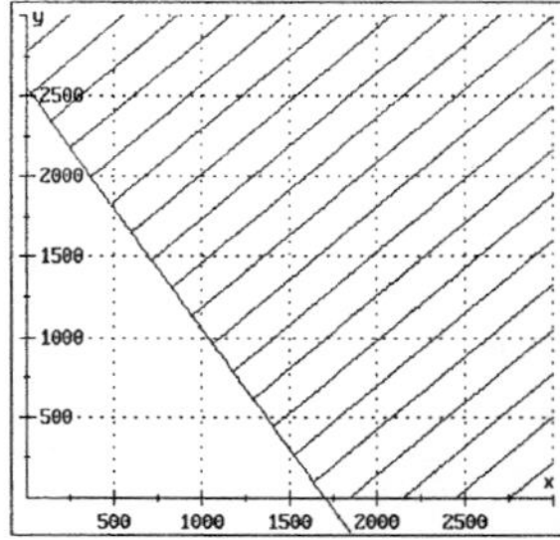

47. **Answers may vary.**

48. **Answers may vary.**

49. An inequality such as $x + (-x) < 10$ is an identity.

50. An inequality such as $x + (-x) > 10$ has no solutions.

Exercise 4.4 (page 253)

1.
$$\begin{aligned} A &= p + prt \\ A - p &= prt \\ \frac{A-p}{pt} &= r, \text{ or } r = \frac{A-p}{pt} \end{aligned}$$

2.
$$\begin{aligned} C &= \frac{5}{9}(F - 32) \\ 9C &= 5(F - 32) \\ 9C &= 5F - 160 \\ 9C + 160 &= 5F \\ \frac{9C + 160}{5} &= F \\ \tfrac{9}{5}C + 32 &= F, \text{ or } F = \tfrac{9}{5}C + 32 \end{aligned}$$

3.
$$\begin{aligned} z &= \frac{x - \mu}{\sigma} \\ z\sigma &= x - \mu \\ z\sigma + \mu &= x, \text{ or } x = z\sigma + \mu \end{aligned}$$

4.
$$\begin{aligned} P &= 2l + 2w \\ P - 2l &= 2w \\ \frac{P - 2l}{2} &= w, \text{ or } w = \frac{P - 2l}{2} \end{aligned}$$

5.
$$\begin{aligned} l &= a + (n-1)d \\ l - a &= (n-1)d \\ \frac{l-a}{n-1} &= d, \text{ or } d = \frac{l-a}{n-1} \end{aligned}$$

6.
$$\begin{aligned} z &= \frac{x - \mu}{\sigma} \\ z\sigma &= x - \mu \\ \mu &= x - z\sigma \end{aligned}$$

7. intersect

8. solid

9. $\begin{cases} y < 3x + 2 \\ y < -2x + 3 \end{cases}$

10. $\begin{cases} y \le x - 2 \\ y \ge 2x + 1 \end{cases}$

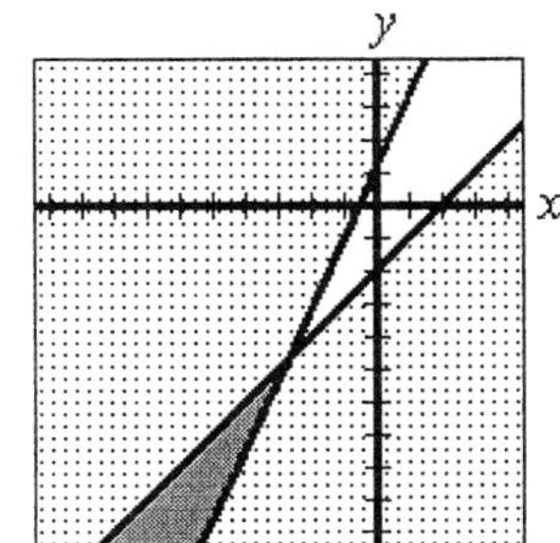

11. $\begin{cases} 3x + 2y > 6 \\ x + 3y \le 2 \end{cases}$

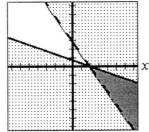

12. $\begin{cases} x + y < 2 \\ x + y \leq 1 \end{cases}$

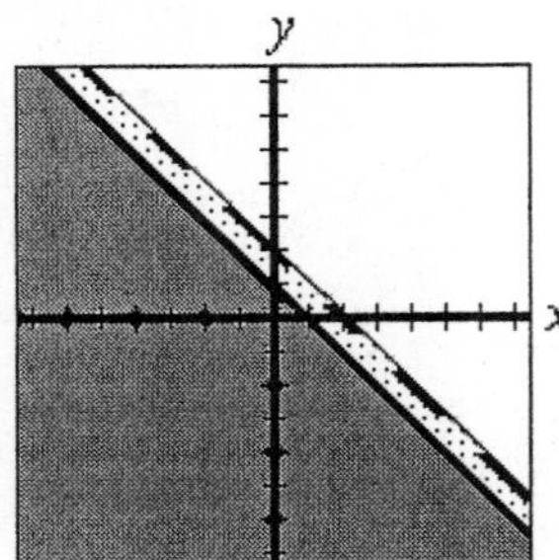

13. $\begin{cases} 3x + y \leq 1 \\ -x + 2y \geq 6 \end{cases}$

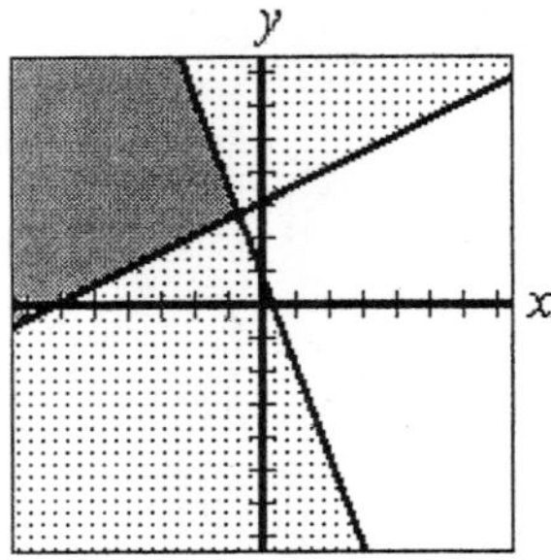

14. $\begin{cases} x + 2y < 3 \\ 2x + 4y < 8 \end{cases}$

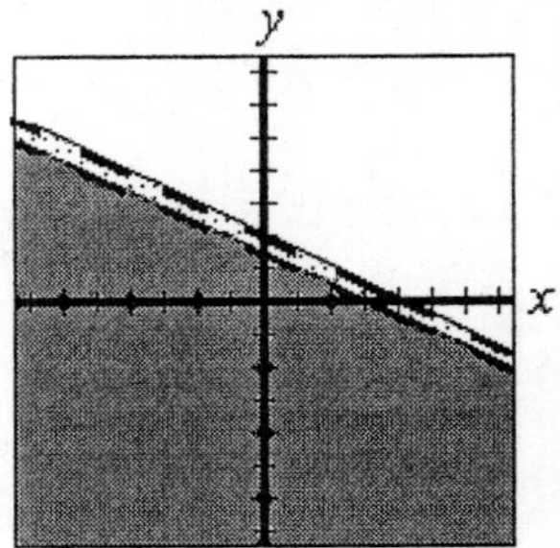

15. $\begin{cases} 2x - y > 4 \\ y < \ -x^2 + 2 \end{cases}$

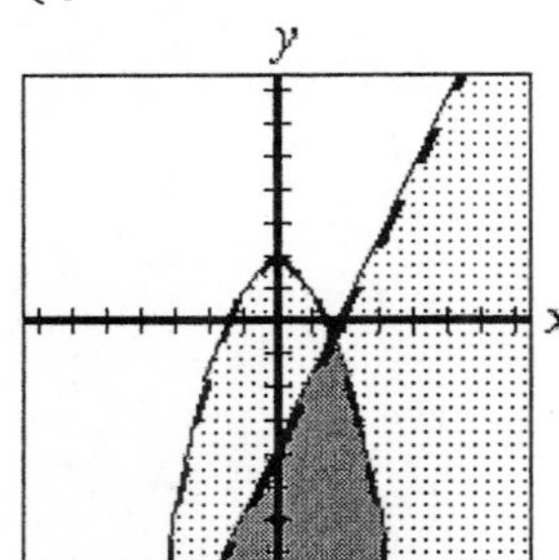

16. $\begin{cases} x \leq y^2 \\ y \geq x \end{cases}$

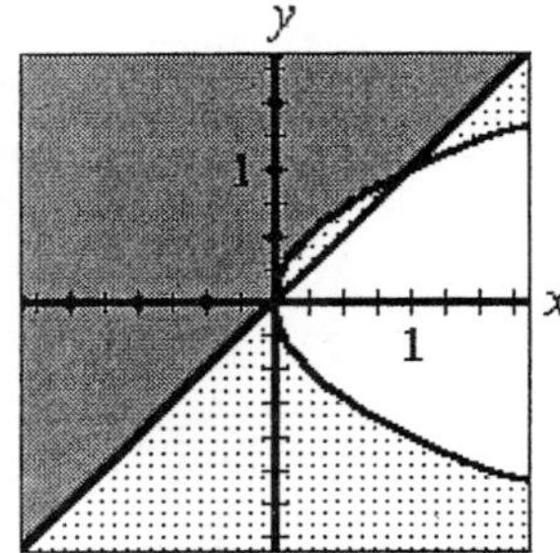

17. $\begin{cases} y > x^2 - 4 \\ y < -x^2 + 4 \end{cases}$

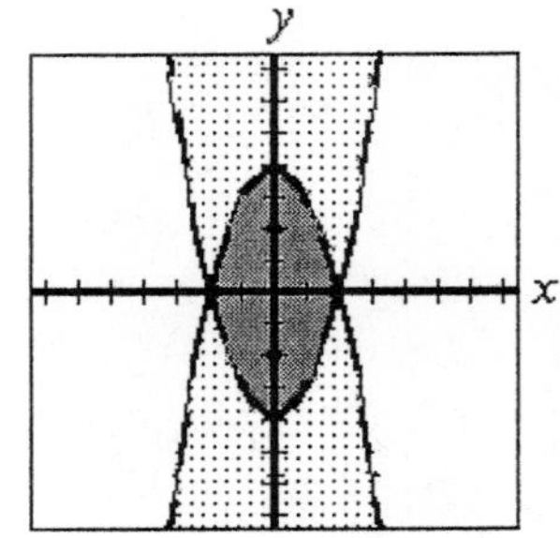

18. $\begin{cases} x \geq y^2 \\ y \geq x^2 \end{cases}$

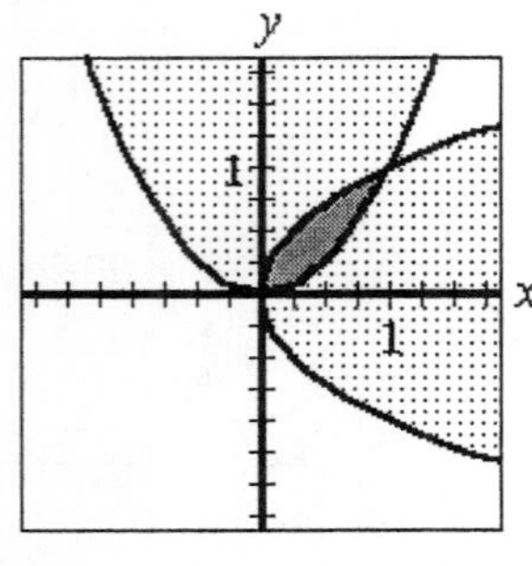

19. $\begin{cases} 2x + 3y \leq 6 \\ 3x + y \leq 1 \\ x \leq 0 \end{cases}$

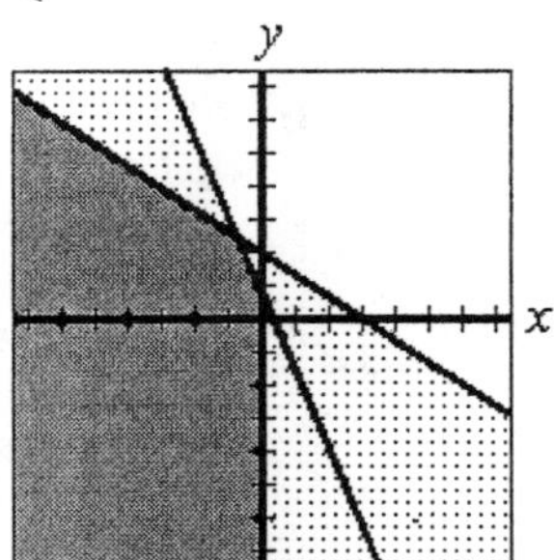

20. $\begin{cases} 2x + y \leq 2 \\ y \geq x \\ x \geq 0 \end{cases}$

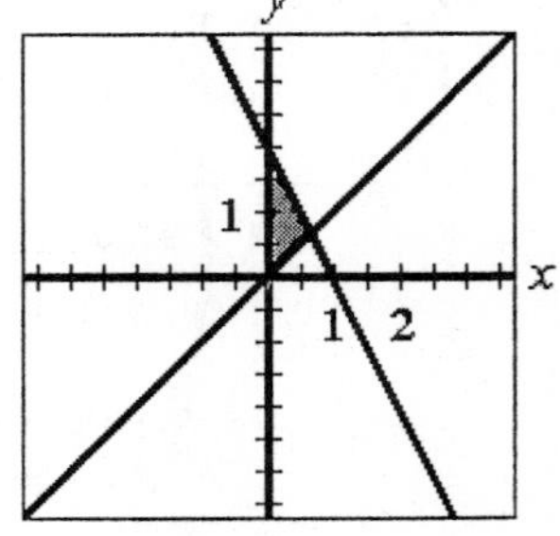

21. $\begin{cases} x - y < 4 \\ y \le 0 \\ x \ge 0 \end{cases}$

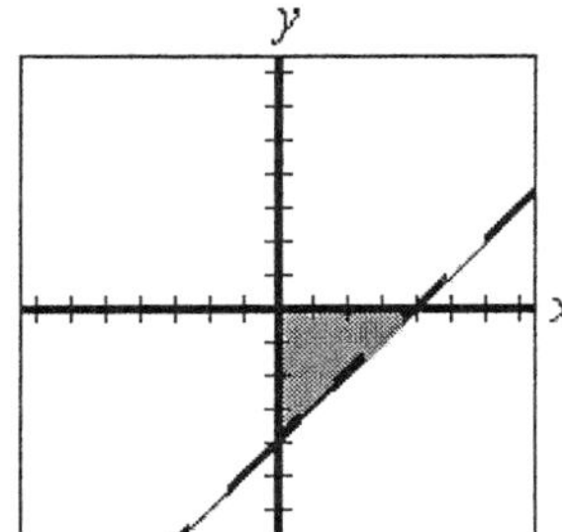

22. 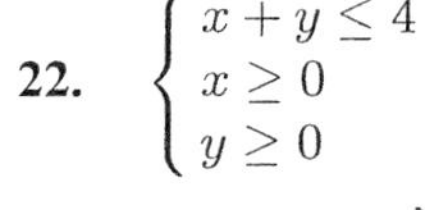
$\begin{cases} x + y \le 4 \\ x \ge 0 \\ y \ge 0 \end{cases}$

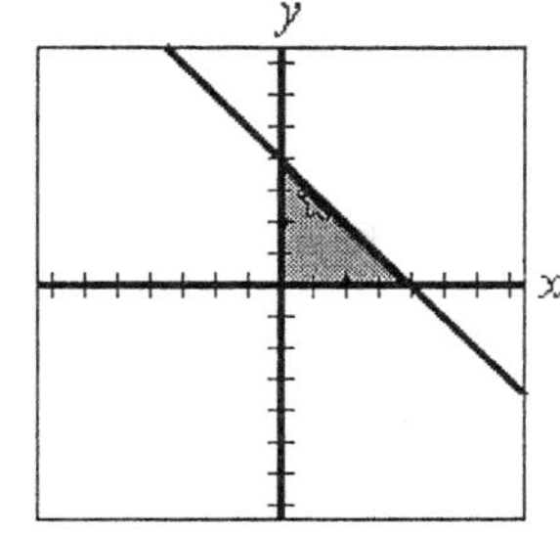

23. 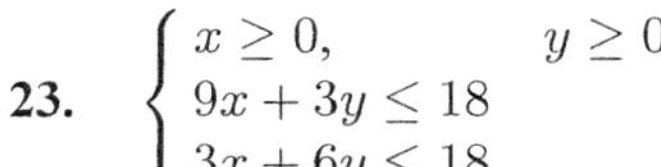
$\begin{cases} x \ge 0, \qquad y \ge 0 \\ 9x + 3y \le 18 \\ 3x + 6y \le 18 \end{cases}$

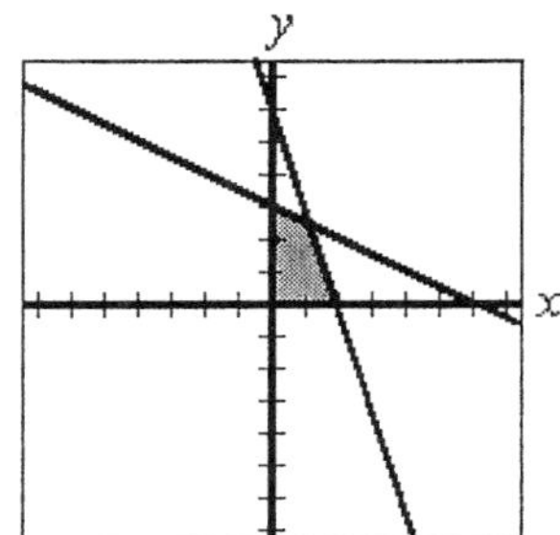

24. $\begin{cases} x + y \ge 1, \quad x - y \le 1 \\ x - y \ge 0, \quad x \le 2 \end{cases}$

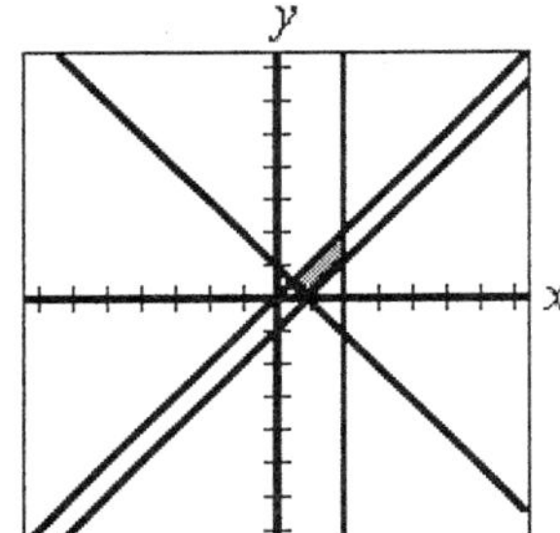

25. See #9.

26. See #17.

27.

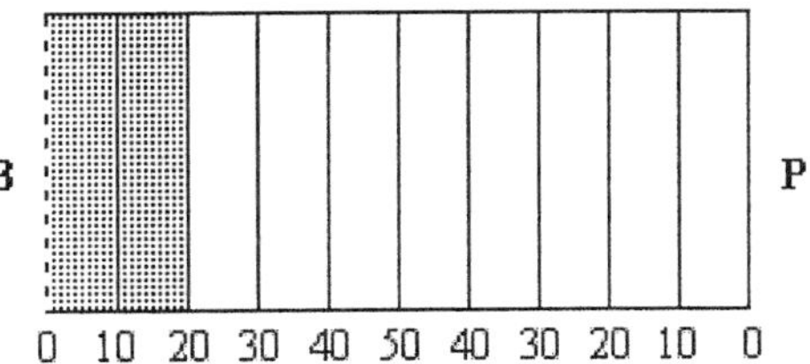

28.

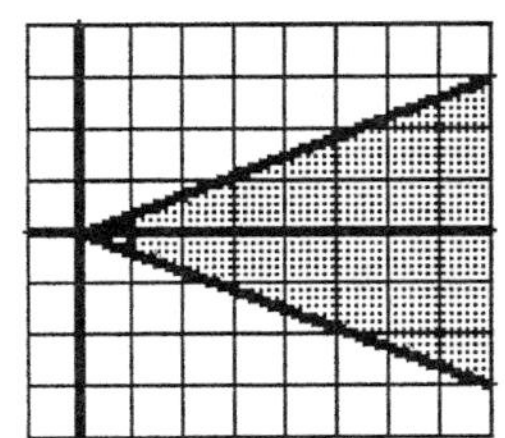

29. Let $x =$ the number of \$10 discs and let $y =$ the number of \$15 discs.

$$\begin{cases} 10x + 15y \geq 30, & x \geq 0 \\ 10x + 15y \leq 60, & y \geq 0 \end{cases}$$

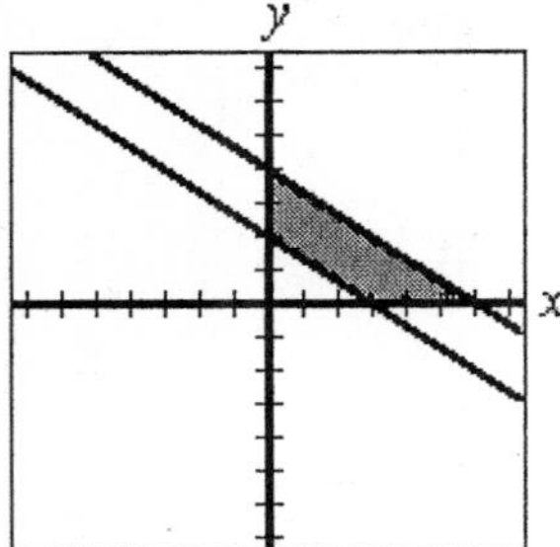

Two solutions are $(1, 2)$ and $(4, 1)$. A customer could buy one \$10 and two \$15 discs, or four \$10 and one \$15 disc.

30. Let $x =$ the # of aluminum boats and let $y =$ the # of fiberglass boats.

$$\begin{cases} 800x + 600y \geq 2400, & x \geq 0 \\ 800x + 600y \leq 4800, & y \geq 0 \end{cases}$$

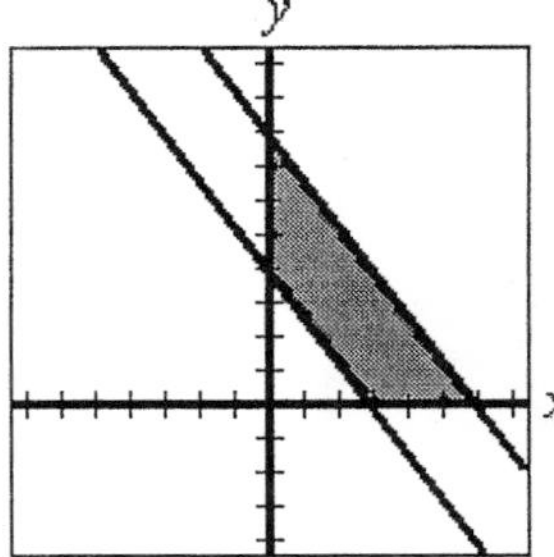

Two solutions are $(4, 1)$ and $(1, 4)$. Northland could buy 4 aluminum and 1 fiberglass, or 1 aluminum and 4 fiberglass.

31. Let $x =$ the # of desk chairs, and let $y =$ the # of side chairs.

$$\begin{cases} 150x + 100y \leq 900, & y > x \\ x \geq 0, & y \geq 0 \end{cases}$$

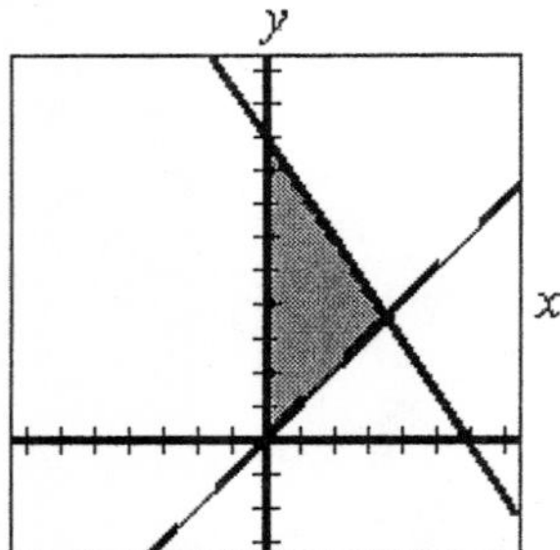

Two solutions are $(2, 4)$ and $(1, 5)$. Best could buy 2 desk chairs and 4 side chairs, or 1 desk chair and 5 side chairs.

32. Let $x =$ the # of air cleaners, and let $y =$ the # of humidifiers.

$$\begin{cases} 500x + 200y \leq 2000, & y > x \\ x \geq 0, & y \geq 0 \end{cases}$$

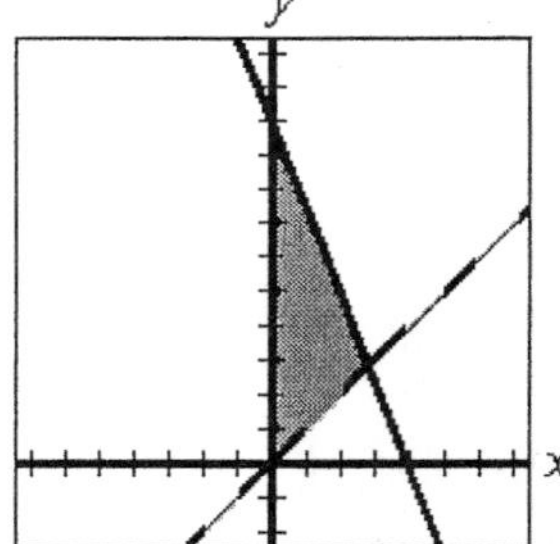

Two solutions are $(1, 2)$ and $(2, 3)$. The company could buy 1 cleaner and 2 humidifiers, or 2 cleaners and 3 humidifiers.

33. **Answers may vary.**

34. **Answers may vary.**

35. No.

36. The graph of $y \geq |x|$ is shown here. Graphing $y \leq k$ on the same axes, and taking only the points which satisfy both inequalities, will result in a triangular region with a vertex at the origin, and a horizontal side through $y = k$. The area will then be $A = \frac{1}{2}bh = \frac{1}{2}(2k)k = k^2$. $k^2 = 25 \Rightarrow \boxed{k = 5}$.

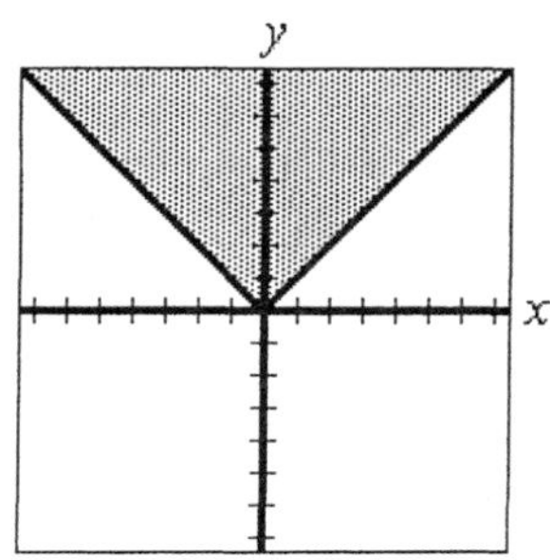

Exercise 4.5 (page 264)

1. $m = \dfrac{\Delta y}{\Delta x} = \dfrac{4-7}{-2-5} = \dfrac{-3}{-7} = \dfrac{3}{7}$

2.
$$\begin{aligned} y - y_1 &= m(x - x_1) \\ y - 7 &= \tfrac{3}{7}(x - 5) \\ 7y - 49 &= 3(x - 5) \\ 7y - 49 &= 3x - 15 \\ -3x + 7y &= 34 \\ 3x - 7y &= -34 \end{aligned}$$

3.
$$\begin{aligned} y - y_1 &= m(x - x_1) \\ y - 7 &= \tfrac{3}{7}(x - 5) \\ y - 7 &= \tfrac{3}{7}x - \tfrac{15}{7} \\ y &= \tfrac{3}{7}x + \tfrac{34}{7} \end{aligned}$$

4.
$$\begin{aligned} y - y_1 &= m(x - x_1) \\ y - 0 &= \tfrac{3}{7}(x - 0) \\ y &= \tfrac{3}{7}x \end{aligned}$$

5. constraints

6. feasible

7. objective

8. corner; edge

9.

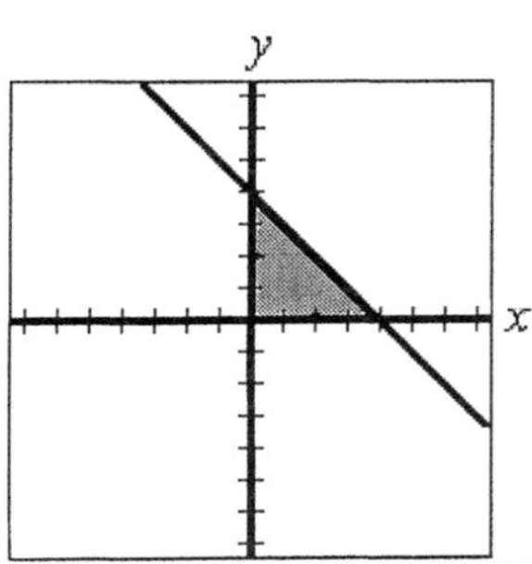

Vertex	$P = 2x + 3y$	Maximum?
$(0, 0)$	$= 0$	No
$(4, 0)$	$= 8$	No
$(0, 4)$	$= 12$	YES

P has a maximum value of 12 at $(0, 4)$.

10.

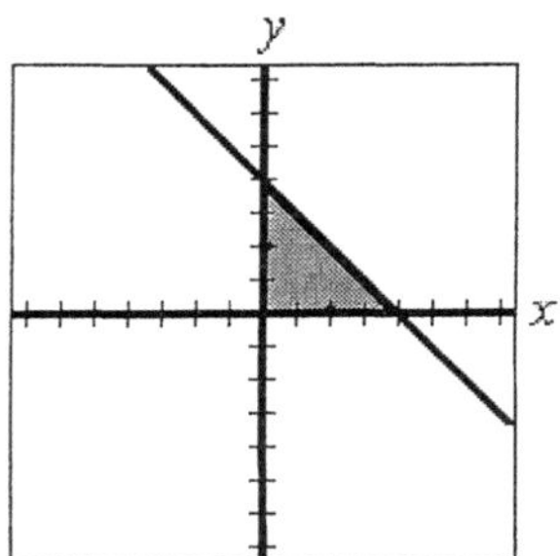

Vertex	$P = 3x + 2y$	Maximum?
$(0, 0)$	$= 0$	No
$(4, 0)$	$= 12$	YES
$(0, 4)$	$= 8$	No

P has a maximum value of 12 at $(4, 0)$.

11.

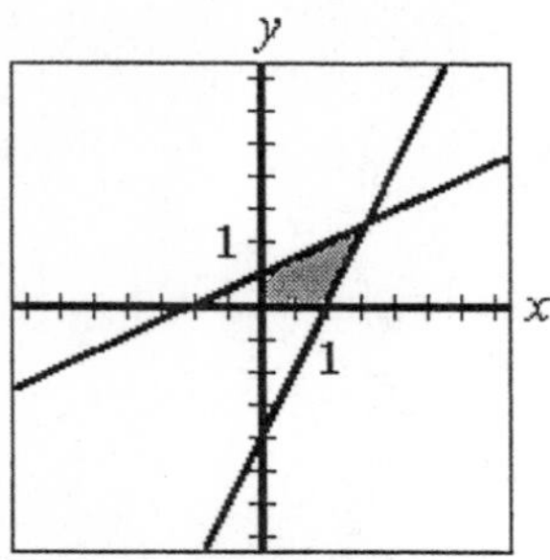

Vertex	$P = y + \frac{1}{2}x$	Maximum?
$(0, 0)$	$= 0$	No
$(1, 0)$	$= \frac{1}{2}$	No
$\left(\frac{5}{3}, \frac{4}{3}\right)$	$= \frac{13}{6}$	YES
$\left(0, \frac{1}{2}\right)$	$= \frac{1}{2}$	No

P has a maximum value of $\frac{13}{6}$ at $\left(\frac{5}{3}, \frac{4}{3}\right)$.

12.

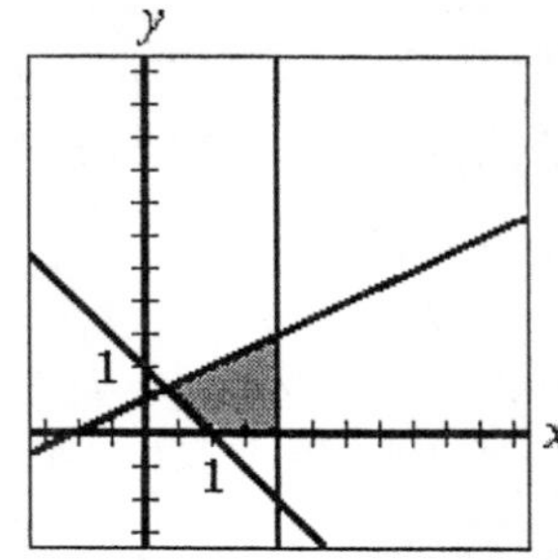

Vertex	$P = 4y - x$	Maximum?
$(1, 0)$	$= -1$	No
$(2, 0)$	$= -2$	No
$\left(2, \frac{3}{2}\right)$	$= 4$	YES
$\left(\frac{1}{3}, \frac{2}{3}\right)$	$= \frac{7}{3}$	No

P has a maximum value of 4 at $\left(2, \frac{3}{2}\right)$.

13.

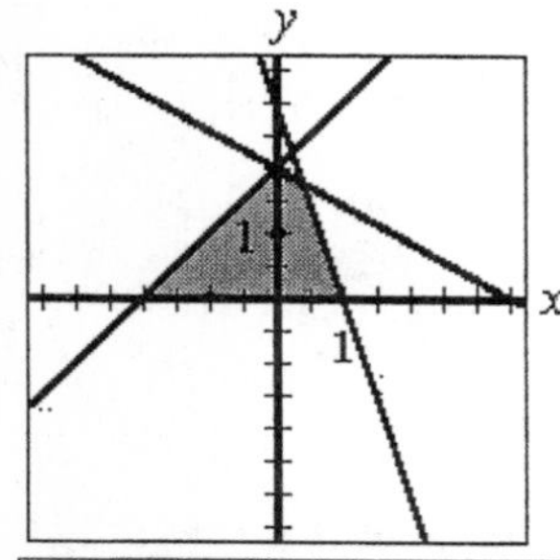

Vertex	$P = 2x + y$	Maximum?
$(1, 0)$	$= 0$	No
$\left(\frac{3}{7}, \frac{12}{7}\right)$	$= \frac{18}{7}$	YES
$(0, 2)$	$= 2$	No
$(-2, 0)$	$= -4$	No

P has a maximum value of $\frac{18}{7}$ at $\left(\frac{3}{7}, \frac{12}{7}\right)$.

14.

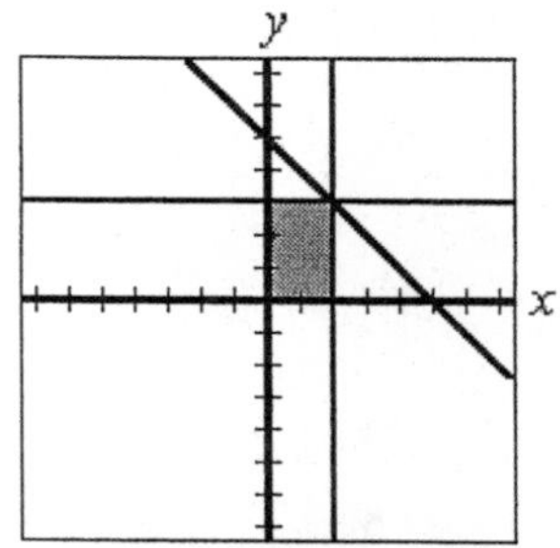

Vertex	$P = x - 2y$	Maximum?
$(0, 0)$	$= 0$	No
$(2, 0)$	$= 2$	YES
$(2, 3)$	$= -4$	No
$(0, 3)$	$= -6$	No

P has a maximum value of 2 at $(2, 0)$.

15.

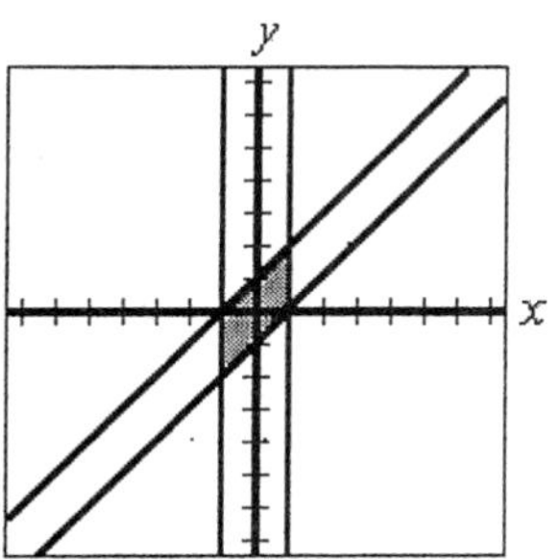

Vertex	$P = 3x - 2y$	Maximum?
$(1,0)$	$= 3$	YES
$(1,2)$	$= -1$	No
$(-1,0)$	$= -3$	No
$(-1,-2)$	$= 1$	No

P has a maximum value of 3 at $(1,0)$.

16.

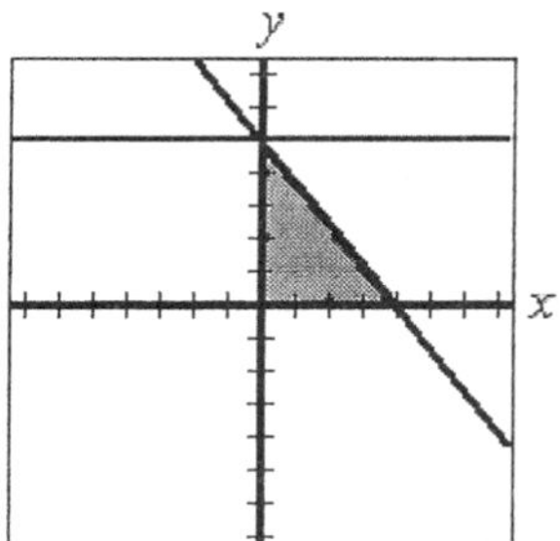

Vertex	$P = x - y$	Maximum?
$(0,0)$	$= 0$	No
$(4,0)$	$= 4$	YES
$(0,5)$	$= -5$	No

P has a maximum value of 4 at $(4,0)$.

17.

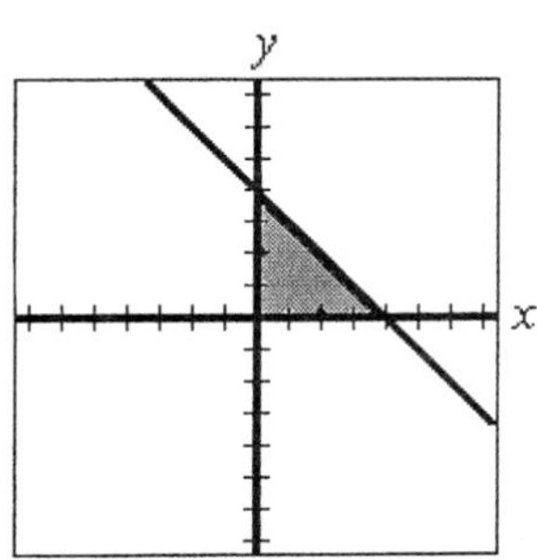

Vertex	$P = 5x + 12y$	Minimum?
$(4,0)$	$= 20$	No
$(0,4)$	$= 48$	No
$(0,0)$	$= 0$	YES

P has a minimum value of 0 at $(0,0)$.

18.

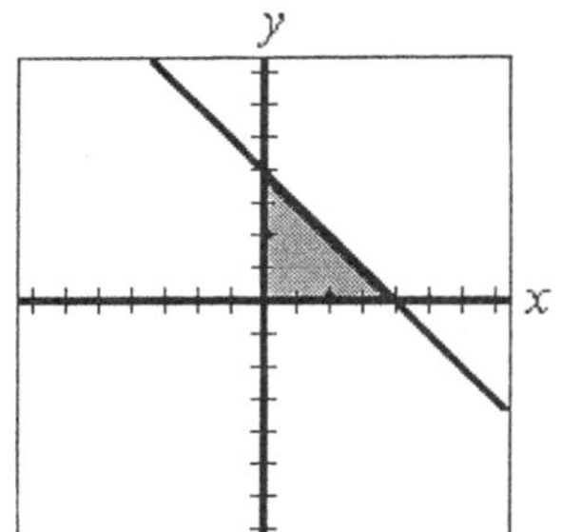

Vertex	$P = 3x + 6y$	Minimum?
$(4,0)$	$= 12$	No
$(0,4)$	$= 24$	No
$(0,0)$	$= 0$	YES

P has a minimum value of 0 at $(0,0)$.

19.

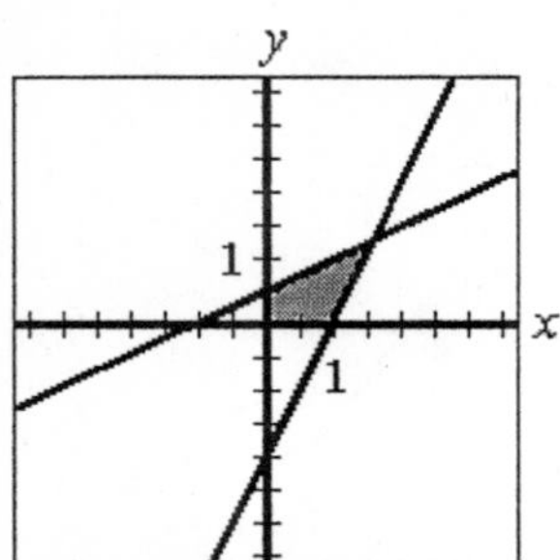

Vertex	$P = 3y + x$	Minimum?
$(0,0)$	$= 0$	YES
$(1,0)$	$= 1$	No
$\left(\frac{5}{3}, \frac{4}{3}\right)$	$= \frac{17}{3}$	No
$\left(0, \frac{1}{2}\right)$	$= \frac{3}{2}$	No

P has a minimum value of 0 at $(0,0)$.

20.

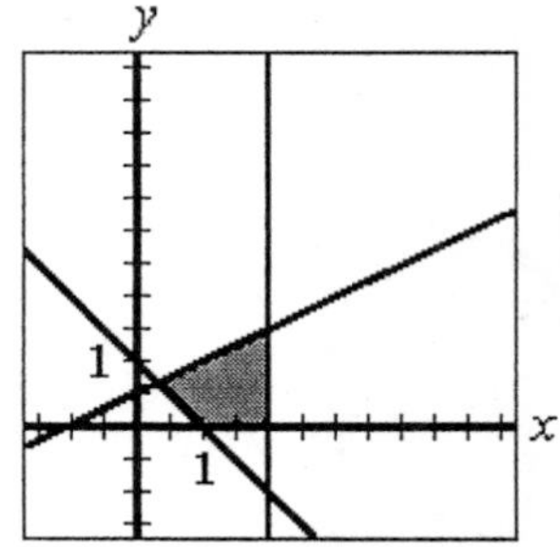

Vertex	$P = 5y + x$	Minimum?
$(1,0)$	$= 1$	YES
$(2,0)$	$= 2$	No
$\left(2, \frac{3}{2}\right)$	$= \frac{19}{2}$	No
$\left(\frac{1}{3}, \frac{2}{3}\right)$	$= \frac{11}{3}$	No

P has a minimum value of 1 at $(1,0)$.

21.

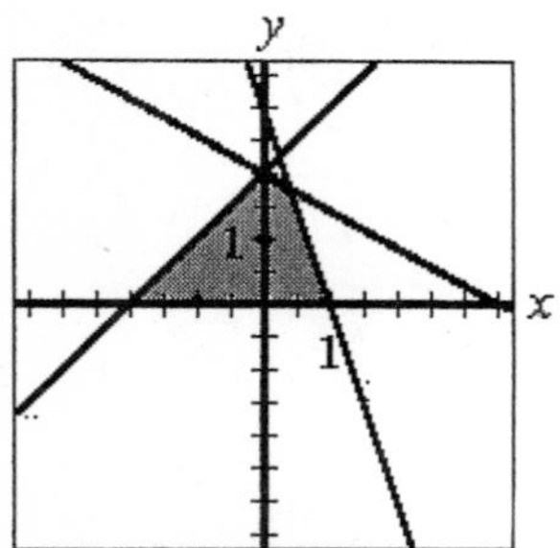

Vertex	$P = 6x + 2y$	Minimum?
$(1,0)$	$= 6$	No
$\left(\frac{3}{7}, \frac{12}{7}\right)$	$= 6$	No
$(0,2)$	$= 4$	No
$(-2,0)$	$= -12$	YES

P has a minimum value of -12 at $(-2,0)$.

22.

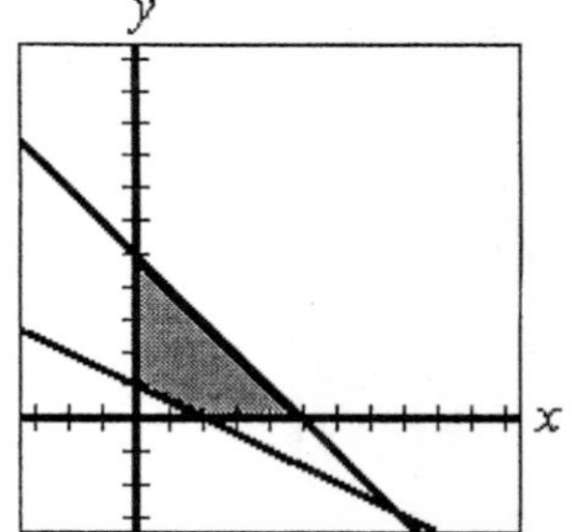

Vertex	$P = 2y - x$	Minimum?
$(2,0)$	$= -2$	No
$(5,0)$	$= -5$	YES
$(0,5)$	$= 10$	No
$(0,1)$	$= 2$	No

P has a minimum value of -5 at $(5,0)$.

23.

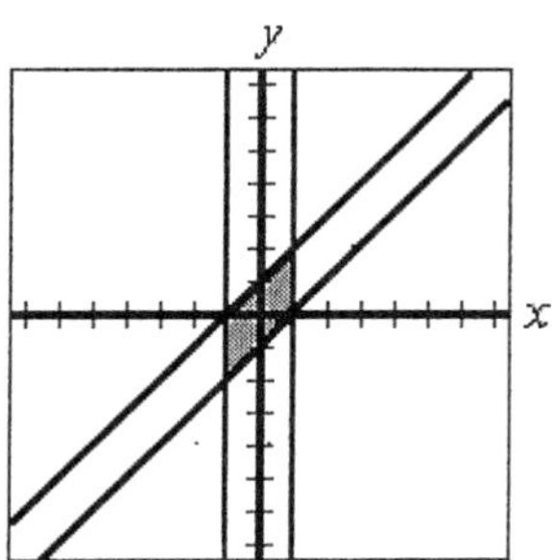

Vertex	$P = 2x - 2y$	Minimum?
$(1, 0)$	$= 2$	No
$(1, 2)$	$= -2$	YES
$(-1, 0)$	$= -2$	YES
$(-1, -2)$	$= 2$	No

P has a minimum value of -2 at $(1, 2)$ and at $(-1, 0)$.

24.

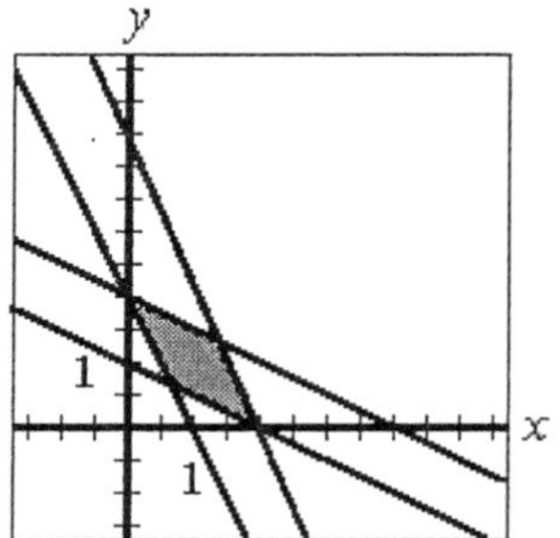

Vertex	$P = y - 2x$	Minimum?
$(\frac{2}{3}, \frac{2}{3})$	$= -\frac{2}{3}$	No
$(2, 0)$	$= -4$	YES
$(\frac{4}{3}, \frac{4}{3})$	$= -\frac{4}{3}$	No
$(0, 2)$	$= 2$	No

P has a minimum value of -4 at $(2, 0)$.

25. Let $x =$ the number of tables made, and let $y =$ the number of chairs made.

$$\begin{cases} x \geq 0 & y \geq 0 \\ 2x + 3y \leq 42 & \text{(Tom's hours)} \\ 6x + 2y \leq 42 & \text{(Carlos' hours)} \end{cases}$$

Profit $= P = 100x + 80y$.

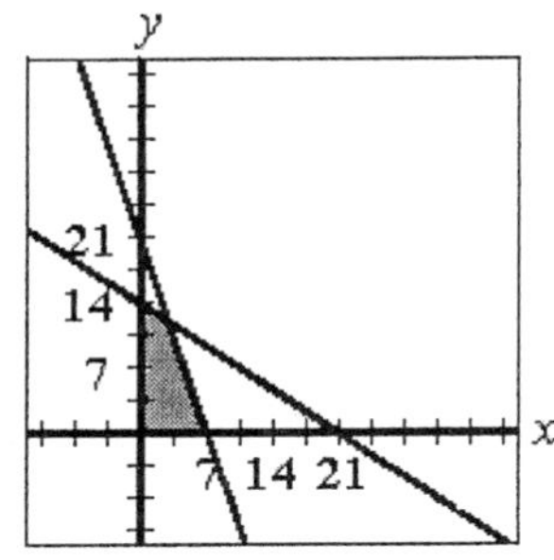

Vertex	$P = 100x + 80y$
$(0, 0)$	$= 0$
$(7, 0)$	$= 700$
$(3, 12)$	$= 1260$
$(0, 14)$	$= 1120$

The maximum income of \$1260 results when they make 3 tables and 12 chairs.

26. Let $x =$ the number of snowmen made, and let $y =$ the number of Santas made.

$$\begin{cases} x \geq 0 & y \geq 0 \\ 4x + 3y \leq 20 & \text{(Nina's hours)} \\ 2x + 4y \leq 20 & \text{(Rob's hours)} \end{cases}$$

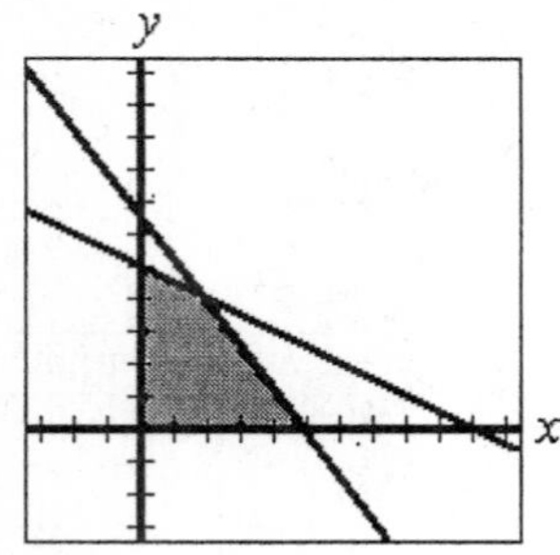

Profit $= P = 80x + 64y$.

Vertex	$P = 80x + 64y$
$(0,0)$	$= 0$
$(5,0)$	$= 400$
$(2,4)$	$= 416$
$(0,5)$	$= 320$

The maximum income of $416 results when they make 2 snowmen and 4 Santas.

27. Let $x =$ the number of IBMs stocked, and let $y =$ the number of Macintoshes stocked.

$$\begin{cases} 20 \leq x \leq 30 \\ 30 \leq y \leq 50 \\ x + y \leq 60 & \text{(Total stock)} \end{cases}$$

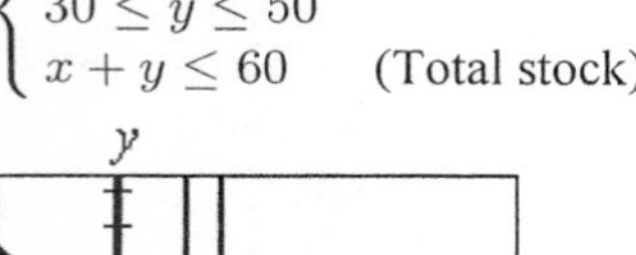

Let the objective function be the total commissions: $P = 50x + 40y$.

Vertex	$P = 50x + 40y$
$(20,30)$	$= 2200$
$(30,30)$	$= 2700$
$(20,40)$	$= 2600$

The maximum amount of commissions is $2700, which results from stocking 30 of each type.

28. Let $x =$ the number of grams of supplement A, and let $y =$ the number of grams of supplement B.

$$\begin{cases} x \geq 0 & y \geq 0 \\ 3x + 2y \geq 16 & \text{(Vitamin C)} \\ 2x + 6y \geq 34 & \text{(Vitamin B)} \end{cases}$$

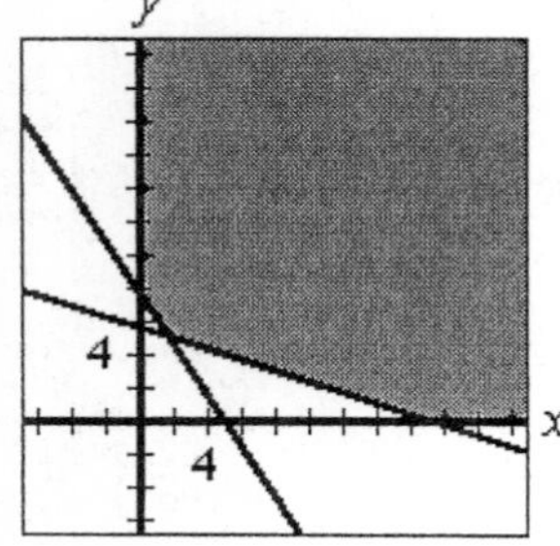

Let the objective function be the total cost: $P = 3x + 4y$.

Vertex	$P = 3x + 4y$
$(17,0)$	$= 51$
$(2,5)$	$= 26$
$(0,8)$	$= 32$

The minimum cost is 26 cents, which results from using 2 grams of supplement A and 5 grams of supplement B.

29. Let $x =$ the number of VCRs made, and let $y =$ the number of TVs made.

$$\begin{cases} x \geq 0 & y \geq 0 \\ 3x + 4y \leq 180 & \text{(electronics hours)} \\ 2x + 3y \leq 120 & \text{(assembly hours)} \\ 2x + y \leq 60 & \text{(finishing hours)} \end{cases}$$

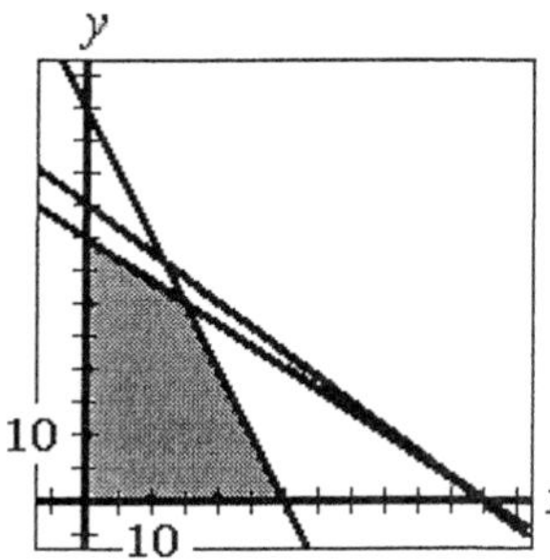

Let the objective function be the profit: $P = 40x + 32y$.

Vertex	$P = 40x + 32y$
$(0, 0)$	$= 0$
$(0, 30)$	$= 960$
$(15, 30)$	$= 1560$
$(0, 40)$	$= 1280$

The maximum profit of \$1560 results when they make 15 VCRs and 30 TVs.

30. Let $x =$ the number of fast (2.66 GHz) chips, and let $y =$ the number of slow (1.66 GHz) chips.

$$\begin{cases} 0 \leq x \leq 50 \\ 0 \leq y \leq 100 \\ 6x + 3y \leq 360 & \text{(total labor hours)} \end{cases}$$

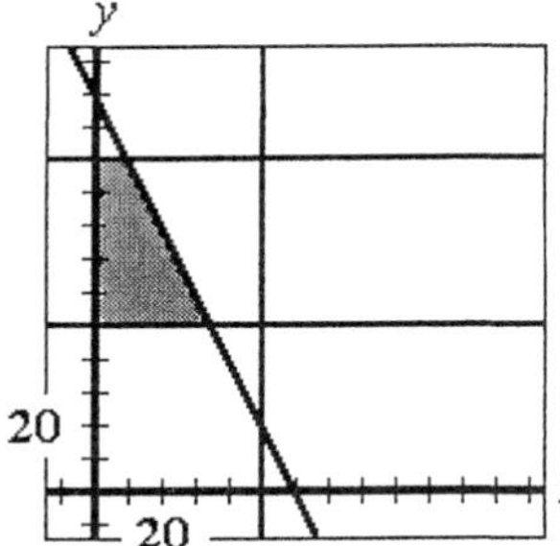

Let the objective function be the profit: $P = 20x + 27y$.

Vertex	$P = 20x + 27y$
$(0, 50)$	$= 1350$
$(35, 50)$	$= 2050$
$(10, 100)$	$= 2900$
$(0, 100)$	$= 2700$

The maximum profit of \$2900 results when they make ten 2.66 GHz chips and one hundred 1.66 GHz chips.

31. Let $x =$ the amount in stocks, and let $y =$ the amount in bonds.

$$\begin{cases} x \geq 100000 \\ y \geq 50000 \\ x + y \leq 200000 & \text{(total amount)} \end{cases}$$

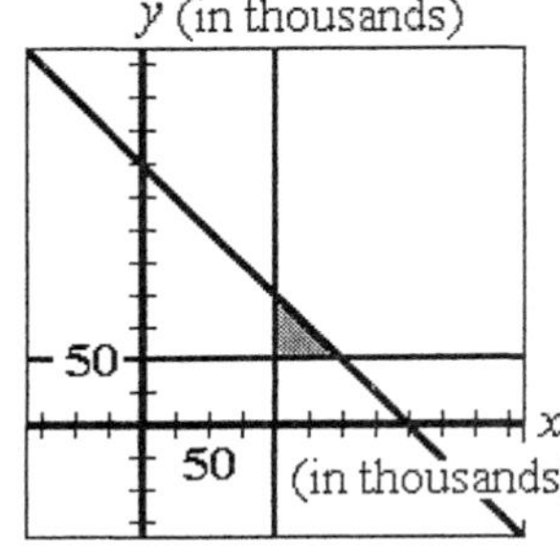

Let the objective function be the income: $P = 0.09x + 0.07y$.

Vertex	$P = 0.09x + 0.07y$
$(100000, 50000)$	$= 12500$
$(150000, 50000)$	$= 17000$
$(100000, 100000)$	$= 16000$

The maximum income of \$17,000 results when she invests \$150,000 in stocks and \$50,000 in bonds.

32. Let $x =$ the number of acres of soybeans, and let $y =$ the number of acres of flowers.

$$\begin{cases} x \geq 0 & y \geq 0 \\ 8x + 12y \leq 100000 & \text{(\# workers)} \\ x \geq 3y & \text{(requirement)} \\ 250x + 300y \leq 3000000 & \text{(budget)} \end{cases}$$

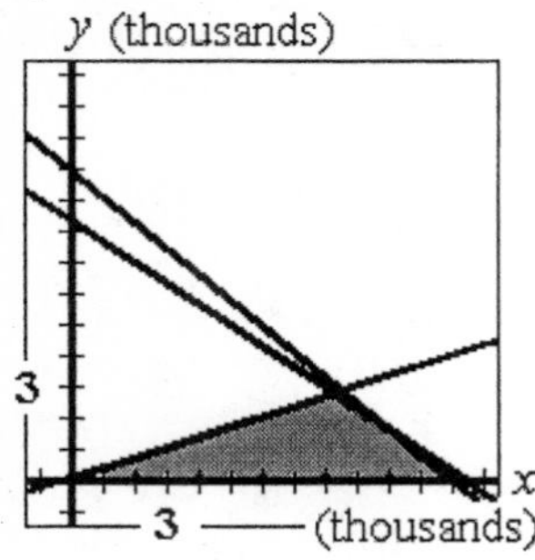

Let the objective function be the profit: $P = 1600x + 2000y$.

Vertex	$P = 1600x + 2000y$
$(0, 0)$	$= 0$
$(12000, 0)$	$= 19200000$
$(10000, 1667)$	$= 19334000$
$(8333, 2778)$	$= 18888800$

The maximum profit of approximately \$19,334,000 results when they plant 10,000 acres of soybeans and 1667 acres of flowers.

33-36. Answers may vary.

Chapter 4 Summary (page 267)

1.
$$\begin{aligned} 5(x-2) &\leq 5 \\ 5x - 10 + 10 &\leq 5 + 10 \\ 5x &\leq 15 \\ x &\leq 3 \end{aligned}$$

solution set: $(-\infty, 3]$

3

2.
$$\begin{aligned} 3x + 4 &> 10 \\ 3x + 4 - 4 &> 10 - 4 \\ 3x &> 6 \\ x &> 2 \end{aligned}$$

solution set: $(2, \infty)$

2

3.
$$\begin{aligned} \frac{1}{3}x - 2 &\geq \frac{1}{2}x + 2 \\ 6\left(\frac{1}{3}x - 2\right) &\geq 6\left(\frac{1}{2}x + 2\right) \\ 2x - 12 + 12 &\geq 3x + 12 + 12 \\ 2x - 3x &\geq 3x - 3xy + 24 \\ -x &\geq 24 \\ x &\leq -24 \end{aligned}$$

solution set: $(-\infty, -24]$

-24

4.
$$\begin{aligned} \frac{7}{4}(x+3) &< \frac{3}{8}(x-3) \\ 8 \cdot \frac{7}{4}(x+3) &< 8 \cdot \frac{3}{8}(x-3) \\ 14(x+3) &< 3(x-3) \\ 14x + 42 &< 3x - 9 \\ 14x &< 3x - 51 \\ 11x &< -51 \\ x &< -\tfrac{51}{11} \end{aligned}$$

solution set: $\left(-\infty, -\frac{51}{11}\right)$

$-\frac{51}{11}$

5. $3 < 3x + 4 < 10$

$3 - 4 < 3x + 4 - 4 < 10 - 4$

$-1 < 3x < 6$

$-\frac{1}{3} < \frac{3x}{3} < \frac{6}{3}$

$-\frac{1}{3} < x < 2 \Rightarrow$ solution set: $(-\frac{1}{3}, 2) \Rightarrow$

6. $4x > 3x + 2 > x - 3$

$4x > 3x + 2$ and $3x + 2 > x - 3$

$4x - 3x > 3x - 3x + 2$ $\quad$ $3x + 2 - 2 > x - 3 - 2$

$x > 2$ $\quad$ $3x - x > x - x - 5$

$x > 2$ $\quad$ $2x > -5$

$x > -\frac{5}{2}$

If $x > 2$ **and** $x > -\frac{5}{2}$, $x > 2$. The solution set is $(2, \infty)$.

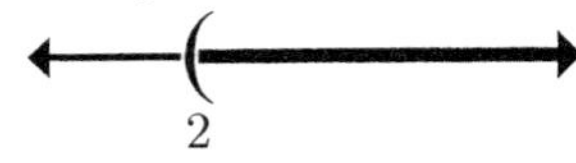

7. $-5 \le 2x - 3 < 5$

$-5 + 3 \le 2x - 3 + 2 < 5 + 3$

$-2 \le 2x < 8$

$-1 \le x < 4 \Rightarrow$ solution set: $[-1, 4) \Rightarrow$

8. Let $x =$ the amount invested at 7%.

$0.06(10000) + 0.07x \ge 2000$

$600 + 0.07x \ge 2000$

$0.07x \ge 1400$

$x \ge 20000 \Rightarrow$ She must invest at least \$20,000.

9. $|-7| = 7$ **10.** $|8| = 8$ **11.** $-|7| = -7$ **12.** $-|-12| = -12$

13. $f(x) = |x + 1| - 3$; Shift $f(x) = |x|$ left 1 and down 3.

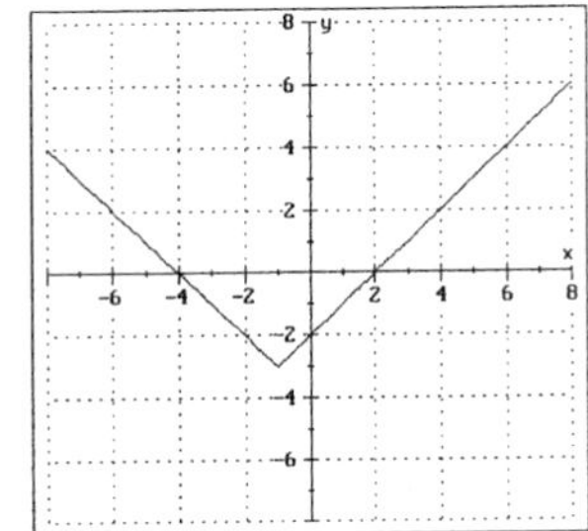

14. $f(x) = |x - 2| + 1$; Shift $f(x) = |x|$ right 2 and up 1.

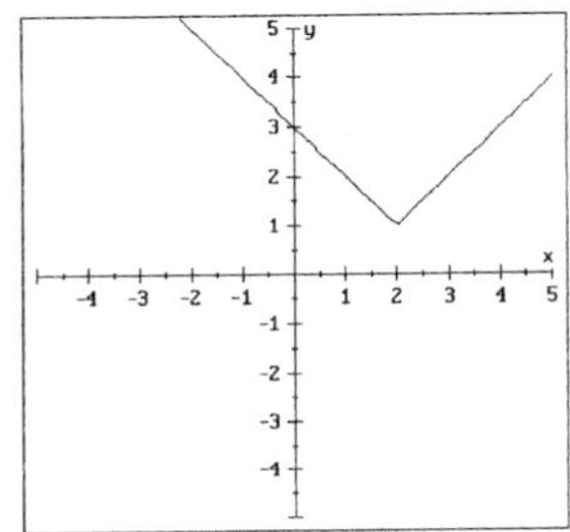

15.
$$|3x+1|=10$$
$$3x+1=10 \quad \textbf{or} \quad 3x+1=-10$$
$$3x=9 \qquad\qquad 3x=-11$$
$$x=3 \qquad\qquad x=-\tfrac{11}{3}$$

16.
$$\left|\tfrac{3}{2}x-4\right|=9$$
$$\tfrac{3}{2}x-4=9 \quad \textbf{or} \quad \tfrac{3}{2}x-4=-9$$
$$\tfrac{3}{2}x=13 \qquad\qquad \tfrac{3}{2}x=-5$$
$$x=\tfrac{26}{3} \qquad\qquad x=-\tfrac{10}{3}$$

17.
$$\left|\frac{2-x}{3}\right|=4$$
$$\frac{2-x}{3}=4 \quad \textbf{or} \quad \frac{2-x}{3}=-4$$
$$2-x=12 \qquad\qquad 2-x=-12$$
$$-x=10 \qquad\qquad -x=-14$$
$$x=-10 \qquad\qquad x=14$$

18.
$$|3x+2|=|2x-3|$$
$$3x+2=2x-3 \quad \textbf{or} \quad 3x+2=-(2x-3)$$
$$x=-5 \qquad\qquad 3x+2=-2x+3$$
$$5x=1$$
$$x=\tfrac{1}{5}$$

19.
$$|5x-4|=|4x-5|$$
$$5x-4=4x-5 \quad \textbf{or} \quad 5x-4=-(4x-5)$$
$$x=-1 \qquad\qquad 5x-4=-4x+5$$
$$9x=9$$
$$x=1$$

20.
$$\left|\frac{3-2x}{2}\right|=\left|\frac{3x-2}{3}\right|$$
$$\frac{3-2x}{2}=\frac{3x-2}{3} \quad \textbf{or} \quad \frac{3-2x}{2}=-\frac{3x-2}{3}$$
$$3(3-2x)=2(3x-2) \qquad\qquad 3(3-2x)=-2(3x-2)$$
$$9-6x=6x-4 \qquad\qquad 9-6x=-6x+4$$
$$-12x=-13 \qquad\qquad 0=-5$$
$$x=\tfrac{13}{12} \qquad\qquad \text{(no solution from this part)}$$

21.
$$|2x+7|<3$$
$$-3<2x+7<3$$
$$-3-7<2x+7-7<3-7$$
$$-10<2x<-4$$
$$-5<x<-2$$
solution set: $(-5,-2)$

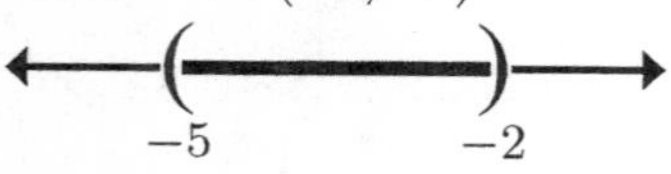

22.
$$|5-3x|\le 14$$
$$-14\le 5-3x\le 14$$
$$-14-5\le 5-5-3x\le 14-5$$
$$-19\le -3x\le 9$$
$$\frac{-19}{-3}\ge\frac{-3x}{-3}\ge\frac{9}{-3}$$
$$\frac{19}{3}\ge x\ge -3, \text{ or } -3\le x\le\frac{19}{3}$$
solution set: $\left[-3, \tfrac{19}{3}\right]$

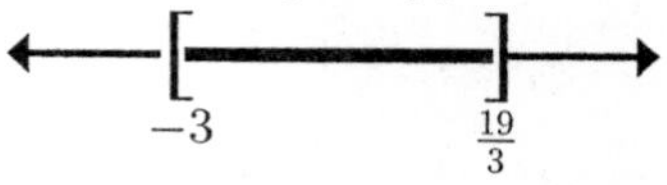

23. $\left|\frac{2}{3}x + 14\right| < 0$

Since an absolute value can never be negative, there is no solution.

24. $\left|\dfrac{1-5x}{3}\right| > 7$

$$\frac{1-5x}{3} < -7 \quad \textbf{or} \quad \frac{1-5x}{3} > 7$$

$$1-5x < -21 \qquad 1-5x > 21$$

$$-5x < -22 \qquad -5x > 20$$

$$x > \tfrac{22}{5} \qquad x < -4$$

solution set: $(-\infty, -4) \cup \left(\frac{22}{5}, \infty\right)$

-4 $\frac{22}{5}$

25. $|3x - 8| \geq 4$

$$3x - 8 \leq -4 \quad \textbf{or} \quad 3x - 8 \geq 4$$

$$3x \leq 4 \qquad 3x \geq 12$$

$$x \leq \tfrac{4}{3} \qquad x \geq 4$$

solution set: $\left(-\infty, \frac{4}{3}\right] \cup [4, \infty)$

$\frac{4}{3}$ 4

26. $\left|\frac{3}{2}x - 14\right| \geq 0$

Since an absolute value is always at least 0, this inequality is true for all real #s. Solution set: $(-\infty, \infty)$

0

27. $2x + 3y > 6$

28. $y \leq 4 - x$

29. $-2 < x < 4$

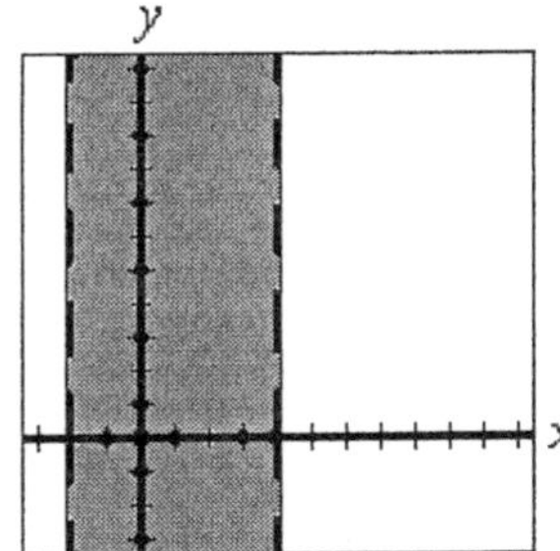

30. $y \leq -2$ or $y > 1$

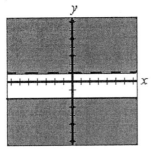

31. $\begin{cases} y \geq x + 1 \\ 3x + 2y < 6 \end{cases}$

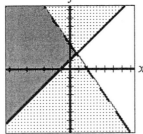

32. $\begin{cases} y \geq x^2 - 4 \\ y < x + 3 \end{cases}$

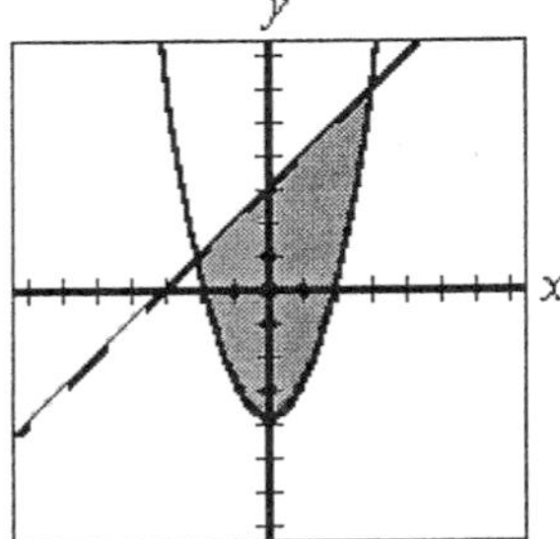

33.

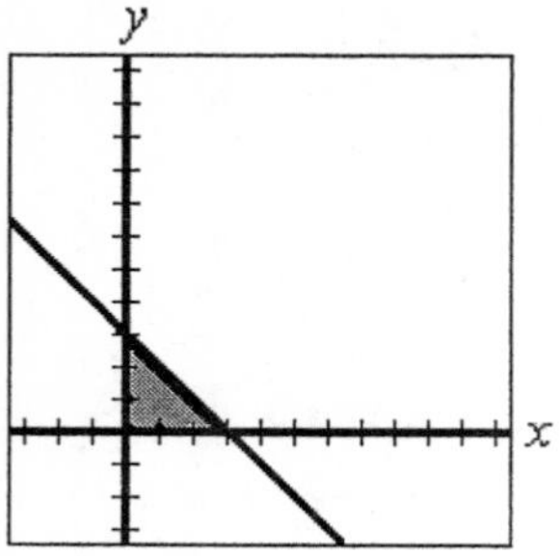

Vertex	$P = 2x + y$	Maximum?
$(0, 0)$	$= 0$	No
$(3, 0)$	$= 6$	YES
$(0, 3)$	$= 3$	No

P has a maximum value of 6 at $(3, 0)$.

34. Let $x =$ the number of bags of fertilizer X, and let $y =$ the number of bags of fertilizer Y.

$$\begin{cases} x \geq 0 & y \geq 0 \\ 6x + 10y \leq 20000 & \text{(Nitrogen)} \\ 8x + 6y \leq 16400 & \text{(Phosphorus)} \\ 6x + 4y \leq 12000 & \text{(Potash)} \end{cases}$$

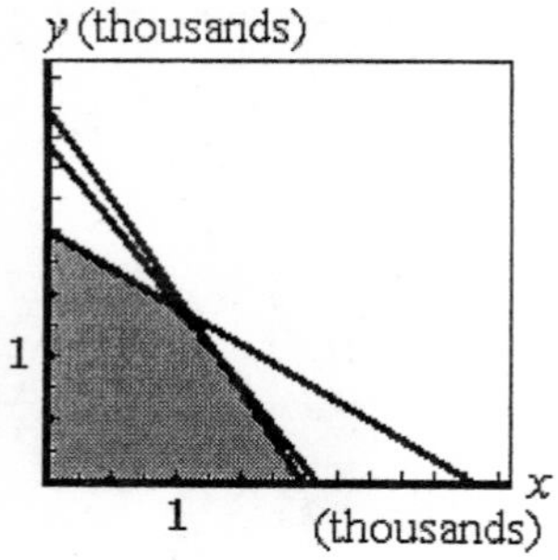

Let the objective function be the total profit: $P = 6x + 5y$.

Vertex	$P = 6x + 5y$
$(2000, 0)$	$= 12000$
$(1600, 600)$	$= 12600$
$(1000, 1400)$	$= 13000$
$(0, 2000)$	$= 10000$

The maximum profit is \$13,000, which results from using 1000 bags of fertilizer X and 1400 bags of fertilizer Y.

Chapter 4 Test (page 270)

1.
$$\begin{aligned} -2(2x + 3) &\geq 14 \\ -4x - 6 + 6 &\geq 14 + 6 \\ -4x &\geq 20 \\ \frac{-4x}{-4} &\leq \frac{20}{-4} \\ x &\leq -5 \end{aligned}$$

solution set: $(-\infty, -5]$

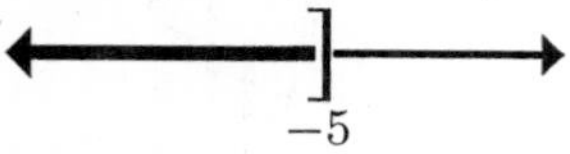

2.
$$\begin{aligned} &-2 < \frac{x-4}{3} < 4 \\ &-6 < x - 4 < 12 \\ &-6 + 4 < x - 4 + 4 \leq 12 + 4 \\ &-2 < x < 16 \end{aligned}$$

solution set: $(-2, 16)$

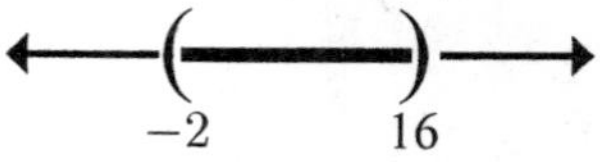

3. $|5 - 8| = |-3| = 3$

4. $|4\pi - 4| = 4\pi - 4 \ (4\pi - 4 > 0)$

5. $f(x) = |x+1| - 4$; Shift $f(x) = |x|$ left 1 and down 4.

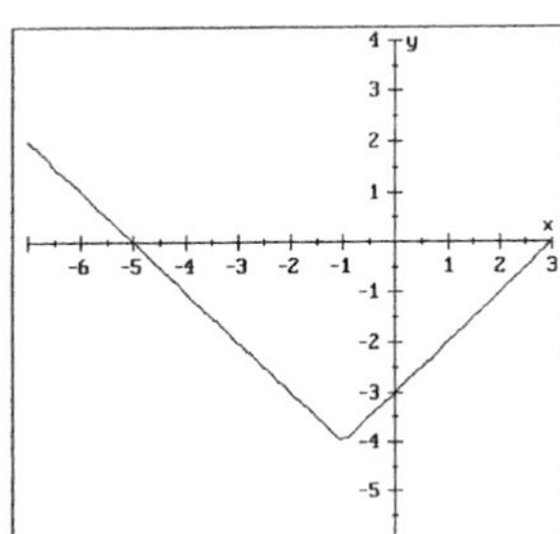

6. $f(x) = |x-2| + 3$; Shift $f(x) = |x|$ right 2 and up 3.

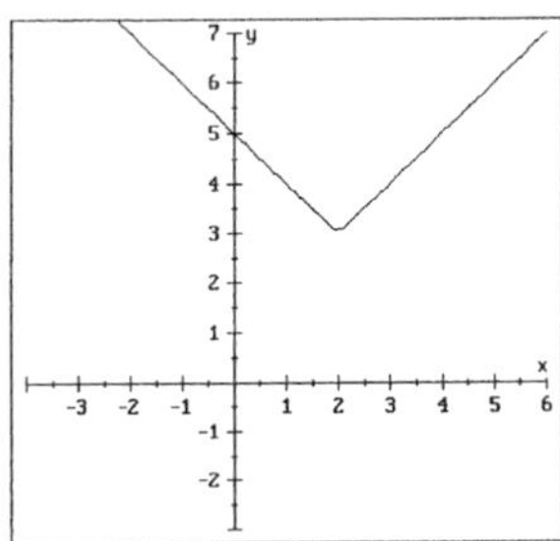

7.
$$|2x+3| = 11$$
$$2x+3 = 11 \quad \textbf{or} \quad 2x+3 = -11$$
$$2x = 8 \qquad\qquad 2x = -14$$
$$x = 4 \qquad\qquad x = -7$$

8.
$$|4-3x| = 19$$
$$4-3x = 19 \quad \textbf{or} \quad 4-3x = -19$$
$$-3x = 15 \qquad\qquad -3x = -23$$
$$x = -5 \qquad\qquad x = \tfrac{23}{3}$$

9.
$$|3x+4| = |x+12|$$
$$3x+4 = x+12 \quad \textbf{or} \quad 3x+4 = -(x+12)$$
$$2x = 8 \qquad\qquad 3x+4 = -x-12$$
$$x = 4 \qquad\qquad 4x = -16$$
$$x = -4$$

10.
$$|3-2x| = |2x+3|$$
$$3-2x = 2x+3 \quad \textbf{or} \quad 3-2x = -(2x+3)$$
$$-4x = 0 \qquad\qquad 3-2x = -2x-3$$
$$x = 0 \qquad\qquad 0 = -6$$
(no solution from this part)

11. $|x+3| \le 4$
$$-4 \le x+3 \le 4$$
$$-4-3 \le x+3-3 \le 4-3$$
$$-7 \le x \le 1$$
solution set: $[-7, 1]$

$\longleftarrow [\!\!\!-\!\!-\!\!-\!\!-]\longrightarrow$ (closed interval from -7 to 1)

12.
$$|2x-4| > 22$$
$$2x-4 < -22 \quad \textbf{or} \quad 2x-4 > 22$$
$$2x < -18 \qquad\qquad 2x > 26$$
$$x < -9 \qquad\qquad x > 13$$
solution set: $(-\infty, -9) \cup (13, \infty)$

$-9 \qquad 13$

13.

$$|4-2x|>2$$
$$4-2x<-2 \quad \textbf{or} \quad 4-2x>2$$
$$-2x<-6 \qquad\qquad -2x>-2$$
$$x>3 \qquad\qquad x<1$$

solution set: $(-\infty,1)\cup(3,\infty)$

14.

$$|2x-4|\le 2$$
$$-2\le 2x-4\le 2$$
$$-2+4\le 2x-4+4\le 2+4$$
$$2\le 2x\le 6$$
$$1\le x\le 3$$

solution set: $[1,3]$

15. $3x+2y\ge 6$

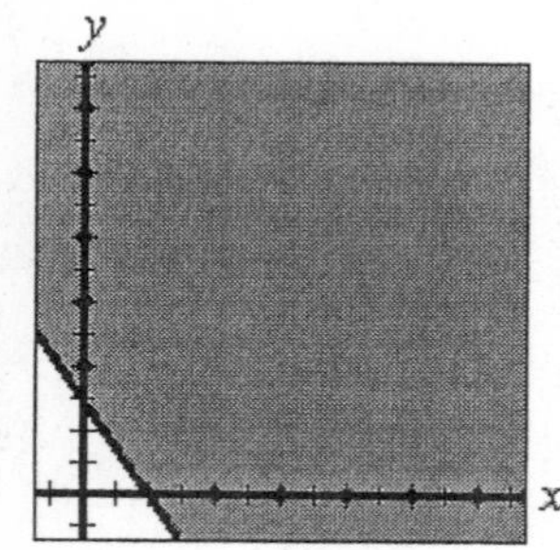

16. $-2\le y<5$

17. $\begin{cases} 2x-3y\ge 6 \\ y\le -x+1 \end{cases}$

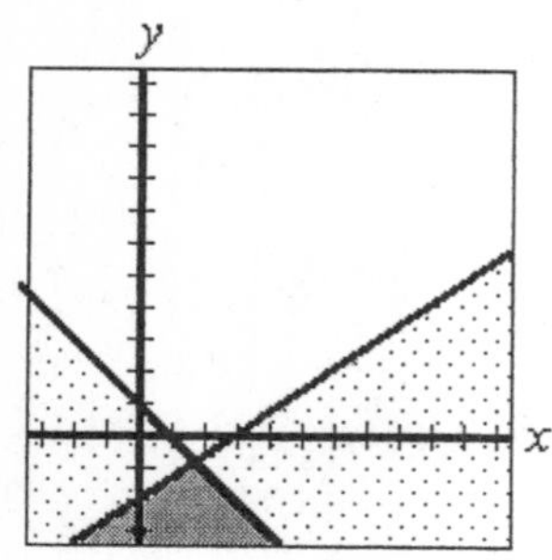

18.

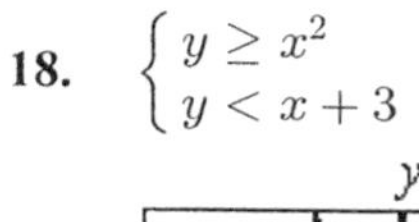

$\begin{cases} y\ge x^2 \\ y<x+3 \end{cases}$

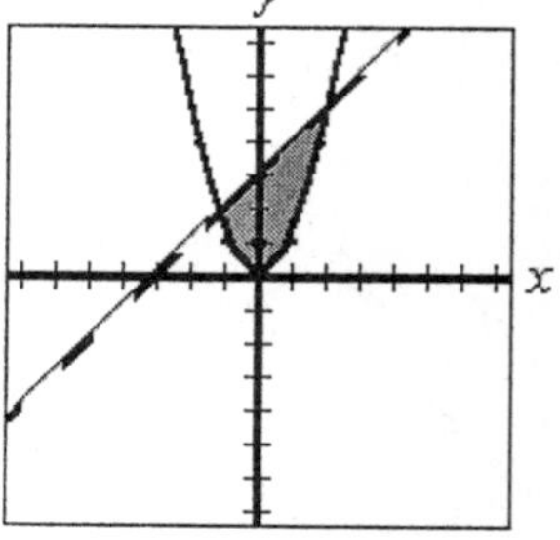

19.

Vertex	$P=3x-y$	Maximum?
$(0,1)$	$=-1$	No
$(1,1)$	$=2$	YES
$(1,2)$	$=1$	No
$(\frac{1}{3},2)$	$=-1$	No

P has a maximum value of 2 at $(1,1)$.

Cumulative Review Exercises (page 271)

1.

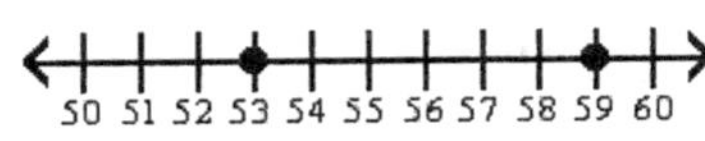

2. The additive inverse of -5 is 5.

3. $x - xy = 2 - 2(-4) = 2 + 8 = 10$

4. $\dfrac{x^2 - y^2}{3x + y} = \dfrac{2^2 - (-4)^2}{3(2) + (-4)} = \dfrac{4 - 16}{6 - 4} = \frac{-12}{2} = -6$

5. $(x^2x^3)^2 = (x^5)^2 = x^{10}$

6. $(x^2)^3(x^4)^2 = x^6x^8 = x^{14}$

7. $\left(\dfrac{x^3}{x^5}\right)^{-2} = \left(\dfrac{x^5}{x^3}\right)^2 = (x^2)^2 = x^4$

8. $\dfrac{a^2b^n}{a^nb^2} = a^{2-n}b^{n-2}$

9. $32{,}600{,}000 = 3.26 \times 10^7$

10. $0.000012 = 1.2 \times 10^{-5}$

11.
$$\begin{aligned} 3x - 6 &= 20 \\ 3x &= 26 \\ x &= \frac{26}{3} \end{aligned}$$

12.
$$\begin{aligned} 6(x - 1) &= 2(x + 3) \\ 6x - 6 &= 2x + 6 \\ 4x &= 12 \\ x &= 3 \end{aligned}$$

13.
$$\begin{aligned} \frac{5b}{2} - 10 &= \frac{b}{3} + 3 \\ 6\left(\frac{5b}{2} - 10\right) &= 6\left(\frac{b}{3} + 3\right) \\ 3(5b) - 60 &= 2b + 18 \\ 15b - 60 &= 2b + 18 \\ 13b &= 78 \\ b &= 6 \end{aligned}$$

14.
$$\begin{aligned} 2a - 5 &= -2a + 4(a - 2) + 1 \\ 2a - 5 &= -2a + 4a - 8 + 1 \\ 2a - 5 &= 2a - 7 \\ -5 &= -7 \end{aligned}$$
contradiction

15.
$$\begin{aligned} 3x + 2y &= 12 \\ 2y &= -3x + 12 \\ y &= -\tfrac{3}{2}x + 6 \\ m &= -\tfrac{3}{2} \end{aligned} \qquad \begin{aligned} 2x - 3y &= 5 \\ -3y &= -2x + 5 \\ y &= \tfrac{2}{3}x - \tfrac{5}{3} \\ m &= \tfrac{2}{3} \end{aligned}$$
perpendicular

16.
$$\begin{aligned} 3x &= y + 4 \\ -y &= -3x + 4 \\ y &= 3x - 4 \\ m &= 3 \end{aligned} \qquad \begin{aligned} y &= 3(x - 4) - 1 \\ y &= 3x - 12 - 1 \\ y &= 3x - 13 \\ m &= 3 \end{aligned}$$
parallel

17. $3x + y = 8$
$y = -3x + 8$
$m = -3$
Use $m = \frac{1}{3}$ (the perpendicular slope):
$y - y_1 = m(x - x_1)$
$y - 3 = \frac{1}{3}(x + 2)$
$y = \frac{1}{3}x + \frac{11}{3}$

18. $A = \frac{1}{2}h(b_1 + b_2)$
$2A = h(b_1 + b_2)$
$\frac{2A}{b_1 + b_2} = h$, or $h = \frac{2A}{b_1 + b_2}$

19. $f(2) = 3(2)^2 - 2 = 3(4) - 2$
$= 12 - 2 = 10$

20. $f(-2) = 3(-2)^2 - (-2) = 3(4) + 2$
$= 12 + 2 = 14$

21. $\begin{cases} 2x + y = 5 \\ x - 2y = 0 \end{cases}$

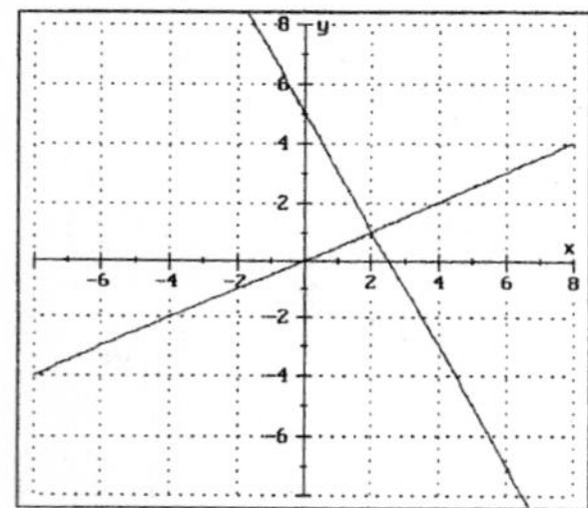

solution: $(2, 1)$

22. $\begin{cases} 3x + y = 4 & (1) \\ 2x - 3y = -1 & (2) \end{cases}$
Substitute $y = -3x + 4$ from (1) into (2):
$2x - 3\mathbf{y} = -1$
$2x - 3(\mathbf{-3x + 4}) = -1$
$2x + 9x - 12 = -1$
$11x = 11$
$x = 1$
Substitute this and solve for y:
$y = -3x + 4$
$y = -3(1) + 4$
$y = 1$
Solution: $(1, 1)$

23.
$$\begin{array}{rl} x + 2y = -2 & \\ \underline{2x - y = 6} & \Rightarrow \times 2 \end{array} \quad \begin{array}{rcl} x + 2y &=& -2 \\ \underline{4x - 2y} &=& \underline{12} \\ 5x &=& 10 \\ x &=& 2 \end{array}$$
Substitute and solve for y:
$x + 2y = -2$
$2 + 2y = -2$
$2y = -4$
$y = -2$
The solution is $(2, -2)$.

24.
$$\begin{array}{ll} \frac{x}{10} + \frac{y}{5} = \frac{1}{2} \Rightarrow \times 10 & x + 2y = 5 \\ \frac{x}{2} - \frac{y}{5} = \frac{13}{10} \Rightarrow \times 10 & 5x - 2y = 13 \\ \hline & 6x = 18 \\ & x = 3 \end{array}$$
Substitute and solve for y:
$x + 2y = 5$
$3 + 2y = 5$
$2y = 2$
$y = 1$
The solution is $(3, 1)$.

25.

$$\begin{array}{ll} (1) & x+y+z=1 \\ (2) & 2x-y-z=-4 \\ (3) & x-2y+z=4 \end{array} \qquad \begin{array}{lrl} (1) & x+y+z & = 1 \\ (2) & 2x-y-z & = -4 \\ \hline (4) & 3x & = -3 \\ & x & = -1 \end{array} \qquad \begin{array}{lrl} (2) & 2x-y-z & = -4 \\ (3) & x-2y+z & = 4 \\ \hline (5) & 3x-3y & = 0 \end{array}$$

$$\begin{aligned} 3x-3y &= 0 \\ 3(-1)-3y &= 0 \\ -3-3y &= 0 \\ -3y &= 3 \\ y &= -1 \end{aligned} \qquad \begin{aligned} x+y+z &= 1 \\ -1+(-1)+z &= 1 \\ -2+z &= 1 \\ z &= 3 \end{aligned}$$

Solution: $\boxed{(-1,-1,3)}$

26.

$$\begin{array}{ll} (1) & x+2y+3z=1 \\ (2) & 3x+2y+z=-1 \\ (3) & 2x+3y+z=-2 \end{array} \qquad \begin{array}{lrl} (1) & x+2y+3z & = 1 \\ -3\cdot(2) & -9x-6y-3z & = 3 \\ \hline (4) & -8x-4y & = 4 \end{array} \qquad \begin{array}{lrl} (1) & x+2y+3z & = 1 \\ -3\cdot(3) & -6x-9y-3z & = 6 \\ \hline (5) & -5x-7y & = 7 \end{array}$$

$$\begin{array}{l} -8x-4y=4 \Rightarrow \times 7 \\ \underline{-5x-7y=7} \Rightarrow \times(-4) \end{array} \qquad \begin{array}{rcr} -56x-28y & = & 28 \\ 20x+28y & = & -28 \\ \hline -36x & = & 0 \\ x & = & 0 \end{array} \qquad \begin{aligned} -5x-7y &= 7 \\ -5(0)-7y &= 7 \\ y &= -1 \end{aligned}$$

$$\begin{aligned} 2x+3y+z &= -2 \\ 2(0)+3(-1)+z &= -2 \\ -3+z &= -2 \\ z &= 1 \end{aligned}$$

$\boxed{\text{The solution is } (0,-1,1).}$

27. $\begin{vmatrix} 3 & -2 \\ 1 & -1 \end{vmatrix} = 3(-1)-(-2)(1) = -3-(-2) = -3+2 = -1$

28. $\begin{vmatrix} 2 & 3 & -1 \\ -1 & -1 & 2 \\ 4 & 1 & -1 \end{vmatrix} = 2\begin{vmatrix} -1 & 2 \\ 1 & -1 \end{vmatrix} - 3\begin{vmatrix} -1 & 2 \\ 4 & -1 \end{vmatrix} + (-1)\begin{vmatrix} -1 & -1 \\ 4 & 1 \end{vmatrix} = 2(-1)-3(-7)-1(3) = 16$

29. $x = \dfrac{\begin{vmatrix} -1 & -3 \\ -7 & 4 \end{vmatrix}}{\begin{vmatrix} 4 & -3 \\ 3 & 4 \end{vmatrix}} = \dfrac{-4-(21)}{16-(-9)} = \dfrac{-25}{25} = -1;\ y = \dfrac{\begin{vmatrix} 4 & -1 \\ 3 & -7 \end{vmatrix}}{\begin{vmatrix} 4 & -3 \\ 3 & 4 \end{vmatrix}} = \dfrac{-28-(-3)}{25} = \dfrac{-25}{25} = -1$

solution: $(-1,-1)$

30. $D = \begin{vmatrix} 1 & -2 & -1 \\ 3 & 1 & -1 \\ 2 & -1 & 1 \end{vmatrix} = 1\begin{vmatrix} 1 & -1 \\ -1 & 1 \end{vmatrix} - (-2)\begin{vmatrix} 3 & -1 \\ 2 & 1 \end{vmatrix} + (-1)\begin{vmatrix} 3 & 1 \\ 2 & -1 \end{vmatrix} = 1(0)+2(5)-1(-5)$

$= 15$

continued on next page

30. continued

$$N_x = \begin{vmatrix} -2 & -2 & -1 \\ 6 & 1 & -1 \\ -1 & -1 & 1 \end{vmatrix} = -2\begin{vmatrix} 1 & -1 \\ -1 & 1 \end{vmatrix} - (-2)\begin{vmatrix} 6 & -1 \\ -1 & 1 \end{vmatrix} + (-1)\begin{vmatrix} 6 & 1 \\ -1 & -1 \end{vmatrix} = 15$$

$$N_y = \begin{vmatrix} 1 & -2 & -1 \\ 3 & 6 & -1 \\ 2 & -1 & 1 \end{vmatrix} = 1\begin{vmatrix} 6 & -1 \\ -1 & 1 \end{vmatrix} - (-2)\begin{vmatrix} 3 & -1 \\ 2 & 1 \end{vmatrix} + (-1)\begin{vmatrix} 3 & 6 \\ 2 & -1 \end{vmatrix} = 30$$

$$N_z = \begin{vmatrix} 1 & -2 & -2 \\ 3 & 1 & 6 \\ 2 & -1 & -1 \end{vmatrix} = 1\begin{vmatrix} 1 & 6 \\ -1 & -1 \end{vmatrix} - (-2)\begin{vmatrix} 3 & 6 \\ 2 & -1 \end{vmatrix} + (-2)\begin{vmatrix} 3 & 1 \\ 2 & -1 \end{vmatrix} = -15$$

$$x = \frac{N_x}{D} = \frac{15}{15} = 1;\ y = \frac{N_y}{D} = \frac{30}{15} = 2;\ z = \frac{N_z}{D} = \frac{-15}{15} = -1 \Rightarrow \text{solution: } (1, 2, -1)$$

31.
$$\begin{aligned} -3(x-4) &\geq x - 32 \\ -3x + 12 &\geq x - 32 \\ -4x &\geq -44 \\ x &\leq 11 \end{aligned}$$

32.
$$\begin{aligned} -8 < -3x + 1 < 10 \\ -9 < -3x < 9 \\ 3 > x > -3, \text{ or } -3 < x < 3 \end{aligned}$$

33.
$$|4x - 3| = 9$$
$$\begin{aligned} 4x - 3 &= 9 \quad \textbf{or} \quad & 4x - 3 &= -9 \\ 4x &= 12 & 4x &= -6 \\ x &= 3 & x &= -\tfrac{3}{2} \end{aligned}$$

34.
$$|2x - 1| = |3x + 4|$$
$$\begin{aligned} 2x - 1 &= 3x + 4 \quad \textbf{or} \quad & 2x - 1 &= -(3x + 4) \\ -x &= 5 & 2x - 1 &= -3x - 4 \\ x &= -5 & 5x &= -3 \\ & & x &= -\tfrac{3}{5} \end{aligned}$$

35.
$$\begin{aligned} |3x - 2| &\leq 4 \\ -4 \leq 3x &- 2 \leq 4 \\ -2 \leq 3x &\leq 6 \\ -\frac{2}{3} \leq x &\leq 2 \end{aligned}$$

36.

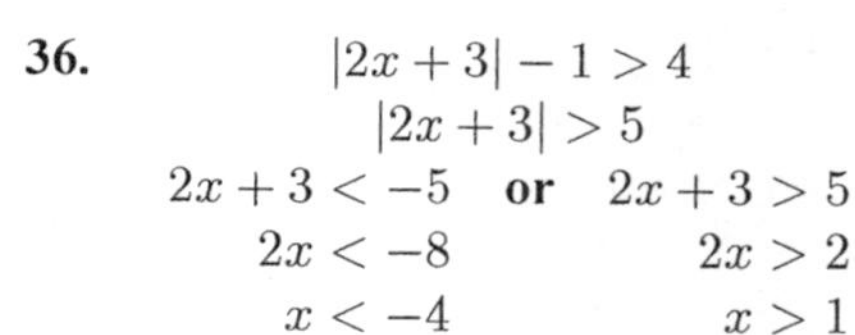

$$|2x + 3| - 1 > 4$$
$$|2x + 3| > 5$$
$$\begin{aligned} 2x + 3 &< -5 \quad \textbf{or} \quad & 2x + 3 &> 5 \\ 2x &< -8 & 2x &> 2 \\ x &< -4 & x &> 1 \end{aligned}$$

37. $2x - 3y \leq 12$

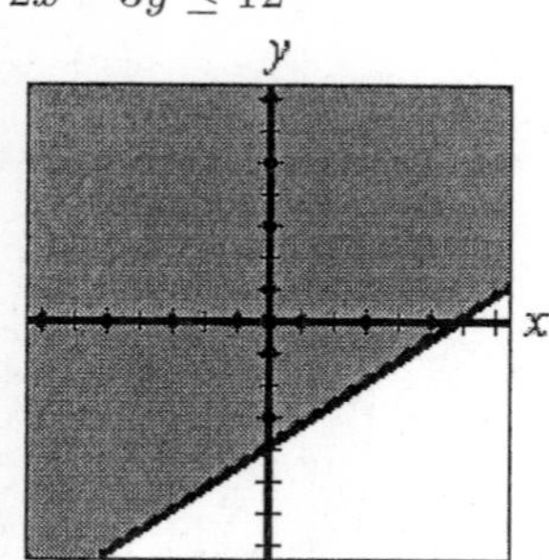

38. $3 > x \geq -2$, or $-2 \leq x < 3$

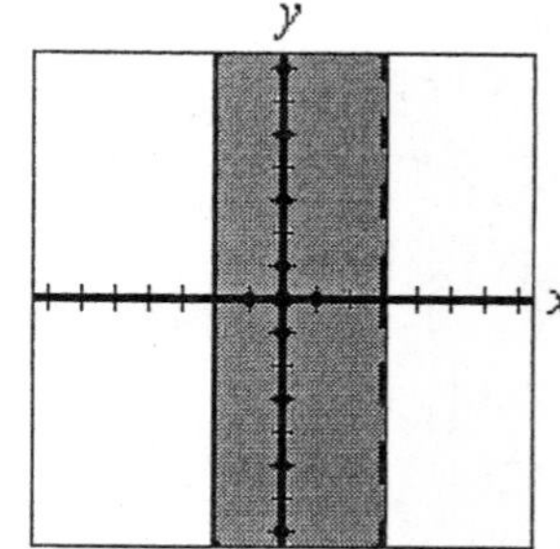

39. $\begin{cases} 3x - 2y < 6 \\ y < -x + 2 \end{cases}$

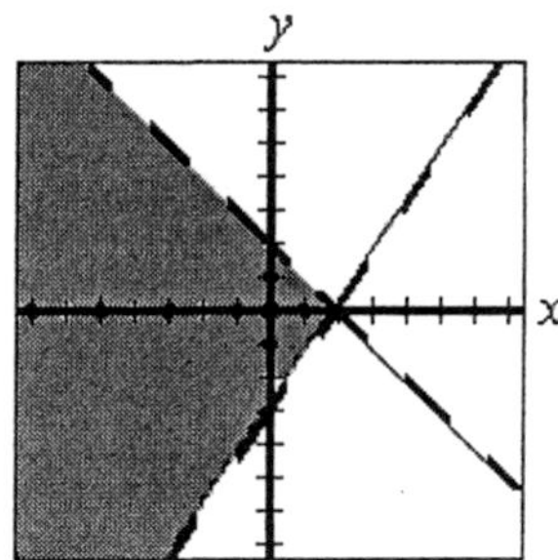

40. $\begin{cases} y < x + 2 \\ 3x + y \leq 6 \end{cases}$

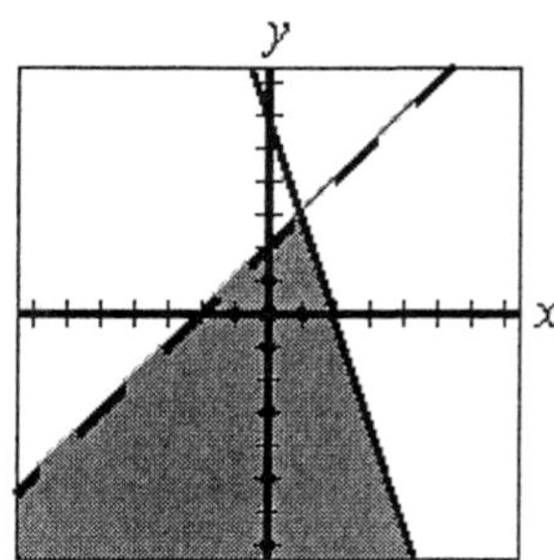

41. Let $x =$ the number of mice used, and let $y =$ the number of rats used.

Form the following system of constraints:

$$\begin{cases} x \geq 0 \\ y \geq 0 \\ 12x + 8y \leq 240 \quad \text{(Maze 1)} \\ 10x + 15y \leq 300 \quad \text{(Maze 2)} \end{cases}$$

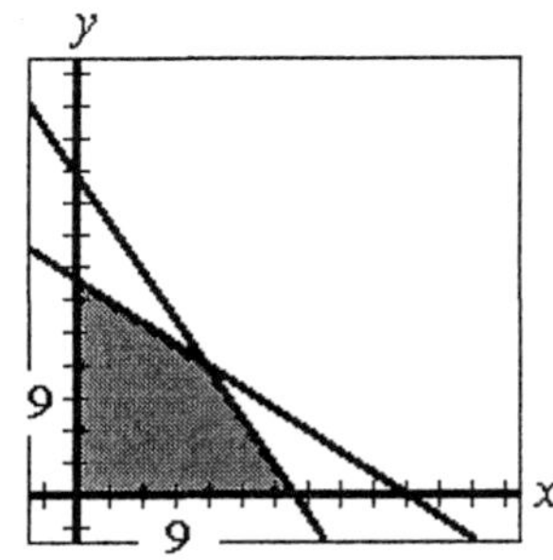

Let the objective function be the total number of animals: $P = x + y$.

Vertex	$P = x + y$
$(0, 0)$	$= 0$
$(20, 0)$	$= 20$
$(12, 12)$	$= 24$
$(0, 20)$	$= 20$

The maximum number of animals that can be used in both mazes is 24, resulting from using 12 mice and 12 rats.

Exercise 5.1 (page 281)

1. $a^3a^2 = a^5$

2. $\dfrac{b^3b^3}{b^4} = \dfrac{b^6}{b^4} = b^2$

3. $\dfrac{3(y^3)^{10}}{y^3y^4} = \dfrac{3y^{30}}{y^7} = 3y^{23}$

4. $\dfrac{4x^{-4}x^5}{2x^{-6}} = \dfrac{2x}{x^{-6}} = 2x^7$

5. $114{,}000{,}000 = 1.14 \times 10^8$

6. $0.0000001 = 1 \times 10^{-7}$

7. sum; whole

8. one

9. binomial

10. trinomial

11. one

12. $P(x)$

13. monomial

14. binomial

15. trinomial

16. binomial

17. binomial

18. none of these

19. monomial

20. monomial

21. deg $= 2$

22. deg $= 17$

23. deg $= 8$ **24.** deg $= 10$ **25.** deg $= 3 + 3 + 4 = 10$

26. deg $= 1$ **27.** deg $= 0$ **28.** deg $= 12$

29. $3x - 2x^4 + 7 - 5x^2 = -2x^4 - 5x^2 + 3x + 7$

30. $-x^2 + 3x^5 - 7x + 3x^3 = 3x^5 + 3x^3 - x^2 - 7x$

31. $a^2x - ax^3 + 7a^3x^5 - 5a^3x^2 = 7a^3x^5 - ax^3 - 5a^3x^2 + a^2x$

32. $4x^2y^7 - 3x^5y^2 + 4x^3y^3 - 2x^4y^6 + 5x^6 = 5x^6 - 3x^5y^2 - 2x^4y^6 + 4x^3y^3 + 4x^2y^7$

33. $4y^2 - 2y^5 + 7y - 5y^3 = 7y + 4y^2 - 5y^3 - 2y^5$

34. $y^3 + 3y^2 + 8y^4 - 2 = -2 + 3y^2 + y^3 + 8y^4$

35. $5x^3y^6 + 2x^4y - 5x^3y^3 + x^5y^7 - 2y^4 = 2x^4y - 5x^3y^3 - 2y^4 + 5x^3y^6 + x^5y^7$

36. $-x^3y^2 + x^2y^3 - 2x^3y + x^7y^6 - 3x^6 = -3x^6 - 2x^3y - x^3y^2 + x^2y^3 + x^7y^6$

37. $P(0) = 2(0)^2 + 0 + 2 = 2(0) + 2 = 2$ **38.** $P(1) = 2(1)^2 + 1 + 2 = 2(1) + 3 = 5$

39. $P(-2) = 2(-2)^2 + (-2) + 2 = 2(4) + 0 = 8$ **40.** $P(-3) = 2(-3)^2 + (-3) + 2 = 2(9) - 1 = 17$

41. $h = f(t) = f(0) = -16(0)^2 + 64(0) = 0 + 0 = 0$ ft

42. $h = f(t) = f(1) = -16(1)^2 + 64(1) = -16(1) + 64 = -16 + 64 = 48$ ft

43. $h = f(t) = f(2) = -16(2)^2 + 64(2) = -16(4) + 128 = -64 + 128 = 64$ ft

44. $h = f(t) = f(4) = -16(4)^2 + 64(4) = -16(16) + 256 = -256 + 256 = 0$ ft

45. $x^2 + y^2 = 2^2 + (-3)^2 = 4 + 9 = 13$ **46.** $x^3 + y^3 = 2^3 + (-3)^3 = 8 + (-27) = -19$

47. $x^3 - y^3 = 2^3 - (-3)^3 = 8 - (-27) = 35$ **48.** $x^2 - y^2 = 2^2 - (-3)^2 = 4 - 9 = -5$

49. $3x^2y + xy^3 = 3(2)^2(-3) + 2(-3)^3 = 3(4)(-3) + 2(-27) = -36 - 54 = -90$

50. $8xy - xy^2 = 8(2)(-3) - (2)(-3)^2 = -48 - 2(9) = -48 - 18 = -66$

51. $-2xy^2 + x^2y = -2(2)(-3)^2 + (2)^2(-3) = -2(2)(9) + 4(-3) = -36 - 12 = -48$

52. $-x^3y - x^2y^2 = -(2)^3(-3) - (2)^2(-3)^2 = -8(-3) - (4)(9) = 24 - 36 = -12$

SECTION 5.1

Problems 53-58 are to be solved using a calculator. The keystrokes needed to solve each problem using a TI-83 graphing calculator appear in each solution. There may be other solutions. Keystrokes for other calculators may be slightly different.

53. $x^2y = (3.7)^2(-2.5) \Rightarrow$ [3] [.] [7] [x^2] [×] [(-)] [2] [.] [5] [ENTER] $\{-34.225\}$

54. $xyz^2 = (3.7)(-2.5)(8.9)^2 \Rightarrow$ [3] [.] [7] [×] [(-)] [2] [.] [5] [×] [8] [.] [9] [x^2] [ENTER] $\{-732.6925\}$

55. $\dfrac{x^2}{z^2} = \dfrac{(3.7)^2}{(8.9)^2} \Rightarrow$ [3] [.] [7] [x^2] [÷] [8] [.] [9] [x^2] [ENTER] $\{0.17283171\}$

56. $\dfrac{z^3}{y^2} = \dfrac{(8.9)^3}{(-2.5)^2} \Rightarrow$ [8] [.] [9] [^] [3] [÷] [(] [(-)] [2] [.] [5] [)] [x^2] [ENTER] $\{112.79504\}$

57. $\dfrac{x+y+z}{xyz} = \dfrac{3.7+(-2.5)+8.9}{(3.7)(-2.5)(8.9)} \Rightarrow$ [(] [3] [.] [7] [+] [(-)] [2] [.] [5] [+] [8] [.] [9] [)] [÷] [(] [3] [.] [7] [×] [(-)] [2] [.] [5] [×] [8] [.] [9] [)] [ENTER] $\{-0.12268448\}$

58. $\dfrac{x+yz}{xy+z} = \dfrac{3.7+(-2.5)(8.9)}{(3.7)(-2.5)+8.9} \Rightarrow$ [(] [3] [.] [7] [+] [(-)] [2] [.] [5] [×] [8] [.] [9] [)] [÷] [(] [3] [.] [7] [×] [(-)] [2] [.] [5] [+] [8] [.] [9] [)] [ENTER] $\{53\}$

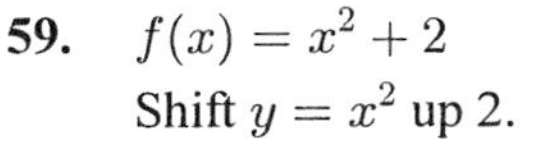

59. $f(x) = x^2 + 2$

Shift $y = x^2$ up 2.

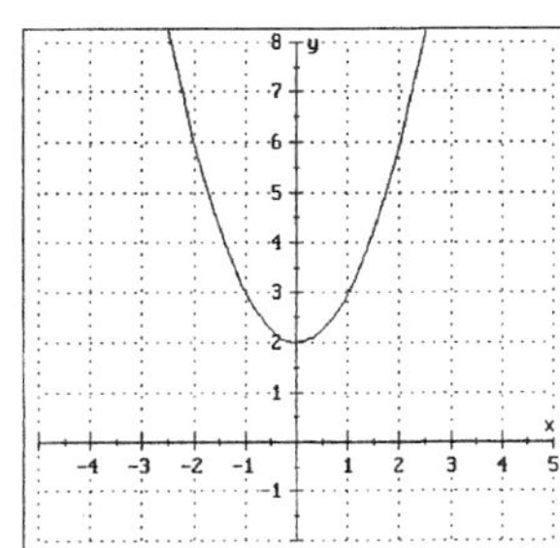

60. $f(x) = x^3 - 2$

Shift $y = x^3$ down 2.

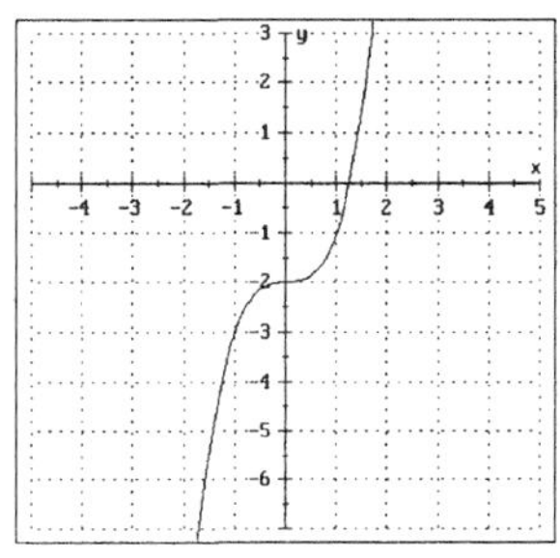

61. $f(x) = -x^3$

Reflect $y = x^3$ about x-axis.

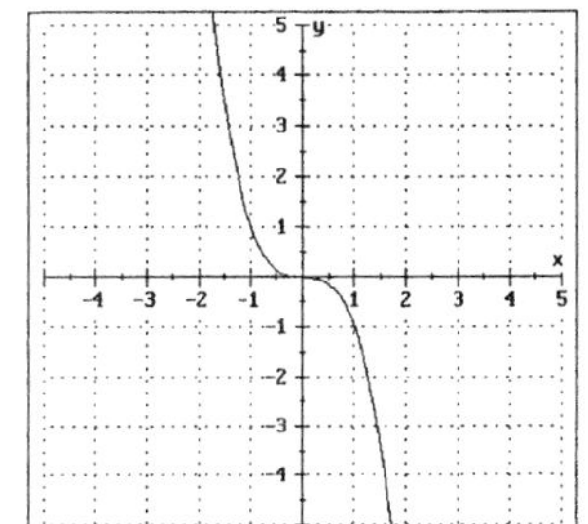

62. $f(x) = -x^2 + 1$

Reflect $y = x^2$ about x-axis and shift up 1.

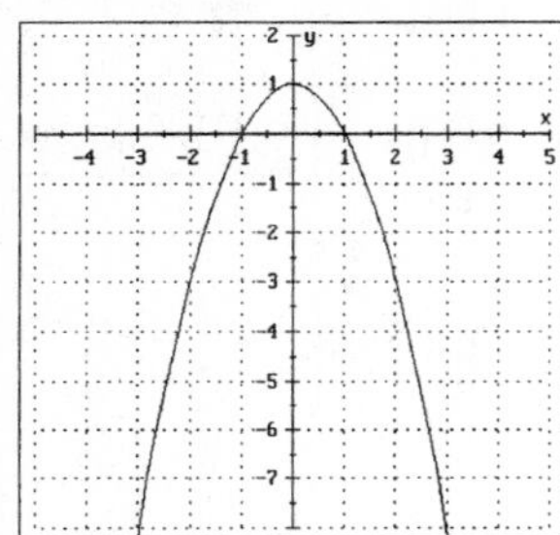

63. $f(x) = -x^3 + x$

x	y
-2	6
-1	0
$-\frac{1}{2}$	$-\frac{3}{8}$
0	0

x	y
$\frac{1}{2}$	$\frac{3}{8}$
1	0
2	-6

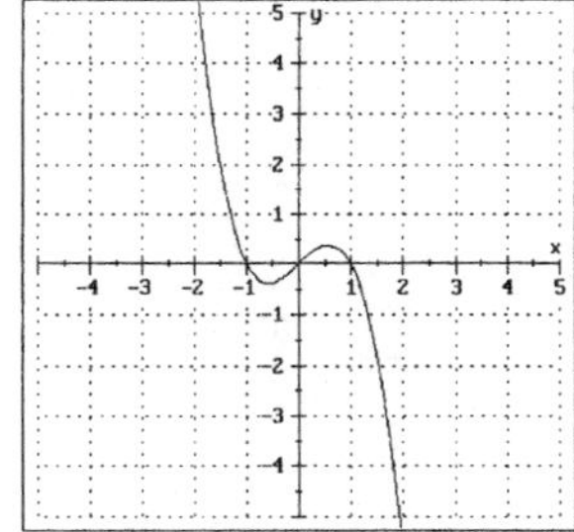

64. $f(x) = x^3 - x$

x	y
-2	-6
-1	0
$-\frac{1}{2}$	$\frac{3}{8}$
0	0

x	y
$\frac{1}{2}$	$-\frac{3}{8}$
1	0
2	6

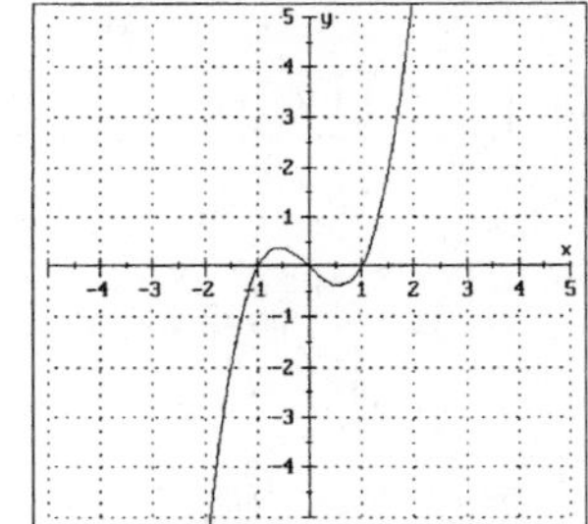

65. $f(x) = x^2 - 2x + 1$

x	y
-1	4
0	1
1	0

x	y
2	1
3	4

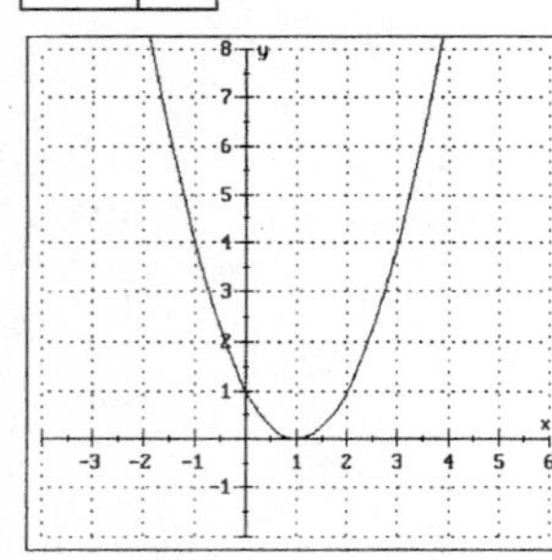

66. $f(x) = x^2 - 2x - 3$

x	y
-1	0
0	-3
1	-4

x	y
2	-3
3	0

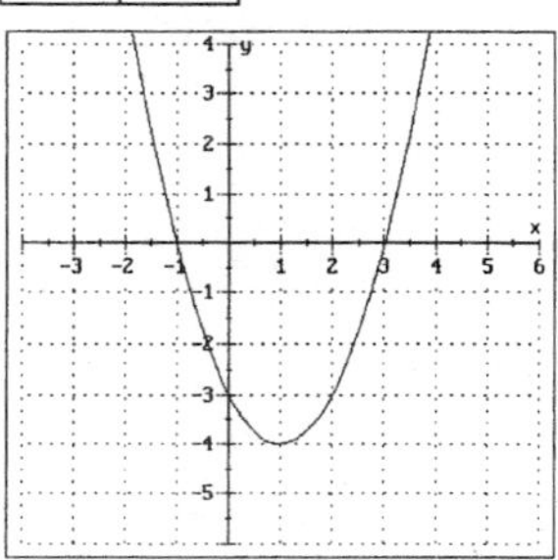

67. $f(x) = 2.75x^2 - 4.7x + 1.5$

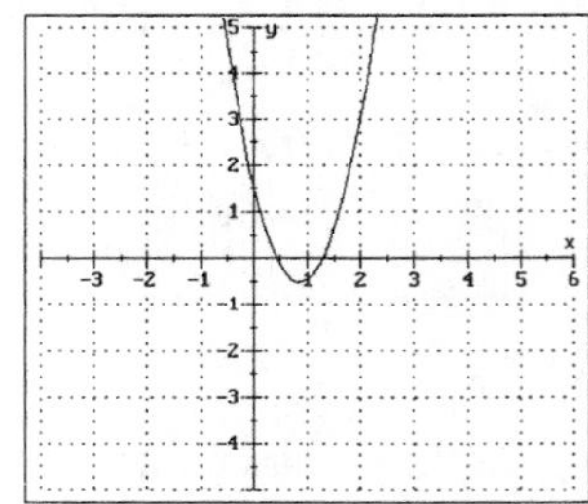

68. $f(x) = -2.5x^2 + 1.7x + 3.2$

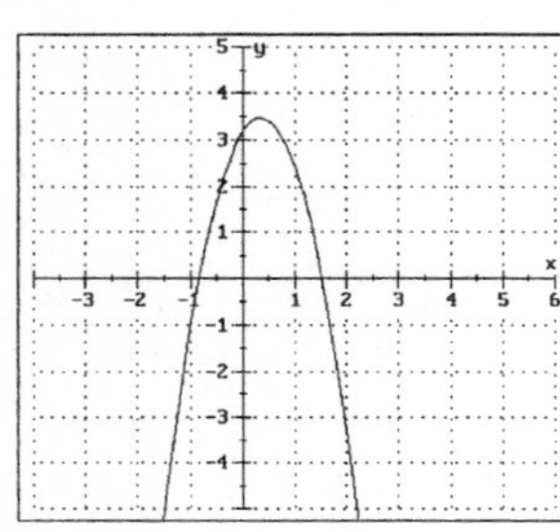

69. $f(x) = 0.25x^2 - 0.5x - 2.5$

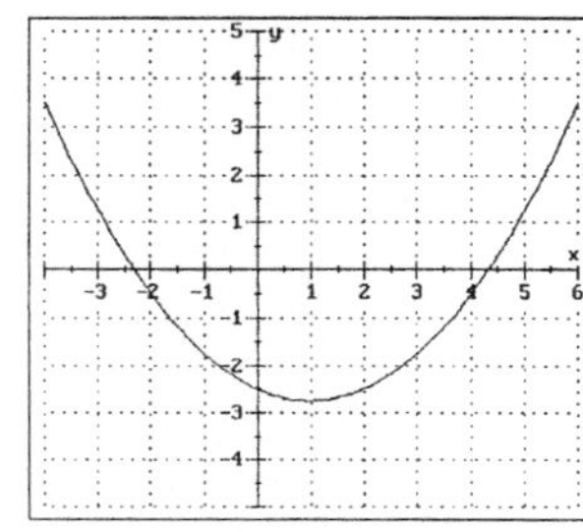

70. $f(x) = 0.37x^2 - 1.4x - 1.5$

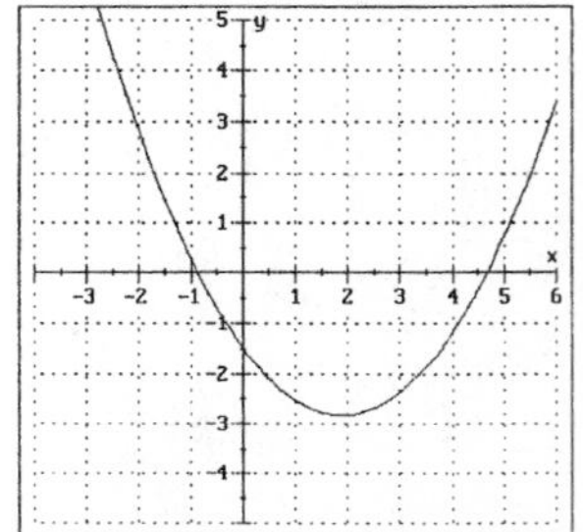

71. $f(1) = -16(1)^2 + 32(1) + 4 = -16(1) + 32 + 4 = -16 + 32 + 4 = 20$; 20 feet

72. $f(2) = -16(2)^2 + 32(2) + 4 = -16(4) + 64 + 4 = -64 + 64 + 4 = 4$; 4 feet

73. $d = f(v) = f(30) = 0.04(30)^2 + 0.9(30) = 0.04(900) + 27 = 36 + 27 = 63$ ft

74. $d = f(v) = f(50) = 0.04(50)^2 + 0.9(50) = 0.04(2500) + 45 = 100 + 45 = 145$ ft

75. $d = f(v) = f(60) = 0.04(60)^2 + 0.9(60) = 0.04(3600) + 54 = 144 + 54 = 198$ ft

76. $d = f(v) = f(70) = 0.04(70)^2 + 0.9(70) = 0.04(4900) + 63 = 196 + 63 = 259$ ft

77. $c = f(x) = f(1) = -2(1)^2 + 12(1) = -2(1) + 12 = -2 + 12 = 10 \text{ in.}^2$

78. $c = f(x) = f(2) = -2(2)^2 + 12(2) = -2(4) + 24 = -8 + 24 = 16 \text{ in.}^2$

79. $c = f(x) = f(3) = -2(3)^2 + 12(3) = -2(9) + 36 = -18 + 36 = 18 \text{ in.}^2$

80. $c = f(x) = f(4) = -2(4)^2 + 12(4) = -2(16) + 48 = -32 + 48 = 16 \text{ in.}^2$

81. **a.** $f(0) = 0.001(0)^3 - 0.12(0)^2 + 3.6(0) + 10 = 0 - 0 + 0 + 10 = 10$ m

b. $f(20) = 0.001(20)^3 - 0.12(20)^2 + 3.6(20) + 10$
$= 0.001(8000) - 0.12(400) + 72 + 10 = 8 - 48 + 72 + 10 = 42$ m

c. $f(40) = 0.001(40)^3 - 0.12(40)^2 + 3.6(40) + 10$
$= 0.001(64000) - 0.12(1600) + 144 + 10 = 64 - 192 + 144 + 10 = 26$ m

d. $f(60) = 0.001(60)^3 - 0.12(60)^2 + 3.6(60) + 10$
$= 0.001(216000) - 0.12(3600) + 216 + 10 = 216 - 432 + 216 + 10 = 10$ m

82. $C(4) = -0.0625(4)^4 + (4)^3 + 16(4) = -0.0625(256) + 64 + 64 = -16 + 128 = 112$ calls

83. **Answers may vary.**

84. **Answers may vary.**

85. $P(2) + P(3) = [2^2 - 5(2)] + [3^2 - 5(3)] = [4 - 10] + [9 - 15] = -6 + (-6) = -12$
$P(2 + 3) = P(5) = 5^2 - 5(5) = 0 \Rightarrow$ They are not equal.

86. $P(2) - P(3) = [2^3 - 3(2)] - [3^3 - 3(3)] = [8 - 6] - [27 - 9] = (2) - (-18) = 20$
$P(2 - 3) = P(-1) = (-1)^3 - 3(-1) = -1 + 3 = 2 \Rightarrow$ They are not equal.

87. $P(P(0)) = P(0^2 - 2(0) - 3) = P(-3) = (-3)^2 - 2(-3) - 3 = 9 + 6 - 3 = 12$

88. $P(P(-1)) = P(2(-1)^2 - (-1) - 5) = P(2(1) + 1 - 5) = P(-2) = 2(-2)^2 - (-2) - 5 = 5$

89. **Answers may vary.**

90. **Answers may vary.**

Exercise 5.2 (page 287)

1. $2x+3 \le 11$
$2x \le 8$
$x \le 4$
solution set: $(-\infty, 4]$

2. $\frac{2}{3}x+5 > 11$
$\frac{2}{3}x > 6$
$2x > 18$
$x > 9$
solution set: $(9, \infty)$

3. $|x-4| < 5$
$-5 < x-4 < 5$
$-1 < x < 9$
solution set: $(-1, 9)$

4. $|2x+1| \ge 7$
$2x+1 \le -7$ **or** $2x+1 \ge 7$
$2x \le -8 \qquad 2x \ge 6$
$x \le -4 \qquad x \ge 3$
solution set: $(-\infty, -4] \cup [3, \infty)$

5. exponents

6. unlike

7. coefficients

8. negative (or opposite)

9. like terms, $3x+7x=10x$

10. unlike terms

11. unlike terms

12. like terms, $3mn+5mn=8mn$

13. like terms, $3r^2t^3-8r^2t^3=-5r^2t^3$

14. like terms, $9u^2v+10u^2v=19u^2v$

15. unlike terms

16. unlike terms

17. $8x+4x=12x$

18. $-2y+16y=14y$

19. $5x^3y^2z-3x^3y^2z=2x^3y^2z$

20. $8wxy-12wxy=-4wxy$

21. $-2x^2y^3+3xy^4-5x^2y^3=-7x^2y^3+3xy^4$

22. $3ab^4-4a^2b^2-2ab^4+2a^2b^2=ab^4-2a^2b^2$

23. $(3x^2y)^2+2x^4y^2-x^4y^2=9x^4y^2+2x^4y^2-x^4y^2=10x^4y^2$

24. $(5x^2y^4)^3-(5x^3y^6)^2=125x^6y^{12}-25x^6y^{12}=100x^6y^{12}$

25. $(2a+3)+(3a+5)=2a+3+3a+5$
$=5a+8$

26. $(5y-2)+(-2y+3)=5y-2-2y+3$
$=3y+1$

27. $(4t+3)-(2t+5)=4t+3-2t-5$
$=2t-2$

28. $(6z-5)-(-2z+1)=6z-5+2z-1$
$=8z-6$

29. $(3x^2+2x+1)+(-2x^2-7x+5)=3x^2+2x+1-2x^2-7x+5=x^2-5x+6$

30. $(-2a^2 - 5a - 7) + (-3a^2 + 7a + 1) = -2a^2 - 5a - 7 - 3a^2 + 7a + 1 = -5a^2 + 2a - 6$

31. $(-a^2 + 2a + 3) - (4a^2 - 2a - 1) = -a^2 + 2a + 3 - 4a^2 + 2a + 1 = -5a^2 + 4a + 4$

32. $(x^2 - 3x + 8) - (3x^2 + x + 3) = x^2 - 3x + 8 - 3x^2 - x - 3 = -2x^2 - 4x + 5$

33. $(7y^3 + 4y^2 + y + 3) + (-8y^3 - y + 3) = 7y^3 + 4y^2 + y + 3 - 8y^3 - y + 3 = -y^3 + 4y^2 + 6$

34. $(6x^3 + 3x - 2) - (2x^3 + 3x^2 + 5) = 6x^3 + 3x - 2 - 2x^3 - 3x^2 - 5 = 4x^3 - 3x^2 + 3x - 7$

35. $$\begin{aligned}(3x^2 + 4x - 3) + (2x^2 - 3x - 1) - (x^2 + x + 7) &= 3x^2 + 4x - 3 + 2x^2 - 3x - 1 - x^2 - x - 7\\ &= 4x^2 - 11\end{aligned}$$

36. $$\begin{aligned}&(-2x^2 + 6x + 5) - (-4x^2 - 7x + 2) - (4x^2 + 10x + 5)\\ &\qquad = -2x^2 + 6x + 5 + 4x^2 + 7x - 2 - 4x^2 - 10x - 5\\ &\qquad = -2x^2 + 3x - 2\end{aligned}$$

37. $$\begin{aligned}&(3x^3 - 2x + 3) + (4x^3 + 3x^2 - 2) + (-4x^3 - 3x^2 + x + 12)\\ &\qquad = 3x^3 - 2x + 3 + 4x^3 + 3x^2 - 2 - 4x^3 - 3x^2 + x + 12\\ &\qquad = 3x^3 - x + 13\end{aligned}$$

38. $$\begin{aligned}&(x^4 - 3x^2 + 4) + (-2x^4 - x^3 + 3x^2) + (3x^2 + 2x + 1)\\ &\qquad = x^4 - 3x^2 + 4 - 2x^4 - x^3 + 3x^2 + 3x^2 + 2x + 1\\ &\qquad = -x^4 - x^3 + 3x^2 + 2x + 5\end{aligned}$$

39. $$\begin{aligned}&(3y^2 - 2y + 4) + [(2y^2 - 3y + 2) - (y^2 + 4y + 3)]\\ &\qquad = 3y^2 - 2y + 4 + [2y^2 - 3y + 2 - y^2 - 4y - 3]\\ &\qquad = 3y^2 - 2y + 4 + [y^2 - 7y - 1]\\ &\qquad = 3y^2 - 2y + 4 + y^2 - 7y - 1 = 4y^2 - 9y + 3\end{aligned}$$

40. $$\begin{aligned}&(-t^2 - t - 1) - [(t^2 + 3t - 1) - (-2t^2 + 4)]\\ &\qquad = -t^2 - t - 1 - [t^2 + 3t - 1 + 2t^2 - 4]\\ &\qquad = -t^2 - t - 1 - [3t^2 + 3t - 5]\\ &\qquad = -t^2 - t - 1 - 3t^2 - 3t + 5 = -4t^2 - 4t + 4\end{aligned}$$

41. $$\begin{array}{rrrrr} & 3x^3 & -\,2x^2 & +\,4x & -\,3\\ & -2x^3 & +\,3x^2 & +\,3x & -\,2\\ + & 5x^3 & -\,7x^2 & +\,7x & -\,12\\ \hline & 6x^3 & -\,6x^2 & +\,14x & -\,17\end{array}$$

42. $$\begin{array}{rrrrr} & 7a^3 & & +\,3a & +\,7\\ & -2a^3 & +\,4a^2 & & -\,13\\ + & 3a^3 & -\,3a^2 & +\,4a & +\,5\\ \hline & 8a^3 & +\,a^2 & +\,7a & -\,1\end{array}$$

43.
$$\begin{array}{r} -2y^4-2y^3+\ 4y^2-\ 3y+10 \\ -3y^4+7y^3-\quad y^2+14y-\ 3 \\ -3y^3-\ 5y^2-\ 5y+\ 7 \\ +\ \underline{-4y^4+\ y^3-13y^2+14y-\ 2} \\ -9y^4+3y^3-15y^2+20y+12 \end{array}$$

44.
$$\begin{array}{r} 17t^4+3t^3-\ 2t^2-\ 3t+\ 4 \\ -12t^4-2t^3+\ 3t^2-\ 5t-17 \\ -2t^4-7t^3+\ 4t^2+12t-\ 5 \\ +\ \underline{5t^4+\ t^3+\ 5t^2-13t+12} \\ 8t^4-5t^3+10t^2-\ 9t-\ 6 \end{array}$$

45.
$$\begin{array}{r} 3x^2-4x+17 \\ -\ \underline{2x^2+4x-\ 5} \\ 3x^2-4x+17 \\ +\ \underline{-2x^2-4x+\ 5} \\ x^2-8x+22 \end{array}$$

46.
$$\begin{array}{r} -2y^2-\ 4y+3 \\ -\ \underline{3y^2+10y-5} \\ -2y^2-\ 4y+3 \\ +\ \underline{-3y^2-10y+5} \\ -5y^2-14y+8 \end{array}$$

47.
$$\begin{array}{r} -5y^3+\ 4y^2-11y+\ 3 \\ -\ \underline{-2y^3-14y^2+17y-32} \\ -5y^3+\ 4y^2-11y+\ 3 \\ +\ \underline{2y^3+14y^2-17y+32} \\ -3y^3+18y^2-28y+35 \end{array}$$

48.
$$\begin{array}{r} 17x^4-\ 3x^2-65x-12 \\ -\ \underline{23x^4+14x^2+\ 3x-23} \\ 17x^4-\ 3x^2-65x-12 \\ +\ \underline{-23x^4-14x^2-\ 3x+23} \\ -6x^4-17x^2-68x+11 \end{array}$$

49. $3(x+2)+2(x-5)=3x+6+2x-10$
$=5x-4$

50. $-2(x-4)+5(x+1)=-2x+8+5x+5$
$=3x+13$

51. $-6(t-4)-5(t-1)=-6t+24-5t+5$
$=-11t+29$

52. $4(a+5)-3(a-1)=4a+20-3a+3$
$=a+23$

53. $2(x^3+x^2)+3(2x^3-x^2)=2x^3+2x^2+6x^3-3x^2=8x^3-x^2$

54. $3(y^2+2y)-4(y^2-4)=3y^2+6y-4y^2+16=-y^2+6y+16$

55. $-3(2m-n)+2(m-3n)=-6m+3n+2m-6n=-4m-3n$

56. $5(p-2q)-4(2p+q)=5p-10q-8p-4q=-3p-14q$

57. $-5(2x^3+7x^2+4x)-2(3x^3-4x^2-4x)=-10x^3-35x^2-20x-6x^3+8x^2+8x$
$=-16x^3-27x^2-12x$

58. $-3(3a^2+4b^3+7)+4(5a^2-2b^3+3)=-9a^2-12b^3-21+20a^2-8b^3+12$
$=11a^2-20b^3-9$

59. $4(3z^2-4z+5)+6(-2z^2-3z+4)-2(4z^2+3z-5)$
$=12z^2-16z+20-12z^2-18z+24-8z^2-6z+10$
$=-8z^2-40z+54$

60. $-3(4x^3 - 2x^2 + 4) - 4(3x^3 + 4x^2 + 3x) + 5(3x - 4)$
$= -12x^3 + 6x^2 - 12 - 12x^3 - 16x^2 - 12x + 15x - 20$
$= -24x^3 - 10x^2 + 3x - 32$

61. $5(2a^2 + 4a - 2) - 2(-3a^2 - a + 12) - 2(a^2 + 3a - 5)$
$= 10a^2 + 20a - 10 + 6a^2 + 2a - 24 - 2a^2 - 6a + 10$
$= 14a^2 + 16a - 24$

62. $-2(2b^2 - 3b + 3) + 3(3b^2 + 2b - 8) - (3b^2 - b + 4)$
$= -4b^2 + 6b - 6 + 9b^2 + 6b - 24 - 3b^2 + b - 4$
$= 2b^2 + 13b - 34$

63. $x + (x + 1) + (x + 2) = x + x + 1 + x + 2 = 3x + 3$

64. $2(x + 1) + 2(x + 4) = 2x + 2 + 2x + 8 = 4x + 10$

65. $(2x^2 + 5x + 1) - (x^2 - 5) = 2x^2 + 5x + 1 - x^2 + 5 = x^2 + 5x + 6$ meters

66. $(3y^2 - 2y + 1) + (2y^2 + y + 1) = 3y^2 - 2y + 1 + 2y^2 + y + 1 = 5y^2 - y + 2$ feet

67. $f(x) = 1100x + 125000$; $f(10) = 1100(10) + 125000 = 11000 + 125000 =$ \$136,000

68. **a.** $f(x) = 1400x + 150000$
b. $f(12) = 1400(12) + 150000 = 16800 + 150000 = 166800$

69. $y = (1100x + 125000) + (1400x + 150000) = 2500x + 275000$

70. **a.** $f(20) = 1100(20) + 125000 = 22000 + 125000 = 147000$
$f(20) = 1400(20) + 150000 = 28000 + 150000 = 178000$
$147000 + 178000 =$ \$325,000
b. $y = 2500(20) + 275000 = 50000 + 275000 =$ \$325,000

71. $y = -2100x + 16600$

72. $y = -2700x + 19200$

73. $y = (-2100x + 16600) + (-2700x + 19200) = -4800x + 35800$

74. 1st car: $y = -2100(3) + 16600 = -6300 + 16600 = 10300$
2nd car: $y = -2700(3) + 19200 = -8100 + 19200 = 11100$
$10300 + 11100 =$ \$21,400
OR
$y = -4800(3) + 35800 = -14400 + 35800 =$ \$21,400

75. **Answers may vary.**

76. **Answers may vary.**

77. $[(-2x^2 - x + 7) + (5x^2 + 3x - 1)] - (3x^2 + 4x - 3)$
$= [-2x^2 - x + 7 + 5x^2 + 3x - 1] - 3x^2 - 4x + 3$
$= -2x^2 - x + 7 + 5x^2 + 3x - 1 - 3x^2 - 4x + 3$
$= -2x + 9$

78. $[(x^2 + x + 2) + (2x^3 - x + 9)] - (8x^3 + 2x^2 - 1)$
$= [x^2 + x + 2 + 2x^3 - x + 9] - 8x^3 - 2x^2 + 1$
$= x^2 + x + 2 + 2x^3 - x + 9 - 8x^3 - 2x^2 + 1$
$= -6x^3 - x^2 + 12$

79. $[(2x^2 - 4x + 3) - (8x^2 + 5x - 3)] + (-2x^2 + 7x - 4)$
$= [2x^2 - 4x + 3 - 8x^2 - 5x + 3] - 2x^2 + 7x - 4$
$= 2x^2 - 4x + 3 - 8x^2 - 5x + 3 - 2x^2 + 7x - 4$
$= -8x^2 - 2x + 2$

80. $[(7x^3 - 4x) - (x^2 + 2)] + (5 + 3x) = [7x^3 - 4x - x^2 - 2] + 5 + 3x$
$= 7x^3 - 4x - x^2 - 2 + 5 + 3x$
$= 7x^3 - x^2 - x + 3$

Exercise 5.3 (page 297)

1. $|3a - b| = |3(-2) - 4| = |-6 - 4|$
$= |-10| = 10$

2. $|ab - b^2| = |-2(4) - 4^2| = |-8 - 16|$
$= |-24| = 24$

3. $-|a^2b - b^0| = -|(-2)^2(4) - 1| = -|4(4) - 1| = -|16 - 1| = -|15| = -15$

4. $\left|\frac{a^3b^2 + ab}{2(ab)^2 - a^3}\right| = \left|\frac{(-2)^3(4)^2 + (-2)(4)}{2[(-2)(4)]^2 - (-2)^3}\right| = \left|\frac{-8(16) - 8}{2(-8)^2 - (-8)}\right| = \left|\frac{-128 - 8}{2(64) + 8}\right| = \left|\frac{-136}{136}\right|$
$= |-1| = 1$

5. The current value of ABC $=$ \$126.5 per share, and the current value of WD $=$ \$73.5 per share. Total value $= 200(126.5) + 350(73.5) = 25300 + 25725 =$ \$51,025.

6. $5{,}870{,}000{,}000{,}000 = 5.87 \times 10^{12}$

7. variable

8. monomial

9. term

10. First; Outer; Inner; Last

11. $x^2 + 2xy + y^2$

12. $x^2 - 2xy + y^2$

13. $x^2 - y^2$

14. perfect square

15. $(2a^2)(-3ab) = -6a^3b$

16. $(-3x^2y)(3xy) = -9x^3y^2$

17. $(-3ab^2c)(5ac^2) = -15a^2b^2c^3$

18. $(-2m^2n)(-4mn^3) = 8m^3n^4$

19. $(4a^2b)(-5a^3b^2)(6a^4) = -120a^9b^3$

20. $(2x^2y^3)(4xy^5)(-5y^6) = -40x^3y^{14}$

21. $(5x^3y^2)^4\left(\frac{1}{5}x^{-2}\right)^2 = (625x^{12}y^8)\left(\frac{1}{25}x^{-4}\right) = 25x^8y^8$

22. $(4a^{-2}b^{-1})^2(2a^3b^4)^4 = (16a^{-4}b^{-2})(16a^{12}b^{16}) = 256a^8b^{14}$

23. $(-5xx^2)(-3xy)^4 = (-5x^3)(81x^4y^4) = -405x^7y^4$

24. $(-2a^2ab^2)^3(-3ab^2b^2) = (-2a^3b^2)^3(-3ab^4) = (-8a^9b^6)(-3ab^4) = 24a^{10}b^{10}$

25. $3(x+2) = 3x + 3(2) = 3x + 6$

26. $-5(a+b) = -5a + (-5)b$
$= -5a - 5b$

27. $-a(a-b) = -a(a) + (-a)(-b)$
$= -a^2 + ab$

28. $y^2(y-1) = y^2(y) + y^2(-1) = y^3 - y^2$

29. $3x(x^2+3x) = 3x(x^2) + 3x(3x)$
$= 3x^3 + 9x^2$

30. $-2x(3x^2-2) = -2x(3x^2) + (-2x)(-2)$
$= -6x^3 + 4x$

31. $-2x(3x^2-3x+2) = -2x(3x^2) + (-2x)(-3x) + (-2x)(2) = -6x^3 + 6x^2 - 4x$

32. $3a(4a^2+3a-4) = 3a(4a^2) + 3a(3a) + 3a(-4) = 12a^3 + 9a^2 - 12a$

33. $5a^2b^3(2a^4b - 5a^0b^3) = 5a^2b^3(2a^4b) + 5a^2b^3(-5a^0b^3) = 10a^6b^4 - 25a^2b^6$

34. $-2a^3b(3a^0b^4 - 2a^2b^3) = -2a^3b(3a^0b^4) + (-2a^3b)(-2a^2b^3) = -6a^3b^5 + 4a^5b^4$

35. $7rst(r^2+s^2-t^2) = 7rst(r^2) + 7rst(s^2) + 7rst(-t^2) = 7r^3st + 7rs^3t - 7rst^3$

36. $3x^2yz(x^2-2y+3z^2) = 3x^2yz(x^2) + 3x^2yz(-2y) + 3x^2yz(3z^2) = 3x^4yz - 6x^2y^2z + 9x^2yz^3$

37. $4m^2n(-3mn)(m+n) = -12m^3n^2(m+n) = -12m^4n^2 - 12m^3n^3$

38. $-3a^2b^3(2b)(3a+b) = -6a^2b^4(3a+b) = -18a^3b^4 - 6a^2b^5$

39. $(x+2)(x+3) = x^2 + 3x + 2x + 6$
$= x^2 + 5x + 6$

40. $(y-3)(y+4) = y^2 + 4y - 3y - 12$
$= y^2 + y - 12$

41. $(z-7)(z-2) = z^2 - 2z - 7z + 14$
$= z^2 - 9z + 14$

42. $(x+3)(x-5) = x^2 - 5x + 3x - 15$
$= x^2 - 2x - 15$

43. $(2a+1)(a-2) = 2a^2 - 4a + a - 2$
$= 2a^2 - 3a - 2$

44. $(3b-1)(2b-1) = 6b^2 - 3b - 2b + 1$
$= 6b^2 - 5b + 1$

45. $(3t-2)(2t+3) = 6t^2+9t-4t-6$
$= 6t^2+5t-6$

46. $(p+3)(3p-4) = 3p^2-4p+9p-12$
$= 3p^2+5p-12$

47. $(3y-z)(2y-z) = 6y^2-3yz-2yz+z^2 = 6y^2-5yz+z^2$

48. $(2m+n)(3m+n) = 6m^2+2mn+3mn+n^2 = 6m^2+5mn+n^2$

49. $(2x-3y)(x+2y) = 2x^2+4xy-3xy-6y^2 = 2x^2+xy-6y^2$

50. $(3y+2z)(y-3z) = 3y^2-9yz+2yz-6z^2 = 3y^2-7yz-6z^2$

51. $(3x+y)(3x-3y) = 9x^2-9xy+3xy-3y^2 = 9x^2-6xy-3y^2$

52. $(2x-y)(3x+2y) = 6x^2+4xy-3xy-2y^2 = 6x^2+xy-2y^2$

53. $(4a-3b)(2a+5b) = 8a^2+20ab-6ab-15b^2 = 8a^2+14ab-15b^2$

54. $(3a+2b)(2a-7b) = 6a^2-21ab+4ab-14b^2 = 6a^2-17ab-14b^2$

55. $(x+2)^2 = (x+2)(x+2) = x^2+2x+2x+4 = x^2+4x+4$

56. $(x-3)^2 = (x-3)(x-3) = x^2-3x-3x+9 = x^2-6x+9$

57. $(a-4)^2 = (a-4)(a-4) = a^2-4a-4a+16 = a^2-8a+16$

58. $(y+5)^2 = (y+5)(y+5) = y^2+5y+5y+25 = y^2+10y+25$

59. $(2a+b)^2 = (2a+b)(2a+b) = 4a^2+2ab+2ab+b^2 = 4a^2+4ab+b^2$

60. $(a-2b)^2 = (a-2b)(a-2b) = a^2-2ab-2ab+4b^2 = a^2-4ab+4b^2$

61. $(2x-y)^2 = (2x-y)(2x-y) = 4x^2-2xy-2xy+y^2 = 4x^2-4xy+y^2$

62. $(3m+4n)(3m+4n) = 9m^2+12mn+12mn+16n^2 = 9m^2+24mn+16n^2$

63. $(x+2)(x-2) = x^2-2x+2x-4$
$= x^2-4$

64. $(z+3)(z-3) = z^2-3z+3z-9$
$= z^2-9$

65. $(a+b)(a-b) = a^2-ab+ab-b^2$
$= a^2-b^2$

66. $(p+q)(p-q) = p^2-pq+pq-q^2$
$= p^2-q^2$

67. $(2x+3y)(2x-3y) = 4x^2-6xy+6xy-9y^2 = 4x^2-9y^2$

68. $(3a+4b)(3a-4b) = 9a^2-12ab+12ab-16b^2 = 9a^2-16b^2$

69. $(x-y)(x^2+xy+y^2) = x(x^2+xy+y^2) - y(x^2+xy+y^2)$
$= x^3 + x^2y + xy^2 - x^2y - xy^2 - y^3 = x^3 - y^3$

70. $(x+y)(x^2-xy+y^2) = x(x^2-xy+y^2) + y(x^2-xy+y^2)$
$= x^3 - x^2y + xy^2 + x^2y - xy^2 + y^3 = x^3 + y^3$

71. $(3y+1)(2y^2+3y+2) = 3y(2y^2+3y+2) + 1(2y^2+3y+2)$
$= 6y^3 + 9y^2 + 6y + 2y^2 + 3y + 2 = 6y^3 + 11y^2 + 9y + 2$

72. $(a+2)(3a^2+4a-2) = a(3a^2+4a-2) + 2(3a^2+4a-2)$
$= 3a^3 + 4a^2 - 2a + 6a^2 + 8a - 4 = 3a^3 + 10a^2 + 6a - 4$

73. $(2a-b)(4a^2+2ab+b^2) = 2a(4a^2+2ab+b^2) - b(4a^2+2ab+b^2)$
$= 8a^3 + 4a^2b + 2ab^2 - 4a^2b - 2ab^2 - b^3 = 8a^3 - b^3$

74. $(x-3y)(x^2+3xy+9y^2) = x(x^2+3xy+9y^2) - 3y(x^2+3xy+9y^2)$
$= x^3 + 3x^2y + 9xy^2 - 3x^2y - 9xy^2 - 27y^3 = x^3 - 27y^3$

75. $(2x-1)[2x^2-3(x+2)] = (2x-1)(2x^2-3x-6) = 2x(2x^2-3x-6) - 1(2x^2-3x-6)$
$= 4x^3 - 6x^2 - 12x - 2x^2 + 3x + 6$
$= 4x^3 - 8x^2 - 9x + 6$

76. $(x+1)^2[x^2-2(x+2)] = (x+1)(x+1)(x^2-2x-4)$
$= (x^2+2x+1)(x^2-2x-4)$
$= x^2(x^2-2x-4) + 2x(x^2-2x-4) + 1(x^2-2x-4)$
$= x^4 - 2x^3 - 4x^2 + 2x^3 - 4x^2 - 8x + x^2 - 2x - 4$
$= x^4 - 7x^2 - 10x - 4$

77. $(a+b)(a-b)(a-3b) = (a^2-b^2)(a-3b) = a^3 - 3a^2b - ab^2 + 3b^3$

78. $(x-y)(x+2y)(x-2y) = (x-y)(x^2-4y^2) = x^3 - x^2y - 4xy^2 + 4y^3$

79. $(a+b)^3 = (a+b)(a+b)(a+b) = \left(a^2+2ab+b^2\right)(a+b)$
$= a^2(a+b) + 2ab(a+b) + b^2(a+b)$
$= a^3 + a^2b + 2a^2b + 2ab^2 + ab^2 + b^3$
$= a^3 + 3a^2b + 3ab^2 + b^3$

80. $(2m-n)^3 = (2m-n)(2m-n)(2m-n) = \left(4m^2-4mn+n^2\right)(2m-n)$
$= 4m^2(2m-n) - 4mn(2m-n) + n^2(2m-n)$
$= 8m^3 - 4m^2n - 8m^2n + 4mn^2 + 2mn^2 - n^3$
$= 8m^3 - 12m^2n + 6mn^2 - n^3$

81. $(2p+q)(p-2q)^2 = (2p+q)(p-2q)(p-2q) = (2p+q)\left(p^2-4pq+4q^2\right)$
$= 2p\left(p^2-4pq+4q^2\right)+q\left(p^2-4pq+4q^2\right)$
$= 2p^3-8p^2q+8pq^2+p^2q-4pq^2+4q^3$
$= 2p^3-7p^2q+4pq^2+4q^3$

82. $(a-b)(2a+b)^2 = (a-b)(2a+b)(2a+b) = (a-b)\left(4a^2+4ab+b^2\right)$
$= a\left(4a^2+4ab+b^2\right)-b\left(4a^2+4ab+b^2\right)$
$= 4a^3+4a^2b+ab^2-4a^2b-4ab^2-b^3$
$= 4a^3-3ab^2-b^3$

83. $x^3(2x^2+x^{-2}) = 2x^5+x$

84. $x^{-4}(2x^{-3}-5x^2) = 2x^{-7}-5x^{-2} = \dfrac{2}{x^7}-\dfrac{5}{x^2}$

85. $x^3y^{-6}z^{-2}(3x^{-2}y^2z-x^3y^{-4}) = 3xy^{-4}z^{-1}-x^6y^{-10}z^{-2} = \dfrac{3x}{y^4z}-\dfrac{x^6}{y^{10}z^2}$

86. $ab^{-2}c^{-3}(a^{-4}bc^3+a^{-3}b^4c^3) = a^{-3}b^{-1}c^0+a^{-2}b^2c^0 = \dfrac{1}{a^3b}+\dfrac{b^2}{a^2}$

87. $(x^{-1}+y)(x^{-1}-y) = x^{-2}-x^{-1}y+x^{-1}y-y^2 = \dfrac{1}{x^2}-y^2$

88. $(x^{-1}-y)(x^{-1}-y) = x^{-2}-x^{-1}y-x^{-1}y+y^2 = x^{-2}-2x^{-1}y+y^2 = \dfrac{1}{x^2}-\dfrac{2y}{x}+y^2$

89. $(2x^{-3}+y^3)(2x^3-y^{-3}) = 4x^0-2x^{-3}y^{-3}+2x^3y^3-y^0 = 4-\dfrac{2}{x^3y^3}+2x^3y^3-1$
$= 2x^3y^3-\dfrac{2}{x^3y^3}+3$

90. $(5x^{-4}-4y^2)(5x^2-4y^{-4}) = 25x^{-2}-20x^{-4}y^{-4}-20x^2y^2+16y^{-2}$
$= \dfrac{25}{x^2}-\dfrac{20}{x^4y^4}-20x^2y^2+\dfrac{16}{y^2}$

91. $x^n(x^{2n}-x^n) = x^{3n}-x^{2n}$

92. $a^{2n}(a^n+a^{2n}) = a^{3n}+a^{4n}$

93. $(x^n+1)(x^n-1) = x^{2n}-x^n+x^n-1 = x^{2n}-1$

94. $(x^n-a^n)(x^n+a^n) = x^{2n}+x^na^n-x^na^n-a^{2n} = x^{2n}-a^{2n}$

95. $(x^n-y^n)(x^n-y^{-n}) = x^{2n}-x^ny^{-n}-x^ny^n+y^0 = x^{2n}-\dfrac{x^n}{y^n}-x^ny^n+1$

96. $(x^n+y^n)(x^n+y^{-n}) = x^{2n}+x^ny^{-n}+x^ny^n+y^0 = x^{2n}+\dfrac{x^n}{y^n}+x^ny^n+1$

97. $(x^{2n} + y^{2n})(x^{2n} - y^{2n}) = x^{4n} - x^{2n}y^{2n} + x^{2n}y^{2n} - y^{4n} = x^{4n} - y^{4n}$

98. $(a^{3n} - b^{3n})(a^{3n} + b^{3n}) = a^{6n} + a^{3n}b^{3n} - a^{3n}b^{3n} - b^{6n} = a^{6n} - b^{6n}$

99. $(x^n + y^n)(x^n + 1) = x^{2n} + x^n + x^n y^n + y^n$

100. $(1 - x^n)(x^{-n} - 1) = x^{-n} - 1 - x^0 + x^n = \dfrac{1}{x^n} - 1 - 1 + x^n = x^n + \dfrac{1}{x^n} - 2$

101. $$\begin{aligned} 3x(2x+4) - 3x^2 &= 6x^2 + 12x - 3x^2 \\ &= 3x^2 + 12x \end{aligned}$$

102. $$\begin{aligned} 2y - 3y(y^2+4) &= 2y - 3y^3 - 12y \\ &= -3y^3 - 10y \end{aligned}$$

103. $$\begin{aligned} 3pq - p(p-q) &= 3pq - p^2 + pq \\ &= -p^2 + 4pq \end{aligned}$$

104. $$\begin{aligned} -4rs(r-2) + 4rs &= -4r^2s + 8rs + 4rs \\ &= -4r^2s + 12rs \end{aligned}$$

105. $$\begin{aligned} 2m(m-n) - (m+n)(m-2n) &= 2m^2 - 2mn - (m^2 - 2mn + mn - 2n^2) \\ &= 2m^2 - 2mn - m^2 + 2mn - mn + 2n^2 \\ &= m^2 - mn + 2n^2 \end{aligned}$$

106. $$\begin{aligned} -3y(2y+z) + (2y-z)(3y+2z) &= -6y^2 - 3yz + 6y^2 + 4yz - 3yz - 2z^2 \\ &= -2yz - 2z^2 \end{aligned}$$

107. $$\begin{aligned} (x+3)(x-3) + (2x-1)(x+2) &= x^2 - 3x + 3x - 9 + 2x^2 + 4x - x - 2 \\ &= 3x^2 + 3x - 11 \end{aligned}$$

108. $$\begin{aligned} (2b+3)(b-1) - (b+2)(3b-1) &= 2b^2 - 2b + 3b - 3 - (3b^2 - b + 6b - 2) \\ &= 2b^2 + b - 3 - 3b^2 + b - 6b + 2 \\ &= -b^2 - 4b - 1 \end{aligned}$$

109. $$\begin{aligned} (3x-4)^2 - (2x+3)^2 &= (3x-4)(3x-4) - (2x+3)(2x+3) \\ &= 9x^2 - 12x - 12x + 16 - (4x^2 + 6x + 6x + 9) \\ &= 9x^2 - 24x + 16 - 4x^2 - 6x - 6x - 9 \\ &= 5x^2 - 36x + 7 \end{aligned}$$

110. $$\begin{aligned} (3y+1)^2 + (2y-4)^2 &= (3y+1)(3y+1) + (2y-4)(2y-4) \\ &= 9y^2 + 3y + 3y + 1 + 4y^2 - 8y - 8y + 16 \\ &= 13y^2 - 10y + 17 \end{aligned}$$

111. $$\begin{aligned} 3(x-3y)^2 + 2(3x+y)^2 &= 3(x-3y)(x-3y) + 2(3x+y)(3x+y) \\ &= 3(x^2 - 3xy - 3xy + 9y^2) + 2(9x^2 + 3xy + 3xy + y^2) \\ &= 3x^2 - 9xy - 9xy + 27y^2 + 18x^2 + 6xy + 6xy + 2y^2 \\ &= 21x^2 - 6xy + 29y^2 \end{aligned}$$

112. $2(x-y^2)^2 - 3(y^2+2x)^2 = 2(x-y^2)(x-y^2) - 3(y^2+2x)(y^2+2x)$
$= 2(x^2 - xy^2 - xy^2 + y^4) - 3(y^4 + 2xy^2 + 2xy^2 + 4x^2)$
$= 2x^2 - 2xy^2 - 2xy^2 + 2y^4 - 3y^4 - 6xy^2 - 6xy^2 - 12x^2$
$= -10x^2 - 16xy^2 - y^4$

113. $5(2y-z)^2 + 4(y+2z)^2 = 5(2y-z)(2y-z) + 4(y+2z)(y+2z)$
$= 5(4y^2 - 2yz - 2yz + z^2) + 4(y^2 + 2yz + 2yz + 4z^2)$
$= 20y^2 - 10yz - 10yz + 5z^2 + 4y^2 + 8yz + 8yz + 16z^2$
$= 24y^2 - 4yz + 21z^2$

114. $3(x+2z)^2 - 2(2x-z)^2 = 3(x+2z)(x+2z) - 2(2x-z)(2x-z)$
$= 3(x^2 + 2xz + 2xz + 4z^2) - 2(4x^2 - 2xz - 2xz + z^2)$
$= 3x^2 + 6xz + 6xz + 12z^2 - 8x^2 + 4xz + 4xz - 2z^2$
$= -5x^2 + 20xz + 10z^2$

115. $(3.21x - 7.85)(2.87x + 4.59) = 9.2127x^2 + 14.7339x - 22.5295x - 36.0315$
$= 9.2127x^2 - 7.7956x - 36.0315$

116. $(7.44y + 56.7)(-2.1y - 67.3) = -15.624y^2 - 500.712y - 119.07y - 3815.91$
$= -15.624y^2 - 619.782y - 3815.91$

117. $(-17.3y + 4.35)^2 = (-17.3y + 4.35)(-17.3y + 4.35)$
$= 299.29y^2 - 75.255y - 75.255y + 18.9225$
$= 299.29y^2 - 150.51y + 18.9225$

118. $(-0.31x + 29.3)(-81x - 0.2) = 25.11x^2 + 0.062x - 2373.3x - 5.86$
$= 25.11x^2 - 2373.238x - 5.86$

119. **a.** Area of large square $= x^2$
b. Area I $= (x-y)^2$
c. Area II $= y(x-y) = xy - y^2$
d. Area III $= y(x-y) = xy - y^2$
e. Area IV $= y^2$
f. Area I = Area of large square − Area II − Area III − Area IV
$(x-y)^2 = x^2 - (xy - y^2) - (xy - y^2) - y^2$
$(x-y)^2 = x^2 - xy + y^2 - xy + y^2 - y^2$
$(x-y)^2 = x^2 - 2xy + y^2$

120. **a.** Area of $ABCD = (x+y)(x-y)$
b. Area I $= x(x-y) = x^2 - xy$
c. Area II $= y(x-y) = xy - y^2$
d. Area of $ABCD$ = Area I + Area II
$(x+y)(x-y) = x^2 - xy + xy - y^2$
$(x+y)(x-y) = x^2 - y^2$

121. $x = -\frac{1}{5}(375) + 90 = -75 + 90 = 15$

122. $20 = -\frac{1}{5}p + 90$
$\frac{1}{5}p = 70$
$p = \$350$

123. $r = p\left(-\frac{1}{5}p + 90\right) = -\frac{1}{5}p^2 + 90p$

124. $r = 400\left(-\frac{1}{5}(400) + 90\right) = 400(-80 + 90) = 400(10) = \$4000$

125. $A = lw = (2x - 3)(2x - 3) = 4x^2 - 6x - 6x + 9 = (4x^2 - 12x + 9)\text{ ft}^2$

126. $A = lw = (x - 2)(x + 4) = x^2 + 4x - 2x - 8 = (x^2 + 2x - 8)\text{ ft}^2$

127. $A = \frac{1}{2}bh = \frac{1}{2}(b + 5)(b - 2) = \frac{1}{2}(b^2 - 2b + 5b - 10) = \frac{1}{2}(b^2 + 3b - 10)\text{ in.}^2$

128. $$\begin{aligned} V = lwh = (x+2)(x+2)(x+2) &= \left(x^2 + 2x + 2x + 4\right)(x+2) \\ &= \left(x^2 + 4x + 4\right)(x+2) \\ &= x^2(x+2) + 4x(x+2) + 4(x+2) \\ &= x^3 + 2x^2 + 4x^2 + 8x + 4x + 8 = \left(x^3 + 6x^2 + 12x + 8\right)\text{ ft}^3 \end{aligned}$$

129. Answers may vary.

130. Answers may vary.

131. $0.35 \times 10^7 + 1.96 \times 10^7 = (0.35 + 1.96) \times 10^7 = 2.31 \times 10^7$

132. $$\begin{aligned} 1.435 \times 10^8 + 2.11 \times 10^7 = 1.435 \times 10^8 + 0.211 \times 10^8 &= (1.435 + 0.211) \times 10^8 \\ &= 1.646 \times 10^8 \end{aligned}$$

Exercise 5.4 (page 306)

1. $(a + 4)(a - 4) = a^2 - 16$

2. $(2b + 3)(2b - 3) = 4b^2 - 9$

3. $(4r^2 + 3s)(4r^2 - 3s) = 16r^4 - 9s^2$

4. $(5a + 2b^3)(5a - 2b^3) = 25a^2 - 4b^6$

5. $(m + 4)(m^2 - 4m + 16) = m^3 - 4m^2 + 16m + 4m^2 - 16m + 64 = m^3 + 64$

6. $(p - q)(p^2 + pq + q^2) = p^3 + p^2q + pq^2 - p^2q - pq^2 - q^3 = p^3 - q^3$

7. factoring

8. prime-factored

9. greatest common factor

10. prime; irreducible

11. $6 = 2 \cdot 3$

12. $10 = 2 \cdot 5$

13. $135 = 3^3 \cdot 5$

14. $98 = 2 \cdot 7^2$

15. $128 = 2^7$

16. $357 = 3 \cdot 7 \cdot 17$

17. $325 = 5^2 \cdot 13$

18. $288 = 2^5 \cdot 3^2$

19. $36 = 2^2 \cdot 3^2$; $48 = 2^4 \cdot 3$; gcf $= 2^2 \cdot 3 = 12$

20. $45 = 3^2 \cdot 5$; $75 = 3 \cdot 5^2$; gcf $= 3 \cdot 5 = 15$

21. $42 = 2 \cdot 3 \cdot 7$; $36 = 2^2 \cdot 3^2$; $98 = 2 \cdot 7^2$
gcf $= 2$

22. $16 = 2^4$; $40 = 2^3 \cdot 5$; $60 = 2^2 \cdot 3 \cdot 5$
gcf $= 2^2 = 4$

23. $4a^2b = 2^2 \cdot a^2b$; $8a^3c = 2^3 \cdot a^3c$
gcf $= 2^2 \cdot a^2 = 4a^2$

24. $6x^3y^2z = 2 \cdot 3 \cdot x^3y^2z$; $9xyz^2 = 3^2xyz^2$
gcf $= 3 \cdot x \cdot y \cdot z = 3xyz$

25. $18x^4y^3z^2 = 2 \cdot 3^2 \cdot x^4y^3z^2$; $-12xy^2z^3 = -1 \cdot 2^2 \cdot 3xy^2z^2$; gcf $= 2 \cdot 3 \cdot x \cdot y^2 \cdot z^2 = 6xy^2z^2$

26. $6x^2y^3 = 2 \cdot 3 \cdot x^2y^2$; $24xy^3 = 2^3 \cdot 3 \cdot xy^3$; $40x^2y^2z^3 = 2^3 \cdot 5 \cdot x^2y^2z^3$; gcf $= 2 \cdot xy^2 = 2xy^2$

27. $3a - 12 = 3(a - \underline{4})$

28. $5t + 25 = 5(t + \underline{5})$

29. $8z^2 + 2z = 2z(4z + \underline{1})$

30. $9t^3 - 3t^2 = 3t^2(3t - \underline{1})$

31. $2x + 8 = \mathbf{2} \cdot x + \mathbf{2} \cdot 4 = \mathbf{2}(x + 4)$

32. $3y - 9 = \mathbf{3} \cdot y - \mathbf{3} \cdot 3 = \mathbf{3}(y - 3)$

33. $2x^2 - 6x = \boldsymbol{2x} \cdot x - \boldsymbol{2x} \cdot 3 = \boldsymbol{2x}(x - 3)$

34. $3y^3 + 3y^2 = \boldsymbol{3y^2} \cdot y + \boldsymbol{3y^2} \cdot 1 = \boldsymbol{3y^2}(y + 1)$

35. $5xy + 12ab^2 \Rightarrow$ prime

36. $7x^2 + 14x = \boldsymbol{7x} \cdot x + \boldsymbol{7x} \cdot 2 = \boldsymbol{7x}(x + 2)$

37. $15x^2y - 10x^2y^2 = \boldsymbol{5x^2y} \cdot 3 - \boldsymbol{5x^2y} \cdot 2y$
$= \boldsymbol{5x^2y}(3 - 2y)$

38. $11m^3n^2 - 12x^2y \Rightarrow$ prime

39. $63x^3y^2 + 81x^2y^4 = \boldsymbol{9x^2y^2} \cdot 7x + \boldsymbol{9x^2y^2} \cdot 9y^2 = \boldsymbol{9x^2y^2}(7x + 9y^2)$

40. $33a^3b^4c - 16xyz \Rightarrow$ prime

41. $14r^2s^3 + 15t^6 \Rightarrow$ prime

42. $13ab^2c^3 - 26a^3b^2c = \boldsymbol{13ab^2c} \cdot c^2 - \boldsymbol{13ab^2c} \cdot 2a^2 = \boldsymbol{13ab^2c}(c^2 - 2a^2)$

43. $27z^3 + 12z^2 + 3z = \boldsymbol{3z} \cdot 9z^2 + \boldsymbol{3z} \cdot 4z + \boldsymbol{3z} \cdot 1 = \boldsymbol{3z}(9z^2 + 4z + 1)$

44. $25t^6 - 10t^3 + 5t^2 = \boldsymbol{5t^2}(5t^4) - \boldsymbol{5t^2}(2t) + \boldsymbol{5t^2}(1) = \boldsymbol{5t^2}(5t^4 - 2t + 1)$

45. $24s^3 - 12s^2t + 6st^2 = \boldsymbol{6s}(4s^2) - \boldsymbol{6s}(2st) + \boldsymbol{6s}(t^2) = \boldsymbol{6s}(4s^2 - 2st + t^2)$

46. $18y^2z^2 + 12y^2z^3 - 24y^4z^3 = \boldsymbol{6y^2z^2}(3) + \boldsymbol{6y^2z^2}(2z) - \boldsymbol{6y^2z^2}(4y^2z) = \boldsymbol{6y^2z^2}(3 + 2z - 4y^2z)$

47. $45x^{10}y^3 - 63x^7y^7 + 81x^{10}y^{10} = \boldsymbol{9x^7y^3}(5x^3) - \boldsymbol{9x^7y^3}(7y^4) + \boldsymbol{9x^7y^3}(9x^3y^7)$
$= \boldsymbol{9x^7y^3}(5x^3 - 7y^4 + 9x^3y^7)$

48. $48u^6v^6 - 16u^4v^4 - 3u^6v^3 = \boldsymbol{u^4v^3}(48u^2v^3) - \boldsymbol{u^4v^3}(16v) - \boldsymbol{u^4v^3}(3u^2)$
$= \boldsymbol{u^4v^3}(48u^2v^3 - 16v - 3u^2)$

49. $25x^3 - 14y^3 + 36x^3y^3 \Rightarrow$ prime

50. $9m^4n^3p^2 + 18m^2n^3p^4 - 27m^3n^4p = \boldsymbol{9m^2n^3p}(m^2p) + \boldsymbol{9m^2n^3p}(2p^3) - \boldsymbol{9m^2n^3p}(3mn)$
$= \boldsymbol{9m^2n^3p}(m^2p + 2p^3 - 3mn)$

51. $-3a - 6 = \mathbf{(-3)}(a) + \mathbf{(-3)}(2) = \mathbf{-3}(a+2)$

52. $-6b + 12 = \mathbf{(-6)}(b) + \mathbf{(-6)}(-2) = \mathbf{-6}(b-2)$

53. $-3x^2 - x = \boldsymbol{(-x)}(3x) + \boldsymbol{(-x)}(1) = \boldsymbol{-x}(3x+1)$

54. $-4a^3 + a^2 = \boldsymbol{(-a^2)}(4a) + \boldsymbol{(-a^2)}(-1) = \boldsymbol{-a^2}(4a-1)$

55. $-6x^2 - 3xy = \boldsymbol{(-3x)}(2x) + \boldsymbol{(-3x)}(y) = \boldsymbol{-3x}(2x+y)$

56. $-15y^3 + 25y^2 = \boldsymbol{(-5y^2)}(3y) + \boldsymbol{(-5y^2)}(-5) = \boldsymbol{-5y^2}(3y-5)$

57. $-18a^2b - 12ab^2 = \boldsymbol{(-6ab)}(3a) + \boldsymbol{(-6ab)}(2b) = \boldsymbol{-6ab}(3a+2b)$

58. $-21t^5 + 28t^3 = \boldsymbol{(-7t^3)}(3t^2) + \boldsymbol{(-7t^3)}(-4) = \boldsymbol{-7t^3}(3t^2-4)$

59. $-63u^3v^6z^9 + 28u^2v^7z^2 - 21u^3v^3z^4 = -7u^2v^3z^2(9uv^3z^7 - 4v^4 + 3uz^2)$

60. $-56x^4y^3z^2 - 72x^3y^4z^5 + 80xy^2z^3 = -8xy^2z^2(7x^3y + 9x^2y^2z^3 - 10z)$

61. $x^{n+2} + x^{n+3} = x^2(x^{n+2-2} + x^{n+3-2}) = x^2(x^n + x^{n+1})$

62. $y^{n+3} + y^{n+5} = y^3(y^{n+3-3} + y^{n+5-3}) = y^3(y^n + y^{n+2})$

63. $2y^{n+2} - 3y^{n+3} = y^n(2y^{n+2-n} - 3y^{n+3-n}) = y^n(2y^2 - 3y^3)$

64. $4x^{n+3} - 5x^{n+5} = x^n(4x^{n+3-n} - 5x^{n+5-n}) = x^n(4x^3 - 5x^5)$

65. $x^4 - 5x^6 = x^{-2}\left(x^{4-(-2)} - 5x^{6-(-2)}\right) = x^{-2}\left(x^6 - 5x^8\right)$

66. $7y^4 + y = y^{-4}\left(7y^{4-(-4)} + y^{1-(-4)}\right) = y^{-4}\left(7y^8 + y^5\right)$

67. $t^5 + 4t^{-6} = t^{-3}\left(t^{5-(-3)} + 4t^{-6-(-3)}\right) = t^{-3}(t^8 + 4t^{-3})$

68. $6p^3 - p^{-2} = p^{-5}\left(6p^{3-(-5)} - p^{-2-(-5)}\right) = p^{-5}(6p^8 - p^3)$

69. $8y^{2n} + 12 + 16y^{-2n} = 4y^{-2n}\left(2y^{2n-(-2n)} + 3y^{0-(-2n)} + 4y^{-2n-(-2n)}\right) = 4y^{-2n}(2y^{4n} + 3y^{2n} + 4)$

70. $21x^{6n} + 7x^{3n} + 14 = 7x^{-3n}\left(3x^{6n-(-3n)} + x^{3n-(-3n)} + 2x^{0-(-3n)}\right) = 7x^{-3n}(3x^{9n} + x^{6n} + 2x^{3n})$

71. $4\boldsymbol{(x+y)} + t\boldsymbol{(x+y)} = \boldsymbol{(x+y)}(4+t)$

72. $5\boldsymbol{(a-b)} - t\boldsymbol{(a-b)} = \boldsymbol{(a-b)}(5-t)$

73. $\boldsymbol{(a-b)}r - \boldsymbol{(a-b)}s = \boldsymbol{(a-b)}(r-s)$

74. $\boldsymbol{(x+y)}u + \boldsymbol{(x+y)}v = \boldsymbol{(x+y)}(u+v)$

75. $3\boldsymbol{(m+n+p)} + x\boldsymbol{(m+n+p)} = \boldsymbol{(m+n+p)}(3+x)$

76. $x\boldsymbol{(x-y-z)} + y\boldsymbol{(x-y-z)} = \boldsymbol{(x-y-z)}(x+y)$

77. $(x+y)\boldsymbol{(x+y)}+z\boldsymbol{(x+y)}=\boldsymbol{(x+y)}[(x+y)+z]=(x+y)(x+y+z)$

78. $(a-b)^2+(a-b)=(a-b)\boldsymbol{(a-b)}+1\boldsymbol{(a-b)}=\boldsymbol{(a-b)}[(a-b)+1]=(a-b)(a-b+1)$

79. $(u+v)^2-(u+v)=(u+v)\boldsymbol{(u+v)}-1\boldsymbol{(u+v)}=\boldsymbol{(u+v)}[(u+v)-1]=(u+v)(u+v-1)$

80. $$\begin{aligned} a(x-y)-(x-y)^2 &= a\boldsymbol{(x-y)}-(x-y)\boldsymbol{(x-y)}=\boldsymbol{(x-y)}[a-(x-y)] \\ &= (x-y)(a-x+y) \end{aligned}$$

81. $-a\boldsymbol{(x+y)}+b\boldsymbol{(x+y)}=\boldsymbol{(x+y)}(-a+b)=-(x+y)(a-b)$

82. $-b\boldsymbol{x(a-b)}-c\boldsymbol{x(a-b)}=\boldsymbol{x(a-b)}(-b-c)=-x(a-b)(b+c)$

83. $$\begin{aligned} ax+bx+ay+by &= x(a+b)+y(a+b) \\ &= (a+b)(x+y) \end{aligned}$$

84. $$\begin{aligned} ar-br+as-bs &= r(a-b)+s(a-b) \\ &= (a-b)(r+s) \end{aligned}$$

85. $$\begin{aligned} x^2+yx+2x+2y &= x(x+y)+2(x+y) \\ &= (x+y)(x+2) \end{aligned}$$

86. $$\begin{aligned} 2c+2d-cd-d^2 &= 2(c+d)-d(c+d) \\ &= (c+d)(2-d) \end{aligned}$$

87. $3c-cd+3d-c^2=3c+3d-c^2-cd=3(c+d)-c(c+d)=(c+d)(3-c)$

88. $x^2+4y-xy-4x=x^2-4x-xy+4y=x(x-4)-y(x-4)=(x-4)(x-y)$

89. $a^2-4b+ab-4a=a^2+ab-4a-4b=a(a+b)-4(a+b)=(a+b)(a-4)$

90. $7u+v^2-7v-uv=v^2-7v-uv+7u=v(v-7)-u(v-7)=(v-7)(v-u)$

91. $$\begin{aligned} ax+bx-a-b &= x(a+b)-1(a+b) \\ &= (a+b)(x-1) \end{aligned}$$

92. $$\begin{aligned} x^2y-ax-xy+a &= x(xy-a)-1(xy-a) \\ &= (xy-a)(x-1) \end{aligned}$$

93. $x^2+xy+xz+xy+y^2+zy=x(x+y+z)+y(z+y+z)=(x+y+z)(x+y)$

94. $ab-b^2-bc+ac-bc-c^2=b(a-b-c)+c(a-b-c)=(a-b-c)(b+c)$

95. $$\begin{aligned} mpx+mqx+npx+nqx &= x(mp+mq+np+nq)=x[m(p+q)+n(p+q)] \\ &= x(p+q)(m+n) \end{aligned}$$

96. $$\begin{aligned} abd-abe+acd-ace &= a(bd-be+cd-ce)=a[b(d-e)+c(d-e)] \\ &= a(d-e)(b+c) \end{aligned}$$

97. $$\begin{aligned} x^2y+xy^2+2xyz+xy^2+y^3+2y^2z &= y(x^2+xy+2xz+xy+y^2+2yz) \\ &= y[x(x+y+2z)+y(x+y+2z)] \\ &= y(x+y+2z)(x+y) \end{aligned}$$

98. $a^3 - 2a^2b + a^2c - a^2b + 2ab^2 - abc = a(a^2 - 2ab + ac - ab + 2b^2 - bc)$
$$= a[a(a - 2b + c) - b(a - 2b + c)]$$
$$= a(a - 2b + c)(a - b)$$

99. $2n^4p - 2n^2 - n^3p^2 + np + 2mn^3p - 2mn = n(2n^3p - 2n - n^2p^2 + p + 2mn^2p - 2m)$
$$= n(2n^3p - n^2p^2 + 2mn^2p - 2n + p - 2m)$$
$$= n[n^2p(2n - p + 2m) - 1(2n - p + 2m)]$$
$$= n(2n - p + 2m)(n^2p - 1)$$

100. $a^2c^3 + ac^2 + a^3c^2 - 2a^2bc^2 - 2bc^2 + c^3 = c^2(a^2c + a + a^3 - 2a^2b - 2b + c)$
$$= c^2(a^2c + a^3 - 2a^2b + c + a - 2b)$$
$$= c^2[a^2(c + a - 2b) + 1(c + a - 2b)]$$
$$= c^2(c + a - 2b)(a^2 + 1)$$

101.
$$r_1r_2 = rr_2 + rr_1$$
$$r_1r_2 - rr_1 = rr_2$$
$$r_1(r_2 - r) = rr_2$$
$$r_1 = \frac{rr_2}{r_2 - r}$$

102.
$$r_1r_2 = rr_2 + rr_1$$
$$r_1r_2 = r(r_2 + r_1)$$
$$\frac{r_1r_2}{r_2 + r_1} = r, \text{ or } r = \frac{r_1r_2}{r_2 + r_1}$$

103.
$$d_1d_2 = fd_2 + fd_1$$
$$d_1d_2 = f(d_2 + d_1)$$
$$\frac{d_1d_2}{d_2 + d_1} = f, \text{ or } f = \frac{d_1d_2}{d_2 + d_1}$$

104.
$$d_1d_2 = fd_2 + fd_1$$
$$d_1d_2 - fd_1 = fd_2$$
$$d_1(d_2 - f) = fd_2$$
$$d_1 = \frac{fd_2}{d_2 - f}$$

105.
$$b^2x^2 + a^2y^2 = a^2b^2$$
$$b^2x^2 = a^2b^2 - a^2y^2$$
$$b^2x^2 = a^2(b^2 - y^2)$$
$$\frac{b^2x^2}{b^2 - y^2} = a^2, \text{ or } a^2 = \frac{b^2x^2}{b^2 - y^2}$$

106.
$$b^2x^2 + a^2y^2 = a^2b^2$$
$$a^2y^2 = a^2b^2 - b^2x^2$$
$$a^2y^2 = b^2(a^2 - x^2)$$
$$\frac{a^2y^2}{a^2 - x^2} = b^2, \text{ or } b^2 = \frac{a^2y^2}{a^2 - x^2}$$

107.
$$S(1 - r) = a - lr$$
$$S - Sr = a - lr$$
$$S - a = Sr - lr$$
$$S - a = r(S - l)$$
$$\frac{S - a}{S - l} = r, \text{ or } r = \frac{S - a}{S - l}$$

108.
$$Sn = (n - 2)180°$$
$$Sn = 180°n - 360°$$
$$360° = 180°n - Sn$$
$$360° = n(180° - S)$$
$$\frac{360°}{180° - S} = n, \text{ or } n = \frac{360°}{180° - S}$$

109.
$$H(a+b)=2ab$$
$$Ha+Hb=2ab$$
$$Hb=2ab-Ha$$
$$Hb=a(2b-H)$$
$$\frac{Hb}{2b-H}=a, \text{ or } a=\frac{Hb}{2b-H}$$

110.
$$H(a+b)=2ab$$
$$Ha+Hb=2ab$$
$$Ha=2ab-Hb$$
$$Ha=b(2a-H)$$
$$\frac{Ha}{2a-H}=b, \text{ or } b=\frac{Ha}{2a-H}$$

111.
$$3xy-x=2y+3$$
$$3xy-2y=x+3$$
$$y(3x-2)=x+3$$
$$y=\frac{x+3}{3x-2}$$

112.
$$x(5y+3)=y-1$$
$$5xy+3x=y-1$$
$$3x+1=y-5xy$$
$$3x+1=y(1-5x)$$
$$\frac{3x+1}{1-5x}=y$$

113. $2x^3+5x^2-2x+8=x(2x^2+5x-2)+8=x[x(2x+5)-2]+8$

114. $4x^4+5x^2-3x+9=x(4x^3+5x-3)+9=x[x(4x^2+5)-3]+9$

115. **a.** $(2x^2)(6x)=12x^3$ in.^{2}
b. $(5x)(4x)=20x^2$ in.^{2}
c. $12x^3-20x^2=4x^2(3x-5)$ in.^{2}

116. $x^3+4x^2+5x+20=x^2(x+4)+5(x+4)$
$=(x+4)(x^2+5)$
The width is $x+4$ feet.

117. **Answers may vary.**

118. **Answers may vary.**

119. **Answers may vary.**

120. Divisors of 28: $1,2,4,7,14,28 \Rightarrow$ Sum $=1+2+4+7+14+28=56=2\cdot 28$

121. $14=2\cdot 7$; $45=3^2\cdot 5$; gcf $=1 \Rightarrow$ relatively prime

122. $24=2^3\cdot 3$; $63=3^2\cdot 7$; $112=2^4\cdot 7$; gcf $=1 \Rightarrow$ relatively prime

123. $60=2^2\cdot 3\cdot 5$; $28=2^2\cdot 7$; $36=2^2\cdot 3^2$; gcf $=2^2 \Rightarrow$ not relatively prime

124. $55=5\cdot 11$; $49=7^2$; $78=2\cdot 3\cdot 13$; gcf $=1 \Rightarrow$ relatively prime

125. $12x^2y=2^2\cdot 3\cdot x^2y$; $5ab^3=5\cdot ab^3$; $35x^2b^3=5\cdot 7\cdot x^2b^3$; gcf $=1 \Rightarrow$ relatively prime

126. $18uv=2\cdot 3^2\cdot uv$; $25rs=5^2\cdot rs$; $12rsuv=2^2\cdot 3\cdot rsuv$; gcf $=1 \Rightarrow$ relatively prime

Exercise 5.5 (page 314)

1.
$$(x+1)(x+1)=x^2+x+x+1$$
$$=x^2+2x+1$$

2.
$$(2m-3)(m-2)=2m^2-4m-3m+6$$
$$=2m^2-7m+6$$

3. $(2m+n)(2m+n)=4m^2+2mn+2mn+n^2=4m^2+4mn+n^2$

SECTION 5.5

4. $(3m-2n)(3m-2n) = 9m^2 - 6mn - 6mn + 4n^2 = 9m^2 - 12mn + 4n^2$

5. $(a+4)(a+3) = a^2 + 3a + 4a + 12$
$= a^2 + 7a + 12$

6. $(3b+2)(2b-5) = 6b^2 - 15b + 4b - 10$
$= 6b^2 - 11b - 10$

7. $(4r-3s)(2r-s) = 8r^2 - 4rs - 6rs + 3s^2 = 8r^2 - 10rs + 3s^2$

8. $(5a-2b)(3a+4b) = 15a^2 + 20ab - 6ab - 8b^2 = 15a^2 + 14ab - 8b^2$

9. $1, 4, 9, 16, 25, 36, 49, 64, 81, 100$

10. $(p-q)$

11. cannot

12. $1, 8, 27, 64, 125, 216, 343, 512, 729, 1000$

13. $(p^2 - pq + q^2)$

14. $(p^2 + pq + q^2)$

15. $x^2 - 4 = x^2 - 2^2 = (x+2)(x-2)$

16. $y^2 - 9 = y^2 - 3^2 = (y+3)(y-3)$

17. $t^2 - 225 = t^2 - 15^2 = (t+15)(t-15)$

18. $p^2 - 400 = p^2 - 20^2 = (p+20)(p-20)$

19. $9y^2 - 64 = (3y)^2 - 8^2 = (3y+8)(3y-8)$

20. $16x^4 - 81y^2 = (4x^2)^2 - (9y)^2$
$= (4x^2 + 9y)(4x^2 - 9y)$

21. $x^2 + 25 \Rightarrow$ prime (sum of two squares)

22. $144a^2 - b^4 = (12a)^2 - (b^2)^2$
$= (12a + b^2)(12a - b^2)$

23. $625a^2 - 169b^4 = (25a)^2 - (13b^2)^2$
$= (25a + 13b^2)(25a - 13b^2)$

24. $4y^2 + 9z^4 \Rightarrow$ prime (sum of two squares)

25. $81a^4 - 49b^2 = (9a^2)^2 - (7b)^2$
$= (9a^2 + 7b)(9a^2 - 7b)$

26. $64r^6 - 121s^2 = (8r^3)^2 - (11s)^2$
$= (8r^3 + 11s)(8r^3 - 11s)$

27. $36x^4y^2 - 49z^4 = (6x^2y)^2 - (7z^2)^2 = (6x^2y + 7z^2)(6x^2y - 7z^2)$

28. $4a^2b^4c^6 - 9d^8 = (2ab^2c^3)^2 - (3d^4)^2 = (2ab^2c^3 + 3d^4)(2ab^2c^3 - 3d^4)$

29. $(x+y)^2 - z^2 = [(x+y)+z][(x+y)-z]$
$= (x+y+z)(x+y-z)$

30. $a^2 - (b-c)^2 = [a+(b-c)][a-(b-c)]$
$= (a+b-c)(a-b+c)$

31. $(a-b)^2 - c^2 = [(a-b)+c][(a-b)-c] = (a-b+c)(a-b-c)$

32. $(m+n)^2 - p^4 = [(m+n)+p^2][(m+n)-p^2] = (m+n+p^2)(m+n-p^2)$

33. $x^4 - y^4 = (x^2+y^2)(x^2-y^2) = (x^2+y^2)(x+y)(x-y)$

34. $16a^4 - 81b^4 = (4a^2+9b^2)(4a^2-9b^2) = (4a^2+9b^2)(2a+3b)(2a-3b)$

35. $256x^4y^4 - z^8 = (16x^2y^2 + z^4)(16x^2y^2 - z^4) = (16x^2y^2 + z^4)(4xy + z^2)(4xy - z^2)$

36. $225a^4 - 16b^8c^{12} = (15a^2 + 4b^4c^6)(15a^2 - 4b^4c^6)$

37. $2x^2 - 288 = 2(x^2 - 144)$
$= 2(x + 12)(x - 12)$

38. $8x^2 - 72 = 8(x^2 - 9)$
$= 8(x + 3)(x - 3)$

39. $2x^3 - 32x = 2x(x^2 - 16)$
$= 2x(x + 4)(x - 4)$

40. $3x^3 - 243x = 3x(x^2 - 81)$
$= 3x(x + 9)(x - 9)$

41. $5x^3 - 125x = 5x(x^2 - 25)$
$= 5x(x + 5)(x - 5)$

42. $6x^4 - 216x^2 = 6x^2(x^2 - 36)$
$= 6x^2(x + 6)(x - 6)$

43. $r^2s^2t^2 - t^2x^4y^2 = t^2(r^2s^2 - x^4y^2) = t^2(rs + x^2y)(rs - x^2y)$

44. $16a^4b^3c^4 - 64a^2bc^6 = 16a^2bc^4(a^2b^2 - 4c^2) = 16a^2bc^4(ab + 2c)(ab - 2c)$

45. $r^3 + s^3 = (r + s)(r^2 - rs + s^2)$

46. $t^3 - v^3 = (t - v)(t^2 + tv + v^2)$

47. $p^3 - q^3 = (p - q)(p^2 + pq + q^2)$

48. $m^3 + n^3 = (m + n)(m^2 - mn + n^2)$

49. $x^3 - 8y^3 = x^3 - (2y)^3 = (x - 2y)[x^2 + x(2y) + (2y)^2] = (x - 2y)(x^2 + 2xy + 4y^2)$

50. $27a^3 + b^3 = (3a)^3 + b^3 = (3a + b)[(3a)^2 - 3ab + b^2] = (3a + b)(9a^2 - 3ab + b^2)$

51. $64a^3 - 125b^6 = (4a)^3 - (5b^2)^3 = (4a - 5b^2)[(4a)^2 + (4a)(5b^2) + (5b^2)^2]$
$= (4a - 5b^2)(16a^2 + 20ab^2 + 25b^4)$

52. $8x^6 + 125y^3 = (2x^2)^3 + (5y)^3 = (2x^2 + 5y)[(2x^2)^2 - (2x^2)(5y) + (5y)^2]$
$= (2x^2 + 5y)(4x^4 - 10x^2y + 25y^2)$

53. $125x^3y^6 + 216z^9 = (5xy^2)^3 + (6z^3)^3 = (5xy^2 + 6z^3)[(5xy^2)^2 - (5xy^2)(6z^3) + (6z^3)^2]$
$= (5xy^2 + 6z^3)(25x^2y^4 - 30xy^2z^3 + 36z^6)$

54. $1000a^6 - 343b^3c^6 = (10a^2)^3 - (7bc^2)^3 = (10a^2 - 7bc^2)[(10a^2)^2 + (10a^2)(7bc^2) + (7bc^2)^2]$
$= (10a^2 - 7bc^2)(100a^4 + 70a^2bc^2 + 49b^2c^4)$

55. $x^6 + y^6 = (x^2)^3 + (y^2)^3 = (x^2 + y^2)[(x^2)^2 - x^2y^2 + (y^2)^2] = (x^2 + y^2)(x^4 - x^2y^2 + y^4)$

56. $x^9 + y^9 = (x^3)^3 + (y^3)^3 = (x^3 + y^3)[(x^3)^2 - x^3y^3 + (y^3)^2]$
$= (x + y)(x^2 - xy + y^2)(x^6 - x^3y^3 + y^6)$

57. $5x^3 + 625 = 5(x^3 + 125) = 5(x^3 + 5^3) = 5(x + 5)(x^2 - 5x + 25)$

58. $2x^3 - 128 = 2(x^3 - 64) = 2(x^3 - 4^3) = 2(x - 4)(x^2 + 4x + 16)$

59. $4x^5 - 256x^2 = 4x^2(x^3 - 64) = 4x^2(x^3 - 4^3) = 4x^2(x - 4)(x^2 + 4x + 16)$

60. $2x^6 + 54x^3 = 2x^3(x^3 + 27) = 2x^3(x^3 + 3^3) = 2x^3(x + 3)(x^2 - 3x + 9)$

61. $$\begin{aligned} 128u^2v^3 - 2t^3u^2 = 2u^2(64v^3 - t^3) = 2u^2[(4v)^3 - t^3] &= 2u^2(4v - t)[(4v)^2 + 4vt + t^2] \\ &= 2u^2(4v - t)(16v^2 + 4vt + t^2) \end{aligned}$$

62. $$\begin{aligned} 56rs^2t^3 + 7rs^2v^6 = 7rs^2(8t^3 + v^6) = 7rs^2[(2t)^3 + (v^2)^3] &= 7rs^2(2t + v^2)[(2t)^2 - 2tv^2 + (v^2)^2] \\ &= 7rs^2(2t + v^2)(4t^2 - 2tv^2 + v^4) \end{aligned}$$

63. $(a + b)x^3 + 27(a + b) = (a + b)(x^3 + 27) = (a + b)(x^3 + 3^3) = (a + b)(x + 3)(x^2 - 3x + 9)$

64. $(c - d)r^3 - (c - d)s^3 = (c - d)(r^3 - s^3) = (c - d)(r - s)(r^2 + rs + s^2)$

65. $x^{2m} - y^{4n} = (x^m)^2 - (y^{2n})^2 = (x^m + y^{2n})(x^m - y^{2n})$

66. $$\begin{aligned} a^{4m} - b^{8n} = (a^{2m})^2 - (b^{4n})^2 = (a^{2m} + b^{4n})(a^{2m} - b^{4n}) &= (a^{2m} + b^{4n})[(a^m)^2 - (b^{2n})^2] \\ &= (a^{2m} + b^{4n})(a^m + b^{2n})(a^m - b^{2n}) \end{aligned}$$

67. $100a^{4m} - 81b^{2n} = (10a^{2m})^2 - (9b^n)^2 = (10a^{2m} + 9b^n)(10a^{2m} - 9b^n)$

68. $25x^{8m} - 36y^{4n} = (5x^{4m})^2 - (6y^{2n})^2 = (5x^{4m} + 6y^{2n})(5x^{4m} - 6y^{2n})$

69. $x^{3n} - 8 = (x^n)^3 - 2^3 = (x^n - 2)[(x^n)^2 + 2x^n + 2^2] = (x^n - 2)(x^{2n} + 2x^n + 4)$

70. $a^{3m} + 64 = (a^m)^3 + 4^3 = (a^m + 4)[(a^m)^2 - 4a^m + 4^2] = (a^m + 4)(a^{2m} - 4a^m + 16)$

71. $a^{3m} + b^{3n} = (a^m)^3 + (b^n)^3 = (a^m + b^n)[(a^m)^2 - a^mb^n + (b^n)^2] = (a^m + b^n)(a^{2m} - a^mn^n + b^{2n})$

72. $$\begin{aligned} x^{6m} - y^{3n} = (x^{2m})^3 - (y^n)^3 &= (x^{2m} - y^n)[(x^{2m})^2 + x^{2m}y^n + (y^n)^2] \\ &= (x^{2m} - y^n)(x^{4m} + x^{2m}y^n + y^{2n}) \end{aligned}$$

73. $$\begin{aligned} 2x^{6m} + 16y^{3m} = 2(x^{6m} + 8y^{3m}) &= 2[(x^{2m})^3 + (2y^m)^3] \\ &= 2(x^{2m} + 2y^m)[(x^{2m})^2 - 2x^{2m}y^m + (2y^m)^2] \\ &= 2(x^{2m} + 2y^m)(x^{4m} - 2x^{2m}y^m + 4y^{2m}) \end{aligned}$$

74. $$\begin{aligned} 24 + 3c^{3m} = 3(8 + c^{3m}) = 3[2^3 + (c^m)^3] &= 3(2 + c^m)[2^2 - 2c^m + (c^m)^2] \\ &= 3(2 + c^m)(4 - 2c^m + c^{2m}) \end{aligned}$$

75. $$\begin{aligned} a^2 - b^2 + a + b = (a^2 - b^2) + (a + b) &= (a + b)(a - b) + 1(a + b) \\ &= (a + b)(a - b + 1) \end{aligned}$$

76. $$\begin{aligned} x^2 - y^2 - x - y = (x^2 - y^2) - (x + y) &= (x + y)(x - y) - 1(x + y) \\ &= (x + y)(x - y - 1) \end{aligned}$$

77. $a^2 - b^2 + 2a - 2b = (a^2 - b^2) + 2(a - b) = (a + b)(a - b) + 2(a - b)$
$= (a - b)(a + b + 2)$

78. $m^2 - n^2 + 3m + 3n = (m^2 - n^2) + 3(m + n) = (m + n)(m - n) + 3(m + n)$
$= (m + n)(m - n + 3)$

79. $2x + y + 4x^2 - y^2 = (2x + y) + (4x^2 - y^2) = (2x + y)(1) + (2x + y)(2x - y)$
$= (2x + y)(1 + 2x - y)$

80. $m - 2n + m^2 - 4n^2 = (m - 2n) + (m^2 - 4n^2) = (m - 2n)(1) + (m + 2n)(m - 2n)$
$= (m - 2n)(1 + m + 2n)$

81. $0.5gt_1^2 - 0.5gt_2^2 = 0.5g(t_1^2 - t_2^2)$
$= 0.5g(t_1 + t_2)(t_1 - t_2)$

82. $\pi R^2 - \pi r^2 = \pi(R^2 - r^2)$
$= \pi(R + r)(R - r)$

83. $V = \frac{4}{3}\pi r_1^3 - \frac{4}{3}\pi r_2^3 = \frac{4}{3}\pi(r_1^3 - r_2^3) = \frac{4}{3}\pi(r_1 - r_2)(r_1^2 + r_1r_2 + r_2^2)$

84. $h = 144 - 16t^2 = 16(9 - t^2) = 16(3^2 - t^2) = 16(3 + t)(3 - t)$

85. Answers may vary.

86. Answers may vary.

87. $x^{32} - y^{32} = (x^{16} + y^{16})(x^{16} - y^{16}) = (x^{16} + y^{16})(x^8 + y^8)(x^8 - y^8)$
$= (x^{16} + y^{16})(x^8 + y^8)(x^4 + y^4)(x^4 - y^4)$
$= (x^{16} + y^{16})(x^8 + y^8)(x^4 + y^4)(x^2 + y^2)(x^2 - y^2)$
$= (x^{16} + y^{16})(x^8 + y^8)(x^4 + y^4)(x^2 + y^2)(x + y)(x - y)$

88. In the fifth line of the proof, both sides of the equation are divided by $x - y$. However, in the first line, it is stated that $x = y$. This means that $x - y = 0$. Thus, in the fifth line of the proof, both sides of the equation are divided by 0, which is not valid.

Exercise 5.6 (page 325)

1. $\frac{2 + x}{11} = 3$
$2 + x = 33$
$x = 31$

2. $\frac{3y - 12}{2} = 9$
$3y - 12 = 18$
$3y = 30$
$y = 10$

3. $\frac{2}{3}(5t - 3) = 38$
$2(5t - 3) = 114$
$10t - 6 = 114$
$10t = 120$
$t = 12$

4. $3(p+2)=4p$
$3p+6=4p$
$-p=-6$
$p=6$

5. $11r+6(3-r)=3$
$11r+18-6r=3$
$5r=-15$
$r=-3$

6. $2q^2-9=q(q+3)+q^2$
$2q^2-9=q^2+3q+q^2$
$2q^2-9=2q^2+3q$
$-9=3q$
$-3=q$

7. $2xy+y^2$

8. $2xy+y^2$

9. x^2-y^2

10. b^2-4ac; integer

11. $x+2$

12. $x-2$

13. $x-3$

14. $x+3$

15. $2a+1$

16. $3p-4$

17. $2m+3n$

18. $4r-s$

19. $x^2+2x+1=(x+1)(x+1)=(x+1)^2$

20. $y^2-2y+1=(y-1)(y-1)=(y-1)^2$

21. $a^2-18a+81=(a-9)(a-9)=(a-9)^2$

22. $b^2+12b+36=(b+6)(b+6)=(b+6)^2$

23. $4y^2+4y+1=(2y+1)(2y+1)$
$=(2y+1)^2$

24. $9x^2+6x+1=(3x+1)(3x+1)$
$=(3x+1)^2$

25. $9b^2-12b+4=(3b-2)(3b-2)$
$=(3b-2)^2$

26. $4a^2-12a+9=(2a-3)(2a-3)$
$=(2a-3)^2$

27. $9z^2+24z+16=(3z+4)(3z+4)$
$=(3z+4)^2$

28. $16z^2-24z+9=(4z-3)(4z-3)$
$=(4z-3)^2$

29. $x^2+9x+8=(x+1)(x+8)$

30. $y^2+7y+6=(y+1)(y+6)$

31. $x^2-7x+10=(x-5)(x-2)$

32. $c^2-7c+12=(c-3)(c-4)$

33. $b^2+8b+18 \Rightarrow$ prime

34. $x^2-12x+35=(x-5)(x-7)$

35. $x^2-x-30=(x-6)(x+5)$

36. $a^2+4a-45=(a+9)(a-5)$

37. $a^2+5a-50=(a+10)(a-5)$

38. $b^2+9b-36=(b+12)(b-3)$

39. $y^2-4y-21=(y-7)(y+3)$

40. $x^2+4x-28 \Rightarrow$ prime

41. $3x^2+12x-63=3(x^2+4x-21)$
$=3(x+7)(x-3)$

42. $2y^2+4y-48=2(y^2+2y-24)$
$=2(y+6)(y-4)$

43. $a^2b^2-13ab^2+22b^2=b^2(a^2-13a+22)=b^2(a-11)(a-2)$

44. $a^2b^2x^2-18a^2b^2x+81a^2b^2=a^2b^2(x^2-18x+81)=a^2b^2(x-9)(x-9)$

45. $b^2x^2 - 12bx^2 + 35x^2 = x^2(b^2 - 12b + 35)$
$= x^2(b-5)(b-7)$

46. $c^3x^2 + 11c^3x - 42c^3 = c^3(x^2 + 11x - 42)$
$= c^3(x+14)(x-3)$

47. $-a^2 + 4a + 32 = -(a^2 - 4a - 32)$
$= -(a-8)(a+4)$

48. $-x^2 - 2x + 15 = -(x^2 + 2x - 15)$
$= -(x+5)(x-3)$

49. $-3x^2 + 15x - 18 = -3(x^2 - 5x + 6)$
$= -3(x-2)(x-3)$

50. $-2y^2 - 16y + 40 = -2(y^2 + 8y - 20)$
$= -2(y+10)(y-2)$

51. $-4x^2 + 4x + 80 = -4(x^2 - x - 20)$
$= -4(x-5)(x+4)$

52. $-5a^2 + 40a - 75 = -5(a^2 - 8a + 15)$
$= -5(a-5)(a-3)$

53. $6y^2 + 7y + 2 = (2y+1)(3y+2)$

54. $6x^2 - 11x + 3 = (2x-3)(3x-1)$

55. $8a^2 + 6a - 9 = (4a-3)(2a+3)$

56. $15b^2 + 4b - 4 = (5b-2)(3b+2)$

57. $6x^2 - 5x - 4 = (3x-4)(2x+1)$

58. $18y^2 - 3y - 10 = (6y-5)(3y+2)$

59. $5x^2 + 4x + 1 \Rightarrow$ prime

60. $6z^2 + 17z + 12 = (3z+4)(2z+3)$

61. $8x^2 - 10x + 3 = (4x-3)(2x-1)$

62. $4a^2 + 20a + 3 \Rightarrow$ prime

63. $a^2 - 3ab - 4b^2 = (a-4b)(a+b)$

64. $b^2 + 2bc - 80c^2 = (b+10c)(b-8c)$

65. $2y^2 + yt - 6t^2 = (2y-3t)(y+2t)$

66. $3x^2 - 10xy - 8y^2 = (3x+2y)(x-4y)$

67. $3x^3 - 10x^2 + 3x = x(3x^2 - 10x + 3)$
$= x(3x-1)(x-3)$

68. $3t^3 - 3t^2 + t = t(3t^2 - 3t + 1)$

69. $-3a^2 + ab + 2b^2 = -(3a^2 - ab - 2b^2)$
$= -(3a+2b)(a-b)$

70. $-2x^2 + 3xy + 5y^2 = -(2x^2 - 3xy - 5y^2)$
$= -(2x-5y)(x+y)$

71. $-4x^2 - 9 + 12x = -4x^2 + 12x - 9 = -(4x^2 - 12x + 9) = -(2x-3)(2x-3) = -(2x-3)^2$

72. $6x + 4 + 9x^2 = 9x^2 + 6x + 4 \Rightarrow$ prime

73. $5a^2 + 45b^2 - 30ab = 5a^2 - 30ab + 45b^2 = 5(a^2 - 6ab + 9b^2) = 5(a-3b)(a-3b) = 5(a-3b)^2$

74. $-90x^2 + 2 - 8x = -90x^2 - 8x + 2 = -2(45x^2 + 4x - 1) = -2(9x-1)(5x+1)$

75. $8x^2z + 6xyz + 9y^2z = z(8x^2 + 6xy + 9y^2)$ NOTE: $8x^2 + 6xy + 9y^2$ is a prime trinomial.

76. $x^3 - 60xy^2 + 7x^2y = x(x^2 - 60y^2 + 7xy) = x(x^2 + 7xy - 60y^2) = x(x+12y)(x-5y)$

77. $21x^4 - 10x^3 - 16x^2 = x^2(21x^2 - 10x - 16) = x^2(7x-8)(3x+2)$

78. $16x^3 - 50x^2 + 36x = 2x(8x^2 - 25x + 18) = 2x(8x - 9)(x - 2)$

79. $x^4 + 8x^2 + 15 = (x^2 + 5)(x^2 + 3)$

80. $x^4 + 11x^2 + 24 = (x^2 + 3)(x^2 + 8)$

81. $y^4 - 13y^2 + 30 = (y^2 - 10)(y^2 - 3)$

82. $y^4 - 13y^2 + 42 = (y^2 - 6)(y^2 - 7)$

83. $a^4 - 13a^2 + 36 = (a^2 - 4)(a^2 - 9) = (a + 2)(a - 2)(a + 3)(a - 3)$

84. $b^4 - 17b^2 + 16 = (b^2 - 16)(b^2 - 1) = (b + 4)(b - 4)(b + 1)(b - 1)$

85. $z^4 - z^2 - 12 = (z^2 - 4)(z^2 + 3) = (z + 2)(z - 2)(z^2 + 3)$

86. $c^4 - 8c^2 - 9 = (c^2 - 9)(c^2 + 1) = (c + 3)(c - 3)(c^2 + 1)$

87. $4x^3 + x^6 + 3 = x^6 + 4x^3 + 3 = (x^3 + 1)(x^3 + 3) = (x + 1)(x^2 - x + 1)(x^3 + 3)$

88. $a^6 - 2 + a^3 = a^6 + a^3 - 2 = (a^3 + 2)(a^3 - 1) = (a^3 + 2)(a - 1)(a^2 + a + 1)$

89. $x^{2n} + 2x^n + 1 = (x^n + 1)(x^n + 1) = (x^n + 1)^2$

90.
$$\begin{aligned} x^{4n} - 2x^{2n} + 1 = (x^{2n} - 1)(x^{2n} - 1) &= [(x^n)^2 - 1^2][(x^n)^2 - 1^2] \\ &= (x^n + 1)(x^n - 1)(x^n + 1)(x^n - 1) \\ &= (x^n + 1)^2(x^n - 1)^2 \end{aligned}$$

91. $2a^{6n} - 3a^{3n} - 2 = (2a^{3n} + 1)(a^{3n} - 2)$

92. $b^{2n} - b^n - 6 = (b^n - 3)(b^n + 2)$

93. $x^{4n} + 2x^{2n}y^{2n} + y^{4n} = (x^{2n} + y^{2n})(x^{2n} + y^{2n}) = (x^{2n} + y^{2n})^2$

94. $y^{6n} + 2y^{3n}z + z^2 = (y^{3n} + z)(y^{3n} + z) = (y^{3n} + z)^2$

95. $6x^{2n} + 7x^n - 3 = (3x^n - 1)(2x^n + 3)$

96. $12y^{4n} + 10y^{2n} + 2 = 2(6y^{4n} + 5y^{2n} + 1) = 2(3y^{2n} + 1)(2y^{2n} + 1)$

97. $(x + 1)^2 + 2(x + 1) + 1 = [(x + 1) + 1][(x + 1) + 1] = (x + 2)(x + 2) = (x + 2)^2$

98. $(a + b)^2 - 2(a + b) + 1 = [(a + b) - 1][(a + b) - 1] = (a + b - 1)(a + b - 1) = (a + b - 1)^2$

99. $(a + b)^2 - 2(a + b) - 24 = [(a + b) - 6][(a + b) + 4] = (a + b - 6)(a + b + 4)$

100. $(x - y)^2 + 3(x - y) - 10 = [(x - y) + 5][(x - y) - 2] = (x - y + 5)(x - y - 2)$

101. $6(x + y)^2 - 7(x + y) - 20 = [(3(x + y) + 4][2(x + y) - 5] = (3x + 3y + 4)(2x + 2y - 5)$

102. $2(x - z)^2 + 9(x - z) + 4 = [2(x - z) + 1][(x - z) + 4] = (2x - 2z + 1)(x - z + 4)$

SECTION 5.6

103. $x^2 + 4x + 4 - y^2 = (x^2 + 4x + 4) - y^2 = (x+2)(x+2) - y^2 = (x+2)^2 - y^2$
$= [(x+2) + y][(x+2) - y]$
$= (x + 2 + y)(x + 2 - y)$

104. $x^2 - 6x + 9 - 4y^2 = (x^2 - 6x + 9) - 4y^2 = (x-3)(x-3) - 4y^2$
$= (x-3)^2 - (2y)^2$
$= [(x-3) + 2y][(x-3) - 2y]$
$= (x - 3 + 2y)(x - 3 - 2y)$

105. $x^2 + 2x + 1 - 9z^2 = (x^2 + 2x + 1) - 9z^2 = (x+1)(x+1) - 9z^2$
$= (x+1)^2 - (3z)^2$
$= [(x+1) + 3z][(x+1) - 3z]$
$= (x + 1 + 3z)(x + 1 - 3z)$

106. $x^2 + 10x + 25 - 16z^2 = (x^2 + 10x + 25) - 16z^2 = (x+5)(x+5) - 16z^2$
$= (x+5)^2 - (4z)^2$
$= [(x+5) + 4z][(x+5) - 4z]$
$= (x + 5 + 4z)(x + 5 - 4z)$

107. $c^2 - 4a^2 + 4ab - b^2 = c^2 - (4a^2 - 4ab + b^2) = c^2 - (2a - b)(2a - b)$
$= c^2 - (2a - b)^2$
$= [c + (2a - b)][c - (2a - b)]$
$= (c + 2a - b)(c - 2a + b)$

108. $4c^2 - a^2 - 6ab - 9b^2 = 4c^2 - (a^2 + 6ab + 9b^2) = 4c^2 - (a + 3b)(a + 3b)$
$= (2c)^2 - (a + 3b)^2$
$= [2c + (a + 3b)][2c - (a + 3b)]$
$= (2c + a + 3b)(2c - a - 3b)$

109. $a^2 - b^2 + 8a + 16 = (a^2 + 8a + 16) - b^2 = (a+4)(a+4) - b^2 = (a+4)^2 - b^2$
$= [(a+4) + b][(a+4) - b]$
$= (a + 4 + b)(a + 4 - b)$

110. $a^2 + 14a - 25b^2 + 49 = (a^2 + 14a + 49) - 25b^2 = (a+7)(a+7) - 25b^2$
$= (a+7)^2 - (5b)^2$
$= [(a+7) + 5b][(a+7) - 5b]$
$= (a + 7 + 5b)(a + 7 - 5b)$

111. $4x^2 - z^2 + 4xy + y^2 = (4x^2 + 4xy + y^2) - z^2 = (2x + y)(2x + y) - z^2$
$= (2x + y)^2 - z^2$
$= [(2x + y) + z][(2x + y) - z]$
$= (2x + y + z)(2x + y - z)$

112. $x^2 - 4xy - 4z^2 + 4y^2 = (x^2 - 4xy + 4y^2) - 4z^2 = (x - 2y)(x - 2y) - 4z^2$
$$= (x - 2y)^2 - (2z)^2$$
$$= [(x - 2y) + 2z][(x - 2y) - 2z]$$
$$= (x - 2y + 2z)(x - 2y - 2z)$$

113. $a^2 - 17a + 16$: $a = 1, b = -17, c = 16 \Rightarrow$ key # $= ac = 1(16) = 16$.
Find two factors of 16 which add to equal -17: -1 and -16.
Rewrite and factor: $a^2 - 17a + 16 = a^2 - a - 16a + 16$
$$= a(a - 1) - 16(a - 1) = (a - 1)(a - 16)$$

114. $b^2 - 4b - 21$: $a = 1, b = -4, c = -21 \Rightarrow$ key # $= ac = 1(-21) = -21$.
Find two factors of -21 which add to equal -4: $+3$ and -7.
Rewrite and factor: $b^2 - 4b - 21 = b^2 + 3b - 7b - 21$
$$= b(b + 3) - 7(b + 3) = (b + 3)(b - 7)$$

115. $2u^2 + 5u + 3$: $a = 2, b = 5, c = 3 \Rightarrow$ key # $= ac = 2(3) = 6$.
Find two factors of 6 which add to equal 5: $+2$ and $+3$.
Rewrite and factor: $2u^2 + 5u + 3 = 2u^2 + 2u + 3u + 3$
$$= 2u(u + 1) + 3(u + 1) = (u + 1)(2u + 3)$$

116. $6y^2 + 5y - 6$: $a = 6, b = 5, c = -6 \Rightarrow$ key # $= ac = 6(-6) = -36$.
Find two factors of -36 which add to equal 5: $+9$ and -4.
Rewrite and factor: $6y^2 + 5y - 6 = 6y^2 + 9y - 4y - 6$
$$= 3y(2y + 3) - 2(2y + 3) = (2y + 3)(3y - 2)$$

117. $20r^2 - 7rs - 6s^2$: $a = 20, b = -7, c = -6 \Rightarrow$ key # $= ac = 20(-6) = -120$.
Find two factors of -120 which add to equal -7: -15 and $+8$.
Rewrite and factor: $20r^2 - 7rs - 6s^2 = 20r^2 - 15rs + 8rs - 6s^2$
$$= 5r(4r - 3s) + 2s(4r - 3s) = (4r - 3s)(5r + 2s)$$

118. $6s^2 + st - 12t^2$: $a = 6, b = 1, c = -12 \Rightarrow$ key # $= ac = 6(-12) = -72$.
Find two factors of -72 which add to equal 1: -8 and $+9$.
Rewrite and factor: $6s^2 + st - 12t^2 = 6s^2 - 8st + 9st - 12t^2$
$$= 2s(3s - 4t) + 3t(3s - 4t) = (3s - 4t)(2s + 3t)$$

119. $20u^2 + 19uv + 3v^2$: $a = 20, b = 19, c = 3 \Rightarrow$ key # $= ac = 20(3) = 60$.
Find two factors of 60 which add to equal 19: $+15$ and $+4$.
Rewrite and factor: $20u^2 + 19uv + 3v^2 = 20u^2 + 15uv + 4uv + 3v^2$
$$= 5u(4u + 3v) + v(4u + 3v) = (4u + 3v)(5u + v)$$

120. $12m^2 + mn - 6n^2$: $a = 12, b = 1, c = -6 \Rightarrow$ key # $= ac = 12(-6) = -72$.
Find two factors of -72 which add to equal 1: $+9$ and -8.
Rewrite and factor: $12m^2 + mn - 6n^2 = 12m^2 + 9mn - 8mn - 6n^2$
$$= 3m(4m + 3n) - 2n(4m + 3n) = (4m + 3n)(3m - 2n)$$

121. $x^2 + 6x + 9 = (x+3)(x+3)$
One side is $x + 3$.

122. $25x^2 - 40x + 16 = (5x-4)(5x-4)$
One side is $5x - 4$.

123. $4x^2 + 20x - 11 = (2x+11)(2x-1)$
length $= 2x + 11$ in., width $= 2x - 1$ in.
difference $= (2x + 11) - (2x - 1)$
$= 2x + 11 - 2x + 1 = 12$
The difference is 12 inches.

124. $72x^2 + 120x - 400 = 8(9x^2 + 15x - 50)$
$= 8(3x-5)(3x+10)$
height $= 3x - 5$ ft, length $= 3x + 10$ ft

125. **Answers may vary.**

126. **Answers may vary.**

127. $x^2 - q^2 = x^2 + 0x - q^2$: $a = 1, b = 0, c = -q^2 \Rightarrow b^2 - 4ac = 0^2 - 4(1)(-q^2) = 4q^2 = (2q)^2$.
Since $4q^2$ is a perfect square, the binomial is factorable, and the test works.

128. $ax^2 + ax + a$: $a = a, b = a, c = a \Rightarrow b^2 - 4ac = a^2 - 4a(a) = a^2 - 4a^2 = -3a^2$.
$-3a^2$ is a nonnegative perfect square only for $a = 0$. The test does not work. The test actually determines the factorability of $x^2 + x + 1$. NOTE: $ax^2 + ax + a = a(x^2 + x + 1)$.

Exercise 5.7 (page 330)

1. $(3a^2 + 4a - 2) + (4a^2 - 3a - 5) = 3a^2 + 4a - 2 + 4a^2 - 3a - 5 = 7a^2 + a - 7$

2. $(-4b^2 - 3b - 2) - (3b^2 - 2b + 5) = -4b^2 - 3b - 2 - 3b^2 + 2b - 5 = -7b^2 - b - 7$

3. $5(2y^2 - 3y + 3) - 2(3y^2 - 2y + 6) = 10y^2 - 15y + 15 - 6y^2 + 4y - 12 = 4y^2 - 11y + 3$

4. $4(3x^2 + 3x + 3) + 3(x^2 - 3x - 4) = 12x^2 + 12x + 12 + 3x^2 - 9x - 12 = 15x^2 + 3x$

5. $(m+4)(m-2) = m^2 - 2m + 4m - 8 = m^2 + 2m - 8$

6. $(3p + 4q)(2p - 3q) = 6p^2 - 9pq + 8pq - 12q^2 = 6p^2 - pq - 12q^2$

7. common factors

8. difference; cubes; difference

9. trinomial

10. grouping

11. $x^2 + 8x + 16 = (x+4)(x+4) = (x+4)^2$

12. $20 + 11x - 3x^2 = -3x^2 + 11x + 20 = -(3x^2 - 11x - 20) = -(3x+4)(x-5)$

13. $8x^3y^3 - 27 = (2xy)^3 - 3^3 = (2xy - 3)[(2xy)^2 + (2xy)(3) + 3^2] = (2xy - 3)(4x^2y^2 + 6xy + 9)$

14. $3x^2y + 6xy^2 - 12xy = 3xy(x + 2y - 4)$

15. $xy - ty + xs - ts = y(x - t) + s(x - t)$
$= (x - t)(y + s)$

16. $bc + b + cd + d = b(c+1) + d(c+1)$
$= (c+1)(b+d)$

17. $25x^2 - 16y^2 = (5x)^2 - (4y)^2$
$= (5x+4y)(5x-4y)$

18. $27x^9 - y^3 = (3x^3)^3 - y^3 = (3x^3 - y)[(3x^3)^2 + 3x^3y + y^2] = (3x^3 - y)(9x^6 + 3x^3y + y^2)$

19. $12x^2 + 52x + 35 = (6x+5)(2x+7)$

20. $12x^2 + 14x - 6 = 2(6x^2 + 7x - 3)$
$= 2(3x-1)(2x+3)$

21. $6x^2 - 14x + 8 = 2(3x^2 - 7x + 4)$
$= 2(3x-4)(x-1)$

22. $12x^2 - 12 = 12(x^2 - 1)$
$= 12(x+1)(x-1)$

23. $56x^2 - 15x + 1 = (8x-1)(7x-1)$

24. $7x^2 - 57x + 8 = (7x-1)(x-8)$

25. $4x^2y^2 + 4xy^2 + y^2 = y^2(4x^2 + 4x + 1)$
$= y^2(2x+1)(2x+1)$

26. $100z^2 - 81t^2 = (10z)^2 - (9t)^2$
$= (10z+9t)(10z-9t)$

27. $x^3 + (a^2y)^3 = (x + a^2y)[x^2 - xa^2y + (a^2y)^2] = (x + a^2y)(x^2 - a^2xy + a^4y^2)$

28. $4x^2y^2z^2 - 26x^2y^2z^3 = 2x^2y^2z^2(2 - 13z)$

29. $2x^3 - 54 = 2(x^3 - 27) = 2(x^3 - 3^3) = 2(x-3)(x^2 + 3x + 3^2) = 2(x-3)(x^2 + 3x + 9)$

30. $4(xy)^3 + 256 = 4((xy)^3 + 64) = 4[(xy)^3 + 4^3] = 4(xy+4)[(xy)^2 - xy(4) + 4^2]$
$= 4(xy+4)(x^2y^2 - 4xy + 16)$

31. $ae + bf + af + be = ae + af + be + bf = a(e+f) + b(e+f) = (e+f)(a+b)$

32. $a^2x^2 + b^2y^2 + b^2x^2 + a^2y^2 = a^2x^2 + a^2y^2 + b^2x^2 + b^2y^2 = a^2(x^2+y^2) + b^2(x^2+y^2)$
$= (x^2+y^2)(a^2+b^2)$

33. $2(x+y)^2 + (x+y) - 3 = [2(x+y)+3][(x+y)-1] = (2x+2y+3)(x+y-1)$

34. $(x-y)^3 + 125 = (x-y)^3 + 5^3 = [(x-y)+5][(x-y)^2 - 5(x-y) + 5^2]$
$= (x-y+5)[(x-y)^2 - 5(x-y) + 25]$

35. $625x^4 - 256y^4 = (25x^2)^2 - (16y^2)^2 = (25x^2 + 16y^2)(25x^2 - 16y^2)$
$= (25x^2 + 16y^2)(5x+4y)(5x-4y)$

36. $2(a-b)^2 + 5(a-b) + 3 = [2(a-b)+3][(a-b)+1] = (2a - 2b + 3)(a - b + 1)$

37. $36x^4 - 36 = 36(x^4 - 1) = 36(x^2+1)(x^2-1) = 36(x^2+1)(x+1)(x-1)$

38. $6x^2 - 63 - 13x = 6x^2 - 13x - 63 = (3x+7)(2x-9)$

39. $2x^6 + 2y^6 = 2(x^6 + y^6) = 2[(x^2)^3 + (y^2)^3] = 2(x^2+y^2)(x^4 - x^2y^2 + y^4)$

40. $x^4 - x^4y^4 = x^4(1-y^4) = x^4(1+y^2)(1-y^2) = x^4(1+y^2)(1+y)(1-y)$
$= -x^4\left(y^2+1\right)(y+1)(y-1)$

41. $a^4 - 13a^2 + 36 = (a^2-4)(a^2-9) = (a+2)(a-2)(a+3)(a-3)$

42. $x^4 - 17x^2 + 16 = (x^2-16)(x^2-1) = (x+4)(x-4)(x+1)(x-1)$

43. $x^2 + 6x + 9 - y^2 = (x+3)^2 - y^2 = (x+3+y)(x+3-y)$

44. $x^2 + 10x + 25 - y^8 = (x+5)^2 - (y^4)^2 = (x+5+y^4)(x+5-y^4)$

45. $4x^2 + 4x + 1 - 4y^2 = (2x+1)^2 - (2y)^2 = (2x+1+2y)(2x+1-2y)$

46. $9x^2 - 6x + 1 - 25y^2 = (3x-1)^2 - (5y)^2 = (3x-1+5y)(3x-1-5y)$

47. $x^2 - y^2 - 2y - 1 = x^2 - \left(y^2+2y+1\right) = x^2 - (y+1)^2 = [x+(y+1)][x-(y+1)]$
$= (x+y+1)(x-y-1)$

48. $a^2 - b^2 + 4b - 4 = a^2 - \left(b^2-4b+4\right) = a^2 - (b-2)^2 = [a+(b-2)][a-(b-2)]$
$= (a+b-2)(a-b+2)$

49. $x^5 + x^2 - x^3 - 1 = x^2(x^3+1) - 1(x^3+1) = (x^3+1)(x^2-1)$
$= (x+1)(x^2-x+1)(x+1)(x-1)$

50. $x^5 - x^2 - 4x^3 + 4 = x^2(x^3-1) - 4(x^3-1) = (x^3-1)(x^2-4)$
$= (x-1)(x^2+x+1)(x+2)(x-2)$

51. $x^5 - 9x^3 + 8x^2 - 72 = x^3(x^2-9) + 8(x^2-9) = (x^2-9)(x^3+8)$
$= (x+3)(x-3)(x+2)(x^2-2x+4)$

52. $x^5 - 4x^3 - 8x^2 + 32 = x^3(x^2-4) - 8(x^2-4) = (x^2-4)(x^3-8)$
$= (x+2)(x-2)(x-2)(x^2+2x+4)$
$= (x+2)(x-2)^2\left(x^2+2x+4\right)$

53. $2x^5z - 2x^2y^3z - 2x^3y^2z + 2y^5z = 2z(x^5 - x^2y^3 - x^3y^2 + y^5)$
$= 2z[x^2(x^3-y^3) - y^2(x^3-y^3)]$
$= 2z(x^3-y^3)(x^2-y^2)$
$= 2z(x-y)(x^2+xy+y^2)(x+y)(x-y)$
$= 2z(x+y)(x-y)^2\left(x^2+xy+y^2\right)$

54. $x^2y^3 - 4x^2y - 9y^3 + 36y = y(x^2y^2 - 4x^2 - 9y^2 + 36) = y[x^2(y^2-4) - 9(y^2-4)]$
$= y(y^2-4)(x^2-9)$
$= y(y+2)(y-2)(x+3)(x-3)$

55. $x^{2m} - x^m - 6 = (x^m - 3)(x^m + 2)$

56. $a^{2n} - b^{2n} = (a^n)^2 - (b^n)^2$
$= (a^n + b^n)(a^n - b^n)$

57. $a^{3n} - b^{3n} = (a^n)^3 - (b^n)^3$
$= (a^n - b^n)(a^{2n} + a^n b^n + b^{2n})$

58. $x^{3m} + y^{3m} = (x^m)^3 + (y^m)^3$
$= (x^m + y^m)(x^{2m} - x^m y^m + y^{2m})$

59. $x^{-2} + 2x^{-1} + 1 = (x^{-1} + 1)(x^{-1} + 1) = (x^{-1} + 1)^2 = \left(\frac{1}{x} + 1\right)^2$

60. $4a^{-2} - 12a^{-1} + 9 = (2a^{-1} - 3)(2a^{-1} - 3) = (2a^{-1} - 3)^2 = \left(\frac{2}{a} - 3\right)^2$

61. $6x^{-2} - 5x^{-1} - 6 = (3x^{-1} + 2)(2x^{-1} - 3) = \left(\frac{3}{x} + 2\right)\left(\frac{2}{x} - 3\right)$

62. $x^{-4} - y^{-4} = (x^{-2} + y^{-2})(x^{-2} - y^{-2}) = (x^{-2} + y^{-2})(x^{-1} + y^{-1})(x^{-1} - y^{-1})$
$= \left(\frac{1}{x^2} + \frac{1}{y^2}\right)\left(\frac{1}{x} + \frac{1}{y}\right)\left(\frac{1}{x} - \frac{1}{y}\right)$

63. **Answers may vary.**

64. **Answers may vary.**

65. **Answers may vary.**

66. $2x^2 + 7x + 3 = (2x + 1)(x + 3)$; $x^2 - 2x - 15 = (x - 5)(x + 3)$; gcf $= x + 3$

67. $x^4 + x^2 + 1 = x^4 + x^2 + x^2 + 1 - x^2 = x^4 + 2x^2 + 1 - x^2 = (x^2 + 1)^2 - x^2$
$= (x^2 + 1 + x)(x^2 + 1 - x)$

68. $x^4 + 7x^2 + 16 = x^4 + 7x^2 + x^2 + 16 - x^2 = x^4 + 8x^2 + 16 - x^2 = (x^2 + 4)^2 - x^2$
$= (x^2 + 4 + x)(x^2 + 4 - x)$

Exercise 5.8 (page 338)

1. 2, 3, 5, 7

2. 8, 9, 10, 12, 14, 15, 16

3. $V = \frac{4}{3}\pi r^3 = \frac{4}{3}\pi(21.23)^3 = 40{,}081.00 \text{ cm}^3$

4. $V = \frac{1}{3}\pi r^2 h = \frac{1}{3}\pi(12.33)^2(14.7) = 2340.30 \text{ m}^3$

5. $ax^2 + bx + c = 0$

6. $ab = 0$

7.
$4x^2 + 8x = 0$
$4x(x + 2) = 0$
$4x = 0$ **or** $x + 2 = 0$
$x = 0$ $\quad x = -2$

8.
$x^2 - 9 = 0$
$(x + 3)(x - 3) = 0$
$x + 3 = 0$ **or** $x - 3 = 0$
$x = -3$ $\quad x = 3$

9.
$$\begin{aligned} y^2 - 16 &= 0 \\ (y+4)(y-4) &= 0 \end{aligned}$$
$$y+4=0 \quad \textbf{or} \quad y-4=0$$
$$y=-4 \qquad\qquad y=4$$

10.
$$\begin{aligned} 5y^2 - 10y &= 0 \\ 5y(y-2) &= 0 \end{aligned}$$
$$5y=0 \quad \textbf{or} \quad y-2=0$$
$$y=0 \qquad\qquad y=2$$

11.
$$\begin{aligned} x^2 + x &= 0 \\ x(x+1) &= 0 \end{aligned}$$
$$x=0 \quad \textbf{or} \quad x+1=0$$
$$x=-1$$

12.
$$\begin{aligned} x^2 - 3x &= 0 \\ x(x-3) &= 0 \end{aligned}$$
$$x=0 \quad \textbf{or} \quad x-3=0$$
$$x=3$$

13.
$$\begin{aligned} 5y^2 - 25 &= 0 \\ 5y(y-5) &= 0 \end{aligned}$$
$$5y=0 \quad \textbf{or} \quad y-5=0$$
$$y=0 \qquad\qquad y=5$$

14.
$$\begin{aligned} y^2 - 36 &= 0 \\ (y+6)(y-6) &= 0 \end{aligned}$$
$$y+6=0 \quad \textbf{or} \quad y-6=0$$
$$y=-6 \qquad\qquad y=6$$

15.
$$\begin{aligned} z^2 + 8z + 15 &= 0 \\ (z+5)(z+3) &= 0 \end{aligned}$$
$$z+5=0 \quad \textbf{or} \quad z+3=0$$
$$z=-5 \qquad\qquad z=-3$$

16.
$$\begin{aligned} w^2 + 7w + 12 &= 0 \\ (w+4)(w+3) &= 0 \end{aligned}$$
$$w+4=0 \quad \textbf{or} \quad w+3=0$$
$$w=-4 \qquad\qquad w=-3$$

17.
$$\begin{aligned} y^2 - 7y + 6 &= 0 \\ (y-6)(y-1) &= 0 \end{aligned}$$
$$y-6=0 \quad \textbf{or} \quad y-1=0$$
$$y=6 \qquad\qquad y=1$$

18.
$$\begin{aligned} n^2 - 5n + 6 &= 0 \\ (n-2)(n-3) &= 0 \end{aligned}$$
$$n-2=0 \quad \textbf{or} \quad n-3=0$$
$$n=2 \qquad\qquad n=3$$

19.
$$\begin{aligned} y^2 - 7y + 12 &= 0 \\ (y-4)(y-3) &= 0 \end{aligned}$$
$$y-4=0 \quad \textbf{or} \quad y-3=0$$
$$y=4 \qquad\qquad y=3$$

20.
$$\begin{aligned} x^2 - 3x + 2 &= 0 \\ (x-2)(x-1) &= 0 \end{aligned}$$
$$x-2=0 \quad \textbf{or} \quad x-1=0$$
$$x=2 \qquad\qquad x=1$$

21.
$$\begin{aligned} x^2 + 6x + 8 &= 0 \\ (x+4)(x+2) &= 0 \end{aligned}$$
$$x+4=0 \quad \textbf{or} \quad x+2=0$$
$$x=-4 \qquad\qquad x=-2$$

22.
$$\begin{aligned} x^2 + 9x + 20 &= 0 \\ (x+5)(x+4) &= 0 \end{aligned}$$
$$x+5=0 \quad \textbf{or} \quad x+4=0$$
$$x=-5 \qquad\qquad x=-4$$

23.
$$\begin{aligned} 3m^2 + 10m + 3 &= 0 \\ (3m+1)(m+3) &= 0 \end{aligned}$$
$$3m+1=0 \quad \textbf{or} \quad m+3=0$$
$$3m=-1 \qquad\qquad m=-3$$
$$m=-\tfrac{1}{3}$$

24.
$$\begin{aligned} 2r^2 + 5r + 3 &= 0 \\ (2r+3)(r+1) &= 0 \end{aligned}$$
$$2r+3=0 \quad \textbf{or} \quad r+1=0$$
$$2r=-3 \qquad\qquad r=-1$$
$$r=-\tfrac{3}{2}$$

25.
$$\begin{aligned} 2y^2 - 5y + 2 &= 0 \\ (2y-1)(y-2) &= 0 \\ 2y - 1 = 0 \quad &\textbf{or} \quad y - 2 = 0 \\ 2y = 1 \quad & \quad y = 2 \\ y = \tfrac{1}{2} & \end{aligned}$$

26.
$$\begin{aligned} 2x^2 - 3x + 1 &= 0 \\ (2x-1)(x-1) &= 0 \\ 2x - 1 = 0 \quad &\textbf{or} \quad x - 1 = 0 \\ 2x = 1 \quad & \quad x = 1 \\ x = \tfrac{1}{2} & \end{aligned}$$

27.
$$\begin{aligned} 2x^2 - x - 1 &= 0 \\ (2x+1)(x-1) &= 0 \\ 2x + 1 = 0 \quad &\textbf{or} \quad x - 1 = 0 \\ 2x = -1 \quad & \quad x = 1 \\ x = -\tfrac{1}{2} & \end{aligned}$$

28.
$$\begin{aligned} 2x^2 - 3x - 5 &= 0 \\ (2x-5)(x+1) &= 0 \\ 2x - 5 = 0 \quad &\textbf{or} \quad x + 1 = 0 \\ 2x = 5 \quad & \quad x = -1 \\ x = \tfrac{5}{2} & \end{aligned}$$

29.
$$\begin{aligned} 3s^2 - 5s - 2 &= 0 \\ (3s+1)(s-2) &= 0 \\ 3s + 1 = 0 \quad &\textbf{or} \quad s - 2 = 0 \\ 3s = -1 \quad & \quad s = 2 \\ s = -\tfrac{1}{3} & \end{aligned}$$

30.
$$\begin{aligned} 8t^2 + 10t - 3 &= 0 \\ (4t-1)(2t+3) &= 0 \\ 4t - 1 = 0 \quad &\textbf{or} \quad 2t + 3 = 0 \\ 4t = 1 \quad & \quad 2t = -3 \\ t = \tfrac{1}{4} \quad & \quad t = -\tfrac{3}{2} \end{aligned}$$

31.
$$\begin{aligned} x(x-6) + 9 &= 0 \\ x^2 - 6x + 9 &= 0 \\ (x-3)(x-3) &= 0 \\ x - 3 = 0 \quad &\textbf{or} \quad x - 3 = 0 \\ x = 3 \quad & \quad x = 3 \end{aligned}$$

32.
$$\begin{aligned} x^2 + 8(x+2) &= 0 \\ x^2 + 8x + 16 &= 0 \\ (x+4)(x+4) &= 0 \\ x + 4 = 0 \quad &\textbf{or} \quad x + 4 = 0 \\ x = -4 \quad & \quad x = -4 \end{aligned}$$

33.
$$\begin{aligned} 8a^2 &= 3 - 10a \\ 8a^2 + 10a - 3 &= 0 \\ (4a-1)(2a+3) &= 0 \\ 4a - 1 = 0 \quad &\textbf{or} \quad 2a + 3 = 0 \\ 4a = 1 \quad & \quad 2a = -3 \\ a = \tfrac{1}{4} \quad & \quad a = -\tfrac{3}{2} \end{aligned}$$

34.
$$\begin{aligned} 5z^2 &= 6 - 13z \\ 5z^2 + 13z - 6 &= 0 \\ (5z-2)(z+3) &= 0 \\ 5z - 2 = 0 \quad &\textbf{or} \quad z + 3 = 0 \\ 5z = 2 \quad & \quad z = -3 \\ z = \tfrac{2}{5} & \end{aligned}$$

35.
$$\begin{aligned} b(6b-7) &= 10 \\ 6b^2 - 7b - 10 &= 0 \\ (6b+5)(b-2) &= 0 \\ 6b + 5 = 0 \quad &\textbf{or} \quad b - 2 = 0 \\ 6b = -5 \quad & \quad b = 2 \\ b = -\tfrac{5}{6} & \end{aligned}$$

36.
$$\begin{aligned} 2y(4y+3) &= 9 \\ 8y^2 + 6y &= 9 \\ 8y^2 + 6y - 9 &= 0 \\ (4y-3)(2y+3) &= 0 \\ 4y - 3 = 0 \quad &\textbf{or} \quad 2y + 3 = 0 \\ 4y = 3 \quad & \quad 2y = -3 \\ y = \tfrac{3}{4} \quad & \quad y = -\tfrac{3}{2} \end{aligned}$$

37.
$$\begin{aligned} \frac{3a^2}{2} &= \frac{1}{2} - a \\ 3a^2 &= 1 - 2a \\ 3a^2 + 2a - 1 &= 0 \\ (3a-1)(a+1) &= 0 \end{aligned}$$
$$3a - 1 = 0 \quad \textbf{or} \quad a + 1 = 0$$
$$3a = 1 \qquad a = -1$$
$$a = \tfrac{1}{3}$$

38.
$$\begin{aligned} x^2 &= \frac{1}{2}(x+1) \\ 2x^2 &= x + 1 \\ 2x^2 - x - 1 &= 0 \\ (2x+1)(x-1) &= 0 \end{aligned}$$
$$2x + 1 = 0 \quad \textbf{or} \quad x - 1 = 0$$
$$2x = -1 \qquad x = 1$$
$$x = -\tfrac{1}{2}$$

39.
$$\begin{aligned} \frac{1}{2}x^2 - \frac{5}{4}x &= -\frac{1}{2} \\ 2x^2 - 5x &= -2 \\ 2x^2 - 5x + 2 &= 0 \\ (2x-1)(x-2) &= 0 \end{aligned}$$
$$2x - 1 = 0 \quad \textbf{or} \quad x - 2 = 0$$
$$2x = 1 \qquad x = 2$$
$$x = \tfrac{1}{2}$$

40.
$$\begin{aligned} \frac{1}{4}x^2 + \frac{3}{4}x &= 1 \\ x^2 + 3x &= 4 \\ x^2 + 3x - 4 &= 0 \\ (x+4)(x-1) &= 0 \end{aligned}$$
$$x + 4 = 0 \quad \textbf{or} \quad x - 1 = 0$$
$$x = -4 \qquad x = 1$$

41.
$$\begin{aligned} x\left(3x + \frac{22}{5}\right) &= 1 \\ 3x^2 + \frac{22}{5}x &= 1 \\ 15x^2 + 22x &= 5 \\ 15x^2 + 22x - 5 &= 0 \\ (5x-1)(3x+5) &= 0 \end{aligned}$$
$$5x - 1 = 0 \quad \textbf{or} \quad 3x + 5 = 0$$
$$5x = 1 \qquad 3x = -5$$
$$x = \tfrac{1}{5} \qquad x = -\tfrac{5}{3}$$

42.
$$\begin{aligned} x\left(\frac{x}{11} - \frac{1}{7}\right) &= \frac{6}{77} \\ \frac{x^2}{11} - \frac{1}{7}x &= \frac{6}{77} \\ 7x^2 - 11x &= 6 \\ 7x^2 - 11x - 6 &= 0 \\ (7x+3)(x-2) &= 0 \end{aligned}$$
$$7x + 3 = 0 \quad \textbf{or} \quad x - 2 = 0$$
$$7x = -3 \qquad x = 2$$
$$x = -\tfrac{3}{7}$$

43.
$$\begin{aligned} x^3 + x^2 &= 0 \\ x^2(x+1) &= 0 \end{aligned}$$
$$x^2 = 0 \quad \textbf{or} \quad x + 1 = 0$$
$$x = 0 \qquad x = -1$$

44.
$$\begin{aligned} 2x^4 + 8x^3 &= 0 \\ 2x^3(x+4) &= 0 \end{aligned}$$
$$2x^3 = 0 \quad \textbf{or} \quad x + 4 = 0$$
$$x = 0 \qquad x = -4$$

45.
$$\begin{aligned} y^3 - 49y &= 0 \\ y(y^2 - 49) &= 0 \\ y(y+7)(y-7) &= 0 \end{aligned}$$
$$y = 0 \quad \textbf{or} \quad y + 7 = 0 \quad \textbf{or} \quad y - 7 = 0$$
$$y = -7 \qquad y = 7$$

46.

$$2z^3 - 200z = 0$$
$$2z(z^2 - 100) = 0$$
$$2z(z + 10)(z - 10) = 0$$

$2z = 0$ **or** $z + 10 = 0$ **or** $z - 10 = 0$

$z = 0$ $\quad z = -10$ $\quad z = 10$

47.

$$x^3 - 4x^2 - 21x = 0$$
$$x(x^2 - 4x - 21) = 0$$
$$x(x + 3)(x - 7) = 0$$

$x = 0$ **or** $x + 3 = 0$ **or** $x - 7 = 0$

$x = -3$ $\quad x = 7$

48.

$$x^3 + 8x^2 - 9x = 0$$
$$x(x^2 + 8x - 9) = 0$$
$$x(x + 9)(x - 1) = 0$$

$x = 0$ **or** $x + 9 = 0$ **or** $x - 1 = 0$

$x = -9$ $\quad x = 1$

49.

$$z^4 - 13z^2 + 36 = 0$$
$$(z^2 - 4)(z^2 - 9) = 0$$
$$(z + 2)(z - 2)(z + 3)(z - 3) = 0$$

$z + 2 = 0$ **or** $z - 2 = 0$ **or** $z + 3 = 0$ **or** $z - 3 = 0$

$z = -2$ $\quad z = 2$ $\quad z = -3$ $\quad z = 3$

50.

$$y^4 - 10y^2 + 9 = 0$$
$$(y^2 - 1)(y^2 - 9) = 0$$
$$(y + 1)(y - 1)(y + 3)(y - 3) = 0$$

$y + 1 = 0$ **or** $y - 1 = 0$ **or** $y + 3 = 0$ **or** $y - 3 = 0$

$y = -1$ $\quad y = 1$ $\quad y = -3$ $\quad y = 3$

51.

$$3a(a^2 + 5a) = -18a$$
$$3a^3 + 15a^2 = -18a$$
$$3a^3 + 15a^2 + 18a = 0$$
$$3a(a^2 + 5a + 6) = 0$$
$$3a(a + 2)(a + 3) = 0$$

$3a = 0$ **or** $a + 2 = 0$ **or** $a + 3 = 0$

$a = 0$ $\quad a = -2$ $\quad a = -3$

52.

$$7t^3 = 2t\left(t + \frac{5}{2}\right)$$
$$7t^3 = 2t^2 + 5t$$
$$7t^3 - 2t^2 - 5t = 0$$
$$t(7t^2 - 2t - 5) = 0$$
$$t(7t + 5)(t - 1) = 0$$

$t = 0$ **or** $7t + 5 = 0$ **or** $t - 1 = 0$

$7t = -5$ $\quad t = 1$

$t = -\frac{5}{7}$

53.
$$\begin{aligned} \frac{x^2(6x+37)}{35} &= x \\ x^2(6x+37) &= 35x \\ 6x^3+37x^2 &= 35x \\ 6x^3+37x^2-35x &= 0 \\ x(6x^2+37x-35) &= 0 \\ x(6x-5)(x+7) &= 0 \end{aligned}$$
$$x=0 \quad \textbf{or} \quad 6x-5=0 \quad \textbf{or} \quad x+7=0$$
$$6x=5 \qquad x=-7$$
$$x=\tfrac{5}{6}$$

54.
$$\begin{aligned} x^2 &= -\frac{4x^3(3x+5)}{3} \\ 3x^2 &= -[4x^3(3x+5)] \\ 3x^2 &= -12x^4-20x^3 \\ 12x^4+20x^3+3x^2 &= 0 \\ x^2(12x^2+20x+3) &= 0 \\ x^2(6x+1)(2x+3) &= 0 \end{aligned}$$
$$x^2=0 \quad \textbf{or} \quad 6x+1=0 \quad \textbf{or} \quad 2x+3=0$$
$$x=0 \qquad 6x=-1 \qquad 2x=-3$$
$$x=-\tfrac{1}{6} \qquad x=-\tfrac{3}{2}$$

55.
$$\begin{aligned} f(x) &= 0 \\ x^2-49 &= 0 \\ (x+7)(x-7) &= 0 \end{aligned}$$
$$x+7=0 \quad \textbf{or} \quad x-7=0$$
$$x=-7 \qquad x=7$$

56.
$$\begin{aligned} f(x) &= 0 \\ x^2+11x &= 0 \\ x(x+11) &= 0 \end{aligned}$$
$$x=0 \quad \textbf{or} \quad x+11=0$$
$$x=-11$$

57.
$$\begin{aligned} f(x) &= 0 \\ 2x^2+5x-3 &= 0 \\ (2x-1)(x+3) &= 0 \end{aligned}$$
$$2x-1=0 \quad \textbf{or} \quad x+3=0$$
$$2x=1 \qquad x=-3$$
$$x=\tfrac{1}{2}$$

58.
$$\begin{aligned} f(x) &= 0 \\ 3x^2-x-2 &= 0 \\ (3x+2)(x-1) &= 0 \end{aligned}$$
$$3x+2=0 \quad \textbf{or} \quad x-1=0$$
$$3x=-2 \qquad x=1$$
$$x=-\tfrac{2}{3}$$

59.
$$\begin{aligned} f(x) &= 0 \\ 5x^3+3x^2-2x &= 0 \\ x(5x^2+3x-2) &= 0 \\ x(5x-2)(x+1) &= 0 \end{aligned}$$
$$x=0 \quad \textbf{or} \quad 5x-2=0 \quad \textbf{or} \quad x+1=0$$
$$5x=2 \qquad x=-1$$
$$x=\tfrac{2}{5}$$

60.

$$\begin{aligned} f(x) &= 0 \\ x^4 - 26x^2 + 25 &= 0 \\ (x^2 - 25)(x^2 - 1) &= 0 \\ (x+5)(x-5)(x+1)(x-1) &= 0 \end{aligned}$$

$x + 5 = 0$ **or** $x - 5 = 0$ **or** $x + 1 = 0$ **or** $x - 1 = 0$

$x = -5$ $\quad x = 5$ $\quad x = -1$ $\quad x = 1$

61.

$$\begin{aligned} x^3 + 3x^2 - x - 3 &= 0 \\ x^2(x+3) - 1(x+3) &= 0 \\ (x+3)(x^2-1) &= 0 \\ (x+3)(x+1)(x-1) &= 0 \end{aligned}$$

$x + 3 = 0$ **or** $x + 1 = 0$ **or** $x - 1 = 0$

$x = -3$ $\quad x = -1$ $\quad x = 1$

62.

$$\begin{aligned} x^3 - x^2 - 4x + 4 &= 0 \\ x^2(x-1) - 4(x-1) &= 0 \\ (x-1)(x^2-4) &= 0 \\ (x-1)(x+2)(x-2) &= 0 \end{aligned}$$

$x - 1 = 0$ **or** $x + 2 = 0$ **or** $x - 2 = 0$

$x = 1$ $\quad x = -2$ $\quad x = 2$

63.

$$\begin{aligned} 2r^3 + 3r^2 - 18r - 27 &= 0 \\ r^2(2r+3) - 9(2r+3) &= 0 \\ (2r+3)(r^2-9) &= 0 \\ (2r+3)(r+3)(r-3) &= 0 \end{aligned}$$

$2r + 3 = 0$ **or** $r + 3 = 0$ **or** $r - 3 = 0$

$2r = -3$ $\quad r = -3$ $\quad r = 3$

$r = -\frac{3}{2}$

64.

$$\begin{aligned} 3s^3 - 2s^2 - 3s + 2 &= 0 \\ s^2(3s-2) - 1(3s-2) &= 0 \\ (3s-2)(s^2-1) &= 0 \\ (3s-2)(s+1)(s-1) &= 0 \end{aligned}$$

$3s - 2 = 0$ **or** $s + 1 = 0$ **or** $s - 1 = 0$

$3s = 2$ $\quad s = -1$ $\quad s = 1$

$s = \frac{2}{3}$

65.

$$\begin{aligned} 3y^3 + y^2 &= 4(3y+1) \\ 3y^3 + y^2 &= 12y + 4 \\ 3y^3 + y^2 - 12y - 4 &= 0 \\ y^2(3y+1) - 4(3y+1) &= 0 \\ (3y+1)(y^2-4) &= 0 \\ (3y+1)(y+2)(y-2) &= 0 \end{aligned}$$

$3y + 1 = 0$ **or** $y + 2 = 0$ **or** $y - 2 = 0$

$3y = -1$ $\quad y = -2$ $\quad y = 2$

$y = -\frac{1}{3}$

66.

$$\begin{aligned} w^3 + 16 &= w(w+16) \\ w^3 + 16 &= w^2 + 16w \\ w^3 - w^2 - 16w + 16 &= 0 \\ w^2(w-1) - 16(w-1) &= 0 \\ (w-1)(w^2-16) &= 0 \\ (w-1)(w+4)(w-4) &= 0 \end{aligned}$$

$w - 1 = 0$ **or** $w + 4 = 0$ **or** $w - 4 = 0$

$w = 1$ $\quad w = -4$ $\quad w = 4$

67. Let $x =$ the first even integer and $x + 2 =$ the second.

$$\boxed{\text{First}} \cdot \boxed{\text{Second}} = 288$$

$$\begin{aligned} x(x+2) &= 288 \\ x^2 + 2x - 288 &= 0 \\ (x+18)(x-16) &= 0 \end{aligned}$$

$x + 18 = 0$ **or** $x - 16 = 0$

$x = -18$ $\quad x = 16$

The integers are -18 and -16, or 16 and 18.

68. Let $x =$ the first odd integer and $x + 2 =$ the second.

$$\boxed{\text{First}} \cdot \boxed{\text{Second}} = 143$$

$$\begin{aligned} x(x+2) &= 143 \\ x^2 + 2x - 143 &= 0 \\ (x+13)(x-11) &= 0 \end{aligned}$$

$x + 13 = 0$ **or** $x - 11 = 0$

$x = -13$ $\quad x = 11$

The integers are -13 and -11, or 11 and 13.

69. Let $x =$ the first positive integer and $x+1 =$ the second.

$$\boxed{\text{First}^2} + \boxed{\text{Second}^2} = 85$$
$$x^2 + (x+1)^2 = 85$$
$$x^2 + (x+1)(x+1) = 85$$
$$x^2 + x^2 + 2x + 1 = 85$$
$$2x^2 + 2x - 84 = 0$$
$$2(x^2 + x - 42) = 0$$
$$2(x+7)(x-6) = 0$$
$$x+7=0 \quad \textbf{or} \quad x-6=0$$
$$x=-7 \qquad\qquad x=6$$

(not positive)

The integers are 6 and 7.

70. Let $x =$ the first positive integer, $x+1 =$ the second and $x+2 =$ the third.

$$\boxed{\text{First}^2} + \boxed{\text{Second}^2} + \boxed{\text{Third}^2} = 77$$
$$x^2 + (x+1)^2 + (x+2)^2 = 77$$
$$x^2 + (x+1)(x+1) + (x+2)(x+2) = 77$$
$$x^2 + x^2 + 2x + 1 + x^2 + 4x + 4 = 77$$
$$3x^2 + 6x - 72 = 0$$
$$3(x^2 + 2x - 24) = 0$$
$$3(x+6)(x-4) = 0$$
$$x+6=0 \quad \textbf{or} \quad x-4=0$$
$$x=-6 \qquad\qquad x=4$$

(not positive)

The integers are $4, 5$ and 6.

71.

$$\boxed{\text{Length}} \cdot \boxed{\text{Width}} = \boxed{\text{Area}}$$
$$(w+4)w = 96$$
$$w^2 + 4w = 96$$
$$w^2 + 4w - 96 = 0$$
$$(w+12)(w-8) = 0$$
$$w+12=0 \quad \textbf{or} \quad w-8=0$$
$$x=-12 \qquad\qquad w=8$$

(not positive)

The dimensions are 8 m by 12 m, so the perimeter is 40 meters.

72. Let $x =$ one side and $3x =$ the other side.

$$\boxed{\text{Length}} \cdot \boxed{\text{Width}} = \boxed{\text{Area}}$$
$$3x \cdot x = 147$$
$$3x^2 - 147 = 0$$
$$3(x^2 - 49) = 0$$
$$3(x+7)(x-7) = 0$$
$$x+7=0 \quad \textbf{or} \quad x-7=0$$
$$x=-7 \qquad\qquad x=7$$

(not positive)

The dimensions are 7 cm by 21cm.

73.

$$\boxed{\text{Length}} \cdot \boxed{\text{Width}} = \boxed{\text{Area}}$$
$$(2x-5)x = 375$$
$$2x^2 - 5x = 375$$
$$2x^2 - 5x - 375 = 0$$
$$(2x+25)(x-15) = 0$$
$$2x+25=0 \quad \textbf{or} \quad x-15=0$$
$$2x=-25 \qquad\qquad x=15$$

(x is not positive)

The dimensions are 15 ft by 25 ft.

74.

$$\frac{1}{2} \cdot \boxed{\text{Base}} \cdot \boxed{\text{Height}} = \boxed{\text{Area}}$$
$$\frac{1}{2}(2h+3)h = 162$$
$$(2h+3)h = 324$$
$$2h^2 + 3h - 324 = 0$$
$$(2h+27)(h-12) = 0$$
$$2h+27=0 \quad \textbf{or} \quad h-12=0$$
$$2h=-27 \qquad\qquad h=12$$

(h is not positive)

The height is 12 cm.

75. The length of the mural is $18 - 2w$, while the width is $11 - 2w$.

$$\boxed{\text{Length}} \cdot \boxed{\text{Width}} = \boxed{\text{Area}}$$
$$(18-2w)(11-2w) = 60$$
$$198 - 36w - 22w + 4w^2 = 60$$
$$4w^2 - 58w + 138 = 0$$
$$2(2w^2 - 29w + 69) = 0$$
$$2(2w-23)(w-3) = 0$$
$$2w - 23 = 0 \quad \textbf{or} \quad w - 3 = 0$$
$$2w = 23 \qquad\qquad w = 3$$
$$w = \tfrac{23}{2}$$
(makes width < 0)

The dimensions are 12 ft by 5 ft.

76. The area of the yard is $48 \cdot 100 = 4800$ square feet. The area of the garden will then be 1200 square feet. Let the dimensions of the garden be x feet and $x + 40$ feet.

$$\boxed{\text{Length}} \cdot \boxed{\text{Width}} = \boxed{\text{Area}}$$
$$x(x+40) = 1200$$
$$x^2 + 40x = 1200$$
$$x^2 + 40x - 1200 = 0$$
$$(x+60)(x-20) = 0$$
$$x + 60 = 0 \quad \textbf{or} \quad x - 20 = 0$$
$$x = -60 \qquad\qquad x = 20$$
(x is not positive)

The dimensions will be 20 feet by 60 feet, so the perimeter will be 160 feet.

77. Let $w =$ the width of the room and let $2w =$ the length of the room.

$$\boxed{\text{Length}} \cdot \boxed{\text{Width}} = \boxed{\text{Area}}$$
$$(2w-12)w = 560$$
$$2w^2 - 12w - 560 = 0$$
$$2(w^2 - 6w - 280) = 0$$
$$2(w+14)(w-20) = 0$$
$$w + 14 = 0 \quad \textbf{or} \quad w - 20 = 0$$
$$w = -14 \qquad\qquad w = 20$$
(impossible)

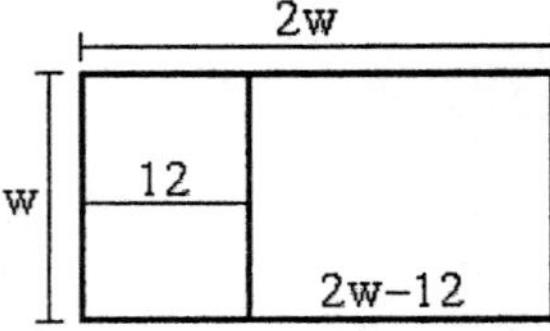

The dimensions are 20 feet by 40 feet.

78. Let x = the old length of a side and $x + 4$ = the new length.

$$9 \cdot \boxed{\text{Old area}} = \boxed{\text{New area}}$$
$$9x^2 = (x+4)^2$$
$$9x^2 = x^2 + 8x + 16$$
$$8x^2 - 8x - 16 = 0$$
$$8(x^2 - x - 2) = 0$$
$$8(x+1)(x-2) = 0$$
$$x + 1 = 0 \quad \textbf{or} \quad x - 2 = 0$$
$$x = -1 \qquad\qquad x = 2$$
(impossible)

The original perimeter was 8 inches.

79.
$$h = vt - 16t^2$$
$$0 = 160t - 16t^2$$
$$0 = 16t(10 - t)$$
$$16t = 0 \quad \textbf{or} \quad 10 - t = 0$$
$$t = 0 \qquad\qquad 10 = t$$

The object will hit the ground after 10 sec.

80.
$$h = vt - 16t^2$$
$$0 = 208 - 16t^2$$
$$0 = 16t(13 - t)$$
$$16t = 0 \quad \textbf{or} \quad 13 - t = 0$$
$$t = 0 \qquad\qquad 13 = t$$

The object will hit the ground after 13 seconds.

81.
$$h = vt - 16t^2$$
$$3344 = 480t - 16t^2$$
$$16t^2 - 480t + 3344 = 0$$
$$16(t^2 - 30t + 209) = 0$$
$$16(t - 11)(t - 19) = 0$$
$$t - 11 = 0 \quad \textbf{or} \quad t - 19 = 0$$
$$t = 11 \qquad\qquad t = 19$$

The cannonball will be at that height after 11 and 19 seconds.

82.
$$h = vt - 16t^2$$
$$192 = 128t - 16t^2$$
$$16t^2 - 128t + 192 = 0$$
$$16(t^2 - 8t + 12) = 0$$
$$16(t - 2)(t - 6) = 0$$
$$t - 2 = 0 \quad \textbf{or} \quad t - 6 = 0$$
$$t = 2 \qquad\qquad t = 6$$

The stone will be at that height after 2 and 6 seconds.

83. Let w = the width and $2w + 20$ = the length.

$$\boxed{\text{Length}} \cdot \boxed{\text{Width}} = \boxed{\text{Area}}$$
$$(2w + 20)w = 6000$$
$$2w^2 + 20w - 6000 = 0$$
$$2(w^2 + 10w - 3000) = 0$$
$$2(w - 50)(w + 60) = 0$$
$$w - 50 = 0 \quad \textbf{or} \quad w + 60 = 0$$
$$w = 50 \qquad\qquad w = -60$$
(impossible)

The width is 50 meters.

84. Let $w =$ the width and $w + 4 =$ the length.

$\boxed{\text{Length}} \cdot \boxed{\text{Width}} = \boxed{\text{Area}}$

$$(w+4)w = 285$$
$$w^2 + 4w - 285 = 0$$
$$(w+19)(w-15) = 0$$
$$w + 19 = 0 \quad \text{or} \quad w - 15 = 0$$
$$w = -19 \qquad\qquad w = 15$$
(impossible)

The dimensions are 15 feet by 19 feet, so the perimeter is 68 feet.

85. Let $x =$ the width of the pool and $2x - 10 =$ the length of the pool.

$\boxed{\text{Length}} \cdot \boxed{\text{Width}} = \boxed{\text{Area of pool}}$

$$(2x-10)x = 1500$$
$$2x^2 - 10x - 1500 = 0$$
$$2(x^2 - 5x - 750) = 0$$
$$2(x+25)(x-30) = 0$$
$$x + 25 = 0 \quad \text{or} \quad x - 30 = 0$$
$$x = -25 \qquad\qquad x = 30$$
(impossible)

The dimensions of the pool are 30 feet by 50 feet. Then let $w =$ the width of the walkway.

$30w + 30w + 50w + 50w + 4w^2 = \boxed{\text{Area of walkway}}$

$$4w^2 + 160w = 516$$
$$4w^2 + 160w - 516 = 0$$
$$4(w^2 + 40w - 129) = 0$$
$$4(w+43)(w-3) = 0$$
$$w + 43 = 0 \quad \text{or} \quad w - 3 = 0$$
$$w = -43 \qquad\qquad w = 3$$
(impossible)

The walkway should be 3 feet wide.

86.
$$A = \frac{h(B+b)}{2}$$
$$44 = \frac{h(18+h)}{2}$$
$$88 = h^2 + 18h$$
$$0 = h^2 + 18h - 88$$
$$0 = (h+22)(h-4)$$
$$h + 22 = 0 \quad \text{or} \quad h - 4 = 0$$
$$h = -22 \qquad\qquad h = 4 \text{ ft}$$
(impossible)

87. $x^2 - 4x + 7$

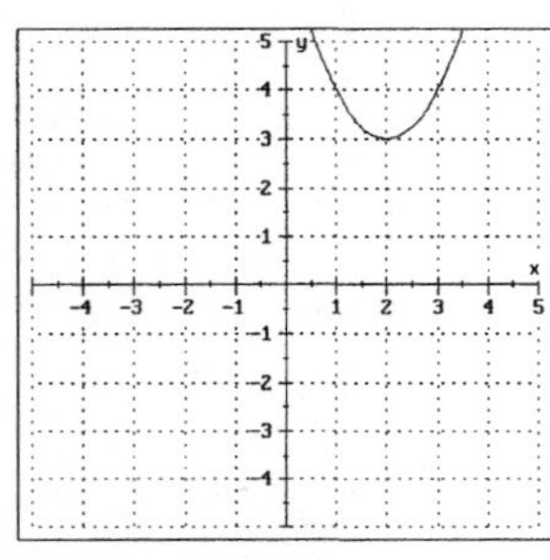

x-intercepts: none $\Rightarrow$ no solution

88. $2x^2 - 7x + 4$

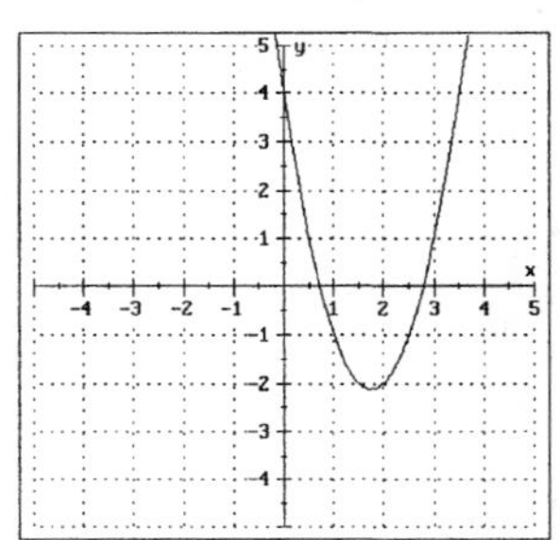

x-intercepts: 2.78, 0.72

89. $-3x^3 - 2x^2 + 5 = 0$

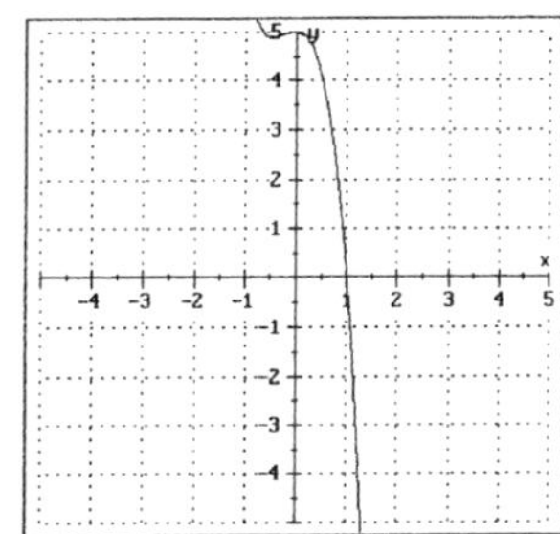

x-intercept: 1.00

90. $-2x^3 - 3x - 5 = 0$

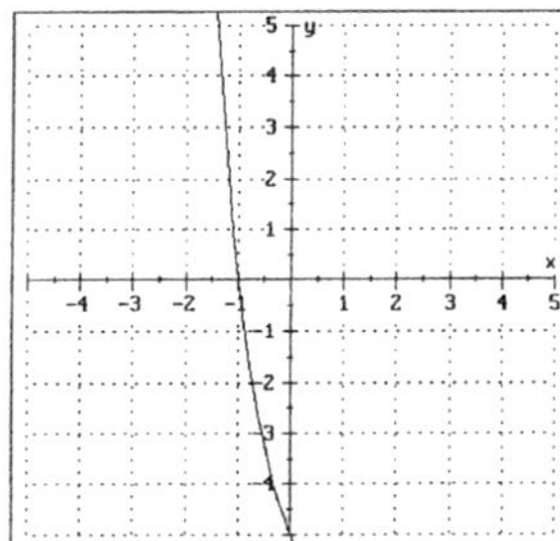

x-intercept: -1.00

91. **Answers may vary.**

92. **Answers may vary.**

93. $(x-3)(x-5) = 0$
$x^2 - 8x + 15 = 0$

94. $(x+2)(x-6) = 0$
$x^2 - 4x - 12 = 0$

95. $(x-0)(x+5) = 0$
$x^2 + 5x = 0$

96. $\left(x - \frac{1}{2}\right)\left(x - \frac{1}{3}\right) = 0$
$x^2 - \frac{1}{3}x - \frac{1}{2}x + \frac{1}{6} = 0$
$6x^2 - 2x - 3x + 1 = 0$
$6x^2 - 5x + 1 = 0$

Chapter 5 Summary (page 342)

1. degree $= 5$

2. degree $= 4 + 4 = 8$

3. $P(0) = -0^2 + 4(0) + 6 = 6$

4. $P(1) = -1^2 + 4(1) + 6$
$= -1 + 4 + 6 = 9$

5. $P(-t) = -(-t)^2 + 4(-t) + 6$
$= -t^2 - 4t + 6$

6. $P(z) = -z^2 + 4z + 6$

7. $f(x) = x^3 - 1$
Shift $f(x) = x^3$ down 1.

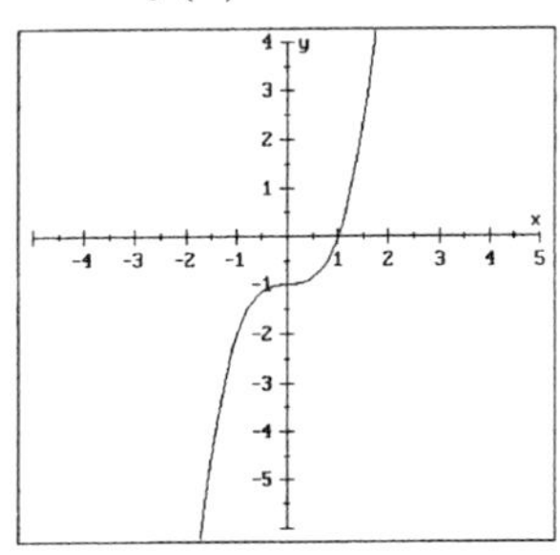

8. $f(x) = x^2 - 2x$

x	y
-1	3
0	0

x	y
1	-1
2	0

x	y
3	3

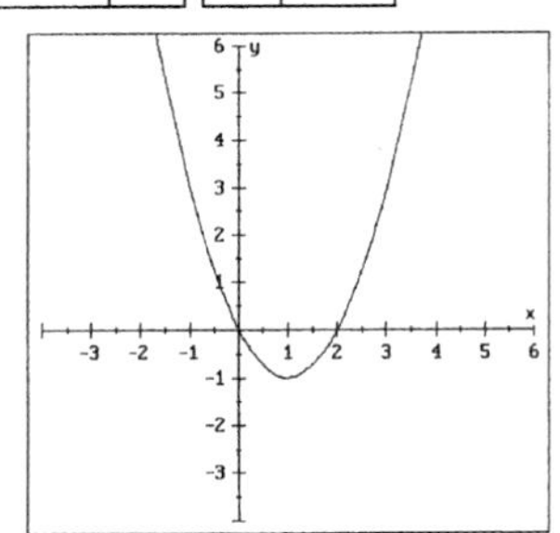

9. $(3x^2+4x+9)+(2x^2-2x+7)=3x^2+4x+9+2x^2-2x+7=5x^2+2x+16$

10. $(4x^3+4x^2+7)-(-2x^3-x-2)=4x^3+4x^2+7+2x^3+x+2=6x^3+4x^2+x+9$

11. $(2x^2-5x+9)-(x^2-3)-(-3x^2+4x-7)$
$=2x^2-5x+9-x^2+3+3x^2-4x+7=4x^2-9x+19$

12. $2(7x^3-6x^2+4x-3)-3(7x^3+6x^2+4x-3)$
$=14x^3-12x^2+8x-6-21x^3-18x^2-12x+9=-7x^3-30x^2-4x+3$

13. $(8a^2b^2)(-2abc)=-16a^3b^3c$

14. $(-3xy^2z)(2xz^3)=-6x^2y^2z^4$

15. $2xy^2(x^3y-4xy^5)=2x^4y^3-8x^2y^7$

16. $a^2b(a^2+2ab+b^2)=a^4b+2a^3b^2+a^2b^3$

17. $(8x-5)(2x+3)=16x^2+24x-10x-15=16x^2+14x-15$

18. $(3x+2)(2x-4)=6x^2-12x+4x-8=6x^2-8x-8$

19. $(2x-5y)^2=(2x-5y)(2x-5y)=4x^2-10xy-10xy+25y^2=4x^2-20xy+25y^2$

20. $(5x-4)(3x-2)=15x^2-10x-12x+8=15x^2-22x+8$

21. $(3x^2-2)(x^2-x+2)=3x^4-3x^3+6x^2-2x^2+2x-4=3x^4-3x^3+4x^2+2x-4$

22. $(2a-b)(a+b)(a-2b)=\left(2a^2+ab-b^2\right)(a-2b)=\left(2a^2+ab-b^2\right)a+\left(2a^2+ab-b^2\right)(-2b)$
$=2a^3+a^2b-ab^2-4a^2b-2ab^2+2b^3$
$=2a^3-3a^2b-3ab^2+2b^3$

23. $4x+8=4(x+2)$

24. $3x^2-6x=3x(x-2)$

25. $5x^2y^3-10xy^2=5xy^2(xy-2)$

26. $7a^4b^2+49a^3b=7a^3b(ab+7)$

27. $-8x^2y^3z^4-12x^4y^3z^2=-4x^2y^3z^2(2z^2+3x^2)$

28. $12a^6b^4c^2+15a^2b^4c^6=3a^2b^4c^2(4a^4+5c^4)$

29. $27x^3y^3z^3+81x^4y^5z^2-90x^2y^3z^7=9x^2y^3z^2(3xz+9x^2y^2-10z^5)$

30. $-36a^5b^4c^2+60a^7b^5c^3-24a^2b^3c^7=-12a^2b^3c^2(3a^3b-5a^5b^2c+2c^5)$

31. $x^{2n}+x^n=x^n(x^{2n-n}+1)=x^n(x^n+1)$

32. $y^{2n}-y^{4n}=y^{2n}(1-y^{4n-2n})=y^{2n}(1-y^{2n})$

33. $x^{-4}-x^{-2}=x^{-2}(x^{-4-(-2)}-1)$
$=x^{-2}(x^{-2}-1)$

34. $a^6+1=a^{-3}(a^{6-(-3)}+a^{0-(-3)})$
$=a^{-3}(a^9+a^3)$

35. $5x^2(x+y)^3-15x^3(x+y)^4=5x^2(x+y)^3[1-3x(x+y)]=5x^2(x+y)^3(1-3x^2-3xy)$

36. $-49a^3b^2(a-b)^4+63a^2b^4(a-b)^3=-7a^2b^2(a-b)^3[7a(a-b)-9b^2]$
$=-7a^2b^2(a-b)^3(7a^2-7ab-9b^2)$

37. $xy+2y+4x+8=y(x+2)+4(x+2)$
$=(x+2)(y+4)$

38. $ac+bc+3a+3b=c(a+b)+3(a+b)$
$=(a+b)(c+3)$

39. $x^4+4y+4x^2+x^2y=x^4+x^2y+4x^2+4y=x^2(x^2+y)+4(x^2+y)=(x^2+y)(x^2+4)$

40. $a^5+b^2c+a^2c+a^3b^2=a^5+a^3b^2+a^2c+b^2c=a^3(a^2+b^2)+c(a^2+b^2)=(a^2+b^2)(a^3+c)$

41.
$$S=2wh+2wl+2lh$$
$$S-2wl=2wh+2lh$$
$$S-2wl=h(2w+2l)$$
$$\frac{S-2wl}{2w+2l}=h\text{, or } h=\frac{S-2wl}{2w+2l}$$

42.
$$S=2wh+2wl+2lh$$
$$S-2wh=2wl+2lh$$
$$S-2wh=l(2w+2h)$$
$$\frac{S-2wh}{2w+2h}=l\text{, or } l=\frac{S-2wh}{2w+2h}$$

43. $z^2-16=z^2-4^2=(z+4)(z-4)$

44. $y^2-121=y^2-11^2=(y+11)(y-11)$

45. $x^2y^4-64z^6=(xy^2)^2-(8z^3)^2$
$=(xy^2+8z^3)(xy^2-8z^3)$

46. $a^2b^2+c^2 \Rightarrow$ prime

75. $(x+z)^2-t^2=[(x+z)+t][(x+z)-t]$
$=(x+z+t)(x+z-t)$

48. $c^2-(a+b)^2=[c+(a+b)][c-(a+b)]$
$=(c+a+b)(c-a-b)$

49. $2x^4-98=2(x^4-49)$
$=2(x^2+7)(x^2-7)$

50. $3x^6-300x^2=3x^2(x^4-100)$
$=3x^2(x^2+10)(x^2-10)$

51. $x^3+343=x^3+7^3$
$=(x+7)(x^2-7x+49)$

52. $a^3-125=a^3-5^3$
$=(a-5)(a^2+5a+25)$

53. $8y^3-512=8(y^3-64)=8(y^3-4^3)=8(y-4)(y^2+4y+16)$

54. $4x^3y+108yz^3=4y(x^3+27z^3)=4y\left[x^3+(3z)^3\right]=4y(x+3z)(x^2-3xz+9z^2)$

55. $x^2+10x+25=(x+5)(x+5)$

56. $a^2-14a+49=(a-7)(a-7)$

57. $y^2+21y+20=(y+20)(y+1)$

58. $z^2-11z+30=(z-5)(z-6)$

59. $-x^2-3x+28=-(x^2+3x-28)$
$=-(x+7)(x-4)$

60. $y^2-5y-24=(y-8)(y+3)$

61. $4a^2-5a+1=(4a-1)(a-1)$

62. $3b^2+2b+1 \Rightarrow$ prime

63. $7x^2+x+2 \Rightarrow$ prime

64. $-15x^2+14x+8=-(15x^2-14x-8)$
$=-(5x+2)(3x-4)$

65. $y^3 + y^2 - 2y = y(y^2 + y - 2)$
$= y(y+2)(y-1)$

66. $2a^4 + 4a^3 - 6a^2 = 2a^2(a^2 + 2a - 3)$
$= 2a^2(a+3)(a-1)$

67. $-3x^2 - 9x - 6 = -3(x^2 + 3x + 2)$
$= -3(x+2)(x+1)$

68. $8x^2 - 4x - 24 = 4(2x^2 - x - 6)$
$= 4(2x+3)(x-2)$

69. $15x^2 - 57xy - 12y^2 = 3(5x^2 - 19xy - 4y^2)$
$= 3(5x+y)(x-4y)$

70. $30x^2 + 65xy + 10y^2 = 5(6x^2 + 13xy + 2y^2)$
$= 5(6x+y)(x+2y)$

71. $24x^2 - 23xy - 12y^2 = (8x+3y)(3x-4y)$

72. $14x^2 + 13xy - 12y^2 = (2x+3y)(7x-4y)$

73. $x^3 + 5x^2 - 6x = x(x^2 + 5x - 6)$
$= x(x-1)(x+6)$

74. $3x^2y - 12xy - 63y = 3y(x^2 - 4x - 21)$
$= 3y(x-7)(x+3)$

75. $z^2 - 4 + zx - 2x = (z^2 - 4) + zx - 2x = (z+2)(z-2) + x(z-2) = (z-2)(z+2+x)$

76. $x^2 + 2x + 1 - p^2 = (x^2 + 2x + 1) - p^2 = (x+1)^2 - p^2 = (x+1+p)(x+1-p)$

77. $x^2 + 4x + 4 - 4p^4 = (x^2 + 4x + 4) - 4p^4 = (x+2)^2 - (2p^2)^2$
$= (x+2+2p^2)(x+2-2p^2)$

78. $y^2 + 3y + 2 + 2x + xy = (y^2 + 3y + 2) + 2x + xy = (y+1)(y+2) + x(2+y)$
$= (y+2)(y+1+x)$

79. $x^{2m} + 2x^m - 3 = (x^m + 3)(x^m - 1)$

80. $x^{-2} - x^{-1} - 2 = (x^{-1} - 2)(x^{-1} + 1)$
$= \left(\frac{1}{x} - 2\right)\left(\frac{1}{x} + 1\right)$

81.
$4x^2 - 3x = 0$
$x(4x-3) = 0$
$x = 0$ **or** $4x - 3 = 0$
$4x = 3$
$x = \frac{3}{4}$

82.
$x^2 - 36 = 0$
$(x+6)(x-6) = 0$
$x + 6 = 0$ **or** $x - 6 = 0$
$x = -6$ $\quad x = 6$

83.
$12x^2 + 4x - 5 = 0$
$(2x-1)(6x+5) = 0$
$2x - 1 = 0$ **or** $6x + 5 = 0$
$2x = 1$ $\quad 6x = -5$
$x = \frac{1}{2}$ $\quad x = -\frac{5}{6}$

84.
$7y^2 - 37y + 10 = 0$
$(7y-2)(y-5) = 0$
$7y - 2 = 0$ **or** $y - 5 = 0$
$7y = 2$ $\quad y = 5$
$y = \frac{2}{7}$

85.
$$t^2(15t-2)=8t$$
$$15t^3-2t^2=8t$$
$$15t^3-2t^2-8t=0$$
$$t(15t^2-2t-8)=0$$
$$t(3t+2)(5t-4)=0$$
$t=0$ **or** $3t+2=0$ **or** $5t-4=0$
$3t=-2$ $\quad$ $5t=4$
$t=-\frac{2}{3}$ $\quad$ $t=\frac{4}{5}$

86.
$$3u^3=u(19u+14)$$
$$3u^3=19u^2+14u$$
$$3u^3-19u^2-14u=0$$
$$u(3u^2-19u-14)=0$$
$$u(3u+2)(u-7)=0$$
$u=0$ **or** $3u+2=0$ **or** $u-7=0$
$3u=-2$ $\quad$ $u=7$
$u=-\frac{2}{3}$

87. Let h = the height and $h+3$ = the width
$$lwh=V$$
$$12(h+3)h=840$$
$$12h^2+36h-840=0$$
$$3(4h^2+12h-280)=0$$
$$3(4h+40)(h-7)=0$$
$4h+40=0$ **or** $h-7=0$
$4h=-40$ $\quad$ $h=7$
$h=-10$
(impossible) $\quad$ The height is 7 cm.

88. Let x = one side of the base and $x+3$ = the other side of the base.
$$\frac{Bh}{3}=V$$
$$\frac{x(x+3)9}{3}=1020$$
$$\frac{9x^2+27x}{3}=1020$$
$$9x^2+27x=3060$$
$$9x^2+27x-3060=0$$
$$9(x^2+3x-340)=0$$
$$9(x+20)(x-17)=0$$
$x+20=0$ **or** $x-17=0$
$x=-20$ $\quad$ $x=17$
(impossible) $\quad$ The dimensions are 17 m by 20 m.

Chapter 5 Test (page 347)

1. degree $=5$

2. degree $=9+4=13$

3. $P(2)=-3(2)^2+2(2)-1=-3(4)+4-1=-12+4-1=-9$

4. $P(-1)=-3(-1)^2+2(-1)-1=-3(1)-2-1=-3-2-1=-6$

5. $f(x)=x^2+2x$

x	y
-3	3
-2	0
-1	-1

x	y
0	0
1	3

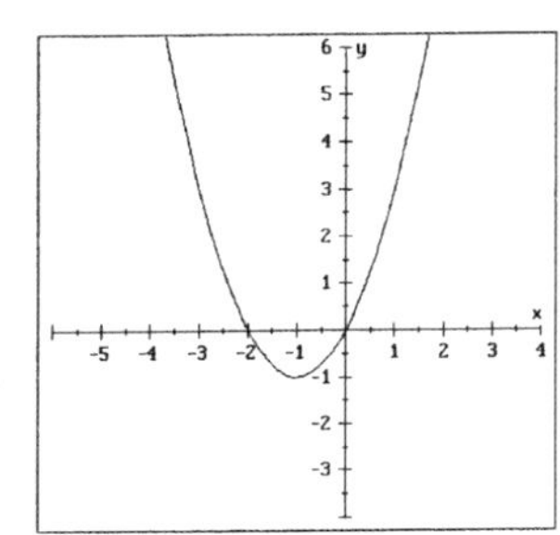

6. $(2y^2+4y+3)+(3y^2-3y-4)=2y^2+4y+3+3y^2-3y-4=5y^2+y-1$

7. $(-3u^2 + 2u - 7) - (u^2 + 7) = -3u^2 + 2u - 7 - u^2 - 7 = -4u^2 + 2u - 14$

8. $3(2a^2 - 4a + 2) - 4(-a^2 - 3a - 4) = 6a^2 - 12a + 6 + 4a^2 + 12a + 16 = 10a^2 + 22$

9. $-2(2x^2 - 2) + 3(x^2 + 5x - 2) = -4x^2 + 4 + 3x^2 + 15x - 6 = -x^2 + 15x - 2$

10. $(3x^3y^2z)(-2xy^{-1}z^3) = -6x^4yz^4$

11. $-5a^2b(3ab^3 - 2ab^4) = -15a^3b^4 + 10a^3b^5$

12. $(z+4)(z-4) = z^2 - 4z + 4z - 16$
$= z^2 - 16$

13. $(3x-2)(4x+3) = 12x^2 + 9x - 8x - 6$
$= 12x^2 + x - 6$

14. $(u-v)^2(2u+v) = (u^2 - 2uv + v^2)(2u+v) = (u^2 - 2uv + v^2)(2u) + (u^2 - 2uv + v^2)(v)$
$= 2u^3 - 4u^2v + 2uv^2 + u^2v - 2uv^2 + v^3$
$= 2u^3 - 3u^2v + v^3$

15. $3xy^2 + 6x^2y = 3xy(y + 2x)$

16. $12a^3b^2c - 3a^2b^2c^2 + 6abc^3 = 3abc(4a^2b - abc + 2c^2)$

17. $x^2y^{n+2} + y^n = y^n(x^2y^{n+2-n} + y^{n-n}) = y^n(x^2y^2 + 1)$

18. $a^nb^n - ab^{-n} = b^n(a^nb^{n-n} - ab^{-n-n}) = b^n(a^n - ab^{-2n})$

19. $(u-v)r + (u-v)s = (u-v)(r+s)$

20. $ax - xy + ay - y^2 = x(a-y) + y(a-y)$
$= (a-y)(x+y)$

21. $x^2 - 49 = x^2 - 7^2 = (x+7)(x-7)$

22. $2x^2 - 32 = 2(x^2 - 16) = 2(x+4)(x-4)$

23. $4y^4 - 64 = 4(y^4 - 16) = 4(y^2+4)(y^2-4) = 4(y^2+4)(y+2)(y-2)$

24. $b^3 + 125 = b^3 + 5^3$
$= (b+5)(b^2 - 5b + 25)$

25. $b^3 - 27 = b^3 - 3^3$
$= (b-3)(b^2 + 3b + 9)$

26. $3u^3 - 24 = 3(u^3 - 8) = 3(u^3 - 2^3) = 3(u-2)(u^2 + 2u + 4)$

27. $a^2 - 5a - 6 = (a-6)(a+1)$

28. $6b^2 + b - 2 = (3b+2)(2b-1)$

29. $6u^2 + 9u - 6 = 3(2u^2 + 3u - 2)$
$= 3(2u-1)(u+2)$

30. $20r^2 - 15r - 5 = 5(4r^2 - 3r - 1)$
$= 5(4r+1)(r-1)$

31. $x^{2n} + 2x^n + 1 = (x^n+1)(x^n+1) = (x^n+1)^2$

32. $x^2 + 6x + 9 - y^2 = (x^2 + 6x + 9) - y^2 = (x+3)^2 - y^2 = (x+3+y)(x+3-y)$

33.
$$\begin{aligned} r_1r_2 - r_2r &= r_1r \\ r_1r_2 &= r_1r + r_2r \\ r_1r_2 &= r(r_1 + r_2) \\ \frac{r_1r_2}{r_1 + r_2} &= r, \text{ or } r = \frac{r_1r_2}{r_1 + r_2} \end{aligned}$$

34.
$$\begin{aligned} x^2 - 5x - 6 &= 0 \\ (x+1)(x-6) &= 0 \\ x + 1 = 0 \quad &\textbf{or} \quad x - 6 = 0 \\ x = -1 \quad & \quad\quad x = 6 \end{aligned}$$

35. Let x and $x + 1$ represent the integers.
$$\begin{aligned} x(x+1) &= 156 \\ x^2 + x - 156 &= 0 \\ (x+13)(x-12) &= 0 \\ x + 13 = 0 \quad &\textbf{or} \quad x - 12 = 0 \\ x = -13 \quad & \quad\quad x = 12 \end{aligned}$$
(impossible)

The integers are 12 and 13 (sum of 25).

36.
$$\begin{aligned} \boxed{\text{Area of border}} &= 70 \\ 2w(1) + 2w(1) + w(1) + w(1) + 4(1) &= 70 \\ 6w + 4 &= 70 \\ 6w &= 66 \\ w &= 11 \end{aligned}$$
The dimensions are 11 ft by 22 ft.

Exercise 6.1 (page 356)

1. $3x^2 - 9x = 3x(x-3)$

2. $6t^2 - 5t - 6 = (3t+2)(2t-3)$

3. $27x^6 + 64y^3 = (3x^2)^3 + (4y)^3 = (3x^2 + 4y)(9x^4 - 12x^2y + 16y^2)$

4. $x^2 + ax + 2x + 2a = x(x+a) + 2(x+a) = (x+a)(x+2)$

5. rational

6. 0

7. asymptote

8. $ad = bc$

9. a

10. 1; 0

11. $\frac{a}{b}$; 0

12. -1

13.
$$\begin{aligned} t = f(r) &= \frac{600}{r} \\ f(30) &= \frac{600}{30} \\ &= 20 \end{aligned}$$
$t = 20$ hours

14.
$$\begin{aligned} t = f(r) &= \frac{600}{r} \\ f(40) &= \frac{600}{40} \\ &= 15 \end{aligned}$$
$t = 15$ hours

15.
$$\begin{aligned} t = f(r) &= \frac{600}{r} \\ f(50) &= \frac{600}{50} \\ &= 12 \end{aligned}$$
$t = 12$ hours

16.
$$\begin{aligned} t = f(r) &= \frac{600}{r} \\ f(60) &= \frac{600}{60} \\ &= 10 \end{aligned}$$
$t = 10$ hours

17.
$$\begin{aligned} C = f(p) &= \frac{50{,}000p}{100 - p} \\ f(10) &= \frac{50{,}000(10)}{100 - 10} \\ &= \frac{500{,}000}{90} \\ &\approx 5555.56 \end{aligned}$$
$C = \$5555.56$

18.
$$\begin{aligned} C = f(p) &= \frac{50{,}000p}{100 - p} \\ f(30) &= \frac{50{,}000(30)}{100 - 30} \\ &= \frac{1{,}500{,}000}{70} \\ &\approx 21{,}428.57 \end{aligned}$$
$C = \$21{,}428.57$

19.
$$\begin{aligned} C = f(p) &= \frac{50{,}000p}{100 - p} \\ f(50) &= \frac{50{,}000(50)}{100 - 50} \\ &= \frac{2{,}500{,}000}{50} \\ &= 50{,}000.00 \end{aligned}$$
$C = \$50{,}000$

20. $C = f(p) = \frac{50{,}000p}{100 - p}$; $f(80) = \frac{50{,}000(80)}{100 - 80} = \frac{4{,}000{,}000}{20} = 200{,}000.00$; $C = \$200{,}000$

21. $c = f(x) = 1.25x + 700$

22. Refer to **#21**. $\overline{c} = f(x) = \dfrac{1.25x + 700}{x}$

23. Refer to **#21**. $c = f(500) = 1.25(500) + 700 = \$1325$

24. Refer to **#22**. $\overline{c} = f(500) = \dfrac{1.25(500) + 700}{500} = \dfrac{1325}{500} = \$2.65$

25. Refer to **#22**. $\overline{c} = f(1000) = \dfrac{1.25(1000) + 700}{1000} = \dfrac{1950}{1000} = \$1.95$

26. Refer to **#22**. $\overline{c} = f(2000) = \dfrac{1.25(2000) + 700}{2000} = \dfrac{3200}{2000} = \$1.60$

27. $c = f(n) = 0.09n + 7.50$

28. Refer to **#27**. $\overline{c} = f(n) = \dfrac{0.09n + 7.50}{n}$

29. Refer to **#27**. $c = f(775) = 0.09(775) + 7.50 = 69.75 + 7.50 = \$77.25$

30. Refer to **#28**. $\overline{c} = f(775) = \dfrac{0.09(775) + 7.50}{775} = \dfrac{77.25}{775} \approx \$0.0997 = 9.97¢$

31. Refer to **#28**. $\overline{c} = f(1000) = \dfrac{0.09(1000) + 7.50}{1000} = \dfrac{97.50}{1000} = \$0.0975 = 9.75¢$

32. Refer to **#28**. $\overline{c} = f(1200) = \dfrac{0.09(1200) + 7.50}{1200} = \dfrac{115.5}{1200} = \$0.09625 = 9.625¢$

33. $f(t) = \dfrac{t^2 + 2t}{2t + 2}$

$f(15) = \dfrac{15^2 + 2(15)}{2(15) + 2}$

$= \dfrac{225 + 30}{30 + 2} = \dfrac{255}{32} \approx 7.96875$

It will take them almost 8 days.

34. $f(t) = \dfrac{t^2 + 2t}{2t + 2}$

Crew 2: 20 days $\Rightarrow$ Crew 1: 18 days

$f(18) = \dfrac{18^2 + 2(18)}{2(18) + 2}$

$= \dfrac{324 + 36}{36 + 2} = \dfrac{360}{38} \approx 9.4737$

It will take them about 9.5 days.

35. $f(t) = \dfrac{t^2 + 3t}{2t + 3}$

Small pipe: 7 hrs $\Rightarrow$ Large pipe: 4 hrs

$f(4) = \dfrac{4^2 + 3(4)}{2(4) + 3}$

$= \dfrac{16 + 12}{8 + 3} = \dfrac{28}{11} = 2.5455$

It will take the pipes about 2.55 hours.

36. $f(t) = \dfrac{t^2 + 3t}{2t + 3}$

$f(8) = \dfrac{8^2 + 3(8)}{2(8) + 3}$

$= \dfrac{64 + 24}{16 + 3} = \dfrac{88}{19} = 4.6316$

It will take the pipes about 4.63 hours.

37. $f(x) = \dfrac{x}{x-2}$

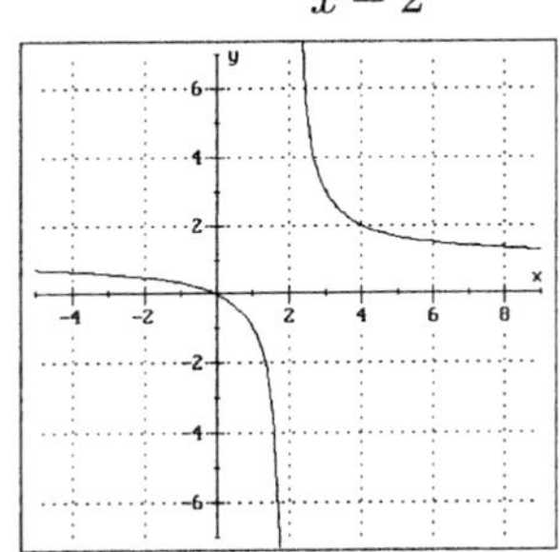

domain: $(-\infty, 2) \cup (2, \infty)$

38. $f(x) = \dfrac{x+2}{x}$

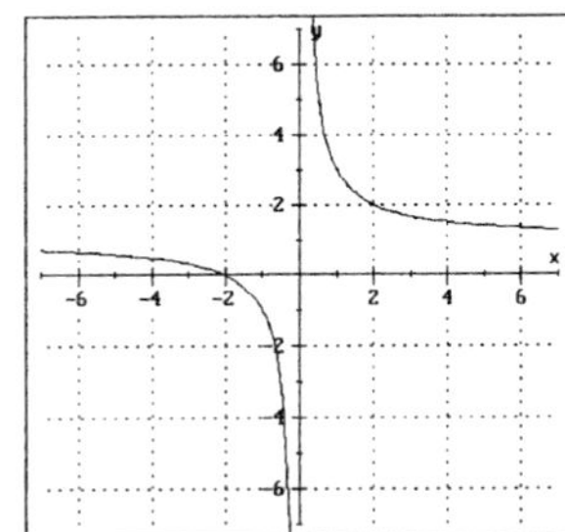

domain: $(-\infty, 0) \cup (0, \infty)$

39. $f(x) = \dfrac{x+1}{x^2-4}$

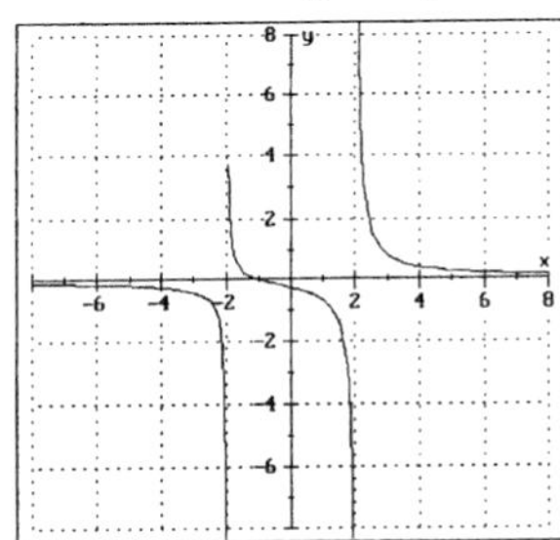

domain: $(-\infty, -2) \cup (-2, 2) \cup (2, \infty)$

40. $f(x) = \dfrac{x-2}{x^2-3x-4}$

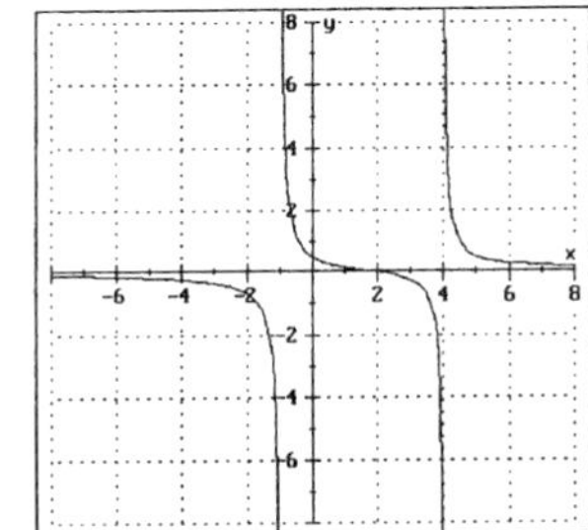

domain: $(-\infty, -1) \cup (-1, 4) \cup (4, \infty)$

41. $\dfrac{12}{18} = \dfrac{\cancel{6} \cdot 2}{\cancel{6} \cdot 3} = \dfrac{2}{3}$

42. $\dfrac{25}{55} = \dfrac{\cancel{5} \cdot 5}{\cancel{5} \cdot 11} = \dfrac{5}{11}$

43. $-\dfrac{112}{36} = -\dfrac{\cancel{4} \cdot 28}{\cancel{4} \cdot 9} = -\dfrac{28}{9}$

44. $-\dfrac{49}{21} = -\dfrac{\cancel{7} \cdot 7}{\cancel{7} \cdot 3} = -\dfrac{7}{3}$

45. $\dfrac{288}{312} = \dfrac{\cancel{24} \cdot 12}{\cancel{24} \cdot 13} = \dfrac{12}{13}$

46. $\dfrac{144}{72} = \dfrac{\cancel{72} \cdot 2}{\cancel{72} \cdot 1} = \dfrac{2}{1} = 2$

47. $-\dfrac{244}{74} = -\dfrac{\cancel{2} \cdot 122}{\cancel{2} \cdot 37} = -\dfrac{122}{37}$

48. $-\dfrac{512}{236} = -\dfrac{\cancel{4} \cdot 128}{\cancel{4} \cdot 59} = -\dfrac{128}{59}$

49. $\dfrac{12x^3}{3x} = \dfrac{12}{3} \cdot \dfrac{x^3}{x} = 4x^2$

50. $-\dfrac{15a^2}{25a^3} = -\dfrac{15}{25} \cdot \dfrac{a^2}{a^3} = -\dfrac{3}{5a}$

51. $\dfrac{-24x^3y^4}{18x^4y^3} = -\dfrac{24}{18} \cdot \dfrac{x^3y^4}{x^4y^3} = -\dfrac{4y}{3x}$

52. $\dfrac{15a^5b^4}{21b^3c^2} = \dfrac{15}{21} \cdot \dfrac{a^5b^4}{b^3c^2} = \dfrac{5a^5b}{7c^2}$

53. $\dfrac{(3x^3)^2}{9x^4} = \dfrac{9x^6}{9x^4} = x^2$

54. $\dfrac{8(x^2y^3)^3}{2(xy^2)^2} = \dfrac{8x^6y^9}{2x^2y^4} = 4x^4y^5$

55. $-\dfrac{11x(x-y)}{22(x-y)} = -\dfrac{x}{2}$

56. $\dfrac{x(x-2)^2}{(x-2)^3} = \dfrac{x}{x-2}$

57. $\dfrac{9y^2(y-z)}{21y(y-z)^2} = \dfrac{3y}{7(y-z)}$

58. $\dfrac{-3ab^2(a-b)}{9ab(b-a)} = \dfrac{-3ab^2(a-b)}{-9ab(a-b)} = \dfrac{b}{3}$

59. $\dfrac{(a-b)(c-d)}{(c-d)(a-b)} = 1$

60. $\dfrac{(p+q)(p-r)}{(r-p)(p+q)} = \dfrac{(p+q)(p-r)}{-(p-r)(p+q)} = -1$

61. $\dfrac{x+y}{x^2-y^2} = \dfrac{x+y}{(x+y)(x-y)} = \dfrac{1}{x-y}$

62. $\dfrac{x-y}{x^2-y^2} = \dfrac{x-y}{(x+y)(x-y)} = \dfrac{1}{x+y}$

63. $\dfrac{x^2-x-2}{2x+4} = \dfrac{(x-2)(x+1)}{2(x+2)}$; lowest terms

64. $\dfrac{y-xy}{xy-x} = \dfrac{y(1-x)}{x(y-1)}$; lowest terms

65. $\dfrac{5x-10}{x^2-4x+4} = \dfrac{5(x-2)}{(x-2)(x-2)} = \dfrac{5}{x-2}$

66. $\dfrac{2a^2+5a+3}{6a+9} = \dfrac{(2a+3)(a+1)}{3(2a+3)} = \dfrac{a+1}{3}$

67. $\dfrac{12-3x^2}{x^2-x-2} = \dfrac{3(4-x^2)}{(x-2)(x+1)} = \dfrac{3(2+x)(2-x)}{(x-2)(x+1)} = \dfrac{-3(x+2)(x-2)}{(x-2)(x+1)} = \dfrac{-3(x+2)}{x+1}$

68. $\dfrac{x^2+2x-15}{x^2-25} = \dfrac{(x+5)(x-3)}{(x+5)(x-5)} = \dfrac{x-3}{x-5}$

69. $\dfrac{3x+6y}{x+2y} = \dfrac{3(x+2y)}{x+2y} = 3$

70. $\dfrac{x^2+y^2}{x+y} \Rightarrow$ lowest terms

71. $\dfrac{x^3+8}{x^2-2x+4} = \dfrac{(x+2)(x^2-2x+4)}{x^2-2x+4}$
$= x+2$

72. $\dfrac{x^2+3x+9}{x^3-27} = \dfrac{x^2+3x+9}{(x-3)(x^2+3x+9)} = \dfrac{1}{x-3}$

73. $\dfrac{x^2+2x+1}{x^2+4x+3} = \dfrac{(x+1)(x+1)}{(x+1)(x+3)} = \dfrac{x+1}{x+3}$

74. $\dfrac{6x^2+x-2}{8x^2+2x-3} = \dfrac{(3x+2)(2x-1)}{(4x+3)(2x-1)} = \dfrac{3x+2}{4x+3}$

75. $\dfrac{3m-6n}{3n-6m} = \dfrac{3(m-2n)}{3(n-2m)} = \dfrac{m-2n}{n-2m}$

76. $\dfrac{ax+by+ay+bx}{a^2-b^2} = \dfrac{ax+ay+bx+by}{a^2-b^2} = \dfrac{a(x+y)+b(x+y)}{a^2-b^2} = \dfrac{(x+y)(a+b)}{(a+b)(a-b)} = \dfrac{x+y}{a-b}$

77. $\dfrac{4x^2+24x+32}{16x^2+8x-48} = \dfrac{4(x^2+6x+8)}{8(2x^2+x-6)} = \dfrac{4(x+4)(x+2)}{8(2x-3)(x+2)} = \dfrac{x+4}{2(2x-3)}$

78. $\dfrac{a^2-4}{a^3-8} = \dfrac{(a+2)(a-2)}{(a-2)(a^2+2a+4)} = \dfrac{a+2}{a^2+2a+4}$

79. $\dfrac{3x^2-3y^2}{x^2+2y+2x+yx} = \dfrac{3(x^2-y^2)}{x^2+2x+yx+2y} = \dfrac{3(x+y)(x-y)}{(x+2)(x+y)} = \dfrac{3(x-y)}{x+2}$

80. $\dfrac{x^2+x-30}{x^2-x-20} = \dfrac{(x+6)(x-5)}{(x-5)(x+4)} = \dfrac{x+6}{x+4}$

81. $\dfrac{4x^2+8x+3}{6+x-2x^2} = \dfrac{4x^2+8x+3}{-(2x^2-x-6)} = \dfrac{(2x+3)(2x+1)}{-(2x+3)(x-2)} = \dfrac{2x+1}{-(x-2)} = \dfrac{2x+1}{2-x}$

82. $\dfrac{6x^2+13x+6}{6-5x-6x^2} = \dfrac{6x^2+13x+6}{-(6x^2+5x-6)} = \dfrac{(3x+2)(2x+3)}{-(3x-2)(2x+3)} = \dfrac{3x+2}{-(3x-2)} = -\dfrac{3x+2}{3x-2}$

83. $\dfrac{a^3+27}{4a^2-36} = \dfrac{a^3+27}{4(a^2-9)} = \dfrac{(a+3)(a^2-3a+9)}{4(a+3)(a-3)} = \dfrac{a^2-3a+9}{4(a-3)}$

84. $\dfrac{a-b}{b^2-a^2} = \dfrac{a-b}{(b+a)(b-a)} = \dfrac{a-b}{-(a+b)(a-b)} = -\dfrac{1}{a+b}$

85. $\dfrac{2x^2-3x-9}{2x^2+3x-9} = \dfrac{(2x+3)(x-3)}{(2x-3)(x+3)}$

$\Rightarrow$ lowest terms

86. $\dfrac{6x^2-7x-5}{2x^2+5x+2} = \dfrac{(3x-5)(2x+1)}{(2x+1)(x+2)}$

$= \dfrac{3x-5}{x+2}$

87. $\dfrac{(m+n)^3}{m^2+2mn+n^2} = \dfrac{(m+n)^3}{(m+n)(m+n)} = \dfrac{(m+n)^3}{(m+n)^2} = m+n$

88. $\dfrac{x^3-27}{3x^2-8x-3} = \dfrac{(x-3)(x^2+3x+9)}{(3x+1)(x-3)} = \dfrac{x^2+3x+9}{3x+1}$

89. $\dfrac{m^3-mn^2}{mn^2+m^2n-2m^3} = \dfrac{m(m^2-n^2)}{m(n^2+mn-2m^2)} = \dfrac{m(m+n)(m-n)}{m(n+2m)(n-m)} = \dfrac{m(m+n)(m-n)}{-m(n+2m)(m-n)}$

$= -\dfrac{m+n}{2m+n}$

90. $\dfrac{p^3+p^2q-2pq^2}{pq^2+p^2q-2p^3} = \dfrac{p(p^2+pq-2q^2)}{p(q^2+pq-2p^2)} = \dfrac{p(p+2q)(p-q)}{p(q+2p)(q-p)} = \dfrac{p(p+2q)(p-q)}{-p(2p+q)(p-q)} = -\dfrac{p+2q}{2p+q}$

$= \dfrac{-p-2q}{q+2p}$

91. $\dfrac{x^4-y^4}{(x^2+2xy+y^2)(x^2+y^2)} = \dfrac{(x^2+y^2)(x^2-y^2)}{(x+y)^2(x^2+y^2)} = \dfrac{(x^2+y^2)(x+y)(x-y)}{(x+y)^2(x^2+y^2)} = \dfrac{x-y}{x+y}$

92. $\dfrac{-4x-4+3x^2}{4x^2-2-7x} = \dfrac{3x^2-4x-4}{4x^2-7x-2} = \dfrac{(3x+2)(x-2)}{(4x+1)(x-2)} = \dfrac{3x+2}{4x+1}$

93. $\dfrac{4a^2-9b^2}{2a^2-ab-6b^2} = \dfrac{(2a+3b)(2a-3b)}{(2a+3b)(a-2b)} = \dfrac{2a-3b}{a-2b}$

94. $$\frac{x^2+2xy}{x+2y+x^2-4y^2}=\frac{x(x+2y)}{(x+2y)1+(x+2y)(x-2y)}=\frac{x(x+2y)}{(x+2y)(1+x-2y)}=\frac{x}{1+x-2y}$$

95. $$\begin{aligned}\frac{x-y}{x^3-y^3-x+y}=\frac{x-y}{(x^3-y^3)-(x-y)}&=\frac{x-y}{(x-y)(x^2+xy+y^2)-1(x-y)}\\&=\frac{x-y}{(x-y)(x^2+xy+y^2-1)}\\&=\frac{1}{x^2+xy+y^2-1}\end{aligned}$$

96. $$\begin{aligned}\frac{2x^2+2x-12}{x^3+3x^2-4x-12}=\frac{2(x^2+x-6)}{x^2(x+3)-4(x+3)}=\frac{2(x+3)(x-2)}{(x+3)(x^2-4)}&=\frac{2(x+3)(x-2)}{(x+3)(x+2)(x-2)}\\&=\frac{2}{x+2}\end{aligned}$$

97. $$\frac{px-py+qx-qy}{px+qx+py+qy}=\frac{p(x-y)+q(x-y)}{x(p+q)+y(p+q)}=\frac{(x-y)(p+q)}{(p+q)(x+y)}=\frac{x-y}{x+y}$$

98. $$\frac{6xy-4x-9y+6}{6y^2-13y+6}=\frac{2x(3y-2)-3(3y-2)}{(3y-2)(2y-3)}=\frac{(3y-2)(2x-3)}{(3y-2)(2y-3)}=\frac{2x-3}{2y-3}$$

99. $$\frac{(x^2-1)(x+1)}{(x^2-2x+1)^2}=\frac{(x+1)(x-1)(x+1)}{[(x-1)^2]^2}=\frac{(x+1)^2(x-1)}{(x-1)^4}=\frac{(x+1)^2}{(x-1)^3}$$

100. $$\frac{(x^2+2x+1)(x^2-2x+1)}{(x^2-1)^2}=\frac{(x+1)^2(x-1)^2}{(x+1)(x-1)(x+1)(x-1)}=1$$

101. $$\begin{aligned}\frac{(2x^2+3xy+y^2)(3a+b)}{(x+y)(2xy+2bx+y^2+by)}=\frac{(2x+y)(x+y)(3a+b)}{(x+y)[2x(y+b)+y(y+b)]}&=\frac{(2x+y)(x+y)(3a+b)}{(x+y)(y+b)(2x+y)}\\&=\frac{3a+b}{y+b}\end{aligned}$$

102. $$\begin{aligned}\frac{(x-1)(6ax+9x+4a+6)}{(3x+2)(2ax-2a+3x-3)}&=\frac{(x-1)[3x(2a+3)+2(2a+3)]}{(3x+2)[2a(x-1)+3(x-1)]}\\&=\frac{(x-1)(2a+3)(3x+2)}{(3x+2)(x-1)(2a+3)}=1\end{aligned}$$

103. a. $$\begin{aligned}C=f(p)&=\frac{50{,}000p}{100-p}\\f(40)&=\frac{50{,}000(40)}{100-40}\\&=\frac{2{,}000{,}000}{60}\\&\approx 33333.33\end{aligned}$$
$C=\$33333.33$

b. $$\begin{aligned}C=f(p)&=\frac{50{,}000p}{100-p}\\f(70)&=\frac{50{,}000(70)}{100-70}\\&=\frac{3{,}500{,}000}{30}\\&\approx 116666.67\end{aligned}$$
$C=\$116666.67$

104. a. $\bar{c} = f(700) = \frac{1.25(700)+800}{700} = \frac{1675}{700} \approx \$2.39$

b. $\bar{c} = f(2500) = \frac{1.25(2500)+800}{2500} = \frac{3925}{2500} = \$1.57$

105. Answers may vary.

106. Answers may vary.

107. $\frac{a-3b}{2b-a} = \frac{-(3b-a)}{-(a-2b)} = \frac{3b-a}{a-2b}$. The two answers are the same.

108. The work is not correct. You must treat the factor of 3 as a **common** factor of the numerator. Thus, to simplify the fraction, you must factor 3 from the **entire** numerator:

$$\frac{3x^2+6}{3y} = \frac{3(x^2+\mathbf{2})}{3y} = \frac{x^2+2}{y}$$

109. You can divide out the 4's in parts a and d.

110. You can divide out the 3's in parts a and c.

Exercise 6.2 (page 365)

1. $-2a^2(3a^3-a^2) = -6a^5+2a^4$

2. $(2t-1)^2 = (2t-1)(2t-1) = 4t^2-4t+1$

3. $(m^n+2)(m^n-2) = m^{2n}-4$

4. $(3b^{-n}+c)(b^{-n}-c) = 3b^{-2n}-2b^{-n}c-c^2$
$= \frac{3}{b^{2n}} - \frac{2c}{b^n} - c^2$

5. $\frac{ac}{bd}$

6. $\frac{ad}{bc}$

7. 0

8. 1

9. $\frac{3}{4}\cdot\frac{5}{3}\cdot\frac{8}{7} = \frac{\cancel{3}}{\cancel{4}}\cdot\frac{5}{\cancel{3}}\cdot\frac{\cancel{4}\cdot 2}{7} = \frac{10}{7}$

10. $-\frac{5}{6}\cdot\frac{3}{7}\cdot\frac{14}{25} = -\frac{\cancel{5}}{\cancel{2}\cdot\cancel{3}}\cdot\frac{\cancel{3}}{\cancel{7}}\cdot\frac{\cancel{2}\cdot\cancel{7}}{\cancel{5}\cdot 5} = -\frac{1}{5}$

11. $-\frac{6}{11}\div\frac{36}{55} = -\frac{6}{11}\cdot\frac{55}{36} = -\frac{\cancel{6}}{\cancel{11}}\cdot\frac{\cancel{11}\cdot 5}{\cancel{6}\cdot 6}$
$= -\frac{5}{6}$

12. $\frac{17}{12}\div\frac{34}{3} = \frac{17}{12}\cdot\frac{3}{34} = \frac{\cancel{17}}{\cancel{3}\cdot 4}\cdot\frac{\cancel{3}}{\cancel{17}\cdot 2} = \frac{1}{8}$

13. $\frac{x^2y^2}{cd}\cdot\frac{c^{-2}d^2}{x} = x^{2-1}y^2c^{-2-1}d^{2-1}$
$= xy^2c^{-3}d = \frac{xy^2d}{c^3}$

14. $\frac{a^{-2}b^2}{x^{-1}y}\cdot\frac{a^4b^4}{x^2y^3} = \frac{a^2b^6}{xy^4}$

15. $\frac{-x^2y^{-2}}{x^{-1}y^{-3}}\div\frac{x^{-3}y^2}{x^4y^{-1}} = \frac{-x^2y^{-2}}{x^{-1}y^{-3}}\cdot\frac{x^4y^{-1}}{x^{-3}y^2} = \frac{-x^6y^{-3}}{x^{-4}y^{-1}} = -x^{6-(-4)}y^{-3-(-1)} = -x^{10}y^{-2} = -\frac{x^{10}}{y^2}$

16. $\frac{(a^3)^2}{b^{-1}}\div\frac{(a^3)^{-2}}{b^{-1}} = \frac{a^6}{b^{-1}}\cdot\frac{b^{-1}}{a^{-6}} = a^{6-(-6)}b^{-1-(-1)} = a^{12}b^0 = a^{12}$

17. $\dfrac{x^2+2x+1}{x}\cdot\dfrac{x^2-x}{x^2-1}=\dfrac{(x+1)(x+1)}{x}\cdot\dfrac{x(x-1)}{(x+1)(x-1)}=x+1$

18. $\dfrac{a+6}{a^2-16}\cdot\dfrac{3a-12}{3a+18}=\dfrac{a+6}{(a+4)(a-4)}\cdot\dfrac{3(a-4)}{3(a+6)}=\dfrac{1}{a+4}$

19. $\dfrac{2x^2-x-3}{x^2-1}\cdot\dfrac{x^2+x-2}{2x^2+x-6}=\dfrac{(2x-3)(x+1)}{(x+1)(x-1)}\cdot\dfrac{(x+2)(x-1)}{(2x-3)(x+2)}=1$

20. $\dfrac{9x^2+3x-20}{3x^2-7x+4}\cdot\dfrac{3x^2-5x+2}{9x^2+18x+5}=\dfrac{(3x+5)(3x-4)}{(3x-4)(x-1)}\cdot\dfrac{(3x-2)(x-1)}{(3x+5)(3x+1)}=\dfrac{3x-2}{3x+1}$

21. $\dfrac{x^2-16}{x^2-25}\div\dfrac{x+4}{x-5}=\dfrac{x^2-16}{x^2-25}\cdot\dfrac{x-5}{x+4}=\dfrac{(x+4)(x-4)}{(x+5)(x-5)}\cdot\dfrac{x-5}{x+4}=\dfrac{x-4}{x+5}$

22. $\dfrac{a^2-9}{a^2-49}\div\dfrac{a+3}{a+7}=\dfrac{a^2-9}{a^2-49}\cdot\dfrac{a+7}{a+3}=\dfrac{(a+3)(a-3)}{(a+7)(a-7)}\cdot\dfrac{a+7}{a+3}=\dfrac{a-3}{a-7}$

23. $\dfrac{a^2+2a-35}{12x}\div\dfrac{ax-3x}{a^2+4a-21}=\dfrac{a^2+2a-35}{12x}\cdot\dfrac{a^2+4a-21}{ax-3x}$

$$=\frac{(a+7)(a-5)}{12x}\cdot\frac{(a+7)(a-3)}{x(a-3)}=\frac{(a+7)^2(a-5)}{12x^2}$$

24. $\dfrac{x^2-4}{2b-bx}\div\dfrac{x^2+4x+4}{2b+bx}=\dfrac{x^2-4}{2b-bx}\cdot\dfrac{2b+bx}{x^2+4x+4}=\dfrac{(x+2)(x-2)}{b(2-x)}\cdot\dfrac{b(2+x)}{(x+2)(x+2)}=-1$

25. $\dfrac{3t^2-t-2}{6t^2-5t-6}\cdot\dfrac{4t^2-9}{2t^2+5t+3}=\dfrac{(3t+2)(t-1)}{(3t+2)(2t-3)}\cdot\dfrac{(2t+3)(2t-3)}{(2t+3)(t+1)}=\dfrac{t-1}{t+1}$

26. $\dfrac{2p^2-5p-3}{p^2-9}\cdot\dfrac{2p^2+5p-3}{2p^2+5p+2}=\dfrac{(2p+1)(p-3)}{(p+3)(p-3)}\cdot\dfrac{(2p-1)(p+3)}{(2p+1)(p+2)}=\dfrac{2p-1}{p+2}$

27. $\dfrac{3n^2+5n-2}{12n^2-13n+3}\div\dfrac{n^2+3n+2}{4n^2+5n-6}=\dfrac{3n^2+5n-2}{12n^2-13n+3}\cdot\dfrac{4n^2+5n-6}{n^2+3n+2}$

$$=\frac{(3n-1)(n+2)}{(3n-1)(4n-3)}\cdot\frac{(4n-3)(n+2)}{(n+2)(n+1)}=\frac{n+2}{n+1}$$

28. $\dfrac{8y^2-14y-15}{6y^2-11y-10}\div\dfrac{4y^2-9y-9}{3y^2-7y-6}=\dfrac{8y^2-14y-15}{6y^2-11y-10}\cdot\dfrac{3y^2-7y-6}{4y^2-9y-9}$

$$=\frac{(4y+3)(2y-5)}{(3y+2)(2y-5)}\cdot\frac{(3y+2)(y-3)}{(4y+3)(y-3)}=1$$

29. $(x+1)\cdot\dfrac{1}{x^2+2x+1}=\dfrac{x+1}{1}\cdot\dfrac{1}{(x+1)(x+1)}=\dfrac{1}{x+1}$

30. $$\frac{x^2-4}{x} \div (x+2) = \frac{x^2-4}{x} \div \frac{x+2}{1} = \frac{x^2-4}{x} \cdot \frac{1}{x+2} = \frac{(x+2)(x-2)}{x} \cdot \frac{1}{x+2} = \frac{x-2}{x}$$

31. $$(x^2-x-2) \cdot \frac{x^2+3x+2}{x^2-4} = \frac{x^2-x-2}{1} \cdot \frac{x^2+3x+2}{x^2-4} = \frac{(x-2)(x+1)}{1} \cdot \frac{(x+2)(x+1)}{(x+2)(x-2)} = (x+1)^2$$

32. $$(2x^2-9x-5) \cdot \frac{x}{2x^2+x} = \frac{2x^2-9x-5}{1} \cdot \frac{x}{2x^2+x} = \frac{(2x+1)(x-5)}{1} \cdot \frac{x}{x(2x+1)} = x-5$$

33. $$(2x^2-15x+25) \div \frac{2x^2-3x-5}{x+1} = \frac{2x^2-15x+25}{1} \cdot \frac{x+1}{2x^2-3x-5} = \frac{(2x-5)(x-5)}{1} \cdot \frac{x+1}{(2x-5)(x+1)} = x-5$$

34. $$(x^2-6x+9) \div \frac{x^2-9}{x+3} = \frac{x^2-6x+9}{1} \cdot \frac{x+3}{x^2-9} = \frac{(x-3)(x-3)}{1} \cdot \frac{x+3}{(x+3)(x-3)} = x-3$$

35. $$\frac{x^3+y^3}{x^3-y^3} \div \frac{x^2-xy+y^2}{x^2+xy+y^2} = \frac{x^3+y^3}{x^3-y^3} \cdot \frac{x^2+xy+y^2}{x^2-xy+y^2} = \frac{(x+y)(x^2-xy+y^2)}{(x-y)(x^2+xy+y^2)} \cdot \frac{x^2+xy+y^2}{x^2-xy+y^2} = \frac{x+y}{x-y}$$

36. $$\frac{x^2-6x+9}{4-x^2} \div \frac{x^2-9}{x^2-8x+12} = \frac{x^2-6x+9}{4-x^2} \cdot \frac{x^2-8x+12}{x^2-9} = \frac{(x-3)(x-3)}{(2+x)(2-x)} \cdot \frac{(x-6)(x-2)}{(x+3)(x-3)} = -\frac{(x-3)(x-6)}{(x+2)(x+3)}$$

37. $$\frac{m^2-n^2}{2x^2+3x-2} \cdot \frac{2x^2+5x-3}{n^2-m^2} = \frac{(m+n)(m-n)}{(2x-1)(x+2)} \cdot \frac{(2x-1)(x+3)}{(n+m)(n-m)} = -\frac{x+3}{x+2}$$

38. $$\frac{x^2-y^2}{2x^2+2xy+x+y} \cdot \frac{2x^2-5x-3}{yx-3y-x^2+3x} = \frac{(x+y)(x-y)}{2x(x+y)+1(x+y)} \cdot \frac{(2x+1)(x-3)}{y(x-3)-x(x-3)} = \frac{(x+y)(x-y)}{(x+y)(2x+1)} \cdot \frac{(2x+1)(x-3)}{(x-3)(y-x)} = -1$$

39. $$\frac{ax+ay+bx+by}{x^3-27} \cdot \frac{x^2+3x+9}{xc+xd+yc+yd} = \frac{a(x+y)+b(x+y)}{(x-3)(x^2+3x+9)} \cdot \frac{x^2+3x+9}{x(c+d)+y(c+d)} = \frac{(x+y)(a+b)}{(x-3)(x^2+3x+9)} \cdot \frac{x^2+3x+9}{(c+d)(x+y)} = \frac{a+b}{(x-3)(c+d)}$$

40. $$\frac{x^2+3x+yx+3y}{x^2-9} \cdot \frac{x-3}{x+3} = \frac{x(x+3)+y(x+3)}{(x+3)(x-3)} \cdot \frac{x-3}{x+3} = \frac{(x+3)(x+y)}{(x+3)(x-3)} \cdot \frac{x-3}{x+3} = \frac{x+y}{x+3}$$

41. $\dfrac{x^2-x-6}{x^2-4}\cdot\dfrac{x^2-x-2}{9-x^2}=\dfrac{(x-3)(x+2)}{(x+2)(x-2)}\cdot\dfrac{(x-2)(x+1)}{(3+x)(3-x)}=-\dfrac{x+1}{x+3}$

42. $$\begin{aligned}\frac{2x^2-7x-4}{20-x-x^2}\div\frac{2x^2-9x-5}{x^2-25}&=-\frac{2x^2-7x-4}{x^2+x-20}\cdot\frac{x^2-25}{2x^2-9x-5}\\&=-\frac{(2x+1)(x-4)}{(x+5)(x-4)}\cdot\frac{(x+5)(x-5)}{(2x+1)(x-5)}=-1\end{aligned}$$

43. $$\begin{aligned}\frac{2x^2+3xy+y^2}{y^2-x^2}\div\frac{6x^2+5xy+y^2}{2x^2-xy-y^2}&=\frac{2x^2+3xy+y^2}{y^2-x^2}\cdot\frac{2x^2-xy-y^2}{6x^2+5xy+y^2}\\&=\frac{(2x+y)(x+y)}{(y+x)(y-x)}\cdot\frac{(2x+y)(x-y)}{(2x+y)(3x+y)}=-\frac{2x+y}{3x+y}\end{aligned}$$

44. $\dfrac{p^3-q^3}{q^2-p^2}\cdot\dfrac{q^2+pq}{p^3+p^2q+pq^2}=\dfrac{(p-q)(p^2+pq+q^2)}{(q+p)(q-p)}\cdot\dfrac{q(q+p)}{p(p^2+pq+q^2)}=-\dfrac{q}{p}$

45. $\dfrac{3x^2y^2}{6x^3y}\cdot\dfrac{-4x^7y^{-2}}{18x^{-2}y}\div\dfrac{36x}{18y^{-2}}=\dfrac{3x^2y^2}{6x^3y}\cdot\dfrac{-4x^7y^{-2}}{18x^{-2}y}\cdot\dfrac{18y^{-2}}{36x}=\dfrac{-2^3\cdot3^3x^9y^{-2}}{2^4\cdot3^5x^2y^2}=-\dfrac{x^7}{18y^4}$

46. $\dfrac{9ab^3}{7xy}\cdot\dfrac{14xy^2}{27z^3}\div\dfrac{18a^2b^2x}{3z^2}=\dfrac{9ab^3}{7xy}\cdot\dfrac{14xy^2}{27z^3}\cdot\dfrac{3z^2}{18a^2b^2x}=\dfrac{2\cdot3^3\cdot7ab^3xy^2z^2}{2\cdot3^5\cdot7a^2b^2x^2yz^3}=\dfrac{by}{9axz}$

47. $$\begin{aligned}(4x+12)\cdot\frac{x^2}{2x-6}\div\frac{2}{x-3}&=\frac{4x+12}{1}\cdot\frac{x^2}{2x-6}\cdot\frac{x-3}{2}=\frac{4(x+3)}{1}\cdot\frac{x^2}{2(x-3)}\cdot\frac{x-3}{2}\\&=x^2(x+3)\end{aligned}$$

48. $$\begin{aligned}(4x^2-9)\div\frac{2x^2+5x+3}{x+2}\div(2x-3)&=\frac{4x^2-9}{1}\div\frac{2x^2+5x+3}{x+2}\div\frac{2x-3}{1}\\&=\frac{4x^2-9}{1}\cdot\frac{x+2}{2x^2+5x+3}\cdot\frac{1}{2x-3}\\&=\frac{(2x+3)(2x-3)}{1}\cdot\frac{x+2}{(2x+3)(x+1)}\cdot\frac{1}{2x-3}\\&=\frac{x+2}{x+1}\end{aligned}$$

49. $$\begin{aligned}\frac{2x^2-2x-4}{x^2+2x-8}\cdot\frac{3x^2+15x}{x+1}\div\frac{4x^2-100}{x^2-x-20}&=\frac{2x^2-2x-4}{x^2+2x-8}\cdot\frac{3x^2+15x}{x+1}\cdot\frac{x^2-x-20}{4x^2-100}\\&=\frac{2(x-2)(x+1)}{(x+4)(x-2)}\cdot\frac{3x(x+5)}{x+1}\cdot\frac{(x-5)(x+4)}{4(x+5)(x-5)}\\&=\frac{3x}{2}\end{aligned}$$

50. $$\frac{6a^2-7a-3}{a^2-1} \div \frac{4a^2-12a+9}{a^2-1} \cdot \frac{2a^2-a-3}{3a^2-2a-1} = \frac{6a^2-7a-3}{a^2-1} \cdot \frac{a^2-1}{4a^2-12a+9} \cdot \frac{2a^2-a-3}{3a^2-2a-1}$$
$$= \frac{(3a+1)(2a-3)}{a^2-1} \cdot \frac{a^2-1}{(2a-3)(2a-3)} \cdot \frac{(2a-3)(a+1)}{(3a+1)(a-1)} = \frac{a+1}{a-1}$$

51. $$\frac{2t^2+5t+2}{t^2-4t+16} \div \frac{t+2}{t^3+64} \div \frac{2t^3+9t^2+4t}{t+1} = \frac{2t^2+5t+2}{t^2-4t+16} \cdot \frac{t^3+64}{t+2} \cdot \frac{t+1}{2t^3+9t^2+4t}$$
$$= \frac{(2t+1)(t+2)}{t^2-4t+16} \cdot \frac{(t+4)(t^2-4t+16)}{t+2} \cdot \frac{t+1}{t(2t+1)(t+4)} = \frac{t+1}{t}$$

52. $$\frac{a^6-b^6}{a^4-a^3b} \cdot \frac{a^3}{a^4+a^2b^2+b^4} \div \frac{1}{a} = \frac{a^6-b^6}{a^4-a^3b} \cdot \frac{a^3}{a^4+a^2b^2+b^4} \cdot \frac{a}{1}$$
$$= \frac{(a^2-b^2)(a^4+a^2b^2+b^4)}{a^3(a-b)} \cdot \frac{a^3}{a^4+a^2b^2+b^4} \cdot \frac{a}{1}$$
$$= \frac{(a+b)(a-b)}{a^3(a-b)} \cdot \frac{a^3}{1} \cdot \frac{a}{1} = a(a+b)$$

53. $$\frac{x^4-3x^2-4}{x^4-1} \cdot \frac{x^2+3x+2}{x^2+4x+4} = \frac{(x^2-4)(x^2+1)}{(x^2+1)(x^2-1)} \cdot \frac{(x+1)(x+2)}{(x+2)(x+2)}$$
$$= \frac{(x+2)(x-2)}{(x+1)(x-1)} \cdot \frac{(x+1)(x+2)}{(x+2)(x+2)} = \frac{x-2}{x-1}$$

54. $$\frac{x^3+2x^2+4x+8}{y^2-1} \cdot \frac{y^2+2y+1}{x^4-16} = \frac{x^2(x+2)+4(x+2)}{(y+1)(y-1)} \cdot \frac{(y+1)(y+1)}{(x^2+4)(x^2-4)}$$
$$= \frac{(x+2)(x^2+4)}{(y+1)(y-1)} \cdot \frac{(y+1)(y+1)}{(x^2+4)(x+2)(x-2)} = \frac{y+1}{(y-1)(x-2)}$$

55. $$(x^2-x-6) \div (x-3) \div (x-2) = \frac{x^2-x-6}{1} \div \frac{x-3}{1} \div \frac{x-2}{1}$$
$$= \frac{(x-3)(x+2)}{1} \cdot \frac{1}{x-3} \cdot \frac{1}{x-2} = \frac{x+2}{x-2}$$

56. $$(x^2-x-6) \div [(x-3) \div (x-2)] = \frac{x^2-x-6}{1} \div \left[\frac{x-3}{1} \div \frac{x-2}{1}\right]$$
$$= \frac{x^2-x-6}{1} \div \left[\frac{x-3}{1} \cdot \frac{1}{x-2}\right]$$
$$= \frac{(x-3)(x+2)}{1} \cdot \frac{x-2}{x-3} = (x+2)(x-2)$$

57. $$\frac{3x^2-2x}{3x+2} \div (3x-2) \div \frac{3x}{3x-3} = \frac{3x^2-2x}{3x+2} \cdot \frac{1}{3x-2} \cdot \frac{3x-3}{3x}$$
$$= \frac{x(3x-2)}{3x+2} \cdot \frac{1}{3x-2} \cdot \frac{3(x-1)}{3x} = \frac{x-1}{3x+2}$$

58. $(2x^2-3x-2)\div\dfrac{2x^2-x-1}{x-2}\div(x-1)=\dfrac{2x^2-3x-2}{1}\cdot\dfrac{x-2}{2x^2-x-1}\cdot\dfrac{1}{x-1}$

$$=\frac{(2x+1)(x-2)}{1}\cdot\frac{x-2}{(2x+1)(x-1)}\cdot\frac{1}{x-1}$$

$$=\frac{(x-2)^2}{(x-1)^2}$$

59. $\dfrac{2x^2+5x-3}{x^2+2x-3}\div\left(\dfrac{x^2+2x-35}{x^2-6x+5}\div\dfrac{x^2-9x+14}{2x^2-5x+2}\right)=$

$$=\frac{2x^2+5x-3}{x^2+2x-3}\div\left(\frac{x^2+2x-35}{x^2-6x+5}\cdot\frac{2x^2-5x+2}{x^2-9x+14}\right)$$

$$=\frac{2x^2+5x-3}{x^2+2x-3}\div\left(\frac{(x+7)(x-5)}{(x-5)(x-1)}\cdot\frac{(2x-1)(x-2)}{(x-7)(x-2)}\right)$$

$$=\frac{2x^2+5x-3}{x^2+2x-3}\div\frac{(x+7)(2x-1)}{(x-1)(x-7)}=\frac{(2x-1)(x+3)}{(x+3)(x-1)}\cdot\frac{(x-1)(x-7)}{(x+7)(2x-1)}=\frac{x-7}{x+7}$$

60. $\dfrac{x^2-4}{x^2-x-6}\div\left(\dfrac{x^2-x-2}{x^2-8x+15}\cdot\dfrac{x^2-3x-10}{x^2+3x+2}\right)$

$$=\frac{x^2-4}{x^2-x-6}\div\left(\frac{(x-2)(x+1)}{(x-3)(x-5)}\cdot\frac{(x-5)(x+2)}{(x+2)(x+1)}\right)$$

$$=\frac{x^2-4}{x^2-x-6}\div\frac{x-2}{x-3}=\frac{(x+2)(x-2)}{(x-3)(x+2)}\cdot\frac{x-3}{x-2}=1$$

61. $\dfrac{x^2-x-12}{x^2+x-2}\div\dfrac{x^2-6x+8}{x^2-3x-10}\cdot\dfrac{x^2-3x+2}{x^2-2x-15}=\dfrac{x^2-x-12}{x^2+x-2}\cdot\dfrac{x^2-3x-10}{x^2-6x+8}\cdot\dfrac{x^2-3x+2}{x^2-2x-15}$

$$=\frac{(x-4)(x+3)}{(x+2)(x-1)}\cdot\frac{(x-5)(x+2)}{(x-4)(x-2)}\cdot\frac{(x-2)(x-1)}{(x-5)(x+3)}=1$$

62. $\dfrac{4x^2-10x+6}{x^4-3x^3}\div\dfrac{2x-3}{2x^3}\cdot\dfrac{x-3}{2x-2}=\dfrac{2(2x^2-5x+3)}{x^3(x-3)}\cdot\dfrac{2x^3}{2x-3}\cdot\dfrac{x-3}{2(x-1)}$

$$=\frac{2(2x-3)(x-1)}{x^3(x-3)}\cdot\frac{2x^3}{2x-3}\cdot\frac{x-3}{2(x-1)}=2$$

63. $\dfrac{x-3}{x^3+4}$ is in lowest terms.

$$\left(\frac{x-3}{x^3+4}\right)^2=\frac{(x-3)(x-3)}{(x^3+4)(x^3+4)}=\frac{x^2-6x+9}{x^6+8x^3+16}$$

64. $\dfrac{2t^2+t}{t-1}=\dfrac{t(2t+1)}{t-1}$ is in lowest terms.

$$\left(\frac{2t^2+t}{t-1}\right)^2=\frac{(2t^2+t)(2t^2+t)}{(t-1)(t-1)}=\frac{4t^4+4t^3+t^2}{t^2-2t+1}$$

65. $\dfrac{2m^2-m-3}{x^2-1} = \dfrac{(2m-3)(m+1)}{(x+1)(x-1)}$ is in lowest terms.

$$\left(\frac{2m^2-m-3}{x^2-1}\right)^2 = \frac{(2m^2-m-3)(2m^2-m-3)}{(x^2-1)(x^2-1)}$$
$$= \frac{4m^4-2m^3-6m^2-2m^3+m^2+3m-6m^2+3m+9}{x^4-2x^2+1}$$
$$= \frac{4m^4-4m^3-11m^2+6m+9}{x^4-2x^2+1}$$

66. $\dfrac{-k-3}{x^2-x+1}$ is in lowest terms.

$$\left(\frac{-k-3}{x^2-x+1}\right)^2 = \frac{(-k-3)(-k-3)}{(x^2-x+1)(x^2-x+1)}$$
$$= \frac{k^2+6k+9}{x^4-x^3+x^2-x^3+x^2-x+x^2-x+1}$$
$$= \frac{k^2+6k+9}{x^4-2x^3+3x^2-2x+1}$$

67. $A = \dfrac{1}{2}bh = \dfrac{1}{2}\cdot\dfrac{b^2-4}{b+3}\cdot\dfrac{b^2-9}{b+2} = \dfrac{1}{2}\cdot\dfrac{(b+2)(b-2)}{b+3}\cdot\dfrac{(b+3)(b-3)}{b+2} = \dfrac{(b-2)(b-3)}{2}$ cm^2

68. $V = lwh = \dfrac{x^2+3x+2}{x+4}\cdot\dfrac{2x+8}{x^2+4x+4}\cdot\dfrac{x+2}{x^2+3x} = \dfrac{(x+2)(x+1)}{x+4}\cdot\dfrac{2(x+4)}{(x+2)(x+2)}\cdot\dfrac{x+2}{x(x+3)}$
$$= \frac{2(x+1)}{4(x+3)}$$

69. $d = rt = \dfrac{k^2-k-6}{k-4}\cdot\dfrac{k^2-16}{k^2-2k-3} = \dfrac{(k-3)(k+2)}{k-4}\cdot\dfrac{(k+4)(k-4)}{(k-3)(k+1)} = \dfrac{(k+2)(k+4)}{k+1}$ miles

70. $d = rt = \dfrac{k_1^2+3k_1+2}{k_1-3}\cdot\dfrac{k_1^2-3k_1}{k_1+1} = \dfrac{(k_1+1)(k_1+2)}{k_1-3}\cdot\dfrac{k_1(k_1-3)}{k_1+1} = k_1(k_1+2)$ miles

71. **Answers may vary.**

72. **Answers may vary.**

73. $\dfrac{x^2}{y}\ \boxed{\div}\ \dfrac{x}{y^2}\ \boxed{\times}\ \dfrac{x^2}{y^2} = \dfrac{x^3}{y}$

74. $\dfrac{x^2}{y}\ \boxed{\div}\ \dfrac{x}{y^2}\ \boxed{\div}\ \dfrac{x^2}{y^2} = \dfrac{y^3}{x}$

Exercise 6.3 (page 374)

1. $(-1, 4] \Rightarrow$ 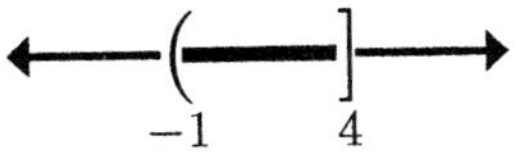

2. $(-\infty, -5] \cup [4, \infty)$

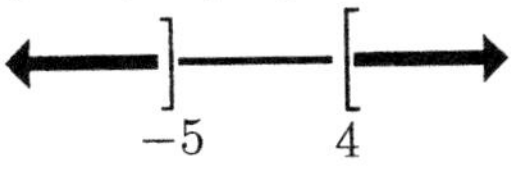

3. $$\begin{aligned} P &= 2l + 2w \\ P - 2l &= 2w \\ \frac{P-2l}{2} &= w, \text{ or } w = \frac{P-2l}{2} \end{aligned}$$

4. $$\begin{aligned} S &= \frac{a - lr}{1-r} \\ S(1-r) &= a - lr \\ S - Sr + lr &= a, \text{ or } a = S - Sr + lr \end{aligned}$$

5. $\dfrac{a+c}{b}$

6. $\dfrac{a-c}{b}$

7. subtract; keep

8. add; common

9. LCD

10. factor; highest

11. $\dfrac{3}{4}+\dfrac{7}{4}=\dfrac{10}{4}=\dfrac{5}{2}$

12. $\dfrac{5}{11}+\dfrac{2}{11}=\dfrac{7}{11}$

13. $\dfrac{10}{33}-\dfrac{21}{33}=\dfrac{-11}{33}=-\dfrac{1}{3}$

14. $\dfrac{8}{15}-\dfrac{2}{15}=\dfrac{6}{15}=\dfrac{2}{5}$

15. $\dfrac{3}{4y}+\dfrac{8}{4y}=\dfrac{11}{4y}$

16. $\dfrac{5}{3z^2}-\dfrac{6}{3z^2}=\dfrac{-1}{3z^2}=-\dfrac{1}{3z^2}$

17. $\dfrac{3}{a+b}-\dfrac{a}{a+b}=\dfrac{3-a}{a+b}$

18. $\dfrac{x}{x+4}+\dfrac{5}{x+4}=\dfrac{x+5}{x+4}$

19. $$\begin{aligned} \frac{3x}{2x+2}+\frac{x+4}{2x+2}=\frac{4x+4}{2x+2} &= \frac{4(x+1)}{2(x+1)} \\ &= 2 \end{aligned}$$

20. $$\begin{aligned} \frac{4y}{y-4}-\frac{16}{y-4}=\frac{4y-16}{y-4} &= \frac{4(y-4)}{y-4} \\ &= 4 \end{aligned}$$

21. $\dfrac{3x}{x-3}-\dfrac{9}{x-3}=\dfrac{3x-9}{x-3}=\dfrac{3(x-3)}{x-3}=3$

22. $\dfrac{9x}{x-y}-\dfrac{9y}{x-y}=\dfrac{9x-9y}{x-y}=\dfrac{9(x-y)}{x-y}=9$

23. $$\begin{aligned} \frac{5x}{x+1}+\frac{3}{x+1}-\frac{2x}{x+1} &= \frac{3x+3}{x+1} \\ &= \frac{3(x+1)}{x+1}=3 \end{aligned}$$

24. $\dfrac{4}{a+4}-\dfrac{2a}{a+4}+\dfrac{3a}{a+4}=\dfrac{a+4}{a+4}=1$

25. $\dfrac{3(x^2+x)}{x^2-5x+6}+\dfrac{-3(x^2-x)}{x^2-5x+6}=\dfrac{3x^2+3x}{x^2-5x+6}+\dfrac{-3x^2+3x}{x^2-5x+6}=\dfrac{6x}{(x-3)(x-2)}$

26. $\dfrac{2x+4}{x^2+13x+12}-\dfrac{x+3}{x^2+13x+12}=\dfrac{x+1}{(x+12)(x+1)}=\dfrac{1}{x+12}$

27. $8=2^3$; $12=2^2\cdot 3$; $18=2\cdot 3^2 \Rightarrow \text{LCD}=2^3\cdot 3^2=72$

28. $10=2\cdot 5$; $15=3\cdot 5$; $28=2^2\cdot 7 \Rightarrow \text{LCD}=2^2\cdot 3\cdot 5\cdot 7=420$

29. $x^2+3x=x(x+3)$; $x^2-9=(x+3)(x-3) \Rightarrow \text{LCD}=x(x+3)(x-3)$

30. $3y^2-6y=3y(y-2)$; $3y(y-4)=3y(y-4) \Rightarrow \text{LCD}=3y(y-2)(y-4)$

31. $x^3+27=(x+3)(x^2-3x+9)$; $x^2+6x+9=(x+3)^2 \Rightarrow \text{LCD}=(x+3)^2(x^2-3x+9)$

32. $x^3 - 8 = (x-2)(x^2+2x+4)$; $x^2 - 4x + 4 = (x-2)^2 \Rightarrow \text{LCD} = (x-2)^2(x^2+2x+4)$

33. $2x^2 + 5x + 3 = (2x+3)(x+1)$
$4x^2 + 12x + 9 = (2x+3)^2$
$x^2 + 2x + 1 = (x+1)^2$
$\text{LCD} = (2x+3)^2(x+1)^2$

34. $2x^2 + 5x + 3 = (2x+3)(x+1)$
$4x^2 + 12x + 9 = (2x+3)^2$
$4x + 6 = 2(2x+3)$
$\text{LCD} = 2(2x+3)^2(x+1)$

35. $\frac{1}{2} + \frac{1}{3} = \frac{1 \cdot 3}{2 \cdot 3} + \frac{1 \cdot 2}{3 \cdot 2} = \frac{3}{6} + \frac{2}{6} = \frac{5}{6}$

36. $\frac{5}{6} + \frac{2}{7} = \frac{5 \cdot 7}{6 \cdot 7} + \frac{2 \cdot 6}{7 \cdot 6} = \frac{35}{42} + \frac{12}{42} = \frac{47}{42}$

37. $\frac{7}{15} - \frac{17}{25} = \frac{7 \cdot 5}{15 \cdot 5} - \frac{17 \cdot 3}{25 \cdot 3} = \frac{35}{75} - \frac{51}{75} = -\frac{16}{75}$

38. $\frac{8}{9} - \frac{5}{12} = \frac{8 \cdot 4}{9 \cdot 4} - \frac{5 \cdot 3}{12 \cdot 3} = \frac{32}{36} - \frac{15}{36} = \frac{17}{36}$

39. $\frac{a}{2} + \frac{2a}{5} = \frac{a \cdot 5}{2 \cdot 5} + \frac{2a \cdot 2}{5 \cdot 2} = \frac{5a}{10} + \frac{4a}{10} = \frac{9a}{10}$

40. $\frac{b}{6} + \frac{3a}{4} = \frac{b \cdot 2}{6 \cdot 2} + \frac{3a \cdot 3}{4 \cdot 3} = \frac{2b}{12} + \frac{9a}{12} = \frac{2b + 9a}{12}$

41. $\frac{3a}{2} - \frac{4b}{7} = \frac{3a \cdot 7}{2 \cdot 7} - \frac{4b \cdot 2}{7 \cdot 2} = \frac{21a}{14} - \frac{8b}{14} = \frac{21a - 8b}{14}$

42. $\frac{2m}{3} - \frac{4n}{5} = \frac{2m \cdot 5}{3 \cdot 5} - \frac{4n \cdot 3}{5 \cdot 3} = \frac{10m}{15} - \frac{12n}{15} = \frac{10m - 12n}{15}$

43. $\frac{3}{4x} + \frac{2}{3x} = \frac{3 \cdot 3}{4x \cdot 3} + \frac{2 \cdot 4}{3x \cdot 4} = \frac{9}{12x} + \frac{8}{12x} = \frac{17}{12x}$

44. $\frac{2}{5a} + \frac{3}{2b} = \frac{2 \cdot 2b}{5a \cdot 2b} + \frac{3 \cdot 5a}{2b \cdot 5a} = \frac{4b}{10ab} + \frac{15a}{10ab} = \frac{4b + 15a}{10ab}$

45. $\frac{3a}{2b} - \frac{2b}{3a} = \frac{3a \cdot 3a}{2b \cdot 3a} - \frac{2b \cdot 2b}{3a \cdot 2b} = \frac{9a^2}{6ab} - \frac{4b^2}{6ab} = \frac{9a^2 - 4b^2}{6ab}$

46. $\frac{5m}{2n} - \frac{3n}{4m} = \frac{5m \cdot 2m}{2n \cdot 2m} - \frac{3n \cdot n}{4m \cdot n} = \frac{10m^2}{4mn} - \frac{3n^2}{4mn} = \frac{10m^2 - 3n^2}{4mn}$

47. $\frac{a+b}{3} + \frac{a-b}{7} = \frac{(a+b)7}{3(7)} + \frac{(a-b)3}{7(3)} = \frac{7a+7b}{21} + \frac{3a-3b}{21} = \frac{10a+4b}{21}$

48. $\frac{x-y}{2} + \frac{x+y}{3} = \frac{(x-y)3}{2(3)} + \frac{(x+y)2}{3(2)} = \frac{3x-3y}{6} + \frac{2x+2y}{6} = \frac{5x-y}{6}$

49. $\frac{3}{x+2} + \frac{5}{x-4} = \frac{3(x-4)}{(x+2)(x-4)} + \frac{5(x+2)}{(x-4)(x+2)} = \frac{3x-12}{(x+2)(x-4)} + \frac{5x+10}{(x+2)(x-4)}$
$= \frac{8x-2}{(x+2)(x-4)} = \frac{2(4x-1)}{(x+2)(x-4)}$

50. $$\frac{2}{a+4}-\frac{6}{a+3}=\frac{2(a+3)}{(a+4)(a+3)}-\frac{6(a+4)}{(a+3)(a+4)}=\frac{2a+6}{(a+4)(a+3)}-\frac{6a+24}{(a+4)(a+3)}$$
$$=\frac{-4a-18}{(a+4)(a+3)}=-\frac{2(2a+9)}{(a+4)(a+3)}$$

51. $$\frac{x+2}{x+5}-\frac{x-3}{x+7}=\frac{(x+2)(x+7)}{(x+5)(x+7)}-\frac{(x-3)(x+5)}{(x+7)(x+5)}=\frac{x^2+9x+14}{(x+5)(x+7)}-\frac{x^2+2x-15}{(x+5)(x+7)}$$
$$=\frac{7x+29}{(x+5)(x+7)}$$

52. $$\frac{7}{x+3}+\frac{4x}{x+6}=\frac{7(x+6)}{(x+3)(x+6)}+\frac{4x(x+3)}{(x+6)(x+3)}=\frac{7x+42}{(x+3)(x+6)}+\frac{4x^2+12x}{(x+3)(x+6)}$$
$$=\frac{4x^2+19x+42}{(x+3)(x+6)}$$

53. $$x+\frac{1}{x}=\frac{x}{1}+\frac{1}{x}=\frac{x(x)}{1(x)}+\frac{1}{x}=\frac{x^2}{x}+\frac{1}{x}=\frac{x^2+1}{x}$$

54. $$2-\frac{1}{x+1}=\frac{2}{1}-\frac{1}{x+1}=\frac{2(x+1)}{1(x+1)}-\frac{1}{x+1}=\frac{2x+2}{x+1}-\frac{1}{x+1}=\frac{2x+1}{x+1}$$

55. $$\frac{x+8}{x-3}-\frac{x-14}{3-x}=\frac{x+8}{x-3}-\frac{-x+14}{x-3}=\frac{2x-6}{x-3}=\frac{2(x-3)}{x-3}=2$$

56. $$\frac{3-x}{2-x}+\frac{x-1}{x-2}=\frac{-3+x}{x-2}+\frac{x-1}{x-2}=\frac{2x-4}{x-2}=\frac{2(x-2)}{x-2}=2$$

57. $$\frac{2a+1}{3a+2}-\frac{a-4}{2-3a}=\frac{2a+1}{3a+2}-\frac{-a+4}{3a-2}=\frac{(2a+1)(3a-2)}{(3a+2)(3a-2)}+\frac{(a-4)(3a+2)}{(3a-2)(3a+2)}$$
$$=\frac{6a^2-a-2}{(3a+2)(3a-2)}+\frac{3a^2-10a-8}{(3a+2)(3a-2)}$$
$$=\frac{9a^2-11a-10}{(3a+2)(3a-2)}$$

58. $$\frac{4}{x-2}+\frac{5}{4-x^2}=\frac{4}{x-2}+\frac{-5}{x^2-4}=\frac{4}{x-2}+\frac{-5}{(x+2)(x-2)}$$
$$=\frac{4(x+2)}{(x-2)(x+2)}+\frac{-5}{(x+2)(x-2)}$$
$$=\frac{4x+8}{(x-2)(x+2)}+\frac{-5}{(x-2)(x+2)}$$
$$=\frac{4x+3}{(x-2)(x+2)}$$

59. $$\begin{aligned}\frac{x}{x^2+5x+6}+\frac{x}{x^2-4}&=\frac{x}{(x+2)(x+3)}+\frac{x}{(x+2)(x-2)}\\&=\frac{x(x-2)}{(x+2)(x+3)(x-2)}+\frac{x(x+3)}{(x+2)(x-2)(x+3)}\\&=\frac{x^2-2x}{(x+2)(x+3)(x-2)}+\frac{x^2+3x}{(x+2)(x+3)(x-2)}\\&=\frac{2x^2+x}{(x+2)(x+3)(x-2)}\end{aligned}$$

60. $$\begin{aligned}\frac{x}{3x^2-2x-1}+\frac{4}{3x^2+10x+3}&=\frac{x}{(3x+1)(x-1)}+\frac{4}{(3x+1)(x+3)}\\&=\frac{x(x+3)}{(3x+1)(x-1)(x+3)}+\frac{4(x-1)}{(3x+1)(x+3)(x-1)}\\&=\frac{x^2+3x}{(3x+1)(x-1)(x+3)}+\frac{4x-4}{(3x+1)(x-1)(x+3)}\\&=\frac{x^2+7x-4}{(3x+1)(x-1)(x+3)}\end{aligned}$$

61. $$\begin{aligned}\frac{4}{x^2-2x-3}-\frac{x}{3x^2-7x-6}&=\frac{4}{(x-3)(x+1)}-\frac{x}{(3x+2)(x-3)}\\&=\frac{4(3x+2)}{(x-3)(x+1)(3x+2)}-\frac{x(x+1)}{(3x+2)(x-3)(x+1)}\\&=\frac{12x+8}{(x-3)(x+1)(3x+2)}-\frac{x^2+x}{(x-3)(x+1)(3x+2)}\\&=\frac{-x^2+11x+8}{(x-3)(x+1)(3x+2)}\end{aligned}$$

62. $$\begin{aligned}\frac{2a}{a^2-2a-8}+\frac{3}{a^2-5a+4}&=\frac{2a}{(a-4)(a+2)}+\frac{3}{(a-4)(a-1)}\\&=\frac{2a(a-1)}{(a-4)(a+2)(a-1)}+\frac{3(a+2)}{(a-4)(a-1)(a+2)}\\&=\frac{2a^2-2a}{(a-4)(a+2)(a-1)}+\frac{3a+6}{(a-4)(a+2)(a-1)}\\&=\frac{2a^2+a+6}{(a-4)(a+2)(a-1)}\end{aligned}$$

63. $$\frac{8}{x^2-9}+\frac{2}{x-3}-\frac{6}{x}=\frac{8}{(x+3)(x-3)}+\frac{2}{x-3}-\frac{6}{x}$$
$$=\frac{8x}{x(x+3)(x-3)}+\frac{2x(x+3)}{x(x+3)(x-3)}-\frac{6(x+3)(x-3)}{x(x+3)(x-3)}$$
$$=\frac{8x}{x(x+3)(x-3)}+\frac{2x^2+6x}{x(x+3)(x-3)}-\frac{6(x^2-9)}{x(x+3)(x-3)}$$
$$=\frac{8x}{x(x+3)(x-3)}+\frac{2x^2+6x}{x(x+3)(x-3)}-\frac{6x^2-54}{x(x+3)(x-3)}$$
$$=\frac{-4x^2+14x+54}{x(x+3)(x-3)}$$

64. $$\frac{x}{x^2-4}-\frac{x}{x+2}+\frac{2}{x}=\frac{x}{(x+2)(x-2)}-\frac{x}{x+2}+\frac{2}{x}$$
$$=\frac{x(x)}{x(x+2)(x-2)}-\frac{x(x)(x-2)}{x(x+2)(x-2)}+\frac{2(x+2)(x-2)}{x(x+2)(x-2)}$$
$$=\frac{x^2}{x(x+2)(x-2)}-\frac{x^2(x-2)}{x(x+2)(x-2)}+\frac{2(x^2-4)}{x(x+2)(x-2)}$$
$$=\frac{x^2}{x(x+2)(x-2)}-\frac{x^3-2x^2}{x(x+2)(x-2)}+\frac{2x^2-8}{x(x+2)(x-2)}$$
$$=\frac{-x^3+5x^2-8}{x(x+2)(x-2)}$$

65. $$\frac{x}{x+1}-\frac{x}{1-x^2}+\frac{1}{x}=\frac{x}{x+1}+\frac{x}{x^2-1}+\frac{1}{x}$$
$$=\frac{x}{x+1}+\frac{x}{(x+1)(x-1)}+\frac{1}{x}$$
$$=\frac{x(x)(x-1)}{x(x+1)(x-1)}+\frac{x(x)}{x(x+1)(x-1)}+\frac{1(x+1)(x-1)}{x(x+1)(x-1)}$$
$$=\frac{x^2(x-1)}{x(x+1)(x-1)}+\frac{x^2}{x(x+1)(x-1)}+\frac{x^2-1}{x(x+1)(x-1)}$$
$$=\frac{x^3-x^2}{x(x+1)(x-1)}+\frac{x^2}{x(x+1)(x-1)}+\frac{x^2-1}{x(x+1)(x-1)}$$
$$=\frac{x^3+x^2-1}{x(x+1)(x-1)}$$

66. $$\frac{y}{y-2} - \frac{2}{y+2} - \frac{-8}{4-y^2} = \frac{y}{y-2} - \frac{2}{y+2} - \frac{8}{y^2-4}$$
$$= \frac{y}{y-2} - \frac{2}{y+2} - \frac{8}{(y+2)(y-2)}$$
$$= \frac{y(y+2)}{(y+2)(y-2)} - \frac{2(y-2)}{(y+2)(y-2)} - \frac{8}{(y+2)(y-2)}$$
$$= \frac{y^2+2y}{(y+2)(y-2)} - \frac{2y-4}{(y+2)(y-2)} - \frac{8}{(y+2)(y-2)}$$
$$= \frac{y^2-4}{(y+2)(y-2)} = \frac{(y+2)(y-2)}{(y+2)(y-2)} = 1$$

67. $$2x+3+\frac{1}{x+1} = \frac{2x+3}{1} + \frac{1}{x+1} = \frac{(2x+3)(x+1)}{1(x+1)} + \frac{1}{x+1}$$
$$= \frac{2x^2+5x+3}{x+1} + \frac{1}{x+1} = \frac{2x^2+5x+4}{x+1}$$

68. $$x+1+\frac{1}{x-1} = \frac{x+1}{1} + \frac{1}{x-1} = \frac{(x+1)(x-1)}{1(x-1)} + \frac{1}{x-1}$$
$$= \frac{x^2-1}{x-1} + \frac{1}{x-1} = \frac{x^2}{x-1}$$

69. $$1+x-\frac{x}{x-5} = \frac{x+1}{1} - \frac{x}{x-5} = \frac{(x+1)(x-5)}{1(x-5)} - \frac{x}{x-5}$$
$$= \frac{x^2-4x-5}{x-5} - \frac{x}{x-5} = \frac{x^2-5x-5}{x-5}$$

70. $$2-x+\frac{3}{x-9} = \frac{2-x}{1} + \frac{3}{x-9} = \frac{(2-x)(x-9)}{1(x-9)} + \frac{3}{x-9}$$
$$= \frac{-x^2+11x-18}{x-9} + \frac{3}{x-9} = \frac{-x^2+11x-15}{x-9}$$

71. $$\frac{3x}{x-1} - 2x - x^2 = \frac{3x}{x-1} - \frac{x^2+2x}{1} = \frac{3x}{x-1} - \frac{(x^2+2x)(x-1)}{1(x-1)}$$
$$= \frac{3x}{x-1} - \frac{x^3+x^2-2x}{x-1} = \frac{-x^3-x^2+5x}{x-1}$$

72. $$\frac{23}{x-1} + 4x - 5x^2 = \frac{23}{x-1} + \frac{4x-5x^2}{1} = \frac{23}{x-1} + \frac{(4x-5x^2)(x-1)}{1(x-1)}$$
$$= \frac{23}{x-1} + \frac{-5x^3+9x^2-4x}{x-1} = \frac{-5x^3+9x^2-4x+23}{x-1}$$

73. $$\frac{y+4}{y^2+7y+12}-\frac{y-4}{y+3}+\frac{47}{y+4}=\frac{y+4}{(y+4)(y+3)}-\frac{y-4}{y+3}+\frac{47}{y+4}$$
$$=\frac{y+4}{(y+4)(y+3)}-\frac{(y-4)(y+4)}{(y+4)(y+3)}+\frac{47(y+3)}{(y+4)(y+3)}$$
$$=\frac{y+4}{(y+4)(y+3)}-\frac{y^2-16}{(y+4)(y+3)}+\frac{47y+141}{(y+4)(y+3)}$$
$$=\frac{-y^2+48y+161}{(y+4)(y+3)}$$

74. $$\frac{x+3}{2x^2-5x+2}-\frac{3x-1}{x^2-x-2}=\frac{x+3}{(2x-1)(x-2)}-\frac{3x-1}{(x-2)(x+1)}$$
$$=\frac{(x+3)(x+1)}{(2x-1)(x-2)9x+1)}-\frac{(3x-1)(2x-1)}{(2x-1)(x-2)(x+1)}$$
$$=\frac{x^2+4x+3}{(2x-1)(x-2)(x+1)}-\frac{6x^2-5x+1}{(2x-1)(x-2)(x+1)}$$
$$=\frac{-5x^2+9x+2}{(2x-1)(x-2)(x+1)}$$
$$=\frac{-(5x^2-9x-2)}{(2x-1)(x-2)(x+1)}$$
$$=\frac{-(5x+1)(x-2)}{(2x-1)(x-2)(x+1)}=-\frac{5x+1}{(2x-1)(x+1)}$$

75. $$\frac{3}{x+1}-\frac{2}{x-1}+\frac{x+3}{x^2-1}=\frac{3}{x+1}-\frac{2}{x-1}+\frac{x+3}{(x+1)(x-1)}$$
$$=\frac{3(x-1)}{(x+1)(x-1)}-\frac{2(x+1)}{(x+1)(x-1)}+\frac{x+3}{(x+1)(x-1)}$$
$$=\frac{3x-3}{(x+1)(x-1)}-\frac{2x+2}{(x+1)(x-1)}+\frac{x+3}{(x+1)(x-1)}$$
$$=\frac{2x-2}{(x+1)(x-1)}=\frac{2(x-1)}{(x+1)(x-1)}=\frac{2}{x+1}$$

76. $$\frac{2}{x-2}+\frac{3}{x+2}-\frac{x-1}{x^2-4}=\frac{2}{x-2}+\frac{3}{x+2}-\frac{x-1}{(x+2)(x-2)}$$
$$=\frac{2(x+2)}{(x+2)(x-2)}+\frac{3(x-2)}{(x+2)(x-2)}-\frac{x-1}{(x+2)(x-2)}$$
$$=\frac{2x+4}{(x+2)(x-2)}+\frac{3x-6}{(x+2)(x-2)}-\frac{x-1}{(x+2)(x-2)}$$
$$=\frac{4x-1}{(x+2)(x-2)}$$

77. $$\frac{x-2}{x^2-3x}+\frac{2x-1}{x^2+3x}-\frac{2}{x^2-9}=\frac{x-2}{x(x-3)}+\frac{2x-1}{x(x+3)}-\frac{2}{(x+3)(x-3)}$$
$$=\frac{(x-2)(x+3)}{x(x-3)(x+3)}+\frac{(2x-1)(x-3)}{x(x-3)(x+3)}-\frac{2x}{x(x-3)(x+3)}$$
$$=\frac{x^2+x-6}{x(x-3)(x+3)}+\frac{2x^2-7x+3}{x(x-3)(x+3)}-\frac{2x}{x(x-3)(x+3)}$$
$$=\frac{3x^2-8x-3}{x(x-3)(x+3)}=\frac{(3x+1)(x-3)}{x(x-3)(x+3)}=\frac{3x+1}{x(x+3)}$$

78. $$\frac{2}{x-1}-\frac{2x}{x^2-1}-\frac{x}{x^2+2x+1}=\frac{2}{x-1}-\frac{2x}{(x+1)(x-1)}-\frac{x}{(x+1)^2}$$
$$=\frac{2(x+1)^2}{(x-1)(x+1)^2}-\frac{2x(x+1)}{(x-1)(x+1)^2}-\frac{x(x-1)}{(x-1)(x+1)^2}$$
$$=\frac{2x^2+4x+2}{(x-1)(x+1)^2}-\frac{2x^2+2x}{(x-1)(x+1)^2}-\frac{x^2-x}{(x-1)(x+1)^2}$$
$$=\frac{-x^2+3x+2}{(x-1)(x+1)^2}$$

79. $$\frac{5}{x^2-25}-\frac{3}{2x^2-9x-5}+1=\frac{5}{(x+5)(x-5)}-\frac{3}{(2x+1)(x-5)}+\frac{1}{1}$$
$$=\frac{5(2x+1)}{(x+5)(x-5)(2x+1)}-\frac{3(x+5)}{(x+5)(x-5)(2x+1)}+\frac{(x+5)(x-5)(2x+1)}{(x+5)(x-5)(2x+1)}$$
$$=\frac{10x+5}{(x+5)(x-5)(2x+1)}-\frac{3x+15}{(x+5)(x-5)(2x+1)}+\frac{2x^3+x^2-50x-25}{(x+5)(x-5)(2x+1)}$$
$$=\frac{2x^3+x^2-43x-35}{(x+5)(x-5)(2x+1)}$$

80. $$\frac{3x}{2x-1}+\frac{x+1}{3x+2}+\frac{2x}{6x^3+x^2-2x}=\frac{3x}{2x-1}+\frac{x+1}{3x+2}+\frac{2x}{x(3x+2)(2x-1)}$$
$$=\frac{3x(x)(3x+2)}{x(2x-1)(3x+2)}+\frac{x(x+1)(2x-1)}{x(2x-1)(3x+2)}+\frac{2x}{x(2x-1)(3x+2)}$$
$$=\frac{9x^3+6x^2}{x(2x-1)(3x+2)}+\frac{2x^3+x^2-x}{x(2x-1)(3x+2)}+\frac{2x}{x(2x-1)(3x+2)}$$
$$=\frac{11x^3+7x^2+x}{x(2x-1)(3x+2)}=\frac{x(11x^2+7x+1)}{x(2x-1)(3x+2)}=\frac{11x^2+7x+1}{(2x-1)(3x+2)}$$

81. $$\frac{3x}{x-3}+\frac{4}{x-2}-\frac{5x}{x^3-5x^2+6x}=\frac{3x}{x-3}+\frac{4}{x-2}-\frac{5x}{x(x-3)(x-2)}$$
$$=\frac{3x(x)(x-2)}{x(x-3)(x-2)}+\frac{4x(x-3)}{x(x-3)(x-2)}-\frac{5x}{x(x-3)(x-2)}$$
$$=\frac{3x^3-6x^2}{x(x-3)(x-2)}+\frac{4x^2-12x}{x(x-3)(x-2)}-\frac{5x}{x(x-3)(x-2)}$$
$$=\frac{3x^3-2x^2-17x}{x(x-3)(x-2)}=\frac{x(3x^2-2x-17)}{x(x-3)(x-2)}=\frac{3x^2-2x-17}{(x-3)(x-2)}$$

82. $$\frac{2x-1}{x^2+x-6}-\frac{3x-5}{x^2-2x-15}+\frac{2x-3}{x^2-7x+10}$$
$$=\frac{2x-1}{(x+3)(x-2)}-\frac{3x-5}{(x-5)(x+3)}+\frac{2x-3}{(x-2)(x-5)}$$
$$=\frac{(2x-1)(x-5)}{(x+3)(x-2)(x-5)}-\frac{(3x-5)(x-2)}{(x+3)(x-2)(x-5)}+\frac{(2x-3)(x+3)}{(x+3)(x-2)(x-5)}$$
$$=\frac{2x^2-11x+5}{(x+3)(x-2)(x-5)}-\frac{3x^2-11x+10}{(x+3)(x-2)(x-5)}+\frac{2x^2+3x-9}{(x+3)(x-2)(x-5)}$$
$$=\frac{x^2+3x-14}{(x+3)(x-2)(x-5)}$$

83. $$2+\frac{4a}{a^2-1}-\frac{2}{a+1}=\frac{2}{1}+\frac{4a}{(a+1)(a-1)}-\frac{2}{(a+1)}$$
$$=\frac{2(a+1)(a-1)}{(a+1)(a-1)}+\frac{4a}{(a+1)(a-1)}-\frac{2(a-1)}{(a+1)(a-1)}$$
$$=\frac{2a^2-2}{(a+1)(a-1)}+\frac{4a}{(a+1)(a-1)}-\frac{2a-2}{(a+1)(a-1)}$$
$$=\frac{2a^2+2a}{(a+1)(a-1)}=\frac{2a(a+1)}{(a+1)(a-1)}=\frac{2a}{a-1}$$

84. $$\frac{a}{a-1}-\frac{a+1}{2a-2}+a=\frac{a}{a-1}-\frac{a+1}{2(a-1)}+\frac{a}{1}$$
$$=\frac{2a}{2(a-1)}-\frac{a+1}{2(a-1)}+\frac{2a(a-1)}{2(a-1)}$$
$$=\frac{2a}{2(a-1)}-\frac{a+1}{2(a-1)}+\frac{2a^2-2a}{2(a-1)}$$
$$=\frac{2a^2-a-1}{2(a-1)}=\frac{(2a+1)(a-1)}{2(a-1)}=\frac{2a+1}{2}$$

85. $$\frac{x+5}{2x^2-2}+\frac{x}{2x+2}-\frac{3}{x-1}=\frac{x+5}{2(x+1)(x-1)}+\frac{x}{2(x+1)}-\frac{3}{x-1}$$
$$=\frac{x+5}{2(x+1)(x-1)}+\frac{x(x-1)}{2(x+1)(x-1)}-\frac{3(2)(x+1)}{2(x+1)(x-1)}$$
$$=\frac{x+5}{2(x+1)(x-1)}+\frac{x^2-x}{2(x+1)(x-1)}-\frac{6x+6}{2(x+1)(x-1)}$$
$$=\frac{x^2-6x-1}{2(x+1)(x-1)}$$

86. $$\frac{a}{2-a}+\frac{3}{a-2}-\frac{3a-2}{a^2-4}=\frac{-a}{a-2}+\frac{3}{a-2}-\frac{3a-2}{a^2-4}=\frac{-a+3}{a-2}-\frac{3a-2}{(a+2)(a-2)}$$
$$=\frac{(-a+3)(a+2)}{(a+2)(a-2)}-\frac{3a-2}{(a+2)(a-2)}$$
$$=\frac{-a^2+a+6}{(a+2)(a-2)}-\frac{3a-2}{(a+2)(a-2)}$$
$$=\frac{-a^2-2a+8}{(a+2)(a-2)}$$
$$=\frac{-(a+4)(a-2)}{(a+2)(a-2)}=-\frac{a+4}{a+2}$$

87. $$\frac{a}{a-b}+\frac{b}{a+b}+\frac{a^2+b^2}{b^2-a^2}=\frac{-a}{b-a}+\frac{b}{b+a}+\frac{a^2+b^2}{(b+a)(b-a)}$$
$$=\frac{-a(b+a)}{(b+a)(b-a)}+\frac{b(b-a)}{(b+a)(b-a)}+\frac{a^2+b^2}{(b+a)(b-a)}$$
$$=\frac{-ab-a^2}{(b+a)(b-a)}+\frac{b^2-ab}{(b+a)(b-a)}+\frac{a^2+b^2}{(b+a)(b-a)}$$
$$=\frac{2b^2-2ab}{(b+a)(b-a)}=\frac{2b(b-a)}{(b+a)(b-a)}=\frac{2b}{b+a}$$

88. $$\frac{1}{x+y}-\frac{1}{x-y}-\frac{2y}{y^2-x^2}=\frac{1}{y+x}+\frac{1}{y-x}-\frac{2y}{(y+x)(y-x)}$$
$$=\frac{1(y-x)}{(y+x)(y-x)}+\frac{1(y+x)}{(y+x)(y-x)}-\frac{2y}{(y+x)(y-x)}$$
$$=\frac{y-x}{(y+x)(y-x)}+\frac{y+x}{(y+x)(y-x)}-\frac{2y}{(y+x)(y-x)}$$
$$=\frac{2y-2y}{(y+x)(y-x)}=\frac{0}{(y+x)(y-x)}=0$$

89. $$\frac{7n^2}{m-n}+\frac{3m}{n-m}-\frac{3m^2-n}{m^2-2mn+n^2}=\frac{7n^2}{m-n}-\frac{3m}{m-n}-\frac{3m^2-n}{(m-n)^2}$$
$$=\frac{7n^2(m-n)}{(m-n)^2}-\frac{3m(m-n)}{(m-n)^2}-\frac{3m^2-n}{(m-n)^2}$$
$$=\frac{7mn^2-7n^3}{(m-n)^2}-\frac{3m^2-3mn}{(m-n)^2}-\frac{3m^2-n}{(m-n)^2}$$
$$=\frac{7mn^2-7n^3-6m^2+3mn+n}{(m-n)^2}$$

90. $$\frac{3b}{2a-b}+\frac{2a-1}{b-2a}-\frac{3a^2+b}{b^2-4ab+4a^2}=\frac{-3b}{b-2a}+\frac{2a-1}{b-2a}-\frac{3a^2+b}{(b-2a)^2}$$
$$=\frac{-3b(b-2a)}{(b-2a)^2}+\frac{(2a-1)(b-2a)}{(b-2a)^2}-\frac{3a^2+b}{(b-2a)^2}$$
$$=\frac{-3b^2+6ab}{(b-2a)^2}+\frac{-4a^2+2ab-b+2a}{(b-2a)^2}-\frac{3a^2+b}{(b-2a)^2}$$
$$=\frac{-7a^2+8ab+2a-2b-3b^2}{(b-2a)^2}$$

91. $$\frac{m+1}{m^2+2m+1}+\frac{m-1}{m^2-2m+1}+\frac{2}{m^2-1}=\frac{m+1}{(m+1)^2}+\frac{m-1}{(m-1)^2}+\frac{2}{(m+1)(m-1)}$$
$$=\frac{1}{m+1}+\frac{1}{m-1}+\frac{2}{(m+1)(m-1)}$$
$$=\frac{1(m-1)}{(m+1)(m-1)}+\frac{1(m+1)}{(m+1)(m-1)}+\frac{2}{(m+1)(m-1)}$$
$$=\frac{2m+2}{(m+1)(m-1)}=\frac{2(m+1)}{(m+1)(m-1)}=\frac{2}{m-1}$$

92. $$\frac{a+2}{a^2+3a+2}+\frac{a-1}{a^2-1}+\frac{3}{a+1}=\frac{a+2}{(a+2)(a+1)}+\frac{a-1}{(a+1)(a-1)}+\frac{3}{a+1}$$
$$=\frac{1}{a+1}+\frac{1}{a+1}+\frac{3}{a+1}=\frac{5}{a+1}$$

93. $$\left(\frac{1}{x-1}+\frac{1}{1-x}\right)^2=\left(\frac{1}{x-1}+\frac{-1}{x-1}\right)^2=\left(\frac{0}{x-1}\right)^2=0^2=0$$

94. $$\left(\frac{1}{a-1}-\frac{1}{1-a}\right)^2=\left(\frac{1}{a-1}+\frac{1}{a-1}\right)^2=\left(\frac{2}{a-1}\right)^2=\frac{4}{(a-1)^2}$$

95. $$\left(\frac{x}{x-3}+\frac{3}{3-x}\right)^3=\left(\frac{x}{x-3}+\frac{-3}{x-3}\right)^3=\left(\frac{x-3}{x-3}\right)^3=1^3=1$$

96. $$\left(\frac{2y}{y+4}+\frac{8}{y+4}\right)^3=\left(\frac{2y+8}{y+4}\right)^3=\left(\frac{2(y+4)}{y+4}\right)^3=2^3=8$$

97. $\frac{a}{b}+\frac{c}{d}=\frac{ad}{bd}+\frac{cb}{db}=\frac{ad+bc}{bd}$

98. $\frac{a}{b}-\frac{c}{d}=\frac{ad}{bd}-\frac{cb}{db}=\frac{ad-bc}{bd}$

99. $$\begin{aligned} h=\frac{9x-2}{2}+x+\frac{3x+6}{x} &= \frac{(9x-2)x}{2x}+\frac{x(2x)}{1(2x)}+\frac{(3x+6)2}{x(2)} \\ &= \frac{9x^2-2x}{2x}+\frac{2x^2}{2x}+\frac{6x+12}{12}=\frac{11x^2+4x+12}{2x}\text{ ft} \end{aligned}$$

100. $$\begin{aligned} P=2l+2w=2\cdot\frac{3x+2}{x}+2\cdot\frac{5x+8}{x+1} &= \frac{2(3x+2)(x+1)}{x(x+1)}+\frac{2(5x+8)(x)}{(x+1)x} \\ &= \frac{2(3x^2+5x+2)+2x(5x+8)}{x(x+1)} \\ &= \frac{6x^2+10x+4+10x^2+16x}{x(x+1)} \\ &= \frac{16x^2+26x+4}{x(x+1)}=\frac{2(8x^2+13x+2)}{x(x+1)}\text{ in.} \end{aligned}$$

101. $w=\frac{7x+5}{2}-\frac{8x-6}{x}=\frac{(7x+5)x}{2x}-\frac{(8x-6)2}{2x}=\frac{7x^2+5x}{2x}-\frac{16x-12}{2x}=\frac{7x^2-11x+12}{2x}$

102. $\frac{x+2}{4}-\frac{x+3}{6}=\frac{3(x+2)}{3(4)}-\frac{2(x+3)}{2(6)}=\frac{3x+6-(2x+6)}{12}=\frac{3x+6-2x-6}{12}=\frac{x}{12}$ more

103. **Answers may vary.**

104. **Answers may vary.**

105. In the second line, the **whole** numerator must be subtracted.
The second line should have $\frac{8x+2-3x-8}{5}$.

106. In the third line, the problem uses the false conclusions $(x+y)^2=x^2+y^2$ and $(x-y)^2=x^2-y^2$.

Exercise 6.4 (page 384)

1. $$\begin{aligned} \frac{8(a-5)}{3} &= 2(a-4) \\ 8(a-5) &= 6(a-4) \\ 8a-40 &= 6a-24 \\ 2a &= 16 \\ a &= 8 \end{aligned}$$

2.
$$
\begin{aligned}
\frac{3t^2}{5}+\frac{7t}{10}&=\frac{3t+6}{5}\\
10\left(\frac{3t^2}{5}+\frac{7t}{10}\right)&=10\cdot\frac{3t+6}{5}\\
2(3t^2)+7t&=2(3t+6)\\
6t^2+7t&=6t+12\\
6t^2+t-12&=0\\
(2t+3)(3t-4)&=0
\end{aligned}
$$
$$
\begin{array}{lcl}
2t+3=0 & \textbf{or} & 3t-4=0\\
2t=-3 & & 3t=4\\
t=-\tfrac{3}{2} & & t=\tfrac{4}{3}
\end{array}
$$

3.
$$
\begin{aligned}
a^4-13a^2+36&=0\\
(a^2-4)(a^2-9)&=0\\
(a+2)(a-2)(a+3)(a-3)&=0
\end{aligned}
$$
$$
\begin{array}{lclclcl}
a+2=0 & \textbf{or} & a-2=0 & \textbf{or} & a+3=0 & \textbf{or} & a-3=0\\
a=-2 & & a=2 & & a=-3 & & a=3
\end{array}
$$

4.
$$|2x-1|=9$$
$$
\begin{array}{lcl}
2x-1=9 & \textbf{or} & 2x-1=-9\\
2x=10 & & 2x=-8\\
x=5 & & x=-4
\end{array}
$$

5. complex

6. $\div$

7. $\dfrac{\frac{1}{2}}{\frac{3}{4}}=\dfrac{1}{2}\div\dfrac{3}{4}=\dfrac{1}{2}\cdot\dfrac{4}{3}=\dfrac{2}{3}$

8. $-\dfrac{\frac{3}{4}}{\frac{1}{2}}=-\dfrac{3}{4}\div\dfrac{1}{2}=-\dfrac{3}{4}\cdot\dfrac{2}{1}=-\dfrac{3}{2}$

9. $\dfrac{-\frac{2}{3}}{\frac{6}{9}}=-\dfrac{2}{3}\div\dfrac{6}{9}=-\dfrac{2}{3}\cdot\dfrac{9}{6}=-1$

10. $\dfrac{\frac{11}{18}}{\frac{22}{27}}=\dfrac{11}{18}\div\dfrac{22}{27}=\dfrac{11}{18}\cdot\dfrac{27}{22}=\dfrac{3}{4}$

11. $\dfrac{\frac{1}{2}+\frac{1}{3}}{\frac{1}{4}}=\dfrac{\frac{3}{6}+\frac{2}{6}}{\frac{1}{4}}=\dfrac{\frac{5}{6}}{\frac{1}{4}}=\dfrac{5}{6}\div\dfrac{1}{4}=\dfrac{5}{6}\cdot\dfrac{4}{1}=\dfrac{10}{3}$

12. $\dfrac{\frac{1}{4}-\frac{1}{5}}{\frac{1}{3}}=\dfrac{\frac{5}{20}-\frac{4}{20}}{\frac{1}{3}}=\dfrac{\frac{1}{20}}{\frac{1}{3}}=\dfrac{1}{20}\div\dfrac{1}{3}=\dfrac{1}{20}\cdot\dfrac{3}{1}=\dfrac{3}{20}$

13. $\dfrac{\frac{1}{2}-\frac{2}{3}}{\frac{2}{3}+\frac{1}{2}}=\dfrac{\frac{3}{6}-\frac{4}{6}}{\frac{4}{6}+\frac{3}{6}}=\dfrac{-\frac{1}{6}}{\frac{7}{6}}=-\dfrac{1}{6}\div\dfrac{7}{6}=-\dfrac{1}{6}\cdot\dfrac{6}{7}=-\dfrac{1}{7}$

14. $\dfrac{\frac{2}{3}+\frac{4}{5}}{\frac{2}{5}-\frac{1}{3}}=\dfrac{\frac{10}{15}+\frac{12}{15}}{\frac{6}{15}-\frac{5}{15}}=\dfrac{\frac{22}{15}}{\frac{1}{15}}=\dfrac{22}{15}\div\dfrac{1}{15}=\dfrac{22}{15}\cdot\dfrac{15}{1}=22$

15. $\dfrac{\frac{4x}{y}}{\frac{6xz}{y^2}}=\dfrac{4x}{y}\div\dfrac{6xz}{y^2}=\dfrac{4x}{y}\cdot\dfrac{y^2}{6xz}=\dfrac{2y}{3z}$

16. $\dfrac{\frac{5t^4}{9x}}{\frac{2t}{18x}}=\dfrac{5t^4}{9x}\div\dfrac{2t}{18x}=\dfrac{5t^4}{9x}\cdot\dfrac{18x}{2t}=\dfrac{5t^3}{1}=5t^3$

17. $\dfrac{5ab^2}{\frac{ab}{25}}=5ab^2\div\dfrac{ab}{25}=\dfrac{5ab^2}{1}\cdot\dfrac{25}{ab}=\dfrac{125b}{1}=125b$

18. $\dfrac{\frac{6a^2b}{4t}}{3a^2b^2}=\dfrac{6a^2b}{4t}\div 3a^2b^2=\dfrac{6a^2b}{4t}\cdot\dfrac{1}{3a^2b^2}=\dfrac{1}{2bt}$

19. $\dfrac{\frac{x-y}{xy}}{\frac{y-x}{x}}=\dfrac{x-y}{xy}\div\dfrac{y-x}{x}=\dfrac{x-y}{xy}\cdot\dfrac{x}{y-x}=\dfrac{x-y}{xy}\cdot\dfrac{x}{-(x-y)}=-\dfrac{1}{y}$

20. $\dfrac{\frac{x^2+5x+6}{3xy}}{\frac{x^2-9}{6xy}}=\dfrac{x^2+5x+6}{3xy}\div\dfrac{x^2-9}{6xy}=\dfrac{x^2+5x+6}{3xy}\cdot\dfrac{6xy}{x^2-9}$

$$=\dfrac{(x+3)(x+2)}{3xy}\cdot\dfrac{6xy}{(x+3)(x-3)}=\dfrac{2(x+2)}{x-3}$$

21. $\dfrac{\frac{1}{x}-\frac{1}{y}}{xy}=\dfrac{\left(\frac{1}{x}-\frac{1}{y}\right)\cdot xy}{(xy)\cdot xy}=\dfrac{\frac{1}{x}\cdot xy-\frac{1}{y}\cdot xy}{x^2y^2}=\dfrac{y-x}{x^2y^2}$

22. $\dfrac{xy}{\frac{1}{x}-\frac{1}{y}}=\dfrac{(xy)\cdot xy}{\left(\frac{1}{x}-\frac{1}{y}\right)\cdot xy}=\dfrac{x^2y^2}{\frac{1}{x}\cdot xy-\frac{1}{y}\cdot xy}=\dfrac{x^2y^2}{y-x}$

23. $\dfrac{\frac{1}{a}+\frac{1}{b}}{\frac{1}{a}}=\dfrac{\left(\frac{1}{a}+\frac{1}{b}\right)\cdot ab}{\frac{1}{a}\cdot ab}=\dfrac{\frac{1}{a}\cdot ab+\frac{1}{b}\cdot ab}{b}=\dfrac{b+a}{b}$

24. $\dfrac{\frac{1}{b}}{\frac{1}{a}-\frac{1}{b}}=\dfrac{\frac{1}{b}\cdot ab}{\left(\frac{1}{a}-\frac{1}{b}\right)\cdot ab}=\dfrac{a}{\frac{1}{a}\cdot ab-\frac{1}{b}\cdot ab}=\dfrac{a}{b-a}$

25. $\dfrac{1+\frac{x}{y}}{1-\frac{x}{y}}=\dfrac{\left(1+\frac{x}{y}\right)\cdot y}{\left(1-\frac{x}{y}\right)\cdot y}=\dfrac{1\cdot y+\frac{x}{y}\cdot y}{1\cdot y-\frac{x}{y}\cdot y}=\dfrac{y+x}{y-x}$

26. $\dfrac{\frac{x}{y}+1}{1-\frac{x}{y}} = \dfrac{\left(\frac{x}{y}+1\right)\cdot y}{\left(1-\frac{x}{y}\right)\cdot y} = \dfrac{\frac{x}{y}\cdot y+1\cdot y}{1\cdot y-\frac{x}{y}\cdot y} = \dfrac{x+y}{y-x}$

27. $\dfrac{\frac{y}{x}-\frac{x}{y}}{\frac{1}{x}+\frac{1}{y}} = \dfrac{\left(\frac{y}{x}-\frac{x}{y}\right)\cdot xy}{\left(\frac{1}{x}+\frac{1}{y}\right)\cdot xy} = \dfrac{\frac{y}{x}\cdot xy-\frac{x}{y}\cdot xy}{\frac{1}{x}\cdot xy+\frac{1}{y}\cdot xy} = \dfrac{y^2-x^2}{y+x} = \dfrac{(y+x)(y-x)}{y+x} = y-x$

28. $\dfrac{\frac{y}{x}-\frac{x}{y}}{\frac{1}{y}-\frac{1}{x}} = \dfrac{\left(\frac{y}{x}-\frac{x}{y}\right)\cdot xy}{\left(\frac{1}{y}-\frac{1}{x}\right)\cdot xy} = \dfrac{\frac{y}{x}\cdot xy-\frac{x}{y}\cdot xy}{\frac{1}{y}\cdot xy-\frac{1}{x}\cdot xy} = \dfrac{y^2-x^2}{x-y} = \dfrac{(y+x)(y-x)}{x-y} = -(y+x)$

29. $\dfrac{\frac{1}{a}-\frac{1}{b}}{\frac{a}{b}-\frac{b}{a}} = \dfrac{\left(\frac{1}{a}-\frac{1}{b}\right)\cdot ab}{\left(\frac{a}{b}-\frac{b}{a}\right)\cdot ab} = \dfrac{\frac{1}{a}\cdot ab-\frac{1}{b}\cdot ab}{\frac{a}{b}\cdot ab-\frac{b}{a}\cdot ab} = \dfrac{b-a}{a^2-b^2} = \dfrac{b-a}{(a+b)(a-b)} = -\dfrac{1}{a+b}$

30. $\dfrac{\frac{1}{a}+\frac{1}{b}}{\frac{a}{b}-\frac{b}{a}} = \dfrac{\left(\frac{1}{a}+\frac{1}{b}\right)\cdot ab}{\left(\frac{a}{b}-\frac{b}{a}\right)\cdot ab} = \dfrac{\frac{1}{a}\cdot ab+\frac{1}{b}\cdot ab}{\frac{a}{b}\cdot ab-\frac{b}{a}\cdot ab} = \dfrac{b+a}{a^2-b^2} = \dfrac{b+a}{(a+b)(a-b)} = \dfrac{1}{a-b}$

31. $\dfrac{x+1-\frac{6}{x}}{\frac{1}{x}} = \dfrac{\left(x+1-\frac{6}{x}\right)\cdot x}{\frac{1}{x}\cdot x} = \dfrac{x^2+x-6}{1} = x^2+x-6$

32. $\dfrac{x-1-\frac{2}{x}}{\frac{x}{3}} = \dfrac{\left(x-1-\frac{2}{x}\right)\cdot 3x}{\frac{x}{3}\cdot 3x} = \dfrac{3x^2-3x-6}{x^2} = \dfrac{3(x^2-x-2)}{x^2} = \dfrac{3(x-2)(x+1)}{x^2}$

33. $\dfrac{5xy}{1+\frac{1}{xy}} = \dfrac{(5xy)\cdot xy}{\left(1+\frac{1}{xy}\right)\cdot xy} = \dfrac{5x^2y^2}{xy+1}$

34. $\dfrac{3a}{a+\frac{1}{a}} = \dfrac{(3a)\cdot a}{\left(a+\frac{1}{a}\right)\cdot a} = \dfrac{3a^2}{a^2+1}$

35. $\dfrac{1+\frac{6}{x}+\frac{8}{x^2}}{1+\frac{1}{x}-\frac{12}{x^2}} = \dfrac{\left(1+\frac{6}{x}+\frac{8}{x^2}\right)\cdot x^2}{\left(1+\frac{1}{x}-\frac{12}{x^2}\right)\cdot x^2} = \dfrac{x^2+6x+8}{x^2+x-12} = \dfrac{(x+4)(x+2)}{(x+4)(x-3)} = \dfrac{x+2}{x-3}$

36. $\dfrac{1-x-\frac{2}{x}}{\frac{6}{x^2}+\frac{1}{x}-1} = \dfrac{\left(1-x-\frac{2}{x}\right)\cdot x^2}{\left(\frac{6}{x^2}+\frac{1}{x}-1\right)\cdot x^2} = \dfrac{x^2-x^3-2x}{6+x-x^2} = \dfrac{-x(x^2-x+2)}{-(x^2-x-6)} = \dfrac{x(x^2-x+2)}{(x-3)(x+2)}$

37. $$\frac{\frac{1}{a+1}+1}{\frac{3}{a-1}+1}=\frac{\left(\frac{1}{a+1}+1\right)(a+1)(a-1)}{\left(\frac{3}{a-1}+1\right)(a+1)(a-1)}=\frac{1(a-1)+1(a+1)(a-1)}{3(a+1)+1(a+1)(a-1)}=\frac{a-1+a^2-1}{3a+3+a^2-1}$$
$$=\frac{a^2+a-2}{a^2+3a+2}=\frac{(a+2)(a-1)}{(a+2)(a+1)}=\frac{a-1}{a+1}$$

38. $$\frac{2+\frac{3}{x+1}}{\frac{1}{x}+x+x^2}=\frac{\left(2+\frac{3}{x+1}\right)(x)(x+1)}{\left(\frac{1}{x}+x+x^2\right)(x)(x+1)}=\frac{2x(x+1)+3x}{1(x+1)+x^2(x+1)+x^3(x+1)}$$
$$=\frac{2x^2+2x+3x}{(x+1)(x^3+x^2+1)}=\frac{2x^2+5x}{(x+1)(x^3+x^2+1)}$$

39. $$\frac{x^{-1}+y^{-1}}{x}=\frac{\frac{1}{x}+\frac{1}{y}}{x}=\frac{\left(\frac{1}{x}+\frac{1}{y}\right)(xy)}{x(xy)}=\frac{y+x}{x^2y}$$

40. $$\frac{x^{-1}-y^{-1}}{y}=\frac{\frac{1}{x}-\frac{1}{y}}{y}=\frac{\left(\frac{1}{x}-\frac{1}{y}\right)(xy)}{y(xy)}=\frac{y-x}{xy^2}$$

41. $$\frac{y}{x^{-1}-y^{-1}}=\frac{y}{\frac{1}{x}-\frac{1}{y}}=\frac{y(xy)}{\left(\frac{1}{x}-\frac{1}{y}\right)(xy)}=\frac{xy^2}{y-x}$$

42. $$\frac{x^{-1}+y^{-1}}{(x+y)^{-1}}=\frac{\frac{1}{x}+\frac{1}{y}}{\frac{1}{x+y}}=\frac{\left(\frac{1}{x}+\frac{1}{y}\right)(xy)(x+y)}{\left(\frac{1}{x+y}\right)(xy)(x+y)}=\frac{(y+x)(x+y)}{xy}=\frac{(x+y)^2}{xy}$$

43. $$\frac{x^{-1}+y^{-1}}{x^{-1}-y^{-1}}=\frac{\frac{1}{x}+\frac{1}{y}}{\frac{1}{x}-\frac{1}{y}}=\frac{\left(\frac{1}{x}+\frac{1}{y}\right)(xy)}{\left(\frac{1}{x}-\frac{1}{y}\right)(xy)}=\frac{y+x}{y-x}$$

44. $$\frac{(x+y)^{-1}}{x^{-1}+y^{-1}}=\frac{\frac{1}{x+y}}{\frac{1}{x}+\frac{1}{y}}=\frac{\left(\frac{1}{x+y}\right)(xy)(x+y)}{\left(\frac{1}{x}+\frac{1}{y}\right)(xy)(x+y)}=\frac{xy}{(y+x)(x+y)}=\frac{xy}{(x+y)^2}$$

45. $$\frac{x+y}{x^{-1}+y^{-1}}=\frac{x+y}{\frac{1}{x}+\frac{1}{y}}=\frac{(x+y)(xy)}{\left(\frac{1}{x}+\frac{1}{y}\right)(xy)}=\frac{(x+y)xy}{y+x}=xy$$

46. $$\frac{x-y}{x^{-1}-y^{-1}}=\frac{x-y}{\frac{1}{x}-\frac{1}{y}}=\frac{(x-y)(xy)}{\left(\frac{1}{x}-\frac{1}{y}\right)(xy)}=\frac{(x-y)xy}{y-x}=-xy$$

47. $$\frac{x-y^{-2}}{y-x^{-2}}=\frac{x-\frac{1}{y^2}}{y-\frac{1}{x^2}}=\frac{\left(x-\frac{1}{y^2}\right)(x^2y^2)}{\left(y-\frac{1}{x^2}\right)(x^2y^2)}=\frac{x^3y^2-x^2}{x^2y^3-y^2}=\frac{x^2(xy^2-1)}{y^2(x^2y-1)}$$

48. $$\frac{x^{-2}-y^{-2}}{x^{-1}-y^{-1}}=\frac{\frac{1}{x^2}-\frac{1}{y^2}}{\frac{1}{x}-\frac{1}{y}}=\frac{\left(\frac{1}{x^2}-\frac{1}{y^2}\right)(x^2y^2)}{\left(\frac{1}{x}-\frac{1}{y}\right)(x^2y^2)}=\frac{y^2-x^2}{xy^2-x^2y}=\frac{(y+x)(y-x)}{xy(y-x)}=\frac{x+y}{xy}$$

49. $$\frac{1+\frac{a}{b}}{1-\frac{a}{1-\frac{a}{b}}}=\frac{1+\frac{a}{b}}{1-\frac{a(b)}{\left(1-\frac{a}{b}\right)(b)}}=\frac{1+\frac{a}{b}}{1-\frac{ab}{b-a}}=\frac{\left(1+\frac{a}{b}\right)(b)(b-a)}{\left(1-\frac{ab}{b-a}\right)(b)(b-a)}=\frac{b(b-a)+a(b-a)}{b(b-a)-ab(b)}$$

$$=\frac{b^2-ab+ab-a^2}{b^2-ab-ab^2}$$

$$=\frac{b^2-a^2}{b(b-a-ab)}$$

$$=\frac{(b+a)(b-a)}{b(b-a-ab)}$$

50. $$\frac{1+\frac{2}{1+\frac{a}{b}}}{1-\frac{a}{b}}=\frac{1+\frac{2(b)}{\left(1+\frac{a}{b}\right)(b)}}{1-\frac{a}{b}}=\frac{1+\frac{2b}{b+a}}{1-\frac{a}{b}}=\frac{\left(1+\frac{2b}{b+a}\right)(b)(b+a)}{\left(1-\frac{a}{b}\right)(b)(b+a)}=\frac{b(b+a)+2b^2}{b(b+a)-a(b+a)}$$

$$=\frac{b^2+ab+2b^2}{b^2+ab-ab-a^2}$$

$$=\frac{3b^2+ab}{b^2-a^2}$$

$$=\frac{b(3b+a)}{(b+a)(b-a)}$$

51. $$\frac{x-\frac{1}{x}}{1+\frac{1}{\frac{1}{x}}}=\frac{x-\frac{1}{x}}{1+\frac{1(x)}{\frac{1}{x}(x)}}=\frac{x-\frac{1}{x}}{1+\frac{x}{1}}=\frac{x-\frac{1}{x}}{1+x}=\frac{\left(x-\frac{1}{x}\right)x}{(1+x)x}=\frac{x^2-1}{x(x+1)}=\frac{(x+1)(x-1)}{x(x+1)}$$

$$=\frac{x-1}{x}$$

52. $$\frac{\frac{a^2+3a+4}{ab}}{2+\frac{3+a}{\frac{2}{a}}} = \frac{\frac{a^2+3a+4}{ab}}{2+\frac{(3+a)(a)}{\frac{2}{a}(a)}} = \frac{\frac{a^2+3a+4}{ab}}{2+\frac{a^2+3a}{2}} = \frac{\left(\frac{a^2+3a+4}{ab}\right)(2ab)}{\left(2+\frac{a^2+3a}{2}\right)(2ab)} = \frac{2(a^2+3a+4)}{4ab+ab(a^2+3a)}$$
$$= \frac{2(a^2+3a+4)}{4ab+a^3b+3a^2b}$$
$$= \frac{2(a^2+3a+4)}{ab(a^2+3a+4)}$$
$$= \frac{2}{ab}$$

53. $$\frac{b}{b+\frac{2}{2+\frac{1}{2}}} = \frac{b}{b+\frac{2(2)}{\left(2+\frac{1}{2}\right)(2)}} = \frac{b}{b+\frac{4}{4+1}} = \frac{b}{b+\frac{4}{5}} = \frac{b(5)}{\left(b+\frac{4}{5}\right)(5)} = \frac{5b}{5b+4}$$

54. $$\frac{2y}{y-\frac{y}{3-\frac{1}{2}}} = \frac{2y}{y-\frac{y(2)}{\left(3-\frac{1}{2}\right)(2)}} = \frac{2y}{y-\frac{2y}{6-1}} = \frac{2y}{y-\frac{2y}{5}} = \frac{2y(5)}{\left(y-\frac{2y}{5}\right)(5)} = \frac{10y}{5y-2y} = \frac{10y}{3y} = \frac{10}{3}$$

55. $$a+\frac{a}{1+\frac{a}{a+1}} = a+\frac{a(a+1)}{\left(1+\frac{a}{a+1}\right)(a+1)} = a+\frac{a(a+1)}{a+1+a} = a+\frac{a^2+a}{2a+1} = \frac{a(2a+1)}{2a+1}+\frac{a^2+a}{2a+1}$$
$$= \frac{2a^2+a+a^2+a}{2a+1}$$
$$= \frac{3a^2+2a}{2a+1}$$

56. $$b+\frac{b}{1-\frac{b+1}{b}} = b+\frac{b(b)}{\left(1-\frac{b+1}{b}\right)(b)} = b+\frac{b^2}{b-(b+1)} = b+\frac{b^2}{-1} = b-b^2$$

57. $$\frac{x-\frac{1}{1-\frac{x}{2}}}{\frac{3}{x+\frac{2}{3}}-x} = \frac{x-\frac{1(2)}{\left(1-\frac{x}{2}\right)2}}{\frac{3(3)}{\left(x+\frac{2}{3}\right)3}-x} = \frac{x-\frac{2}{2-x}}{\frac{9}{3x+2}-x} = \frac{\left(x-\frac{2}{2-x}\right)(2-x)(3x+2)}{\left(\frac{9}{3x+2}-x\right)(2-x)(3x+2)}$$
$$= \frac{x(2-x)(3x+2)-2(3x+2)}{9(2-x)-x(2-x)(3x+2)}$$
$$= \frac{(2x-x^2)(3x+2)-2(3x+2)}{9(2-x)-(3x^2+2x)(2-x)}$$
$$= \frac{(3x+2)(-x^2+2x-2)}{(2-x)(-3x^2-2x+9)}$$

58. $$\frac{\frac{2x}{x-\frac{1}{x}}-\frac{1}{x}}{2x+\frac{2x}{1-\frac{1}{x}}}=\frac{\frac{2x(x)}{\left(x-\frac{1}{x}\right)x}-\frac{1}{x}}{2x+\frac{2x(x)}{\left(1-\frac{1}{x}\right)x}}=\frac{\frac{2x^2}{x^2-1}-\frac{1}{x}}{2x+\frac{2x^2}{x-1}}=\frac{\frac{2x^2}{(x+1)(x-1)}-\frac{1}{x}}{2x+\frac{2x^2}{x-1}}$$

$$=\frac{\left(\frac{2x^2}{(x+1)(x-1)}-\frac{1}{x}\right)(x)(x+1)(x-1)}{\left(2x+\frac{2x^2}{x-1}\right)(x)(x+1)(x-1)}$$

$$=\frac{2x^2(x)-1(x+1)(x-1)}{2x(x)(x+1)(x-1)+2x^2(x)(x+1)}$$

$$=\frac{2x^3-x^2+1}{(2x^3-2x^2)(x+1)+2x^3(x+1)}$$

$$=\frac{2x^3-x^2+1}{(4x^3-2x^2)(x+1)}=\frac{2x^3-x^2+1}{2x^2(2x-1)(x+1)}$$

59. $$\frac{2x+\frac{1}{2-\frac{x}{2}}}{\frac{4}{\frac{x}{2}-2}-x}=\frac{2x+\frac{1(2)}{(2-\frac{x}{2})2}}{\frac{4(2)}{(\frac{x}{2}-2)2}-x}=\frac{2x+\frac{2}{4-x}}{\frac{8}{x-4}-x}=\frac{\left(2x+\frac{2}{4-x}\right)(x-4)}{\left(\frac{8}{x-4}-x\right)(x-4)}$$

$$=\frac{2x^2-8x-2}{8-x^2+4x}=\frac{2(x^2-4x-1)}{-x^2+4x+8}$$

60. $$\frac{3x-\frac{1}{3-\frac{x}{2}}}{\frac{3}{\frac{x}{2}-3}+x}=\frac{3x-\frac{1(2)}{(3-\frac{x}{2})2}}{\frac{3(2)}{(\frac{x}{2}-3)2}+x}=\frac{3x-\frac{2}{6-x}}{\frac{6}{x-6}+x}=\frac{\left(3x-\frac{2}{6-x}\right)(x-6)}{\left(\frac{6}{x-6}+x\right)(x-6)}$$

$$=\frac{3x^2-18x+2}{6+x^2-6x}=\frac{3x^2-18x+2}{x^2-6x+6}$$

61. $$\frac{\frac{1}{x^2+3x+2}+\frac{1}{x^2+x-2}}{\frac{3x}{x^2-1}-\frac{x}{x+2}}=\frac{\frac{1}{(x+2)(x+1)}+\frac{1}{(x+2)(x-1)}}{\frac{3x}{(x+1)(x-1)}-\frac{x}{x+2}}$$

$$=\frac{\left(\frac{1}{(x+2)(x+1)}+\frac{1}{(x+2)(x-1)}\right)(x+2)(x+1)(x-1)}{\left(\frac{3x}{(x+1)(x-1)}-\frac{x}{x+2}\right)(x+2)(x+1)(x-1)}$$

$$=\frac{x-1+x+1}{3x(x+2)-x(x+1)(x-1)}$$

$$=\frac{2x}{3x^2+6x-x^3+x}$$

$$=\frac{2x}{-x^3+3x^2+7x}=\frac{2x}{-x(x^2-3x-7)}=\frac{-2}{x^2-3x-7}$$

62. $$\frac{\frac{1}{x^2-1}-\frac{2}{x^2+4x+3}}{\frac{2}{x^2+2x-3}+\frac{1}{x+3}}=\frac{\frac{1}{(x+1)(x-1)}-\frac{2}{(x+1)(x+3)}}{\frac{2}{(x+3)(x-1)}+\frac{1}{x+3}}$$
$$=\frac{\left(\frac{1}{(x+1)(x-1)}-\frac{2}{(x+1)(x+3)}\right)(x+1)(x-1)(x+3)}{\left(\frac{2}{(x+3)(x-1)}+\frac{1}{x+3}\right)(x+1)(x-1)(x+3)}$$
$$=\frac{x+3-2(x-1)}{2(x+1)+(x+1)(x-1)}=\frac{-x+5}{x^2+2x+1}$$

63. $$\frac{1}{\frac{1}{k_1}+\frac{1}{k_2}}=\frac{1k_1k_2}{\left(\frac{1}{k_1}+\frac{1}{k_2}\right)k_1k_2}=\frac{k_1k_2}{k_2+k_1}$$

64. $$\frac{1}{\frac{1}{R_1}+\frac{1}{R_2}}=\frac{1R_1R_2}{\left(\frac{1}{R_1}+\frac{1}{R_2}\right)R_1R_2}=\frac{R_1R_2}{R_2+R_1}$$

65. $$\text{average}=\frac{\frac{k}{2}+\frac{k}{3}+\frac{k}{2}}{3}=\frac{\frac{3k}{6}+\frac{2k}{6}+\frac{3k}{6}}{3}=\frac{\frac{8k}{6}}{\frac{3}{1}}=\frac{8k}{6}\div\frac{3}{1}=\frac{8k}{6}\cdot\frac{1}{3}=\frac{8k}{18}=\frac{4k}{9}$$

66. $$\frac{d_1+d_2}{\frac{d_1}{s_1}+\frac{d_2}{s_2}}=\frac{(d_1+d_2)s_1s_2}{\left(\frac{d_1}{s_1}+\frac{d_2}{s_2}\right)s_1s_2}=\frac{s_1s_2(d_1+d_2)}{d_1s_2+d_2s_1}$$

67. Answers may vary.

68. Answers may vary.

69. $$(x^{-1}y^{-1})(x^{-1}+y^{-1})^{-1}=\frac{1}{x}\cdot\frac{1}{y}\cdot\frac{1}{\frac{1}{x}+\frac{1}{y}}=\frac{1}{xy}\cdot\frac{1(xy)}{\left(\frac{1}{x}+\frac{1}{y}\right)xy}=\frac{1}{xy}\cdot\frac{xy}{y+x}=\frac{1}{y+x}$$

70. $$[(x^{-1}+1)^{-1}+1]^{-1}=\frac{1}{(x^{-1}+1)^{-1}+1}=\frac{1}{\frac{1}{\frac{1}{x}+1}+1}=\frac{1}{\frac{x}{1+x}+1}=\frac{1+x}{x+1+x}=\frac{x+1}{2x+1}$$

Exercise 6.5 (page 394)

1. $$(m^2n^{-3})^{-2}=m^{-4}n^6=\frac{n^6}{m^4}$$

2. $$\frac{a^{-1}}{a^{-1}+1}=\frac{a^{-1}\cdot a^1}{(a^{-1}+1)\cdot a^1}=\frac{a^0}{a^0+a^1}$$
$$=\frac{1}{1+a}$$

3. $$\frac{a^0+2a^0-3a^0}{(a-b)^0}=\frac{1+2-3}{1}=0$$

4. $$(4x^{-2}+3)(2x-4)=8x^{-1}-16x^{-2}+6x-12=\frac{8}{x}-\frac{16}{x^2}+6x-12$$

5. rational

6. extraneous

7.

$$\frac{1}{4}+\frac{9}{x}=1$$
$$\left(\frac{1}{4}+\frac{9}{x}\right)4x=1(4x)$$
$$x+36=4x$$
$$-3x=-36$$
$$x=12$$

The answer checks.

8.

$$\frac{1}{3}-\frac{10}{x}=-3$$
$$\left(\frac{1}{3}-\frac{10}{x}\right)3x=-3(3x)$$
$$x-30=-9x$$
$$10x=30$$
$$x=3$$

The answer checks.

9.

$$\frac{34}{x}-\frac{3}{2}=-\frac{13}{20}$$
$$\left(\frac{34}{x}-\frac{3}{2}\right)20x=-\left(\frac{13}{20}\right)20x$$
$$680-30x=-13x$$
$$-17x=-680$$
$$x=40$$

The answer checks.

10.

$$\frac{1}{2}+\frac{7}{x}=2+\frac{1}{x}$$
$$\left(\frac{1}{2}+\frac{7}{x}\right)2x=\left(2+\frac{1}{x}\right)2x$$
$$x+14=4x+2$$
$$-3x=-12$$
$$x=4$$

The answer checks.

11.

$$\frac{3}{y}+\frac{7}{2y}=13$$
$$\left(\frac{3}{y}+\frac{7}{2y}\right)2y=(13)2y$$
$$6+7=26y$$
$$-26y=-13$$
$$y=\frac{-13}{-26}=\frac{1}{2}$$

The answer checks.

12.

$$\frac{2}{x}+\frac{1}{2}=\frac{7}{2x}$$
$$\left(\frac{2}{x}+\frac{1}{2}\right)2x=\left(\frac{7}{2x}\right)2x$$
$$4+x=7$$
$$x=3$$

The answer checks.

13.

$$\frac{x+1}{x}-\frac{x-1}{x}=0$$
$$\left(\frac{x+1}{x}-\frac{x-1}{x}\right)x=(0)x$$
$$x+1-(x-1)=0$$
$$x+1-x+1=0$$
$$2\neq 0$$

There is no solution.

14.

$$\frac{2}{x}+\frac{1}{2}=\frac{9}{4x}-\frac{1}{2x}$$
$$\left(\frac{2}{x}+\frac{1}{2}\right)4x=\left(\frac{9}{4x}-\frac{1}{2x}\right)4x$$
$$8+2x=9-2$$
$$2x=-1$$
$$x=-\frac{1}{2}$$

The answer checks.

15.
$$\frac{7}{5x}-\frac{1}{2}=\frac{5}{6x}+\frac{1}{3}$$
$$\left(\frac{7}{5x}-\frac{1}{2}\right)30x=\left(\frac{5}{6x}+\frac{1}{3}\right)30x$$
$$42-15x=25+10x$$
$$-25x=-17$$
$$x=\frac{-17}{-25}=\frac{17}{25}$$
The answer checks.

16.
$$\frac{x-3}{x-1}-\frac{2x-4}{x-1}=0$$
$$\left(\frac{x-3}{x-1}-\frac{2x-4}{x-1}\right)(x-1)=0(x-1)$$
$$x-3-(2x-4)=0$$
$$x-3-2x+4=0$$
$$-x+1=0$$
$$1=x$$
The answer does not check.
There is no solution.

17.
$$\frac{y-3}{y+2}=3-\frac{1-2y}{y+2}$$
$$\frac{y-3}{y+2}\cdot(y+2)=\left(3-\frac{1-2y}{y+2}\right)(y+2)$$
$$y-3=3(y+2)-(1-2y)$$
$$y-3=3y+6-1+2y$$
$$y-3=5y+5$$
$$-4y=8$$
$$y=-2$$
The answer does not check.
There is no solution.

18.
$$\frac{3a-5}{a-1}-2=\frac{2a}{1-a}$$
$$\left(\frac{3a-5}{a-1}-2\right)(a-1)=\frac{-2a}{a-1}\cdot(a-1)$$
$$3a-5-2(a-1)=-2a$$
$$3a-5-2a+2=-2a$$
$$a-3=-2a$$
$$3a=3$$
$$a=1$$
The answer does not check.
There is no solution.

19.
$$\frac{3-5y}{2+y}=\frac{3+5y}{2-y}$$
$$\left(\frac{3-5y}{2+y}\right)(2+y)(2-y)=\left(\frac{3+5y}{2-y}\right)(2+y)(2-y)$$
$$(3-5y)(2-y)=(3+5y)(2+y)$$
$$5y^2-13y+6=5y^2+13y+6$$
$$-26y=0$$
$$y=0$$
The answer checks.

20.
$$\frac{x}{x-2}=1+\frac{1}{x-3}$$
$$\left(\frac{x}{x-2}\right)(x-2)(x-3)=\left(1+\frac{1}{x-3}\right)(x-2)(x-3)$$
$$x(x-3)=(x-2)(x-3)+x-2$$
$$x^2-3x=x^2-5x+6+x-2$$
$$x^2-3x=x^2-4x+4$$
$$x=4$$
The answer checks.

21.

$$\frac{a+2}{a+1} = \frac{a-4}{a-3}$$
$$\left(\frac{a+2}{a+1}\right)(a+1)(a-3) = \left(\frac{a-4}{a-3}\right)(a+1)(a-3)$$
$$(a+2)(a-3) = (a-4)(a+1)$$
$$a^2 - a - 6 = a^2 - 3a - 4$$
$$2a = 2$$
$$a = 1$$

The answer checks.

22.

$$\frac{z+2}{z+8} - \frac{z-3}{z-2} = 0$$
$$\left(\frac{z+2}{z+8} - \frac{z-3}{z-2}\right)(z+8)(z-2) = 0(z+8)(z-2)$$
$$(z+2)(z-2) - (z-3)(z+8) = 0$$
$$z^2 - 4 - (z^2 + 5z - 24) = 0$$
$$z^2 - 4 - z^2 - 5z + 24 = 0$$
$$-5z = -20$$
$$z = 4$$

The answer checks.

23.

$$\frac{x+2}{x+3} - 1 = \frac{1}{3 - 2x - x^2}$$
$$\frac{x+2}{x+3} - 1 = \frac{-1}{x^2 + 2x - 3}$$
$$\frac{x+2}{x+3} - 1 = \frac{-1}{(x+3)(x-1)}$$
$$\left(\frac{x+2}{x+3} - 1\right)(x+3)(x-1) = \left(\frac{-1}{(x+3)(x-1)}\right)(x+3)(x-1)$$
$$(x+2)(x-1) - (x+3)(x-1) = -1$$
$$x^2 + x - 2 - (x^2 + 2x - 3) = -1$$
$$x^2 + x - 2 - x^2 - 2x + 3 = -1$$
$$-x = -2$$
$$x = 2$$

The answer checks.

24.

$$\frac{x-3}{x-2} - \frac{1}{x} = \frac{x-3}{x}$$
$$\left(\frac{x-3}{x-2} - \frac{1}{x}\right)(x)(x-2) = \left(\frac{x-3}{x}\right)(x)(x-2)$$
$$(x-3)x - 1(x-2) = (x-3)(x-2)$$
$$x^2 - 3x - x + 2 = x^2 - 5x + 6$$
$$x^2 - 4x + 2 = x^2 - 5x + 6$$
$$x = 4$$

The answer checks.

25.

$$\begin{aligned}
\frac{x}{x+2} &= 1 - \frac{3x+2}{x^2+4x+4} \\
\frac{x}{x+2} &= 1 - \frac{3x+2}{(x+2)(x+2)} \\
\left(\frac{x}{x+2}\right)(x+2)(x+2) &= \left(1 - \frac{3x+2}{(x+2)(x+2)}\right)(x+2)(x+2) \\
x(x+2) &= (x+2)(x+2) - (3x+2) \\
x^2 + 2x &= x^2 + 4x + 4 - 3x - 2 \\
x^2 + 2x &= x^2 + x + 2 \\
x &= 2 \qquad \text{The answer checks.}
\end{aligned}$$

26.

$$\begin{aligned}
\frac{3+2a}{a^2+6+5a} + \frac{2-5a}{a^2-4} &= \frac{2-3a}{a^2-6+a} \\
\frac{2a+3}{a^2+5a+6} + \frac{-5a+2}{a^2-4} &= \frac{-3a+2}{a^2+a-6} \\
\frac{2a+3}{(a+3)(a+2)} + \frac{-5a+2}{(a+2)(a-2)} &= \frac{-3a+2}{(a+3)(a-2)} \\
\left(\frac{2a+3}{(a+3)(a+2)} + \frac{-5a+2}{(a+2)(a-2)}\right)(a+3)(a+2)(a-2) & \\
&= \left(\frac{-3a+2}{(a+3)(a-2)}\right)(a+3)(a+2)(a-2) \\
(2a+3)(a-2) + (-5a+2)(a+3) &= (-3a+2)(a+2) \\
2a^2 - a - 6 + -5a^2 - 13a + 6 &= -3a^2 - 4a + 4 \\
-3a^2 - 14a &= -3a^2 - 4a + 4 \\
-10a &= 4 \\
a &= \tfrac{4}{-10} = -\tfrac{2}{5} \qquad \text{The answer checks.}
\end{aligned}$$

27.

$$\begin{aligned}
\frac{2}{x-2} + \frac{1}{x+1} &= \frac{1}{x^2-x-2} \\
\left(\frac{2}{x-2} + \frac{1}{x+1}\right)(x-2)(x+1) &= \left(\frac{1}{(x-2)(x+1)}\right)(x-2)(x+1) \\
2(x+1) + 1(x-2) &= 1 \\
2x + 2 + x - 2 &= 1 \\
3x &= 1 \\
x &= \tfrac{1}{3} \qquad \text{The answer checks.}
\end{aligned}$$

28.

$$
\begin{aligned}
\frac{5}{y-1}+\frac{3}{y-3} &= \frac{8}{y-2}\\
\left(\frac{5}{y-1}+\frac{3}{y-3}\right)(y-1)(y-3)(y-2) &= \left(\frac{8}{y-2}\right)(y-1)(y-3)(y-2)\\
5(y-3)(y-2)+3(y-1)(y-2) &= 8(y-1)(y-3)\\
5(y^2-5y+6)+3(y^2-3y+2) &= 8(y^2-4y+3)\\
5y^2-25y+30+3y^2-9y+6 &= 8y^2-32y+24\\
8y^2-34y+36 &= 8y^2-32y+24\\
-2y &= -12\\
y &= 6 \qquad \text{The answer checks.}
\end{aligned}
$$

29.

$$
\begin{aligned}
\frac{3}{a-2}-\frac{1}{a-1} &= \frac{7}{(a-2)(a-1)}\\
\left(\frac{3}{a-2}-\frac{1}{a-1}\right)(a-2)(a-1) &= \left(\frac{7}{(a-2)(a-1)}\right)(a-2)(a-1)\\
3(a-1)-1(a-2) &= 7\\
3a-3-a+2 &= 7\\
2a-1 &= 7\\
2a &= 8\\
a &= 4 \qquad \text{The answer checks.}
\end{aligned}
$$

30.

$$
\begin{aligned}
\frac{5}{x+6}-\frac{3}{x-4} &= \frac{2}{x+3}\\
\left(\frac{5}{x+6}-\frac{3}{x-4}\right)(x+6)(x-4)(x+3) &= \left(\frac{2}{x+3}\right)(x+6)(x-4)(x+3)\\
5(x-4)(x+3)-3(x+6)(x+3) &= 2(x+6)(x-4)\\
5\left(x^2-x-12\right)-3\left(x^2+9x+18\right) &= 2\left(x^2+2x-24\right)\\
5x^2-5x-60-3x^2-27x-54 &= 2x^2+4x-48\\
2x^2-32x-114 &= 2x^2+4x-48\\
-36x &= 66\\
x &= \frac{66}{-36} = -\frac{11}{6} \qquad \text{The answer checks.}
\end{aligned}
$$

31.
$$\begin{aligned}
\frac{a-1}{a+3}-\frac{1-2a}{3-a}&=\frac{2-a}{a-3}\\
\frac{a-1}{a+3}-\frac{2a-1}{a-3}&=\frac{2-a}{a-3}\\
\left(\frac{a-1}{a+3}-\frac{2a-1}{a-3}\right)(a+3)(a-3)&=\left(\frac{2-a}{a-3}\right)(a+3)(a-3)\\
(a-1)(a-3)-(2a-1)(a+3)&=(2-a)(a+3)\\
a^2-4a+3-(2a^2+5a-3)&=-a^2-a+6\\
a^2-4a+3-2a^2-5a+3&=-a^2-a+6\\
-a^2-9a+6&=-a^2-a+6\\
-8a&=0\\
a&=0
\end{aligned}$$
The answer checks.

32.
$$\begin{aligned}
\frac{5}{2z^2+z-3}-\frac{2}{2z+3}&=\frac{z+1}{z-1}-1\\
\left(\frac{5}{(2z+3)(z-1)}-\frac{2}{2z+3}\right)(2z+3)(z-1)&=\left(\frac{z+1}{z-1}-1\right)(2z+3)(z-1)\\
5-2(z-1)&=(z+1)(2z+3)-(2z+3)(z-1)\\
5-2z+2&=2z^2+5z+3-(2z^2+z-3)\\
-2z+7&=2z^2+5z+3-2z^2-z+3\\
-2z+7&=4z+6\\
-6z&=-1\\
z&=\frac{-1}{-6}=\frac{1}{6}
\end{aligned}$$
The answer checks.

33.
$$\begin{aligned}
\frac{5}{x+4}+\frac{1}{x+4}&=x-1\\
\left(\frac{5}{x+4}+\frac{1}{x+4}\right)(x+4)&=(x-1)(x+4)\\
5+1&=x^2+3x-4\\
0&=x^2+3x-10\\
0&=(x+5)(x-2)
\end{aligned}$$

$x+5=0$ **or** $x-2=0$

$x=-5$ $\quad$ $x=2$ $\quad$ Both answers check.

34.
$$\frac{2}{x-1}+\frac{x-2}{3}=\frac{4}{x-1}$$
$$\left(\frac{2}{x-1}+\frac{x-2}{3}\right)(3)(x-1)=\left(\frac{4}{x-1}\right)(3)(x-1)$$
$$6+(x-2)(x-1)=12$$
$$6+x^2-3x+2=12$$
$$x^2-3x-4=0$$
$$(x+1)(x-4)=0$$
$x+1=0$ **or** $x-4=0$
$x=-1$ $x=4$ Both answers check.

35.
$$\frac{3}{x+1}-\frac{x-2}{2}=\frac{x-2}{x+1}$$
$$\left(\frac{3}{x+1}-\frac{x-2}{2}\right)(2)(x+1)=\left(\frac{x-2}{x+1}\right)(2)(x+1)$$
$$6-(x-2)(x+1)=2x-4$$
$$6-(x^2-x-2)=2x-4$$
$$6-x^2+x+2=2x-4$$
$$0=x^2+x-12$$
$$0=(x+4)(x-3)$$
$x+4=0$ **or** $x-3=0$
$x=-4$ $x=3$ Both answers check.

36.
$$\frac{x-4}{x-3}+\frac{x-2}{x-3}=x-3$$
$$\left(\frac{x-4}{x-3}+\frac{x-2}{x-3}\right)(x-3)=(x-3)(x-3)$$
$$x-4+x-2=x^2-6x+9$$
$$2x-6=x^2-6x+9$$
$$0=x^2-8x+15$$
$$0=(x-5)(x-3)$$
$x-5=0$ **or** $x-3=0$
$x=5$ $x=3$ $x=5$ checks, but $x=3$ does not and is not a solution.

37.

$$\frac{2}{x-3}+\frac{3}{4}=\frac{17}{2x}$$

$$\left(\frac{2}{x-3}+\frac{3}{4}\right)(4x)(x-3)=\left(\frac{17}{2x}\right)(4x)(x-3)$$

$$8x+3x(x-3)=34(x-3)$$

$$8x+3x^2-9x=34x-102$$

$$3x^2-35x+102=0$$

$$(3x-17)(x-6)=0$$

$3x-17=0$ **or** $x-6=0$

$3x=17$ $\quad x=6$

$x=\frac{17}{3}$ Both answers check.

38.

$$\frac{30}{y-2}+\frac{24}{y-5}=13$$

$$\left(\frac{30}{y-2}+\frac{24}{y-5}\right)(y-2)(y-5)=13(y-2)(y-5)$$

$$30(y-5)+24(y-2)=13(y^2-7y+10)$$

$$30y-150+24y-48=13y^2-91y+130$$

$$0=13y^2-145y+328$$

$$0=(13y-41)(y-8)$$

$13y-41=0$ **or** $y-8=0$

$13y=41$ $\quad y=8$

$y=\frac{41}{13}$ Both answers check.

39.

$$\frac{x+4}{x+7}-\frac{x}{x+3}=\frac{3}{8}$$

$$\left(\frac{x+4}{x+7}-\frac{x}{x+3}\right)(8)(x+7)(x+3)=\left(\frac{3}{8}\right)(8)(x+7)(x+3)$$

$$8(x+4)(x+3)-8x(x+7)=3(x+7)(x+3)$$

$$8(x^2+7x+12)-8x^2-56x=3(x^2+10x+21)$$

$$8x^2+56x+96-8x^2-56x=3x^2+30x+63$$

$$96=3x^2+30x+63$$

$$0=3x^2+30x-33$$

$$0=3(x^2+10x-11)$$

$$0=3(x+11)(x-1)$$

$x+11=0$ **or** $x-1=0$

$x=-11$ $\quad x=1$ Both answers check.

40.

$$\frac{5}{x+4} - \frac{1}{3} = \frac{x-1}{x}$$
$$\left(\frac{5}{x+4} - \frac{1}{3}\right)(3x)(x+4) = \left(\frac{x-1}{x}\right)(3x)(x+4)$$
$$15x - x(x+4) = 3(x-1)(x+4)$$
$$15x - x^2 - 4x = 3(x^2 + 3x - 4)$$
$$-x^2 + 11x = 3x^2 + 9x - 12$$
$$0 = 4x^2 - 2x - 12$$
$$0 = 2(2x^2 - x - 6)$$
$$0 = 2(2x+3)(x-2)$$

$2x + 3 = 0$ **or** $x - 2 = 0$

$2x = -3$ $\quad x = 2$

$x = -\frac{3}{2}$

Both answers check.

41.

$$S = \frac{a}{1-r}$$
$$S(1-r) = \left(\frac{a}{1-r}\right)(1-r)$$
$$S - Sr = a$$
$$S - a = Sr$$
$$\frac{S-a}{S} = r, \text{ or } r = \frac{S-a}{S}$$

42.

$$\frac{1}{p} + \frac{1}{q} = \frac{1}{f}$$
$$\left(\frac{1}{p} + \frac{1}{q}\right)pqf = \left(\frac{1}{f}\right)pqf$$
$$qf + pf = pq$$
$$f(q+p) = pq$$
$$f = \frac{pq}{q+p}$$

43.

$$\frac{1}{p} + \frac{1}{q} = \frac{1}{f}$$
$$\left(\frac{1}{p} + \frac{1}{q}\right)pqf = \left(\frac{1}{f}\right)pqf$$
$$qf + pf = pq$$
$$qf = pq - pf$$
$$qf = p(q-f)$$
$$\frac{qf}{q-f} = p, \text{ or } p = \frac{qf}{q-f}$$

44.

$$\frac{1}{R} = \frac{1}{r_1} + \frac{1}{r_2}$$
$$\left(\frac{1}{R}\right)Rr_1r_2 = \left(\frac{1}{r_1} + \frac{1}{r_2}\right)Rr_1r_2$$
$$r_1r_2 = Rr_2 + Rr_1$$
$$r_1r_2 = R(r_2 + r_1)$$
$$\frac{r_1r_2}{r_2+r_1} = R, \text{ or } R = \frac{r_1r_2}{r_2+r_1}$$

45.

$$S = \frac{a-lr}{1-r}$$
$$S(1-r) = \left(\frac{a-lr}{1-r}\right)(1-r)$$
$$S - Sr = a - lr$$
$$S - a = Sr - lr$$
$$S - a = r(S-l)$$
$$\frac{S-a}{S-l} = r, \text{ or } r = \frac{S-a}{S-l}$$

46.

$$H = \frac{2ab}{a+b}$$
$$H(a+b) = \left(\frac{2ab}{a+b}\right)(a+b)$$
$$Ha + Hb = 2ab$$
$$Hb = 2ab - Ha$$
$$Hb = a(2b - H)$$
$$\frac{Hb}{2b-H} = a, \text{ or } a = \frac{Hb}{2b-H}$$

47.

$$\begin{aligned}\frac{1}{R} &= \frac{1}{r_1} + \frac{1}{r_2} + \frac{1}{r_3}\\ \left(\frac{1}{R}\right)Rr_1r_2r_3 &= \left(\frac{1}{r_1} + \frac{1}{r_2} + \frac{1}{r_3}\right)Rr_1r_2r_3\\ r_1r_2r_3 &= Rr_2r_3 + Rr_1r_3 + Rr_1r_2\\ r_1r_2r_3 &= R(r_2r_3 + r_1r_3 + r_1r_2)\\ \frac{r_1r_2r_3}{r_2r_3 + r_1r_3 + r_1r_2} &= R, \text{ or } R = \frac{r_1r_2r_3}{r_2r_3 + r_1r_3 + r_1r_2}\end{aligned}$$

48.

$$\begin{aligned}\frac{1}{R} &= \frac{1}{r_1} + \frac{1}{r_2} + \frac{1}{r_3}\\ \left(\frac{1}{R}\right)Rr_1r_2r_3 &= \left(\frac{1}{r_1} + \frac{1}{r_2} + \frac{1}{r_3}\right)Rr_1r_2r_3\\ r_1r_2r_3 &= Rr_2r_3 + Rr_1r_3 + Rr_1r_2\\ r_1r_2r_3 - Rr_1r_3 - Rr_1r_2 &= Rr_2r_3\\ r_1(r_2r_3 - Rr_3 - Rr_2) &= Rr_2r_3\\ r_1 &= \frac{Rr_2r_3}{r_2r_3 - Rr_3 - Rr_2} = \frac{-Rr_2r_3}{Rr_3 + Rr_2 - r_2r_3}\end{aligned}$$

49.

$$\begin{aligned}\frac{1}{f} &= \frac{1}{s_1} + \frac{1}{s_2}\\ \left(\frac{1}{f}\right)fs_1s_2 &= \left(\frac{1}{s_1} + \frac{1}{s_2}\right)fs_1s_2\\ s_1s_2 &= fs_2 + fs_1\\ s_1s_2 &= f(s_2 + s_1)\\ \frac{s_1s_2}{s_2 + s_1} &= f, \text{ or } f = \frac{s_1s_2}{s_2 + s_1}\end{aligned}$$

$$f = \frac{s_1s_2}{s_2 + s_1} = \frac{(60 \text{ in.})(5 \text{ in.})}{60 \text{ in.} + 5 \text{ in.}} = \frac{300 \text{ in.}^2}{65 \text{ in.}} = \frac{60}{13} \text{ in.} = 4\frac{8}{13} \text{ inches}$$

50.

$$\begin{aligned}\frac{1}{f} &= 0.6\left(\frac{1}{r_1} + \frac{1}{r_2}\right)\\ \left(\frac{1}{f}\right)fr_1r_2 &= 0.6\left(\frac{1}{r_1} + \frac{1}{r_2}\right)fr_1r_2\\ r_1r_2 &= 0.6fr_2 + 0.6fr_1\\ r_1r_2 &= f(0.6r_2 + 0.6r_1)\\ \frac{r_1r_2}{0.6r_2 + 0.6r_1} &= f, \text{ or } f = \frac{r_1r_2}{0.6(r_2 + r_1)}\end{aligned}$$

$$f = \frac{r_1r_2}{0.6(r_2 + r_1)} = \frac{(8 \text{ cm})(8 \text{ cm})}{0.6(8 \text{ cm} + 8 \text{ cm})} = \frac{64 \text{ cm}^2}{0.6(16 \text{ cm})} = \frac{64 \text{ cm}^2}{9.6 \text{ cm}} = \frac{640 \text{ cm}^2}{96 \text{ cm}} = \frac{20}{3} \text{ cm}$$

51. The first painter paints $\frac{1}{5}$ of a house in 1 day, while the second paints $\frac{1}{3}$ of a house in 1 day. Let $x =$ the number of days it takes them to paint the house together. Then they can paint $\frac{1}{x}$ of a house together in 1 day.

$$\boxed{\text{Amount 1st paints in 1 day}} + \boxed{\text{Amount 2nd paints in 1 day}} = \boxed{\text{Amount painted by both in 1 day}}$$

$$\frac{1}{5} + \frac{1}{3} = \frac{1}{x}$$
$$\left(\frac{1}{5} + \frac{1}{3}\right)15x = \frac{1}{x} \cdot 15x$$
$$3x + 5x = 15$$
$$8x = 15$$
$$x = \frac{15}{8} = 1\frac{7}{8}$$

They can paint the house together in $1\frac{7}{8}$ days.

52. The first reader reads $\frac{1}{8}$ of the pages in 1 hour, while the second reads $\frac{1}{10}$ of the pages in 1 hour. Let $x =$ the number of hours it takes them to read the pages together. Then they can read $\frac{1}{x}$ of the pages together in 1 hour.

$$\boxed{\text{Amount 1st reads in 1 hour}} + \boxed{\text{Amount 2nd reads in 1 hour}} = \boxed{\text{Amount read by both in 1 hour}}$$

$$\frac{1}{8} + \frac{1}{10} = \frac{1}{x}$$
$$\left(\frac{1}{8} + \frac{1}{10}\right)40x = \frac{1}{x} \cdot 40x$$
$$5x + 4x = 40$$
$$9x = 40$$
$$x = \frac{40}{9} = 4\frac{4}{9} \text{ hours}$$

They can meet the deadline.

53. The first belt can move $\frac{1}{10}$ of the 1000 bushels in 1 minute, while the second can move $\frac{1}{14}$ of the 1000 bushels in 1 minute. Let $x =$ the number of minutes for them to move the 1000 bushels together. Then $\frac{1}{x} =$ the amount of the 1000 bushels they can move together in 1 minute.

$$\boxed{\text{Amount 1st moves in 1 minute}} + \boxed{\text{Amount 2nd moves in 1 minute}} = \boxed{\text{Amount moved together in 1 minute}}$$

$$\frac{1}{10} + \frac{1}{14} = \frac{1}{x}$$
$$\left(\frac{1}{10} + \frac{1}{14}\right)(70x) = \frac{1}{x} \cdot 70x$$
$$7x + 5x = 70$$
$$12x = 70$$
$$x = \frac{70}{12} = \frac{35}{6} = 5\frac{5}{6}$$

It will take them $5\frac{5}{6}$ minutes to move 1000 bushels working together.

54. The first crew can finish $\frac{1}{12}$ of the roof in 1 hour, while the second can finish $\frac{1}{10}$ of the roof in 1 hour. Let $x =$ the number of hours needed for the crews to finish the roof together. Then $\frac{1}{x} =$ the amount of the roof they can finish together in 1 hour.

Amount 1st finishes in 1 hour	+	Amount 2nd finishes in 1 hour	=	Amount finished together in 1 hour

$$\frac{1}{12} + \frac{1}{10} = \frac{1}{x}$$
$$\left(\frac{1}{12} + \frac{1}{10}\right)(60x) = \frac{1}{x} \cdot 60x$$
$$5x + 6x = 60$$
$$11x = 60$$
$$x = \frac{60}{11} = 5\frac{5}{11}$$

It will take them $5\frac{5}{11}$ hours, so they will not finish before the predicted rain.

55. The first drain can drain $\frac{1}{3}$ of the pool in 1 day, while the second can drain $\frac{1}{2}$ of the pool in 1 day. Let $x =$ the number of days needed for them to drain the pool together. Then $\frac{1}{x} =$ the amount of the pool they can drain together in 1 day.

Amount 1st drains in 1 day	+	Amount 2nd drains in 1 day	=	Amount drained together in 1 day

$$\frac{1}{3} + \frac{1}{2} = \frac{1}{x}$$
$$\left(\frac{1}{3} + \frac{1}{2}\right)6x = \frac{1}{x} \cdot 6x$$
$$2x + 3x = 6$$
$$5x = 6$$
$$x = \frac{6}{5} = 1\frac{1}{5}$$

It will take the two drains $1\frac{1}{5}$ days to drain the pool.

56. The first pipe can fill $\frac{1}{9}$ of the pool in 1 hour. Let $x =$ the number of hours needed for the second pipe to fill the pool alone. Then it can fill $\frac{1}{x}$ of the pool alone. Since both pipes can fill the pool in 3 hours, they fill $\frac{1}{3}$ of the pool in 1 hour.

Amount 1st fills in 1 hour	+	Amount 2nd fills in 1 hour	=	Amount filled together in 1 hour

$$\frac{1}{9} + \frac{1}{x} = \frac{1}{3}$$
$$\left(\frac{1}{9} + \frac{1}{x}\right)(9x) = \frac{1}{3} \cdot 9x$$
$$x + 9 = 3x$$
$$9 = 2x, \text{ or } x = \frac{9}{2} = 4\frac{1}{2}$$

It takes the second pipe $4\frac{1}{2}$ hours to fill the pool alone.

57. The first pipe fills $\frac{1}{3}$ of the pond in 1 week, while the second fills $\frac{1}{5}$ of the pond in a week and evaporation empties $\frac{1}{10}$ of the pond in a week. Let $x =$ the number of weeks needed to fill the pond with both pipes, considering evaporation. Then $\frac{1}{x}$ of the pond is filled in 1 week.

$$\boxed{\text{Amount 1st pipe fills in 1 week}} + \boxed{\text{Amount 2nd pipe fills in 1 week}} - \boxed{\text{Amount emptied in 1 week}} = \boxed{\text{Total amount filled in 1 week}}$$

$$\begin{aligned} \frac{1}{3} + \frac{1}{5} - \frac{1}{10} &= \frac{1}{x} \\ \left(\frac{1}{3} + \frac{1}{5} - \frac{1}{10}\right)30x &= \frac{1}{x} \cdot 30x \\ 10x + 6x - 3x &= 30 \\ 13x &= 30 \\ x &= \frac{30}{13} = 2\frac{4}{13} \end{aligned}$$

It will take $2\frac{4}{13}$ weeks to fill the pond.

58. Sally can clean $\frac{1}{6}$ of the house in 1 hour, while her father can clean $\frac{1}{4}$ of the house in 1 hour and her brother can mess up $\frac{1}{8}$ of the house in 1 hour. Let $x =$ the number of hours needed for both to clean up the house, considering her brother. Then $\frac{1}{x} =$ the amount of the house cleaned in 1 hr.

$$\boxed{\text{Amount Sally cleans in 1 hour}} + \boxed{\text{Amount her father cleans in 1 hour}} - \boxed{\text{Amount her brother messes up in 1 hour}} = \boxed{\text{Total amount cleaned in 1 hour}}$$

$$\begin{aligned} \frac{1}{6} + \frac{1}{4} - \frac{1}{8} &= \frac{1}{x} \\ \left(\frac{1}{6} + \frac{1}{4} - \frac{1}{8}\right)24x &= \frac{1}{x} \cdot 24x \\ 4x + 6x - 3x &= 24 \\ 7x &= 24 \\ x &= \frac{24}{7} = 3\frac{3}{7} \end{aligned}$$

It will take $3\frac{3}{7}$ hours to clean the house.

59. Let $w =$ the rate at which he walks. Then $w + 5 =$ the rate at which he bicycles.

	Rate	Time	Dist.
Walk	w	$\frac{24}{w}$	24
Bicycle	$w+5$	$\frac{24}{w+5}$	24

$$\boxed{\text{Time walking}} + \boxed{\text{Time bicycling}} = \boxed{\text{Total time}}$$

$$\frac{24}{w} + \frac{24}{w+5} = 11$$

$$\left(\frac{24}{w} + \frac{24}{w+5}\right)(w)(w+5) = 11w(w+5)$$

$$24(w+5) + 24w = 11w^2 + 55w$$

$$0 = 11w^2 + 7w - 120$$

$$0 = (11w + 40)(w - 3)$$

$11w + 40 = 0$ **or** $w - 3 = 0$

$w = -\frac{40}{11}$ $\quad$ $w = 3$ $\quad$ $w = 3$ is the only answer that makes sense.

He walks at a rate of 3 miles per hour.

60. Let $r =$ the rate of the slower train. Then $r + 10 =$ the rate of the faster train.

	Rate	Time	Dist.
Slower	r	$\frac{315}{r}$	315
Faster	$r+10$	$\frac{315}{r+10}$	315

$$\boxed{\text{Faster time}} + 2 = \boxed{\text{Slower time}}$$

$$\frac{315}{r+10} + 2 = \frac{315}{r}$$

$$\left(\frac{315}{r+10} + 2\right)(r)(r+10) = \frac{315}{r} \cdot r(r+10)$$

$$315r + 2r(r+10) = 315(r+10)$$

$$2r^2 + 20r - 3150 = 0$$

$$2(r+45)(r-35) = 0$$

$r + 45 = 0$ **or** $r - 35 = 0$

$r = -45$ $\quad$ $r = 35$ $\quad$ $r = 35$ is the only answer that makes sense.

The trains have rates of 35 and 45 miles per hour.

61. Let $r =$ the first rate. Then $r - 20 =$ the second rate.

	Rate	Time	Dist.
First	r	$\frac{120}{r}$	120
Second	$r - 20$	$\frac{120}{r-20}$	120

$$\boxed{\text{First time}} + \boxed{\text{Second time}} = 5$$

$$\frac{120}{r} + \frac{120}{r-20} = 5$$

$$\left(\frac{120}{r} + \frac{120}{r-20}\right)(r)(r-20) = 5r(r-20)$$

$$120(r-20) + 120r = 5r^2 - 100r$$

$$0 = 5r^2 - 340r + 2400$$

$$0 = 5(r-60)(r-8)$$

$r - 60 = 0$ **or** $r - 8 = 0$

$r = 60$ $\quad$ $r = 8$ $\quad$ $r = 60$ is the only answer that makes sense.

The train traveled at rates of 60 and 40 miles per hour.

62. Let $t =$ the time from Rockford to St. Louis. Then $t - 3 =$ the time from Rockford to Chicago.

	Rate	Time	Dist.
to St. Louis	$\frac{275}{t}$	t	275
to Chicago	$\frac{110}{t-3}$	$t - 3$	110

$$\boxed{\text{First rate}} = \boxed{\text{Second rate}}$$

$$\frac{275}{t} = \frac{110}{t-3}$$

$$\left(\frac{275}{t}\right)(t)(t-3) = \left(\frac{110}{t-3}\right)(t)(t-3)$$

$$275(t-3) = 110t$$

$$275t - 825 = 110t$$

$$165t = 825$$

$$t = \frac{825}{165} = 5$$

The driver who went to Chicago was on the road for 2 hours.

63. Let $c =$ the speed of the current.

	Rate	Time	Dist.
Downstream	$12+c$	$\frac{45}{12+c}$	45
Upstream	$12-c$	$\frac{27}{12-c}$	27

$$\boxed{\text{Time downstream}} = \boxed{\text{Time upstream}}$$

$$\frac{45}{12+c} = \frac{27}{12-c}$$

$$\frac{45}{12+c}(12+c)(12-c) = \frac{27}{12-c}(12+c)(12-c)$$

$$45(12-c) = 27(12+c)$$

$$540 - 45c = 324 + 27c$$

$$-72c = -216$$

$$c = \frac{-216}{-72} = 3$$

The speed of the current is 3 miles per hour.

64. Let $c =$ the speed of the current.

	Rate	Time	Dist.
Downstream	$3+c$	$\frac{10}{3+c}$	10
Upstream	$3-c$	$\frac{10}{3-c}$	10

$$\boxed{\text{Time downstream}} + \boxed{\text{Time upstream}} = 12$$

$$\frac{10}{3+c} + \frac{10}{3-c} = 12$$

$$\left(\frac{10}{3+c} + \frac{10}{3-c}\right)(3+c)(3-c) = 12(3+c)(3-c)$$

$$10(3-c) + 10(3+c) = 12(9-c^2)$$

$$30 - 10c + 30 + 10c = -12c^2 + 108$$

$$12c^2 - 48 = 0$$

$$12(c+2)(c-2) = 0$$

$c + 2 = 0$ **or** $c - 2 = 0$

$c = -2$ $\qquad$ $c = 2$ $\quad$ $c = 2$ is the only answer that makes sense.

The speed of the current is 2 miles per hour.

65. Let $w =$ the speed of the wind.

	Rate	Time	Dist.
Downwind	$340+w$	$\frac{200}{340+w}$	200
Upwind	$340-w$	$\frac{140}{340-w}$	140

$$\boxed{\text{Time downwind}} = \boxed{\text{Time upwind}}$$

$$\frac{200}{340+w} = \frac{140}{340-w}$$

$$\left(\frac{200}{340+w}\right)(340+w)(340-w) = \left(\frac{140}{340-w}\right)(340+w)(340-w)$$

$$200(340-w) = 140(340+w)$$

$$68000 - 200w = 47600 + 140w$$

$$-340w = -20400$$

$$w = 60$$

The speed of the wind is 60 miles per hour.

66. Let $p =$ the speed of the plane in still air.

	Rate	Time	Dist.
Downwind	$p+40$	$\frac{650}{p+40}$	650
Upwind	$p-40$	$\frac{475}{p-40}$	475

$$\boxed{\text{Time downwind}} = \boxed{\text{Time upwind}}$$

$$\frac{650}{p+40} = \frac{475}{p-40}$$

$$\left(\frac{650}{p+40}\right)(p+40)(p-40) = \left(\frac{475}{p-40}\right)(p+40)(p-40)$$

$$650(p-40) = 475(p+40)$$

$$650p - 26000 = 475p + 19000$$

$$175p = 45000$$

$$p \approx 257$$

The speed of the plane is about 257 miles per hour.

67. Let x = the number purchased.
Then the unit cost $= \frac{224}{x}$.

$$\boxed{\text{New unit cost}} \cdot \boxed{\text{New \# motors}} = 224$$

$$\left(\frac{224}{x} - 4\right)(x+1) = 224$$

$$\left(\frac{224}{x} - 4\right)(x+1)x = 224x$$

$$224(x+1) - 4x(x+1) = 224x$$

$$224x + 224 - 4x^2 - 4x = 224x$$

$$-4x^2 - 4x + 224 = 0$$

$$-4(x+8)(x-7) = 0$$

$x + 8 = 0$ **or** $x - 7 = 0$

$x = -8$ $\quad x = 7$

He originally bought 7 motors.

68. Let x = the number purchased.
Then the unit cost $= \frac{1800}{x}$.

$$\boxed{\text{New unit cost}} \cdot \boxed{\text{New \# ovens}} = 1800$$

$$\left(\frac{1800}{x} + 25\right)(x-1) = 1800$$

$$\left(\frac{1800}{x} + 25\right)(x-1)x = 1800x$$

$$1800(x-1) + 25x(x-1) = 1800x$$

$$1800x - 1800 + 25x^2 - 25x = 1800x$$

$$25x^2 - 25x - 1800 = 0$$

$$25(x+8)(x-9) = 0$$

$x + 8 = 0$ **or** $x - 9 = 0$

$x = -8$ $\quad x = 9$

She originally bought 9 ovens.

69. Let x = the number of days.
Then the daily cost $= \frac{1200}{x}$.

$$\boxed{\text{New daily cost}} \cdot \boxed{\text{New \# days}} = 1200$$

$$\left(\frac{1200}{x} - 20\right)(x+3) = 1200$$

$$\left(\frac{1200}{x} - 20\right)(x+3)x = 1200x$$

$$1200(x+3) - 20x(x+3) = 1200x$$

$$1200x + 3600 - 20x^2 - 60x = 1200x$$

$$-20x^2 - 60x + 3600 = 0$$

$$-20(x+15)(x-12) = 0$$

$x + 15 = 0$ **or** $x - 12 = 0$

$x = -15$ $\quad x = 12$

Her original vacation was 12 days.

70.

$$k = \frac{1}{\frac{1}{k_1} + \frac{1}{k_2}}$$

$$k = \frac{1k_1k_2}{\left(\frac{1}{k_1} + \frac{1}{k_2}\right)k_1k_2}$$

$$k = \frac{k_1k_2}{k_2 + k_1}$$

$$k(k_2 + k_1) = k_1k_2$$

$$kk_2 + kk_1 = k_1k_2$$

$$kk_2 = k_1k_2 - kk_1$$

$$kk_2 = k_1(k_2 - k)$$

$$\frac{kk_2}{k_2 - k} = k_1$$

$$k_1 = \frac{kk_2}{k_2 - k}$$

$$= \frac{1{,}900{,}000(4{,}200{,}000)}{4{,}200{,}000 - 1{,}900{,}000}$$

$$= \frac{7.98 \times 10^{12}}{2.3 \times 10^6}$$

$$\approx 3.5 \times 10^6$$

k_1 should have a stiffness of about 3,500,000 in. lb/rad.

71. **Answers may vary.**

72. **Answers may vary.**

73. $\dfrac{x}{x-3}+\dfrac{3}{x-3}=\dfrac{6}{x-3}$

74.
$$(x-1)^{-1}-x^{-1}=6^{-1}$$
$$\frac{1}{x-1}-\frac{1}{x}=\frac{1}{6}$$
$$\left(\frac{1}{x-1}-\frac{1}{x}\right)6x(x-1)=\frac{1}{6}\cdot 6x(x-1)$$
$$6x-6(x-1)=x(x-1)$$
$$0=x^2-x-6$$
$$0=(x+2)(x-3)$$

$x+2=0$ **or** $x-3=0$

$x=-2$ $\quad$ $x=3$

Both solutions check.

Exercise 6.6 (page 402)

1. $2(x^2+4x-1)+3(2x^2-2x+2)=2x^2+8x-2+6x^2-6x+6=8x^2+2x+4$

2. $3(2a^2-3a+2)-4(2a^2+4a-7)=6a^2-9a+6-8a^2-16a+28=-2a^2-25a+34$

3. $-2(3y^3-2y+7)-3(y^2+2y-4)+4(y^3+2y-1)$
$=-6y^3+4y-14-3y^2-6y+12+4y^3+8y-4=-2y^3-3y^2+6y-6$

4. $3(4y^3+3y-2)+2(3y^2-y+3)-5(2y^3-y^2-2)$
$=12y^3+9y-6+6y^2-2y+6-10y^3+5y^2+10=2y^3+11y^2+7y+10$

5. $\dfrac{1}{b}$

6. algorithm

7. quotient

8. $\dfrac{6}{3a-2}$

9. $\dfrac{4x^2y^3}{8x^5y^2}=\dfrac{y}{2x^3}$

10. $\dfrac{25x^4y^7}{5xy^9}=\dfrac{5x^3}{y^2}$

11. $\dfrac{33a^{-2}b^2}{44a^2b^{-2}}=\dfrac{3b^4}{4a^4}$

12. $\dfrac{-63a^4b^{-3}}{81a^{-3}b^3}=-\dfrac{7a^7}{9b^6}$

13. $\dfrac{45x^{-2}y^{-3}t^0}{-63x^{-1}y^4t^2}=-\dfrac{5}{7xy^7t^2}$

14. $\dfrac{112a^0b^2c^{-3}}{48a^4b^0c^4}=\dfrac{7b^2}{3a^4c^7}$

15. $\dfrac{-65a^{2n}b^nc^{3n}}{-15a^nb^{-n}c}=\dfrac{13a^nb^{2n}c^{3n-1}}{3}$

16. $\dfrac{-32x^{-3n}y^{-2n}z}{40x^{-2}y^{-n}z^{n+1}}=-\dfrac{4}{5x^{3n-2}y^nz^n}$

17. $\dfrac{4x^2-x^3}{6x}=\dfrac{4x^2}{6x}-\dfrac{x^3}{6x}=\dfrac{2x}{3}-\dfrac{x^2}{6}$

18. $\dfrac{5y^4+45y^3}{15y^2}=\dfrac{5y^4}{15y^2}+\dfrac{45y^3}{15y^2}=\dfrac{y^2}{3}+3y$

19. $\dfrac{4x^2y^3+x^3y^2}{6xy}=\dfrac{4x^2y^3}{6xy}+\dfrac{x^3y^2}{6xy}$
$=\dfrac{2xy^2}{3}+\dfrac{x^2y}{6}$

20. $\dfrac{3a^3y^2-18a^4y^3}{27a^2y^2}=\dfrac{3a^3y^2}{27a^2y^2}-\dfrac{18a^4y^3}{27a^2y^2}$
$=\dfrac{a}{9}-\dfrac{2a^2y}{3}$

21. $\dfrac{24x^6y^7 - 12x^5y^{12} + 36xy}{48x^2y^3} = \dfrac{24x^6y^7}{48x^2y^3} - \dfrac{12x^5y^{12}}{48x^2y^3} + \dfrac{36xy}{48x^2y^3} = \dfrac{x^4y^4}{2} - \dfrac{x^3y^9}{4} + \dfrac{3}{4xy^2}$

22. $\dfrac{9x^4y^3 + 18x^2y - 27xy^4}{9x^3y^3} = \dfrac{9x^4y^3}{9x^3y^3} + \dfrac{18x^2y}{9x^3y^3} - \dfrac{27xy^4}{9x^3y^3} = x + \dfrac{2}{xy^2} - \dfrac{3y}{x^2}$

23. $\dfrac{3a^{-2}b^3 - 6a^2b^{-3} + 9a^{-2}}{12a^{-1}b} = \dfrac{3a^{-2}b^3}{12a^{-1}b} - \dfrac{6a^2b^{-3}}{12a^{-1}b} + \dfrac{9a^{-2}}{12a^{-1}b} = \dfrac{b^2}{4a} - \dfrac{a^3}{2b^4} + \dfrac{3}{4ab}$

24. $\dfrac{4x^3y^{-2} + 8x^{-2}y^2 - 12y^4}{12x^{-1}y^{-1}} = \dfrac{4x^3y^{-2}}{12x^{-1}y^{-1}} + \dfrac{8x^{-2}y^2}{12x^{-1}y^{-1}} - \dfrac{12y^4}{12x^{-1}y^{-1}} = \dfrac{x^4}{3y} + \dfrac{2y^3}{3x} - xy^5$

25. $\dfrac{x^ny^n - 3x^{2n}y^{2n} + 6x^{3n}y^{3n}}{x^ny^n} = \dfrac{x^ny^n}{x^ny^n} - \dfrac{3x^{2n}y^{2n}}{x^ny^n} + \dfrac{6x^{3n}y^{3n}}{x^ny^n} = 1 - 3x^ny^n + 6x^{2n}y^{2n}$

26. $\dfrac{2a^n - 3a^nb^{2n} - 6b^{4n}}{a^nb^{n-1}} = \dfrac{2a^n}{a^nb^{n-1}} - \dfrac{3a^nb^{2n}}{a^nb^{n-1}} - \dfrac{6b^{4n}}{a^nb^{n-1}} = \dfrac{2}{b^{n-1}} - 3b^{n+1} - \dfrac{6b^{3n+1}}{a^n}$

27.
$$\begin{array}{r} x+2 \\ x+3\,\overline{)\,x^2+5x+6} \\ \underline{x^2+3x} \\ 2x+6 \\ \underline{2x+6} \\ 0 \end{array}$$

28.
$$\begin{array}{r} x-2 \\ x-3\,\overline{)\,x^2-5x+6} \\ \underline{x^2-3x} \\ -2x+6 \\ \underline{-2x+6} \\ 0 \end{array}$$

29.
$$\begin{array}{r} x+7 \\ x+3\,\overline{)\,x^2+10x+21} \\ \underline{x^2+3x} \\ 7x+21 \\ \underline{7x+21} \\ 0 \end{array}$$

30.
$$\begin{array}{r} x+3 \\ x+7\,\overline{)\,x^2+10x+21} \\ \underline{x^2+7x} \\ 3x+21 \\ \underline{3x+21} \\ 0 \end{array}$$

31.
$$\begin{array}{r} 3x-5+\frac{3}{2x+3} \\ 2x+3\,\overline{)\,6x^2-x-12} \\ \underline{6x^2+9x} \\ -10x-12 \\ \underline{-10x-15} \\ 3 \end{array}$$

32.
$$\begin{array}{r} 3x+4 \\ 2x-3\,\overline{)\,6x^2-x-12} \\ \underline{6x^2-9x} \\ 8x-12 \\ \underline{8x-12} \\ 0 \end{array}$$

33.
$$\begin{array}{r} 3x^2+x+2+\frac{8}{x-1} \\ x-1\,\overline{)\,3x^3-2x^2+x+6} \\ \underline{3x^3-3x^2} \\ x^2+x \\ \underline{x^2-x} \\ 2x+6 \\ \underline{2x-2} \\ 8 \end{array}$$

34.
$$\begin{array}{r} 4a^2-3a+0+\frac{7}{a+1} \\ a+1\,\overline{)\,4a^3+a^2-3a+7} \\ \underline{4a^3+4a^2} \\ -3a^2-3a \\ \underline{-3a^2-3a} \\ 0a+7 \\ \underline{0a+0} \\ 7 \end{array}$$

35.
$$\begin{array}{r} 2x^2+5x+3+\frac{4}{3x-2} \\ 3x-2\,\overline{)\,6x^3+11x^2-x-2} \\ \underline{6x^3-4x^2} \\ 15x^2-x \\ \underline{15x^2-10x} \\ 9x-2 \\ \underline{9x-6} \\ 4 \end{array}$$

36.
$$\begin{array}{r} 3x^2+x-2+\frac{16}{2x+3} \\ 2x+3\,\overline{)\,6x^3+11x^2-x+10} \\ \underline{6x^3+9x^2} \\ 2x^2-x \\ \underline{2x^2+3x} \\ -4x+10 \\ \underline{-4x-6} \\ 16 \end{array}$$

37.
$$\begin{array}{rl} & 3x^2 + 4x + 3 \\ 2x-3 & \overline{)6x^3 - x^2 - 6x - 9} \\ & \underline{6x^3 - 9x^2} \\ & 8x^2 - 6x \\ & \underline{8x^2 - 12x} \\ & 6x - 9 \\ & \underline{6x - 9} \\ & 0 \end{array}$$

38.
$$\begin{array}{rl} & 4x^2 - x - 1 \\ 4x+5 & \overline{)16x^3 + 16x^2 - 9x - 5} \\ & \underline{16x^3 + 20x^2} \\ & -4x^2 - 9x \\ & \underline{-4x^2 - 5x} \\ & -4x - 5 \\ & \underline{-4x - 5} \\ & 0 \end{array}$$

39.
$$\begin{array}{rl} & a + 1 \\ a+1 & \overline{)a^2 + 2a + 1} \\ & \underline{a^2 + a} \\ & a + 1 \\ & \underline{a + 1} \\ & 0 \end{array}$$

40.
$$\begin{array}{rl} & 3a + 5 \\ 2a-3 & \overline{)6a^2 + a - 15} \\ & \underline{6a^2 - 9a} \\ & 10a - 15 \\ & \underline{10a - 15} \\ & 0 \end{array}$$

41.
$$\begin{array}{rl} & 2y + 2 \\ 5y-2 & \overline{)10y^2 + 6y - 4} \\ & \underline{10y^2 - 4y} \\ & 10y - 4 \\ & \underline{10y - 4} \\ & 0 \end{array}$$

42.
$$\begin{array}{rl} & x - 8y \\ x-2y & \overline{)x^2 - 10xy + 16y^2} \\ & \underline{x^2 - 2xy} \\ & -8xy + 16y^2 \\ & \underline{-8xy + 16y^2} \\ & 0 \end{array}$$

43.
$$\begin{array}{rl} & 6x - 12 \\ x-1 & \overline{)6x^2 - 18x + 12} \\ & \underline{6x^2 - 6x} \\ & -12x + 12 \\ & \underline{-12x + 12} \\ & 0 \end{array}$$

44.
$$\begin{array}{rl} & 3x^2 + 7x + 3 - \frac{9}{2x+3} \\ 2x+3 & \overline{)6x^3 + 23x^2 + 27x + 0} \\ & \underline{6x^3 + 9x^2} \\ & 14x^2 + 27x \\ & \underline{14x^2 + 21x} \\ & 6x + 0 \\ & \underline{6x + 9} \\ & -9 \end{array}$$

45.
$$\begin{array}{rl} & 3x^2 - x + 2 \\ 3x-2 & \overline{)9x^3 - 9x^2 + 8x - 4} \\ & \underline{9x^3 - 6x^2} \\ & -3x^2 + 8x \\ & \underline{-3x^2 + 2x} \\ & 6x - 4 \\ & \underline{6x - 4} \\ & 0 \end{array}$$

46.
$$\begin{array}{rl} & 2x^2 + 3x - 1 \\ 4x-3 & \overline{)8x^3 + 6x^2 - 13x + 3} \\ & \underline{8x^3 - 6x^2} \\ & 12x^2 - 13x \\ & \underline{12x^2 - 9x} \\ & -4x + 3 \\ & \underline{-4x + 3} \\ & 0 \end{array}$$

47.
$$\begin{array}{rl} & 4x^3 - 3x^2 + 3x + 1 \\ 4x+3 & \overline{)16x^4 + 0x^3 + 3x^2 + 13x + 3} \\ & \underline{16x^4 + 12x^3} \\ & -12x^3 + 3x^2 \\ & \underline{-12x^3 - 9x^2} \\ & 12x^2 + 13x \\ & \underline{12x^2 + 9x} \\ & 4x + 3 \\ & \underline{4x + 3} \\ & 0 \end{array}$$

48.
$$\begin{array}{rl} & 3x^2 - x + 2 \\ 3x+2 & \overline{)9x^3 + 3x^2 + 4x + 4} \\ & \underline{9x^3 + 6x^2} \\ & -3x^2 + 4x \\ & \underline{-3x^2 - 2x} \\ & 6x + 4 \\ & \underline{6x + 4} \\ & 0 \end{array}$$

49.

$$
\begin{array}{rr}
 & a^2 + a + 1 + \frac{2}{a-1} \\
a - 1 & \overline{)\,a^3 + 0a^2 + 0a + 1} \\
 & \underline{a^3 - a^2 } \\
 & a^2 + 0a \\
 & \underline{a^2 - a } \\
 & a + 1 \\
 & \underline{a - 1} \\
 & 2
\end{array}
$$

50.

$$
\begin{array}{rr}
 & 9a^2 + 6ab + 4b^2 \\
3a - 2b & \overline{)\,27a^3 + 0a^2b + 0ab^2 - 8b^3} \\
 & \underline{27a^3 - 18a^2b } \\
 & 18a^2b + 0ab^2 \\
 & \underline{18a^2b - 12ab^2 } \\
 & 12ab^2 - 8b^3 \\
 & \underline{12ab^2 - 8b^3} \\
 & 0
\end{array}
$$

51.

$$
\begin{array}{rr}
 & 5a^2 - 3a - 4 \\
3a - 4 & \overline{)\,15a^3 - 29a^2 + 0a + 16} \\
 & \underline{15a^3 - 20a^2 } \\
 & -9a^2 + 0a \\
 & \underline{-9a^2 + 12a } \\
 & -12a + 16 \\
 & \underline{-12a + 16} \\
 & 0
\end{array}
$$

52.

$$
\begin{array}{rr}
 & 2x^2 - 3x + 4 \\
2x - 3 & \overline{)\,4x^3 - 12x^2 + 17x - 12} \\
 & \underline{4x^3 - 6x^2 } \\
 & -6x^2 + 17x \\
 & \underline{-6x^2 + 9x } \\
 & 8x - 12 \\
 & \underline{8x - 12} \\
 & 0
\end{array}
$$

53.

$$
\begin{array}{rr}
 & 6y - 12 \\
y - 2 & \overline{)\,6y^2 - 24y + 24} \\
 & \underline{6y^2 - 12y } \\
 & -12y + 24 \\
 & \underline{-12y + 24} \\
 & 0
\end{array}
$$

54.

$$
\begin{array}{rr}
 & a - 18 \\
-a + 3 & \overline{)\,-a^2 + 21a - 54} \\
 & \underline{-a^2 + 3a } \\
 & 18a - 54 \\
 & \underline{18a - 54} \\
 & 0
\end{array}
$$

55.

$$
\begin{array}{rr}
 & 16x^4 - 8x^3y + 4x^2y^2 - 2xy^3 + y^4 \\
2x + y & \overline{)\,32x^5 + 0x^4y + 0x^3y^2 + 0x^2y^3 + 0xy^4 + y^5} \\
 & \underline{32x^5 + 16x^4y } \\
 & -16x^4y + 0x^3y^2 \\
 & \underline{-16x^4y - 8x^3y^2 } \\
 & 8x^3y^2 + 0x^2y^3 \\
 & \underline{8x^3y^2 + 4x^2y^3 } \\
 & -4x^2y^3 + 0xy^4 \\
 & \underline{-4x^2y^3 - 2xy^4 } \\
 & 2xy^4 + y^5 \\
 & \underline{2xy^4 + y^5} \\
 & 0
\end{array}
$$

56.

$$
\begin{array}{rr}
 & 27x^3 + 9x^2y + 3xy^2 + y^3 \\
3x - y & \overline{)\,81x^4 + 0x^3y + 0x^2y^2 + 0xy^3 - y^4} \\
 & \underline{81x^4 - 27x^3y } \\
 & 27x^3y + 0x^2y^2 \\
 & \underline{27x^3y - 9x^2y^2 } \\
 & 9x^2y^2 + 0xy^3 \\
 & \underline{9x^2y^2 - 3xy^3 } \\
 & 3xy^3 - y^4 \\
 & \underline{3xy^3 - y^4} \\
 & 0
\end{array}
$$

57.

$$\begin{array}{r|l} & x^4 + x^2 + 4 \\ \cline{2-2} x^2 - 2 & x^6 - x^4 + 2x^2 - 8 \\ & \underline{x^6 - 2x^4} \\ & \quad x^4 + 2x^2 \\ & \quad \underline{x^4 - 2x^2} \\ & \qquad 4x^2 - 8 \\ & \qquad \underline{4x^2 - 8} \\ & \qquad\qquad 0 \end{array}$$

58.

$$\begin{array}{r|l} & x^4 - x^2 - 3 \\ \cline{2-2} x^2 + 3 & x^6 + 2x^4 - 6x^2 - 9 \\ & \underline{x^6 + 3x^4} \\ & \quad - x^4 - 6x^2 \\ & \quad \underline{- x^4 - 3x^2} \\ & \qquad - 3x^2 - 9 \\ & \qquad \underline{- 3x^2 - 9} \\ & \qquad\qquad 0 \end{array}$$

59.

$$\begin{array}{r|l} & x^2 + x + 1 \\ \cline{2-2} x^2 + x + 2 & x^4 + 2x^3 + 4x^2 + 3x + 2 \\ & \underline{x^4 + x^3 + 2x^2} \\ & \quad x^3 + 2x^2 + 3x \\ & \quad \underline{x^3 + x^2 + 2x} \\ & \qquad x^2 + x + 2 \\ & \qquad \underline{x^2 + x + 2} \\ & \qquad\qquad 0 \end{array}$$

60.

$$\begin{array}{r|l} & x^2 + 2x + 3 \\ \cline{2-2} 2x^2 - x - 1 & 2x^4 + 3x^3 + 3x^2 - 5x - 3 \\ & \underline{2x^4 - x^3 - x^2} \\ & \quad 4x^3 + 4x^2 - 5x \\ & \quad \underline{4x^3 - 2x^2 - 2x} \\ & \qquad 6x^2 - 3x - 3 \\ & \qquad \underline{6x^2 - 3x - 3} \\ & \qquad\qquad 0 \end{array}$$

61.

$$\begin{array}{r|l} & x^2 + x + 2 \\ \cline{2-2} x^2 + 0x + 3 & x^4 + x^3 + 5x^2 + 3x + 6 \\ & \underline{x^4 + 0x^3 + 3x^2} \\ & \quad x^3 + 2x^2 + 3x \\ & \quad \underline{x^3 + 0x^2 + 3x} \\ & \qquad 2x^2 + 0x + 6 \\ & \qquad \underline{2x^2 + 0x + 6} \\ & \qquad\qquad 0 \end{array}$$

62.

$$\begin{array}{r|l} & x^2 - 2 + \dfrac{-x^2 + 7x + 4}{x^3 + 2x + 1} \\ \cline{2-2} x^3 + 0x^2 + 2x + 1 & x^5 + 0x^4 + 0x^3 + 0x^2 + 3x + 2 \\ & \underline{x^5 + 0x^4 + 2x^3 + x^2} \\ & \qquad - 2x^3 - x^2 + 3x + 2 \\ & \qquad \underline{- 2x^3 + 0x^2 - 4x - 2} \\ & \qquad\qquad - x^2 + 7x + 4 \end{array}$$

63.

$$\begin{array}{r|l} & 9.8x + 16.4 - \frac{36.5}{x-2} \\ \cline{2-2} x - 2 & 9.8x^2 - 3.2x - 69.3 \\ & \underline{9.8x^2 - 19.6x} \\ & \qquad 16.4x - 69.3 \\ & \qquad \underline{16.4x - 32.8} \\ & \qquad\qquad - 36.5 \end{array}$$

64.

$$\begin{array}{r|l} & -8.9x - 28.732 - \frac{122.9584}{2.5x-3.7} \\ \cline{2-2} 2.5x - 3.7 & -22.25x^2 - 38.9x - 16.65 \\ & \underline{-22.25x^2 + 32.93x} \\ & \qquad - 71.83x - 16.65 \\ & \qquad \underline{- 71.83x + 106.3084} \\ & \qquad\qquad - 122.9584 \end{array}$$

65.

$$A = lw \qquad \frac{A}{w} = l \qquad \frac{9x^2+21x+10}{3x+2} = l$$

$$\begin{array}{r|l} & 3x + 5 \\ \hline 3x+2 & 9x^2 + 21x + 10 \\ & 9x^2 + 6x \\ \hline & \quad 15x + 10 \\ & \quad 15x + 10 \\ \hline & \qquad 0 \end{array}$$

The length is $3x + 5$.

66.

$$A = \tfrac{1}{2}bh \qquad 2 \cdot \frac{A}{b} = h \qquad 2 \cdot \frac{10x^2+x-2}{5x-2} = l$$

$$\begin{array}{r|l} & 2x + 1 \\ \hline 5x-2 & 10x^2 + x - 2 \\ & 10x^2 - 4x \\ \hline & \quad 5x - 2 \\ & \quad 5x - 2 \\ \hline & \qquad 0 \end{array}$$

The height is $2(2x + 1)$, or $4x + 2$.

67. Dog sled

$$rt = d$$

$$r = \frac{d}{t} = \frac{12x^2+13x-14}{4x+7}$$

$$\begin{array}{r|l} & 3x - 2 \\ \hline 4x+7 & 12x^2 + 13x - 14 \\ & 12x^2 + 21x \\ \hline & \quad -8x - 14 \\ & \quad -8x - 14 \\ \hline & \qquad 0 \end{array}$$

$$r = 3x - 2$$

Snowshoes

$$rt = d$$

$$t = \frac{d}{r} = \frac{3x^2+19x+20}{3x+4}$$

$$\begin{array}{r|l} & x + 5 \\ \hline 3x+4 & 3x^2 + 19x + 20 \\ & 3x^2 + 4x \\ \hline & \quad 15x + 20 \\ & \quad 15x + 20 \\ \hline & \qquad 0 \end{array}$$

$$t = x + 5$$

68. Cashews

$$pn = v$$

$$n = \frac{v}{p} = \frac{x^4+4x^2+16}{x^2+2x+4}$$

$$\begin{array}{r|l} & x^2 - 2x + 4 \\ \hline x^2+2x+4 & x^4 + 0x^3 + 4x^2 + 0x + 16 \\ & x^4 + 2x^3 + 4x^2 \\ \hline & \quad -2x^3 + 0x^2 + 0x \\ & \quad -2x^3 - 4x^2 - 8x \\ \hline & \qquad 4x^2 + 8x + 16 \\ & \qquad 4x^2 + 8x + 16 \\ \hline & \qquad\qquad 0 \end{array}$$

The number of pounds is $x^2 - 2x + 4$.

Sunflower seeds

$$pn = v$$

$$p = \frac{v}{n} = \frac{x^4-x^2-42}{x^2+6}$$

$$\begin{array}{r|l} & x^2 - 7 \\ \hline x^2+6 & x^4 - x^2 - 42 \\ & x^4 + 6x^2 \\ \hline & \quad -7x^2 - 42 \\ & \quad -7x^2 - 42 \\ \hline & \qquad 0 \end{array}$$

The price per pound is $x^2 - 7$.

69. Answers may vary.

70. Answers may vary.

71.

$$\begin{array}{r|l} & 5x + 7 \\ \hline 2x-3 & 10x^2 - x - 21 \\ & 10x^2 - 15x \\ \hline & \quad 14x - 21 \\ & \quad 14x - 21 \\ \hline & \qquad 0 \end{array}$$

$0 \Rightarrow$ It is a factor.

72.
$$\begin{array}{r|l} & x^4 + x^3 + x^2 + x + 1 \\ \hline x-1 & x^5 + 0x^4 + 0x^3 + 0x^2 + 0x - 1 \\ & \underline{x^5 - x^4} \\ & x^4 + 0x^3 \\ & \underline{x^4 - x^3} \\ & x^3 + 0x^2 \\ & \underline{x^3 - x^2} \\ & x^2 + 0x \\ & \underline{x^2 - x} \\ & x - 1 \\ & \underline{x - 1} \\ & 0 \end{array}$$
$0 \Rightarrow$ It is a factor.

Exercise 6.7 (page 409)

1. $f(1) = 3(1)^2 + 2(1) - 1 = 4$

2. $f(-2) = 3(-2)^2 + 2(-2) - 1 = 7$

3. $f(2a) = 3(2a)^2 + 2(2a) - 1$
$= 12a^2 + 4a - 1$

4. $f(-t) = 3(-t)^2 + 2(-t) - 1$
$= 3t^2 - 2t - 1$

5. $2(x^2 + 4x - 1) + 3(2x^2 - 2x + 2) = 2x^2 + 8x - 2 + 6x^2 - 6x + 6 = 8x^2 + 2x + 4$

6. $-2(3y^3 - 2y + 7) - 3(y^2 + 2y - 4) + 4(y^3 + 2y - 1)$
$= -6y^3 + 4y - 14 - 3y^2 - 6y + 12 + 4y^3 + 8y - 4 = -2y^3 - 3y^2 + 6y - 6$

7. $P(r)$

8. $x - r$

9. $\begin{array}{r|rrr} 1 & 1 & 1 & -2 \\ & & 1 & 2 \\ \hline & 1 & 2 & 0 \end{array} \Rightarrow \boxed{x+2}$

10. $\begin{array}{r|rrr} 2 & 1 & 1 & -6 \\ & & 2 & 6 \\ \hline & 1 & 3 & 0 \end{array} \Rightarrow \boxed{x+3}$

11. $\begin{array}{r|rrr} 4 & 1 & -7 & 12 \\ & & 4 & -12 \\ \hline & 1 & -3 & 0 \end{array} \Rightarrow \boxed{x-3}$

12. $\begin{array}{r|rrr} 5 & 1 & -6 & 5 \\ & & 5 & -5 \\ \hline & 1 & -1 & 0 \end{array} \Rightarrow \boxed{x-1}$

13. $\begin{array}{r|rrr} -4 & 1 & 6 & 8 \\ & & -4 & -8 \\ \hline & 1 & 2 & 0 \end{array} \Rightarrow \boxed{x+2}$

14. $\begin{array}{r|rrr} -3 & 1 & -2 & -15 \\ & & -3 & 15 \\ \hline & 1 & -5 & 0 \end{array} \Rightarrow \boxed{x-5}$

15. $\begin{array}{r|rrr} -2 & 1 & -5 & 14 \\ & & -2 & 14 \\ \hline & 1 & -7 & 28 \end{array} \Rightarrow \boxed{x-7+\frac{28}{x+2}}$

16. $\begin{array}{r|rrr} -6 & 1 & 13 & 42 \\ & & -6 & -42 \\ \hline & 1 & 7 & 0 \end{array} \Rightarrow \boxed{x+7}$

17.
$$\begin{array}{r|rrrr} 3 & 3 & -10 & 5 & -6 \\ & & 9 & -3 & 6 \\ \hline & 3 & -1 & 2 & 0 \end{array}$$
$$\Rightarrow \boxed{3x^2 - x + 2}$$

18.
$$\begin{array}{r|rrrr} 3 & 2 & -9 & 10 & -3 \\ & & 6 & -9 & 3 \\ \hline & 2 & -3 & 1 & 0 \end{array}$$
$$\Rightarrow \boxed{2x^2 - 3x + 1}$$

19.
$$\begin{array}{r|rrrr} 2 & 2 & 0 & -5 & -6 \\ & & 4 & 8 & 6 \\ \hline & 2 & 4 & 3 & 0 \end{array}$$
$$\Rightarrow \boxed{2x^2 + 4x + 3}$$

20.
$$\begin{array}{r|rrrr} -2 & 4 & 5 & 0 & -1 \\ & & -8 & 6 & -12 \\ \hline & 4 & -3 & 6 & -13 \end{array}$$
$$\Rightarrow \boxed{4x^2 - 3x + 6 - \frac{13}{x+2}}$$

21.
$$\begin{array}{r|rrrr} -1 & 6 & 5 & 0 & 4 \\ & & -6 & 1 & -1 \\ \hline & 6 & -1 & 1 & 3 \end{array}$$
$$\Rightarrow \boxed{6x^2 - x + 1 + \frac{3}{x+1}}$$

22.
$$\begin{array}{r|rrr} 4 & -3 & 1 & 4 \\ & & -12 & -44 \\ \hline & -3 & -11 & -40 \end{array}$$
$$\Rightarrow \boxed{-3x - 11 - \frac{40}{x-4}}$$

23.
$$\begin{array}{r|rrr} 0.2 & 7.2 & -2.1 & 0.5 \\ & & 1.44 & -0.132 \\ \hline & 7.2 & -0.66 & 0.368 \end{array}$$
$$\Rightarrow \boxed{7.2x - 0.66 + \frac{0.368}{x-0.2}}$$

24.
$$\begin{array}{r|rrr} 0.4 & 8.1 & 3.2 & -5.7 \\ & & 3.24 & 2.576 \\ \hline & 8.1 & 6.44 & -3.124 \end{array}$$
$$\Rightarrow \boxed{8.1x + 6.44 - \frac{3.124}{x-0.4}}$$

25.
$$\begin{array}{r|rrr} -1.7 & 2.7 & 1.0 & -5.2 \\ & & -4.59 & 6.103 \\ \hline & 2.7 & -3.59 & 0.903 \end{array}$$
$$\Rightarrow \boxed{2.7x - 3.59 + \frac{0.903}{x+1.7}}$$

26.
$$\begin{array}{r|rrr} -2.5 & 1.3 & -0.5 & -2.3 \\ & & -3.25 & 9.375 \\ \hline & 1.3 & -3.75 & 7.075 \end{array}$$
$$\Rightarrow \boxed{1.3x - 3.75 + \frac{7.075}{x+2.5}}$$

27.
$$\begin{array}{r|rrrr} -57 & 9 & 0 & 0 & -25 \\ & & -513 & 29241 & -1666737 \\ \hline & 9 & -513 & 29241 & -1666762 \end{array}$$
$$\Rightarrow \boxed{9x^2 - 513x + 29{,}241 - \frac{1{,}666{,}762}{x+57}}$$

28.
$$\begin{array}{r|rrrr} 2.3 & 0.5 & 0.0 & 1.0 & 0.0 \\ & & 1.15 & 2.645 & 8.3835 \\ \hline & 0.5 & 1.15 & 3.645 & 8.3835 \end{array}$$
$$\Rightarrow \boxed{0.5x^2 + 1.15x + 3.645 + \frac{8.3835}{x-2.3}}$$

29. $P(1) = 2(1)^3 - 4(1)^2 + 2(1) - 1 = \boxed{-1}$
$$\begin{array}{r|rrrr} 1 & 2 & -4 & 2 & -1 \\ & & 2 & -2 & 0 \\ \hline & 2 & -2 & 0 & \boxed{-1} \end{array}$$

30. $P(2) = 2(2)^3 - 4(2)^2 + 2(2) - 1 = \boxed{3}$
$$\begin{array}{r|rrrr} 2 & 2 & -4 & 2 & -1 \\ & & 4 & 0 & 4 \\ \hline & 2 & 0 & 2 & \boxed{3} \end{array}$$

31. $P(-2) = 2(-2)^3 - 4(-2)^2 + 2(-2) - 1 = \boxed{-37}$
$$\begin{array}{r|rrrr} -2 & 2 & -4 & 2 & -1 \\ & & -4 & 16 & -36 \\ \hline & 2 & -8 & 18 & \boxed{-37} \end{array}$$

32. $P(-1) = 2(-1)^3 - 4(-1)^2 + 2(-1) - 1 = \boxed{-9}$
$$\begin{array}{r|rrrr} -1 & 2 & -4 & 2 & -1 \\ & & -2 & 6 & -8 \\ \hline & 2 & -6 & 8 & \boxed{-9} \end{array}$$

33. $P(3) = 2(3)^3 - 4(3)^2 + 2(3) - 1 = \boxed{23}$

$$\begin{array}{r|rrrr} 3 & 2 & -4 & 2 & -1 \\ & & 6 & 6 & 24 \\ \hline & 2 & 2 & 8 & \boxed{23} \end{array}$$

34. $P(-4) = 2(-4)^3 - 4(-4)^2 + 2(-4) - 1 = \boxed{-201}$

$$\begin{array}{r|rrrr} -4 & 2 & -4 & 2 & -1 \\ & & -8 & 48 & -200 \\ \hline & 2 & -12 & 50 & \boxed{-201} \end{array}$$

35. $P(0) = 2(0)^3 - 4(0)^2 + 2(0) - 1 = \boxed{-1}$

$$\begin{array}{r|rrrr} 0 & 2 & -4 & 2 & -1 \\ & & 0 & 0 & 0 \\ \hline & 2 & -4 & 2 & \boxed{-1} \end{array}$$

36. $P(4) = 2(4)^3 - 4(4)^2 + 2(4) - 1 = \boxed{71}$

$$\begin{array}{r|rrrr} 4 & 2 & -4 & 2 & -1 \\ & & 8 & 16 & 72 \\ \hline & 2 & 4 & 18 & \boxed{71} \end{array}$$

37. $Q(-1) = (-1)^4 - 3(-1)^3 + 2(-1)^2 + (-1) - 3 = \boxed{2}$

$$\begin{array}{r|rrrrr} -1 & 1 & -3 & 2 & 1 & -3 \\ & & -1 & 4 & -6 & 5 \\ \hline & 1 & -4 & 6 & -5 & \boxed{2} \end{array}$$

38. $Q(1) = (1)^4 - 3(1)^3 + 2(1)^2 + (1) - 3 = \boxed{-2}$

$$\begin{array}{r|rrrrr} 1 & 1 & -3 & 2 & 1 & -3 \\ & & 1 & -2 & 0 & 1 \\ \hline & 1 & -2 & 0 & 1 & \boxed{-2} \end{array}$$

39. $Q(2) = (2)^4 - 3(2)^3 + 2(2)^2 + (2) - 3 = \boxed{-1}$

$$\begin{array}{r|rrrrr} 2 & 1 & -3 & 2 & 1 & -3 \\ & & 2 & -2 & 0 & 2 \\ \hline & 1 & -1 & 0 & 1 & \boxed{-1} \end{array}$$

40. $Q(-2) = (-2)^4 - 3(-2)^3 + 2(-2)^2 + (-2) - 3 = \boxed{43}$

$$\begin{array}{r|rrrrr} -2 & 1 & -3 & 2 & 1 & -3 \\ & & -2 & 10 & -24 & 46 \\ \hline & 1 & -5 & 12 & -23 & \boxed{43} \end{array}$$

41. $Q(3) = (3)^4 - 3(3)^3 + 2(3)^2 + (3) - 3 = \boxed{18}$

$$\begin{array}{r|rrrrr} 3 & 1 & -3 & 2 & 1 & -3 \\ & & 3 & 0 & 6 & 21 \\ \hline & 1 & 0 & 2 & 7 & \boxed{18} \end{array}$$

42. $Q(0) = (0)^4 - 3(0)^3 + 2(0)^2 + (0) - 3 = \boxed{-3}$

$$\begin{array}{r|rrrrr} 0 & 1 & -3 & 2 & 1 & -3 \\ & & 0 & 0 & 0 & 0 \\ \hline & 1 & -3 & 2 & 1 & \boxed{-3} \end{array}$$

43. $Q(-3) = (-3)^4 - 3(-3)^3 + 2(-3)^2 + (-3) - 3 = \boxed{174}$

$$\begin{array}{r|rrrrr} -3 & 1 & -3 & 2 & 1 & -3 \\ & & -3 & 18 & -60 & 177 \\ \hline & 1 & -6 & 20 & -59 & \boxed{174} \end{array}$$

44. $Q(-4) = (-4)^4 - 3(-4)^3 + 2(-4)^2 + (-4) - 3 = \boxed{473}$

$$\begin{array}{r|rrrrr} -4 & 1 & -3 & 2 & 1 & -3 \\ & & -4 & 28 & -120 & 476 \\ \hline & 1 & -7 & 30 & -119 & \boxed{473} \end{array}$$

45.
$$\begin{array}{r|rrrr} 2 & 1 & -4 & 1 & -2 \\ & & 2 & -4 & -6 \\ \hline & 1 & -2 & -3 & \boxed{-8} \end{array}$$

46.
$$\begin{array}{r|rrrr} 1 & 1 & -3 & 1 & 1 \\ & & 1 & -2 & -1 \\ \hline & 1 & -2 & -1 & \boxed{0} \end{array}$$

47.
$$\begin{array}{r|rrrr} 3 & 2 & 0 & 1 & 2 \\ & & 6 & 18 & 57 \\ \hline & 2 & 6 & 19 & \boxed{59} \end{array}$$

48.
$$\begin{array}{r|rrrr} -2 & 1 & 1 & 0 & 1 \\ & & -2 & 2 & -4 \\ \hline & 1 & -1 & 2 & \boxed{-3} \end{array}$$

49.
$$\begin{array}{r|rrrrr} -2 & 1 & -2 & 1 & -3 & 2 \\ & & -2 & 8 & -18 & 42 \\ \hline & 1 & -4 & 9 & -21 & \boxed{44} \end{array}$$

50.
$$\begin{array}{r|rrrrrr} -1 & 1 & 3 & 0 & -1 & 0 & 1 \\ & & -1 & -2 & 2 & -1 & 1 \\ \hline & 1 & 2 & -2 & 1 & -1 & \boxed{2} \end{array}$$

51.
$$\begin{array}{r|rrrrrr} -\frac{1}{2} & 3 & 0 & 0 & 0 & 0 & 1 \\ & & -\frac{3}{2} & \frac{3}{4} & -\frac{3}{8} & \frac{3}{16} & -\frac{3}{32} \\ \hline & 3 & -\frac{3}{2} & \frac{3}{4} & -\frac{3}{8} & \frac{3}{16} & \boxed{\frac{29}{32}} \end{array}$$

52.
$$\begin{array}{r|rrrrrrrr} 2 & 5 & 0 & 0 & -7 & 0 & 1 & 0 & 1 \\ & & 10 & 20 & 40 & 66 & 132 & 266 & 532 \\ \hline & 5 & 10 & 20 & 33 & 66 & 133 & 266 & \boxed{533} \end{array}$$

53.
$$\begin{array}{r|rrrr} 3 & 1 & -3 & 5 & -15 \\ & & 3 & 0 & 15 \\ \hline & 1 & 0 & 5 & \boxed{0} \end{array} \Rightarrow \text{factor}$$

54.
$$\begin{array}{r|rrrr} -1 & 1 & 2 & -2 & -3 \\ & & -1 & -1 & 3 \\ \hline & 1 & 1 & -3 & \boxed{0} \end{array} \Rightarrow \text{factor}$$

55.
$$\begin{array}{r|rrr} -2 & 3 & -7 & 4 \\ & & -6 & 26 \\ \hline & 3 & -13 & \boxed{30} \end{array} \Rightarrow \text{not a factor}$$

56.
$$\begin{array}{r|rrrr} 0 & 7 & -5 & -8 & 0 \\ & & 0 & 0 & 0 \\ \hline & 7 & -5 & -8 & \boxed{0} \end{array} \Rightarrow \text{factor}$$

57.
$$\begin{array}{r|rrrrrrr} 2 & 1 & 0 & 0 & 0 & 0 & 0 & 0 \\ & & 2 & 4 & 8 & 16 & 32 & 64 \\ \hline & 1 & 2 & 4 & 8 & 16 & 32 & \boxed{64} \end{array}$$

58.
$$\begin{array}{r|rrrrrr} -3 & 1 & 0 & 0 & 0 & 0 & 0 \\ & & -3 & 9 & -27 & 81 & -243 \\ \hline & 1 & -3 & 9 & -27 & 81 & \boxed{-243} \end{array}$$

59. **Answers may vary.**

60. **Answers may vary.**

61. remainder $= P(1) = 1^{100} - 1^{99} + 1^{98} - 1^{97} + \cdots + 1^2 - 1 + 1 = 1$

62. remainder $= P(-1) = (-1)^{100} - (-1)^{99} + (-1)^{98} - (-1)^{97} + \cdots + (-1)^2 - (-1) + 1 = 101$

Exercise 6.8 (page 420)

1. $(x^2x^3)^2 = (x^5)^2 = x^{10}$

2. $\left(\dfrac{a^3a^5}{a^{-2}}\right)^3 = \left(\dfrac{a^8}{a^{-2}}\right)^3 = (a^{10})^3 = a^{30}$

3. $\dfrac{b^0 - 2b^0}{b^0} = \dfrac{1-2}{1} = \dfrac{-1}{1} = -1$

4. $\left(\dfrac{2r^{-2}r^{-3}}{4r^{-5}}\right)^{-3} = \left(\dfrac{2r^{-5}}{4r^{-5}}\right)^{-3} = \left(\dfrac{1}{2}\right)^{-3}$
$= 2^3 = 8$

5. $35{,}000 = 3.5 \times 10^4$

6. $0.00035 = 3.5 \times 10^{-4}$

7. $2.5 \times 10^{-3} = 0.0025$

8. $2.5 \times 10^4 = 25{,}000$

9. unit costs; rates

10. proportion

11. extremes; means

12. similar

13. direct

14. inverse

15. rational

16. linear

17. joint

18. combined

19. direct

20. neither

21. neither

22. inverse

23.
$$\frac{x}{5} = \frac{15}{25}$$
$$25x = 75$$
$$x = 3$$

24.
$$\frac{4}{y} = \frac{6}{27}$$
$$108 = 6y$$
$$18 = y$$

25.
$$\frac{r-2}{3} = \frac{r}{5}$$
$$5(r-2) = 3r$$
$$5r - 10 = 3r$$
$$2r = 10$$
$$r = 5$$

26.
$$\frac{x+1}{x-1} = \frac{6}{4}$$
$$4(x+1) = 6(x-1)$$
$$4x + 4 = 6x - 6$$
$$-2x = -10$$
$$x = 5$$

27.
$$\frac{3}{n} = \frac{2}{n+1}$$
$$3(n+1) = 2n$$
$$3n + 3 = 2n$$
$$n = -3$$

28.
$$\frac{4}{x+3} = \frac{3}{5}$$
$$20 = 3(x+3)$$
$$20 = 3x + 9$$
$$11 = 3x$$
$$\frac{11}{3} = x$$

29.
$$\begin{aligned}\frac{5}{5z+3} &= \frac{2z}{2z^2+6}\\ 5(2z^2+6) &= 2z(5z+3)\\ 10z^2+30 &= 10z^2+6z\\ 30 &= 6z\\ 5 &= z\end{aligned}$$

30.
$$\begin{aligned}\frac{9t+6}{t(t+3)} &= \frac{7}{t+3}\\ (9t+6)(t+3) &= 7(t^2+3t)\\ 9t^2+33t+18 &= 7t^2+21t\\ 2t^2+12t+18 &= 0\\ 2(t+3)(t+3) &= 0\end{aligned}$$
$t=-3 \Rightarrow$ extraneous, so no solution.

31.
$$\begin{aligned}\frac{2}{c} &= \frac{c-3}{2}\\ 4 &= c^2-3c\\ 0 &= c^2-3c-4\\ 0 &= (c+1)(c-4)\end{aligned}$$
$c+1=0$ **or** $c-4=0$

$c=-1$ $\quad$ $c=4$

32.
$$\begin{aligned}\frac{y}{4} &= \frac{4}{y}\\ y^2 &= 16\\ y^2-16 &= 0\\ (y+4)(y-4) &= 0\end{aligned}$$
$y+4=0$ **or** $y-4=0$

$y=-4$ $\quad$ $y=4$

33.
$$\begin{aligned}\frac{2}{3x} &= \frac{6x}{36}\\ 72 &= 18x^2\\ 0 &= 18x^2-72\\ 0 &= 18(x+2)(x-2)\end{aligned}$$
$x+2=0$ **or** $x-2=0$

$x=-2$ $\quad$ $x=2$

34.
$$\begin{aligned}\frac{2}{x+6} &= \frac{-2x}{5}\\ 10 &= -2x^2-12x\\ 2x^2+12x+10 &= 0\\ 2(x+5)(x+1)\end{aligned}$$
$x+5=0$ **or** $x+1=0$

$x=-5$ $\quad$ $x=-1$

35.
$$\begin{aligned}\frac{2(x+3)}{3} &= \frac{4(x-4)}{5}\\ 10(x+3) &= 12(x-4)\\ 10x+30 &= 12x-48\\ -2x &= -78\\ x &= 39\end{aligned}$$

36.
$$\begin{aligned}\frac{x+4}{5} &= \frac{3(x-2)}{3}\\ 3x+12 &= 15(x-2)\\ 3x+12 &= 15x-30\\ -12x &= -42\\ x &= \frac{-42}{-12} = \frac{7}{2}\end{aligned}$$

37.
$$\begin{aligned}\frac{1}{x+3} &= \frac{-2x}{x+5}\\ x+5 &= -2x^2-6x\\ 2x^2+7x+5 &= 0\\ (2x+5)(x+1)\end{aligned}$$
$2x+5=0$ **or** $x+1=0$

$x=-\frac{5}{2}$ $\quad$ $x=-1$

38.
$$\begin{aligned}\frac{x-1}{x+1} &= \frac{2}{3x}\\ 3x^2-3x &= 2x+2\\ 3x^2-5x-2 &= 0\\ (3x+1)(x-2) &= 0\end{aligned}$$
$3x+1=0$ **or** $x-2=0$

$x=-\frac{1}{3}$ $\quad$ $x=2$

39.
$$\frac{a-4}{a+2} = \frac{a-5}{a+1}$$
$$(a-4)(a+1) = (a+2)(a-5)$$
$$a^2 - 3a - 4 = a^2 - 3a - 10$$
$$-3a - 4 = -3a - 10$$
$$-4 \neq -10$$
no solution

40.
$$\frac{z+2}{z+6} = \frac{z-4}{z-2}$$
$$(z+2)(z-2) = (z+6)(z-4)$$
$$z^2 - 4 = z^2 + 2z - 24$$
$$-4 = 2z - 24$$
$$20 = 2z$$
$$10 = z$$

41. $A = kp^2$

42. $z = \frac{k}{t^3}$

43. $v = \frac{k}{r^3}$

44. $r = ks^2$

45. $B = kmn$

46. $C = kxyz$

47. $P = \frac{ka^2}{j^3}$

48. $M = \frac{kxz^2}{n^3}$

49. L varies jointly with m and n.

50. P varies directly with m and inversely with n.

51. E varies jointly with a and the square of b.

52. U varies jointly with r, the square of s, and t.

53. X varies directly with x^2 and inversely with y^2.

54. Z varies directly with w and inversely with the product of x and y.

55. R varies directly with L and inversely with d^2.

56. e varies jointly with P and L and inversely with A.

57. Let $c =$ the cost of 5 shirts.
$$\frac{2}{25} = \frac{5}{c}$$
$$2c = 125$$
$$c = 62.5$$
5 shirts will cost \$62.50.

58. Let $b =$ the number of bottles for 10 gallons.
$$\frac{4}{2} = \frac{b}{10}$$
$$40 = 2b$$
$$20 = b$$
20 bottles of ketchup will be needed.

59. Let $g =$ gallons of gas for 315 miles.
$$\frac{42}{1} = \frac{315}{g}$$
$$42g = 315$$
$$g = \frac{315}{42} = 7.5$$
7.5 gallons of gas are needed.

60. Let $L =$ the length of a real engine.
$$\frac{L \text{ in.}}{9 \text{ in.}} = \frac{87 \text{ ft}}{1 \text{ ft}}$$
$$L = 783$$
The real length is 783 inches, or $65\frac{1}{4}$ feet.

61. Let $w =$ the width if it were a real house.
$$\frac{1 \text{in.}}{1 \text{ ft}} = \frac{32 \text{ in.}}{w \text{ ft}}$$
$$w = 32$$
The width would be 32 feet.

62. Let $t =$ the number of teachers.
$$\frac{3}{50} = \frac{t}{2700}$$
$$8100 = 50t$$
$$162 = t$$
162 teachers will be needed.

63. Let h = the actual height of the building.

$$\frac{7 \text{ in.}}{280 \text{ ft}} = \frac{2 \text{ in.}}{h \text{ ft}}$$
$$7h = 560$$
$$h = 80$$

The building is 80 feet tall.

64. 6 gallons $= 6 \cdot 128$ oz. $= 768$ oz.

$$\frac{50}{1} \stackrel{?}{=} \frac{768}{16}$$
$$16 \cdot 50 \stackrel{?}{=} 768$$
$$800 \stackrel{?}{=} 768$$

The directions are not exactly correct. However, they are almost correct.

65. Let d = the dosage required.

$$\frac{0.006}{1} = \frac{d}{30}$$
$$0.18 = d$$

The dosage is 0.18 grams.

66. Let m = the mass of the child.

$$\frac{0.025}{1} = \frac{1.125}{m}$$
$$0.025m = 1.125$$
$$m = 45$$

The mass of the child is 45 kilograms.

67. Let h = the height of the tree.

$$\frac{6}{4} = \frac{h}{28}$$
$$168 = 4h$$
$$42 = h$$

The tree is 42 feet tall.

68.

$$\frac{5}{7} = \frac{h}{30}$$
$$150 = 7h$$
$$h = \frac{150}{7} = 21\frac{3}{7}$$

The height of the flagpole is $21\frac{3}{7}$ feet.

69.

$$\frac{20}{32} = \frac{w}{75}$$
$$1500 = 32w$$
$$w = \frac{1500}{32} = 46\frac{7}{8}$$

The width of the river is $46\frac{7}{8}$ feet.

70.

$$\frac{150}{1000} = \frac{x}{5280}$$
$$792000 = 1000x$$
$$792 = x$$

The plane will ascend 792 feet.

71.

$$\frac{1350}{1} = \frac{x}{5}$$
$$6750 = x$$

The plane will descend 6750 feet.

72.

$$\frac{100}{300} = \frac{x}{2640}$$
$$264000 = 300x$$
$$880 = x$$

The hill is 880 feet tall.

73.

$$A = kr^2$$
$$A = \pi r^2$$
$$A = \pi(6 \text{ in.})^2$$
$$\boxed{A = 36\pi \text{ in.}^2}$$

74.

$$s = kt^2$$
$$1024 = k(8)^2$$
$$1024 = 64k$$
$$16 = k$$

$$s = 16t^2$$
$$s = 16(10)^2$$
$$s = 16(100)$$
$$\boxed{s = 1600 \text{ ft}}$$

75.

$$d = kg$$
$$288 = k(12)$$
$$24 = k$$

$$d = 24g$$
$$d = 24(18)$$
$$\boxed{d = 432 \text{ mi}}$$

76.

$$h = ka$$
$$144 = k(8)$$
$$18 = k$$

$$h = 18a$$
$$1152 = 18a$$
$$\boxed{64 \text{ acres} = a}$$

77.
$$t = \frac{k}{n} \qquad 10 = \frac{k}{25} \qquad 250 = k$$
$$t = \frac{250}{n} \qquad t = \frac{250}{10} \qquad \boxed{t = 25 \text{ days}}$$

78.
$$l = \frac{k}{w} \qquad 12 = \frac{k}{18} \qquad 216 = k$$
$$l = \frac{216}{w} \qquad 16 = \frac{216}{w} \qquad 16w = 216 \qquad \boxed{w = 13.5 \text{ ft}}$$

79.
$$V = \frac{k}{P} \qquad 20 = \frac{k}{6} \qquad 120 = k$$
$$V = \frac{120}{P} \qquad V = \frac{120}{10} \qquad \boxed{V = 12 \text{ in.}^3}$$

80.
$$V = \frac{k}{a} \qquad 7000 = \frac{k}{3} \qquad 21000 = k$$
$$V = \frac{21000}{a} \qquad V = \frac{21000}{7} \qquad \boxed{V = \$3000}$$

81.
$$f = \frac{k}{l} \qquad 256 = \frac{k}{2} \qquad 512 = k$$
$$f = \frac{512}{l} \qquad f = \frac{512}{6} \qquad \boxed{f = 85\tfrac{1}{3}}$$

82. $A_1 = klw$
$A_2 = k(3l)(3w) = 9klw = 9A_1$
The area is multiplied by 9.

83. $V_1 = klwh$
$V_2 = k(2l)(3w)(2h) = 12klwh = 12V_1$
The volume is multiplied by 12.

84.
$$c = kth \qquad 1800 = k(4)(6) \qquad 1800 = 24k \qquad 75 = k$$
$$c = 75th \qquad c = 75(10)(12) \qquad \boxed{c = \$9000}$$

85. $g = khr^2$
$g = 23.5hr^2$
$g = 23.5(20)(7.5)^2$
$\boxed{g = 26{,}437.5 \text{ gallons}}$

86.
$$l = \frac{kxy}{z} \qquad 30 = \frac{k(15)(5)}{10} \qquad 30 = \frac{75k}{10} \qquad 30 = 7.5k \Rightarrow \boxed{4 = k}$$

87. $V = kC$
$6 = k(2)$
$3 = k \Rightarrow$ The resistance is 3 ohms.

88. $P = kC^2$
$P = 5(3)^2$
$\boxed{P = 45 \text{ watts}}$

89.
$$D = \frac{k}{wd^3} \qquad 1.1 = \frac{k}{4(4)^3} \qquad 1.1 = \frac{k}{256} \qquad 281.6 = k$$
$$D = \frac{281.6}{wd^3} \qquad D = \frac{281.6}{2(8)^3} \qquad D = \frac{281.6}{1024} \qquad \boxed{D = 0.275 \text{ in.}}$$

90. Refer to **#89.**
$$D = \frac{281.6}{wd^3}$$
$$D = \frac{281.6}{8(2)^3} = \frac{281.6}{64} = \boxed{4.4 \text{ in.}}$$

91.
$$P = \frac{kT}{V} \qquad 1 = \frac{k(273)}{1} \qquad \frac{1}{273} = k$$

$$P = \frac{\frac{1}{273}T}{V} \qquad 1 = \frac{\frac{1}{273}T}{2} \qquad 2 = \frac{1}{273}T \qquad \boxed{546\text{ K} = T}$$

92.
$$T = \frac{ks^2}{r} \qquad 32 = \frac{k(8)^2}{6} \qquad 32 = \frac{k(64)}{6} \qquad 192 = 64k \qquad 3 = k$$

$$T = \frac{3s^2}{r} \qquad T = \frac{3(4)^2}{3} \qquad \boxed{T = 16\text{ lb}}$$

93-96. Answers may vary.

97. This is not direct variation. For this to be direct variation, one temperature would have to be a constant multiple of the other.

98. This is not inverse variation. For this to be inverse variation, the product of the two amounts would have to be constant.

Chapter 6 Summary (page 425)

1. $f(3) = \frac{3(3)+2}{3} = \frac{11}{3}$

2. $f(100) = \frac{3(100)+2}{100} = \frac{302}{100}$

3. 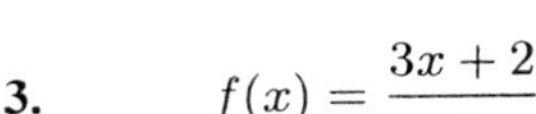
$f(x) = \frac{3x+2}{x}$

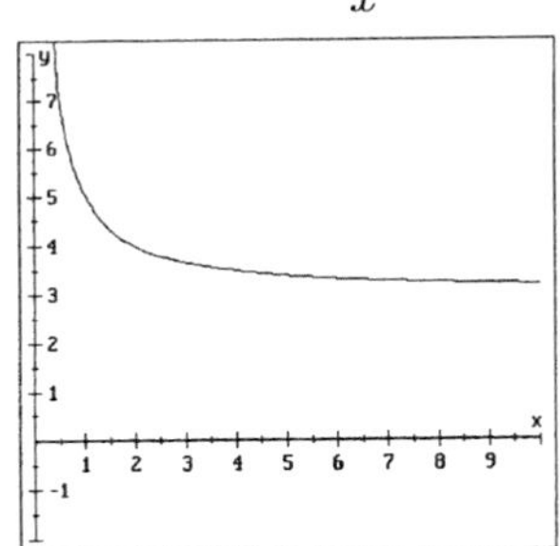

Horizontal: $y = 3$
Vertical: $x = 0$

4. $\frac{248x^2y}{576xy^2} = \frac{8 \cdot 31x^2y}{8 \cdot 72xy^2} = \frac{31x}{72y}$

5. $\frac{212m^3n}{588m^2n^3} = \frac{4 \cdot 53m^3n}{4 \cdot 147m^2n^3} = \frac{53m}{147n^2}$

6. $\frac{x^2-49}{x^2+14x+49} = \frac{(x+7)(x-7)}{(x+7)(x+7)} = \frac{x-7}{x+7}$

7. $\frac{x^2+6x+36}{x^3-216} = \frac{x^2+6x+36}{(x-6)(x^2+6x+36)} = \frac{1}{x-6}$

8. $\frac{x^2-2x+4}{2x^3+16} = \frac{x^2-2x+4}{2(x^3+8)} = \frac{x^2-2x+4}{2(x+2)(x^2-2x+4)} = \frac{1}{2(x+2)} = \frac{1}{2x+4}$

9. $\frac{x-y}{y-x} = \frac{x-y}{-1(x-y)} = -1$

10. $\frac{2m-2n}{n-m} = \frac{2(m-n)}{-1(m-n)} = -2$

11. $$\frac{m^3+m^2n-2mn^2}{2m^3-mn^2-m^2n}=\frac{m(m^2+mn-2n^2)}{m(2m^2-mn-n^2)}=\frac{m(m+2n)(m-n)}{m(2m+n)(m-n)}=\frac{m+2n}{2m+n}$$

12. $$\frac{ac-ad+bc-bd}{d^2-c^2}=\frac{a(c-d)+b(c-d)}{(d+c)(d-c)}=\frac{(c-d)(a+b)}{(d+c)(d-c)}=-\frac{a+b}{c+d}=\frac{-a-b}{c+d}$$

13. $$\frac{x^2+4x+4}{x^2-x-6}\cdot\frac{x^2-9}{x^2+5x+6}=\frac{(x+2)(x+2)}{(x+2)(x-3)}\cdot\frac{(x+3)(x-3)}{(x+2)(x+3)}=1$$

14. $$\frac{x^3-64}{x^2+4x+16}\div\frac{x^2-16}{x+4}=\frac{x^3-64}{x^2+4x+16}\cdot\frac{x+4}{x^2-16}=\frac{(x-4)(x^2+4x+16)}{x^2+4x+16}\cdot\frac{x+4}{(x+4)(x-4)}$$
$$=1$$

15. $$\frac{x^2+3x+2}{x^2-x-6}\cdot\frac{3x^2-3x}{x^2-3x-4}\div\frac{x^2+3x+2}{x^2-2x-8}$$
$$=\frac{x^2+3x+2}{x^2-x-6}\cdot\frac{3x^2-3x}{x^2-3x-4}\cdot\frac{x^2-2x-8}{x^2+3x+2}$$
$$=\frac{(x+2)(x+1)}{(x-3)(x+2)}\cdot\frac{3x(x-1)}{(x-4)(x+1)}\cdot\frac{(x-4)(x+2)}{(x+2)(x+1)}$$
$$=\frac{3x(x-1)}{(x-3)(x+1)}$$

16. $$\frac{x^2-x-6}{x^2-3x-10}\div\frac{x^2-x}{x^2-5x}\cdot\frac{x^2-4x+3}{x^2-6x+9}=\frac{x^2-x-6}{x^2-3x-10}\cdot\frac{x^2-5x}{x^2-x}\cdot\frac{x^2-4x+3}{x^2-6x+9}$$
$$=\frac{(x-3)(x+2)}{(x-5)(x+2)}\cdot\frac{x(x-5)}{x(x-1)}\cdot\frac{(x-3)(x-1)}{(x-3)(x-3)}=1$$

17. $$\frac{5y}{x-y}-\frac{3}{x-y}=\frac{5y-3}{x-y}$$

18. $$\frac{3x-1}{x^2+2}+\frac{3(x-2)}{x^2+2}=\frac{3x-1+3x-6}{x^2+2}$$
$$=\frac{6x-7}{x^2+2}$$

19. $$\frac{3}{x+2}+\frac{2}{x+3}=\frac{3(x+3)}{(x+2)(x+3)}+\frac{2(x+2)}{(x+2)(x+3)}=\frac{3x+9+2x+4}{(x+2)(x+3)}=\frac{5x+13}{(x+2)(x+3)}$$

20. $$\frac{4x}{x-4}-\frac{3}{x+3}=\frac{4x(x+3)}{(x-4)(x+3)}-\frac{3(x-4)}{(x-4)(x+3)}=\frac{4x^2+12x-3x+12}{(x-4)(x+3)}=\frac{4x^2+9x+12}{(x-4)(x+3)}$$

21. $$\frac{2x}{x+1}+\frac{3x}{x+2}+\frac{4x}{x^2+3x+2}=\frac{2x}{x+1}+\frac{3x}{x+2}+\frac{4x}{(x+2)(x+1)}$$
$$=\frac{2x(x+2)}{(x+1)(x+2)}+\frac{3x(x+1)}{(x+1)(x+2)}+\frac{4x}{(x+2)(x+1)}$$
$$=\frac{2x^2+4x+3x^2+3x+4x}{(x+1)(x+2)}=\frac{5x^2+11x}{(x+1)(x+2)}$$

22. $$\frac{5x}{x-3}+\frac{5}{x^2-5x+6}+\frac{x+3}{x-2}=\frac{5x}{x-3}+\frac{5}{(x-3)(x-2)}+\frac{x+3}{x-2}$$
$$=\frac{5x(x-2)}{(x-3)(x-2)}+\frac{5}{(x-3)(x-2)}+\frac{(x+3)(x-3)}{(x-3)(x-2)}$$
$$=\frac{5x^2-10x+5+x^2-9}{(x-3)(x-2)}$$
$$=\frac{6x^2-10x-4}{(x-3)(x-2)}=\frac{2(3x+1)(x-2)}{(x-3)(x-2)}=\frac{2(3x+1)}{x-3}$$

23. $$\frac{3(x+2)}{x^2-1}-\frac{2}{x+1}+\frac{4(x+3)}{x^2-2x+1}=\frac{3(x+2)}{(x+1)(x-1)}-\frac{2}{(x+1)}+\frac{4(x+3)}{(x-1)(x-1)}$$
$$=\frac{3(x+2)(x-1)}{(x+1)(x-1)(x-1)}-\frac{2(x-1)(x-1)}{(x+1)(x-1)(x-1)}+\frac{4(x+3)(x+1)}{(x+1)(x-1)(x-1)}$$
$$=\frac{3(x^2+x-2)-2(x^2-2x+1)+4(x^2+4x+3)}{(x+1)(x-1)(x-1)}$$
$$=\frac{3x^2+3x-6-2x^2+4x-2+4x^2+16x+12}{(x+1)(x-1)(x-1)}=\frac{5x^2+23x+4}{(x+1)(x-1)(x-1)}$$

24. $$\frac{-2(3+x)}{x^2+6x+9}+\frac{3(x+2)}{x^2-6x+9}-\frac{1}{x^2-9}$$
$$=\frac{-2(3+x)}{(x+3)(x+3)}+\frac{3(x+2)}{(x-3)(x-3)}-\frac{1}{(x+3)(x-3)}$$
$$=\frac{-2}{x+3}+\frac{3(x+2)}{(x-3)(x-3)}-\frac{1}{(x+3)(x-3)}$$
$$=\frac{-2(x-3)(x-3)}{(x+3)(x-3)(x-3)}+\frac{3(x+2)(x+3)}{(x+3)(x-3)(x-3)}-\frac{1(x-3)}{(x+3)(x-3)(x-3)}$$
$$=\frac{-2(x^2-6x+9)+3(x^2+5x+6)-x+3}{(x+3)(x-3)(x-3)}=\frac{x^2+26x+3}{(x+3)(x-3)(x-3)}$$

25. $$\frac{\frac{3}{x}-\frac{2}{y}}{xy}=\frac{\left(\frac{3}{x}-\frac{2}{y}\right)xy}{xy(xy)}=\frac{3y-2x}{x^2y^2}$$

26. $$\frac{\frac{1}{x}+\frac{2}{y}}{\frac{2}{x}-\frac{1}{y}}=\frac{\left(\frac{1}{x}+\frac{2}{y}\right)xy}{\left(\frac{2}{x}-\frac{1}{y}\right)xy}=\frac{y+2x}{2y-x}$$

27. $$\frac{2x+3+\frac{1}{x}}{x+2+\frac{1}{x}}=\frac{\left(2x+3+\frac{1}{x}\right)x}{\left(x+2+\frac{1}{x}\right)x}=\frac{2x^2+3x+1}{x^2+2x+1}=\frac{(2x+1)(x+1)}{(x+1)(x+1)}=\frac{2x+1}{x+1}$$

28. $$\frac{6x+13+\frac{6}{x}}{6x+5-\frac{6}{x}}=\frac{\left(6x+13+\frac{6}{x}\right)x}{\left(6x+5-\frac{6}{x}\right)x}=\frac{6x^2+13x+6}{6x^2+5x-6}=\frac{(2x+3)(3x+2)}{(2x+3)(3x-2)}=\frac{3x+2}{3x-2}$$

29. $\dfrac{1-\frac{1}{x}-\frac{2}{x^2}}{1+\frac{4}{x}+\frac{3}{x^2}} = \dfrac{\left(1-\frac{1}{x}-\frac{2}{x^2}\right)x^2}{\left(1+\frac{4}{x}+\frac{3}{x^2}\right)x^2} = \dfrac{x^2-x-2}{x^2+4x+3} = \dfrac{(x-2)(x+1)}{(x+3)(x+1)} = \dfrac{x-2}{x+3}$

30. $\dfrac{x^{-1}+1}{x+1} = \dfrac{\frac{1}{x}+1}{x+1} = \dfrac{\left(\frac{1}{x}+1\right)x}{(x+1)x} = \dfrac{1+x}{(x+1)x} = \dfrac{1}{x}$

31. $\dfrac{x^{-1}-y^{-1}}{x^{-1}+y^{-1}} = \dfrac{\frac{1}{x}-\frac{1}{y}}{\frac{1}{x}+\frac{1}{y}} = \dfrac{\left(\frac{1}{x}-\frac{1}{y}\right)xy}{\left(\frac{1}{x}+\frac{1}{y}\right)xy} = \dfrac{y-x}{y+x}$

32. $\dfrac{(x-y)^{-2}}{x^{-2}-y^{-2}} = \dfrac{\frac{1}{(x-y)^2}}{\frac{1}{x^2}-\frac{1}{y^2}} = \dfrac{\left(\frac{1}{(x-y)^2}\right)x^2y^2(x-y)^2}{\left(\frac{1}{x^2}-\frac{1}{y^2}\right)x^2y^2(x-y)^2} = \dfrac{x^2y^2}{y^2(x-y)^2-x^2(x-y)^2}$

$$= \dfrac{x^2y^2}{(x-y)^2(y^2-x^2)}$$

33.
$$\begin{aligned} \frac{4}{x}-\frac{1}{10} &= \frac{7}{2x} \\ \left(\frac{4}{x}-\frac{1}{10}\right)10x &= \frac{7}{2x}\cdot 10x \\ 40-x &= 35 \\ 5 &= x \end{aligned}$$
The answer checks.

34.
$$\begin{aligned} \frac{2}{x+5}-\frac{1}{6} &= \frac{1}{x+4} \\ \left(\frac{2}{x+5}-\frac{1}{6}\right)6(x+5)(x+4) &= \left(\frac{1}{x+4}\right)6(x+5)(x+4) \\ 12(x+4)-(x+5)(x+4) &= 6(x+5) \\ 12x+48-x^2-9x-20 &= 6x+30 \\ 0 &= x^2+3x+2 \\ 0 &= (x+2)(x+1) \end{aligned}$$

$x+2=0$ **or** $x+1=0$

$x=-2$ $\quad$ $x=-1$ $\quad$ Both answers check.

35.

$$\begin{aligned}
\frac{2(x-5)}{x-2} &= \frac{6x+12}{4-x^2} \\
\frac{2(x-5)}{x-2} &= \frac{6(x+2)}{(2+x)(2-x)} \\
\frac{2(x-5)}{x-2} &= \frac{-6(x+2)}{(x+2)(x-2)} \\
2(x-5)(x+2) &= -6(x+2) \\
2x^2-6x-20 &= -6x-12 \\
2x^2-8 &= 0 \\
2(x+2)(x-2) &= 0 \\
x = 2, \text{ or } x &= -2
\end{aligned}$$

Neither answer checks, so there is no solution.

36.

$$\begin{aligned}
\frac{7}{x+9} - \frac{x+2}{2} &= \frac{x+4}{x+9} \\
\left(\frac{7}{x+9} - \frac{x+2}{2}\right)2(x+9) &= \left(\frac{x+4}{x+9}\right)2(x+9) \\
14-(x+2)(x+9) &= 2(x+4) \\
14-x^2-11x-18 &= 2x+8 \\
0 &= x^2+13x+12 \\
0 &= (x+12)(x+1)
\end{aligned}$$

$x+12=0$ **or** $x+1=0$

$x=-12$ $\quad$ $x=-1$ $\quad$ Both answers check.

37.

$$\begin{aligned}
\frac{x^2}{a^2} - \frac{y^2}{b^2} &= 1 \\
\left(\frac{x^2}{a^2} - \frac{y^2}{b^2}\right)a^2b^2 &= a^2b^2 \\
x^2b^2 - y^2a^2 &= a^2b^2 \\
x^2b^2 - a^2b^2 &= y^2a^2 \\
\frac{x^2b^2 - a^2b^2}{a^2} &= y^2
\end{aligned}$$

38.

$$\begin{aligned}
H &= \frac{2ab}{a+b} \\
H(a+b) &= 2ab \\
Ha + Hb &= 2ab \\
Ha &= 2ab - Hb \\
Ha &= b(2a-H) \\
\frac{Ha}{2a-H} &= b
\end{aligned}$$

39. Let $r =$ the usual rate. Then $r - 10 =$ the slower rate.

	Rate	Time	Dist.
Usual	r	$\frac{200}{r}$	200
Slower	$r - 10$	$\frac{200}{r-10}$	200

$$\boxed{\text{Usual time}} + 1 = \boxed{\text{Slower time}}$$

$$\frac{200}{r} + 1 = \frac{200}{r-10}$$

$$\left(\frac{200}{r} + 1\right)(r)(r-10) = \frac{200}{r-10} \cdot r(r-10)$$

$$200(r-10) + r(r-10) = 200r$$

$$r^2 - 10r - 2000 = 0$$

$$(r+40)(r-50) = 0$$

$$r + 40 = 0 \quad \textbf{or} \quad r - 50 = 0$$

$$r = -40 \qquad\qquad r = 50$$

$r = 50$ is the only answer that makes sense. The usual rate is 50 miles per hour.

40. Let $r =$ the usual rate. Then $r + 40 =$ the faster rate.

	Rate	Time	Dist.
Usual	r	$\frac{600}{r}$	600
Faster	$r + 40$	$\frac{600}{r+40}$	600

$$\boxed{\text{Usual time}} - \frac{1}{2} = \boxed{\text{Faster time}}$$

$$\frac{600}{r} - \frac{1}{2} = \frac{600}{r+40}$$

$$\left(\frac{600}{r} - \frac{1}{2}\right)(2r)(r+40) = \frac{600}{r+40} \cdot 2r(r+40)$$

$$1200(r+40) + r(r+40) = 1200r$$

$$r^2 + 40r - 48000 = 0$$

$$(r+240)(r-200) = 0$$

$$r + 240 = 0 \quad \textbf{or} \quad r - 200 = 0$$

$$r = -240 \qquad\qquad r = 200$$

$r = 200$ is the only answer that makes sense. The usual rate is 200 miles per hour.

41. The first pipe can drain $\frac{1}{24}$ of the tank in 1 hour, while the second can drain $\frac{1}{36}$ of the pool in 1 hour. Let $x =$ the number of hours needed for them to drain the tank together. Then $\frac{1}{x} =$ the amount of the tank they can drain together in 1 hour.

$$\boxed{\text{Amount 1st drains in 1 hour}} + \boxed{\text{Amount 2nd drains in 1 hour}} = \boxed{\text{Amount drained together in 1 hour}}$$

$$\frac{1}{24} + \frac{1}{36} = \frac{1}{x}$$
$$\left(\frac{1}{24} + \frac{1}{36}\right)72x = \frac{1}{x} \cdot 72x$$
$$3x + 2x = 72$$
$$5x = 72$$
$$x = \frac{72}{5} = 14\frac{2}{5}$$

It will take the two pipes $14\frac{2}{5}$ hours to drain the tank.

42. The first man can side $\frac{1}{14}$ of the house in 1 day. Let $x =$ the number of days needed for the second man to side the house alone. Then he can side $\frac{1}{x}$ of the house alone. Since both men can side the house in 8 days, they side $\frac{1}{8}$ of the house in 1 day.

$$\boxed{\text{Amount 1st sides in 1 day}} + \boxed{\text{Amount 2nd sides in 1 day}} = \boxed{\text{Amount sided together in 1 day}}$$

$$\frac{1}{14} + \frac{1}{x} = \frac{1}{8}$$
$$\left(\frac{1}{14} + \frac{1}{x}\right)(56x) = \frac{1}{8} \cdot 56x$$
$$4x + 56 = 7x$$
$$56 = 3x, \text{ or } x = \frac{56}{3} = 18\frac{2}{3}$$

It takes the second man $18\frac{2}{3}$ days to side the house alone.

43. $\dfrac{-5x^6y^3}{10x^3y^6} = -\dfrac{x^3}{2y^3}$

44. $\dfrac{30x^3y^2 - 15x^2y - 10xy^2}{-10xy} = \dfrac{30x^3y^2}{-10xy} + \dfrac{-15x^2y}{-10xy} + \dfrac{-10xy^2}{-10xy} = -3x^2y + \dfrac{3x}{2} + y$

45.

$$\begin{array}{r|l} & x + 5y \\ \hline 3x - 2y & 3x^2 + 13xy - 10y^2 \\ & \underline{3x^2 - 2xy} \\ & \quad 15xy - 10y^2 \\ & \quad \underline{15xy - 10y^2} \\ & \qquad 0 \end{array}$$

46.

$$\begin{array}{r|l} & x^2 + 2x - 1 + \frac{6}{2x+3} \\ \hline 2x + 3 & 2x^3 + 7x^2 + 4x + 3 \\ & \underline{2x^3 + 3x^2} \\ & \quad 4x^2 + 4x \\ & \quad \underline{4x^2 + 6x} \\ & \qquad -2x + 3 \\ & \qquad \underline{-2x - 3} \\ & \qquad\qquad 6 \end{array}$$

47.
$$\begin{array}{r|rrrr} 5 & 1 & -3 & -8 & -10 \\ & & 5 & 10 & 10 \\ \hline & 1 & 2 & 2 & \boxed{0} \end{array} \Rightarrow \text{factor}$$

48.
$$\begin{array}{r|rrrr} -5 & 1 & 4 & -5 & 5 \\ & & -5 & 5 & 0 \\ \hline & 1 & -1 & 0 & \boxed{5} \end{array} \Rightarrow \text{not a factor}$$

49.
$$\frac{x+1}{8} = \frac{4x-2}{24}$$
$$24(x+1) = 8(4x-2)$$
$$24x + 24 = 32x - 16$$
$$-8x = -40$$
$$x = 5$$

50.
$$\frac{1}{x+6} = \frac{x+10}{12}$$
$$12 = (x+6)(x+10)$$
$$12 = x^2 + 16x + 60$$
$$0 = x^2 + 16x + 48$$
$$0 = (x+12)(x+4)$$
$$x + 12 = 0 \quad \textbf{or} \quad x + 4 = 0$$
$$x = -12 \qquad\qquad x = -4$$

51. Let h = the height of the tree.
$$\frac{44}{2.5} = \frac{h}{4}$$
$$176 = 2.5h$$
$70.4 = h \Rightarrow$ The tree is 70.4 feet tall.

52.
$$x = ky \qquad x = 6y$$
$$12 = k(2) \qquad x = 6(12)$$
$$6 = k \qquad \boxed{x = 72}$$

53.
$$x = \frac{k}{y} \qquad x = \frac{72}{y}$$
$$24 = \frac{k}{3} \qquad 12 = \frac{72}{y}$$
$$72 = k \qquad 12y = 72$$
$$\boxed{y = 6}$$

54.
$$x = kyz$$
$$24 = k(3)(4)$$
$$24 = 12k$$
$$\boxed{2 = k}$$

55.
$$x = \frac{kt}{y}$$
$$2 = \frac{k(8)}{64}$$
$$128 = 8k$$
$$\boxed{16 = k}$$

56.
$$T = kv \qquad T = kv$$
$$1575 = k(90{,}000) \qquad T = 0.0175(312{,}000)$$
$$\frac{1575}{90{,}000} = k \qquad \boxed{T = \$5460}$$
$$0.0175 = k$$

Chapter 6 Test (page 429)

1. $\dfrac{-12x^2y^3z^2}{18x^3y^4z^2} = -\dfrac{2}{3xy}$

2. $\dfrac{2x+4}{x^2-4} = \dfrac{2(x+2)}{(x+2)(x-2)} = \dfrac{2}{x-2}$

3. $\dfrac{3y-6z}{2z-y} = \dfrac{3(y-2z)}{2z-y} = -3$

4. $\dfrac{2x^2+7x+3}{4x+12} = \dfrac{(2x+1)(x+3)}{4(x+3)} = \dfrac{2x+1}{4}$

5. $f(10) = \dfrac{5(10)-2}{10} = \dfrac{48}{10} = \dfrac{24}{5}$

6. $\dfrac{x^2y^{-2}}{x^3z^2} \cdot \dfrac{x^2z^4}{y^2z} = \dfrac{x^4y^{-2}z^4}{x^3y^2z^3} = \dfrac{xz}{y^4}$

7. $\dfrac{(x+1)(x+2)}{10} \cdot \dfrac{5}{x+2} = \dfrac{x+1}{2}$

8. $\dfrac{u^2+5u+6}{u^2-4} \cdot \dfrac{u^2-5u+6}{u^2-9} = \dfrac{(u+2)(u+3)}{(u+2)(u-2)} \cdot \dfrac{(u-2)(u-3)}{(u+3)(u-3)} = 1$

9. $$\begin{aligned}\frac{x^3+y^3}{4} \div \frac{x^2-xy+y^2}{2x+2y} = \frac{x^3+y^3}{4} \cdot \frac{2x+2y}{x^2-xy+y^2} &= \frac{(x+y)(x^2-xy+y^2)}{4} \cdot \frac{2(x+y)}{x^2-xy+y^2} \\ &= \frac{(x+y)^2}{2}\end{aligned}$$

10. $$\begin{aligned}\frac{xu+2u+3x+6}{u^2-9} \cdot \frac{2u-6}{x^2+3x+2} &= \frac{u(x+2)+3(x+2)}{(u+3)(u-3)} \cdot \frac{2(u-3)}{(x+2)(x+1)} \\ &= \frac{(x+2)(u+3)}{(u+3)(u-3)} \cdot \frac{2(u-3)}{(x+2)(x+1)} = \frac{2}{x+1}\end{aligned}$$

11. $\dfrac{a^2+7a+12}{a+3} \div \dfrac{16-a^2}{a-4} = \dfrac{a^2+7a+12}{a+3} \cdot \dfrac{a-4}{16-a^2} = \dfrac{(a+4)(a+3)}{a+3} \cdot \dfrac{a-4}{(4+a)(4-a)} = -1$

12. $\dfrac{3t}{t+3} + \dfrac{9}{t+3} = \dfrac{3t+9}{t+3} = \dfrac{3(t+3)}{t+3)} = 3$

13. $\dfrac{3w}{w-5} + \dfrac{w+10}{5-w} = \dfrac{3w}{w-5} - \dfrac{w+10}{w-5} = \dfrac{2w-10}{w-5} = \dfrac{2(w-5)}{w-5} = 2$

14. $\dfrac{2}{r} + \dfrac{r}{s} = \dfrac{2s}{rs} + \dfrac{rr}{rs} = \dfrac{2s+r^2}{rs}$

15. $$\begin{aligned}\frac{x+2}{x+1} - \frac{x+1}{x+2} = \frac{(x+2)(x+2)}{(x+1)(x+2)} - \frac{(x+1)(x+1)}{(x+1)(x+2)} &= \frac{x^2+4x+4-(x^2+2x+1)}{(x+1)(x+2)} \\ &= \frac{x^2+4x+4-x^2-2x-1}{(x+1)(x+2)} \\ &= \frac{2x+3}{(x+1)(x+2)}\end{aligned}$$

16. $\dfrac{\frac{2u^2w^3}{v^2}}{\frac{4uw^4}{uv}} = \dfrac{2u^2w^3}{v^2} \div \dfrac{4uw^4}{uv} = \dfrac{2u^2w^3}{v^2} \cdot \dfrac{uv}{4uw^4} = \dfrac{2u^3w^3v}{4uv^2w^4} = \dfrac{u^2}{2vw}$

17. $\dfrac{\frac{x}{y}+\frac{1}{2}}{\frac{x}{2}-\frac{1}{y}} = \dfrac{\left(\frac{x}{y}+\frac{1}{2}\right)2y}{\left(\frac{x}{2}-\frac{1}{y}\right)2y} = \dfrac{2x+y}{xy-2}$

18.
$$\begin{aligned}
\frac{2}{x-1}+\frac{5}{x+2}&=\frac{11}{x+2}\\
\left(\frac{2}{x-1}+\frac{5}{x+2}\right)(x-1)(x+2)&=\left(\frac{11}{x+2}\right)(x-1)(x+2)\\
2(x+2)+5(x-1)&=11(x-1)\\
2x+4+5x-5&=11x-11\\
7x-1&=11x-11\\
-4x&=-10\\
x&=\frac{-10}{-4}=\frac{5}{2}
\end{aligned}$$
The answer checks.

19.
$$\begin{aligned}
\frac{u-2}{u-3}+3&=u+\frac{u-4}{3-u}\\
\frac{u-2}{u-3}+3&=u+\frac{4-u}{u-3}\\
\left(\frac{u-2}{u-3}+3\right)(u-3)&=\left(u+\frac{4-u}{u-3}\right)(u-3)\\
u-2+3(u-3)&=u(u-3)+4-u\\
4u-11&=u^2-4u+4\\
0&=u^2-8u+15\\
0&=(u-5)(u-3)
\end{aligned}$$
$u-5=0$ **or** $u-3=0$

$u=5$ $\qquad u=3$ $\qquad u=3$ does not check and is not a solution. $u=5$ is a solution.

20.
$$\begin{aligned}
\frac{x^2}{a^2}+\frac{y^2}{b^2}&=1\\
\left(\frac{x^2}{a^2}+\frac{y^2}{b^2}\right)a^2b^2&=a^2b^2\\
x^2b^2+y^2a^2&=a^2b^2\\
x^2b^2&=a^2b^2-y^2a^2\\
x^2b^2&=a^2(b^2-y^2)\\
\frac{x^2b^2}{b^2-y^2}&=a^2
\end{aligned}$$

21.
$$\begin{aligned}
\frac{1}{r}&=\frac{1}{r_1}+\frac{1}{r_2}\\
\frac{1}{r}\cdot rr_1r_2&=\left(\frac{1}{r_1}+\frac{1}{r_2}\right)rr_1r_2\\
r_1r_2&=rr_2+rr_1\\
r_1r_2-rr_2&=rr_1\\
r_2(r_1-r)&=rr_1\\
r_2&=\frac{rr_1}{r_1-r}
\end{aligned}$$

22. Let $r =$ the usual rate (in nautical miles per day). Then $r + 11 =$ the faster rate.

	Rate	Time	Dist.
Usual	r	$\frac{440}{r}$	440
Faster	$r+11$	$\frac{440}{r+11}$	440

$$\boxed{\text{Usual time}} - 2 = \boxed{\text{Faster time}}$$

$$\frac{440}{r} - 2 = \frac{440}{r+11}$$

$$\left(\frac{440}{r} - 2\right)(r)(r+11) = \frac{440}{r+11} \cdot r(r+11)$$

$$440(r+11) - 2r(r+11) = 440r$$

$$-2r^2 - 22r + 4840 = 0$$

$$-2(r+55)(r-44) = 0$$

$r + 55 = 0$ **or** $r - 44 = 0$

$r = -55$ $\qquad r = 44$ $\qquad r = 44$ is the only answer that makes sense.

The usual rate is 44 nautical miles per day.

The usual time is $\frac{440}{44} = 10$ days.

23. Let $r =$ the usual rate. Then $r + 0.04 =$ the higher rate.

Interest = Principal · Rate (· Time) ⇒ Principal $= \dfrac{\text{Interest}}{\text{Rate}}$

$$\boxed{\text{Original principal}} - 2000 = \boxed{\text{New principal}}$$

$$\frac{300}{r} - 2000 = \frac{300}{r+0.04}$$

$$\left(\frac{300}{r} - 2000\right)r(r+0.04) = \left(\frac{300}{r+0.04}\right)r(r+0.04)$$

$$300(r+0.04) - 2000r(r+0.04) = 300r$$

$$-2000r^2 - 80r + 12 = 0$$

$$-4(500r^2 + 20r - 3) = 0$$

$$-4(50r-3)(10r+1) = 0$$

$50r - 3 = 0$ **or** $10r + 1 = 0$

$r = 0.06$ $\qquad r = -0.10$ $\qquad r = 0.06$ is the only answer that makes sense.

She would invest \$5000 at 6% or \$3000 at 10%.

24. $$\frac{18x^2y^3 - 12x^3y^2 + 9xy}{-3xy^4} = \frac{18x^2y^3}{-3xy^4} + \frac{-12x^3y^2}{-3xy^4} + \frac{9xy}{-3xy^4} = -\frac{6x}{y} + \frac{4x^2}{y^2} - \frac{3}{y^3}$$

25.

$$\begin{array}{r|l} & 3x^2 + 4x + 2 \\ \hline 2x-1 & 6x^3 + 5x^2 + 0x - 2 \\ & \underline{6x^3 - 3x^2} \\ & \quad 8x^2 + 0x \\ & \quad \underline{8x^2 - 4x} \\ & \qquad 4x - 2 \\ & \qquad \underline{4x - 2} \\ & \qquad\quad 0 \end{array}$$

26.

$$\begin{array}{r|l} & x^2 - 5x + 10 \\ \hline x+1 & x^3 - 4x^2 + 5x + 3 \\ & \underline{x^3 + x^2} \\ & \quad -5x^2 + 5x \\ & \quad \underline{-5x^2 - 5x} \\ & \qquad 10x + 3 \\ & \qquad \underline{10x + 10} \\ & \qquad\quad -7 \end{array}$$

27.

$$\begin{array}{r|rrrr} 2 & 4 & 3 & 2 & -1 \\ & & 8 & 22 & 48 \\ \hline & 4 & 11 & 24 & \boxed{47} \end{array}$$

28. Let h = the height of the tree.

$$\frac{12}{2} = \frac{h}{3}$$
$$36 = 2h$$
$$18 = h$$

The tree is 18 feet tall.

29.

$$\frac{3}{x-2} = \frac{x+3}{2x}$$
$$6x = (x+3)(x-2)$$
$$6x = x^2 + x - 6$$
$$0 = x^2 - 5x - 6$$
$$0 = (x-6)(x+1)$$
$$x - 6 = 0 \quad \textbf{or} \quad x + 1 = 0$$
$$x = 6 \qquad\qquad x = -1$$

30.

$$V = \frac{k}{t} \qquad V = \frac{1100}{t}$$
$$55 = \frac{k}{20} \qquad 75 = \frac{1100}{t}$$
$$1100 = k \qquad 75t = 1100$$
$$t = \tfrac{1100}{75}$$
$$\boxed{t = \tfrac{44}{3}}$$

Cumulative Review Exercises (page 430)

1. $a^3b^2a^5b^2 = a^8b^4$

2. $\dfrac{a^3b^6}{a^7b^2} = \dfrac{b^4}{a^4}$

3. $\left(\dfrac{2a^2}{3b^4}\right)^{-4} = \left(\dfrac{3b^4}{2a^2}\right)^4 = \dfrac{81b^{16}}{16a^8}$

4.

$$\left(\frac{x^{-2}y^3}{x^2x^3y^4}\right)^{-3} = \left(\frac{x^2x^3y^4}{x^{-2}y^3}\right)^3$$
$$= (x^7y)^3 = x^{21}y^3$$

5. $4.25 \times 10^4 = 42{,}500$

6. $7.12 \times 10^{-4} = 0.000712$

7.

$$\frac{a+2}{5} - \frac{8}{5} = 4a - \frac{a+9}{2}$$
$$\left(\frac{a+2}{5} - \frac{8}{5}\right)10 = \left(4a - \frac{a+9}{2}\right)10$$
$$2(a+2) - 16 = 40a - 5(a+9)$$
$$2a + 4 - 16 = 40a - 5a - 45$$
$$33 = 33a$$
$$1 = a$$

8.

$$\frac{3x-4}{6} - \frac{x-2}{2} = \frac{-2x-3}{3}$$
$$\left(\frac{3x-4}{6} - \frac{x-2}{2}\right)6 = \left(\frac{-2x-3}{3}\right)6$$
$$3x - 4 - 3(x-2) = 2(-2x-3)$$
$$3x - 4 - 3x + 6 = -4x - 6$$
$$4x = -8$$
$$x = -2$$

9. $m = \dfrac{\Delta y}{\Delta x} = \dfrac{5-10}{-2-4} = \dfrac{-5}{-6} = \dfrac{5}{6}$

10. $3x + 4y = 13$
$4y = -3x + 13$
$y = -\frac{3}{4}x + \frac{13}{4}$
$m = -\frac{3}{4}$

11. $y = 3x + 2 \Rightarrow m = 3$
A parallel line will also have $m = 3$.

12. $y = 3x + 2 \Rightarrow m = 3$
A perpendicular line will have $m = -\frac{1}{3}$.

13. $f(0) = 0^2 - 2(0) = 0$

14. $f(-2) = (-2)^2 - 2(-2) = 8$

15. $f\left(\frac{2}{5}\right) = \left(\frac{2}{5}\right)^2 - 2\left(\frac{2}{5}\right) = \frac{4}{25} - \frac{4}{5} = -\frac{16}{25}$

16. $f(t-1) = (t-1)^2 - 2(t-1) = t^2 - 2t + 1 - 2t + 2 = t^2 - 4t + 3$

17. $y = \dfrac{kxz}{r}$

18. Since the graph does not pass the vertical line test, it is not the graph of a function.

19.
$$x - 2 \le 3x + 1 \le 5x - 4$$

$x - 2 \le 3x + 1$ and $3x + 1 \le 5x - 4$

$x - 2 - 1 \le 3x + 1 - 1$ | $3x + 1 + 4 \le 5x - 4 + 4$

$x - x - 3 \le 3x - x$ | $3x - 3x + 5 \le 5x - 3x$

$-3 \le 2x$ | $5 \le 2x$

$\dfrac{-3}{2} \le x$ | $\dfrac{5}{2} \le x$

$x \ge -\dfrac{3}{2}$ | $x \ge \dfrac{5}{2}$

If $x \ge -\dfrac{3}{2}$ **and** $x \ge \dfrac{5}{2}$, then $x \ge \dfrac{5}{2}$. The solution set is $\left[\dfrac{5}{2}, \infty\right)$

$\frac{5}{2}$

CUMULATIVE REVIEW EXERCISES

20.
$$\left|\frac{3a}{5}-2\right|+1 \geq \frac{6}{5}$$
$$\left|\frac{3a}{5}-2\right| \geq \frac{1}{5}$$
$$\frac{3a}{5}-2 \leq -\frac{1}{5} \quad \text{or} \quad \frac{3a}{5}-2 \geq \frac{1}{5}$$
$$\frac{3a}{5} \leq \frac{9}{5} \qquad \frac{3a}{5} \geq \frac{11}{5}$$
$$3a \leq 9 \qquad 3a \geq 11$$
$$a \leq 3 \qquad a \geq \frac{11}{3}$$
solution set: $(-\infty, 3] \cup \left[\frac{11}{3}, \infty\right)$

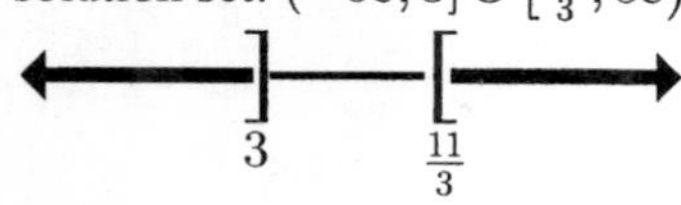

21. trinomial

22. degree $= 3 + 4 = 7$

23. $f(-2) = -3(-2)^3 + (-2) - 4 = 18$

24. $y = f(x) = 2x^2 - 3$

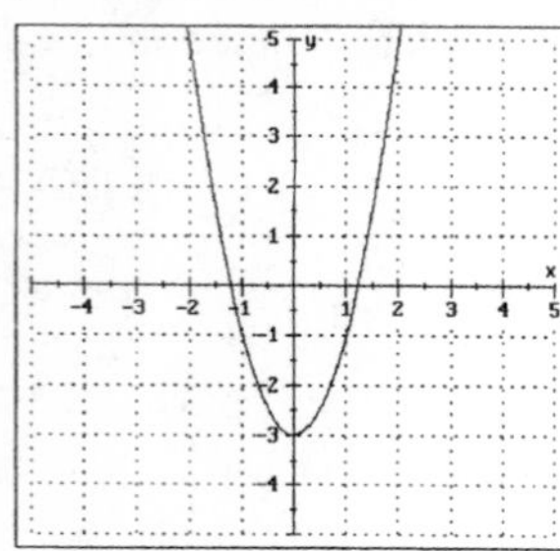

25. $(3x^2 - 2x + 7) + (-2x^2 + 2x + 5) + (3x^2 - 4x + 2)$
$= 3x^2 - 2x + 7 - 2x^2 + 2x + 5 + 3x^2 - 4x + 2 = 4x^2 - 4x + 14$

26. $(-5x^2 + 3x + 4) - (-2x^2 + 3x + 7) = -5x^2 + 3x + 4 + 2x^2 - 3x - 7 = -3x^2 - 3$

27. $(3x + 4)(2x - 5) = 6x^2 - 15x + 8x - 20$
$= 6x^2 - 7x - 20$

28. $(2x^n - 1)(x^n + 2) = 2x^{2n} + 4x^n - x^n - 2$
$= 2x^{2n} + 3x^n - 2$

29. $3r^2s^3 - 6rs^4 = 3rs^3(r - 2s)$

30. $5(x - y) - a(x - y) = (x - y)(5 - a)$

31. $xu + yv + xv + yu = xu + xv + yu + yv = x(u + v) + y(u + v) = (u + v)(x + y)$

32. $81x^4 - 16y^4 = (9x^2 + 4y^2)(9x^2 - 4y^2) = (9x^2 + 4y^2)(3x + 2y)(3x - 2y)$

33. $8x^3 - 27y^6 = (2x)^3 - (3y^2)^3 = (2x - 3y^2)(4x^2 + 6xy^2 + 9y^4)$

34. $6x^2 + 5x - 6 = (2x + 3)(3x - 2)$

35. $9x^2 - 30x + 25 = (3x - 5)(3x - 5)$

36. $15x^2 - x - 6 = (5x + 3)(3x - 2)$

37. $27a^3 + 8b^3 = (3a + 2b)(9a^2 - 6ab + 4b^2)$

CUMULATIVE REVIEW EXERCISES

38. $6x^2 + x - 35 = (3x - 7)(2x + 5)$

39. $x^2 + 10x + 25 - y^4 = (x + 5)^2 - y^4 = (x + 5 + y^2)(x + 5 - y^2)$

40. $y^2 - x^2 + 4x - 4 = y^2 - (x^2 - 4x + 4) = y^2 - (x - 2)^2 = (y + x - 2)(y - x + 2)$

41.
$$\begin{aligned} x^3 - 4x &= 0 \\ x(x^2 - 4) &= 0 \\ x(x + 2)(x - 2) &= 0 \end{aligned}$$
$x = 0$ **or** $x + 2 = 0$ **or** $x - 2 = 0$

$x = -2 \qquad x = 2$

42.
$$\begin{aligned} 6x^2 + 7 &= -23x \\ 6x^2 + 23x + 7 &= 0 \\ (2x + 7)(3x + 1) &= 0 \end{aligned}$$
$2x + 7 = 0$ **or** $3x + 1 = 0$

$x = -\frac{7}{2} \qquad x = -\frac{1}{3}$

43. $\dfrac{2x^2y + xy - 6y}{3x^2y + 5xy - 2y} = \dfrac{y(2x^2 + x - 6)}{y(3x^2 + 5x - 2)} = \dfrac{y(2x - 3)(x + 2)}{y(3x - 1)(x + 2)} = \dfrac{2x - 3}{3x - 1}$

44.
$$\begin{aligned} \frac{x^2 - 4}{x^2 + 9x + 20} \div \frac{x^2 + 5x + 6}{x^2 + 4x - 5} \cdot \frac{x^2 + 3x - 4}{(x - 1)^2} &= \frac{x^2 - 4}{x^2 + 9x + 20} \cdot \frac{x^2 + 4x - 5}{x^2 + 5x + 6} \cdot \frac{x^2 + 3x - 4}{(x - 1)^2} \\ &= \frac{(x + 2)(x - 2)}{(x + 4)(x + 5)} \cdot \frac{(x + 5)(x - 1)}{(x + 2)(x + 2)} \cdot \frac{(x + 4)(x - 1)}{(x - 1)(x - 1)} \\ &= \frac{x - 2}{x + 3} \end{aligned}$$

45.
$$\begin{aligned} \frac{2}{x + y} + \frac{3}{x - y} - \frac{x - 3y}{x^2 - y^2} &= \frac{2}{x + y} + \frac{3}{x - y} - \frac{x - 3y}{(x + y)(x - y)} \\ &= \frac{2(x - y)}{(x + y)(x - y)} + \frac{3(x + y)}{(x + y)(x - y)} - \frac{x - 3y}{(x + y)(x - y)} \\ &= \frac{2x - 2y + 3x + 3y - x + 3y}{(x + y)(x - y)} \\ &= \frac{4x + 4y}{(x + y)(x - y)} = \frac{4(x + y)}{(x + y)(x - y)} = \frac{4}{x - y} \end{aligned}$$

46. $\dfrac{\frac{a}{b} + b}{a - \frac{b}{a}} = \dfrac{\left(\frac{a}{b} + b\right)ab}{\left(a - \frac{b}{a}\right)ab} = \dfrac{a^2 + ab^2}{a^2b - b^2}$

47.
$$\begin{aligned} \frac{5x - 3}{x + 2} &= \frac{5x + 3}{x - 2} \\ (5x - 3)(x - 2) &= (5x + 3)(x + 2) \\ 5x^2 - 13x + 6 &= 5x^2 = 13x + 6 \\ -26x &= 0 \\ x &= 0 \end{aligned}$$
The answer checks.

48.

$$\frac{3}{x-2}+\frac{x^2}{(x+3)(x-2)}=\frac{x+4}{x+3}$$
$$\left(\frac{3}{x-2}+\frac{x^2}{(x+3)(x-2)}\right)(x+3)(x-2)=\left(\frac{x+4}{x+3}\right)(x+3)(x-2)$$
$$3(x+3)+x^2=(x+4)(x-2)$$
$$x^2+3x+9=x^2+2x-8$$
$$x=-17$$

49.

$$\begin{array}{r|l} & x+4 \\ \hline x+5 & x^2+9x+20 \\ & \underline{x^2+5x} \\ & \quad 4x+20 \\ & \quad \underline{4x+20} \\ & \qquad 0 \end{array}$$

50.

$$\begin{array}{r|l} & -x^2+x+5+\frac{8}{x-1} \\ \hline x-1 & -x^3+2x^2+4x+3 \\ & \underline{-x^3+x^2} \\ & \quad x^2+4x \\ & \quad \underline{x^2-x} \\ & \qquad 5x+3 \\ & \qquad \underline{5x-5} \\ & \qquad\quad 8 \end{array}$$

Exercise 7.1 (page 443)

1. $\frac{x^2+7x+12}{x^2-16}=\frac{(x+4)(x+3)}{(x+4)(x-4)}=\frac{x+3}{x-4}$

2. $\frac{a^3-b^3}{b^2-a^2}=\frac{(a-b)(a^2+ab+b^2)}{(b+a)(b-a)}=-\frac{a^2+ab+b^2}{a+b}$

3. $\frac{x^2-x-6}{x^2-2x-3}\cdot\frac{x^2-1}{x^2+x-2}=\frac{(x-3)(x+2)}{(x-3)(x+1)}\cdot\frac{(x+1)(x-1)}{(x+2)(x-1)}=1$

4. $\frac{x^2-3x-4}{x^2-5x+6}\div\frac{x^2-2x-3}{x^2-x-2}=\frac{(x-4)(x+1)}{(x-3)(x-2)}\cdot\frac{(x-2)(x+1)}{(x-3)(x+1)}=\frac{(x-4)(x+1)}{(x-3)^2}$

5. $\frac{3}{m+1}+\frac{3m}{m-1}=\frac{3(m-1)}{(m+1)(m-1)}+\frac{3m(m+1)}{(m+1)(m-1)}=\frac{3m^2+6m-3}{(m+1)(m-1)}=\frac{3(m^2+2m-1)}{(m+1)(m-1)}$

6. $\frac{2x+3}{3x-1}-\frac{x-4}{2x+1}=\frac{(2x+3)(2x+1)}{(3x-1)(2x+1)}-\frac{(x-4)(3x-1)}{(3x-1)(2x+1)}=\frac{x^2+21x-1}{(3x-1)(2x+1)}$

7. $(5x^2)^2$ **8.** $6^2=36$ **9.** positive **10.** $|x|$

11. 3; up **12.** 5; left **13.** x **14.** x

15. odd **16.** even **17.** 0 **18.** standard

19. $3x^2$ **20.** x **21.** a^2+b^3 **22.** $\frac{x}{y}$

23. $\sqrt{121} = \sqrt{11^2} = 11$ **24.** $\sqrt{144} = \sqrt{12^2} = 12$ **25.** $-\sqrt{64} = -\sqrt{8^2} = -8$

26. $-\sqrt{1} = -\sqrt{1^2} = -1$ **27.** $\sqrt{\frac{1}{9}} = \sqrt{\left(\frac{1}{3}\right)^2} = \frac{1}{3}$ **28.** $-\sqrt{\frac{4}{25}} = -\sqrt{\left(\frac{2}{5}\right)^2} = -\frac{2}{5}$

29. $-\sqrt{\frac{25}{49}} = -\sqrt{\left(\frac{5}{7}\right)^2} = -\frac{5}{7}$ **30.** $\sqrt{\frac{49}{81}} = \sqrt{\left(\frac{7}{9}\right)^2} = \frac{7}{9}$

31. $\sqrt{-25}$: not a real number **32.** $\sqrt{0.25} = 0.5$ **33.** $\sqrt{0.16} = 0.4$

34. $\sqrt{-49}$: not a real number **35.** $\sqrt{(-4)^2} = \sqrt{16} = 4$ **36.** $\sqrt{(-9)^2} = \sqrt{81} = 9$

37. $\sqrt{-36}$: not a real number **38.** $-\sqrt{-4}$: not a real number **39.** $\sqrt{12} \approx 3.4641$

40. $\sqrt{340} \approx 18.4391$ **41.** $\sqrt{679.25} \approx 26.0624$ **42.** $\sqrt{0.0063} \approx 0.0794$

43. $\sqrt{4x^2} = \sqrt{(2x)^2} = |2x| = 2|x|$ **44.** $\sqrt{16y^4} = \sqrt{(4y^2)^2} = |4y^2| = 4y^2$

45. $\sqrt{9a^4} = \sqrt{(3a^2)^2} = |3a^2| = 3a^2$ **46.** $\sqrt{16b^2} = \sqrt{(4b)^2} = |4b| = 4|b|$

47. $\sqrt{(t+5)^2} = |t+5|$ **48.** $\sqrt{(a+6)^2} = |a+6|$

49. $\sqrt{(-5b)^2} = |-5b| = 5|b|$ **50.** $\sqrt{(-8c)^2} = |-8c| = 8|c|$

51. $\sqrt{a^2+6a+9} = \sqrt{(a+3)^2} = |a+3|$ **52.** $\sqrt{x^2+10x+25} = \sqrt{(x+5)^2} = |x+5|$

53. $\sqrt{t^2+24t+144} = \sqrt{(t+12)^2}$
$= |t+12|$

54. $\sqrt{m^2+30m+225} = \sqrt{(m+15)^2}$
$= |m+15|$

55. $\sqrt[3]{1} = \sqrt[3]{1^3} = 1$ **56.** $\sqrt[3]{-8} = \sqrt[3]{(-2)^3} = -2$ **57.** $\sqrt[3]{-125} = \sqrt[3]{(-5)^3} = -5$

58. $\sqrt[3]{512} = \sqrt[3]{8^3} = 8$ **59.** $\sqrt[3]{-\frac{8}{27}} = \sqrt[3]{\left(-\frac{2}{3}\right)^3} = -\frac{2}{3}$ **60.** $\sqrt[3]{\frac{125}{216}} = \sqrt[3]{\left(\frac{5}{6}\right)^3} = \frac{5}{6}$

61. $\sqrt[3]{0.064} = 0.4$ **62.** $\sqrt[3]{0.001} = 0.1$

63. $\sqrt[3]{8a^3} = \sqrt[3]{(2a)^3} = 2a$ **64.** $\sqrt[3]{-27x^6} = \sqrt[3]{(-3x^2)^3} = -3x^2$

65. $\sqrt[3]{-1000p^3q^3} = \sqrt[3]{(-10pq)^3} = -10pq$ **66.** $\sqrt[3]{343a^6b^3} = \sqrt[3]{(7a^2b)^3} = 7a^2b$

67. $\sqrt[3]{-\frac{1}{8}m^6n^3} = \sqrt[3]{\left(-\frac{1}{2}m^2n\right)^3} = -\frac{1}{2}m^2n$ **68.** $\sqrt[3]{\frac{27}{1000}a^6b^6} = \sqrt[3]{\left(\frac{3}{10}a^2b^2\right)^3} = \frac{3}{10}a^2b^2$

69. $\sqrt[3]{0.008z^9} = \sqrt[3]{(0.2z^3)^3} = 0.2z^3$ **70.** $\sqrt[3]{0.064s^9t^6} = \sqrt[3]{(0.4s^3t^2)^3} = 0.4s^3t^2$

71. $\sqrt[4]{81} = \sqrt[4]{3^4} = 3$

72. $\sqrt[6]{64} = \sqrt[6]{2^6} = 2$

73. $-\sqrt[5]{243} = -\sqrt[5]{3^5} = -3$

74. $-\sqrt[4]{625} = -\sqrt[4]{5^4} = -5$

75. $\sqrt[5]{-32} = \sqrt[5]{(-2)^5} = -2$

76. $\sqrt[6]{729} = \sqrt[6]{3^6} = 3$

77. $\sqrt[4]{\frac{16}{625}} = \sqrt[4]{\left(\frac{2}{5}\right)^4} = \frac{2}{5}$

78. $\sqrt[5]{-\frac{243}{32}} = \sqrt[5]{\left(-\frac{3}{2}\right)^5} = -\frac{3}{2}$

79. $-\sqrt[5]{-\frac{1}{32}} = -\sqrt[5]{\left(-\frac{1}{2}\right)^5} = -\left(-\frac{1}{2}\right) = \frac{1}{2}$

80. $\sqrt[6]{-729} \Rightarrow$ not a real number

81. $\sqrt[4]{-256} \Rightarrow$ not a real number

82. $-\sqrt[4]{\frac{81}{256}} = -\sqrt[4]{\left(\frac{3}{4}\right)^4} = -\frac{3}{4}$

83. $\sqrt[4]{16x^4} = \sqrt[4]{(2x)^4} = |2x| = 2|x|$

84. $\sqrt[5]{32a^5} = \sqrt[5]{(2a)^5} = 2a$

85. $\sqrt[3]{8a^3} = \sqrt[3]{(2a)^3} = 2a$

86. $\sqrt[6]{64x^6} = \sqrt[6]{(2x)^6} = |2x| = 2|x|$

87. $\sqrt[4]{\frac{1}{16}x^4} = \sqrt[4]{\left(\frac{1}{2}x\right)^4} = \left|\frac{1}{2}x\right| = \frac{1}{2}|x|$

88. $\sqrt[4]{\frac{1}{81}x^8} = \sqrt[4]{\left(\frac{1}{3}x^2\right)^4} = \left|\frac{1}{3}x^2\right| = \frac{1}{3}x^2$

89. $\sqrt[4]{x^{12}} = \sqrt[4]{(x^3)^4} = |x^3|$

90. $\sqrt[8]{x^{24}} = \sqrt[8]{(x^3)^8} = |x^3|$

91. $\sqrt[5]{-x^5} = \sqrt[5]{(-x)^5} = -x$

92. $\sqrt[3]{-x^6} = \sqrt[3]{(-x^2)^3} = -x^2$

93. $\sqrt[3]{-27a^6} = \sqrt[3]{(-3a^2)^3} = -3a^2$

94. $\sqrt[5]{-32x^5} = \sqrt[5]{(-2x)^5} = -2x$

95. $\sqrt[25]{(x+2)^{25}} = x + 2$

96. $\sqrt[44]{(x+4)^{44}} = |x + 4|$

97. $\sqrt[8]{0.00000001x^{16}y^8} = \sqrt[8]{(0.1x^2y)^8} = |0.1x^2y| = 0.1x^2|y|$

98. $\sqrt[5]{0.00032x^{10}y^5} = \sqrt[5]{(0.2x^2y)^5} = 0.2x^2y$

99. $f(4) = \sqrt{4-4} = \sqrt{0} = 0$

100. $f(8) = \sqrt{8-4} = \sqrt{4} = 2$

101. $f(20) = \sqrt{20-4} = \sqrt{16} = 4$

102. $f(29) = \sqrt{29-4} = \sqrt{25} = 5$

103. $g(9) = \sqrt{9-8} = \sqrt{1} = 1$

104. $g(17) = \sqrt{17-8} = \sqrt{9} = 3$

105. $g(8.25) = \sqrt{8.25-8} = \sqrt{0.25} = 0.5$

106. $g(8.64) = \sqrt{8.64-8} = \sqrt{0.64} = 0.8$

107. $f(4) = \sqrt{4^2+1} = \sqrt{17} \approx 4.1231$

108. $f(6) = \sqrt{6^2+1} = \sqrt{37} \approx 6.0828$

109. $f(2.35) = \sqrt{(2.35)^2+1} = \sqrt{6.5225} \approx 2.5539$

110. $f(21.57) = \sqrt{(21.57)^2+1} = \sqrt{466.2649} \approx 21.5932$

111. $f(x) = \sqrt{x+4}$; Shift $y = \sqrt{x}$ left 4.

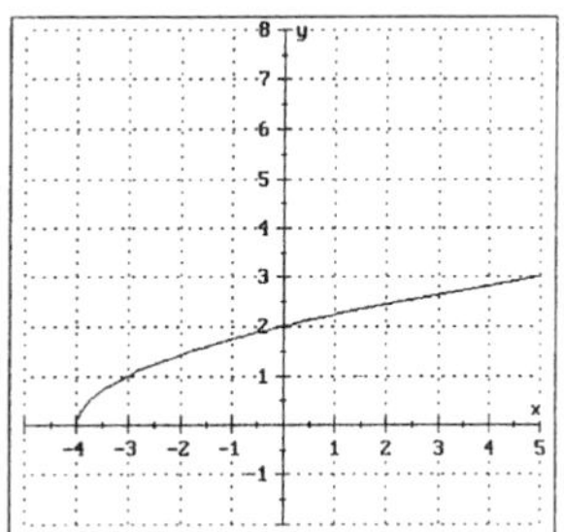

$D = [-4, \infty)$; $R = [0, \infty)$

112. $f(x) = -\sqrt{x-2}$; Reflect $y = \sqrt{x}$ about the x-axis and shift right 2.

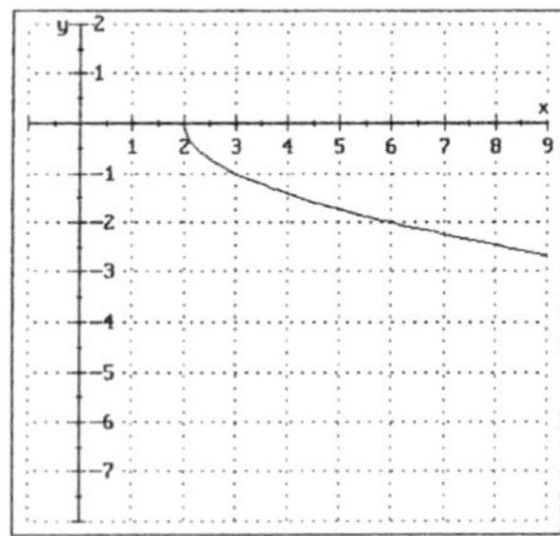

$D = [2, \infty)$; $R = (-\infty, 0]$

113. $f(x) = -\sqrt{x} - 3$; Reflect $y = \sqrt{x}$ about the x-axis and shift down 3.

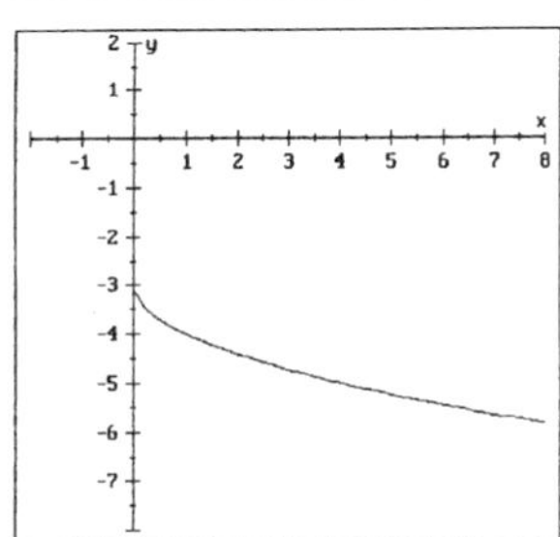

$D = [0, \infty)$; $R = (-\infty, -3]$

114. $f(x) = \sqrt[3]{x} - 1$; Shift $y = \sqrt[3]{x}$ down 1.

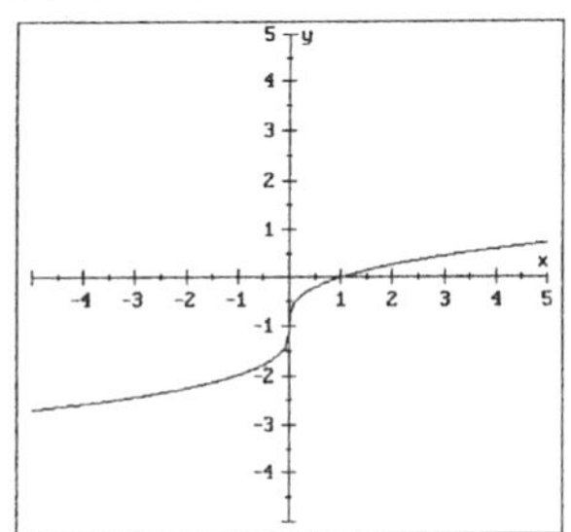

$D = (-\infty, \infty)$; $R = (-\infty, \infty)$

115. mean $= \dfrac{2+5+5+6+7}{5} = \dfrac{25}{5} = 5$

Original term	Mean	Difference (term−mean)	Square of difference
2	5	−3	9
5	5	0	0
5	5	0	0
6	5	1	1
7	5	2	4

st. dev. $= \sqrt{\dfrac{9+0+0+1+4}{5}} \approx 1.67$

116. mean $= \dfrac{3+6+7+9+11+12}{6} = \dfrac{48}{6} = 8$

Original term	Mean	Difference (term−mean)	Square of difference
3	8	−5	25
6	8	−2	4
7	8	−1	1
9	8	1	1
11	8	3	9
12	8	4	16

st. dev. $= \sqrt{\dfrac{25+4+1+1+9+16}{6}} \approx 3.06$

117. $s_{\bar{x}} = \dfrac{s}{\sqrt{N}} = \dfrac{65}{\sqrt{30}} \approx 11.8673$

118. $\sigma_{\bar{x}} = \dfrac{\sigma}{\sqrt{N}} = \dfrac{12.7}{\sqrt{32}} = 2.2451$

119. $r = \sqrt{\dfrac{A}{\pi}} = \sqrt{\dfrac{9\pi}{\pi}} = \sqrt{9} = 3$ units

120. $d = \sqrt{2s^2} = \sqrt{2(90)^2} = \sqrt{16200}$
≈ 127.28 feet

121. $t = \frac{\sqrt{s}}{4} = \frac{\sqrt{256}}{4} = \frac{16}{4} = 4$ seconds

122. $s = k\sqrt{l} = 3.24\sqrt{400} = 3.24(20)$
$= 64.8$ mph

123. $I = \sqrt{\frac{P}{18}} = \sqrt{\frac{980}{18}} \approx \sqrt{54.44} \approx 7.4$ amps

124. $p = \frac{590}{\sqrt{t}} = \frac{590}{\sqrt{71}} \approx \frac{590}{8.43} \approx 70$ beats/minute

125. Answers may vary.

126. Answers may vary.

127. $\sqrt{x^2 - 4x + 4} = \sqrt{(x-2)^2} = |x - 2|$.
$|x - 2| = x - 2$ when $x - 2 \geq 0$, or $x \geq 2$.

128. $\sqrt{x^2} = |x|$, and $|x| = x$ if $x \geq 0$.
If $x < 0$, then $\sqrt{x^2} \neq x$.

Exercise 7.2 (page 450)

1. $(4x + 2)(3x - 5) = 12x^2 - 14x - 10$

2. $(3y - 5)(2y + 3) = 6y^2 - y - 15$

3. $(5t + 4s)(3t - 2s) = 15t^2 + 2ts - 8s^2$

4. $(4r - 3)(2r^2 + 3r - 4) = 8r^3 + 12r^2 - 16r - 6r^2 - 9r + 12 = 8r^3 + 6r^2 - 25r + 12$

5. hypotenuse

6. legs

7. $a^2 + b^2 = c^2$

8. sum; legs

9. distance

10. $\sqrt{(x_2 - x_1)^2 + (y_2 - y_1)^2}$

11.
$$\begin{aligned} c^2 &= a^2 + b^2 \\ c^2 &= 6^2 + 8^2 \\ c^2 &= 36 + 64 \\ c^2 &= 100 \\ c &= \sqrt{100} \\ c &= 10 \text{ ft} \end{aligned}$$

12.
$$\begin{aligned} c^2 &= a^2 + b^2 \\ 26^2 &= 10^2 + b^2 \\ 676 &= 100 + b^2 \\ 576 &= b^2 \\ \sqrt{576} &= b \\ 24 \text{ cm} &= b \end{aligned}$$

13.
$$\begin{aligned} c^2 &= a^2 + b^2 \\ 82^2 &= a^2 + 18^2 \\ 6724 &= a^2 + 324 \\ 6400 &= a^2 \\ \sqrt{6400} &= a \\ 80 \text{ m} &= a \end{aligned}$$

14.
$$\begin{aligned} c^2 &= a^2 + b^2 \\ 25^2 &= a^2 + 7^2 \\ 625 &= a^2 + 49 \\ 576 &= a^2 \\ \sqrt{576} &= a \\ 24 \text{ ft} &= a \end{aligned}$$

15.
$$\begin{aligned} c^2 &= a^2 + b^2 \\ 50^2 &= 14^2 + b^2 \\ 2500 &= 196 + b^2 \\ 2304 &= b^2 \\ \sqrt{2304} &= b \\ 48 \text{ in.} &= b \end{aligned}$$

16.
$$\begin{aligned} c^2 &= a^2 + b^2 \\ c^2 &= 8^2 + 15^2 \\ c^2 &= 64 + 225 \\ c^2 &= 289 \\ c &= \sqrt{289} \\ c &= 17 \text{ cm} \end{aligned}$$

17. Let $x =$ the length of the diagonal.

$$\begin{aligned} 7^2 + 7^2 &= x^2 \\ 49 + 49 &= x^2 \\ 98 &= x^2 \\ \sqrt{98} &= x \\ 9.9 \text{ cm} &\approx x \end{aligned}$$

18. Let $x =$ the length of the diagonal.

$$\begin{aligned} 7^2 + \left(\sqrt{98}\right)^2 &= x^2 \\ 49 + 98 &= x^2 \\ 147 &= x^2 \\ \sqrt{147} &= x \\ 12.1 \text{ cm} &\approx x \end{aligned}$$

19. $$\begin{aligned} d = \sqrt{(x_2 - x_1)^2 + (y_2 - y_1)^2} = \sqrt{(0-3)^2 + [0-(-4)]^2} = \sqrt{(-3)^2 + (4)^2} &= \sqrt{9+16} \\ &= \sqrt{25} = 5 \end{aligned}$$

20. $$\begin{aligned} d = \sqrt{(x_2 - x_1)^2 + (y_2 - y_1)^2} = \sqrt{[0-(-6)]^2 + (0-8)^2} = \sqrt{6^2 + (-8)^2} &= \sqrt{36+64} \\ &= \sqrt{100} = 10 \end{aligned}$$

21. $$\begin{aligned} d = \sqrt{(x_2 - x_1)^2 + (y_2 - y_1)^2} = \sqrt{(2-5)^2 + (4-8)^2} = \sqrt{(-3)^2 + (-4)^2} &= \sqrt{9+16} \\ &= \sqrt{25} = 5 \end{aligned}$$

22. $$\begin{aligned} d = \sqrt{(x_2 - x_1)^2 + (y_2 - y_1)^2} = \sqrt{(5-8)^2 + (9-13)^2} = \sqrt{(-3)^2 + (-4)^2} &= \sqrt{9+16} \\ &= \sqrt{25} = 5 \end{aligned}$$

23. $$\begin{aligned} d = \sqrt{(x_2 - x_1)^2 + (y_2 - y_1)^2} = \sqrt{(-2-3)^2 + (-8-4)^2} = \sqrt{(-5)^2 + (-12)^2} &= \sqrt{25+144} \\ &= \sqrt{169} = 13 \end{aligned}$$

24. $$\begin{aligned} d = \sqrt{(x_2 - x_1)^2 + (y_2 - y_1)^2} = \sqrt{(-5-7)^2 + (-2-3)^2} = \sqrt{(-12)^2 + (-5)^2} &= \sqrt{144+25} \\ &= \sqrt{169} = 13 \end{aligned}$$

25. $$\begin{aligned} d = \sqrt{(x_2 - x_1)^2 + (y_2 - y_1)^2} = \sqrt{(6-12)^2 + (8-16)^2} = \sqrt{(-6)^2 + (-8)^2} &= \sqrt{36+64} \\ &= \sqrt{100} = 10 \end{aligned}$$

26. $$\begin{aligned} d = \sqrt{(x_2 - x_1)^2 + (y_2 - y_1)^2} = \sqrt{(10-2)^2 + [4-(-2)]^2} = \sqrt{8^2 + 6^2} &= \sqrt{64+36} \\ &= \sqrt{100} = 10 \end{aligned}$$

27. $$\begin{aligned} d = \sqrt{(x_2 - x_1)^2 + (y_2 - y_1)^2} = \sqrt{[-3-(-5)]^2 + [5-(-5)]^2} = \sqrt{2^2 + 10^2} &= \sqrt{4+100} \\ &= \sqrt{104} \approx 10.2 \end{aligned}$$

28. $$\begin{aligned} d = \sqrt{(x_2 - x_1)^2 + (y_2 - y_1)^2} = \sqrt{(2-4)^2 + [-3-(-8)]^2} = \sqrt{(-2)^2 + 5^2} &= \sqrt{4+25} \\ &= \sqrt{29} \approx 5.4 \end{aligned}$$

29. Let the points be represented by $A(5,1)$, $B(7,0)$ and $C(3,0)$. Find the length of $\overline{AB}$ and $\overline{AC}$:

$\overline{AB}: \sqrt{(5-7)^2+(1-0)^2} = \sqrt{(-2)^2+1^2} = \sqrt{5}$

$\overline{AC}: \sqrt{(5-3)^2+(1-0)^2} = \sqrt{(2)^2+1^2} = \sqrt{5}$

Since $\overline{AB}$ and $\overline{AC}$ have the same length, $(5,1)$ is equidistant from $(7,0)$ and $(3,0)$.

30. Let the points be represented by $A(2,3)$, $B(-3,4)$ and $C(1,-2)$. Find the length of each side:

$\overline{AB}: \sqrt{[2-(-3)]^2+(3-4)^2} = \sqrt{5^2+(-1)^2} = \sqrt{26}$

$\overline{AC}: \sqrt{(2-1)^2+[3-(-2)]^2} = \sqrt{1^2+5^2} = \sqrt{26}$

$\overline{BC}: \sqrt{(-3-1)^2+[4-(-2)]^2} = \sqrt{(-4)^2+6^2} = \sqrt{52}$

Note that $(\text{length of } \overline{AB})^2 + (\text{length of } \overline{AC})^2 = (\text{length of } \overline{BC})^2$

$$\left(\sqrt{26}\right)^2 + \left(\sqrt{26}\right)^2 = \left(\sqrt{52}\right)^2$$

$$26+26=52$$

Thus, $\overline{BC}$ is the hypotenuse of a right triangle.

31. Let the points be represented by $A(-2,4)$, $B(2,8)$ and $C(6,4)$. Find the length of each side:

$\overline{AB}: \sqrt{(-2-2)^2+(4-8)^2} = \sqrt{(-4)^2+(-4)^2} = \sqrt{32}$

$\overline{AC}: \sqrt{(-2-6)^2+(4-4)^2} = \sqrt{(-8)^2+0^2} = \sqrt{64}$

$\overline{BC}: \sqrt{(2-6)^2+(8-4)^2} = \sqrt{(-4)^2+4^2} = \sqrt{32}$

Since $\overline{AB}$ and $\overline{BC}$ have the same length, the triangle is isosceles.

32. Let the points be represented by $A(-2,13)$, $B(-8,9)$ and $C(-2,5)$. Find the length of each side:

$\overline{AB}: \sqrt{[-2-(-8)]^2+(13-9)^2} = \sqrt{6^2+4^2} = \sqrt{36+16} = \sqrt{52}$

$\overline{AC}: \sqrt{[-2-(-2)]^2+(13-5)^2} = \sqrt{0^2+8^2} = \sqrt{64}$

$\overline{BC}: \sqrt{[-8-(-2)]^2+(9-5)^2} = \sqrt{(-6)^2+4^2} = \sqrt{36+16} = \sqrt{52}$

Since $\overline{AB}$ and $\overline{BC}$ have the same length, the triangle is isosceles.

33.
$$d^2 = 5^2+12^2$$
$$d^2 = 25+144$$
$$d^2 = 169$$
$$d = 13 \text{ ft}$$

34.
$$15^2+h^2 = 17^2$$
$$225+h^2 = 289$$
$$h^2 = 64$$
$$h = 8 \text{ ft}$$

35. Let $x =$ distance to 2nd
$$x^2 = 90^2+90^2$$
$$x^2 = 8100+8100$$
$$x^2 = 16200$$
$$x \approx 127 \text{ ft}$$

36. Refer to **#35**.

distance $= 127.3 - 60.5 \approx 67$ ft

37. Refer to the diagram provided. The 3rd baseman is at B, so $\overline{BC}$ has a length of 10 ft. Let $\overline{AC}$ and $\overline{AB}$ both have a length x.

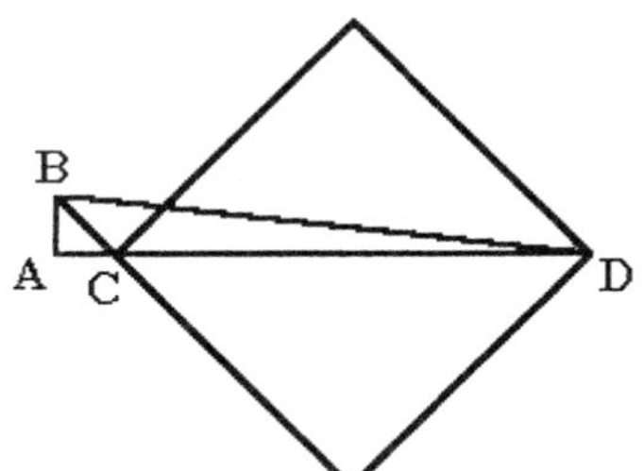

$$x^2 + x^2 = 10^2$$
$$2x^2 = 100$$
$$x^2 = 50$$
$$x = \sqrt{50} \approx 7.1 \text{ ft}$$

From **#35**, the length of $\overline{CD} \approx 127.3$ ft, so $\overline{AD}$ has a length of about $127.3 + 7.1 = 134.4$ ft. Let y = the length of $\overline{BD}$.

$$y^2 = (7.1)^2 + (134.4)^2$$
$$y^2 = 18113.77$$
$$y = \sqrt{18113.77} \approx 135 \text{ ft}$$

38. Refer to the diagram provided. The shortstop is at C, so the length of $\overline{AC}$ is 60 ft. Let x represent the length of $\overline{AB}$ and $\overline{BC}$.

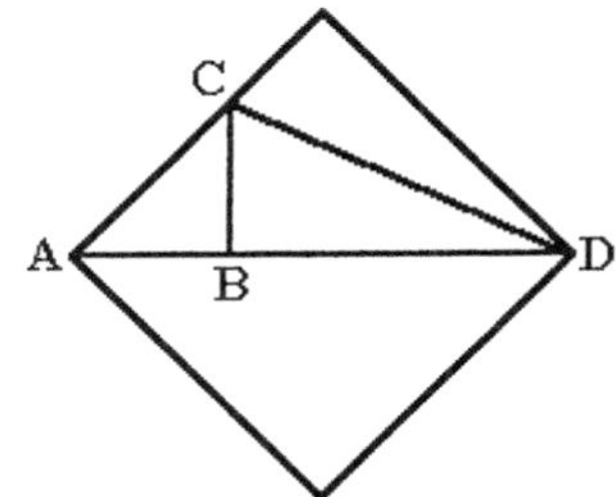

$$x^2 + x^2 = 60^2$$
$$2x^2 = 3600$$
$$x^2 = 1800$$
$$x = \sqrt{1800} \approx 42.4 \text{ ft}$$

From **#37**, the length of $\overline{AD}$ is 127.3 ft, so the length of $\overline{BD}$ is $127.3 - 42.4 = 84.9$ ft. Let x = the length of $\overline{CD}$.

$$x^2 = (84.9)^2 + (42.4)^2$$
$$x^2 = 9005.77 \Rightarrow x = \sqrt{9005.77} \approx 94.9 \text{ ft.}$$

39. $d = \sqrt{a^2 + b^2 + c^2} = \sqrt{12^2 + 24^2 + 17^2} = \sqrt{1009} = 31.76 \Rightarrow$ The racket will not fit.

40. $d = \sqrt{a^2 + b^2 + c^2} = \sqrt{21^2 + 21^2 + 3^2} = \sqrt{891} = 29.8 \Rightarrow$ The femur will not fit.

41. $d = \sqrt{a^2 + b^2 + c^2} = \sqrt{21^2 + 21^2 + 21^2} = \sqrt{1323} = 36.4 \Rightarrow$ The femur will fit.

42.
$$37^2 = 9^2 + h^2$$
$$1369 = 81 + h^2$$
$$1288 = h^2$$
$$35.9 = h$$
The ladder will reach.

43. Let x = direct distance from A to D.
$$x^2 = 52^2 + (105 + 60)^2$$
$$x^2 = 52^2 + 165^2$$
$$x^2 = 29929$$
$$x = 173 \text{ yd}$$

44. Let x = direct distance from A to E.
$x^2 = (42+26)^2 + (300-15)^2$
$x^2 = 68^2 + 285^2$
$x^2 = 85849$
$x = 293$ yd
savings $= 383 - 293 = 90$ yd

45. Let x = half the length of the stretched wire.
$x^2 = 20^2 + 1^2$
$x^2 = 401$
$x = 20.025$ ft
The stretched wire has a length of 40.05 ft.
It has been stretched by 0.05 ft.

46. $s = \sqrt{A} = \sqrt{49} = 7$ ft
perimeter $= 4s = 4(7) = 28$ ft

47. $A = 6\sqrt[3]{V^2} = 6\sqrt[3]{8^2} = 6\sqrt[3]{64}$
$= 6(4) = 24\text{ cm}^2$

48. Area of one grain $= 6\sqrt[3]{V^2} = 6\sqrt[3]{(6\times 10^{-6})^2} = 6\sqrt[3]{36\times 10^{-12}} = 1.98\times 10^{-3}$
Area of one cup $= (1.5\times 10^6)(1.98\times 10^{-3}) = 2.97\times 10^3$, or about 3000 in.^{2}

49. **Answers may vary.**

50. **Answers may vary.**

51. $I = \frac{703w}{h^2} = \frac{703(104)}{(54.1)^2} = \frac{73112}{2926.81} \approx 25$

52. $I = \frac{703w}{h^2} = \frac{703(220)}{72^2} = \frac{154{,}660}{5184}$
≈ 29.83
Higher risk

Exercise 7.3 (page 460)

1.
$$\begin{aligned} 5x - 4 &< 11 \\ 5x &< 15 \\ x &< 3 \end{aligned}$$

2.
$$\begin{aligned} 2(3t-5) &\ge 8 \\ 6t - 10 &\ge 8 \\ 6t &\ge 18 \\ t &\ge 3 \end{aligned}$$

3.
$$\begin{aligned} \frac{4}{5}(r-3) &> \frac{2}{3}(r+2) \\ 15\cdot\frac{4}{5}(r-3) &> 15\cdot\frac{2}{3}(r+2) \\ 12(r-3) &> 10(r+2) \\ 2r &> 56 \\ r &> 28 \end{aligned}$$

4.
$$\begin{aligned} -4 < 2x - 4 &\le 8 \\ 0 < 2x &\le 12 \\ 0 < x &\le 6 \end{aligned}$$

5. Let x = pints of water added (0% alcohol).

[Alcohol at start] + [Alcohol added] = [Alcohol at end]

$$\begin{aligned} 0.20(5) + 0(x) &= 0.15(5+x) \\ 1 + 0 &= 0.75 + 0.15x \\ 0.25 &= 0.15x \\ x = \frac{0.25}{0.15} &= \frac{5}{3} \end{aligned}$$
$\Rightarrow 1\frac{2}{3}$ pints of water should be added.

6. Original # of boxes $= x$; # sold $= x - 4$

Original cost $= \frac{70}{x}$; New cost $= \frac{70}{x} + 2$

$$\boxed{\text{Number sold}} \cdot \boxed{\text{Charge per box}} = 70$$

$$(x-4)\left(\frac{70}{x} + 2\right) = 70$$

$$70 + 2x - \frac{280}{x} - 8 = 70$$

$$2x - \frac{280}{x} - 8 = 0$$

$$2x - \frac{280}{x} - 8 = 0$$

$$\left(2x - \frac{280}{x} - 8\right)x = 0(x)$$

$$2x^2 - 8x - 280 = 0$$

$$2(x-14)(x+10) = 0$$

$x + 10 = 0$ **or** $x - 14 = 0$

$x = -10$ $\quad$ $x = 14$

The grocer sold $14 - 4 = 10$ boxes.

7. $a \cdot a \cdot a \cdot a$ **8.** a^{m+n} **9.** a^{mn} **10.** $a^n b^n$

11. $\dfrac{a^n}{b^n}$ **12.** $1; 0$ **13.** $\dfrac{1}{a^n}; 0$ **14.** a^{m-n}

15. $\left(\dfrac{b}{a}\right)^n$ **16.** $\sqrt[n]{x}$ **17.** $|x|$ **18.** $\left(\sqrt[n]{x}\right)^m$

19. $7^{1/3} = \sqrt[3]{7}$ **20.** $26^{1/2} = \sqrt{26}$ **21.** $8^{1/5} = \sqrt[5]{8}$ **22.** $13^{1/7} = \sqrt[7]{13}$

23. $(3x)^{1/4} = \sqrt[4]{3x}$ **24.** $(4ab)^{1/6} = \sqrt[6]{4ab}$ **25.** $\left(\frac{1}{2}x^3y\right)^{1/4} = \sqrt[4]{\frac{1}{2}x^3y}$

26. $\left(\frac{3}{4}a^2b^2\right)^{1/5} = \sqrt[5]{\frac{3}{4}a^2b^2}$ **27.** $(4a^2b^3)^{1/5} = \sqrt[5]{4a^2b^3}$ **28.** $(5pq^2)^{1/3} = \sqrt[3]{5pq^2}$

29. $(x^2+y^2)^{1/2} = \sqrt{x^2+y^2}$ **30.** $(x^3+y^3)^{1/3} = \sqrt[3]{x^3+y^3}$ **31.** $\sqrt{11} = 11^{1/2}$

32. $\sqrt[3]{12} = 12^{1/3}$ **33.** $\sqrt[4]{3a} = (3a)^{1/4}$ **34.** $\sqrt[7]{12xy} = (12xy)^{1/7}$

35. $3\sqrt[5]{a} = 3a^{1/5}$ **36.** $4\sqrt[3]{p} = 4p^{1/3}$ **37.** $\sqrt[6]{\frac{1}{7}abc} = \left(\frac{1}{7}abc\right)^{1/6}$

38. $\sqrt[7]{\frac{3}{8}p^2q} = \left(\frac{3}{8}p^2q\right)^{1/7}$ **39.** $\sqrt[5]{\frac{1}{2}mn} = \left(\frac{1}{2}mn\right)^{1/5}$ **40.** $\sqrt[8]{\frac{2}{7}p^2q} = \left(\frac{2}{7}p^2q\right)^{1/8}$

41. $\sqrt[3]{a^2-b^2} = (a^2-b^2)^{1/3}$ **42.** $\sqrt{x^2+y^2} = (x^2+y^2)^{1/2}$ **43.** $4^{1/2} = \sqrt{4} = 2$

44. $64^{1/2} = \sqrt{64} = 8$ **45.** $27^{1/3} = \sqrt[3]{27} = 3$ **46.** $125^{1/3} = \sqrt[3]{125} = 5$

47. $16^{1/4} = \sqrt[4]{16} = 2$ **48.** $625^{1/4} = \sqrt[4]{625} = 5$ **49.** $32^{1/5} = \sqrt[5]{32} = 2$

50. $0^{1/5} = \sqrt[5]{0} = 0$ **51.** $\left(\frac{1}{4}\right)^{1/2} = \sqrt{\frac{1}{4}} = \frac{1}{2}$ **52.** $\left(\frac{1}{16}\right)^{1/2} = \sqrt{\frac{1}{16}} = \frac{1}{4}$

53. $\left(\frac{1}{8}\right)^{1/3} = \sqrt[3]{\frac{1}{8}} = \frac{1}{2}$ **54.** $\left(\frac{1}{16}\right)^{1/4} = \sqrt[4]{\frac{1}{16}} = \frac{1}{2}$ **55.** $-16^{1/4} = -\sqrt[4]{16} = -2$

56. $-125^{1/3} = -\sqrt[3]{125} = -5$ **57.** $(-27)^{1/3} = \sqrt[3]{-27} = -3$ **58.** $(-125)^{1/3} = \sqrt[3]{-125} = -5$

59. $(-64)^{1/2} = \sqrt{-64}$; not a real number

60. $(-243)^{1/5} = \sqrt[5]{-243} = -3$

61. $0^{1/3} = \sqrt[3]{0} = 0$

62. $(-216)^{1/2} = \sqrt{-216}$; not a real number

63. $(25y^2)^{1/2} = \left[(5y)^2\right]^{1/2} = |5y| = 5|y|$

64. $(-27x^3)^{1/3} = \left[(-3x)^3\right]^{1/3} = -3x$

65. $(16x^4)^{1/4} = \left[(2x)^4\right]^{1/4} = |2x| = 2|x|$

66. $(-16x^4)^{1/2} \Rightarrow$ not a real number

67. $(243x^5)^{1/5} = \left[(3x)^5\right]^{1/5} = 3x$

68. $\left[(x+1)^4\right]^{1/4} = |x+1|$

69. $(-64x^8)^{1/4} \Rightarrow$ not a real number

70. $\left[(x+5)^3\right]^{1/3} = x+5$

71. $36^{3/2} = \left(36^{1/2}\right)^3 = 6^3 = 216$

72. $27^{2/3} = \left(27^{1/3}\right)^2 = 3^2 = 9$

73. $81^{3/4} = \left(81^{1/4}\right)^3 = 3^3 = 27$

74. $100^{3/2} = \left(100^{1/2}\right)^3 = 10^3 = 1000$

75. $144^{3/2} = \left(144^{1/2}\right)^3 = 12^3 = 1728$

76. $1000^{2/3} = \left(1000^{1/3}\right)^2 = 10^2 = 100$

77. $\left(\frac{1}{8}\right)^{2/3} = \left[\left(\frac{1}{8}\right)^{1/3}\right]^2 = \left(\frac{1}{2}\right)^2 = \frac{1}{4}$

78. $\left(\frac{4}{9}\right)^{3/2} = \left[\left(\frac{4}{9}\right)^{1/2}\right]^3 = \left(\frac{2}{3}\right)^3 = \frac{8}{27}$

79. $\left(25x^4\right)^{3/2} = \left[\left(25x^4\right)^{1/2}\right]^3 = \left(5x^2\right)^3 = 125x^6$

80. $\left(27a^3b^3\right)^{2/3} = \left[\left(27a^3b^3\right)^{1/3}\right]^2 = (3ab)^2 = 9a^2b^2$

81. $\left(\frac{8x^3}{27}\right)^{2/3} = \left[\left(\frac{8x^3}{27}\right)^{1/3}\right]^2 = \left(\frac{2x}{3}\right)^2 = \frac{4x^2}{9}$

82. $\left(\frac{27}{64y^6}\right)^{2/3} = \left[\left(\frac{27}{64y^6}\right)^{1/3}\right]^2 = \left(\frac{3}{4y^2}\right)^2 = \frac{9}{16y^4}$

Problems 83-86 are to be solved using a calculator. The keystrokes needed to solve each problem using a TI-83 graphing calculator appear in each solution. There may be other solutions. Keystrokes for other calculators may be slightly different.

83. [1] [5] [^] [(] [1] [÷] [3] [)] [ENTER] {2.4662...} ⇒ 2.47

84. [5] [0] [.] [5] [^] [(] [1] [÷] [4] [)] [ENTER] {2.66577...} ⇒ 2.67

85. [1] [.] [0] [4] [5] [^] [(] [1] [÷] [5] [)] [ENTER] {1.0088...} ⇒ 1.01

86. [(] [(-)] [1] [0] [0] [0] [)] [^] [(] [2] [÷] [5] [)] [ENTER] {15.8489...} ⇒ 15.85

87. $4^{-1/2} = \frac{1}{4^{1/2}} = \frac{1}{2}$

88. $8^{-1/3} = \frac{1}{8^{1/3}} = \frac{1}{2}$

89. $4^{-3/2} = \frac{1}{4^{3/2}} = \frac{1}{(4^{1/2})^3} = \frac{1}{2^3} = \frac{1}{8}$

90. $25^{-5/2} = \frac{1}{25^{5/2}} = \frac{1}{(25^{1/2})^5} = \frac{1}{5^5} = \frac{1}{3125}$

91. $(16x^2)^{-3/2} = \frac{1}{(16x^2)^{3/2}} = \frac{1}{\left[(16x^2)^{1/2}\right]^3} = \frac{1}{(4x)^3} = \frac{1}{64x^3}$

92. $(81c^4)^{-3/2} = \frac{1}{(81c^4)^{3/2}} = \frac{1}{\left[(81c^4)^{1/2}\right]^3} = \frac{1}{(9c^2)^3} = \frac{1}{729c^6}$

93. $(-27y^3)^{-2/3} = \frac{1}{(-27y^3)^{2/3}} = \frac{1}{\left[(-27y^3)^{1/3}\right]^2} = \frac{1}{(-3y)^2} = \frac{1}{9y^2}$

94. $(-8z^9)^{-2/3} = \frac{1}{(-8z^9)^{2/3}} = \frac{1}{\left[(-8z^9)^{1/3}\right]^2} = \frac{1}{(-2z^3)^2} = \frac{1}{4z^6}$

95. $(-32p^5)^{-2/5} = \frac{1}{(-32p^5)^{2/5}} = \frac{1}{\left[(-32p^5)^{1/5}\right]^2} = \frac{1}{(-2p)^2} = \frac{1}{4p^2}$

96. $(16q^6)^{-5/2} = \frac{1}{(16q^6)^{5/2}} = \frac{1}{\left[(16q^6)^{1/2}\right]^5} = \frac{1}{(4q^3)^5} = \frac{1}{1024q^{15}}$

97. $\left(\frac{1}{4}\right)^{-3/2} = \left(\frac{4}{1}\right)^{3/2} = 4^{3/2} = (4^{1/2})^3 = 2^3 = 8$

98. $\left(\frac{4}{25}\right)^{-3/2} = \left(\frac{25}{4}\right)^{3/2} = \left[\left(\frac{25}{4}\right)^{1/2}\right]^3 = \left(\frac{5}{2}\right)^3 = \frac{125}{8}$

99. $\left(\frac{27}{8}\right)^{-4/3} = \left(\frac{8}{27}\right)^{4/3} = \left[\left(\frac{8}{27}\right)^{1/3}\right]^4 = \left(\frac{2}{3}\right)^4 = \frac{16}{81}$

100. $\left(\frac{25}{49}\right)^{-3/2} = \left(\frac{49}{25}\right)^{3/2} = \left[\left(\frac{49}{25}\right)^{1/2}\right]^3 = \left(\frac{7}{5}\right)^3 = \frac{343}{125}$

101. $\left(-\frac{8x^3}{27}\right)^{-1/3} = \left(-\frac{27}{8x^3}\right)^{1/3} = -\frac{3}{2x}$

102. $\left(\frac{16}{81y^4}\right)^{-3/4} = \left(\frac{81y^4}{16}\right)^{3/4} = \left[\left(\frac{81y^4}{16}\right)^{1/4}\right]^3 = \left(\frac{3y}{2}\right)^3 = \frac{27y^3}{8}$

Problems 103-106 are to be solved using a calculator. The keystrokes needed to solve each problem using a TI-83 graphing calculator appear in each solution. There may be other solutions. Keystrokes for other calculators may be slightly different.

103. [1] [7] [^] [(] [(-)] [1] [÷] [2] [)] [ENTER] $\{0.2425...\} \Rightarrow 0.24$

104. [2] [.] [4] [5] [^] [(] [(-)] [2] [÷] [3] [)] [ENTER] $\{0.5502...\} \Rightarrow 0.55$

105. [(] [(-)] [.] [2] [5] [)] [^] [(] [(-)] [1] [÷] [5] [)] [ENTER] $\{-1.3195...\} \Rightarrow -1.32$

106. [(] [(-)] [1] [7] [.] [1] [)] [^] [(] [(-)] [3] [÷] [7] [)] [ENTER] $\{-0.296...\} \Rightarrow -0.30$

107. $5^{4/9}5^{4/9} = 5^{4/9+4/9} = 5^{8/9}$

108. $4^{2/5}4^{2/5} = 4^{2/5+2/5} = 4^{4/5}$

109. $\left(4^{1/5}\right)^3 = 4^{(1/5)\cdot 3} = 4^{3/5}$

110. $\left(3^{1/3}\right)^5 = 3^{(1/3)\cdot 5} = 3^{5/3}$

111. $\dfrac{9^{4/5}}{9^{3/5}} = 9^{4/5-3/5} = 9^{1/5}$

112. $\dfrac{7^{2/3}}{7^{1/2}} = 7^{2/3-1/2} = 7^{1/6}$

113. $\dfrac{7^{1/2}}{7^0} = 7^{1/2-0} = 7^{1/2}$

114. $5^{1/3}5^{-5/3} = 5^{1/3-5/3} = 5^{-4/3} = \dfrac{1}{5^{4/3}}$

115. $6^{-2/3}6^{-4/3} = 6^{-6/3} = 6^{-2} = \dfrac{1}{36}$

116. $\dfrac{3^{4/3}3^{1/3}}{3^{2/3}} = \dfrac{3^{5/3}}{3^{2/3}} = 3^{3/3} = 3^1 = 3$

117. $\dfrac{2^{5/6}2^{1/3}}{2^{1/2}} = \dfrac{2^{7/6}}{2^{1/2}} = 2^{4/6} = 2^{2/3}$

118. $\dfrac{5^{1/3}5^{1/2}}{5^{1/3}} = \dfrac{5^{5/6}}{5^{1/3}} = 5^{3/6} = 5^{1/2}$

119. $a^{2/3}a^{1/3} = a^{3/3} = a^1 = a$

120. $b^{3/5}b^{1/5} = b^{4/5}$

121. $\left(a^{2/3}\right)^{1/3} = a^{(2/3)(1/3)} = a^{2/9}$

122. $\left(t^{4/5}\right)^{10} = t^{(4/5)\cdot 10} = t^{40/5} = t^8$

123. $\left(a^{1/2}b^{1/3}\right)^{3/2} = a^{3/4}b^{1/2}$

124. $\left(a^{3/5}b^{3/2}\right)^{2/3} = a^{2/5}b^1 = a^{2/5}b$

125. $\left(mn^{-2/3}\right)^{-3/5} = m^{-3/5}n^{2/5} = \dfrac{n^{2/5}}{m^{3/5}}$

126. $\left(r^{-2}s^3\right)^{1/3} = r^{-2/3}s = \dfrac{s}{r^{2/3}}$

127. $\dfrac{\left(4x^3y\right)^{1/2}}{(9xy)^{1/2}} = \dfrac{2x^{3/2}y^{1/2}}{3x^{1/2}y^{1/2}} = \dfrac{2x}{3}$

128. $\dfrac{(27x^3y)^{1/3}}{(8xy^2)^{2/3}} = \dfrac{3xy^{1/3}}{4x^{2/3}y^{4/3}} = \dfrac{3x^{1/3}}{4y}$

129. $\left(27x^{-3}\right)^{-1/3} = (27)^{-1/3}x = \dfrac{1}{3}x$

130. $\left(16a^{-2}\right)^{-1/2} = (16)^{-1/2}a = \dfrac{1}{4}a$

131. $y^{1/3}\left(y^{2/3} + y^{5/3}\right) = y^{3/3} + y^{6/3} = y + y^2$

132. $y^{2/5}\left(y^{-2/5} + y^{3/5}\right) = y^0 + y^{5/5} = 1 + y$

133. $x^{3/5}\left(x^{7/5} - x^{2/5} + 1\right) = x^{10/5} - x^{5/5} + x^{3/5} = x^2 - x + x^{3/5}$

134. $x^{4/3}\left(x^{2/3} + 3x^{5/3} - 4\right) = x^{6/3} + 3x^{9/3} - 4x^{4/3} = x^2 + 3x^3 - 4x^{4/3}$

135. $(x^{1/2}+2)(x^{1/2}-2) = x^{2/2} - 2x^{1/2} + 2x^{1/2} - 4 = x - 4$

136. $(x^{1/2}+y^{1/2})(x^{1/2}-y^{1/2}) = x^{2/2} - x^{1/2}y^{1/2} + x^{1/2}y^{1/2} - y^{2/2} = x - y$

137. $(x^{2/3}-x)(x^{2/3}+x) = x^{4/3} + x^{5/3} - x^{5/3} - x^2 = x^{4/3} - x^2$

138. $(x^{1/3}+x^2)(x^{1/3}-x^2) = x^{2/3} - x^{7/3} + x^{7/3} - x^4 = x^{2/3} - x^4$

139. $$\begin{aligned}(x^{2/3}+y^{2/3})^2 = (x^{2/3}+y^{2/3})(x^{2/3}+y^{2/3}) &= x^{4/3} + x^{2/3}y^{2/3} + x^{2/3}y^{2/3} + y^{4/3} \\ &= x^{4/3} + 2x^{2/3}y^{2/3} + y^{4/3}\end{aligned}$$

140. $$\begin{aligned}(a^{1/2}-b^{2/3})^2 = (a^{1/2}-b^{2/3})(a^{1/2}-b^{2/3}) &= a^{2/2} - a^{1/2}b^{2/3} - a^{1/2}b^{2/3} + b^{4/3} \\ &= a - 2a^{1/2}b^{2/3} + b^{4/3}\end{aligned}$$

141. $$\begin{aligned}(a^{3/2}-b^{3/2})^2 = (a^{3/2}-b^{3/2})(a^{3/2}-b^{3/2}) &= a^{6/2} - a^{3/2}b^{3/2} - a^{3/2}b^{3/2} + b^{6/2} \\ &= a^3 - 2a^{3/2}b^{3/2} + b^3\end{aligned}$$

142. $$\begin{aligned}(x^{-1/2}-x^{1/2})^2 = (x^{-1/2}-x^{1/2})(x^{-1/2}-x^{1/2}) &= x^{-2/2} - x^{1/2}x^{-1/2} - x^{1/2}x^{-1/2} + x^{2/2} \\ &= x^{-1} - 1 - 1 + x = \frac{1}{x} - 2 + x\end{aligned}$$

143. $\sqrt[6]{p^3} = (p^3)^{1/6} = p^{3/6} = p^{1/2} = \sqrt{p}$

144. $\sqrt[8]{q^2} = (q^2)^{1/8} = q^{1/4} = \sqrt[4]{q}$

145. $\sqrt[4]{25b^2} = (5^2b^2)^{1/4} = 5^{1/2}b^{1/2} = \sqrt{5b}$

146. $\sqrt[9]{-8x^6} = (-8x^6)^{1/9} = [(-2)^3]^{1/9}(x^6)^{1/9} = (-2)^{1/3}x^{2/3} = -\sqrt[3]{2x^2}$

147. Answers may vary.

148. Answers may vary.

149. $16^{2/4} = 2^2 = 4; 16^{1/2} = 4$

150. $(8)^{1\frac{1}{3}} = 8^{4/3} = 2^4 = 16$
$(25)^{2\frac{1}{2}} = 25^{5/2} = 5^5 = 3125$

Exercise 7.4 (page 469)

1. $3x^2y^3(-5x^3y^{-4}) = -15x^5y^{-1} = \dfrac{-15x^5}{y}$

2. $-2a^2b^{-2}(4a^{-2}b^4 - 2a^2b + 3a^3b^2) = -8a^0b^2 + 4a^4b^{-1} - 6a^5b^0 = -8b^2 + \dfrac{4a^4}{b} - 6a^5$

3. $$\begin{aligned}(3t+2)^2 &= (3t+2)(3t+2) \\ &= 9t^2 + 12t + 4\end{aligned}$$

4. $(5r-3s)(5r+2s) = 25r^2 - 5rs - 6s^2$

SECTION 7.4

5.
$$\begin{array}{r} 3p + \quad 4 + \frac{-5}{2p-5} \\ 2p-5 \,\overline{)\,6p^2 - \quad 7p - \quad 25} \\ \underline{6p^2 - 15p \qquad\quad} \\ 8p - \quad 25 \\ \underline{8p - \quad 20} \\ -5 \end{array}$$

6.
$$\begin{array}{r} 2m^2 - \quad mn + \quad\; n^2 \\ 3m+n \,\overline{)\,6m^3 - \quad m^2n + 2mn^2 + n^3} \\ \underline{6m^3 + 2m^2n \qquad\qquad\qquad} \\ -3m^2n + 2mn^2 \qquad \\ \underline{-3m^2n - \quad mn^2 \qquad} \\ 3mn^2 + n^3 \\ \underline{3mn^2 + n^3} \\ 0 \end{array}$$

7. $\sqrt[n]{a}\sqrt[n]{b}$

8. $\dfrac{\sqrt[n]{a}}{\sqrt[n]{b}}$

9. $\sqrt{6}\sqrt{6} = \sqrt{36} = 6$

10. $\sqrt{11}\sqrt{11} = \sqrt{121} = 11$

11. $\sqrt{t}\sqrt{t} = \sqrt{t^2} = t$

12. $-\sqrt{z}\sqrt{z} = -\sqrt{z^2} = -z$

13. $\sqrt[3]{5x^2}\sqrt[3]{25x} = \sqrt[3]{125x^3} = 5x$

14. $\sqrt[4]{25a}\sqrt[4]{25a^3} = \sqrt[4]{625a^4} = 5a$

15. $\dfrac{\sqrt{500}}{\sqrt{5}} = \sqrt{\dfrac{500}{5}} = \sqrt{100} = 10$

16. $\dfrac{\sqrt{128}}{\sqrt{2}} = \sqrt{\dfrac{128}{2}} = \sqrt{64} = 8$

17. $\dfrac{\sqrt{98x^3}}{\sqrt{2x}} = \sqrt{\dfrac{98x^3}{2x}} = \sqrt{49x^2} = 7x$

18. $\dfrac{\sqrt{75y^5}}{\sqrt{3y}} = \sqrt{\dfrac{75y^5}{3y}} = \sqrt{25y^4} = 5y^2$

19. $\dfrac{\sqrt{180ab^4}}{\sqrt{5ab^2}} = \sqrt{\dfrac{180ab^4}{5ab^2}} = \sqrt{36b^2} = 6b$

20. $\dfrac{\sqrt{112ab^3}}{\sqrt{7ab}} = \sqrt{\dfrac{112ab^3}{7ab}} = \sqrt{16b^2} = 4b$

21. $\dfrac{\sqrt[3]{48}}{\sqrt[3]{6}} = \sqrt[3]{\dfrac{48}{6}} = \sqrt[3]{8} = 2$

22. $\dfrac{\sqrt[3]{64}}{\sqrt[3]{8}} = \sqrt[3]{\dfrac{64}{8}} = \sqrt[3]{8} = 2$

23. $\dfrac{\sqrt[3]{189a^4}}{\sqrt[3]{7a}} = \sqrt[3]{\dfrac{189a^4}{7a}} = \sqrt[3]{27a^3} = 3a$

24. $\dfrac{\sqrt[3]{243x^7}}{\sqrt[3]{9x}} = \sqrt[3]{\dfrac{243x^7}{9x}} = \sqrt[3]{27x^6} = 3x^2$

25. $\sqrt{20} = \sqrt{4 \cdot 5} = \sqrt{4}\sqrt{5} = 2\sqrt{5}$

26. $\sqrt{8} = \sqrt{4 \cdot 2} = \sqrt{4}\sqrt{2} = 2\sqrt{2}$

27. $-\sqrt{200} = -\sqrt{100 \cdot 2} = -\sqrt{100}\sqrt{2}$
$= -10\sqrt{2}$

28. $-\sqrt{250} = -\sqrt{25 \cdot 10} = -\sqrt{25}\sqrt{10}$
$= -5\sqrt{10}$

29. $\sqrt[3]{80} = \sqrt[3]{8 \cdot 10} = \sqrt[3]{8}\sqrt[3]{10} = 2\sqrt[3]{10}$

30. $\sqrt[3]{270} = \sqrt[3]{27 \cdot 10} = \sqrt[3]{27}\sqrt[3]{10} = 3\sqrt[3]{10}$

31. $\sqrt[3]{-81} = \sqrt[3]{-27 \cdot 3} = \sqrt[3]{-27}\sqrt[3]{3} = -3\sqrt[3]{3}$

32. $\sqrt[3]{-72} = \sqrt[3]{-8 \cdot 9} = \sqrt[3]{-8}\sqrt[3]{9} = -2\sqrt[3]{9}$

33. $\sqrt[4]{32} = \sqrt[4]{16 \cdot 2} = \sqrt[4]{16}\sqrt[4]{2} = 2\sqrt[4]{2}$

34. $\sqrt[4]{48} = \sqrt[4]{16 \cdot 3} = \sqrt[4]{16}\sqrt[4]{3} = 2\sqrt[4]{3}$

35. $\sqrt[5]{96} = \sqrt[5]{32 \cdot 3} = \sqrt[5]{32}\sqrt[5]{3} = 2\sqrt[5]{3}$

36. $\sqrt[7]{256} = \sqrt[7]{128 \cdot 2} = \sqrt[7]{128}\sqrt[7]{2} = 2\sqrt[7]{2}$

37. $\sqrt{\frac{7}{9}} = \frac{\sqrt{7}}{\sqrt{9}} = \frac{\sqrt{7}}{3}$

38. $\sqrt{\frac{3}{4}} = \frac{\sqrt{3}}{\sqrt{4}} = \frac{\sqrt{3}}{2}$

39. $\sqrt[3]{\frac{7}{64}} = \frac{\sqrt[3]{7}}{\sqrt[3]{64}} = \frac{\sqrt[3]{7}}{4}$

40. $\sqrt[3]{\frac{4}{125}} = \frac{\sqrt[3]{4}}{\sqrt[3]{125}} = \frac{\sqrt[3]{4}}{5}$

41. $\sqrt[4]{\frac{3}{10{,}000}} = \frac{\sqrt[4]{3}}{\sqrt[4]{10{,}000}} = \frac{\sqrt[4]{3}}{10}$

42. $\sqrt[5]{\frac{4}{243}} = \frac{\sqrt[5]{4}}{\sqrt[5]{243}} = \frac{\sqrt[5]{4}}{3}$

43. $\sqrt[5]{\frac{3}{32}} = \frac{\sqrt[5]{3}}{\sqrt[5]{32}} = \frac{\sqrt[5]{3}}{2}$

44. $\sqrt[6]{\frac{5}{64}} = \frac{\sqrt[6]{5}}{\sqrt[6]{64}} = \frac{\sqrt[6]{5}}{2}$

45. $\sqrt{50x^2} = \sqrt{25x^2 \cdot 2} = \sqrt{25x^2}\sqrt{2} = 5x\sqrt{2}$

46. $\sqrt{75a^2} = \sqrt{25a^2 \cdot 3} = \sqrt{25a^2}\sqrt{3} = 5a\sqrt{3}$

47. $\sqrt{32b} = \sqrt{16 \cdot 2b} = \sqrt{16}\sqrt{2b} = 4\sqrt{2b}$

48. $\sqrt{80c} = \sqrt{16 \cdot 5c} = \sqrt{16}\sqrt{5c} = 4\sqrt{5c}$

49. $-\sqrt{112a^3} = -\sqrt{16a^2 \cdot 7a} = -\sqrt{16a^2}\sqrt{7a} = -4a\sqrt{7a}$

50. $\sqrt{147a^5} = \sqrt{49a^4 \cdot 3a} = \sqrt{49a^4}\sqrt{3a} = 7a^2\sqrt{3a}$

51. $\sqrt{175a^2b^3} = \sqrt{25a^2b^2 \cdot 7b} = \sqrt{25a^2b^2}\sqrt{7b} = 5ab\sqrt{7b}$

52. $\sqrt{128a^3b^5} = \sqrt{64a^2b^4 \cdot 2ab} = \sqrt{64a^2b^4}\sqrt{2ab} = 8ab^2\sqrt{2ab}$

53. $-\sqrt{300xy} = -\sqrt{100 \cdot 3xy} = -\sqrt{100}\sqrt{3xy} = -10\sqrt{3xy}$

54. $\sqrt{200x^2y} = \sqrt{100x^2 \cdot 2y} = \sqrt{100x^2}\sqrt{2y}$
$= 10x\sqrt{2y}$

55. $\sqrt[3]{-54x^6} = \sqrt[3]{-27x^6 \cdot 2} = \sqrt[3]{-27x^6}\sqrt[3]{2}$
$= -3x^2\sqrt[3]{2}$

56. $-\sqrt[3]{-81a^3} = -\sqrt[3]{-27a^3 \cdot 3} = -\sqrt[3]{-27a^3}\sqrt[3]{3} = -(-3a)\sqrt[3]{3} = 3a\sqrt[3]{3}$

57. $\sqrt[3]{16x^{12}y^3} = \sqrt[3]{8x^{12}y^3 \cdot 2} = \sqrt[3]{8x^{12}y^3}\sqrt[3]{2}$
$= 2x^4y\sqrt[3]{2}$

58. $\sqrt[3]{40a^3b^6} = \sqrt[3]{8a^3b^6 \cdot 5} = \sqrt[3]{8a^3b^6}\sqrt[3]{5}$
$= 2ab^2\sqrt[3]{5}$

59. $\sqrt[4]{32x^{12}y^4} = \sqrt[4]{16x^{12}y^4 \cdot 2} = \sqrt[4]{16x^{12}y^4}\sqrt[4]{2}$
$= 2x^3y\sqrt[4]{2}$

60. $\sqrt[5]{64x^{10}y^5} = \sqrt[5]{32x^{10}y^5 \cdot 2} = \sqrt[5]{32x^{10}y^5}\sqrt[5]{2}$
$= 2x^2y\sqrt[5]{2}$

61. $\sqrt{\frac{z^2}{16x^2}} = \frac{\sqrt{z^2}}{\sqrt{16x^2}} = \frac{z}{4x}$

62. $\sqrt{\frac{b^4}{64a^8}} = \frac{\sqrt{b^4}}{\sqrt{64a^8}} = \frac{b^2}{8a^4}$

63. $\sqrt[4]{\frac{5x}{16z^4}} = \frac{\sqrt[4]{5x}}{\sqrt[4]{16z^4}} = \frac{\sqrt[4]{5x}}{2z}$

64. $\sqrt[3]{\frac{11a^2}{125b^6}} = \frac{\sqrt[3]{11a^2}}{\sqrt[3]{125b^6}} = \frac{\sqrt[3]{11a^2}}{5b^2}$

65. $4\sqrt{2x}+6\sqrt{2x}=10\sqrt{2x}$

66. $6\sqrt[3]{5y}+3\sqrt[3]{5y}=9\sqrt[3]{5y}$

67. $8\sqrt[5]{7a^2}-7\sqrt[5]{7a^2}=\sqrt[5]{7a^2}$

68. $10\sqrt[6]{12xyz}-\sqrt[6]{12xyz}=9\sqrt[6]{12xyz}$

69. $\sqrt{3}+\sqrt{27}=\sqrt{3}+\sqrt{9}\sqrt{3}$
$=\sqrt{3}+3\sqrt{3}=4\sqrt{3}$

70. $\sqrt{8}+\sqrt{32}=\sqrt{4}\sqrt{2}+\sqrt{16}\sqrt{2}$
$=2\sqrt{2}+4\sqrt{2}=6\sqrt{2}$

71. $\sqrt{2}-\sqrt{8}=\sqrt{2}-\sqrt{4}\sqrt{2}$
$=\sqrt{2}-2\sqrt{2}=-\sqrt{2}$

72. $\sqrt{20}-\sqrt{125}=\sqrt{4}\sqrt{5}-\sqrt{25}\sqrt{5}$
$=2\sqrt{5}-5\sqrt{5}=-3\sqrt{5}$

73. $\sqrt{98}-\sqrt{50}=\sqrt{49}\sqrt{2}-\sqrt{25}\sqrt{2}=7\sqrt{2}-5\sqrt{2}=2\sqrt{2}$

74. $\sqrt{72}-\sqrt{200}=\sqrt{36}\sqrt{2}-\sqrt{100}\sqrt{2}=6\sqrt{2}-10\sqrt{2}=-4\sqrt{2}$

75. $3\sqrt{24}+\sqrt{54}=3\sqrt{4}\sqrt{6}+\sqrt{9}\sqrt{6}=3(2)\sqrt{6}+3\sqrt{6}=6\sqrt{6}+3\sqrt{6}=9\sqrt{6}$

76. $\sqrt{18}+2\sqrt{50}=\sqrt{9}\sqrt{2}+2\sqrt{25}\sqrt{2}=3\sqrt{2}+2(5)\sqrt{2}=3\sqrt{2}+10\sqrt{2}=13\sqrt{2}$

77. $\sqrt[3]{24}+\sqrt[3]{3}=\sqrt[3]{8}\sqrt[3]{3}+\sqrt[3]{3}=2\sqrt[3]{3}+\sqrt[3]{3}=3\sqrt[3]{3}$

78. $\sqrt[3]{16}+\sqrt[3]{128}=\sqrt[3]{8}\sqrt[3]{2}+\sqrt[3]{64}\sqrt[3]{2}=2\sqrt[3]{2}+4\sqrt[3]{2}=6\sqrt[3]{2}$

79. $\sqrt[3]{32}-\sqrt[3]{108}=\sqrt[3]{8}\sqrt[3]{4}-\sqrt[3]{27}\sqrt[3]{4}=2\sqrt[3]{4}-3\sqrt[3]{4}=-\sqrt[3]{4}$

80. $\sqrt[3]{80}-\sqrt[3]{10{,}000}=\sqrt[3]{8}\sqrt[3]{10}-\sqrt[3]{1000}\sqrt[3]{10}=2\sqrt[3]{10}-10\sqrt[3]{10}=-8\sqrt[3]{10}$

81. $2\sqrt[3]{125}-5\sqrt[3]{64}=2(5)-5(4)=10-20$
$=-10$

82. $3\sqrt[3]{27}+12\sqrt[3]{216}=3(3)+12(6)=9+72$
$=81$

83. $14\sqrt[4]{32}-15\sqrt[4]{162}=14\sqrt[4]{16}\sqrt[4]{2}-15\sqrt[4]{81}\sqrt[4]{2}=14(2)\sqrt[4]{2}-15(3)\sqrt[4]{2}=28\sqrt[4]{2}-45\sqrt[4]{2}$
$=-17\sqrt[4]{2}$

84. $23\sqrt[4]{768}+\sqrt[4]{48}=23\sqrt[4]{256}\sqrt[4]{3}+\sqrt[4]{16}\sqrt[4]{3}=23(4)\sqrt[4]{3}+2\sqrt[4]{3}=92\sqrt[4]{3}+2\sqrt[4]{3}=94\sqrt[4]{3}$

85. $3\sqrt[4]{512}+2\sqrt[4]{32}=3\sqrt[4]{256}\sqrt[4]{2}+2\sqrt[4]{16}\sqrt[4]{2}=3(4)\sqrt[4]{2}+2(2)\sqrt[4]{2}=12\sqrt[4]{2}+4\sqrt[4]{2}=16\sqrt[4]{2}$

86. $4\sqrt[4]{243}-\sqrt[4]{48}=4\sqrt[4]{81}\sqrt[4]{3}-\sqrt[4]{16}\sqrt[4]{3}=4(3)\sqrt[4]{3}-2\sqrt[4]{3}=12\sqrt[4]{3}-2\sqrt[4]{3}=10\sqrt[4]{3}$

87. $\sqrt{98}-\sqrt{50}-\sqrt{72}=\sqrt{49}\sqrt{2}-\sqrt{25}\sqrt{2}-\sqrt{36}\sqrt{2}=7\sqrt{2}-5\sqrt{2}-6\sqrt{2}=-4\sqrt{2}$

88. $\sqrt{20}+\sqrt{125}-\sqrt{80}=\sqrt{4}\sqrt{5}+\sqrt{25}\sqrt{5}-\sqrt{16}\sqrt{5}=2\sqrt{5}+5\sqrt{5}-4\sqrt{5}=3\sqrt{5}$

89. $\sqrt{18}+\sqrt{300}-\sqrt{243}=\sqrt{9}\sqrt{2}+\sqrt{100}\sqrt{3}-\sqrt{81}\sqrt{3}=3\sqrt{2}+10\sqrt{3}-9\sqrt{3}=3\sqrt{2}+\sqrt{3}$

90. $\sqrt{80}-\sqrt{128}+\sqrt{288}=\sqrt{16}\sqrt{5}-\sqrt{64}\sqrt{2}+\sqrt{144}\sqrt{2}=4\sqrt{5}-8\sqrt{2}+12\sqrt{2}$
$=4\sqrt{5}+4\sqrt{2}$

91. $2\sqrt[3]{16}-\sqrt[3]{54}-3\sqrt[3]{128}=2\sqrt[3]{8}\sqrt[3]{2}-\sqrt[3]{27}\sqrt[3]{2}-3\sqrt[3]{64}\sqrt[3]{2}=2(2)\sqrt[3]{2}-3\sqrt[3]{2}-3(4)\sqrt[3]{2}$
$=4\sqrt[3]{2}-3\sqrt[3]{2}-12\sqrt[3]{2}=-11\sqrt[3]{2}$

92. $\sqrt[4]{48}-\sqrt[4]{243}-\sqrt[4]{768}=\sqrt[4]{16}\sqrt[4]{3}-\sqrt[4]{81}\sqrt[4]{3}-\sqrt[4]{256}\sqrt[4]{3}=2\sqrt[4]{3}-3\sqrt[4]{3}-4\sqrt[4]{3}=-5\sqrt[4]{3}$

93. $\sqrt{25y^2z}-\sqrt{16y^2z}=\sqrt{25y^2}\sqrt{z}-\sqrt{16y^2}\sqrt{z}=5y\sqrt{z}-4y\sqrt{z}=y\sqrt{z}$

94. $\sqrt{25yz^2}+\sqrt{9yz^2}=\sqrt{25z^2}\sqrt{y}+\sqrt{9z^2}\sqrt{y}=5z\sqrt{y}+3z\sqrt{y}=8z\sqrt{y}$

95. $\sqrt{36xy^2}+\sqrt{49xy^2}=\sqrt{36y^2}\sqrt{x}+\sqrt{49y^2}\sqrt{x}=6y\sqrt{x}+7y\sqrt{x}=13y\sqrt{x}$

96. $3\sqrt{2x}-\sqrt{8x}=3\sqrt{2x}-\sqrt{4}\sqrt{2x}=3\sqrt{2x}-2\sqrt{2x}=\sqrt{2x}$

97. $2\sqrt[3]{64a}+2\sqrt[3]{8a}=2\sqrt[3]{64}\sqrt[3]{a}+2\sqrt[3]{8}\sqrt[3]{a}=2(4)\sqrt[3]{a}+2(2)\sqrt[3]{a}=8\sqrt[3]{a}+4\sqrt[3]{a}=12\sqrt[3]{a}$

98. $3\sqrt[4]{x^4y}-2\sqrt[4]{x^4y}=3\sqrt[4]{x^4}\sqrt[4]{y}-2\sqrt[4]{x^4}\sqrt[4]{y}=3x\sqrt[4]{y}-2x\sqrt[4]{y}=x\sqrt[4]{y}$

99. $\sqrt{y^5}-\sqrt{9y^5}-\sqrt{25y^5}=\sqrt{y^4}\sqrt{y}-\sqrt{9y^4}\sqrt{y}-\sqrt{25y^4}\sqrt{y}=y^2\sqrt{y}-3y^2\sqrt{y}-5y^2\sqrt{y}$
$=-7y^2\sqrt{y}$

100. $\sqrt{8y^7}+\sqrt{32y^7}-\sqrt{2y^7}=\sqrt{4y^6}\sqrt{2y}+\sqrt{16y^6}\sqrt{2y}-\sqrt{y^6}\sqrt{2y}$
$=2y^3\sqrt{2y}+4y^3\sqrt{2y}-y^3\sqrt{2y}=5y^3\sqrt{2y}$

101. $\sqrt[5]{x^6y^2}+\sqrt[5]{32x^6y^2}+\sqrt[5]{x^6y^2}=\sqrt[5]{x^5}\sqrt[5]{xy^2}+\sqrt[5]{32x^5}\sqrt[5]{xy^2}+\sqrt[5]{x^5}\sqrt[5]{xy^2}$
$=x\sqrt[5]{xy^2}+2x\sqrt[5]{xy^2}+x\sqrt[5]{xy^2}=4x\sqrt[5]{xy^2}$

102. $\sqrt[3]{xy^4}+\sqrt[3]{8xy^4}-\sqrt[3]{27xy^4}=\sqrt[3]{y^3}\sqrt[3]{xy}+\sqrt[3]{8y^3}\sqrt[3]{xy}-\sqrt[3]{27y^3}\sqrt[3]{xy}$
$=y\sqrt[3]{xy}+2y\sqrt[3]{xy}-3y\sqrt[3]{xy}=0$

103. $\sqrt{x^2+2x+1}+\sqrt{x^2+2x+1}=\sqrt{(x+1)^2}+\sqrt{(x+1)^2}=x+1+x+1=2x+2$

104. $\sqrt{4x^2+12x+9}+\sqrt{9x^2+6x+1}=\sqrt{(2x+3)^2}+\sqrt{(3x+1)^2}=2x+3+3x+1=5x+4$

105. $x=2.00$;
$h=2\sqrt{2}\approx 2.83$

106. $x=y=\frac{3}{\sqrt{2}}\approx 2.12$

107. $h=2(5)=10.00$;
$x=5\sqrt{3}\approx 8.66$

108. $x=\frac{7}{\sqrt{3}}\approx 4.04$;
$h\approx 2(4.04)\approx 8.08$

109. $x=\frac{9.37}{2}\approx 4.69$;
$y\approx 4.69\sqrt{3}\approx 8.11$

110. $x=\frac{12.26}{\sqrt{3}}\approx 7.08$;
$h\approx 2(7.08)\approx 14.16$

111. $x = y = \frac{17.12}{\sqrt{2}} \approx 12.11$

112. $x = 32.10; h = 32.10\sqrt{2} \approx 45.40$

113. $x = 5\sqrt{3} \approx 8.66$ mm
$h = 2x = 2\left(5\sqrt{3}\right) = 10\sqrt{3} \approx 17.32$ mm

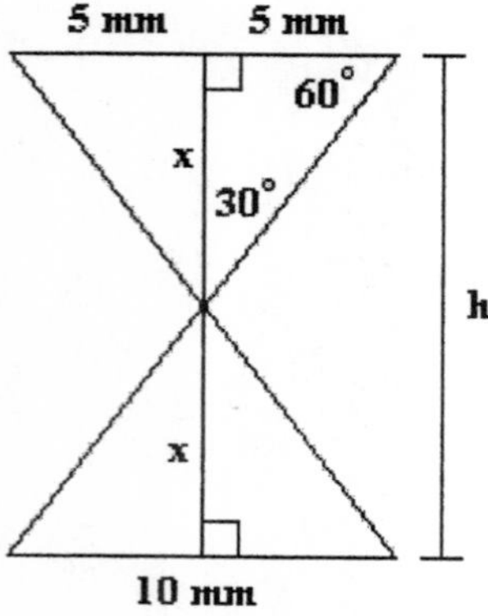

114. $a = \dfrac{12}{2} = 6; y = 6\sqrt{3}$
$b = \dfrac{28}{2} = 14; x = 14\sqrt{3}$
$h = 6\sqrt{3} + 14\sqrt{3} = 20\sqrt{3}$ in. ≈ 34.64 in.

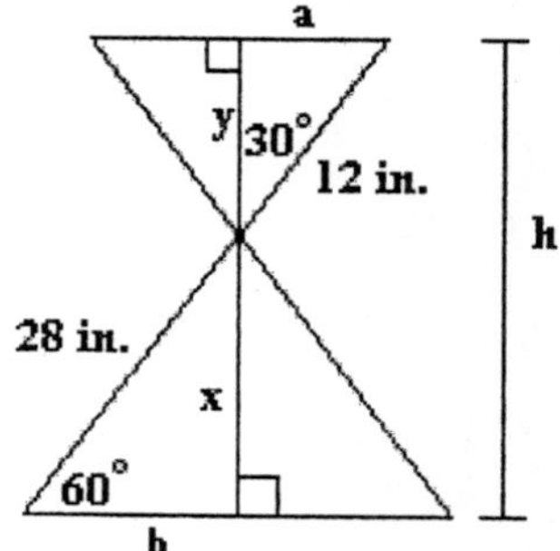

115. Answers may vary.

116. Answers may vary.

117. $\sqrt{a+b} = \sqrt{a} + \sqrt{b}$ if either a or b equals 0 and the other is nonnegative.

118.
$$\begin{aligned}\sqrt{3} + \sqrt{3^2} + \sqrt{3^3} + \sqrt{3^4} + \sqrt{3^5} &= \sqrt{3} + 3 + \sqrt{3^2}\sqrt{3} + 3^2 + \sqrt{3^4}\sqrt{3} \\ &= \sqrt{3} + 3 + 3\sqrt{3} + 9 + 9\sqrt{3} = 12 + 13\sqrt{3}\end{aligned}$$

Exercise 7.5 (page 478)

1.
$$\begin{aligned}\frac{2}{3-a} &= 1 \\ 2 &= 3 - a \\ a &= 1\end{aligned}$$

2.
$$\begin{aligned}5(s-4) &= -5(s-4) \\ 5s - 20 &= -5s + 20 \\ 10s &= 40 \\ s &= 4\end{aligned}$$

3.
$$\begin{aligned}\frac{8}{b-2} + \frac{3}{2-b} &= -\frac{1}{b} \\ \frac{8}{b-2} + \frac{-3}{b-2} &= -\frac{1}{b} \\ \frac{5}{b-2} &= \frac{-1}{b} \\ 5b &= -b + 2 \\ 6b &= 2 \\ b &= \frac{2}{6} = \frac{1}{3}\end{aligned}$$

4.
$$\begin{aligned} \frac{2}{x-2}+\frac{1}{x+1} &= \frac{1}{(x+1)(x-2)} \\ (x+1)(x-2)\left[\frac{2}{x-2}+\frac{1}{x+1}\right] &= (x+1)(x-2)\cdot\frac{1}{(x+1)(x-2)} \\ 2(x+1)+1(x-2) &= 1 \\ 2x+2+x-2 &= 1 \\ 3x &= 1 \\ x &= \frac{1}{3} \end{aligned}$$

5. $2; \sqrt{7}; \sqrt{5}$

6. distributive

7. FOIL

8. $\sqrt{x}-1$

9. conjugate

10. $\sqrt{5}-2$

11. $\sqrt{2}\sqrt{8}=\sqrt{16}=4$

12. $\sqrt{3}\sqrt{27}=\sqrt{81}=9$

13. $\sqrt{5}\sqrt{10}=\sqrt{50}=\sqrt{25}\sqrt{2}=5\sqrt{2}$

14. $\sqrt{7}\sqrt{35}=\sqrt{245}=\sqrt{49}\sqrt{5}=7\sqrt{5}$

15. $2\sqrt{3}\sqrt{6}=2\sqrt{18}=2\sqrt{9}\sqrt{2}=2(3)\sqrt{2}$
$=6\sqrt{2}$

16. $3\sqrt{11}\sqrt{33}=3\sqrt{363}=3\sqrt{121}\sqrt{3}$
$=3(11)\sqrt{3}=33\sqrt{3}$

17. $\sqrt[3]{5}\sqrt[3]{25}=\sqrt[3]{125}=5$

18. $\sqrt[3]{7}\sqrt[3]{49}=\sqrt[3]{343}=7$

19. $\left(3\sqrt[3]{9}\right)\left(2\sqrt[3]{3}\right)=6\sqrt[3]{27}=6(3)=18$

20. $\left(2\sqrt[3]{16}\right)\left(-\sqrt[3]{4}\right)=-2\sqrt[3]{64}=-2(4)=-8$

21. $\sqrt[3]{2}\sqrt[3]{12}=\sqrt[3]{24}=\sqrt[3]{8}\sqrt[3]{3}=2\sqrt[3]{3}$

22. $\sqrt[3]{3}\sqrt[3]{18}=\sqrt[3]{54}=\sqrt[3]{27}\sqrt[3]{2}=3\sqrt[3]{2}$

23. $\sqrt{ab^3}\sqrt{ab}=\sqrt{a^2b^4}=ab^2$

24. $\sqrt{8x}\sqrt{2x^3y}=\sqrt{16x^4y}=\sqrt{16x^4}\sqrt{y}$
$=4x^2\sqrt{y}$

25. $\sqrt{5ab}\sqrt{5a}=\sqrt{25a^2b}=\sqrt{25a^2}\sqrt{b}$
$=5a\sqrt{b}$

26. $\sqrt{15rs^2}\sqrt{10r}=\sqrt{150r^2s^2}$
$=\sqrt{25r^2s^2}\sqrt{6}=5rs\sqrt{6}$

27. $\sqrt[3]{5r^2s}\sqrt[3]{2r}=\sqrt[3]{10r^3s}=\sqrt[3]{r^3}\sqrt[3]{10s}$
$=r\sqrt[3]{10s}$

28. $\sqrt[3]{3xy^2}\sqrt[3]{9x^3}=\sqrt[3]{27x^4y^2}=\sqrt[3]{27x^3}\sqrt[3]{xy^2}$
$=3x\sqrt[3]{xy^2}$

29. $\sqrt[3]{a^5b}\sqrt[3]{16ab^5}=\sqrt[3]{16a^6b^6}=\sqrt[3]{8a^6b^6}\sqrt[3]{2}$
$=2a^2b^2\sqrt[3]{2}$

30. $\sqrt[3]{3x^4y}\sqrt[3]{18x}=\sqrt[3]{54x^5y}=\sqrt[3]{27x^3}\sqrt[3]{2x^2y}$
$=3x\sqrt[3]{2x^2y}$

31. $\sqrt{x(x+3)}\sqrt{x^3(x+3)}=\sqrt{x^4(x+3)^2}$
$=x^2(x+3)$

32. $\sqrt{y^2(x+y)}\sqrt{(x+y)^3}=\sqrt{y^2(x+y)^4}$
$=y(x+y)^2$

33. $\sqrt[3]{6x^2(y+z)^2}\sqrt[3]{18x(y+z)} = \sqrt[3]{108x^3(y+z)^3} = \sqrt[3]{27x^3(y+z)^3}\sqrt[3]{4} = 3x(y+z)\sqrt[3]{4}$

34. $\sqrt[3]{9x^2y(z+1)^2}\sqrt[3]{6xy^2(z+1)} = \sqrt[3]{54x^3y^3(z+1)^3} = \sqrt[3]{27x^3y^3(z+1)^3}\sqrt[3]{2} = 3xy(z+1)\sqrt[3]{2}$

35. $3\sqrt{5}\left(4-\sqrt{5}\right) = 12\sqrt{5} - 3\sqrt{25} = 12\sqrt{5} - 3(5) = 12\sqrt{5} - 15$

36. $2\sqrt{7}\left(3\sqrt{7}-1\right) = 6\sqrt{49} - 2\sqrt{7} = 6(7) - 2\sqrt{7} = 42 - 2\sqrt{7}$

37. $3\sqrt{2}\left(4\sqrt{3}+2\sqrt{7}\right) = 12\sqrt{6} + 6\sqrt{14}$

38. $-\sqrt{3}\left(\sqrt{7}-\sqrt{5}\right) = -\sqrt{21} + \sqrt{15}$

39. $-2\sqrt{5x}\left(4\sqrt{2x} - 3\sqrt{3}\right) = -8\sqrt{10x^2} + 6\sqrt{15x} = -8x\sqrt{10} + 6\sqrt{15x}$

40. $3\sqrt{7t}\left(2\sqrt{7t} + 3\sqrt{3t^2}\right) = 6\sqrt{49t^2} + 9\sqrt{21t^3} = 6(7t) + 9t\sqrt{21t} = 42t + 9t\sqrt{21t}$

41. $\left(\sqrt{2}+1\right)\left(\sqrt{2}-3\right) = \sqrt{4} - 3\sqrt{2} + \sqrt{2} - 3 = 2 - 2\sqrt{2} - 3 = -1 - 2\sqrt{2}$

42. $\left(2\sqrt{3}+1\right)\left(\sqrt{3}-1\right) = 2\sqrt{9} - 2\sqrt{3} + \sqrt{3} - 1 = 6 - \sqrt{3} - 1 = 5 - \sqrt{3}$

43. $(4\sqrt{x}+3)(2\sqrt{x}-5) = 8\sqrt{x^2} - 20\sqrt{x} + 6\sqrt{x} - 15 = 8x - 14\sqrt{x} - 15$

44. $(7\sqrt{y}+2)(3\sqrt{y}-5) = 21\sqrt{y^2} - 35\sqrt{y} + 6\sqrt{y} - 10 = 21y - 29\sqrt{y} - 10$

45. $\left(\sqrt{5z}+\sqrt{3}\right)\left(\sqrt{5z}+\sqrt{3}\right) = \sqrt{25z^2} + \sqrt{15z} + \sqrt{15z} + \sqrt{9} = 5z + 2\sqrt{15z} + 3$

46. $\left(\sqrt{3p}-\sqrt{2}\right)\left(\sqrt{3p}+\sqrt{2}\right) = \sqrt{9p^2} + \sqrt{6p} - \sqrt{6p} - \sqrt{4} = 3p - 2$

47. $\left(\sqrt{3x}-\sqrt{2y}\right)\left(\sqrt{3x}+\sqrt{2y}\right) = \sqrt{9x^2} + \sqrt{6xy} - \sqrt{6xy} - \sqrt{4y^2} = 3x - 2y$

48. $\left(\sqrt{3m}+\sqrt{2n}\right)\left(\sqrt{3m}+\sqrt{2n}\right) = \sqrt{9m^2} + \sqrt{6mn} + \sqrt{6mn} + \sqrt{4n^2} = 3m + 2\sqrt{6mn} + 2n$

49. $\left(2\sqrt{3a}-\sqrt{b}\right)\left(\sqrt{3a}+3\sqrt{b}\right) = 2\sqrt{9a^2} + 6\sqrt{3ab} - \sqrt{3ab} - 3\sqrt{b^2} = 6a + 5\sqrt{3ab} - 3b$

50. $(5\sqrt{p}-\sqrt{3q})(\sqrt{p}+2\sqrt{3q}) = 5\sqrt{p^2} + 10\sqrt{3pq} - \sqrt{3pq} - 2\sqrt{9q^2} = 5p + 9\sqrt{3pq} - 6q$

51. $\left(3\sqrt{2r}-2\right)^2 = \left(3\sqrt{2r}-2\right)\left(3\sqrt{2r}-2\right) = 9\sqrt{4r^2} - 6\sqrt{2r} - 6\sqrt{2r} + 4 = 18r - 12\sqrt{2r} + 4$

52. $\left(2\sqrt{3t}+5\right)^2 = \left(2\sqrt{3t}+5\right)\left(2\sqrt{3t}+5\right) = 4\sqrt{9t^2}+10\sqrt{3t}+10\sqrt{3t}+25$
$= 12t+20\sqrt{3t}+25$

53. $-2\left(\sqrt{3x}+\sqrt{3}\right)^2 = -2\left(\sqrt{3x}+\sqrt{3}\right)\left(\sqrt{3x}+\sqrt{3}\right)$
$= -2\left(\sqrt{9x^2}+\sqrt{9x}+\sqrt{9x}+\sqrt{9}\right)$
$= -2(3x+3\sqrt{x}+3\sqrt{x}+3) = -2(3x+6\sqrt{x}+3) = -6x-12\sqrt{x}-6$

54. $3\left(\sqrt{5x}-\sqrt{3}\right)^2 = 3\left(\sqrt{5x}-\sqrt{3}\right)\left(\sqrt{5x}-\sqrt{3}\right) = 3\left(\sqrt{25x^2}-\sqrt{15x}-\sqrt{15x}+\sqrt{9}\right)$
$= 3\left(5x-2\sqrt{15x}+3\right) = 15x-6\sqrt{15x}+9$

55. $\sqrt{\frac{1}{7}} = \frac{\sqrt{1}}{\sqrt{7}} = \frac{1\sqrt{7}}{\sqrt{7}\sqrt{7}} = \frac{\sqrt{7}}{7}$

56. $\sqrt{\frac{5}{3}} = \frac{\sqrt{5}}{\sqrt{3}} = \frac{\sqrt{5}\sqrt{3}}{\sqrt{3}\sqrt{3}} = \frac{\sqrt{15}}{3}$

57. $\sqrt{\frac{2}{3}} = \frac{\sqrt{2}}{\sqrt{3}} = \frac{\sqrt{2}\sqrt{3}}{\sqrt{3}\sqrt{3}} = \frac{\sqrt{6}}{3}$

58. $\sqrt{\frac{3}{2}} = \frac{\sqrt{3}}{\sqrt{2}} = \frac{\sqrt{3}\sqrt{2}}{\sqrt{2}\sqrt{2}} = \frac{\sqrt{6}}{2}$

59. $\frac{\sqrt{5}}{\sqrt{8}} = \frac{\sqrt{5}\sqrt{2}}{\sqrt{8}\sqrt{2}} = \frac{\sqrt{10}}{\sqrt{16}} = \frac{\sqrt{10}}{4}$

60. $\frac{\sqrt{3}}{\sqrt{50}} = \frac{\sqrt{3}\sqrt{2}}{\sqrt{50}\sqrt{2}} = \frac{\sqrt{6}}{\sqrt{100}} = \frac{\sqrt{6}}{10}$

61. $\frac{\sqrt{8}}{\sqrt{2}} = \sqrt{\frac{8}{2}} = \sqrt{4} = 2$

62. $\frac{\sqrt{27}}{\sqrt{3}} = \sqrt{\frac{27}{3}} = \sqrt{9} = 3$

63. $\frac{1}{\sqrt[3]{2}} = \frac{1\sqrt[3]{4}}{\sqrt[3]{2}\sqrt[3]{4}} = \frac{\sqrt[3]{4}}{\sqrt[3]{8}} = \frac{\sqrt[3]{4}}{2}$

64. $\frac{2}{\sqrt[3]{6}} = \frac{2\sqrt[3]{36}}{\sqrt[3]{6}\sqrt[3]{36}} = \frac{2\sqrt[3]{36}}{\sqrt[3]{216}} = \frac{2\sqrt[3]{36}}{6} = \frac{\sqrt[3]{36}}{3}$

65. $\frac{3}{\sqrt[3]{9}} = \frac{3\sqrt[3]{3}}{\sqrt[3]{9}\sqrt[3]{3}} = \frac{3\sqrt[3]{3}}{\sqrt[3]{27}} = \frac{3\sqrt[3]{3}}{3} = \sqrt[3]{3}$

66. $\frac{2}{\sqrt[3]{a}} = \frac{2\sqrt[3]{a^2}}{\sqrt[3]{a}\sqrt[3]{a^2}} = \frac{2\sqrt[3]{a^2}}{\sqrt[3]{a^3}} = \frac{2\sqrt[3]{a^2}}{a}$

67. $\frac{\sqrt[3]{2}}{\sqrt[3]{9}} = \frac{\sqrt[3]{2}\sqrt[3]{3}}{\sqrt[3]{9}\sqrt[3]{3}} = \frac{\sqrt[3]{6}}{\sqrt[3]{27}} = \frac{\sqrt[3]{6}}{3}$

68. $\frac{\sqrt[3]{9}}{\sqrt[3]{54}} = \sqrt[3]{\frac{9}{54}} = \sqrt[3]{\frac{1}{6}} = \frac{1\sqrt[3]{36}}{\sqrt[3]{6}\sqrt[3]{36}} = \frac{\sqrt[3]{36}}{6}$

69. $\frac{\sqrt{8x^2y}}{\sqrt{xy}} = \sqrt{\frac{8x^2y}{xy}} = \sqrt{8x} = 2\sqrt{2x}$

70. $\frac{\sqrt{9xy}}{\sqrt{3x^2y}} = \sqrt{\frac{9xy}{3x^2y}} = \sqrt{\frac{3}{x}} = \frac{\sqrt{3}\sqrt{x}}{\sqrt{x}\sqrt{x}}$
$= \frac{\sqrt{3x}}{x}$

71. $\dfrac{\sqrt{10xy^2}}{\sqrt{2xy^3}} = \sqrt{\dfrac{10xy^2}{2xy^3}} = \sqrt{\dfrac{5}{y}} = \dfrac{\sqrt{5}\sqrt{y}}{\sqrt{y}\sqrt{y}} = \dfrac{\sqrt{5y}}{y}$

72. $\dfrac{\sqrt{5ab^2c}}{\sqrt{10abc}} = \sqrt{\dfrac{5ab^2c}{10abc}} = \sqrt{\dfrac{b}{2}} = \dfrac{\sqrt{b}\sqrt{2}}{\sqrt{2}\sqrt{2}} = \dfrac{\sqrt{2b}}{2}$

73. $\dfrac{\sqrt[3]{4a^2}}{\sqrt[3]{2ab}} = \sqrt[3]{\dfrac{4a^2}{2ab}} = \sqrt[3]{\dfrac{2a}{b}} = \dfrac{\sqrt[3]{2a}\sqrt[3]{b^2}}{\sqrt[3]{b}\sqrt[3]{b^2}} = \dfrac{\sqrt[3]{2ab^2}}{b}$

74. $\dfrac{\sqrt[3]{9x}}{\sqrt[3]{3xy}} = \sqrt[3]{\dfrac{9x}{3xy}} = \sqrt[3]{\dfrac{3}{y}} = \dfrac{\sqrt[3]{3}\sqrt[3]{y^2}}{\sqrt[3]{y}\sqrt[3]{y^2}} = \dfrac{\sqrt[3]{3y^2}}{y}$

75. $\dfrac{1}{\sqrt[4]{4}} = \dfrac{1\sqrt[4]{4}}{\sqrt[4]{4}\sqrt[4]{4}} = \dfrac{\sqrt[4]{4}}{\sqrt[4]{16}} = \dfrac{\sqrt[4]{4}}{2}$

76. $\dfrac{1}{\sqrt[5]{2}} = \dfrac{1\sqrt[5]{16}}{\sqrt[5]{2}\sqrt[5]{16}} = \dfrac{\sqrt[5]{16}}{\sqrt[5]{32}} = \dfrac{\sqrt[5]{16}}{2}$

77. $\dfrac{1}{\sqrt[5]{16}} = \dfrac{1\sqrt[5]{2}}{\sqrt[5]{16}\sqrt[5]{2}} = \dfrac{\sqrt[5]{2}}{\sqrt[5]{32}} = \dfrac{\sqrt[5]{2}}{2}$

78. $\dfrac{4}{\sqrt[4]{32}} = \dfrac{4}{\sqrt[4]{16}\sqrt[4]{2}} = \dfrac{4}{2\sqrt[4]{2}} = \dfrac{2\sqrt[4]{8}}{\sqrt[4]{2}\sqrt[4]{8}} = \dfrac{2\sqrt[4]{8}}{2} = \sqrt[4]{8}$

79. $\dfrac{1}{\sqrt{2}-1} = \dfrac{1\left(\sqrt{2}+1\right)}{\left(\sqrt{2}-1\right)\left(\sqrt{2}+1\right)} = \dfrac{\sqrt{2}+1}{\sqrt{4}-1} = \dfrac{\sqrt{2}+1}{2-1} = \dfrac{\sqrt{2}+1}{1} = \sqrt{2}+1$

80. $\dfrac{3}{\sqrt{3}-1} = \dfrac{3\left(\sqrt{3}+1\right)}{\left(\sqrt{3}-1\right)\left(\sqrt{3}+1\right)} = \dfrac{3\left(\sqrt{3}+1\right)}{\sqrt{9}-1} = \dfrac{3\left(\sqrt{3}+1\right)}{3-1} = \dfrac{3\left(\sqrt{3}+1\right)}{2}$

81. $\dfrac{\sqrt{2}}{\sqrt{5}+3} = \dfrac{\sqrt{2}\left(\sqrt{5}-3\right)}{\left(\sqrt{5}+3\right)\left(\sqrt{5}-3\right)} = \dfrac{\sqrt{2}\left(\sqrt{5}-3\right)}{\sqrt{25}-9} = \dfrac{\sqrt{2}\left(\sqrt{5}-3\right)}{5-9} = \dfrac{\sqrt{2}\left(\sqrt{5}-3\right)}{-4}$

$= \dfrac{\sqrt{10}-3\sqrt{2}}{-4}$

$= \dfrac{3\sqrt{2}-\sqrt{10}}{4}$

82. $\dfrac{\sqrt{3}}{\sqrt{3}-2} = \dfrac{\sqrt{3}\left(\sqrt{3}+2\right)}{\left(\sqrt{3}-2\right)\left(\sqrt{3}+2\right)} = \dfrac{\sqrt{3}\left(\sqrt{3}+2\right)}{\sqrt{9}-4} = \dfrac{\sqrt{3}\left(\sqrt{3}+2\right)}{3-4} = \dfrac{\sqrt{3}\left(\sqrt{3}+2\right)}{-1}$

$= \dfrac{\sqrt{9}+2\sqrt{3}}{-1}$

$= -\left(3+2\sqrt{3}\right)$

$= -3-2\sqrt{3}$

83. $\dfrac{\sqrt{3}+1}{\sqrt{3}-1}=\dfrac{\left(\sqrt{3}+1\right)\left(\sqrt{3}+1\right)}{\left(\sqrt{3}-1\right)\left(\sqrt{3}+1\right)}=\dfrac{\sqrt{9}+\sqrt{3}+\sqrt{3}+1}{\sqrt{9}-1}=\dfrac{4+2\sqrt{3}}{2}=\dfrac{2\left(2+\sqrt{3}\right)}{2}$

$=2+\sqrt{3}$

84. $\dfrac{\sqrt{2}-1}{\sqrt{2}+1}=\dfrac{\left(\sqrt{2}-1\right)\left(\sqrt{2}-1\right)}{\left(\sqrt{2}+1\right)\left(\sqrt{2}-1\right)}=\dfrac{\sqrt{4}-\sqrt{2}-\sqrt{2}+1}{\sqrt{4}-1}=\dfrac{3-2\sqrt{2}}{1}=3-2\sqrt{2}$

85. $\dfrac{\sqrt{7}-\sqrt{2}}{\sqrt{2}+\sqrt{7}}=\dfrac{\left(\sqrt{7}-\sqrt{2}\right)\left(\sqrt{2}-\sqrt{7}\right)}{\left(\sqrt{2}+\sqrt{7}\right)\left(\sqrt{2}-\sqrt{7}\right)}=\dfrac{\sqrt{14}-7-2+\sqrt{14}}{\sqrt{4}-\sqrt{49}}=\dfrac{2\sqrt{14}-9}{-5}=\dfrac{9-2\sqrt{14}}{5}$

86. $\dfrac{\sqrt{3}+\sqrt{2}}{\sqrt{3}-\sqrt{2}}=\dfrac{\left(\sqrt{3}+\sqrt{2}\right)\left(\sqrt{3}+\sqrt{2}\right)}{\left(\sqrt{3}-\sqrt{2}\right)\left(\sqrt{3}+\sqrt{2}\right)}=\dfrac{3+\sqrt{6}+\sqrt{6}+2}{3-2}=5+2\sqrt{6}$

87. $\dfrac{2}{\sqrt{x}+1}=\dfrac{2(\sqrt{x}-1)}{(\sqrt{x}+1)(\sqrt{x}-1)}=\dfrac{2(\sqrt{x}-1)}{\sqrt{x^2}-1}=\dfrac{2(\sqrt{x}-1)}{x-1}$

88. $\dfrac{3}{\sqrt{x}-2}=\dfrac{3(\sqrt{x}+2)}{(\sqrt{x}-2)(\sqrt{x}+2)}=\dfrac{3(\sqrt{x}+2)}{\sqrt{x^2}-4}=\dfrac{3(\sqrt{x}+2)}{x-4}$

89. $\dfrac{x}{\sqrt{x}-4}=\dfrac{x(\sqrt{x}+4)}{(\sqrt{x}-4)(\sqrt{x}+4)}=\dfrac{x(\sqrt{x}+4)}{\sqrt{x^2}-16}=\dfrac{x(\sqrt{x}+4)}{x-16}$

90. $\dfrac{2x}{\sqrt{x}+1}=\dfrac{2x(\sqrt{x}-1)}{(\sqrt{x}+1)(\sqrt{x}-1)}=\dfrac{2x(\sqrt{x}-1)}{\sqrt{x^2}-1}=\dfrac{2x(\sqrt{x}-1)}{x-1}$

91. $\dfrac{2z-1}{\sqrt{2z}-1}=\dfrac{(2z-1)\left(\sqrt{2z}+1\right)}{\left(\sqrt{2z}-1\right)\left(\sqrt{2z}+1\right)}=\dfrac{(2z-1)\left(\sqrt{2z}+1\right)}{\sqrt{4z^2}-1}=\dfrac{(2z-1)\left(\sqrt{2z}+1\right)}{2z-1}$

$=\sqrt{2z}+1$

92. $\dfrac{3t-1}{\sqrt{3t}+1}=\dfrac{(3t-1)\left(\sqrt{3t}-1\right)}{\left(\sqrt{3t}+1\right)\left(\sqrt{3t}-1\right)}=\dfrac{(3t-1)\left(\sqrt{3t}-1\right)}{\sqrt{9t^2}-1}=\dfrac{(3t-1)\left(\sqrt{3t}-1\right)}{3t-1}=\sqrt{3t}-1$

93. $\dfrac{\sqrt{x}-\sqrt{y}}{\sqrt{x}+\sqrt{y}}=\dfrac{(\sqrt{x}-\sqrt{y})(\sqrt{x}-\sqrt{y})}{(\sqrt{x}+\sqrt{y})(\sqrt{x}-\sqrt{y})}=\dfrac{\sqrt{x^2}-\sqrt{xy}-\sqrt{xy}+\sqrt{y^2}}{\sqrt{x^2}-\sqrt{y^2}}=\dfrac{x-2\sqrt{xy}+y}{x-y}$

94. $\dfrac{\sqrt{x}+\sqrt{y}}{\sqrt{x}-\sqrt{y}}=\dfrac{(\sqrt{x}+\sqrt{y})(\sqrt{x}+\sqrt{y})}{(\sqrt{x}-\sqrt{y})(\sqrt{x}+\sqrt{y})}=\dfrac{\sqrt{x^2}+\sqrt{xy}+\sqrt{xy}+\sqrt{y^2}}{\sqrt{x^2}-\sqrt{y^2}}=\dfrac{x+2\sqrt{xy}+y}{x-y}$

95. $\dfrac{\sqrt{3}+1}{2}=\dfrac{\left(\sqrt{3}+1\right)\left(\sqrt{3}-1\right)}{2\left(\sqrt{3}-1\right)}=\dfrac{\sqrt{9}-1}{2\left(\sqrt{3}-1\right)}=\dfrac{2}{2\left(\sqrt{3}-1\right)}=\dfrac{1}{\sqrt{3}-1}$

96. $\dfrac{\sqrt{5}-1}{2}=\dfrac{\left(\sqrt{5}-1\right)\left(\sqrt{5}+1\right)}{2\left(\sqrt{5}+1\right)}=\dfrac{\sqrt{25}-1}{2\left(\sqrt{5}+1\right)}=\dfrac{4}{2\left(\sqrt{5}+1\right)}=\dfrac{2}{\sqrt{5}+1}$

97. $\dfrac{\sqrt{x}+3}{x}=\dfrac{(\sqrt{x}+3)(\sqrt{x}-3)}{x(\sqrt{x}-3)}=\dfrac{\sqrt{x^2}-9}{x(\sqrt{x}-3)}=\dfrac{x-9}{x(\sqrt{x}-3)}$

98. $\dfrac{2+\sqrt{x}}{5x}=\dfrac{(2+\sqrt{x})(2-\sqrt{x})}{5x(2-\sqrt{x})}=\dfrac{4-\sqrt{x^2}}{5x(2-\sqrt{x})}=\dfrac{4-x}{5x(2-\sqrt{x})}$

99. $\dfrac{\sqrt{x}+\sqrt{y}}{\sqrt{x}}=\dfrac{(\sqrt{x}+\sqrt{y})(\sqrt{x}-\sqrt{y})}{\sqrt{x}(\sqrt{x}-\sqrt{y})}=\dfrac{\sqrt{x^2}-\sqrt{y^2}}{\sqrt{x^2}-\sqrt{xy}}=\dfrac{x-y}{x-\sqrt{xy}}$

100. $\dfrac{\sqrt{x}-\sqrt{y}}{\sqrt{x}+\sqrt{y}}=\dfrac{(\sqrt{x}-\sqrt{y})(\sqrt{x}+\sqrt{y})}{(\sqrt{x}+\sqrt{y})(\sqrt{x}+\sqrt{y})}=\dfrac{\sqrt{x^2}-\sqrt{y^2}}{\sqrt{x^2}+\sqrt{xy}+\sqrt{xy}+\sqrt{y^2}}=\dfrac{x-y}{x+2\sqrt{xy}+y}$

101. $r=\sqrt{\dfrac{A}{\pi}}=\dfrac{\sqrt{A}}{\sqrt{\pi}}=\dfrac{\sqrt{A}\sqrt{\pi}}{\sqrt{\pi}\sqrt{\pi}}=\dfrac{\sqrt{A\pi}}{\pi}$

102. $p(t)=\dfrac{9}{\sqrt{t}}=\dfrac{9\sqrt{t}}{\sqrt{t}\sqrt{t}}=\dfrac{9\sqrt{t}}{t}$

103. If the area of the aperture is again cut in half, the area will equal $9\pi/4$ cm^2.

$$\begin{aligned} A&=\pi r^2\\ \frac{9\pi}{4}&=\pi\left(\frac{d}{2}\right)^2\\ \frac{9\pi}{4}&=\frac{\pi d^2}{4}\\ 36\pi&=4\pi d^2\\ \frac{36\pi}{4\pi}&=d^2\\ 9&=d^2\\ 3&=d \end{aligned}$$

$$\begin{aligned} f\text{-number}&=\frac{f}{d}\\ &=\frac{12}{3}\\ &=4 \end{aligned}$$

$\boxed{f/4}$

104. If the area of the aperture is again cut in half, the area will equal $9\pi/8$ cm^2.

$$\begin{aligned} A&=\pi r^2\\ \frac{9\pi}{8}&=\pi\left(\frac{d}{2}\right)^2\\ \frac{9\pi}{8}&=\frac{\pi d^2}{4}\\ 36\pi&=8\pi d^2\\ \frac{36\pi}{8\pi}&=d^2\\ \frac{9}{2}&=d^2\\ \sqrt{\frac{9}{2}}&=d\\ 2.1213&\approx d \end{aligned}$$

$$\begin{aligned} f\text{-number}&=\frac{f}{d}\\ &=\frac{12}{2.1213}\\ &\approx 5.7 \end{aligned}$$

$\boxed{f/5.7}$

105. Answers may vary.

106. Answers may vary.

107. $\frac{\sqrt{x}-3}{4}=\frac{(\sqrt{x}-3)(\sqrt{x}+3)}{4(\sqrt{x}+3)}=\frac{x-9}{4(\sqrt{x}+3)}$

108. $\frac{2\sqrt{3x}+4}{\sqrt{3x}-1}=\frac{\left(2\sqrt{3x}+4\right)\left(2\sqrt{3x}-4\right)}{\left(\sqrt{3x}-1\right)\left(2\sqrt{3x}-4\right)}=\frac{4\sqrt{9x^2}-16}{2\sqrt{9x^2}-4\sqrt{3x}-2\sqrt{3x}+4}$

$$=\frac{12x-16}{6x-6\sqrt{3x}+4}$$
$$=\frac{2(6x-8)}{2\left(3x-3\sqrt{3x}+2\right)}$$
$$=\frac{6x-8}{3x-3\sqrt{3x}+2}$$

Exercise 7.6 (page 487)

1. $f(0)=3(0)^2-4(0)+2=2$

2. $f(-3)=3(-3)^2-4(-3)+2=41$

3. $f(2)=3(2)^2-4(2)+2=6$

4. $f\left(\frac{1}{2}\right)=3\left(\frac{1}{2}\right)^2-4\left(\frac{1}{2}\right)+2=\frac{3}{4}$

5. $x^n=y^n$

6. isolate

7. square

8. cube

9. extraneous

10. check; extraneous

11.
$$\begin{aligned}\sqrt{5x-6}&=2\\ \left(\sqrt{5x-6}\right)^2&=2^2\\ 5x-6&=4\\ 5x&=10\\ x&=2\end{aligned}$$
The answer checks.

12.
$$\begin{aligned}\sqrt{7x-10}&=12\\ \left(\sqrt{7x-10}\right)^2&=12^2\\ 7x-10&=144\\ 7x&=154\\ x&=22\end{aligned}$$
The answer checks.

13.
$$\begin{aligned}\sqrt{6x+1}+2&=7\\ \sqrt{6x+1}&=5\\ \left(\sqrt{6x+1}\right)^2&=5^2\\ 6x+1&=25\\ 6x&=24\\ x&=4\end{aligned}$$
The answer checks.

14.
$$\begin{aligned}\sqrt{6x+13}-2&=5\\ \sqrt{6x+13}&=7\\ \left(\sqrt{6x+13}\right)^2&=7^2\\ 6x+13&=49\\ 6x&=36\\ x&=6\end{aligned}$$
The answer checks.

15.
$$
\begin{aligned}
2\sqrt{4x+1} &= \sqrt{x+4} \\
\left(2\sqrt{4x+1}\right)^2 &= \left(\sqrt{x+4}\right)^2 \\
4(4x+1) &= x+4 \\
16x+4 &= x+4 \\
15x &= 0 \\
x &= 0
\end{aligned}
$$
The answer checks.

16.
$$
\begin{aligned}
\sqrt{3(x+4)} &= \sqrt{5x-12} \\
\left(\sqrt{3x+12}\right)^2 &= \left(\sqrt{5x-12}\right)^2 \\
3x+12 &= 5x-12 \\
-2x &= -24 \\
x &= 12
\end{aligned}
$$
The answer checks.

17.
$$
\begin{aligned}
\sqrt[3]{7n-1} &= 3 \\
\left(\sqrt[3]{7n-1}\right)^3 &= 3^3 \\
7n-1 &= 27 \\
7n &= 28 \\
n &= 4
\end{aligned}
$$
The answer checks.

18.
$$
\begin{aligned}
\sqrt[3]{12m+4} &= 4 \\
\left(\sqrt[3]{12m+4}\right)^3 &= 4^3 \\
12m+4 &= 64 \\
12m &= 60 \\
m &= 5
\end{aligned}
$$
The answer checks.

19.
$$
\begin{aligned}
\sqrt[4]{10p+1} &= \sqrt[4]{11p-7} \\
\left(\sqrt[4]{10p+1}\right)^4 &= \left(\sqrt[4]{11p-7}\right)^4 \\
10p+1 &= 11p-7 \\
-p &= -8 \\
p &= 8
\end{aligned}
$$
The answer checks.

20.
$$
\begin{aligned}
\sqrt[4]{10y+2} &= 2\sqrt[4]{2} \\
\left(\sqrt[4]{10y+2}\right)^4 &= \left(2\sqrt[4]{2}\right)^4 \\
10y+2 &= 16(2) \\
10y+2 &= 32 \\
10y &= 30 \\
y &= 3
\end{aligned}
$$
The answer checks.

21.
$$
\begin{aligned}
x &= \frac{\sqrt{12x-5}}{2} \\
2x &= \sqrt{12x-5} \\
(2x)^2 &= \left(\sqrt{12x-5}\right)^2 \\
4x^2 &= 12x-5
\end{aligned}
$$
$$4x^2 - 12x + 5 = 0$$
$$(2x-1)(2x-5)$$
$$
\begin{aligned}
2x-1 &= 0 \quad \textbf{or} \quad & 2x-5 &= 0 \\
2x &= 1 & 2x &= 5 \\
x &= \tfrac{1}{2} & x &= \tfrac{5}{2}
\end{aligned}
$$
Both answers check.

22.
$$
\begin{aligned}
x &= \frac{\sqrt{16x-12}}{2} \\
2x &= \sqrt{16x-12} \\
(2x)^2 &= \left(\sqrt{16x-12}\right)^2 \\
4x^2 &= 16x-12
\end{aligned}
$$
$$4x^2 - 16x + 12 = 0$$
$$4(x-3)(x-1) = 0$$
$$
\begin{aligned}
x-3 &= 0 \quad \textbf{or} \quad & x-1 &= 0 \\
x &= 3 & x &= 1
\end{aligned}
$$
Both answers check.

23.
$$\begin{aligned}\sqrt{x+2} &= \sqrt{4-x}\\ \left(\sqrt{x+2}\right)^2 &= \left(\sqrt{4-x}\right)^2\\ x+2 &= 4-x\\ 2x &= 2\\ x &= 1\end{aligned}$$
The answer checks.

24.
$$\begin{aligned}\sqrt{6-x} &= \sqrt{2x+3}\\ \left(\sqrt{6-x}\right)^2 &= \left(\sqrt{2x+3}\right)^2\\ 6-x &= 2x+3\\ -3x &= -3\\ x &= 1\end{aligned}$$
The answer checks.

25.
$$\begin{aligned}2\sqrt{x} &= \sqrt{5x-16}\\ \left(2\sqrt{x}\right)^2 &= \left(\sqrt{5x-16}\right)^2\\ 4x &= 5x-16\\ -x &= -16\\ x &= 16\end{aligned}$$
The answer checks.

26.
$$\begin{aligned}3\sqrt{x} &= \sqrt{3x+12}\\ \left(3\sqrt{x}\right)^2 &= \left(\sqrt{3x+12}\right)^2\\ 9x &= 3x+12\\ 6x &= 12\\ x &= 2\end{aligned}$$
The answer checks.

27.
$$\begin{aligned}r-9 &= \sqrt{2r-3}\\ (r-9)^2 &= \left(\sqrt{2r-3}\right)^2\\ r^2-18r+81 &= 2r-3\\ r^2-20r+84 &= 0\\ (r-14)(r-6) &= 0\end{aligned}$$
$r-14=0$ **or** $r-6=0$

$r=14$ $\quad$ $r=6$

solution $\quad$ not a solution

28.
$$\begin{aligned}-s-3 &= 2\sqrt{5-s}\\ (-s-3)^2 &= \left(2\sqrt{5-s}\right)^2\\ s^2+6s+9 &= 4(5-s)\\ s^2+6s+9 &= 20-4s\\ s^2+10s-11 &= 0\\ (s+11)(s-1) &= 0\end{aligned}$$
$s+11=0$ **or** $s-1=0$

$s=-11$ $\quad$ $s=1$

solution $\quad$ not a solution

29.
$$\begin{aligned}\sqrt{-5x+24} &= 6-x\\ \left(\sqrt{-5x+24}\right)^2 &= (6-x)^2\\ -5x+24 &= 36-12x+x^2\\ 0 &= x^2-7x+12\\ 0 &= (x-3)(x-4)\end{aligned}$$
$x-3=0$ **or** $x-4=0$

$x=3$ $\quad$ $x=4$

solution $\quad$ solution

30.
$$\begin{aligned}\sqrt{-x+2} &= x-2\\ \left(\sqrt{-x+2}\right)^2 &= (x-2)^2\\ -x+2 &= x^2-4x+4\\ 0 &= x^2-3x+2\\ 0 &= (x-2)(x-1)\end{aligned}$$
$x-2=0$ **or** $x-1=0$

$x=2$ $\quad$ $x=1$

solution $\quad$ not a solution

31.
$$\sqrt{y+2} = 4 - y$$
$$\left(\sqrt{y+2}\right) = (4-y)^2$$
$$y + 2 = 16 - 8y + y^2$$
$$0 = y^2 - 9y + 14$$
$$0 = (y-2)(y-7)$$
$y - 2 = 0$ **or** $y - 7 = 0$
$y = 2$ $\quad$ $y = 7$
solution $\quad$ not a solution

32.
$$\sqrt{22y+86} = y + 9$$
$$\left(\sqrt{22y+86}\right)^2 = (y+9)^2$$
$$22y + 86 = y^2 + 18y + 81$$
$$0 = y^2 - 4y - 5$$
$$0 = (y-5)(y+1)$$
$y - 5 = 0$ **or** $y + 1 = 0$
$y = 5$ $\quad$ $y = -1$
solution $\quad$ solution

33.
$$\sqrt{x}\sqrt{x+16} = 15$$
$$\left(\sqrt{x}\sqrt{x+16}\right)^2 = 15^2$$
$$x(x+16) = 225$$
$$x^2 + 16x - 225 = 0$$
$$(x-9)(x+25) = 0$$
$x - 9 = 0$ **or** $x + 25 = 0$
$x = 9$ $\quad$ $x = -25$
solution $\quad$ not a solution

34.
$$\sqrt{x}\sqrt{x+6} = 4$$
$$\left(\sqrt{x}\sqrt{x+6}\right)^2 = 4^2$$
$$x(x+6) = 16$$
$$x^2 + 6x - 16 = 0$$
$$(x-2)(x+8) = 0$$
$x - 2 = 0$ **or** $x + 8 = 0$
$x = 2$ $\quad$ $x = -8$
solution $\quad$ not a solution

35.
$$\sqrt[3]{x^3 - 7} = x - 1$$
$$\left(\sqrt[3]{x^3 - 7}\right)^3 = (x-1)^3$$
$$x^3 - 7 = x^3 - 3x^2 + 3x - 1$$
$$3x^2 - 3x - 6 = 0$$
$$3(x-2)(x+1) = 0$$
$x - 2 = 0$ **or** $x + 1 = 0$
$x = 2$ $\quad$ $x = -1$
solution $\quad$ solution

36.
$$\sqrt[3]{x^3 + 56} - 2 = x$$
$$\sqrt[3]{x^3 + 56} = x + 2$$
$$\left(\sqrt[3]{x^3 + 56}\right)^3 = (x+2)^3$$
$$x^3 + 56 = x^3 + 6x^2 + 12x + 8$$
$$0 = 6x^2 + 12x - 48$$
$$0 = 6(x-2)(x+4)$$
$x - 2 = 0$ **or** $x + 4 = 0$
$x = 2$ $\quad$ $x = -4$
solution $\quad$ solution

37.
$$\sqrt[4]{x^4 + 4x^2 - 4} = -x$$
$$\left(\sqrt[4]{x^4 + 4x^2 - 4}\right)^4 = (-x)^4$$
$$x^4 + 4x^2 - 4 = x^4$$
$$4x^2 - 4 = 0$$
$$4(x+1)(x-1) = 0$$
$x + 1 = 0$ **or** $x - 1 = 0$
$x = -1$ $\quad$ $x = 1$
solution $\quad$ not a solution

38.
$$\sqrt[4]{8x - 8} + 2 = 0$$
$$\sqrt[4]{8x - 8} = -2$$
$$\left(\sqrt[4]{8x - 8}\right)^4 = (-2)^4$$
$$8x - 8 = 16$$
$$8x = 24$$
$$x = 3$$
The answer does not check. $\Rightarrow$ no solution

39.
$$\begin{aligned} \sqrt[4]{12t+4}+2&=0\\ \sqrt[4]{12t+4}&=-2\\ \left(\sqrt[4]{12t+4}\right)^4&=(-2)^4\\ 12t+4&=16\\ 12t&=12\\ t&=1 \end{aligned}$$
The answer does not check. $\Rightarrow$ no solution

40.
$$\begin{aligned} u&=\sqrt[4]{u^4-6u^2+24}\\ u^4&=\left(\sqrt[4]{u^4-6u^2+24}\right)\\ u^4&=u^4-6u^2+24\\ 6u^2-24&=0\\ 6(u+2)(u-2)&=0 \end{aligned}$$
$u+2=0$ **or** $u-2=0$
$u=-2$ $\qquad u=2$
not a solution $\qquad$ solution

41.
$$\begin{aligned} \sqrt{2y+1}&=1-2\sqrt{y}\\ \left(\sqrt{2y+1}\right)^2&=\left(1-2\sqrt{y}\right)^2\\ 2y+1&=1-4\sqrt{y}+4y\\ 4\sqrt{y}&=2y\\ \left(4\sqrt{y}\right)^2&=(2y)^2\\ 16y&=4y^2\\ 0&=4y^2-16y\\ 0&=4y(y-4) \end{aligned}$$
$4y=0$ **or** $y-4=0$
$y=0$ $\qquad y=4$
solution $\qquad$ not a solution

42.
$$\begin{aligned} \sqrt{u}+3&=\sqrt{u-3}\\ \left(\sqrt{u}+3\right)^2&=\left(\sqrt{u-3}\right)^2\\ u+6\sqrt{u}+9&=u-3\\ 6\sqrt{u}&=-12\\ \left(6\sqrt{u}\right)^2&=(-12)^2\\ 36u&=144\\ u&=4 \end{aligned}$$
The answer does not check. $\Rightarrow$ no solution

43.
$$\begin{aligned} \sqrt{y+7}+3&=\sqrt{y+4}\\ \left(\sqrt{y+7}+3\right)^2&=\left(\sqrt{y+4}\right)^2\\ y+7+6\sqrt{y+7}+9&=y+4\\ 6\sqrt{y+7}&=-12\\ \left(6\sqrt{y+7}\right)^2&=(-12)^2\\ 36(y+7)&=144\\ y+7&=4\\ y&=-3 \end{aligned}$$
The answer does not check. $\Rightarrow$ no solution

44.
$$\begin{aligned} 1+\sqrt{z}&=\sqrt{z+3}\\ \left(1+\sqrt{z}\right)^2&=\left(\sqrt{z+3}\right)^2\\ 1+2\sqrt{z}+z&=z+3\\ 2\sqrt{z}&=2\\ \left(2\sqrt{z}\right)^2&=2^2\\ 4z&=4\\ z&=1 \end{aligned}$$
The answer checks.

45.
$$\sqrt{v}+\sqrt{3}=\sqrt{v+3}$$
$$\left(\sqrt{v}+\sqrt{3}\right)^2=\left(\sqrt{v+3}\right)^2$$
$$v+2\sqrt{3v}+3=v+3$$
$$2\sqrt{3v}=0$$
$$\left(2\sqrt{3v}\right)^2=0^2$$
$$12v=0$$
$$v=0$$
The answer checks.

46.
$$\sqrt{x}+2=\sqrt{x+4}$$
$$\left(\sqrt{x}+2\right)^2=\left(\sqrt{x+4}\right)^2$$
$$x+4\sqrt{x}+4=x+4$$
$$4\sqrt{x}=0$$
$$\left(4\sqrt{x}\right)^2=0^2$$
$$16x=0$$
$$x=0$$
The answer checks.

47.
$$2+\sqrt{u}=\sqrt{2u+7}$$
$$\left(2+\sqrt{u}\right)^2=\left(\sqrt{2u+7}\right)^2$$
$$4+4\sqrt{u}+u=2u+7$$
$$4\sqrt{u}=u+3$$
$$\left(4\sqrt{u}\right)^2=(u+3)^2$$
$$16u=u^2+6u+9$$
$$0=u^2-10u+9$$
$$0=(u-9)(u-1)$$
$u-9=0$ **or** $u-1=0$

$u=9$ $\qquad$ $u=1$

solution $\qquad$ solution

48.
$$5r+4=\sqrt{5r+20}+4r$$
$$r+4=\sqrt{5r+20}$$
$$(r+4)^2=\left(\sqrt{5r+20}\right)^2$$
$$r^2+8r+16=5r+20$$
$$r^2+3r-4=0$$
$$(r+4)(r-1)=0$$
$r+4=0$ **or** $r-1=0$

$r=-4$ $\qquad$ $r=1$

solution $\qquad$ solution

49.
$$\sqrt{6t+1}-3\sqrt{t}=-1$$
$$\sqrt{6t+1}=3\sqrt{t}-1$$
$$\left(\sqrt{6t+1}\right)^2=\left(3\sqrt{t}-1\right)^2$$
$$6t+1=9t-6\sqrt{t}+1$$
$$6\sqrt{t}=3t$$
$$\left(6\sqrt{t}\right)^2=(3t)^2$$
$$36t=9t^2$$
$$0=9t^2-36t$$
$$0=9t(t-4)$$
$9t=0$ **or** $t-4=0$

$t=0$ $\qquad$ $t=4$

not a solution $\qquad$ solution

50.
$$\sqrt{4s+1}-\sqrt{6s}=-1$$
$$\sqrt{4s+1}=\sqrt{6s}-1$$
$$\left(\sqrt{4s+1}\right)^2=\left(\sqrt{6s}-1\right)^2$$
$$4s+1=6s-2\sqrt{6s}+1$$
$$2\sqrt{6s}=2s$$
$$\left(2\sqrt{6s}\right)^2=(2s)^2$$
$$24s=4s^2$$
$$0=4s^2-24s$$
$$0=4s(s-6)$$
$4s=0$ **or** $s-6=0$

$s=0$ $\qquad$ $s=6$

not a solution $\qquad$ solution

51.
$$\begin{aligned}
\sqrt{2x+5}+\sqrt{x+2} &= 5\\
\sqrt{2x+5} &= 5-\sqrt{x+2}\\
\left(\sqrt{2x+5}\right)^2 &= \left(5-\sqrt{x+2}\right)^2\\
2x+5 &= 25-10\sqrt{x+2}+x+2\\
10\sqrt{x+2} &= -x+22\\
\left(10\sqrt{x+2}\right)^2 &= (-x+22)^2\\
100(x+2) &= x^2-44x+484\\
0 &= x^2-144x+284\\
0 &= (x-142)(x-2)
\end{aligned}$$

$x-142=0$ **or** $x-2=0$

$x=142$ $\quad$ $x=2$

not a solution $\quad$ solution

52.
$$\begin{aligned}
\sqrt{2x+5}+\sqrt{2x+1}+4 &= 0\\
\sqrt{2x+5} &= -4-\sqrt{2x+1}\\
\left(\sqrt{2x+5}\right)^2 &= \left(-4-\sqrt{2x+1}\right)^2\\
2x+5 &= 16+8\sqrt{2x+1}+2x+1\\
-12 &= 8\sqrt{2x+1}\\
(-12)^2 &= \left(8\sqrt{2x+1}\right)^2
\end{aligned}$$

$$\begin{aligned}
(-12)^2 &= \left(8\sqrt{2x+1}\right)^2\\
144 &= 64(2x+1)\\
144 &= 128x+64\\
80 &= 128x\\
x &= \frac{80}{128}=\frac{5}{8}
\end{aligned}$$

The answer does not check.
no solution

53.
$$\begin{aligned}
\sqrt{z-1}+\sqrt{z+2} &= 3\\
\sqrt{z-1} &= 3-\sqrt{z+2}\\
\left(\sqrt{z-1}\right)^2 &= \left(3-\sqrt{z+2}\right)^2\\
z-1 &= 9-6\sqrt{z+2}+z+2\\
6\sqrt{z+2} &= 12\\
\sqrt{z+2} &= 2\\
\left(\sqrt{z+2}\right)^2 &= 2^2\\
z+2 &= 4
\end{aligned}$$

$z=2$: The answer checks.

54. $\sqrt{16v+1}+\sqrt{8v+1}=12$

$$\begin{aligned}
\sqrt{16v+1} &= 12-\sqrt{8v+1}\\
\left(\sqrt{16v+1}\right)^2 &= \left(12-\sqrt{8v+1}\right)^2\\
16v+1 &= 144-24\sqrt{8v+1}+8v+1\\
24\sqrt{8v+1} &= -8v+144\\
3\sqrt{8v+1} &= -v+18\\
\left(3\sqrt{8v+1}\right)^2 &= (-v+18)^2\\
9(8v+1) &= v^2-36v+324\\
0 &= v^2-108v+315\\
0 &= (v-3)(v-105)
\end{aligned}$$

$v-3=0$ **or** $v-105=0$

$v=3$ $\quad$ $v=105$

solution $\quad$ not a solution

55. $\sqrt{x-5}-\sqrt{x+3}=4$

$$\begin{aligned}
\sqrt{x-5} &= \sqrt{x+3}+4\\
\left(\sqrt{x-5}\right)^2 &= \left(\sqrt{x+3}+4\right)^2\\
x-5 &= x+3+8\sqrt{x+3}+16\\
-24 &= 8\sqrt{x+3}\\
-3 &= \sqrt{x+3}\\
(-3)^2 &= \left(\sqrt{x+3}\right)^2\\
9 &= x+3\\
6 &= x
\end{aligned}$$

The answer does not check. ⇒ no solution

56. $\sqrt{x+8}-\sqrt{x-4}=-2$

$$\begin{aligned}
\sqrt{x+8} &= \sqrt{x-4}-2\\
\left(\sqrt{x+8}\right)^2 &= \left(\sqrt{x-4}-2\right)^2\\
x+8 &= x-4-4\sqrt{x-4}+4\\
4\sqrt{x-4} &= -8\\
\sqrt{x-4} &= -2\\
\left(\sqrt{x-4}\right)^2 &= (-2)^2\\
x-4 &= 4\\
x &= 8
\end{aligned}$$

The answer does not check. ⇒ no solution

57.

$$\begin{aligned}
\sqrt{x+1}+\sqrt{3x} &= \sqrt{5x+1} \\
\left(\sqrt{x+1}+\sqrt{3x}\right)^2 &= \left(\sqrt{5x+1}\right)^2 \\
x+1+2\sqrt{3x(x+1)}+3x &= 5x+1 \\
2\sqrt{3x^2+3x} &= x \\
\left(2\sqrt{3x^2+3x}\right)^2 &= x^2 \\
12x^2+12x &= x^2 \\
11x^2+12x &= 0 \\
x(11x+12) &= 0
\end{aligned}$$

$x = 0$ **or** $11x+12=0$

$x = -\frac{12}{11}$

solution — not a solution

58.

$$\begin{aligned}
\sqrt{3x}-\sqrt{x+1} &= \sqrt{x-2} \\
\left(\sqrt{3x}-\sqrt{x+1}\right)^2 &= \left(\sqrt{x-2}\right)^2 \\
3x-2\sqrt{3x(x+1)}+x+1 &= x-2 \\
3x+3 &= 2\sqrt{3x^2+3x} \\
(3x+3)^2 &= \left(2\sqrt{3x^2+3x}\right)^2 \\
9x^2+18x+9 &= 12x^2+12x \\
0 &= 3x^2-6x-9 \\
0 &= 3(x-3)(x+1)
\end{aligned}$$

$x-3=0$ **or** $x+1=0$

$x=3$ — $x=-1$

solution — not a solution

59.

$$\begin{aligned}
\sqrt{\sqrt{a}+\sqrt{a+8}} &= 2 \\
\left(\sqrt{\sqrt{a}+\sqrt{a+8}}\right)^2 &= 2^2 \\
\sqrt{a}+\sqrt{a+8} &= 4 \\
\left(\sqrt{a}+\sqrt{a+8}\right)^2 &= 4^2 \\
a+2\sqrt{a(a+8)}+a+8 &= 16 \\
2\sqrt{a^2+8a} &= -2a+8 \\
\left(2\sqrt{a^2+8a}\right)^2 &= (-2a+8)^2 \\
4a^2+32a &= 4a^2-32a+64 \\
64a &= 64
\end{aligned}$$

$a = 1$: The answer checks.

60.

$$\sqrt{\sqrt{2y}-\sqrt{y-1}}=1$$
$$\left(\sqrt{\sqrt{2y}-\sqrt{y-1}}\right)^2=1^2$$
$$\sqrt{2y}-\sqrt{y-1}=1$$
$$\left(\sqrt{2y}-\sqrt{y-1}\right)^2=1^2$$
$$2y-2\sqrt{2y(y-1)}+y-1=1$$
$$3y-2=2\sqrt{2y^2-2y}$$
$$(3y-2)^2=\left(2\sqrt{2y^2-2y}\right)^2$$
$$9y^2-12y+4=8y^2-8y$$
$$y^2-4y+4=0$$
$$(y-2)(y-2)=0$$
$y=2$: The answer checks.

61.

$$\frac{6}{\sqrt{x+5}}=\sqrt{x}$$
$$\left(\frac{6}{\sqrt{x+5}}\right)^2=\left(\sqrt{x}\right)^2$$
$$\frac{36}{x+5}=x$$
$$36=x^2+5x$$
$$0=x^2+5x-36$$
$$0=(x+9)(x-4)$$
$x+9=0$ **or** $x-4=0$

$x=-9$ not a solution; $x=4$ solution

62.

$$\frac{\sqrt{2x}}{\sqrt{x+2}}=\sqrt{x-1}$$
$$\left(\frac{\sqrt{2x}}{\sqrt{x+2}}\right)^2=\left(\sqrt{x-1}\right)^2$$
$$\frac{2x}{x+2}=x-1$$
$$2x=x^2+x-2$$
$$0=x^2-x-2$$
$$0=(x-2)(x+1)$$
$x-2=0$ **or** $x+1=0$

$x=2$ solution; $x=-1$ not a solution

63.

$$\sqrt{x+2}+\sqrt{2x-3}=\sqrt{11-x}$$
$$\left(\sqrt{x+2}+\sqrt{2x-3}\right)^2=\left(\sqrt{11-x}\right)^2$$
$$x+2+2\sqrt{(x+2)(2x-3)}+2x-3=11-x$$
$$2\sqrt{2x^2+x-6}=-4x+12$$
$$\left(2\sqrt{2x^2+x-6}\right)^2=(-4x+12)^2$$
$$8x^2+4x-24=16x^2-96x+144$$
$$0=8x^2-100x+168$$
$$0=4(2x-21)(x-2)$$
$2x-21=0$ **or** $x-2=0$

$x=\frac{21}{2}$ not a solution; $x=2$ solution

64.

$$\sqrt{8-x}-\sqrt{3x-8}=\sqrt{x-4}$$
$$\left(\sqrt{8-x}-\sqrt{3x-8}\right)^2=\left(\sqrt{x-4}\right)^2$$
$$8-x-2\sqrt{(8-x)(3x-8)}+3x-8=x-4$$
$$x+4=2\sqrt{-3x^2+32x-64}$$
$$(x+4)^2=\left(2\sqrt{-3x^2+32x-64}\right)^2$$
$$x^2+8x+16=-12x^2+128x-256$$
$$13x^2-120x+272=0$$
$$(13x-68)(x-4)=0$$

$13x-68=0$ **or** $x-4=0$

$x=\frac{68}{13}$ $\quad$ $x=4$

not a solution $\quad$ solution

65.

$$v=\sqrt{2gh}$$
$$v^2=\left(\sqrt{2gh}\right)^2$$
$$v^2=2gh$$
$$\frac{v^2}{2g}=\frac{2gh}{2g}$$
$$\frac{v^2}{2g}=h\text{, or }h=\frac{v^2}{2g}$$

66.

$$d=1.4\sqrt{h}$$
$$d^2=\left(1.4\sqrt{h}\right)^2$$
$$d^2=1.96h$$
$$\frac{d^2}{1.96}=\frac{1.96h}{1.96}$$
$$\frac{d^2}{1.96}=h\text{, or }h=\frac{d^2}{1.96}$$

67.

$$T=2\pi\sqrt{\frac{l}{32}}$$
$$T^2=\left(2\pi\sqrt{\frac{l}{32}}\right)^2$$
$$T^2=4\pi^2\cdot\frac{l}{32}$$
$$32(T^2)=32\left(4\pi^2\cdot\frac{l}{32}\right)$$
$$32T^2=4\pi^2 l$$
$$\frac{32T^2}{4\pi^2}=\frac{4\pi^2 l}{4\pi^2}$$
$$\frac{8T^2}{\pi^2}=l\text{, or }l=\frac{8T^2}{\pi^2}$$

68.

$$d=\sqrt[3]{\frac{12V}{\pi}}$$
$$d^3=\left(\sqrt[3]{\frac{12V}{\pi}}\right)^3$$
$$d^3=\frac{12V}{\pi}$$
$$\pi(d^3)=\pi\left(\frac{12V}{\pi}\right)$$
$$\pi d^3=12V$$
$$\frac{\pi d^3}{12}=\frac{12V}{12}$$
$$\frac{\pi d^3}{12}=V\text{, or }V=\frac{\pi d^3}{12}$$

69.

$$\begin{aligned} r &= \sqrt[3]{\frac{A}{P}} - 1 \\ r+1 &= \sqrt[3]{\frac{A}{P}} \\ (r+1)^3 &= \left(\sqrt[3]{\frac{A}{P}}\right)^3 \\ (r+1)^3 &= \frac{A}{P} \\ P(r+1)^3 &= P\left(\frac{A}{P}\right) \\ P(r+1)^3 &= A, \text{ or } A = P(r+1)^3 \end{aligned}$$

70.

$$\begin{aligned} r &= \sqrt[3]{\frac{A}{P}} - 1 \\ r+1 &= \sqrt[3]{\frac{A}{P}} \\ (r+1)^3 &= \left(\sqrt[3]{\frac{A}{P}}\right)^3 \\ (r+1)^3 &= \frac{A}{P} \\ P(r+1)^3 &= P\left(\frac{A}{P}\right) \\ P(r+1)^3 &= A \\ P &= \frac{A}{(r+1)^3} \end{aligned}$$

71.

$$\begin{aligned} L_A &= L_B\sqrt{1 - \frac{v^2}{c^2}} \\ L_A^2 &= \left(L_B\sqrt{1 - \frac{v^2}{c^2}}\right)^2 \\ L_A^2 &= L_B^2\left(1 - \frac{v^2}{c^2}\right) \\ \frac{L_A^2}{L_B^2} &= 1 - \frac{v^2}{c^2} \\ \frac{v^2}{c^2} &= 1 - \frac{L_A^2}{L_B^2} \\ c^2\left(\frac{v^2}{c^2}\right) &= c^2\left(1 - \frac{L_A^2}{L_B^2}\right) \\ v^2 &= c^2\left(1 - \frac{L_A^2}{L_B^2}\right) \end{aligned}$$

72.

$$\begin{aligned} R_1 &= \sqrt{\frac{A}{\pi} - R_2^2} \\ R_1^2 &= \left(\sqrt{\frac{A}{\pi} - R_2^2}\right)^2 \\ R_1^2 &= \frac{A}{\pi} - R_2^2 \\ R_1^2 + R_2^2 &= \frac{A}{\pi} \\ \pi\left(R_1^2 + R_2^2\right) &= \pi\left(\frac{A}{\pi}\right) \\ \pi R_1^2 + \pi R_2^2 &= A, \text{ or } A = \pi R_1^2 + \pi R_2^2 \end{aligned}$$

73.

$$\begin{aligned} s &= 1.45\sqrt{r} \\ 65 &= 1.45\sqrt{r} \\ 65^2 &= \left(1.45\sqrt{r}\right)^2 \\ 4225 &= 2.1025r \\ 2010 \text{ ft} &\approx r \end{aligned}$$

74.

$$\begin{aligned} d &= 1.4\sqrt{h} \\ 25 &= 1.4\sqrt{h} \\ 25^2 &= \left(1.4\sqrt{h}\right)^2 \\ 625 &= 1.96h \\ 319 \text{ ft} &\approx h \end{aligned}$$

75.
$$v = \sqrt[3]{\frac{P}{0.02}}$$
$$v = \sqrt[3]{\frac{500}{0.02}}$$
$$v = \sqrt[3]{25000}$$
$$v \approx 29 \text{ mph}$$

76.
$$l = \sqrt{f^2 + h^2}$$
$$10 = \sqrt{f^2 + 6^2}$$
$$10^2 = \left(\sqrt{f^2 + 36}\right)^2$$
$$100 = f^2 + 36$$
$$64 = f^2$$
$$8 \text{ ft} = f$$

77. $r = 1 - \sqrt[n]{\frac{T}{C}} = 1 - \sqrt[5]{\frac{9000}{22000}} \approx 1 - \sqrt[5]{0.40909} \approx 1 - 0.836 \approx 0.164 \approx 16\%$

78. $r = \sqrt[n]{\frac{V}{P}} - 1 = \sqrt[5]{\frac{1338.23}{1000}} - 1 = \sqrt[5]{1.33923} - 1 \approx 1.060 - 1 \approx 0.060 \approx 6\%$

79. Graph $y = \sqrt{5x}$ and $y = \sqrt{100 - 3x^2}$:

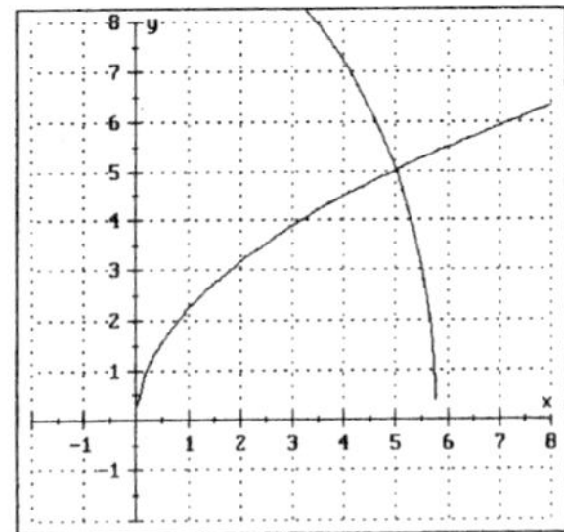

equilibrium price: $x = \$5$

80. Graph $y = \sqrt{23x}$ and $y = \sqrt{312 - 2x^2}$:

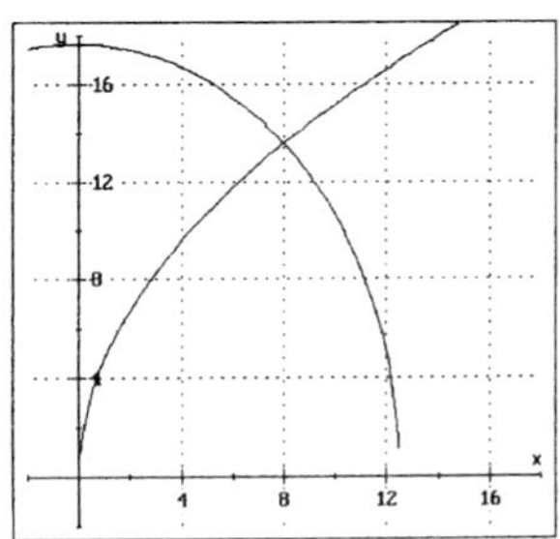

equilibrium price: $x = \$8$

81.
$$r = \sqrt[4]{\frac{8kl}{\pi R}}$$
$$r^4 = \left(\sqrt[4]{\frac{8kl}{\pi R}}\right)^4$$
$$r^4 = \frac{8kl}{\pi R}$$
$$\pi R r^4 = 8kl$$
$$R = \frac{8kl}{\pi r^4}$$

82.
$$s = \sqrt[3]{\frac{P}{0.02}}$$
$$s^3 = \left(\sqrt[3]{\frac{P}{0.02}}\right)^3$$
$$s^3 = \frac{P}{0.02}$$
$$0.02s^3 = P$$

83. **Answers may vary.**

84. **Answers may vary.**

85.
$$\sqrt[3]{2x} = \sqrt{x}$$
$$\left(\sqrt[3]{2x}\right)^2 = \left(\sqrt{x}\right)^2$$
$$\left[\left(\sqrt[3]{2x}\right)^2\right]^3 = x^3$$
$$4x^2 = x^3$$
$$0 = x^3 - 4x^2$$
$$0 = x^2(x-4)$$
$x = 0$ or $x = 4$

86.
$$\sqrt[4]{x} = \sqrt{\frac{x}{4}}$$
$$\left(\sqrt[4]{x}\right)^2 = \left(\sqrt{\frac{x}{4}}\right)^2$$
$$\left[\left(\sqrt[4]{x}\right)^2\right]^4 = \left(\frac{x}{4}\right)^4$$
$$x^2 = \frac{x^4}{256}$$
$$0 = x^4 - 256x^2$$
$$0 = x^2(x^2 - 256)$$
$$0 = x^2(x+16)(x-16)$$
$x = 0$ or $x = 16$ ($x = -16$ does not check.)

Exercise 7.7 (page 498)

1. $\frac{x^2 - x - 6}{9 - x^2} \cdot \frac{x^2 + x - 6}{x^2 - 4} = \frac{(x-3)(x+2)}{(3+x)(3-x)} \cdot \frac{(x+3)(x-2)}{(x+2)(x-2)} = -1$

2.
$$\frac{3x+4}{x-2} + \frac{x-4}{x+2} = \frac{(3x+4)(x+2)}{(x-2)(x+2)} + \frac{(x-4)(x-2)}{(x+2)(x-2)}$$
$$= \frac{3x^2 + 6x + 4x + 8 + x^2 - 2x - 4x + 8}{(x+2)(x-2)} = \frac{4x^2 + 4x + 16}{(x+2)(x-2)}$$

3. Let w = the speed of the wind.

	Rate	Time	Dist.
Downwind	$200 + w$	$\frac{330}{200+w}$	330
Upwind	$200 - w$	$\frac{330}{200-w}$	330

$$\frac{330}{200+w} + \frac{330}{200-w} = \frac{10}{3}$$
$$\left(\frac{330}{200+w} + \frac{330}{200-w}\right)3(200+w)(200-w) = \frac{10}{3} \cdot 3(200+w)(200-w)$$
$$330(3)(200-w) + 330(3)(200+w) = 10(40{,}000 - w^2)$$
$$198{,}000 - 990w + 198{,}000 + 990w = 400{,}000 - 10w^2$$
$$10w^2 = 4000$$
$$w^2 = 400$$
$$w = 20 \text{ mph}$$

4. Let $r =$ the increase in the rate.

	Rate	Time	Dist.
First rate	50	$\frac{135}{50} = \frac{27}{10}$	135
Faster rate	$50 + r$	$\frac{135}{50+r}$	135

$$\frac{135}{50+r} = \frac{27}{10} - \frac{1}{2}$$
$$\frac{135}{50+r} = \frac{22}{10}$$
$$1350 = 22(50+r)$$
$$1350 = 1100 + 22r$$
$$250 = 22r$$
$$r = \tfrac{250}{22} = 11.4 \text{ mph faster}$$

5. imaginary

6. i

7. -1

8. $-i$

9. 1

10. $\sqrt{a}\sqrt{b}$

11. $\dfrac{\sqrt{a}}{\sqrt{b}}$

12. complex

13. 5; 7

14. c; d

15. conjugates

16. $\sqrt{a^2 + b^2}$

17. $\sqrt{-9} = \sqrt{-1 \cdot 9} = \sqrt{i^2 \cdot 3^2} = 3i$

18. $\sqrt{-16} = \sqrt{-1 \cdot 16} = \sqrt{i^2 \cdot 4^2} = 4i$

19. $\sqrt{-36} = \sqrt{-1 \cdot 36} = \sqrt{i^2 \cdot 6^2} = 6i$

20. $\sqrt{-81} = \sqrt{-1 \cdot 81} = \sqrt{i^2 \cdot 9^2} = 9i$

21. $\sqrt{-7} = \sqrt{-1 \cdot 7} = \sqrt{i^2 \cdot 7} = i\sqrt{7}$

22. $\sqrt{-11} = \sqrt{-1 \cdot 11} = \sqrt{i^2 \cdot 11} = i\sqrt{11}$

23. $3 + 7i \stackrel{?}{=} \sqrt{9} + (5+2)i$
$3 + 7i \stackrel{?}{=} 3 + 7i$
They are equal.

24. $\sqrt{4} + \sqrt{25}i \stackrel{?}{=} 2 - (-5)i$
$2 + 5i \stackrel{?}{=} 2 + 5i$
They are equal.

25. $8 + 5i \stackrel{?}{=} 2^3 + \sqrt{25}i^3$
$8 + 5i \stackrel{?}{=} 8 + 5(-i)$
$8 + 5i \stackrel{?}{=} 8 - 5i$
They are not equal.

26. $4 - 7i \stackrel{?}{=} -4i^2 + 7i^3$
$4 - 7i \stackrel{?}{=} -4(-1) + 7(-i)$
$4 - 7i \stackrel{?}{=} 4 - 7i$
They are equal.

27. $\sqrt{4} + \sqrt{-4} \stackrel{?}{=} 2 - 2i$
$2 + \sqrt{-1 \cdot 4} \stackrel{?}{=} 2 - 2i$
$2 + 2i = 2 - 2i$
They are not equal.

28. $\sqrt{-9} - i \stackrel{?}{=} 4i$
$\sqrt{-1 \cdot 9} - i \stackrel{?}{=} 4i$
$3i - i \stackrel{?}{=} 4i$
$2i \stackrel{?}{=} 4i$
They are not equal.

29. $(3+4i)+(5-6i) = 3+4i+5-6i$
$= 8-2i$

30. $(5+3i)-(6-9i) = 5+3i-6+9i$
$= -1+12i$

31. $(7-3i)-(4+2i) = 7-3i-4-2i$
$= 3-5i$

32. $(8+3i)+(-7-2i) = 8+3i-7-2i$
$= 1+i$

33. $(8+5i)+(7+2i) = 8+5i+7+2i$
$= 15+7i$

34. $(-7+9i)-(-2-8i) = -7+9i+2+8i$
$= -5+17i$

35. $(1+i)-2i+(5-7i) = 1+i-2i+5-7i = 6-8i$

36. $(-9+i)-5i+(2+7i) = -9+i-5i+2+7i = -7+3i$

37. $(5+3i)-(3-5i)+\sqrt{-1} = 5+3i-3+5i+i = 2+9i$

38. $(8+7i)-(-7-\sqrt{-64})+(3-i) = 8+7i+7+8i+3-i = 18+14i$

39. $(-8-\sqrt{3}i)-(7-3\sqrt{3}i) = -8-\sqrt{3}i-7+3\sqrt{3}i = -15+2\sqrt{3}i$

40. $(2+2\sqrt{2}i)+(-3-\sqrt{2}i) = 2+2\sqrt{2}i-3-\sqrt{2}i = -1+\sqrt{2}i$

41. $3i(2-i) = 6i-3i^2 = 6i-3(-1) = 6i+3 = 3+6i$

42. $-4i(3+4i) = -12i-16i^2 = -12i-16(-1) = -12i+16 = 16-12i$

43. $-5i(5-5i) = -25i+25i^2 = -25i+25(-1) = -25i-25 = -25-25i$

44. $2i(7+2i) = 14i+4i^2 = 14i+4(-1) = 14i-4 = -4+14i$

45. $(2+i)(3-i) = 6-2i+3i-i^2 = 6+i-(-1) = 6+i+1 = 7+i$

46. $(4-i)(2+i) = 8+4i-2i-i^2 = 8+2i-(-1) = 8+2i+1 = 9+2i$

47. $(2-4i)(3+2i) = 6+4i-12i-8i^2 = 6-8i-8(-1) = 6-8i+8 = 14-8i$

48. $(3-2i)(4-3i) = 12-9i-8i+6i^2 = 12-17i+6(-1) = 12-17i-6 = 6-17i$

49. $(2+\sqrt{2}i)(3-\sqrt{2}i) = 6-2\sqrt{2}i+3\sqrt{2}i-2i^2 = 6+\sqrt{2}i-2(-1) = 8+\sqrt{2}i$

50. $(5+\sqrt{3}i)(2-\sqrt{3}i) = 10-5\sqrt{3}i+2\sqrt{3}i-3i^2 = 10-3\sqrt{3}i-3(-1) = 13-3\sqrt{3}i$

51. $(8-\sqrt{-1})(-2-\sqrt{-16}) = (8-i)(-2-4i) = -16-32i+2i+4i^2 = -16-30i-4$
$= -20-30i$

52. $(-1+\sqrt{-4})(2+\sqrt{-9}) = (-1+2i)(2+3i) = -2-3i+4i+6i^2 = -2+i-6 = -8+i$

53. $(2+i)^2 = (2+i)(2+i) = 4+2i+2i+i^2 = 4+4i-1 = 3+4i$

54. $(3-2i)^2 = (3-2i)(3-2i) = 9-6i-6i+4i^2 = 9-12i-4 = 5-12i$

55. $(2+3i)^2 = (2+3i)(2+3i) = 4+6i+6i+9i^2 = 4+12i-9 = -5+12i$

56. $(1-3i)^2 = (1-3i)(1-3i) = 1-3i-3i+9i^2 = 1-6i-9 = -8-6i$

57. $i(5+i)(3-2i) = i(15-10i+3i-2i^2) = i(15-7i+2) = i(17-7i) = 17i-7i^2 = 7+17i$

58. $i(-3-2i)(1-2i) = i(-3+6i-2i+4i^2) = i(-3+4i-4) = i(-7+4i) = -7i+4i^2$
$= -4-7i$

59. $(2+i)(2-i)(1+i) = (4-2i+2i-i^2)(1+i) = 5(1+i) = 5+5i$

60. $(3+2i)(3-2i)(i+1) = (9-6i+6i-4i^2)(1+i) = 13(1+i) = 13+13i$

61. $(3+i)[(3-2i)+(2+i)] = (3+i)(5-i) = 15-3i+5i-i^2 = 16+2i$

62. $(2-3i)[(5-2i)-(2i+1)] = (2-3i)(4-4i) = 8-8i-12i+12i^2 = -4-20i$

63. $\dfrac{1}{i} = \dfrac{1i^3}{ii^3} = \dfrac{i^3}{i^4} = \dfrac{i^3}{1} = i^3 = -i = 0-i$

64. $\dfrac{1}{i^3} = \dfrac{i}{i^3 i} = \dfrac{i}{i^4} = \dfrac{i}{1} = i = 0+i$

65. $\dfrac{4}{5i^3} = \dfrac{4i}{5i^3 i} = \dfrac{4i}{5i^4} = \dfrac{4i}{5(1)} = \dfrac{4}{5}i = 0+\dfrac{4}{5}i$

66. $\dfrac{3}{2i} = \dfrac{3i^3}{2ii^3} = \dfrac{3i^3}{2i^4} = \dfrac{3(-i)}{2} = 0-\dfrac{3}{2}i$

67. $\dfrac{3i}{8\sqrt{-9}} = \dfrac{3i}{8(3i)} = \dfrac{1}{8} = \dfrac{1}{8}+0i$

68. $\dfrac{5i^3}{2\sqrt{-4}} = \dfrac{5i^3}{2(2i)} = \dfrac{5i^3}{4i} = \dfrac{5i^2}{4} = -\dfrac{5}{4}+0i$

69. $\dfrac{-3}{5i^5} = \dfrac{-3i^3}{5i^5 i^3} = \dfrac{-3(-i)}{5i^8} = \dfrac{3i}{5} = 0+\dfrac{3}{5}i$

70. $\dfrac{-4}{6i^7} = \dfrac{-4i}{6i^7 i} = \dfrac{-4i}{6i^8} = 0-\dfrac{2}{3}i$

71. $\dfrac{5}{2-i} = \dfrac{5(2+i)}{(2-i)(2+i)} = \dfrac{5(2+i)}{4-i^2} = \dfrac{5(2+i)}{5} = 2+i$

72. $\dfrac{26}{3-2i} = \dfrac{26(3+2i)}{(3-2i)(3+2i)} = \dfrac{26(3+2i)}{9-4i^2} = \dfrac{26(3+2i)}{13} = 2(3+2i) = 6+4i$

73. $\dfrac{13i}{5+i} = \dfrac{13i(5-i)}{(5+i)(5-i)} = \dfrac{65i-13i^2}{25-i^2} = \dfrac{13+65i}{26} = \dfrac{13}{26}+\dfrac{65}{26}i = \dfrac{1}{2}+\dfrac{5}{2}i$

74. $\dfrac{2i}{5+3i} = \dfrac{2i(5-3i)}{(5+3i)(5-3i)} = \dfrac{10i-6i^2}{25-9i^2} = \dfrac{10i+6}{34} = \dfrac{6}{34}+\dfrac{10}{34}i = \dfrac{3}{17}+\dfrac{5}{17}i$

75. $$\frac{-12}{7-\sqrt{-1}}=\frac{-12}{7-i}=\frac{-12(7+i)}{(7-i)(7+i)}=\frac{-84-12i}{49-i^2}=\frac{-84-12i}{50}=\frac{-84}{50}-\frac{12}{50}i=-\frac{42}{25}-\frac{6}{25}i$$

76. $$\frac{4}{3+\sqrt{-1}}=\frac{4}{3+i}=\frac{4(3-i)}{(3+i)(3-i)}=\frac{12-4i}{9-i^2}=\frac{12-4i}{10}=\frac{12}{10}-\frac{4}{10}i=\frac{6}{5}-\frac{2}{5}i$$

77. $$\frac{5i}{6+2i}=\frac{5i(6-2i)}{(6+2i)(6-2i)}=\frac{30i-10i^2}{36-4i^2}=\frac{10+30i}{40}=\frac{10}{40}+\frac{30}{40}i=\frac{1}{4}+\frac{3}{4}i$$

78. $$\frac{-4i}{2-6i}=\frac{-4i(2+6i)}{(2-6i)(2+6i)}=\frac{-8i-24i^2}{4-36i^2}=\frac{24-8i}{40}=\frac{24}{40}-\frac{8}{40}i=\frac{3}{5}-\frac{1}{5}i$$

79. $$\frac{3-2i}{3+2i}=\frac{(3-2i)(3-2i)}{(3+2i)(3-2i)}=\frac{9-6i-6i+4i^2}{9-4i^2}=\frac{5-12i}{13}=\frac{5}{13}-\frac{12}{13}i$$

80. $$\frac{2+3i}{2-3i}=\frac{(2+3i)(2+3i)}{(2-3i)(2+3i)}=\frac{4+6i+6i+9i^2}{4-9i^2}=\frac{-5+12i}{13}=-\frac{5}{13}+\frac{12}{13}i$$

81. $$\frac{3+2i}{3+i}=\frac{(3+2i)(3-i)}{(3+i)(3-i)}=\frac{9-3i+6i-2i^2}{9-i^2}=\frac{11+3i}{10}=\frac{11}{10}+\frac{3}{10}i$$

82. $$\frac{2-5i}{2+5i}=\frac{(2-5i)(2-5i)}{(2+5i)(2-5i)}=\frac{4-10i-10i+25i^2}{4-25i^2}=\frac{-21-20i}{29}=-\frac{21}{29}-\frac{20}{29}i$$

83. $$\frac{\sqrt{5}-\sqrt{3}i}{\sqrt{5}+\sqrt{3}i}=\frac{\left(\sqrt{5}-\sqrt{3}i\right)\left(\sqrt{5}-\sqrt{3}i\right)}{\left(\sqrt{5}+\sqrt{3}i\right)\left(\sqrt{5}-\sqrt{3}i\right)}=\frac{5-\sqrt{15}i-\sqrt{15}i+3i^2}{5-3i^2}=\frac{2-2\sqrt{15}i}{8}$$
$$=\frac{2}{8}-\frac{2\sqrt{15}}{8}i$$
$$=\frac{1}{4}-\frac{\sqrt{15}}{4}i$$

84. $$\frac{\sqrt{3}+\sqrt{2}i}{\sqrt{3}-\sqrt{2}i}=\frac{\left(\sqrt{3}+\sqrt{2}i\right)\left(\sqrt{3}+\sqrt{2}i\right)}{\left(\sqrt{3}-\sqrt{2}i\right)\left(\sqrt{3}+\sqrt{2}i\right)}=\frac{3+\sqrt{6}i+\sqrt{6}i+2i^2}{3-2i^2}=\frac{1+2\sqrt{6}i}{5}=\frac{1}{5}+\frac{2\sqrt{6}}{5}i$$

85. $$\left(\frac{i}{3+2i}\right)^2=\frac{i^2}{(3+2i)^2}=\frac{-1}{(3+2i)(3+2i)}=\frac{-1}{9+12i+4i^2}=\frac{-1}{5+12i}=\frac{-1(5-12i)}{(5+12i)(5-12i)}$$
$$=\frac{-5+12i}{25-144i^2}$$
$$=-\frac{5}{169}+\frac{12}{169}i$$

86. $\left(\frac{5+i}{2+i}\right)^2 = \frac{(5+i)^2}{(2+i)^2} = \frac{(5+i)(5+i)}{(2+i)(2+i)} = \frac{25+10i+i^2}{4+4i+i^2} = \frac{24+10i}{3+4i} = \frac{(24+10i)(3-4i)}{(3+4i)(3-4i)}$

$= \frac{72-96i+30i-40i^2}{9-16i^2}$

$= \frac{112-66i}{25} = \frac{112}{25} - \frac{66}{25}i$

87. $\frac{i(3-i)}{3+i} = \frac{(3i-i^2)(3-i)}{(3+i)(3-i)} = \frac{(1+3i)(3-i)}{9-i^2} = \frac{3-i+9i-3i^2}{10} = \frac{6+8i}{10} = \frac{3}{5} + \frac{4}{5}i$

88. $\frac{5+3i}{i(3-5i)} = \frac{5+3i}{3i-5i^2} = \frac{5+3i}{5+3i} = 1$

89. $\frac{(2-5i)-(5-2i)}{5-i} = \frac{2-5i-5+2i}{5-i} = \frac{-3-3i}{5-i} = \frac{(-3-3i)(5+i)}{(5-i)(5+i)} = \frac{-15-3i-15i-3i^2}{25-i^2}$

$= \frac{-12-18i}{26}$

$= \frac{-12}{26} - \frac{18}{26}i$

$= -\frac{6}{13} - \frac{9}{13}i$

90. $\frac{5i}{(5+2i)+(2+i)} = \frac{5i}{7+3i} = \frac{5i(7-3i)}{(7+3i)(7-3i)} = \frac{35i-15i^2}{49-9i^2} = \frac{15+35i}{58} = \frac{15}{58} + \frac{35}{58}i$

91. $i^{21} = i^{20}i^1 = (i^4)^5 i = 1^5 i = i$

92. $i^{19} = i^{16}i^3 = (i^4)^4 i^3 = 1^4 i^3 = i^3 = -i$

93. $i^{27} = i^{24}i^3 = (i^4)^6 i^3 = 1^6 i^3 = i^3 = -i$

94. $i^{22} = i^{20}i^2 = (i^4)^5 i^2 = 1^5 i^2 = i^2 = -1$

95. $i^{100} = (i^4)^{25} = 1^{25} = 1$

96. $i^{42} = i^{40}i^2 = (i^4)^{10} i^2 = 1^{10} i^2 = i^2 = -1$

97. $i^{97} = i^{96}i^1 = (i^4)^{24} i = 1^{24} i = i$

98. $i^{200} = (i^4)^{50} = 1^{50} = 1$

99. $|6+8i| = \sqrt{6^2+8^2} = \sqrt{36+64} = \sqrt{100} = 10$

100. $|12+5i| = \sqrt{12^2+5^2} = \sqrt{144+25} = \sqrt{169} = 13$

101. $|12-5i| = \sqrt{12^2+(-5)^2} = \sqrt{144+25} = \sqrt{169} = 13$

102. $|3-4i| = \sqrt{3^2+(-4)^2} = \sqrt{9+16} = \sqrt{25} = 5$

103. $|5+7i| = \sqrt{5^2+7^2} = \sqrt{25+49} = \sqrt{74}$

104. $|6-5i| = \sqrt{6^2+(-5)^2} = \sqrt{36+25} = \sqrt{61}$

105. $\left|\frac{3}{5}-\frac{4}{5}i\right|=\sqrt{\left(\frac{3}{5}\right)^2+\left(\frac{4}{5}\right)^2}=\sqrt{\frac{9}{25}+\frac{16}{25}}=\sqrt{\frac{25}{25}}=\sqrt{1}=1$

106. $\left|\frac{5}{13}+\frac{12}{13}i\right|=\sqrt{\left(\frac{5}{13}\right)^2+\left(\frac{12}{13}\right)^2}=\sqrt{\frac{25}{169}+\frac{144}{169}}=\sqrt{\frac{169}{169}}=\sqrt{1}=1$

107.
$$\begin{aligned} x^2-2x+26&=0\\ (1-5i)^2-2(1-5i)+26&=0\\ (1-5i)(1-5i)-2+10i+26&=0\\ 1-10i+25i^2+24+10i&=0\\ 1-10i-25+24+10i&=0\\ 0&=0 \end{aligned}$$

108.
$$\begin{aligned} x^2-6x+13&=0\\ (3-2i)^2-6(3-2i)+13&=0\\ (3-2i)(3-2i)-18+12i+13&=0\\ 9-12i+4i^2-5+12i&=0\\ 9-12i-4-5+12i&=0\\ 0&=0 \end{aligned}$$

109.
$$\begin{aligned} x^4-3x^2-4&=0\\ i^4-3i^2-4&=0\\ 1-3(-1)-4&=0\\ 1+3-4&=0\\ 0&=0 \end{aligned}$$

110.
$$\begin{aligned} x^2+x+1&=0\\ (2+i)^2+(2+i)+1&=0\\ 4+4i+i^2+2+i+1&=0\\ 4+4i-1+2+i+1&=0\\ 6+5i&\neq 0 \end{aligned}$$

111. $V=IR=(2-3i)(2+i)=4+2i-6i-3i^2=4-4i-3(-1)=4-4i+3=7-4i$ volts

112. $V=IR\Rightarrow R=\frac{V}{I}=\frac{18+i}{3-2i}=\frac{(18+i)(3+2i)}{(3-2i)(3+2i)}=\frac{54+39i+2i^2}{9-4i^2}=\frac{52+39i}{13}=4+3i$ ohms

113. $Z=\frac{V}{I}=\frac{1.7+0.5i}{0.5i}=\frac{(1.7+0.5i)i}{(0.5i)i}=\frac{1.7i+0.5i^2}{0.5i^2}=\frac{-0.5+1.7i}{-0.5}=1-3.4i$

114. $Z=\frac{V}{I}=\frac{1.6-0.4i}{-0.2i}=\frac{(1.6-0.4i)i}{(-0.2i)i}=\frac{1.6i-0.4i^2}{-0.2i^2}=\frac{0.4+1.6i}{0.2}=2+8i$

115. Answers may vary.

116. Answers may vary.

117. $\frac{3-i}{2}=\frac{(3-i)(3+i)}{2(3+i)}=\frac{9-i^2}{2(3+i)}=\frac{10}{2(3+i)}=\frac{5}{3+i}$

118. $\frac{2+3i}{2-3i}=\frac{(2+3i)(2-3i)}{(2-3i)(2-3i)}=\frac{4-9i^2}{4-12i+9i^2}=\frac{13}{-5-12i}$

Chapter 7 Summary (page 501)

1. $\sqrt{49}=\sqrt{7^2}=7$

2. $-\sqrt{121}=-\sqrt{11^2}=-11$

3. $-\sqrt{36}=-\sqrt{6^2}=-6$

4. $\sqrt{225}=\sqrt{15^2}=15$

5. $\sqrt[3]{-27}=\sqrt[3]{(-3)^3}=-3$

6. $-\sqrt[3]{216}=-\sqrt[3]{6^3}=-6$

7. $\sqrt[4]{625} = \sqrt[4]{5^4} = 5$

8. $\sqrt[5]{-32} = \sqrt[5]{(-2)^5} = -2$

9. $\sqrt{25x^2} = \sqrt{5^2x^2} = |5x| = 5|x|$

10. $\sqrt{x^2+4x+4} = \sqrt{(x+2)^2} = |x+2|$

11. $\sqrt[3]{27a^6b^3} = 3a^2b$

12. $\sqrt[4]{256x^8y^4} = |4x^2y| = 4x^2|y|$

13. $y = f(x) = \sqrt{x+2}$; Shift $y = \sqrt{x}$ left 2.

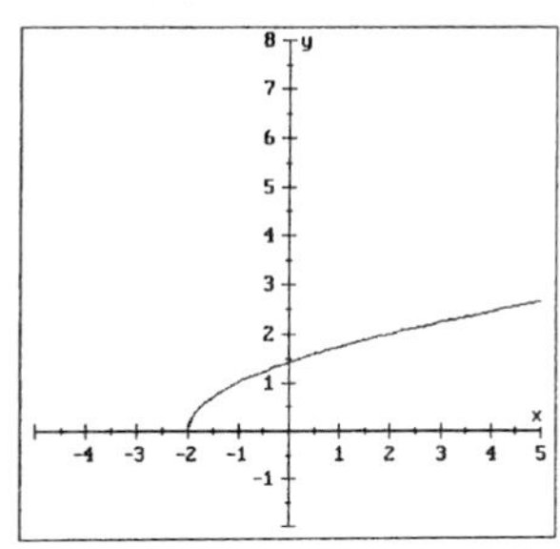

14. $y = f(x) = -\sqrt{x-1}$; Reflect $y = \sqrt{x}$ about the x-axis and shift right 1.

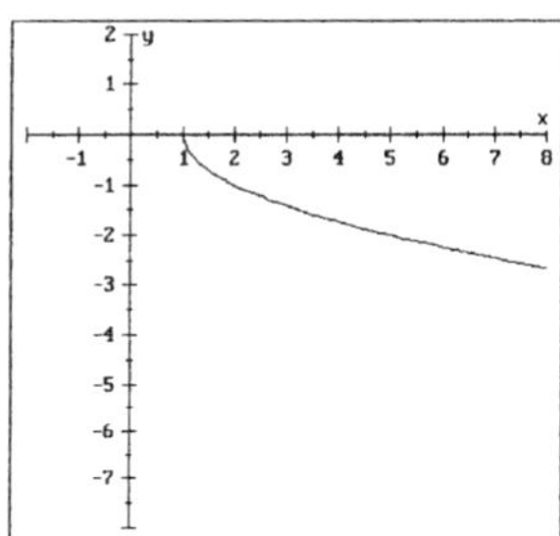

15. $y = f(x) = -\sqrt{x}+2$; Reflect $y = \sqrt{x}$ about the x-axis and shift up 2.

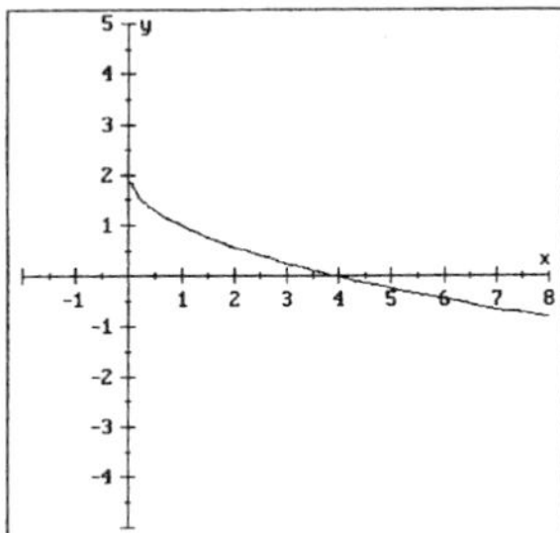

16. $y = f(x) = -\sqrt[3]{x}+3$; Reflect $y = \sqrt[3]{x}$ about the x-axis and shift up 3.

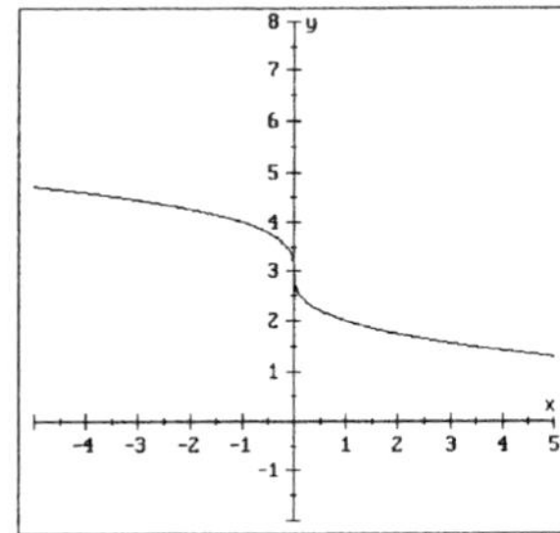

17. mean $= \frac{4+8+12+16+20}{5} = \frac{60}{5} = 12$

18.

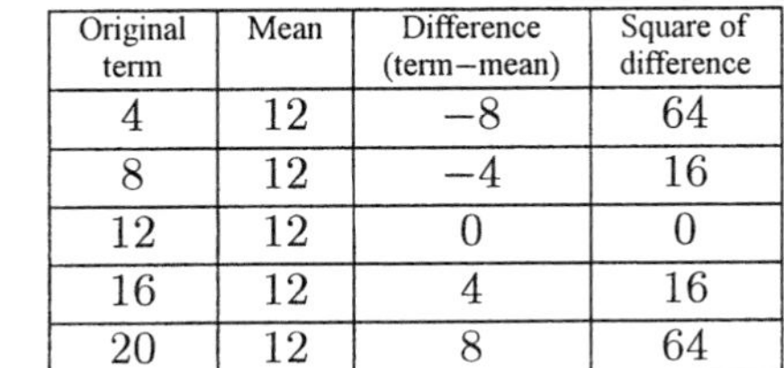

Original term	Mean	Difference (term−mean)	Square of difference
4	12	−8	64
8	12	−4	16
12	12	0	0
16	12	4	16
20	12	8	64

$$\begin{aligned} s &= \sqrt{\frac{64+16+0+16+64}{5}} \\ &= \sqrt{\frac{160}{5}} \\ &= \sqrt{32} \\ &\approx 5.7 \end{aligned}$$

19. $d = 1.4\sqrt{h}$

$d = 1.4\sqrt{4.7} \approx 3.0$ miles

20.

$$\begin{aligned} d &= 1.4\sqrt{h} \\ 4 &= 1.4\sqrt{h} \\ 16 &= 1.96h \\ 8.2\text{ ft} &= h \end{aligned}$$

21. Let x = one-half of d.

$125^2 = x^2 + 117^2$
$15625 = x^2 + 13689$
$1936 = x^2$
$44 = x$
$d = 2x = 2(44) = 88$ yd

22. Let x = one-half of d.

$8900^2 = x^2 + 3900^2$
$79{,}210{,}000 = x^2 + 15{,}210{,}000$
$64{,}000{,}000 = x^2$
$8000 = x$
$d = 2x = 2(8000) = 16{,}000$ yd

23. $d = \sqrt{(x_2 - x_1)^2 + (y_2 - y_1)^2} = \sqrt{(0-5)^2 + [0-(-12)]^2} = \sqrt{(-5)^2 + 12^2} = \sqrt{169} = 13$

24. $d = \sqrt{(x_2 - x_1)^2 + (y_2 - y_1)^2} = \sqrt{(-4-(-2))^2 + (6-8)^2} = \sqrt{(-2)^2 + (-2)^2} = \sqrt{8} \approx 2.83$

25. $25^{1/2} = 5$

26. $-36^{1/2} = -6$

27. $9^{3/2} = \left(9^{1/2}\right)^3 = 3^3 = 27$

28. $16^{3/2} = \left(16^{1/2}\right)^3 = 4^3 = 64$

29. $(-8)^{1/3} = -2$

30. $-8^{2/3} = -(8^{1/3})^2 = -2^2 = -4$

31. $8^{-2/3} = \frac{1}{8^{2/3}} = \frac{1}{(8^{1/3})^2} = \frac{1}{2^2} = \frac{1}{4}$

32. $8^{-1/3} = \frac{1}{8^{1/3}} = \frac{1}{2}$

33. $-49^{5/2} = -(49^{1/2})^5 = -7^5 = -16{,}807$

34. $\frac{1}{25^{5/2}} = \frac{1}{(25^{1/2})^5} = \frac{1}{5^5} = \frac{1}{3125}$

35. $\left(\frac{1}{4}\right)^{-3/2} = 4^{3/2} = (4^{1/2})^3 = 2^3 = 8$

36. $\left(\frac{4}{9}\right)^{-3/2} = \left(\frac{9}{4}\right)^{3/2} = \left[\left(\frac{9}{4}\right)^{1/2}\right]^3 = \left(\frac{3}{2}\right)^3 = \frac{27}{8}$

37. $(27x^3y)^{1/3} = 3xy^{1/3}$

38. $(81x^4y^2)^{1/4} = 3xy^{1/2}$

39. $(25x^3y^4)^{3/2} = 125x^{9/2}y^6$

40. $(8u^2v^3)^{-2/3} = \frac{1}{4u^{4/3}v^2}$

41. $5^{1/4}5^{1/2} = 5^{1/4+1/2} = 5^{3/4}$

42. $a^{3/7}a^{2/7} = a^{3/7+2/7} = a^{5/7}$

43. $u^{1/2}\left(u^{1/2} - u^{-1/2}\right) = u^{2/2} - u^0 = u - 1$

44. $v^{2/3}\left(v^{1/3} + v^{4/3}\right) = v^{3/3} + v^{6/3} = v + v^2$

45. $\left(x^{1/2} + y^{1/2}\right)^2 = \left(x^{1/2} + y^{1/2}\right)\left(x^{1/2} + y^{1/2}\right) = x^{2/2} + x^{1/2}y^{1/2} + x^{1/2}y^{1/2} + y^{2/2}$
$= x + 2x^{1/2}y^{1/2} + y$

46. $\left(a^{2/3} + b^{2/3}\right)\left(a^{2/3} - b^{2/3}\right) = a^{4/3} - a^{2/3}b^{2/3} + a^{2/3}b^{2/3} - b^{4/3} = a^{4/3} - b^{4/3}$

47. $\sqrt[6]{5^2} = 5^{2/6} = 5^{1/3} = \sqrt[3]{5}$

48. $\sqrt[8]{x^4} = x^{4/8} = x^{1/2} = \sqrt{x}$

49. $\sqrt[9]{27a^3b^6} = \left(3^3a^3b^6\right)^{1/9} = 3^{3/9}a^{3/9}b^{6/9} = 3^{1/3}a^{1/3}b^{2/3} = \sqrt[3]{3ab^2}$

50. $\sqrt[4]{25a^2b^2} = 5^{2/4}a^{2/4}b^{2/4} = 5^{1/2}a^{1/2}b^{1/2} = \sqrt{5ab}$

51. $\sqrt{240} = \sqrt{16 \cdot 15} = \sqrt{16}\sqrt{15} = 4\sqrt{15}$

52. $\sqrt[3]{54} = \sqrt[3]{27 \cdot 2} = \sqrt[3]{27}\sqrt[3]{2} = 3\sqrt[3]{2}$

53. $\sqrt[4]{32} = \sqrt[4]{16 \cdot 2} = \sqrt[4]{16}\sqrt[4]{2} = 2\sqrt[4]{2}$

54. $\sqrt[5]{96} = \sqrt[5]{32 \cdot 3} = \sqrt[5]{32}\sqrt[5]{3} = 2\sqrt[5]{3}$

55. $\sqrt{8x^3} = \sqrt{4x^2}\sqrt{2x} = 2x\sqrt{2x}$

56. $\sqrt{18x^4y^3} = \sqrt{9x^4y^2}\sqrt{2y} = 3x^2y\sqrt{2y}$

57. $\sqrt[3]{16x^5y^4} = \sqrt[3]{8x^3y^3}\sqrt[3]{2x^2y} = 2xy\sqrt[3]{2x^2y}$

58. $\sqrt[3]{54x^7y^3} = \sqrt[3]{27x^6y^3}\sqrt[3]{2x} = 3x^2y\sqrt[3]{2x}$

59. $\dfrac{\sqrt{32x^3}}{\sqrt{2x}} = \sqrt{\dfrac{32x^3}{2x}} = \sqrt{16x^2} = 4x$

60. $\dfrac{\sqrt[3]{16x^5}}{\sqrt[3]{2x^2}} = \sqrt[3]{\dfrac{16x^5}{2x^2}} = \sqrt[3]{8x^3} = 2x$

61. $\sqrt[3]{\dfrac{2a^2b}{27x^3}} = \dfrac{\sqrt[3]{2a^2b}}{3x}$

62. $\sqrt{\dfrac{17xy}{64a^4}} = \dfrac{\sqrt{17xy}}{8a^2}$

63. $\sqrt{2} + \sqrt{8} = \sqrt{2} + \sqrt{4}\sqrt{2} = \sqrt{2} + 2\sqrt{2}$
$= 3\sqrt{2}$

64. $\sqrt{20} - \sqrt{5} = \sqrt{4}\sqrt{5} - \sqrt{5} = 2\sqrt{5} - \sqrt{5}$
$= \sqrt{5}$

65. $2\sqrt[3]{3} - \sqrt[3]{24} = 2\sqrt[3]{3} - \sqrt[3]{8}\sqrt[3]{3} = 2\sqrt[3]{3} - 2\sqrt[3]{3} = 0$

66. $\sqrt[4]{32} + 2\sqrt[4]{162} = \sqrt[4]{16}\sqrt[4]{2} + 2\sqrt[4]{81}\sqrt[4]{2} = 2\sqrt[4]{2} + 2(3)\sqrt[4]{2} = 2\sqrt[4]{2} + 6\sqrt[4]{2} = 8\sqrt[4]{2}$

67. $2x\sqrt{8} + 2\sqrt{200x^2} + \sqrt{50x^2} = 2x\sqrt{4}\sqrt{2} + 2\sqrt{100x^2}\sqrt{2} + \sqrt{25x^2}\sqrt{2}$
$= 2x(2)\sqrt{2} + 2(10x)\sqrt{2} + 5x\sqrt{2}$
$= 4x\sqrt{2} + 20x\sqrt{2} + 5x\sqrt{2} = 29x\sqrt{2}$

68. $3\sqrt{27a^3} - 2a\sqrt{3a} + 5\sqrt{75a^3} = 3\sqrt{9a^2}\sqrt{3a} - 2a\sqrt{3a} + 5\sqrt{25a^2}\sqrt{3a}$
$= 3(3a)\sqrt{3a} - 2a\sqrt{3a} + 5(5a)\sqrt{3a}$
$= 9a\sqrt{3a} - 2a\sqrt{3a} + 25a\sqrt{3a} = 32a\sqrt{3a}$

69. $\sqrt[3]{54} - 3\sqrt[3]{16} + 4\sqrt[3]{128} = \sqrt[3]{27}\sqrt[3]{2} - 3\sqrt[3]{8}\sqrt[3]{2} + 4\sqrt[3]{64}\sqrt[3]{2}$
$= 3\sqrt[3]{2} - 3(2)\sqrt[3]{2} + 4(4)\sqrt[3]{2} = 3\sqrt[3]{2} - 6\sqrt[3]{2} + 16\sqrt[3]{2} = 13\sqrt[3]{2}$

70. $2\sqrt[4]{32x^5} + 4\sqrt[4]{162x^5} - 5x\sqrt[4]{512x} = 2\sqrt[4]{16x^4}\sqrt[4]{2x} + 4\sqrt[4]{81x^4}\sqrt[4]{2x} - 5x\sqrt[4]{256}\sqrt[4]{2x}$
$= 2(2x)\sqrt[4]{2x} + 4(3x)\sqrt[4]{2x} - 5x(4)\sqrt[4]{2x}$
$= 4x\sqrt[4]{2x} + 12x\sqrt[4]{2x} - 20x\sqrt[4]{2x} = -4x\sqrt[4]{2x}$

71. hypotenuse $= 7\sqrt{2}$ m

72. shorter leg $= \dfrac{1}{2}\left(12\sqrt{3}\right) = 6\sqrt{3}$ cm
longer leg $= \sqrt{3}\left(6\sqrt{3}\right) = 6(3) = 18$ cm

73. $x = 5\sqrt{2} \approx 7.07$ in.

74. $x = \sqrt{3}\left(\frac{1}{2} \cdot 10\right) = 5\sqrt{3} \approx 8.66$ cm

75. $\left(2\sqrt{5}\right)\left(3\sqrt{2}\right) = 6\sqrt{10}$

76. $2\sqrt{6}\sqrt{216} = 2\sqrt{6}\sqrt{36}\sqrt{6} = 2(6)(6) = 72$

77. $\sqrt{9x}\sqrt{x} = \sqrt{9x^2} = 3x$

78. $\sqrt[3]{3}\sqrt[3]{9} = \sqrt[3]{27} = 3$

79. $-\sqrt[3]{2x^2}\sqrt[3]{4x} = -\sqrt[3]{8x^3}$
$= -2x$

80. $-\sqrt[4]{256x^5y^{11}}\sqrt[4]{625x^9y^3} = -\sqrt[4]{256x^4y^8}\sqrt[4]{xy^3}\sqrt[4]{625x^8}\sqrt[4]{xy^3} = -4xy^2(5x^2)\sqrt[4]{x^2y^6}$
$= -20x^3y^2\sqrt[4]{y^4}\sqrt[4]{x^2y^2}$
$= -20x^3y^3\sqrt[4]{x^2y^2} = -20x^3y^3\sqrt{xy}$

81. $\sqrt{2}\left(\sqrt{8} - 3\right) = \sqrt{16} - 3\sqrt{2} = 4 - 3\sqrt{2}$

82. $\sqrt{2}\left(\sqrt{2} + 3\right) = \sqrt{4} + 3\sqrt{2} = 2 + 3\sqrt{2}$

83. $\sqrt{5}\left(\sqrt{2} - 1\right) = \sqrt{10} - \sqrt{5}$

84. $\sqrt{3}\left(\sqrt{3} + \sqrt{2}\right) = \sqrt{9} + \sqrt{6} = 3 + \sqrt{6}$

85. $\left(\sqrt{2} + 1\right)\left(\sqrt{2} - 1\right) = \sqrt{4} - \sqrt{2} + \sqrt{2} - 1 = 1$

86. $\left(\sqrt{3} + \sqrt{2}\right)\left(\sqrt{3} + \sqrt{2}\right) = \sqrt{9} + \sqrt{6} + \sqrt{6} + \sqrt{4} = 5 + 2\sqrt{6}$

87. $(\sqrt{x} + \sqrt{y})(\sqrt{x} - \sqrt{y}) = \sqrt{x^2} - \sqrt{xy} + \sqrt{xy} - \sqrt{y^2} = x - y$

88. $(2\sqrt{u} + 3)(3\sqrt{u} - 4) = 6\sqrt{u^2} - 8\sqrt{u} + 9\sqrt{u} - 12 = 6u + \sqrt{u} - 12$

89. $\dfrac{1}{\sqrt{3}} = \dfrac{1\sqrt{3}}{\sqrt{3}\sqrt{3}} = \dfrac{\sqrt{3}}{3}$

90. $\dfrac{\sqrt{3}}{\sqrt{5}} = \dfrac{\sqrt{3}\sqrt{5}}{\sqrt{5}\sqrt{5}} = \dfrac{\sqrt{15}}{5}$

91. $\dfrac{x}{\sqrt{xy}} = \dfrac{x\sqrt{xy}}{\sqrt{xy}\sqrt{xy}} = \dfrac{x\sqrt{xy}}{xy} = \dfrac{\sqrt{xy}}{y}$

92. $\dfrac{\sqrt[3]{uv}}{\sqrt[3]{u^5v^7}} = \dfrac{\sqrt[3]{uv}\sqrt[3]{uv^2}}{\sqrt[3]{u^5v^7}\sqrt[3]{uv^2}} = \dfrac{\sqrt[3]{u^2v^3}}{\sqrt[3]{u^6v^9}} = \dfrac{v\sqrt[3]{u^2}}{u^2v^3} = \dfrac{\sqrt[3]{u^2}}{u^2v^2}$

93. $\dfrac{2}{\sqrt{2} - 1} = \dfrac{2\left(\sqrt{2} + 1\right)}{\left(\sqrt{2} - 1\right)\left(\sqrt{2} + 1\right)} = \dfrac{2\left(\sqrt{2} + 1\right)}{\sqrt{4} - 1} = \dfrac{2\left(\sqrt{2} + 1\right)}{2 - 1} = \dfrac{2\left(\sqrt{2} + 1\right)}{1} = 2\left(\sqrt{2} + 1\right)$

94. $$\frac{\sqrt{2}}{\sqrt{3}-1}=\frac{\sqrt{2}\left(\sqrt{3}+1\right)}{\left(\sqrt{3}-1\right)\left(\sqrt{3}+1\right)}=\frac{\sqrt{2}\left(\sqrt{3}+1\right)}{\sqrt{9}-1}=\frac{\sqrt{2}\left(\sqrt{3}+1\right)}{3-1}=\frac{\sqrt{2}\left(\sqrt{3}+1\right)}{2}=\frac{\sqrt{6}+\sqrt{2}}{2}$$

95. $$\frac{2x-32}{\sqrt{x}+4}=\frac{(2x-32)\left(\sqrt{x}-4\right)}{\left(\sqrt{x}+4\right)\left(\sqrt{x}-4\right)}=\frac{2(x-16)\left(\sqrt{x}-4\right)}{\sqrt{x^2}-16}=\frac{2(x-16)\left(\sqrt{x}-4\right)}{x-16}=2\left(\sqrt{x}-4\right)$$

96. $$\frac{\sqrt{a}+1}{\sqrt{a}-1}=\frac{\left(\sqrt{a}+1\right)\left(\sqrt{a}+1\right)}{\left(\sqrt{a}-1\right)\left(\sqrt{a}+1\right)}=\frac{\sqrt{a^2}+\sqrt{a}+\sqrt{a}+1}{\sqrt{a^2}-1}=\frac{a+2\sqrt{a}+1}{a-1}$$

97. $$\frac{\sqrt{3}}{5}=\frac{\sqrt{3}\sqrt{3}}{5\sqrt{3}}=\frac{3}{5\sqrt{3}}$$

98. $$\frac{\sqrt[3]{9}}{3}=\frac{\sqrt[3]{9}\sqrt[3]{3}}{3\sqrt[3]{3}}=\frac{\sqrt[3]{27}}{3\sqrt[3]{3}}=\frac{3}{3\sqrt[3]{3}}=\frac{1}{\sqrt[3]{3}}$$

99. $$\frac{3-\sqrt{x}}{2}=\frac{\left(3-\sqrt{x}\right)\left(3+\sqrt{x}\right)}{2\left(3+\sqrt{x}\right)}=\frac{9-\sqrt{x^2}}{2\left(3+\sqrt{x}\right)}=\frac{9-x}{2\left(3+\sqrt{x}\right)}$$

100. $$\frac{\sqrt{a}-\sqrt{b}}{\sqrt{a}}=\frac{\left(\sqrt{a}-\sqrt{b}\right)\left(\sqrt{a}+\sqrt{b}\right)}{\sqrt{a}\left(\sqrt{a}+\sqrt{b}\right)}=\frac{\sqrt{a^2}-\sqrt{b^2}}{\sqrt{a^2}+\sqrt{ab}}=\frac{a-b}{a+\sqrt{ab}}$$

101. $$\begin{aligned}\sqrt{y+3}&=\sqrt{2y-19}\\ \left(\sqrt{y+3}\right)^2&=\left(\sqrt{2y-19}\right)^2\\ y+3&=2y-19\\ -y&=-22\\ y&=22\end{aligned}$$
The answer checks.

102. $$\begin{aligned}u&=\sqrt{25u-144}\\ u^2&=\left(\sqrt{25u-144}\right)^2\\ u^2&=25u-144\end{aligned}$$
$u^2-25u+144=0$
$(u-9)(u-16)=0$
$u=9$ or $u=16$; Both answers check.

103. $$\begin{aligned}r&=\sqrt{12r-27}\\ r^2&=\left(\sqrt{12r-27}\right)^2\\ r^2&=12r-27\end{aligned}$$
$r^2-12r+27=0$
$(r-9)(r-3)=0$
$r=9$ or $r=3$; Both answers check.

104. $$\begin{aligned}\sqrt{z+1}+\sqrt{z}&=2\\ \sqrt{z+1}&=2-\sqrt{z}\\ \left(\sqrt{z+1}\right)^2&=\left(2-\sqrt{z}\right)^2\\ z+1&=4-4\sqrt{z}+z\\ 4\sqrt{z}&=3\\ \left(4\sqrt{z}\right)^2&=3^2\\ 16z&=9\\ z&=\tfrac{9}{16}\end{aligned}$$
$z=\frac{9}{16}$; The answer checks.

105. $\sqrt{2x+5}-\sqrt{2x}=1$

$$\begin{aligned}\sqrt{2x+5}&=1+\sqrt{2x}\\ \left(\sqrt{2x+5}\right)^2&=\left(1+\sqrt{2x}\right)^2\\ 2x+5&=1+2\sqrt{2x}+2x\\ 4&=2\sqrt{2x}\\ 4^2&=\left(2\sqrt{2x}\right)^2\\ 16&=8x\\ 2&=x\text{; The answer checks.}\end{aligned}$$

106. $\sqrt[3]{x^3+8}=x+2$

$$\begin{aligned}\left(\sqrt[3]{x^3+8}\right)^3&=(x+2)^3\\ x^3+8&=x^3+6x^2+12x+8\\ 0&=6x^2+12x\\ 0&=6x(x+2)\end{aligned}$$

$x=0$ or $x=-2$; Both answers check.

107. $(5+4i)+(7-12i)=5+4i+7-12i=12-8i$

108. $(-6-40i)-(-8+28i)=-6-40i+8-28i=2-68i$

109. $(-32+\sqrt{-144})-(64+\sqrt{-81})=-32+\sqrt{144i^2}-64-\sqrt{81i^2}=-32+12i-64-9i$
$=-96+3i$

110. $(-8+\sqrt{-8})+(6-\sqrt{-32})=-8+\sqrt{4i^2}\sqrt{2}+6-\sqrt{16i^2}\sqrt{2}$
$=-8+2i\sqrt{2}+6-4i\sqrt{2}=-2-2\sqrt{2}i$

111. $(2-7i)(-3+4i)=-6+8i+21i-28i^2=-6+29i+28=22+29i$

112. $(-5+6i)(2+i)=-10-5i+12i+6i^2=-10+7i-6=-16+7i$

113. $(5-\sqrt{-27})(-6+\sqrt{-12})=(5-3i\sqrt{3})(-6+2i\sqrt{3})=-30+10i\sqrt{3}+18i\sqrt{3}-6i^2(3)$
$=-30+28i\sqrt{3}+18=-12+28\sqrt{3}i$

114. $(2+\sqrt{-128})(3-\sqrt{-98})=(2+8i\sqrt{2})(3-7i\sqrt{2})=6-14i\sqrt{2}+24i\sqrt{2}-56i^2(2)$
$=6+10i\sqrt{2}+112=118+10\sqrt{2}i$

115. $\frac{3}{4i}=\frac{3}{4i}\cdot\frac{i}{i}=\frac{3i}{4i^2}=\frac{3i}{-4}=-\frac{3}{4}i=0-\frac{3}{4}i$

116. $\frac{-2}{5i^3}=\frac{-2}{5i^3}\cdot\frac{i}{i}=\frac{-2i}{5i^4}=\frac{-2i}{5}=0-\frac{2}{5}i$

117. $\frac{6}{2+i}=\frac{6}{2+i}\cdot\frac{2-i}{2-i}=\frac{6(2-i)}{4-i^2}=\frac{6(2-i)}{5}=\frac{12-6i}{5}=\frac{12}{5}-\frac{6}{5}i$

118. $\frac{7}{3-i}=\frac{7}{3-i}\cdot\frac{3+i}{3+i}=\frac{7(3+i)}{9-i^2}=\frac{7(3+i)}{10}=\frac{21+7i}{10}=\frac{21}{10}+\frac{7}{10}i$

119. $\frac{4+i}{4-i}=\frac{4+i}{4-i}\cdot\frac{4+i}{4+i}=\frac{16+8i+i^2}{16-i^2}=\frac{15+8i}{17}=\frac{15}{17}+\frac{8}{17}i$

120. $\frac{3-i}{3+i} = \frac{3-i}{3+i} \cdot \frac{3-i}{3-i} = \frac{9-6i+i^2}{9-i^2} = \frac{8-6i}{10} = \frac{8}{10} - \frac{6}{10}i = \frac{4}{5} - \frac{3}{5}i$

121. $\frac{3}{5+\sqrt{-4}} = \frac{3}{5+2i} = \frac{3}{5+2i} \cdot \frac{5-2i}{5-2i} = \frac{3(5-2i)}{25-4i^2} = \frac{3(5-2i)}{29} = \frac{15-6i}{29} = \frac{15}{29} - \frac{6}{29}i$

122. $\frac{2}{3-\sqrt{-9}} = \frac{2}{3-3i} = \frac{2}{3-3i} \cdot \frac{3+3i}{3+3i} = \frac{2(3+3i)}{9-9i^2} = \frac{2(3+3i)}{18} = \frac{3+3i}{9} = \frac{3}{9} + \frac{3}{9}i = \frac{1}{3} + \frac{1}{3}i$

123. $|9+12i| = \sqrt{9^2+12^2} = \sqrt{81+144} = \sqrt{225} = 15 = 15+0i$

124. $|24-10i| = \sqrt{24^2+(-10)^2} = \sqrt{576+100} = \sqrt{676} = 26 = 26+0i$

125. $i^{12} = (i^4)^3 = 1^3 = 1$

126. $i^{583} = i^{580}i^3 = (i^4)^{145}i^3 = 1^{145}(-i) = -i$

Chapter 7 Test (page 507)

1. $\sqrt{49} = \sqrt{7^2} = 7$

2. $\sqrt[3]{64} = \sqrt[3]{4^3} = 4$

3. $\sqrt{4x^2} = \sqrt{(2x)^2} = |2x| = 2|x|$

4. $\sqrt[3]{8x^3} = \sqrt[3]{(2x)^3} = 2x$

5. $f(x) = \sqrt{x-2}$; Shift $y = \sqrt{x}$ right 2.

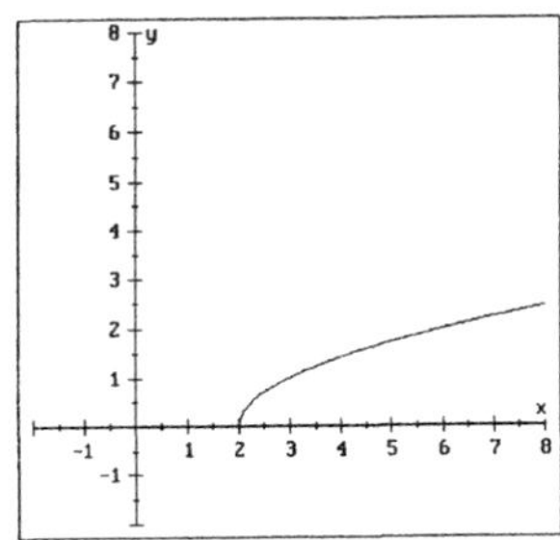

domain $= [2, \infty)$, range $= [0, \infty)$

6. $f(x) = \sqrt[3]{x}+3$; Shift $y = \sqrt[3]{x}$ up 3.

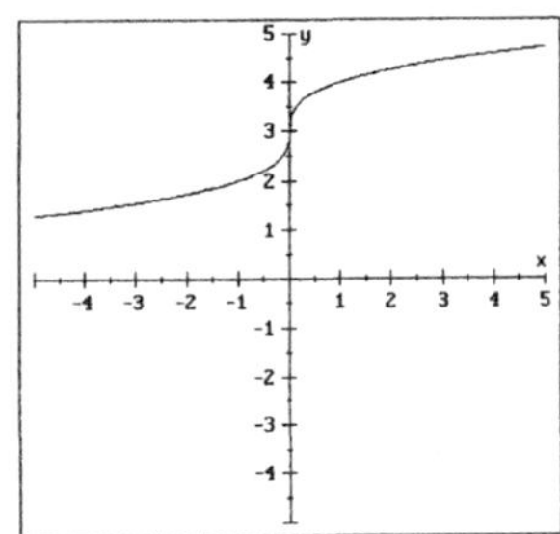

domain $= (-\infty, \infty)$, range $= (-\infty, \infty)$

7.
$$\begin{aligned} 53^2 &= 45^2 + h^2 \\ 2809 &= 2025 + h^2 \\ 784 &= h^2 \\ 28 \text{ in.} &= h \end{aligned}$$

8.
$$\begin{aligned} 2^2 &= \left(\frac{w}{2}\right)^2 + (1.9)^2 \\ 4 &= \frac{w^2}{4} + 3.61 \\ 0.39 &= \frac{w^2}{4} \\ 1.56 &= w^2 \\ 1.25 \text{ meters} &= w \end{aligned}$$

9. $d = \sqrt{(x_2-x_1)^2 + (y_2-y_1)^2} = \sqrt{(6-0)^2+(8-0)^2} = \sqrt{6^2+8^2} = \sqrt{100} = 10$

CHAPTER 7 TEST

10. $d=\sqrt{(x_2-x_1)^2+(y_2-y_1)^2}=\sqrt{(-2-22)^2+(5-12)^2}=\sqrt{(-24)^2+(-7)^2}=\sqrt{625}=25$

11. $16^{1/4}=2$

12. $27^{2/3}=\left(27^{1/3}\right)^2=3^2=9$

13. $36^{-3/2}=\dfrac{1}{36^{3/2}}=\dfrac{1}{\left(36^{1/2}\right)^3}=\dfrac{1}{6^3}=\dfrac{1}{216}$

14. $\left(-\dfrac{8}{27}\right)^{-2/3}=\left(-\dfrac{27}{8}\right)^{2/3}=\left[\left(-\dfrac{27}{8}\right)^{1/3}\right]^2=\left(-\dfrac{3}{2}\right)^2=\dfrac{9}{4}$

15. $\dfrac{2^{5/3}2^{1/6}}{2^{1/2}}=\dfrac{2^{10/6}2^{1/6}}{2^{3/6}}=2^{10/6+1/6-3/6}=2^{8/6}=2^{4/3}$

16. $\dfrac{(8x^3y)^{1/2}(8xy^5)^{1/2}}{(x^3y^6)^{1/3}}=\dfrac{8^{1/2}x^{3/2}y^{1/2}8^{1/2}x^{1/2}y^{5/2}}{x^{3/3}y^{6/3}}=\dfrac{8^{2/2}x^{4/2}y^{6/2}}{xy^2}=\dfrac{8x^2y^3}{xy^2}=8xy$

17. $\sqrt{48}=\sqrt{16}\sqrt{3}=4\sqrt{3}$

18. $\sqrt{250x^3y^5}=\sqrt{25x^2y^4}\sqrt{10xy}=5xy^2\sqrt{10xy}$

19. $\dfrac{\sqrt[3]{24x^{15}y^4}}{\sqrt[3]{y}}=\sqrt[3]{\dfrac{24x^{15}y^4}{y}}=\sqrt[3]{24x^{15}y^3}=\sqrt[3]{8x^{15}y^3}\sqrt[3]{3}=2x^5y\sqrt[3]{3}$

20. $\sqrt{\dfrac{3a^5}{48a^7}}=\sqrt{\dfrac{1}{16a^2}}=\dfrac{1}{4a}$

21. $\sqrt{12x^2}=\sqrt{4x^2}\sqrt{3}=2|x|\sqrt{3}$

22. $\sqrt{8x^6}=\sqrt{4x^6}\sqrt{2}=2|x^3|\sqrt{2}$

23. $\sqrt[3]{81x^3}=\sqrt[3]{27x^3}\sqrt[3]{3}=3x\sqrt[3]{3}$

24. $\sqrt{18x^4y^9}=\sqrt{9x^4y^8}\sqrt{2y}=|3x^2y^4|\sqrt{2y}=\quad 3x^2y^4\sqrt{2y}$

25. $\sqrt{12}-\sqrt{27}=\sqrt{4}\sqrt{3}-\sqrt{9}\sqrt{3}=2\sqrt{3}-3\sqrt{3}=-\sqrt{3}$

26. $$\begin{aligned}2\sqrt[3]{40}-\sqrt[3]{5000}+4\sqrt[3]{625}=2\sqrt[3]{8}\sqrt[3]{5}-\sqrt[3]{1000}\sqrt[3]{5}+4\sqrt[3]{125}\sqrt[3]{5}&=2(2)\sqrt[3]{5}-10\sqrt[3]{5}+4(5)\sqrt[3]{5}\\&=4\sqrt[3]{5}-10\sqrt[3]{5}+20\sqrt[3]{5}\\&=14\sqrt[3]{5}\end{aligned}$$

27. $$\begin{aligned}2\sqrt{48y^5}-3y\sqrt{12y^3}=2\sqrt{16y^4}\sqrt{3y}-3y\sqrt{4y^2}\sqrt{3y}&=2(4y^2)\sqrt{3y}-3y(2y)\sqrt{3y}\\&=8y^2\sqrt{3y}-6y^2\sqrt{3y}=2y^2\sqrt{3y}\end{aligned}$$

28. $\sqrt[4]{768z^5}+z\sqrt[4]{48z}=\sqrt[4]{256z^4}\sqrt[4]{3z}+z\sqrt[4]{16}\sqrt[4]{3z}=4z\sqrt[4]{3z}+2z\sqrt[4]{3z}=6z\sqrt[4]{3z}$

29. $-2\sqrt{xy}\left(3\sqrt{x}+\sqrt{xy^3}\right)=-6\sqrt{x^2y}-2\sqrt{x^2y^4}=-6x\sqrt{y}-2xy^2$

CHAPTER 7 TEST

30. $\left(3\sqrt{2}+\sqrt{3}\right)\left(2\sqrt{2}-3\sqrt{3}\right)=6\sqrt{4}-9\sqrt{6}+2\sqrt{6}-3\sqrt{9}=12-7\sqrt{6}-9=3-7\sqrt{6}$

31. $\dfrac{1}{\sqrt{5}}=\dfrac{1\sqrt{5}}{\sqrt{5}\sqrt{5}}=\dfrac{\sqrt{5}}{5}$

32. $\dfrac{3t-1}{\sqrt{3t}-1}=\dfrac{(3t-1)\left(\sqrt{3t}+1\right)}{\left(\sqrt{3t}-1\right)\left(\sqrt{3t}+1\right)}=\dfrac{(3t-1)\left(\sqrt{3t}+1\right)}{\sqrt{9t^2}-1}=\dfrac{(3t-1)\left(\sqrt{3t}+1\right)}{3t-1}=\sqrt{3t}+1$

33. $\dfrac{\sqrt{3}}{\sqrt{7}}=\dfrac{\sqrt{3}\sqrt{3}}{\sqrt{7}\sqrt{3}}=\dfrac{3}{\sqrt{21}}$

34. $\dfrac{\sqrt{a}+\sqrt{b}}{\sqrt{a}-\sqrt{b}}=\dfrac{\left(\sqrt{a}+\sqrt{b}\right)\left(\sqrt{a}-\sqrt{b}\right)}{\left(\sqrt{a}-\sqrt{b}\right)\left(\sqrt{a}-\sqrt{b}\right)}=\dfrac{\sqrt{a^2}-\sqrt{b^2}}{\sqrt{a^2}-\sqrt{ab}-\sqrt{ab}+\sqrt{b^2}}=\dfrac{a-b}{a-2\sqrt{ab}+b}$

35.
$$\begin{aligned}\sqrt[3]{6n+4}-4&=0\\ \sqrt[3]{6n+4}&=4\\ \left(\sqrt[3]{6n+4}\right)^3&=4^3\\ 6n+4&=64\\ 6n&=60\\ n&=10\end{aligned}$$
The answer checks.

36.
$$\begin{aligned}1-\sqrt{u}&=\sqrt{u-3}\\ \left(1-\sqrt{u}\right)^2&=\left(\sqrt{u-3}\right)^2\\ 1-2\sqrt{u}+u&=u-3\\ 4&=2\sqrt{u}\\ 4^2&=\left(2\sqrt{u}\right)^2\\ 16&=4u\\ 4&=u\end{aligned}$$
The answer does not check. ⇒ no solution

37. $(2+4i)+(-3+7i)=2+4i-3+7i=-1+11i$

38. $(3-\sqrt{-9})-(-1+\sqrt{-16})=3-3i+1-4i=4-7i$

39. $2i(3-4i)=6i-8i^2=6i+8=8+6i$

40. $(3+2i)(-4-i)=-12-3i-8i-2i^2=-12-11i+2=-10-11i$

41. $\dfrac{1}{i\sqrt{2}}=\dfrac{1}{i\sqrt{2}}\cdot\dfrac{i\sqrt{2}}{i\sqrt{2}}=\dfrac{i\sqrt{2}}{2i^2}=-\dfrac{\sqrt{2}i}{2}=0-\dfrac{\sqrt{2}}{2}i$

42. $\dfrac{2+i}{3-i}=\dfrac{2+i}{3-i}\cdot\dfrac{3+i}{3+i}=\dfrac{6+5i+i^2}{9-i^2}=\dfrac{5+5i}{10}=\dfrac{1}{2}+\dfrac{1}{2}$

Exercise 8.1 (page 519)

1. $$\begin{aligned} \frac{t+9}{2}+\frac{t+2}{5} &= \frac{8}{5}+4t \\ 5(t+9)+2(t+2) &= 2(8)+40t \\ 5t+45+2t+4 &= 16+40t \\ 33 &= 33t \\ 1 &= t \end{aligned}$$

2. $$\begin{aligned} \frac{1-5x}{2x}+4 &= \frac{x+3}{x} \\ 1-5x+4(2x) &= 2(x+3) \\ 1-5x+8x &= 2x+6 \\ x &= 5 \end{aligned}$$

3. $$\begin{aligned} 3(t-3)+3t &\le 2(t+1)+t+1 \\ 3t-9+3t &\le 2t+2+t+1 \\ 3t &\le 12 \\ t &\le 4 \end{aligned}$$

4. $$\begin{aligned} -2(y+4)-3y+8 &\ge 3(2y-3)-y \\ -2y-8-3y+8 &\ge 6y-9-y \\ 9 &\ge 10y \\ \tfrac{9}{10} &\ge y, \text{ or } y \le \tfrac{9}{10} \end{aligned}$$

5. $x=\sqrt{c}; x=-\sqrt{c}$

6. $6; 9; 9; x^2+6x+9$

7. positive or negative

8. $A=P(1+r)^t$

9. $$\begin{aligned} &6x^2+12x=0 \\ &6x(x+2)=0 \\ &6x=0 \quad \text{or} \quad x+2=0 \\ &x=0 \qquad\qquad x=-2 \end{aligned}$$

10. $$\begin{aligned} &5x^2+11x=0 \\ &x(5x+11)=0 \\ &x=0 \quad \text{or} \quad 5x+11=0 \\ &\qquad\qquad\qquad x=-\tfrac{11}{5} \end{aligned}$$

11. $$\begin{aligned} &2y^2-50=0 \\ &2(y+5)(y-5)=0 \\ &y+5=0 \quad \text{or} \quad y-5=0 \\ &y=-5 \qquad\qquad y=5 \end{aligned}$$

12. $$\begin{aligned} &4y^2-64=0 \\ &4(y+4)(y-4)=0 \\ &y+4=0 \quad \text{or} \quad y-4=0 \\ &y=-4 \qquad\qquad y=4 \end{aligned}$$

13. $$\begin{aligned} &r^2+6r+8=0 \\ &(r+2)(r+4)=0 \\ &r+2=0 \quad \text{or} \quad r+4=0 \\ &r=-2 \qquad\qquad r=-4 \end{aligned}$$

14. $$\begin{aligned} &x^2+9x+20=0 \\ &(x+4)(x+5)=0 \\ &x+4=0 \quad \text{or} \quad x+5=0 \\ &x=-4 \qquad\qquad x=-5 \end{aligned}$$

15. $$\begin{aligned} &7x-6=x^2 \\ &0=x^2-7x+6 \\ &0=(x-6)(x-1) \\ &x-6=0 \quad \text{or} \quad x-1=0 \\ &x=6 \qquad\qquad x=1 \end{aligned}$$

16. $$\begin{aligned} &5t-6=t^2 \\ &0=t^2-5t+6 \\ &0=(t-2)(t-3) \\ &t-2=0 \quad \text{or} \quad t-3=0 \\ &t=2 \qquad\qquad t=3 \end{aligned}$$

17.
$$2z^2 - 5z + 2 = 0$$
$$(2z-1)(z-2) = 0$$
$$2z - 1 = 0 \quad \textbf{or} \quad z - 2 = 0$$
$$z = \tfrac{1}{2} \qquad\qquad z = 2$$

18.
$$2x^2 - x - 1 = 0$$
$$(2x+1)(x-1) = 0$$
$$2x + 1 = 0 \quad \textbf{or} \quad x - 1 = 0$$
$$x = -\tfrac{1}{2} \qquad\qquad x = 1$$

19.
$$6s^2 + 11s - 10 = 0$$
$$(2s+5)(3s-2) = 0$$
$$2s + 5 = 0 \quad \textbf{or} \quad 3s - 2 = 0$$
$$s = -\tfrac{5}{2} \qquad\qquad s = \tfrac{2}{3}$$

20.
$$3x^2 + 10x - 8 = 0$$
$$(x+4)(3x-2) = 0$$
$$x + 4 = 0 \quad \textbf{or} \quad 3x - 2 = 0$$
$$x = -4 \qquad\qquad x = \tfrac{2}{3}$$

21.
$$x^2 = 36$$
$$x = \pm\sqrt{36} = \pm 6$$

22.
$$x^2 = 144$$
$$x = \pm\sqrt{144} = \pm 12$$

23.
$$z^2 = 5$$
$$z = \pm\sqrt{5}$$

24.
$$u^2 = 24$$
$$u = \pm\sqrt{24} = \pm 2\sqrt{6}$$

25.
$$3x^2 - 16 = 0$$
$$3x^2 = 16$$
$$x^2 = \frac{16}{3}$$
$$x = \pm\sqrt{\frac{16}{3}} = \pm\frac{4}{\sqrt{3}} = \pm\frac{4\sqrt{3}}{3}$$

26.
$$5x^2 - 49 = 0$$
$$5x^2 = 49$$
$$x^2 = \frac{49}{5}$$
$$x = \pm\sqrt{\frac{49}{5}} = \pm\frac{7}{\sqrt{5}} = \pm\frac{7\sqrt{5}}{5}$$

27.
$$(y+1)^2 = 1$$
$$y + 1 = \pm\sqrt{1}$$
$$y + 1 = \pm 1$$
$$y = -1 \pm 1$$
$$y = 0 \text{ or } y = -2$$

28.
$$(y-1)^2 = 4$$
$$y - 1 = \pm\sqrt{4}$$
$$y - 1 = \pm 2$$
$$y = 1 \pm 2$$
$$y = 3 \text{ or } y = -1$$

29.
$$(s-7)^2 - 9 = 0$$
$$(s-7)^2 = 9$$
$$s - 7 = \pm\sqrt{9}$$
$$s - 7 = \pm 3$$
$$s = 7 \pm 3$$
$$s = 10 \text{ or } s = 4$$

30.
$$(t+4)^2 = 16$$
$$t + 4 = \pm\sqrt{16}$$
$$t + 4 = \pm 4$$
$$t = -4 \pm 4$$
$$t = 0 \text{ or } t = -8$$

31.
$$(x+5)^2 - 3 = 0$$
$$(x+5)^2 = 3$$
$$x + 5 = \pm\sqrt{3}$$
$$x = -5 \pm \sqrt{3}$$

32.
$$(x+3)^2 - 7 = 0$$
$$(x+3)^2 = 7$$
$$x + 3 = \pm\sqrt{7}$$
$$x = -3 \pm \sqrt{7}$$

33.
$$(x-2)^2 - 5 = 0$$
$$(x-2)^2 = 5$$
$$x - 2 = \pm\sqrt{5}$$
$$x = 2 \pm \sqrt{5}$$

34.
$$(x-5)^2 - 11 = 0$$
$$(x-5)^2 = 11$$
$$x - 5 = \pm\sqrt{11}$$
$$x = 5 \pm \sqrt{11}$$

35.
$$p^2 + 16 = 0$$
$$p^2 = -16$$
$$p = \pm\sqrt{-16}$$
$$p = \pm 4i$$

36.
$$\begin{aligned} q^2+25&=0\\ q^2&=-25\\ q&=\pm\sqrt{-25}\\ q&=\pm 5i \end{aligned}$$

37.
$$\begin{aligned} 4m^2+81&=0\\ 4m^2&=-81\\ m^2&=-\frac{81}{4}\\ m&=\pm\sqrt{-\frac{81}{4}}\\ m&=\pm\frac{9}{2}i \end{aligned}$$

38.
$$\begin{aligned} 9n^2+121&=0\\ 9n^2&=-121\\ n^2&=-\frac{121}{9}\\ n&=\pm\sqrt{-\frac{121}{9}}\\ n&=\pm\frac{11}{3}i \end{aligned}$$

39.
$$\begin{aligned} 2d^2&=3h\\ d^2&=\frac{3h}{2}\\ d&=\sqrt{\frac{3h}{2}}\\ d&=\frac{\sqrt{3h}}{\sqrt{2}}\cdot\frac{\sqrt{2}}{\sqrt{2}}\\ d&=\frac{\sqrt{6h}}{2} \end{aligned}$$

40.
$$\begin{aligned} 2x^2&=d^2\\ \sqrt{2x^2}&=d\\ x\sqrt{2}&=d, \text{ or } d=x\sqrt{2} \end{aligned}$$

41.
$$\begin{aligned} E&=mc^2\\ \frac{E}{m}&=c^2\\ \sqrt{\frac{E}{m}}&=c\\ \frac{\sqrt{E}}{\sqrt{m}}\cdot\frac{\sqrt{m}}{\sqrt{m}}&=c\\ \frac{\sqrt{Em}}{m}&=c \end{aligned}$$

42.
$$\begin{aligned} S&=\frac{1}{2}gt^2\\ 2S&=gt^2\\ \frac{2S}{g}&=t^2\\ \sqrt{\frac{2S}{g}}&=t\\ \frac{\sqrt{2S}}{\sqrt{g}}\cdot\frac{\sqrt{g}}{\sqrt{g}}&=t\\ \frac{\sqrt{2Sg}}{g}&=t \end{aligned}$$

43.
$$\begin{aligned} x^2+2x-8&=0\\ x^2+2x&=8\\ x^2+2x+1&=8+1\\ (x+1)^2&=9\\ x+1&=\pm 3\\ x&=-1\pm 3\\ x=2 &\text{ or } x=-4 \end{aligned}$$

44.
$$\begin{aligned} x^2+6x+5&=0\\ x^2+6x&=-5\\ x^2+6x+9&=-5+9\\ (x+3)^2&=4\\ x+3&=\pm 2\\ x&=-3\pm 2\\ x=-5 &\text{ or } x=-1 \end{aligned}$$

45.
$$\begin{aligned} x^2-6x+8&=0\\ x^2-6x&=-8\\ x^2-6x+9&=-8+9\\ (x-3)^2&=1\\ x-3&=\pm 1\\ x&=3\pm 1\\ x=4 &\text{ or } x=2 \end{aligned}$$

46.
$$\begin{aligned} x^2+8x+15&=0\\ x^2+8x&=-15\\ x^2+8x+16&=-15+16\\ (x+4)^2&=1\\ x+4&=\pm 1\\ x&=-4\pm 1\\ x=-5 &\text{ or } x=-3 \end{aligned}$$

47.
$$\begin{aligned} x^2+5x+4&=0\\ x^2+5x&=-4\\ x^2+5x+\frac{25}{4}&=-4+\frac{25}{4}\\ \left(x+\frac{5}{2}\right)^2&=\frac{9}{4}\\ x+\frac{5}{2}&=\pm\frac{3}{2}\\ x&=-\frac{5}{2}\pm\frac{3}{2} \end{aligned}$$
$x=-\frac{2}{2}=-1$ or $x=-\frac{8}{2}=-4$

48.
$$\begin{aligned} x^2-11x+30&=0\\ x^2-11x&=-30\\ x^2-11x+\frac{121}{4}&=-30+\frac{121}{4}\\ \left(x-\frac{11}{2}\right)^2&=\frac{1}{4}\\ x-\frac{11}{2}&=\pm\frac{1}{2}\\ x&=\frac{11}{2}\pm\frac{1}{2} \end{aligned}$$
$x=\frac{12}{2}=6$ or $x=\frac{10}{2}=5$

49.
$$\begin{aligned} x+1&=2x^2\\ -2x^2+x&=-1\\ x^2-\frac{1}{2}x&=\frac{1}{2}\\ x^2-\frac{1}{2}x+\frac{1}{16}&=\frac{1}{2}+\frac{1}{16}\\ \left(x-\frac{1}{4}\right)^2&=\frac{9}{16}\\ x-\frac{1}{4}&=\pm\frac{3}{4}\\ x&=\frac{1}{4}\pm\frac{3}{4} \end{aligned}$$
$x=\frac{4}{4}=1$ or $x=-\frac{2}{4}=-\frac{1}{2}$

50.
$$\begin{aligned} -2&=2x^2-5x\\ 2x^2-5x&=-2\\ x^2-\frac{5}{2}x&=-1\\ x^2-\frac{5}{2}x+\frac{25}{16}&=-1+\frac{25}{16}\\ \left(x-\frac{5}{4}\right)^2&=\frac{9}{16}\\ x-\frac{5}{4}&=\pm\frac{3}{4}\\ x&=\frac{5}{4}\pm\frac{3}{4} \end{aligned}$$
$x=\frac{8}{4}=2$ or $x=\frac{2}{4}=\frac{1}{2}$

51.
$$\begin{aligned} 6x^2+11x+3&=0\\ 6x^2+11x&=-3\\ x^2+\frac{11}{6}x&=-\frac{1}{2}\\ x^2+\frac{11}{6}x+\frac{121}{144}&=-\frac{1}{2}+\frac{121}{144}\\ \left(x+\frac{11}{12}\right)^2&=\frac{49}{144}\\ x+\frac{11}{12}&=\pm\frac{7}{12}\\ x&=-\frac{11}{12}\pm\frac{7}{12} \end{aligned}$$
$x=-\frac{4}{12}=-\frac{1}{3}$ or $x=-\frac{18}{12}=-\frac{3}{2}$

52.
$$\begin{aligned} 6x^2+x-2&=0\\ 6x^2+x&=2\\ x^2+\frac{1}{6}x&=\frac{1}{3}\\ x^2+\frac{1}{6}x+\frac{1}{144}&=\frac{1}{3}+\frac{1}{144}\\ \left(x+\frac{1}{12}\right)^2&=\frac{49}{144}\\ x+\frac{1}{12}&=\pm\frac{7}{12}\\ x&=-\frac{1}{12}\pm\frac{7}{12} \end{aligned}$$
$x=\frac{6}{12}=\frac{1}{2}$ or $x=-\frac{8}{12}=-\frac{2}{3}$

53.
$$\begin{aligned} 9 - 6r &= 8r^2 \\ -8r^2 - 6r &= -9 \\ r^2 + \frac{3}{4}r &= \frac{9}{8} \\ r^2 + \frac{3}{4}r + \frac{9}{64} &= \frac{9}{8} + \frac{9}{64} \\ \left(r + \frac{3}{8}\right)^2 &= \frac{81}{64} \\ r + \frac{3}{8} &= \pm\frac{9}{8} \\ r &= -\frac{3}{8} \pm \frac{9}{8} \end{aligned}$$
$r = \frac{6}{8} = \frac{3}{4}$ or $r = -\frac{12}{8} = -\frac{3}{2}$

54.
$$\begin{aligned} 11m - 10 &= 3m^2 \\ -3m^2 + 11m &= 10 \\ m^2 - \frac{11}{3}m &= -\frac{10}{3} \\ m^2 - \frac{11}{3}m + \frac{121}{36} &= -\frac{10}{3} + \frac{121}{36} \\ \left(m - \frac{11}{6}\right)^2 &= \frac{1}{36} \\ m - \frac{11}{6} &= \pm\frac{1}{6} \\ m &= \frac{11}{6} \pm \frac{1}{6} \end{aligned}$$
$m = \frac{12}{6} = 2$ or $m = \frac{10}{6} = \frac{5}{3}$

55.
$$\begin{aligned} \frac{7x+1}{5} &= -x^2 \\ 7x + 1 &= -5x^2 \\ 5x^2 + 7x &= -1 \\ x^2 + \frac{7}{5}x &= -\frac{1}{5} \\ x^2 + \frac{7}{5}x + \frac{49}{100} &= -\frac{1}{5} + \frac{49}{100} \\ \left(x + \frac{7}{10}\right)^2 &= \frac{29}{100} \\ x + \frac{7}{10} &= \pm\frac{\sqrt{29}}{10} \\ x &= -\frac{7}{10} \pm \frac{\sqrt{29}}{10} \end{aligned}$$

56.
$$\begin{aligned} \frac{3x^2}{8} &= \frac{1}{8} - x \\ 3x^2 &= 1 - 8x \\ 3x^2 + 8x &= 1 \\ x^2 + \frac{8}{3}x &= \frac{1}{3} \\ x^2 + \frac{8}{3}x + \frac{16}{9} &= \frac{1}{3} + \frac{16}{9} \\ \left(x + \frac{4}{3}\right)^2 &= \frac{19}{9} \\ x + \frac{4}{3} &= \pm\frac{\sqrt{19}}{3} \\ x &= -\frac{4}{3} \pm \frac{\sqrt{19}}{3} \end{aligned}$$

57.
$$\begin{aligned} p^2 + 2p + 2 &= 0 \\ p^2 + 2p &= -2 \\ p^2 + 2p + 1 &= -2 + 1 \\ (p+1)^2 &= -1 \\ p + 1 &= \pm\sqrt{-1} \\ p + 1 &= \pm i \\ p &= -1 \pm i \end{aligned}$$

58.
$$\begin{aligned} x^2 - 6x + 10 &= 0 \\ x^2 - 6x &= -10 \\ x^2 - 6x + 9 &= -10 + 9 \\ (x-3)^2 &= -1 \\ x - 3 &= \pm\sqrt{-1} \\ x - 3 &= \pm i \\ x &= 3 \pm i \end{aligned}$$

59.
$$\begin{aligned} y^2+8y+18 &= 0 \\ y^2+8y &= -18 \\ y^2+8y+16 &= -18+16 \\ (y+4)^2 &= -2 \\ y+4 &= \pm\sqrt{-2} \\ y &= -4\pm i\sqrt{2} \end{aligned}$$

60.
$$\begin{aligned} t^2+t+3 &= 0 \\ t^2+t &= -3 \\ t^2+t+\frac{1}{4} &= -3+\frac{1}{4} \\ \left(t+\frac{1}{2}\right)^2 &= -\frac{11}{4} \\ t+\frac{1}{2} &= \pm\sqrt{-\frac{11}{4}} \\ t &= -\frac{1}{2}\pm\frac{\sqrt{11}}{2}i \end{aligned}$$

61.
$$\begin{aligned} 3m^2-2m+3 &= 0 \\ m^2-\frac{2}{3}m &= -1 \\ m^2-\frac{2}{3}m+\frac{1}{9} &= -1+\frac{1}{9} \\ \left(m-\frac{1}{3}\right)^2 &= -\frac{8}{9} \\ m-\frac{1}{3} &= \pm\sqrt{-\frac{8}{9}} \\ m &= \frac{1}{3}\pm\frac{\sqrt{8}}{3}i \\ m &= \frac{1}{3}\pm\frac{2\sqrt{2}}{3}i \end{aligned}$$

62.
$$\begin{aligned} 4p^2+2p+3 &= 0 \\ p^2+\frac{1}{2}p &= -\frac{3}{4} \\ p^2+\frac{1}{2}p+\frac{1}{16} &= -\frac{3}{4}+\frac{1}{16} \\ \left(p+\frac{1}{4}\right)^2 &= -\frac{11}{16} \\ p+\frac{1}{4} &= \pm\sqrt{-\frac{11}{16}} \\ p &= -\frac{1}{4}\pm\frac{\sqrt{11}}{4}i \end{aligned}$$

63.
$$\begin{aligned} f(x) &= 0 \\ 2x^2+x-5 &= 0 \\ x^2+\frac{1}{2}x &= \frac{5}{2} \\ x^2+\frac{1}{2}x+\frac{1}{16} &= \frac{5}{2}+\frac{1}{16} \\ \left(x+\frac{1}{4}\right)^2 &= \frac{41}{16} \\ x+\frac{1}{4} &= \pm\sqrt{\frac{41}{16}} \\ x+\frac{1}{4} &= \pm\frac{\sqrt{41}}{4} \\ x &= -\frac{1}{4}\pm\frac{\sqrt{41}}{4} \end{aligned}$$

64.
$$\begin{aligned} f(x) &= 0 \\ 3x^2-2x-4 &= 0 \\ x^2-\frac{2}{3}x &= \frac{4}{3} \\ x^2-\frac{2}{3}x+\frac{1}{9} &= \frac{4}{3}+\frac{1}{9} \\ \left(x-\frac{1}{3}\right)^2 &= \frac{13}{9} \\ x-\frac{1}{3} &= \pm\sqrt{\frac{13}{9}} \\ x-\frac{1}{3} &= \pm\frac{\sqrt{13}}{3} \\ x &= \frac{1}{3}\pm\frac{\sqrt{13}}{3} \end{aligned}$$

65.
$$\begin{aligned} f(x) &= 0 \\ x^2 + x - 3 &= 0 \\ x^2 + x &= 3 \\ x^2 + x + \frac{1}{4} &= 3 + \frac{1}{4} \\ \left(x + \frac{1}{2}\right)^2 &= \frac{13}{4} \\ x + \frac{1}{2} &= \pm\sqrt{\frac{13}{4}} \\ x &= -\frac{1}{2} \pm \frac{\sqrt{13}}{2} \end{aligned}$$

66.
$$\begin{aligned} f(x) &= 0 \\ x^2 + 2x - 4 &= 0 \\ x^2 + 2x &= 4 \\ x^2 + 2x + 1 &= 4 + 1 \\ (x+1)^2 &= 5 \\ x + 1 &= \pm\sqrt{5} \\ x &= -1 \pm \sqrt{5} \end{aligned}$$

67.
$$\begin{aligned} s &= 16t^2 \\ 256 &= 16t^2 \\ 16 &= t^2 \\ \pm\sqrt{16} &= t \\ \pm 4 &= t \end{aligned}$$

$t = 4$ is the only answer that makes sense, so it will take 4 seconds.

68.
$$\begin{aligned} l &= \frac{32t^2}{4\pi^2} \\ 5 &= \frac{32t^2}{4\pi^2} \\ 20\pi^2 &= 32t^2 \\ \frac{20\pi^2}{32} &= t^2 \\ \pm\sqrt{\frac{20\pi^2}{32}} &= t \end{aligned}$$

t must be positive, so $t = \sqrt{\frac{20\pi^2}{32}} \approx 2.48$ sec.

69.
$$\begin{aligned} s^2 &= 10.5l \\ s^2 &= 10.5(500) \\ s^2 &= 5250 \\ s &= \pm\sqrt{5250} \\ s &\approx \pm 72.5 \end{aligned}$$

s must be positive, so the speed was about 72.5 mph.

70.
$$\begin{aligned} p^2 &= \frac{348100}{t} \\ p^2 &= \frac{348100}{64} \\ p &= \pm\sqrt{\frac{348100}{64}} \\ p &= \pm 73.75 \end{aligned}$$

p must be positive, so the rate is 73.75.

71.
$$\begin{aligned} A &= P(1+r)^t \\ 9193.60 &= 8500(1+r)^2 \\ \frac{9193.60}{8500} &= (1+r)^2 \\ 1.0816 &= (1+r)^2 \\ \pm\sqrt{1.0816} &= \sqrt{(1+r)^2} \\ \pm 1.04 &= 1 + r \\ -1 \pm 1.04 &= r \end{aligned}$$

$r = 0.04$ or $r = -2.04$; r must be positive, so $r = 0.04$, or 4%.

72.
$$\begin{aligned} A &= P(1+r)^t \\ 14045 &= 12500(1+r)^2 \\ \frac{14045}{12500} &= (1+r)^2 \\ 1.1236 &= (1+r)^2 \\ \pm\sqrt{1.1236} &= \sqrt{(1+r)^2} \\ \pm 1.06 &= 1 + r \\ -1 \pm 1.06 &= r \end{aligned}$$

$r = 0.06$ or $r = -2.06$; r must be positive, so $r = 0.06$, or 6%.

73.
$$\begin{aligned} A &= 100 \\ lw &= 100 \\ (1.9x)(x) &= 100 \\ 1.9x^2 &= 100 \\ x^2 &= \frac{100}{1.9} \\ x &= \pm\sqrt{\frac{100}{1.9}} \approx \pm 7.25 \end{aligned}$$
x must be positive, so $x = 7\frac{1}{4}$ ft.
$1.9x \approx 1.9(7.25) \approx 13.75 = 13\frac{3}{4}$ ft.

74. The object is dropped from a height of 48 ft, so $s = 48$. When the object has a height of $h = 5$ it will hit the woman.
$$\begin{aligned} h &= s - 16t^2 \\ 5 &= 48 - 16t^2 \\ 16t^2 &= 43 \\ t^2 &= \frac{43}{16} \\ t &= \pm\sqrt{\frac{43}{16}} \approx \pm 1.6 \end{aligned}$$
t must be positive, so the woman has about 1.6 seconds to move.

75. **Answers may vary.**

76. **Answers may vary.**

77. $\left(\frac{1}{2}\sqrt{3}\right)^2 = \left(\frac{\sqrt{3}}{2}\right)^2 = \frac{3}{4}$

78.
$$\begin{aligned} x^2 + \sqrt{3}x - \tfrac{1}{4} &= 0 \\ x^2 + \sqrt{3}x + \tfrac{3}{4} &= \tfrac{1}{4} + \tfrac{3}{4} \\ \left(x + \tfrac{\sqrt{3}}{2}\right)^2 &= 1 \\ x + \tfrac{\sqrt{3}}{2} &= \pm 1 \\ x &= -\tfrac{\sqrt{3}}{2} \pm 1 \end{aligned}$$

Exercise 8.2 (page 526)

1.
$$\begin{aligned} Ax + By &= C \\ By &= -Ax + C \\ B &= \frac{-Ax + C}{y} \end{aligned}$$

2.
$$\begin{aligned} R &= \frac{kL}{d^2} \\ Rd^2 &= kL \\ \frac{Rd^2}{k} &= L, \text{ or } L = \frac{Rd^2}{k} \end{aligned}$$

3. $\sqrt{24} = \sqrt{4}\sqrt{6} = 2\sqrt{6}$

4. $\sqrt{288} = \sqrt{144}\sqrt{2} = 12\sqrt{2}$

5. $\frac{3}{\sqrt{3}} = \frac{3}{\sqrt{3}} \cdot \frac{\sqrt{3}}{\sqrt{3}} = \frac{3\sqrt{3}}{3} = \sqrt{3}$

6. $\frac{1}{2-\sqrt{3}} = \frac{1}{2-\sqrt{3}} \cdot \frac{2+\sqrt{3}}{2+\sqrt{3}} = \frac{2+\sqrt{3}}{4+2\sqrt{3}-2\sqrt{3}-3} = \frac{2+\sqrt{3}}{1} = 2+\sqrt{3}$

7. 3; −2; 6

8. $x = \frac{-b \pm \sqrt{b^2 - 4ac}}{2a}$

9. $x^2 + 3x + 2 = 0$

$a = 1, b = 3, c = 2$

$$\begin{aligned} x &= \frac{-b \pm \sqrt{b^2 - 4ac}}{2a} \\ &= \frac{-3 \pm \sqrt{3^2 - 4(1)(2)}}{2(1)} \\ &= \frac{-3 \pm \sqrt{9-8}}{2} \\ &= \frac{-3 \pm \sqrt{1}}{2} = \frac{-3 \pm 1}{2} \end{aligned}$$

$x = -\frac{2}{2} = -1$ or $x = -\frac{4}{2} = -2$

10. $x^2 - 3x + 2 = 0$

$a = 1, b = -3, c = 2$

$$\begin{aligned} x &= \frac{-b \pm \sqrt{b^2 - 4ac}}{2a} \\ &= \frac{3 \pm \sqrt{(-3)^2 - 4(1)(2)}}{2(1)} \\ &= \frac{3 \pm \sqrt{9-8}}{2} \\ &= \frac{3 \pm \sqrt{1}}{2} = \frac{3 \pm 1}{2} \end{aligned}$$

$x = \frac{4}{2} = 2$ or $x = \frac{2}{2} = 1$

11. $x^2 - 2x - 15 = 0$

$a = 1, b = -2, c = -15$

$$\begin{aligned} x &= \frac{-b \pm \sqrt{b^2 - 4ac}}{2a} \\ &= \frac{2 \pm \sqrt{(-2)^2 - 4(1)(-15)}}{2(1)} \\ &= \frac{2 \pm \sqrt{4+60}}{2} \\ &= \frac{2 \pm \sqrt{64}}{2} = \frac{2 \pm 8}{2} \end{aligned}$$

$x = \frac{10}{2} = 5$ or $x = -\frac{6}{2} = -3$

12. $x^2 - 2x - 35 = 0$

$a = 1, b = -2, c = -35$

$$\begin{aligned} x &= \frac{-b \pm \sqrt{b^2 - 4ac}}{2a} \\ &= \frac{2 \pm \sqrt{(-2)^2 - 4(1)(-35)}}{2(1)} \\ &= \frac{2 \pm \sqrt{4+140}}{2} \\ &= \frac{2 \pm \sqrt{144}}{2} = \frac{2 \pm 12}{2} \end{aligned}$$

$x = \frac{14}{2} = 7$ or $x = -\frac{10}{2} = -5$

13. $x^2 + 12x = -36$

$x^2 + 12x + 36 = 0$

$a = 1, b = 12, c = 36$

$$\begin{aligned} x &= \frac{-b \pm \sqrt{b^2 - 4ac}}{2a} \\ &= \frac{-12 \pm \sqrt{12^2 - 4(1)(36)}}{2(1)} \\ &= \frac{-12 \pm \sqrt{144-144}}{2} \\ &= \frac{-12 \pm \sqrt{0}}{2} = \frac{-12 \pm 0}{2} \end{aligned}$$

$x = -\frac{12}{2} = -6$ or $x = -\frac{12}{2} = -6$

14. $y^2 - 18y = -81$

$y^2 - 18y + 81 = 0$

$a = 1, b = -18, c = 81$

$$\begin{aligned} y &= \frac{-b \pm \sqrt{b^2 - 4ac}}{2a} \\ &= \frac{18 \pm \sqrt{(-18)^2 - 4(1)(81)}}{2(1)} \\ &= \frac{18 \pm \sqrt{324-324}}{2} \\ &= \frac{18 \pm \sqrt{0}}{2} = \frac{18 \pm 0}{2} \end{aligned}$$

$y = \frac{18}{2} = 9$ or $y = \frac{18}{2} = 9$

15. $2x^2 - x - 3 = 0$

$a = 2, b = -1, c = -3$

$$x = \frac{-b \pm \sqrt{b^2 - 4ac}}{2a}$$
$$= \frac{1 \pm \sqrt{(-1)^2 - 4(2)(-3)}}{2(2)}$$
$$= \frac{1 \pm \sqrt{1 + 24}}{4}$$
$$= \frac{1 \pm \sqrt{25}}{4} = \frac{1 \pm 5}{4}$$

$x = \frac{6}{4} = \frac{3}{2}$ or $x = -\frac{4}{4} = -1$

16. $3x^2 - 10x + 8 = 0$

$a = 3, b = -10, c = 8$

$$x = \frac{-b \pm \sqrt{b^2 - 4ac}}{2a}$$
$$= \frac{10 \pm \sqrt{(-10)^2 - 4(3)(8)}}{2(3)}$$
$$= \frac{10 \pm \sqrt{100 - 96}}{6}$$
$$= \frac{10 \pm \sqrt{4}}{6} = \frac{10 \pm 2}{6}$$

$x = \frac{12}{6} = 2$ or $x = \frac{8}{6} = \frac{4}{3}$

17. $6x^2 - x - 1 = 0$

$a = 6, b = -1, c = -1$

$$x = \frac{-b \pm \sqrt{b^2 - 4ac}}{2a}$$
$$= \frac{1 \pm \sqrt{(-1)^2 - 4(6)(-1)}}{2(6)}$$
$$= \frac{1 \pm \sqrt{1 + 24}}{12}$$
$$= \frac{1 \pm \sqrt{25}}{12} = \frac{1 \pm 5}{12}$$

$x = \frac{6}{12} = \frac{1}{2}$ or $x = -\frac{4}{12} = -\frac{1}{3}$

18. $2x^2 + 5x - 3 = 0$

$a = 2, b = 5, c = -3$

$$x = \frac{-b \pm \sqrt{b^2 - 4ac}}{2a}$$
$$= \frac{-5 \pm \sqrt{5^2 - 4(2)(-3)}}{2(2)}$$
$$= \frac{-5 \pm \sqrt{25 + 24}}{4}$$
$$= \frac{-5 \pm \sqrt{49}}{4} = \frac{-5 \pm 7}{4}$$

$x = \frac{2}{4} = \frac{1}{2}$ or $x = -\frac{12}{4} = -3$

19. $15x^2 - 14x = 8$

$15x^2 - 14x - 8 = 0$

$a = 15, b = -14, c = -8$

$$x = \frac{-b \pm \sqrt{b^2 - 4ac}}{2a}$$
$$= \frac{14 \pm \sqrt{(-14)^2 - 4(15)(-8)}}{2(15)}$$
$$= \frac{14 \pm \sqrt{196 + 480}}{30}$$
$$= \frac{14 \pm \sqrt{676}}{30} = \frac{14 \pm 26}{30}$$

$x = \frac{40}{30} = \frac{4}{3}$ or $x = -\frac{12}{30} = -\frac{2}{5}$

20. $4x^2 = -5x + 6$

$4x^2 + 5x - 6 = 0$

$a = 4, b = 5, c = -6$

$$x = \frac{-b \pm \sqrt{b^2 - 4ac}}{2a}$$
$$= \frac{-5 \pm \sqrt{5^2 - 4(4)(-6)}}{2(4)}$$
$$= \frac{-5 \pm \sqrt{25 + 96}}{8}$$
$$= \frac{-5 \pm \sqrt{121}}{8} = \frac{-5 \pm 11}{8}$$

$x = \frac{6}{8} = \frac{3}{4}$ or $x = -\frac{16}{8} = -2$

21.
$$8u = -4u^2 - 3$$
$$4u^2 + 8u + 3 = 0$$
$$a = 4, b = 8, c = 3$$
$$u = \frac{-b \pm \sqrt{b^2 - 4ac}}{2a} = \frac{-8 \pm \sqrt{8^2 - 4(4)(3)}}{2(4)} = \frac{-8 \pm \sqrt{64 - 48}}{8} = \frac{-8 \pm \sqrt{16}}{8} = \frac{-8 \pm 4}{8}$$
$$u = \tfrac{-4}{8} = -\tfrac{1}{2} \text{ or } u = \tfrac{-12}{8} = -\tfrac{3}{2}$$

22.
$$4t + 3 = 4t^2$$
$$4t^2 - 4t - 3 = 0$$
$$a = 4, b = -4, c = -3$$
$$t = \frac{-b \pm \sqrt{b^2 - 4ac}}{2a} = \frac{4 \pm \sqrt{(-4)^2 - 4(4)(-3)}}{2(4)} = \frac{4 \pm \sqrt{16 + 48}}{8} = \frac{4 \pm \sqrt{64}}{8} = \frac{4 \pm 8}{8}$$
$$t = \tfrac{12}{8} = \tfrac{3}{2} \text{ or } t = \tfrac{-4}{8} = -\tfrac{1}{2}$$

23.
$$16y^2 + 8y - 3 = 0$$
$$a = 16, b = 8, c = -3$$
$$y = \frac{-b \pm \sqrt{b^2 - 4ac}}{2a} = \frac{-8 \pm \sqrt{8^2 - 4(16)(-3)}}{2(16)} = \frac{-8 \pm \sqrt{64 + 192}}{32} = \frac{-8 \pm \sqrt{256}}{32} = \frac{-8 \pm 16}{32}$$
$$y = \tfrac{8}{32} = \tfrac{1}{4} \text{ or } y = \tfrac{-24}{32} = -\tfrac{3}{4}$$

24.
$$16x^2 + 16x + 3 = 0$$
$$a = 16, b = 16, c = 3$$
$$x = \frac{-b \pm \sqrt{b^2 - 4ac}}{2a} = \frac{-16 \pm \sqrt{16^2 - 4(16)(3)}}{2(16)} = \frac{-16 \pm \sqrt{256 - 192}}{32} = \frac{-16 \pm \sqrt{64}}{32} = \frac{-16 \pm 8}{32}$$
$$x = \tfrac{-8}{32} = -\tfrac{1}{4} \text{ or } x = \tfrac{-24}{32} = -\tfrac{3}{4}$$

25.
$$5x^2 + 5x + 1 = 0$$
$$a = 5, b = 5, c = 1$$
$$x = \frac{-b \pm \sqrt{b^2 - 4ac}}{2a} = \frac{-5 \pm \sqrt{5^2 - 4(5)(1)}}{2(5)} = \frac{-5 \pm \sqrt{25 - 20}}{10} = \frac{-5 \pm \sqrt{5}}{10} = \frac{-5}{10} \pm \frac{\sqrt{5}}{10} = -\frac{1}{2} \pm \frac{\sqrt{5}}{10}$$

26.
$$4w^2 + 6w + 1 = 0$$
$$a = 4, b = 6, c = 1$$
$$w = \frac{-b \pm \sqrt{b^2 - 4ac}}{2a} = \frac{-6 \pm \sqrt{6^2 - 4(4)(1)}}{2(4)} = \frac{-6 \pm \sqrt{36 - 16}}{8} = \frac{-6 \pm \sqrt{20}}{8} = -\frac{6}{8} \pm \frac{\sqrt{4}\sqrt{5}}{8} = -\frac{3}{4} \pm \frac{2\sqrt{5}}{8} = -\frac{3}{4} \pm \frac{\sqrt{5}}{4}$$

27.
$$\begin{aligned} \frac{x^2}{2}+\frac{5}{2}x &= -1 \\ x^2+5x &= -2 \\ x^2+5x+2 &= 0 \\ a=1, b=5, c=2 & \\ x &= \frac{-b\pm\sqrt{b^2-4ac}}{2a} \\ &= \frac{-5\pm\sqrt{5^2-4(1)(2)}}{2(1)} \\ &= \frac{-5\pm\sqrt{25-8}}{2} \\ &= \frac{-5\pm\sqrt{17}}{2} = -\frac{5}{2}\pm\frac{\sqrt{17}}{2} \end{aligned}$$

28.
$$\begin{aligned} -3x &= \frac{x^2}{2}+2 \\ -6x &= x^2+4 \\ x^2+6x+4 &= 0 \\ a=1, b=6, c=4 & \\ x &= \frac{-b\pm\sqrt{b^2-4ac}}{2a} \\ &= \frac{-6\pm\sqrt{6^2-4(1)(4)}}{2(1)} \\ &= \frac{-6\pm\sqrt{36-16}}{2} \\ &= \frac{-6\pm\sqrt{20}}{2} \\ &= \frac{-6\pm2\sqrt{5}}{2} = -\frac{6}{2}\pm\frac{2\sqrt{5}}{2} = -3\pm\sqrt{5} \end{aligned}$$

29.
$$\begin{aligned} 2x^2-1 &= 3x \\ 2x^2-3x-1 &= 0 \\ a=2, b=-3, c=-1 & \\ x &= \frac{-b\pm\sqrt{b^2-4ac}}{2a} \\ &= \frac{3\pm\sqrt{(-3)^2-4(2)(-1)}}{2(2)} \\ &= \frac{3\pm\sqrt{9+8}}{4} \\ &= \frac{3\pm\sqrt{17}}{4} = \frac{3}{4}\pm\frac{\sqrt{17}}{4} \end{aligned}$$

30.
$$\begin{aligned} -9x &= 2-3x^2 \\ 3x^2-9x-2 &= 0 \\ a=3, b=-9, c=-2 & \\ x &= \frac{-b\pm\sqrt{b^2-4ac}}{2a} \\ &= \frac{9\pm\sqrt{(-9)^2-4(3)(-2)}}{2(3)} \\ &= \frac{9\pm\sqrt{81+24}}{6} \\ &= \frac{9\pm\sqrt{105}}{6} = \frac{9}{6}\pm\frac{\sqrt{105}}{6} = \frac{3}{2}\pm\frac{\sqrt{105}}{6} \end{aligned}$$

31.
$$\begin{aligned} x^2+2x+2 &= 0 \\ a=1, b=2, c=2 & \\ x &= \frac{-b\pm\sqrt{b^2-4ac}}{2a} \\ &= \frac{-2\pm\sqrt{2^2-4(1)(2)}}{2(1)} \\ &= \frac{-2\pm\sqrt{4-8}}{2} \\ &= \frac{-2\pm\sqrt{-4}}{2} \\ &= \frac{-2\pm\sqrt{-1\cdot4}}{2} \\ &= \frac{-2\pm2i}{2} = \frac{-2}{2}\pm\frac{2i}{2} = -1\pm i \end{aligned}$$

32.
$$\begin{aligned} x^2+3x+3 &= 0 \\ a=1, b=3, c=3 & \\ x &= \frac{-b\pm\sqrt{b^2-4ac}}{2a} \\ &= \frac{-3\pm\sqrt{3^2-4(1)(3)}}{2(1)} \\ &= \frac{-3\pm\sqrt{9-12}}{2} \\ &= \frac{-3\pm\sqrt{-3}}{2} \\ &= \frac{-3\pm\sqrt{-1\cdot3}}{2} \\ &= \frac{-3\pm i\sqrt{3}}{2} = -\frac{3}{2}\pm\frac{\sqrt{3}}{2}i \end{aligned}$$

33. $2x^2 + x + 1 = 0$

$a = 2, b = 1, c = 1$

$$\begin{aligned} x &= \frac{-b \pm \sqrt{b^2 - 4ac}}{2a} \\ &= \frac{-1 \pm \sqrt{1^2 - 4(2)(1)}}{2(2)} \\ &= \frac{-1 \pm \sqrt{1 - 8}}{4} \\ &= \frac{-1 \pm \sqrt{-7}}{4} \\ &= \frac{-1 \pm \sqrt{-1 \cdot 7}}{4} \\ &= \frac{-1 \pm i\sqrt{7}}{4} = -\frac{1}{4} \pm \frac{\sqrt{7}}{4}i \end{aligned}$$

34. $3x^2 + 2x + 1 = 0$

$a = 3, b = 2, c = 1$

$$\begin{aligned} x &= \frac{-b \pm \sqrt{b^2 - 4ac}}{2a} \\ &= \frac{-2 \pm \sqrt{2^2 - 4(3)(1)}}{2(3)} \\ &= \frac{-2 \pm \sqrt{4 - 12}}{6} \\ &= \frac{-2 \pm \sqrt{-8}}{6} \\ &= \frac{-2 \pm \sqrt{-1 \cdot 4 \cdot 2}}{6} \\ &= \frac{-2 \pm 2i\sqrt{2}}{6} \\ &= -\frac{2}{6} \pm \frac{2\sqrt{2}}{6}i = -\frac{1}{3} \pm \frac{\sqrt{2}}{3}i \end{aligned}$$

35. $3x^2 - 4x = -2$

$3x^2 - 4x + 2 = 0$

$a = 3, b = -4, c = 2$

$$\begin{aligned} x &= \frac{-b \pm \sqrt{b^2 - 4ac}}{2a} \\ &= \frac{4 \pm \sqrt{(-4)^2 - 4(3)(2)}}{2(3)} \\ &= \frac{4 \pm \sqrt{16 - 24}}{6} \\ &= \frac{4 \pm \sqrt{-8}}{6} \\ &= \frac{4 \pm \sqrt{-1 \cdot 4 \cdot 2}}{6} \\ &= \frac{4 \pm 2i\sqrt{2}}{6} \\ &= \frac{4}{6} \pm \frac{2\sqrt{2}}{6}i = \frac{2}{3} \pm \frac{\sqrt{2}}{3}i \end{aligned}$$

36. $2x^2 + 3x = -3$

$2x^2 + 3x + 3 = 0$

$a = 2, b = 3, c = 3$

$$\begin{aligned} x &= \frac{-b \pm \sqrt{b^2 - 4ac}}{2a} \\ &= \frac{-3 \pm \sqrt{3^2 - 4(2)(3)}}{2(2)} \\ &= \frac{-3 \pm \sqrt{9 - 24}}{4} \\ &= \frac{-3 \pm \sqrt{-15}}{4} \\ &= \frac{-3 \pm \sqrt{-1 \cdot 15}}{4} \\ &= \frac{-3 \pm i\sqrt{15}}{4} = -\frac{3}{4} \pm \frac{\sqrt{15}}{4}i \end{aligned}$$

37.

$$\begin{aligned}
3x^2 - 2x &= -3 \\
3x^2 - 2x + 3 &= 0 \\
a = 3, b &= -2, c = 3 \\
x &= \frac{-b \pm \sqrt{b^2 - 4ac}}{2a} \\
&= \frac{2 \pm \sqrt{(-2)^2 - 4(3)(3)}}{2(3)} \\
&= \frac{2 \pm \sqrt{4 - 36}}{6} \\
&= \frac{2 \pm \sqrt{-32}}{6} \\
&= \frac{2 \pm \sqrt{-1 \cdot 16 \cdot 2}}{6} \\
&= \frac{2 \pm 4i\sqrt{2}}{6} \\
&= \frac{2}{6} \pm \frac{4\sqrt{2}}{6}i = \frac{1}{3} \pm \frac{2\sqrt{2}}{3}i
\end{aligned}$$

38.

$$\begin{aligned}
5x^2 &= 2x - 1 \\
5x^2 - 2x + 1 &= 0 \\
a = 5, b &= -2, c = 1 \\
x &= \frac{-b \pm \sqrt{b^2 - 4ac}}{2a} \\
&= \frac{2 \pm \sqrt{(-2)^2 - 4(5)(1)}}{2(5)} \\
&= \frac{2 \pm \sqrt{4 - 20}}{10} \\
&= \frac{2 \pm \sqrt{-16}}{10} \\
&= \frac{2 \pm \sqrt{-1 \cdot 16}}{10} \\
&= \frac{2 \pm 4i}{10} = \frac{2}{10} \pm \frac{4i}{10} = \frac{1}{5} \pm \frac{2}{5}i
\end{aligned}$$

39.

$$\begin{aligned}
f(x) &= 0 \\
4x^2 + 4x - 19 &= 0 \\
a = 4, b &= 4, c = -19 \\
x &= \frac{-b \pm \sqrt{b^2 - 4ac}}{2a} \\
&= \frac{-4 \pm \sqrt{4^2 - 4(4)(-19)}}{2(4)} \\
&= \frac{-4 \pm \sqrt{16 + 304}}{8} \\
&= \frac{-4 \pm \sqrt{320}}{8} \\
&= -\frac{4}{8} \pm \frac{8\sqrt{5}}{8} = -\frac{1}{2} \pm \sqrt{5}
\end{aligned}$$

40.

$$\begin{aligned}
f(x) &= 0 \\
9x^2 + 12x - 8 &= 0 \\
a = 9, b &= 12, c = -8 \\
x &= \frac{-b \pm \sqrt{b^2 - 4ac}}{2a} \\
&= \frac{-12 \pm \sqrt{12^2 - 4(9)(-8)}}{2(9)} \\
&= \frac{-12 \pm \sqrt{144 + 288}}{18} \\
&= \frac{-12 \pm \sqrt{432}}{18} \\
&= -\frac{2}{3} \pm \frac{12\sqrt{3}}{18} = -\frac{2}{3} \pm \frac{2\sqrt{3}}{3}
\end{aligned}$$

41.

$$\begin{aligned} f(x) &= 0 \\ 3x^2 + 2x + 2 &= 0 \\ a = 3, b &= 2, c = 2 \\ x &= \frac{-b \pm \sqrt{b^2 - 4ac}}{2a} \\ &= \frac{-2 \pm \sqrt{2^2 - 4(3)(2)}}{2(3)} \\ &= \frac{-2 \pm \sqrt{4 - 24}}{6} \\ &= \frac{-2 \pm \sqrt{-20}}{6} \\ &= -\frac{2}{6} \pm \frac{2i\sqrt{5}}{6} = -\frac{1}{3} \pm \frac{\sqrt{5}}{3} i \end{aligned}$$

42.

$$\begin{aligned} f(x) &= 0 \\ 4x^2 + x + 1 &= 0 \\ a = 4, b &= 1, c = 1 \\ x &= \frac{-b \pm \sqrt{b^2 - 4ac}}{2a} \\ &= \frac{-1 \pm \sqrt{1^2 - 4(4)(1)}}{2(4)} \\ &= \frac{-1 \pm \sqrt{1 - 16}}{8} \\ &= \frac{-1 \pm \sqrt{-15}}{8} \\ &= -\frac{1}{8} \pm \frac{\sqrt{15}}{8} i \end{aligned}$$

43. $0.7x^2 - 3.5x - 25 = 0$

$a = 0.7, b = -3.5, c = -25$

$$\begin{aligned} x &= \frac{-b \pm \sqrt{b^2 - 4ac}}{2a} \\ &= \frac{3.5 \pm \sqrt{(-3.5)^2 - 4(0.7)(-25)}}{2(0.7)} \\ &= \frac{3.5 \pm \sqrt{12.25 + 70}}{1.4} \\ &= \frac{3.5 \pm \sqrt{82.25}}{1.4} \\ &= \frac{3.5 \pm 9.069}{1.4} \end{aligned}$$

$x = 8.98$ or $x = -3.98$

44. $-4.5x^2 + 0.2x + 3.75 = 0$

$a = -4.5, b = 0.2, c = 3.75$

$$\begin{aligned} x &= \frac{-b \pm \sqrt{b^2 - 4ac}}{2a} \\ &= \frac{-0.2 \pm \sqrt{(0.2)^2 - 4(-4.5)(3.75)}}{2(-4.5)} \\ &= \frac{-0.2 \pm \sqrt{0.04 + 67.5}}{-9} \\ &= \frac{-0.2 \pm \sqrt{67.54}}{-9} \\ &= \frac{-0.2 \pm 8.218}{-9} \end{aligned}$$

$x = -0.89$ or $x = 0.94$

45.
$$C = \frac{N^2 - N}{2}$$
$$2C = N^2 - N$$
$$N^2 - N - 2C = 0$$
$$a = 1, b = -1, c = -2C$$
$$N = \frac{-b \pm \sqrt{b^2 - 4ac}}{2a}$$
$$= \frac{1 \pm \sqrt{(-1)^2 - 4(1)(-2C)}}{2(1)}$$
$$= \frac{1 \pm \sqrt{1 + 8C}}{2}$$

46.
$$A = 2\pi r^2 + 2\pi rh$$
$$2\pi r^2 + 2\pi rh - A = 0$$
$$a = 2\pi, b = 2\pi h, c = -A$$
$$r = \frac{-b \pm \sqrt{b^2 - 4ac}}{2a}$$
$$= \frac{-2\pi h \pm \sqrt{(2\pi h)^2 - 4(2\pi)(-A)}}{2(2\pi)}$$
$$= \frac{-2\pi h \pm \sqrt{4\pi^2 h^2 + 8\pi A}}{4\pi}$$
$$= \frac{-2\pi h \pm 2\sqrt{\pi^2 h^2 + 2\pi A}}{4\pi}$$
$$= \frac{-\pi h + \sqrt{\pi^2 h^2 + 2\pi A}}{2\pi}$$ [choose only + in numerator to make r positive]

47. Let x and $x + 2$ represent the integers.
$$x(x + 2) = 288$$
$$x^2 + 2x - 288 = 0$$
$$(x + 18)(x - 16) = 0$$
$$x = -18 \text{ or } x = 16$$
Since the integers are positive, they must be 16 and 18.

48. Let x and $x + 2$ represent the integers.
$$x(x + 2) = 143$$
$$x^2 + 2x - 143 = 0$$
$$(x + 13)(x - 11) = 0$$
$$x = -13 \text{ or } x = 11$$
Since the integers are negative, they must be -13 and -11.

49. Let x and $x + 1$ represent the integers.
$$x^2 + (x + 1)^2 = 85$$
$$x^2 + x^2 + 2x + 1 = 85$$
$$2x^2 + 2x - 84 = 0$$
$$2(x + 7)(x - 6) = 0$$
$$x = -7 \text{ or } x = 6$$
Since the integers are positive, they must be 6 and 7.

50. Let x, $x + 1$ and $x + 2$ represent the integers.
$$x^2 + (x + 1)^2 + (x + 2)^2 = 77$$
$$x^2 + x^2 + 2x + 1 + x^2 + 4x + 4 = 77$$
$$3x^2 + 6x - 72 = 0$$
$$3(x + 6)(x - 4) = 0$$
$$x = -6 \text{ or } x = 4$$
Since the integers are positive, they must be 4, 5 and 6.

51. $(x - 3)(x - 5) = 0$
$$x^2 - 8x + 15 = 0$$

52. $(x + 4)(x - 6) = 0$
$$x^2 - 2x - 24 = 0$$

53. $(x - 2)(x - 3)(x + 4) = 0$
$$(x^2 - 5x + 6)(x + 4) = 0$$
$$x^3 - x^2 - 14x + 24 = 0$$

54. $(x - 3)(x + 3)(x - 4)(x + 4) = 0$
$$(x^2 - 9)(x^2 - 16) = 0$$
$$x^4 - 25x^2 + 144 = 0$$

55. length · width = Area
$(x+4)x = 96$
$x^2 + 4x - 96 = 0$
$(x+12)(x-8) = 0$
$x = -12$ or $x = 8$
Since the width is positive, the dimensions are 8 ft by 12 ft.

56. length · width = Area
$(2x-3)x = 77$
$2x^2 - 3x = 77$
$2x^2 - 3x - 77 = 0$
$(2x+11)(x-7) = 49$
$x = -\frac{11}{2}$ or $x = 7$
Since the width is positive, the dimensions are 7 ft by 11 ft.

57. Let $s =$ the length of a side.
Area = perimeter
$s^2 = 4s$
$s^2 - 4s = 0$
$s(s-4) = 0$
$s = 0$ or $s = 4$
Since the length cannot be 0, the length of a side is 4 units.

58. Let w represent the width.
Then $w + 2$ represents the length.
Area = Perimeter + 11
$w(w+2) = 2w + 2(w+2) + 11$
$w^2 + 2w = 4w + 15$
$w^2 - 2w - 15 = 0$
$(w+3)(w-5) = 0$
$w = -3$ or $w = 5$
Since the width is positive, the dimensions are 5 in. by 7 in., for a perimeter of 24 in.

59. Let b represent the base.
Then $3b + 5$ represents the height.
$\frac{1}{2}$base · height = Area
$\frac{1}{2}b(3b+5) = 6$
$b(3b+5) = 12$
$3b^2 + 5b - 12 = 0$
$(3b-4)(b+3) = 0$
$b = \frac{4}{3}$ or $b = -3$
Since the base is positive, it must be $\frac{4}{3}$ cm.

60. Let b represent the base.
Then $2b + 4$ represents the height.
$\frac{1}{2}$base · height = Area
$\frac{1}{2}b(2b+4) = 15$
$b(2b+4) = 30$
$2b^2 + 4b - 30 = 0$
$2(b+5)(b-3) = 0$
$b = -5$ or $b = 3$
Since the base is positive, it must be 3 m, and the height must be 10 m.

61. Let $r =$ the slower rate. Then $r + 20 =$ the faster rate.

	Rate	Time	Dist.
Slower	r	$\frac{150}{r}$	150
Faster	$r+20$	$\frac{150}{r+20}$	150

$$\boxed{\text{Faster time}} + 2 = \boxed{\text{Slower time}}$$

$$\frac{150}{r+20} + 2 = \frac{150}{r}$$

$$\left(\frac{150}{r+20} + 2\right)(r)(r+20) = \frac{150}{r} \cdot r(r+20)$$

$$150r + 2r(r+20) = 150(r+20)$$

$$2r^2 + 40r - 3000 = 0$$

$$2(r+50)(r-30) = 0$$

$r = -50$ or $r = 30$ $\quad r = 30$ is the only answer that makes sense.
Her original speed was 30 mph.

62. Let $r =$ the faster rate. Then $r - 4 =$ the slower rate.

	Rate	Time	Dist.
Slower	$r-4$	$\frac{160}{r-4}$	160
Faster	r	$\frac{160}{r}$	160

$$\boxed{\text{Slower time}} - 2 = \boxed{\text{Faster time}}$$

$$\frac{160}{r-4} - 2 = \frac{160}{r}$$

$$\left(\frac{160}{r-4} - 2\right)(r)(r-4) = \frac{160}{r} \cdot r(r-4)$$

$$160r - 2r(r-4) = 160(r-4)$$

$$-2r^2 + 8r + 640 = 0$$

$$-2(r+16)(r-20) = 0$$

$r = -16$ or $r = 20$ $\quad r = 20$ is the only answer that makes sense.
His original speed was 20 mph.

63. Let $x =$ the number of 10¢ increases. Then the ticket price will be $4 + 0.10x$, while the projected attendance will be $300 - 5x$, for total receipts of $(4 + 0.10x)(300 - 5x)$.

$$\text{Total} = 1248$$

$$(4 + 0.10x)(300 - 5x) = 1248$$

$$1200 + 10x - 0.5x^2 = 1248$$

$$-0.5x^2 + 10x - 48 = 0$$

$$x^2 - 20x + 96 = 0$$

$$(x-12)(x-8) = 0$$

$x = 12$ or $x = 8 \Rightarrow 4 + 0.10(12) = 5.20$; $4 + 0.10(8) = 4.80$

The ticket price would be either \$5.20 or \$4.80.

64. Let $x =$ the number of 5¢ increases. Then the fare will be $25 + 5x$, while the projected # of passengers will be $3000 - 80x$, for total receipts of $(25 + 5x)(3000 - 80x)$.

$$\begin{aligned}
\text{Total} &= 99400 \\
(25 + 5x)(3000 - 80x) &= 99400 \\
75{,}000 + 13{,}000x - 400x^2 &= 99400 \\
-400x^2 + 13{,}000x - 24{,}400 &= 0 \\
2x^2 - 65x + 122 &= 0 \\
(2x - 61)(x - 2) &= 0
\end{aligned}$$

$x = 30.5$ or $x = 2 \Rightarrow$ The smallest increase would be 10¢.

65. Let $x =$ the number of additional subscribers. Then the profit per subscriber will be $20 + 0.01x$, for a total profit of $(20 + 0.01x)(3000 + x)$.

$$\begin{aligned}
\text{Total profit} &= 120000 \\
(20 + 0.01x)(3000 + x) &= 120000 \\
60{,}000 + 50x + 0.01x^2 &= 120000 \\
0.01x^2 + 50x - 60{,}000 &= 0 \\
x^2 + 5000x - 6{,}000{,}000 &= 0 \\
(x + 6000)(x - 1000) &= 0
\end{aligned}$$

$x = -6000$ (impossible) or $x = 1000 \Rightarrow$ The total number of subscribers would be 4000.

66.

$$\begin{aligned}
1000(1 + r)^2 + 2000(1 + r) &= 3368.10 \\
1000(1 + 2r + r^2) + 2000 + 2000r &= 3368.10 \\
1000r^2 + 4000r + 3000 &= 3368.10 \\
1000r^2 + 4000r - 368.10 &= 0 \\
r^2 + 4r - 0.3681 &= 0
\end{aligned}$$

$a = 1, b = 4, c = -0.3681$

$$\begin{aligned}
r &= \frac{-b \pm \sqrt{b^2 - 4ac}}{2a} \\
&= \frac{-4 \pm \sqrt{4^2 - 4(1)(-0.3681)}}{2(1)} \\
&= \frac{-4 \pm \sqrt{16 + 1.4724}}{2} = \frac{-4 \pm \sqrt{17.4724}}{2} = \frac{-4 \pm 4.18}{2} = \frac{0.18}{2} \text{ or } \frac{-8.18}{2} \text{ (impossible)}
\end{aligned}$$

$r = 0.09$, or 9%

67. Let w = the constant width.

$$\begin{aligned} \text{Frame area} &= \text{Picture area} \\ (12+2w)(10+2w) - 12(10) &= 12(10) \\ 120 + 44w + 4w^2 - 240 &= 0 \\ 4w^2 + 44w - 120 &= 0 \\ 4(w^2 + 11w - 30) &= 0 \\ w^2 + 11w - 30 &= 0 \Rightarrow a = 1, b = 11, c = -30 \end{aligned}$$

$$w = \frac{-11 \pm \sqrt{11^2 - 4(1)(-30)}}{2(1)}$$

$$= \frac{-11 \pm \sqrt{121 + 120}}{2} = \frac{-11 \pm \sqrt{241}}{2} = \frac{-11 \pm 15.52}{2} = \frac{4.52}{2} \text{ or } \frac{-26.52}{2} \text{ (impossible)}$$

$w = 2.26$ in.

68. Let s = the original length of a side.

$$\begin{aligned} \text{Volume} &= 200 \\ (s-4)(s-4)\cdot 2 &= 200 \\ (s-4)(s-4) &= 100 \\ s^2 - 8s + 16 - 100 &= 0 \\ s^2 - 8s - 84 &= 0 \\ (s-14)(s+6) &= 0 \end{aligned}$$

$s = 14$ or $s = -6$ (impossible) $\Rightarrow$ The original size should be 14 in. by 14 in.

69. $P = 0.03x^2 - 1.37x + 82.51$

$75 = 0.03x^2 - 1.37x + 82.51$

$0 = 0.03x^2 - 1.37x + 7.51$

$a = 0.03$, $b = -1.37$, $c = 7.51$

$$x = \frac{-b \pm \sqrt{b^2 - 4ac}}{2a}$$

$$= \frac{1.37 \pm \sqrt{(-1.37)^2 - 4(0.03)(7.51)}}{2(0.03)} = \frac{1.37 \pm \sqrt{0.9757}}{0.06} \approx 39.3 \text{ or } 6.4$$

Since $x \leq 30$, the only answer that works is 6.4. The model indicates the desired result happened in early 1976.

70. $B = 0.0596x^2 - 0.3811x + 14.2709$
$15 = 0.0596x^2 - 0.3811x + 14.2709$
$0 = 0.0596x^2 - 0.3811x - 0.7291$
$a = 0.0596,\ b = -0.3811,\ c = -0.7291$

$$x = \frac{-b \pm \sqrt{b^2 - 4ac}}{2a}$$
$$= \frac{0.3811 \pm \sqrt{(-0.3811)^2 - 4(0.0596)(-0.7291)}}{2(0.0596)} = \frac{0.3811 \pm \sqrt{0.31905465}}{0.1192} \approx 7.94 \text{ or } -1.54$$

Since $x \geq 0$, the only answer that works is 7.9. This is about 8 years after 1995, or 2003.

71. Let $[H^+]$ (and then $[A^-]$) $= x$ and $[HA] = 0.1 - x$.

$$\frac{[H^+][A^-]}{[HA]} = 4 \times 10^{-4}$$
$$\frac{x^2}{0.1 - x} = 4 \times 10^{-4}$$
$$x^2 = 4 \times 10^{-5} - (4 \times 10^{-4})x$$
$$x^2 + (4 \times 10^{-4})x - 4 \times 10^{-5} = 0$$
$$x = \frac{-b \pm \sqrt{b^2 - 4ac}}{2a} = \frac{-4 \times 10^{-4} \pm \sqrt{(4 \times 10^{-4})^2 - 4(1)(-4 \times 10^{-5})}}{2(1)}$$
$$\approx \frac{-4 \times 10^{-4} \pm 0.012655}{2} \approx \frac{0.012255}{2} \text{ or } -\frac{0.013055}{2} \text{ (impossible)}$$

The concentration is about $0.00613\text{ M} = 6.13 \times 10^{-3}$ M.

72. Let $[H^+]$ (and then $[HS^-]$) $= x$ and $[HHS] = 0.1 - x$.

$$\frac{[H^+][HS^-]}{[HHS]} = 1.0 \times 10^{-7}$$
$$\frac{x^2}{0.1 - x} = 1.0 \times 10^{-7}$$
$$x^2 = 1.0 \times 10^{-8} - (1.0 \times 10^{-7})x$$
$$x^2 + (1 \times 10^{-7})x - 1 \times 10^{-8} = 0$$
$$x = \frac{-b \pm \sqrt{b^2 - 4ac}}{2a} = \frac{-1 \times 10^{-7} \pm \sqrt{(1 \times 10^{-7})^2 - 4(1)(-1 \times 10^{-8})}}{2(1)}$$
$$\approx \frac{-1 \times 10^{-7} \pm 0.0002}{2} \approx \frac{0.0001999}{2} \text{ or } -\frac{0.0002001}{2} \text{ (impossible)}$$

The concentration is about 9.995×10^{-5} M.

73. **Answers may vary.**

74. **Answers may vary.**

75. $x^2 + 2\sqrt{2}x - 6 = 0$

$a = 1, b = 2\sqrt{2}, c = -6$

$$x = \frac{-b \pm \sqrt{b^2 - 4ac}}{2a} = \frac{-2\sqrt{2} \pm \sqrt{\left(2\sqrt{2}\right)^2 - 4(1)(-6)}}{2(1)} = \frac{-2\sqrt{2} \pm \sqrt{8 + 24}}{2} = \frac{-2\sqrt{2} \pm \sqrt{32}}{2} = \frac{-2\sqrt{2} \pm 4\sqrt{2}}{2}$$

$x = \frac{2\sqrt{2}}{2} = \sqrt{2}$ or $x = \frac{-6\sqrt{2}}{2} = -3\sqrt{2}$

76. $\sqrt{2}x^2 + x - \sqrt{2} = 0$

$a = \sqrt{2}, b = 1, c = -\sqrt{2}$

$$x = \frac{-b \pm \sqrt{b^2 - 4ac}}{2a} = \frac{-1 \pm \sqrt{(1)^2 - 4(\sqrt{2})(-\sqrt{2})}}{2(\sqrt{2})} = \frac{-1 \pm \sqrt{1 + 8}}{2\sqrt{2}} = \frac{-1 \pm \sqrt{9}}{2\sqrt{2}} = \frac{-1 \pm 3}{2\sqrt{2}} = \frac{(-1 \pm 3)}{2\sqrt{2}} \cdot \frac{\sqrt{2}}{\sqrt{2}} = \frac{(-1 \pm 3)\sqrt{2}}{4}$$

$x = \frac{2\sqrt{2}}{4} = \frac{\sqrt{2}}{2}$ or $x = \frac{-4\sqrt{2}}{4} = -\sqrt{2}$

77. $x^2 - 3ix - 2 = 0$

$a = 1, b = -3i, c = -2$

$$x = \frac{-b \pm \sqrt{b^2 - 4ac}}{2a} = \frac{3i \pm \sqrt{(-3i)^2 - 4(1)(-2)}}{2(1)} = \frac{3i \pm \sqrt{9i^2 + 8}}{2} = \frac{3i \pm \sqrt{-1}}{2} = \frac{3i \pm i}{2}$$

$x = \frac{4i}{2} = 2i$ or $x = \frac{2i}{2} = i$

78. $ix^2 + 3x - 2i = 0$

$a = i, b = 3, c = -2i$

$$x = \frac{-b \pm \sqrt{b^2 - 4ac}}{2a} = \frac{-3 \pm \sqrt{3^2 - 4(i)(-2i)}}{2(i)} = \frac{-3 \pm \sqrt{9 + 8i^2}}{2i} = \frac{-3 \pm \sqrt{1}}{2i} = \frac{-3 \pm 1}{2i} = \frac{-3 \pm 1}{2i} \cdot \frac{i}{i} = \frac{(-3 \pm 1)i}{-2}$$

$x = \frac{-2i}{-2} = i$ or $x = \frac{-4i}{-2} = 2i$

Exercise 8.3 (page 536)

1.

$$\frac{1}{4} + \frac{1}{t} = \frac{1}{2t}$$
$$\left(\frac{1}{4} + \frac{1}{t}\right)4t = \frac{1}{2t} \cdot 4t$$
$$t + 4 = 2$$
$$t = -2$$

2.

$$\frac{p-3}{3p} + \frac{1}{2p} = \frac{1}{4}$$
$$\left(\frac{p-3}{3p} + \frac{1}{2p}\right)(12p) = \frac{1}{4} \cdot 12p$$
$$4(p-3) + 6 = 3p$$
$$4p - 12 + 6 = 3p$$
$$p = 6$$

3. $m = \frac{\Delta y}{\Delta x} = \frac{-4-5}{-2-3} = \frac{-9}{-5} = \frac{9}{5}$

4. Use the slope from problem #3.

$$\begin{aligned} y - y_1 &= m(x - x_1) \\ y - 5 &= \frac{9}{5}(x-3) \\ 5(y-5) &= 5 \cdot \frac{9}{5}(x-3) \\ 5y - 25 &= 9x - 27 \\ -9x + 5y &= -2, \text{ or } 9x - 5y = 2 \end{aligned}$$

5. $b^2 - 4ac$

6. conjugates

7. rational; unequal

8. $-\frac{b}{a}; \frac{c}{a}$

9. $4x^2 - 4x + 1 = 0; a = 4, b = -4, c = 1$

$$\begin{aligned} b^2 - 4ac &= (-4)^2 - 4(4)(1) \\ &= 16 - 16 = 0 \end{aligned}$$

The solutions are rational and equal.

10. $6x^2 - 5x - 6 = 0; a = 6, b = -5, c = -6$

$$\begin{aligned} b^2 - 4ac &= (-5)^2 - 4(6)(-6) \\ &= 25 + 144 = 169 \end{aligned}$$

The solutions are rational and unequal.

11. $5x^2 + x + 2 = 0; a = 5, b = 1, c = 2$

$$\begin{aligned} b^2 - 4ac &= 1^2 - 4(5)(2) \\ &= 1 - 40 = -39 \end{aligned}$$

The solutions are complex conjugates.

12. $3x^2 + 10x - 2 = 0; a = 3, b = 10, c = -2$

$$\begin{aligned} b^2 - 4ac &= 10^2 - 4(3)(-2) \\ &= 100 + 24 = 124 \end{aligned}$$

The solutions are irrational and unequal.

13.

$$\begin{aligned} 2x^2 &= 4x - 1 \\ 2x^2 - 4x + 1 &= 0; a = 2, b = -4, c = 1 \\ b^2 - 4ac &= (-4)^2 - 4(2)(1) \\ &= 16 - 8 = 8 \end{aligned}$$

The solutions are irrational and unequal.

14.

$$\begin{aligned} 9x^2 &= 12x - 4 \\ 9x^2 - 12x + 4 &= 0; a = 9, b = -12, c = 4 \\ b^2 - 4ac &= (-12)^2 - 4(9)(4) \\ &= 144 - 144 = 0 \end{aligned}$$

The solutions are rational and equal.

15.

$$\begin{aligned} x(2x-3) &= 20 \\ 2x^2 - 3x - 20 &= 0; a = 2, b = -3, c = -20 \\ b^2 - 4ac &= (-3)^2 - 4(2)(-20) \\ &= 9 + 160 = 169 \end{aligned}$$

The solutions are rational and unequal.

16.

$$\begin{aligned} x(x-3) &= -10 \\ x^2 - 3x + 10 &= 0; a = 1, b = -3, c = 10 \\ b^2 - 4ac &= (-3)^2 - 4(1)(10) \\ &= 9 - 40 = -31 \end{aligned}$$

The solutions are complex conjugates.

17. $x^2 + kx + 9 = 0; a = 1, b = k, c = 9$

Set the discriminant equal to 0:

$$\begin{aligned} b^2 - 4ac &= 0 \\ k^2 - 4(1)(9) &= 0 \\ k^2 - 36 &= 0 \\ k^2 &= 36 \\ k &= \pm 6 \end{aligned}$$

18. $kx^2 - 12x + 4 = 0; a = k, b = -12, c = 4$

Set the discriminant equal to 0:

$$\begin{aligned} b^2 - 4ac &= 0 \\ (-12)^2 - 4(k)(4) &= 0 \\ 144 - 16k &= 0 \\ 144 &= 16k \\ 9 &= k \end{aligned}$$

19. $9x^2 + 4 = -kx$

$9x^2 + kx + 4 = 0;\ a = 9, b = k, c = 4$

Set the discriminant equal to 0:

$$\begin{aligned} b^2 - 4ac &= 0 \\ k^2 - 4(9)(4) &= 0 \\ k^2 - 144 &= 0 \\ k^2 &= 144 \\ k &= \pm 12 \end{aligned}$$

20. $9x^2 - kx + 25 = 0;\ a = 9, b = -k, c = 25$

Set the discriminant equal to 0:

$$\begin{aligned} b^2 - 4ac &= 0 \\ (-k)^2 - 4(9)(25) &= 0 \\ k^2 - 900 &= 0 \\ k^2 &= 900 \\ k &= \pm 30 \end{aligned}$$

21. $(k-1)x^2 + (k-1)x + 1 = 0$

$a = k-1, b = k-1, c = 1$

Set the discriminant equal to 0:

$$\begin{aligned} b^2 - 4ac &= 0 \\ (k-1)^2 - 4(k-1)(1) &= 0 \\ k^2 - 2k + 1 - 4k + 4 &= 0 \\ k^2 - 6k + 5 &= 0 \\ (k-5)(k-1) &= 0 \end{aligned}$$

$k - 5 = 0$ **or** $k - 1 = 0$

$k = 5$ $\quad$ $k = 1$

doesn't work

22. $(k+3)x^2 + 2kx + 4 = 0$

$a = k+3, b = 2k, c = 4$

Set the discriminant equal to 0:

$$\begin{aligned} b^2 - 4ac &= 0 \\ (2k)^2 - 4(k+3)(4) &= 0 \\ 4k^2 - 16(k+3) &= 0 \\ 4k^2 - 16k - 48 &= 0 \\ 4(k^2 - 4k - 12) &= 0 \\ 4(k-6)(k+2) &= 0 \end{aligned}$$

$k - 6 = 0$ **or** $k + 2 = 0$

$k = 6$ $\quad$ $k = -2$

23. $(k+4)x^2 + 2kx + 9 = 0$

$a = k+4, b = 2k, c = 9$

Set the discriminant equal to 0:

$$\begin{aligned} b^2 - 4ac &= 0 \\ (2k)^2 - 4(k+4)(9) &= 0 \\ 4k^2 - 36(k+4) &= 0 \\ 4k^2 - 36k - 144 &= 0 \\ 4(k^2 - 9k - 36) &= 0 \\ 4(k-12)(k+3) &= 0 \end{aligned}$$

$k - 12 = 0$ **or** $k + 3 = 0$

$k = 12$ $\quad$ $k = -3$

24. $(k+15)x^2 + (k-30)x + 4 = 0$

$a = k+15, b = k-30, c = 4$

Set the discriminant equal to 0:

$$\begin{aligned} b^2 - 4ac &= 0 \\ (k-30)^2 - 4(k+15)(4) &= 0 \\ k^2 - 60k + 900 - 16(k+15) &= 0 \\ k^2 - 60k + 900 - 16k - 240 &= 0 \\ k^2 - 76k + 660 &= 0 \\ (k-66)(k-10) &= 0 \end{aligned}$$

$k - 66 = 0$ **or** $k - 10 = 0$

$k = 66$ $\quad$ $k = 10$

25. $1492x^2 + 1776x - 1984 = 0$

$a = 1492, b = 1776, c = -1984$

$$\begin{aligned} b^2 - 4ac &= (1776)^2 - 4(1492)(-1984) \\ &= 3{,}154{,}176 + 11{,}840{,}512 \\ &= 14{,}994{,}688 \end{aligned}$$

The solutions are real numbers.

26. $1776x^2 - 1492x + 1984 = 0$

$a = 1776, b = -1492, c = 1984$

$$\begin{aligned} b^2 - 4ac &= (-1492)^2 - 4(1776)(1984) \\ &= 2{,}226{,}064 - 14{,}094{,}336 \\ &= -11{,}868{,}272 \end{aligned}$$

The solutions are not real numbers.

27.
$$\begin{aligned} 3x^2 + 4x &= k \\ 3x^2 + 4x - k &= 0 \end{aligned}$$
$a = 3, b = 4, c = -k$
Set the discriminant less than 0:
$$\begin{aligned} b^2 - 4ac &< 0 \\ 4^2 - 4(3)(-k) &< 0 \\ 16 + 12k &< 0 \\ 12k &< -16 \\ k &< -\frac{16}{12}, \text{ or } k < -\frac{4}{3} \end{aligned}$$

28.
$$\begin{aligned} kx^2 - 4x &= 7 \\ kx^2 - 4x - 7 &= 0 \end{aligned}$$
$a = k, b = -4, c = -7$
Set the discriminant less than 0:
$$\begin{aligned} b^2 - 4ac &< 0 \\ (-4)^2 - 4(k)(-7) &< 0 \\ 16 + 28k &< 0 \\ 28k &< -16 \\ k &< -\frac{16}{28}, \text{ or } k < -\frac{4}{7} \end{aligned}$$

29.
$$\begin{aligned} x^4 - 17x^2 + 16 &= 0 \\ (x^2 - 16)(x^2 - 1) &= 0 \end{aligned}$$
$$\begin{aligned} x^2 - 16 &= 0 & \textbf{or} \quad x^2 - 1 &= 0 \\ x^2 &= 16 & x^2 &= 1 \\ x &= \pm 4 & x &= \pm 1 \end{aligned}$$

30.
$$\begin{aligned} x^4 - 10x^2 + 9 &= 0 \\ (x^2 - 9)(x^2 - 1) &= 0 \end{aligned}$$
$$\begin{aligned} x^2 - 9 &= 0 & \textbf{or} \quad x^2 - 1 &= 0 \\ x^2 &= 9 & x^2 &= 1 \\ x &= \pm 3 & x &= \pm 1 \end{aligned}$$

31.
$$\begin{aligned} x^4 - 3x^2 &= -2 \\ x^4 - 3x^2 + 2 &= 0 \\ (x^2 - 2)(x^2 - 1) &= 0 \end{aligned}$$
$$\begin{aligned} x^2 - 2 &= 0 & \textbf{or} \quad x^2 - 1 &= 0 \\ x^2 &= 2 & x^2 &= 1 \\ x &= \pm\sqrt{2} & x &= \pm 1 \end{aligned}$$

32.
$$\begin{aligned} x^4 - 29x^2 &= -100 \\ x^4 - 29x^2 + 100 &= 0 \\ (x^2 - 25)(x^2 - 4) &= 0 \end{aligned}$$
$$\begin{aligned} x^2 - 25 &= 0 & \textbf{or} \quad x^2 - 4 &= 0 \\ x^2 &= 25 & x^2 &= 4 \\ x &= \pm 5 & x &= \pm 2 \end{aligned}$$

33.
$$\begin{aligned} x^4 &= 6x^2 - 5 \\ x^4 - 6x^2 + 5 &= 0 \\ (x^2 - 5)(x^2 - 1) &= 0 \end{aligned}$$
$$\begin{aligned} x^2 - 5 &= 0 & \textbf{or} \quad x^2 - 1 &= 0 \\ x^2 &= 5 & x^2 &= 1 \\ x &= \pm\sqrt{5} & x &= \pm 1 \end{aligned}$$

34.
$$\begin{aligned} x^4 &= 8x^2 - 7 \\ x^4 - 8x^2 + 7 &= 0 \\ (x^2 - 7)(x^2 - 1) &= 0 \end{aligned}$$
$$\begin{aligned} x^2 - 7 &= 0 & \textbf{or} \quad x^2 - 1 &= 0 \\ x^2 &= 7 & x^2 &= 1 \\ x &= \pm\sqrt{7} & x &= \pm 1 \end{aligned}$$

35.
$$\begin{aligned} 2x^4 - 10x^2 &= -8 \\ 2x^4 - 10x^2 + 8 &= 0 \\ 2(x^2 - 4)(x^2 - 1) &= 0 \end{aligned}$$
$$\begin{aligned} x^2 - 4 &= 0 & \textbf{or} \quad x^2 - 1 &= 0 \\ x^2 &= 4 & x^2 &= 1 \\ x &= \pm 2 & x &= \pm 1 \end{aligned}$$

36.
$$\begin{aligned} 3x^4 + 12 &= 15x^2 \\ 3x^4 - 15x^2 + 12 &= 0 \\ 3(x^2 - 4)(x^2 - 1) &= 0 \end{aligned}$$
$$\begin{aligned} x^2 - 4 &= 0 & \textbf{or} \quad x^2 - 1 &= 0 \\ x^2 &= 4 & x^2 &= 1 \\ x &= \pm 2 & x &= \pm 1 \end{aligned}$$

37.
$$\begin{aligned} 2x^4 + 24 &= 26x^2 \\ 2x^4 - 26x^2 + 24 &= 0 \\ 2(x^2 - 12)(x^2 - 1) &= 0 \end{aligned}$$
$$\begin{array}{lcl} x^2 - 12 = 0 & \textbf{or} & x^2 - 1 = 0 \\ x^2 = 12 & & x^2 = 1 \\ x = \pm\sqrt{12} & & x = \pm 1 \\ \quad = \pm 2\sqrt{3} & & \end{array}$$

38.
$$\begin{aligned} 4x^4 &= -9 + 13x^2 \\ 4x^4 - 13x^2 + 9 &= 0 \\ (4x^2 - 9)(x^2 - 1) &= 0 \end{aligned}$$
$$\begin{array}{lcl} 4x^2 - 9 = 0 & \textbf{or} & x^2 - 1 = 0 \\ x^2 = \frac{9}{4} & & x^2 = 1 \\ x = \pm\frac{3}{2} & & x = \pm 1 \end{array}$$

39.
$$\begin{aligned} t^4 + 3t^2 &= 28 \\ t^4 + 3t^2 - 28 &= 0 \\ (t^2 + 7)(t^2 - 4) &= -0 \end{aligned}$$
$$\begin{array}{lcl} t^2 + 7 = 0 & \textbf{or} & t^2 - 4 = 0 \\ t^2 = -7 & & t^2 = 4 \\ t = \pm\sqrt{-7} & & t = \pm 2 \\ \quad = \pm i\sqrt{7} & & \end{array}$$

40.
$$\begin{aligned} t^4 + 4t^2 - 5 &= 0 \\ (t^2 + 5)(t^2 - 1) &= 0 \end{aligned}$$
$$\begin{array}{lcl} t^2 + 5 = 0 & \textbf{or} & t^2 - 1 = 0 \\ t^2 = -5 & & t^2 = 1 \\ t = \pm\sqrt{-5} & & t = \pm 1 \\ \quad = \pm i\sqrt{5} & & \end{array}$$

41.
$$\begin{aligned} x - 6\sqrt{x} + 8 &= 0 \\ (\sqrt{x} - 2)(\sqrt{x} - 4) &= 0 \end{aligned}$$
$$\begin{array}{lcl} \sqrt{x} - 2 = 0 & \textbf{or} & \sqrt{x} - 4 = 0 \\ \sqrt{x} = 2 & & \sqrt{x} = 4 \\ \left(\sqrt{x}\right)^2 = 2^2 & & \left(\sqrt{x}\right)^2 = 4^2 \\ x = 4 & & x = 16 \\ \text{Solution} & & \text{Solution} \end{array}$$

42.
$$\begin{aligned} x - 5\sqrt{x} + 4 &= 0 \\ (\sqrt{x} - 1)(\sqrt{x} - 4) &= 0 \end{aligned}$$
$$\begin{array}{lcl} \sqrt{x} - 1 = 0 & \textbf{or} & \sqrt{x} - 4 = 0 \\ \sqrt{x} = 1 & & \sqrt{x} = 4 \\ \left(\sqrt{x}\right)^2 = 1^2 & & \left(\sqrt{x}\right)^2 = 4^2 \\ x = 1 & & x = 16 \\ \text{Solution} & & \text{Solution} \end{array}$$

43.
$$\begin{aligned} 2x - \sqrt{x} &= 3 \\ 2x - \sqrt{x} - 3 &= 0 \\ (2\sqrt{x} - 3)(\sqrt{x} + 1) &= 0 \end{aligned}$$
$$\begin{array}{lcl} 2\sqrt{x} - 3 = 0 & \textbf{or} & \sqrt{x} + 1 = 0 \\ 2\sqrt{x} = 3 & & \sqrt{x} = -1 \\ \left(2\sqrt{x}\right)^2 = 3^2 & & \left(\sqrt{x}\right)^2 = (-1)^2 \\ 4x = 9 & & x = 1 \\ x = \frac{9}{4} & & \text{Does not check} \\ \text{Solution} & & \end{array}$$

44.
$$\begin{aligned} 3x - 4 &= -4\sqrt{x} \\ 3x + 4\sqrt{x} - 4 &= 0 \\ (3\sqrt{x} - 2)(\sqrt{x} + 2) &= 0 \end{aligned}$$
$$\begin{array}{lcl} 3\sqrt{x} - 2 = 0 & \textbf{or} & \sqrt{x} + 2 = 0 \\ 3\sqrt{x} = 2 & & \sqrt{x} = -2 \\ \left(3\sqrt{x}\right)^2 = 2^2 & & \left(\sqrt{x}\right)^2 = (-2)^2 \\ 9x = 4 & & x = 4 \\ x = \frac{4}{9} & & \text{Does not check} \\ \text{Solution} & & \end{array}$$

45.

$$2x + x^{1/2} - 3 = 0$$
$$\left(2x^{1/2} + 3\right)\left(x^{1/2} - 1\right) = 0$$

$2x^{1/2} + 3 = 0$ **or** $x^{1/2} - 1 = 0$

$2x^{1/2} = -3$ $\qquad x^{1/2} = 1$

$x^{1/2} = -\frac{3}{2}$ $\qquad \left(x^{1/2}\right)^2 = 1^2$

$\left(x^{1/2}\right)^2 = \left(-\frac{3}{2}\right)^2$ $\qquad x = 1$

$x = \frac{9}{4}$ $\qquad$ Solution

Does not check

46.

$$2x - x^{1/2} - 1 = 0$$
$$\left(2x^{1/2} + 1\right)\left(x^{1/2} - 1\right) = 0$$

$2x^{1/2} + 1 = 0$ **or** $x^{1/2} - 1 = 0$

$2x^{1/2} = -1$ $\qquad x^{1/2} = 1$

$x^{1/2} = -\frac{1}{2}$ $\qquad \left(x^{1/2}\right)^2 = 1^2$

$\left(x^{1/2}\right)^2 = \left(-\frac{1}{2}\right)^2$ $\qquad x = 1$

$x = \frac{1}{4}$ $\qquad$ Solution

Does not check

47.

$$3x + 5x^{1/2} + 2 = 0$$
$$\left(3x^{1/2} + 2\right)\left(x^{1/2} + 1\right) = 0$$

$3x^{1/2} + 2 = 0$ **or** $x^{1/2} + 1 = 0$

$3x^{1/2} = -2$ $\qquad x^{1/2} = -1$

$x^{1/2} = -\frac{2}{3}$ $\qquad \left(x^{1/2}\right)^2 = (-1)^2$

$\left(x^{1/2}\right)^2 = \left(-\frac{2}{3}\right)^2$ $\qquad x = 1$

$x = \frac{4}{9}$ $\qquad$ Does not check

Does not check

48.

$$3x - 4x^{1/2} + 1 = 0$$
$$\left(3x^{1/2} - 1\right)\left(x^{1/2} - 1\right) = 0$$

$3x^{1/2} - 1 = 0$ **or** $x^{1/2} - 1 = 0$

$3x^{1/2} = 1$ $\qquad x^{1/2} = 1$

$x^{1/2} = \frac{1}{3}$ $\qquad \left(x^{1/2}\right)^2 = 1^2$

$\left(x^{1/2}\right)^2 = \left(\frac{1}{3}\right)^2$ $\qquad x = 1$

$x = \frac{1}{9}$ $\qquad$ Solution

Solution

49.

$$x^{2/3} + 5x^{1/3} + 6 = 0$$
$$\left(x^{1/3} + 2\right)\left(x^{1/3} + 3\right) = 0$$

$x^{1/3} + 2 = 0$ **or** $x^{1/3} + 3 = 0$

$x^{1/3} = -2$ $\qquad x^{1/3} = -3$

$\left(x^{1/3}\right)^3 = (-2)^3$ $\qquad \left(x^{1/3}\right)^3 = (-3)^3$

$x = -8$ $\qquad x = -27$

Solution $\qquad$ Solution

50.

$$x^{2/3} - 7x^{1/3} + 12 = 0$$
$$\left(x^{1/3} - 4\right)\left(x^{1/3} - 3\right) = 0$$

$x^{1/3} - 4 = 0$ **or** $x^{1/3} - 3 = 0$

$x^{1/3} = 4$ $\qquad x^{1/3} = 3$

$\left(x^{1/3}\right)^3 = (4)^3$ $\qquad \left(x^{1/3}\right)^3 = (3)^3$

$x = 64$ $\qquad x = 27$

Solution $\qquad$ Solution

51.

$$x^{2/3} - 2x^{1/3} - 3 = 0$$
$$(x^{1/3} - 3)(x^{1/3} + 1) = 0$$
$$x^{1/3} - 3 = 0 \quad \textbf{or} \quad x^{1/3} + 1 = 0$$
$$x^{1/3} = 3 \qquad x^{1/3} = -1$$
$$(x^{1/3})^3 = (3)^3 \qquad (x^{1/3})^3 = (-1)^3$$
$$x = 27 \qquad x = -1$$

Solution Solution

52.

$$x^{2/3} + 4x^{1/3} - 5 = 0$$
$$(x^{1/3} + 5)(x^{1/3} - 1) = 0$$
$$x^{1/3} + 5 = 0 \quad \textbf{or} \quad x^{1/3} - 1 = 0$$
$$x^{1/3} = -5 \qquad x^{1/3} = 1$$
$$(x^{1/3})^3 = (-5)^3 \qquad (x^{1/3})^3 = (1)^3$$
$$x = -125 \qquad x = 1$$

Solution Solution

53.

$$x + 5 + \frac{4}{x} = 0$$
$$x\left(x + 5 + \frac{4}{x}\right) = x(0)$$
$$x^2 + 5x + 4 = 0$$
$$(x + 4)(x + 1) = 0$$
$$x + 4 = 0 \quad \textbf{or} \quad x + 1 = 0$$
$$x = -4 \qquad x = -1$$

54.

$$x - 4 + \frac{3}{x} = 0$$
$$x\left(x - 4 + \frac{3}{x}\right) = x(0)$$
$$x^2 - 4x + 3 = 0$$
$$(x - 3)(x - 1) = 0$$
$$x - 3 = 0 \quad \textbf{or} \quad x - 1 = 0$$
$$x = 3 \qquad x = 1$$

55.

$$x + 1 = \frac{20}{x}$$
$$x + 1 - \frac{20}{x} = 0$$
$$x\left(x + 1 - \frac{20}{x}\right) = x(0)$$
$$x^2 + x - 20 = 0$$
$$(x + 5)(x - 4) = 0$$
$$x + 5 = 0 \quad \textbf{or} \quad x - 4 = 0$$
$$x = -5 \qquad x = 4$$

56.

$$x + \frac{15}{x} = 8$$
$$x - 8 + \frac{15}{x} = 0$$
$$x\left(x - 8 + \frac{15}{x}\right) = x(0)$$
$$x^2 - 8x + 15 = 0$$
$$(x - 5)(x - 3) = 0$$
$$x - 5 = 0 \quad \textbf{or} \quad x - 3 = 0$$
$$x = 5 \qquad x = 3$$

57.

$$\frac{1}{x - 1} + \frac{3}{x + 1} = 2$$
$$\left(\frac{1}{x - 1} + \frac{3}{x + 1}\right)(x - 1)(x + 1) = 2(x + 1)(x - 1)$$
$$1(x + 1) + 3(x - 1) = 2(x^2 - 1)$$
$$x + 1 + 3x - 3 = 2x^2 - 2$$
$$0 = 2x^2 - 4x$$
$$0 = 2x(x - 2) \qquad 2x = 0 \quad \textbf{or} \quad x - 2 = 0$$
$$x = 0 \qquad x = 2$$

58.

$$\frac{6}{x-2} - \frac{12}{x-1} = -1$$
$$\left(\frac{6}{x-2} - \frac{12}{x-1}\right)(x-1)(x-2) = -1(x-1)(x-2)$$
$$6(x-1) - 12(x-2) = -1(x^2 - 3x + 2)$$
$$6x - 6 - 12x + 24 = -x^2 + 3x - 2$$
$$x^2 - 9x + 20 = 0$$
$$(x-4)(x-5) = 0 \qquad x - 4 = 0 \quad \textbf{or} \quad x - 5 = 0$$
$$x = 4 \qquad\qquad x = 5$$

59.

$$\frac{1}{x+2} + \frac{24}{x+3} = 13$$
$$\left(\frac{1}{x+2} + \frac{24}{x+3}\right)(x+2)(x+3) = 13(x+2)(x+3)$$
$$1(x+3) + 24(x+2) = 13(x^2 + 5x + 6)$$
$$x + 3 + 24x + 48 = 13x^2 + 65x + 78$$
$$0 = 13x^2 + 40x + 27$$
$$0 = (13x+27)(x+1) \qquad 13x + 27 = 0 \quad \textbf{or} \quad x + 1 = 0$$
$$13x = -27 \qquad\qquad x = -1$$
$$x = -\tfrac{27}{13}$$

60.

$$\frac{3}{x} + \frac{4}{x+1} = 2$$
$$\left(\frac{3}{x} + \frac{4}{x+1}\right)(x)(x+1) = 2(x)(x+1)$$
$$3(x+1) + 4(x) = 2(x^2 + x)$$
$$3x + 3 + 4x = 2x^2 + 2x$$
$$0 = 2x^2 - 5x - 3$$
$$0 = (2x+1)(x-3) \qquad 2x + 1 = 0 \quad \textbf{or} \quad x - 3 = 0$$
$$2x = -1 \qquad\qquad x = 3$$
$$x = -\tfrac{1}{2}$$

61.

$$\begin{aligned}
x + \frac{2}{x-2} &= 0 \\
\left(x + \frac{2}{x-2}\right)(x-2) &= 0(x-2) \\
x(x-2) + 2 &= 0 \\
x^2 - 2x + 2 &= 0 \\
x^2 - 2x &= -2 \\
x^2 - 2x + 1 &= -2 + 1 \\
(x-1)^2 &= -1 \\
x - 1 &= \pm\sqrt{-1} \\
x &= 1 \pm i
\end{aligned}$$

62.

$$\begin{aligned}
x + \frac{x+5}{x-3} &= 0 \\
\left(x + \frac{x+5}{x-3}\right)(x-3) &= 0(x-3) \\
x(x-3) + x + 5 &= 0 \\
x^2 - 2x + 5 &= 0 \\
x^2 - 2x &= -5 \\
x^2 - 2x + 1 &= -5 + 1 \\
(x-1)^2 &= -4 \\
x - 1 &= \pm\sqrt{-4} \\
x &= 1 \pm 2i
\end{aligned}$$

63.

$$\begin{aligned}
x^{-4} - 2x^{-2} + 1 &= 0 \\
(x^{-2} - 1)(x^{-2} - 1) &= 0
\end{aligned}$$

$x^{-2} - 1 = 0$ **or** $x^{-2} - 1 = 0$

$$\begin{aligned}
x^{-2} &= 1 & x^{-2} &= 1 \\
\frac{1}{x^2} &= 1 & \frac{1}{x^2} &= 1 \\
1 &= x^2 & 1 &= x^2 \\
\pm 1 &= x & \pm 1 &= x
\end{aligned}$$

64.

$$\begin{aligned}
4x^{-4} + 1 &= 5x^{-2} \\
4x^{-4} - 5x^{-2} + 1 &= 0 \\
(4x^{-2} - 1)(x^{-2} - 1) &= 0
\end{aligned}$$

$4x^{-2} - 1 = 0$ **or** $x^{-2} - 1 = 0$

$$\begin{aligned}
4x^{-2} &= 1 & x^{-2} &= 1 \\
\frac{4}{x^2} &= 1 & \frac{1}{x^2} &= 1 \\
4 &= x^2 & 1 &= x^2 \\
\pm 2 &= x & \pm 1 &= x
\end{aligned}$$

65.

$$\begin{aligned}
8a^{-2} - 10a^{-1} - 3 &= 0 \\
(2a^{-1} - 3)(4a^{-1} + 1) &= 0
\end{aligned}$$

$2a^{-1} - 3 = 0$ **or** $4a^{-1} + 1 = 0$

$$\begin{aligned}
2a^{-1} &= 3 & 4a^{-1} &= -1 \\
\frac{2}{a} &= 3 & \frac{4}{a} &= -1 \\
2 &= 3a & 4 &= -a \\
\tfrac{2}{3} &= a & -4 &= a
\end{aligned}$$

66.

$$\begin{aligned}
2y^{-2} - 5y^{-1} &= 3 \\
2y^{-2} - 5y^{-1} - 3 &= 0 \\
(y^{-1} - 3)(2y^{-1} + 1) &= 0
\end{aligned}$$

$y^{-1} - 3 = 0$ **or** $2y^{-1} + 1 = 0$

$$\begin{aligned}
y^{-1} &= 3 & 2y^{-1} &= -1 \\
\frac{1}{y} &= 3 & \frac{2}{y} &= -1 \\
1 &= 3y & 2 &= -y \\
\tfrac{1}{3} &= y & -2 &= y
\end{aligned}$$

67. Let $y = m + 1$.

$$8(m+1)^{-2} - 30(m+1)^{-1} + 7 = 0$$
$$8y^{-2} - 30y^{-1} + 7 = 0$$
$$(4y^{-1} - 1)(2y^{-1} - 7) = 0$$

$$\begin{aligned} 4y^{-1} - 1 &= 0 \\ 4y^{-1} &= 1 \\ \frac{4}{y} &= 1 \\ 4 &= y \\ 4 &= m + 1 \\ 3 &= m \end{aligned} \quad \textbf{or} \quad \begin{aligned} 2y^{-1} - 7 &= 0 \\ 2y^{-1} &= 7 \\ \frac{2}{y} &= 7 \\ 2 &= 7y \\ \tfrac{2}{7} &= y \\ \tfrac{2}{7} &= m + 1 \\ -\tfrac{5}{7} &= m \end{aligned}$$

68. Let $y = p - 2$.

$$2(p-2)^{-2} + 3(p-2)^{-1} - 5 = 0$$
$$2y^{-2} + 3y^{-1} - 5 = 0$$
$$(y^{-1} - 1)(2y^{-1} + 5) = 0$$

$$\begin{aligned} y^{-1} - 1 &= 0 \\ y^{-1} &= 1 \\ \frac{1}{y} &= 1 \\ 1 &= y \\ 1 &= p - 2 \\ 3 &= p \end{aligned} \quad \textbf{or} \quad \begin{aligned} 2y^{-1} + 5 &= 0 \\ 2y^{-1} &= -5 \\ \frac{2}{y} &= -5 \\ 2 &= -5y \\ -\tfrac{2}{5} &= y \\ -\tfrac{2}{5} &= p - 2 \\ \tfrac{8}{5} &= p \end{aligned}$$

69.
$$\begin{aligned} x^2 + y^2 &= r^2 \\ x^2 &= r^2 - y^2 \\ x &= \pm\sqrt{r^2 - y^2} \end{aligned}$$

70.
$$\begin{aligned} x^2 + y^2 &= r^2 \\ y^2 &= r^2 - x^2 \\ y &= \pm\sqrt{r^2 - x^2} \end{aligned}$$

71.
$$\begin{aligned} I &= \frac{k}{d^2} \\ Id^2 &= k \\ d^2 &= \frac{k}{I} \\ d &= \pm\sqrt{\frac{k}{I}} = \pm\frac{\sqrt{kI}}{I} \end{aligned}$$

72.
$$\begin{aligned} V &= \frac{1}{3}\pi r^2 h \\ 3V &= \pi r^2 h \\ \frac{3V}{\pi h} &= r^2 \\ \pm\sqrt{\frac{3V}{\pi h}} &= r, \text{ or } r = \pm\frac{\sqrt{3V\pi h}}{\pi h} \end{aligned}$$

73. $xy^2 + 3xy + 7 = 0;\ a = x, b = 3x, c = 7$

$$y = \frac{-b \pm \sqrt{b^2 - 4ac}}{2a} = \frac{-3x \pm \sqrt{(3x)^2 - 4(x)(7)}}{2x} = \frac{-3x \pm \sqrt{9x^2 - 28x}}{2x}$$

74. $kx = ay - x^2$

$x^2 + kx - ay = 0;\ a = 1, b = k, c = -ay$

$$x = \frac{-b \pm \sqrt{b^2 - 4ac}}{2a} = \frac{-k \pm \sqrt{k^2 - 4(1)(-ay)}}{2(1)} = \frac{-k \pm \sqrt{k^2 + 4ay}}{2}$$

75. $\sigma = \sqrt{\dfrac{\Sigma x^2}{N} - \mu^2}$

$$\sigma^2 = \frac{\Sigma x^2}{N} - \mu^2$$

$$\mu^2 = \frac{\Sigma x^2}{N} - \sigma^2$$

76. $\sigma = \sqrt{\dfrac{\Sigma x^2}{N} - \mu^2}$

$$\sigma^2 = \frac{\Sigma x^2}{N} - \mu^2$$

$$N\sigma^2 = \Sigma x^2 - N\mu^2$$

$$N\sigma^2 + N\mu^2 = \Sigma x^2$$

$$N(\sigma^2 + \mu^2) = \Sigma x^2 \Rightarrow N = \frac{\Sigma x^2}{\sigma^2 + \mu^2}$$

77. $12x^2 - 5x - 2 = 0;\ a = 12, b = -5, c = -2$

$(4x + 1)(3x - 2) = 0$

$4x + 1 = 0$ **or** $3x - 2 = 0$

$4x = -1$ $\qquad 3x = 2$

$x = -\frac{1}{4}$ $\qquad x = \frac{2}{3}$

$$-\frac{b}{a} = -\frac{-5}{12} = \frac{5}{12} \qquad \frac{c}{a} = \frac{-2}{12} = -\frac{1}{6}$$

$$-\frac{1}{4} + \frac{2}{3} = -\frac{3}{12} + \frac{8}{12} = \frac{5}{12} \qquad \left(-\frac{1}{4}\right)\left(\frac{2}{3}\right) = -\frac{1}{6}$$

78. $8x^2 - 2x - 3 = 0;\ a = 8, b = -2, c = -3$

$(4x - 3)(2x + 1) = 0$

$4x - 3 = 0$ **or** $2x + 1 = 0$

$4x = 3$ $\qquad 2x = -1$

$x = \frac{3}{4}$ $\qquad x = -\frac{1}{2}$

$$-\frac{b}{a} = -\frac{-2}{8} = \frac{1}{4} \qquad \frac{c}{a} = \frac{-3}{8} = -\frac{3}{8}$$

$$-\frac{1}{2} + \frac{3}{4} = -\frac{2}{4} + \frac{3}{4} = \frac{1}{4} \qquad \left(-\frac{1}{2}\right)\left(\frac{3}{4}\right) = -\frac{3}{8}$$

79. $2x^2 + 5x + 1 = 0;\ a = 2, b = 5, c = 1;\ -\dfrac{b}{a} = -\dfrac{5}{2};\ \dfrac{c}{a} = \dfrac{1}{2}$

$$x = \frac{-b \pm \sqrt{b^2 - 4ac}}{2a} = \frac{-5 \pm \sqrt{5^2 - 4(2)(1)}}{2(2)} = \frac{-5 \pm \sqrt{17}}{4} = -\frac{5}{4} \pm \frac{\sqrt{17}}{4}$$

$$\frac{-5 + \sqrt{17}}{4} + \frac{-5 - \sqrt{17}}{4} = \frac{-10}{4} = -\frac{5}{2}$$

$$\left(\frac{-5 + \sqrt{17}}{4}\right)\left(\frac{-5 - \sqrt{17}}{4}\right) = \frac{25 + 5\sqrt{17} - 5\sqrt{17} - 17}{16} = \frac{8}{16} = \frac{1}{2}$$

80. $3x^2+9x+1=0;\ a=3, b=9, c=1;\ -\frac{b}{a}=-\frac{9}{3}=-3;\ \frac{c}{a}=\frac{1}{3}$

$$x=\frac{-b\pm\sqrt{b^2-4ac}}{2a}=\frac{-9\pm\sqrt{9^2-4(3)(1)}}{2(3)}=\frac{-9\pm\sqrt{69}}{6}=-\frac{3}{2}\pm\frac{\sqrt{69}}{6}$$

$$\frac{-9+\sqrt{69}}{6}+\frac{-9-\sqrt{69}}{6}=\frac{-18}{6}=-3$$

$$\left(\frac{-9+\sqrt{69}}{6}\right)\left(\frac{-9-\sqrt{69}}{6}\right)=\frac{81+9\sqrt{69}-9\sqrt{69}-69}{36}=\frac{12}{36}=\frac{1}{3}$$

81. $3x^2-2x+4=0;\ a=3, b=-2, c=4;\ -\frac{b}{a}=-\frac{-2}{3}=\frac{2}{3};\ \frac{c}{a}=\frac{4}{3}$

$$x=\frac{-b\pm\sqrt{b^2-4ac}}{2a}=\frac{-(-2)\pm\sqrt{(-2)^2-4(3)(4)}}{2(3)}=\frac{2\pm\sqrt{-44}}{6}=\frac{1}{3}\pm\frac{i\sqrt{11}}{3}$$

$$\frac{1+i\sqrt{11}}{3}+\frac{1-i\sqrt{11}}{3}=\frac{2}{3}$$

$$\left(\frac{1+i\sqrt{11}}{3}\right)\left(\frac{1-i\sqrt{11}}{3}\right)=\frac{1-i\sqrt{11}+i\sqrt{11}-11i^2}{9}=\frac{1+11}{9}=\frac{12}{9}=\frac{4}{3}$$

82. $2x^2-x+4=0;\ a=2, b=-1, c=4;\ -\frac{b}{a}=-\frac{-1}{2}=\frac{1}{2};\ \frac{c}{a}=\frac{4}{2}=2$

$$x=\frac{-b\pm\sqrt{b^2-4ac}}{2a}=\frac{-(-1)\pm\sqrt{(-1)^2-4(2)(4)}}{2(2)}=\frac{1\pm\sqrt{-31}}{4}=\frac{1}{4}\pm\frac{i\sqrt{31}}{4}$$

$$\frac{1+i\sqrt{31}}{4}+\frac{1-i\sqrt{31}}{4}=\frac{2}{4}=\frac{1}{2}$$

$$\left(\frac{1+i\sqrt{31}}{4}\right)\left(\frac{1-i\sqrt{31}}{4}\right)=\frac{1-i\sqrt{31}+i\sqrt{31}-31i^2}{16}=\frac{1+31}{16}=\frac{32}{16}=2$$

83. $x^2+2x+5=0;\ a=1, b=2, c=5;\ -\frac{b}{a}=-\frac{2}{1}=-2;\ \frac{c}{a}=\frac{5}{1}=5$

$$x=\frac{-b\pm\sqrt{b^2-4ac}}{2a}=\frac{-2\pm\sqrt{2^2-4(1)(5)}}{2(1)}=\frac{-2\pm\sqrt{-16}}{2}=\frac{-2\pm 4i}{2}=-1\pm 2i$$

$(-1+2i)+(-1-2i)=-2;\ (-1+2i)(-1-2i)=1+2i-2i-4i^2=1+4=5$

84. $x^2-4x+13=0;\ a=1, b=-4, c=13;\ -\frac{b}{a}=-\frac{-4}{1}=4;\ \frac{c}{a}=\frac{13}{1}=13$

$$x=\frac{-b\pm\sqrt{b^2-4ac}}{2a}=\frac{-(-4)\pm\sqrt{(-4)^2-4(1)(13)}}{2(1)}=\frac{4\pm\sqrt{-36}}{2}=\frac{4\pm 6i}{2}=2\pm 3i$$

$(2+3i)+(2-3i)=4;\ (2+3i)(2-3i)=4-6i+6i-9i^2=4+9=13$

85. **Answers may vary.**

86. **Answers may vary.**

87. No

88. Yes

Exercise 8.4 (page 549)

1. $$\begin{aligned} 3x + 5 &= 5x - 15 \\ -2x &= -20 \\ x &= 10 \end{aligned}$$

2. $$\begin{aligned} 14x - 10 + 22x + 10 &= 180 \\ 36x &= 180 \\ x &= 5 \end{aligned}$$

3. Let t = the time of the second train.
Then $t + 3$ = the time of the first train.

	Rate	Time	Dist.
First	30	$t+3$	$30(t+3)$
Second	55	t	$55t$

1st distance = 2nd distance

$$\begin{aligned} 30(t+3) &= 55t \\ 30t + 90 &= 55t \\ -25t &= -90 \\ t &= \tfrac{-90}{-25} = \tfrac{18}{5} = 3\tfrac{3}{5} \text{ hours} \end{aligned}$$

4. Let x = amount at 7%. Then the amount at 8% = 25,000 − x.

$$\begin{aligned} 0.07x + 0.08(25{,}000 - x) &= 1900 \\ 7x + 8(25{,}000 - x) &= 190{,}000 \\ 7x + 200{,}000 - 8x &= 190{,}000 \\ -x &= -10{,}000 \\ x &= 10{,}000 \end{aligned}$$

\$10,000 is at 7%, and **\$15,000 is at 8%**.

5. $f(x) = ax^2 + bx + c$; $a \neq 0$

6. parabolas

7. vertex

8. axis

9. upward

10. downward

11. to the right

12. to the left

13. upward

14. variance

15. $f(x) = x^2$
vertex: $(0, 0)$; opens U

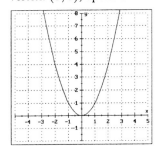

16. $f(x) = -x^2$
vertex: $(0, 0)$; opens D

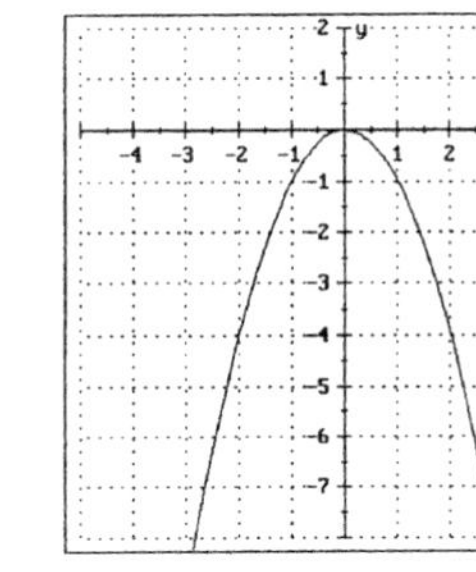

17. $f(x) = x^2 + 2$
vertex: $(0, 2)$; opens U

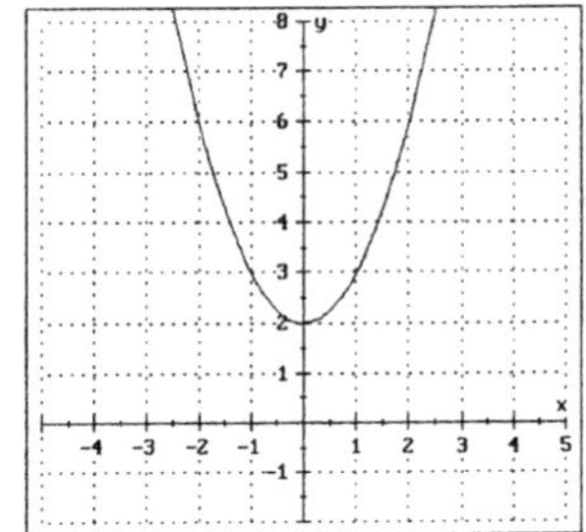

18. $f(x) = x^2 - 3$
vertex: $(0, -3)$; opens U

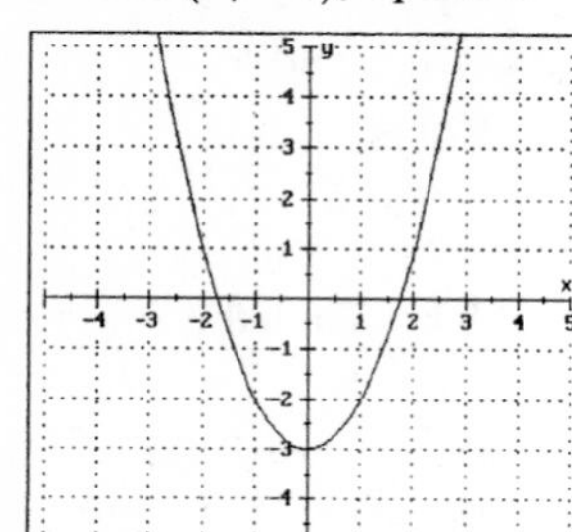

19. $f(x) = -(x-2)^2$
vertex: $(2, 0)$; opens D

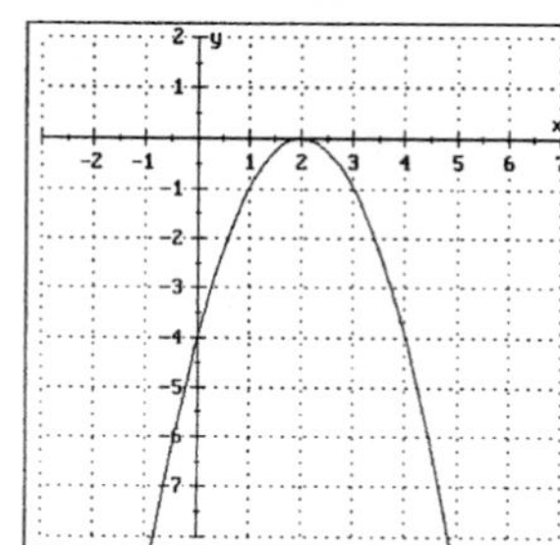

20. $f(x) = (x+2)^2$
vertex: $(-2, 0)$; opens U

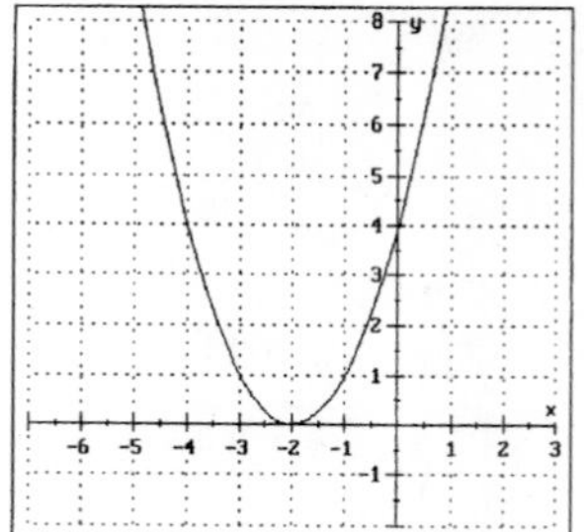

21. $f(x) = (x-3)^2 + 2$
vertex: $(3, 2)$; opens U

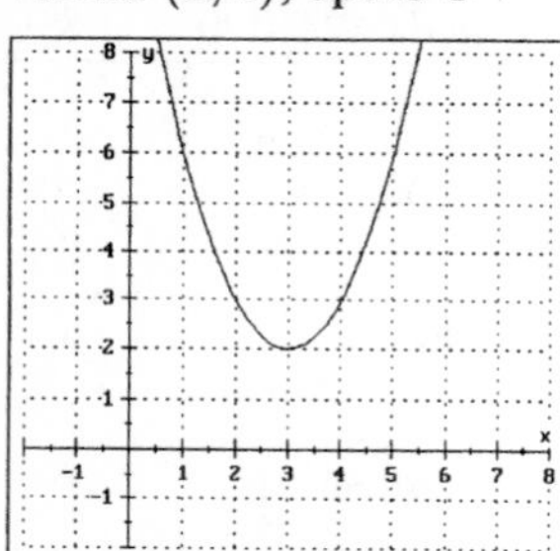

22. $f(x) = (x+1)^2 - 2$
vertex: $(-1, -2)$; opens U

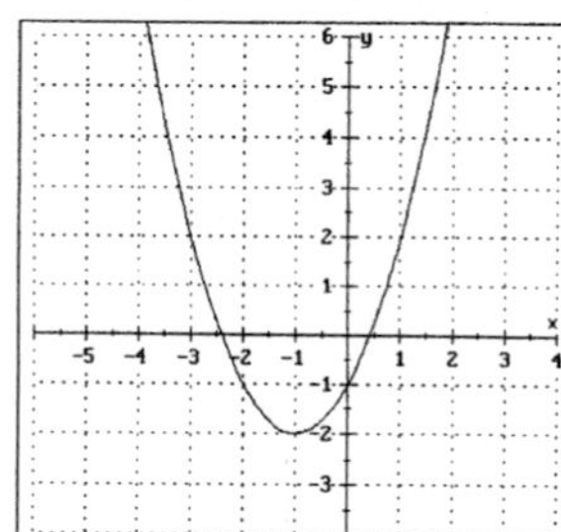

23. $f(x) = x^2 + x - 6$
$f(x) = \left(x + \frac{1}{2}\right)^2 - \frac{25}{4}$
vertex: $\left(-\frac{1}{2}, -\frac{25}{4}\right)$; opens U

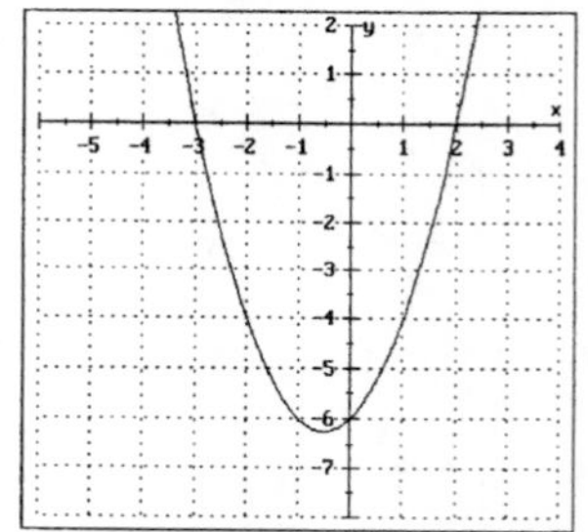

24. $f(x) = x^2 - x - 6$
$f(x) = \left(x - \frac{1}{2}\right)^2 - \frac{25}{4}$
vertex: $\left(\frac{1}{2}, -\frac{25}{4}\right)$; opens U

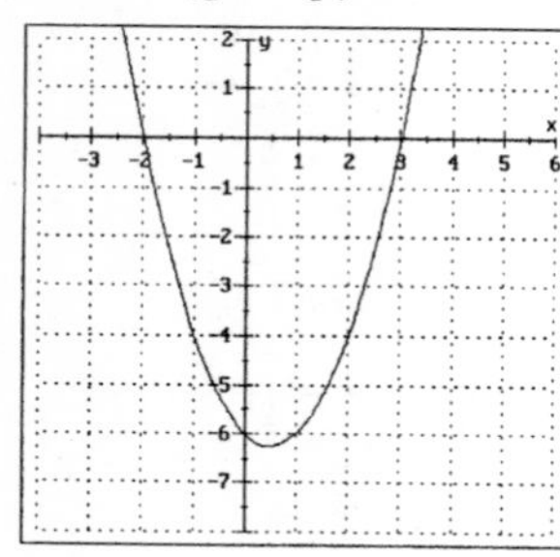

25. $f(x) = -2x^2 + 4x + 1$
$f(x) = -2(x-1)^2 + 3$
vertex: $(1, 3)$; opens D

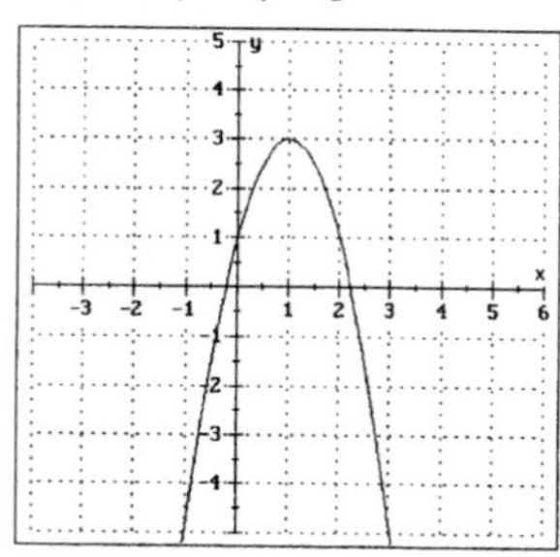

26. $f(x) = -2x^2 + 4x + 3$
$f(x) = -2(x-1)^2 + 5$
vertex: $(1, 5)$; opens D

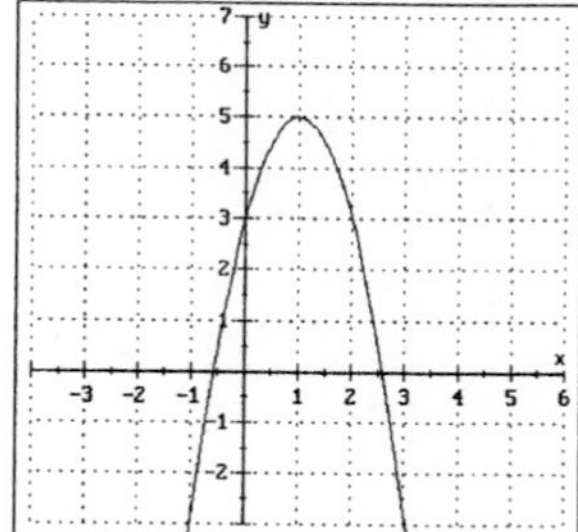

27. $f(x) = 3x^2 - 12x + 10$
$f(x) = 3(x-2)^2 - 2$
vertex: $(2, -2)$; opens U

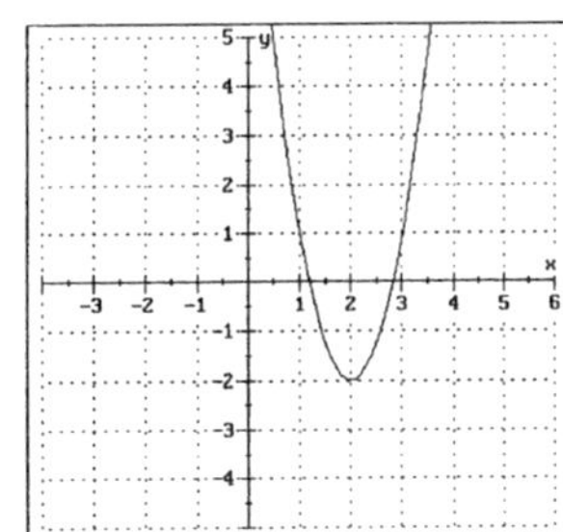

28. $f(x) = 3x^2 - 12x + 9$
$f(x) = 3(x-2)^2 - 3$
vertex: $(2, -3)$; opens U

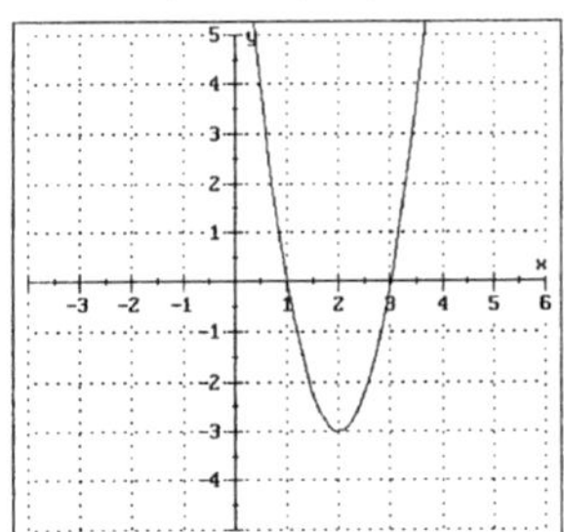

29. $y = (x-1)^2 + 2; V(1, 2)$; axis: $x = 1$

30. $y = 2(x-2)^2 - 1; V(2, -1)$; axis: $x = 2$

31. $y = 2(x+3)^2 - 4;$
$V(-3, -4)$; axis: $x = -3$

32. $y = -3(x+1)^2 + 3; V(-1, 3)$; axis: $x = -1$

33. $y = -3x^2 \Rightarrow y = -3(x-0)^2 + 0$
$V(0, 0)$; axis: $x = 0$

34. $y = 3x^2 - 3 \Rightarrow y = 3(x-0)^2 - 3$
$V(0, -3)$; axis: $x = 0$

35. $y = 2x^2 - 4x$
$y = 2(x^2 - 2x)$
$y = 2(x^2 - 2x + 1) - 2$
$y = 2(x-1)^2 - 2$
$V(1, -2)$; axis: $x = 1$

36. $y = 3x^2 + 6x$
$y = 3(x^2 + 2x)$
$y = 3(x^2 + 2x + 1) - 3$
$y = 3(x+1)^2 - 3$
$V(-1, -3)$; axis: $x = -1$

37. $y = -4x^2 + 16x + 5$
$y = -4(x^2 - 4x) + 5$
$y = -4(x^2 - 4x + 4) + 5 + 16$
$y = -4(x-2)^2 + 21$
$V(2, 21)$; axis: $x = 2$

38. $y = 5x^2 + 20x + 25$
$y = 5(x^2 + 4x) + 25$
$y = 5(x^2 + 4x + 4) + 25 - 20$
$y = 5(x+2)^2 + 5$
$V(-2, 5)$; axis: $x = -2$

39. $y - 7 = 6x^2 - 5x$

$$y = 6\left(x^2 - \frac{5}{6}x\right) + 7$$
$$y = 6\left(x^2 - \frac{5}{6}x + \frac{25}{144}\right) + 7 - \frac{25}{24}$$
$$y = 6\left(x - \frac{5}{12}\right)^2 + \frac{143}{24}$$

$V\left(\frac{5}{12}, \frac{143}{24}\right)$; axis: $x = \frac{5}{12}$

40. $y - 2 = 3x^2 + 4x$

$$y = 3\left(x^2 + \frac{4}{3}x\right) + 2$$
$$y = 3\left(x^2 + \frac{4}{3}x + \frac{4}{9}\right) + 2 - \frac{4}{3}$$
$$y = 3\left(x + \frac{2}{3}\right)^2 + \frac{2}{3}$$

$V\left(-\frac{2}{3}, \frac{2}{3}\right)$; axis: $x = -\frac{2}{3}$

41. $y - 2 = (x-5)^2$
$y = (x-5)^2 + 2 \Rightarrow V(5,2)$

42. $y = ax^2$
$y = a(x-0)^2 + 0 \Rightarrow V(0,0)$

43. $y = 2x^2 - x + 1$

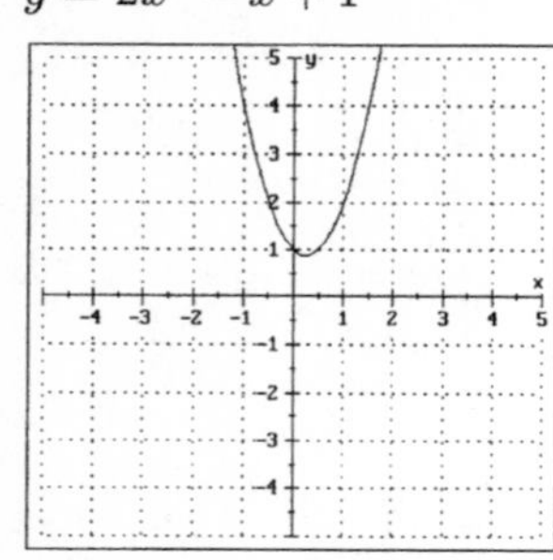

$V(0.25, 0.88)$

44. $y = x^2 + 5x - 6$

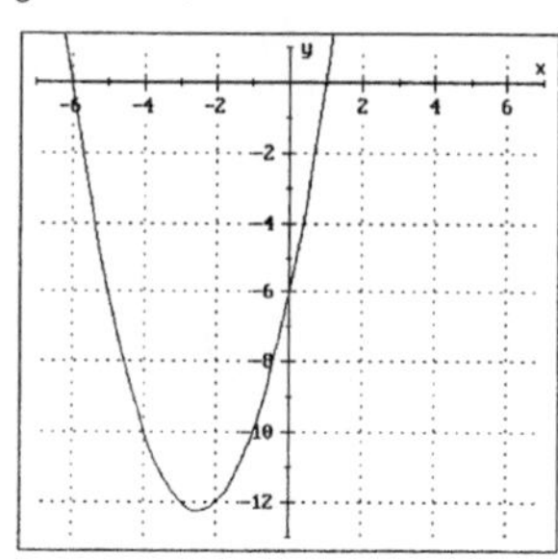

$V(-2.5, -12.25)$

45. $y = 7 + x - x^2$

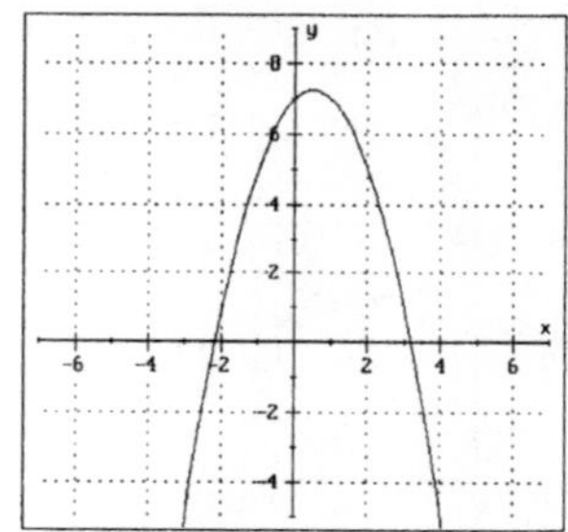

$V(0.5, 7.25)$

46. $y = 2x^2 - 3x + 2$

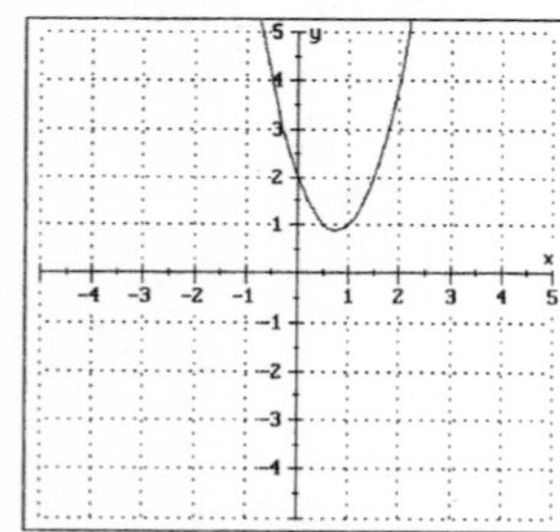

$V(0.75, 0.88)$

47. $y = x^2 + x - 6$

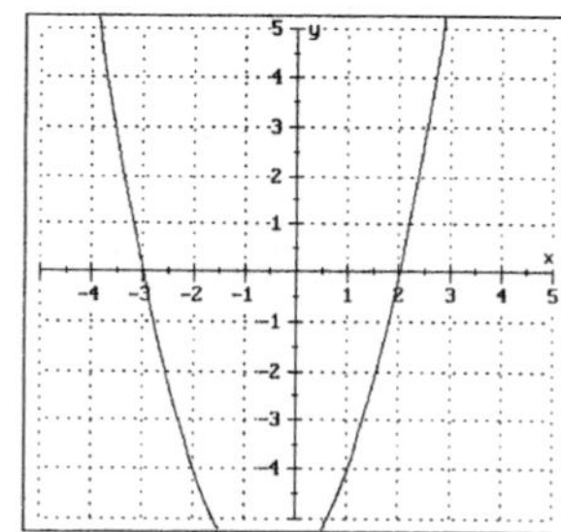

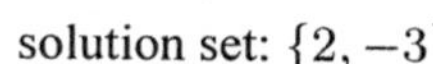

solution set: $\{2, -3\}$

48. $y = 2x^2 - 5x - 3$

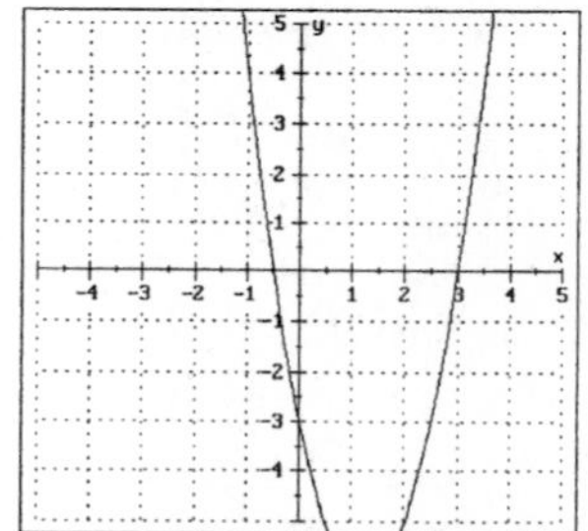

solution set: $\{3, -0.5\}$

49. $y = 0.5x^2 - 0.7x - 3$

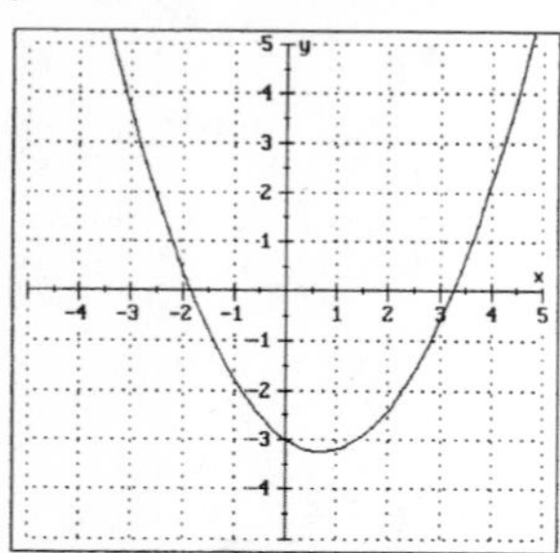

solution set: $\{-1.85, 3.25\}$

50. $y = 2x^2 - 0.5x - 2$

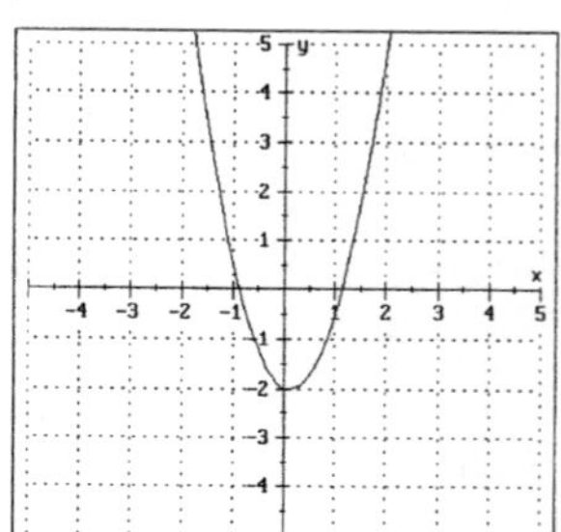

solution set: $\{-0.88, 1.13\}$

51. Since the graph of the height equation is a parabola, the max. height occurs at the vertex.
$s = 48t - 16t^2$
$s = -16(t^2 - 3t)$
$s = -16\left(t^2 - 3t + \frac{9}{4}\right) + 36$
$s = -16\left(t - \frac{3}{2}\right)^2 + 36$
$V\left(\frac{3}{2}, 36\right) \Rightarrow$ max. height $= 36$ ft

To find the time it takes for the ball to return to earth, set $s = 0$ and solve for t.
$s = 48t - 16t^2$
$0 = 48t - 16t^2$
$0 = 16t(3 - t)$
$t = 0$ or $t = 3$
The ball returns to earth after 3 seconds.
(0 seconds is when the ball is originally thrown.)

52. Since the graph of the height equation is a parabola, the max. height occurs at the vertex.
$s = -16t^2 + 32t + 48$
$s = -16(t^2 - 2t) + 48$
$s = -16(t^2 - 2t + 1) + 48 + 16$
$s = -16(t - 1)^2 + 64$
$V(1, 64) \Rightarrow$ max. height $= 64$ ft

To find the time it takes for the ball to return to earth, set $s = 0$ and solve for t.
$s = -16t^2 + 32t + 48$
$0 = -16t^2 + 32t + 48$
$0 = -16(t - 3)(t + 2)$
$t = 3$ or $t = -2$
The ball returns to earth after 3 seconds.
(-2 seconds is before the ball is thrown.)

53. Let $w =$ the width of the rectangle.
Then $100 - w =$ the length.
$A = w(100 - w)$
$A = -w^2 + 100w$
$A = -(w^2 - 100w + 2500) + 2500$
$A = -(w - 50)^2 + 2500$
dim: 50 ft by 50 ft; area $= 2500$ ft^2

54. Let $w =$ the length of the 2 congruent sides.
Then $1000 - 2w =$ the other side.
$A = w(1000 - 2w)$
$A = -2w^2 + 1000w$
$A = -2(w^2 - 500w + 62500) + 125000$
$A = -2(w - 250)^2 + 125000$
dim: 250 ft by 500 ft

55. Let $w =$ the width of the rectangle.
Then $150 - w =$ the length.
$A = w(150 - w)$
$A = -w^2 + 150w$
$A = -(w^2 - 150w + 5625) + 5625$
$A = -(w - 75)^2 + 5625$
dim: 75 ft by 75 ft; area $= 5625$ ft^2

56. Since the equation is a quadratic function, the minimum cost will occur at the vertex.
$C(n) = 2.2n^2 - 66n + 655$
$C(n) = 2.2\left(n^2 - 30n\right) + 655$
$C(n) = 2.2\left(n^2 - 30n + 225\right) + 655 - 495$
$C(n) = 2.2(n - 15)^2 + 160$
The minimum cost is \$160 for 15 minutes.

57. Graph $y = H = 3.3x^2 - 59.4x + 281.3$ and find the y-coordinate of the vertex:

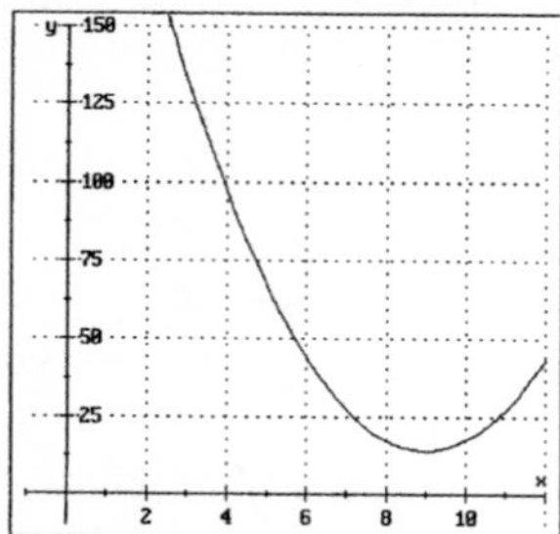

The minimum level was 14 feet.

58. Graph $y = E = 0.058x^2 - 1.162x + 50.604$ and find the y-coordinate of the vertex:

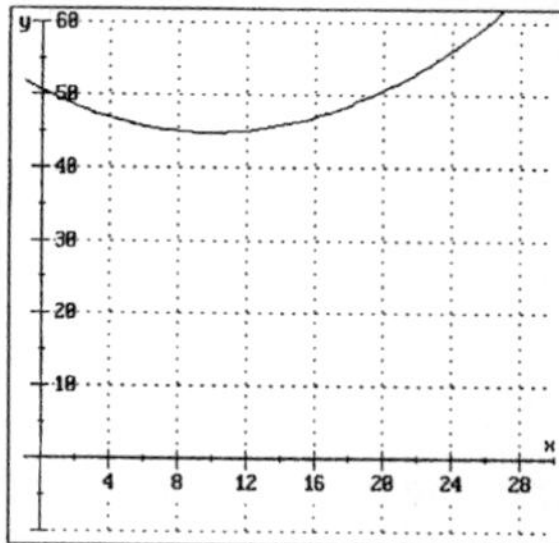

The lowest enrollment was about 44.8 million.

59. Graph $y = R = -\frac{x^2}{1000} + 10x$ and find the x-coordinate of the vertex:

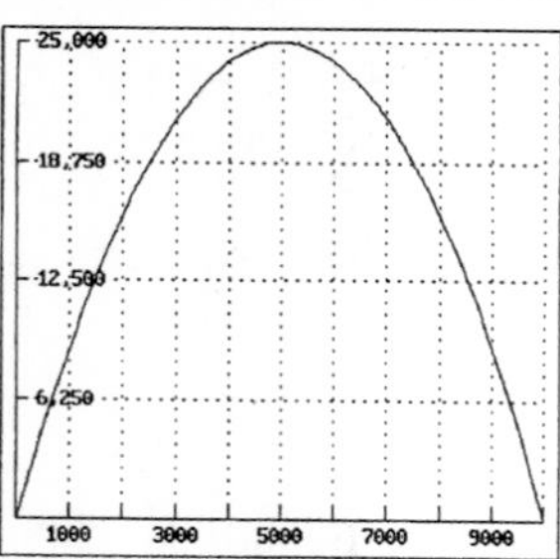

5000 stereos should be sold.

60. The maximum revenue will be the y-coord. of the vertex of the parabola graphed in **#59.** The maximum revenue is $25,000.

61. Graph $y = R = -\frac{x^2}{728} + 9x$ and find the x- and y- coordinates of the vertex:

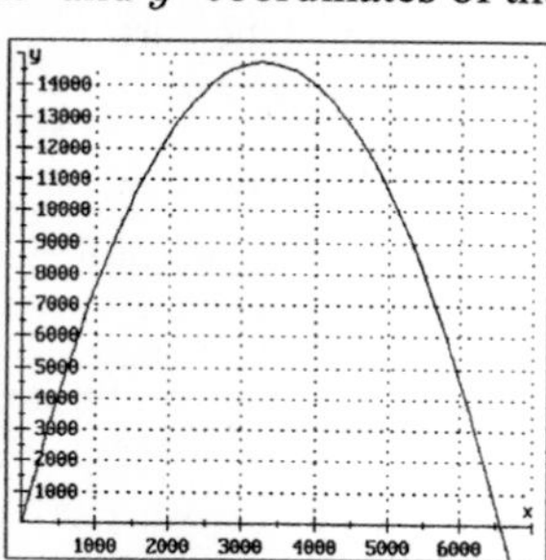

Max. revenue = $14,742; # radios = 3276

62. Graph $y = R = -\frac{x^2}{5} + 80x - 1000$ and find the x- and y-coordinates of the vertex:

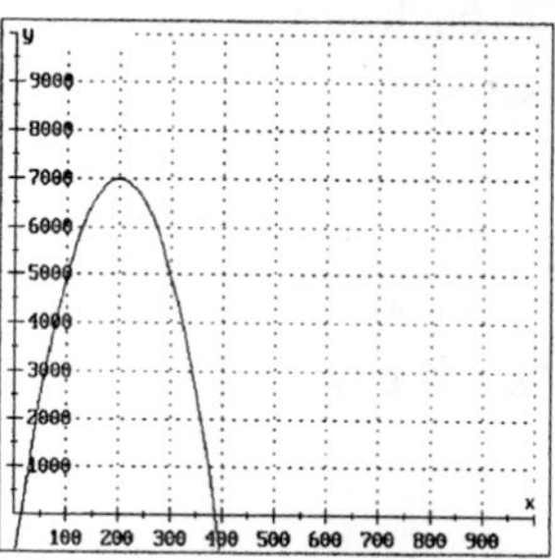

Max. revenue = $7000; # stereos = 200

63. Let $x =$ the number of \$1 increases to the price. Then the sales will be $4000 - 100x$, and the revenue will be $(30 + x)(4000 - 100x)$. Find the vertex of the parabola $y = (30 + x)(4000 - 100x)$.

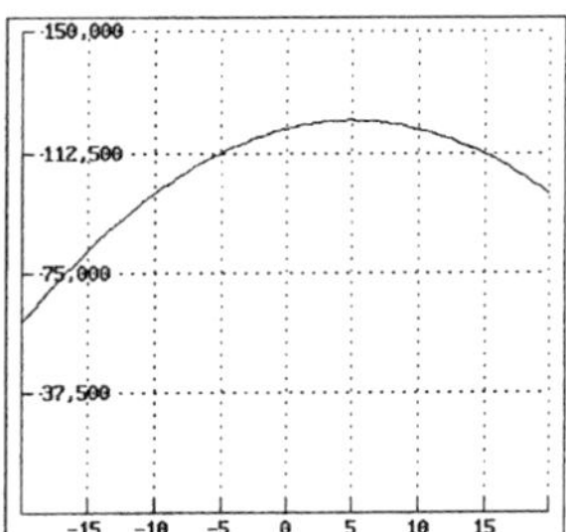

The price should increase \$5, to a total of \$35.

64. Let $x =$ the number of \$1 decreases to the price. Then the sales will be $525 + 75x$, and the revenue will be $(57 - x)(525 + 75x)$. Find the vertex of the parabola $y = (57 - x)(525 + 75x)$.

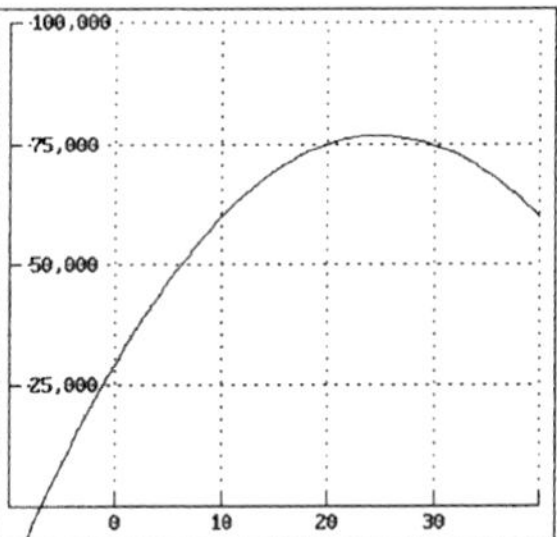

The price should decr. \$25, to a total of \$32.

65. Graph $y = 50x(1 - x)$ and find the x-coordinate(s) when $y = 9.375$.

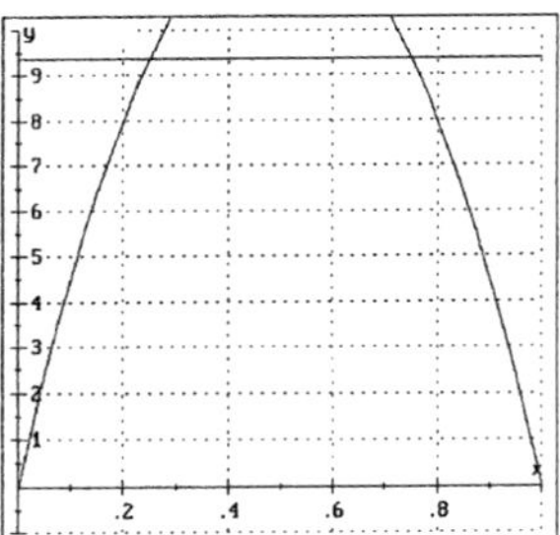

$p = 0.25$ or $p = 0.75$

66. Graph $y = 75x(1 - x)$ and find the x-coordinate(s) when $y = 12$.

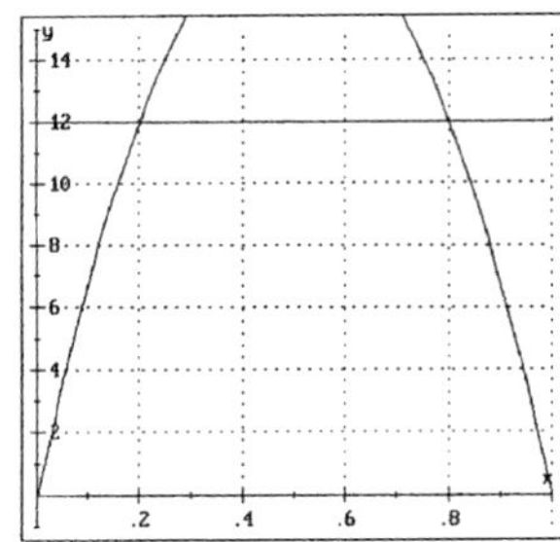

$p = 0.20$ or $p = 0.80$

67. **Answers may vary.**

68. **Answers may vary.**

69. Graph $y = x^2 + x + 1$ to find x-intercept(s):

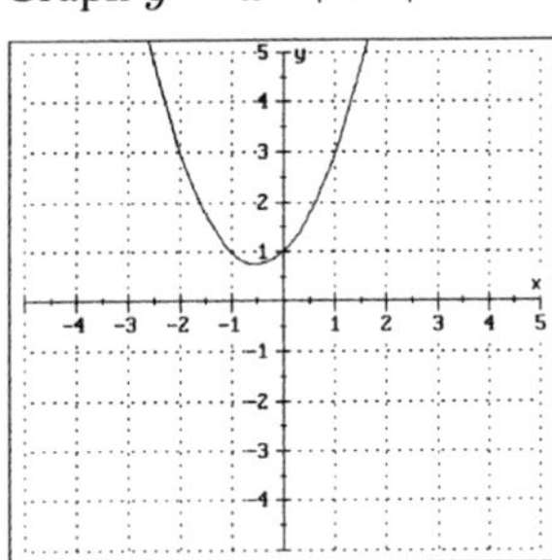

There are no x-intercepts, which means there is no solution to the equation.

70.

$$y = ax^2 + bx + c$$

$$y = a\left(x^2 + \frac{b}{a}x\right) + c$$

$$y = a\left(x^2 + \frac{b}{a}x + \frac{b^2}{4a^2}\right) + c - \frac{b^2}{4a}$$

$$y = a\left(a + \frac{b}{2a}\right)^2 + \left(c - \frac{b^2}{4a}\right)$$

Vertex: $\left(-\frac{b}{2a}, c - \frac{b^2}{4a}\right)$

Exercise 8.5 (page 559)

1. $y = kx$

2. $y = \frac{k}{t}$

3. $t = kxy$

4. $d = \frac{kt}{u^2}$

5. $y = 3x - 4$
$m = 3$

6. $\frac{2x - y}{5} = 8$
$2x - y = 40$
$2x - 40 = y$: $m = 2$

7. greater

8. less

9. undefined

10. sign

11. $x^2 - 5x + 4 < 0$
$(x - 4)(x - 1) < 0$
$x - 4$ −−−−−−−−0 ++++
$x - 1$ −−−0++++++++++ 
←—(——)—→ at 1, 4
solution set: $(1, 4)$

12. $x^2 - 3x - 4 > 0$
$(x - 4)(x + 1) > 0$
$x - 4$ −−−−−−−−−0++++
$x + 1$ −−− 0++++++++++ 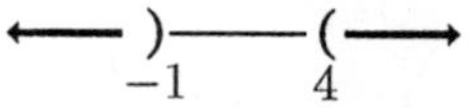
←—)——(—→ at −1, 4
solution set: $(-\infty, -1) \cup (4, \infty)$

13. $x^2 - 8x + 15 > 0$
$(x - 5)(x - 3) > 0$
$x - 5$ −−−−−−−−−0++++
$x - 3$ −−− 0++++++++++ 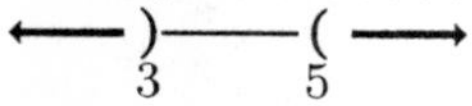
←—)——(—→ at 3, 5
solution set: $(-\infty, 3) \cup (5, \infty)$

14. $x^2 + 2x - 8 < 0$
$(x - 2)(x + 4) < 0$
$x - 2$ −−−−−−−−− 0++++
$x + 4$ −−− 0++++++++++ 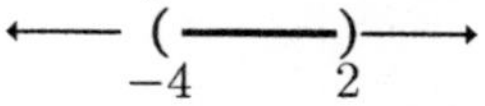
←—(——)—→ at −4, 2
solution set: $(-4, 2)$

15. $x^2 + x - 12 \le 0$
$(x - 3)(x + 4) \le 0$
$x - 3$ −−−−−−−−− 0++++
$x + 4$ −−− 0++++++++++ 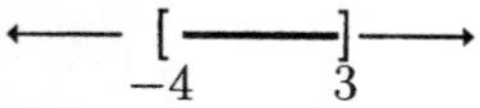
←—[——]—→ at −4, 3
solution set: $[-4, 3]$

16. $x^2 + 7x + 12 \ge 0$
$(x + 3)(x + 4) \ge 0$
$x + 3$ −−−−−−−−−0 ++++
$x + 4$ −−− 0++++++ ++++ 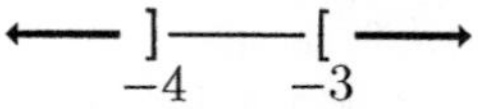
←—]——[—→ at −4, −3
solution set: $(-\infty, -4] \cup [-3, \infty)$

17. $x^2 + 2x \ge 15$
$x^2 + 2x - 15 \ge 0$
$(x - 3)(x + 5) \ge 0$
$x - 3$ −−−−−−−−−0++++
$x + 5$ −−− 0++++++++++ 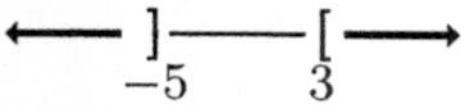
←—]——[—→ at −5, 3
solution set: $(-\infty, -5] \cup [3, \infty)$

18. $x^2 - 8x \le -15$
$x^2 - 8x + 15 \le 0$
$(x - 5)(x - 3) \le 0$
$x - 5$ −−−−−−−−0 ++++
$x - 3$ −−−0++++++++++ 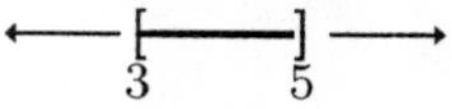
←—[——]—→ at 3, 5
solution set: $[3, 5]$

19.
$$x^2 + 8x < -16$$
$$x^2 + 8x + 16 < 0$$
$$(x + 4)(x + 4) < 0$$

$x + 4$ ---- 0+++++++
$x + 4$ ---- 0+++++++
←——————→
−4

Since the product is never negative, there is no solution.

20.
$$x^2 + 6x \geq -9$$
$$x^2 + 6x + 9 \geq 0$$
$$(x + 3)(x + 3) \geq 0$$

$x + 3$ -----0 +++++++
$x + 3$ -----0 +++++++
←——————→

solution set: $(-\infty, \infty)$

21.
$$x^2 \geq 9$$
$$x^2 - 9 \geq 0$$
$$(x - 3)(x + 3) \geq 0$$

$x - 3$ ---------0++++
$x + 3$ --- 0+++++++++++
←——]——[——→
−3 3

solution set: $(-\infty, -3] \cup [3, \infty)$

22.
$$x^2 \geq 16$$
$$x^2 - 16 \geq 0$$
$$(x - 4)(x + 4) \geq 0$$

$x - 4$ ---------0++++
$x + 4$ --- 0+++++++++++
←——]——[——→
−4 4

solution set: $(-\infty, -4] \cup [4, \infty)$

23.
$$2x^2 - 50 < 0$$
$$2(x - 5)(x + 5) < 0$$

$x - 5$ --------- 0++++
$x + 5$ --- 0++++++ ++++
←—— (——)——→
−5 5

solution set: $(-5, 5)$

24.
$$3x^2 - 243 < 0$$
$$3(x - 9)(x + 9) < 0$$

$x - 9$ --------- 0++++
$x + 9$ --- 0++++++ ++++
←—— (——)——→
−9 9

solution set: $(-9, 9)$

25.
$$\frac{1}{x} < 2$$
$$\frac{1}{x} - 2 < 0$$
$$\frac{1}{x} - \frac{2x}{x} < 0$$
$$\frac{1 - 2x}{x} < 0$$

$1 - 2x$ +++++++++ 0---
x ---0+++++++++++
←——)——(——→
0 $\frac{1}{2}$

solution set: $(-\infty, 0) \cup \left(\frac{1}{2}, \infty\right)$

26.
$$\frac{1}{x} > 3$$
$$\frac{1}{x} - 3 > 0$$
$$\frac{1}{x} - \frac{3x}{x} > 0$$
$$\frac{1 - 3x}{x} > 0$$

$1 - 3x$ ++++++++++0---
x ---0++++++++++
←——(——)——→
0 $\frac{1}{3}$

solution set: $\left(0, \frac{1}{3}\right)$

27.

$$\frac{4}{x} \geq 2$$
$$\frac{4}{x} - 2 \geq 0$$
$$\frac{4}{x} - \frac{2x}{x} \geq 0$$
$$\frac{4-2x}{x} \geq 0$$

$4-2x$ ++++++++++0 – – –
x – – – 0 +++++++++++

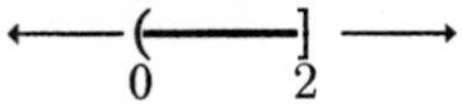

solution set: $(0, 2]$

28.

$$-\frac{6}{x} < 12$$
$$-\frac{6}{x} - 12 < 0$$
$$\frac{-6}{x} - \frac{12x}{x} < 0$$
$$\frac{-6-12x}{x} < 0$$

$-6-12x$ +++ 0 – – – – – – – –
x – – – – – – – – –0++++

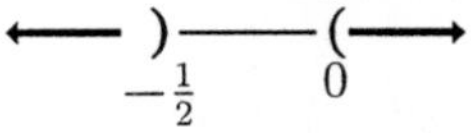

solution set: $\left(-\infty, -\frac{1}{2}\right) \cup (0, \infty)$

29.

$$-\frac{5}{x} < 3$$
$$-\frac{5}{x} - 3 < 0$$
$$\frac{-5}{x} - \frac{3x}{x} < 0$$
$$\frac{-5-3x}{x} < 0$$

$-5-3x$ +++ 0 – – – – – – – –
x – – – – – – – – –0++++

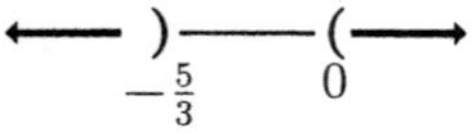

solution set: $\left(-\infty, -\frac{5}{3}\right) \cup (0, \infty)$

30.

$$\frac{4}{x} \geq 8$$
$$\frac{4}{x} - 8 \geq 0$$
$$\frac{4}{x} - \frac{8x}{x} \geq 0$$
$$\frac{4-8x}{x} \geq 0$$

$4-8x$ ++++++++++0– – –
x – – – 0++++++++++

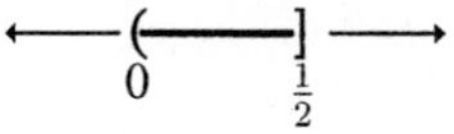

solution set: $\left(0, \frac{1}{2}\right]$

31.

$$\frac{x^2-x-12}{x-1} < 0$$
$$\frac{(x-4)(x+3)}{x-1} < 0$$

$x-4$ – – – – – – – – – – – 0++++
$x-1$ – – – – – – – – 0+++ +++++
$x+3$ – – – 0++++++ ++++ +++++

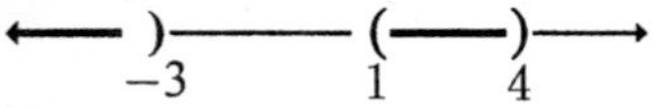

solution set: $(-\infty, -3) \cup (1, 4)$

32.

$$\frac{x^2+x-6}{x-4} \geq 0$$
$$\frac{(x-2)(x+3)}{x-4} \geq 0$$

$x-4$ – – – – – – – – – – – 0++++
$x-2$ – – – – – – – – 0++++++++
$x+3$ – – – 0+++++++++++++++

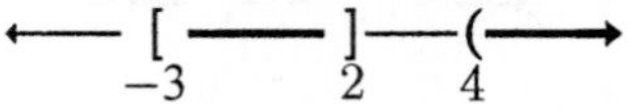

solution set: $[-3, 2] \cup (4, \infty)$

33. $\frac{x^2+x-20}{x+2} \geq 0$

$\frac{(x-4)(x+5)}{x+2} \geq 0$

$x-4$ ––––––––– –––– 0++++
$x+2$ ––––––––– 0++++ +++++
$x+5$ ––– 0+++++++++++ +++++

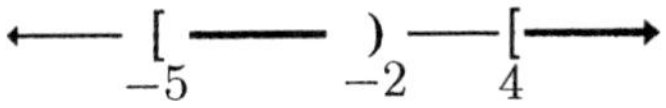

$\longleftarrow$ [——) —[$\longrightarrow$ at -5, -2, 4

solution set: $[-5,-2) \cup [4,\infty)$

34. $\frac{x^2-10x+25}{x+5} < 0$

$\frac{(x-5)(x-5)}{x+5} < 0$

$x-5$ ––––––––– 0++++
$x-5$ ––––––––– 0++++
$x+5$ ––– 0++++++ +++++

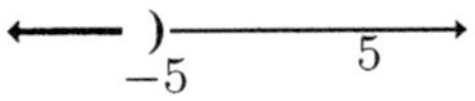

$\longleftarrow$) $\longrightarrow$ at -5, 5

solution set: $(-\infty,-5)$

35. $\frac{x^2-4x+4}{x+4} < 0$

$\frac{(x-2)(x-2)}{x+4} < 0$

$x-2$ ––––––––– 0++++
$x-2$ ––––––––– 0++++
$x+4$ ––– 0++++++ +++++

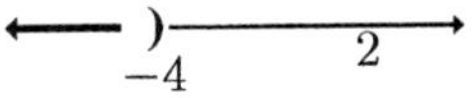

$\longleftarrow$) $\longrightarrow$ at -4, 2

solution set: $(-\infty,-4)$

36. $\frac{2x^2-5x+2}{x+2} > 0$

$\frac{(2x-1)(x-2)}{x+2} > 0$

$x-2$ ––––––––– –––– 0++++
$2x-1$ ––––––––– 0+++++++++
$x+2$ ––– 0++++++++++++++++

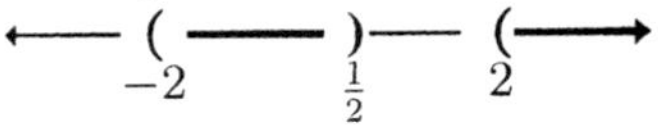

$\longleftarrow$ (——) — ($\longrightarrow$ at -2, $\frac{1}{2}$, 2

solution set: $\left(-2,\frac{1}{2}\right) \cup (2,\infty)$

37. $\frac{6x^2-5x+1}{2x+1} > 0$

$\frac{(2x-1)(3x-1)}{2x+1} > 0$

$2x-1$ ––––––––– –––– 0++++
$3x-1$ ––––––––– 0+++++++++
$2x+1$ ––– 0++++++++++++++++

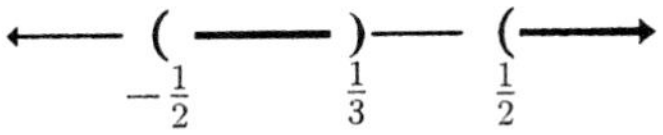

$\longleftarrow$ (——) — ($\longrightarrow$ at $-\frac{1}{2}$, $\frac{1}{3}$, $\frac{1}{2}$

solution set: $\left(-\frac{1}{2},\frac{1}{3}\right) \cup \left(\frac{1}{2},\infty\right)$

38. $\frac{6x^2+11x+3}{3x-1} < 0$

$\frac{(3x+1)(2x+3)}{3x-1} < 0$

$3x-1$ –––––––––– ––––– 0++++
$3x+1$ –––––––––– 0++++++++++
$2x+3$ ––– 0+++++++++++++++++++

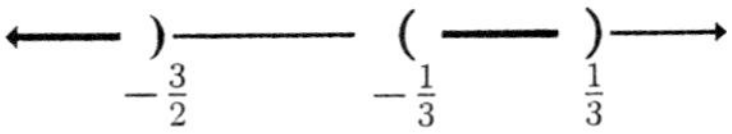

$\longleftarrow$) —— (——) $\longrightarrow$ at $-\frac{3}{2}$, $-\frac{1}{3}$, $\frac{1}{3}$

solution set: $\left(-\infty,-\frac{3}{2}\right) \cup \left(-\frac{1}{3},\frac{1}{3}\right)$

39.

$$\frac{3}{x-2} < \frac{4}{x}$$
$$\frac{3}{x-2} - \frac{4}{x} < 0$$
$$\frac{3x}{x(x-2)} - \frac{4(x-2)}{x(x-2)} < 0$$
$$\frac{-x+8}{x(x-2)} < 0$$

$-x+8$ +++++++++++++++++0– – –
$x-2$ – – – – – – – – 0++++++++++
x – – – 0+++++ ++++++++++

⟵(——)— (⟶ at 0, 2, 8

solution set: $(0, 2) \cup (8, \infty)$

40.

$$\frac{-6}{x+1} \geq \frac{1}{x}$$
$$\frac{-6}{x+1} - \frac{1}{x} \geq 0$$
$$\frac{-6x}{x(x+1)} - \frac{x+1}{x(x+1)} \geq 0$$
$$\frac{-7x-1}{x(x+1)} \geq 0$$

x – – – – – – – – – – – – – – – 0++++
$-7x-1$ +++++++++++++ 0– – – – – – –
$x+1$ – – – 0+++++++++++++++++++++

⟵)——— [——)⟶ at -1, $-\frac{1}{7}$, 0

solution set: $(-\infty, -1) \cup \left[-\frac{1}{7}, 0\right)$

41.

$$\frac{-5}{x+2} \geq \frac{4}{2-x}$$
$$\frac{-5}{x+2} - \frac{4}{2-x} \geq 0$$
$$\frac{-5(2-x)}{(x+2)(2-x)} - \frac{4(x+2)}{(x+2)(2-x)} \geq 0$$
$$\frac{x-18}{(x+2)(2-x)} \geq 0$$

$x-18$ – – – – – – – – – – – – – – – 0++++
$2-x$ +++++++++++++ 0– – – – – – –
$x+2$ – – – 0++++++++++++++++++

⟵)——— (——] ⟶ at -2, 2, 18

solution set: $(-\infty, -2) \cup (2, 18]$

42.

$$\frac{-6}{x-3} < \frac{5}{3-x}$$
$$\frac{-6}{x-3} - \frac{5}{3-x} < 0$$
$$\frac{-6}{x-3} + \frac{5}{x-3} < 0$$
$$\frac{-1}{x-3} < 0$$
$$\frac{1}{x-3} > 0$$

$x-3$ – – – – – 0+++++++

⟵——(——⟶ at 3

solution set: $(3, \infty)$

43.

$$\frac{7}{x-3} \geq \frac{2}{x+4}$$
$$\frac{7}{x-3} - \frac{2}{x+4} \geq 0$$
$$\frac{7(x+4)}{(x-3)(x+4)} - \frac{2(x-3)}{(x-3)(x+4)} \geq 0$$
$$\frac{5x+34}{(x-3)(x+4)} \geq 0$$

$x-3$ – – – – – – – – – – – – 0++++
$x+4$ – – – – – – – – 0+++++ +++++
$5x+34$ – – – 0++++++ +++++ +++++

⟵ [——) —(⟶ at $-\frac{34}{5}$, -4, 3

solution set: $\left[-\frac{34}{5}, -4\right) \cup (3, \infty)$

44.

$$\frac{-5}{x-4} < \frac{3}{x+1}$$
$$\frac{-5}{x-4} - \frac{3}{x+1} < 0$$
$$\frac{-5(x+1)}{(x-4)(x+1)} - \frac{3(x-4)}{(x-4)(x+1)} < 0$$
$$\frac{-8x+7}{(x-4)(x+1)} < 0$$

$x-4$ – – – – – – – – – – – – 0++++
$-8x+7$ +++++++++++ 0– – – – – – –
$x+1$ – – – 0+++++++++++++++++

⟵ (——)— (⟶ at -1, $\frac{7}{8}$, 4

solution set: $\left(-1, \frac{7}{8}\right) \cup (4, \infty)$

45.

$$\frac{x}{x+4} \le \frac{1}{x+1}$$
$$\frac{x}{x+4} - \frac{1}{x+1} \le 0$$
$$\frac{x(x+1)}{(x+4)(x+1)} - \frac{1(x+4)}{(x+4)(x+1)} \le 0$$
$$\frac{x^2-4}{(x+4)(x+1)} \le 0$$
$$\frac{(x+2)(x-2)}{(x+4)(x+1)} \le 0$$

```
x − 2  ----------- -----0+++
x + 1  ----------- 0++++ ++++
x + 2  ------0++++ +++++ ++++
x + 4  -- 0+++++++++ +++++ ++++
←— ( —— ] —— ( ——]—→
   −4     −2    −1   2
```

solution set: $(-4,-2] \cup (-1,2]$

46.

$$\frac{x}{x+9} \ge \frac{1}{x+1}$$
$$\frac{x}{x+9} - \frac{1}{x+1} \ge 0$$
$$\frac{x(x+1)}{(x+9)(x+1)} - \frac{1(x+9)}{(x+9)(x+1)} \ge 0$$
$$\frac{x^2-9}{(x+9)(x+1)} \ge 0$$
$$\frac{(x+3)(x-3)}{(x+9)(x+1)} \ge 0$$

```
x − 3  ------------ -----0+++
x + 1  ------------ 0++++ ++++
x + 3  -------0++++ +++++ ++++
x + 9  --- 0+++++++++ +++++ ++++
←—— ) —— [ —— ) ——[—→
    −9     −3    −1   3
```

solution set: $(-\infty,-9) \cup [-3,-1) \cup [3,\infty)$

47.

$$\frac{x}{x+16} > \frac{1}{x+1}$$
$$\frac{x}{x+16} - \frac{1}{x+1} > 0$$
$$\frac{x(x+1)}{(x+16)(x+1)} - \frac{1(x+16)}{(x+16)(x+1)} > 0$$
$$\frac{x^2-16}{(x+16)(x+1)} > 0$$
$$\frac{(x+4)(x-4)}{(x+16)(x+1)} > 0$$

```
x − 4   ------- ---- ----0+++
x + 1   ------- ---- 0++++ ++++
x + 4   ------- 0++++ +++++ ++++
x + 16  --- 0++++ +++++ +++++ ++++
←—— ) —— ( —— ) ——(—→
   −16    −4    −1   4
```

solution set: $(-\infty,-16) \cup (-4,-1) \cup (4,\infty)$

48.

$$\frac{x}{x+25} < \frac{1}{x+1}$$
$$\frac{x}{x+25} - \frac{1}{x+1} < 0$$
$$\frac{x(x+1)}{(x+25)(x+1)} - \frac{1(x+25)}{(x+25)(x+1)} < 0$$
$$\frac{x^2-25}{(x+25)(x+1)} < 0$$
$$\frac{(x+5)(x-5)}{(x+25)(x+1)} < 0$$

```
x − 5   ------ ---- ----0+++
x + 1   ------ ---- 0++++ ++++
x + 5   ------ 0++++ +++++ ++++
x + 25  -- 0++++ +++++ +++++ ++++
←— ( —— ) —— ( ——)—→
  −25    −5    −1   5
```

solution set: $(-25,-5) \cup (-1,5)$

49. $(x+2)^2 > 0$

$(x+2)(x+2) > 0$

$x+2$ - - - - - 0+++++++

$x+2$ - - - - - 0+++++++

⟵) (⟶

-2

solution set: $(-\infty, -2) \cup (-2, \infty)$

50. $(x-3)^2 < 0$

$(x-3)(x-3) < 0$

$x-3$ - - - - 0+++++++

$x-3$ - - - - 0+++++++

⟵ 3 ⟶

Since the product is never negative, there is no solution.

51. $x^2 - 2x - 3 < 0$

Graph $y = x^2 - 2x - 3$ and find the x-coordinates of points below the x-axis.

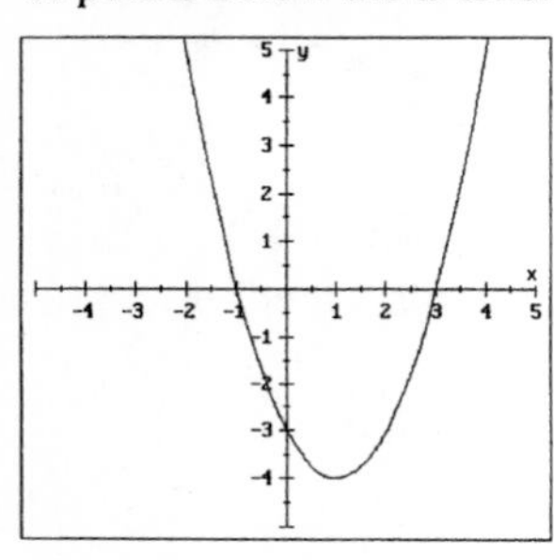

$(-1, 3)$

52. $x^2 + x - 6 > 0$

Graph $y = x^2 + x - 6$ and find the x-coordinates of points above the x-axis.

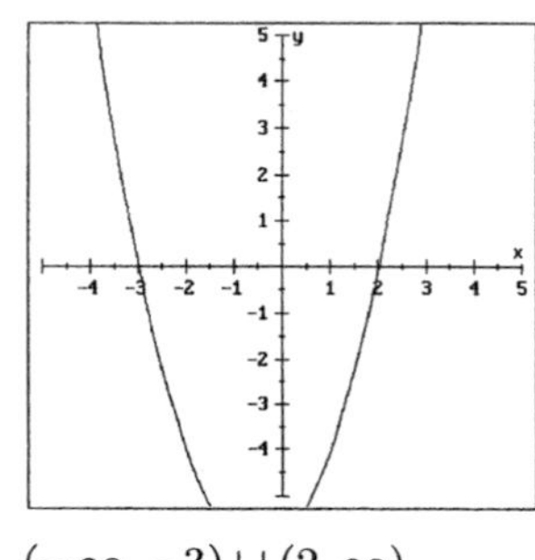

$(-\infty, -3) \cup (2, \infty)$

53. $\frac{x+3}{x-2} > 0$

Graph $y = (x+3)/(x-2)$ and find the x-coordinates of points above the x-axis.

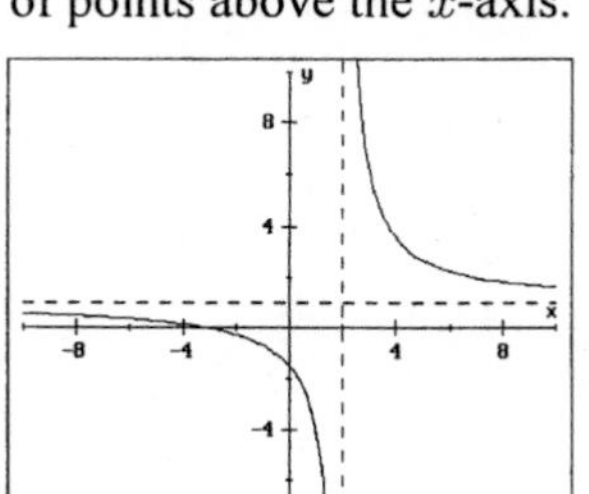

$(-\infty, -3) \cup (2, \infty)$

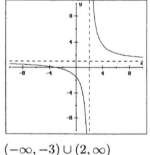

54. $\frac{3}{x} < 2 \Rightarrow \frac{3}{x} - 2 < 0$

Graph $y = (3/x) - 2$ and find the x-coordinates of points below the x-axis.

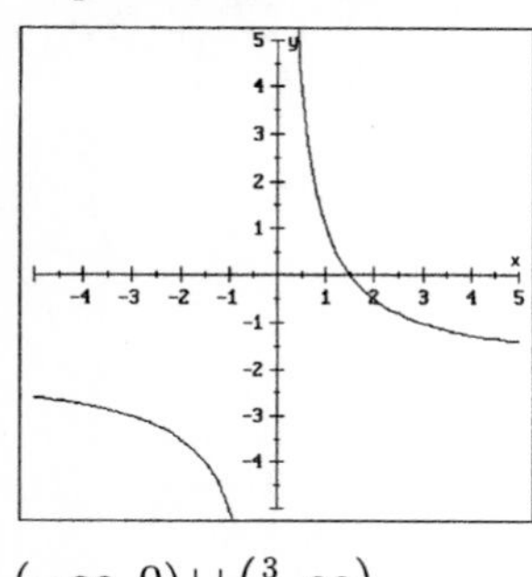

$(-\infty, 0) \cup \left(\frac{3}{2}, \infty\right)$

55. $y < x^2 + 1$

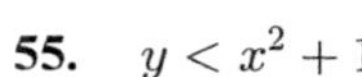

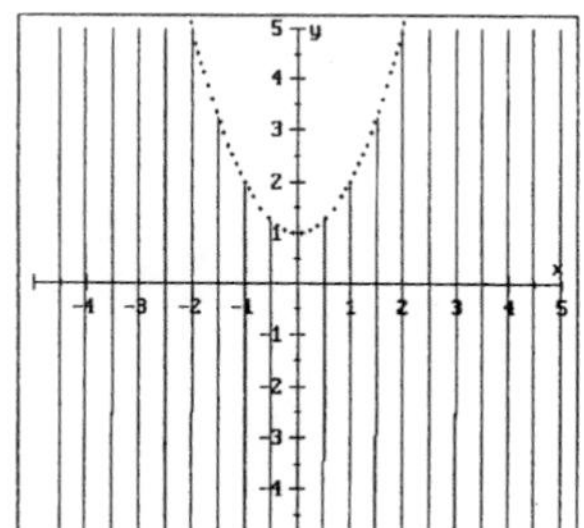

56. $y > x^2 - 3$

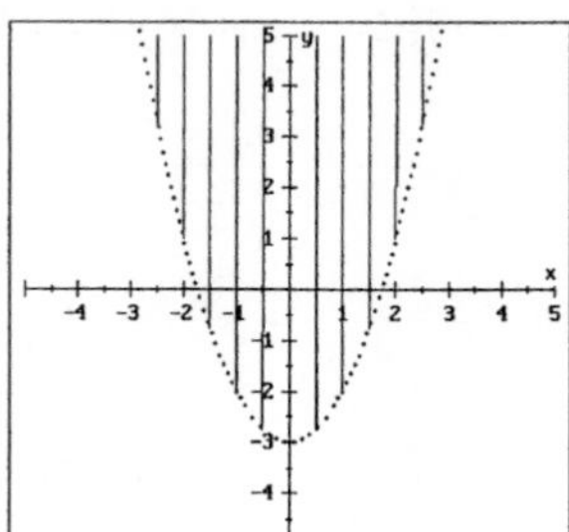

57. $y \le x^2 + 5x + 6$

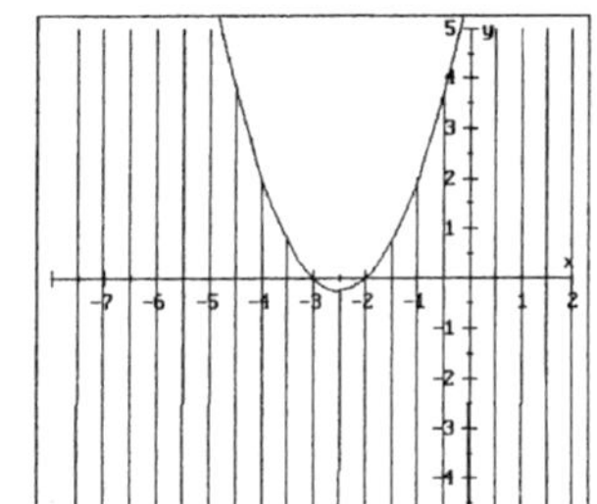

58. $y \ge x^2 + 5x + 4$

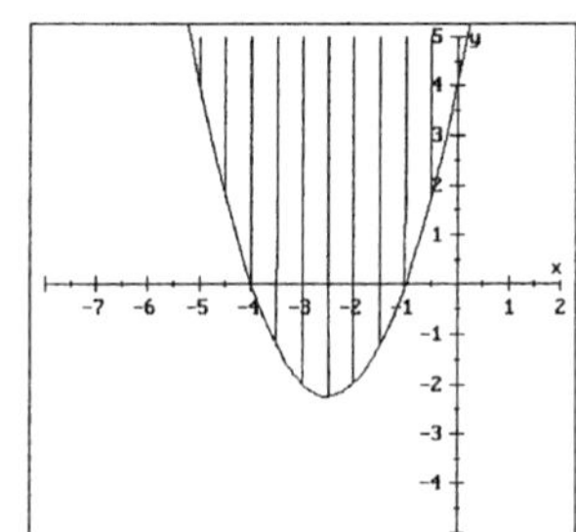

59. $y \ge (x-1)^2$

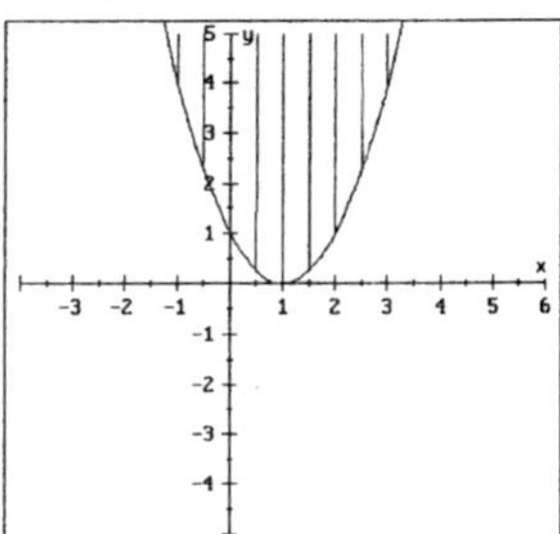

60. $y \le (x+2)^2$

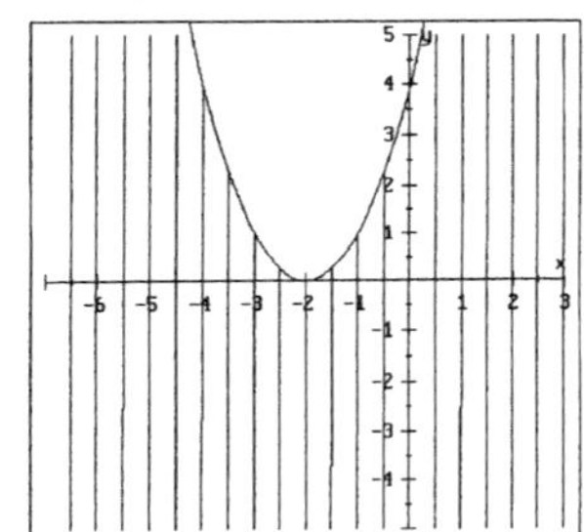

61. $-x^2 - y + 6 > -x$
$-x^2 + x + 6 > y$

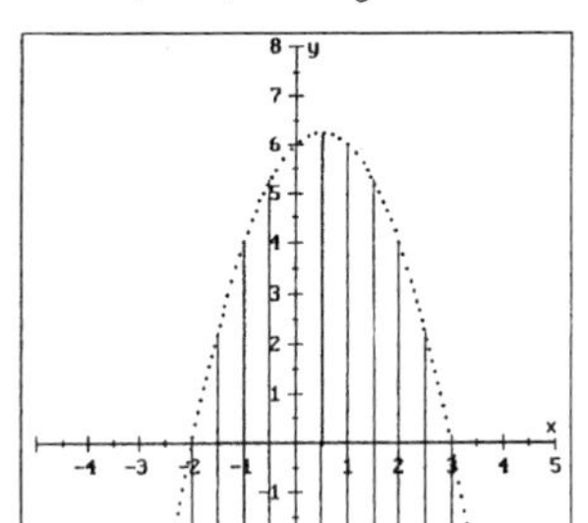

62. $y > (x+3)(x-2)$

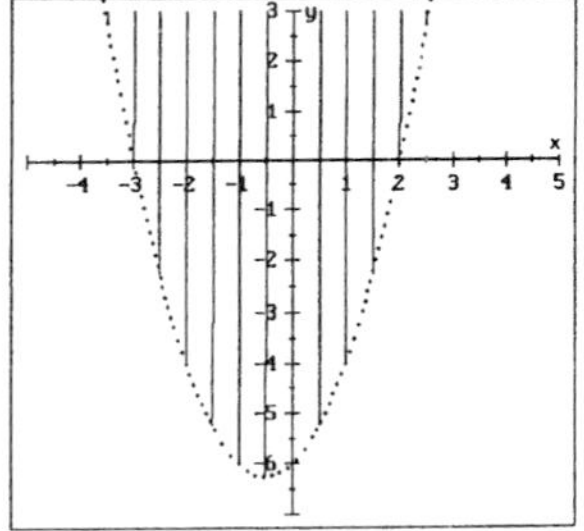

63. $y < |x+4|$

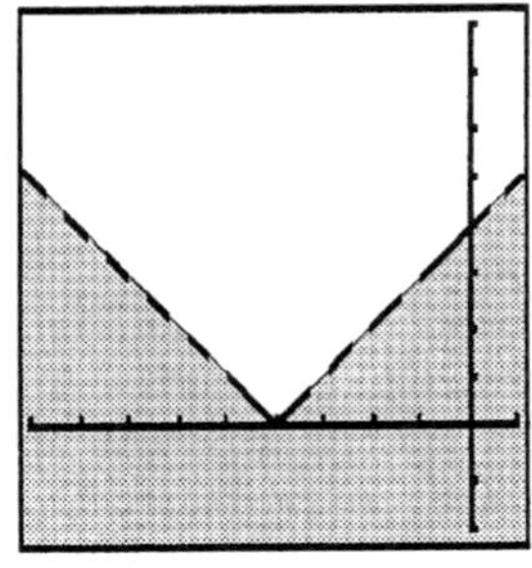

64. $y \ge |x-3|$

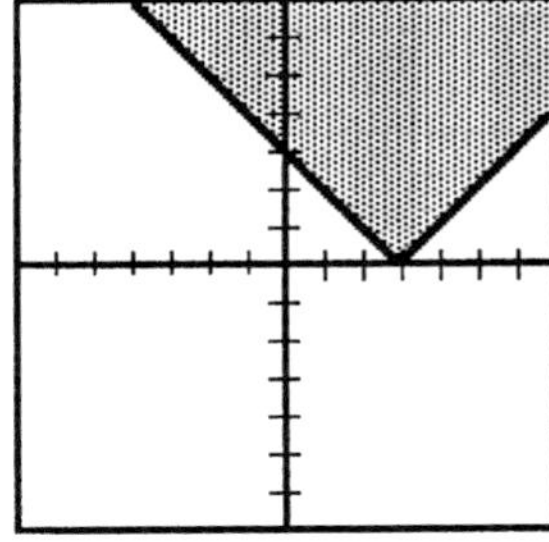

65. $y \le -|x| + 2$

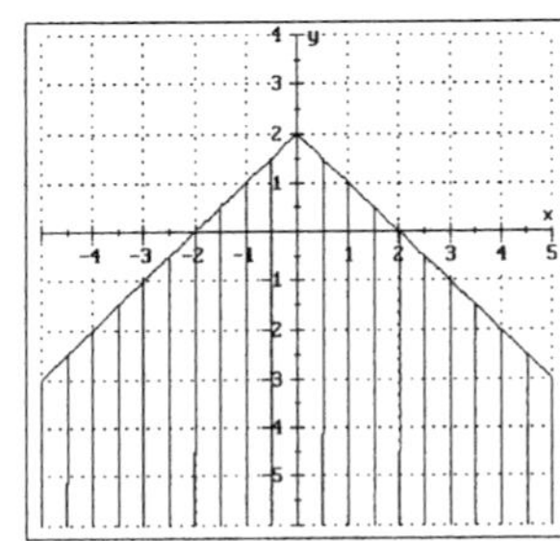

66. $y > |x| - 2$

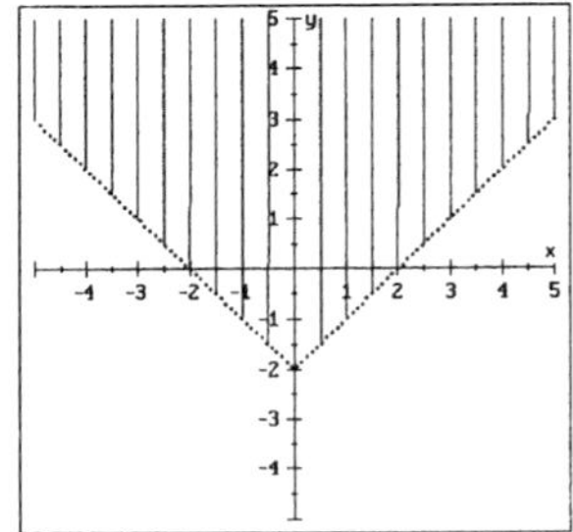

67. **Answers may vary.**

68. **Answers may vary.**

69. It will be positive if 4, 2 or 0 factors are negative.

70. It will be negative if 3 or 1 factors are negative.

Exercise 8.6 (page 567)

1. $\dfrac{3x^2+x-14}{4-x^2} = \dfrac{(3x+7)(x-2)}{(2+x)(2-x)} = -\dfrac{3x+7}{x+2}$

2. $\dfrac{2x^3+14x^2}{3+2x-x^2} \cdot \dfrac{x^2-3x}{x} = -\dfrac{2x^3+14x^2}{x^2-2x-3} \cdot \dfrac{x^2-3x}{x} = -\dfrac{2x^2(x+7)}{(x-3)(x+1)} \cdot \dfrac{x(x-3)}{x}$
$= -\dfrac{2x^2(x+7)}{x+1}$

3. $\dfrac{8+2x-x^2}{12+x-3x^2} \div \dfrac{3x^2+5x-2}{3x-1} = \dfrac{x^2-2x-8}{3x^2-x-12} \cdot \dfrac{3x-1}{3x^2+5x-2}$
$= \dfrac{(x-4)(x+2)}{3x^2-x-12} \cdot \dfrac{3x-1}{(3x-1)(x+2)} = \dfrac{x-4}{3x^2-x-12}$

4. $\dfrac{x-1}{1+\frac{x}{x-2}} = \dfrac{x-1}{1+\frac{x}{x-2}} \cdot \dfrac{x-2}{x-2} = \dfrac{x^2-3x+2}{x-2+x} = \dfrac{x^2-3x+2}{2x-2} = \dfrac{(x-2)(x-1)}{2(x-1)} = \dfrac{x-2}{2}$

5. $f(x)+g(x)$

6. $f(x)-g(x)$

7. $f(x)g(x)$

8. $\dfrac{f(x)}{g(x)}$

9. domain

10. $f(g(x))$

11. $f(x)$

12. $f(x)$

13. $f+g = f(x)+g(x) = 3x+4x = 7x$
domain $= (-\infty, \infty)$

14. $f-g = f(x)-g(x) = 3x-4x = -x$
domain $= (-\infty, \infty)$

15. $f \cdot g = f(x) \cdot g(x) = 3x \cdot 4x = 12x^2$
domain $= (-\infty, \infty)$

16. $f/g = \dfrac{f(x)}{g(x)} = \dfrac{3x}{4x} = \dfrac{3}{4}$ (for $x \neq 0$)
domain $= (-\infty, 0) \cup (0, \infty)$

17. $g-f = g(x)-f(x) = 4x-3x = x$
domain $= (-\infty, \infty)$

18. $g+f = g(x)+f(x) = 4x+3x = 7x$
domain $= (-\infty, \infty)$

19. $g/f = \dfrac{g(x)}{f(x)} = \dfrac{4x}{3x} = \dfrac{4}{3}$ (for $x \neq 0$)
domain $= (-\infty, 0) \cup (0, \infty)$

20. $g \cdot f = g(x) \cdot f(x) = 4x \cdot 3x = 12x^2$
domain $= (-\infty, \infty)$

21. $f+g = f(x)+g(x) = 2x+1+x-3 = 3x-2$; domain $= (-\infty, \infty)$

22. $f - g = f(x) - g(x) = (2x+1) - (x-3) = 2x + 1 - x + 3 = x + 4$; domain $= (-\infty, \infty)$

23. $f \cdot g = f(x) \cdot g(x) = (2x+1)(x-3) = 2x^2 - 5x - 3$; domain $= (-\infty, \infty)$

24. $f/g = \dfrac{f(x)}{g(x)} = \dfrac{2x+1}{x-3}$; domain $= (-\infty, 3) \cup (3, \infty)$

25. $g - f = g(x) - f(x) = (x-3) - (2x+1) = x - 3 - 2x - 1 = -x - 4$; domain $= (-\infty, \infty)$

26. $g + f = g(x) + f(x) = x - 3 + 2x + 1 = 3x - 2$; domain $= (-\infty, \infty)$

27. $g/f = \dfrac{g(x)}{f(x)} = \dfrac{x-3}{2x+1}$; domain $= \left(-\infty, -\dfrac{1}{2}\right) \cup \left(-\dfrac{1}{2}, \infty\right)$

28. $g \cdot f = g(x) \cdot f(x) = (x-3)(2x+1) = 2x^2 - 5x - 3$; domain $= (-\infty, \infty)$

29. $f - g = f(x) - g(x) = (3x-2) - (2x^2+1) = 3x - 2 - 2x^2 - 1 = -2x^2 + 3x - 3$
domain $= (-\infty, \infty)$

30. $f + g = f(x) + g(x) = 3x - 2 + 2x^2 + 1 = 2x^2 + 3x - 1$; domain $= (-\infty, \infty)$

31. $f/g = \dfrac{f(x)}{g(x)} = \dfrac{3x-2}{2x^2+1}$; domain $= (-\infty, \infty)$

32. $f \cdot g = f(x) \cdot g(x) = (3x-2)(2x^2+1) = 6x^3 - 4x^2 + 3x - 2$; domain $= (-\infty, \infty)$

33. $f - g = f(x) - g(x) = (x^2-1) - (x^2-4) = x^2 - 1 - x^2 + 4 = 3$; domain $= (-\infty, \infty)$

34. $f + g = f(x) + g(x) = x^2 - 1 + x^2 - 4 = 2x^2 - 5$; domain $= (-\infty, \infty)$

35. $g/f = \dfrac{g(x)}{f(x)} = \dfrac{x^2-4}{x^2-1} = \dfrac{(x+2)(x-2)}{(x+1)(x-1)}$; domain $= (-\infty, -1) \cup (-1, 1) \cup (1, \infty)$

36. $g \cdot f = g(x) \cdot f(x) = (x^2-4)(x^2-1) = x^4 - 5x^2 + 4$; domain $= (-\infty, \infty)$

37. $(f \circ g)(2) = f(g(2)) = f(2^2 - 1) = f(3) = 2(3) + 1 = 7$

38. $(g \circ f)(2) = g(f(2)) = g(2(2) + 1) = g(5) = 5^2 - 1 = 24$

39. $(g \circ f)(-3) = g(f(-3)) = g(2(-3) + 1) = g(-5) = (-5)^2 - 1 = 24$

40. $(f \circ g)(-3) = f(g(-3)) = f((-3)^2 - 1) = f(8) = 2(8) + 1 = 17$

41. $(f \circ g)(0) = f(g(0)) = f(0^2 - 1) = f(-1) = 2(-1) + 1 = -1$

42. $(g \circ f)(0) = g(f(0)) = g(2(0) + 1) = g(1) = 1^2 - 1 = 0$

43. $(f \circ g)\left(\frac{1}{2}\right) = f\left(g\left(\frac{1}{2}\right)\right) = f\left(\left(\frac{1}{2}\right)^2 - 1\right) = f\left(-\frac{3}{4}\right) = 2\left(-\frac{3}{4}\right) + 1 = -\frac{3}{2} + 1 = -\frac{1}{2}$

44. $(g \circ f)\left(\frac{1}{3}\right) = g\left(f\left(\frac{1}{3}\right)\right) = g\left(2\left(\frac{1}{3}\right) + 1\right) = g\left(\frac{5}{3}\right) = \left(\frac{5}{3}\right)^2 - 1 = \frac{25}{9} - 1 = \frac{16}{9}$

45. $(f \circ g)(x) = f(g(x)) = f(x^2 - 1) = 2(x^2 - 1) + 1 = 2x^2 - 2 + 1 = 2x^2 - 1$

46. $(g \circ f)(x) = g(f(x)) = g(2x + 1) = (2x + 1)^2 - 1 = 4x^2 + 4x + 1 - 1 = 4x^2 + 4x$

47. $(g \circ f)(2x) = g(f(2x)) = g(2(2x) + 1) = g(4x + 1) = (4x + 1)^2 - 1 = 16x^2 + 8x + 1 - 1$
$= 16x^2 + 8x$

48. $(f \circ g)(2x) = f(g(2x)) = f\left((2x)^2 - 1\right) = f\left(4x^2 - 1\right) = 2\left(4x^2 - 1\right) + 1 = 8x^2 - 2 + 1$
$= 8x^2 - 1$

49. $(f \circ g)(4) = f(g(4)) = f(4^2 + 4) = f(20) = 3(20) - 2 = 58$

50. $(g \circ f)(4) = g(f(4)) = g(3(4) - 2) = g(10) = 10^2 + 10 = 110$

51. $(g \circ f)(-3) = g(f(-3)) = g(3(-3) - 2) = g(-11) = (-11)^2 + (-11) = 110$

52. $(f \circ g)(-3) = f(g(-3)) = f\left((-3)^2 + (-3)\right) = f(6) = 3(6) - 2 = 16$

53. $(g \circ f)(0) = g(f(0)) = g(3(0) - 2) = g(-2) = (-2)^2 + (-2) = 2$

54. $(f \circ g)(0) = f(g(0)) = f(0^2 + 0) = f(0) = 3(0) - 2 = -2$

55. $(g \circ f)(x) = g(f(x)) = g(3x - 2) = (3x - 2)^2 + 3x - 2 = 9x^2 - 12x + 4 + 3x - 2$
$= 9x^2 - 9x + 2$

56. $(f \circ g)(x) = f(g(x)) = f(x^2 + x) = 3(x^2 + x) - 2 = 3x^2 + 3x - 2$

57. $\frac{f(x+h) - f(x)}{h} = \frac{2(x+h) + 3 - (2x+3)}{h} = \frac{2x + 2h + 3 - 2x - 3}{h} = \frac{2h}{h} = 2$

58. $\frac{f(x+h) - f(x)}{h} = \frac{3(x+h) - 5 - (3x-5)}{h} = \frac{3x + 3h - 5 - 3x + 5}{h} = \frac{3h}{h} = 3$

59. $\frac{f(x+h) - f(x)}{h} = \frac{(x+h)^2 - x^2}{h} = \frac{x^2 + 2xh + h^2 - x^2}{h} = \frac{2xh + h^2}{h} = 2x + h$

60. $\frac{f(x+h) - f(x)}{h} = \frac{(x+h)^2 - 1 - (x^2 - 1)}{h} = \frac{x^2 + 2xh + h^2 - 1 - x^2 + 1}{h}$
$= \frac{2xh + h^2}{h} = 2x + h$

61. $$\frac{f(x+h)-f(x)}{h}=\frac{2(x+h)^2-1-(2x^2-1)}{h}=\frac{2x^2+4xh+2h^2-1-2x^2+1}{h}=\frac{4xh+2h^2}{h}=4x+2h$$

62. $$\frac{f(x+h)-f(x)}{h}=\frac{3(x+h)^2-3x^2}{h}=\frac{3x^2+6xh+3h^2-3x^2}{h}=\frac{6xh+3h^2}{h}=6x+3h$$

63. $$\frac{f(x+h)-f(x)}{h}=\frac{(x+h)^2+(x+h)-(x^2+x)}{h}=\frac{x^2+2xh+h^2+x+h-x^2-x}{h}=\frac{2xh+h^2+h}{h}=2x+h+1$$

64. $$\frac{f(x+h)-f(x)}{h}=\frac{(x+h)^2-(x+h)-(x^2-x)}{h}=\frac{x^2+2xh+h^2-x-h-x^2+x}{h}=\frac{2xh+h^2-h}{h}=2x+h-1$$

65. $$\frac{f(x+h)-f(x)}{h}=\frac{(x+h)^2+3(x+h)-4-(x^2+3x-4)}{h}=\frac{x^2+2xh+h^2+3x+3h-4-x^2-3x+4}{h}=\frac{2xh+h^2+3h}{h}=2x+h+3$$

66. $$\frac{f(x+h)-f(x)}{h}=\frac{(x+h)^2-4(x+h)+3-(x^2-4x+3)}{h}=\frac{x^2+2xh+h^2-4x-4h+3-x^2+4x-3}{h}=\frac{2xh+h^2-4h}{h}=2x+h-4$$

67. $$\frac{f(x+h)-f(x)}{h}=\frac{2(x+h)^2+3(x+h)-7-(2x^2+3x-7)}{h}=\frac{2x^2+4xh+2h^2+3x+3h-7-2x^2-3x+7}{h}=\frac{4xh+2h^2+3h}{h}=4x+2h+3$$

68. $$\frac{f(x+h)-f(x)}{h}=\frac{3(x+h)^2-2(x+h)+4-(3x^2-2x+4)}{h}=\frac{3x^2+6xh+3h^2-2x-2h+4-3x^2+2x-4}{h}=\frac{6xh+3h^2-2h}{h}=6x+3h-2$$

69. $$\frac{f(x)-f(a)}{x-a}=\frac{(2x+3)-(2a+3)}{x-a}=\frac{2x+3-2a-3}{x-a}=\frac{2x-2a}{x-a}=\frac{2(x-a)}{x-a}=2$$

70. $$\frac{f(x)-f(a)}{x-a}=\frac{(3x-5)-(3a-5)}{x-a}=\frac{3x-5-3a+5}{x-a}=\frac{3x-3a}{x-a}=\frac{3(x-a)}{x-a}=3$$

71. $$\frac{f(x)-f(a)}{x-a}=\frac{x^2-a^2}{x-a}=\frac{(x+a)(x-a)}{x-a}=x+a$$

72. $$\frac{f(x)-f(a)}{x-a}=\frac{(x^2-1)-(a^2-1)}{x-a}=\frac{x^2-1-a^2+1}{x-a}=\frac{x^2-a^2}{x-a}=\frac{(x+a)(x-a)}{x-a}=x+a$$

73. $$\begin{aligned}\frac{f(x)-f(a)}{x-a}=\frac{(2x^2-1)-(2a^2-1)}{x-a}=\frac{2x^2-1-2a^2+1}{x-a}&=\frac{2x^2-2a^2}{x-a}\\&=\frac{2(x+a)(x-a)}{x-a}\\&=2(x+a)=2x+2a\end{aligned}$$

74. $$\frac{f(x)-f(a)}{x-a}=\frac{3x^2-3a^2}{x-a}=\frac{3(x+a)(x-a)}{x-a}=3(x+a)=3x+3a$$

75. $$\begin{aligned}\frac{f(x)-f(a)}{x-a}=\frac{(x^2+x)-(a^2+a)}{x-a}=\frac{x^2+x-a^2-a}{x-a}&=\frac{x^2-a^2+x-a}{x-a}\\&=\frac{(x+a)(x-a)+1(x-a)}{x-a}\\&=\frac{(x-a)(x+a+1)}{x-a}=x+a+1\end{aligned}$$

76. $$\begin{aligned}\frac{f(x)-f(a)}{x-a}=\frac{(x^2-x)-(a^2-a)}{x-a}=\frac{x^2-x-a^2+a}{x-a}&=\frac{x^2-a^2-x+a}{x-a}\\&=\frac{(x+a)(x-a)-1(x-a)}{x-a}\\&=\frac{(x-a)(x+a-1)}{x-a}=x+a-1\end{aligned}$$

77. $$\begin{aligned}\frac{f(x)-f(a)}{x-a}=\frac{(x^2+3x-4)-(a^2+3a-4)}{x-a}&=\frac{x^2+3x-4-a^2-3a+4}{x-a}\\&=\frac{x^2-a^2+3x-3a}{x-a}\\&=\frac{(x+a)(x-a)+3(x-a)}{x-a}\\&=\frac{(x-a)(x+a+3)}{x-a}=x+a+3\end{aligned}$$

78. $$\frac{f(x)-f(a)}{x-a}=\frac{(x^2-4x+3)-(a^2-4a+3)}{x-a}=\frac{x^2-4x+3-a^2+4a-3}{x-a}$$
$$=\frac{x^2-a^2-4x+4a}{x-a}$$
$$=\frac{(x+a)(x-a)-4(x-a)}{x-a}$$
$$=\frac{(x-a)(x+a-4)}{x-a}=x+a-4$$

79. $$\frac{f(x)-f(a)}{x-a}=\frac{(2x^2+3x-7)-(2a^2+3a-7)}{x-a}=\frac{2x^2+3x-7-2a^2-3a+7}{x-a}$$
$$=\frac{2x^2-2a^2+3x-3a}{x-a}$$
$$=\frac{2(x+a)(x-a)+3(x-a)}{x-a}$$
$$=\frac{(x-a)(2(x+a)+3)}{x-a}$$
$$=2(x+a)+3=2x+2a+3$$

80. $$\frac{f(x)-f(a)}{x-a}=\frac{(3x^2-2x+4)-(3a^2-2a+4)}{x-a}=\frac{3x^2-2x+4-3a^2+2a-4}{x-a}$$
$$=\frac{3x^2-3a^2-2x+2a}{x-a}$$
$$=\frac{3(x+a)(x-a)-2(x-a)}{x-a}$$
$$=\frac{(x-a)(3(x+a)-2)}{x-a}$$
$$=3(x+a)-2=3x+3a-2$$

81. $(f\circ g)(x)=f(g(x))=f(2x-5)=(2x-5)+1=2x-4$
$(g\circ f)(x)=g(f(x))=g(x+1)=2(x+1)-5=2x+2-5=2x-3$

82. $(f\circ g)(x)=f(g(x))=f(3x^2-2)=(3x^2-2)^2+1=9x^4-12x^2+4+1=9x^4-12x^2+5$
$(g\circ f)(x)=g(f(x))=g(x^2+1)=3(x^2+1)^2-2=3(x^4+2x^2+1)-2=3x^4+6x^2+1$

83. $f(a)=a^2+2a-3;\ f(h)=h^2+2h-3\Rightarrow f(a)+f(h)=a^2+h^2+2a+2h-6$
$f(a+h)=(a+h)^2+2(a+h)-3=a^2+2ah+h^2+2a+2h-3$
$=a^2+h^2+2ah+2a+2h-3$

84. $g(a)=2a^2+10;\ g(h)=2h^2+10\Rightarrow g(a)+g(h)=2a^2+2h^2+20$
$g(a+h)=2(a+h)^2+10=2(a^2+2ah+h^2)+10=2a^2+2h^2+4ah+10$

85. $$\frac{f(x+h)-f(x)}{h}=\frac{(x+h)^3-1-(x^3-1)}{h}=\frac{x^3+3x^2h+3xh^2+h^3-1-x^3+1}{h}$$
$$=\frac{3x^2h+3xh^2+h^3}{h}$$
$$=\frac{h(3x^2+3xh+h^2)}{h}=3x^2+3xh+h^2$$

86. $$\frac{f(x+h)-f(x)}{h}=\frac{(x+h)^3+2-(x^3+2)}{h}=\frac{x^3+3x^2h+3xh^2+h^3+2-x^3-2}{h}$$
$$=\frac{3x^2h+3xh^2+h^3}{h}=3x^2+3xh+h^2$$

87. $F(t)=2700-200t$; $C(F)=\frac{5}{9}(F-32)$
$C(F(t))=C(2700-200t)=\frac{5}{9}(2700-200t-32)=\frac{5}{9}(2668-200t)$

88. $C(t)=34+\frac{1}{6}t$; $F(C)=\frac{9}{5}C+32$; $F(C(t))=F(34+\frac{1}{6}t)=\frac{9}{5}(34+\frac{1}{6}t)+32$

89. **Answers may vary.**

90. **Answers may vary.**

91. It is associative. Examples will vary.

92. It is not distributive. Examples will vary.

Exercise 8.7 (page 576)

1. $3-\sqrt{-64}=3-\sqrt{64i^2}=3-8i$

2. $(2-3i)+(4+5i)=2-3i+4+5i$
$=6+2i$

3. $(3+4i)(2-3i)=6-9i+8i-12i^2=6-i-12(-1)=6-i+12=18-i$

4. $$\frac{6+7i}{3-4i}=\frac{6+7i}{3-4i}\cdot\frac{3+4i}{3+4i}=\frac{18+24i+21i+28i^2}{9-16i^2}=\frac{-10+45i}{25}=-\frac{10}{25}+\frac{45}{25}i=-\frac{2}{5}+\frac{9}{5}i$$

5. $|6-8i|=\sqrt{6^2+(-8)^2}=\sqrt{36+64}=\sqrt{100}=10$

6. $$\left|\frac{2+i}{3-i}\right|=\left|\frac{2+i}{3-i}\cdot\frac{3+i}{3+i}\right|=\left|\frac{5+5i}{10}\right|=\left|\frac{1}{2}+\frac{1}{2}i\right|=\sqrt{\left(\frac{1}{2}\right)^2+\left(\frac{1}{2}\right)^2}=\sqrt{\frac{1}{2}}=\frac{1}{\sqrt{2}}=\frac{\sqrt{2}}{2}$$

7. one-to-one

8. horizontal

9. 2

10. the inverse of f; f inverse

11. x

12. $y=x$

13. Each input has a different output.
one-to-one

14. The inputs $x=2$ and $x=-2$ have the same output. not one-to-one

15. The inputs $x=2$ and $x=-2$ have the same output. not one-to-one

16. Each input has a different output.
one-to-one

17. one-to-one **18.** one-to-one **19.** one-to-one **20.** one-to-one

21. not one-to-one **22.** not one-to-one **23.** one-to-one **24.** one-to-one

25. inverse $= \{(2,3),(1,2),(0,1)\}$. Since each x-coordinate is paired with only one y-coordinate, the inverse relation **is a function.**

26. inverse $= \{(1,4),(1,5),(1,6),(1,7)\}$. Since $x = 1$ is paired with more than one y-coordinate, the inverse relation **is not a function.**

27. inverse $= \{(2,1),(3,2),(3,1),(5,1)\}$. Since $x = 3$ is paired with more than one y-coordinate, the inverse relation **is not a function.**

28. inverse $= \{(-1,-1),(0,0),(1,1),(2,2)\}$. Since each x-coordinate is paired with only one y-coordinate, the inverse relation **is a function.**

29. inverse $= \{(1,1),(4,2),(9,3),(16,4)\}$. Since each x-coordinate is paired with only one y-coordinate, the inverse relation **is a function.**

30. inverse $= \{(1,1),(1,2),(1,3),(1,4)\}$. Since $x = 1$ is paired with more than one y-coordinate, the inverse relation **is not a function.**

31.

$$\begin{aligned} f(x) &= 3x+1 \\ y &= 3x+1 \\ x &= 3y+1 \\ x-1 &= 3y \\ \frac{x-1}{3} &= y \\ \frac{1}{3}x - \frac{1}{3} &= y \\ f^{-1}(x) &= \tfrac{1}{3}x - \tfrac{1}{3} \end{aligned}$$

$f \circ f^{-1}$

$$\begin{aligned} f[f^{-1}(x)] &= f\left(\tfrac{1}{3}x - \tfrac{1}{3}\right) \\ &= 3 \cdot \left(\tfrac{1}{3}x - \tfrac{1}{3}\right) + 1 \\ &= x - 1 + 1 \\ &= x \end{aligned}$$

$f^{-1} \circ f$

$$\begin{aligned} f^{-1}[f(x)] &= f^{-1}(3x+1) \\ &= \tfrac{1}{3}(3x+1) - \tfrac{1}{3} \\ &= x + \tfrac{1}{3} - \tfrac{1}{3} \\ &= x \end{aligned}$$

32.

$$\begin{aligned} y+1 &= 5x \\ y = f(x) &= 5x - 1 \\ y+1 &= 5x \\ x &= 5y-1 \\ \frac{x+1}{5} &= y \\ \frac{1}{5}x + \frac{1}{5} &= y \\ f^{-1}(x) &= \tfrac{1}{5}x + \tfrac{1}{5} \end{aligned}$$

$f \circ f^{-1}$

$$\begin{aligned} f[f^{-1}(x)] &= f\left(\tfrac{1}{5}x + \tfrac{1}{5}\right) \\ &= 5 \cdot \left(\tfrac{1}{5}x + \tfrac{1}{5}\right) - 1 \\ &= x + 1 - 1 \\ &= x \end{aligned}$$

$f^{-1} \circ f$

$$\begin{aligned} f^{-1}[f(x)] &= f^{-1}(5x-1) \\ &= \tfrac{1}{5}(5x-1) + \tfrac{1}{5} \\ &= x - \tfrac{1}{5} + \tfrac{1}{5} \\ &= x \end{aligned}$$

33.

$$\begin{aligned} x+4 &= 5y \\ y = f(x) &= \frac{x+4}{5} \\ x &= \frac{y+4}{5} \\ 5x &= y+4 \\ 5x-4 &= y \\ f^{-1}(x) &= 5x-4 \end{aligned}$$

$f \circ f^{-1}$

$$\begin{aligned} f\left[f^{-1}(x)\right] &= f(5x-4) \\ &= \frac{(5x-4)+4}{5} \\ &= \frac{5x}{5} \\ &= x \end{aligned}$$

$f^{-1} \circ f$

$$\begin{aligned} f^{-1}[f(x)] &= f^{-1}\left(\frac{x+4}{5}\right) \\ &= 5 \cdot \left(\frac{x+4}{5}\right) - 4 \\ &= x+4-4 \\ &= x \end{aligned}$$

34.

$$\begin{aligned} x &= 3y+1 \\ x-1 &= 3y \\ y = f(x) &= \frac{x-1}{3} \\ x &= \frac{y-1}{3} \\ 3x &= y-1 \\ 3x+1 &= y \\ f^{-1}(x) &= 3x+1 \end{aligned}$$

$f \circ f^{-1}$

$$\begin{aligned} f\left[f^{-1}(x)\right] &= f(3x+1) \\ &= \frac{(3x+1)-1}{3} \\ &= \frac{3x}{3} \\ &= x \end{aligned}$$

$f^{-1} \circ f$

$$\begin{aligned} f^{-1}[f(x)] &= f^{-1}\left(\frac{x-1}{3}\right) \\ &= 3 \cdot \left(\frac{x-1}{3}\right) + 1 \\ &= x-1+1 \\ &= x \end{aligned}$$

35.

$$\begin{aligned} f(x) &= \frac{x-4}{5} \\ y &= \frac{x-4}{5} \\ x &= \frac{y-4}{5} \\ 5x &= y-4 \\ 5x+4 &= y \\ f^{-1}(x) &= 5x+4 \end{aligned}$$

$f \circ f^{-1}$

$$\begin{aligned} f\left[f^{-1}(x)\right] &= f(5x+4) \\ &= \frac{(5x+4)-4}{5} \\ &= \frac{5x}{5} \\ &= x \end{aligned}$$

$f^{-1} \circ f$

$$\begin{aligned} f^{-1}[f(x)] &= f^{-1}\left(\frac{x-4}{5}\right) \\ &= 5 \cdot \left(\frac{x-4}{5}\right) + 4 \\ &= x-4+4 \\ &= x \end{aligned}$$

36.

$$\begin{aligned} f(x) &= \frac{2x+6}{3} \\ y &= \frac{2x+6}{3} \\ x &= \frac{2y+6}{3} \\ 3x &= 2y+6 \\ 3x-6 &= 2y \\ \tfrac{3}{2}x-3 &= y \\ f^{-1}(x) &= \tfrac{3}{2}x-3 \end{aligned}$$

$f \circ f^{-1}$

$$\begin{aligned} f\left[f^{-1}(x)\right] &= f\left(\tfrac{3}{2}x-3\right) \\ &= \frac{2\left(\frac{3}{2}x-3\right)+6}{3} \\ &= \frac{3x-6+6}{3} \\ &= \frac{3x}{3} \\ &= x \end{aligned}$$

$f^{-1} \circ f$

$$\begin{aligned} f^{-1}[f(x)] &= f^{-1}\left(\frac{2x+6}{3}\right) \\ &= \frac{3}{2}\left(\frac{2x+6}{3}\right) - 3 \\ &= \frac{2x+6}{2} - 3 \\ &= x+3-3 \\ &= x \end{aligned}$$

37.

$$\begin{aligned} 4x - 5y &= 20 \\ 5y &= 4x - 20 \\ y = f(x) &= \tfrac{4}{5}x - 4 \\ x &= \tfrac{4}{5}y - 4 \\ x + 4 &= \tfrac{4}{5}y \\ \tfrac{5}{4}(x+4) &= y \\ f^{-1}(x) &= \tfrac{5}{4}x + 5 \end{aligned}$$

$f \circ f^{-1}$

$$\begin{aligned} f[f^{-1}(x)] &= f\left(\tfrac{5}{4}x + 5\right) \\ &= \tfrac{4}{5}\left(\tfrac{5}{4}x + 5\right) - 4 \\ &= x + 4 - 4 \\ &= x \end{aligned}$$

$f^{-1} \circ f$

$$\begin{aligned} f^{-1}[f(x)] &= f^{-1}\left(\tfrac{4}{5}x - 4\right) \\ &= \tfrac{5}{4}\left(\tfrac{4}{5}x - 4\right) + 5 \\ &= x - 5 + 5 \\ &= x \end{aligned}$$

38.

$$\begin{aligned} 3x + 5y &= 15 \\ 5y &= -3x + 15 \\ y = f(x) &= -\tfrac{3}{5}x + 3 \\ x &= -\tfrac{3}{5}y + 3 \\ \tfrac{3}{5}y &= -x + 3 \\ y &= \tfrac{5}{3}(-x + 3) \\ f^{-1}(x) &= -\tfrac{5}{3}x + 5 \end{aligned}$$

$f \circ f^{-1}$

$$\begin{aligned} f[f^{-1}(x)] &= f\left(-\tfrac{5}{3}x + 5\right) \\ &= -\tfrac{3}{5}\left(-\tfrac{5}{3}x + 5\right) + 3 \\ &= x - 3 + 3 \\ &= x \end{aligned}$$

$f^{-1} \circ f$

$$\begin{aligned} f^{-1}[f(x)] &= f^{-1}\left(-\tfrac{3}{5}x + 3\right) \\ &= -\tfrac{5}{3}\left(-\tfrac{3}{5}x + 3\right) + 5 \\ &= x - 5 + 5 \\ &= x \end{aligned}$$

39.

$$\begin{aligned} y &= 4x + 3 \\ x &= 4y + 3 \\ x - 3 &= 4y \\ \frac{x-3}{4} &= y \end{aligned}$$

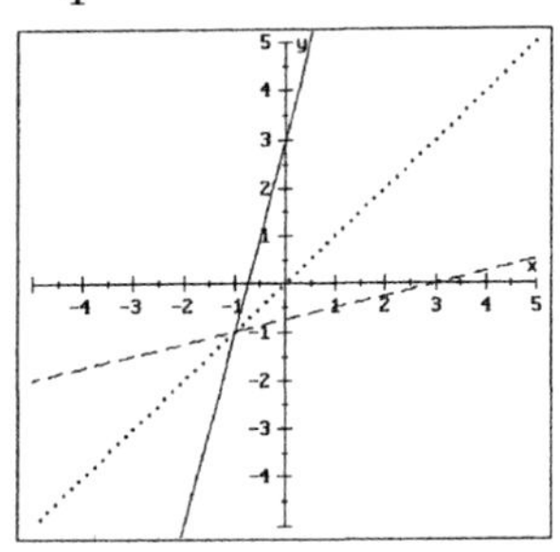

40.

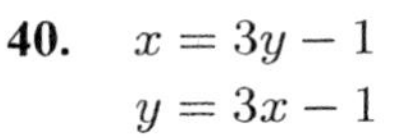

$$\begin{aligned} x &= 3y - 1 \\ y &= 3x - 1 \end{aligned}$$

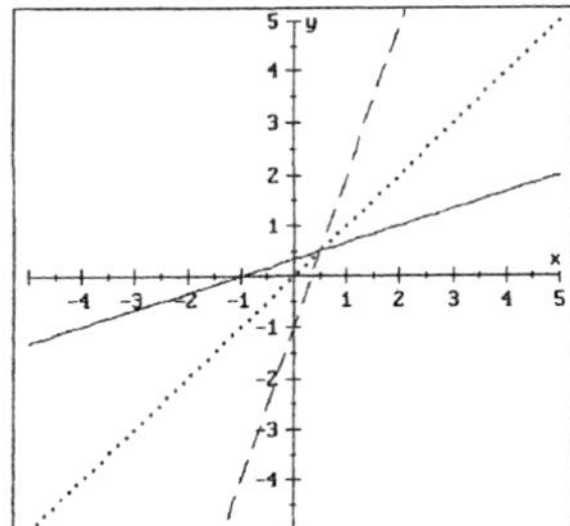

41. $x = \dfrac{y-2}{3}$

$y = \dfrac{x-2}{3}$

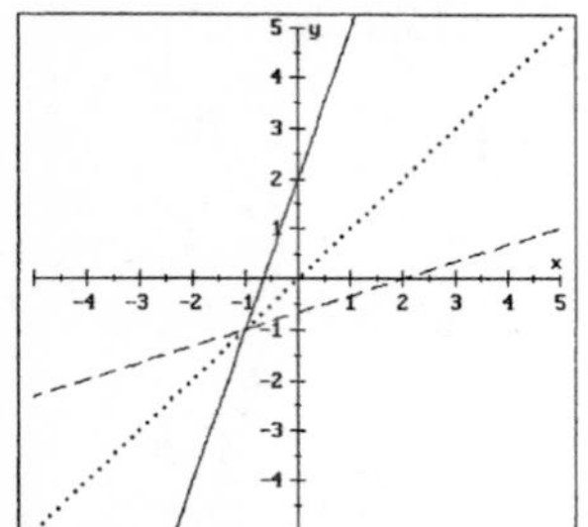

42. $y = \dfrac{x+3}{4}$

$x = \dfrac{y+3}{4}$

$4x = y + 3$

$4x - 3 = y$

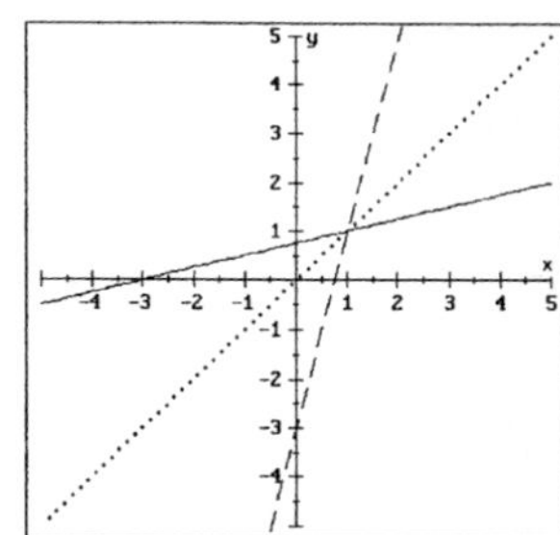

43. $3x - y = 5$

$3y - x = 5$

$3y = x + 5$

$y = \dfrac{x+5}{3}$

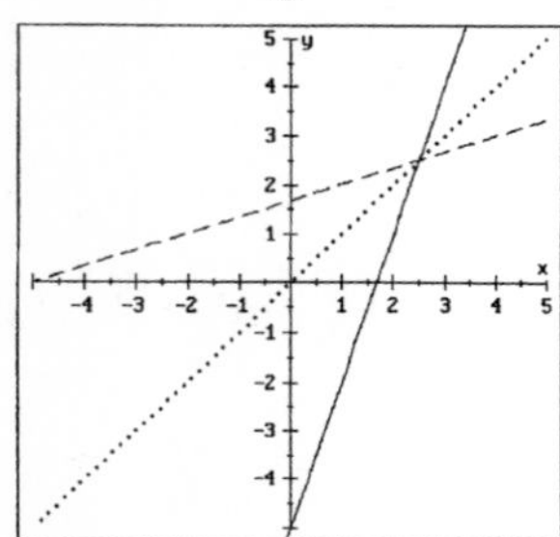

44. $2x + 3y = 9$

$2y + 3x = 9$

$2y = -3x + 9$

$y = \dfrac{-3x+9}{2}$

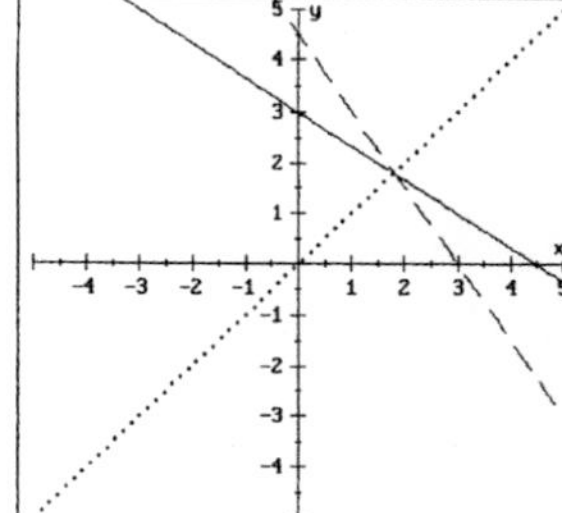

45. $3(x + y) = 2x + 4$

$3(y + x) = 2y + 4$

$3y + 3x = 2y + 4$

$y = 4 - 3x$

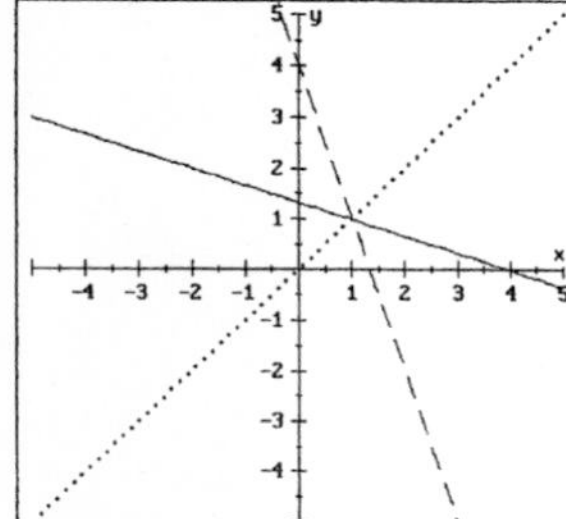

46. $-4(y - 1) + x = 2$

$-4(x - 1) + y = 2$

$-4x + 4 + y = 2$

$y = 4x - 2$

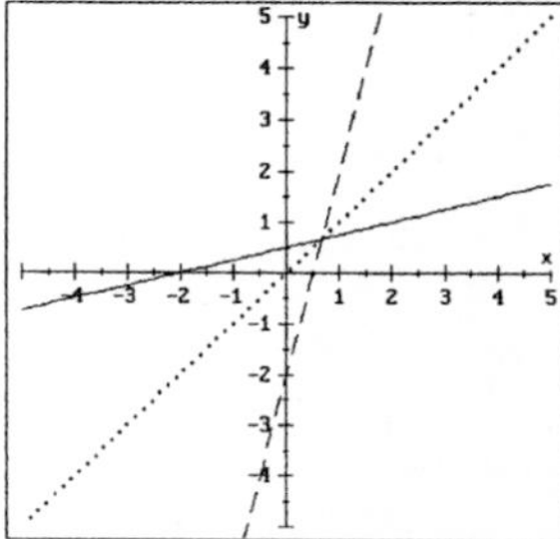

47.
$$\begin{aligned} y &= x^2 + 4 \\ x &= y^2 + 4 \\ x - 4 &= y^2 \\ \pm\sqrt{x-4} &= y \end{aligned}$$
The relation **is not** a function.

48.
$$\begin{aligned} y &= x^2 + 5 \\ x &= y^2 + 5 \\ x - 5 &= y^2 \\ \pm\sqrt{x-5} &= y \end{aligned}$$
The relation **is not** a function.

49.
$$\begin{aligned} y &= x^3 \\ x &= y^3 \\ \sqrt[3]{x} &= y \end{aligned}$$
The relation **is** a function.

50.
$$\begin{aligned} xy &= 4 \\ yx &= 4 \\ y &= \frac{4}{x} \end{aligned}$$
The relation **is** a function.

51.
$$\begin{aligned} y &= |x| \\ x &= |y| \end{aligned}$$
The relation **is not** a function.

52.
$$\begin{aligned} y &= \sqrt[3]{x} \\ x &= \sqrt[3]{y} \\ x^3 &= \left(\sqrt[3]{y}\right)^3 \\ x^3 &= y \end{aligned}$$
The relation **is** a function.

53.
$$\begin{aligned} y &= 2x^3 - 3 \\ x &= 2y^3 - 3 \\ x + 3 &= 2y^3 \\ \frac{x+3}{2} &= y^3 \\ \sqrt[3]{\frac{x+3}{2}} &= \sqrt[3]{y^3}, \text{ or } y = f^{-1}(x) = \sqrt[3]{\frac{x+3}{2}} \end{aligned}$$

54.
$$\begin{aligned} y &= \frac{3}{x^3} - 1 \\ x &= \frac{3}{y^3} - 1 \\ x + 1 &= \frac{3}{y^3} \\ y^3(x+1) &= 3 \\ y^3 &= \frac{3}{x+1} \\ y &= f^{-1}(x) = \sqrt[3]{\frac{3}{x+1}} \end{aligned}$$

55. $y = x^2 + 1$

inverse: $x = y^2 + 1$

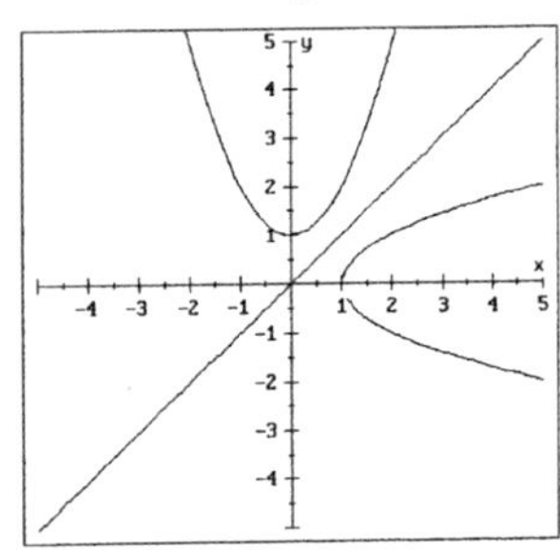

56. $y = \frac{1}{4}x^2 - 3$

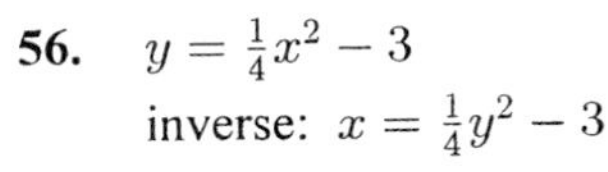

inverse: $x = \frac{1}{4}y^2 - 3$

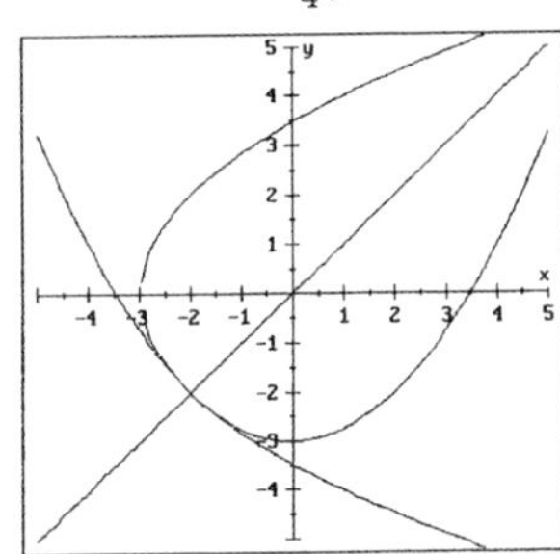

57. $y = \sqrt{x}$
inverse: $x = \sqrt{y}$

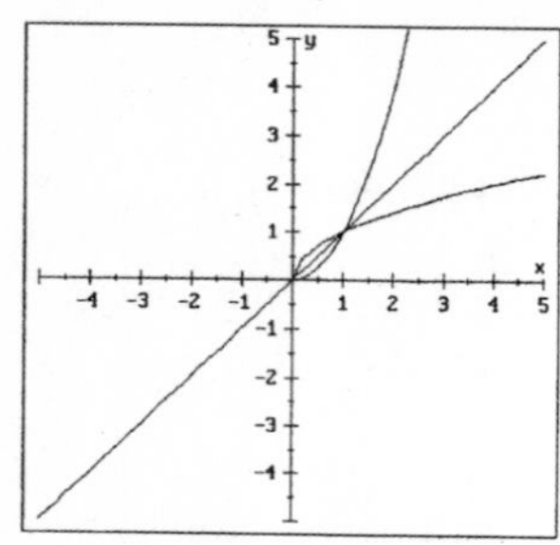

58. $y = |x|$
inverse: $x = |y|$

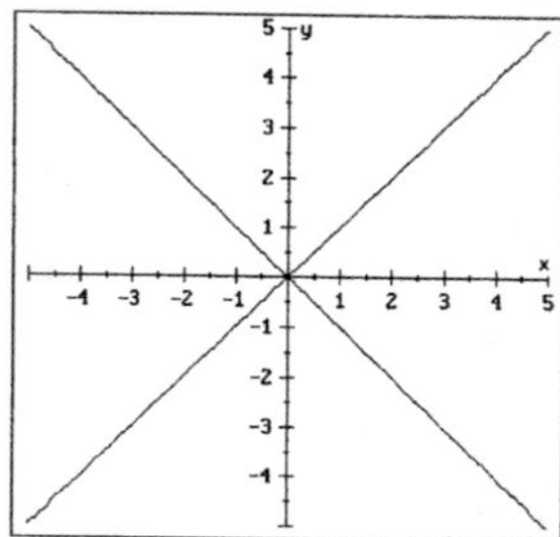

59. **Answers may vary.**

60. **Answers may vary.**

61.
$$\begin{aligned} y &= \frac{x+1}{x-1} \\ x &= \frac{y+1}{y-1} \\ x(y-1) &= y+1 \\ xy - x &= y+1 \\ xy - y &= x+1 \\ y(x-1) &= x+1 \\ y &= \frac{x+1}{x-1} \end{aligned}$$

62.
$$\begin{aligned} (f \circ f^{-1})(x) &= f\left(f^{-1}(x)\right) \\ &= f\left(\frac{x+1}{x-1}\right) \\ &= \frac{\frac{x+1}{x-1}+1}{\frac{x+1}{x-1}-1} \\ &= \frac{\frac{x+1}{x-1}+1}{\frac{x+1}{x-1}-1} \cdot \frac{x-1}{x-1} \\ &= \frac{x+1+x-1}{x+1-(x-1)} \\ &= \frac{2x}{2} = x \end{aligned}$$

Chapter 8 Summary (page 580)

1.
$$\begin{aligned} 12x^2 + x - 6 &= 0 \\ (4x+3)(3x-2) &= 0 \end{aligned}$$
$4x + 3 = 0$ **or** $3x - 2 = 0$
$x = -\frac{3}{4}$ $\qquad x = \frac{2}{3}$

2.
$$\begin{aligned} 6x^2 + 17x + 5 &= 0 \\ (2x+5)(3x+1) &= 0 \end{aligned}$$
$2x + 5 = 0$ **or** $3x + 1 = 0$
$x = -\frac{5}{2}$ $\qquad x = -\frac{1}{3}$

3.
$$\begin{aligned} 15x^2 + 2x - 8 &= 0 \\ (3x-2)(5x+4) &= 0 \end{aligned}$$
$3x - 2 = 0$ **or** $5x + 4 = 0$
$x = \frac{2}{3}$ $\qquad x = -\frac{4}{5}$

4.
$$\begin{aligned} (x+2)^2 &= 36 \\ x + 2 &= \pm\sqrt{36} \\ x + 2 &= \pm 6 \\ x &= -2 \pm 6 \end{aligned}$$
$x = 4$ or $x = -8$

5.
$$\begin{aligned} x^2+6x+8&=0\\ x^2+6x&=-8\\ x^2+6x+9&=-8+9\\ (x+3)^2&=1\\ x+3&=\pm 1\\ x&=-3\pm 1\end{aligned}$$
$x=-2$ or $x=-4$

6.
$$\begin{aligned} 2x^2-9x+7&=0\\ x^2-\frac{9}{2}x+\frac{7}{2}&=0\\ x^2-\frac{9}{2}x&=-\frac{7}{2}\\ x^2-\frac{9}{2}x+\frac{81}{16}&=-\frac{56}{16}+\frac{81}{16}\\ \left(x-\frac{9}{4}\right)^2&=\frac{25}{16}\\ x-\frac{9}{4}&=\pm\frac{5}{4}\\ x&=\frac{9}{4}\pm\frac{5}{4}\end{aligned}$$
$x=\frac{7}{2}$ or $x=1$

7.
$$\begin{aligned} 2x^2-x-5&=0\\ x^2-\frac{1}{2}x-\frac{5}{2}&=0\\ x^2-\frac{1}{2}x&=\frac{5}{2}\\ x^2-\frac{1}{2}x+\frac{1}{16}&=\frac{5}{2}+\frac{1}{16}\\ \left(x-\frac{1}{4}\right)^2&=\frac{41}{16}\\ x-\frac{1}{4}&=\pm\frac{\sqrt{41}}{4}\\ x&=\frac{1}{4}\pm\frac{\sqrt{41}}{4}\end{aligned}$$

8.
$x^2-8x-9=0$

$a=1, b=-8, c=-9$
$$\begin{aligned} x&=\frac{-b\pm\sqrt{b^2-4ac}}{2a}\\ &=\frac{-(-8)\pm\sqrt{(-8)^2-4(1)(-9)}}{2(1)}\\ &=\frac{8\pm\sqrt{64+36}}{2}\\ &=\frac{8\pm\sqrt{100}}{2}=\frac{8\pm 10}{2}\end{aligned}$$
$x=\frac{18}{2}=9$ or $x=\frac{-2}{2}=-1$

9.
$x^2-10x=0$

$a=1, b=-10, c=0$
$$\begin{aligned} x&=\frac{-b\pm\sqrt{b^2-4ac}}{2a}\\ &=\frac{-(-10)\pm\sqrt{(-10)^2-4(1)(0)}}{2(1)}\\ &=\frac{10\pm\sqrt{100+0}}{2}\\ &=\frac{10\pm\sqrt{100}}{2}=\frac{10\pm 10}{2}\end{aligned}$$
$x=\frac{20}{2}=10$ or $x=\frac{0}{2}=0$

10.
$2x^2+13x-7=0$

$a=2, b=13, c=-7$
$$\begin{aligned} x&=\frac{-b\pm\sqrt{b^2-4ac}}{2a}\\ &=\frac{-(13)\pm\sqrt{13^2-4(2)(-7)}}{2(2)}\\ &=\frac{-13\pm\sqrt{169+56}}{4}\\ &=\frac{-13\pm\sqrt{225}}{4}=\frac{-13\pm 15}{4}\end{aligned}$$
$x=\frac{2}{4}=\frac{1}{2}$ or $x=\frac{-28}{4}=-7$

11. $3x^2 + 20x - 7 = 0$
$a = 3, b = 20, c = -7$

$$x = \frac{-b \pm \sqrt{b^2 - 4ac}}{2a} = \frac{-20 \pm \sqrt{(20)^2 - 4(3)(-7)}}{2(3)} = \frac{-20 \pm \sqrt{400 + 84}}{6} = \frac{-20 \pm \sqrt{484}}{6} = \frac{-20 \pm 22}{6}$$

$$x = \frac{2}{6} = \frac{1}{3} \text{ or } x = \frac{-42}{6} = -7$$

12. $2x^2 - x - 2 = 0$
$a = 2, b = -1, c = -2$

$$x = \frac{-b \pm \sqrt{b^2 - 4ac}}{2a} = \frac{-(-1) \pm \sqrt{(-1)^2 - 4(2)(-2)}}{2(2)} = \frac{1 \pm \sqrt{1 + 16}}{4} = \frac{1 \pm \sqrt{17}}{4} = \frac{1}{4} \pm \frac{\sqrt{17}}{4}$$

13. $x^2 + x + 2 = 0$
$a = 1, b = 1, c = 2$

$$x = \frac{-b \pm \sqrt{b^2 - 4ac}}{2a} = \frac{-1 \pm \sqrt{1^2 - 4(1)(2)}}{2(1)} = \frac{-1 \pm \sqrt{1 - 8}}{2} = \frac{-1 \pm \sqrt{-7}}{2} = -\frac{1}{2} \pm \frac{\sqrt{7}}{2}i$$

14. Let w represent the original width.
Then $w + 2$ represents the original length.
The new dimensions are then $2w$ and $2(w + 2) = 2w + 4$.

Old Area + 72 = New Area

$$w(w + 2) + 72 = 2w(2w + 4)$$
$$w^2 + 2w + 72 = 4w^2 + 8w$$
$$0 = 3w^2 + 6w - 72$$
$$0 = 3(w + 6)(w - 4)$$

$w = -6$ or $w = 4$

Since the width is positive, the dimensions are 4 cm by 6 cm.

15. Let w represent the original width.
Then $w + 1$ represents the original length.
The new dimensions are then $2w$ and $3(w + 1) = 3w + 3$.

Old Area + 30 = New Area

$$w(w + 1) + 30 = 2w(3w + 3)$$
$$w^2 + w + 30 = 6w^2 + 6w$$
$$0 = 5w^2 + 5w - 30$$
$$0 = 5(w + 3)(w - 2)$$

$w = -3$ or $w = 2$

Since the width is positive, the dimensions are 2 ft by 3 ft.

16. When the rocket hits the ground, $h = 0$:

$h = 112t - 16t^2$

$0 = 112t - 16t^2$

$0 = 16t(7 - t)$

$t = 0$ or $t = 7$

It hits the ground after 7 seconds.

17. The maximum height occurs at the vertex:

$h = 112t - 16t^2$

$h = -16t^2 + 112t$

$h = -16(t^2 - 7t)$

$h = -16\left(t^2 - 7t + \frac{49}{4}\right) + 196$

$h = -16\left(t^2 - 7t + \frac{49}{4}\right) + 196$

$h = -16\left(t - \frac{7}{2}\right)^2 + 196$

Vertex: $\left(\frac{7}{2}, 196\right) \Rightarrow$ max. height $= 196$ ft

18. $3x^2 + 4x - 3 = 0$

$a = 3, b = 4, c = -3$

$b^2 - 4ac = 4^2 - 4(3)(-3)$

$= 16 + 36 = 52$

irrational unequal solutions

19. $4x^2 - 5x + 7 = 0$

$a = 4, b = -5, c = 7$

$b^2 - 4ac = (-5)^2 - 4(4)(7)$

$= 25 - 112 = -87$

complex conjugate solutions

20. $(k-8)x^2 + (k+16)x = -49$

$(k-8)x^2 + (k+16)x + 49 = 0$

$a = k - 8, b = k + 16, c = 49$

Set the discriminant equal to 0:

$$b^2 - 4ac = 0$$

$$(k+16)^2 - 4(k-8)(49) = 0$$

$$k^2 + 32k + 256 - 196k + 1568 = 0$$

$$k^2 - 164k + 1824 = 0$$

$$(k-12)(k-152) = 0$$

$k - 12 = 0$ **or** $k - 152 = 0$

$k = 12$ $\quad$ $k = 152$

21. $3x^2 + 4x = k + 1$

$3x^2 + 4x - k - 1 = 0$

$a = 3, b = 4, c = -k - 1$

Set the discriminant ≥ 0:

$$b^2 - 4ac \geq 0$$

$$4^2 - 4(3)(-k-1) \geq 0$$

$$16 + 12k + 12 \geq 0$$

$$12k \geq -28$$

$$k \geq -\frac{28}{12}$$

$$k \geq -\frac{7}{3}$$

22. $x - 13x^{1/2} + 12 = 0$

$\left(x^{1/2} - 12\right)\left(x^{1/2} - 1\right) = 0$

$x^{1/2} - 12 = 0$ **or** $x^{1/2} - 1 = 0$

$x^{1/2} = 12$ $\quad$ $x^{1/2} = 1$

$\left(x^{1/2}\right)^2 = (12)^2$ $\quad$ $\left(x^{1/2}\right)^2 = 1^2$

$x = 144$ $\quad$ $x = 1$

Solution. $\quad$ Solution.

23. $a^{2/3} + a^{1/3} - 6 = 0$

$\left(a^{1/3} - 2\right)\left(a^{1/3} + 3\right) = 0$

$a^{1/3} - 2 = 0$ **or** $a^{1/3} + 3 = 0$

$a^{1/3} = 2$ $\quad$ $a^{1/3} = -3$

$\left(a^{1/3}\right)^3 = (2)^3$ $\quad$ $\left(a^{1/3}\right)^3 = (-3)^3$

$a = 8$ $\quad$ $a = -27$

Solution. $\quad$ Solution.

24.

$$\frac{1}{x+1} - \frac{1}{x} = -\frac{1}{x+1}$$

$$\left(\frac{1}{x+1} - \frac{1}{x}\right)(x)(x+1) = -\frac{1}{x+1}(x)(x+1)$$

$$1(x) - 1(x+1) = -x$$

$$x - x - 1 = -x$$

$$-1 = -x$$

$$1 = x$$

25.
$$\frac{6}{x+2}+\frac{6}{x+1}=5$$
$$\left(\frac{6}{x+2}+\frac{6}{x+1}\right)(x+2)(x+1)=5(x+2)(x+1)$$
$$6(x+1)+6(x+2)=5(x^2+3x+2)$$
$$6x+6+6x+12=5x^2+15x+10$$
$$0=5x^2+3x-8$$
$$0=(5x+8)(x-1) \qquad 5x+8=0 \quad \textbf{or} \quad x-1=0$$
$$x=-\tfrac{8}{5} \qquad\qquad x=1$$

26. $3x^2-14x+3=0$

$\text{sum}=-\frac{b}{a}=-\frac{-14}{3}=\frac{14}{3}$

27. $3x^2-14x+3=0$

$\text{product}=\frac{c}{a}=\frac{3}{3}=1$

28. $y=2x^2-3$

$y=2(x-0)^2-3$

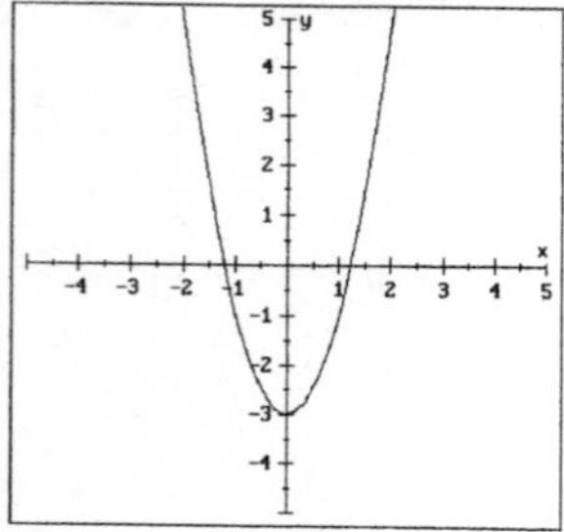

vertex: $(0,-3)$

29. $y=-2x^2-1$

$y=-2(x-0)^2-1$

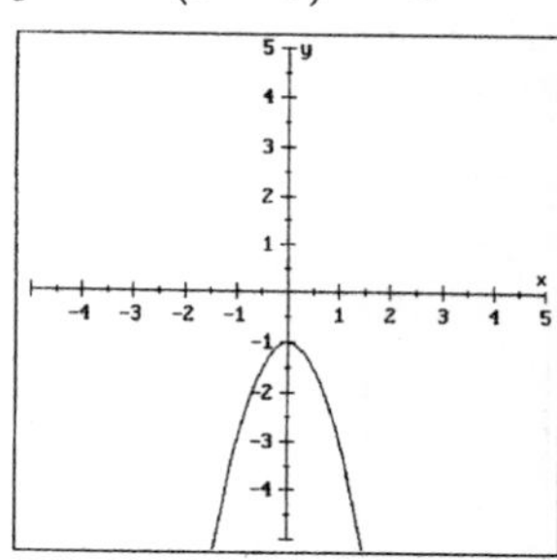

vertex: $(0,-1)$

30. $y=-4(x-2)^2+1$

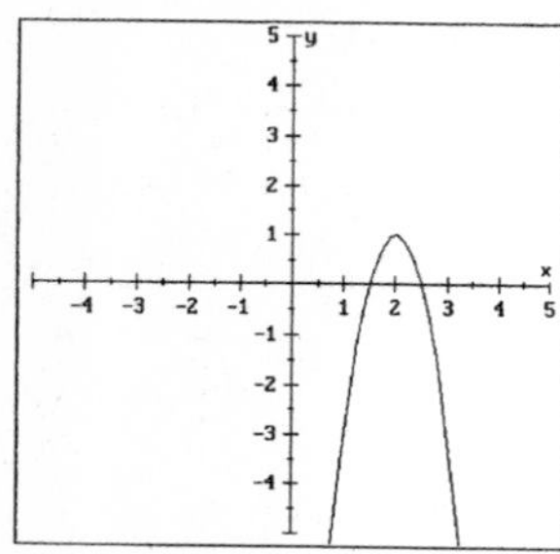

vertex: $(2,1)$

31.
$$\begin{aligned} y &= 5x^2+10x-1 \\ &= 5\left(x^2+2x\right)-1 \\ &= 5\left(x^2+2x+1\right)-1-5 \\ &= 5(x+1)^2-6 \end{aligned}$$

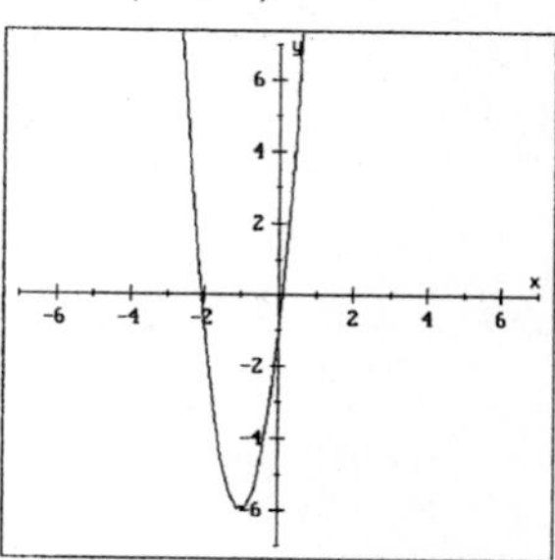

vertex: $(-1,-6)$

32. $y = 3x^2 - 12x - 5 = 3(x^2 - 4x) - 5 = 3(x^2 - 4x + 4) - 5 - 12 = 3(x - 2)^2 - 17$
vertex: $(2, -17)$

33.
$$x^2 + 2x - 35 > 0$$
$$(x + 7)(x - 5) > 0$$
$x - 5$ – – – – – – – – – 0 ++++
$x + 7$ – – – 0++++++ ++++
$-7 \qquad 5$
solution set: $(-\infty, -7) \cup (5, \infty)$

34.
$$x^2 + 7x - 18 < 0$$
$$(x - 2)(x + 9) < 0$$
$x - 2$ – – – – – – – – – 0++++
$x + 9$ – – – 0++++++++++
$-9 \qquad 2$
solution set: $(-9, 2)$

35.
$$\frac{3}{x} \le 5$$
$$\frac{3}{x} - 5 \le 0$$
$$\frac{3}{x} - \frac{5x}{x} \le 0$$
$$\frac{3 - 5x}{x} \le 0$$
$3 - 5x$ +++++++++ 0 – – –
x – – – 0+++++++++++
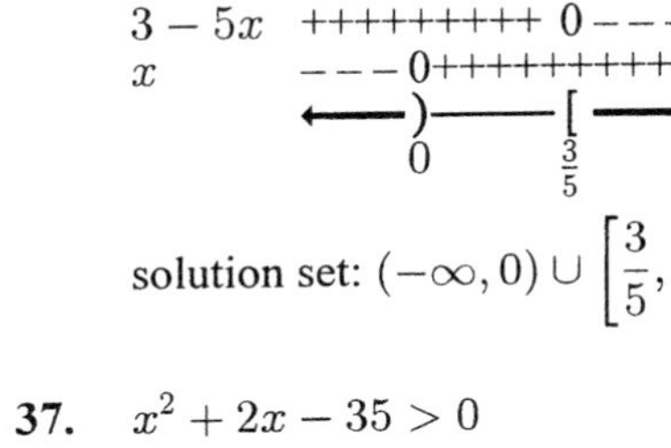
$0 \qquad \frac{3}{5}$

solution set: $(-\infty, 0) \cup \left[\frac{3}{5}, \infty\right)$

36.
$$\frac{2x^2 - x - 28}{x - 1} > 0$$
$$\frac{(2x + 7)(x - 4)}{x - 1} > 0$$
$x - 4$ – – – – – – – – – – – – – 0++++
$x - 1$ – – – – – – – – – 0+++++++++
$2x + 7$ – – – 0+++++++++++++++++
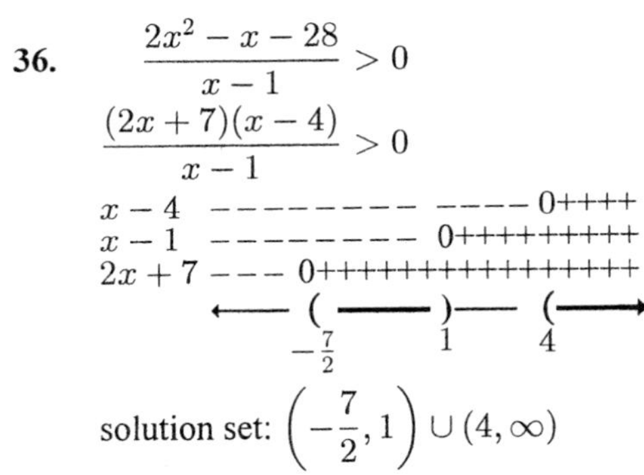
$-\frac{7}{2} \qquad 1 \qquad 4$

solution set: $\left(-\frac{7}{2}, 1\right) \cup (4, \infty)$

37. $x^2 + 2x - 35 > 0$
Graph $y = x^2 + 2x - 35$
and find the x-coordinates
of points above the x-axis.

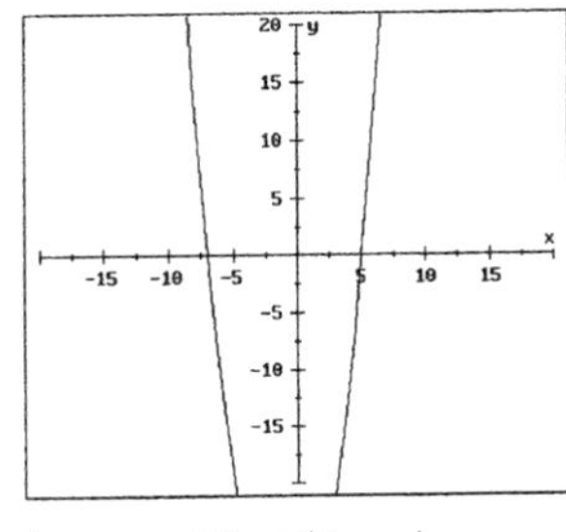

$(-\infty, -7) \cup (5, \infty)$

38. $x^2 + 7x - 18 < 0$
Graph $y = x^2 + 7x - 18$
and find the x-coordinates
of points below the x-axis.

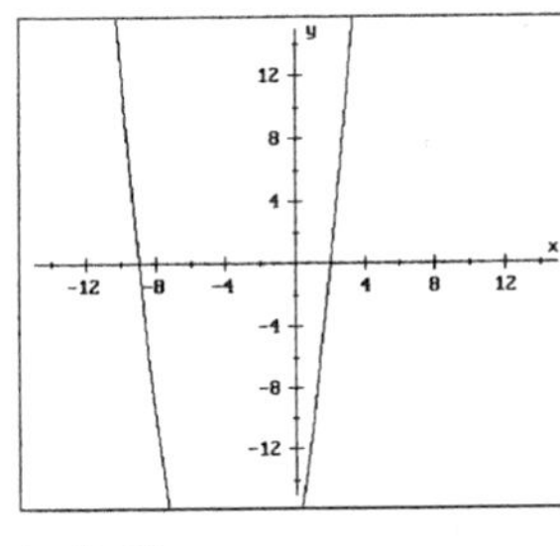

$(-9, 2)$

39. $\frac{3}{x} \le 5 \Rightarrow \frac{3}{x} - 5 \le 0$
Graph $y = (3/x) - 5$
and find the x-coordinates
of points below or on the x-axis.

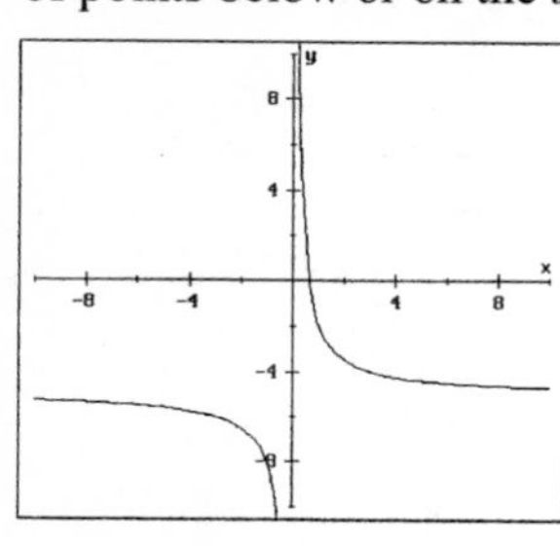

$(-\infty, 0) \cup \left[\frac{3}{5}, \infty\right)$

40. $\frac{2x^2-x-28}{x-1} > 0$
Graph $y = (2x^2 - x - 28)/(x - 1)$
and find the x-coordinates
of points above the x-axis.

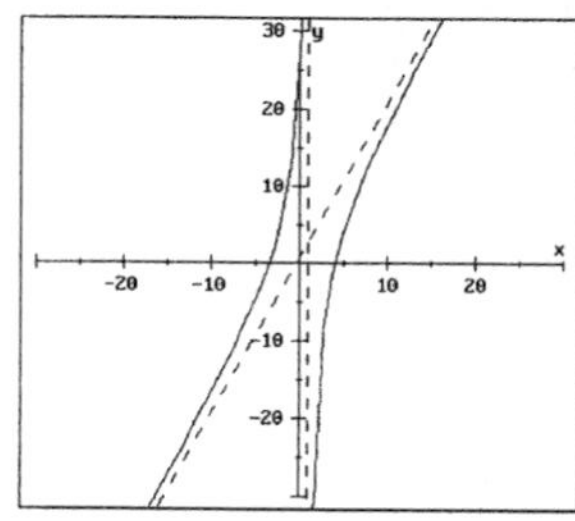

$\left(-\frac{7}{2}, 1\right) \cup (4, \infty)$

41. $y < \frac{1}{2}x^2 - 1$

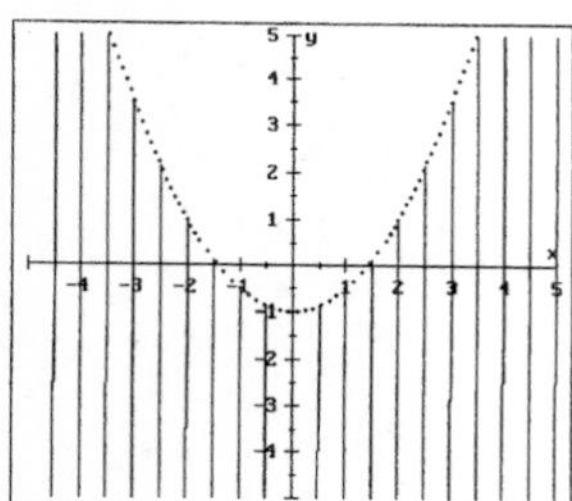

42. $y \ge -|x|$

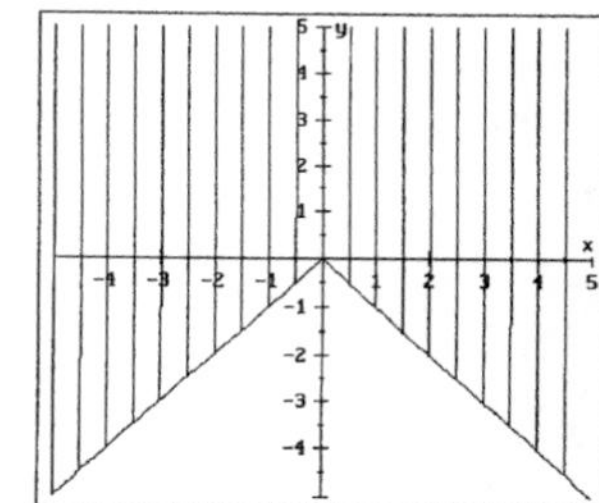

43. $f + g = f(x) + g(x) = 2x + x + 1$
$= 3x + 1$

44. $f - g = f(x) - g(x) = 2x - (x + 1)$
$= x - 1$

45. $f \cdot g = f(x)g(x) = 2x(x + 1) = 2x^2 + 2x$

46. $f/g = \frac{f(x)}{g(x)} = \frac{2x}{x+1}$

47. $(f \circ g)(2) = f(g(2)) = f(2 + 1)$
$= f(3) = 2(3) = 6$

48. $(g \circ f)(-1) = g(f(-1)) = g(2(-1))$
$= g(-2)$
$= -2 + 1 = -1$

49. $(f \circ g)(x) = f(g(x)) = f(x + 1) = 2(x + 1)$

50. $(g \circ f)(x) = g(f(x)) = g(2x) = 2x + 1$

51. $f(x) = 2(x-3)$

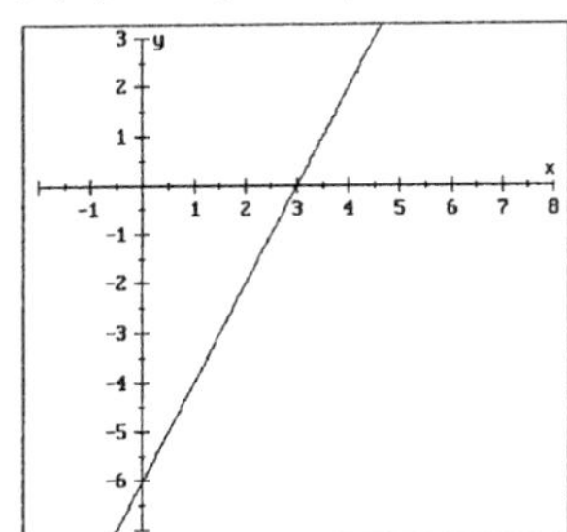

one-to-one

52. $f(x) = x(2x-3)$

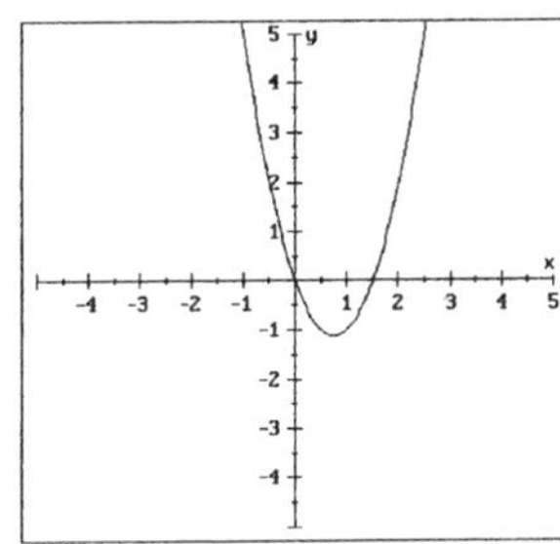

not one-to-one

53. $f(x) = -3(x-2)^2 + 5$

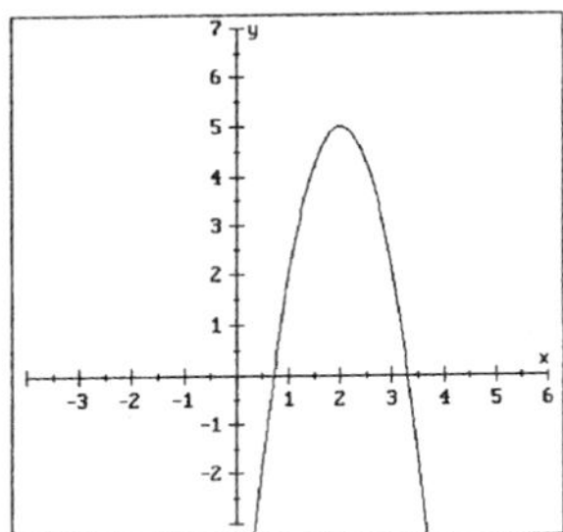

not one-to-one

54. $f(x) = |x|$

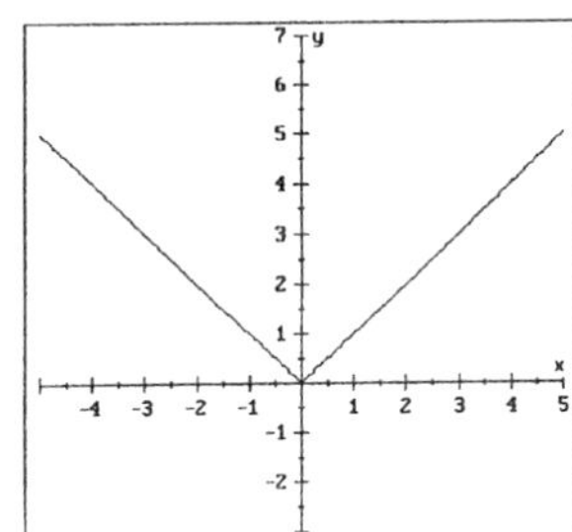

not one-to-one

55.
$$\begin{aligned} y &= 6x - 3 \\ x &= 6y - 3 \\ x + 3 &= 6y \\ \tfrac{x+3}{6} &= y, \text{ or } y = f^{-1}(x) = \tfrac{x+3}{6} \end{aligned}$$

56.
$$\begin{aligned} y &= 4x + 5 \\ x &= 4y + 5 \\ x - 5 &= 4y \\ \tfrac{x-5}{4} &= y, \text{ or } y = f^{-1}(x) = \tfrac{x-5}{4} \end{aligned}$$

57.
$$\begin{aligned} y &= 2x^2 - 1 \\ x &= 2y^2 - 1 \\ x + 1 &= 2y^2 \\ \tfrac{x+1}{2} &= y^2 \\ \sqrt{\tfrac{x+1}{2}} &= y, \text{ or } y = f^{-1}(x) = \sqrt{\tfrac{x+1}{2}} \end{aligned}$$

58.
$$\begin{aligned} y &= |x| \\ x &= |y| \end{aligned}$$

Chapter 8 Test (page 584)

1.
$$\begin{aligned} x^2 + 3x - 18 &= 0 \\ (x+6)(x-3) &= 0 \end{aligned}$$
$x + 6 = 0$ **or** $x - 3 = 0$
$x = -6$ $\quad$ $x = 3$

2.
$$\begin{aligned} x(6x + 19) &= -15 \\ 6x^2 + 19x + 15 &= 0 \\ (2x+3)(3x+5) &= 0 \end{aligned}$$
$2x + 3 = 0$ **or** $3x + 5 = 0$
$x = -\frac{3}{2}$ $\quad$ $x = -\frac{5}{3}$

3. $\left(\frac{1}{2}\cdot 24\right)^2 = 12^2 = 144$

4. $\left(\frac{1}{2}\cdot(-50)\right)^2 = (-25)^2 = 625$

5.
$$\begin{aligned} x^2+4x+1&=0\\ x^2+4x&=-1\\ x^2+4x+4&=-1+4\\ (x+2)^2&=3\\ x+2&=\pm\sqrt{3}\\ x&=-2\pm\sqrt{3} \end{aligned}$$

6.
$$\begin{aligned} x^2-5x-3&=0\\ x^2-5x&=3\\ x^2-5x+\frac{25}{4}&=3+\frac{25}{4}\\ \left(x-\frac{5}{2}\right)^2&=\frac{37}{4}\\ x-\frac{5}{2}&=\pm\sqrt{\frac{37}{4}}\\ x&=\frac{5}{2}\pm\frac{\sqrt{37}}{2} \end{aligned}$$

7.
$$2x^2+5x+1=0$$
$$a=2, b=5, c=1$$
$$\begin{aligned} x&=\frac{-b\pm\sqrt{b^2-4ac}}{2a}\\ &=\frac{-5\pm\sqrt{5^2-4(2)(1)}}{2(2)}\\ &=\frac{-5\pm\sqrt{25-8}}{4}\\ &=\frac{-5\pm\sqrt{17}}{4}=-\frac{5}{4}\pm\frac{\sqrt{17}}{4} \end{aligned}$$

8.
$$x^2-x+3=0$$
$$a=1, b=-1, c=3$$
$$\begin{aligned} x&=\frac{-b\pm\sqrt{b^2-4ac}}{2a}\\ &=\frac{-(-1)\pm\sqrt{(-1)^2-4(1)(3)}}{2(1)}\\ &=\frac{1\pm\sqrt{1-12}}{2}\\ &=\frac{1\pm\sqrt{-11}}{2}=\frac{1}{2}\pm\frac{\sqrt{11}}{2}i \end{aligned}$$

9.
$$3x^2+5x+17=0$$
$$a=3, b=5, c=17$$
$$\begin{aligned} b^2-4ac&=5^2-4(3)(17)\\ &=25-208=-183 \end{aligned}$$
nonreal solutions

10.
$$4x^2-2kx+k-1=0$$
$$a=4, b=-2k, c=k-1$$
Set the discriminant equal to 0:
$$\begin{aligned} b^2-4ac&=0\\ (-2k)^2-4(4)(k-1)&=0\\ 4k^2-16k+16&=0\\ 4(k-2)(k-2)&=0 \end{aligned}$$
$k-2=0$ **or** $k-2=0$
$k=2$ $\qquad$ $k=2$

11. Let $x=$ the length of the shorter leg.
Then $x+14=$ the other length.
$$\begin{aligned} x^2+(x+14)^2&=26^2\\ x^2+x^2+28x+196&=676\\ 2x^2+28x-480&=0\\ 2(x+24)(x-10)&=0 \end{aligned}$$
$x=-24$ or $x=10$
The shorter leg is 10 inches long.

12.
$$\begin{aligned} 2y-3y^{1/2}+1&=0\\ \left(2y^{1/2}-1\right)\left(y^{1/2}-1\right)&=0 \end{aligned}$$
$2y^{1/2}-1=0$ **or** $y^{1/2}-1=0$
$2y^{1/2}=1$ $\qquad$ $y^{1/2}=1$
$\left(y^{1/2}\right)^2=\left(\frac{1}{2}\right)^2$ $\qquad$ $\left(y^{1/2}\right)^2=1^2$
$y=\frac{1}{4}$ $\qquad$ $y=1$
Solution $\qquad$ Solution

13. $y = \frac{1}{2}x^2 - 4 = \frac{1}{2}(x-0)^2 - 4$

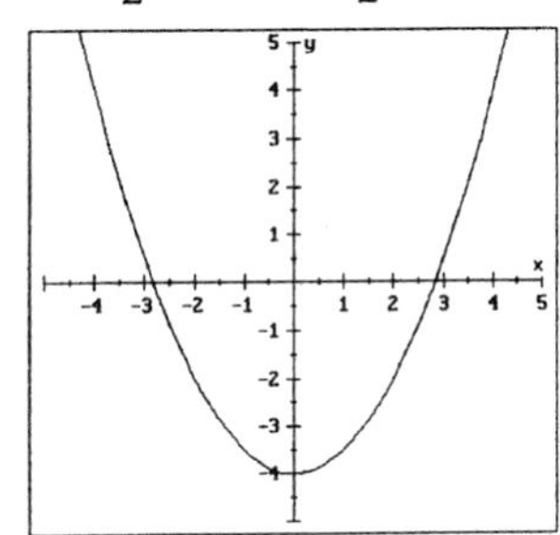

vertex: $(0, -4)$

14. $y = -2x^2 + 8x - 7$

$= -2(x^2 - 4x) - 7$

$= -2(x^2 - 4x + 4) - 7 + 8$

$= -2(x-2)^2 + 1$

Vertex: $(2, 1)$

15. $y \le -x^2 + 3$

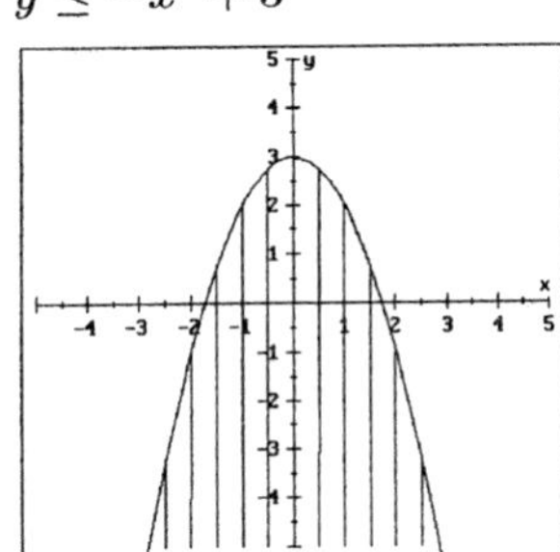

16. $x^2 - 2x - 8 > 0$

$(x+2)(x-4) > 0$

$x - 4$ −−−−−−−−−0 ++++

$x + 2$ −−− 0++++++ ++++

⟵) ——— (⟶ at −2 and 4

solution set: $(-\infty, -2) \cup (4, \infty)$

17. $\frac{x-2}{x+3} \le 0$

$x - 2$ −−−−−−−−0 ++++

$x + 3$ −−−0++++++ ++++

⟵ (———] ⟶ at −3 and 2

solution set: $(-3, 2]$

18. $g + f = g(x) + f(x) = x - 1 + 4x$

$= 5x - 1$

19. $f - g = f(x) - g(x) = 4x - (x-1)$

$= 3x + 1$

20. $g \cdot f = g(x)f(x) = (x-1)4x = 4x^2 - 4x$

21. $g/f = \frac{g(x)}{f(x)} = \frac{x-1}{4x}$

22. $(g \circ f)(1) = g(f(1)) = g(4(1))$

$= g(4) = 4 - 1 = 3$

23. $(f \circ g)(0) = f(g(0)) = f(0-1)$

$= f(-1) = 4(-1) = -4$

24. $(f \circ g)(-1) = f(g(-1)) = f(-1-1) = f(-2) = 4(-2) = -8$

25. $(g \circ f)(-2) = g(f(-2)) = g(4(-2)) = g(-8) = -8 - 1 = -9$

26. $(f \circ g)(x) = f(g(x)) = f(x-1) = 4(x-1)$

27. $(g \circ f)(x) = g(f(x)) = g(4x) = 4x - 1$

28.
$$\begin{aligned} 3x + 2y &= 12 \\ 3y + 2x &= 12 \\ 3y &= -2x + 12 \\ y &= \frac{-2x + 12}{3} \end{aligned}$$

29.
$$\begin{aligned} y &= 3x^2 + 4 \\ x &= 3y^2 + 4 \\ x - 4 &= 3y^2 \\ \frac{x-4}{3} &= y^2 \\ -\sqrt{\frac{x-4}{3}} &= y \end{aligned}$$

Cumulative Review Exercises (page 585)

1. $y = f(x) = 2x^2 - 3$
domain $= (-\infty, \infty)$
range $= [-3, \infty)$

2. $y = f(x) = -|x - 4|$
domain $= (-\infty, \infty)$
range $= (-\infty, 0]$

3.
$$\begin{aligned} y - y_1 &= m(x - x_1) \\ y + 4 &= 3(x + 2) \\ y &= 3x + 2 \end{aligned}$$

4.
$$\begin{aligned} 2x + 3y &= 6 \\ 3y &= -2x + 6 \\ y &= -\frac{2}{3}x + 2 \\ y - y_1 &= m(x - x_1) \\ y + 2 &= -\frac{2}{3}(x - 0) \\ y &= -\frac{2}{3}x - 2 \end{aligned}$$

5. $(2a^2 + 4a - 7) - 2(3a^2 - 4a) = 2a^2 + 4a - 7 - 6a^2 + 8a = -4a^2 + 12a - 7$

6. $(3x + 2)(2x - 3) = 6x^2 - 9x + 4x - 6 = 6x^2 - 5x - 6$

7. $x^4 - 16y^4 = (x^2 + 4y^2)(x^2 - 4y^2) = (x^2 + 4y^2)(x + 2y)(x - 2y)$

8. $15x^2 - 2x - 8 = (5x - 4)(3x + 2)$

9.
$$\begin{aligned} x^2 - 5x - 6 &= 0 \\ (x - 6)(x + 1) &= 0 \end{aligned}$$
$x - 6 = 0$ **or** $x + 1 = 0$
$x = 6$ $\quad\quad x = -1$

10.
$$\begin{aligned} 6a^3 - 2a &= a^2 \\ 6a^3 - a^2 - 2a &= 0 \\ a(6a^2 - a - 2) &= 0 \\ a(3a - 2)(2a + 1) &= 0 \end{aligned}$$
$a = 0$ **or** $3a - 2 = 0$ **or** $2a + 1 = 0$
$a = \frac{2}{3}$ $\quad\quad a = -\frac{1}{2}$

11. $\sqrt{25x^4} = 5x^2$

12. $\sqrt{48t^3} = \sqrt{16t^2}\sqrt{3t} = 4t\sqrt{3t}$

CUMULATIVE REVIEW EXERCISES

13. $\sqrt[3]{-27x^3} = -3x$

14. $\sqrt[3]{\dfrac{128x^4}{2x}} = \sqrt[3]{64x^3} = 4x$

15. $8^{-1/3} = \dfrac{1}{8^{1/3}} = \dfrac{1}{2}$

16. $64^{2/3} = \left(64^{1/3}\right)^2 = 4^2 = 16$

17. $\dfrac{y^{2/3}y^{5/3}}{y^{1/3}} = \dfrac{y^{7/3}}{y^{1/3}} = y^{6/3} = y^2$

18. $\dfrac{x^{5/3}x^{1/2}}{x^{3/4}} = \dfrac{x^{13/6}}{x^{3/4}} = x^{17/12}$

19. $f(x) = \sqrt{x-2}$; Shift $y = \sqrt{x}$ right 2.

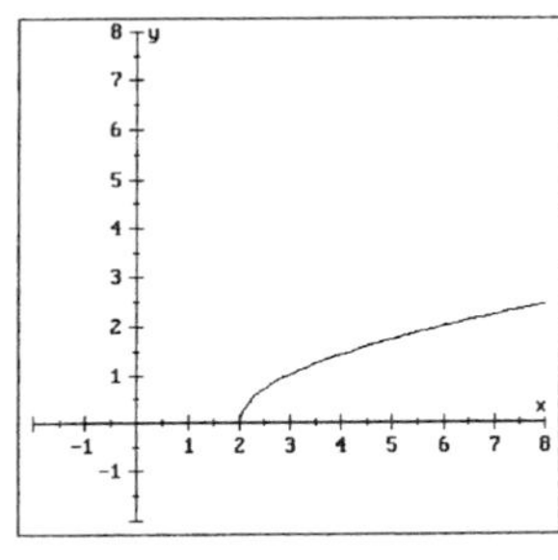

$D = [2, \infty)$, $R = [0, \infty)$

20. $f(x) = -\sqrt{x+2}$; Reflect $y = \sqrt{x}$ about the x-axis and shift left 2.

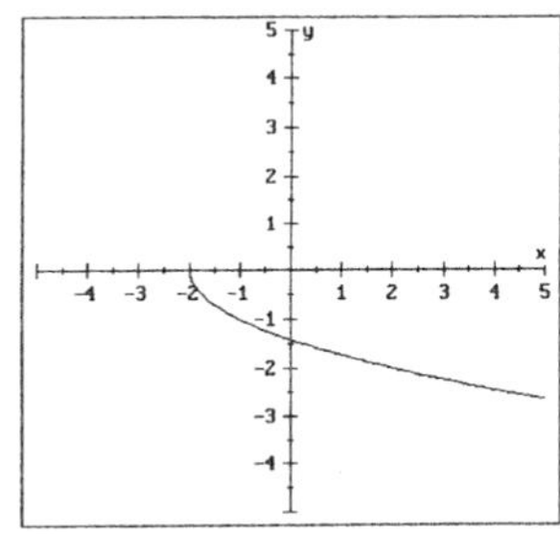

$D = [-2, \infty)$, $R = (-\infty, 0]$

21. $\left(x^{2/3} - x^{1/3}\right)\left(x^{2/3} + x^{1/3}\right) = x^{4/3} + x^{3/3} - x^{3/3} - x^{2/3} = x^{4/3} - x^{2/3}$

22. $\left(x^{-1/2} + x^{1/2}\right)^2 = \left(x^{-1/2} + x^{1/2}\right)\left(x^{-1/2} + x^{1/2}\right) = x^{-2/2} + x^0 + x^0 + x^{2/2} = x + 2 + \frac{1}{x}$

23. $\sqrt{50} - \sqrt{8} + \sqrt{32} = \sqrt{25}\sqrt{2} - \sqrt{4}\sqrt{2} + \sqrt{16}\sqrt{2} = 5\sqrt{2} - 2\sqrt{2} + 4\sqrt{2} = 7\sqrt{2}$

24.
$$\begin{aligned} -3\sqrt[4]{32} - 2\sqrt[4]{162} + 5\sqrt[4]{48} &= -3\sqrt[4]{16}\sqrt[4]{2} - 2\sqrt[4]{81}\sqrt[4]{2} + 5\sqrt[4]{16}\sqrt[4]{3} \\ &= -3(2)\sqrt[4]{2} - 2(3)\sqrt[4]{2} + 5(2)\sqrt[4]{3} \\ &= -6\sqrt[4]{2} - 6\sqrt[4]{2} + 10\sqrt[4]{3} = -12\sqrt[4]{2} + 10\sqrt[4]{3} \end{aligned}$$

25. $3\sqrt{2}(2\sqrt{3} - 4\sqrt{12}) = 6\sqrt{6} - 12\sqrt{24} = 6\sqrt{6} - 12\sqrt{4}\sqrt{6} = 6\sqrt{6} - 24\sqrt{6} = -18\sqrt{6}$

26. $\dfrac{5}{\sqrt[3]{x}} = \dfrac{5}{\sqrt[3]{x}} \cdot \dfrac{\sqrt[3]{x^2}}{\sqrt[3]{x^2}} = \dfrac{5\sqrt[3]{x^2}}{\sqrt[3]{x^3}} = \dfrac{5\sqrt[3]{x^2}}{x}$

27. $\dfrac{\sqrt{x}+2}{\sqrt{x}-1} = \dfrac{\sqrt{x}+2}{\sqrt{x}-1} \cdot \dfrac{\sqrt{x}+1}{\sqrt{x}+1} = \dfrac{x + 3\sqrt{x} + 2}{x - 1}$

28. $\sqrt[6]{x^3y^3} = (x^3y^3)^{1/6} = x^{3/6}y^{3/6} = x^{1/2}y^{1/2} = \sqrt{xy}$

29.

$$
\begin{aligned}
5\sqrt{x+2} &= x+8 \\
\left(5\sqrt{x+2}\right)^2 &= (x+8)^2 \\
25(x+2) &= x^2+16x+64 \\
25x+50 &= x^2+16x+64 \\
0 &= x^2-9x+14 \\
0 &= (x-7)(x-2)
\end{aligned}
$$

$x=7$ or $x=2$ (Both check.)

30.

$$
\begin{aligned}
\sqrt{x}+\sqrt{x+2} &= 2 \\
\sqrt{x} &= 2-\sqrt{x+2} \\
\left(\sqrt{x}\right)^2 &= \left(2-\sqrt{x+2}\right)^2 \\
x &= 4-4\sqrt{x+2}+x+2 \\
4\sqrt{x+2} &= 6 \\
\left(4\sqrt{x+2}\right)^2 &= 6^2 \\
16(x+2) &= 36 \\
16x+32 &= 36 \\
16x &= 4 \\
x &= \frac{4}{16} = \frac{1}{4}
\end{aligned}
$$

31. hypotenuse $= 3\sqrt{2}$ in.

32. hypotenuse $= 2 \cdot \dfrac{3}{\sqrt{3}} = \dfrac{6\sqrt{3}}{3} = 2\sqrt{3}$ in.

33. $d = \sqrt{(-2-4)^2+(6-14)^2} = \sqrt{(-6)^2+(-8)^2} = \sqrt{36+64} = \sqrt{100} = 10$

34. $\left(\frac{1}{2}\cdot 6\right)^2 = 3^2 = 9$

35.

$$
\begin{aligned}
2x^2+x-3 &= 0 \\
x^2+\frac{1}{2}x-\frac{3}{2} &= 0 \\
x^2+\frac{1}{2}x &= \frac{3}{2} \\
x^2+\frac{1}{2}x+\frac{1}{16} &= \frac{3}{2}+\frac{1}{16} \\
\left(x+\frac{1}{4}\right)^2 &= \frac{25}{16} \\
x+\frac{1}{4} &= \pm\frac{5}{4} \\
x &= -\frac{1}{4}\pm\frac{5}{4}
\end{aligned}
$$

$x = \dfrac{4}{4} = 1$ or $x = -\dfrac{6}{4} = -\dfrac{3}{2}$

36. $3x^2+4x-1=0$

$a=3, b=4, c=-1$

$$
\begin{aligned}
x &= \frac{-b\pm\sqrt{b^2-4ac}}{2a} \\
&= \frac{-4\pm\sqrt{4^2-4(3)(-1)}}{2(3)} \\
&= \frac{-4\pm\sqrt{16+12}}{6} \\
&= \frac{-4\pm\sqrt{28}}{6} \\
&= \frac{-4\pm 2\sqrt{7}}{6} = -\frac{2}{3}\pm\frac{\sqrt{7}}{3}
\end{aligned}
$$

CUMULATIVE REVIEW EXERCISES

37. $y = \frac{1}{2}x^2 + 5 = \frac{1}{2}(x-0)^2 + 5$

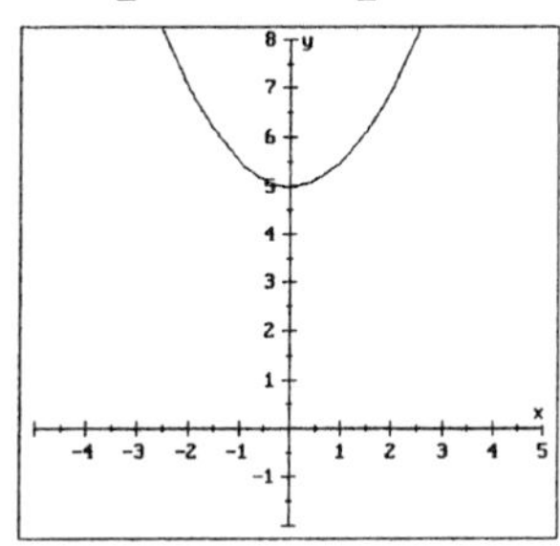

vertex: $(0, 5)$

38. $y \le -x^2 + 3$

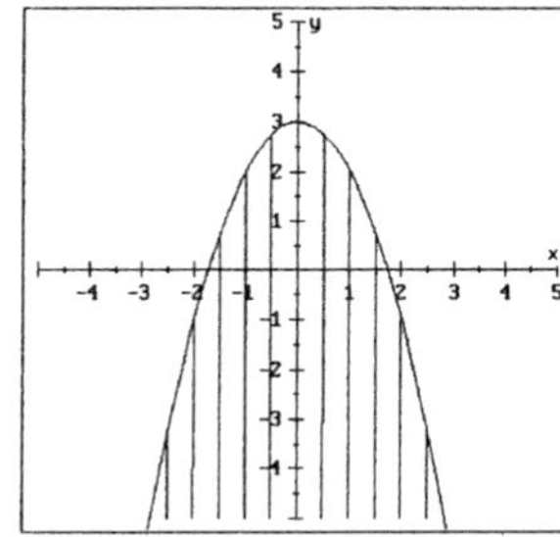

vertex: $(0, 3)$

39. $(3+5i) + (4-3i) = 3 + 5i + 4 - 3i = 7 + 2i$

40. $(7-4i) - (12+3i) = 7 - 4i - 12 - 3i = -5 - 7i$

41. $(2-3i)(2+3i) = 4 + 6i - 6i - 9i^2 = 4 + 9 = 13 = 13 + 0i$

42. $(3+i)(3-3i) = 9 - 9i + 3i - 3i^2 = 9 - 6i + 3 = 12 - 6i$

43. $(3-2i) - (4+i)^2 = 3 - 2i - (16 + 8i + i^2) = 3 - 2i - (15 + 8i) = 3 - 2i - 15 - 8i$
$= -12 - 10i$

44. $\frac{5}{3-i} = \frac{5}{3-i} \cdot \frac{3+i}{3+i} = \frac{5(3+i)}{9-i^2} = \frac{5(3+i)}{10} = \frac{3+i}{2} = \frac{3}{2} + \frac{1}{2}i$

45. $|3+2i| = \sqrt{3^2 + 2^2} = \sqrt{9+4} = \sqrt{13}$

46. $|5-6i| = \sqrt{5^2 + (-6)^2} = \sqrt{25+36} = \sqrt{61}$

47.
$$2x^2 + 4x = k$$
$$2x^2 + 4x - k = 0$$
$$a = 2, b = 4, c = -k$$
Set the discriminant equal to 0:
$$b^2 - 4ac = 0$$
$$4^2 - 4(2)(-k) = 0$$
$$16 + 8k = 0$$
$$8k = -16$$
$$k = -2$$

48.
$$a - 7a^{1/2} + 12 = 0$$
$$\left(a^{1/2} - 3\right)\left(a^{1/2} - 4\right) = 0$$
$a^{1/2} - 3 = 0$ **or** $a^{1/2} - 4 = 0$

$a^{1/2} = 3$ | $a^{1/2} = 4$

$\left(a^{1/2}\right)^2 = 3^2$ | $\left(a^{1/2}\right)^2 = 4^2$

$a = 9$ | $a = 16$

Solution | Solution

49. $x^2 - x - 6 > 0$
$(x+2)(x-3) > 0$

$x-3$ ‒ ‒ ‒ ‒ ‒ ‒ ‒ ‒ ‒ 0 ++++
$x+2$ ‒ ‒ ‒ 0+++++++ ++++

⟵) —— (⟶
−2 3

solution set: $(-\infty, -2) \cup (3, \infty)$

50. $x^2 - x - 6 \le 0$
$(x+2)(x-3) \le 0$

$x-3$ ‒ ‒ ‒ ‒ ‒ ‒ ‒ ‒ 0 ++++
$x+2$ ‒ ‒ ‒ 0+++++++ ++++

⟵ [——] ⟶
−2 3

solution set: $[-2, 3]$

51. $f(-1) = 3(-1)^2 + 2 = 3(1) + 2 = 3 + 2 = 5$

52. $(g \circ f)(2) = g(f(2)) = g(3(2)^2 + 2) = g(14) = 2(14) - 1 = 27$

53. $(f \circ g)(x) = f(g(x)) = f(2x - 1) = 3(2x - 1)^2 + 2 = 3(4x^2 - 4x + 1) + 2 = 12x^2 - 12x + 5$

54. $(g \circ f)(x) = g(f(x)) = g(3x^2 + 2) = 2(3x^2 + 2) - 1 = 6x^2 + 3$

55.
$$y = 3x + 2$$
$$x = 3y + 2$$
$$x - 2 = 3y$$
$$\frac{x-2}{3} = y, \text{ or } y = f^{-1}(x) = \frac{x-2}{3}$$

56.
$$y = x^3 + 4$$
$$x = y^3 + 4$$
$$x - 4 = y^3$$
$$\sqrt[3]{x-4} = y, \text{ or } y = f^{-1}(x) = \sqrt[3]{x-4}$$

Exercise 9.1 (page 598)

`1.
$$3x + 2x - 20 = 180$$
$$5x = 200$$
$$x = 40$$

2. $m(\angle 1) = 2x - 20 = 2(40) - 20 = 60°$

3. $m(\angle 2) = 3x = 3(40) = 120°$

4. $m(\angle 3) = m(\angle 1) = 60°$

5. exponential

6. domain

7. $(0, \infty)$

8. $(0, 1); (1, 3)$

9. increasing

10. decreasing

11. $P\left(1 + \frac{r}{k}\right)^{kt}$

12. $PV(1 + i)^n$

Problems 13-16 are to be solved using a calculator. The keystrokes needed to solve each problem using a TI-83 graphing calculator appear in each solution. There may be other solutions. Keystrokes for other calculators may be slightly different.

13. $2^{\sqrt{2}} \Rightarrow$ [2] [^] [√] [2] [ENTER]
$\{2.6651\}$

14. $7^{\sqrt{2}} \Rightarrow$ [7] [^] [√] [2] [ENTER]
$\{15.6729\}$

15. $5^{\sqrt{5}} \Rightarrow$ [5] [^] [√] [5] [ENTER]
$\{36.5548\}$

16. $6^{\sqrt{3}} \Rightarrow$ [6] [^] [√] [3] [ENTER]
$\{22.2740\}$

17. $\left(2^{\sqrt{3}}\right)^{\sqrt{3}} = 2^{(\sqrt{3})(\sqrt{3})} = 2^3 = 8$

18. $3^{\sqrt{2}}3^{\sqrt{18}} = 3^{\sqrt{2}+\sqrt{18}} = 3^{\sqrt{2}+3\sqrt{2}} = 3^{4\sqrt{2}}$

19. $7^{\sqrt{3}}7^{\sqrt{12}} = 7^{\sqrt{3}+\sqrt{12}} = 7^{\sqrt{3}+2\sqrt{3}} = 7^{3\sqrt{3}}$

20. $\left(3^{\sqrt{5}}\right)^{\sqrt{5}} = 3^{(\sqrt{5})(\sqrt{5})} = 3^5 = 243$

21. $y = f(x) = 3^x$
through $(0,1)$ and $(1,3)$

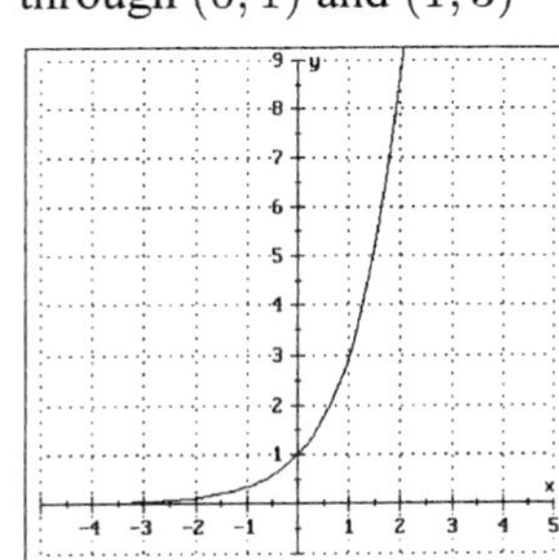

22. $y = f(x) = 5^x$
through $(0,1)$ and $(1,5)$

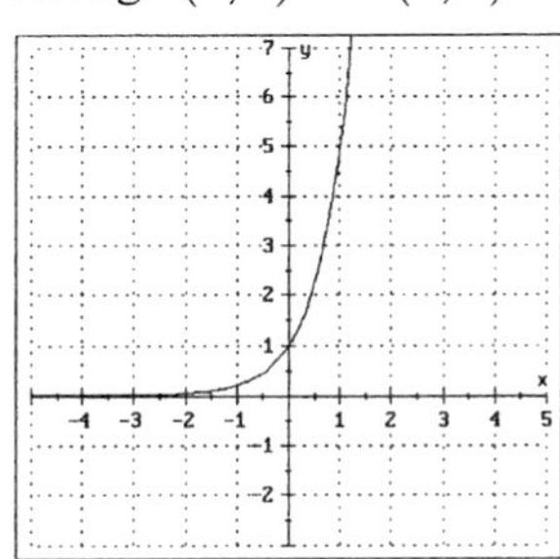

23. $y = f(x) = \left(\frac{1}{3}\right)^x$
through $(0,1)$ and $\left(1, \frac{1}{3}\right)$

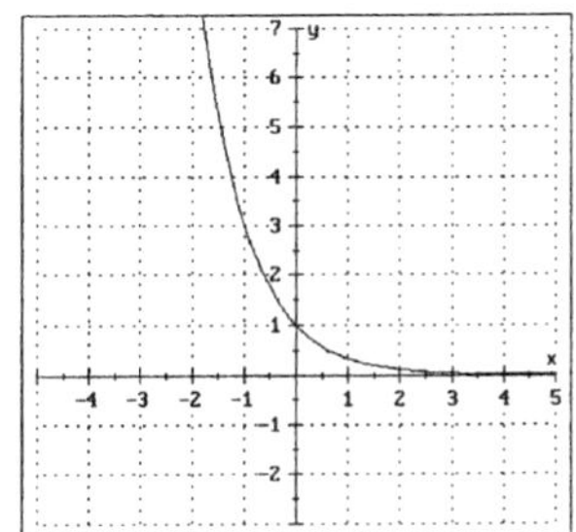

24. $y = f(x) = \left(\frac{1}{5}\right)^x$
through $(0,1)$ and $\left(1, \frac{1}{5}\right)$

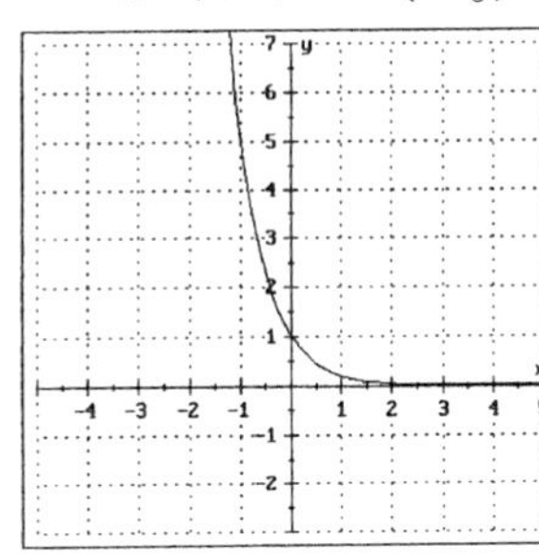

25. $f(x) = 3^x - 2$
Shift $y = 3^x$ down 2.

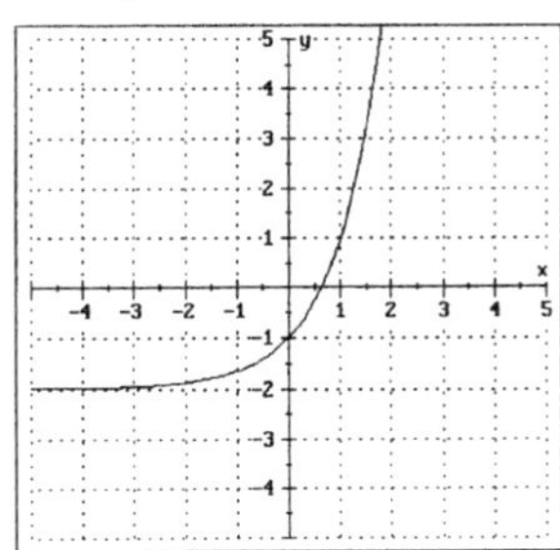

26. $y = f(x) = 2^x + 1$
Shift $y = 2^x$ up 1.

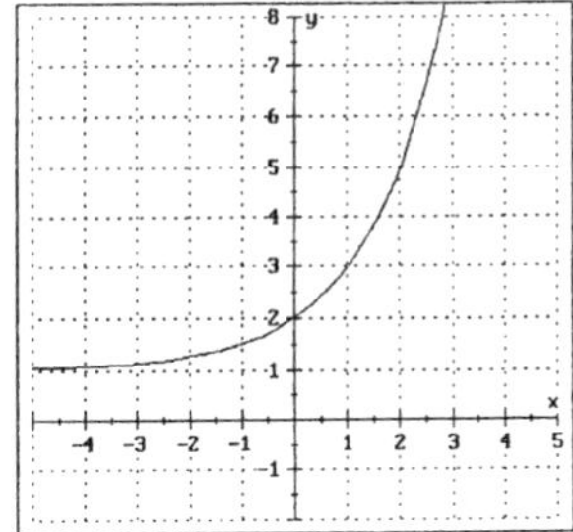

27. $f(x) = 3^{x-1}$
Shift $y = 3^x$ right 1.

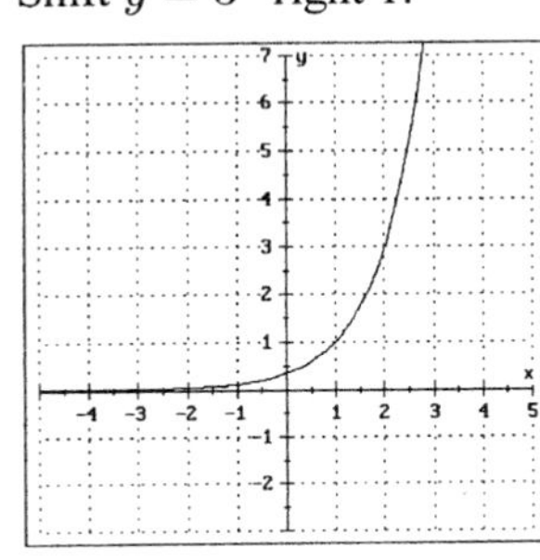

28. $y = f(x) = 2^{x+1}$
Shift $y = 2^x$ left 1.

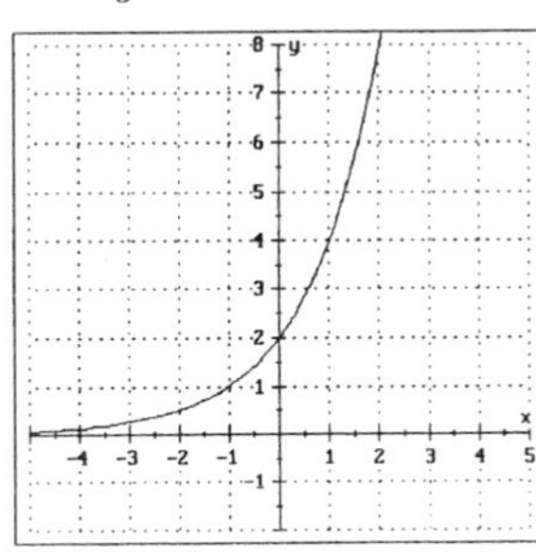

29. $y = b^x$
$\frac{1}{2} = b^1$
$\frac{1}{2} = b$

30. $y = b^x$
$7 = b^1$
$7 = b$

31. $y = b^x$
$2 = b^0$
no such value of b

32. $y = b^x$
$3 = b^1$
$3 = b$

33. $y = b^x$
$2 = b^1$
$2 = b$

34. $y = b^x$
$\frac{1}{3} = b^{-1}$
$3 = b$

35. $y = b^x$
$9 = b^2$
$3 = b$

36. The graph has y-coordinates less than zero. This is impossible. No such value exists.

37. $y = f(x) = \dfrac{1}{2}\left(3^{x/2}\right)$

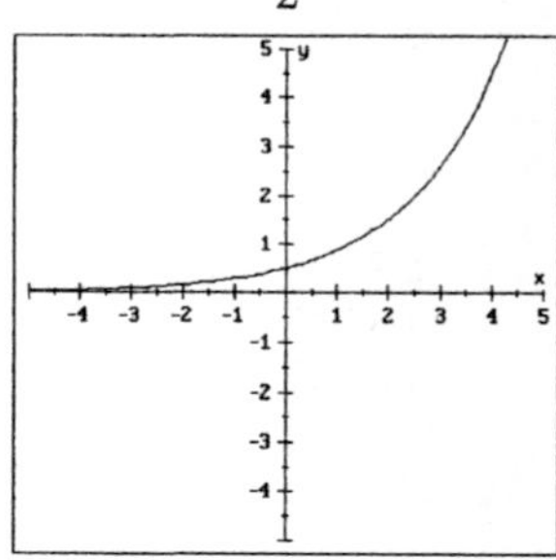

increasing

38. $y = f(x) = -3\left(2^{x/3}\right)$

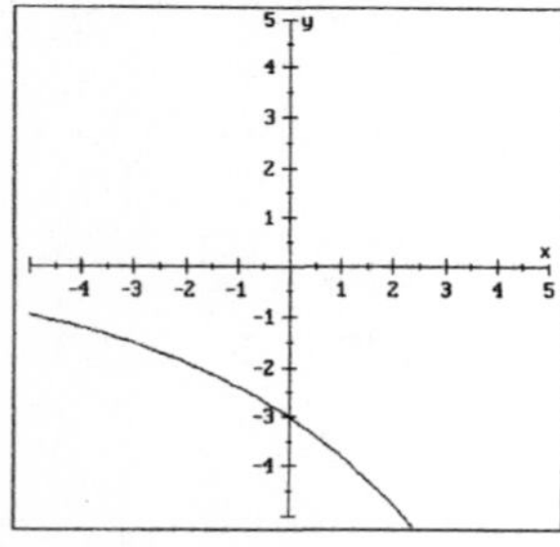

decreasing

39. $y = f(x) = 2\left(3^{-x/2}\right)$

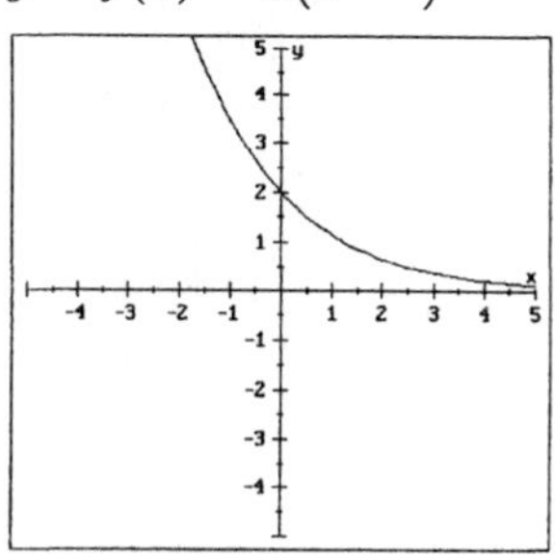

decreasing

40. $y = f(x) = -\dfrac{1}{4}\left(2^{-x/2}\right)$

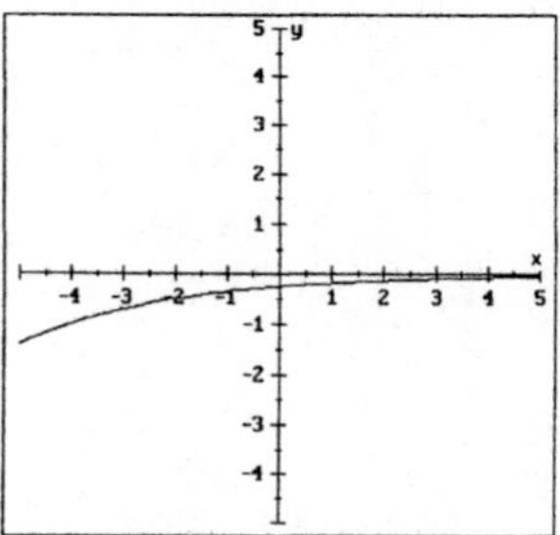

increasing

41. $S(n) = 5.74(1.39)^n = 5.74(1.39)^5$
≈ 29.8
There were about 29.8 million users in 1995.

42. $S(n) = 5.74(1.39)^n = 5.74(1.39)^{10}$
≈ 154.5
There were about 154.5 million users in 2000.

43. $A = A_0\left(\dfrac{2}{3}\right)^t$

$A = A_0\left(\dfrac{2}{3}\right)^5$

$A = \dfrac{32}{243}A_0$

44. $P = (6 \times 10^6)(2.3)^t = (6 \times 10^6)(2.3)^4$
$\approx 1.679046 \times 10^8$

45. $C = (3 \times 10^{-4})(0.7)^t = (3 \times 10^{-4})(0.7)^5 \approx 5.0421 \times 10^{-5}$ coulombs

46. $P = 3745(0.93)^t = 3745(0.93)^{6.75} \approx 2295$ people

47. $$\begin{aligned} A &= P\left(1+\frac{r}{k}\right)^{kt} \\ &= 10{,}000\left(1+\frac{0.08}{4}\right)^{4(10)} \\ &= 10{,}000(1.02)^{40} \\ &\approx \$22{,}080.40 \end{aligned}$$

48. $$\begin{aligned} A &= P\left(1+\frac{r}{k}\right)^{kt} \\ &= 10{,}000\left(1+\frac{0.08}{12}\right)^{12(10)} \\ &= 10{,}000(1.0066666667)^{120} \\ &\approx \$22{,}196.40 \end{aligned}$$

49. $$\begin{aligned} A &= P\left(1+\frac{r}{k}\right)^{kt} \\ &= 1000\left(1+\frac{0.05}{4}\right)^{4(5)} \\ &= 1000(1.0125)^{20} \\ &\approx \$1282.040 \end{aligned}$$

$$\begin{aligned} A &= P\left(1+\frac{r}{k}\right)^{kt} \\ &= 1000\left(1+\frac{0.055}{4}\right)^{4(5)} \\ &= 1000(1.01375)^{20} \\ &\approx \$1314.07 \end{aligned}$$

$$\begin{aligned} \text{difference} &= \$1314.07 - \$1282.04 \\ &= \$32.03 \end{aligned}$$

50. For each of the plans, deposit \$1000 for 10 years:

FIDELITY

$$\begin{aligned} A &= P\left(1+\frac{r}{k}\right)^{kt} \\ &= 1000\left(1+\frac{0.0525}{12}\right)^{12(10)} \\ &= 1000(1.004375)^{120} \\ &\approx \$1688.52 \end{aligned}$$

UNION

$$\begin{aligned} A &= P\left(1+\frac{r}{k}\right)^{kt} \\ &= 1000\left(1+\frac{0.0535}{1}\right)^{1(10)} \\ &= 1000(1.0535)^{10} \\ &\approx \$1684.01 \end{aligned}$$

Fidelity provides the better investment option.

51. $$\begin{aligned} A &= P\left(1+\frac{r}{k}\right)^{kt} \\ &= 1\left(1+\frac{0.05}{1}\right)^{1(300)} \\ &= 1(1.05)^{300} \\ &\approx \$2{,}273{,}996.13 \end{aligned}$$

52. $$\begin{aligned} A &= P\left(1+\frac{r}{k}\right)^{kt} \\ &= 10{,}000\left(1+\frac{0.06}{4}\right)^{4(20)} \\ &= 10{,}000(1.015)^{80} \\ &\approx \$32{,}906.63 \end{aligned}$$

$$\begin{aligned} A &= P\left(1+\frac{r}{k}\right)^{kt} \\ &= 10{,}000\left(1+\frac{0.06}{365}\right)^{365(20)} \\ &= 10{,}000(1.000164384)^{7300} \\ &\approx \$33{,}197.90 \end{aligned}$$

$$\begin{aligned} \text{diff.} &= \$33{,}197.90 - \$32{,}906.63 \\ &= \$291.27 \end{aligned}$$

53. $$\begin{aligned} A &= P\left(1 + \frac{r}{k}\right)^{kt} \\ &= 4700\left(1 + \frac{-0.25}{1}\right)^{1(5)} \\ &= 4700(0.75)^5 \\ &\approx \$1115.33 \end{aligned}$$

54. $$\begin{aligned} A &= P\left(1 + \frac{r}{k}\right)^{kt} \\ &= 15{,}000{,}000\left(1 + \frac{0.06}{1}\right)^{1(193)} \\ &= 15{,}000{,}000(1.06)^{193} \\ &\approx \$1.14847952 \times 10^{12} \\ &\approx \frac{1.14847952 \times 10^{12}}{827{,}000} \approx \$1{,}388{,}730 \end{aligned}$$

55. **Answers may vary.**

56. **Answers may vary.**

57. If the base were 0, then the function would not be defined for $x = 0 \Rightarrow y = 0^0$.

58. If the base were negative, say -4, then the function would not be real for $x = \frac{1}{2} \Rightarrow y = (-4)^{1/2}$.

Exercise 9.2 (page 606)

1. $\sqrt{240x^5} = \sqrt{16x^4}\sqrt{15x} = 4x^2\sqrt{15x}$

2. $$\begin{aligned} \sqrt[3]{-125x^5y^4} &= \sqrt[3]{-125x^3y^3}\sqrt[3]{x^2y} \\ &= -5xy\sqrt[3]{x^2y} \end{aligned}$$

3. $$\begin{aligned} 4\sqrt{48y^3} - 3y\sqrt{12y} &= 4\sqrt{16y^2}\sqrt{3y} - 3y\sqrt{4}\sqrt{3y} = 4(4y)\sqrt{3y} - 3y(2)\sqrt{3y} \\ &= 16y\sqrt{3y} - 6y\sqrt{3y} = 10y\sqrt{3y} \end{aligned}$$

4. $\sqrt[4]{48z^5} + \sqrt[4]{768z^5} = \sqrt[4]{16z^4}\sqrt[4]{3z} + \sqrt[4]{256z^4}\sqrt[4]{3z} = 2z\sqrt[4]{3z} + 4z\sqrt[4]{3z} = 6z\sqrt[4]{3z}$

5. 2.72

6. Pe^{rt}

7. increasing

8. $(1, e)$

9. $A = Pe^{rt}$

10. population; linearly

11. $y = f(x) = e^x + 1$
Shift $y = e^x$ up 1.

12. $y = f(x) = e^x - 2$
Shift $y = e^x$ down 2.

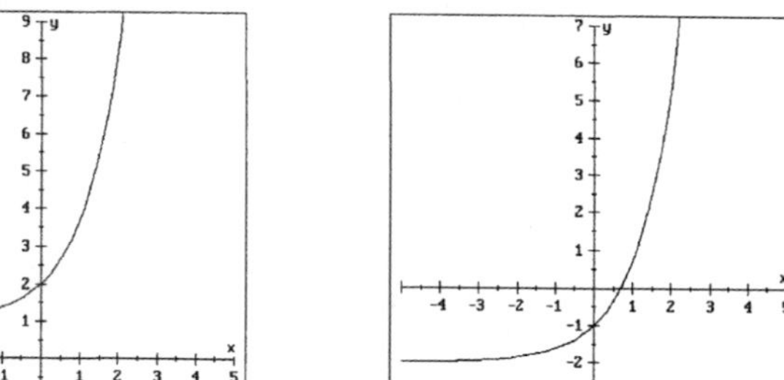

13. $y = f(x) = e^{(x+3)}$
Shift $y = e^x$ left 3.

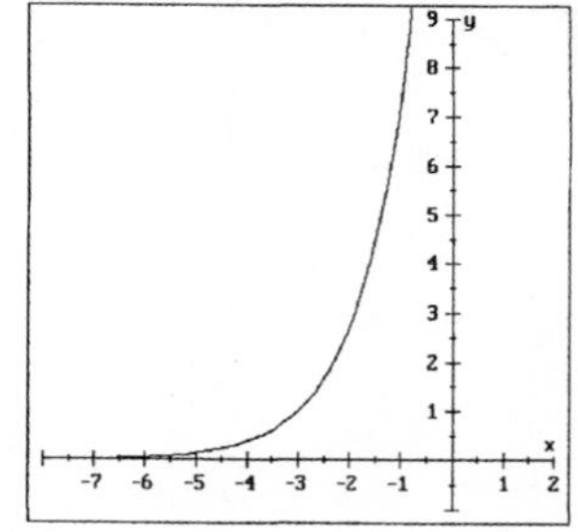

14. $y = f(x) = e^{(x-5)}$
Shift $y = e^x$ right 5.

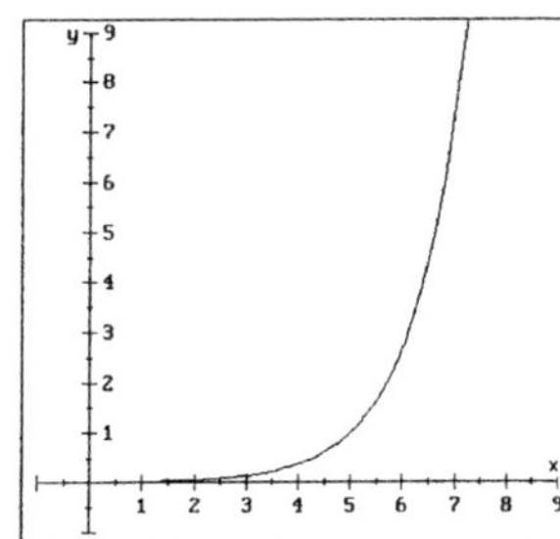

15. $y = f(x) = -e^x$; Reflect $y = e^x$ about the x-axis.

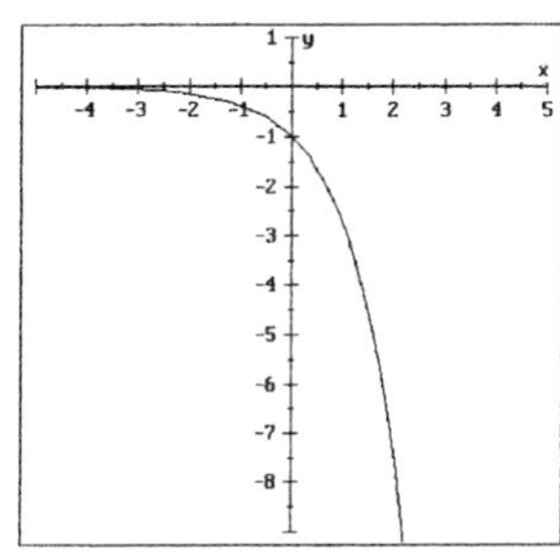

16. $y = f(x) = -e^x + 1$
Reflect $y = e^x$ about the x-axis and shift up 1.

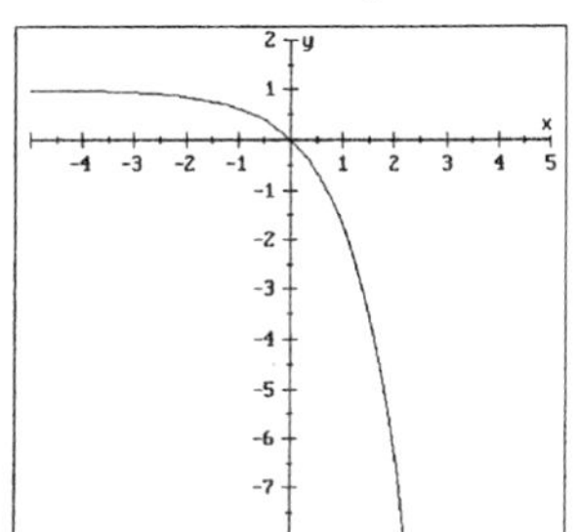

17. $y = f(x) = 2e^x$; Stretch $y = e^x$ vertically by a factor of 2.

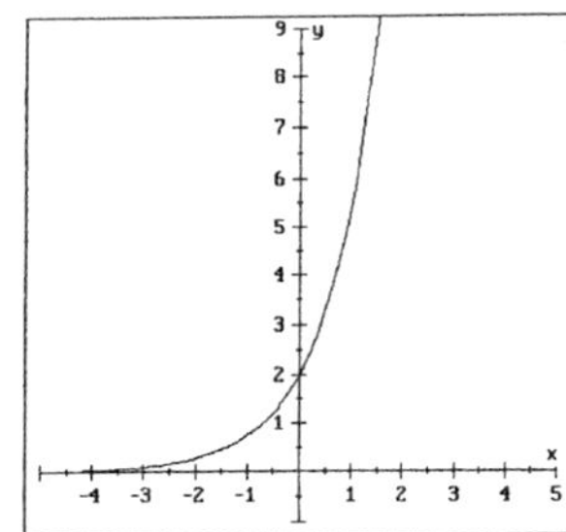

18. $y = f(x) = \frac{1}{2}e^x$
Compress $y = e^x$ vertically by a factor of 2.

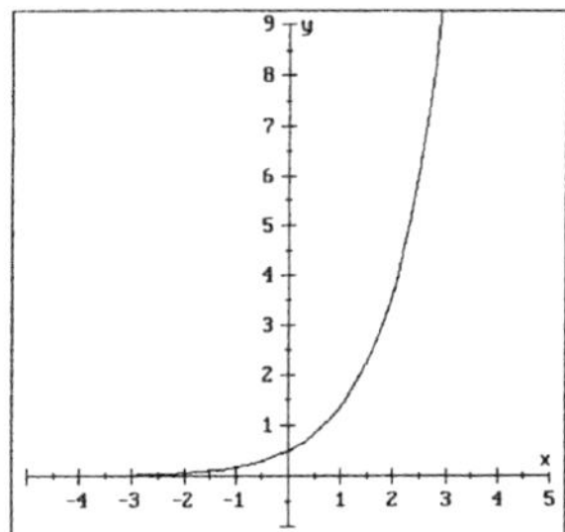

19. The graph should be increasing. The graph could not look like this.

20. The range must be $(0, \infty)$. The graph could not look like this.

21. The graph should go through the point $(0, 1)$. The graph could not look like this.

22. The graph could look like this.

23.
$$\begin{aligned} A &= Pe^{rt} \\ &= 5000e^{0.06(12)} \\ &= 5000e^{0.72} \\ &\approx \$10{,}272.17 \end{aligned}$$

24.
$$\begin{aligned} A &= Pe^{rt} \\ &= 6000e^{0.07(35)} \\ &= 6000e^{2.45} \\ &\approx \$69{,}530.08 \end{aligned}$$

25.
$$\begin{aligned} A &= Pe^{rt} \\ 12000 &= Pe^{0.07(9)} \\ 12000 &= Pe^{0.63} \\ 12000 &\approx P(1.8776106) \\ P &\approx \frac{12000}{1.8776106} \\ P &\approx \$6{,}391.10 \end{aligned}$$

26. $A = Pe^{rt} = 8000e^{0.08(-6)} = 8000e^{-0.48} \approx \$4{,}950.27$

27. $A = Pe^{rt}$
$= 5000e^{0.085(5)}$
$= 5000e^{0.425}$
$\approx \$7{,}647.95$ (continuous)

$A = P\left(1 + \frac{r}{k}\right)^{kt}$
$= 5000\left(1 + \frac{0.085}{1}\right)^{1(5)}$
$= 5000(1.085)^5$
$\approx \$7{,}518.28$ (annual)

28. $A = Pe^{rt}$
$= 30000e^{0.08(20)}$
$= 30000e^{1.6}$
$\approx \$148{,}590.97$ (continuous)

$A = P\left(1 + \frac{r}{k}\right)^{kt}$
$= 30000\left(1 + \frac{0.08}{1}\right)^{1(20)}$
$= 30000(1.08)^{20}$
$\approx \$139{,}828.71$ (annual)

29. $A = Pe^{rt}$
$= 6e^{0.019(30)}$
$= 6e^{0.57}$
≈ 10.6 billion people

30. $A = Pe^{rt}$
$= 6e^{0.019(40)}$
$= 6e^{0.76}$
≈ 12.8 billion people

31. $A = Pe^{rt}$
$= 6e^{0.019(50)}$
$= 6e^{0.95}$
$\approx 6(2.6)$
It will increase by a factor of about 2.6.

32. $P = 173e^{0.03t}$
$= 173e^{0.03(30)}$
$= 173e^{0.9}$
≈ 426

33. $A = 8000e^{-0.008t}$
$= 8000e^{-0.008(20)}$
$= 8000e^{-0.16}$
≈ 6817

34. $P = P_0e^{0.27t}$
$= 2e^{0.27(10)}$
$= 2e^{2.7}$
≈ 30 cases

35. $A = A_0e^{-0.087t}$
$= 50e^{-0.087(30)}$
$= 50e^{-2.61}$
≈ 3.68 grams

36. $A = A_0e^{-0.013t}$
$= 25e^{-0.013(45)}$
$= 25e^{-0.585}$
≈ 13.93 grams

37. $A = A_0e^{-0.00000693t}$
$= 2500e^{-0.00000693(100)}$
$= 2500e^{-0.000693}$
≈ 2498.27 grams

38. $A = A_0e^{-0.0000284t}$
$= 75e^{-0.0000284(50{,}000)}$
$= 75e^{-1.42}$
≈ 18.13 grams

39. $P = 0.3\left(1 - e^{-0.05t}\right)$
$= 0.3\left(1 - e^{-0.05(15)}\right)$
$= 0.3\left(1 - e^{-0.75}\right)$
$\approx 0.3(1 - 0.47237)$
$\approx 0.3(0.52763)$
≈ 0.16

40. $x = 0.08\left(1 - e^{-0.1t}\right)$
$= 0.08\left(1 - e^{-0.1(30)}\right)$
$= 0.08\left(1 - e^{-3}\right)$
$\approx 0.08(1 - 0.049787)$
$\approx 0.08(0.950213)$
≈ 0.076

41. $x = 0.08\left(1 - e^{-0.1t}\right)$
$= 0.08\left(1 - e^{-0.1(0)}\right)$
$= 0.08\left(1 - e^{0}\right)$
$= 0.08(1 - 1)$
$= 0.08(0) = 0$

42. $v = 50\left(1 - e^{-0.2t}\right)$
$= 50\left(1 - e^{-0.2(0)}\right)$
$= 50\left(1 - e^{0}\right)$
$= 50(1 - 1)$
$= 50(0) = 0$ mps

43. $v = 50\left(1 - e^{-0.2t}\right)$
$= 50\left(1 - e^{-0.2(20)}\right)$
$= 50\left(1 - e^{-4}\right)$
$\approx 50(1 - 0.01832)$
$\approx 50(0.98168) \approx 49$ mps

44. $v = 50\left(1 - e^{-0.2t}\right)$
$= 50\left(1 - e^{-0.2(2)}\right)$
$= 50\left(1 - e^{-0.4}\right)$
$\approx 50(1 - 0.67032)$
$\approx 50(0.32968) \approx 16.5$ mps

$v = 50\left(1 - e^{-0.3t}\right)$
$= 50\left(1 - e^{-0.3(2)}\right)$
$= 50\left(1 - e^{-0.6}\right)$
$\approx 50(1 - 0.54881)$
$\approx 50(0.45119) \approx 22.6$ mps $\Rightarrow$ faster

45. $A = Pe^{rt}$
$= 4570e^{-0.06(6.5)}$
$\approx \$3094.15$

46. $A = Pe^{rt}$
$= 7500e^{-0.02(8.25)}$
$\approx \$6359.20$

47. $y = 1000e^{0.02x}$
$y = 31x + 2000$

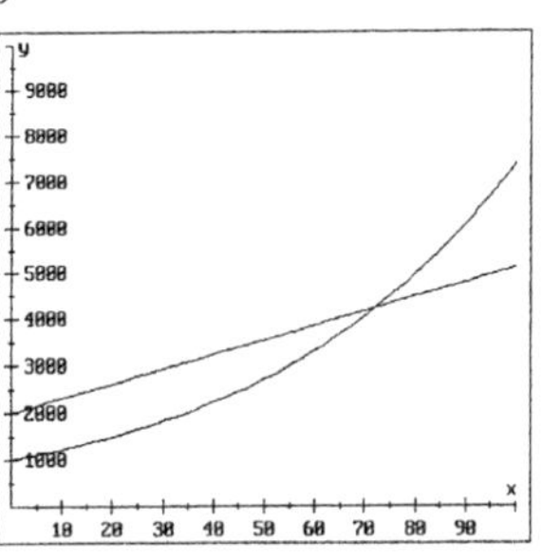

about 72 years

48. $y = 1000e^{0.01x}$
$y = 30.625x + 2000$

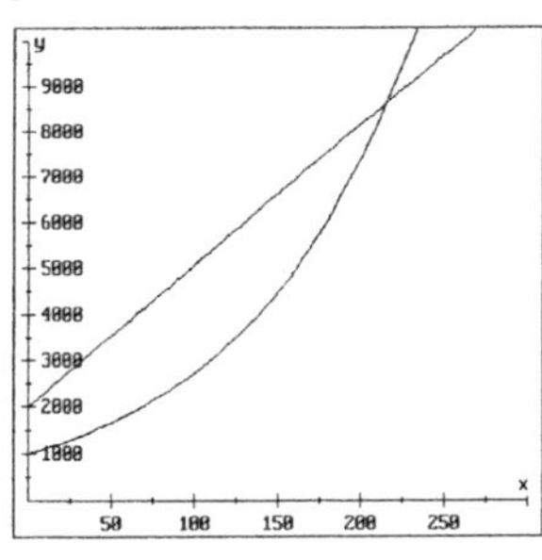

about 215 years

49. **Answers may vary.**

50. **Answers may vary.**

51. $e \approx 2.7182;\ 1 + 1 + \frac{1}{2} + \frac{1}{2\cdot 3} + \frac{1}{2\cdot 3\cdot 4} + \frac{1}{2\cdot 3\cdot 4\cdot 5} \approx 2.7167$

52. $y = f(x) = \dfrac{e^{x} + e^{-x}}{2}$

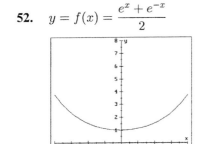

53. $e^{t+5} = ke^{t}$
$e^{t} \cdot e^{5} = ke^{t}$
$e^{5}e^{t} = ke^{t}$
$k = e^{5}$

54. $e^{5t} = k^{t}$
$\left(e^{5}\right)^{t} = k^{t}$
$k = e^{5}$

Exercise 9.3 (page 616)

1.
$$\begin{aligned}\sqrt[3]{6x+4}&=4\\6x+4&=4^3\\6x+4&=64\\6x&=60\\x&=10\end{aligned}$$

2.
$$\begin{aligned}\sqrt{3x-4}&=\sqrt{-7x+2}\\3x-4&=-7x+2\\10x&=6\\x&=\tfrac{6}{10}=\tfrac{3}{5}\end{aligned}$$
$\frac{3}{5}$ does not check $\Rightarrow$ no solution

3.
$$\begin{aligned}\sqrt{a+1}-1&=3a\\\sqrt{a+1}&=3a+1\\a+1&=(3a+1)^2\\a+1&=9a^2+6a+1\\0&=9a^2+5a\\0&=a(9a+5)\end{aligned}$$
$a=0$ or $a=-\frac{5}{9}$ ($-\frac{5}{9}$ does not check.)

4.
$$\begin{aligned}3-\sqrt{t-3}&=\sqrt{t}\\\left(3-\sqrt{t-3}\right)^2&=t\\9-6\sqrt{t-3}+t-3&=t\\6&=6\sqrt{t-3}\\1&=\sqrt{t-3}\\1&=t-3\\4&=t\end{aligned}$$

5. $x=b^y$ **6.** $(0,\infty)$ **7.** range **8.** x

9. inverse **10.** logarithmic **11.** exponent **12.** asymptote

13. $(b,1)$; $(1,0)$ **14.** x **15.** $20\log\frac{E_O}{E_I}$ **16.** $\log\frac{A}{P}$

17. $\log_3 27=3\Rightarrow 3^3=27$

18. $\log_8 8=1\Rightarrow 8^1=8$

19. $\log_{1/2}\frac{1}{4}=2\Rightarrow\left(\frac{1}{2}\right)^2=\frac{1}{4}$

20. $\log_{1/5}1=0\Rightarrow\left(\frac{1}{5}\right)^0=1$

21. $\log_4\frac{1}{64}=-3\Rightarrow 4^{-3}=\frac{1}{64}$

22. $\log_6\frac{1}{36}=-2\Rightarrow 6^{-2}=\frac{1}{36}$

23. $\log_{1/2}\frac{1}{8}=3\Rightarrow\left(\frac{1}{2}\right)^3=\frac{1}{8}$

24. $\log_{1/5}1=0\Rightarrow\left(\frac{1}{5}\right)^0=1$

25. $6^2=36\Rightarrow\log_6 36=2$

26. $10^3=1000\Rightarrow\log_{10}1000=3$

27. $5^{-2}=\frac{1}{25}\Rightarrow\log_5\frac{1}{25}=-2$

28. $3^{-3}=\frac{1}{27}\Rightarrow\log_3\frac{1}{27}=-3$

29. $\left(\frac{1}{2}\right)^{-5}=32\Rightarrow\log_{1/2}32=-5$

30. $\left(\frac{1}{3}\right)^{-3}=27\Rightarrow\log_{1/3}27=-3$

31. $x^y=z\Rightarrow\log_x z=y$

32. $m^n=p\Rightarrow\log_m p=n$

33. $\log_2 16 = x \Rightarrow 2^x = 16 \Rightarrow x = 4$

34. $\log_3 9 = x \Rightarrow 3^x = 9 \Rightarrow x = 2$

35. $\log_4 16 = x \Rightarrow 4^x = 16 \Rightarrow x = 2$

36. $\log_6 216 = x \Rightarrow 6^x = 216 \Rightarrow x = 3$

37. $\log_{1/2} \frac{1}{8} = x \Rightarrow \left(\frac{1}{2}\right)^x = \frac{1}{8} \Rightarrow x = 3$

38. $\log_{1/3} \frac{1}{81} = x \Rightarrow \left(\frac{1}{3}\right)^x = \frac{1}{81} \Rightarrow x = 4$

39. $\log_9 3 = x \Rightarrow 9^x = 3 \Rightarrow x = \frac{1}{2}$

40. $\log_{125} 5 = x \Rightarrow 125^x = 5 \Rightarrow x = \frac{1}{3}$

41. $\log_{1/2} 8 = x \Rightarrow \left(\frac{1}{2}\right)^x = 8 \Rightarrow x = -3$

42. $\log_{1/2} 16 = x \Rightarrow \left(\frac{1}{2}\right)^x = 16 \Rightarrow x = -4$

43. $\log_7 x = 2 \Rightarrow 7^2 = x \Rightarrow x = 49$

44. $\log_5 x = 0 \Rightarrow 5^0 = x \Rightarrow x = 1$

45. $\log_6 x = 1 \Rightarrow 6^1 = x \Rightarrow x = 6$

46. $\log_2 x = 4 \Rightarrow 2^4 = x \Rightarrow x = 16$

47. $\log_{25} x = \frac{1}{2} \Rightarrow 25^{1/2} = x \Rightarrow x = 5$

48. $\log_4 x = \frac{1}{2} \Rightarrow 4^{1/2} = x \Rightarrow x = 2$

49. $\log_5 x = -2 \Rightarrow 5^{-2} = x \Rightarrow x = \frac{1}{25}$

50. $\log_3 x = -2 \Rightarrow 3^{-2} = x \Rightarrow x = \frac{1}{9}$

51. $\log_{36} x = -\frac{1}{2} \Rightarrow 36^{-1/2} = x \Rightarrow x = \frac{1}{6}$

52. $\log_{27} x = -\frac{1}{3} \Rightarrow 27^{-1/3} = x \Rightarrow x = \frac{1}{3}$

53. $\log_{100} \frac{1}{1000} = x \Rightarrow 100^x = \frac{1}{1000} \Rightarrow x = -\frac{3}{2}$

54. $\log_{5/2} \frac{4}{25} = x \Rightarrow \left(\frac{5}{2}\right)^x = \frac{4}{25} \Rightarrow x = -2$

55. $\log_{27} 9 = x \Rightarrow 27^x = 9 \Rightarrow x = \frac{2}{3}$

56. $\log_{12} x = 0 \Rightarrow 12^0 = x \Rightarrow x = 1$

57. $\log_x 5^3 = 3 \Rightarrow x^3 = 5^3 \Rightarrow x = 5$

58. $\log_x 5 = 1 \Rightarrow x^1 = 5 \Rightarrow x = 5$

59. $\log_x \frac{9}{4} = 2 \Rightarrow x^2 = \frac{9}{4} \Rightarrow x = \frac{3}{2}$

60. $\log_x \frac{\sqrt{3}}{3} = \frac{1}{2} \Rightarrow x^{1/2} = \frac{\sqrt{3}}{3} \Rightarrow x = \frac{1}{3}$

61. $\log_x \frac{1}{64} = -3 \Rightarrow x^{-3} = \frac{1}{64} \Rightarrow x = 4$

62. $\log_x \frac{1}{100} = -2 \Rightarrow x^{-2} = \frac{1}{100} \Rightarrow x = 10$

63. $\log_{2\sqrt{2}} x = 2 \Rightarrow (2\sqrt{2})^2 = x \Rightarrow x = 8$

64. $\log_4 8 = x \Rightarrow 4^x = 8 \Rightarrow x = \frac{3}{2}$

65. $2^{\log_2 4} = x \Rightarrow x = 4$

66. $3^{\log_3 5} = x \Rightarrow x = 5$

67. $x^{\log_4 6} = 6 \Rightarrow x = 4$

68. $x^{\log_3 8} = 8 \Rightarrow x = 3$

69. $\log 10^3 = x \Rightarrow 10^x = 10^3 \Rightarrow x = 3$

70. $\log 10^{-2} = x \Rightarrow 10^x = 10^{-2} \Rightarrow x = -2$

71. $10^{\log x} = 100 \Rightarrow \log x = 2 \Rightarrow x = 100$

72. $10^{\log x} = \frac{1}{10} \Rightarrow \log x = -1 \Rightarrow x = \frac{1}{10}$

73. $\log 8.25 \approx 0.9165$

74. $\log 0.77 \approx -0.1135$

75. $\log 0.00867 \approx -2.0620$

76. $\log 375.876 \approx 2.5750$

77. $\log y = 1.4023 \Rightarrow y = 25.25$

78. $\log y = 2.6490 \Rightarrow y = 445.66$

79. $\log y = 4.24 \Rightarrow y = 17{,}378.01$

80. $\log y = 0.926 \Rightarrow y = 8.43$

81. $\log y = -3.71 \Rightarrow y = 0.00$

82. $\log y = -0.28 \Rightarrow y = 0.52$

83. $\log y = \log 8 \Rightarrow \log y = 0.9030 \Rightarrow y = 8$

84. $\log y = \log 7 \Rightarrow \log y = 0.8451 \Rightarrow y = 7$

85. $y = f(x) = \log_3 x$
through $(1, 0)$ and $(3, 1)$

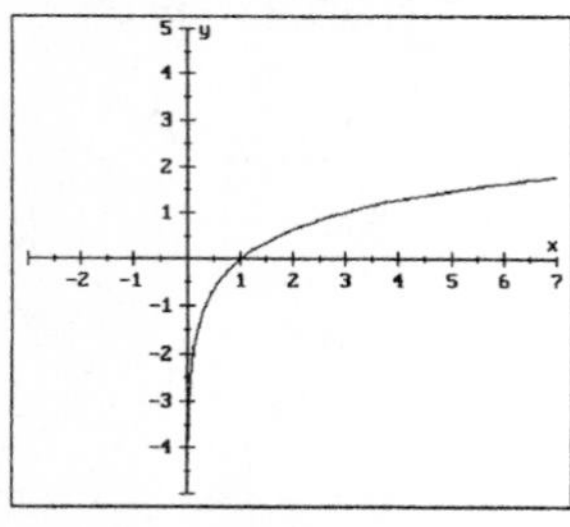

increasing

86. $y = f(x) = \log_{1/3} x$
through $(1, 0)$ and $\left(\frac{1}{3}, 1\right)$

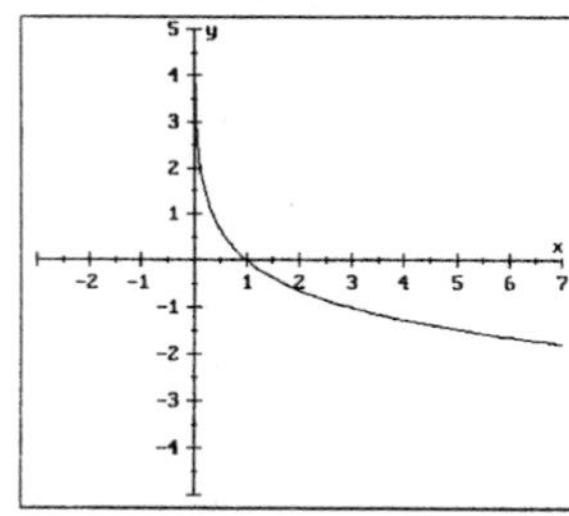

decreasing

87. $y = f(x) = \log_{1/2} x$
through $(1, 0)$ and $\left(\frac{1}{2}, 1\right)$

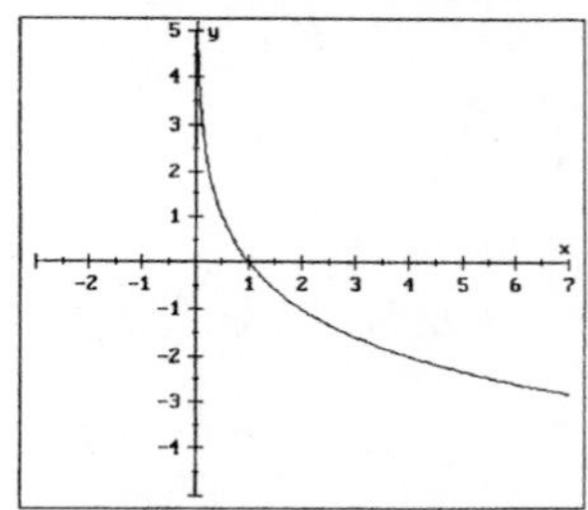

decreasing

88. $y = f(x) = \log_4 x$
through $(1, 0)$ and $(4, 1)$

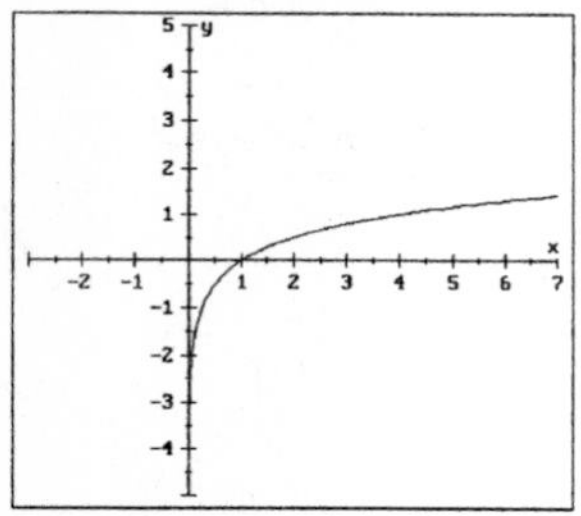

increasing

89. $y = f(x) = 3 + \log_3 x$
Shift $y = \log_3 x$ up 3.

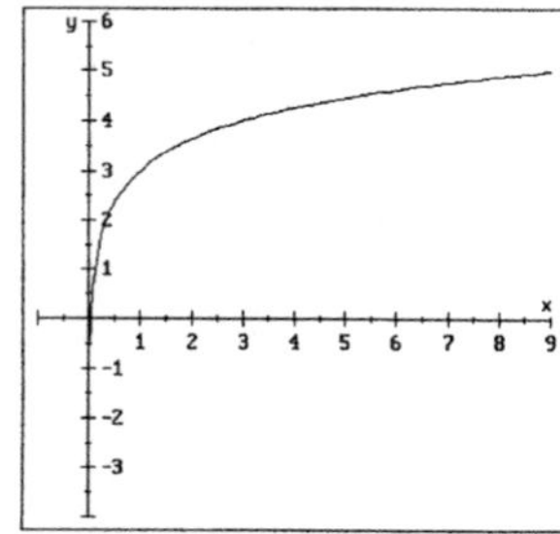

90. $y = f(x) = \log_{1/3} x - 1$
Shift $y = \log_{1/3} x$ down 1.

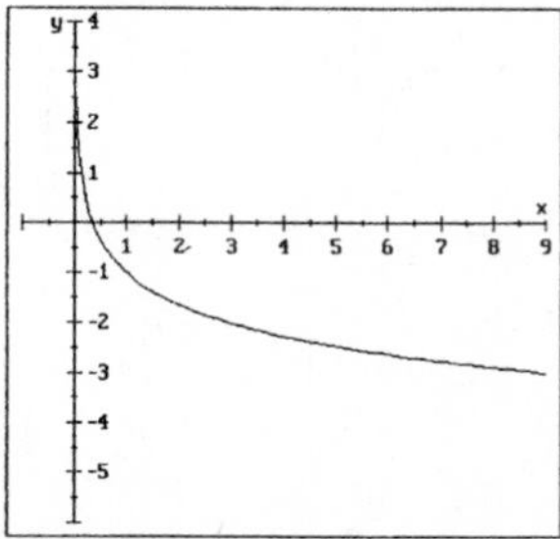

91. $y = f(x) = \log_{1/2}(x-2)$
Shift $y = \log_{1/2} x$ right 2.

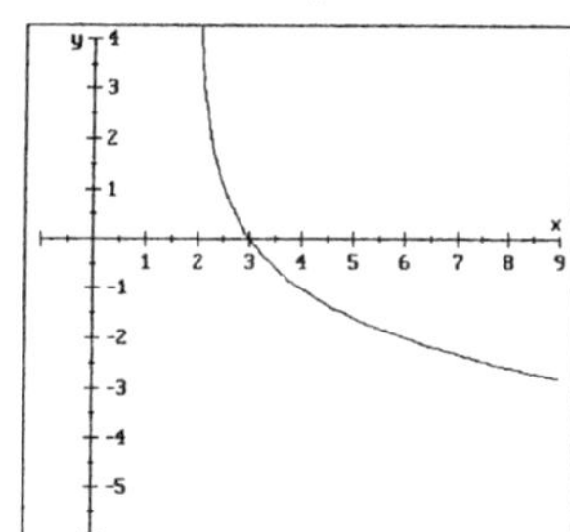

92. $y = f(x) = \log_4(x+2)$
Shift $y = \log_4 x$ left 2.

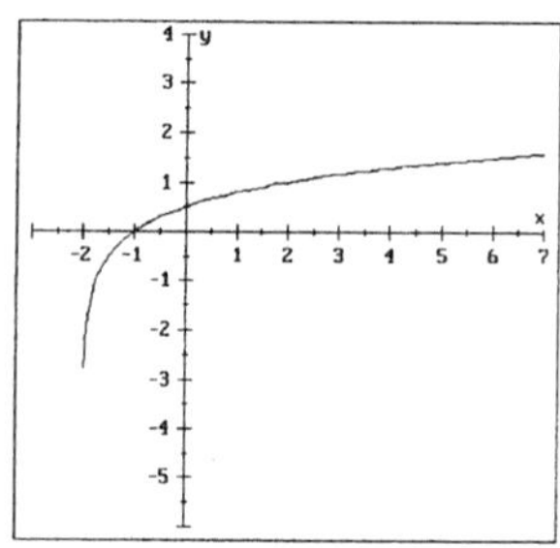

93. $y = f(x) = 2^x$
$y = g(x) = \log_2 x$

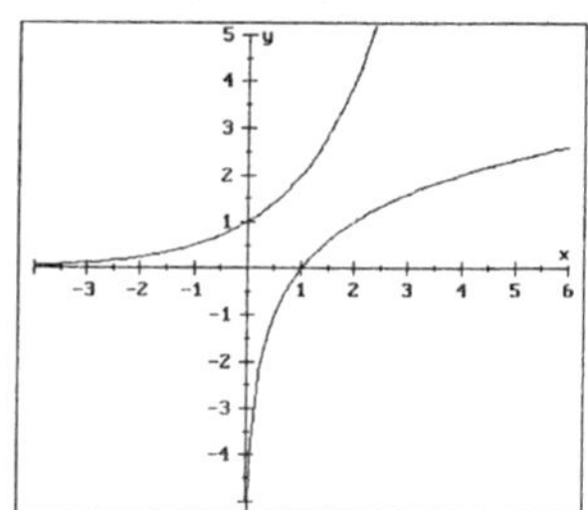

94. $y = f(x) = \left(\frac{1}{2}\right)^x$
$y = g(x) = \log_{1/2} x$

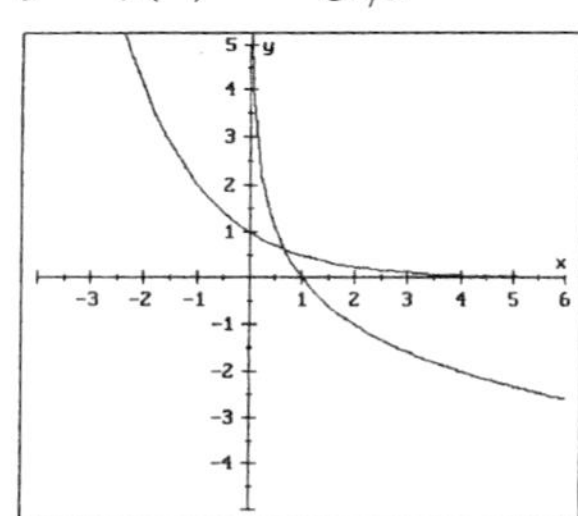

95. $y = f(x) = \left(\frac{1}{4}\right)^x$
$y = g(x) = \log_{1/4} x$

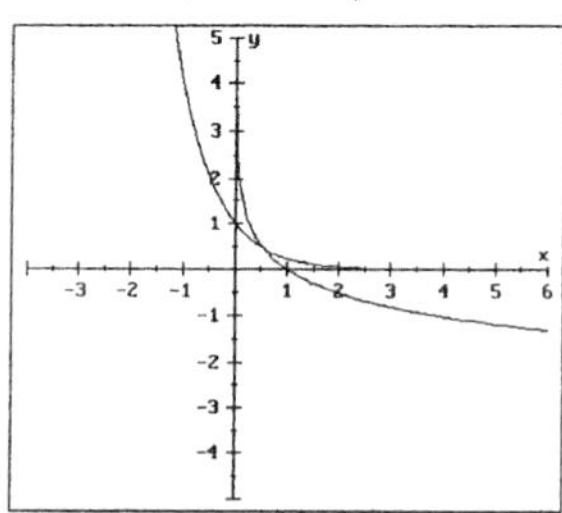

96. $y = f(x) = 4^x$
$y = g(x) = \log_4 x$

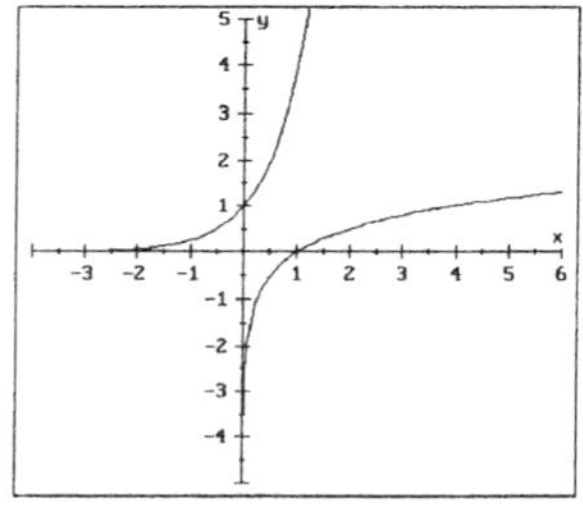

97. $\log_b 2 = 0 \Rightarrow b^0 = 2$
No such value exists.

98. $\log_b(-1) = 0 \Rightarrow b^0 = -1$
No such value exists.

99. $\log_b 9 = 2 \Rightarrow b^2 = 9 \Rightarrow b = 3$

100. $\log_b \frac{1}{2} = 1 \Rightarrow b^1 = \frac{1}{2} \Rightarrow b = \frac{1}{2}$

101. dB gain $= 20 \log \frac{E_O}{E_I} = 20 \log \frac{20}{0.71} = 20 \log 28.169 \approx 29.0$ dB

102. dB gain $= 20 \log \frac{E_O}{E_I} = 20 \log \frac{2.8}{0.05} = 20 \log 56 \approx 35.0$ dB

103. dB gain $= 20 \log \frac{E_O}{E_I} = 20 \log \frac{30}{0.1} = 20 \log 300 \approx 49.5$ dB

104. dB gain $= 20 \log \frac{E_O}{E_I} = 20 \log \frac{80}{0.12} = 20 \log 666.6667 \approx 56.5$ dB

105. $R = \log \frac{A}{P} = \log \frac{5000}{0.2} = \log 25{,}000$
≈ 4.4

106. $R = \log \frac{A}{P} = \log \frac{80000}{0.08} = \log 1000000$
$= 6$

107. $R = \log \frac{A}{P} = \log \frac{2500}{0.25} = \log 10000$
$= 4$

108. $R = \log \frac{A}{P} \Rightarrow 10^R = \frac{A}{P} \Rightarrow A = P \cdot 10^R$
$P \cdot 10^{R+1} = P \cdot 10^R \cdot 10 = A \cdot 10$
The amplitude must be multiplied by 10.

109. $n = \frac{\log V - \log C}{\log\left(1 - \frac{2}{N}\right)} = \frac{\log 2000 - \log 17000}{\log\left(1 - \frac{2}{5}\right)} \approx \frac{-0.929419}{-0.221849} \approx 4.2$ years old

110. $n = \frac{\log V - \log C}{\log\left(1 - \frac{2}{N}\right)} = \frac{\log 189 - \log 470}{\log\left(1 - \frac{2}{12}\right)} \approx \frac{-0.395363}{-0.079181} \approx 5.0$ years old

111. $n = \frac{\log\left[\frac{Ar}{P} + 1\right]}{\log(1 + r)} = \frac{\log\left[\frac{20{,}000(0.12)}{1000} + 1\right]}{\log(1 + 0.12)} = \frac{\log 3.4}{\log 1.12} \approx 10.8$ years

112. $n = \frac{\log\left[\frac{Ar}{P} + 1\right]}{\log(1 + r)} = \frac{\log\left[\frac{50{,}000(0.08)}{5000} + 1\right]}{\log(1 + 0.08)} = \frac{\log 1.8}{\log 1.08} \approx 7.6$ years

113. Answers may vary.

114. Answers may vary.

115. Answers may vary.

116. $y = f(x) = \log_5 x$
(Other functions are possible.)

117. Answers may vary.

Exercise 9.4 (page 623)

1. $y = 5x + 8 \Rightarrow m = 5$
Use the parallel slope:
$y - y_1 = m(x - x_1)$
$y - 0 = 5(x - 0)$
$y = 5x$

2. $y = mx + b$
$y = 9x + 5$

3. $y = \frac{2}{3}x - 12 \Rightarrow m = \frac{2}{3}$
Use the perpendicular slope:
$y - y_1 = m(x - x_1)$
$y - 2 = -\frac{3}{2}(x - 3)$
$y - 2 = -\frac{3}{2}x + \frac{9}{2}$
$y = -\frac{3}{2}x + \frac{13}{2}$

4. $3x + 2y = 9$
$2y = -3x + 9$
$y = -\frac{3}{2}x + \frac{9}{2} \Rightarrow m = -\frac{3}{2}$
Use the parallel slope:
$y - y_1 = m(x - x_1)$
$y - 5 = -\frac{3}{2}(x - (-3))$
$y - 5 = -\frac{3}{2}x - \frac{9}{2}$
$y = -\frac{3}{2}x + \frac{1}{2}$

5. $x = 5$

6. $y = 5$

7. $\dfrac{2x+3}{4x^2-9} = \dfrac{2x+3}{(2x+3)(2x-3)} = \dfrac{1}{2x-3}$

8. $\dfrac{x+1}{x} + \dfrac{x-1}{x+1} = \dfrac{(x+1)(x+1)}{x(x+1)} + \dfrac{(x-1)x}{(x+1)x} = \dfrac{x^2+2x+1}{x(x+1)} + \dfrac{x^2-x}{x(x+1)} = \dfrac{2x^2+x+1}{x(x+1)}$

9. $\dfrac{x^2+3x+2}{3x+12} \cdot \dfrac{x+4}{x^2-4} = \dfrac{(x+2)(x+1)}{3(x+4)} \cdot \dfrac{x+4}{(x+2)(x-2)} = \dfrac{x+1}{3(x-2)}$

10. $\dfrac{1+\frac{y}{x}}{\frac{y}{x}-1} = \dfrac{\left(1+\frac{y}{x}\right)x}{\left(\frac{y}{x}-1\right)x} = \dfrac{x+y}{y-x}$

11. $\log_e x$

12. $(0, \infty)$

13. $(-\infty, \infty)$

14. y-axis

15. 10

16. e

17. $\dfrac{\ln 2}{r}$

18. undefined

19. $\ln 25.25 \approx 3.2288$

20. $\ln 0.523 \approx -0.6482$

21. $\ln 9.89 \approx 2.2915$

22. $\ln 0.00725 \approx -4.9268$

23. $\log(\ln 2) \approx \log(0.6931) \approx -0.1592$

24. $\ln(\log 28.8) \approx \ln(1.4594) \approx 0.3780$

25. $\ln(\log 0.5) = \ln(-0.3010) \Rightarrow$ impossible

26. $\log(\ln 0.2) \approx \log(-1.609) \Rightarrow$ impossible

27. $\ln y = 2.3015 \Rightarrow y = 9.9892$

28. $\ln y = 1.548 \Rightarrow y = 4.7021$

29. $\ln y = 3.17 \Rightarrow y = 23.8075$

30. $\ln y = 0.837 \Rightarrow y = 2.3094$

31. $\ln y = -4.72 \Rightarrow y = 0.0089$

32. $\ln y = -0.48 \Rightarrow y = 0.6188$

33. $\log y = \ln 6 \Rightarrow \log y \approx 1.7918 \Rightarrow y \approx 61.9098$
(The answer will vary if rounding is used on the calculator.)

34. $\ln y = \log 5 \Rightarrow \ln y \approx 0.6990 \Rightarrow y \approx 2.0117$
(The answer will vary if rounding is used on the calculator.)

35. The graph must be increasing. The graph could not look like this.

36. The graph could look like this.

37. The graph must go through $(1, 0)$. The graph could not look like this.

38. The graph must be increasing. The graph could not look like this.

39. $y = -\ln x$

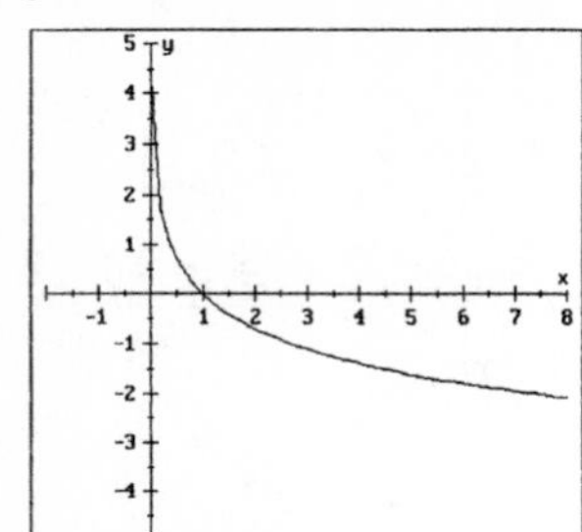

40. $y = \ln x^2$

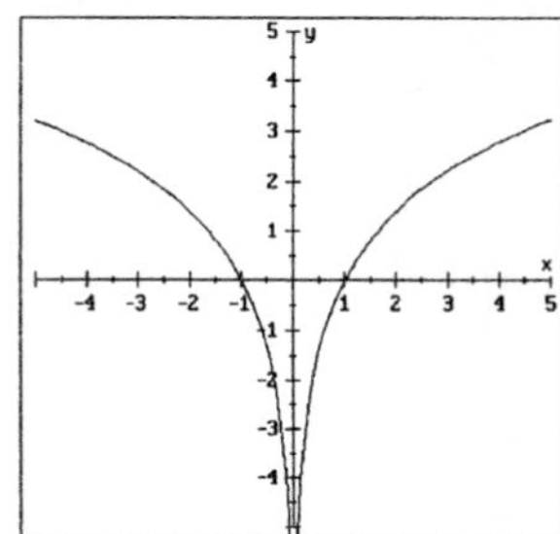

41. $y = \ln(-x)$

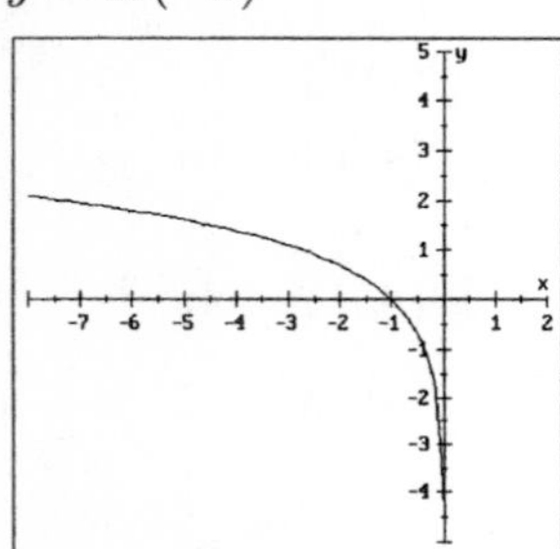

42. $y = \ln\left(\frac{1}{2}x\right)$

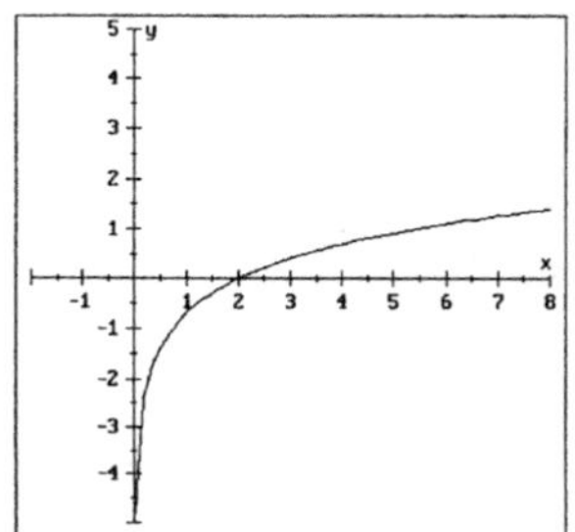

43. $t = \dfrac{\ln 2}{r} = \dfrac{\ln 2}{0.12} \approx 5.8$ years

44. $t = \dfrac{\ln 2}{r} = \dfrac{\ln 2}{0.05} \approx 13.9$ years

45. $t = \dfrac{\ln 3}{r} = \dfrac{\ln 3}{0.12} \approx 9.2$ years

46. $t = \dfrac{\ln 3}{r} = \dfrac{\ln 3}{0.06} \approx 18.3$ years

47. $t = -\dfrac{1}{0.9} \ln \dfrac{50 - T_r}{200 - T_r} = -\dfrac{1}{0.9} \ln \dfrac{50 - 38}{200 - 38} = -\dfrac{1}{0.9} \ln \dfrac{12}{162} \approx -\dfrac{1}{0.9}(-2.6027) \approx 2.9$ hours

48. $t = \dfrac{1}{0.25} \ln \dfrac{98.6 - T_s}{82 - T_s} = \dfrac{1}{0.25} \ln \dfrac{98.6 - 72}{82 - 72} = \dfrac{1}{0.25} \ln \dfrac{26.6}{10} \approx \dfrac{1}{0.25}(0.9783) \approx 3.9$ hours

49. **Answers may vary.**

50. **Answers may vary.**

51. $P = P_0 e^{rt} = P_0 e^{r\frac{\ln 3}{r}} = P_0 e^{\ln 3} = 3P_0$

52. $P = P_0 e^{rt} = P_0 e^{r\frac{\ln 4}{r}} = P_0 e^{\ln 4} = 4P_0$

53. Let $t = \dfrac{\ln 5}{r} \Rightarrow P = P_0 e^{rt} = P_0 e^{r\frac{\ln 5}{r}} = P_0 e^{\ln 5} = 5P_0$

54. $y = \dfrac{1}{1 + e^{-2x}}$

As x gets large, the y-coordinates approach 1, while as x gets large in a negative sense, the y-coordinates approach 0.

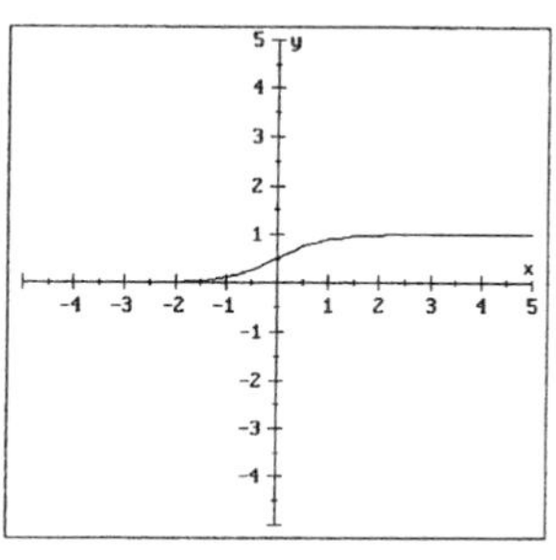

Exercise 9.5 (page 633)

1. $m = \dfrac{\Delta y}{\Delta x} = \dfrac{3-(-4)}{-2-4} = \dfrac{7}{-6} = -\dfrac{7}{6}$

2. $d = \sqrt{(-2-4)^2 + (3-(-4))^2} = \sqrt{(-6)^2 + 7^2} = \sqrt{36+49} = \sqrt{85}$

3. $x = \frac{-2+4}{2} = \frac{2}{2} = 1$

$y = \frac{3+(-4)}{2} = \frac{-1}{2} = -\frac{1}{2}$

midpoint: $\left(1, -\frac{1}{2}\right)$

4. Use the slope $m = -\frac{7}{6}$ from #1.

$y - y_1 = m(x - x_1)$

$y - 3 = -\frac{7}{6}(x+2)$

$y - 3 = -\frac{7}{6}x - \frac{7}{3}$

$y = -\frac{7}{6}x + \frac{2}{3}$

5. 0 **6.** 1 **7.** $M; N$ **8.** x

9. $x; y$ **10.** $-$ **11.** x **12.** x

13. $\neq$ **14.** $=$ **15.** 0 **16.** 1

17. 7 **18.** 8 **19.** 10 **20.** 2

21. 1 **22.** 0 **23.** 0 **24.** 1

25. 7 **26.** 8 **27.** 10 **28.** 2

29. 1 **30.** 0

Problems 31-36 are to be solved using a calculator. The keystrokes needed to solve each problem using a TI-83 graphing calculator appear in each solution. There may be other solutions. Keystrokes for other calculators may be slightly different.

31. [log] [2] [.] [5] [×] [3] [.] [7] [ENTER] {0.96614}

[log] [2] [.] [5] [)] [+] [log] [3] [.] [7] [)] [ENTER] {0.96614}

32. [ln] [1] [1] [.] [3] [÷] [6] [.] [1] [ENTER] {0.61651}
[ln] [1] [1] [.] [3] [)] [−] [ln] [6] [.] [1] [ENTER] {0.61651}

33. [ln] [2] [.] [2] [5] [^] [4] [ENTER] {3.24372}
[4] [ln] [2] [.] [2] [5] [ENTER] {3.24372}

34. [log] [4] [5] [.] [3] [7] [ENTER] {1.65677}
[ln] [4] [5] [.] [3] [7] [)] [÷] [ln] [1] [0] [ENTER] {1.65677}

35. [log] [$\sqrt{\ }$] [2] [4] [.] [3] [ENTER] {0.69280}
[.] [5] [log] [2] [4] [.] [3] [ENTER] {0.69280}

36. [ln] [8] [.] [7] [5] {2.16905}
[log] [8] [.] [7] [5] [)] [÷] [log] [(] [2nd] [e^x] [1] [ENTER] {2.16905}

37. $\log_b xyz = \log_b x + \log_b y + \log_b z$

38. $\log_b 4xz = \log_b 4 + \log_b x + \log_b z$

39. $\log_b \frac{2x}{y} = \log_b 2x - \log_b y = \log_b 2 + \log_b x - \log_b y$

40. $\log_b \frac{x}{yz} = \log_b x - \log_b yz = \log_b x - (\log_b y + \log_b z) = \log_b x - \log_b y - \log_b z$

41. $\log_b x^3y^2 = \log_b x^3 + \log_b y^2 = 3\log_b x + 2\log_b y$

42. $\log_b xy^2z^3 = \log_b x + \log_b y^2 + \log_b z^3 = \log_b x + 2\log_b y + 3\log_b z$

43. $\log_b (xy)^{1/2} = \frac{1}{2}\log_b xy = \frac{1}{2}(\log_b x + \log_b y) = \frac{1}{2}\log_b x + \frac{1}{2}\log_b y$

44. $\log_b x^3y^{1/2} = \log_b x^3 + \log_b y^{1/2} = 3\log_b x + \frac{1}{2}\log_b y$

45. $\log_b x\sqrt{z} = \log_b xz^{1/2} = \log_b x + \log_b z^{1/2} = \log_b x + \frac{1}{2}\log_b z$

46. $\log_b \sqrt{xy} = \log_b (xy)^{1/2} = \frac{1}{2}\log_b xy = \frac{1}{2}(\log_b x + \log_b y) = \frac{1}{2}\log_b x + \frac{1}{2}\log_b y$

47.
$$\begin{aligned}\log_b \frac{\sqrt[3]{x}}{\sqrt[4]{yz}} = \log_b \frac{x^{1/3}}{(yz)^{1/4}} = \log_b x^{1/3} - \log_b (yz)^{1/4} &= \frac{1}{3}\log_b x - \frac{1}{4}\log_b yz \\ &= \frac{1}{3}\log_b x - \frac{1}{4}(\log_b y + \log_b z) \\ &= \frac{1}{3}\log_b x - \frac{1}{4}\log_b y - \frac{1}{4}\log_b z\end{aligned}$$

48. $\log_b \sqrt[4]{\dfrac{x^3y^2}{z^4}} = \log_b \left(\dfrac{x^3y^2}{z^4}\right)^{1/4} = \dfrac{1}{4}\log_b \dfrac{x^3y^2}{z^4} = \dfrac{1}{4}(\log_b x^3 + \log_b y^2 - \log_b z^4)$

$$= \frac{1}{4}(3\log_b x + 2\log_b y - 4\log_b z)$$

$$= \frac{3}{4}\log_b x + \frac{1}{2}\log_b y - \log_b z$$

49. $\log_b (x+1) - \log_b x = \log_b \dfrac{x+1}{x}$

50. $\log_b x + \log_b (x+2) - \log_b 8 = \log_b x(x+2) - \log_b 8 = \log_b \dfrac{x(x+2)}{8}$

51. $2\log_b x + \dfrac{1}{2}\log_b y = \log_b x^2 + \log_b y^{1/2} = \log_b x^2y^{1/2}$

52. $-2\log_b x - 3\log_b y + \log_b z = \log_b x^{-2} + \log_b y^{-3} + \log_b z = \log_b x^{-2}y^{-3}z = \log_b \dfrac{z}{x^2y^3}$

53. $-3\log_b x - 2\log_b y + \dfrac{1}{2}\log_b z = \log_b x^{-3} + \log_b y^{-2} + \log_b z^{1/2} = \log_b x^{-3}y^{-2}z^{1/2} = \log_b \dfrac{z^{1/2}}{x^3y^2}$

54. $3\log_b (x+1) - 2\log_b (x+2) + \log_b x = \log_b (x+1)^3 + \log_b (x+2)^{-2} + \log_b x$

$$= \log_b (x+1)^3(x+2)^{-2}x$$

$$= \log_b \frac{x(x+1)^3}{(x+2)^2}$$

55. $\log_b \left(\dfrac{x}{z} + x\right) - \log_b \left(\dfrac{y}{z} + y\right) = \log_b \dfrac{\frac{x}{z} + x}{\frac{y}{z} + y} = \log_b \dfrac{x + xz}{y + yz} = \log_b \dfrac{x(1+z)}{y(1+z)} = \log_b \dfrac{x}{y}$

56. $\log_b (xy + y^2) - \log_b (xz + yz) + \log_b z = \log_b \dfrac{xy + y^2}{xz + yz} + \log_b z = \log_b \dfrac{y(x+y)}{z(x+y)} + \log_b z$

$$= \log_b \frac{y}{z} + \log_b z$$

$$= \log_b \left(\frac{y}{z} \cdot z\right) = \log_b y$$

57. $\log_b 0 = 1 \Rightarrow b^1 = 0 \Rightarrow b = 0 \Rightarrow$ FALSE $(b \neq 0)$

58. $\log_b (x+y) \neq \log_b x + \log_b y \Rightarrow$ TRUE $(\log_b xy = \log_b x + \log_b y)$

59. $\log_b xy = (\log_b x)(\log_b y) \Rightarrow$ FALSE $(\log_b xy = \log_b x + \log_b y)$

60. $\log_b ab = \log_b a + \log_b b = \log_b a + 1 \Rightarrow$ TRUE

61. $\log_7 7^7 = 7 \Rightarrow 7^7 = 7^7 \Rightarrow$ TRUE

62. $7^{\log_7 7} = 7 \Rightarrow$ TRUE

63. $\dfrac{\log_b A}{\log_b B} = \log_b A - \log_b B \Rightarrow$ FALSE $\left(\log_b \dfrac{A}{B} = \log_b A - \log_b B\right)$

64. $\log_b (A - B) = \dfrac{\log_b A}{\log_b B} \Rightarrow$ FALSE (There is no property that applies to this case.)

65. $3 \log_b \sqrt[3]{a} = 3 \log_b a^{1/3} = \dfrac{1}{3} \cdot 3 \log_b a = \log_b a \Rightarrow$ TRUE

66. $\dfrac{1}{3} \log_b a^3 = 3 \cdot \dfrac{1}{3} \log_b a = \log_b a \Rightarrow$ TRUE

67. $\log_b \dfrac{1}{a} = \log_b 1 - \log_b a = 0 - \log_b a = -\log_b a \Rightarrow$ TRUE

68. $\log_b 2 = \log_2 b \Rightarrow b^{\log_2 b} = 2 \Rightarrow$ FALSE $\left(b^{\log_b 2} = 2\right)$

69. $\log 28 = \log 4 \cdot 7 = \log 4 + \log 7 = 0.6021 + 0.8451 = 1.4472$

70. $\log \dfrac{7}{4} = \log 7 - \log 4 = 0.8451 - 0.6021 = 0.2430$

71. $\log 2.25 = \log \dfrac{9}{4} = \log 9 - \log 4 = 0.9542 - 0.6021 = 0.3521$

72. $\log 36 = \log 4 \cdot 9 = \log 4 + \log 9 = 0.6021 + 0.9542 = 1.5563$

73. $\log \dfrac{63}{4} = \log \dfrac{7 \cdot 9}{4} = \log 7 + \log 9 - \log 4 = 0.8451 + 0.9542 - 0.6021 = 1.1972$

74. $\log \dfrac{4}{63} = \log \dfrac{4}{7 \cdot 9} = \log 4 - \log 7 - \log 9 = 0.6021 - 0.8451 - 0.9542 = -1.1972$

75. $\log 252 = \log 4 \cdot 7 \cdot 9 = \log 4 + \log 7 + \log 9 = 0.6021 + 0.8451 + 0.9542 = 2.4014$

76. $\log 49 = \log 7^2 = 2 \log 7 = 2(0.8451) = 1.6902$

77. $\log 112 = \log 4^2 \cdot 7 = \log 4^2 + \log 7 = 2 \log 4 + \log 7 = 2(0.6021) + 0.8451 = 2.0493$

78. $\log 324 = \log 9^2 \cdot 4 = \log 9^2 + \log 4 = 2 \log 9 + \log 4 = 2(0.9542) + 0.6021 = 2.5105$

79. $\log \dfrac{144}{49} = \log \dfrac{16 \cdot 9}{7^2} = \log \dfrac{4^2 \cdot 9}{7^2} = \log 4^2 + \log 9 - \log 7^2 = 2 \log 4 + \log 9 - 2 \log 7$

$$= 2(0.6021) + 0.9542 - 2(0.8451)$$
$$= 0.4682$$

80. $\log \frac{324}{63} = \log \frac{9^2 \cdot 4}{9 \cdot 7} = \log 9^2 + \log 4 - \log 9 - \log 7 = 2\log 9 + \log 4 - \log 9 - \log 7$
$= \log 9 + \log 4 - \log 7$
$= 0.9542 + 0.6021 - 0.8451 = 0.7112$

81. $\log_3 7 = \frac{\log 7}{\log 3} \approx 1.7712$

82. $\log_7 3 = \frac{\log 3}{\log 7} \approx 0.5646$

83. $\log_{1/3} 3 = \frac{\log 3}{\log \frac{1}{3}} \approx -1.0000$

84. $\log_{1/2} 6 = \frac{\log 6}{\log \frac{1}{2}} \approx -2.5850$

85. $\log_3 8 = \frac{\log 8}{\log 3} \approx 1.8928$

86. $\log_5 10 = \frac{\log 10}{\log 5} \approx 1.4307$

87. $\log_{\sqrt{2}} \sqrt{5} = \frac{\log \sqrt{5}}{\log \sqrt{2}} \approx 2.3219$

88. $\log_\pi e = \frac{\log e}{\log \pi} \approx 0.8736$

89. $\text{pH} = -\log [\text{H}^+] = -\log (1.7 \times 10^{-5})$
≈ 4.77

90. $\text{pH} = -\log [\text{H}^+]$
$13.2 = -\log [\text{H}^+]$
$-13.2 = \log [\text{H}^+]$
$[\text{H}^+] = 6.31 \times 10^{-14}$ gram-ions per liter

91. low pH:
$\text{pH} = -\log [\text{H}^+]$
$6.8 = -\log [\text{H}^+]$
$-6.8 = \log [\text{H}^+]$
$[\text{H}^+] = 1.5849 \times 10^{-7}$

high pH:
$\text{pH} = -\log [\text{H}^+]$
$7.6 = -\log [\text{H}^+]$
$-7.6 = \log [\text{H}^+]$
$[\text{H}^+] = 2.5119 \times 10^{-8}$

92. $\text{pH} = -\log [\text{H}^+] = -\log (6.31 \times 10^{-4}) \approx 3.20$

93. $k \ln 2I = k(\ln 2 + \ln I)$
$= k \ln 2 + k \ln I$
$= k \ln 2 + L$
The loudness increases by $k \ln 2$.

94. $k \ln 3I = k(\ln 3 + \ln I)$
$= k \ln 3 + k \ln I$
$= k \ln 3 + L$
The loudness increases by $k \ln 3$.

95. $L = 3k \ln I = k \cdot 3 \ln I$
$= k \ln I^3$
The intensity must be cubed.

96. $L = 4k \ln I = k \cdot 4 \ln I$
$= k \ln I^4$
The intensity must be raised to the 4th power.

97. **Answers may vary.**

98. **Answers may vary.**

99. $\ln(e^x) = \log_e(e^x) = x$

100.
$$\begin{aligned} \log_b 3x &= 1 + \log_b x \\ \log_b 3 + \log_b x &= 1 + \log_b x \\ \log_b 3 &= 1 \\ b &= 3 \end{aligned}$$

101. Let $\log_{b^2} x = y$. Then
$$\begin{aligned} \left(b^2\right)^y &= x \\ b^{2y} &= x \\ \left(b^{2y}\right)^{1/2} &= x^{1/2} \\ b^y &= x^{1/2} \\ \log_b x^{1/2} &= y \\ \frac{1}{2}\log_b x &= y \end{aligned}$$

102. $e^{x \ln a} = e^{\ln a^x} = a^x$

Exercise 9.6 (page 644)

1.
$$\begin{aligned} 5x^2 - 25x &= 0 \\ 5x(x-5) &= 0 \\ 5x = 0 \quad \textbf{or} \quad x - 5 &= 0 \\ x = 0 \qquad\qquad x &= 5 \end{aligned}$$

2.
$$\begin{aligned} 4y^2 - 25 &= 0 \\ (2y+5)(2y-5) &= 0 \\ 2y + 5 = 0 \quad \textbf{or} \quad 2y - 5 &= 0 \\ y = -\tfrac{5}{2} \qquad\qquad y &= \tfrac{5}{2} \end{aligned}$$

3.
$$\begin{aligned} 3p^2 + 10p &= 8 \\ 3p^2 + 10p - 8 &= 0 \\ (3p-2)(p+4) &= 0 \\ 3p - 2 = 0 \quad \textbf{or} \quad p + 4 &= 0 \\ p = \tfrac{2}{3} \qquad\qquad p &= -4 \end{aligned}$$

4.
$$\begin{aligned} 4t^2 + 1 &= -6t \\ 4t^2 + 6t + 1 = 0 &\Rightarrow a = 4, b = 6, c = 1 \\ x &= \frac{-b \pm \sqrt{b^2 - 4ac}}{2a} \\ &= \frac{-6 \pm \sqrt{6^2 - 4(4)(1)}}{2(4)} \\ &= \frac{-6 \pm \sqrt{36 - 16}}{8} \\ &= \frac{-6 \pm \sqrt{20}}{8} \\ &= -\frac{6}{8} \pm \frac{2\sqrt{5}}{8} = -\frac{3}{4} \pm \frac{\sqrt{5}}{4} \end{aligned}$$

5. exponential

6. logarithmic

7. $A_0 2^{-t/h}$

8. $P_0 e^{kt}$

9.
$$\begin{aligned} 4^x &= 5 \\ \log 4^x &= \log 5 \\ x \log 4 &= \log 5 \\ x &= \frac{\log 5}{\log 4} \\ x &\approx 1.1610 \end{aligned}$$

10.
$$\begin{aligned} 7^x &= 12 \\ \log 7^x &= \log 12 \\ x \log 7 &= \log 12 \\ x &= \frac{\log 12}{\log 7} \\ x &\approx 1.2770 \end{aligned}$$

11.
$$\begin{aligned} e^t &= 50 \\ \ln e^t &= \ln 50 \\ t \ln e &= \ln 50 \\ t &= \ln 50 \\ t &\approx 3.9120 \end{aligned}$$

12.
$$\begin{aligned} e^{-t} &= 0.25 \\ \ln e^{-t} &= \ln 0.25 \\ -t\ln e &= \ln 0.25 \\ t &= -\ln 0.25 \\ t &\approx 1.3863 \end{aligned}$$

13.
$$\begin{aligned} 5 &= 2.1(1.04)^t \\ \frac{5}{2.1} &= (1.04)^t \\ \log\frac{5}{2.1} &= \log(1.04)^t \\ \log\frac{5}{2.1} &= t\log 1.04 \\ \frac{\log\frac{5}{2.1}}{\log 1.04} &= t \\ 22.1184 &\approx t \end{aligned}$$

14.
$$\begin{aligned} 61 &= 1.5(1.02)^t \\ \frac{61}{1.5} &= (1.02)^t \\ \log\frac{61}{1.5} &= \log(1.02)^t \\ \log\frac{61}{1.5} &= t\log 1.02 \\ \frac{\log\frac{61}{1.5}}{\log 1.02} &= t \\ 187.1170 &\approx t \end{aligned}$$

15.
$$\begin{aligned} 13^{x-1} &= 2 \\ \log 13^{x-1} &= \log 2 \\ (x-1)\log 13 &= \log 2 \\ x-1 &= \frac{\log 2}{\log 13} \\ x &= \frac{\log 2}{\log 13}+1 \\ x &\approx 1.2702 \end{aligned}$$

16.
$$\begin{aligned} 5^{x+1} &= 3 \\ \log 5^{x+1} &= \log 3 \\ (x+1)\log 5 &= \log 3 \\ x+1 &= \frac{\log 3}{\log 5} \\ x &= \frac{\log 3}{\log 5}-1 \\ x &\approx -0.3174 \end{aligned}$$

17.
$$\begin{aligned} 2^{x+1} &= 3^x \\ \log 2^{x+1} &= \log 3^x \\ (x+1)\log 2 &= x\log 3 \\ x\log 2+\log 2 &= x\log 3 \\ \log 2 &= x\log 3 - x\log 2 \\ \log 2 &= x(\log 3-\log 2) \\ \frac{\log 2}{\log 3-\log 2} &= x \\ 1.7095 &\approx x \end{aligned}$$

18.
$$\begin{aligned} 5^{x-3} &= 3^{2x} \\ \log 5^{x-3} &= \log 3^{2x} \\ (x-3)\log 5 &= 2x\log 3 \\ x\log 5 - 3\log 5 &= 2x\log 3 \\ x\log 5 - 2x\log 3 &= 3\log 5 \\ x(\log 5 - 2\log 3) &= 3\log 5 \\ x &= \frac{3\log 5}{\log 5 - 2\log 3} \\ x &\approx -8.2144 \end{aligned}$$

19.
$$\begin{aligned} 2^x &= 3^x \\ \log 2^x &= \log 3^x \\ x\log 2 &= x\log 3 \\ 0 &= x\log 3 - x\log 2 \\ 0 &= x(\log 3-\log 2) \\ \frac{0}{\log 3-\log 2} &= x \\ 0 &= x \end{aligned}$$

20.
$$\begin{aligned} 3^{2x} &= 4^x \\ \log 3^{2x} &= \log 4^x \\ 2x\log 3 &= x\log 4 \\ 0 &= x\log 4 - 2x\log 3 \\ 0 &= x(\log 4 - 2\log 3) \\ \frac{0}{\log 4 - 2\log 3} &= x \\ 0 &= x \end{aligned}$$

21.
$$\begin{aligned} 7^{x^2} &= 10 \\ \log 7^{x^2} &= \log 10 \\ x^2 \log 7 &= \log 10 \\ x^2 &= \frac{\log 10}{\log 7} \\ x^2 &\approx 1.1833 \\ x &\approx \pm 1.0878 \end{aligned}$$

22.
$$\begin{aligned} 8^{x^2} &= 11 \\ \log 8^{x^2} &= \log 11 \\ x^2 \log 8 &= \log 11 \\ x^2 &= \frac{\log 11}{\log 8} \\ x^2 &\approx 1.1531 \\ x &\approx \pm 1.0738 \end{aligned}$$

23.
$$\begin{aligned} 8^{x^2} &= 9^x \\ \log 8^{x^2} &= \log 9^x \\ x^2 \log 8 &= x \log 9 \\ x^2 \log 8 - x \log 9 &= 0 \\ x(x \log 8 - \log 9) &= 0 \\ x = 0 \quad \textbf{or} \quad x \log 8 - \log 9 &= 0 \\ x \log 8 &= \log 9 \\ x &= \frac{\log 9}{\log 8} \\ x &\approx 1.0566 \end{aligned}$$

24.
$$\begin{aligned} 5^{x^2} &= 2^{5x} \\ \log 5^{x^2} &= \log 2^{5x} \\ x^2 \log 5 &= 5x \log 2 \\ x^2 \log 5 - 5x \log 2 &= 0 \\ x(x \log 5 - 5 \log 2) &= 0 \\ x = 0 \quad \textbf{or} \quad x \log 5 - 5 \log 2 &= 0 \\ x \log 5 &= 5 \log 2 \\ x &= \frac{5 \log 2}{\log 5} \\ x &\approx 2.1534 \end{aligned}$$

25.
$$\begin{aligned} 2^{x^2-2x} &= 8 \\ 2^{x^2-2x} &= 2^3 \\ x^2 - 2x &= 3 \\ x^2 - 2x - 3 &= 0 \\ (x+1)(x-3) &= 0 \\ x + 1 = 0 \quad &\textbf{or} \quad x - 3 = 0 \\ x = -1 \quad & \quad\quad x = 3 \end{aligned}$$

26.
$$\begin{aligned} 3^{x^2-3x} &= 81 \\ 3^{x^2-3x} &= 3^4 \\ x^2 - 3x &= 4 \\ x^2 - 3x - 4 &= 0 \\ (x+1)(x-4) &= 0 \\ x + 1 = 0 \quad &\textbf{or} \quad x - 4 = 0 \\ x = -1 \quad & \quad\quad x = 4 \end{aligned}$$

27.
$$\begin{aligned} 3^{x^2+4x} &= \frac{1}{81} \\ 3^{x^2+4x} &= 3^{-4} \\ x^2 + 4x &= -4 \\ x^2 + 4x + 4 &= 0 \\ (x+2)(x+2) &= 0 \\ x + 2 = 0 \quad &\textbf{or} \quad x + 2 = 0 \\ x = -2 \quad & \quad\quad x = -2 \end{aligned}$$

28.
$$\begin{aligned} 7^{x^2+3x} &= \frac{1}{49} \\ 7^{x^2+3x} &= 7^{-2} \\ x^2 + 3x &= -2 \\ x^2 + 3x + 2 &= 0 \\ (x+1)(x+2) &= 0 \\ x + 1 = 0 \quad &\textbf{or} \quad x + 2 = 0 \\ x = -1 \quad & \quad\quad x = -2 \end{aligned}$$

29.
$$\begin{aligned} 4^{x+2} - 4^x &= 15 \\ 4^x 4^2 - 4^x &= 15 \\ 16 \cdot 4^x - 4^x &= 15 \\ 15 \cdot 4^x &= 15 \\ 4^x &= 1 \\ x &= 0 \end{aligned}$$

30.
$$\begin{aligned} 3^{x+3} + 3^x &= 84 \\ 3^x 3^3 + 3^x &= 84 \\ 27 \cdot 3^x + 3^x &= 84 \\ 28 \cdot 3^x &= 84 \\ 3^x &= 3 \\ x &= 1 \end{aligned}$$

31.
$$\begin{aligned}
2(3^x) &= 6^{2x} \\
\log 2(3^x) &= \log 6^{2x} \\
\log 2 + \log 3^x &= 2x \log 6 \\
\log 2 + x \log 3 &= 2x \log 6 \\
\log 2 &= 2x \log 6 - x \log 3 \\
\log 2 &= x(2 \log 6 - \log 3) \\
\frac{\log 2}{2 \log 6 - \log 3} &= x \\
0.2789 &\approx x
\end{aligned}$$

32.
$$\begin{aligned}
2(3^{x+1}) &= 3(2^{x-1}) \\
\log 2(3^{x+1}) &= \log 3(2^{x-1}) \\
\log 2 + \log 3^{x+1} &= \log 3 + \log 2^{x-1} \\
\log 2 + (x+1)\log 3 &= \log 3 + (x-1)\log 2 \\
\log 2 + x \log 3 + \log 3 &= \log 3 + x \log 2 - \log 2 \\
2 \log 2 &= x \log 2 - x \log 3 \\
2 \log 2 &= x(\log 2 - \log 3) \\
\frac{2 \log 2}{\log 2 - \log 3} &= x \\
-3.4190 &\approx x
\end{aligned}$$

33. $2^{x+1} = 7 \Rightarrow$ Graph $y = 2^{x+1}$ and $y = 7$.

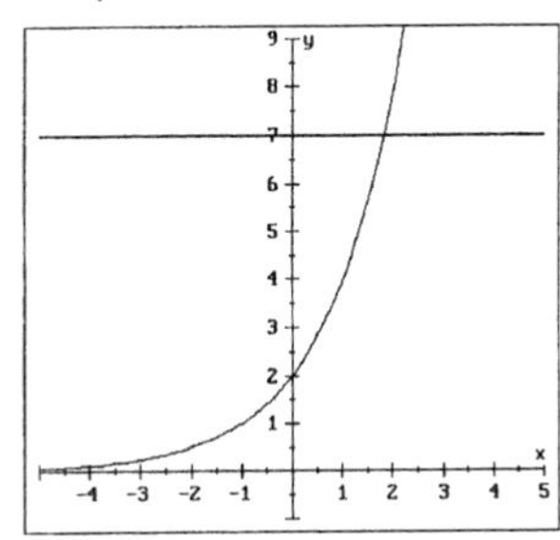

$x \approx 1.8$

34. $3^{x-1} = 2^x \Rightarrow$ Graph $y = 3^{x-1}$ and $y = 2^x$.

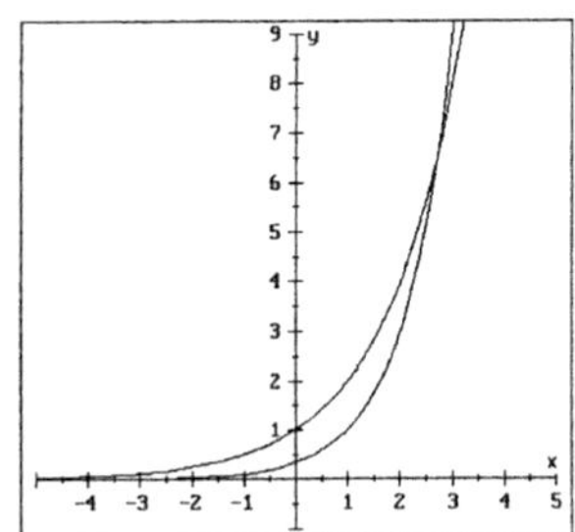

$x \approx 2.7$

35. $2^{x^2-2x} - 8 = 0 \Rightarrow$ Graph $y = 2^{x^2-2x} - 8$ and find any x-intercept(s).

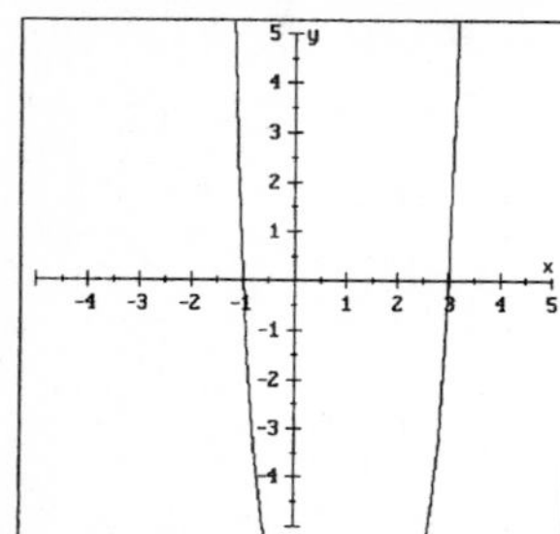

$x = 3$ or $x = -1$

36. $3^x - 10 = 3^{-x}$
Graph $y = 3^x - 10$ and $y = 3^{-x}$.

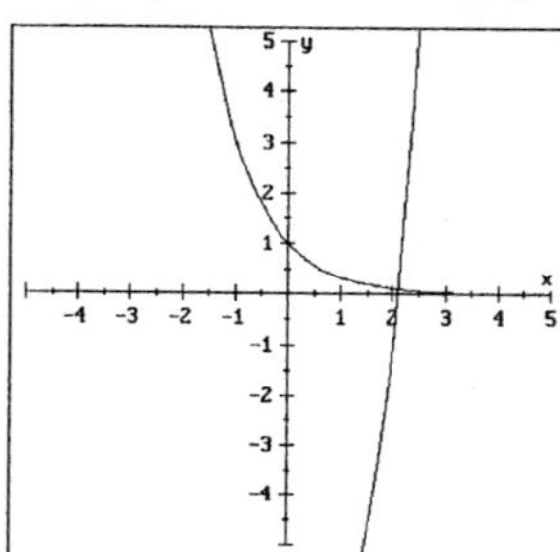

$x \approx 2.1$

37.
$$\begin{aligned} \log 2x &= \log 4 \\ 2x &= 4 \\ x &= 2 \end{aligned}$$

38.
$$\begin{aligned} \log 3x &= \log 9 \\ 3x &= 9 \\ x &= 3 \end{aligned}$$

39.
$$\begin{aligned} \log (3x+1) &= \log (x+7) \\ 3x+1 &= x+7 \\ 2x &= 6 \\ x &= 3 \end{aligned}$$

40.
$$\begin{aligned} \log (x^2+4x) &= \log (x^2+16) \\ x^2+4x &= x^2+16 \\ 4x &= 16 \\ x &= 4 \end{aligned}$$

41.
$$\begin{aligned} \log (3-2x) - \log (x+24) &= 0 \\ \log (3-2x) &= \log (x+24) \\ 3-2x &= x+24 \\ -21 &= 3x \\ -7 &= x \end{aligned}$$

42.
$$\begin{aligned} \log (3x+5) - \log (2x+6) &= 0 \\ \log (3x+5) &= \log (2x+6) \\ 3x+5 &= 2x+6 \\ x &= 1 \end{aligned}$$

43.
$$\begin{aligned} \log \frac{4x+1}{2x+9} &= 0 \\ 10^0 &= \frac{4x+1}{2x+9} \\ 1 &= \frac{4x+1}{2x+9} \\ 2x+9 &= 4x+1 \\ 8 &= 2x \\ 4 &= x \end{aligned}$$

44.
$$\begin{aligned} \log \frac{2-5x}{2(x+8)} &= 0 \\ 10^0 &= \frac{2-5x}{2(x+8)} \\ 1 &= \frac{2-5x}{2x+16} \\ 2x+16 &= 2-5x \\ 7x &= -14 \\ x &= -2 \end{aligned}$$

45.
$$\begin{aligned} \log x^2 &= 2 \\ 10^2 &= x^2 \\ 100 &= x^2 \\ \pm 10 &= x \end{aligned}$$

46.
$$\begin{aligned} \log x^3 &= 3 \\ 10^3 &= x^3 \\ 1000 &= x^3 \\ 10 &= x \end{aligned}$$

47.
$$\begin{aligned} \log x + \log(x-48) &= 2 \\ \log x(x-48) &= 2 \\ 10^2 &= x(x-48) \\ 0 &= x^2 - 48x - 100 \\ 0 &= (x-50)(x+2) \end{aligned}$$
$x - 50 = 0$ **or** $x + 2 = 0$

$x = 50$ $\quad x = -2$

Does not check.

48.
$$\begin{aligned} \log x + \log(x+9) &= 1 \\ \log x(x+9) &= 1 \\ 10^1 &= x(x+9) \\ 0 &= x^2 + 9x - 10 \\ 0 &= (x-1)(x+10) \end{aligned}$$
$x - 1 = 0$ **or** $x + 10 = 0$

$x = 1$ $\quad x = -10$

Does not check.

49.
$$\begin{aligned} \log x + \log(x-15) &= 2 \\ \log x(x-15) &= 2 \\ 10^2 &= x(x-15) \\ 0 &= x^2 - 15x - 100 \\ 0 &= (x-20)(x+5) \end{aligned}$$
$x - 20 = 0$ **or** $x + 5 = 0$

$x = 20$ $\quad x = -5$

Does not check.

50.
$$\begin{aligned} \log x + \log(x+21) &= 2 \\ \log x(x+21) &= 2 \\ 10^2 &= x(x+21) \\ 0 &= x^2 + 21x - 100 \\ 0 &= (x-4)(x+25) \end{aligned}$$
$x - 4 = 0$ **or** $x + 25 = 0$

$x = 4$ $\quad x = -25$

Does not check.

51.
$$\begin{aligned} \log(x+90) &= 3 - \log x \\ \log x + \log(x+90) &= 3 \\ \log x(x+90) &= 3 \\ 10^3 &= x(x+90) \\ 0 &= x^2 + 90x - 1000 \\ 0 &= (x-10)(x+100) \end{aligned}$$
$x - 10 = 0$ **or** $x + 100 = 0$

$x = 10$ $\quad x = -100$

Does not check.

52.
$$\begin{aligned} \log(x-90) &= 3 - \log x \\ \log x + \log(x-90) &= 3 \\ \log x(x-90) &= 3 \\ 10^3 &= x(x-90) \\ 0 &= x^2 - 90x - 1000 \\ 0 &= (x-100)(x+10) \end{aligned}$$
$x - 100 = 0$ **or** $x + 10 = 0$

$x = 100$ $\quad x = -10$

Does not check.

53.
$$\begin{aligned} \log(x-6) - \log(x-2) &= \log\frac{5}{x} \\ \log\frac{x-6}{x-2} &= \log\frac{5}{x} \\ \frac{x-6}{x-2} &= \frac{5}{x} \\ x(x-6) &= 5(x-2) \\ x^2 - 6x &= 5x - 10 \\ x^2 - 11x + 10 &= 0 \\ (x-10)(x-1) &= 0 \end{aligned}$$
$x - 10 = 0$ **or** $x - 1 = 0$

$x = 10$ $\quad x = 1$

Does not check.

54.
$$\begin{aligned} \log(3-2x) - \log(x+9) &= 0 \\ \log(3-2x) &= \log(x+9) \\ 3 - 2x &= x + 9 \\ -6 &= 3x \\ -2 &= x \end{aligned}$$

55.
$$\log x^2 = (\log x)^2$$
$$2 \log x = (\log x)^2$$
$$0 = (\log x)^2 - 2 \log x$$
$$0 = \log x\,(\log x - 2)$$
$$\log x = 0 \quad \textbf{or} \quad \log x - 2 = 0$$
$$x = 1 \qquad \log x = 2$$
$$x = 100$$

56.
$$\log(\log x) = 1$$
$$10^1 = \log x$$
$$10 = \log x$$
$$x = 10^{10}$$

57.
$$\frac{\log(3x-4)}{\log x} = 2$$
$$\log(3x-4) = 2 \log x$$
$$\log(3x-4) = \log x^2$$
$$3x - 4 = x^2$$
$$0 = x^2 - 3x + 4$$
$b^2 - 4ac = (-3)^2 - 4(1)(4) = -7 \Rightarrow$ solutions are nonreal $\Rightarrow$ no solution

58.
$$\frac{\log(8x-7)}{\log x} = 2$$
$$\log(8x-7) = 2 \log x$$
$$\log(8x-7) = \log x^2$$
$$8x - 7 = x^2$$
$$0 = x^2 - 8x + 7$$
$$0 = (x-7)(x-1)$$
$$x - 7 = 0 \quad \textbf{or} \quad x - 1 = 0$$
$$x = 7 \qquad x = 1$$
Does not check.

59.
$$\frac{\log(5x+6)}{2} = \log x$$
$$\log(5x+6) = 2 \log x$$
$$\log(5x+6) = \log x^2$$
$$5x + 6 = x^2$$
$$0 = x^2 - 5x - 6$$
$$0 = (x-6)(x+1)$$
$$x - 6 = 0 \quad \textbf{or} \quad x + 1 = 0$$
$$x = 6 \qquad x = -1$$
Does not check.

60.
$$\frac{1}{2} \log(4x+5) = \log x$$
$$\log(4x+5) = 2 \log x$$
$$\log(4x+5) = \log x^2$$
$$4x + 5 = x^2$$
$$0 = x^2 - 4x - 5$$
$$0 = (x-5)(x+1)$$
$$x - 5 = 0 \quad \textbf{or} \quad x + 1 = 0$$
$$x = 5 \qquad x = -1$$
Does not check.

61.
$$\log_3 x = \log_3 \left(\frac{1}{x}\right) + 4$$
$$\log_3 x = \log_3 \left(\frac{1}{x}\right) + \log_3 81$$
$$\log_3 x = \log_3 \left(\frac{81}{x}\right)$$
$$x = \frac{81}{x}$$
$$x^2 = 81$$
$x = 9$ (-9 does not check.)

62.
$$\log_5(7+x) + \log_5(8-x) - \log_5 2 = 2$$
$$\log_5 \frac{(7+x)(8-x)}{2} = 2$$
$$\frac{(7+x)(8-x)}{2} = 25$$
$$(7+x)(8-x) = 50$$
$$-x^2 + x + 56 = 50$$
$$x^2 - x - 6 = 0$$
$$(x-3)(x+2) = 0$$
$$x - 3 = 0 \quad \textbf{or} \quad x + 2 = 0$$
$$x = 3 \qquad x = -2$$

63.
$$\begin{aligned} 2\log_2 x &= 3 + \log_2(x-2) \\ \log_2 x^2 - \log_2(x-2) &= 3 \\ \log_2 \frac{x^2}{x-2} &= 3 \\ \frac{x^2}{x-2} &= 8 \\ x^2 &= 8(x-2) \\ x^2 &= 8x - 16 \\ x^2 - 8x + 16 &= 0 \\ (x-4)(x-4) &= 0 \end{aligned}$$
$x - 4 = 0$ **or** $x - 4 = 0$
$x = 4$ $\quad x = 4$

64.
$$\begin{aligned} 2\log_3 x - \log_3(x-4) &= 2 + \log_3 2 \\ \log_3 x^2 - \log_3(x-4) - \log_3 2 &= 2 \\ \log_3 \frac{x^2}{2(x-4)} &= 2 \\ \frac{x^2}{2x-8} &= 9 \\ x^2 &= 9(2x-8) \\ x^2 &= 18x - 72 \\ x^2 - 18x + 72 &= 0 \\ (x-6)(x-12) &= 0 \end{aligned}$$
$x - 6 = 0$ **or** $x - 12 = 0$
$x = 6$ $\quad x = 12$

65.
$$\begin{aligned} \log(7y+1) &= 2\log(y+3) - \log 2 \\ \log(7y+1) &= \log(y+3)^2 - \log 2 \\ \log(7y+1) &= \log\frac{y^2+6y+9}{2} \\ 7y+1 &= \frac{y^2+6y+9}{2} \\ 2(7y+1) &= y^2+6y+9 \\ 14y+2 &= y^2+6y+9 \\ 0 &= y^2 - 8y + 7 \\ 0 &= (y-7)(y-1) \end{aligned}$$
$y - 7 = 0$ **or** $y - 1 = 0$
$y = 7$ $\quad y = 1$

66.
$$\begin{aligned} 2\log(y+2) &= \log(y+2) - \log 12 \\ \log(y+2) &= -\log 12 \\ \log(y+2) &= \log 12^{-1} \\ y + 2 &= \frac{1}{12} \\ y &= \frac{1}{12} - 2 = \frac{1}{12} - \frac{24}{12} = -\frac{23}{12} \end{aligned}$$

67. $\log x + \log(x-15) = 2$
Graph $y = \log x + \log(x-15)$ and $y = 2$.

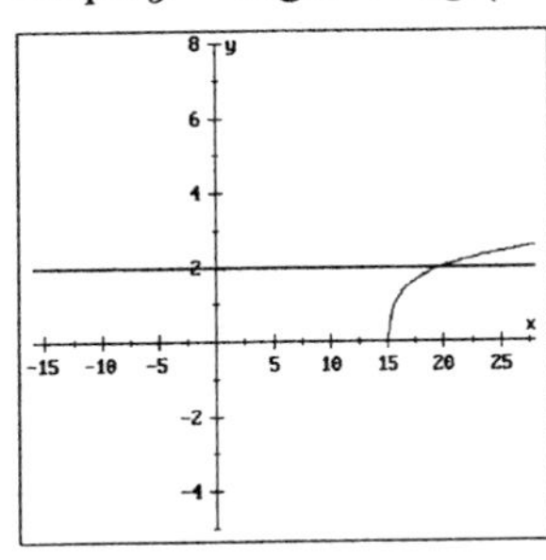

$x = 20$

68. $\log x + \log(x+3) = 1$
Graph $y = \log x + \log(x+3)$ and $y = 1$.

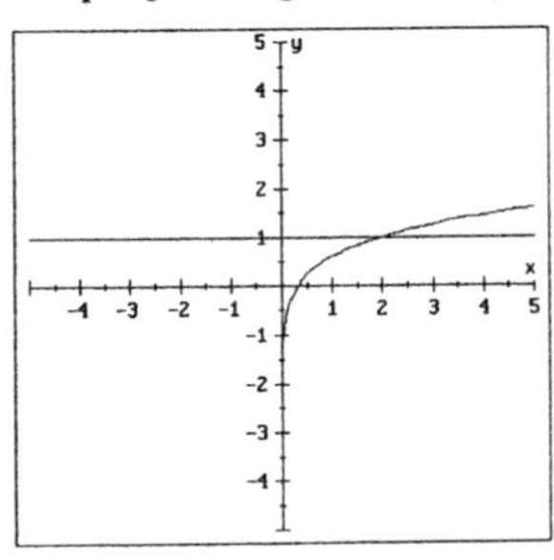

$x = 2$

69. $\ln(2x+5) - \ln 3 = \ln(x-1) \Rightarrow$ Graph $y = \ln(2x+5) - \ln 3$ and $y = \ln(x-1)$.

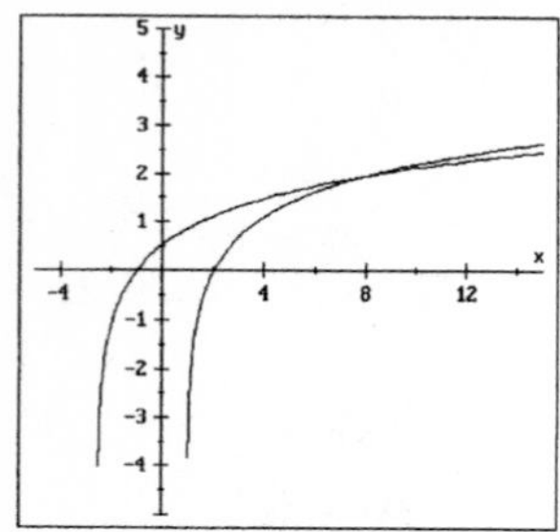

$x = 8$

70. $2\log(x^2+4x) = 1$

Graph $y = 2\log(x^2+4x)$ and $y = 1$.

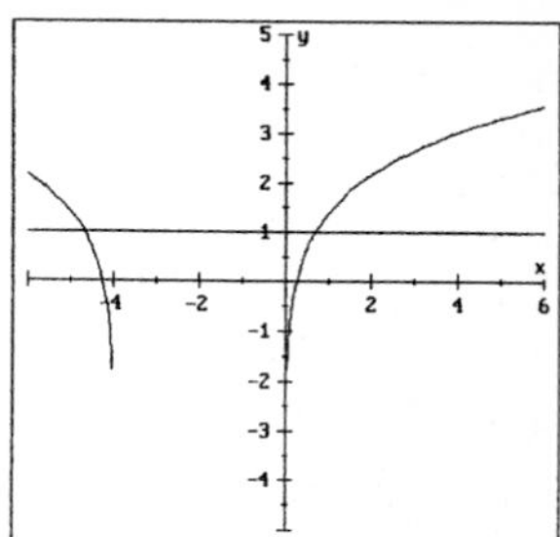

$x \approx -4.7$ or $x \approx 0.7$

71.

$$\begin{aligned} A &= A_0 e^{-0.013t} \\ 0.5A_0 &= A_0 e^{-0.013t} \\ 0.5 &= e^{-0.013t} \\ \ln 0.5 &= \ln e^{-0.013t} \\ \ln 0.5 &= -0.013t \\ \frac{\ln 0.5}{-0.013} &= t \\ 53 \text{ days} &\approx t \end{aligned}$$

72.

$$\begin{aligned} A &= A_0 e^{-0.0000284t} \\ 0.5A_0 &= A_0 e^{-0.0000284t} \\ 0.5 &= e^{-0.0000284t} \\ \ln 0.5 &= \ln e^{-0.0000284t} \\ \ln 0.5 &= -0.0000284t \\ \frac{\ln 0.5}{-0.0000284} &= t \\ 24{,}000 \text{ years} &\approx t \end{aligned}$$

73.

$$\begin{aligned} A &= A_0 2^{-t/h} \\ 0.75A_0 &= A_0 2^{-t/12.4} \\ 0.75 &= 2^{-t/12.4} \\ \log 0.75 &= \log 2^{-t/12.4} \\ \log 0.75 &= -\frac{t}{12.4}\log 2 \\ \frac{\log 0.75}{-\log 2} &= \frac{t}{12.4} \\ -12.4\frac{\log 0.75}{\log 2} &= t \\ 5.1 \text{ years} &\approx t \end{aligned}$$

74.

$$\begin{aligned} A &= A_0 2^{-t/h} \\ 0.80A_0 &= A_0 2^{-2/h} \\ 0.80 &= 2^{-2/h} \\ \log 0.80 &= \log 2^{-2/h} \\ \log 0.80 &= -\frac{2}{h}\log 2 \\ h\log 0.80 &= -2\log 2 \\ h &= \frac{-2\log 2}{\log 0.80} \\ h &\approx 6.2 \text{ years} \end{aligned}$$

75.

$$\begin{aligned} A &= A_0 2^{-t/h} \\ 0.20A_0 &= A_0 2^{-t/18.4} \\ 0.20 &= 2^{-t/18.4} \\ \log 0.20 &= \log 2^{-t/18.4} \\ \log 0.20 &= -\frac{t}{18.4}\log 2 \\ \frac{\log 0.20}{-\log 2} &= \frac{t}{18.4} \\ -18.4\frac{\log 0.20}{\log 2} &= t \\ 42.7 \text{ days} &\approx t \end{aligned}$$

76.

$$\begin{aligned} A &= A_0 2^{-t/h} \\ 1.3A_0 &= A_0 2^{-t/8.4} \\ 1.3 &= 2^{-t/8.4} \\ \log 1.3 &= \log 2^{-t/8.4} \\ \log 1.3 &= -\frac{t}{8.4}\log 2 \\ \frac{\log 1.3}{-\log 2} &= \frac{t}{8.4} \\ -8.4\frac{\log 1.3}{\log 2} &= t \\ -3.2 \text{ hours} &\approx t \Rightarrow 3.2 \text{ hours ago} \end{aligned}$$

77.

$$\begin{aligned} A &= A_0 2^{-t/h} \\ 0.60A_0 &= A_0 2^{-t/5700} \\ 0.60 &= 2^{-t/5700} \\ \log 0.60 &= \log 2^{-t/5700} \\ \log 0.60 &= -\frac{t}{5700}\log 2 \\ \frac{\log 0.60}{-\log 2} &= \frac{t}{5700} \\ -5700\frac{\log 0.60}{\log 2} &= t \\ 4200 \text{ years} &\approx t \end{aligned}$$

78.

$$\begin{aligned} A &= A_0 2^{-t/h} \\ 0.10A_0 &= A_0 2^{-t/5700} \\ 0.10 &= 2^{-t/5700} \\ \log 0.10 &= \log 2^{-t/5700} \\ \log 0.10 &= -\frac{t}{5700}\log 2 \\ \frac{\log 0.10}{-\log 2} &= \frac{t}{5700} \\ -5700\frac{\log 0.10}{\log 2} &= t \\ 19{,}000 \text{ years} &\approx t \end{aligned}$$

79.

$$\begin{aligned} A &= P\left(1+\frac{r}{k}\right)^{kt} \\ 800 &= 500\left(1+\frac{0.085}{2}\right)^{2t} \\ 1.6 &= (1.0425)^{2t} \\ \log 1.6 &= \log (1.0425)^{2t} \\ \log 1.6 &= 2t \log (1.0425) \\ \frac{\log 1.6}{2 \log 1.0425} &= t \\ 5.6 \text{ years} &\approx t \end{aligned}$$

80.

$$\begin{aligned} A &= Pe^{rt} \\ 800 &= 500e^{0.085t} \\ 1.6 &= e^{0.085t} \\ \ln 1.6 &= \ln e^{0.085t} \\ \ln 1.6 &= 0.085t \\ \frac{\ln 1.6}{0.085} &= t \\ 5.5 \text{ years} &\approx t \end{aligned}$$

81.
$$\begin{aligned} A &= P\left(1+\frac{r}{k}\right)^{kt} \\ 2100 &= 1300\left(1+\frac{0.09}{4}\right)^{4t} \\ \frac{2100}{1300} &= (1.0225)^{4t} \\ \log\frac{2100}{1300} &= \log(1.0225)^{4t} \\ \log\frac{21}{13} &= 4t\log(1.0225) \\ \frac{\log\frac{21}{13}}{4\log 1.0225} &= t \\ 5.4\text{ years} &\approx t \end{aligned}$$

82.
$$\begin{aligned} A &= P\left(1+\frac{r}{k}\right)^{kt} \\ 7000 &= 5000\left(1+\frac{r}{1}\right)^{1(5)} \\ \frac{7000}{5000} &= (1+r)^5 \\ \sqrt[5]{\frac{7000}{5000}} &= 1+r \\ 1.0696 &\approx 1+r \\ 0.0696 &\approx r, \text{ or } r \approx 6.96\% \end{aligned}$$

83. $\text{doubling time} = t = \dfrac{\ln 2}{r} = \dfrac{100\ln 2}{100r} \approx \dfrac{70}{100r} = \dfrac{70}{r\text{, written as a }\%}$

84.
$$\begin{aligned} P &= P_0a^t \\ 3P_0 &= P_0a^5 \\ 3 &= a^5 \\ \sqrt[5]{3} &= a \end{aligned}$$

$$\begin{aligned} P &= P_0a^t \\ 2P_0 &= P_0\left(\sqrt[5]{3}\right)^t \\ 2 &= \left(\sqrt[5]{3}\right)^t \\ \log 2 &= \log\left(\sqrt[5]{3}\right)^t \\ \log 2 &= t\log\sqrt[5]{3} \\ \frac{\log 2}{\log\sqrt[5]{3}} &= t \\ 3.2\text{ days} &\approx t \end{aligned}$$

85.
$$\begin{aligned} P &= P_0e^{kt} \\ 2P_0 &= P_0e^{5k} \\ 2 &= e^{5k} \\ \ln 2 &= \ln e^{5k} \\ \ln 2 &= 5k \\ \frac{\ln 2}{5} &= k \end{aligned}$$

$$\begin{aligned} P &= P_0e^{kt} \\ 1{,}000{,}000 &= 30{,}000e^{\frac{\ln 2}{5}t} \\ 33.333 &\approx e^{\frac{\ln 2}{5}t} \\ \ln 33.333 &\approx \ln e^{\frac{\ln 2}{5}t} \\ \ln 33.333 &\approx \frac{\ln 2}{5}t \\ \frac{5\ln 33.333}{\ln 2} &\approx t \\ 25.3\text{ years} &\approx t \end{aligned}$$

86.
$$\begin{aligned} P &= P_0e^{kt} \\ 3P_0 &= P_0e^{15k} \\ 3 &= e^{15k} \\ \ln 3 &= \ln e^{15k} \\ \ln 3 &= 15k \\ \frac{\ln 3}{15} &= k \end{aligned}$$

$$\begin{aligned} P &= P_0e^{kt} \\ 280 &= 140e^{\frac{\ln 3}{15}t} \\ 2 &= e^{\frac{\ln 3}{15}t} \\ \ln 2 &= \ln e^{\frac{\ln 3}{15}t} \\ \ln 2 &= \frac{\ln 3}{15}t \\ \frac{15\ln 2}{\ln 3} &= t \\ 9.5\text{ years} &\approx t \end{aligned}$$

87.
$$\begin{aligned} P &= P_0e^{kt} \\ 2P_0 &= P_0e^{24k} \\ 2 &= e^{24k} \\ \ln 2 &= \ln e^{24k} \\ \ln 2 &= 24k \\ \frac{\ln 2}{24} &= k \end{aligned}$$

$$\begin{aligned} P &= P_0e^{kt} \\ P &= P_0e^{\frac{\ln 2}{24}(36)} \\ P &= P_0e^{\frac{3\ln 2}{2}} \\ P &= P_0(2.828) \end{aligned}$$

It will be about 2.828 times larger.

88.

$$\begin{aligned} I &= I_0 e^{kx} \\ 0.70P_0 &= P_0 e^{6k} \\ 0.7 &= e^{6k} \\ \ln 0.7 &= \ln e^{6k} \\ \ln 0.7 &= 6k \\ \frac{\ln 0.7}{6} &= k \end{aligned}$$

$$\begin{aligned} P &= P_0 e^{kt} \\ 0.20P_0 &= P_0 e^{\frac{\ln 0.7}{6}x} \\ 0.2 &= e^{\frac{\ln 0.7}{6}x} \\ \ln 0.2 &= \ln e^{\frac{\ln 0.7}{6}x} \\ \ln 0.2 &= \frac{\ln 0.7}{6}x \\ \frac{6 \ln 0.2}{\ln 0.7} &= x \\ 27 \text{ meters} &\approx x \end{aligned}$$

89. $n = \frac{1}{\log 2}\left(\log \frac{B}{b}\right) = \frac{1}{\log 2} \log \frac{5 \times 10^6}{500} = \frac{1}{\log 2} \log 10{,}000 \approx 13.3$ generations

90. $n = \frac{1}{\log 2}\left(\log \frac{B}{b}\right) = \frac{1}{\log 2} \log \frac{6 \times 10^7}{800} = \frac{1}{\log 2} \log 75{,}000 \approx 16.2$ generations

91. **Answers may vary.**

92. **Answers may vary.**

93. Since the logarithm of a negative number is not defined (as a real number), the values $x - 3$ and $x^2 + 2$ must be nonnegative. Since $x^2 + 2$ is always greater than 0, the only restriction is that $x - 3 > 0$, or $x > 3$. Thus, x cannot be a solution if $x \leq 3$.

94.

$$\begin{aligned} x^{\log x} &= 10{,}000 \\ \log\left(x^{\log x}\right) &= \log 10{,}000 \\ (\log x)(\log x) &= 4 \\ (\log x)^2 &= 4 \\ \log x &= \pm 2 \end{aligned}$$

$$x = 10^2 = 100, \text{ or } x = 10^{-2} = \frac{1}{100}$$

Chapter 9 Summary (page 648)

1. $5^{\sqrt{2}} \cdot 5^{\sqrt{2}} = 5^{\sqrt{2}+\sqrt{2}} = 5^{2\sqrt{2}}$

2. $\left(2^{\sqrt{5}}\right)^{\sqrt{2}} = 2^{(\sqrt{5})(\sqrt{2})} = 2^{\sqrt{10}}$

3. $y = 3^x$; through $(0, 1)$ and $(1, 3)$

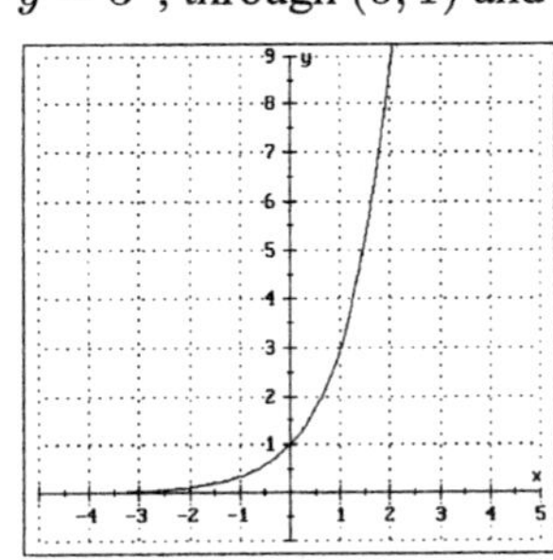

4. 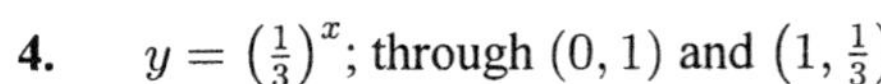

$y = \left(\frac{1}{3}\right)^x$; through $(0, 1)$ and $\left(1, \frac{1}{3}\right)$

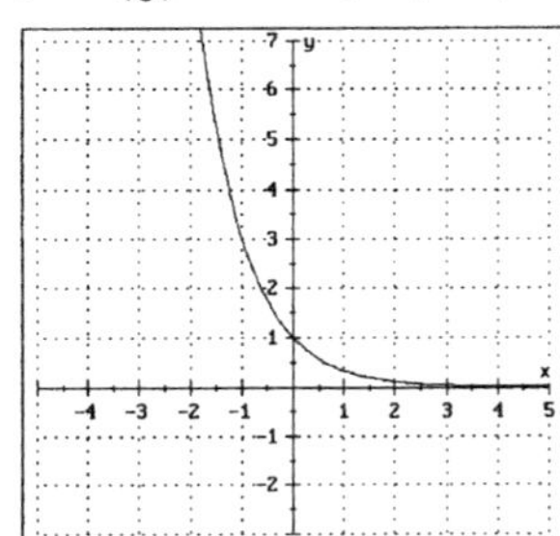

5. The graph will go through $(0,1)$ and $(1,6)$, so $x=1$ and $y=6$.

6. domain $=(-\infty,\infty)$; range $=(0,\infty)$

7. $y=f(x)=\left(\frac{1}{2}\right)^x-2$
Shift $y=\left(\frac{1}{2}\right)^x$ down 2.

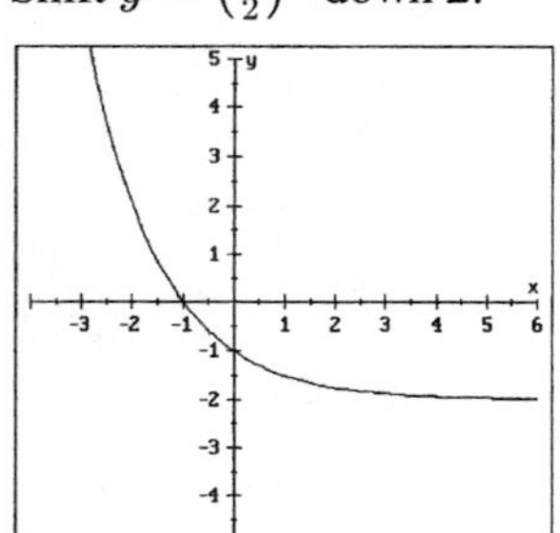

8. $y=f(x)=\left(\frac{1}{2}\right)^{x+2}$
Shift $y=\left(\frac{1}{2}\right)^x$ left 2.

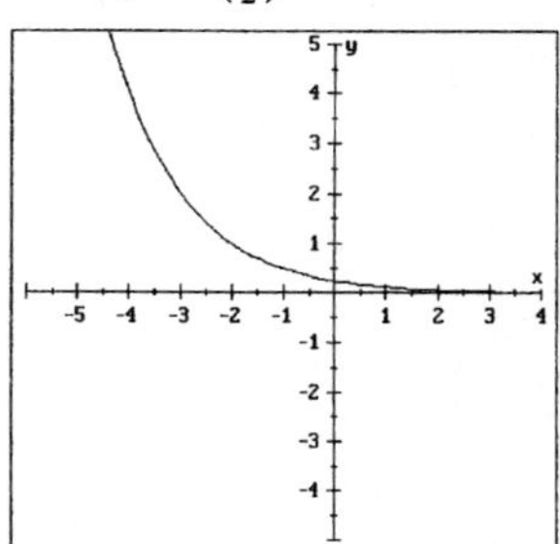

9. $A=P\left(1+\dfrac{r}{k}\right)^{kt}=10500\left(1+\dfrac{0.09}{4}\right)^{4\cdot 60}=10500(1.0225)^{240}\approx \$2{,}189{,}703.45$

10. $A=Pe^{rt}=10500e^{0.09(60)}=10500e^{5.4}\approx \$2{,}324{,}767.37$

11. $y=f(x)=e^x+1$; Shift $y=e^x$ up 1.

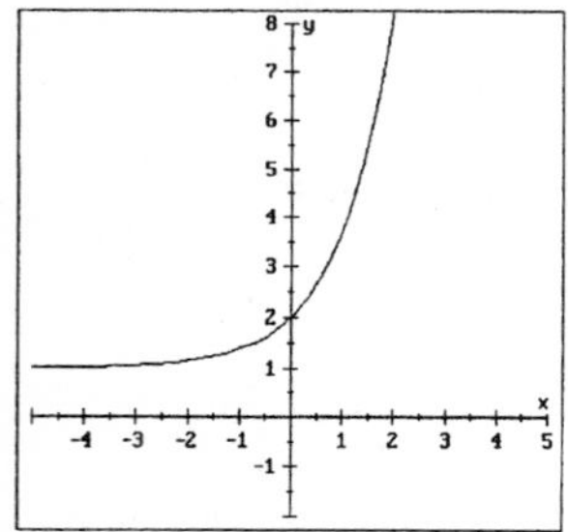

12. $y=f(x)=e^{x-3}$; Shift $y=e^x$ right 3.

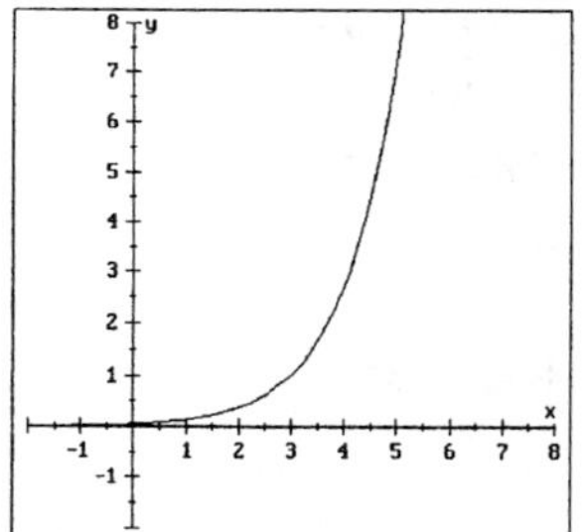

13. $P=P_0e^{kt}=275{,}000{,}000e^{0.015(50)}$
$\approx 582{,}000{,}000$

14. $A=A_0e^{-0.0244t}=50e^{-0.0244(20)}=50e^{-0.488}$
≈ 30.69 g

15. domain $=(0,\infty)$; range $=(-\infty,\infty)$

16. **Answers will vary.**

17. $\log_3 9=2$

18. $\log_9 \dfrac{1}{3}=-\dfrac{1}{2}$

19. $\log_\pi 1=0$

20. $\log_5 0.04=\log_5 \dfrac{1}{25}=-2$

21. $\log_a \sqrt{a}=\log_a a^{1/2}=\dfrac{1}{2}$

22. $\log_a \sqrt[3]{a}=\log_a a^{1/3}=\dfrac{1}{3}$

23. $\log_2 x=5 \Rightarrow 2^5=x \Rightarrow x=32$

24. $\log_{\sqrt{3}} x=4 \Rightarrow \left(\sqrt{3}\right)^4=x \Rightarrow x=9$

25. $\log_{\sqrt{3}} x = 6 \Rightarrow \left(\sqrt{3}\right)^6 = x \Rightarrow x = 27$

26. $\log_{0.1} 10 = x \Rightarrow (0.1)^x = 10$
$\left(\frac{1}{10}\right)^x = 10 \Rightarrow x = -1$

27. $\log_x 2 = -\frac{1}{3} \Rightarrow x^{-1/3} = 2$
$\left(x^{-1/3}\right)^{-3} = 2^{-3} \Rightarrow x = \frac{1}{8}$

28. $\log_x 32 = 5 \Rightarrow x^5 = 32 \Rightarrow x = 2$

29. $\log_{0.25} x = -1 \Rightarrow (0.25)^{-1} = x$
$\left(\frac{1}{4}\right)^{-1} = x \Rightarrow x = 4$

30. $\log_{0.125} x = -\frac{1}{3} \Rightarrow (0.125)^{-1/3} = x$
$\left(\frac{1}{8}\right)^{-1/3} = x \Rightarrow x = 2$

31. $\log_{\sqrt{2}} 32 = x \Rightarrow \left(\sqrt{2}\right)^x = 32$
$\left(2^{1/2}\right)^x = 2^5 \Rightarrow \frac{1}{2}x = 5 \Rightarrow x = 10$

32. $\log_{\sqrt{5}} x = -4 \Rightarrow \left(\sqrt{5}\right)^{-4} = x$
$\left(5^{1/2}\right)^{-4} = x \Rightarrow 5^{-2} = x \Rightarrow x = \frac{1}{25}$

33. $\log_{\sqrt{3}} 9\sqrt{3} = x \Rightarrow \left(\sqrt{3}\right)^x = 9\sqrt{3}$
$\left(3^{1/2}\right)^x = 3^{5/2} \Rightarrow x = 5$

34. 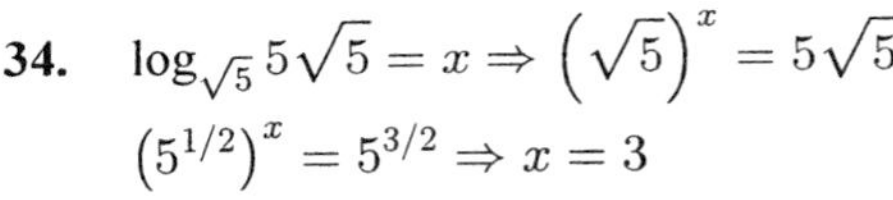
$\log_{\sqrt{5}} 5\sqrt{5} = x \Rightarrow \left(\sqrt{5}\right)^x = 5\sqrt{5}$
$\left(5^{1/2}\right)^x = 5^{3/2} \Rightarrow x = 3$

35. $y = f(x) = \log(x - 2)$
Shift $y = \log x$ right 2.

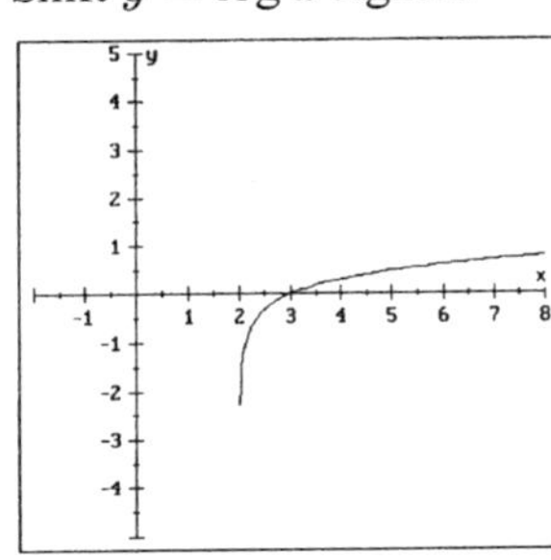

36. $y = f(x) = 3 + \log x$
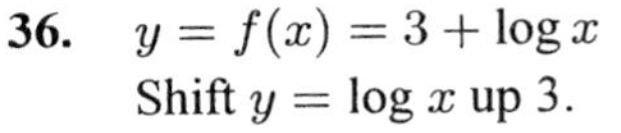
Shift $y = \log x$ up 3.

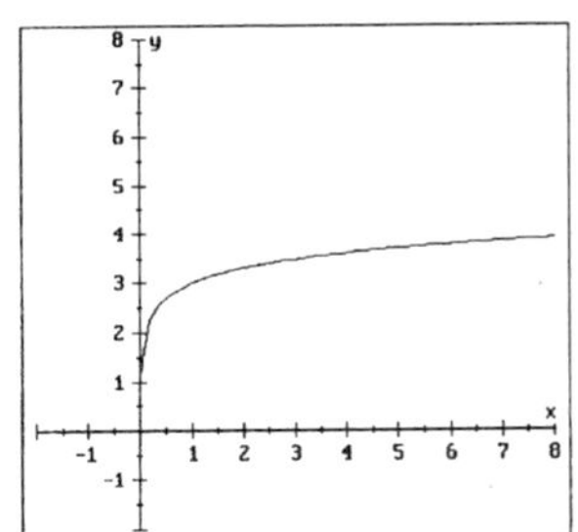

37. $y = 4^x$
$y = \log_4 x$

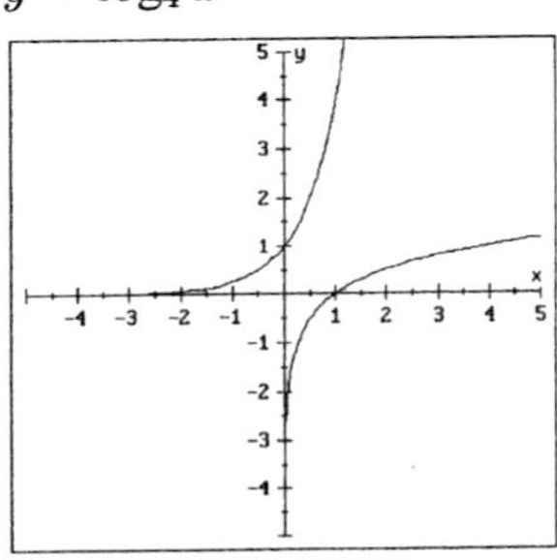

38. 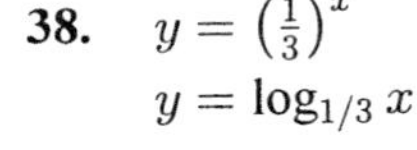
$y = \left(\frac{1}{3}\right)^x$
$y = \log_{1/3} x$

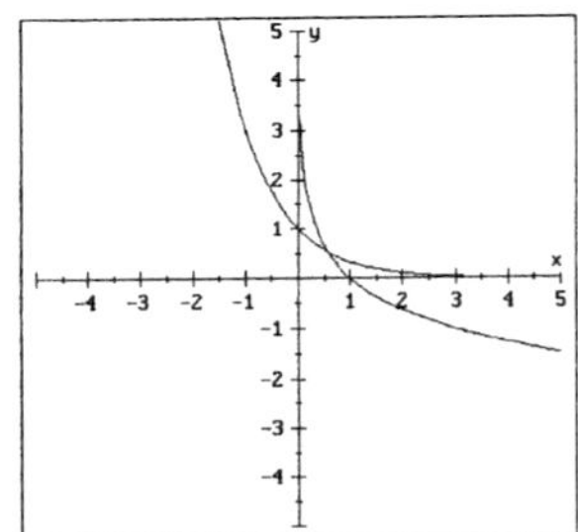

39. dB gain $= 20 \log \frac{E_O}{E_I} = 20 \log \frac{18}{0.04} = 20 \log 450 \approx 53$ dB

40. $R = \log \frac{A}{P} = \log \frac{7500}{0.3} = \log 25{,}000 \approx 4.4$

41. $\ln 452 \approx 6.1137$

42. $\ln(\log 7.85) \approx \ln 0.8949 \approx -0.1111$

43. $\ln x = 2.336 \Rightarrow x = 10.3398$

44. $\ln x = \log 8.8 \Rightarrow x = 2.5715$

45. $y = f(x) = 1 + \ln x$
Shift $y = \ln x$ up 1.

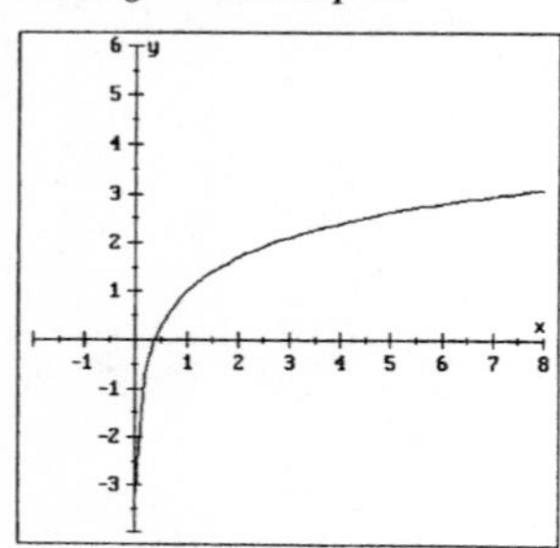

46. $y = f(x) = \ln(x+1)$
Shift $y = \ln x$ left 1.

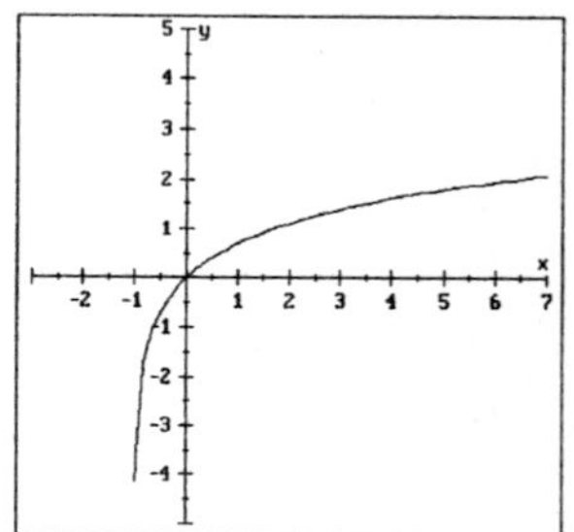

47. $t = \dfrac{\ln 2}{r} = \dfrac{\ln 2}{0.03} \approx 23$ years

48. $\log_7 1 = 0$

49. $\log_7 7 = 1$

50. $\log_7 7^3 = 3$

51. $7^{\log_7 4} = 4$

52. $\ln e^4 = 4$

53. $\ln 1 = 0$

54. $10^{\log_{10} 7} = 7$

55. $e^{\ln 3} = 3$

56. $\log_b b^4 = 4$

57. $\ln e^9 = 9$

58. $\log_b \dfrac{x^2y^3}{z^4} = \log_b x^2 + \log_b y^3 - \log_b z^4 = 2\log_b x + 3\log_b y - 4\log_b z$

59. $$\log_b \sqrt{\frac{x}{yz^2}} = \log_b \left(\frac{x}{yz^2}\right)^{1/2} = \frac{1}{2}\log_b \frac{x}{yz^2} = \frac{1}{2}(\log_b x - \log_b y - \log_b z^2)$$
$$= \frac{1}{2}(\log_b x - \log_b y - 2\log_b z)$$

60. $3\log_b x - 5\log_b y + 7\log_b z = \log_b x^3 - \log_b y^5 + \log_b z^7 = \log_b \dfrac{x^3z^7}{y^5}$

61. $\dfrac{1}{2}\log_b x + 3\log_b y - 7\log_b z = \log_b x^{1/2} + \log_b y^3 - \log_b z^7 = \log_b \dfrac{y^3\sqrt{x}}{z^7}$

62. $\log abc = \log a + \log b + \log c = 0.6 + 0.36 + 2.4 = 3.36$

63. $\log a^2b = \log a^2 + \log b = 2\log a + \log b = 2(0.6) + 0.36 = 1.56$

64. $\log \dfrac{ac}{b} = \log a + \log c - \log b = 0.6 + 2.4 - 0.36 = 2.64$

65. $\log \dfrac{a^2}{c^3b^2} = \log a^2 - \log c^3 - \log b^2 = 2\log a - 3\log c - 2\log b$
$= 2(0.6) - 3(2.4) - 2(0.36) = -6.72$

66. $\log_5 17 = \dfrac{\log 17}{\log 5} \approx 1.7604$

67.
$$\begin{aligned} \text{pH} &= -\log\,[\text{H}^+] \\ 3.1 &= -\log\,[\text{H}^+] \\ -3.1 &= \log\,[\text{H}^+] \\ [\text{H}^+] &= 7.94 \times 10^{-4} \text{ gram-ions per liter} \end{aligned}$$

68.
$$\begin{aligned} k\ln\left(\tfrac{1}{2}I\right) &= k\left(\ln\tfrac{1}{2} + \ln I\right) \\ &= k\ln\tfrac{1}{2} + k\ln I \\ &= k\ln 2^{-1} + L \\ &= -k\ln 2 + L \end{aligned}$$
The loudness decreases by $k \ln 2$.

69.
$$\begin{aligned} 3^x &= 7 \\ \log 3^x &= \log 7 \\ x\log 3 &= \log 7 \\ x &= \frac{\log 7}{\log 3} \approx 1.7712 \end{aligned}$$

70.
$$\begin{aligned} 5^{x+2} &= 625 \\ 5^{x+2} &= 5^4 \\ x+2 &= 4 \\ x &= 2 \end{aligned}$$

71.
$$\begin{aligned} 25 &= 5.5(1.05)^t \\ \frac{25}{5.5} &= (1.05)^t \\ \log\frac{25}{5.5} &= \log\,(1.05)^t \\ \log\frac{25}{5.5} &= t\log 1.05 \\ \frac{\log\frac{25}{5.5}}{\log 1.05} &= t \\ 31.0335 &\approx t \end{aligned}$$

72.
$$\begin{aligned} 4^{2t-1} &= 64 \\ 4^{2t-1} &= 4^3 \\ 2t-1 &= 3 \\ 2t &= 4 \\ t &= 2 \end{aligned}$$

73.
$$\begin{aligned} 2^x &= 3^{x-1} \\ \log 2^x &= \log 3^{x-1} \\ x\log 2 &= (x-1)\log 3 \\ x\log 2 &= x\log 3 - \log 3 \\ \log 3 &= x\log 3 - x\log 2 \\ \log 3 &= x(\log 3 - \log 2) \\ \frac{\log 3}{\log 3 - \log 2} &= x \\ 2.7095 &\approx x \end{aligned}$$

74.
$$\begin{aligned} 2^{x^2+4x} &= \frac{1}{8} \\ 2^{x^2+4x} &= 2^{-3} \\ x^2+4x &= -3 \\ x^2+4x+3 &= 0 \\ (x+3)(x+1) &= 0 \\ x+3 = 0 \quad &\textbf{or} \quad x+1 = 0 \\ x = -3 \quad & \qquad\quad x = -1 \end{aligned}$$

75. $\log x + \log(29 - x) = 2$
$$\log x(29-x) = 2$$
$$10^2 = x(29-x)$$
$$100 = 29x - x^2$$
$$x^2 - 29x + 100 = 0$$
$$(x-25)(x-4) = 0$$
$$x - 25 = 0 \quad \textbf{or} \quad x - 4 = 0$$
$$x = 25 \qquad\qquad x = 4$$

76. $\log_2 x + \log_2(x-2) = 3$
$$\log_2 x(x-2) = 3$$
$$x(x-2) = 2^3$$
$$x^2 - 2x = 8$$
$$x^2 - 2x - 8 = 0$$
$$(x-4)(x+2) = 0$$
$$x - 4 = 0 \quad \textbf{or} \quad x + 2 = 0$$
$$x = 4 \qquad\qquad x = -2$$
Does not check.

77. $\log_2(x+2) + \log_2(x-1) = 2$
$$\log_2(x+2)(x-1) = 2$$
$$(x+2)(x-1) = 2^2$$
$$x^2 + x - 2 = 4$$
$$x^2 + x - 6 = 0$$
$$(x-2)(x+3) = 0$$
$$x - 2 = 0 \quad \textbf{or} \quad x + 3 = 0$$
$$x = 2 \qquad\qquad x = -3$$
Does not check.

78. $\dfrac{\log(7x-12)}{\log x} = 2$
$$\log(7x-12) = 2\log x$$
$$\log(7x-12) = \log x^2$$
$$7x - 12 = x^2$$
$$0 = x^2 - 7x + 12$$
$$0 = (x-4)(x-3)$$
$$x - 4 = 0 \quad \textbf{or} \quad x - 3 = 0$$
$$x = 4 \qquad\qquad x = 3$$

79. $\log x + \log(x-5) = \log 6$
$$\log x(x-5) = \log 6$$
$$x(x-5) = 6$$
$$x^2 - 5x - 6 = 0$$
$$(x-6)(x+1) = 0$$
$$x - 6 = 0 \quad \textbf{or} \quad x + 1 = 0$$
$$x = 6 \qquad\qquad x = -1$$
Does not check.

80. $\log 3 - \log(x-1) = -1$
$$\log\frac{3}{x-1} = -1$$
$$\frac{3}{x-1} = 10^{-1}$$
$$\frac{3}{x-1} = \frac{1}{10}$$
$$30 = x - 1$$
$$31 = x$$

81. $e^{x\ln 2} = 9$
$$e^{\ln 2^x} = 9$$
$$2^x = 9$$
$$\ln 2^x = \ln 9$$
$$x\ln 2 = \ln 9$$
$$x = \frac{\ln 9}{\ln 2} \approx 3.1699$$

82. $\ln x = \ln(x-1)$
$$x = x - 1$$
$$0 = -1$$
There is no solution.

83.
$$\begin{aligned}
\ln x &= \ln(x-1)+1\\
\ln x - \ln(x-1) &= 1\\
\ln\frac{x}{x-1} &= 1\\
\frac{x}{x-1} &= e^1\\
x &= e(x-1)\\
x &= ex-e\\
e &= ex-x\\
e &= x(e-1)\\
\frac{e}{e-1} &= x\\
1.5820 &\approx x
\end{aligned}$$

84.
$$\begin{aligned}
\ln x &= \log_{10} x\\
\ln x &= \frac{\ln x}{\ln 10}\\
\ln x \ln 10 &= \ln x\\
\ln x \ln 10 - \ln x &= 0\\
\ln x(\ln 10 - 1) &= 0\\
\ln x &= 0\\
x &= 1
\end{aligned}$$

85.
$$\begin{aligned}
A &= A_0 2^{-t/h}\\
\frac{2}{3}A_0 &= A_0 2^{-t/5700}\\
\frac{2}{3} &= 2^{-t/5700}\\
\log\frac{2}{3} &= \log 2^{-t/5700}\\
\log\frac{2}{3} &= -\frac{t}{5700}\log 2\\
\frac{\log\frac{2}{3}}{-\log 2} &= \frac{t}{5700}\\
-5700\frac{\log\frac{2}{3}}{\log 2} &= t\\
3300 \text{ years} &\approx t
\end{aligned}$$

Chapter 9 Test (page 653)

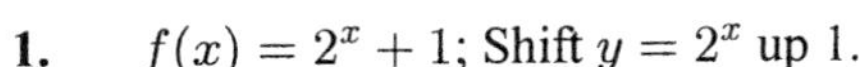

1. $f(x) = 2^x + 1$; Shift $y = 2^x$ up 1.

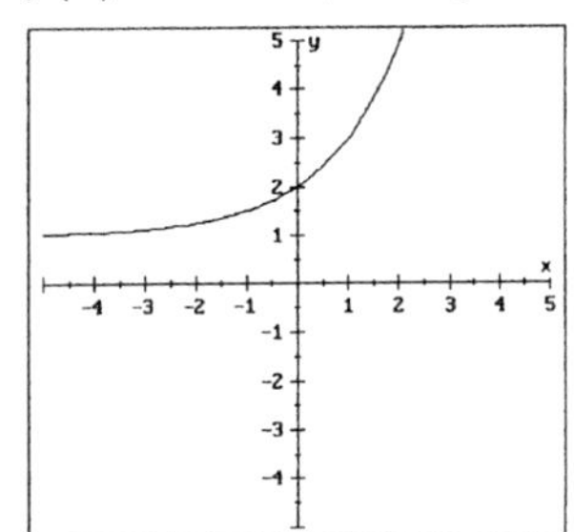

2. $f(x) = 2^{-x}$; Reflect $y = 2^x$ about y-axis.

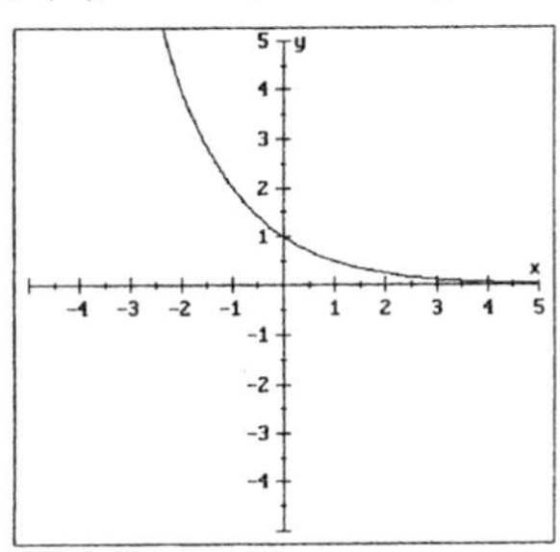

3. $A = A_0(2)^{-t} = 3(2)^{-6} = \frac{3}{2^6} = \frac{3}{64}$ gram

CHAPTER 9 TEST

4. $A = A_0\left(1+\frac{r}{k}\right)^{kt} = 1000\left(1+\frac{0.06}{2}\right)^{2(1)} = 1000(1.03)^2 \approx \$1060.90$

5. $f(x) = e^x$

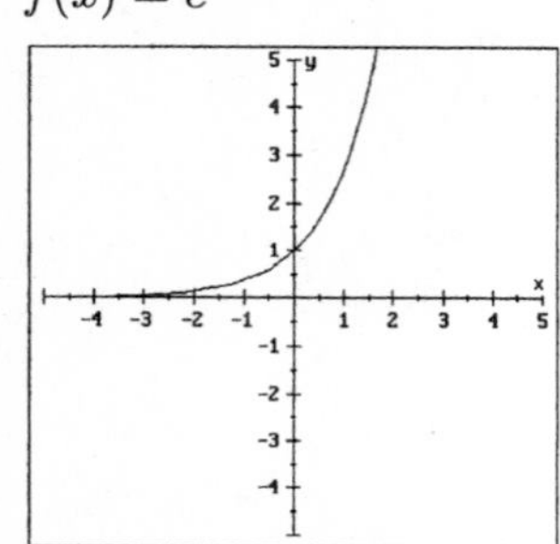

6. $A = A_0e^{rt} = 2000e^{(0.08)10}$
$= 2000e^{0.8}$
$\approx \$4451.08$

7. $\log_4 16 = x \Rightarrow 4^x = 16 \Rightarrow x = 2$

8. $\log_x 81 = 4 \Rightarrow x^4 = 81 \Rightarrow x = 3$

9. $\log_3 x = -3 \Rightarrow 3^{-3} = x \Rightarrow x = \frac{1}{27}$

10. $\log_x 100 = 2 \Rightarrow x^2 = 100 \Rightarrow x = 10$

11. $\log_{3/2}\frac{9}{4} = x \Rightarrow \left(\frac{3}{2}\right)^x = \frac{9}{4} \Rightarrow x = 2$

12. $\log_{2/3} x = -3 \Rightarrow \left(\frac{2}{3}\right)^{-3} = x \Rightarrow x = \frac{27}{8}$

13. $f(x) = -\log_3 x$

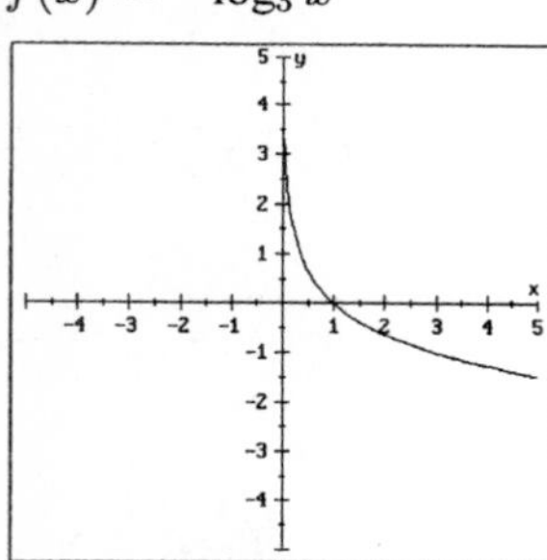

14. $f(x) = \ln x$

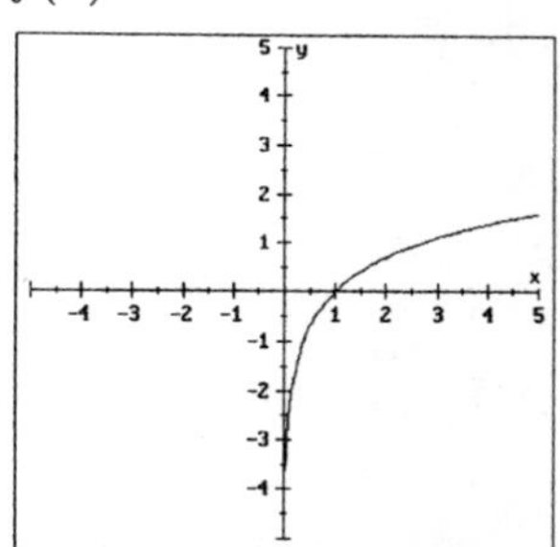

15. $\log a^2bc^3 = \log a^2 + \log b + \log c^3 = 2\log a + \log b + 3\log c$

16. $\ln\sqrt{\frac{a}{b^2c}} = \ln\left(\frac{a}{b^2c}\right)^{1/2} = \frac{1}{2}\ln\frac{a}{b^2c} = \frac{1}{2}(\ln a - \ln b^2 - \ln c) = \frac{1}{2}\ln a - \ln b - \frac{1}{2}\ln c$

17. $\frac{1}{2}\log(a+2) + \log b - 3\log c = \log(a+2)^{1/2} + \log b - \log c^3 = \log\frac{b\sqrt{a+2}}{c^3}$

18. $\frac{1}{3}(\log a - 2\log b) - \log c = \frac{1}{3}(\log a - \log b^2) - \log c = \frac{1}{3}\log\frac{a}{b^2} - \log c = \log\sqrt[3]{\frac{a}{b^2}} - \log c$
$= \log\frac{\sqrt[3]{a}}{c\sqrt[3]{b^2}}$

19. $\log 24 = \log 8 \cdot 3 = \log 2^3 \cdot 3 = \log 2^3 + \log 3 = 3 \log 2 + \log 3 = 3(0.3010) + 0.4771 = 1.3801$

20. $\log \frac{8}{3} = \log \frac{2^3}{3} = \log 2^3 - \log 3 = 3 \log 2 - \log 3 = 3(0.3010) - 0.4771 = 0.4259$

21. $\log_7 3 = \frac{\log 3}{\log 7} \text{or} \frac{\ln 3}{\ln 7}$

22. $\log_\pi e = \frac{\log e}{\log \pi} \text{or} \frac{\ln e}{\ln \pi}$

23. $\log_a ab = \log_a a + \log_a b = 1 + \log_a b \Rightarrow \text{TRUE}$

24. $\frac{\log a}{\log b} = \log a - \log b \Rightarrow \text{FALSE} \ \left(\log \frac{a}{b} = \log a - \log b\right)$

25. $\log a^{-3} = -3 \log a \neq \frac{1}{3 \log a} \Rightarrow \text{FALSE}$

26. $\ln(-x) = -\ln x \Rightarrow$ FALSE (This implies one of the logarithms is negative, which is impossible.)

27. $\text{pH} = -\log [\text{H}^+] = -\log(3.7 \times 10^{-7}) \approx 6.4$

28. $\text{dB gain} = 20 \log \frac{E_O}{E_I} = 20 \log \frac{60}{0.3} = 20 \log 200 \approx 46$

29.
$$\begin{aligned} 5^x &= 3 \\ \log 5^x &= \log 3 \\ x \log 5 &= \log 3 \\ x &= \frac{\log 3}{\log 5} \end{aligned}$$

30.
$$\begin{aligned} 3^{x-1} &= 100^x \\ \log 3^{x-1} &= \log 100^x \\ (x-1)\log 3 &= x \log 100 \\ x \log 3 - \log 3 &= 2x \\ x \log 3 - 2x &= \log 3 \\ x(\log 3 - 2) &= \log 3 \\ x &= \frac{\log 3}{\log 3 - 2} \end{aligned}$$

31.
$$\begin{aligned} \log(5x+2) &= \log(2x+5) \\ 5x+2 &= 2x+5 \\ 3x &= 3 \\ x &= 1 \end{aligned}$$

32.
$$\begin{aligned} \log x + \log(x-9) &= 1 \\ \log x(x-9) &= 1 \\ x(x-9) &= 10 \\ x^2 - 9x - 10 &= 0 \\ (x-10)(x+1) &= 0 \end{aligned}$$

$x - 10 = 0$ **or** $x + 1 = 0$

$x = 10$ $\quad$ $x = -1$

Does not check.

Exercise 10.1 (page 666)

1.
$$|3x - 4| = 11$$
$$3x - 4 = 11 \quad \textbf{or} \quad 3x - 4 = -11$$
$$3x = 15 \qquad 3x = -7$$
$$x = 5 \qquad x = -\tfrac{7}{3}$$

2.
$$\left|\frac{4 - 3x}{5}\right| = 12$$
$$\tfrac{4-3x}{5} = 12 \quad \textbf{or} \quad \tfrac{4-3x}{5} = -12$$
$$4 - 3x = 60 \qquad 4 - 3x = -60$$
$$-3x = 56 \qquad -3x = -64$$
$$x = -\tfrac{56}{3} \qquad x = \tfrac{64}{3}$$

3.
$$|3x + 4| = |5x - 2|$$
$$3x + 4 = 5x - 2 \quad \textbf{or} \quad 3x + 4 = -(5x - 2)$$
$$-2x = -6 \qquad 3x + 4 = -5x + 2$$
$$x = 3 \qquad 8x = -2$$
$$x = -\tfrac{1}{4}$$

4.
$$|6 - 4x| = |x + 2|$$
$$6 - 4x = x + 2 \quad \textbf{or} \quad 6 - 4x = -(x + 2)$$
$$-5x = -4 \qquad 6 - 4x = -x - 2$$
$$x = \tfrac{4}{5} \qquad -3x = -8$$
$$x = \tfrac{8}{3}$$

5. circle; plane

6. radius; center

7. $r^2 < 0$

8. parabola; origin; upward

9. parabola; $(3, 2)$; right

10. parabola; $(-3, 1)$; left

11. $x^2 + y^2 = 9$
C $(0, 0)$; $r = \sqrt{9} = 3$

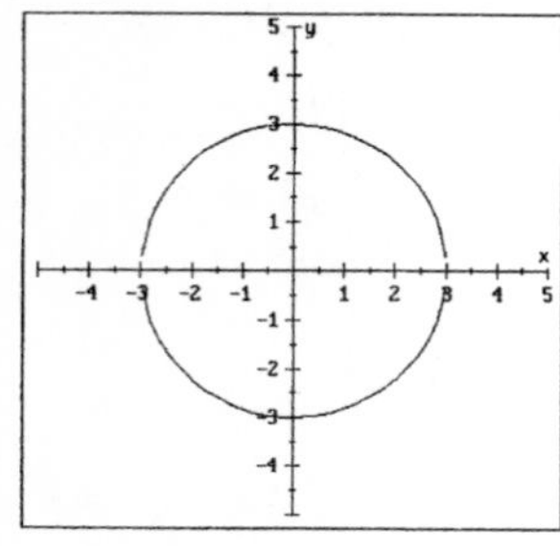

12. $x^2 + y^2 = 16$
C $(0, 0)$; $r = \sqrt{16} = 4$

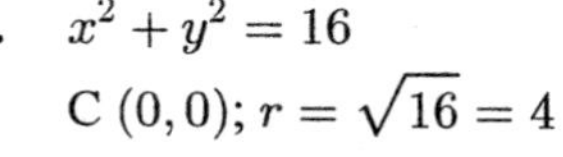

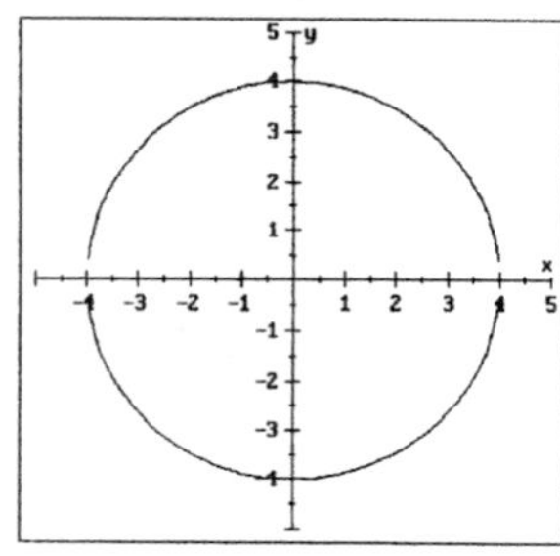

13. $(x - 2)^2 + y^2 = 9$
C $(2, 0)$; $r = \sqrt{9} = 3$

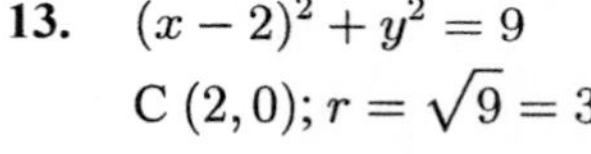

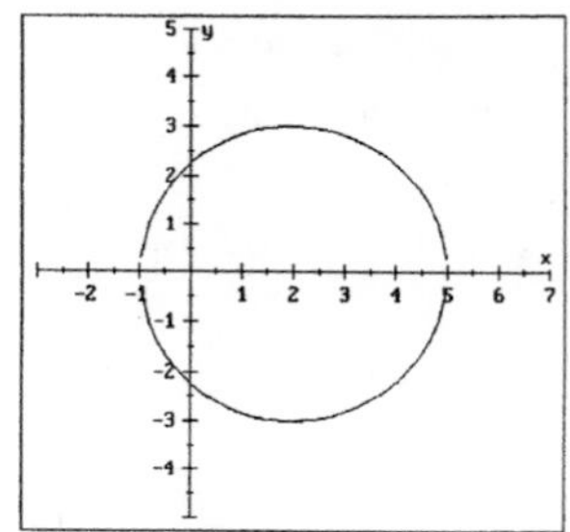

14. $x^2 + (y - 3)^2 = 4$
C $(0, 3)$; $r = \sqrt{4} = 2$

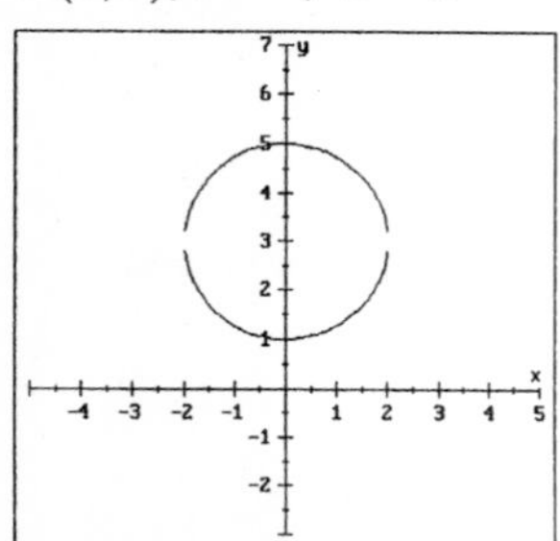

15. $(x - 2)^2 + (y - 4)^2 = 4$
C $(2, 4)$; $r = \sqrt{4} = 2$

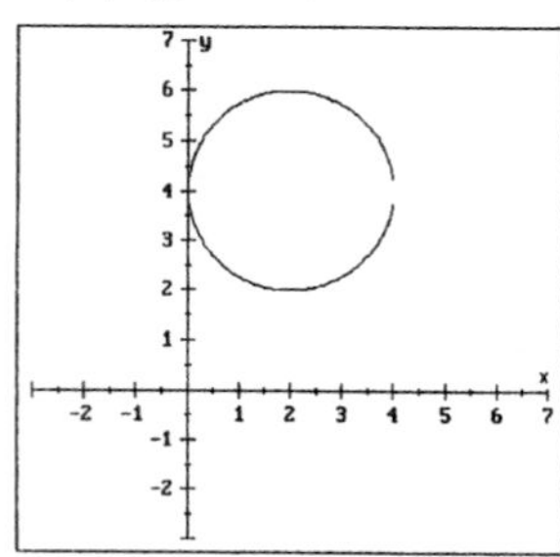

16. $(x - 3)^2 + (y - 2)^2 = 4$
C $(3, 2)$; $r = \sqrt{4} = 2$

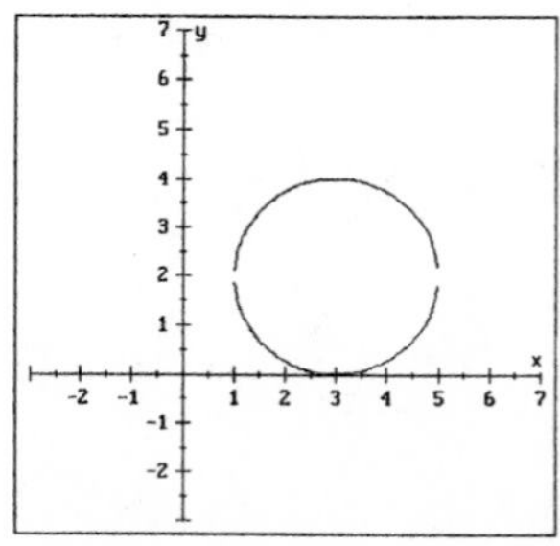

17. $(x+3)^2+(y-1)^2=16$

C $(-3,1)$; $r=\sqrt{16}=4$

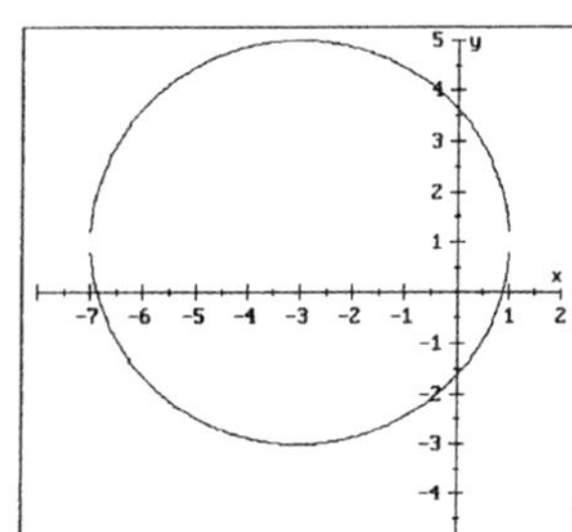

18. $(x-1)^2+(y+4)^2=9$

C $(1,-4)$; $r=\sqrt{9}=3$

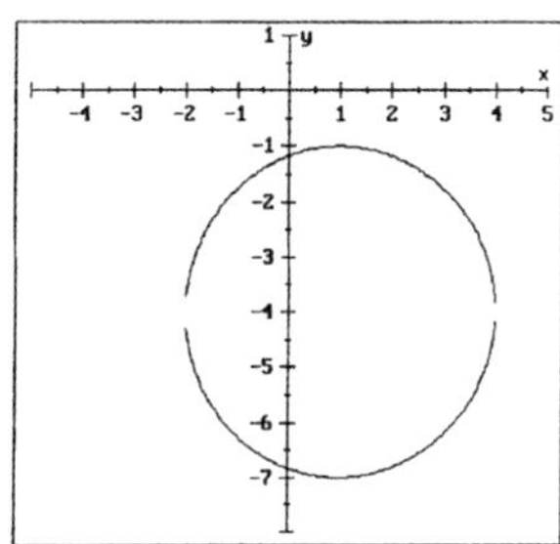

19. $x^2+(y+3)^2=1$

C $(0,-3)$; $r=\sqrt{1}=1$

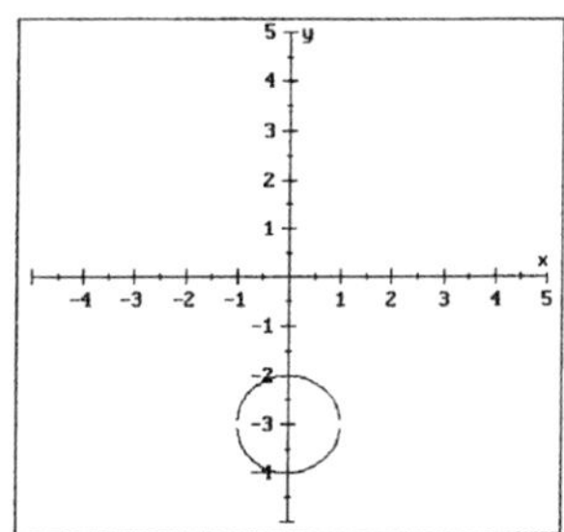

20. $(x+4)^2+y^2=1$

C $(-4,0)$; $r=\sqrt{1}=1$

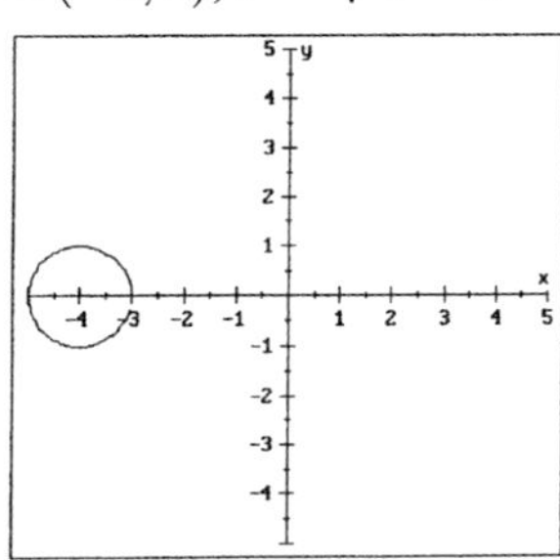

21. $3x^2+3y^2=16$

$$3y^2=16-3x^2$$
$$y^2=\frac{16-3x^2}{3}$$
$$y=\pm\sqrt{\frac{16-3x^2}{3}}$$

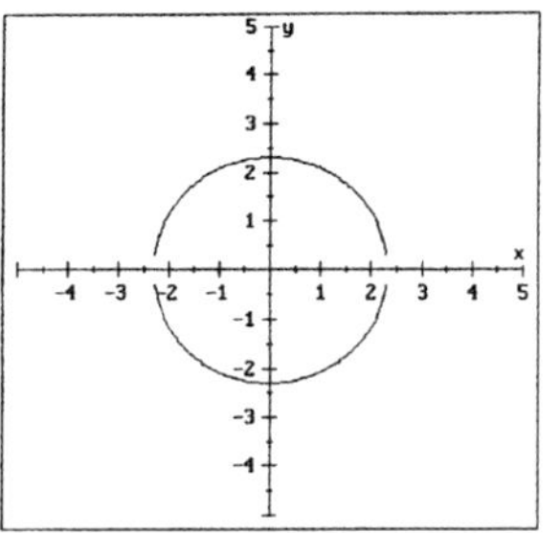

22. $2x^2+2y^2=9$

$$2y^2=9-2x^2$$
$$y^2=\frac{9-2x^2}{2}$$
$$y=\pm\sqrt{\frac{9-2x^2}{2}}$$

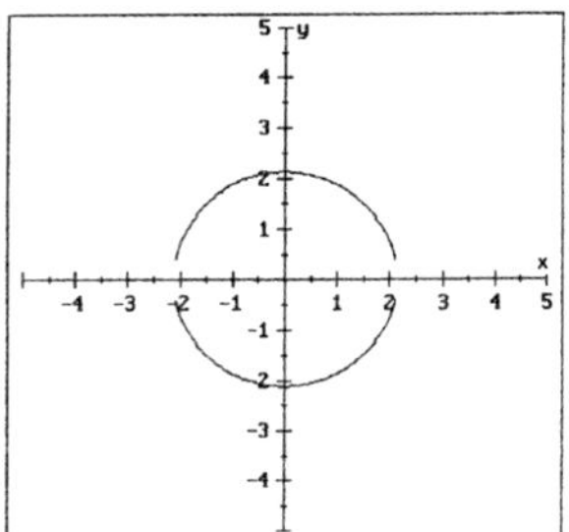

23. $(x+1)^2+y^2=16$

$$y^2=16-(x+1)^2$$
$$y=\pm\sqrt{16-(x+1)^2}$$

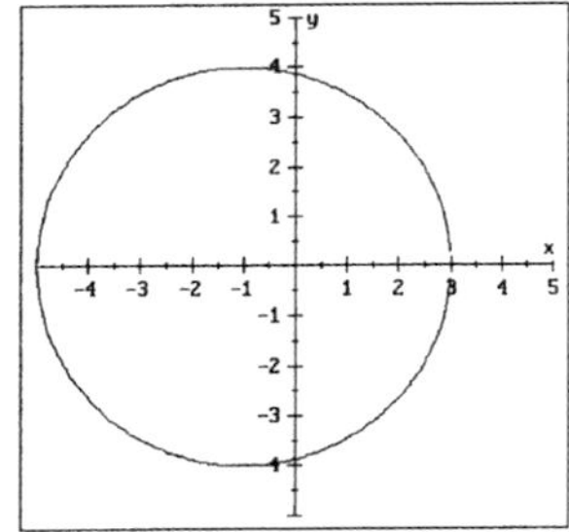

24. $x^2+(y-2)^2=4$

$$(y-2)^2=4-x^2$$
$$y-2=\pm\sqrt{4-x^2}$$
$$y=2\pm\sqrt{4-x^2}$$

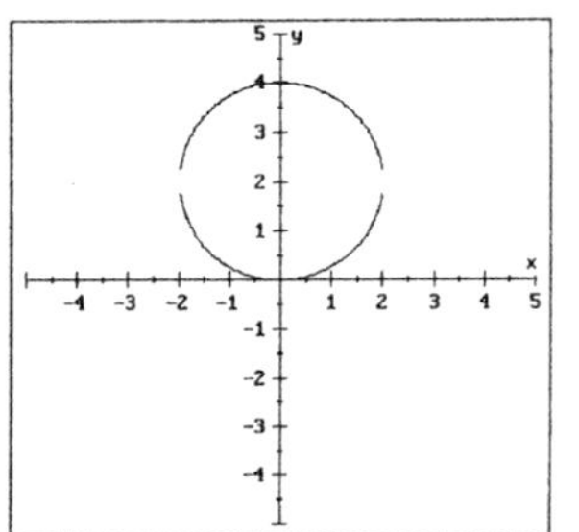

25.
$$(x-h)^2+(y-k)^2=r^2$$
$$(x-0)^2+(y-0)^2=1^2$$
$$x^2+y^2=1$$

26.
$$(x-h)^2+(y-k)^2=r^2$$
$$(x-0)^2+(y-0)^2=4^2$$
$$x^2+y^2=16$$

27.
$$(x-h)^2+(y-k)^2=r^2$$
$$(x-6)^2+(y-8)^2=5^2$$
$$(x-6)^2+(y-8)^2=25$$

28.
$$(x-h)^2+(y-k)^2=r^2$$
$$(x-5)^2+(y-3)^2=2^2$$
$$(x-5)^2+(y-3)^2=4$$

29.
$$(x-h)^2+(y-k)^2=r^2$$
$$(x-(-2))^2+(y-6)^2=12^2$$
$$(x+2)^2+(y-6)^2=144$$

30.
$$(x-h)^2+(y-k)^2=r^2$$
$$(x-5)^2+(y-(-4))^2=6^2$$
$$(x-5)^2+(y+4)^2=36$$

31.
$$(x-h)^2+(y-k)^2=r^2$$
$$(x-0)^2+(y-0)^2=\left(\sqrt{2}\right)^2$$
$$x^2+y^2=2$$

32.
$$(x-h)^2+(y-k)^2=r^2$$
$$(x-0)^2+(y-0)^2=\left(4\sqrt{3}\right)^2$$
$$x^2+y^2=48$$

33.
$$x^2+y^2+2x-8=0$$
$$x^2+2x+y^2=8$$
$$x^2+2x+1+y^2=8+1$$
$$(x+1)^2+y^2=9$$

C $(-1,0)$; $r=\sqrt{9}=3$

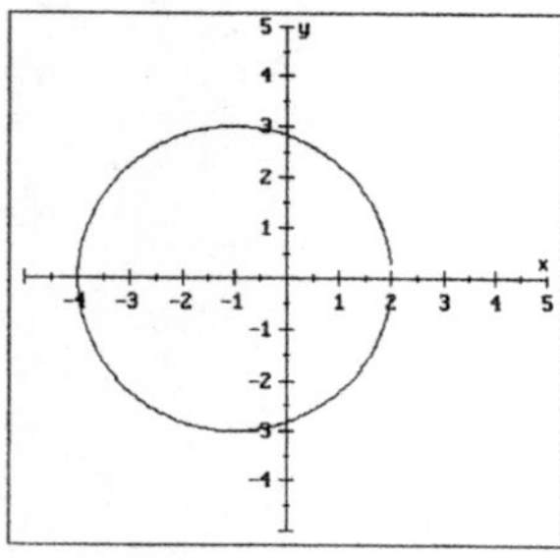

34.
$$x^2+y^2-4y=12$$
$$x^2+y^2-4y+4=12+4$$
$$x^2+(y-2)^2=16$$

C $(0,2)$; $r=\sqrt{16}=4$

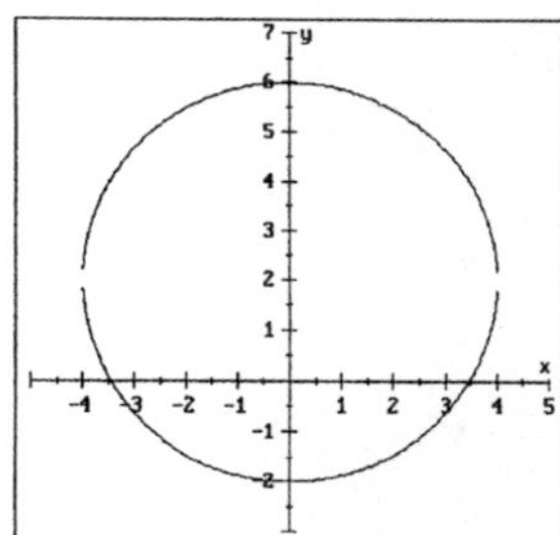

35.
$$9x^2 + 9y^2 - 12y = 5$$
$$x^2 + y^2 - \frac{4}{3}y = \frac{5}{9}$$
$$x^2 + y^2 - \frac{4}{3}y + \frac{4}{9} = \frac{5}{9} + \frac{4}{9}$$
$$x^2 + \left(y - \frac{2}{3}\right)^2 = 1$$
C $\left(0, \frac{2}{3}\right)$; $r = \sqrt{1} = 1$

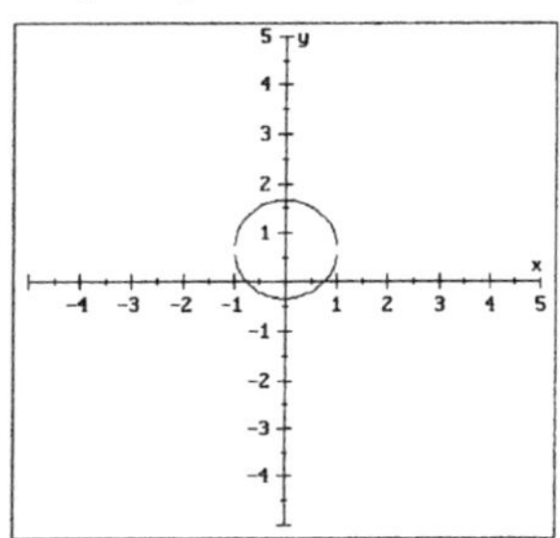

36.
$$4x^2 + 4y^2 + 4y = 15$$
$$x^2 + y^2 + y = \frac{15}{4}$$
$$x^2 + y^2 + y + \frac{1}{4} = \frac{15}{4} + \frac{1}{4}$$
$$x^2 + \left(y + \frac{1}{2}\right)^2 = 4$$
C $\left(0, -\frac{1}{2}\right)$; $r = \sqrt{4} = 2$

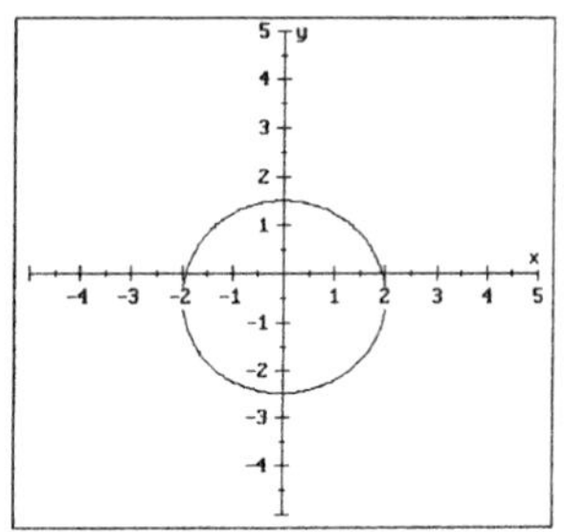

37.
$$x^2 + y^2 - 2x + 4y = -1$$
$$x^2 - 2x + y^2 + 4y = -1$$
$$x^2 - 2x + 1 + y^2 + 4y + 4 = -1 + 1 + 4$$
$$(x-1)^2 + (y+2)^2 = 4$$
C $(1, -2)$; $r = \sqrt{4} = 2$

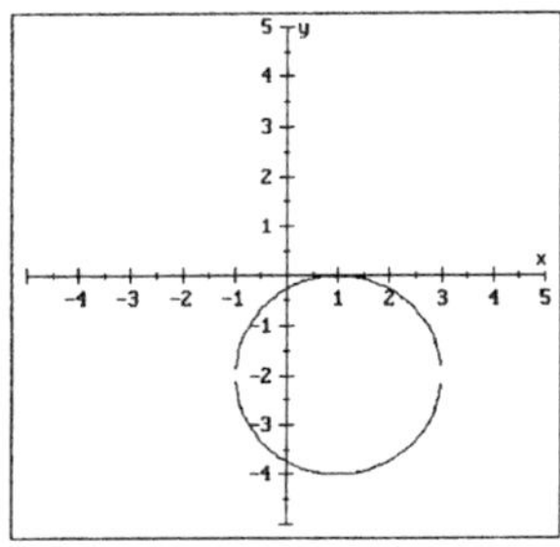

38.
$$x^2 + y^2 + 4x + 2y = 4$$
$$x^2 + 4x + y^2 + 2y = 4$$
$$x^2 + 4x + 4 + y^2 + 2y + 1 = 4 + 4 + 1$$
$$(x+2)^2 + (y+1)^2 = 9$$
C $(-2, -1)$; $r = \sqrt{9} = 3$

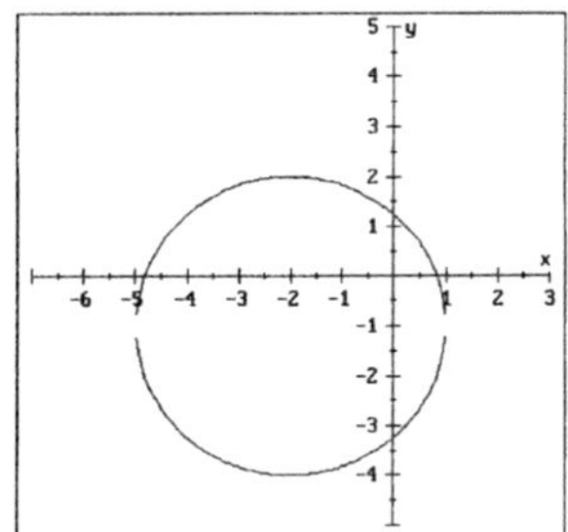

39.
$$x^2 + y^2 + 6x - 4y = -12$$
$$x^2 + 6x + y^2 - 4y = -12$$
$$x^2 + 6x + 9 + y^2 - 4y + 4 = -12 + 9 + 4$$
$$(x+3)^2 + (y-2)^2 = 1$$
C $(-3, 2)$; $r = \sqrt{1} = 1$

40.
$$x^2 + y^2 + 8x + 2y = -13$$
$$x^2 + 8x + y^2 + 2y = -13$$
$$x^2 + 8x + 16 + y^2 + 2y + 1 = -13 + 16 + 1$$
$$(x+4)^2 + (y+1)^2 = 4$$
C $(-4, -1)$; $r = \sqrt{4} = 2$

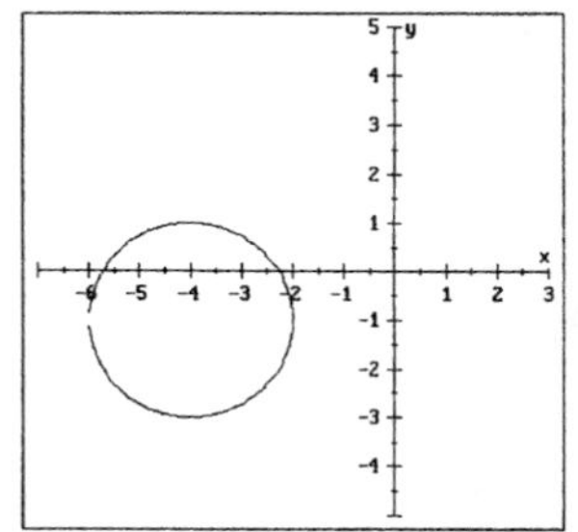

41. $x = y^2$

$x = (y-0)^2 + 0$

V $(0, 0)$; opens R

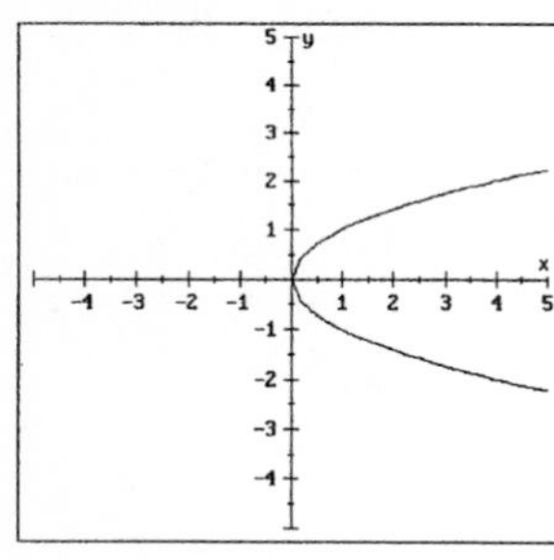

42. $x = -y^2 + 1$

$x = -(y-0)^2 + 1$

V $(1, 0)$; opens L

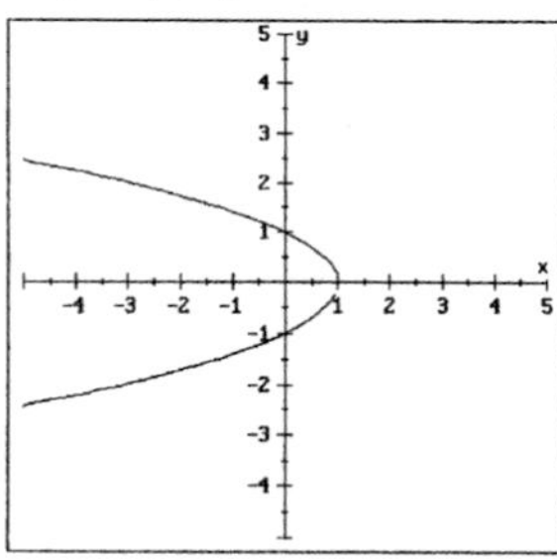

43. $x = -\frac{1}{4}y^2$

$x = -\frac{1}{4}(y-0)^2 + 0$

V $(0, 0)$; opens L

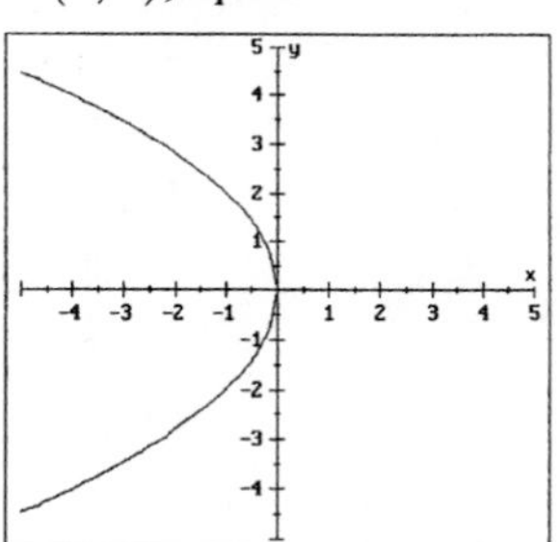

44. $x = 4y^2$

$x = 4(y-0)^2 + 0$

V $(0, 0)$; opens R

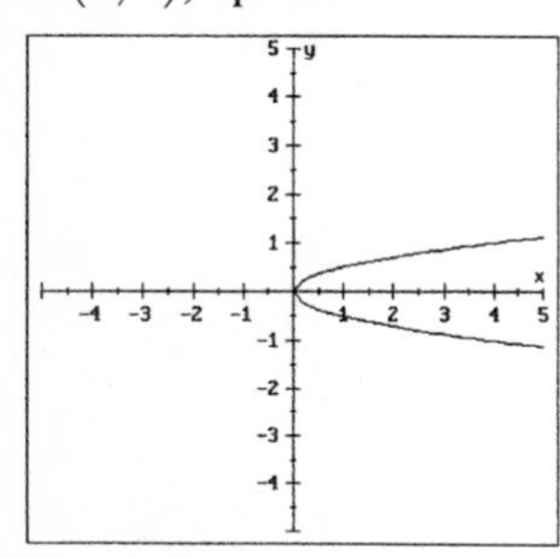

45. $y = x^2 + 4x + 5$

$y = x^2 + 4x + 4 + 5 - 4$

$y = (x+2)^2 + 1$

V $(-2, 1)$; opens U

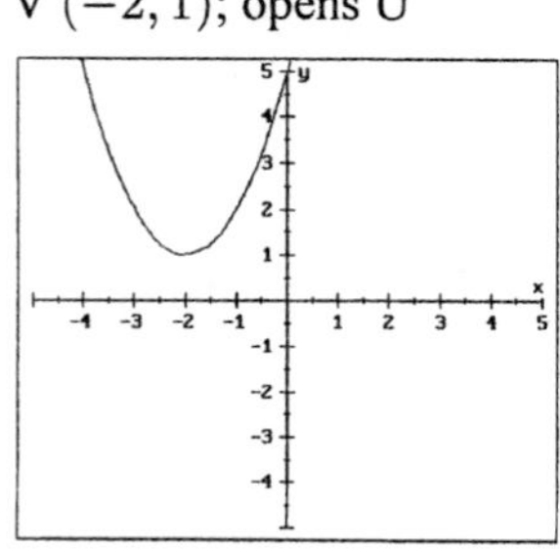

46. $y = -x^2 - 2x + 3$

$y = -(x^2 + 2x) + 3$

$y = -(x^2 + 2x + 1) + 3 + 1$

$y = -(x+1)^2 + 4$

V $(-1, 4)$; opens D

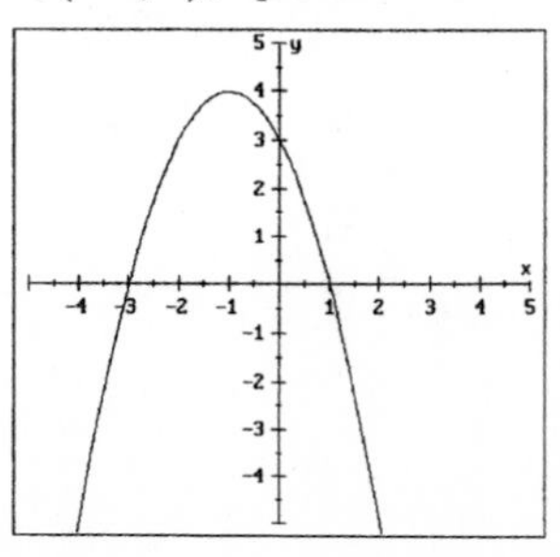

47. $y = -x^2 - x + 1$
$y = -(x^2 + x) + 1$
$y = -\left(x^2 + x + \frac{1}{4}\right) + 1 + \frac{1}{4}$
$y = -\left(x + \frac{1}{2}\right)^2 + \frac{5}{4}$
V $\left(-\frac{1}{2}, \frac{5}{4}\right)$; opens D

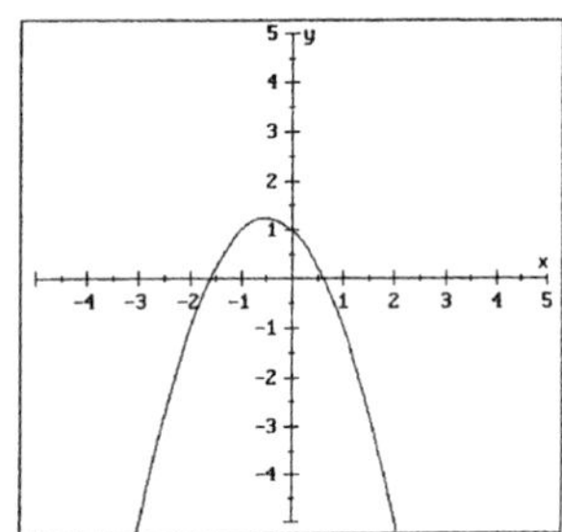

48. $x = \frac{1}{2}y^2 + 2y$
$x = \frac{1}{2}(y^2 + 4y)$
$x = \frac{1}{2}(y^2 + 4y + 4) - 2$
$x = \frac{1}{2}(y + 2)^2 - 2$
V $(-2, -2)$; opens R

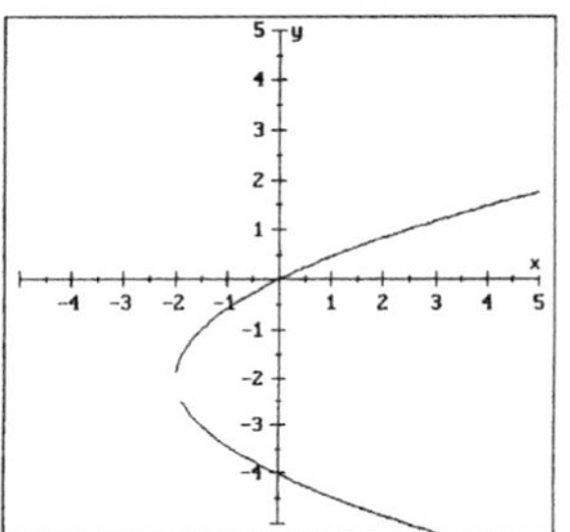

49. $y^2 + 4x - 6y = -1$
$4x = -y^2 + 6y - 1$
$x = -\frac{1}{4}y^2 + \frac{3}{2}y - \frac{1}{4}$
$x = -\frac{1}{4}(y^2 - 6y) - \frac{1}{4}$
$x = -\frac{1}{4}(y^2 - 6y + 9) - \frac{1}{4} + \frac{9}{4}$
$x = -\frac{1}{4}(y - 3)^2 + 2$
V $(2, 3)$; opens L

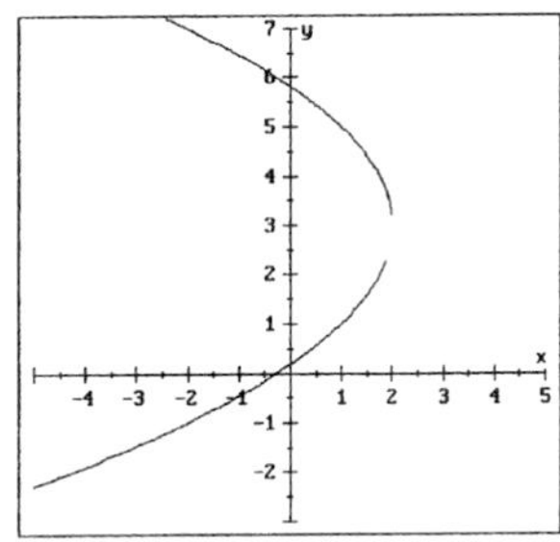

50. $x^2 - 2y - 2x = -7$
$-2y = -x^2 + 2x - 7$
$y = \frac{1}{2}x^2 - x + \frac{7}{2}$
$y = \frac{1}{2}(x^2 - 2x) + \frac{7}{2}$
$y = \frac{1}{2}(x^2 - 2x + 1) + \frac{7}{2} - \frac{1}{2}$
$y = \frac{1}{2}(x - 1)^2 + 3$
V $(1, 3)$; opens U

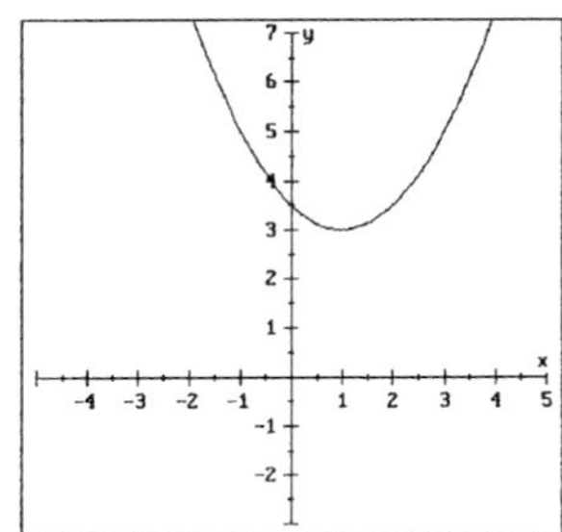

51. $y = 2(x-1)^2 + 3$
V $(1,3)$; opens U

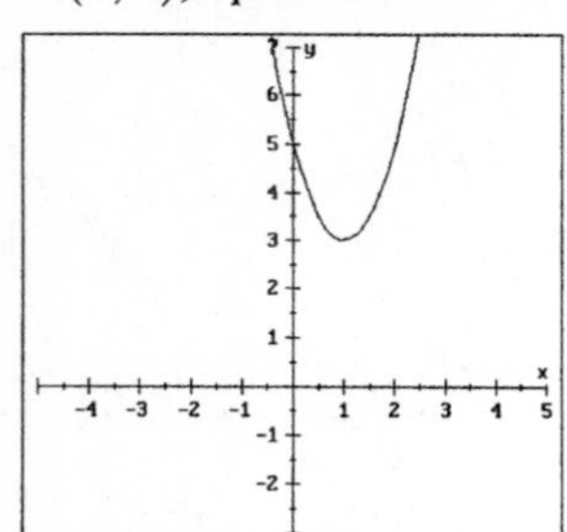

52. $y = -2(x+1)^2 + 2$
V $(-1,2)$; opens D

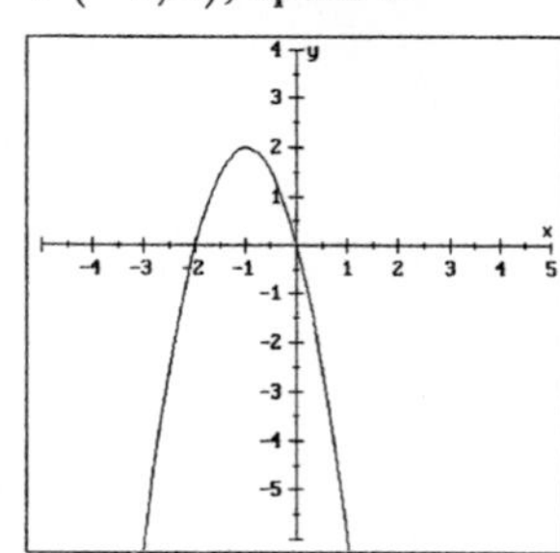

53. $x = 2y^2$

$y^2 = \frac{x}{2}$

$y = \pm\sqrt{\frac{x}{2}}$

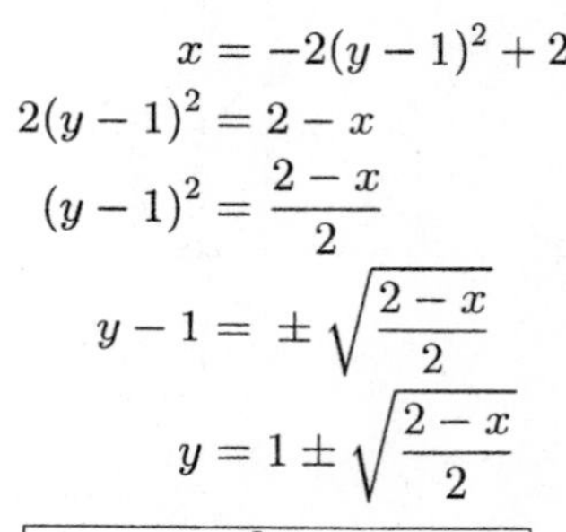

54. $x = y^2 - 4$

$y^2 = x + 4$

$y = \pm\sqrt{x+4}$

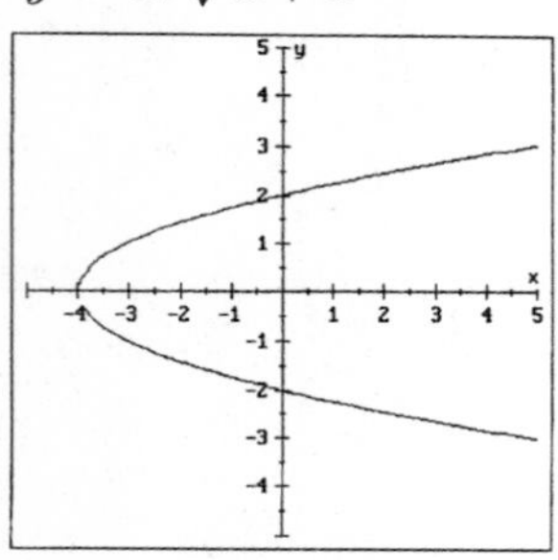

55. $x^2 - 2x + y = 6$

$y = 6 - x^2 + 2x$

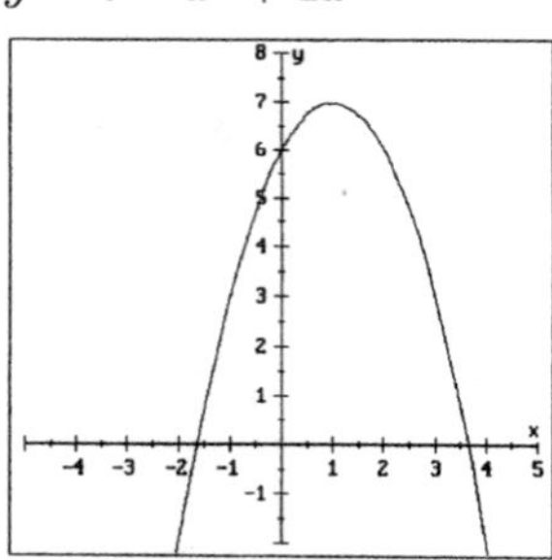

56. $x = -2(y-1)^2 + 2$

$2(y-1)^2 = 2 - x$

$(y-1)^2 = \frac{2-x}{2}$

$y - 1 = \pm\sqrt{\frac{2-x}{2}}$

$y = 1 \pm \sqrt{\frac{2-x}{2}}$

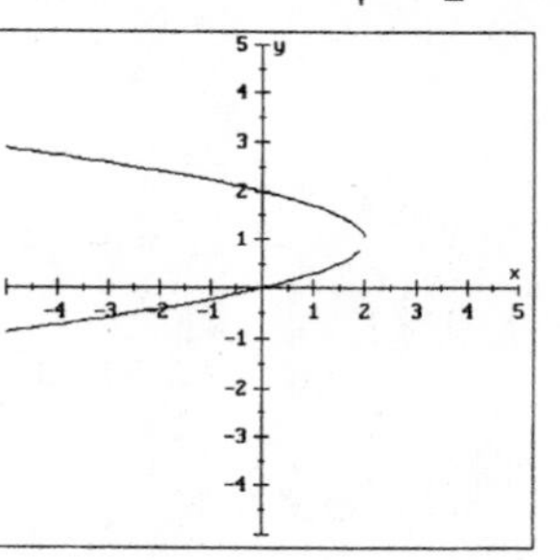

57. The radius of the larger gear is $\sqrt{16} = 4$. Centers: 7 units apart $\Rightarrow$ smaller gear $r = 3$.

$(x-h)^2 + (y-k)^2 = r^2$

$(x-7)^2 + (y-0)^2 = 3^2 \Rightarrow (x-7)^2 + y^2 = 9$

58.

$x^2 + y^2 = 2500$	$(x-10)^2 + y^2 = 900$	smallest $= 50 - 10 - 30 = \underline{10\text{ ft}}$
$(x-0)^2 + (y-0)^2 = 50^2$	$(x-10)^2 + (y-0)^2 = 30^2$	largest $= 50 + 10 - 30 = \underline{30\text{ ft}}$
center: $(0,0)$; radius $= 50$	center: $(10,0)$; radius $= 30$	

59.
$$x^2 + y^2 - 8x - 20y + 16 = 0$$
$$x^2 - 8x + y^2 - 20y = -16$$
$$x^2 - 8x + 16 + y^2 - 20y + 100 = -16 + 16 + 100$$
$$(x-4)^2 + (y-10)^2 = 100$$
center: $(4, 10)$; radius $= 10$
$$x^2 + y^2 + 2x + 4y - 11 = 0$$
$$x^2 + 2x + 1 + y^2 + 4y + 4 = 11 + 1 + 4$$
$$(x+1)^2 + (y+2)^2 = 16$$
center: $(-1, -2)$; radius $= 4$

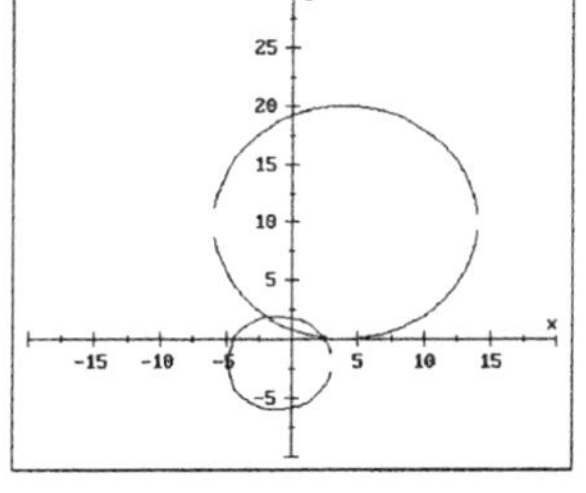

Since the ranges overlap (see graph), they can not be licensed for the same frequency.

60.
$$x^2 + y^2 - 16x - 20y + 155 = 0$$
$$x^2 - 16x + y^2 - 20y = -155$$
$$x^2 - 16x + 64 + y^2 - 20y + 100 = -155 + 64 + 100$$
$$(x-8)^2 + (y-10)^2 = 9$$
center: $(8, 10)$; radius $= 3$

The highways are 11 km and 13 km from the center of town.

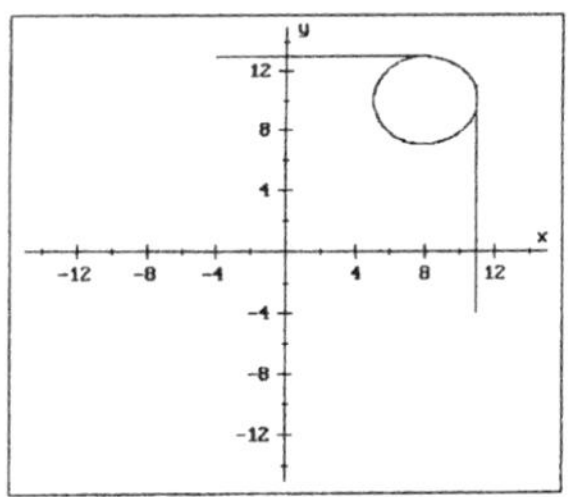

61. Set $y = 0$:
$y = 30x - x^2$
$0 = 30x - x^2$
$0 = x(30 - x)$
$x = 0$ or $x = 30$
It lands 30 feet away.

62. The maximum height is the y-coordinate of the vertex.
$y = 30x - x^2$
$y = -(x^2 - 30x)$
$y = -(x^2 - 30x + 225) + 225$
$y = -(x - 15)^2 + 225$
vertex: $(15, 225) \Rightarrow$ height $= 225$ ft

63. Find the vertex:
$2y^2 - 9x = 18$
$-9x = -2y^2 + 18$
$x = \frac{2}{9}y^2 - 2$
$x = \frac{2}{9}(y - 0)^2 - 2$
vertex: $(-2, 0) \Rightarrow$ distance $= 2$ AU

64. Find the y-coordinate when $x = \pm 4$:
$y = \frac{1}{16}x^2$
$y = \frac{1}{16}(\pm 4)^2$
$y = \frac{1}{16}(16) = 1$ ft deep

65. **Answers may vary.**

66. **Answers may vary.**

67. **Answers may vary.**

68. **Answers may vary.**

Exercise 10.2 (page 678)

1. $3x^{-2}y^2(4x^2+3y^{-2}) = 12x^0y^2 + 9x^{-2}y^0 = 12y^2 + \dfrac{9}{x^2}$

2. $(2a^{-2}-b^{-2})(2a^{-2}+b^{-2}) = 4a^{-4} + 2a^{-2}b^{-2} - 2a^{-2}b^{-2} - b^{-4} = \dfrac{4}{a^4} - \dfrac{1}{b^4}$

3. $\dfrac{x^{-2}+y^{-2}}{x^{-2}-y^{-2}} = \dfrac{x^{-2}+y^{-2}}{x^{-2}-y^{-2}} \cdot \dfrac{x^2y^2}{x^2y^2} = \dfrac{y^2+x^2}{y^2-x^2}$

4. $\dfrac{2x^{-3}-2y^{-3}}{4x^{-3}+4y^{-3}} = \dfrac{2x^{-3}-2y^{-3}}{4x^{-3}+4y^{-3}} \cdot \dfrac{x^3y^3}{x^3y^3} = \dfrac{2y^3-2x^3}{4y^3+4x^3} = \dfrac{2(y^3-x^3)}{4(y^3+x^3)} = \dfrac{y^3-x^3}{2(y^3+x^3)}$

5. ellipse; sum

6. foci

7. center

8. $(\pm a, 0)$; $(0, \pm b)$

9. $(0,0)$

10. (h,k)

11. $\dfrac{x^2}{4} + \dfrac{y^2}{9} = 1$

C $(0,0)$; move 2 horiz. and 3 vert.

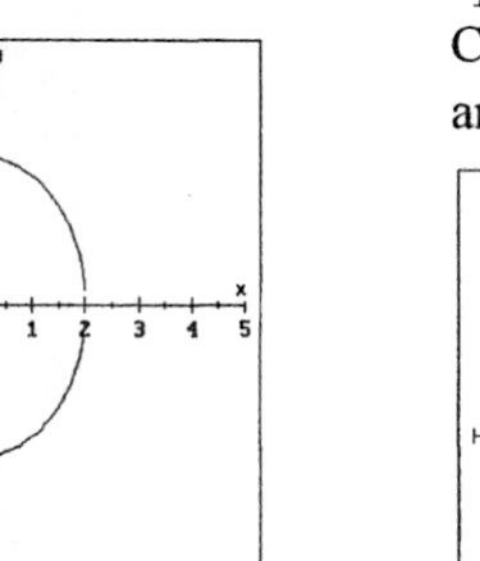

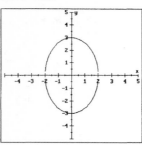

12. $x^2 + \dfrac{y^2}{9} = 1$

$\dfrac{x^2}{1} + \dfrac{y^2}{9} = 1$

C $(0,0)$; move 1 horiz. and 3 vert.

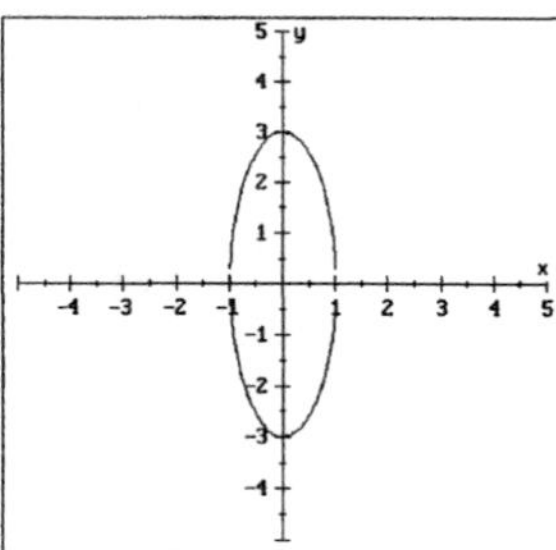

13. $x^2 + 9y^2 = 9$

$\dfrac{x^2}{9} + \dfrac{9y^2}{9} = \dfrac{9}{9}$

$\dfrac{x^2}{9} + \dfrac{y^2}{1} = 1$

C $(0,0)$; move 3 horiz. and 1 vert.

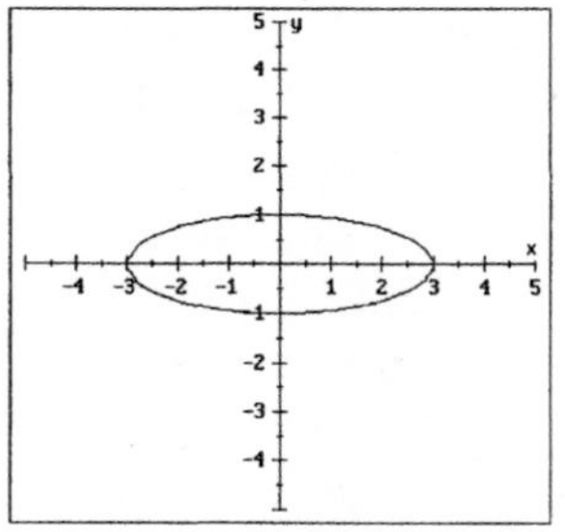

14. $25x^2 + 9y^2 = 225$

$\dfrac{25x^2}{225} + \dfrac{9y^2}{225} = 1$

$\dfrac{x^2}{9} + \dfrac{y^2}{25} = 1$

C $(0,0)$; move 3 horiz. and 5 vert.

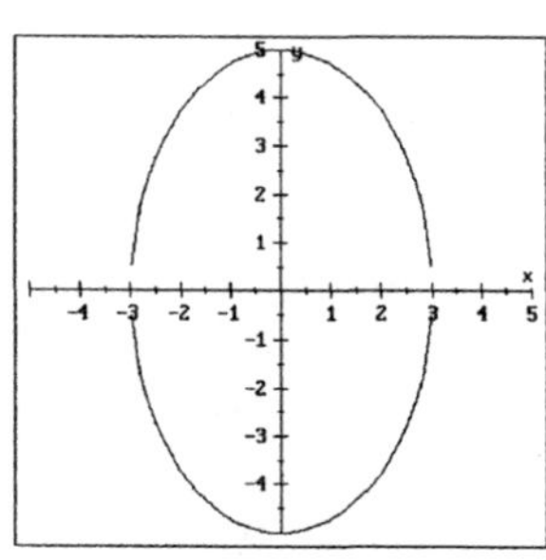

15. $16x^2 + 4y^2 = 64$

$$\frac{16x^2}{64} + \frac{4y^2}{64} = \frac{64}{64}$$

$$\frac{x^2}{4} + \frac{y^2}{16} = 1$$

C $(0,0)$; move 2 horiz. and 4 vert.

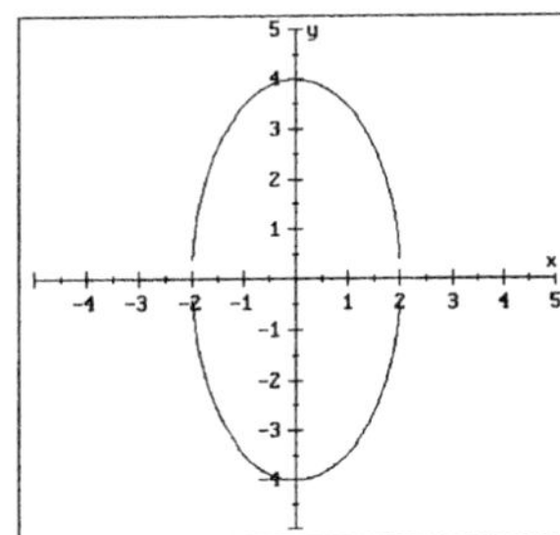

16. $4x^2 + 9y^2 = 36$

$$\frac{4x^2}{36} + \frac{9y^2}{36} = \frac{36}{36}$$

$$\frac{x^2}{9} + \frac{y^2}{4} = 1$$

C $(0,0)$; move 3 horiz. and 2 vert.

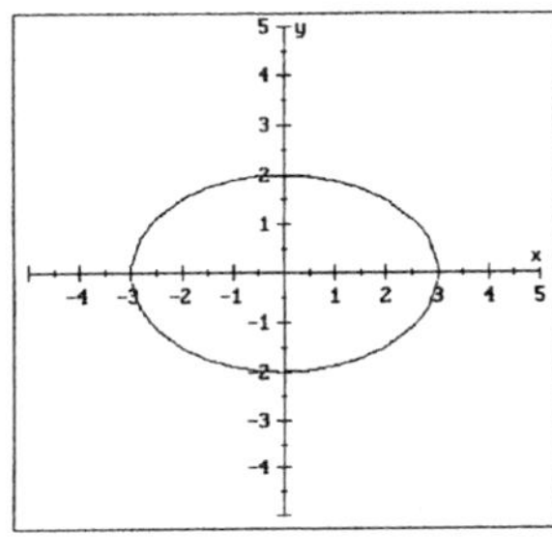

17. $\frac{(x-2)^2}{9} + \frac{(y-1)^2}{4} = 1$

C $(2,1)$; move 3 horiz. and 2 vert.

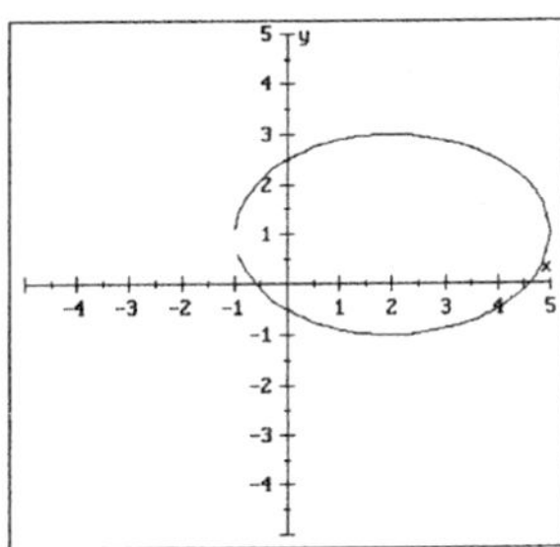

18. $\frac{(x-1)^2}{9} + \frac{(y-3)^2}{4} = 1$

C $(1,3)$; move 3 horiz. and 2 vert.

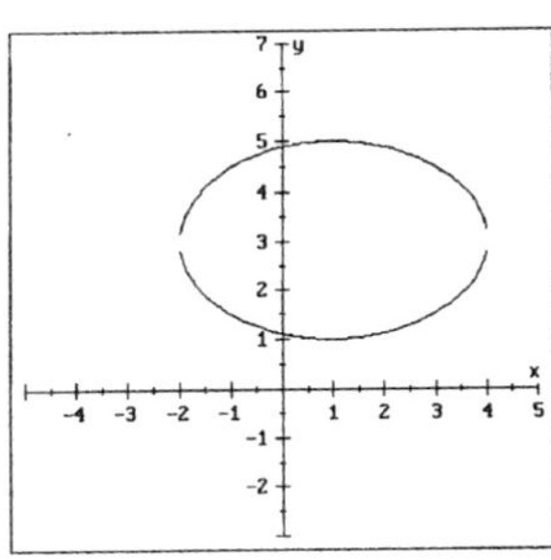

19. $(x+1)^2 + 4(y+2)^2 = 4$

$$\frac{(x+1)^2}{4} + \frac{4(y+2)^2}{4} = \frac{4}{4}$$

$$\frac{(x+1)^2}{4} + \frac{(y+2)^2}{1} = 1$$

C $(-1,-2)$; move 2 horiz. and 1 vert.

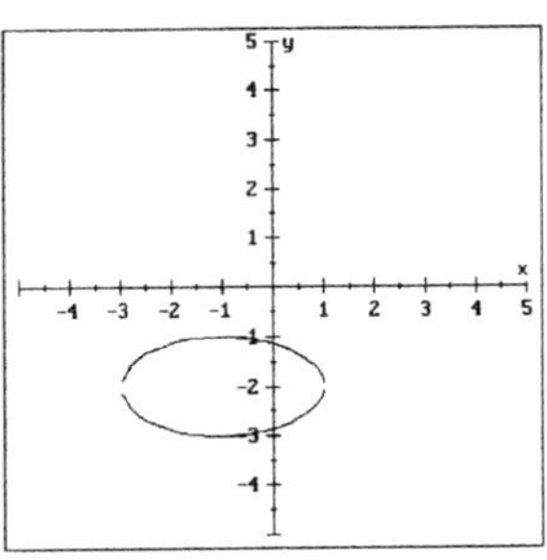

20. $25(x+1)^2 + 9y^2 = 225$

$$\frac{25(x+1)^2}{225} + \frac{9y^2}{225} = \frac{225}{225}$$

$$\frac{(x+1)^2}{9} + \frac{y^2}{25} = 1$$

C $(-1,0)$; move 3 horiz. and 5 vert.

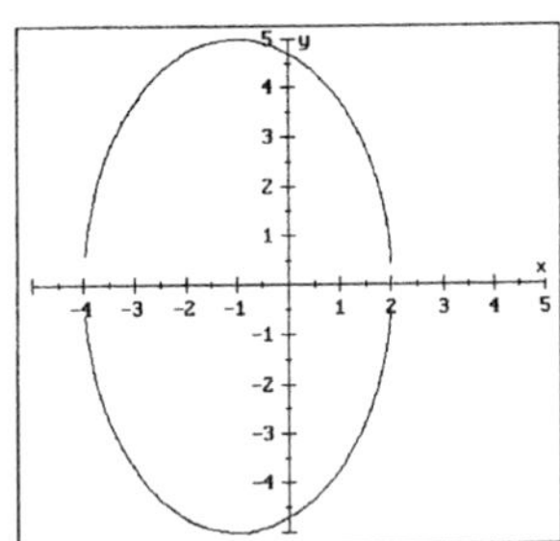

21. $\dfrac{x^2}{9} + \dfrac{y^2}{4} = 1$

$$\frac{y^2}{4} = 1 - \frac{x^2}{9}$$

$$y^2 = 4\left(1 - \frac{x^2}{9}\right)$$

$$y = \pm\sqrt{4\left(1 - \frac{x^2}{9}\right)}$$

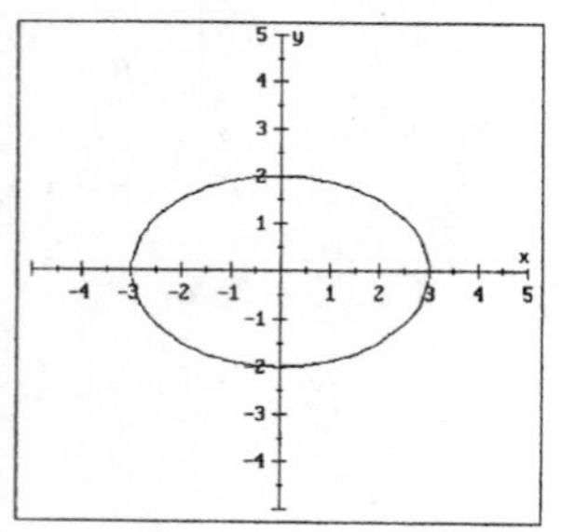

22. $x^2 + 16y^2 = 16$

$$16y^2 = 16 - x^2$$

$$y^2 = \frac{16 - x^2}{16}$$

$$y = \pm\sqrt{\frac{16 - x^2}{16}}$$

23. $\dfrac{x^2}{4} + \dfrac{(y-1)^2}{9} = 1$

$$\frac{(y-1)^2}{9} = 1 - \frac{x^2}{4}$$

$$(y-1)^2 = 9\left(1 - \frac{x^2}{4}\right)$$

$$y - 1 = \pm\sqrt{9\left(1 - \frac{x^2}{4}\right)}$$

$$y = 1 \pm \sqrt{9\left(1 - \frac{x^2}{4}\right)}$$

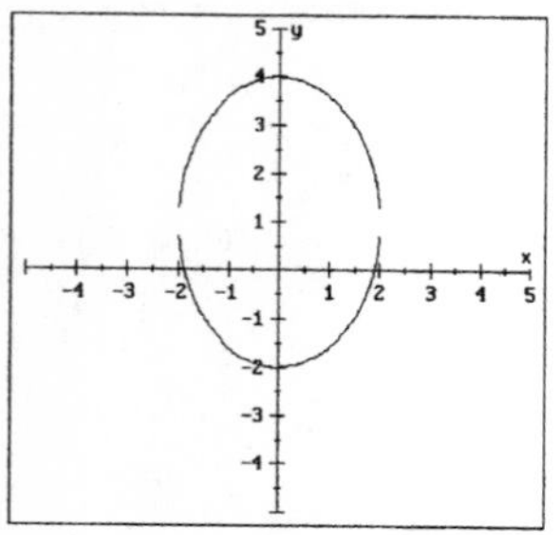

24. $\dfrac{(x+1)^2}{9} + \dfrac{(y-2)^2}{4} = 1$

$$\frac{(y-2)^2}{4} = 1 - \frac{(x+1)^2}{9}$$

$$(y-2)^2 = 4\left(1 - \frac{(x+1)^2}{9}\right)$$

$$y - 2 = \pm\sqrt{4\left(1 - \frac{(x+1)^2}{9}\right)}$$

$$y = 2 \pm \sqrt{4\left(1 - \frac{(x+1)^2}{9}\right)}$$

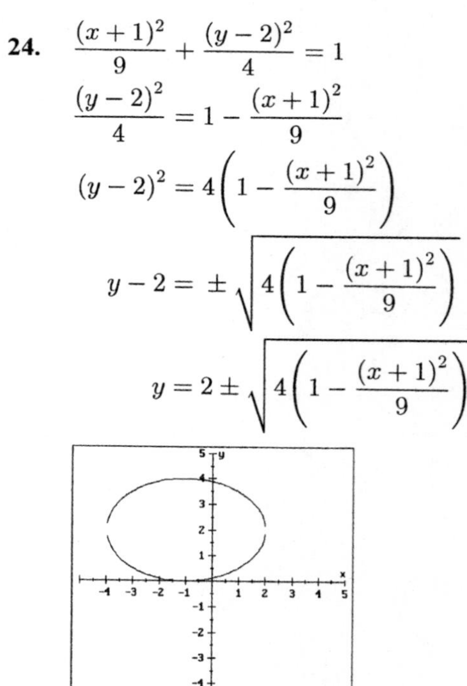

25.

$$x^2 + 4y^2 - 4x + 8y + 4 = 0$$
$$x^2 - 4x + 4\left(y^2 + 2y\right) = -4$$
$$x^2 - 4x + 4 + 4\left(y^2 + 2y + 1\right) = -4 + 4 + 4$$
$$(x-2)^2 + 4(y+1)^2 = 4$$
$$\frac{(x-2)^2}{4} + \frac{(y+1)^2}{1} = 1$$

C $(2, -1)$; move 2 horiz. and 1 vert.

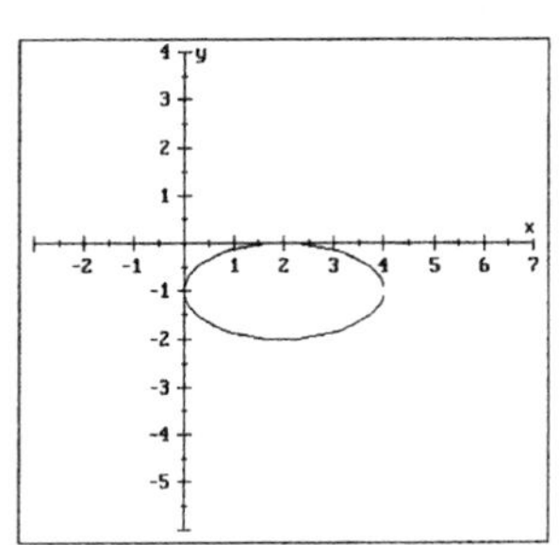

26.

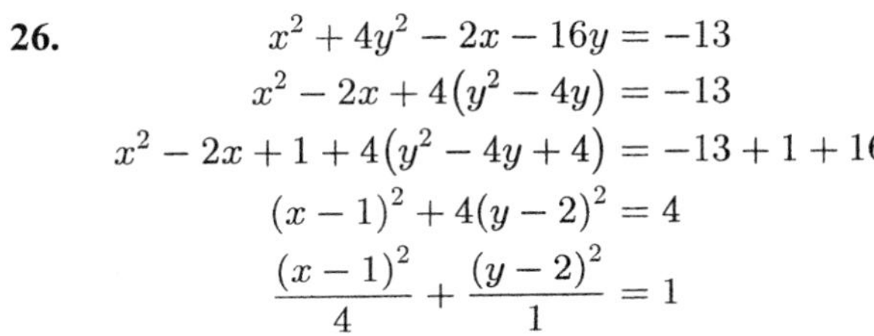

$$x^2 + 4y^2 - 2x - 16y = -13$$
$$x^2 - 2x + 4\left(y^2 - 4y\right) = -13$$
$$x^2 - 2x + 1 + 4\left(y^2 - 4y + 4\right) = -13 + 1 + 16$$
$$(x-1)^2 + 4(y-2)^2 = 4$$
$$\frac{(x-1)^2}{4} + \frac{(y-2)^2}{1} = 1$$

C $(1, 2)$; move 2 horiz. and 1 vert.

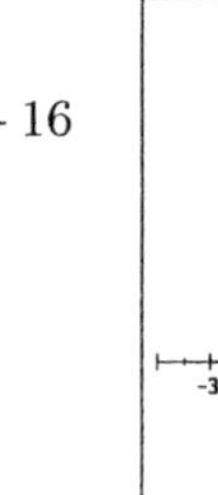

27.

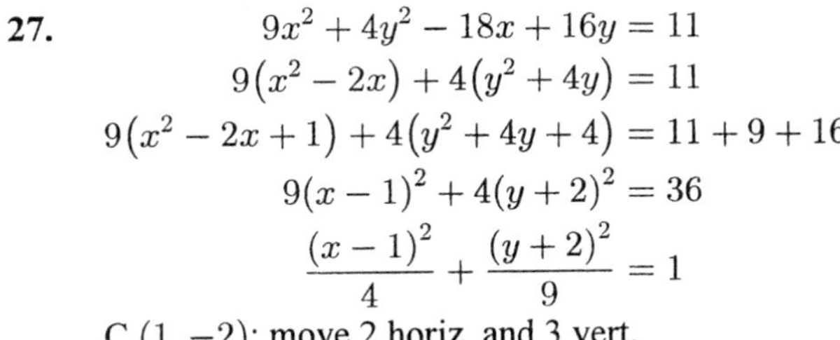

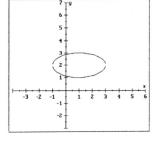

$$9x^2 + 4y^2 - 18x + 16y = 11$$
$$9\left(x^2 - 2x\right) + 4\left(y^2 + 4y\right) = 11$$
$$9\left(x^2 - 2x + 1\right) + 4\left(y^2 + 4y + 4\right) = 11 + 9 + 16$$
$$9(x-1)^2 + 4(y+2)^2 = 36$$
$$\frac{(x-1)^2}{4} + \frac{(y+2)^2}{9} = 1$$

C $(1, -2)$; move 2 horiz. and 3 vert.

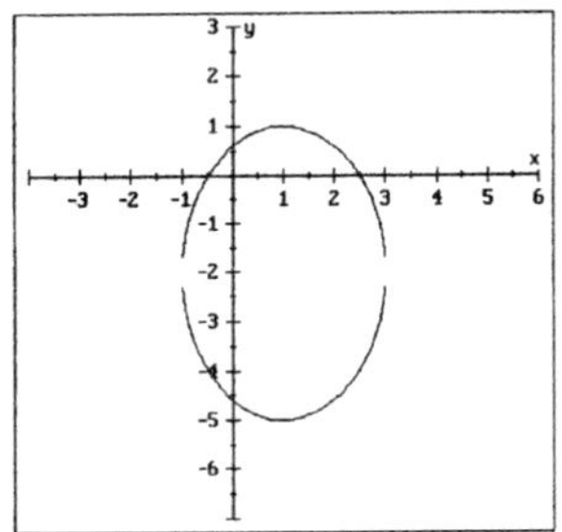

28.

$$16x^2 + 25y^2 - 160x - 200y + 400 = 0$$
$$16\left(x^2 - 10x\right) + 25\left(y^2 - 8y\right) = -400$$
$$16\left(x^2 - 10x + 25\right) + 25\left(y^2 - 8y + 16\right) = -400 + 400 + 400$$
$$16(x-5)^2 + 25(y-4)^2 = 400$$
$$\frac{(x-5)^2}{25} + \frac{(y-4)^2}{16} = 1$$

C $(5, 4)$; move 5 horiz. and 4 vert.

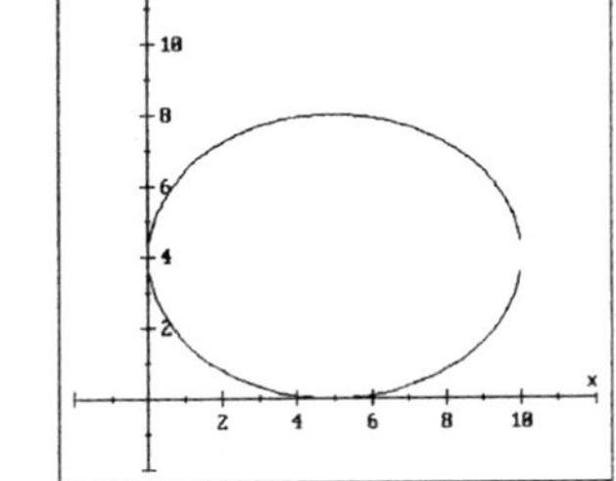

29. $a = 24/2 = 12,\ b = 10/2 = 5$

$$\frac{x^2}{a^2} + \frac{y^2}{b^2} = 1$$
$$\frac{x^2}{12^2} + \frac{y^2}{5^2} = 1$$
$$\frac{x^2}{144} + \frac{y^2}{25} = 1$$

30. $a = 60/2 = 30,\ b = 40/2 = 20$

$$\frac{x^2}{a^2} + \frac{y^2}{b^2} = 1$$
$$\frac{x^2}{30^2} + \frac{y^2}{20^2} = 1$$
$$\frac{x^2}{900} + \frac{y^2}{400} = 1$$

31. Note: $a = 40/2 = 20,\ b = 10$

$$\frac{x^2}{a^2} + \frac{y^2}{b^2} = 1$$
$$\frac{x^2}{400} + \frac{y^2}{100} = 1$$
$$x^2 + 4y^2 = 400$$
$$4y^2 = 400 - x^2$$
$$y^2 = \tfrac{1}{4}(400 - x^2)$$
$$y = \tfrac{1}{2}\sqrt{400 - x^2}$$

32. Let $x = 10$ in Exercise #**31**:

$$y = \tfrac{1}{2}\sqrt{400 - x^2}$$
$$y = \tfrac{1}{2}\sqrt{400 - 10^2}$$
$$y = \tfrac{1}{2}\sqrt{400 - 100}$$
$$y = \tfrac{1}{2}\sqrt{300}$$
$$y = \tfrac{1}{2}\sqrt{100}\sqrt{3}$$
$$y = \tfrac{1}{2}(10)\sqrt{3} = 5\sqrt{3} \text{ ft}$$

33.

$$9x^2 + 16y^2 = 144$$
$$\frac{9x^2}{144} + \frac{16y^2}{144} = \frac{144}{144}$$
$$\frac{x^2}{16} + \frac{y^2}{9} = 1 \Rightarrow \text{area} = \pi ab = \pi(4)(3) = 12\pi \text{ square units}$$

34.

$$4x^2 + 9y^2 = 576$$
$$\frac{4x^2}{576} + \frac{9y^2}{576} = \frac{576}{576}$$
$$\frac{x^2}{144} + \frac{y^2}{64} = 1$$
$$\text{A} = \pi ab = \pi(12)(8) = 96\pi$$

$$9x^2 + 25y^2 = 900$$
$$\frac{9x^2}{900} + \frac{25y^2}{900} = \frac{900}{900}$$
$$\frac{x^2}{100} + \frac{y^2}{36} = 1$$
$$\text{A} = \pi ab = \pi(10)(6) = 60\pi$$

Area of track $= 96\pi - 60\pi = 36\pi$ (square units)

35. **Answers may vary.**

36. **Answers may vary.**

37. It is a circle.

38.

$$x^2 - 2x + y^2 + 4y + 20 = 0$$
$$x^2 - 2x + y^2 + 4y = -20$$
$$x^2 - 2x + 1 + y^2 + 4y + 4 = -20 + 1 + 4$$
$$(x-1)^2 + (y+2)^2 = -15$$

Since the left side of the equation is the sum of two squares, the left side must represent a non-negative number. Since the right side is a negative number, there is no graph.

Exercise 10.3 (page 688)

1. $-6x^4 + 9x^3 - 6x^2 = -3x^2(2x^2 - 3x + 2)$

2. $4a^2 - b^2 = (2a)^2 - b^2 = (2a+b)(2a-b)$

3. $15a^2 - 4ab - 4b^2 = (3a - 2b)(5a + 2b)$

4. $8p^3 - 27q^3 = (2p)^3 - (3q)^3 = (2p-3q)\left((2p)^2 + (2p)(3q) + (3q)^2\right)$
$= (2p-3q)\left(4p^2 + 6pq + 9q^2\right)$

5. hyperbola; difference

6. foci

7. center

8. $(\pm a, 0)$; y-intercepts

9. $(0,0)$

10. (h,k)

11. $\dfrac{x^2}{9} - \dfrac{y^2}{4} = 1$
C $(0,0)$; open horiz.;
move 3 horiz. and 2 vert.

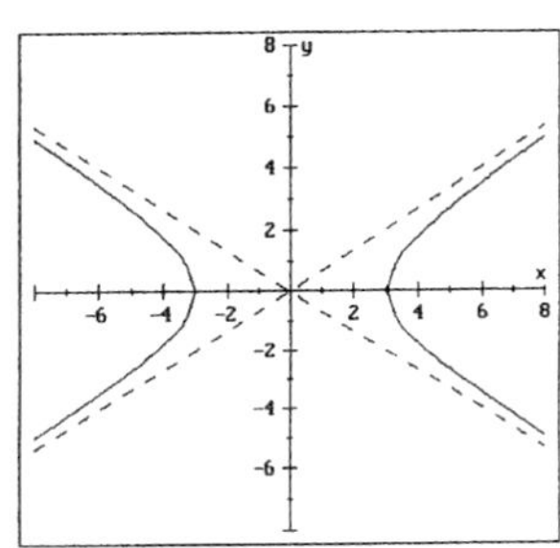

12. $\dfrac{x^2}{4} - \dfrac{y^2}{4} = 1$
C $(0,0)$; open horiz.;
move 2 horiz. and 2 vert.

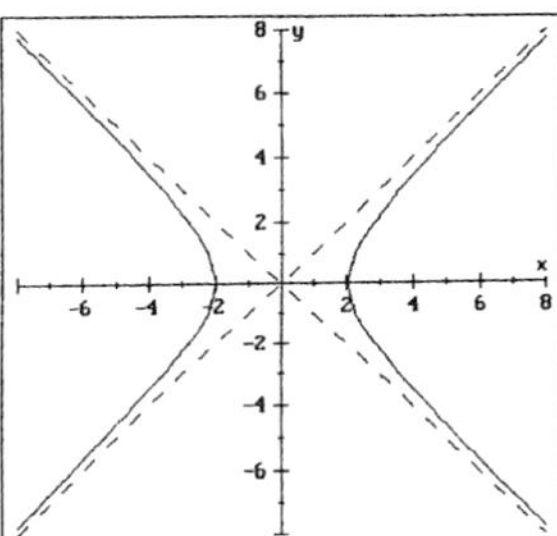

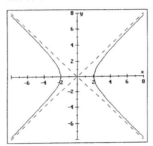

13. $\dfrac{y^2}{4} - \dfrac{x^2}{9} = 1$
C $(0,0)$; open vert.;
move 3 horiz. and 2 vert.

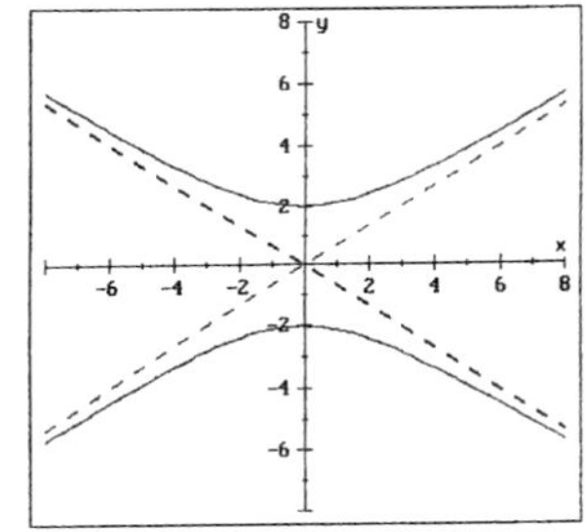

14. $\dfrac{y^2}{4} - \dfrac{x^2}{64} = 1$
C $(0,0)$; open vert.;
move 8 horiz. and 2 vert.

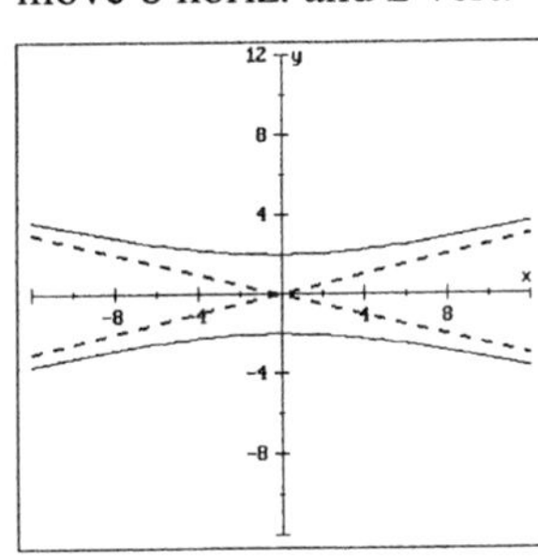

15. $25x^2 - y^2 = 25$
$\dfrac{25x^2}{25} - \dfrac{y^2}{25} = 1$
$\dfrac{x^2}{1} - \dfrac{y^2}{25} = 1$
C $(0,0)$; open horiz.;
move 1 horiz. and 5 vert.

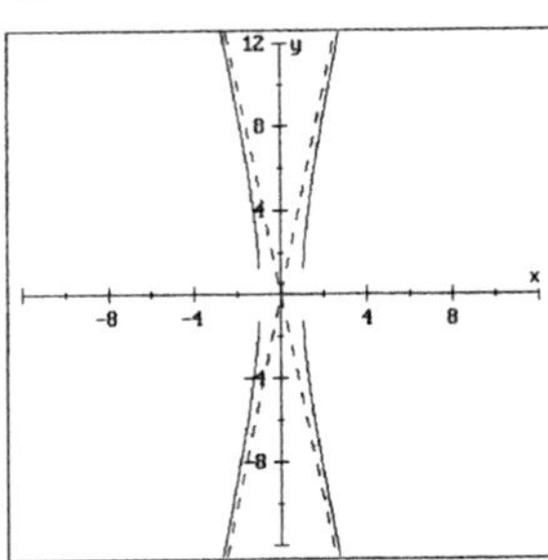

16. $9x^2 - 4y^2 = 36$
$\dfrac{9x^2}{36} - \dfrac{4y^2}{36} = \dfrac{36}{36}$
$\dfrac{x^2}{4} - \dfrac{y^2}{9} = 1$
C $(0,0)$; open horiz.;
move 2 horiz. and 3 vert.

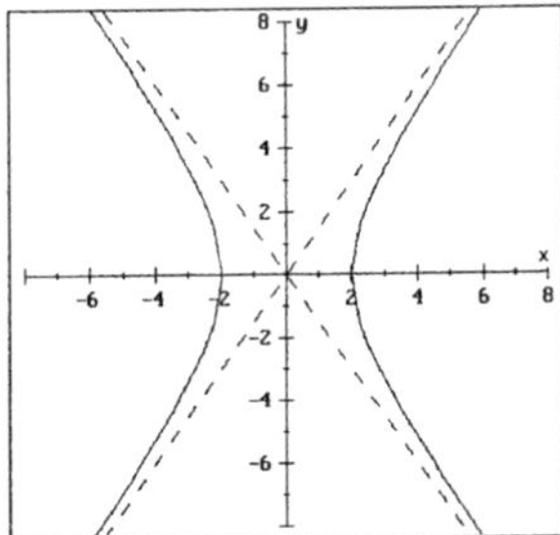

17. $\dfrac{(x-2)^2}{9} - \dfrac{y^2}{16} = 1$
C $(2,0)$; open horiz.;
move 3 horiz. and 4 vert.

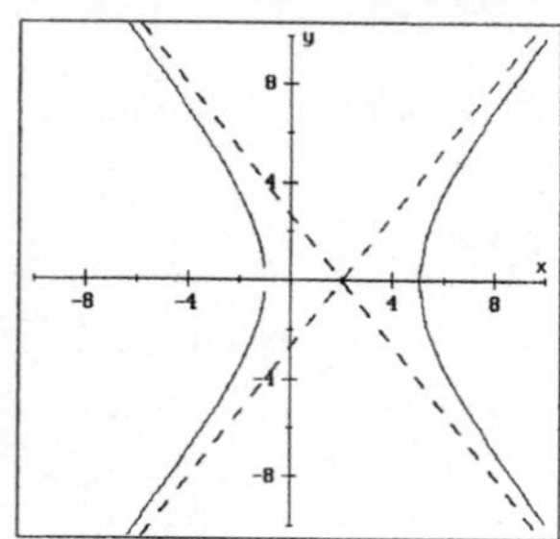

18. $\dfrac{(x+2)^2}{16} - \dfrac{(y-3)^2}{25} = 1$
C $(-2,3)$; open horiz.;
move 4 horiz. and 5 vert.

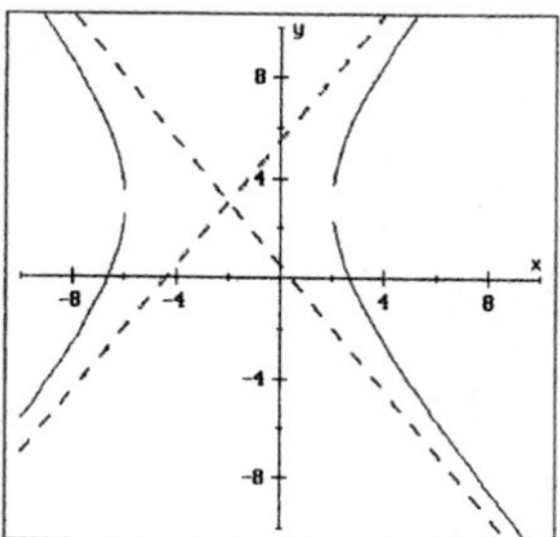

19. $\dfrac{(y+1)^2}{1} - \dfrac{(x-2)^2}{4} = 1$
C $(2,-1)$; open vert.;
move 2 horiz. and 1 vert.

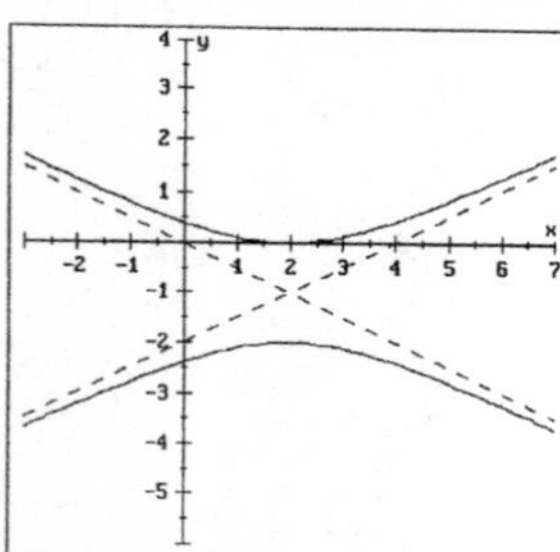

20. $\dfrac{(y-2)^2}{4} - \dfrac{(x+1)^2}{1} = 1$
C $(-1,2)$; open vert.;
move 1 horiz. and 2 vert.

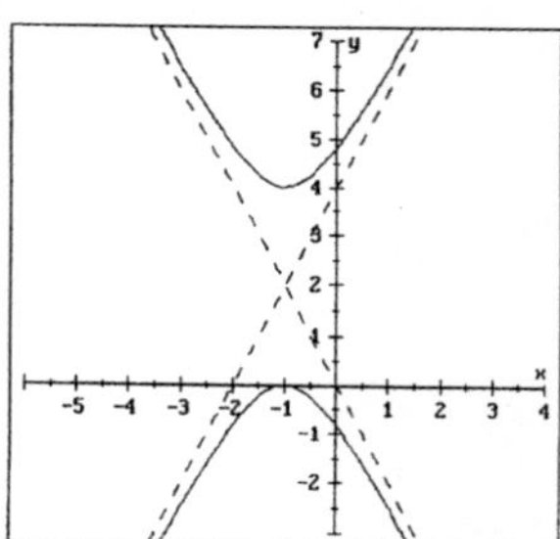

21. $4(x+3)^2 - (y-1)^2 = 4$

$$\frac{4(x+3)^2}{4} - \frac{(y-1)^2}{4} = \frac{4}{4}$$

$$\frac{(x+3)^2}{1} - \frac{(y-1)^2}{4} = 1$$

C $(-3,1)$; open horiz.;
move 1 horiz. and 2 vert.

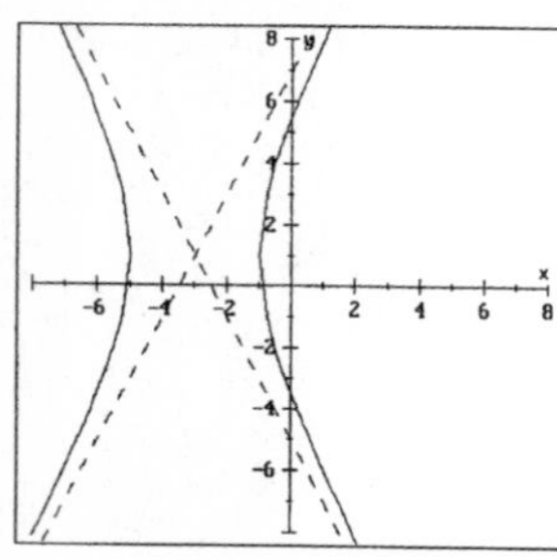

22. $(x+5)^2 - 16y^2 = 16$

$$\frac{(x+5)^2}{16} - \frac{16y^2}{16} = \frac{16}{16}$$

$$\frac{(x+5)^2}{16} - \frac{y^2}{1} = 1$$

C $(-5,0)$; open horiz.;
move 4 horiz. and 1 vert.

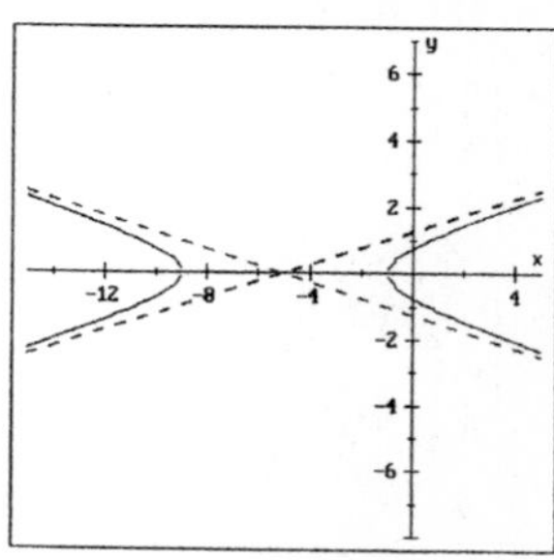

23. $xy = 8$

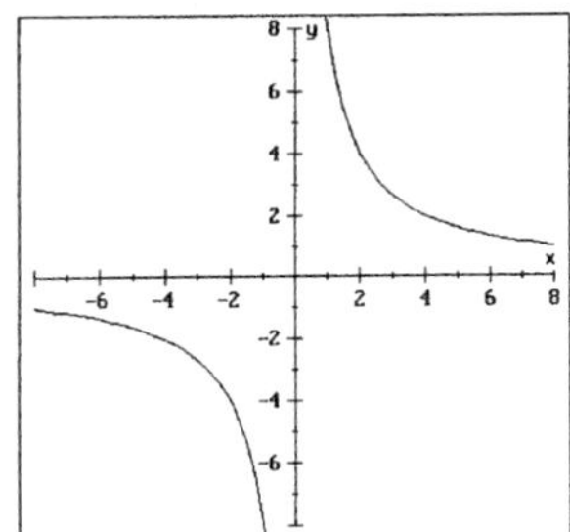

24. $xy = -10$

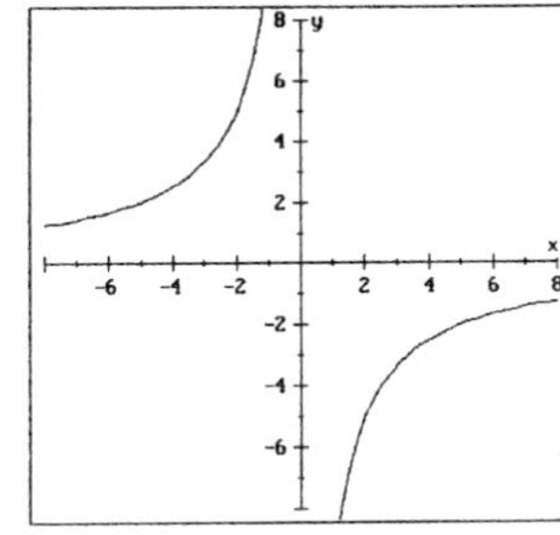

25. $\dfrac{x^2}{9} - \dfrac{y^2}{4} = 1$

$\dfrac{y^2}{4} = \dfrac{x^2}{9} - 1$

$y^2 = 4\left(\dfrac{x^2}{9} - 1\right)$

$y = \pm\sqrt{4\left(\dfrac{x^2}{9} - 1\right)}$

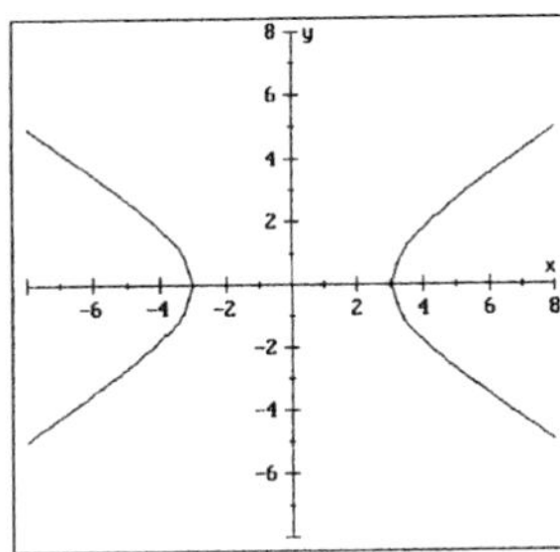

26. $y^2 - 16x^2 = 16$

$y^2 = 16 + 16x^2$

$y = \pm\sqrt{16 + 16x^2}$

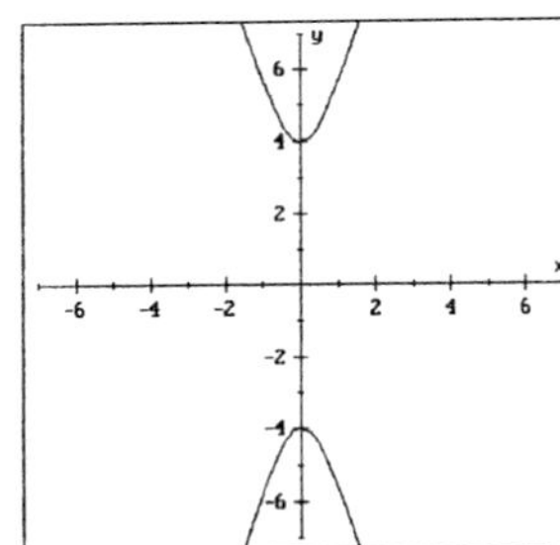

27.
$$\frac{x^2}{4}-\frac{(y-1)^2}{9}=1$$
$$\frac{(y-1)^2}{9}=\frac{x^2}{4}-1$$
$$(y-1)^2=9\left(\frac{x^2}{4}-1\right)$$
$$y-1=\pm\sqrt{9\left(\frac{x^2}{4}-1\right)}$$
$$y=1\pm\sqrt{9\left(\frac{x^2}{4}-1\right)}$$

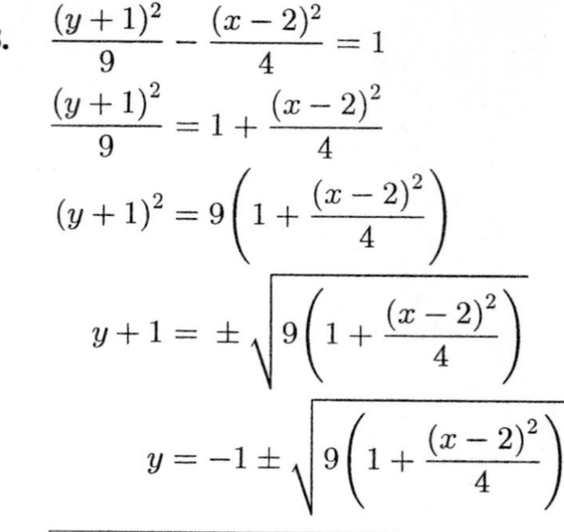

28.
$$\frac{(y+1)^2}{9}-\frac{(x-2)^2}{4}=1$$
$$\frac{(y+1)^2}{9}=1+\frac{(x-2)^2}{4}$$
$$(y+1)^2=9\left(1+\frac{(x-2)^2}{4}\right)$$
$$y+1=\pm\sqrt{9\left(1+\frac{(x-2)^2}{4}\right)}$$
$$y=-1\pm\sqrt{9\left(1+\frac{(x-2)^2}{4}\right)}$$

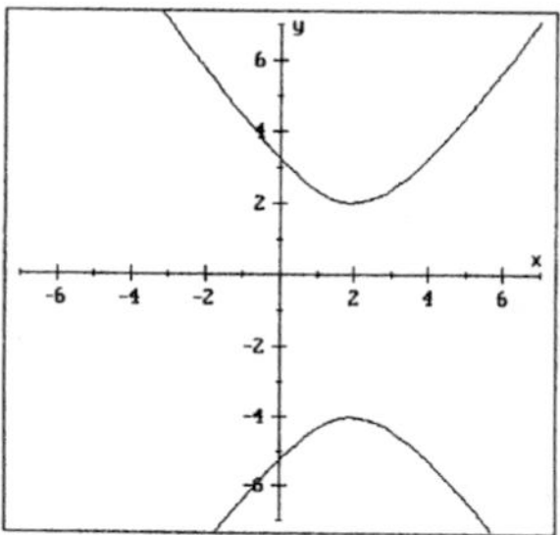

29.
$$4x^2-y^2+8x-4y=4$$
$$4x^2+8x-y^2-4y=4$$
$$4(x^2+2x)-(y^2+4y)=4$$
$$4(x^2+2x+1)-(y^2+4y+4)=4+4-4$$
$$4(x+1)^2-(y+2)^2=4$$
$$\frac{(x+1)^2}{1}-\frac{(y+2)^2}{4}=1$$
C $(-1,-2)$; opens horiz.; move 1 horiz. and 2 vert.

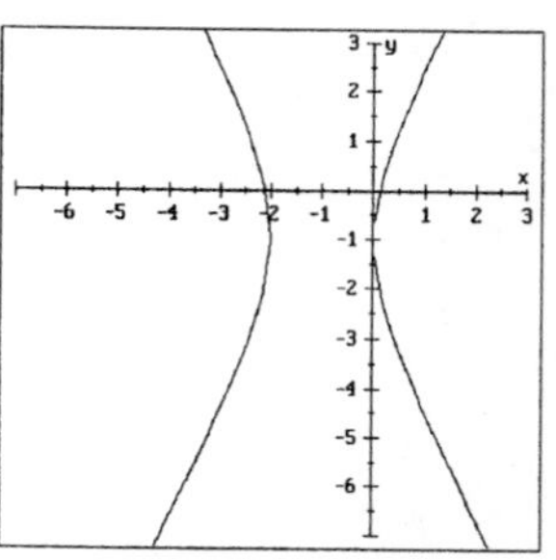

30.
$$x^2-9y^2-4x-54y=86$$
$$x^2-4x-9y^2-54y=86$$
$$(x^2-4x)-9(y^2+6y)=86$$
$$(x^2-4x+4)-9(y^2+6y+9)=86+4-81$$
$$(x-2)^2-9(y+3)^2=9$$
$$\frac{(x-2)^2}{9}-\frac{(y+3)^2}{1}=1$$
C $(2,-3)$; opens horiz.; move 3 horiz. and 1 vert.

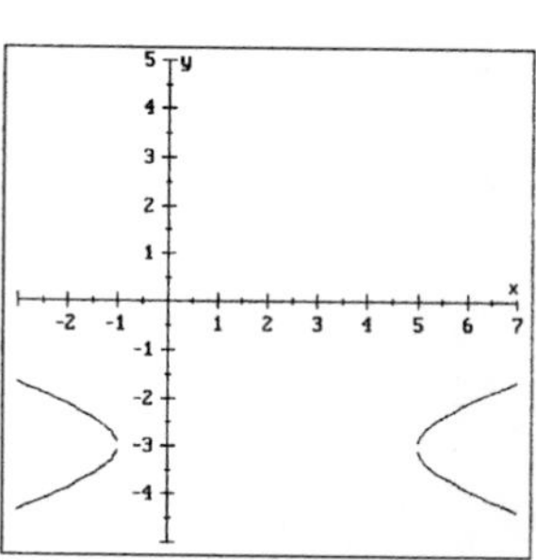

31.
$$
\begin{aligned}
4y^2 - x^2 + 8y + 4x &= 4 \\
4y^2 + 8y - x^2 + 4x &= 4 \\
4\left(y^2 + 2y\right) - \left(x^2 - 4x\right) &= 4 \\
4\left(y^2 + 2y + 1\right) - \left(x^2 - 4x + 4\right) &= 4 + 4 - 4 \\
4(y+1)^2 - (x-2)^2 &= 4 \\
\frac{(y+1)^2}{1} - \frac{(x-2)^2}{4} &= 1
\end{aligned}
$$
C $(2, -1)$; opens vert.; move 2 horiz. and 1 vert.

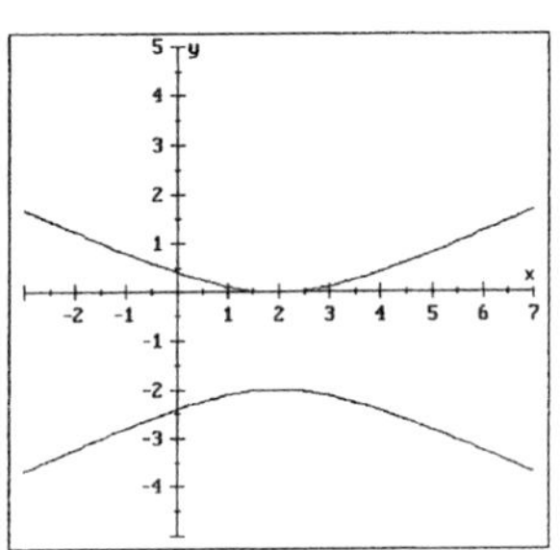

32.
$$
\begin{aligned}
y^2 - 4x^2 - 4y - 8x &= 4 \\
y^2 - 4y - 4x^2 - 8x &= 4 \\
\left(y^2 - 4y\right) - 4\left(x^2 + 2x\right) &= 4 \\
\left(y^2 - 4y + 4\right) - 4\left(x^2 + 2x + 1\right) &= 4 + 4 - 4 \\
(y-2)^2 - 4(x+1)^2 &= 4 \\
\frac{(y-2)^2}{4} - \frac{(x+1)^2}{1} &= 1
\end{aligned}
$$
C $(-1, 2)$; opens vert.; move 1 horiz. and 2 vert.

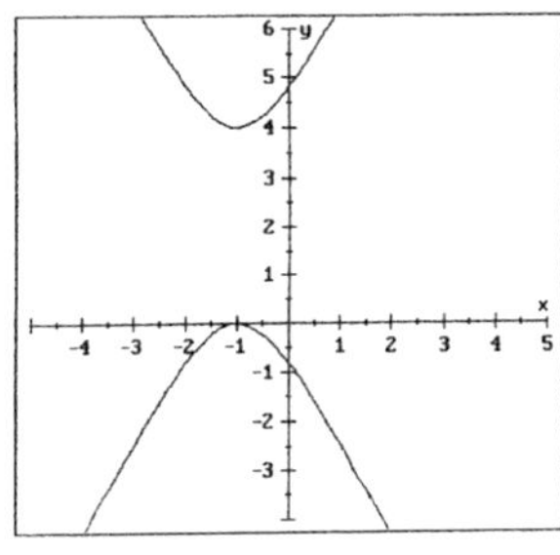

33.
$$
\begin{aligned}
9y^2 - x^2 &= 81 \\
\frac{9y^2}{81} - \frac{x^2}{81} &= \frac{81}{81} \\
\frac{y^2}{9} - \frac{x^2}{81} &= 1
\end{aligned}
$$
distance $= \sqrt{9} = 3$ units

34.
$$
\begin{aligned}
x^2 - 4y^2 &= 576 \\
\frac{x^2}{576} - \frac{4y^2}{576} &= \frac{576}{576} \\
\frac{x^2}{576} - \frac{y^2}{144} &= 1 \\
\frac{x^2}{576} - \frac{5^2}{144} &= 1 \\
\frac{x^2}{576} - \frac{25}{144} &= \frac{144}{144} \\
\frac{x^2}{576} &= \frac{169}{144} \\
144x^2 &= 576(169) \\
x^2 &= \frac{576(169)}{144} = 676 \\
x &= \sqrt{676} = 26 \text{ mi} \quad (\text{and } y = 5 \text{ mi})
\end{aligned}
$$

35. $y^2 - x^2 = 25$

$\frac{y^2}{25} - \frac{x^2}{25} = 1$

vertex: $(0, 5)$

Let $y = 10$:

$10^2 - x^2 = 25$

$-x^2 = -75$

$x = \sqrt{75} = 5\sqrt{3}$

width $= 2(5\sqrt{3}) = 10\sqrt{3}$ miles

36. $x^2 - 4y^2 = 4$

$\frac{x^2}{4} - \frac{4y^2}{4} = \frac{4}{4}$

$\frac{x^2}{4} - \frac{y^2}{1} = 1$

$a = 2 \Rightarrow 2a = 4$

The smallest distance is 4 units.

37. **Answers may vary.**

38. **Answers may vary.**

39. If $a = b$, the rectangle is a square.

40. $x^2 - y^2 = 1$

$y^2 - x^2 = 1$

They have the same center and the same asymptotes.

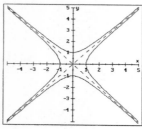

Exercise 10.4 (page 694)

1. $\sqrt{200x^2} - 3\sqrt{98x^2} = \sqrt{100x^2}\sqrt{2} - 3\sqrt{49x^2}\sqrt{2} = 10x\sqrt{2} - 3(7x)\sqrt{2} = -11x\sqrt{2}$

2. $a\sqrt{112a} - 5\sqrt{175a^3} = a\sqrt{16}\sqrt{7a} - 5\sqrt{25a^2}\sqrt{7a} = 4a\sqrt{7a} - 5(5a)\sqrt{7a} = -21a\sqrt{7a}$

3. $\frac{3t\sqrt{2t} - 2\sqrt{2t^3}}{\sqrt{18t} - \sqrt{2t}} = \frac{3t\sqrt{2t} - 2\sqrt{t^2}\sqrt{2t}}{\sqrt{9}\sqrt{2t} - \sqrt{2t}} = \frac{3t\sqrt{2t} - 2t\sqrt{2t}}{3\sqrt{2t} - \sqrt{2t}} = \frac{t\sqrt{2t}}{2\sqrt{2t}} = \frac{t}{2}$

4. $\sqrt[3]{\frac{x}{4}} + \sqrt[3]{\frac{x}{32}} - \sqrt[3]{\frac{x}{500}} = \sqrt[3]{\frac{x}{4} \cdot \frac{2}{2}} + \sqrt[3]{\frac{x}{32} \cdot \frac{2}{2}} - \sqrt[3]{\frac{x}{500} \cdot \frac{2}{2}} = \sqrt[3]{\frac{2x}{8}} + \sqrt[3]{\frac{2x}{64}} - \sqrt[3]{\frac{2x}{1000}}$

$= \frac{\sqrt[3]{2x}}{2} + \frac{\sqrt[3]{2x}}{4} - \frac{\sqrt[3]{2x}}{10}$

$= \frac{10\sqrt[3]{2x}}{20} + \frac{5\sqrt[3]{2x}}{20} - \frac{2\sqrt[3]{2x}}{20}$

$= \frac{13\sqrt[3]{2x}}{20}$

5. graphing; substitution

6. two

7. $\begin{cases} 8x^2 + 32y^2 = 256 \\ x = 2y \end{cases}$

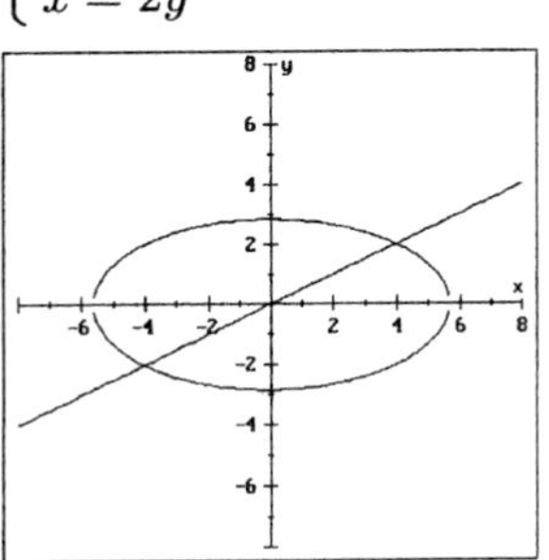

$(-4, -2), (4, 2)$

8. $\begin{cases} x^2 + y^2 = 2 \\ x + y = 2 \end{cases}$

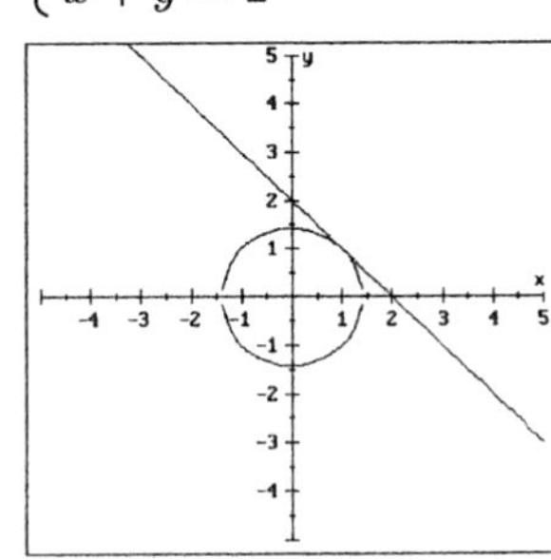

$(1, 1)$

9. $\begin{cases} x^2 + y^2 = 10 \\ y = 3x^2 \end{cases}$

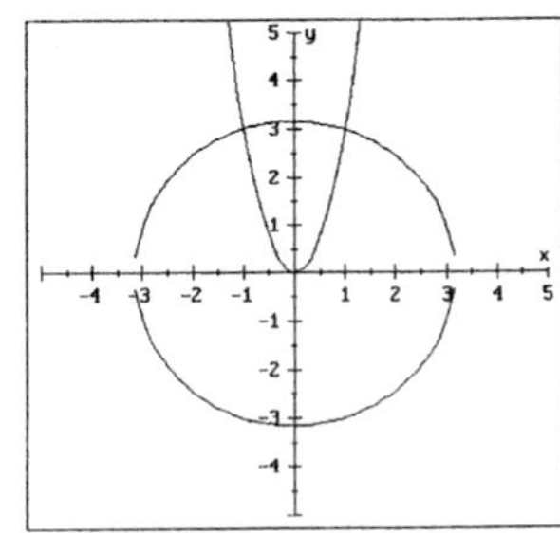

$(-1, 3), (1, 3)$

10. $\begin{cases} x^2 + y^2 = 5 \\ x + y = 3 \end{cases}$

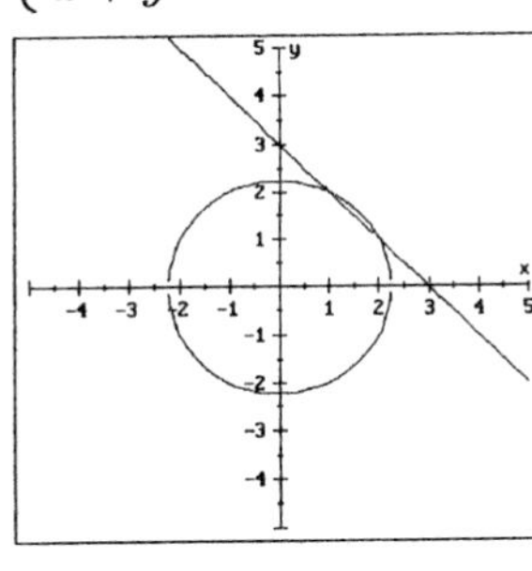

$(1, 2), (2, 1)$

11. $\begin{cases} x^2 + y^2 = 25 \\ 12x^2 + 64y^2 = 768 \end{cases}$

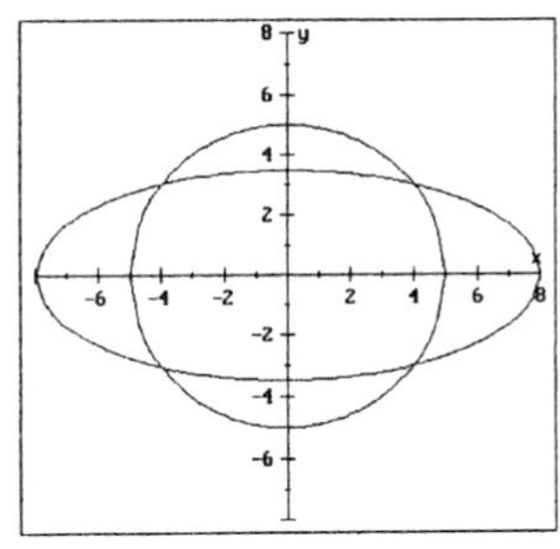

$(-4, 3), (4, 3), (4, -3)$
$(-4, -3)$

12. $\begin{cases} x^2 + y^2 = 13 \\ y = x^2 - 1 \end{cases}$

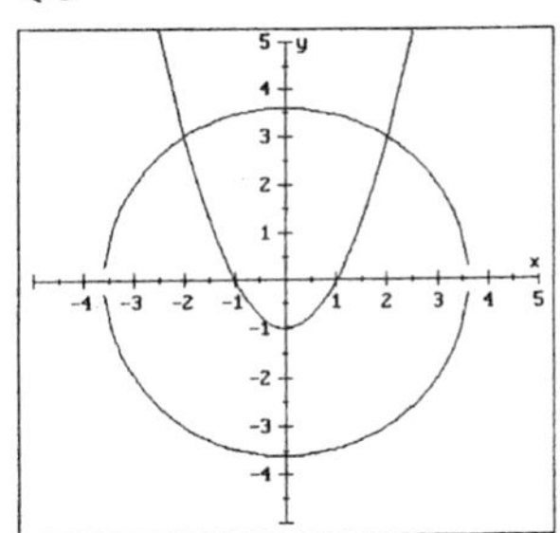

$(-2, 3), (2, 3)$

13. $\begin{cases} x^2 - 13 = -y^2 \\ y = 2x - 4 \end{cases}$

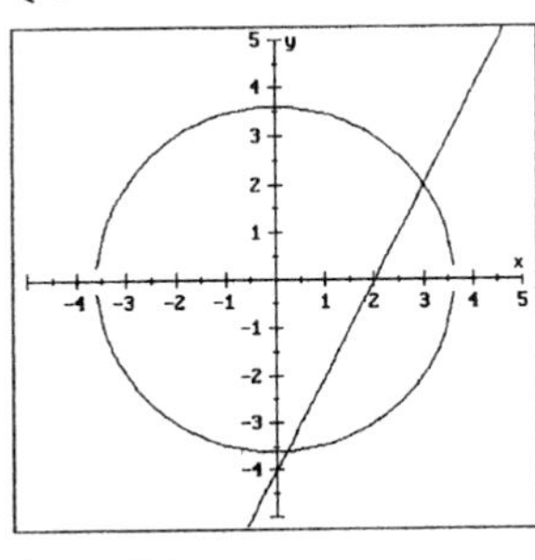

$\left(\frac{1}{5}, -\frac{18}{5}\right), (3, 2)$

14. $\begin{cases} x^2 + y^2 = 20 \\ y = x^2 \end{cases}$

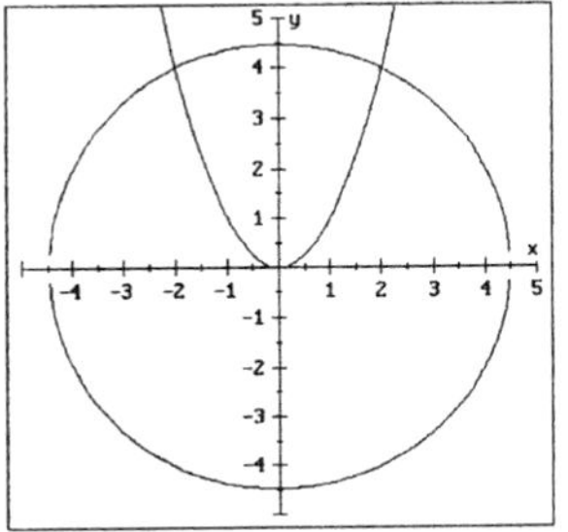

$(-2, 4), (2, 4)$

15. $\begin{cases} x^2 - 6x - y = -5 \\ x^2 - 6x + y = -5 \end{cases}$

$\begin{cases} y = x^2 - 6x + 5 \\ y = -x^2 + 6x - 5 \end{cases}$

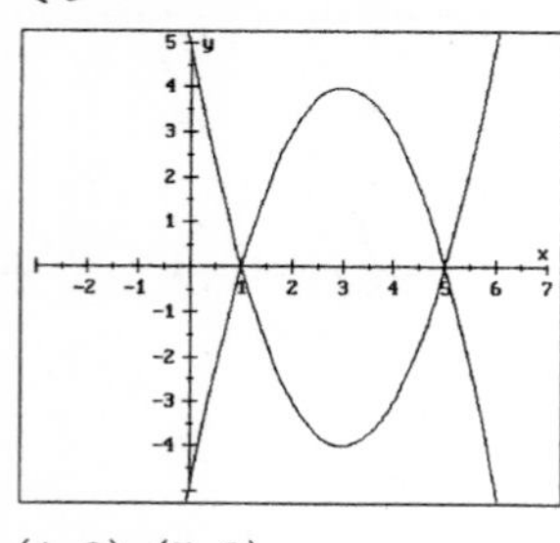

$(1, 0), (5, 0)$

16. $\begin{cases} x^2 - y^2 = -5 \\ 3x^2 + 2y^2 = 30 \end{cases}$

$\begin{cases} y = \pm\sqrt{x^2 + 5} \\ y = \pm\sqrt{\frac{30-3x^2}{2}} \end{cases}$

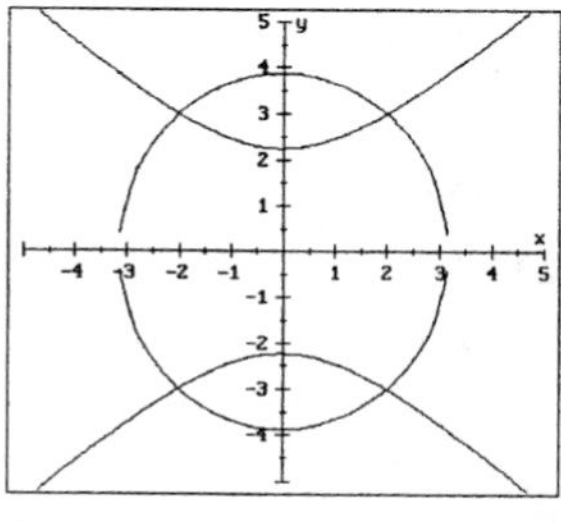

$(-2, 3), (2, 3), (2, -3), (-2, -3)$

17. $\begin{cases} (1) & 25x^2 + 9y^2 = 225 \\ (2) & 5x + 3y = 15 \end{cases}$

Substitute $x = -\frac{3}{5}y + 3$ from (2) into (1):

$$\begin{aligned} 25x^2 + 9y^2 &= 225 \\ 25\left(-\tfrac{3}{5}y + 3\right)^2 + 9y^2 &= 225 \\ 25\left(\tfrac{9}{25}y^2 - \tfrac{18}{5}y + 9\right) + 9y^2 &= 225 \\ 9y^2 - 90y + 225 + 9y^2 &= 225 \\ 18y^2 - 90y &= 0 \\ 18y(y - 5) &= 0 \end{aligned}$$

$18y = 0$ **or** $y - 5 = 0$

$y = 0$ $\quad y = 5$

Substitute these and solve for x:

$$\begin{aligned} 5x + 3y &= 15 & 5x + 3y &= 15 \\ 5x + 3(0) &= 15 & 5x + 3(5) &= 15 \\ 5x &= 15 & 5x &= 0 \\ x &= 3 & x &= 0 \end{aligned}$$

Solutions: $(3, 0), (0, 5)$

18. $\begin{cases} (1) & x^2 + y^2 = 20 \\ (2) & y = x^2 \end{cases}$

Substitute $x^2 = y$ from (2) into (1):

$$\begin{aligned} x^2 + y^2 &= 20 \\ y + y^2 &= 20 \\ y^2 + y - 20 &= 0 \\ (y + 5)(y - 4) &= 0 \end{aligned}$$

$y + 5 = 0$ **or** $y - 4 = 0$

$y = -5$ $\quad y = 4$

Substitute these and solve for x:

$$\begin{aligned} x^2 &= y & x^2 &= y \\ x^2 &= -5 & x^2 &= 4 \\ &\text{complex} & x &= \pm 2 \end{aligned}$$

Solutions: $(2, 4), (-2, 4)$

19. $\begin{cases} (1) & x^2 + y^2 = 2 \\ (2) & x + y = 2 \end{cases}$

Substitute $x = 2 - y$ from (2) into (1):

$$\begin{aligned} x^2 + y^2 &= 2 \\ (2 - y)^2 + y^2 &= 2 \\ 4 - 4y + y^2 + y^2 &= 2 \\ 2y^2 - 4y + 2 &= 0 \\ 2(y - 1)(y - 1) &= 0 \end{aligned}$$

$y - 1 = 0$ **or** $y - 1 = 0$

$y = 1$ $\quad y = 1$

Substitute this and solve for x:

$$\begin{aligned} x &= 2 - y \\ x &= 2 - 1 \\ x &= 1 \end{aligned}$$

Solution: $(1, 1)$

20. $\begin{cases} (1) & x^2+y^2=36 \\ (2) & 49x^2+36y^2=1764 \end{cases}$

Substitute $x^2=36-y^2$ from (1) into (2):

$$\begin{aligned} 49x^2+36y^2 &= 1764 \\ 49(36-y^2)+36y^2 &= 1764 \\ 1764-49y^2+36y^2 &= 1764 \\ -13y^2 &= 0 \\ y &= 0 \end{aligned}$$

Substitute this and solve for x:

$$\begin{aligned} x^2 &= 36-y^2 \\ x^2 &= 36-0 \\ x^2 &= 36 \\ x &= \pm 6 \end{aligned}$$

Solutions: $(6,0),(-6,0)$

21. $\begin{cases} (1) & x^2+y^2=5 \\ (2) & x+y=3 \end{cases}$

Substitute $x=3-y$ from (2) into (1):

$$\begin{aligned} x^2+y^2 &= 5 \\ (3-y)^2+y^2 &= 5 \\ 9-6y+y^2+y^2 &= 5 \\ 2y^2-6y+4 &= 0 \\ 2(y-2)(y-1) &= 0 \end{aligned}$$

$y-2=0$ **or** $y-1=0$

$y=2 \qquad y=1$

Substitute these and solve for x:

$x=3-y \qquad x=3-y$

$x=3-2 \qquad x=3-1$

$x=1 \qquad x=2$

Solutions: $(1,2),(2,1)$

22. $\begin{cases} (1) & x^2-x-y=2 \\ (2) & 4x-3y=0 \end{cases}$

Substitute $y=\frac{4}{3}x$ from (2) into (1):

$$\begin{aligned} x^2-x-y &= 2 \\ x^2-x-\tfrac{4}{3}x &= 2 \\ 3x^2-3x-4x &= 6 \\ 3x^2-7x-6 &= 0 \\ (3x+2)(x-3) &= 0 \end{aligned}$$

$3x+2=0$ **or** $x-3=0$

$x=-\frac{2}{3} \qquad x=3$

Substitute these and solve for y:

$y=\frac{4}{3}x \qquad y=\frac{4}{3}x$

$y=\frac{4}{3}\left(-\frac{2}{3}\right) \qquad y=\frac{4}{3}(3)$

$y=-\frac{8}{9} \qquad y=4$

Solutions: $\left(-\frac{2}{3},-\frac{8}{9}\right),(3,4)$

23. $\begin{cases} (1) & x^2+y^2=13 \\ (2) & y=x^2-1 \end{cases}$

Substitute $x^2=13-y^2$ from (1) into (2):

$$\begin{aligned} y &= x^2-1 \\ y &= 13-y^2-1 \\ y^2+y-12 &= 0 \\ (y+4)(y-3) &= 0 \end{aligned}$$

$y+4=0$ **or** $y-3=0$

$y=-4 \qquad y=3$

Substitute these and solve for x:

$x^2=13-y^2 \qquad x^2=13-y^2$

$x^2=13-16 \qquad x^2=13-9$

$x^2=-4 \qquad x^2=4$

complex $\qquad x=\pm 2$

Solutions: $(2,3),(-2,3)$

24.

$$\begin{aligned} x^2+y^2 &= 25 \Rightarrow (\times 3) & 3x^2+3y^2 &= 75 \\ 2x^2-3y^2 &= 5 \Rightarrow & 2x^2-3y^2 &= 5 \\ & & 5x^2 &= 80 \\ & & x^2 &= 16 \\ & & x &= \pm 4 \end{aligned}$$

Substitute and solve for y:

$x^2+y^2=25 \qquad x^2+y^2=25$

$4^2+y^2=25 \qquad (-4)^2+y^2=25$

$y^2=9 \qquad y^2=9$

$y=\pm 3 \qquad y=\pm 3$

Solutions: $(4,3),(4,-3),(-4,3),(-4,-3)$

25. $\begin{cases} (1) & x^2 + y^2 = 30 \\ (2) & y = x^2 \end{cases}$

Substitute $x^2 = y$ from (2) into (1):

$$x^2 + y^2 = 30$$
$$y + y^2 = 30$$
$$y^2 + y - 30 = 0$$
$$(y - 5)(y + 6) = 0$$

$y - 5 = 0$ **or** $y + 6 = 0$

$y = 5 \qquad y = -6$

Substitute these and solve for x:

$x^2 = y \qquad x^2 = y$

$x^2 = 5 \qquad x^2 = -6$

$x = \pm\sqrt{5} \qquad$ complex

Solutions: $\left(\sqrt{5}, 5\right), \left(-\sqrt{5}, 5\right)$

26. $9x^2 - 7y^2 = 81 \Rightarrow$
$x^2 + y^2 = 9 \Rightarrow (\times 7)$

$$\begin{aligned} 9x^2 - 7y^2 &= 81 \\ 7x^2 + 7y^2 &= 63 \\ \hline 16x^2 &= 144 \\ x^2 &= 9 \\ x &= \pm 3 \end{aligned}$$

Substitute and solve for y:

$x^2 + y^2 = 9 \qquad x^2 + y^2 = 9$

$3^2 + y^2 = 9 \qquad (-3)^2 + y^2 = 9$

$y^2 = 0 \qquad y^2 = 0$

$y = 0 \qquad y = 0$

Solutions: $(3, 0), (-3, 0)$

27.
$$\begin{aligned} x^2 + y^2 &= 13 \\ x^2 - y^2 &= 5 \\ \hline 2x^2 &= 18 \\ x^2 &= 9 \\ x &= \pm 3 \end{aligned}$$

Substitute and solve for y:

$x^2 + y^2 = 13 \qquad x^2 + y^2 = 13$

$3^2 + y^2 = 13 \qquad (-3)^2 + y^2 = 13$

$y^2 = 4 \qquad y^2 = 4$

$y = \pm 2 \qquad y = \pm 2$

Solutions: $(3, 2), (3, -2), (-3, 2), (-3, -2)$

28.
$$\begin{aligned} 2x^2 + y^2 &= 6 \\ x^2 - y^2 &= 3 \\ \hline 3x^2 &= 9 \\ x^2 &= 3 \\ x &= \pm\sqrt{3} \end{aligned}$$

Substitute and solve for y:

$2x^2 + y^2 = 6 \qquad 2x^2 + y^2 = 6$

$2\left(\sqrt{3}\right)^2 + y^2 = 6 \qquad 2\left(-\sqrt{3}\right)^2 + y^2 = 6$

$y^2 = 0 \qquad y^2 = 0$

$y = 0 \qquad y = 0$

Solutions: $\left(\sqrt{3}, 0\right), \left(-\sqrt{3}, 0\right)$

29.
$$\begin{aligned} x^2 + y^2 &= 20 \\ x^2 - y^2 &= -12 \\ \hline 2x^2 &= 8 \\ x^2 &= 4 \\ x &= \pm 2 \end{aligned}$$

Substitute and solve for y:

$x^2 + y^2 = 20 \qquad x^2 + y^2 = 20$

$2^2 + y^2 = 20 \qquad (-2)^2 + y^2 = 20$

$y^2 = 16 \qquad y^2 = 16$

$y = \pm 4 \qquad y = \pm 4$

Solutions: $(2, 4), (2, -4), (-2, 4), (-2, -4)$

30. $\begin{cases} (1) & xy = -\frac{9}{2} \\ (2) & 3x + 2y = 6 \end{cases}$

Substitute $x = -\frac{9}{2y}$ from (1) into (2):

$$3\left(-\tfrac{9}{2y}\right) + 2y = 6$$
$$-\tfrac{27}{2y} + 2y = 6$$
$$-27 + 4y^2 = 12y$$
$$4y^2 - 12y - 27 = 0$$
$$(2y + 3)(2y - 9) = 0$$

$2y + 3 = 0$ **or** $2y - 9 = 0$

$y = -\frac{3}{2}$ $\qquad y = \frac{9}{2}$

Substitute these and solve for x:

$x = -\dfrac{9}{2y}$ $\qquad x = -\dfrac{9}{2y}$

$x = -\dfrac{9}{2\left(-\frac{3}{2}\right)}$ $\qquad x = -\dfrac{9}{2\left(\frac{9}{2}\right)}$

$x = -\frac{9}{-3} = 3$ $\qquad x = -\frac{9}{9}$

Solutions: $\left(3, -\frac{3}{2}\right), \left(-1, \frac{9}{2}\right)$

31. $\begin{cases} (1) & y^2 = 40 - x^2 \\ (2) & y = x^2 - 10 \end{cases}$

Substitute $x^2 = 40 - y^2$ from (1) into (2):

$$y = x^2 - 10$$
$$y = 40 - y^2 - 10$$
$$y^2 + y - 30 = 0$$
$$(y + 6)(y - 5) = 0$$

$y + 6 = 0$ **or** $y - 5 = 0$

$y = -6$ $\qquad y = 5$

Substitute these and solve for x:

$x^2 = 40 - y^2$ $\qquad x^2 = 40 - y^2$

$x^2 = 40 - 36$ $\qquad x^2 = 40 - 25$

$x^2 = 4$ $\qquad x^2 = 15$

$x = \pm 2$ $\qquad x = \pm\sqrt{15}$

$(2, -6), (-2, -6), \left(\sqrt{15}, 5\right), \left(-\sqrt{15}, 5\right)$

32.
$$\begin{array}{r} x^2 - 6x - y = -5 \\ x^2 - 6x + y = -5 \\ \hline 2x^2 - 12x = -10 \end{array}$$
$$2(x - 5)(x - 1) = 0$$

$x - 5 = 0$ **or** $x - 1 = 0$

$x = 5$ $\qquad x = 1$

Substitute and solve for y:

$x^2 - 6x - y = -5$ $\qquad x^2 - 6x - y = -5$

$5^2 - 6(5) - y = -5$ $\qquad 1^2 - 6(1) - y = -5$

$-5 - y = -5$ $\qquad -5 - y = -5$

$y = 0$ $\qquad y = 0$

Solutions: $(5, 0), (1, 0)$

33. $\begin{cases} (1) & y = x^2 - 4 \\ (2) & x^2 - y^2 = -16 \end{cases}$

Substitute $x^2 = y + 4$ from (1) into (2):

$$x^2 - y^2 = -16$$
$$y + 4 - y^2 = -16$$
$$y^2 - y - 20 = 0$$
$$(y + 4)(y - 5) = 0$$

$y + 4 = 0$ **or** $y - 5 = 0$

$y = -4$ $\qquad y = 5$

Substitute these and solve for x:

$x^2 = y + 4$ $\qquad x^2 = y + 4$

$x^2 = -4 + 4$ $\qquad x^2 = 5 + 4$

$x^2 = 0$ $\qquad x^2 = 9$

$x = 0$ $\qquad x = \pm 3$

Solutions: $(0, -4), (3, 5), (-3, 5)$

34. $$\begin{array}{lll} 6x^2+8y^2=182 \Rightarrow (\times 3) & 18x^2+24y^2 &= 546 \\ 8x^2-3y^2=\ \ 24 \Rightarrow (\times 8) & 64x^2-24y^2 &= 192 \\ \hline & 82x^2 &= 748 \\ & x^2 &= 9 \\ & x &= \pm 3 \end{array}$$

Substitute and solve for y:

$$\begin{array}{rr} 6x^2+8y^2=182 & 6x^2+8y^2=182 \\ 6(3)^2+8y^2=182 & 6(-3)^2+8y^2=182 \\ 8y^2=128 & 8y^2=128 \\ y^2=16 & y^2=16 \\ y=\pm 4 & y=\pm 4 \end{array}$$

Solutions: $(3,4),(3,-4),(-3,4),(-3,-4)$

35. $$\begin{array}{lll} x^2-\ y^2=-5 \Rightarrow (\times 2) & 2x^2-2y^2 &= -10 \\ 3x^2+2y^2=\ 30 \Rightarrow & 3x^2+2y^2 &= 30 \\ \hline & 5x^2 &= 20 \\ & x^2 &= 4 \\ & x &= \pm 2 \end{array}$$

Substitute and solve for y:

$$\begin{array}{rr} x^2-y^2=-5 & x^2-y^2=-5 \\ 2^2-y^2=-5 & (-2)^2-y^2=-5 \\ -y^2=-9 & -y^2=-9 \\ y^2=9 & y^2=9 \\ y=\pm 3 & y=\pm 3 \end{array}$$

Solutions: $(2,3),(2,-3),(-2,3),(-2,-3)$

36. $$\begin{array}{ll} \frac{1}{x}+\frac{1}{y} &= 5 \\ \frac{1}{x}-\frac{1}{y} &= -3 \\ \hline \frac{2}{x} &= 2 \\ 2 &= 2x \\ 1 &= x \end{array}$$

Substitute and solve for y:

$$\begin{array}{r} \frac{1}{x}+\frac{1}{y}=5 \\ \frac{1}{1}+\frac{1}{y}=5 \\ \frac{1}{y}=4 \\ y=\frac{1}{4} \end{array}$$

Solution: $\left(1,\frac{1}{4}\right)$

37. $$\begin{array}{lll} \frac{1}{x}+\frac{2}{y}=1 \Rightarrow & \frac{1}{x}+\frac{2}{y} &= 1 \\ \frac{2}{x}-\frac{1}{y}=\frac{1}{3} \Rightarrow (\times 2) & \frac{4}{x}-\frac{2}{y} &= \frac{2}{3} \\ \hline & \frac{5}{x} &= \frac{5}{3} \\ & 15 &= 5x \\ & 3 &= x \end{array}$$

Substitute and solve for y:

$$\begin{array}{r} \frac{1}{x}+\frac{2}{y}=1 \\ \frac{1}{3}+\frac{2}{y}=1 \\ \frac{2}{y}=\frac{2}{3} \\ 6=2y \\ 3=y \end{array}$$

Solution: $(3,3)$

38. $\frac{1}{x}+\frac{3}{y}=4 \Rightarrow$ $\quad \frac{1}{x}+\frac{3}{y}=4$

$\frac{2}{x}-\frac{1}{y}=7 \Rightarrow (\times 3) \quad \frac{6}{x}-\frac{3}{y}=21$

$$\begin{aligned} \frac{7}{x} &= 25 \\ 7 &= 25x \\ \frac{7}{25} &= x \end{aligned}$$

Substitute and solve for y:

$$\begin{aligned} \frac{1}{x}+\frac{3}{y} &= 4 \\ \frac{1}{\frac{7}{25}}+\frac{3}{y} &= 4 \\ \frac{25}{7}+\frac{3}{y} &= 4 \\ 75y+63 &= 84y \\ 63 &= 9y \\ 7 &= y \end{aligned}$$

Solution: $\left(\frac{7}{25}, 7\right)$

39. $\begin{cases} (1) & 3y^2 = xy \\ (2) & 2x^2+xy-84=0 \end{cases}$

From (1): $3y^2 - xy = 0$

$y(3y-x)=0$

$y=0$ or $y=\dfrac{1}{3}x$

Substitute these into (2):

$$\begin{aligned} 2x^2+xy-84 &= 0 \\ 2x^2+x(0)-84 &= 0 \\ 2x^2 &= 84 \\ x^2 &= 42 \\ x &= \pm\sqrt{42} \end{aligned}$$

$$\begin{aligned} 2x^2+xy-84 &= 0 \\ 2x^2+x\left(\tfrac{1}{3}x\right)-84 &= 0 \\ 2x^2+\tfrac{1}{3}x^2 &= 84 \\ 6x^2+x^2 &= 252 \\ 7x^2 &= 252 \\ x^2 &= 36 \\ x &= \pm 6 \end{aligned}$$

(substitute and solve for y)

Solutions: $\left(\sqrt{42}, 0\right), \left(-\sqrt{42}, 0\right), (6,2), (-6,-2)$

40. $x^2+y^2=10 \Rightarrow (\times 3) \quad 3x^2+3y^2=30$

$2x^2-3y^2=5 \Rightarrow \quad 2x^2-3y^2=5$

$$\begin{aligned} 5x^2 &= 35 \\ x^2 &= 7 \\ x &= \pm\sqrt{7} \end{aligned}$$

Substitute and solve for y:

$$\begin{aligned} x^2+y^2 &= 10 \\ 7+y^2 &= 10 \\ y^2 &= 3 \\ y &= \pm\sqrt{3} \end{aligned} \qquad \begin{aligned} x^2+y^2 &= 10 \\ 7+y^2 &= 25 \\ y^2 &= 3 \\ y &= \pm\sqrt{3} \end{aligned}$$

Solutions: $\left(\sqrt{7}, \sqrt{3}\right), \left(\sqrt{7}, -\sqrt{3}\right), \left(-\sqrt{7}, \sqrt{3}\right), \left(-\sqrt{7}, -\sqrt{3}\right)$

41. $\begin{cases} (1) & xy = \frac{1}{6} \\ (2) & y+x=5xy \end{cases}$

Substitute $x=\frac{1}{6y}$ from (1) into (2):

$$\begin{aligned} y+\frac{1}{6y} &= \frac{5y}{6y} \\ 6y^2+1 &= 5y \\ 6y^2-5y+1 &= 0 \\ (2y-1)(3y-1) &= 0 \end{aligned}$$

$2y-1=0$ **or** $3y-1=0$

$y=\frac{1}{2} \qquad y=\frac{1}{3}$

Substitute these and solve for x:

$$\begin{aligned} x &= \frac{1}{6y} \\ x &= \frac{1}{6\left(\frac{1}{2}\right)} \\ x &= \frac{1}{3} \end{aligned} \qquad \begin{aligned} x &= \frac{1}{6y} \\ x &= \frac{1}{6\left(\frac{1}{3}\right)} \\ x &= \frac{1}{2} \end{aligned}$$

Solutions: $\left(\frac{1}{3}, \frac{1}{2}\right), \left(\frac{1}{2}, \frac{1}{3}\right)$

42. $\begin{cases} (1) & xy = \frac{1}{12} \\ (2) & y + x = 7xy \end{cases}$

Substitute $x = \frac{1}{12y}$ from (1) into (2):

$$y + \frac{1}{12y} = \frac{7y}{12y}$$
$$12y^2 + 1 = 7y$$
$$12y^2 - 7y + 1 = 0$$
$$(4y - 1)(3y - 1) = 0$$

$4y - 1 = 0$ **or** $3y - 1 = 0$

$y = \frac{1}{4}$ $\quad$ $y = \frac{1}{3}$

Substitute these and solve for x:

$x = \frac{1}{12y}$ $\quad$ $x = \frac{1}{12y}$

$x = \dfrac{1}{12\left(\frac{1}{4}\right)}$ $\quad$ $x = \dfrac{1}{12\left(\frac{1}{3}\right)}$

$x = \frac{1}{3}$ $\quad$ $x = \frac{1}{4}$

Solutions: $\left(\frac{1}{3}, \frac{1}{4}\right), \left(\frac{1}{4}, \frac{1}{3}\right)$

43. Let the integers be x and y. Then the equations are

$\begin{cases} (1) & xy = 32 \\ (2) & x + y = 12 \end{cases}$

Substitute $x = \frac{32}{y}$ from (1) into (2):

$$\frac{32}{y} + y = 12$$
$$32 + y^2 = 12y$$
$$y^2 - 12y + 32 = 0$$
$$(y - 4)(y - 8) = 0$$

$y - 4 = 0$ **or** $y - 8 = 0$

$y = 4$ $\quad$ $y = 8$

Substitute these and solve for x:

$x = \frac{32}{y} = \frac{32}{4} = 8$ $\quad$ $x = \frac{32}{y} = \frac{32}{8} = 4$

The integers are 8 and 4.

44. Let the numbers be x and y. Then the equations are

$\begin{cases} (1) & x^2 + y^2 = 221 \\ (2) & x + y = 9 \end{cases}$

Substitute $x = 9 - y$ from (2) into (1):

$$x^2 + y^2 = 221$$
$$(9 - y)^2 + y^2 = 221$$
$$81 - 18y + y^2 + y^2 = 221$$
$$2y^2 - 18y - 140 = 0$$
$$2(y - 14)(y + 5) = 0$$

$y - 14 = 0$ **or** $y + 5 = 0$

$y = 14$ $\quad$ $y = -5$

Substitute these and solve for x:

$x = 9 - y$ $\quad$ $x = 9 - y$

$x = 9 - 14$ $\quad$ $x = 9 - (-5)$

$x = -5$ $\quad$ $x = 14$

The numbers are -5 and 14.

45. Let $l =$ the length of the rectangle, and $w =$ the width of the rectangle. Then the equations are:

$\begin{cases} (1) & lw = 63 \\ (2) & 2l + 2w = 32 \end{cases}$

Substitute $l = \frac{63}{w}$ from (1) into (2):

$$2\left(\frac{63}{w}\right) + 2w = 32$$
$$\frac{126}{w} + 2w = 32$$
$$126 + 2w^2 = 32w$$
$$2w^2 - 32y + 126 = 0$$
$$2(w - 7)(w - 9) = 0$$

$w - 7 = 0$ **or** $w - 9 = 0$

$w = 7$ $\quad$ $w = 9$

Substitute these and solve for l:

$l = \frac{63}{w} = \frac{63}{7} = 9$ $\quad$ $l = \frac{63}{w} = \frac{63}{9} = 7$

The dimensions are 7 cm by 9 cm.

46. Let $r =$ Ignacio's rate, and let $p =$ the amount Ignacio invested.
Then Carol invested $p + 500$ at a rate of $r - 0.01$. The equations are

$$\begin{cases} (1)\ pr = 225 \Rightarrow p = \frac{225}{r} \\ (2)\ (p+500)(r-0.01) = 240 \end{cases}$$

Substitute $p = \frac{225}{r}$ from (1) into (2):

$$\left(\tfrac{225}{r} + 500\right)(r - 0.01) = 240$$
$$225 - \tfrac{2.25}{r} + 500r - 5 = 240$$
$$225r - 2.25 + 500r^2 - 5r = 240r$$
$$500r^2 - 20r - 2.25 = 0$$
$$2000r^2 - 80r - 9 = 0$$
$$(100r - 9)(20r + 1) = 0$$

$100r - 9 = 0$ **or** $20r + 1 = 0$

$r = 0.09$ $\qquad r = -0.05$

Substitute $r = 0.09$ and solve for p:

$p = \frac{225}{r} = \frac{225}{0.09} = 2500$

Ignacio invested \$2500 at 9%.

47. Let $r =$ Rania's rate, and let $p =$ the amount Rania invested.
Then Jerome invested $p + 150$ at a rate of $r + 0.015$. The equations are

$$\begin{cases} (1)\ pr = 67.50 \Rightarrow p = \frac{67.5}{r} \\ (2)\ (p+150)(r+0.015) = 94.5 \end{cases}$$

Substitute $p = \frac{67.5}{r}$ from (1) into (2):

$$\left(\tfrac{67.5}{r} + 150\right)(r + 0.015) = 94.5$$
$$67.5 + \tfrac{1.0125}{r} + 150r + 2.25 = 94.5$$
$$67.5r + 1.0125 + 150r^2 + 2.25r = 94.5r$$
$$150r^2 - 24.75r - 1.0125 = 0$$
$$12{,}000r^2 - 1980r + 81 = 0$$
$$(100r - 9)(120r - 9) = 0$$

$100r - 9 = 0$ **or** $120r - 9 = 0$

$r = 0.09$ $\qquad r = 0.075$

Substitute and solve for p:

$p = \frac{67.5}{r} = \frac{67.5}{0.09} = 750$ or $p = \frac{67.5}{r} = \frac{67.5}{0.075} = 900$

Rania invested \$750 at 9% or \$900 at 7.5%.

48. The point on the hill has coordinates $(3y, y)$.

$$y = -\frac{1}{6}x^2 + 2x$$
$$y = -\frac{1}{6}(3y)^2 + 2(3y)$$
$$y = -\frac{1}{6}(9y^2) + 6y$$
$$0 = -\frac{3}{2}y^2 + 5y$$
$$0 = 3y^2 - 10y$$
$$0 = y(3y - 10)$$

$y = 0$ **or** $3y - 10 = 0$

$y = \frac{10}{3}$

$$x = 3y = 3\left(\frac{10}{3}\right) = 10$$

$$d = \sqrt{(10-0)^2 + \left(\frac{10}{3} - 0\right)^2}$$
$$= \sqrt{100 + \frac{100}{9}}$$
$$= \sqrt{\frac{1000}{9}} = \frac{\sqrt{1000}}{3} = \frac{10\sqrt{10}}{3} \text{ miles}$$

49. Let r = Jim's rate and t = Jim's time. Then his brother's rate was $r - 17$ and his time was $t + 1.5$.

$$\begin{cases} (1)\ rt = 306 \Rightarrow t = \frac{306}{r} \\ (2)\ (r-17)(t+1.5) = 306 \end{cases}$$

Substitute $t = \frac{306}{r}$ from (1) into (2):

$$\begin{aligned} (r-17)\left(\tfrac{306}{r} + 1.5\right) &= 306 \\ 306 + 1.5r - \tfrac{5202}{r} - 25.5 &= 306 \\ 306r + 1.5r^2 - 5202 - 25.5r &= 306r \\ 1.5r^2 - 25.5r - 5202 &= 0 \\ 3r^2 - 51r - 10{,}404 &= 0 \\ (3r+153)(r-68) &= 0 \end{aligned}$$

$3r + 153 = 0$ **or** $r - 68 = 0$

$r = -153/3$ $\qquad$ $r = 68$

Substitute and solve for t:

$t = \frac{306}{r} = \frac{306}{68} = 4.5$

Jim drove for 4.5 hours at 68 miles per hour.

50. **Answers may vary.**

51. **Answers may vary.**

52. $0, 1, 2, 3, 4$

53. $0, 1, 2, 3, 4$

Exercise 10.5 (page 701)

1.

$$\begin{aligned} (6x-10) + (3x+10) &= 180 \\ 9x &= 180 \\ x &= 20 \end{aligned}$$

2.

$$\begin{aligned} (4x-20) + (10x-10) &= 180 \\ 14x - 30 - 180 \\ 14x &= 210 \\ x &= 15 \end{aligned}$$

3. domains

4. increasing

5. constant; $f(x)$

6. increase

7. step

8. greatest integer

9. increasing on $(-\infty, 0)$, decreasing on $(0, \infty)$

10. constant on $(-\infty, 0)$, decreasing on $(0, \infty)$

11. decreasing on $(-\infty, 0)$, constant on $(0, 2)$ increasing on $(2, \infty)$

12. constant on $(-\infty, 0)$, increasing on $(0, 2)$ decreasing on $(2, \infty)$

13. $f(x) = \begin{cases} -1 & \text{if } x \le 0 \\ x & \text{if } x > 0 \end{cases}$

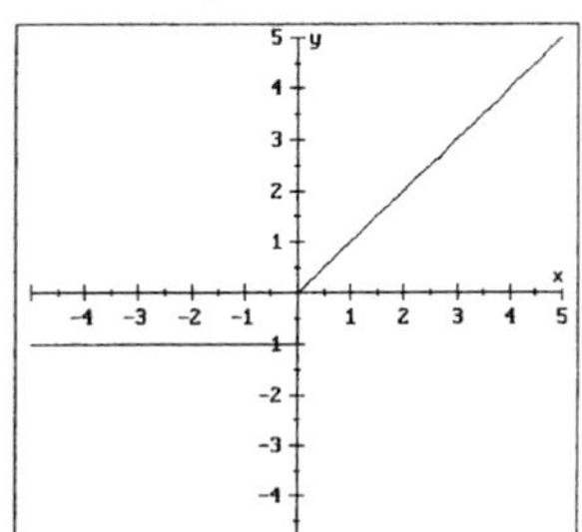

constant on $(-\infty, 0)$
increasing on $(0, \infty)$

14. $f(x) = \begin{cases} -2 & \text{if } x \le 0 \\ x^2 & \text{if } x > 0 \end{cases}$

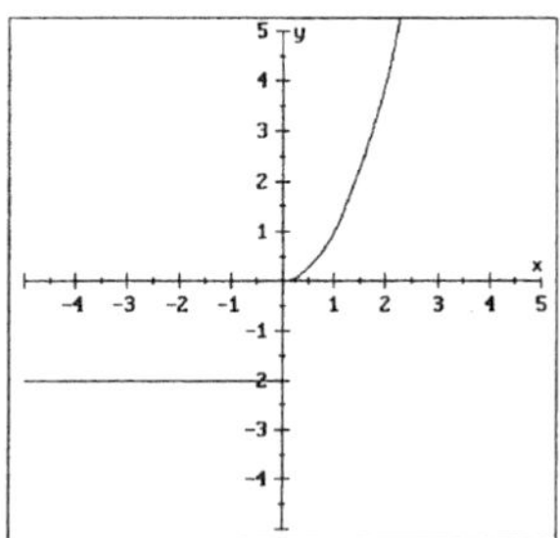

constant on $(-\infty, 0)$
increasing on $(0, \infty)$

15. $f(x) = \begin{cases} -x & \text{if } x \le 0 \\ x & \text{if } 0 < x < 2 \\ -x & \text{if } x \ge 2 \end{cases}$

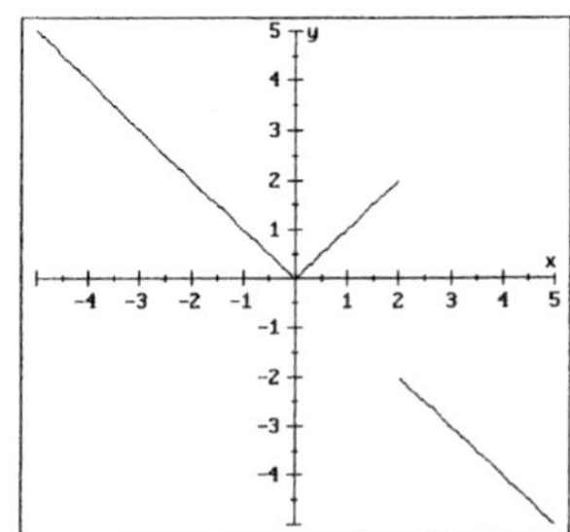

decreasing on $(-\infty, 0)$
increasing on $(0, 2)$
decreasing on $(2, \infty)$

16. $f(x) = \begin{cases} -x & \text{if } x < 0 \\ x^2 & \text{if } 0 \le x \le 1 \\ 1 & \text{if } x > 1 \end{cases}$

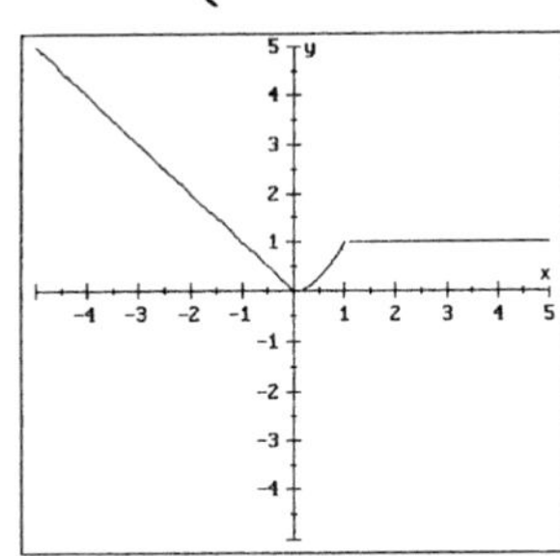

decreasing on $(-\infty, 0)$
increasing on $(0, 1)$
constant on $(1, \infty)$

17. $f(x) = -[[x]]$

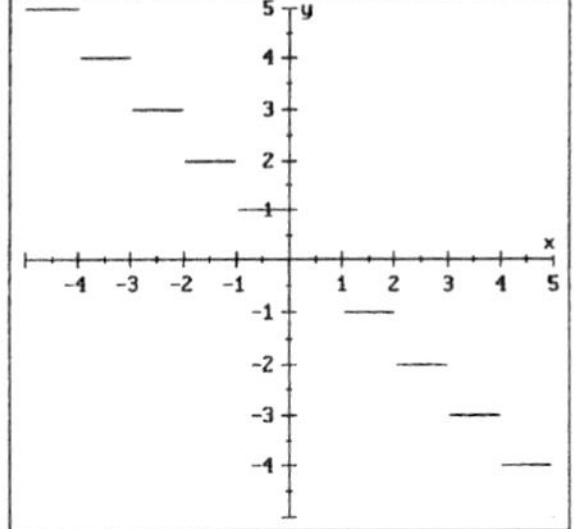

18. $f(x) = [[x]] + 2$

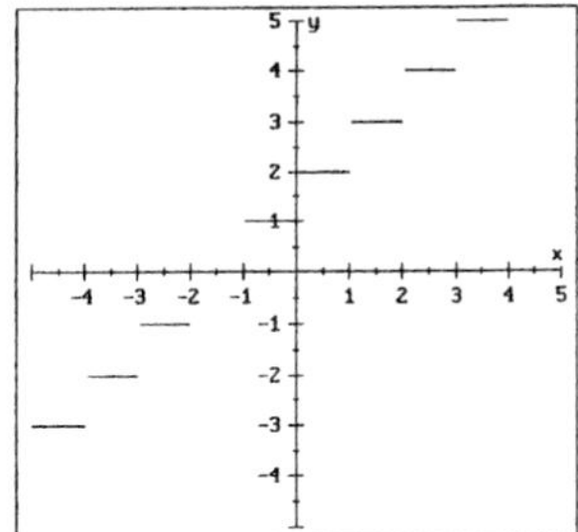

19. $f(x) = 2[[x]]$

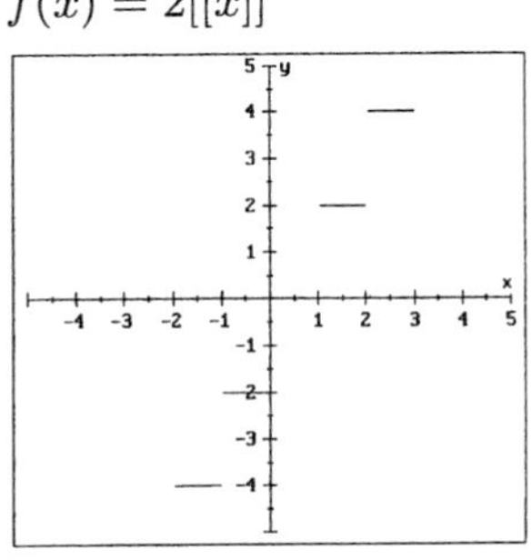

20. $f(x) = \left[\left[\frac{1}{2}x\right]\right]$

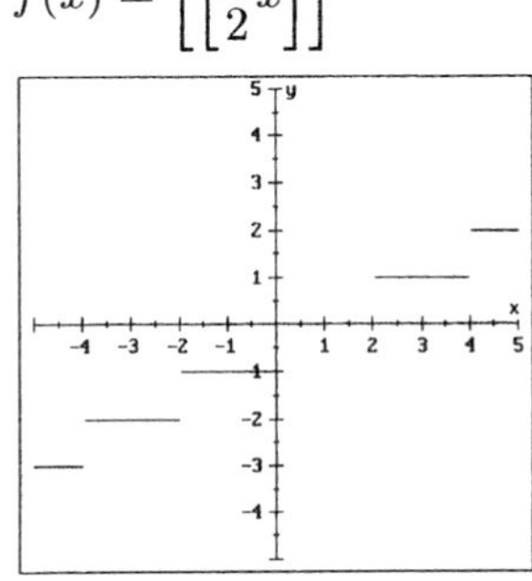

21. $f(x) = \begin{cases} -1 & \text{if } x < 0 \\ 0 & \text{if } x = 0 \\ 1 & \text{if } x > 0 \end{cases}$

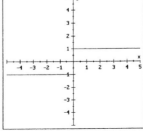

22. $f(x) = \begin{cases} 1 & \text{if } x > 0 \\ 0 & \text{if } x < 0 \end{cases}$

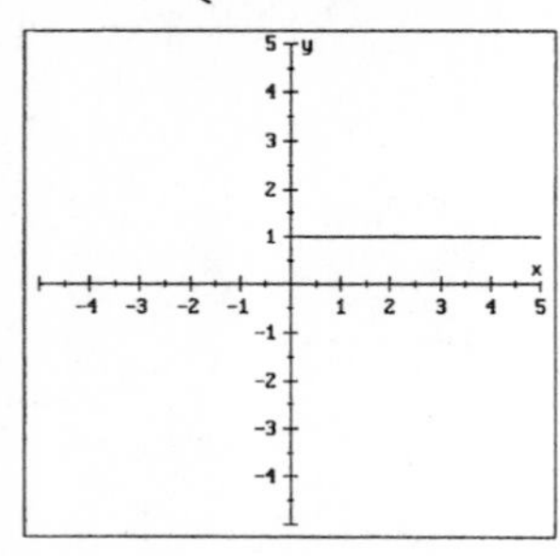

23. Find y when $x = 2.5$.
cost = \$30

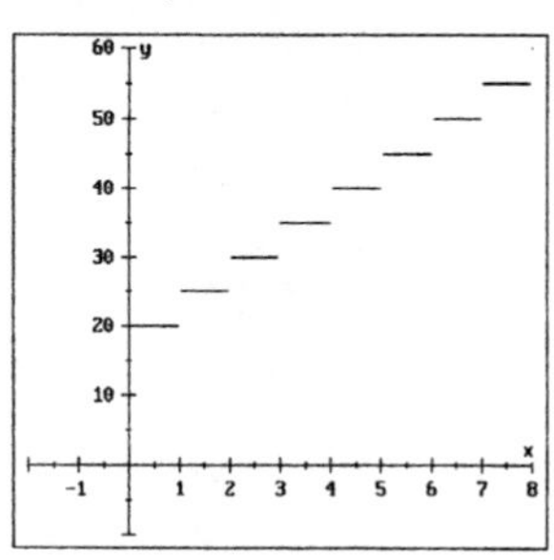

24. Find y when $x = 10.25$.
cost = \$23

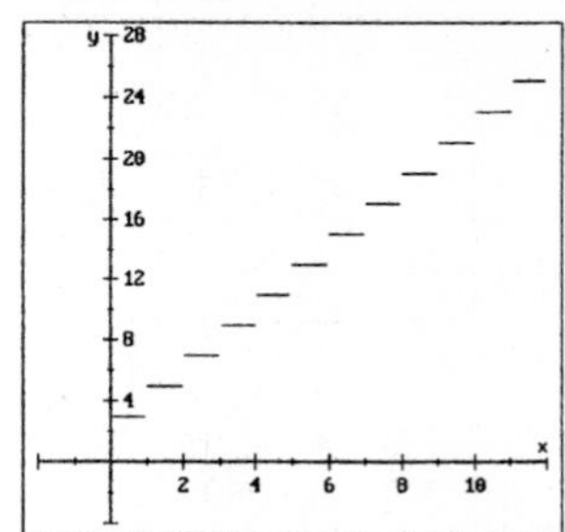

25. After 2 hours, B is cheaper.

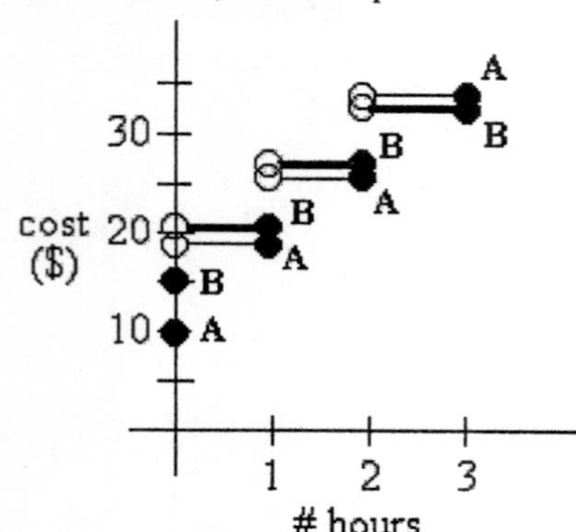

26. The royalties are \$0.70 per book for the first 50,000 books, so $I = 0.7s$ for $0 \le s \le 50{,}000$. The first 50,000 books earn \$35,000 therefore. After that, the royalties are \$1 per book, and the # of books over 50,000 is $s - 50{,}000$. The income on the books over 50,000 is then $1(s - 50{,}000)$. The total income when $s > 50{,}000$ is then $35{,}000 + (s - 50{,}000)$, or $s - 15{,}000$. The function I is described by:

$$I = \begin{cases} 0.7s & \text{if } 0 \le s \le 50{,}000 \\ s - 15{,}000 & \text{if } s > 50{,}000 \end{cases}$$

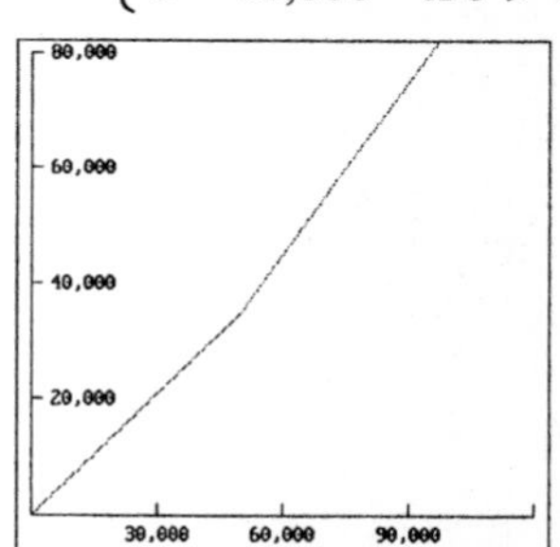

27. **Answers may vary.**

28. **Answers may vary.**

29. $f(x) = \begin{cases} x & \text{if } x < -2 \\ -x & \text{if } x > -2 \end{cases}$

30. $f(x) = \begin{cases} 1 & \text{if } x < 0 \\ x & \text{if } 0 < x < 5 \\ -x & \text{if } x > 5 \end{cases}$

Chapter 10 Summary (page 704)

1. $(x-1)^2 + (y+2)^2 = 9$
C $(1, -2)$; $r = \sqrt{9} = 3$

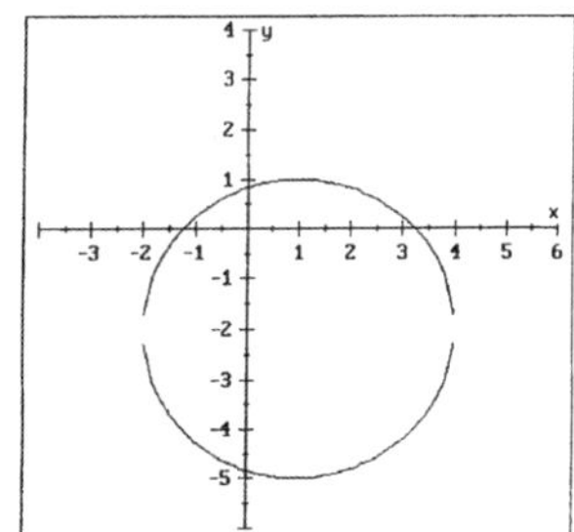

2. $x^2 + y^2 = 16$
C $(0, 0)$; $r = \sqrt{16} = 4$

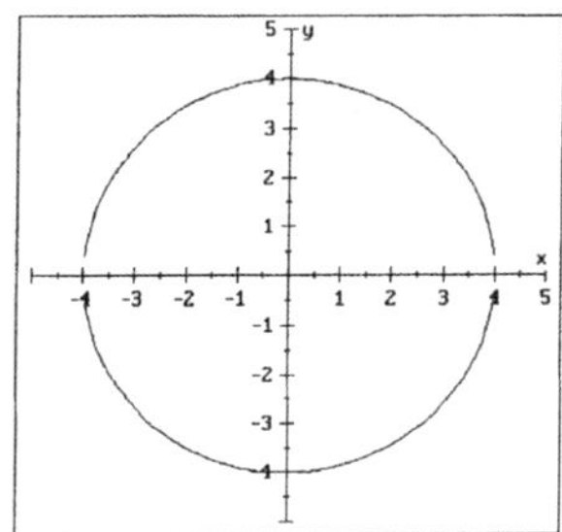

3.
$$x^2 + y^2 + 4x - 2y = 4$$
$$x^2 + 4x + y^2 - 2y = 4$$
$$x^2 + 4x + 4 + y^2 - 2y + 1 = 4 + 4 + 1$$
$$(x+2)^2 + (y-1)^2 = 9$$
C $(-2, 1)$; $r = \sqrt{9} = 3$

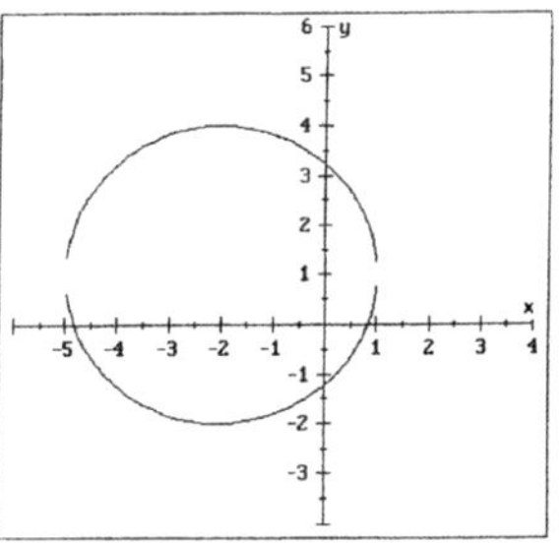

4. $x = -3(y-2)^2 + 5$
V $(5, 2)$; opens L

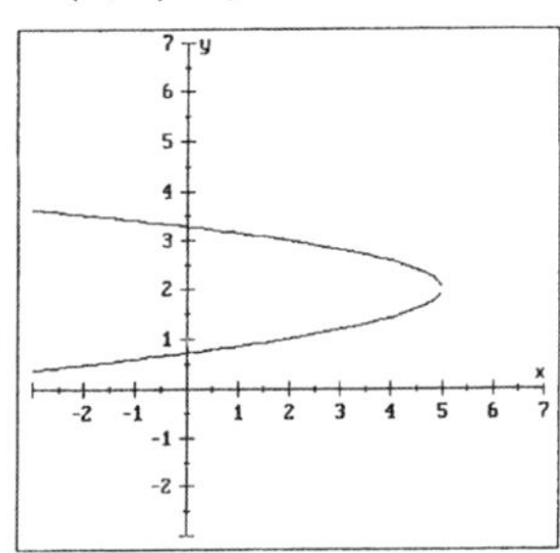

5. $x = 2(y+1)^2 - 2$
V $(-2, -1)$; opens R

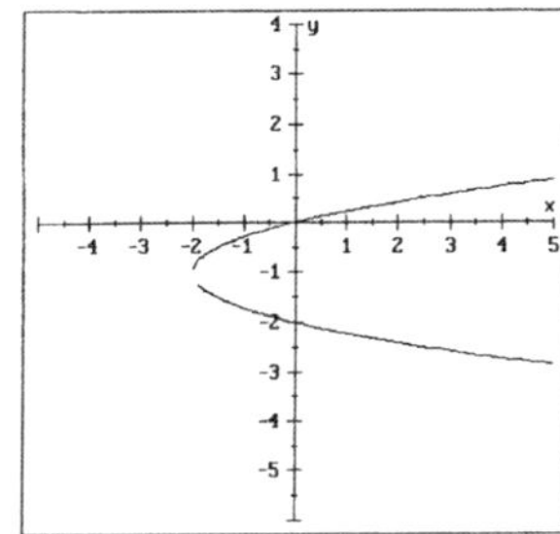

6. $9x^2 + 16y^2 = 144$

$$\frac{9x^2}{144} + \frac{16y^2}{144} = \frac{144}{144}$$

$$\frac{x^2}{16} + \frac{y^2}{9} = 1$$

C $(0,0)$; move 4 horiz. and 3 vert.

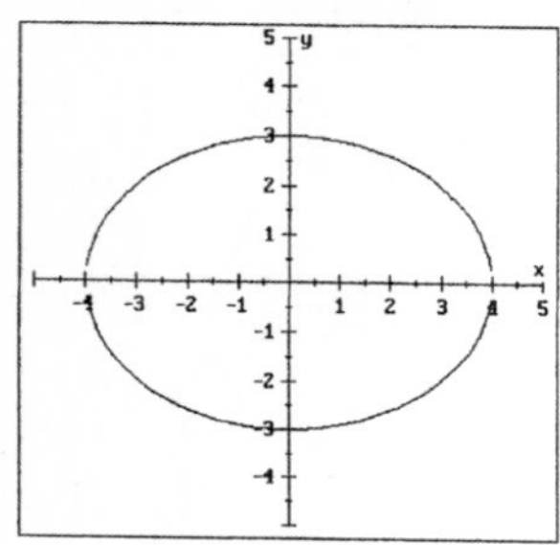

7. $\dfrac{(x-2)^2}{4} + \dfrac{(y-1)^2}{9} = 1$

C $(2,1)$; move 2 horiz. and 3 vert.

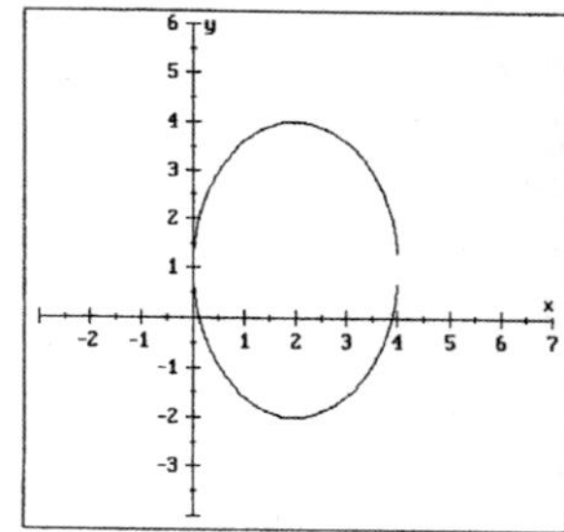

8.

$$4x^2 + 9y^2 + 8x - 18y = 23$$

$$4x^2 + 8x + 9y^2 - 18y = 23$$

$$4(x^2 + 2x) + 9(y^2 - 2y) = 23$$

$$4(x^2 + 2x + 1) + 9(y^2 - 2y + 1) = 23 + 4 + 9$$

$$4(x+1)^2 + 9(y-1)^2 = 36$$

$$\frac{(x+1)^2}{9} + \frac{(y-1)^2}{4} = 1$$

C $(-1,1)$; move 3 horiz. and 2 vert.

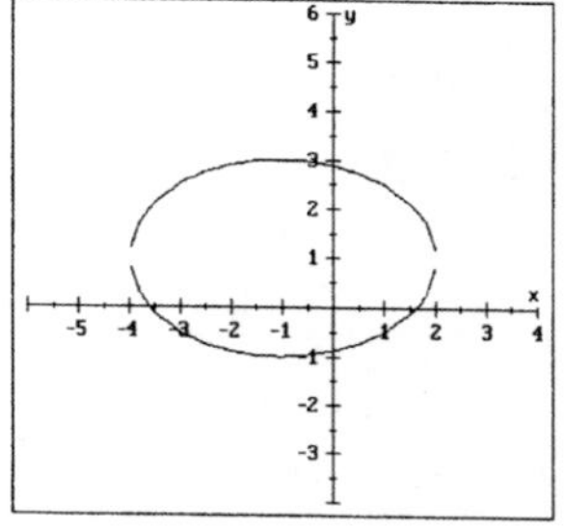

9. $9x^2 - y^2 = -9$

$$\frac{9x^2}{-9} - \frac{y^2}{-9} = \frac{-9}{-9}$$

$$\frac{y^2}{9} - \frac{x^2}{1} = 1$$

C $(0,0)$; opens vert.; move 1 horiz. and 3 vert.

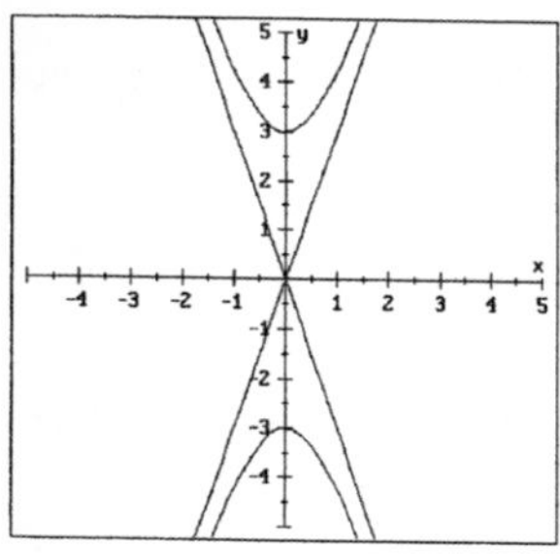

10. 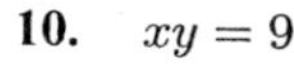 $xy = 9$

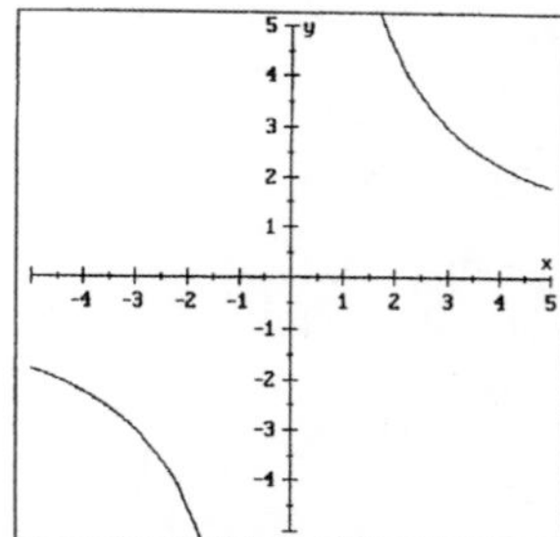

11.
$$\begin{aligned}
4x^2 - 2y^2 + 8x - 8y &= 8 \\
4x^2 + 8x - 2y^2 - 8y &= 8 \\
4(x^2 + 2x) - 2(y^2 + 4y) &= 8 \\
4(x^2 + 2x + 1) - 2(y^2 + 4y + 4) &= 8 \\
4(x+1)^2 - 2(y+2)^2 &= 8 \\
\frac{(x+1)^2}{2} - \frac{(y+2)^2}{4} &= 1 \Rightarrow \text{hyperbola}
\end{aligned}$$

12.
$$\begin{aligned}
9x^2 - 4y^2 - 18x - 8y &= 31 \\
9x^2 - 18x - 4y^2 - 8y &= 31 \\
9(x^2 - 2x) - 4(y^2 + 2y) &= 31 \\
9(x^2 - 2x + 1) - 4(y^2 + 2y + 1) &= 31 + 9 - 4 \\
9(x-1)^2 - 4(y+1)^2 &= 36 \\
\frac{(x-1)^2}{4} - \frac{(y+1)^2}{9} &= 1
\end{aligned}$$

C $(1, -1)$; opens horiz.; move 2 horiz. and 3 vert.

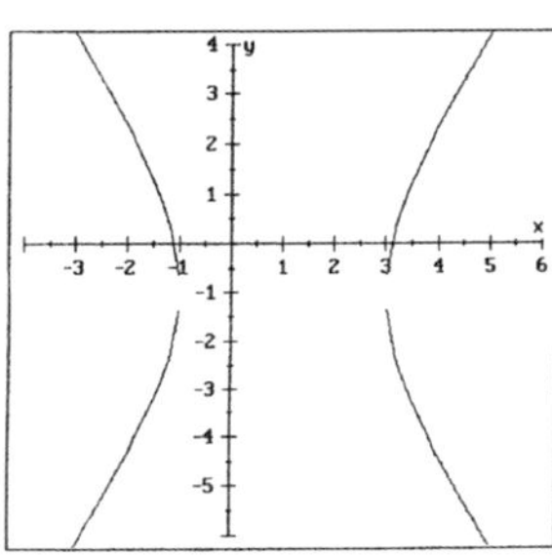

13.
$$\begin{aligned}
3x^2 + y^2 &= 52 \\
x^2 - y^2 &= 12 \\
\hline
4x^2 &= 64 \\
x^2 &= 16 \\
x &= \pm 4
\end{aligned}$$

Substitute and solve for y:

$$\begin{aligned}
x^2 - y^2 &= 12 & x^2 - y^2 &= 12 \\
4^2 - y^2 &= 12 & (-4)^2 - y^2 &= 12 \\
y^2 &= 4 & y^2 &= 4 \\
y &= \pm 2 & y &= \pm 2
\end{aligned}$$

Solutions: $(4,2), (4,-2), (-4,2), (-4,-2)$.

14.
$$\begin{aligned}
\tfrac{x^2}{16} + \tfrac{y^2}{12} = 1 \Rightarrow \times 48 \qquad 3x^2 + 4y^2 &= 48 \\
x^2 - \tfrac{y^2}{3} = 1 \Rightarrow \times (-3) \quad -3x^2 + y^2 &= -3 \\
\hline
5y^2 &= 45 \\
y^2 &= 9 \\
y &= \pm 3
\end{aligned}$$

Substitute and solve for x:

$$\begin{aligned}
3x^2 + 4y^2 &= 48 \\
3x^2 + 4(9) &= 48 \\
3x^2 &= 12 \\
x^2 &= 4 \Rightarrow x = \pm 2
\end{aligned}$$

Solutions: $(2,3), (2,-3), (-2,3), (-2,-3)$

15. increasing on $(-\infty, -2)$; constant on $(-2, 1)$; decreasing on $(1, \infty)$

16. $f(x) = \begin{cases} x & \text{if } x \leq 1 \\ -x^2 & \text{if } x > 1 \end{cases}$

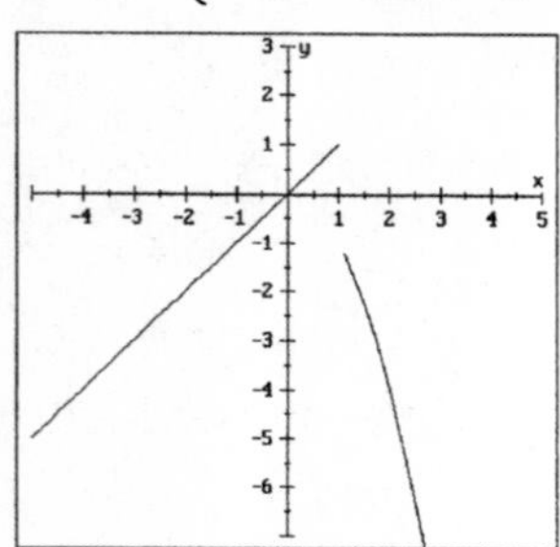

17. $f(x) = 3[[x]]$

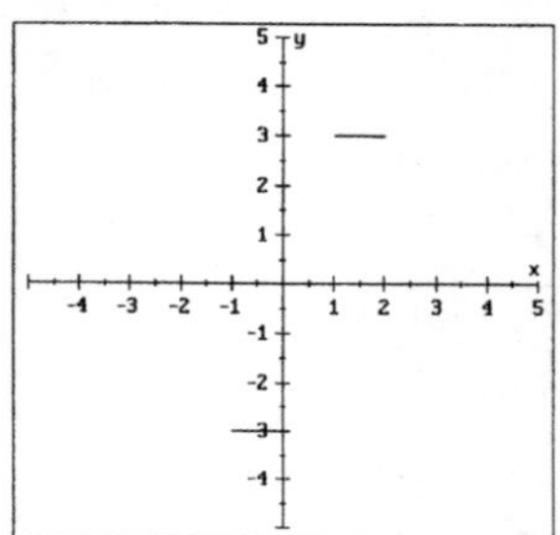

Chapter 10 Test (page 707)

1. $(x-2)^2 + (y+3)^2 = 4$

Center: $(2, -3)$; radius $= 2$

2.

$$x^2 + y^2 + 4x - 6y = 3$$
$$x^2 + 4x + y^2 - 6y = 3$$
$$x^2 + 4x + 4 + y^2 - 6y + 9 = 3 + 4 + 9$$
$$(x+2)^2 + (y-3)^2 = 16$$

Center: $(-2, 3)$; radius $= 4$

3. $(x+1)^2 + (y-2)^2 = 9$

C $(-1, 2)$; $r = \sqrt{9} = 3$

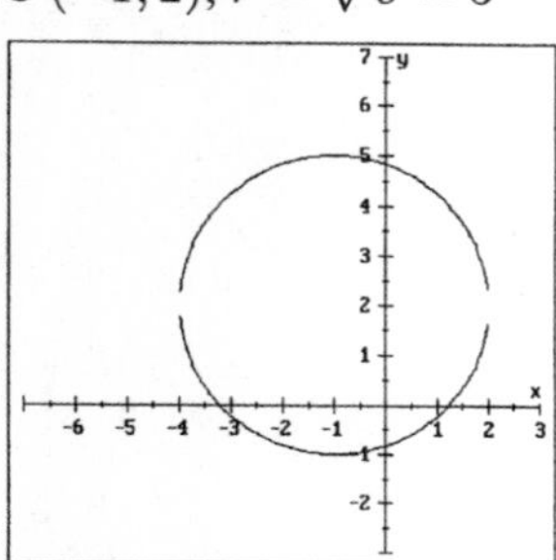

4. $x = (y-2)^2 - 1$

V $(-1, 2)$; opens R

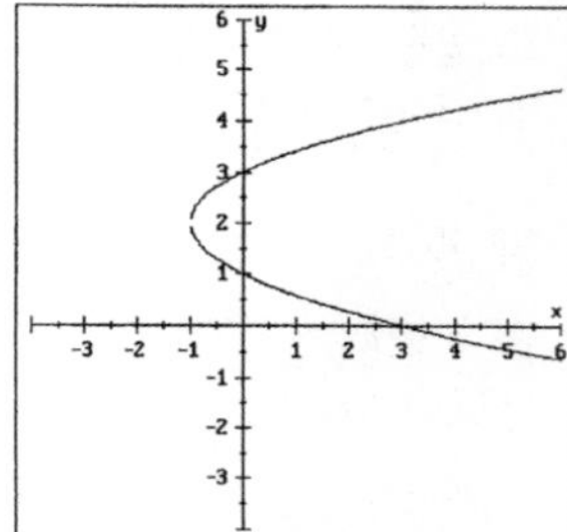

5. $9x^2 + 4y^2 = 36$

$$\frac{9x^2}{36} + \frac{4y^2}{36} = \frac{36}{36}$$

$$\frac{x^2}{4} + \frac{y^2}{9} = 1$$

C $(0, 0)$; move 2 horiz. and 3 vert.

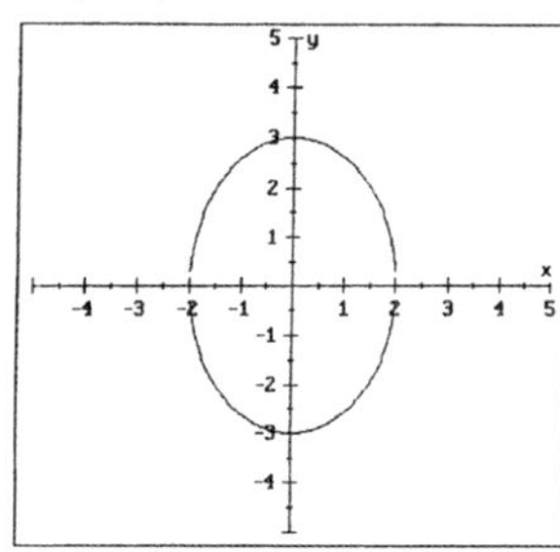

6. $\dfrac{(x-2)^2}{9} - y^2 = 1$

C $(2, 0)$; opens horiz; move 3 horiz and 1 vert

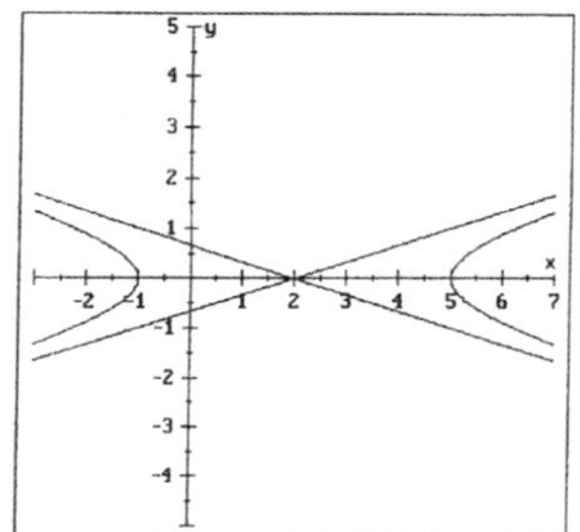

7.

$$4x^2 + y^2 - 24x + 2y = -33$$

$$4(x^2 - 6x) + (y^2 + 2y) = -33$$

$$4(x^2 - 6x + 9) + (y^2 + 2y + 1) = -33 + 36 + 1$$

$$4(x-3)^2 + (y+1)^2 = 4$$

$$\frac{(x-3)^2}{1} + \frac{(y+1)^2}{4} = 1$$

C $(3, -1)$; move 1 horiz. and 2 vert.

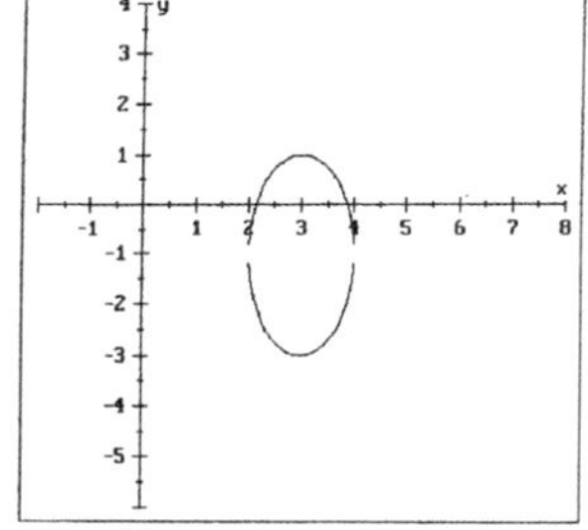

8.

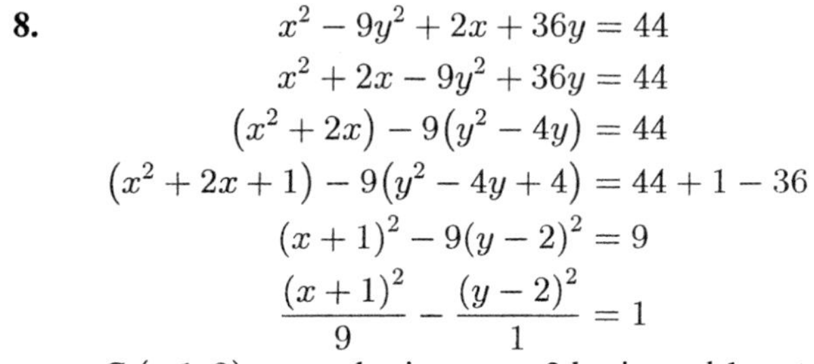

$$x^2 - 9y^2 + 2x + 36y = 44$$

$$x^2 + 2x - 9y^2 + 36y = 44$$

$$(x^2 + 2x) - 9(y^2 - 4y) = 44$$

$$(x^2 + 2x + 1) - 9(y^2 - 4y + 4) = 44 + 1 - 36$$

$$(x+1)^2 - 9(y-2)^2 = 9$$

$$\frac{(x+1)^2}{9} - \frac{(y-2)^2}{1} = 1$$

C $(-1, 2)$; opens horiz.; move 3 horiz. and 1 vert.

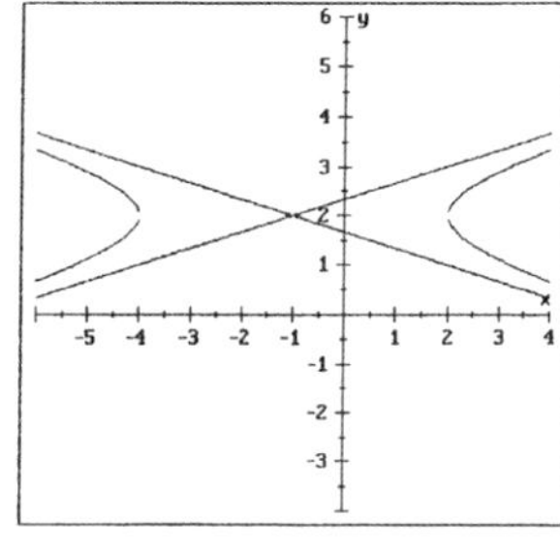

9. $\begin{cases} (1) & 2x - y = -2 \\ (2) & x^2 + y^2 = 16 + 4y \end{cases}$

Substitute $y = 2x + 2$ from (1) into (2):

$$\begin{aligned} x^2 + y^2 &= 16 + 4y \\ x^2 + (2x+2)^2 &= 16 + 4(2x+2) \\ x^2 + 4x^2 + 8x + 4 &= 16 + 8x + 8 \\ 5x^2 - 20 &= 0 \\ 5(x+2)(x-2) &= 0 \end{aligned}$$

$x + 2 = 0$ **or** $x - 2 = 0$
$x = -2$ $\quad x = 2$

Substitute these and solve for y:

$$\begin{aligned} y &= 2x + 2 & y &= 2x + 2 \\ y &= 2(-2) + 2 & y &= 2(2) + 2 \\ y &= -4 + 2 = -2 & y &= 4 + 2 = 6 \end{aligned}$$

Solutions: $(-2, -2), (2, 6)$

10. $\begin{cases} (1) & x^2 + y^2 = 25 \\ (2) & 4x^2 - 9y = 0 \end{cases}$

Substitute $x^2 = 25 - y^2$ from (1) into (2):

$$\begin{aligned} 4x^2 - 9y &= 0 \\ 4(25 - y^2) - 9y &= 0 \\ 100 - 4y^2 - 9y &= 0 \\ -4y^2 - 9y + 100 &= 0 \\ 4y^2 + 9y - 100 &= 0 \\ (y-4)(4y+25) &= 0 \end{aligned}$$

$y - 4 = 0$ **or** $4y + 25 = 0$
$y = 4$ $\quad y = -\frac{25}{4}$

Substitute these and solve for x:

$$\begin{aligned} x^2 &= 25 - y^2 & x^2 &= 25 - y^2 \\ x^2 &= 25 - 4^2 & x^2 &= 25 - \left(-\tfrac{25}{4}\right)^2 \\ x^2 &= 25 - 16 = 9 & x^2 &= 25 - \tfrac{625}{16} = -\tfrac{225}{16} \\ x &= \pm 3 & & x \text{ is nonreal.} \end{aligned}$$

Solutions: $(3, 4), (-3, 4)$

11. increasing: $(-3, 0)$; decreasing: $(0, 3)$

12. $f(x) = \begin{cases} -x^2 & \text{when } x < 0 \\ -x & \text{when } x \geq 0 \end{cases}$

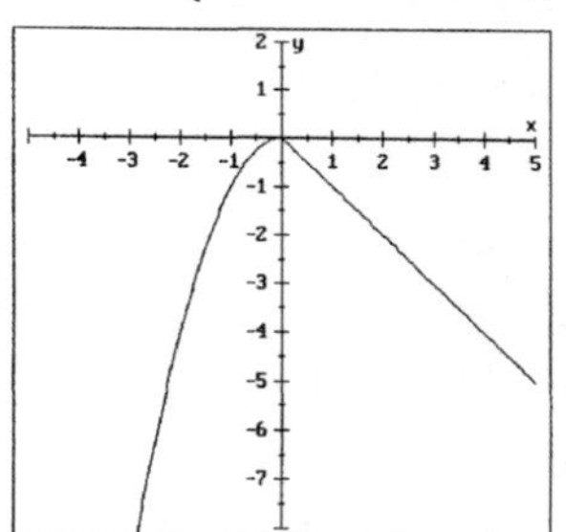

Cumulative Review Exercises (page 708)

1. $(4x - 3y)(3x + y) = 12x^2 + 4xy - 9xy - 3y^2 = 12x^2 - 5xy - 3y^2$

2. $(a^n + 1)(a^n - 3) = a^n a^n - 3a^n + a^n - 3 = a^{2n} - 2a^n - 3$

3. $\dfrac{5a - 10}{a^2 - 4a + 4} = \dfrac{5(a-2)}{(a-2)(a-2)} = \dfrac{5}{a-2}$

4. $\dfrac{a^4 - 5a^2 + 4}{a^2 + 3a + 2} = \dfrac{(a^2-4)(a^2-1)}{(a+2)(a+1)} = \dfrac{(a+2)(a-2)(a+1)(a-1)}{(a+2)(a+1)} = (a-2)(a-1)$
$= a^2 - 3a + 2$

5. $$\frac{a^2-a-6}{a^2-4} \div \frac{a^2-9}{a^2+a-6} = \frac{a^2-a-6}{a^2-4} \cdot \frac{a^2+a-6}{a^2-9} = \frac{(a-3)(a+2)}{(a+2)(a-2)} \cdot \frac{(a+3)(a-2)}{(a+3)(a-3)} = 1$$

6. $$\begin{aligned}\frac{2}{a-2} + \frac{3}{a+2} - \frac{a-1}{a^2-4} &= \frac{2}{a-2} + \frac{3}{a+2} - \frac{a-1}{(a+2)(a-2)} \\ &= \frac{2(a+2)}{(a-2)(a+2)} + \frac{3(a-2)}{(a+2)(a-2)} - \frac{a-1}{(a+2)(a-2)} \\ &= \frac{2(a+2)+3(a-2)-(a-1)}{(a+2)(a-2)} \\ &= \frac{2a+4+3a-6-a+1}{(a+2)(a-2)} = \frac{4a-1}{(a+2)(a-2)}\end{aligned}$$

7. $$\begin{aligned}3x-4y &= 12 \\ -4y &= -3x+12 \\ y &= \frac{3}{4}x-3 \\ m &= \frac{3}{4}\end{aligned} \qquad \begin{aligned} y &= \frac{3}{4}x-5 \\ m &= \frac{3}{4}\end{aligned}$$

Parallel

8. $$\begin{aligned}y &= 3x+4 \\ m &= 3\end{aligned} \qquad \begin{aligned}x &= -3y+4 \\ 3y &= -x+4 \\ y &= -\frac{1}{3}x+\frac{4}{3} \\ m &= -\frac{1}{3}\end{aligned}$$

Perpendicular

9. $$\begin{aligned}y-y_1 &= m(x-x_1) \\ y-5 &= -2(x-0) \\ y-5 &= -2x \\ y &= -2x+5\end{aligned}$$

10. $$\begin{aligned}m = \frac{y_2-y_1}{x_2-x_1} &= \frac{4-(-5)}{-5-8} = \frac{9}{-13} = -\frac{9}{13} \\ y-y_1 &= m(x-x_1) \\ y-(-5) &= -\frac{9}{13}(x-8) \\ y+5 &= -\frac{9}{13}x+\frac{72}{13} \\ y &= -\frac{9}{13}x+\frac{7}{13}\end{aligned}$$

11. $2x-3y<6$

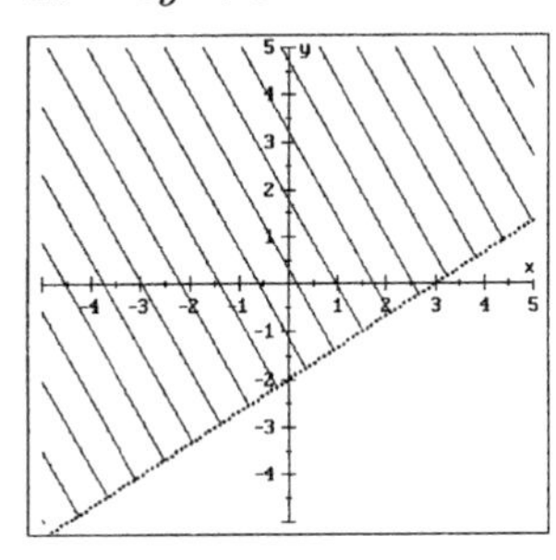

12. $y \geq x^2-4$

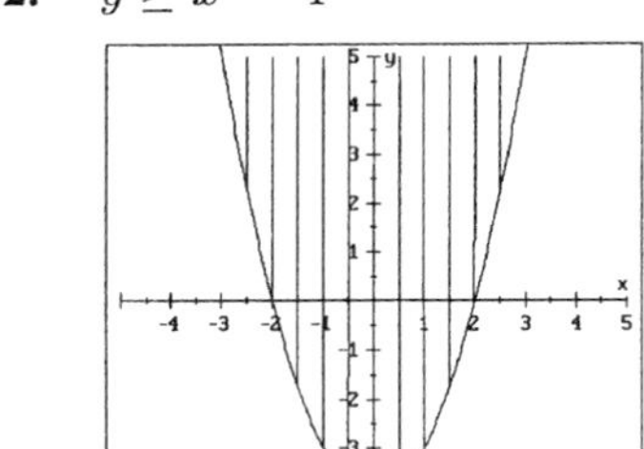

13. $\sqrt{98}+\sqrt{8}-\sqrt{32} = \sqrt{49}\sqrt{2}+\sqrt{4}\sqrt{2}-\sqrt{16}\sqrt{2} = 7\sqrt{2}+2\sqrt{2}-4\sqrt{2} = 5\sqrt{2}$

CUMULATIVE REVIEW EXERCISES

14. $12\sqrt[3]{648x^4} + 3\sqrt[3]{81x^4} = 12\sqrt[3]{216x^3}\sqrt[3]{3x} + 3\sqrt[3]{27x^3}\sqrt[3]{3x} = 12(6x)\sqrt[3]{3x} + 3(3x)\sqrt[3]{3x}$
$= 72x\sqrt[3]{3x} + 9x\sqrt[3]{3x} = 81x\sqrt[3]{3x}$

15.
$$\begin{aligned}
\sqrt{3a+1} &= a-1 \\
\left(\sqrt{3a+1}\right)^2 &= (a-1)^2 \\
3a+1 &= a^2-2a+1 \\
0 &= a^2-5a \\
0 &= a(a-5)
\end{aligned}$$
$a = 0$ **or** $a - 5 = 0$
doesn't check $\quad a = 5$

16.
$$\begin{aligned}
\sqrt{x+3} - \sqrt{3} &= \sqrt{x} \\
\sqrt{x+3} &= \sqrt{x} + \sqrt{3} \\
\left(\sqrt{x+3}\right)^2 &= \left(\sqrt{x}+\sqrt{3}\right)^2 \\
x+3 &= x + 2\sqrt{3x} + 3 \\
0 &= 2\sqrt{3x} \\
0^2 &= \left(2\sqrt{3x}\right)^2 \\
0 &= 4(3x) \\
0 &= 12x \\
0 &= x
\end{aligned}$$

17.
$$\begin{aligned}
6a^2 + 5a - 6 &= 0 \\
(2a+3)(3a-2) &= 0
\end{aligned}$$
$2a + 3 = 0$ **or** $3a - 2 = 0$
$a = -\frac{3}{2} \quad a = \frac{2}{3}$

18.
$$\begin{aligned}
3x^2 + 8x - 1 &= 0 \\
a = 3, b = 8, c &= -1 \\
x &= \frac{-b \pm \sqrt{b^2-4ac}}{2a} \\
&= \frac{-8 \pm \sqrt{8^2 - 4(3)(-1)}}{2(3)} \\
&= \frac{-8 \pm \sqrt{64+12}}{6} \\
&= \frac{-8 \pm \sqrt{76}}{6} \\
&= -\frac{8}{6} \pm \frac{2\sqrt{19}}{6} = -\frac{4}{3} \pm \frac{\sqrt{19}}{3}
\end{aligned}$$

19. $(f \circ g)(x) = f(g(x)) = f(2x+1) = (2x+1)^2 - 2 = 4x^2 + 4x + 1 - 2 = 4x^2 + 4x - 1$

20.
$$\begin{aligned}
y &= 2x^3 - 1 \\
x &= 2y^3 - 1 \\
x + 1 &= 2y^3 \\
\frac{x+1}{2} &= y^3 \\
\sqrt[3]{\frac{x+1}{2}} &= y \\
y = f^{-1}(x) &= \sqrt[3]{\frac{x+1}{2}}
\end{aligned}$$

21. $y = \left(\frac{1}{2}\right)^x$

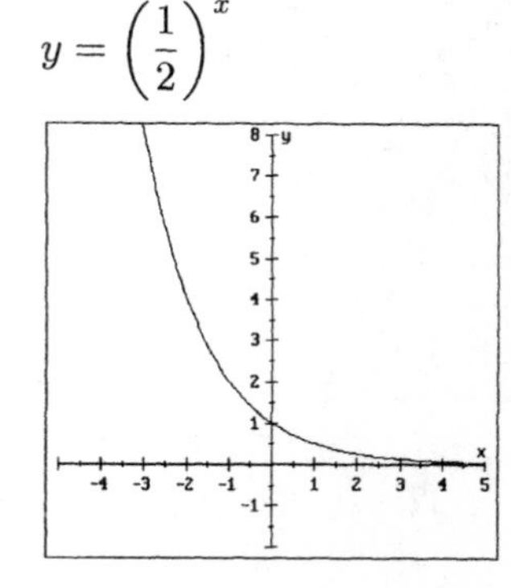

22. $y = \log_2 x \Rightarrow 2^y = x$

23.
$$\begin{aligned} 2^{x+2} &= 3^x \\ \log 2^{x+2} &= \log 3^x \\ (x+2)\log 2 &= x\log 3 \\ x\log 2 + 2\log 2 &= x\log 3 \\ 2\log 2 &= x\log 3 - x\log 2 \\ 2\log 2 &= x(\log 3 - \log 2) \\ \frac{2\log 2}{\log 3 - \log 2} &= x \end{aligned}$$

24.
$$\begin{aligned} 2\log 5 + \log x - \log 4 &= 2 \\ \log 5^2 + \log x - \log 4 &= 2 \\ \log \tfrac{25x}{4} &= 2 \\ 10^2 &= \tfrac{25x}{4} \\ 400 &= 25x \\ 16 &= x \end{aligned}$$

25. $x^2 + (y+1)^2 = 9$

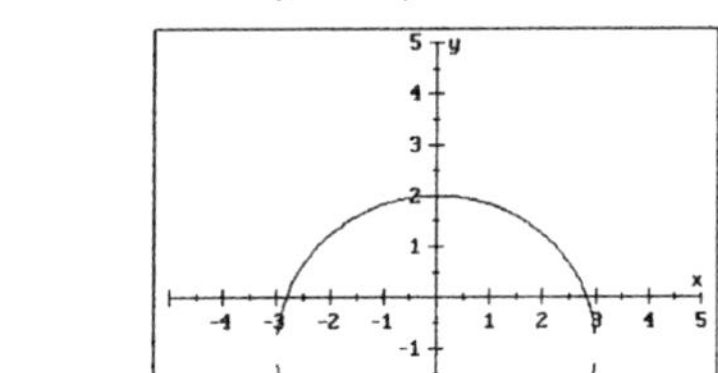

26. $x^2 - 9(y+1)^2 = 9$

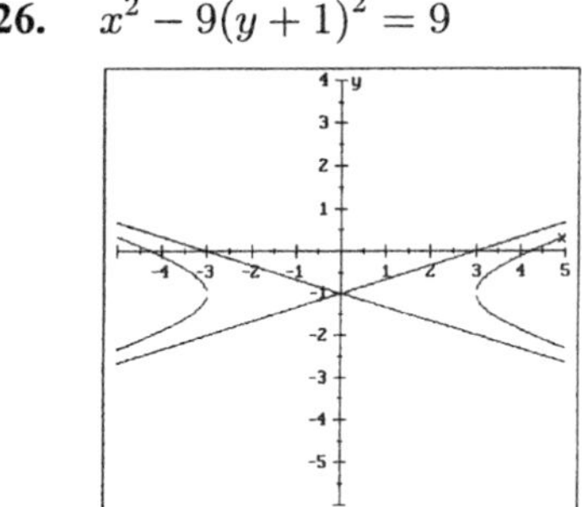

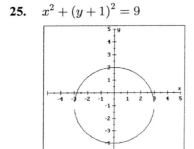

Exercise 11.1 (page 716)

1. $\log_4 16 = x \Rightarrow 4^x = 16 \Rightarrow x = 2$

2. $\log_x 49 = 2 \Rightarrow x^2 = 49 \Rightarrow x = 7$

3. $\log_{25} x = \frac{1}{2} \Rightarrow 25^{1/2} = x \Rightarrow x = 5$

4. $\log_{1/2} \frac{1}{8} = x \Rightarrow \left(\frac{1}{2}\right)^x = \frac{1}{8} \Rightarrow x = 3$

5. one

6. a^{20}

7. Pascal's

8. five factorial

9. 6!

10. $8 \cdot 7!$

11. 1

12. $\dfrac{n!}{2!(n-2)!}a^{n-2}b^2$

13. $3! = 3 \cdot 2 \cdot 1 = 6$

14. $7! = 7 \cdot 6 \cdot 5 \cdot 4 \cdot 3 \cdot 2 \cdot 1 = 5040$

15. $-5! = -1 \cdot 5! = -1 \cdot 5 \cdot 4 \cdot 3 \cdot 2 \cdot 1$
$= -120$

16. $-6! = -1 \cdot 6! = -1 \cdot 6 \cdot 5 \cdot 4 \cdot 3 \cdot 2 \cdot 1$
$= -720$

17. $3! + 4! = 3 \cdot 2 \cdot 1 + 4 \cdot 3 \cdot 2 \cdot 1$
$= 6 + 24 = 30$

18. $2!(3!) = 2 \cdot 1 \cdot 3 \cdot 2 \cdot 1 = 12$

19. $3!(4!) = 3 \cdot 2 \cdot 1 \cdot 4 \cdot 3 \cdot 2 \cdot 1 = 144$

20. $4! + 4! = 4 \cdot 3 \cdot 2 \cdot 1 + 4 \cdot 3 \cdot 2 \cdot 1$
$= 24 + 24 = 48$

21. $8(7!) = 8 \cdot 7 \cdot 6 \cdot 5 \cdot 4 \cdot 3 \cdot 2 \cdot 1 = 40{,}320$

22. $4!(5) = 4 \cdot 3 \cdot 2 \cdot 1 \cdot 5 = 120$

23. $\dfrac{9!}{11!} = \dfrac{9!}{11 \cdot 10 \cdot 9!} = \dfrac{1}{11 \cdot 10} = \dfrac{1}{110}$

24. $\dfrac{13!}{10!} = \dfrac{13 \cdot 12 \cdot 11 \cdot 10!}{10!} = 13 \cdot 12 \cdot 11 = 1716$

25. $\dfrac{49!}{47!} = \dfrac{49 \cdot 48 \cdot 47!}{47!} = 49 \cdot 48 = 2352$

26. $\dfrac{101!}{100!} = \dfrac{101 \cdot 100!}{100!} = 101$

27. $\dfrac{9!}{7!\,0!} = \dfrac{9 \cdot 8 \cdot 7!}{7! \cdot 1} = 9 \cdot 8 = 72$

28. $\dfrac{7!}{5!\,0!} = \dfrac{7 \cdot 6 \cdot 5!}{5! \cdot 1} = 7 \cdot 6 = 42$

29. $\dfrac{5!}{3!(5-3)!} = \dfrac{5!}{3!2!} = \dfrac{5 \cdot 4 \cdot 3!}{3! \cdot 2 \cdot 1} = \dfrac{5 \cdot 4}{2 \cdot 1} = 10$

30. $\dfrac{6!}{4!(6-4)!} = \dfrac{6!}{4!2!} = \dfrac{6 \cdot 5 \cdot 4!}{4! \cdot 2 \cdot 1} = \dfrac{6 \cdot 5}{2 \cdot 1} = 15$

31. $\dfrac{7!}{5!(7-5)!} = \dfrac{7!}{5!2!} = \dfrac{7 \cdot 6 \cdot 5!}{5! \cdot 2 \cdot 1} = \dfrac{7 \cdot 6}{2 \cdot 1} = 21$

32. $\dfrac{8!}{6!(8-6)!} = \dfrac{8!}{6!2!} = \dfrac{8 \cdot 7 \cdot 6!}{6! \cdot 2 \cdot 1} = \dfrac{8 \cdot 7}{2 \cdot 1} = 28$

33. $\dfrac{5!(8-5)!}{4!7!} = \dfrac{5!3!}{4!7!} = \dfrac{5!}{7!} \cdot \dfrac{3!}{4!} = \dfrac{5!}{7 \cdot 6 \cdot 5!} \cdot \dfrac{3!}{4 \cdot 3!} = \dfrac{1}{7 \cdot 6} \cdot \dfrac{1}{4} = \dfrac{1}{168}$

34. $\dfrac{6!7!}{(8-3)!(7-4)!} = \dfrac{6!7!}{5!3!} = \dfrac{7!}{5!} \cdot \dfrac{6!}{3!} = \dfrac{7 \cdot 6 \cdot 5!}{5!} \cdot \dfrac{6 \cdot 5 \cdot 4 \cdot 3!}{3!} = 7 \cdot 6 \cdot 6 \cdot 5 \cdot 4 = 5040$

35. $11! = 39{,}916{,}800$

36. $13! = 6{,}227{,}020{,}800$

37. $20! = 2.432902008 \times 10^{18}$

38. $55! = 1.269640335 \times 10^{73}$

39.
$$\begin{aligned}(x+y)^3 = x^3 + \frac{3!}{1!(3-1)!}x^2y + \frac{3!}{2!(3-2)!}xy^2 + y^3 &= x^3 + \frac{3!}{1!2!}x^2y + \frac{3!}{2!1!}xy^2 + y^3 \\ &= x^3 + \frac{3 \cdot 2!}{1!2!}x^2y + \frac{3 \cdot 2!}{2!1!}xy^2 + y^3 \\ &= x^3 + \frac{3}{1}x^2y + \frac{3}{1}xy^2 + y^3 \\ &= x^3 + 3x^2y + 3xy^2 + y^3\end{aligned}$$

40.
$$\begin{aligned}(x+y)^4 &= x^4 + \frac{4!}{1!(4-1)!}x^3y + \frac{4!}{2!(4-2)!}x^2y^2 + \frac{4!}{3!(4-3)!}xy^3 + y^4 \\ &= x^4 + \frac{4!}{1!3!}x^3y + \frac{4!}{2!2!}x^2y^2 + \frac{4!}{3!1!}xy^3 + y^4 \\ &= x^4 + \frac{4 \cdot 3!}{1!3!}x^3y + \frac{4 \cdot 3 \cdot 2!}{2! \cdot 2 \cdot 1}x^2y^2 + \frac{4 \cdot 3!}{3!1!}xy^3 + y^4 \\ &= x^4 + \frac{4}{1}x^3y + \frac{12}{2}x^2y^2 + \frac{4}{1}xy^3 + y^4 \\ &= x^4 + 4x^3y + 6x^2y^2 + 4xy^3 + y^4\end{aligned}$$

41. $$\begin{aligned}(x-y)^4 &= x^4 + \frac{4!}{1!(4-1)!}x^3(-y) + \frac{4!}{2!(4-2)!}x^2(-y)^2 + \frac{4!}{3!(4-3)!}x(-y)^3 + (-y)^4 \\ &= x^4 + \frac{4!}{1!3!}(-x^3y) + \frac{4!}{2!2!}x^2y^2 + \frac{4!}{3!1!}(-xy^3) + y^4 \\ &= x^4 - \frac{4\cdot 3!}{1!3!}x^3y + \frac{4\cdot 3\cdot 2!}{2!\cdot 2\cdot 1}x^2y^2 - \frac{4\cdot 3!}{3!1!}xy^3 + y^4 \\ &= x^4 - \frac{4}{1}x^3y + \frac{12}{2}x^2y^2 - \frac{4}{1}xy^3 + y^4 \\ &= x^4 - 4x^3y + 6x^2y^2 - 4xy^3 + y^4\end{aligned}$$

42. $$\begin{aligned}(x-y)^3 &= x^3 + \frac{3!}{1!(3-1)!}x^2(-y) + \frac{3!}{2!(3-2)!}x(-y)^2 + (-y)^3 \\ &= x^3 + \frac{3!}{1!2!}(-x^2y) + \frac{3!}{2!1!}xy^2 - y^3 \\ &= x^3 - \frac{3\cdot 2!}{1!2!}x^2y + \frac{3\cdot 2!}{2!1!}xy^2 - y^3 \\ &= x^3 - \frac{3}{1}x^2y + \frac{3}{1}xy^2 - y^3 \\ &= x^3 - 3x^2y + 3xy^2 - y^3\end{aligned}$$

43. $$\begin{aligned}(2x+y)^3 &= (2x)^3 + \frac{3!}{1!(3-1)!}(2x)^2y + \frac{3!}{2!(3-2)!}2xy^2 + y^3 \\ &= 8x^3 + \frac{3!}{1!2!}\cdot 4x^2y + \frac{3!}{2!1!}\cdot 2xy^2 + y^3 \\ &= 8x^3 + \frac{3\cdot 2!}{1!2!}\cdot 4x^2y + \frac{3\cdot 2!}{2!1!}\cdot 2xy^2 + y^3 \\ &= 8x^3 + \frac{3}{1}\cdot 4x^2y + \frac{3}{1}\cdot 2xy^2 + y^3 \\ &= 8x^3 + 12x^2y + 6xy^2 + y^3\end{aligned}$$

44. $$\begin{aligned}(x+2y)^3 &= x^3 + \frac{3!}{1!(3-1)!}x^2(2y) + \frac{3!}{2!(3-2)!}x(2y)^2 + (2y)^3 \\ &= x^3 + \frac{3!}{1!2!}\cdot 2x^2y + \frac{3!}{2!1!}\cdot 4xy^2 + 8y^3 \\ &= x^3 + \frac{3\cdot 2!}{1!2!}\cdot 2x^2y + \frac{3\cdot 2!}{2!1!}\cdot 4xy^2 + 8y^3 \\ &= x^3 + \frac{3}{1}\cdot 2x^2y + \frac{3}{1}\cdot 4xy^2 + 8y^3 \\ &= x^3 + 6x^2y + 12xy^2 + 8y^3\end{aligned}$$

45. $$\begin{aligned}(x-2y)^3 &= x^3 + \frac{3!}{1!(3-1)!}x^2(-2y) + \frac{3!}{2!(3-2)!}x(-2y)^2 + (-2y)^3\\ &= x^3 + \frac{3!}{1!2!}\cdot(-2x^2y) + \frac{3!}{2!1!}\cdot 4xy^2 - 8y^3\\ &= x^3 - \frac{3\cdot 2!}{1!2!}\cdot 2x^2y + \frac{3\cdot 2!}{2!1!}\cdot 4xy^2 - 8y^3\\ &= x^3 - \frac{3}{1}\cdot 2x^2y + \frac{3}{1}\cdot 4xy^2 - 8y^3\\ &= x^3 - 6x^2y + 12xy^2 - 8y^3\end{aligned}$$

46. $$\begin{aligned}(2x-y)^3 &= (2x)^3 + \frac{3!}{1!(3-1)!}(2x)^2(-y) + \frac{3!}{2!(3-2)!}2x(-y)^2 + (-y)^3\\ &= 8x^3 + \frac{3!}{1!2!}\cdot(-4x^2y) + \frac{3!}{2!1!}\cdot 2xy^2 - y^3\\ &= 8x^3 - \frac{3\cdot 2!}{1!2!}\cdot 4x^2y + \frac{3\cdot 2!}{2!1!}\cdot 2xy^2 - y^3\\ &= 8x^3 - \frac{3}{1}\cdot 4x^2y + \frac{3}{1}\cdot 2xy^2 - y^3\\ &= 8x^3 - 12x^2y + 6xy^2 - y^3\end{aligned}$$

47. $$\begin{aligned}(2x+3y)^3 &= (2x)^3 + \frac{3!}{1!(3-1)!}(2x)^2(3y) + \frac{3!}{2!(3-2)!}2x(3y)^2 + (3y)^3\\ &= 8x^3 + \frac{3!}{1!2!}\cdot 4x^2(3y) + \frac{3!}{2!1!}\cdot 2x(9y^2) + 27y^3\\ &= 8x^3 + \frac{3\cdot 2!}{1!2!}\cdot 12x^2y + \frac{3\cdot 2!}{2!1!}\cdot 18xy^2 + 27y^3\\ &= 8x^3 + \frac{3}{1}\cdot 12x^2y + \frac{3}{1}\cdot 18xy^2 + 27y^3\\ &= 8x^3 + 36x^2y + 54xy^2 + 27y^3\end{aligned}$$

48. $$\begin{aligned}(3x-2y)^3 &= (3x)^3 + \frac{3!}{1!(3-1)!}(3x)^2(-2y) + \frac{3!}{2!(3-2)!}(3x)(-2y)^2 + (-2y)^3\\ &= 27x^3 + \frac{3!}{1!2!}\cdot(9x^2)(-2y) + \frac{3!}{2!1!}\cdot 3x(4y^2) - 8y^3\\ &= 27x^3 - \frac{3\cdot 2!}{1!2!}\cdot 18x^2y + \frac{3\cdot 2!}{2!1!}\cdot 12xy^2 - 8y^3\\ &= 27x^3 - \frac{3}{1}\cdot 18x^2y + \frac{3}{1}\cdot 12xy^2 - 8y^3\\ &= 27x^3 - 54x^2y + 36xy^2 - 8y^3\end{aligned}$$

49. $$\left(\frac{x}{2}-\frac{y}{3}\right)^3=\left(\frac{x}{2}\right)^3+\frac{3!}{1!(3-1)!}\left(\frac{x}{2}\right)^2\left(-\frac{y}{3}\right)+\frac{3!}{2!(3-2)!}\left(\frac{x}{2}\right)\left(-\frac{y}{3}\right)^2+\left(-\frac{y}{3}\right)^3$$
$$=\frac{x^3}{8}-\frac{3!}{1!2!}\cdot\frac{x^2}{4}\cdot\frac{y}{3}+\frac{3!}{2!1!}\cdot\frac{x}{2}\cdot\frac{y^2}{9}-\frac{y^3}{27}$$
$$=\frac{x^3}{8}-\frac{3\cdot 2!}{1!2!}\cdot\frac{x^2y}{12}+\frac{3\cdot 2!}{2!1!}\cdot\frac{xy^2}{18}-\frac{y^3}{27}$$
$$=\frac{x^3}{8}-\frac{3}{1}\cdot\frac{x^2y}{12}+\frac{3}{1}\cdot\frac{xy^2}{18}-\frac{y^3}{27}$$
$$=\frac{x^3}{8}-\frac{x^2y}{4}+\frac{xy^2}{6}-\frac{y^3}{27}$$

50. $$\left(\frac{x}{3}+\frac{y}{2}\right)^3=\left(\frac{x}{3}\right)^3+\frac{3!}{1!(3-1)!}\left(\frac{x}{3}\right)^2\left(\frac{y}{2}\right)+\frac{3!}{2!(3-2)!}\left(\frac{x}{3}\right)\left(\frac{y}{2}\right)^2+\left(\frac{y}{2}\right)^3$$
$$=\frac{x^3}{27}+\frac{3!}{1!2!}\cdot\frac{x^2}{9}\cdot\frac{y}{2}+\frac{3!}{2!1!}\cdot\frac{x}{3}\cdot\frac{y^2}{4}+\frac{y^3}{8}$$
$$=\frac{x^3}{27}+\frac{3\cdot 2!}{1!2!}\cdot\frac{x^2y}{18}+\frac{3\cdot 2!}{2!1!}\cdot\frac{xy^2}{12}+\frac{y^3}{8}$$
$$=\frac{x^3}{27}+\frac{3}{1}\cdot\frac{x^2y}{18}+\frac{3}{1}\cdot\frac{xy^2}{12}+\frac{y^3}{8}$$
$$=\frac{x^3}{27}+\frac{x^2y}{6}+\frac{xy^2}{4}+\frac{y^3}{8}$$

51. $$(3+2y)^4=3^4+\frac{4!}{1!(4-1)!}3^3(2y)+\frac{4!}{2!(4-2)!}3^2(2y)^2+\frac{4!}{3!(4-3)!}3(2y)^3+(2y)^4$$
$$=81+\frac{4!}{1!3!}\cdot 27(2y)+\frac{4!}{2!2!}\cdot 9(4y^2)+\frac{4!}{3!1!}\cdot 3(8y^3)+16y^4$$
$$=81+\frac{4\cdot 3!}{1!3!}\cdot 54y+\frac{4\cdot 3\cdot 2!}{2!\cdot 2\cdot 1}\cdot 36y^2+\frac{4\cdot 3!}{3!1!}\cdot 24y^3+16y^4$$
$$=81+\frac{4}{1}\cdot 54y+\frac{12}{2}\cdot 36y^2+\frac{4}{1}\cdot 24y^3+16y^4$$
$$=81+216y+216y^2+96y^3+16y^4$$

52. $$(2x+3)^4=(2x)^4+\frac{4!}{1!(4-1)!}(2x)^3(3)+\frac{4!}{2!(4-2)!}(2x)^2(3)^2+\frac{4!}{3!(4-3)!}(2x)(3)^3+3^4$$
$$=16x^4+\frac{4!}{1!3!}(8x^3)(3)+\frac{4!}{2!2!}(4x^2)(9)+\frac{4!}{3!1!}(2x)(27)+81$$
$$=16x^4+\frac{4\cdot 3!}{1!3!}(24x^3)+\frac{4\cdot 3\cdot 2!}{2!\cdot 2\cdot 1}(36x^2)+\frac{4\cdot 3!}{3!1!}(54x)+81$$
$$=16x^4+\frac{4}{1}(24x^3)+\frac{12}{2}(36x^2)+\frac{4}{1}(54x)+81$$
$$=16x^4+96x^3+216x^2+216x+81$$

53. $\left(\frac{x}{3}-\frac{y}{2}\right)^4$

$$= \left(\frac{x}{3}\right)^4 + \frac{4!}{1!(4-1)!}\left(\frac{x}{3}\right)^3\left(-\frac{y}{2}\right) + \frac{4!}{2!(4-2)!}\left(\frac{x}{3}\right)^2\left(-\frac{y}{2}\right)^2 + \frac{4!}{3!(4-3)!}\left(\frac{x}{3}\right)\left(-\frac{y}{2}\right)^3 + \left(-\frac{y}{2}\right)^4$$

$$= \frac{x^4}{81} - \frac{4!}{1!3!}\cdot\frac{x^3}{27}\cdot\frac{y}{2} + \frac{4!}{2!2!}\cdot\frac{x^2}{9}\cdot\frac{y^2}{4} - \frac{4!}{3!1!}\cdot\frac{x}{3}\cdot\frac{y^3}{8} + \frac{y^4}{16}$$

$$= \frac{x^4}{81} - \frac{4\cdot 3!}{1!3!}\cdot\frac{x^3y}{54} + \frac{4\cdot 3\cdot 2!}{2!2!}\cdot\frac{x^2y^2}{36} - \frac{4\cdot 3!}{3!1!}\cdot\frac{xy^3}{24} + \frac{y^4}{16}$$

$$= \frac{x^4}{81} - \frac{4}{1}\cdot\frac{x^3y}{54} + \frac{12}{2}\cdot\frac{x^2y^2}{36} - \frac{4}{1}\cdot\frac{xy^3}{24} + \frac{y^4}{16}$$

$$= \frac{x^4}{81} - \frac{2x^3y}{27} + \frac{x^2y^2}{6} - \frac{xy^3}{6} + \frac{y^4}{16}$$

54. $\left(\frac{x}{2}+\frac{y}{3}\right)^4$

$$= \left(\frac{x}{2}\right)^4 + \frac{4!}{1!(4-1)!}\left(\frac{x}{2}\right)^3\left(\frac{y}{3}\right) + \frac{4!}{2!(4-2)!}\left(\frac{x}{2}\right)^2\left(\frac{y}{3}\right)^2 + \frac{4!}{3!(4-3)!}\left(\frac{x}{2}\right)\left(\frac{y}{3}\right)^3 + \left(\frac{y}{3}\right)^4$$

$$= \frac{x^4}{16} + \frac{4!}{1!3!}\cdot\frac{x^3}{8}\cdot\frac{y}{3} + \frac{4!}{2!2!}\cdot\frac{x^2}{4}\cdot\frac{y^2}{9} + \frac{4!}{3!1!}\cdot\frac{x}{2}\cdot\frac{y^3}{27} + \frac{y^4}{81}$$

$$= \frac{x^4}{16} + \frac{4\cdot 3!}{1!3!}\cdot\frac{x^3y}{24} + \frac{4\cdot 3\cdot 2!}{2!2!}\cdot\frac{x^2y^2}{36} + \frac{4\cdot 3!}{3!1!}\cdot\frac{xy^3}{54} + \frac{y^4}{81}$$

$$= \frac{x^4}{16} + \frac{4}{1}\cdot\frac{x^3y}{24} + \frac{12}{2}\cdot\frac{x^2y^2}{36} + \frac{4}{1}\cdot\frac{xy^3}{54} + \frac{y^4}{81}$$

$$= \frac{x^4}{16} + \frac{x^3y}{6} + \frac{x^2y^2}{6} + \frac{2xy^3}{27} + \frac{y^4}{81}$$

55.

1
1 1
1 2 1
1 3 3 1
1 4 6 4 1
1 5 10 10 5 1
1 6 15 20 15 6 1
1 7 21 35 35 21 7 1
1 8 28 56 70 56 28 8 1
1 9 36 84 126 126 84 36 9 1

56. $1, 2, 4, 8, 16, 32, 64, 128, 256, 512$; The numbers are consecutive powers of 2.

57. **Answers may vary.**

58. **Answers may vary.**

59. $\frac{n!}{0!(n-0)!} = \frac{n!}{0!n!} = \frac{n!}{1\cdot n!} = \frac{n!}{n!} = 1$

60. $\frac{n!}{n!(n-n)!} = \frac{n!}{n!(0)!} = \frac{n!}{n!\cdot 1} = 1$

61. $1, 1, 2, 3, 5, 8, 13, ...$; Beginning with 2, each number is the sum of the previous two numbers.

Exercise 11.2 (page 720)

1.
$$\begin{aligned} 3x + 2y &= 12 \Rightarrow \\ 2x - \ y &= \ 1 \Rightarrow \times 2 \end{aligned} \qquad \begin{aligned} 3x + 2y &= 12 \\ 4x - 2y &= \ 2 \\ \hline 7x \qquad &= 14 \\ x &= \ 2 \end{aligned}$$

Substitute and solve for y:
$$\begin{aligned} 2x - y &= 1 \\ 2(2) - y &= 1 \\ 4 - y &= 1 \\ -y &= -3 \\ y &= 3 \end{aligned}$$
Solution: $(2, 3)$

2.
$$\begin{cases} (1) & a + b + c = 6 \\ (2) & 2a + b + 3c = 11 \\ (3) & 3a - b - c = 6 \end{cases}$$

Add (1) and (3):
$$\begin{aligned} (1) \quad a + b + c &= \ 6 \\ (3) \quad 3a - b - c &= \ 6 \\ \hline (4) \quad 4a \qquad &= 12 \\ a &= \ 3 \end{aligned}$$

Substitute $a = 3$ into (5) and solve for c:
$$\begin{aligned} 5a + 2c &= 17 \\ 5(3) + 2c &= 17 \\ 2c &= 2 \\ c &= 1 \end{aligned}$$

Add equations (2) and (3)
$$\begin{aligned} (2) \quad 2a + b + 3c &= 11 \\ (3) \quad 3a - b - \ c &= \ 6 \\ \hline (5) \quad 5a \qquad + 2c &= 17 \end{aligned}$$

Substitute $a = 3$ and $c = 1$ and solve for b:
$$\begin{aligned} a + b + c &= 6 \\ 3 + b + 1 &= 6 \\ b &= 2 \end{aligned}$$
Solution: $(3, 2, 1)$

3. $\begin{vmatrix} 2 & -3 \\ 4 & -2 \end{vmatrix} = 2(-2) - (-3)(4) = -4 - (-12) = -4 + 12 = 8$

4. $\begin{vmatrix} 1 & 2 & 3 \\ 4 & 5 & 0 \\ -1 & -2 & 1 \end{vmatrix} = 1\begin{vmatrix} 5 & 0 \\ -2 & 1 \end{vmatrix} - 2\begin{vmatrix} 4 & 0 \\ -1 & 1 \end{vmatrix} + 3\begin{vmatrix} 4 & 5 \\ -1 & -2 \end{vmatrix} = 1(5) - 2(4) + 3(-3) = -12$

5. 3

6. 4

7. 7

8. $3!(9-3)!$

9. In the 2nd term, the exponent on b is 1.
Variables: $a^2b^1 = a^2b$
Coef. $= \dfrac{n!}{r!(n-r)!} = \dfrac{3!}{1!2!} = 3$
Term $= 3a^2b$

10. In the 3rd term, the exponent on b is 2.
Variables: $a^1b^2 = ab^2$
Coef. $= \dfrac{n!}{r!(n-r)!} = \dfrac{3!}{2!1!} = 3$
Term $= 3ab^2$

11. In the 4th term, the exponent on $-y$ is 3.
Variables: $x^1(-y)^3 = -xy^3$
Coef. $= \dfrac{n!}{r!(n-r)!} = \dfrac{4!}{3!1!} = 4$
Term $= 4(-xy^3) = -4xy^3$

12. In the 2nd term, the exponent on $-y$ is 1.
Variables: $x^4(-y)^1 = -x^4y$
Coef. $= \dfrac{n!}{r!(n-r)!} = \dfrac{5!}{1!4!} = 5$
Term $= 5(-x^4y) = -5x^4y$

13. In the 5th term, the exponent on y is 4.
Variables: x^2y^4
Coef. $= \dfrac{n!}{r!(n-r)!} = \dfrac{6!}{4!2!} = 15$
Term $= 15x^2y^4$

14. In the 5th term, the exponent on y is 4.
Variables: x^3y^4
Coef. $= \dfrac{n!}{r!(n-r)!} = \dfrac{7!}{4!3!} = 35$
Term $= 35x^3y^4$

15. In the 3rd term, the exponent on $-y$ is 2.
Variables: $x^6(-y)^2 = x^6y^2$
Coef. $= \dfrac{n!}{r!(n-r)!} = \dfrac{8!}{2!6!} = 28$
Term $= 28x^6y^2$

16. In the 7th term, the exponent on $-y$ is 6.
Variables: $x^3(-y)^6 = x^3y^6$
Coef. $= \dfrac{n!}{r!(n-r)!} = \dfrac{9!}{6!3!} = 84$
Term $= 84x^3y^6$

17. In the 3rd term, the exponent on 3 is 2.
Variables: $x^3(3)^2 = 9x^3$
Coef. $= \dfrac{n!}{r!(n-r)!} = \dfrac{5!}{2!3!} = 10$
Term $= 10(9x^3) = 90x^3$

18. In the 2nd term, the exponent on -2 is 1.
Variables: $x^3(-2)^1 = -2x^3$
Coef. $= \dfrac{n!}{r!(n-r)!} = \dfrac{4!}{1!3!} = 4$
Term $= 4(-2x^3) = -8x^3$

19. In the 3rd term, the exponent on y is 2.
Variables: $(4x)^3y^2 = 64x^3y^2$
Coef. $= \dfrac{n!}{r!(n-r)!} = \dfrac{5!}{2!3!} = 10$
Term $= 10(64x^3y^2) = 640x^3y^2$

20. In the 4th term, the exponent on $4y$ is 3.
Variables: $x^2(4y)^3 = 64x^2y^3$
Coef. $= \dfrac{n!}{r!(n-r)!} = \dfrac{5!}{3!2!} = 10$
Term $= 10(64x^2y^3) = 640x^2y^3$

21. In the 2nd term, the exponent on $-3y$ is 1.
Variables: $x^3(-3y)^1 = -3x^3y$
Coef. $= \dfrac{n!}{r!(n-r)!} = \dfrac{4!}{1!3!} = 4$
Term $= 4(-3x^3y) = -12x^3y$

22. In the 3rd term, the exponent on $-y$ is 2.
Variables: $(3x)^3(-y)^2 = 27x^3y^2$
Coef. $= \dfrac{n!}{r!(n-r)!} = \dfrac{5!}{2!3!} = 10$
Term $= 10(27x^3y^2) = 270x^3y^2$

23. In the 4th term, the exponent on -5 is 3. Variables: $(2x)^4(-5)^3 = (16x^4)(-125) = -2000x^4$
Coef. $= \dfrac{n!}{r!(n-r)!} = \dfrac{7!}{3!4!} = 35$; Term $= 35(-2000x^4) = -70{,}000x^4$

24. In the 6th term, the exponent on 3 is 5. Variables: $(2x)^1(3)^5 = 2x(243) = 486x$
Coef. $= \dfrac{n!}{r!(n-r)!} = \dfrac{6!}{5!1!} = 6$; Term $= 6(486x) = 2916x$

25. In the 5th term, the exponent on $-3y$ is 4. Variables: $(2x)^1(-3y)^4 = 2x(81y^4) = 162xy^4$

Coef. $= \dfrac{n!}{r!(n-r)!} = \dfrac{5!}{4!1!} = 5$; Term $= 5(162xy^4) = 810xy^4$

26. In the 2nd term, the exponent on $-2y$ is 1. Variables: $(3x)^3(-2y)^1 = (27x^3)(-2y) = -54x^3y$

Coef. $= \dfrac{n!}{r!(n-r)!} = \dfrac{4!}{1!3!} = 4$; Term $= 4(-54x^3y) = -216x^3y$

27. In the 3rd term, the exponent on $\sqrt{3}y$ is 2. Variables: $(\sqrt{2}x)^4(\sqrt{3}y)^2 = (4x^4)(3y^2) = 12x^4y^2$

Coef. $= \dfrac{n!}{r!(n-r)!} = \dfrac{6!}{2!4!} = 15$; Term $= 15(12x^4y^2) = 180x^4y^2$

28. In the 2nd term, the exponent on $\sqrt{2}y$ is 1. Variables: $(\sqrt{3}x)^4(\sqrt{2}y)^1 = (9x^4)(\sqrt{2}y) = 9\sqrt{2}x^4y$

Coef. $= \dfrac{n!}{r!(n-r)!} = \dfrac{5!}{1!4!} = 5$; Term $= 5(9\sqrt{2}x^4y) = 45\sqrt{2}x^4y$

29. In the 2nd term, the exponent on $-\dfrac{y}{3}$ is 1. Variables: $\left(\dfrac{x}{2}\right)^3\left(-\dfrac{y}{3}\right)^1 = \left(\dfrac{x^3}{8}\right)\left(-\dfrac{y}{3}\right) = -\dfrac{x^3y}{24}$

Coef. $= \dfrac{n!}{r!(n-r)!} = \dfrac{4!}{1!3!} = 4$; Term $= 4\left(-\dfrac{x^3y}{24}\right) = -\dfrac{x^3y}{6} = -\dfrac{1}{6}x^3y$

30. In the 4th term, the exponent on $\dfrac{y}{2}$ is 3. Variables: $\left(\dfrac{x}{3}\right)^2\left(\dfrac{y}{2}\right)^3 = \left(\dfrac{x^2}{9}\right)\left(\dfrac{y^3}{8}\right) = \dfrac{x^2y^3}{72}$

Coef. $= \dfrac{n!}{r!(n-r)!} = \dfrac{5!}{3!2!} = 10$; Term $= 10\left(\dfrac{x^2y^3}{72}\right) = \dfrac{5x^2y^3}{36} = \dfrac{5}{36}x^2y^3$

31. In the 4th term, the exponent on b is 3.
Variables: $a^{n-3}b^3$

Coef. $= \dfrac{n!}{r!(n-r)!} = \dfrac{n!}{3!(n-3)!}$

Term $= \dfrac{n!}{3!(n-3)!}a^{n-3}b^3$

32. In the 3rd term, the exponent on b is 2.
Variables: $a^{n-2}b^2$

Coef. $= \dfrac{n!}{r!(n-r)!} = \dfrac{n!}{2!(n-2)!}$

Term $= \dfrac{n!}{2!(n-2)!}a^{n-2}b^2$

33. In the 5th term, the exponent on $-b$ is 4.
Variables: $a^{n-4}(-b)^4 = a^{n-4}b^4$

Coef. $= \dfrac{n!}{r!(n-r)!} = \dfrac{n!}{4!(n-4)!}$

Term $= \dfrac{n!}{4!(n-4)!}a^{n-4}b^4$

34. In the 6th term, the exponent on $-b$ is 5.
Variables: $a^{n-5}(-b)^5 = -a^{n-5}b^5$

Coef. $= \dfrac{n!}{r!(n-r)!} = \dfrac{n!}{5!(n-5)!}$

Term $= -\dfrac{n!}{5!(n-5)!}a^{n-5}b^5$

35. In the rth term, the coefficient on b is $r-1$. Variables: $a^{n-(r-1)}b^{r-1} = a^{n-r+1}b^{r-1}$

Coef. $= \dfrac{n!}{r!(n-r)!} = \dfrac{n!}{(r-1)![n-(r-1)]!} = \dfrac{n!}{(r-1)!(n-r+1)!}$

Term $= \dfrac{n!}{(r-1)!(n-r+1)!}a^{n-r+1}b^{r-1}$

36. In the $(r+1)$th term, the coefficient on b is r. Variables: $a^{n-r}b^r$

Coef. $= \dfrac{n!}{r!(n-r)!} = \dfrac{n!}{r!(n-r)!}$; Term $= \dfrac{n!}{r!(n-r)!}a^{n-r}b^r$

37. **Answers may vary.**

38. **Answers may vary.**

39. $\left(x+\dfrac{1}{x}\right)^{10} = \left(x+x^{-1}\right)^{10}$. The constant term occurs when the exponent is 0.

The $(r+1)$th term of $(x+x^{-1})^{10}$ is $\dfrac{10!}{r!(10-r)!}x^{10-r}\left(x^{-1}\right)^r = \dfrac{10!}{r!(10-r)!}x^{10-r}x^{-r}$.

But $\dfrac{10!}{r!(10-r)!}x^{10-r}x^{-r} = \dfrac{10!}{r!(10-r)!}x^{10-2r}$. If $10-2r=0$, then $r=5$.

The term is $\dfrac{10!}{5!(10-5)!}x^{10-5}x^{-5} = \dfrac{10!}{5!5!} = \dfrac{10\cdot 9\cdot 8\cdot 7\cdot 6\cdot 5!}{5!\cdot 5\cdot 4\cdot 3\cdot 2\cdot 1} = 252$

40. $\left(a-\dfrac{1}{a}\right)^{9} = \left(a-a^{-1}\right)^{9}$. The desired term occurs when the exponent is 5.

The $(r+1)$th term of $(a-a^{-1})^9$ is $\dfrac{9!}{r!(9-r)!}a^{9-r}\left(-a^{-1}\right)^r = \dfrac{9!}{r!(9-r)!}a^{9-r}(-a)^{-r}$.

But $\dfrac{9!}{r!(9-r)!}a^{9-r}(-a)^{-r} = (-1)^{-r}\dfrac{9!}{r!(9-r)!}a^{9-r}a^{-r} = (-1)^{-r}\dfrac{9!}{r!(9-r)!}a^{9-2r}$.

If $9-2r=5$, then $r=2$.

The term is $(-1)^{-2}\dfrac{9!}{2!(9-2)!}a^{9-2}a^{-2} = \dfrac{9!}{2!7!}a^5 = \dfrac{9\cdot 8\cdot 7!}{2\cdot 1\cdot 7!} = 36a^5$. Coefficient $= 36$

Exercise 11.3 (page 727)

1. $3(2x^2-4x+7)+4(3x^2+5x-6) = 6x^2-12x+21+12x^2+20x-24 = 18x^2+8x-3$

2.
$$\begin{aligned}(2p+q)\left(3p^2+4pq-3q^2\right) &= 2p\left(3p^2+4pq-3q^2\right)+q\left(3p^2+4pq-3q^2\right)\\ &= 6p^3+8p^2q-6pq^2+3p^2q+4pq^2-3q^3\\ &= 6p^3+11p^2q-2pq^2-3q^3\end{aligned}$$

3.
$$\begin{aligned}\frac{3a+4}{a-2}+\frac{3a-4}{a+2} &= \frac{(3a+4)(a+2)}{(a-2)(a+2)}+\frac{(3a-4)(a-2)}{(a+2)(a-2)}\\ &= \frac{3a^2+10a+8+3a^2-10a+8}{(a+2)(a-2)} = \frac{6a^2+16}{(a+2)(a-2)}\end{aligned}$$

4.

$$\begin{array}{r|l} & 4t^3 \qquad\quad + 4t - 2 \\ \hline 2t-3 & 8t^4 - 12t^3 + 8t^2 - 16t + 6 \\ & \underline{8t^4 - 12t^3} \\ & \qquad\qquad 8t^2 - 16t \\ & \qquad\qquad \underline{8t^2 - 12t} \\ & \qquad\qquad\quad - 4t + 6 \\ & \qquad\qquad\quad \underline{- 4t + 6} \\ & \qquad\qquad\qquad\quad 0 \end{array}$$

5. sequence

6. Fibonacci

7. arithmetic; difference

8. $a_n = a_1 + (n-1)d$

9. arithmetic mean

10. $\dfrac{n(a_1 + a_n)}{2}$

11. series

12. sigma

13. $1 + 2 + 3 + 4 + 5$

14. index

15. $a_1 = 3(1) - 2 = 1$

16. $a_3 = 3(3) - 2 = 7$

17. $a_{25} = 3(25) - 2 = 73$

18. $a_{50} = 3(50) - 2 = 148$

19. $3, 5, 7, 9, 11$

20. $-2, 1, 4, 7, 10$

21. $-5, -8, -11, -14, -17$

22. $8, 3, -2, -7, -12$

23.
$$\begin{aligned} a_n &= a_1 + (n-1)d \\ a_5 &= a_1 + (5-1)d \\ 29 &= 5 + 4d \\ 24 &= 4d \\ 6 &= d \end{aligned}$$
$5, 11, 17, 23, 29$

24.
$$\begin{aligned} a_n &= a_1 + (n-1)d \\ a_6 &= a_1 + (6-1)d \\ 39 &= 4 + 5d \\ 35 &= 5d \\ 7 &= d \end{aligned}$$
$4, 11, 18, 25, 32$

25.
$$\begin{aligned} a_n &= a_1 + (n-1)d \\ a_6 &= a_1 + (6-1)d \\ -39 &= -4 + 5d \\ -35 &= 5d \\ -7 &= d \end{aligned}$$
$-4, -11, -18, -25, -32$

26.
$$\begin{aligned} a_n &= a_1 + (n-1)d \\ a_5 &= a_1 + (5-1)d \\ -37 &= -5 + 4d \\ -32 &= 4d \\ -8 &= d \end{aligned}$$
$-5, -13, -21, -29, -37$

27.
$$\begin{aligned} a_n &= a_1 + (n-1)d \\ a_6 &= a_1 + (6-1)d \\ -83 &= a_1 + 5(7) \\ -83 &= a_1 + 35 \\ -118 &= a_1 \end{aligned}$$
$-118, -111, -104, -97, -90$

28.
$$\begin{aligned} a_n &= a_1 + (n-1)d \\ a_7 &= a_1 + (7-1)d \\ 12 &= a_1 + 6(3) \\ 12 &= a_1 + 18 \\ -6 &= a_1 \end{aligned}$$
$-6, -3, 0, 3, 6$

29.
$$\begin{aligned} a_n &= a_1 + (n-1)d \\ a_7 &= a_1 + (7-1)d \\ 16 &= a_1 + 6(-3) \\ 16 &= a_1 - 18 \\ 34 &= a_1 \end{aligned}$$
$34, 31, 28, 25, 22$

30.
$$\begin{aligned} a_n &= a_1 + (n-1)d \\ a_7 &= a_1 + (7-1)d \\ -12 &= a_1 + 6(-5) \\ -12 &= a_1 - 30 \\ 18 &= a_1 \end{aligned}$$
$18, 13, 8, 3, -2$

31.
$$\begin{aligned} a_n &= a_1 + (n-1)d \\ a_{19} &= a_1 + (19-1)d \\ 131 &= a_1 + 18d \\ a_n &= a_1 + (n-1)d \\ a_{20} &= a_1 + (20-1)d \\ 138 &= a_1 + 19d \end{aligned}$$

$$\begin{aligned} a_1 + 18d = 131 &\Rightarrow \times(-1) & -a_1 - 18d &= -131 \\ \underline{a_1 + 19d = 138} &\Rightarrow & \underline{a_1 + 19d} &\underline{= 138} \\ & & d &= 7 \end{aligned}$$

Substitute and solve for a_1:
$$\begin{aligned} a_1 + 18d &= 131 \\ a_1 + 18(7) &= 131 \\ a_1 + 126 &= 131 \\ a_1 &= 5 \Rightarrow 5, 12, 19, 26, 33 \end{aligned}$$

32.
$$\begin{aligned} a_n &= a_1 + (n-1)d \\ a_{16} &= a_1 + (16-1)d \\ 70 &= a_1 + 15d \\ a_n &= a_1 + (n-1)d \\ a_{18} &= a_1 + (18-1)d \\ 78 &= a_1 + 17d \end{aligned}$$

$$\begin{aligned} a_1 + 15d = 70 &\Rightarrow \times(-1) & -a_1 - 15d &= -70 \\ \underline{a_1 + 17d = 78} &\Rightarrow & \underline{a_1 + 17d} &\underline{= 78} \\ & & 2d &= 8 \\ & & d &= 4 \end{aligned}$$

Substitute and solve for a_1:
$$\begin{aligned} a_1 + 15d &= 70 \\ a_1 + 15(4) &= 70 \\ a_1 + 60 &= 70 \\ a_1 &= 10 \Rightarrow 10, 14, 18, 22, 26 \end{aligned}$$

33.
$$\begin{aligned} a_n &= a_1 + (n-1)d \\ a_{30} &= a_1 + (30-1)d \\ &= 7 + 29(12) \\ &= 7 + 348 = 355 \end{aligned}$$

34.
$$\begin{aligned} a_n &= a_1 + (n-1)d \\ a_{55} &= a_1 + (55-1)d \\ &= -5 + 54(4) \\ &= -5 + 216 = 211 \end{aligned}$$

35.
$$\begin{aligned} a_n &= a_1 + (n-1)d \\ a_2 &= a_1 + (2-1)d \\ -4 &= a_1 + d \\ a_n &= a_1 + (n-1)d \\ a_3 &= a_1 + (3-1)d \\ -9 &= a_1 + 2d \end{aligned}$$

$$\begin{aligned} a_1 + d = -4 &\Rightarrow \times(-1) & -a_1 - d &= 4 \\ \underline{a_1 + 2d = -9} &\Rightarrow & \underline{a_1 + 2d} &\underline{= -9} \\ & & d &= -5 \end{aligned}$$

Substitute and solve for a_1:
$$\begin{aligned} a_1 + d &= -4 \\ a_1 + (-5) &= -4 \\ a_1 &= 1 \end{aligned}$$

Find the desired term:
$$\begin{aligned} a_n &= a_1 + (n-1)d \\ a_{37} &= a_1 + (37-1)d \\ &= 1 + 36(-5) \\ &= 1 - 180 = \boxed{-179} \end{aligned}$$

36. $a_n = a_1 + (n-1)d$
$a_2 = a_1 + (2-1)d$
$6 = a_1 + d$
$a_n = a_1 + (n-1)d$
$a_4 = a_1 + (4-1)d$
$16 = a_1 + 3d$

$a_1 + d = 6 \Rightarrow \times(-1)$ $-a_1 - d = -6$
$a_1 + 3d = 16 \Rightarrow$ $a_1 + 3d = 16$
$2d = 10$
$d = 5$

Substitute and solve for a_1:
$a_1 + d = 6$
$a_1 + 5 = 6$
$a_1 = 1$

Find the desired term:
$a_n = a_1 + (n-1)d$
$a_{40} = a_1 + (40-1)d$
$= 1 + 39(5)$
$= 1 + 195 = \boxed{196}$

37. $a_n = a_1 + (n-1)d$
$a_{27} = a_1 + (27-1)d$
$263 = a_1 + 26(11)$
$263 = a_1 + 286$
$-23 = a_1$

38. $a_n = a_1 + (n-1)d$
$a_{36} = a_1 + (36-1)d$
$-24 = -164 + 35d$
$140 = 35d$
$4 = d$

39. $a_n = a_1 + (n-1)d$
$a_{44} = a_1 + (44-1)d$
$556 = 40 + 43d$
$516 = 43d$
$12 = d$

40. $a_n = a_1 + (n-1)d$
$a_{23} = a_1 + (23-1)d$
$-625 = a_1 + 22(-5)$
$-625 = a_1 - 110$
$-515 = a_1$

41. Form an arithmetic sequence with a 1st term of 2 and a 5th term of 11:
$a_n = a_1 + (n-1)d$
$a_5 = a_1 + (5-1)d$
$11 = 2 + 4d$
$9 = 4d$
$\frac{9}{4} = d$
$2, \boxed{\frac{17}{4}, \frac{13}{2}, \frac{35}{4}}, 11$

42. Form an arithmetic sequence with a 1st term of 5 and a 6th term of 25:
$a_n = a_1 + (n-1)d$
$a_6 = a_1 + (6-1)d$
$25 = 5 + 5d$
$20 = 5d$
$4 = d$
$5, \boxed{9, 13, 17, 21}, 25$

43. Form an arithmetic sequence with a 1st term of 10 and a 6th term of 20:
$a_n = a_1 + (n-1)d$
$a_6 = a_1 + (6-1)d$
$20 = 10 + 5d$
$10 = 5d$
$2 = d$
$10, \boxed{12, 14, 16, 18}, 20$

44. Form an arithmetic sequence with a 1st term of 20 and a 5th term of 30:
$a_n = a_1 + (n-1)d$
$a_5 = a_1 + (5-1)d$
$30 = 20 + 4d$
$10 = 4d$
$\frac{5}{2} = d$
$20, \boxed{\frac{45}{2}, 25, \frac{55}{2}}, 30$

45. Form an arithmetic sequence with a 1st term of 10 and a 3rd term of 19:

$a_n = a_1 + (n-1)d$
$a_3 = a_1 + (3-1)d$
$19 = 10 + 2d$
$9 = 2d$
$\frac{9}{2} = d$
$10, \boxed{\frac{29}{2}}, 19$

46. Form an arithmetic sequence with a 1st term of 5 and a 3rd term of 23:

$a_n = a_1 + (n-1)d$
$a_3 = a_1 + (3-1)d$
$23 = 5 + 2d$
$18 = 2d$
$9 = d$
$5, \boxed{14}, 23$

47. Form an arithmetic sequence with a 1st term of −4.5 and a 3rd term of 7:

$a_n = a_1 + (n-1)d$
$a_3 = a_1 + (3-1)d$
$7 = -4.5 + 2d$
$11.5 = 2d$
$5.75 = d$
$-4.5, \boxed{1.25}, 7$

48. Form an arithmetic sequence with a 1st term of −6.3 and a 3rd term of −5.2:

$a_n = a_1 + (n-1)d$
$a_3 = a_1 + (3-1)d$
$-5.2 = -6.3 + 2d$
$1.1 = 2d$
$0.55 = d$
$-6.3, \boxed{-5.75}, -5.2$

49. $a_1 = 1, d = 3, n = 30$

$a_n = a_1 + (n-1)d = 1 + 29(3) = 88$

$$S_n = \frac{n(a_1 + a_n)}{2} = \frac{30(1+88)}{2} = 1335$$

50. $a_1 = 2, d = 4, n = 28$

$a_n = a_1 + (n-1)d = 2 + 27(4) = 110$

$$S_n = \frac{n(a_1 + a_n)}{2} = \frac{28(2+110)}{2} = 1568$$

51. $a_1 = -5, d = 4, n = 17$

$a_n = a_1 + (n-1)d = -5 + 16(4) = 59$

$$S_n = \frac{n(a_1 + a_n)}{2} = \frac{17(-5+59)}{2} = 459$$

52. $a_1 = -7, d = 6, n = 15$

$a_n = a_1 + (n-1)d = -7 + 14(6) = 77$

$$S_n = \frac{n(a_1 + a_n)}{2} = \frac{15(-7+77)}{2} = 525$$

53. $a_n = a_1 + (n-1)d$
$a_2 = a_1 + (2-1)d$
$7 = a_1 + d$
$a_n = a_1 + (n-1)d$
$a_3 = a_1 + (3-1)d$
$12 = a_1 + 2d$

$$\begin{array}{ll} a_1 + d = 7 \Rightarrow \times(-1) & -a_1 - d = -7 \\ \underline{a_1 + 2d = 12} \Rightarrow & \underline{a_1 + 2d = 12} \\ & d = 5 \end{array}$$

Substitute and solve for a_1:

$a_1 + d = 7$
$a_1 + 5 = 7$
$a_1 = 2, d = 5, n = 12$
$a_n = a_1 + (n-1)d = 2 + 11(5) = 57$

$$S_n = \frac{n(a_1 + a_n)}{2} = \frac{12(2+57)}{2} = 354$$

54. $a_n = a_1 + (n-1)d$
$a_2 = a_1 + (2-1)d$
$5 = a_1 + d$
$a_n = a_1 + (n-1)d$
$a_4 = a_1 + (4-1)d$
$9 = a_1 + 3d$

$$\begin{array}{lrr} a_1 + \ \ d = 5 \Rightarrow \times(-1) & -a_1 - \ \ d = & -5 \\ \underline{a_1 + 3d = 9 \Rightarrow} & a_1 + 3d = & 9 \\ & 2d = & 4 \\ & d = & 2 \end{array}$$

Substitute and solve for a_1:
$a_1 + d = 5$
$a_1 + 2 = 5$
$a_1 = 3, d = 2, n = 16$
$a_n = a_1 + (n-1)d = 3 + 15(2) = 33$
$$S_n = \frac{n(a_1 + a_n)}{2} = \frac{16(3+33)}{2} = 288$$

55. $f(n) = 2n + 1 \Rightarrow f(1) = 3$
$f(n) = 2n + 1 = 31$
$2n = 30$
$n = 15$
$$S_n = \frac{n(a_1 + a_n)}{2} = \frac{15(3+31)}{2} = 255$$

56. $f(n) = 4n + 3 \Rightarrow f(1) = 7$
$f(n) = 4n + 3 = 23$
$4n = 20$
$n = 5$
$$S_n = \frac{n(a_1 + a_n)}{2} = \frac{5(7+23)}{2} = 75$$

57. $a_1 = 1, d = 1, n = 50$
$a_n = a_1 + (n-1)d = 1 + 49(1) = 50$
$$S_n = \frac{n(a_1 + a_n)}{2} = \frac{50(1+50)}{2} = 1275$$

58. $a_1 = 1, d = 1, n = 100$
$a_n = a_1 + (n-1)d = 1 + 99(1) = 100$
$$S_n = \frac{n(a_1 + a_n)}{2} = \frac{100(1+100)}{2} = 5050$$

59. $a_1 = 1, d = 2, n = 50$
$a_n = a_1 + (n-1)d = 1 + 49(2) = 99$
$$S_n = \frac{n(a_1 + a_n)}{2} = \frac{50(1+99)}{2} = 2500$$

60. $a_1 = 2, d = 2, n = 50$
$a_n = a_1 + (n-1)d = 2 + 49(2) = 100$
$$S_n = \frac{n(a_1 + a_n)}{2} = \frac{50(2+100)}{2} = 2550$$

61. $\sum_{k=1}^{4} (3k) = 3(1) + 3(2) + 3(3) + 3(4) = 3 + 6 + 9 + 12$

62. $\sum_{k=1}^{3} (k-9) = (1-9) + (2-9) + (3-9) = (-8) + (-7) + (-6)$

63. $\sum_{k=4}^{6} k^2 = 4^2 + 5^2 + 6^2 = 16 + 25 + 36$

64. $\sum_{k=3}^{5} (-2k) = (-2)(3) + (-2)(4) + (-2)(5) = (-6) + (-8) + (-10)$

65. $\sum_{k=1}^{4} 6k = 6(1) + 6(2) + 6(3) + 6(4) = 6 + 12 + 18 + 24 = 60$

66. $\sum_{k=2}^{5} 3k = 3(2) + 3(3) + 3(4) + 3(5) = 6 + 9 + 12 + 15 = 42$

67. $\sum_{k=3}^{4} (k^2 + 3) = (3^2 + 3) + (4^2 + 3) = 9 + 3 + 16 + 3 = 31$

68. $\sum_{k=2}^{6} (k^2 + 1) = (2^2 + 1) + (3^2 + 1) + (4^2 + 1) + (5^2 + 1) + (6^2 + 1) = 5 + 10 + 17 + 26 + 37$
$= 95$

69. $\sum_{k=4}^{4} (2k + 4) = 2(4) + 4 = 8 + 4 = 12$

70. $\sum_{k=3}^{5} (3k^2 - 7) = \left[3(3)^2 - 7\right] + \left[3(4)^2 - 7\right] + \left[3(5)^2 - 7\right] = 20 + 41 + 68 = 129$

71. $a_1 = 60, d = 50 \Rightarrow 60, 110, 160, 210, 260, 310; n = 121$
$a_n = a_1 + (n-1)d = 60 + (121 - 1)(50) = 60 + 120(50) = \$6060$

72. $a_1 = 9725, d = -275 \Rightarrow 9725, 9450, 9175, 8900, 8625, 8350; n = 17$
$a_n = a_1 + (n-1)d = 9725 + (17-1)(-275) = 9725 - 16(275) = \$5325$

73. $a_1 = 1, d = 1, n = 150, a_n = 150 \Rightarrow S_n = \frac{n(a_1 + a_n)}{2} = \frac{150(1 + 150)}{2} = 11{,}325$ bricks

74. After 1 sec.: $s = 16(1)^2 = 16$; After 2 sec.: $s = 16(2)^2 = 64$; After 3 sec.: $s = 16(3)^2 = 144$
During 2nd second $\Rightarrow$ falls $64 - 16 = 48$ ft; During 3rd second $\Rightarrow$ falls $144 - 64 = 80$ ft

75. After 1 sec.: $s = 16(1)^2 = 16$; After 2 sec.: $s = 16(2)^2 = 64$; After 3 sec.: $s = 16(3)^2 = 144$
During 2nd second $\Rightarrow$ falls $64 - 16 = 48$ ft; During 3rd second $\Rightarrow$ falls $144 - 64 = 80$ ft
The sequence of the amounts fallen during each second is $16, 48, 80 \Rightarrow a_1 = 16, d = 32$
$a_n = a_1 + (n-1)d = 16 + (12-1)(32) = 16 + 11(32) = 368$ ft

76. $180, 360, 540, 720 \Rightarrow a_1 = 180, d = 180$ (a_n represents polygon with $n + 2$ sides)
8 sides $\Rightarrow n = 6 \Rightarrow a_n = a_1 + (n-1)d = 180 + (6-1)(180) = 180 + 5(180) = 1080°$
12 sides $\Rightarrow n = 10 \Rightarrow a_n = a_1 + (n-1)d = 180 + (10-1)(180) = 180 + 9(180) = 1800°$

77. **Answers may vary.**

78. **Answers may vary.**

79. $\sum_{n=1}^{6} \left(\frac{1}{2}n + 1\right)$: $\frac{3}{2}, 2, \frac{5}{2}, 3, \frac{7}{2}, 4$

80. $\frac{3}{2} + 2 + \frac{5}{2} + 3 + \frac{7}{2} + 4 = 9 + \frac{15}{2} = \frac{33}{2} = 16.5$

81. Form an arithmetic sequence with a 1st term of a and a 3rd term of b:

$$a_n = a_1 + (n-1)d$$
$$b = a_1 + (3-1)d$$
$$b = a + 2d$$
$$b - a = 2d$$
$$\frac{b-a}{2} = d \Rightarrow \text{mean} = a_1 + \frac{b-a}{2} = a + \frac{b-a}{2} = \frac{2a}{2} + \frac{b-a}{2} = \frac{a+b}{2}$$

82. Form an arithmetic sequence with a 1st term of a and a 4th term of b:

$$a_n = a_1 + (n-1)d$$
$$b = a + (4-1)d$$
$$b = a + 3d$$
$$b - a = 3d$$
$$\frac{b-a}{3} = d \Rightarrow \text{means: } a_1 + \frac{b-a}{3} = a + \frac{b-a}{3} = \frac{2a+b}{3} \text{ and } \frac{2a+b}{3} + \frac{b-a}{3} = \frac{a+2b}{3};$$
$$\text{sum} = \frac{2a+b}{3} + \frac{a+2b}{3} = \frac{3a+3b}{3} = \frac{3(a+b)}{3} = a+b$$

83. $\sum_{k=1}^{5} 5k = 5(1) + 5(2) + 5(3) + 5(4) + 5(5) = 5(1+2+3+4+5) = 5\sum_{k=1}^{5} k.$

84. $\sum_{k=3}^{6} (k^2 + 3k) = (3^2 + 3(3)) + (4^2 + 3(4)) + (5^2 + 3(5)) + (6^2 + 3(6)) = 140$

$\sum_{k=3}^{6} k^2 = 3^2 + 4^2 + 5^2 + 6^2 = 86; \sum_{k=3}^{6} 3k = 3(3) + 3(4) + 3(5) + 3(6) = 54; 86 + 54 = 140$

85. $\sum_{k=1}^{n} 3 = \sum_{k=1}^{n} 3k^0 = 3(1)^0 + 3(2)^0 + \cdots + 3(n)^0 = 3 + 3 + \cdots + 3 = 3n$

86. $\sum_{k=1}^{3} \frac{k^2}{k} = \frac{3^2}{3} + \frac{2^2}{2} + \frac{1^2}{1} = 3 + 2 + 1 = 6; \sum_{k=1}^{3} k^2 = 1^2 + 2^2 + 3^2 = 14;$

$\sum_{k=1}^{3} k = 1 + 2 + 3 = 6; 6 \neq \frac{14}{6}$

Exercise 11.4 (page 735)

1.

$$x^2 - 5x - 6 \leq 0$$
$$(x-6)(x+1) \leq 0$$

```
x − 6   ---------0++++
x + 1   ---  0+++++++++++
←—— [———————] ——→
    −1       6
```

solution set: $[-1, 6]$

2.

$$a^2 - 7a + 12 \geq 0$$
$$(a-4)(a-3) \geq 0$$

```
a − 4   ---------0 ++++
a − 3   ---  0++++++ ++++
←———] ———————— [———→
    3           4
```

solution set: $(-\infty, 3] \cup [4, \infty)$

3. $\dfrac{x-4}{x+3} > 0$

$x-4$ ---------0++++
$x+3$ ---0++++++++++++

⟵)——(⟶
−3 4

solution set: $(-\infty,-3)\cup(4,\infty)$

4. $\dfrac{t^2+t-20}{t+2} < 0$

$\dfrac{(t+5)(t-4)}{t+2} < 0$

$t-4$ --------------- 0++++
$t+2$ ---------- 0++++++++++
$t+5$ --- 0+++++++++++++++++++++

⟵)—— (——)⟶
−5 −2 4

solution set: $(-\infty,-5)\cup(-2,4)$

5. geometric

6. $a_n = a_1r^{n-1}$

7. common ratio

8. mean

9. $S_n = \dfrac{a_1 - a_1r^n}{1-r}$

10. first

11. $3, 6, 12, 24, 48$

12. $-2, -4, -8, -16, -32$

13. $-5, -1, -\frac{1}{5}, -\frac{1}{25}, -\frac{1}{125}$

14. $8, 4, 2, 1, \frac{1}{2}$

15.
$$\begin{aligned} a_n &= a_1r^{n-1} \\ 32 &= 2r^{3-1} \\ 32 &= 2r^2 \\ 16 &= r^2 \\ \pm 4 &= r, \text{ so } r = 4\ (r>0) \end{aligned}$$
$2, 8, 32, 128, 512$

16.
$$\begin{aligned} a_n &= a_1r^{n-1} \\ 24 &= 3r^{4-1} \\ 24 &= 3r^3 \\ 8 &= r^3 \\ 2 &= r \end{aligned}$$
$3, 6, 12, 24, 48$

17.
$$\begin{aligned} a_n &= a_1r^{n-1} \\ -192 &= -3r^{4-1} \\ -192 &= -3r^3 \\ 64 &= r^3 \\ 4 &= r \end{aligned}$$
$-3, -12, -48, -192, -768$

18.
$$\begin{aligned} a_n &= a_1r^{n-1} \\ 50 &= 2r^{3-1} \\ 50 &= 2r^2 \\ 25 &= r^2 \\ \pm 5 &= r, \text{ so } r = -5\ (r<0) \end{aligned}$$
$2, -10, 50, -250, 1250$

19.
$$\begin{aligned} a_n &= a_1r^{n-1} \\ -4 &= -64r^{5-1} \\ -4 &= -64r^4 \\ \tfrac{1}{16} &= r^4 \\ \pm\tfrac{1}{2} &= r, \text{ so } r = -\tfrac{1}{2}\ (r<0) \end{aligned}$$
$-64, 32, -16, 8, -4$

20.
$$\begin{aligned} a_n &= a_1r^{n-1} \\ -4 &= -64r^{5-1} \\ -4 &= -64r^4 \\ \tfrac{1}{16} &= r^4 \\ \pm\tfrac{1}{2} &= r, \text{ so } r = \tfrac{1}{2}\ (r>0) \end{aligned}$$
$-64, -32, -16, -8, -4$

21.
$$\begin{aligned} a_n &= a_1 r^{n-1} \\ -2 &= -64r^{6-1} \\ -2 &= -64r^5 \\ \tfrac{1}{32} &= r^5 \\ \tfrac{1}{2} &= r \end{aligned}$$
$-64, -32, -16, -8, -4$

22.
$$\begin{aligned} a_n &= a_1 r^{n-1} \\ \tfrac{1}{3} &= -81r^{6-1} \\ \tfrac{1}{3} &= -81r^5 \\ -\tfrac{1}{243} &= r^5 \\ -\tfrac{1}{3} &= r \end{aligned}$$
$-81, 27, -9, 3, -1$

23. If the 3rd term is 50 and the 2nd term is 10, then the common ratio $r = 50 \div 10 = 5$.
$a_1 = a_2 \div r = 10 \div 5 = 2$
$2, 10, 50, 250, 1250$

24. If the 4th term is 81 and the 3rd term is -27, then the common ratio $r = 81 \div (-27) = -3$.
$a_2 = a_3 \div r = -27 \div (-3) = 9$
$a_1 = a_2 \div r = 9 \div (-3) = -3$
$-3, 9, -27, 81, -243$

25. $a_n = a_1 r^{n-1} = 7 \cdot 2^{10-1} = 7 \cdot 2^9 = 7 \cdot 512 = 3584$

26. $a_n = a_1 r^{n-1} = 64 \cdot \left(\frac{1}{2}\right)^{12-1} = 64 \cdot \left(\frac{1}{2}\right)^{11} = 64 \cdot \frac{1}{2048} = \frac{1}{32}$

27.
$$\begin{aligned} a_n &= a_1 r^{n-1} \\ -81 &= a_1(-3)^{8-1} \\ -81 &= a_1(-3)^7 \\ -81 &= -2187a_1 \\ \tfrac{1}{27} &= a_1 \end{aligned}$$

28.
$$\begin{aligned} a_n &= a_1 r^{n-1} \\ 384 &= a_1(2)^{10-1} \\ 384 &= a_1(2)^9 \\ 384 &= 512a_1 \\ \tfrac{3}{4} &= a_1 \end{aligned}$$

29.
$$\begin{aligned} a_n &= a_1 r^{n-1} \\ -1944 &= -8r^{6-1} \\ -1944 &= -8r^5 \\ 243 &= r^5 \\ 3 &= r \end{aligned}$$

30.
$$\begin{aligned} a_n &= a_1 r^{n-1} \\ \tfrac{3}{8} &= 12r^{6-1} \\ \tfrac{3}{8} &= 12r^5 \\ \tfrac{1}{32} &= r^5 \\ \tfrac{1}{2} &= r \end{aligned}$$

31. $a_1 = 2$, $a_5 = 162$
$$\begin{aligned} a_n &= a_1 r^{n-1} \\ a_5 &= a_1 r^{5-1} \\ 162 &= 2r^4 \\ 81 &= r^4 \\ \pm 3 &= r \Rightarrow \text{choose } r = 3 \end{aligned}$$
$2, \boxed{6, 18, 54}, 162$

32. $a_1 = 3$, $a_6 = 96$
$$\begin{aligned} a_n &= a_1 r^{n-1} \\ a_6 &= a_1 r^{6-1} \\ 96 &= 3r^5 \\ 32 &= r^5 \\ 2 &= r \end{aligned}$$
$3, \boxed{6, 12, 24, 48}, 96$

33. $a_1 = -4$, $a_6 = -12500$

$$a_n = a_1 r^{n-1}$$
$$a_6 = a_1 r^{6-1}$$
$$-12500 = -4r^5$$
$$3125 = r^5$$
$$5 = r$$
$$-4, \boxed{-20, -100, -500, -2500}, -12500$$

34. $a_1 = -64$, $a_5 = -1024$

$$a_n = a_1 r^{n-1}$$
$$a_5 = a_1 r^{5-1}$$
$$-1024 = -64r^4$$
$$16 = r^4$$
$$\pm 2 = r \Rightarrow \text{choose } r = -2$$
$$-64, \boxed{128, -256, 512}, -1024$$

35. $a_1 = 2$, $a_3 = 128$

$$a_n = a_1 r^{n-1}$$
$$a_3 = a_1 r^{3-1}$$
$$128 = 2r^2$$
$$64 = r^2$$
$$\pm 8 = r \Rightarrow \text{choose } r = -2$$
$$2, \boxed{-16}, 128$$

36. $a_1 = 3$, $a_3 = 243$

$$a_n = a_1 r^{n-1}$$
$$a_3 = a_1 r^{3-1}$$
$$243 = 3r^2$$
$$81 = r^2$$
$$\pm 9 = r \Rightarrow \text{choose } r = 3$$
$$3, \boxed{27}, 243$$

37. $a_1 = 10$, $a_3 = 20$

$$a_n = a_1 r^{n-1}$$
$$a_3 = a_1 r^{3-1}$$
$$20 = 10r^2$$
$$2 = r^2$$
$$\pm\sqrt{2} = r \Rightarrow \text{choose } r = \sqrt{2}$$
$$10, \boxed{10\sqrt{2}}, 20$$

38. $a_1 = 5$, $a_3 = 15$

$$a_n = a_1 r^{n-1}$$
$$a_3 = a_1 r^{3-1}$$
$$15 = 5r^2$$
$$3 = r^2$$
$$\pm\sqrt{3} = r \Rightarrow \text{choose } r = -\sqrt{3}$$
$$5, \boxed{-5\sqrt{3}}, 15$$

39. $a_1 = -50$, $a_3 = 10$

$$a_n = a_1 r^{n-1}$$
$$a_3 = a_1 r^{3-1}$$
$$10 = -50r^2$$
$$-\tfrac{1}{5} = r^2$$

No such mean exists.

40. $a_1 = -25$, $a_3 = -5$

$$a_n = a_1 r^{n-1}$$
$$a_3 = a_1 r^{3-1}$$
$$-5 = -25r^2$$
$$\tfrac{1}{5} = r^2$$
$$\pm\sqrt{\tfrac{1}{5}} = r \Rightarrow \text{choose } r = \tfrac{\sqrt{5}}{5}$$
$$-25, \boxed{-5\sqrt{5}}, -5$$

41. $a_1 = 2, r = 3, n = 6;\ S_n = \dfrac{a_1 - a_1 r^n}{1 - r} = \dfrac{2 - 2(3)^6}{1 - 3} = \dfrac{2 - 2(729)}{-2} = \dfrac{-1456}{-2} = 728$

42. $a_1 = 2, r = -3, n = 6;\ S_n = \dfrac{a_1 - a_1 r^n}{1 - r} = \dfrac{2 - 2(-3)^6}{1 - (-3)} = \dfrac{2 - 2(729)}{4} = \dfrac{-1456}{4} = -364$

43. $a_1 = 2, r = -3, n = 5;\ S_n = \dfrac{a_1 - a_1 r^n}{1 - r} = \dfrac{2 - 2(-3)^5}{1 - (-3)} = \dfrac{2 - 2(-243)}{4} = \dfrac{488}{4} = 122$

44. $a_1 = 2, r = 3, n = 5;\ S_n = \dfrac{a_1 - a_1 r^n}{1 - r} = \dfrac{2 - 2(3)^5}{1 - 3} = \dfrac{2 - 2(243)}{-2} = \dfrac{-484}{-2} = 242$

45. $a_1 = 3, r = -2, n = 8;\ S_n = \dfrac{a_1 - a_1 r^n}{1 - r} = \dfrac{3 - 3(-2)^8}{1 - (-2)} = \dfrac{3 - 3(256)}{3} = \dfrac{-765}{3} = -255$

46. $a_1 = 3, r = 2, n = 8;\ S_n = \dfrac{a_1 - a_1 r^n}{1 - r} = \dfrac{3 - 3(2)^8}{1 - 2} = \dfrac{3 - 3(256)}{-1} = \dfrac{-765}{-1} = 765$

47. $a_1 = 3, r = 2, n = 7;\ S_n = \dfrac{a_1 - a_1 r^n}{1 - r} = \dfrac{3 - 3(2)^7}{1 - 2} = \dfrac{3 - 3(128)}{-1} = \dfrac{-381}{-1} = 381$

48. $a_1 = 3, r = -2, n = 7;\ S_n = \dfrac{a_1 - a_1 r^n}{1 - r} = \dfrac{3 - 3(-2)^7}{1 - (-2)} = \dfrac{3 - 3(-128)}{3} = \dfrac{387}{3} = 129$

49. If the 3rd term is $\frac{1}{5}$ and the 2nd term is 1, then the common ratio $r = \frac{1}{5} \div 1 = \frac{1}{5}$.

$a_1 = 1 \div \frac{1}{5} = 5, r = \frac{1}{5}, n = 4;\ S_n = \dfrac{a_1 - a_1 r^n}{1 - r} = \dfrac{5 - 5\left(\frac{1}{5}\right)^4}{1 - \frac{1}{5}} = \dfrac{5 - \frac{1}{125}}{\frac{4}{5}} = \dfrac{\frac{624}{125}}{\frac{4}{5}} = \dfrac{156}{25}$

50. If the 3rd term is 4 and the 2nd term is 1, then the common ratio $r = 4 \div 1 = 4$.

$a_1 = 1 \div 4 = \frac{1}{4}, r = 4, n = 5;\ S_n = \dfrac{a_1 - a_1 r^n}{1 - r} = \dfrac{\frac{1}{4} - \frac{1}{4}(4)^5}{1 - 4} = \dfrac{\frac{1}{4} - 256}{-3} = \dfrac{-\frac{1023}{4}}{-3} = \dfrac{341}{4}$

51. If the 4th term is 1 and the 3rd term is -2, then the common ratio $r = 1 \div (-2) = -\frac{1}{2}$.

$a_2 = -2 \div \left(-\frac{1}{2}\right) = 4;\ a_1 = 4 \div \left(-\frac{1}{2}\right) = -8;\ a_1 = -8, r = -\frac{1}{2}, n = 6$

$S_n = \dfrac{a_1 - a_1 r^n}{1 - r} = \dfrac{-8 - (-8)\left(-\frac{1}{2}\right)^6}{1 - \left(-\frac{1}{2}\right)} = \dfrac{-8 + \frac{1}{8}}{\frac{3}{2}} = \dfrac{-\frac{63}{8}}{\frac{3}{2}} = -\dfrac{21}{4}$

52. If the 4th term is 1 and the 3rd term is -3, then the common ratio $r = 1 \div (-3) = -\frac{1}{3}$.

$a_2 = -3 \div \left(-\frac{1}{3}\right) = 9;\ a_1 = 9 \div \left(-\frac{1}{3}\right) = -27;\ a_1 = -27, r = -\frac{1}{3}, n = 5$

$S_n = \dfrac{a_1 - a_1 r^n}{1 - r} = \dfrac{-27 - (-27)\left(-\frac{1}{3}\right)^5}{1 - \left(-\frac{1}{3}\right)} = \dfrac{-27 - \frac{1}{9}}{\frac{4}{3}} = \dfrac{-\frac{244}{9}}{\frac{4}{3}} = -\dfrac{61}{3}$

53. Sequence of population: $500, 500(1.06), 500(1.06)^2, \ldots$

$a_1 = 500, r = 1.06, n = 6 \Rightarrow a_n = a_1 r^{n-1} = 500(1.06)^5 \approx 669$

54. Sequence of population: $98, 98(0.90), 98(0.90)^2, \ldots$

$a_1 = 98, r = 0.90, n = 9 \Rightarrow a_n = a_1 r^{n-1} = 98(0.90)^8 \approx 42$

55. Sequence of amounts: $10000, 10000(0.88), 10000(0.88)^2, \ldots$

$a_1 = 10000, r = 0.88, n = 16 \Rightarrow a_n = a_1 r^{n-1} = 10000(0.88)^{15} \approx \$1{,}469.74$

56. Sequence of amounts: $5000, 5000(1.12), 5000(1.12)^2, \ldots$
$a_1 = 5000, r = 1.12, n = 11 \Rightarrow a_n = a_1 r^{n-1} = 5000(1.12)^{10} \approx \$15{,}529.24$

57. Sequence of values: $70000, 70000(1.06), 70000(1.06)^2, \ldots$
$a_1 = 70000, r = 1.06, n = 13 \Rightarrow a_n = a_1 r^{n-1} = 70000(1.06)^{12} \approx \$140{,}853.75$

58. Sequence of amounts: $5000, 5000(0.91), 5000(0.91)^2, \ldots$
$a_1 = 5000, r = 0.91, n = 6 \Rightarrow a_n = a_1 r^{n-1} = 5000(0.91)^5 \approx \$3{,}120.16$

59. Sequence of areas: $1, \frac{1}{2}, \frac{1}{4}, \ldots$
$a_1 = 1, r = \frac{1}{2}, n = 12 \Rightarrow a_n = a_1 r^{n-1} 1 = 1\left(\frac{1}{2}\right)^{11} = \left(\frac{1}{2}\right)^{11} \approx 0.0005$

60. Sequence of numbers of people in each level of tree: $1, 2, 4, \ldots$
$$a_1 = 1, r = 2, n = 10 \Rightarrow S_n = \frac{a_1 - a_1 r^n}{1 - r} = \frac{1 - 1(2)^{10}}{1 - 2} = \frac{1 - 1024}{-1} = 1024 - 1 = 1023$$

61. Sequence of amounts: $1000(1.03), 1000(1.03)^2, 1000(1.03)^3, \ldots$
$$a_1 = 1030, r = 1.03, n = 4 \Rightarrow S_n = \frac{a_1 - a_1 r^n}{1 - r} = \frac{1030 - 1030(1.03)^4}{1 - 1.03} = \frac{-129.2740743}{-0.03} \approx \$4{,}309.14$$

62. $$a_1 = 500, r = 1.04, n = 4 \Rightarrow S_n = \frac{a_1 - a_1 r^n}{1 - r} = \frac{500 - 500(1.04)^4}{1 - 1.04} = \frac{-84.92928}{-0.04} \approx \$2{,}123.23$$

63. Answers may vary.

64. Answers may vary.

65. Answers may vary.

66. Form a geometric sequence with a 1st term of a and a 3rd term of b:
$$a_n = ar^{n-1}$$
$$b = ar^{3-1}$$
$$\frac{b}{a} = r^2$$
$$\sqrt{\frac{b}{a}} = r \Rightarrow \text{mean} = a \cdot \sqrt{\frac{b}{a}} = a\sqrt{\frac{ba}{a^2}} = a\frac{\sqrt{ab}}{a} = \sqrt{ab}$$

67. arithmetic mean

68. no (see Exercise #39, for example)

69. Answers may vary.

70. Answers may vary.

Exercise 11.5 (page 740)

1. $y = 3x^3 - 4$
function

2. $xy = 12$
function

3. $3x = y^2 + 4$
not a function

4. $x = |y|$
not a function

SECTION 11.5

5. infinite

6. $2+6+18$

7. $S_\infty = \dfrac{a_1}{1-r}$

8. $\dfrac{75}{100} + \dfrac{75}{10{,}000} + \dfrac{75}{1{,}000{,}000} + \cdots$

9. $a_1 = 8, r = \dfrac{1}{2}$
$$S_\infty = \frac{a_1}{1-r} = \frac{8}{1-\frac{1}{2}} = \frac{8}{\frac{1}{2}} = 16$$

10. $a_1 = 12, r = \dfrac{1}{2}$
$$S_\infty = \frac{a_1}{1-r} = \frac{12}{1-\frac{1}{2}} = \frac{12}{\frac{1}{2}} = 24$$

11. $a_1 = 54, r = \dfrac{1}{3}$
$$S_\infty = \frac{a_1}{1-r} = \frac{54}{1-\frac{1}{3}} = \frac{54}{\frac{2}{3}} = 81$$

12. $a_1 = 45, r = \dfrac{1}{3}$
$$S_\infty = \frac{a_1}{1-r} = \frac{45}{1-\frac{1}{3}} = \frac{45}{\frac{2}{3}} = \frac{135}{2}$$

13. $a_1 = 12, r = -\dfrac{1}{2}$
$$S_\infty = \frac{a_1}{1-r} = \frac{12}{1-\left(-\frac{1}{2}\right)} = \frac{12}{\frac{3}{2}} = 8$$

14. $a_1 = 8, r = -\dfrac{1}{2}$
$$S_\infty = \frac{a_1}{1-r} = \frac{8}{1-\left(-\frac{1}{2}\right)} = \frac{8}{\frac{3}{2}} = \frac{16}{3}$$

15. $a_1 = -45, r = -\dfrac{1}{3}$
$$S_\infty = \frac{a_1}{1-r} = \frac{-45}{1-\left(-\frac{1}{3}\right)} = \frac{-45}{\frac{4}{3}} = -\frac{135}{4}$$

16. $a_1 = -54, r = -\dfrac{1}{3}$
$$S_\infty = \frac{a_1}{1-r} = \frac{-54}{1-\left(-\frac{1}{3}\right)} = \frac{-54}{\frac{4}{3}} = -\frac{81}{2}$$

17. $a_1 = \dfrac{9}{2}, r = \dfrac{4}{3} \Rightarrow$ no sum $(|r| > 1)$

18. $a_1 = -112, r = \dfrac{1}{4}$
$$S_\infty = \frac{a_1}{1-r} = \frac{-112}{1-\frac{1}{4}} = \frac{-112}{\frac{3}{4}} = -\frac{448}{3}$$

19. $a_1 = -\dfrac{27}{2}, r = \dfrac{2}{3}$
$$S_\infty = \frac{a_1}{1-r} = \frac{-\frac{27}{2}}{1-\frac{2}{3}} = \frac{-\frac{27}{2}}{\frac{1}{3}} = -\frac{81}{2}$$

20. $a_1 = \dfrac{18}{25}, r = \dfrac{5}{3} \Rightarrow$ no sum $(|r| > 1)$

21. $0.\overline{1} = \frac{1}{10} + \frac{1}{100} + \frac{1}{1000} + \cdots \Rightarrow a_1 = \frac{1}{10}, r = \frac{1}{10} \Rightarrow S_\infty = \dfrac{a_1}{1-r} = \dfrac{\frac{1}{10}}{1-\frac{1}{10}} = \dfrac{\frac{1}{10}}{\frac{9}{10}} = \dfrac{1}{9}$

22. $0.\overline{2} = \frac{2}{10} + \frac{2}{100} + \frac{2}{1000} + \cdots \Rightarrow a_1 = \frac{2}{10}, r = \frac{1}{10} \Rightarrow S_\infty = \dfrac{a_1}{1-r} = \dfrac{\frac{2}{10}}{1-\frac{1}{10}} = \dfrac{\frac{2}{10}}{\frac{9}{10}} = \dfrac{2}{9}$

23. $-0.\overline{3} = -\frac{3}{10} - \frac{3}{100} - \frac{3}{1000} + \cdots \Rightarrow a_1 = -\frac{3}{10}, r = \frac{1}{10} \Rightarrow S_\infty = \dfrac{a_1}{1-r} = \dfrac{-\frac{3}{10}}{1-\frac{1}{10}} = \dfrac{-\frac{3}{10}}{\frac{9}{10}} = -\dfrac{1}{3}$

24. $-0.\overline{4} = -\frac{4}{10} - \frac{4}{100} - \frac{4}{1000} + \cdots \Rightarrow a_1 = -\frac{4}{10}, r = \frac{1}{10} \Rightarrow S_\infty = \frac{a_1}{1-r} = \frac{-\frac{4}{10}}{1-\frac{1}{10}} = \frac{-\frac{4}{10}}{\frac{9}{10}} = -\frac{4}{9}$

25. $0.\overline{12} = \frac{12}{100} + \frac{12}{10{,}000} + \frac{12}{1{,}000{,}000} + \cdots \Rightarrow a_1 = \frac{12}{100}, r = \frac{1}{100}$

$S_\infty = \frac{a_1}{1-r} = \frac{\frac{12}{100}}{1-\frac{1}{100}} = \frac{\frac{12}{100}}{\frac{99}{100}} = \frac{12}{99} = \frac{4}{33}$

26. $0.\overline{21} = \frac{21}{100} + \frac{21}{10{,}000} + \frac{21}{1{,}000{,}000} + \cdots \Rightarrow a_1 = \frac{21}{100}, r = \frac{1}{100}$

$S_\infty = \frac{a_1}{1-r} = \frac{\frac{21}{100}}{1-\frac{1}{100}} = \frac{\frac{21}{100}}{\frac{99}{100}} = \frac{21}{99} = \frac{7}{33}$

27. $0.\overline{75} = \frac{75}{100} + \frac{75}{10{,}000} + \frac{75}{1{,}000{,}000} + \cdots \Rightarrow a_1 = \frac{75}{100}, r = \frac{1}{100}$

$S_\infty = \frac{a_1}{1-r} = \frac{\frac{75}{100}}{1-\frac{1}{100}} = \frac{\frac{75}{100}}{\frac{99}{100}} = \frac{75}{99} = \frac{25}{33}$

28. $0.\overline{57} = \frac{57}{100} + \frac{57}{10{,}000} + \frac{57}{1{,}000{,}000} + \cdots \Rightarrow a_1 = \frac{57}{100}, r = \frac{1}{100}$

$S_\infty = \frac{a_1}{1-r} = \frac{\frac{57}{100}}{1-\frac{1}{100}} = \frac{\frac{57}{100}}{\frac{99}{100}} = \frac{57}{99} = \frac{19}{33}$

29. Distance ball travels down $= 10 + 5 + 2.5 + \cdots = \frac{a_1}{1-r} = \frac{10}{1-\frac{1}{2}} = \frac{10}{\frac{1}{2}} = 20$

Distance ball travels up $= 5 + 2.5 + 1.25 + \cdots = \frac{a_1}{1-r} = \frac{5}{1-\frac{1}{2}} = \frac{5}{\frac{1}{2}} = 10$

Total distance $= 20 + 10 = 30$ m

30. Distance ball travels down $= 12 + 8 + \frac{16}{3} + \cdots = \frac{a_1}{1-r} = \frac{12}{1-\frac{2}{3}} = \frac{12}{\frac{1}{3}} = 36$

Distance ball travels up $= 8 + \frac{16}{3} + \frac{32}{9} + \cdots = \frac{a_1}{1-r} = \frac{8}{1-\frac{2}{3}} = \frac{8}{\frac{1}{3}} = 24$

Total distance $= 36 + 24 = 60$ ft

31. $S_\infty = \frac{a_1}{1-r} = \frac{1000}{1-0.8} = \frac{1000}{0.2}$
$= 5{,}000$ moths

32. $S_\infty = \frac{a_1}{1-r} = \frac{1000}{1-0.9} = \frac{1000}{0.1}$
$= 10{,}000$ moths

33. **Answers may vary.**

34. **Answers may vary.**

35.
$$\begin{aligned} S_\infty &= \frac{a_1}{1-r} \\ 5 &= \frac{1}{1-r} \\ 5(1-r) &= 1 \\ 5-5r &= 1 \\ 4 &= 5r \\ \frac{4}{5} &= r \end{aligned}$$

36.
$$\begin{aligned} S_\infty &= \frac{a_1}{1-r} \\ 9 &= \frac{a_1}{1-(-\frac{2}{3})} \\ 9 &= \frac{a_1}{\frac{5}{3}} \\ \frac{5}{3}(9) &= a_1 \\ 15 &= a_1 \end{aligned}$$

37. $0.\overline{9} = \frac{9}{10} + \frac{9}{100} + \frac{9}{1000} + \cdots \Rightarrow a = \frac{9}{10}, r = \frac{1}{10} \Rightarrow S_\infty = \frac{a_1}{1-r} = \frac{\frac{9}{10}}{1-\frac{1}{10}} = \frac{\frac{9}{10}}{\frac{9}{10}} = \frac{9}{9} = 1$

38. Use #37. $1.\overline{9} = 1 + 0.\overline{9} = 1 + 1 = 2$

39. No. $0.999999 = \frac{999{,}999}{1{,}000{,}000} < 1$

40. $f\left(\frac{1}{2}\right) = 1 + \frac{1}{2} + \left(\frac{1}{2}\right)^2 + \left(\frac{1}{2}\right)^3 + \cdots = 1 + \frac{1}{2} + \frac{1}{4} + \frac{1}{8} + \cdots \Rightarrow a_1 = 1, r = \frac{1}{2}$

$f\left(\frac{1}{2}\right) = \frac{a_1}{1-r} = \frac{1}{1-\frac{1}{2}} = \frac{1}{\frac{1}{2}} = 2$

$f\left(-\frac{1}{2}\right) = 1 + \left(-\frac{1}{2}\right) + \left(-\frac{1}{2}\right)^2 + \left(-\frac{1}{2}\right)^3 + \cdots = 1 - \frac{1}{2} + \frac{1}{4} - \frac{1}{8} + \cdots \Rightarrow a_1 = 1, r = -\frac{1}{2}$

$f\left(-\frac{1}{2}\right) = \frac{a_1}{1-r} = \frac{1}{1-\left(-\frac{1}{2}\right)} = \frac{1}{\frac{3}{2}} = \frac{2}{3}$

Exercise 11.6 (page 749)

1.
$$|2x-3| = 9$$
$$\begin{aligned} 2x-3 &= 9 \\ 2x &= 12 \\ x &= 6 \end{aligned} \quad \textbf{or} \quad \begin{aligned} 2x-3 &= -9 \\ 2x &= -6 \\ x &= -3 \end{aligned}$$

2.
$$\begin{aligned} 2x^2 - x &= 15 \\ 2x^2 - x - 15 &= 0 \\ (2x+5)(x-3) &= 0 \end{aligned}$$
$$\begin{aligned} 2x+5 &= 0 \\ x &= -\tfrac{5}{2} \end{aligned} \quad \textbf{or} \quad \begin{aligned} x-3 &= 0 \\ x &= 3 \end{aligned}$$

3.
$$\begin{aligned} \frac{3}{x-5} &= \frac{8}{x} \\ 3x &= 8(x-5) \\ 3x &= 8x - 40 \\ -5x &= -40 \\ x &= 8 \end{aligned}$$

4.
$$\begin{aligned} \frac{3}{x} &= \frac{x-2}{8} \\ 24 &= x^2 - 2x \\ 0 &= x^2 - 2x - 24 \\ 0 &= (x-6)(x+4) \end{aligned}$$
$$\begin{aligned} x-6 &= 0 \\ x &= 6 \end{aligned} \quad \textbf{or} \quad \begin{aligned} x+4 &= 0 \\ x &= -4 \end{aligned}$$

5. $p \cdot q$

6. permutation

7. $P(n,r)$

8. $P(n,r) = \frac{n!}{(n-r)!}$

9. $n!$

10. 1

11. $\binom{n}{r}$; combinations

12. $C(n,r) = \dfrac{n!}{r!(n-r)!}$

13. 1

14. 1

15. $P(3,3) = \dfrac{3!}{(3-3)!} = \dfrac{3!}{0!} = \dfrac{6}{1} = 6$

16. $P(4,4) = \dfrac{4!}{(4-4)!} = \dfrac{4!}{0!} = \dfrac{24}{1} = 24$

17. $P(5,3) = \dfrac{5!}{(5-3)!} = \dfrac{5!}{2!} = \dfrac{120}{2} = 60$

18. $P(3,2) = \dfrac{3!}{(3-2)!} = \dfrac{3!}{1!} = \dfrac{6}{1} = 6$

19. $P(2,2) \cdot P(3,3) = \dfrac{2!}{(2-2)!} \cdot \dfrac{3!}{(3-3)!} = \dfrac{2!}{0!} \cdot \dfrac{3!}{0!} = \dfrac{2}{1} \cdot \dfrac{6}{1} = 12$

20. $P(3,2) \cdot P(3,3) = \dfrac{3!}{(3-2)!} \cdot \dfrac{3!}{(3-3)!} = \dfrac{3!}{1!} \cdot \dfrac{3!}{0!} = \dfrac{6}{1} \cdot \dfrac{6}{1} = 36$

21. $\dfrac{P(5,3)}{P(4,2)} = \dfrac{\frac{5!}{(5-3)!}}{\frac{4!}{(4-2)!}} = \dfrac{\frac{5!}{2!}}{\frac{4!}{2!}} = \dfrac{\frac{120}{2}}{\frac{24}{2}} = \dfrac{60}{12} = 5$

22. $\dfrac{P(6,2)}{P(5,4)} = \dfrac{\frac{6!}{(6-2)!}}{\frac{5!}{(5-4)!}} = \dfrac{\frac{6!}{4!}}{\frac{5!}{1!}} = \dfrac{\frac{720}{24}}{\frac{120}{1}} = \dfrac{30}{120} = \dfrac{1}{4}$

23. $\dfrac{P(6,2) \cdot P(7,3)}{P(5,1)} = \dfrac{\frac{6!}{(6-2)!} \cdot \frac{7!}{(7-3)!}}{\frac{5!}{(5-1)!}} = \dfrac{\frac{6!}{4!} \cdot \frac{7!}{4!}}{\frac{5!}{4!}} = \dfrac{\frac{720}{24} \cdot \frac{5040}{24}}{\frac{120}{24}} = \dfrac{30 \cdot 210}{5} = 1{,}260$

24. $\dfrac{P(8,3)}{P(5,3) \cdot P(4,3)} = \dfrac{\frac{8!}{(8-3)!}}{\frac{5!}{(5-3)!} \cdot \frac{4!}{(4-3)!}} = \dfrac{\frac{8!}{5!}}{\frac{5!}{2!} \cdot \frac{4!}{1!}} = \dfrac{\frac{40{,}320}{120}}{\frac{120}{2} \cdot \frac{24}{1}} = \dfrac{336}{60 \cdot 24} = \dfrac{336}{1440} = \dfrac{7}{30}$

25. $C(5,3) = \dfrac{5!}{3!(5-3)!} = \dfrac{5!}{3!2!} = \dfrac{120}{6 \cdot 2} = \dfrac{120}{12} = 10$

26. $C(5,4) = \dfrac{5!}{4!(5-4)!} = \dfrac{5!}{4!1!} = \dfrac{120}{24 \cdot 1} = \dfrac{120}{24} = 5$

27. $\dbinom{6}{3} = \dfrac{6!}{3!(6-3)!} = \dfrac{6!}{3!3!} = \dfrac{720}{6 \cdot 6} = \dfrac{720}{36} = 20$

28. $\dbinom{6}{4} = \dfrac{6!}{4!(6-4)!} = \dfrac{6!}{4!2!} = \dfrac{720}{24 \cdot 2} = \dfrac{720}{48} = 15$

29. $\dbinom{5}{4}\dbinom{5}{3} = \dfrac{5!}{4!(5-4)!} \cdot \dfrac{5!}{3!(5-3)!} = \dfrac{5!}{4!1!} \cdot \dfrac{5!}{3!2!} = \dfrac{120}{24 \cdot 1} \cdot \dfrac{120}{6 \cdot 2} = \dfrac{120}{24} \cdot \dfrac{120}{12} = 5 \cdot 10 = 50$

30. $\dbinom{6}{5}\dbinom{6}{4} = \dfrac{6!}{5!(6-5)!} \cdot \dfrac{6!}{4!(6-4)!} = \dfrac{6!}{5!1!} \cdot \dfrac{6!}{4!2!} = \dfrac{720}{120 \cdot 1} \cdot \dfrac{720}{24 \cdot 2} = \dfrac{720}{120} \cdot \dfrac{720}{48} = 6 \cdot 15 = 90$

31. $\dfrac{C(38,37)}{C(19,18)} = \dfrac{\frac{38!}{37!(38-37)!}}{\frac{19!}{18!(19-18)!}} = \dfrac{\frac{38\cdot 37!}{37!1!}}{\frac{19\cdot 18!}{18!1!}} = \dfrac{\frac{38}{1}}{\frac{19}{1}} = \dfrac{38}{19} = 2$

32. $\dfrac{C(25,23)}{C(40,39)} = \dfrac{\frac{25!}{23!(25-23)!}}{\frac{40!}{39!(40-39)!}} = \dfrac{\frac{25\cdot 24\cdot 23!}{23!2!}}{\frac{40\cdot 39!}{39!1!}} = \dfrac{\frac{600}{2}}{\frac{40}{1}} = \dfrac{300}{40} = \dfrac{15}{2}$

33. $C(12,0)C(12,12) = \dfrac{12!}{0!(12-0)!}\cdot\dfrac{12!}{12!(12-12)!} = \dfrac{12!}{0!12!}\cdot\dfrac{12!}{12!0!} = \dfrac{12!}{12!}\cdot\dfrac{12!}{12!} = 1\cdot 1 = 1$

34. $\dfrac{C(8,0)}{C(8,1)} = \dfrac{\frac{8!}{0!(8-0)!}}{\frac{8!}{1!(8-1)!}} = \dfrac{\frac{8!}{0!8!}}{\frac{8!}{1!7!}} = \dfrac{\frac{8!}{8!}}{\frac{8\cdot 7!}{7!}} = \dfrac{1}{8}$

35. $C(n,2) = \dfrac{n!}{2!(n-2)!}$

36. $C(n,3) = \dfrac{n!}{3!(n-3)!}$

37.
$$\begin{aligned}(x+y)^4 &= \binom{4}{0}x^4y^0 + \binom{4}{1}x^3y^1 + \binom{4}{2}x^2y^2 + \binom{4}{3}x^1y^3 + \binom{4}{4}x^0y^4\\ &= \frac{4!}{0!4!}x^4 + \frac{4!}{1!3!}x^3y + \frac{4!}{2!2!}x^2y^2 + \frac{4!}{3!1!}xy^3 + \frac{4!}{4!0!}y^4 = x^4 + 4x^3y + 6x^2y^2 + 4xy^3 + y^4\end{aligned}$$

38.
$$\begin{aligned}(x-y)^2 = \binom{2}{0}x^2(-y)^0 + \binom{2}{1}x^1(-y^1) + \binom{2}{2}x^0(-y)^2 &= \frac{2!}{0!2!}x^2 - \frac{2!}{1!1!}xy + \frac{2!}{2!0!}y^2\\ &= x^2 - 2xy + y^2\end{aligned}$$

39.
$$\begin{aligned}(2x+y)^3 &= \binom{3}{0}(2x)^3y^0 + \binom{3}{1}(2x)^2y^1 + \binom{3}{2}(2x)^1y^2 + \binom{3}{3}(2x)^0y^3\\ &= \frac{3!}{0!3!}\cdot 8x^3 + \frac{3!}{1!2!}\cdot 4x^2y + \frac{3!}{2!1!}\cdot 2xy^2 + \frac{3!}{3!0!}y^3\\ &= 8x^3 + 12x^2y + 6xy^2 + y^3\end{aligned}$$

40.
$$\begin{aligned}&(2x+1)^4\\ &= \binom{4}{0}(2x)^4(1)^0 + \binom{4}{1}(2x)^3(1)^1 + \binom{4}{2}(2x)^2(1)^2 + \binom{4}{3}(2x)^1(1)^3 + \binom{4}{4}(2x)^0(1)^4\\ &= \frac{4!}{0!4!}\cdot 16x^4 + \frac{4!}{1!3!}\cdot 8x^3 + \frac{4!}{2!2!}\cdot 4x^2 + \frac{4!}{3!1!}\cdot 2x + \frac{4!}{4!0!}\cdot 1\\ &= 16x^4 + 32x^3 + 24x^2 + 8x + 1\end{aligned}$$

SECTION 11.6

41. $(3x-2)^4$

$$= \binom{4}{0}(3x)^4(-2)^0 + \binom{4}{1}(3x)^3(-2)^1 + \binom{4}{2}(3x)^2(-2)^2 + \binom{4}{3}(3x)^1(-2)^3 + \binom{4}{4}(3x)^0(-2)^4$$

$$= \frac{4!}{0!4!}\cdot 81x^4(1) + \frac{4!}{1!3!}\cdot 27x^3(-2) + \frac{4!}{2!2!}\cdot 9x^2(4) + \frac{4!}{3!1!}\cdot 3x(-8) + \frac{4!}{4!0!}\cdot 1(16)$$

$$= 81x^4 - 216x^3 + 216x^2 - 96x + 16$$

42. $(3-x^2)^3 = \binom{3}{0}(3)^3(-x^2)^0 + \binom{3}{1}(3)^2(-x^2)^1 + \binom{3}{2}(3)^1(-x^2)^2 + \binom{3}{3}(3)^0(-x^2)^3$

$$= \frac{3!}{0!3!}\cdot 27(1) + \frac{3!}{1!2!}\cdot 9(-x^2) + \frac{3!}{2!1!}\cdot 3(x^4) + \frac{3!}{3!0!}\cdot 1(-x^6)$$

$$= 27 - 27x^2 + 9x^4 - x^6$$

43. $\binom{5}{3}x^2(-5y)^3 = \frac{5!}{3!2!}x^2(-125y^3) = 10x^2(-125y^3) = -1{,}250x^2y^3$

44. $\binom{5}{2}(2x)^3(-y)^2 = \frac{5!}{2!3!}(8x^3)(y^2) = 10(8x^3)(y^2) = 80x^3y^2$

45. $\binom{4}{1}(x^2)^3(-y^3)^1 = \frac{4!}{1!3!}x^6(-y^3) = -4x^6y^3$

46. $\binom{4}{3}(x^3)^1(-y^2)^3 = \frac{4!}{3!1!}x^3(-y^6) = -4x^3y^6$

47. $7\cdot 5 = 35$

48. $5\cdot 3\cdot 4 = 60$

49. $10\cdot 10\cdot 10\cdot 10\cdot 10\cdot 10 = 1{,}000{,}000$

50. $10\cdot 9\cdot 8\cdot 7\cdot 6\cdot 5 = 151{,}200$

51. $9\cdot 9\cdot 8\cdot 7\cdot 6\cdot 5 = 136{,}080$

52. $26\cdot 26\cdot 10\cdot 10\cdot 10\cdot 10 = 6{,}760{,}000$

53. $8\cdot 10\cdot 10\cdot 10\cdot 10\cdot 10\cdot 10 = 8{,}000{,}000$

54. # area codes $= 10\cdot 10\cdot 10 - 2 = 998$
numbers $= 998\cdot 8\cdot 10\cdot 10\cdot 10\cdot 10\cdot 10\cdot 10 = 7.984\times 10^9$

55. $6! = 720$

56. $7! = 5{,}040$

57. $4!\cdot 5! = 24\cdot 120 = 2{,}880$

58. $6!\cdot 4! = 720\cdot 24 = 17{,}280$

59. $25\cdot 24\cdot 23 = 13{,}800$

60. $50\cdot 49\cdot 48 = 117{,}600$

61. $P(10,3) = \frac{10!}{(10-3)!} = \frac{10!}{7!} = 720$

62. $2^{32} = 4{,}294{,}967{,}296$

63. $9 \cdot 10 \cdot 10 \cdot 1 \cdot 1 = 900$

64. 3 letters $= 2 \cdot 26 \cdot 26 = 1352$
4 letters $= 2 \cdot 26 \cdot 26 \cdot 26 = 35{,}152$
Total $= 1{,}352 + 35{,}152 = 36{,}504$

65. $C(14,3) = \frac{14!}{3!(14-3)!} = \frac{14!}{3!11!} = 364$

66. $C(15,3) = \frac{15!}{3!(15-3)!} = \frac{15!}{3!12!} = 455$

67. $C(5,3) = 10 \Rightarrow 5$ persons

68. $C(6,3) = 20 \Rightarrow 6$ persons

69. $C(100,6) = \frac{100!}{6!(100-6)!} = \frac{100!}{6!94!} = \frac{100 \cdot 99 \cdot 98 \cdot 97 \cdot 96 \cdot 95 \cdot 94!}{6!94!} = 1{,}192{,}052{,}400$

70. $C(15,10) \cdot 5 = \frac{15!}{10!(15-10)!} \cdot 5 = \frac{15!}{10!5!} \cdot 5 = \frac{15 \cdot 14 \cdot 13 \cdot 12 \cdot 11 \cdot 10!}{10!5!} \cdot 5 = 15{,}015$

71. $C(3,2) \cdot C(4,2) = \frac{3!}{2!1!} \cdot \frac{4!}{2!2!} = 3 \cdot 6 = 18$

72. $C(5,3) \cdot C(3,2) = \frac{5!}{3!2!} \cdot \frac{3!}{2!1!} = 10 \cdot 3 = 30$

73. $C(12,2) \cdot C(10,3) = \frac{12!}{2!10!} \cdot \frac{10!}{3!7!} = 66 \cdot 120 = 7{,}920$

74. $C(9,5) \cdot C(3,2) = \frac{9!}{5!4!} \cdot \frac{3!}{2!1!} = 126 \cdot 3 = 378$

75. **Answers may vary.**

76. **Answers may vary.**

77. Consider the two people who insist on standing together as one person. Then there are a total of 4 "persons" to be arranged. This can be done in $4! = 24$ ways. However, the two people who are standing together can be arranged in 2 different ways, so there are $24 \cdot 2 = 48$ arrangements.

78. Refer to #77. Since there are 5 people in the line, they can be arranged in $5! = 120$ ways. There are 48 ways the two people COULD stand next to each other, so there are $120 - 48 = 72$ ways in which they are NOT standing next to each other.

Exercise 11.7 (page 755)

1.
$$5^{4x} = \frac{1}{125}$$
$$5^{4x} = 5^{-3}$$
$$4x = -3$$
$$x = -\frac{3}{4}$$

2.
$$8^{-x+1} = \frac{1}{64}$$
$$8^{-x+1} = 8^{-2}$$
$$-x+1 = -2$$
$$-x = -3$$
$$x = 3$$

3.
$$\begin{aligned} 2^{x^2-2x} &= 8 \\ 2^{x^2-2x} &= 2^3 \\ x^2-2x &= 3 \\ x^2-2x-3 &= 0 \\ (x+1)(x-3) &= 0 \\ x+1=0 \quad &\textbf{or} \quad x-3=0 \\ x=-1 \quad & \quad\quad x=3 \end{aligned}$$

4.
$$\begin{aligned} 3^{x^2-3x} &= 81 \\ 3^{x^2-3x} &= 3^4 \\ x^2-3x &= 4 \\ x^2-3x-4 &= 0 \\ (x+1)(x-4) &= 0 \\ x+1=0 \quad &\textbf{or} \quad x-4=0 \\ x=-1 \quad & \quad\quad x=4 \end{aligned}$$

5.
$$\begin{aligned} 3^{x^2+4x} &= \frac{1}{81} \\ 3^{x^2+4x} &= 3^{-4} \\ x^2+4x &= -4 \\ x^2+4x+4 &= 0 \\ (x+2)(x+2) &= 0 \\ x+2=0 \quad &\textbf{or} \quad x+2=0 \\ x=-2 \quad & \quad\quad x=-2 \end{aligned}$$

6.
$$\begin{aligned} 7^{x^2+3x} &= \frac{1}{49} \\ 7^{x^2+3x} &= 7^{-2} \\ x^2+3x &= -2 \\ x^2+3x+2 &= 0 \\ (x+1)(x+2) &= 0 \\ x+1=0 \quad &\textbf{or} \quad x+2=0 \\ x=-1 \quad & \quad\quad x=-2 \end{aligned}$$

7. experiment

8. sample space

9. $\frac{s}{n}$

10. 1

11. 0

12. 0; 1

13. **a.** 6 **b.** 52 **c.** $\frac{6}{52}$; $\frac{3}{26}$

14. **a.** 1 **b.** 270,725 **c.** $\frac{1}{270{,}725}$

15. $\{(1, H), (2, H), (3, H), (4, H), (5, H), (6, H), (1, T), (2, T), (3, T), (4, T), (5, T), (6, T)\}$

16. $\{(H, H, H), (H, H, T), (H, T, H), (H, T, T), (T, H, H), (T, H, T), (T, T, H), (T, T, T)\}$

17. $\{A, B, C, D, E, F, G, H, I, J, K, L, M, N, O, P, Q, R, S, T, U, V, W, X, Y, Z\}$

18. $\{0, 1, 2, 3, 4, 5, 6, 7, 8, 9\}$

19. $\frac{1}{6}$

20. $\frac{2}{6} = \frac{1}{3}$

21. $\frac{4}{6} = \frac{2}{3}$

22. $\frac{3}{6} = \frac{1}{2}$

23. $\frac{19}{42}$

24. $\frac{42}{42} = 1$

25. $\frac{13}{42}$

26. $\frac{9+2}{42} = \frac{11}{42}$

27. $\frac{3}{8}$

28. $\frac{2}{8} = \frac{1}{4}$

29. $\frac{0}{8} = 0$

30. $\frac{1}{8}$

31. rolls of 4: $\{(1, 3), (2, 2), (3, 1)\}$
Probability $= \frac{3}{36} = \frac{1}{12}$

32. $\frac{\text{\# diamonds}}{\text{\# cards}} = \frac{13}{52} = \frac{1}{4}$

33. $\dfrac{\#\text{ red}}{\#\text{ eggs}} = \dfrac{5}{12}$

34. $\dfrac{\#\text{ yellow}}{\#\text{ eggs}} = \dfrac{7}{12}$

35. $\dfrac{\#\text{ ways to get 6 diamonds}}{\#\text{ ways to get 6 cards from the deck of 52}} = \dfrac{\binom{13}{6}}{\binom{52}{6}} = \dfrac{1716}{20{,}358{,}520} = \dfrac{33}{391{,}510}$

36. impossible $\Rightarrow 0$

37. $\dfrac{\#\text{ ways to get 5 clubs}}{\#\text{ ways to get 5 cards from the 26 black cards}} = \dfrac{\binom{13}{5}}{\binom{26}{5}} = \dfrac{1287}{65780} = \dfrac{9}{460}$

38. $\dfrac{\#\text{ face cards}}{\#\text{ cards in deck}} = \dfrac{12}{52} = \dfrac{3}{13}$

39. $SSSS, SSSF, SSFS, SSFF, SFSS, SFSF, SFFS, SFFF,$
$FSSS, FSSF, FSFS, FSFF, FFSS, FFSF, FFFS, FFFF$

40. $SSSS \Rightarrow \dfrac{1}{16}$

41. $SFFF, FSFF, FFSF, FFFS \Rightarrow \dfrac{4}{16} = \dfrac{1}{4}$

42. $SSFF, SFSF, SFFS, FSSF, FSFS, FFSS \Rightarrow \dfrac{6}{16} = \dfrac{3}{8}$

43. $SSSF, SSFS, SFSS, FSSS \Rightarrow \dfrac{4}{16} = \dfrac{1}{4}$

44. $FFFF \Rightarrow \dfrac{1}{16}$

45. 1

46. $\dfrac{176}{282} = \dfrac{88}{141}$

47. $\dfrac{32}{119}$

48. $\dfrac{15}{71}$

49. $\dfrac{\binom{8}{4}}{\binom{10}{4}} = \dfrac{70}{210} = \dfrac{1}{3}$

50. $\dfrac{\binom{5}{3}}{\binom{9}{3}} = \dfrac{10}{84} = \dfrac{5}{42}$

51. **Answers may vary.**

52. **Answers may vary.**

53.
$$\begin{aligned} P(\text{math and art}) &= P(\text{math}) \cdot P(\text{art}|\text{math}) \\ 0.10 &= 0.30 \cdot P(\text{art}|\text{math}) \\ \frac{0.10}{0.30} &= P(\text{art}|\text{math}) \\ \frac{1}{3} &= P(\text{art}|\text{math}) \end{aligned}$$

54.
$$\begin{aligned} P\left(\text{lux and 2}^{\text{nd}}\text{ car}\right) &= P(\text{lux}) \cdot P\left(2^{\text{nd}}|\text{lux}\right) \\ &= 0.2(0.7) \\ &= 0.14 \end{aligned}$$

Chapter 11 Summary (page 758)

1. $(4!)(3!) = 4 \cdot 3 \cdot 2 \cdot 1 \cdot 3 \cdot 2 \cdot 1 = 144$

2. $\dfrac{5!}{3!} = \dfrac{5 \cdot 4 \cdot 3!}{3!} = 5 \cdot 4 = 20$

3. $\dfrac{6!}{2!(6-2)!} = \dfrac{6!}{2!4!} = \dfrac{6 \cdot 5 \cdot 4!}{2 \cdot 1 \cdot 4!} = \dfrac{30}{2} = 15$

4. $\dfrac{12!}{3!(12-3)!} = \dfrac{12!}{3!9!} = \dfrac{12 \cdot 11 \cdot 10 \cdot 9!}{3 \cdot 2 \cdot 1 \cdot 9!} = \dfrac{1320}{6} = 220$

5. $(n-n)! = 0! = 1$

6. $\dfrac{8!}{7!} = \dfrac{8 \cdot 7!}{7!} = 8$

7.
$$\begin{aligned}
(x+y)^5 &= x^5 + \frac{5!}{1!(5-1)!}x^4y + \frac{5!}{2!(5-2)!}x^3y^2 + \frac{5!}{3!(5-3)!}x^2y^3 + \frac{5!}{4!(5-4)!}xy^4 + y^5 \\
&= x^5 + \frac{5!}{1!4!}x^4y + \frac{5!}{2!3!}x^3y^2 + \frac{5!}{3!2!}x^2y^3 + \frac{5!}{4!1!}xy^4 + y^5 \\
&= x^5 + \frac{5 \cdot 4!}{1!4!}x^4y + \frac{5 \cdot 4 \cdot 3!}{2 \cdot 1 \cdot 3!}x^3y^2 + \frac{5 \cdot 4 \cdot 3!}{3! \cdot 2 \cdot 1}x^2y^3 + \frac{5 \cdot 4!}{4! \cdot 1}xy^4 + y^5 \\
&= x^5 + \frac{5}{1}x^4y + \frac{20}{2}x^3y^2 + \frac{20}{2}x^2y^3 + \frac{5}{1}xy^4 + y^5 \\
&= x^5 + 5x^4y + 10x^3y^2 + 10x^2y^3 + 5xy^4 + y^5
\end{aligned}$$

8.
$$\begin{aligned}
(x-y)^4 &= x^4 + \frac{4!}{1!(4-1)!}x^3(-y) + \frac{4!}{2!(4-2)!}x^2(-y)^2 + \frac{4!}{3!(4-3)!}x(-y)^3 + (-y)^4 \\
&= x^4 + \frac{4!}{1!3!}(-x^3y) + \frac{4!}{2!2!}x^2y^2 + \frac{4!}{3!1!}(-xy^3) + y^4 \\
&= x^4 - \frac{4 \cdot 3!}{1!3!}x^3y + \frac{4 \cdot 3 \cdot 2!}{2! \cdot 2 \cdot 1}x^2y^2 - \frac{4 \cdot 3!}{3!1!}xy^3 + y^4 \\
&= x^4 - \frac{4}{1}x^3y + \frac{12}{2}x^2y^2 - \frac{4}{1}xy^3 + y^4 = x^4 - 4x^3y + 6x^2y^2 - 4xy^3 + y^4
\end{aligned}$$

9.
$$\begin{aligned}
(4x-y)^3 &= (4x)^3 + \frac{3!}{1!(3-1)!}(4x)^2(-y) + \frac{3!}{2!(3-2)!}(4x)(-y)^2 + (-y)^3 \\
&= 64x^3 + \frac{3!}{1!2!} \cdot (-16x^2y) + \frac{3!}{2!1!} \cdot 4xy^2 - y^3 \\
&= 64x^3 - \frac{3 \cdot 2!}{1!2!} \cdot 16x^2y + \frac{3 \cdot 2!}{2!1!} \cdot 4xy^2 - y^3 \\
&= 64x^3 - \frac{3}{1} \cdot 16x^2y + \frac{3}{1} \cdot 4xy^2 - y^3 = 64x^3 - 48x^2y + 12xy^2 - y^3
\end{aligned}$$

10.
$$\begin{aligned}
(x+4y)^3 &= x^3 + \frac{3!}{1!(3-1)!}x^2(4y) + \frac{3!}{2!(3-2)!}x(4y)^2 + (4y)^3 \\
&= x^3 + \frac{3!}{1!2!} \cdot 4x^2y + \frac{3!}{2!1!} \cdot 16xy^2 + 64y^3 \\
&= x^3 + \frac{3 \cdot 2!}{1!2!} \cdot 4x^2y + \frac{3 \cdot 2!}{2!1!} \cdot 16xy^2 + 64y^3 \\
&= x^3 + \frac{3}{1} \cdot 4x^2y + \frac{3}{1} \cdot 16xy^2 + 64y^3 = x^3 + 12x^2y + 48xy^2 + 64y^3
\end{aligned}$$

11. In the 3rd term, the exponent on 3 is 2.
Variables: x^2y^2
$$\text{Coef.} = \frac{n!}{r!(n-r)!} = \frac{4!}{2!2!} = 6$$
$\text{Term} = 6x^2y^2$

12. In the 4th term, the exponent on $-y$ is 3.
Variables: $x^2(-y)^3 = -x^2y^3$
$$\text{Coef.} = \frac{n!}{r!(n-r)!} = \frac{5!}{3!2!} = 10$$
$\text{Term} = 10(-x^2y^3) = -10x^2y^3$

13. 2nd term: The exponent on $-4y$ is 1. Variables: $(3x)^2(-4y)^1 = (9x^2)(-4y) = -36x^2y$
$$\text{Coef.} = \frac{n!}{r!(n-r)!} = \frac{3!}{2!1!} = 3;\ \text{Term} = 3(-36x^2y) = -108x^2y$$

14. 3rd term: The exponent on $3y$ is 2. Variables: $(4x)^2(3y)^2 = (16x^2)(9y^2) = 144x^2y^2$
$$\text{Coef.} = \frac{n!}{r!(n-r)!} = \frac{4!}{2!2!} = 6;\ \text{Term} = 6(144x^2y^2) = 864x^2y^2$$

15. $a_n = a_1 + (n-1)d = 7 + (8-1)5 = 7 + 7 \cdot 5 = 42$

16.
$$\begin{aligned} a_n &= a_1 + (n-1)d \\ 242 &= a_1 + (9-1)d \\ 242 &= a_1 + 8d \\ a_n &= a_1 + (n-1)d \\ 212 &= a_1 + (7-1)d \\ 212 &= a_1 + 6d \end{aligned}$$
$$\begin{aligned} a_1 + 8d = 242 &\Rightarrow \times(-1) & -a_1 - 8d &= -242 \\ a_1 + 6d = 212 &\Rightarrow & a_1 + 6d &= 212 \\ & & -2d &= -30 \\ & & d &= 15 \end{aligned}$$
Substitute and solve for a_1:
$$\begin{aligned} a_1 + 6d &= 212 \\ a_1 + 6(15) &= 212 \\ a_1 + 90 &= 212 \\ a_1 &= 122 \Rightarrow 122, 137, 152, 167, 182 \end{aligned}$$

17. Form an arithmetic sequence with a 1st term of 8 and a 4th term of 25:
$$\begin{aligned} a_n &= a_1 + (n-1)d \\ 25 &= 8 + (4-1)d \\ 17 &= 3d \\ \tfrac{17}{3} &= d \Rightarrow 8, \boxed{\tfrac{41}{3}, \tfrac{58}{3}}, 25 \end{aligned}$$

18. $a_1 = 11, d = 7, n = 20$
$a_n = a_1 + (n-1)d = 11 + 19(7) = 144$
$$S_n = \frac{n(a_1 + a_n)}{2} = \frac{20(11 + 144)}{2} = 1{,}550$$

19. $a_1 = 9, d = -\frac{5}{2}, n = 10$
$a_n = a_1 + (n-1)d = 9 + 9\left(-\frac{5}{2}\right) = -\frac{27}{2}$
$$S_n = \frac{n(a_1 + a_n)}{2} = \frac{10(9 - \frac{27}{2})}{2} = -\frac{45}{2}$$

20. $\sum_{k=4}^{6} \frac{1}{2}k = \frac{1}{2}(4) + \frac{1}{2}(5) + \frac{1}{2}(6) = 2 + \frac{5}{2} + 3 = \frac{15}{2}$

21. $\sum_{k=2}^{5} 7k^2 = 7(2)^2 + 7(3)^2 + 7(4)^2 + 7(5)^2 = 28 + 63 + 112 + 175 = 378$

22. $\sum_{k=1}^{4}(3k-4)=(3(1)-4)+(3(2)-4)+(3(3)-4)+(3(4)-4)=-1+2+5+8=14$

23. $\sum_{k=10}^{10} 36k = 36(10) = 360$

24. If the 5th term is $\frac{3}{2}$ and the 4th term is 3, then the common ratio $r = \frac{3}{2} \div 3 = \frac{1}{2}$.

$a_3 = a_4 \div r = 3 \div \frac{1}{2} = 6$

$a_2 = a_3 \div r = 6 \div \frac{1}{2} = 12$

$a_1 = a_2 \div r = 12 \div \frac{1}{2} = 24$

$24, 12, 6, 3, \frac{3}{2}$

25. $$\begin{aligned} a_n &= a_1 r^{n-1} \\ &= \frac{1}{8}(2)^{6-1} \\ &= \frac{1}{8}(32) = 4 \end{aligned}$$

26. $a_1 = -6,\ a_4 = 384$

$$\begin{aligned} a_n &= a_1 r^{n-1} \\ 384 &= -6r^{4-1} \\ -64 &= r^3 \\ -4 &= r \Rightarrow -6, \boxed{24, -96}, 384 \end{aligned}$$

27. $a_1 = 162, r = \frac{1}{3}, n = 7;\ S_n = \frac{a_1 - a_1 r^n}{1-r} = \frac{162 - 162\left(\frac{1}{3}\right)^7}{1-\frac{1}{3}} = \frac{162 - 162\left(\frac{1}{2187}\right)}{-\frac{2}{3}} = \frac{\frac{4372}{27}}{-\frac{2}{3}} = \frac{2186}{9}$

28. $a_1 = \frac{1}{8}, r = -2, n = 8;\ S_n = \frac{a_1 - a_1 r^n}{1-r} = \frac{\frac{1}{8} - \frac{1}{8}(-2)^8}{1-(-2)} = \frac{\frac{1}{8} - \frac{1}{8}(256)}{3} = \frac{-\frac{255}{8}}{3} = -\frac{85}{8}$

29. $a_1 = 25, r = \frac{4}{5};\ S_\infty = \frac{a_1}{1-r} = \frac{25}{1-\frac{4}{5}} = \frac{25}{\frac{1}{5}} = 125$

30. $0.\overline{05} = \frac{5}{100} + \frac{5}{10{,}000} + \frac{5}{1{,}000{,}000} + \cdots \Rightarrow a_1 = \frac{5}{100}, r = \frac{1}{100}$

$S_\infty = \frac{a_1}{1-r} = \frac{\frac{5}{100}}{1-\frac{1}{100}} = \frac{\frac{5}{100}}{\frac{99}{100}} = \frac{5}{99}$

31. $17 \cdot 8 = 136$

32. $P(7,7) = \frac{7!}{(7-7)!} = \frac{7!}{0!} = 7! = 5{,}040$

33. $P(7,0) = \frac{7!}{(7-0)!} = \frac{7!}{7!} = 1$

34. $P(8,6) = \frac{8!}{(8-6)!} = \frac{8!}{2!} = \frac{40{,}320}{2} = 20{,}160$

35. $\frac{P(9,6)}{P(10,7)} = \frac{\frac{9!}{(9-6)!}}{\frac{10!}{(10-7)!}} = \frac{\frac{9!}{3!}}{\frac{10!}{3!}} = \frac{9!}{3!} \cdot \frac{3!}{10!} = \frac{9!}{10!} = \frac{9!}{10 \cdot 9!} = \frac{1}{10}$

36. $C(7,7) = \frac{7!}{7!(7-7)!} = \frac{7!}{7!0!} = \frac{7!}{7!} = 1$

37. $C(7,0) = \frac{7!}{0!(7-0)!} = \frac{7!}{0!7!} = \frac{7!}{7!} = 1$

38. $\binom{8}{6} = \frac{8!}{6!(8-6)!} = \frac{8 \cdot 7 \cdot 6!}{6!2!} = \frac{56}{2} = 28$

39. $\binom{9}{6} = \frac{9!}{6!(9-6)!} = \frac{9 \cdot 8 \cdot 7 \cdot 6!}{6!3!} = \frac{504}{6} = 84$

40. $C(6,3) \cdot C(7,3) = \frac{6!}{3!(6-3)!} \cdot \frac{7!}{3!(7-3)!} = \frac{6!}{3!3!} \cdot \frac{7!}{3!4!} = 20 \cdot 35 = 700$

41. $\frac{C(7,3)}{C(6,3)} = \frac{\frac{7!}{3!(7-3)!}}{\frac{6!}{3!(6-3)!}} = \frac{\frac{7!}{3!4!}}{\frac{6!}{3!3!}} = \frac{35}{20} = \frac{7}{4}$

42. Sequence of amounts: $5000, 5000(0.80), 5000(0.80)^2, \ldots$
$a_1 = 5000, r = 0.80, n = 6 \Rightarrow a_n = a_1 r^{n-1} = 5000(0.80)^5 \approx \$1{,}638.40$

43. Sequence of amounts: $25700, 25700(1.18), 25700(1.18)^2, \ldots$
$a_1 = 25700, r = 1.18, n = 11 \Rightarrow a_n = a_1 r^{n-1} = 25700(1.18)^{10} \approx \$134{,}509.57$

44. $a_1 = 300, d = 75;$
$$\begin{aligned} a_n &= a_1 + (n-1)d \\ 1200 &= 300 + (n-1)(75) \\ 900 &= 75n - 75 \\ 975 &= 75n \Rightarrow n = 13 \Rightarrow \text{in 12 years} \end{aligned}$$

45. $a_1 = 16, d = 32.$
$a_n = a_1 + (n-1)d = 16 + (10-1)(32) = 16 + 9(32) = 304$
$S_n = \frac{n(a_1 + a_n)}{2} = \frac{10(16 + 304)}{2} = 1600 \text{ ft}$

46. $5! = 120$

47. $5! \cdot 3! = 720$

48. $\binom{10}{3} = \frac{10!}{3!7!} = 120$

49. $\binom{5}{2}\binom{6}{2} = 10 \cdot 15 = 150$

50. $\frac{6}{16} = \frac{3}{8}$

51. $\frac{8}{16} = \frac{1}{2}$

52. $\frac{14}{16} = \frac{7}{8}$

53. roll of 11 $\Rightarrow \{(5,6), (6,5)\}$
$\frac{\text{\# roll of 11}}{\text{total \# rolls}} = \frac{2}{36} = \frac{1}{18}$

54. impossible $\Rightarrow 0$

55. $\frac{\text{\# tens}}{\text{total \# cards}} = \frac{4}{52} = \frac{1}{13}$

56. $\dfrac{\overset{\text{\# ways to get 3 aces}}{} \cdot \overset{\text{\# ways to get 2 other cards}}{}}{\text{\# ways to draw 5 cards}} = \dfrac{\binom{4}{3}\binom{48}{2}}{\binom{52}{5}} = \dfrac{4 \cdot 1128}{2{,}598{,}960} = \dfrac{4512}{2{,}598{,}960} = \dfrac{94}{54145}$

57. $\dfrac{\text{\# ways to get 5 spades}}{\text{\# ways to get 5 cards}} = \dfrac{\binom{13}{5}}{\binom{52}{5}} = \dfrac{1287}{2{,}598{,}960} = \dfrac{33}{66{,}640}$

Chapter 11 Test (page 762)

1. $\dfrac{7!}{4!} = \dfrac{7 \cdot 6 \cdot 5 \cdot 4!}{4!} = 7 \cdot 6 \cdot 5 = 210$

2. $0! = 1$

3. In the 2nd term, the exponent on $-y$ is 1.
Variables: $x^4(-y)^1 = -x^4y$
Coef. $= \dfrac{n!}{r!(n-r)!} = \dfrac{5!}{1!4!} = 5$
Term $= 5(-x^4y) = -5x^4y$

4. In the 3rd term, the exponent on $2y$ is 2.
Variables: $(x)^2(2y)^2 = 4x^2y^2$
Coef. $= \dfrac{n!}{r!(n-r)!} = \dfrac{4!}{2!2!} = 6$
Term $= 6(4x^2y^2) = 24x^2y^2$

5. $a_1 = 3, d = 7, n = 10; a_n = a_1 + (n-1)d = 3 + (10-1)(7) = 3 + 9(7) = 66$

6. $a_1 = -2, d = 5, n = 12; a_n = a_1 + (n-1)d = -2 + (12-1)(5) = -2 + 11(5) = 53$
$S_n = \dfrac{n(a_1 + a_n)}{2} = \dfrac{12(-2+53)}{2} = \dfrac{12(51)}{2} = 306$

7. Form an arithmetic sequence with a 1st term of 2 and a 4th term of 98:
$a_n = a_1 + (n-1)d$
$98 = 2 + (4-1)d$
$96 = 3d$
$32 = d \Rightarrow 2, \boxed{34, 66}, 98$

8. $\sum_{k=1}^{3}(2k-3) = (2(1)-3) + (2(2)-3) + (2(3)-3) = -1 + 1 + 3 = 3$

9. $a_1 = -\frac{1}{9}, r = 3, n = 7; a_n = a_1r^{n-1} = -\frac{1}{9}(3)^{7-1} = -\frac{1}{9}(3)^6 = -\frac{1}{9}(729) = -81$

10. $a_1 = \frac{1}{27}, r = 3, n = 6; S_n = \dfrac{a_1 - a_1r^n}{1-r} = \dfrac{\frac{1}{27} - \frac{1}{27}(3)^6}{1-3} = \dfrac{\frac{1}{27} - \frac{1}{27}(729)}{-2} = \dfrac{-\frac{728}{27}}{-2} = \dfrac{364}{27}$

11. $a_1 = 3, a_4 = 648$
$a_n = a_1r^{n-1}$
$648 = 3r^{4-1}$
$216 = r^3$
$6 = r \Rightarrow 3, \boxed{18, 108}, 648$

12. $a_1 = 9, r = \dfrac{1}{3}$
$S_\infty = \dfrac{a_1}{1-r} = \dfrac{9}{1-\frac{1}{3}} = \dfrac{9}{\frac{2}{3}} = \dfrac{27}{2}$

13. $P(5,4) = \dfrac{5!}{(5-4)!} = \dfrac{5!}{1!} = 5! = 120$

14. $P(8,8) = \dfrac{8!}{(8-8)!} = \dfrac{8!}{0!} = 8! = 40{,}320$

15. $C(6,4) = \dfrac{6!}{4!2!} = \dfrac{6 \cdot 5 \cdot 4!}{4! \cdot 2 \cdot 1} = \dfrac{30}{2} = 15$

16. $C(8,3) = \dfrac{8!}{3!5!} = \dfrac{8 \cdot 7 \cdot 6 \cdot 5!}{3 \cdot 2 \cdot 1 \cdot 5!} = \dfrac{56 \cdot 6}{6} = 56$

17. $C(6,0) \cdot P(6,5) = \dfrac{6!}{0!6!} \cdot \dfrac{6!}{1!} = 1 \cdot 6!$
$= 720$

18. $P(8,7) \cdot C(8,7) = \dfrac{8!}{1!} \cdot \dfrac{8!}{7!1!} = 8! \cdot 8$
$= 322{,}560$

19. $\dfrac{P(6,4)}{C(6,4)} = \dfrac{\frac{6!}{2!}}{\frac{6!}{4!2!}} = \dfrac{6!}{2!} \cdot \dfrac{4!2!}{6!} = 4! = 24$

20. $\dfrac{C(9,6)}{P(6,4)} = \dfrac{\frac{9!}{6!3!}}{\frac{6!}{2!}} = \dfrac{84}{360} = \dfrac{7}{30}$

21. $\dbinom{7}{3} = \dfrac{7!}{3!4!} = 35$

22. $\dbinom{5}{1}\dbinom{4}{2} = 5 \cdot 6 = 30$

23. $\dfrac{1}{6}$

24. $\dfrac{8}{52} = \dfrac{2}{13}$

25. $\dfrac{\text{\# ways to get 5 hearts}}{\text{\# ways to get 5 cards}} = \dfrac{\binom{13}{5}}{\binom{52}{5}} = \dfrac{1287}{2{,}598{,}960} = \dfrac{33}{66{,}640}$

26. $\dfrac{\text{\# ways to get 2 heads}}{\text{\# ways to toss 5 times}} = \dfrac{\binom{5}{2}}{2^5} = \dfrac{20}{32} = \dfrac{5}{16}$

Cumulative Review Exercises (page 762)

1. $\begin{cases} 2x + y = 5 \\ x - 2y = 0 \end{cases}$

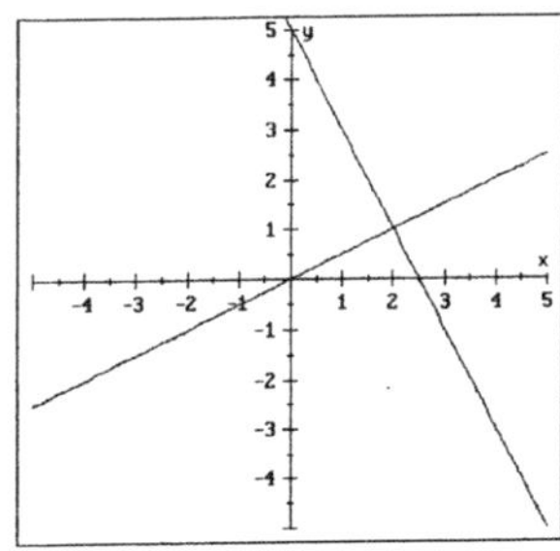

Solution: $(2,1)$

2. $\begin{cases} (1) \quad 3x + y = 4 \\ (2) \quad 2x - 3y = -1 \end{cases}$

Substitute $y = 4 - 3x$ from (1) into (2):

$$\begin{aligned} 2x - 3y &= -1 \\ 2x - 3(4 - 3x) &= -1 \\ 2x - 12 + 9x &= -1 \\ 11x &= 11 \\ x &= 1 \end{aligned}$$

Substitute this and solve for y:

$y = 4 - 3x = 4 - 3(1) = 1$

Solution: $(1,1)$

3. $\begin{aligned} x + 2y &= -2 \\ 2x - y &= 6 \end{aligned} \Rightarrow \times (2)$ $\quad \begin{aligned} x + 2y &= -2 \\ 4x - 2y &= 12 \\ \hline 5x &= 10 \\ x &= 2 \end{aligned}$ $\quad \begin{aligned} x + 2y &= -2 \\ 2 + 2y &= -2 \\ 2y &= -4 \\ y &= -2 \end{aligned}$

Solution: $\boxed{(2,-2)}$

CUMULATIVE REVIEW EXERCISES

4.

$$\begin{array}{rl} \frac{x}{10}+\frac{y}{5}=\frac{1}{2} \Rightarrow \times 10 & x+2y=5 \\ \frac{x}{2}-\frac{y}{5}=\frac{13}{10} \Rightarrow \times 10 & 5x-2y=13 \\ \hline & 6x=18 \\ & x=3 \end{array} \qquad \begin{array}{r} x+2y=5 \\ 3+2y=5 \\ 2y=2 \\ y=1 \Rightarrow \text{Solution: } \boxed{(3,1)} \end{array}$$

5.

$$\begin{vmatrix} 3 & -2 \\ 1 & -1 \end{vmatrix} = 3(-1)-(-2)(1) = -3+2=-1$$

6.

$$y=\frac{\begin{vmatrix} 4 & -1 \\ 3 & -7 \end{vmatrix}}{\begin{vmatrix} 4 & -3 \\ 3 & 4 \end{vmatrix}}=\frac{4(-7)-(-1)(3)}{4(4)-(-3)(3)}=\frac{-25}{25}=-1$$

7.

$$\begin{array}{ll} (1) & x+y+z=1 \\ (2) & 2x-y-z=-4 \\ (3) & x-2y+z=4 \end{array} \quad \begin{array}{ll} (1) & x+y+z=1 \\ (2) & 2x-y-z=-4 \\ \hline (4) & 3x=-3 \\ & x=-1 \end{array} \quad \begin{array}{ll} (2) & 2x-y-z=-4 \\ (3) & x-2y+z=4 \\ \hline (5) & 3x-3y=0 \end{array}$$

$$\begin{array}{r} 3x-3y=0 \\ 3(-1)-3y=0 \\ -3-3y=0 \\ -3y=3 \\ y=-1 \end{array} \qquad \begin{array}{r} x+y+z=1 \\ -1+(-1)+z=1 \\ -2+z=1 \\ z=3 \end{array}$$

$\boxed{\text{The solution is } (-1,-1,3).}$

8.

$$z=\frac{\begin{vmatrix} 1 & 2 & 6 \\ 3 & 2 & 6 \\ 2 & 3 & 6 \end{vmatrix}}{\begin{vmatrix} 1 & 2 & 3 \\ 3 & 2 & 1 \\ 2 & 3 & 1 \end{vmatrix}}=\frac{1\begin{vmatrix} 2 & 6 \\ 3 & 6 \end{vmatrix}-2\begin{vmatrix} 3 & 6 \\ 2 & 6 \end{vmatrix}+6\begin{vmatrix} 3 & 2 \\ 2 & 3 \end{vmatrix}}{1\begin{vmatrix} 2 & 1 \\ 3 & 1 \end{vmatrix}-2\begin{vmatrix} 3 & 1 \\ 2 & 1 \end{vmatrix}+3\begin{vmatrix} 3 & 2 \\ 2 & 3 \end{vmatrix}}=\frac{1(-6)-2(6)+6(5)}{1(-1)-2(1)+3(5)}=\frac{12}{12}=1$$

9.

$$\begin{cases} 3x-2y<6 \\ y<-x+2 \end{cases}$$

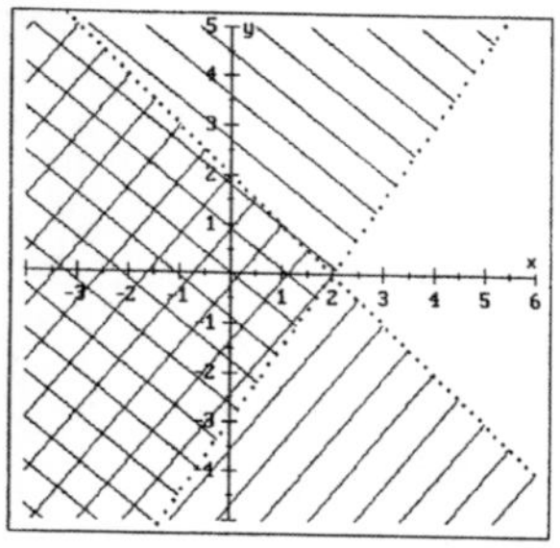

10.

$$\begin{cases} y<x+2 \\ 3x+y\le 6 \end{cases}$$

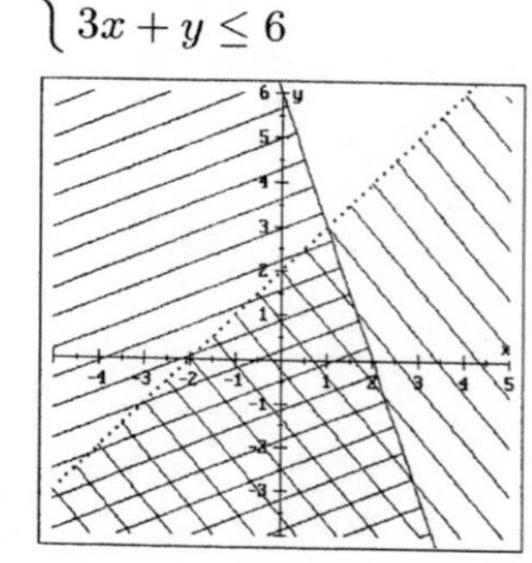

CUMULATIVE REVIEW EXERCISES

11. $y = \left(\frac{1}{2}\right)^x$

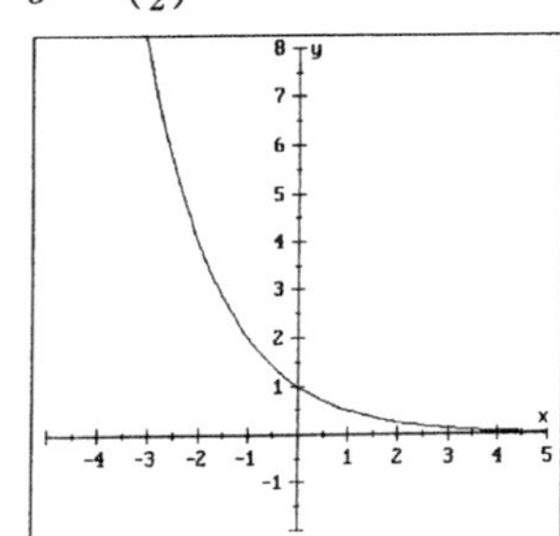

12. $y = \log_2 x \Rightarrow 2^y = x$

13. $\log_x 25 = 2 \Rightarrow x^2 = 25 \Rightarrow x = 5$

14. $\log_5 125 = x \Rightarrow 5^x = 125 \Rightarrow x = 3$

15. $\log_3 x = -3 \Rightarrow 3^{-3} = x \Rightarrow x = \frac{1}{27}$

16. $\log_5 x = 0 \Rightarrow 5^0 = x \Rightarrow x = 1$

17. $y = 2^x$

18. x

19. $\log 98 = \log(14 \cdot 7) = \log 14 + \log 7 = 1.1461 + 0.8451 = 1.9912$

20. $\log 2 = \log \frac{14}{7} = \log 14 - \log 7 = 1.1461 - 0.8451 = 0.3010$

21. $\log 49 = \log 7^2 = 2 \log 7 = 2(0.8451) = 1.6902$

22. $\log \frac{7}{5} = \log \frac{14}{10} = \log 14 - \log 10 = 1.1461 - 1 = 0.1461$

23.

$$\begin{aligned} 2^{x+5} &= 3^x \\ \log 2^{x+2} &= \log 3^x \\ (x+5)\log 2 &= x \log 3 \\ x \log 2 + 5 \log 2 &= x \log 3 \\ 5 \log 2 &= x \log 3 - x \log 2 \\ 5 \log 2 &= x(\log 3 - \log 2) \\ \frac{5 \log 2}{\log 3 - \log 2} &= x \end{aligned}$$

24.

$$\begin{aligned} \log 5 + \log x - \log 4 &= 1 \\ \log \frac{5x}{4} &= 1 \\ 10^1 &= \frac{5x}{4} \\ 40 &= 5x \\ 8 &= x \end{aligned}$$

25. $A = P\left(1 + \frac{r}{k}\right)^{kt} = 9000\left(1 + \frac{-0.12}{1}\right)^{1(9)} \approx \$2{,}848.31$

26. $\log_6 8 = \frac{\log 8}{\log 6} \approx 1.16056$

27. $\frac{6!7!}{5!} = \frac{6 \cdot 5! \cdot 7!}{5!} = 6 \cdot 7! = 30{,}240$

28. $(3a-b)^4$

$$= (3a)^4 + \frac{4!}{1!(4-1)!}(3a)^3(-b) + \frac{4!}{2!(4-2)!}(3a)^2(-b)^2 + \frac{4!}{3!(4-3)!}(3a)(-b)^3 + (-b)^4$$
$$= 81a^4 + \frac{4!}{1!3!}(-27a^3b) + \frac{4!}{2!2!}(9a^2b^2) + \frac{4!}{3!1!}(-3ab^3) + b^4$$
$$= 81a^4 - \frac{4\cdot 3!}{1!3!}(27a^3b) + \frac{4\cdot 3\cdot 2!}{2!\cdot 2\cdot 1}(9a^2b^2) - \frac{4\cdot 3!}{3!1!}(3ab^3) + b^4$$
$$= 81a^4 - \frac{4}{1}(27a^3b) + \frac{12}{2}(9a^2b^2) - \frac{4}{1}(3ab^3) + b^4$$
$$= 81a^4 - 108a^3b + 54a^2b^2 - 12ab^3 + b^4$$

29. In the 7th term, the exponent on $-y$ is 6.
Variables: $(2x)^2(-y)^6 = 4x^2y^6$
Coef. $= \dfrac{n!}{r!(n-r)!} = \dfrac{8!}{6!2!} = 28$
Term $= 28(4x^2y^6) = 112x^2y^6$

30. $a_1 = -11, d = 6, n = 20$
$$a_n = a_1 + (n-1)d$$
$$= -11 + (19)(6)$$
$$= -11 + 114 = 103$$

31. $a_1 = 6, d = 3, n = 20;\ a_n = a_1 + (n-1)d = 6 + (20-1)(3) = 6 + 19(3) = 63$
$$S_n = \frac{n(a_1+a_n)}{2} = \frac{20(6+63)}{2} = \frac{20(69)}{2} = 690$$

32. $a_1 = -3;\ a_4 = 30$:
$$a_n = a_1 + (n-1)d$$
$$30 = -3 + (4-1)d$$
$$33 = 3d$$
$$11 = d \Rightarrow -3, \boxed{8, 19}, 30$$

33. $\sum_{k=1}^{3} 3k^2 = 3(1)^2 + 3(2)^2 + 3(3)^2$
$$= 3 + 12 + 27 = 42$$

34. $\sum_{k=3}^{5}(2k+1) = (2(3)+1) + (2(4)+1) + (2(5)+1) = 7 + 9 + 11 = 27$

35. $a_1 = \frac{1}{27}, r = 3, n = 7;\ a_n = a_1r^{n-1} = \frac{1}{27}(3)^{7-1} = \frac{1}{27}(3)^6 = \frac{1}{27}(729) = 27$

36. $a_1 = \frac{1}{64}, r = 2, n = 10;\ S_n = \dfrac{a_1 - a_1r^n}{1-r} = \dfrac{\frac{1}{64} - \frac{1}{64}(2)^{10}}{1-2} = \dfrac{\frac{1}{64} - \frac{1}{64}(1024)}{-1} = \dfrac{-\frac{1023}{64}}{-1} = \dfrac{1023}{64}$

37. $a_1 = -3,\ a_4 = 192$
$$a_n = a_1r^{n-1}$$
$$192 = -3r^{4-1}$$
$$-64 = r^3$$
$$-4 = r \Rightarrow -3, \boxed{12, -48}, 192$$

38. $a_1 = 9, r = \dfrac{1}{3}$
$$S_\infty = \frac{a_1}{1-r} = \frac{9}{1-\frac{1}{3}} = \frac{9}{\frac{2}{3}} = \frac{27}{2}$$

39. $P(9,3) = \dfrac{9!}{(9-3)!} = \dfrac{9!}{6!} = \dfrac{9\cdot 8\cdot 7\cdot 6!}{6!} = 9\cdot 8\cdot 7 = 504$

40. $C(7,4) = \dfrac{7!}{4!(7-4)!} = \dfrac{7!}{4!3!} = \dfrac{7\cdot 6\cdot 5\cdot 4!}{4!\cdot 3\cdot 2\cdot 1} = \dfrac{210}{6} = 35$

41. $\dfrac{C(8,4)C(8,0)}{P(6,2)} = \dfrac{\frac{8!}{4!4!}\cdot\frac{8!}{0!8!}}{\frac{6!}{4!}} = \dfrac{70\cdot 1}{30} = \dfrac{7}{3}$

42. $C(n,n) = 1$ is smaller than $P(n,n) = n!$.

43. $7! = 5{,}040$

44. $\dbinom{9}{3} = \dfrac{9!}{3!6!} = 84$

45. $\dfrac{12}{52} = \dfrac{3}{13}$

Appendix 1 (page A-4)

1. $y = x^2 - 1$

x-axis	y-axis	origin
$-y = x^2 - 1$	$y = (-x)^2 - 1$	$-y = (-x)^2 - 1$
not equivalent: no symmetry	$y = x^2 - 1$	$-y = x^2 - 1$
	equivalent: symmetry	not equivalent: no symmetry

2. $y = x^3$

x-axis	y-axis	origin
$-y = x^3$	$y = (-x)^3$	$-y = (-x)^3$
not equivalent: no symmetry	$y = -x^3$	$-y = -x^3$
	not equivalent: no symmetry	$y = x^3$
		equivalent: symmetry

3. $y = x^5$

x-axis	y-axis	origin
$-y = x^5$	$y = (-x)^3$	$-y = (-x)^3$
not equivalent: no symmetry	$y = -x^5$	$-y = -x^5$
	not equivalent: no symmetry	$y = x^5$
		equivalent: symmetry

4. $y = x^4$

x-axis	y-axis	origin
$-y = x^4$	$y = (-x)^2$	$-y = (-x)^2$
not equivalent: no symmetry	$y = x^4$	$-y = x^4$
	equivalent: symmetry	not equivalent: no symmetry

5.

$$y = -x^2 + 2$$

x-axis	y-axis	origin
$-y = -x^2 + 2$	$y = (-x)^2 + 2$	$-y = (-x)^2 + 2$
not equivalent: no symmetry	$y = -x^2 + 2$	$-y = x^2 + 2$
	equivalent: **symmetry**	not equivalent: no symmetry

6.

$$y = x^3 + 1$$

x-axis	y-axis	origin
$-y = x^3 + 1$	$y = (-x)^3 + 1$	$-y = (-x)^3 + 1$
not equivalent: no symmetry	$y = -x^3 + 1$	$-y = -x^3 + 1$
	not equivalent: no symmetry	$y = x^3 - 1$
		not equivalent: no symmetry

7.

$$y = x^2 - x$$

x-axis	y-axis	origin
$-y = x^2 - x$	$y = (-x)^2 - (-x)$	$-y = (-x)^2 - (-x)$
not equivalent: no symmetry	$y = x^2 + x$	$-y = x^2 + x$
	not equivalent: no symmetry	not equivalent: no symmetry

8.

$$y^2 = x + 7$$

x-axis	y-axis	origin
$(-y)^2 = x + 7$	$y^2 = -x + 7$	$(-y)^2 = -x + 7$
$y^2 = x + 7$	not equivalent: no symmetry	$y^2 = -x + 7$
equivalent: **symmetry**		not equivalent: no symmetry

9.

$$y = -\lvert x + 2 \rvert$$

x-axis	y-axis	origin
$-y = -\lvert x + 2 \rvert$	$y = -\lvert -x + 2 \rvert$	$-y = -\lvert -x + 2 \rvert$
not equivalent: no symmetry	not equivalent: no symmetry	not equivalent: no symmetry

10.

$$y = \lvert x \rvert - 3$$

x-axis	y-axis	origin
$-y = \lvert x \rvert - 3$	$y = \lvert -x \rvert - 3$	$-y = \lvert -x \rvert - 3$
not equivalent: no symmetry	$y = \lvert -1 \rvert \lvert x \rvert - 3$	$-y = \lvert -1 \rvert \lvert x \rvert - 3$
	$y = \lvert x \rvert - 3$	$-y = \lvert x \rvert - 3$
	equivalent: **symmetry**	not equivalent: no symmetry

11. $|y| = x$

x-axis	y-axis	origin
$\lvert -y\rvert = x$	$\lvert y\rvert = -x$	$\lvert -y\rvert = -x$
$\lvert -1\rvert\lvert y\rvert = x$	not equivalent: no symmetry	$\lvert -1\rvert\lvert y\rvert = -x$
$\lvert y\rvert = x$		$\lvert y\rvert = -x$
equivalent: symmetry		not equivalent: no symmetry

12. $y = 2\sqrt{x}$

x-axis	y-axis	origin
$-y = 2\sqrt{x}$	$y = 2\sqrt{-x}$	$-y = 2\sqrt{-x}$
not equivalent: no symmetry	not equivalent: no symmetry	not equivalent: no symmetry

13. $y = x^4 - 4$

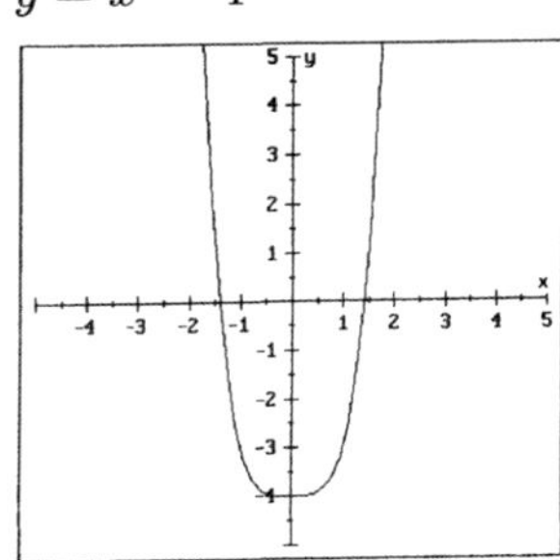

D $(-\infty, \infty)$; R $[-4, \infty)$

14. 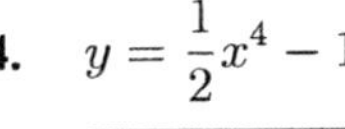 $y = \frac{1}{2}x^4 - 1$

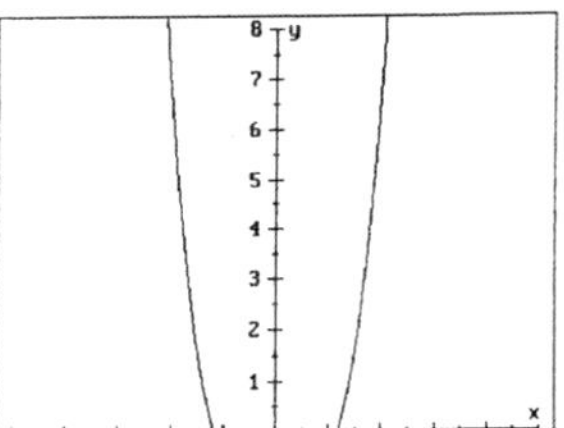

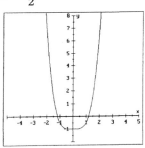

D $(-\infty, \infty)$; R $[-1, \infty)$

15. $y = -x^3$

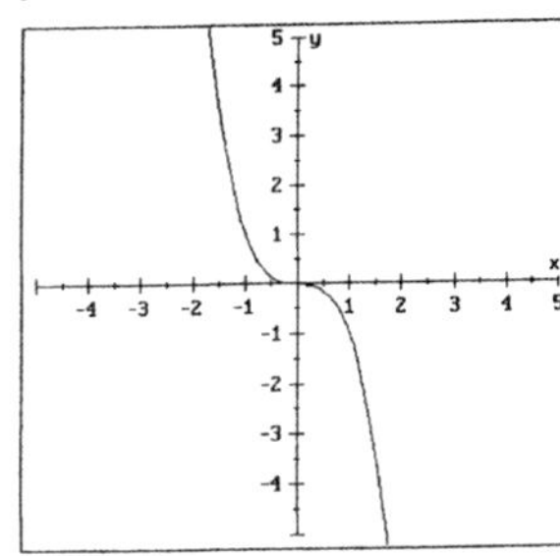

D $(-\infty, \infty)$; R $(-\infty, \infty)$

16. $y = x^3 + 2$

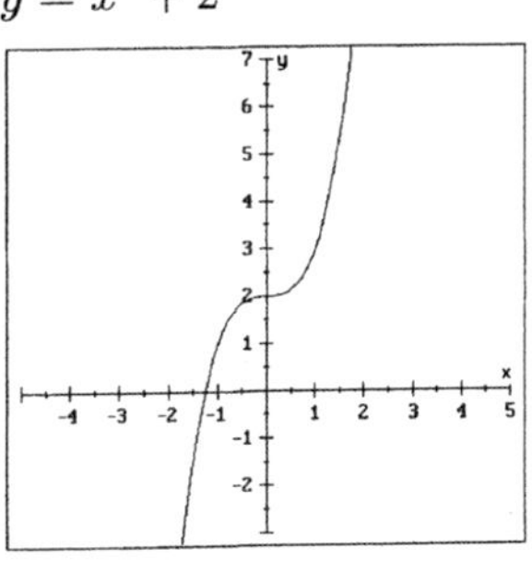

D $(-\infty, \infty)$; R $(-\infty, \infty)$

17. $y = x^4 + x^2$

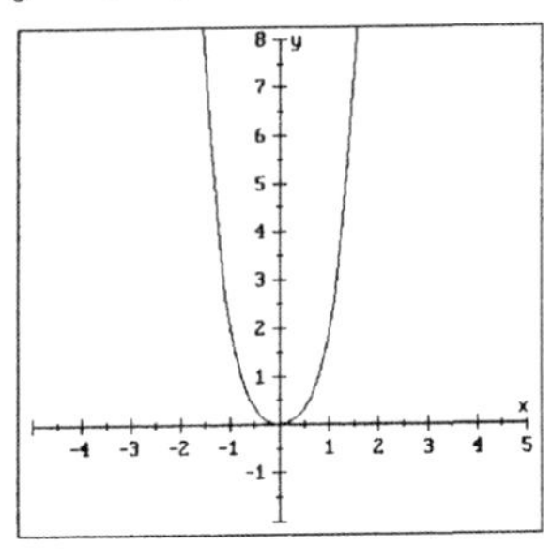

D $(-\infty, \infty)$; R $[0, \infty)$

18. $y = 3 - x^4$

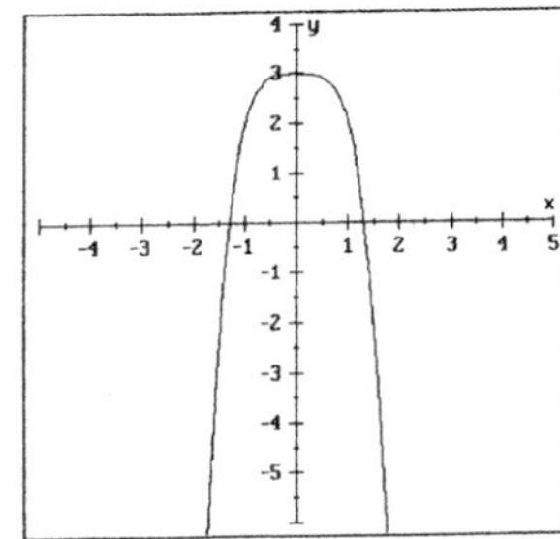

D $(-\infty, \infty)$; R $(-\infty, 3]$

19. $y = x^3 - x$

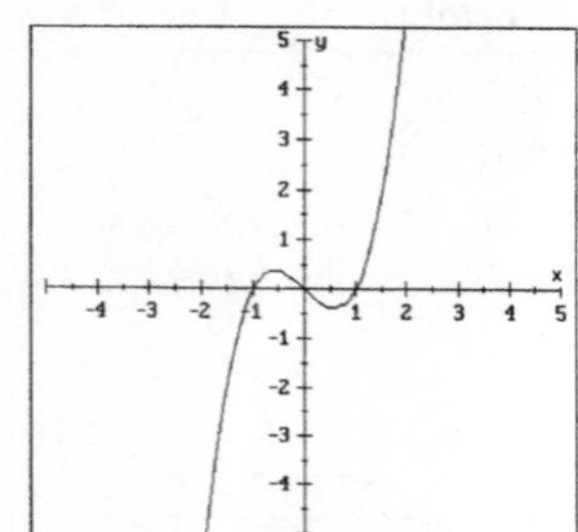

D $(-\infty, \infty)$; R $(-\infty, \infty)$

20. $y = x^3 + x$

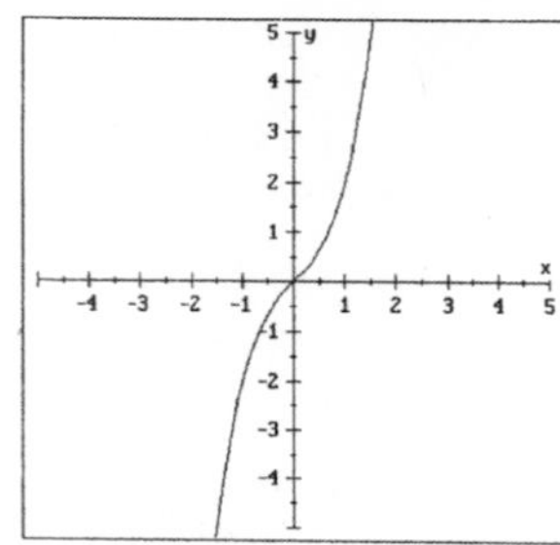

D $(-\infty, \infty)$; R $(-\infty, \infty)$

21. $y = \frac{1}{2}|x| - 1$

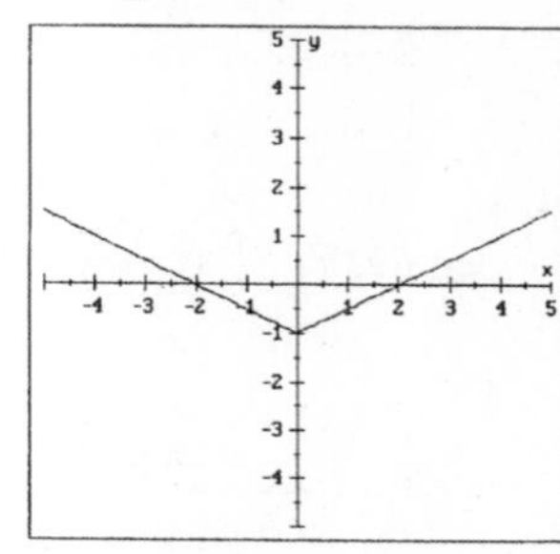

D $(-\infty, \infty)$; R $[-1, \infty)$

22. $y = -|x| + 1$

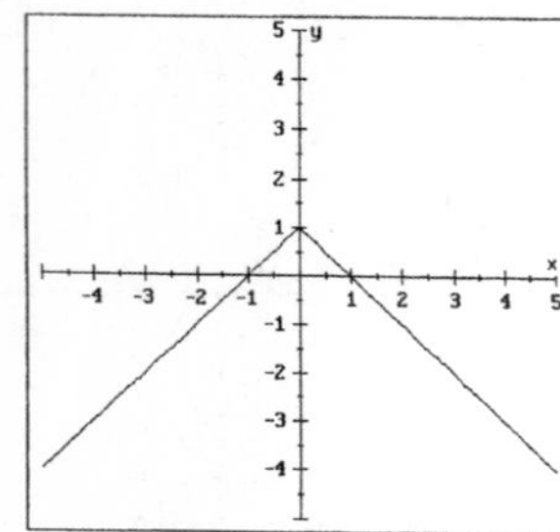

D $(-\infty, \infty)$; R $(-\infty, 1]$

23. $y = -|x + 2|$

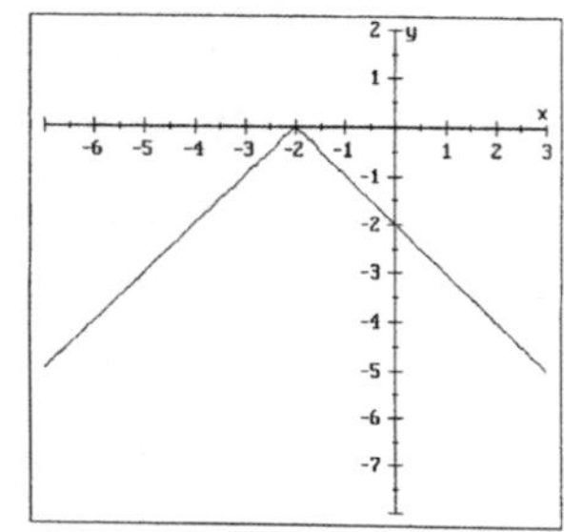

D $(-\infty, \infty)$; R $(-\infty, 0]$

24. $y = |x - 2|$

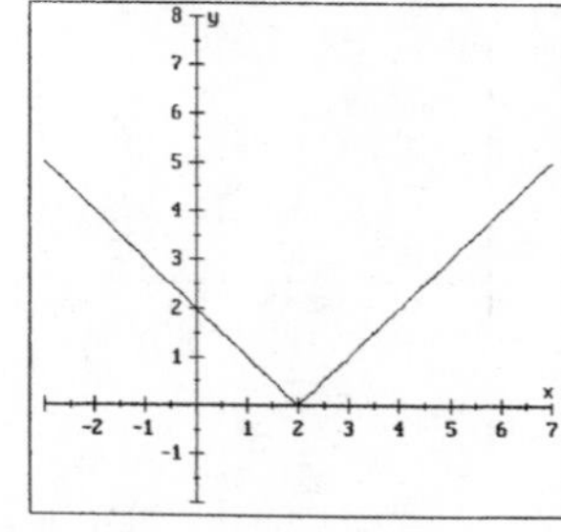

D $(-\infty, \infty)$; R $[0, \infty)$

Sample Final Examination (page A-6)

1. Prime numbers from 40 and 50: 41, 43, 47
Answer: a

2. Commutative property of $\times$: $ab = ba$
Answer: a

3. $\frac{c - ab}{bc} = \frac{6 - 3(-2)}{-2(6)} = \frac{12}{-12} = -1$
Answer: c

4.
$$\frac{1}{2} + \frac{3}{4} \div \frac{5}{6} = \frac{1}{2} + \frac{3}{4} \cdot \frac{6}{5} = \frac{1}{2} + \frac{9}{10} = \frac{5}{10} + \frac{9}{10} = \frac{14}{10} = \frac{7}{5}$$
Answer: d

5. $\left(\frac{a^2}{a^5}\right)^{-5} = \left(\frac{a^5}{a^2}\right)^5 = (a^3)^5 = a^{15}$
Answer: c

6. $0.0000234 = 2.34 \times 10^{-5}$
Answer: a

7. $P(-1) = 2(-1)^2 - (-1) - 1 = 2(1) + 1 - 1 = 2 \Rightarrow$ **Answer: c**

8. $(3x + 2) - (2x - 1) + (x - 3) = 3x + 2 - 2x + 1 + x - 3 = 2x \Rightarrow$ **Answer: b**

9. $$\begin{aligned}(3x-2)(2x+3) &= 6x^2+9x-4x-6\\ &= 6x^2+5x-6\end{aligned}$$

Answer: a

10. $$\begin{array}{r} x-\ \ 2 \\ 2x+1\ \overline{)\,2x^2-3x-2} \\ \underline{2x^2+\ \ x} \\ -4x-2 \\ \underline{-4x-2} \\ 0 \end{array}$$

Answer: c

11. $$\begin{aligned}5x-3 &= -2x+10\\ 7x &= 13\\ x &= \frac{13}{7}\end{aligned}$$

Answer: d

12. Let x and $x+2$ represent the integers.

$$\begin{aligned}x+x+2 &= 44\\ 2x &= 42\\ x &= 21\end{aligned}$$

The integers are 21 and 23. Product $= 483$

Answer: c

13. $$\begin{aligned}2ax-a &= b+x\\ 2ax-x &= a+b\\ x(2a-1) &= a+b\\ x &= \frac{a+b}{2a-1}\end{aligned}$$

Answer: b

14. $$|2x+5| = 13$$

$$\begin{aligned}2x+5 &= 13 \quad \textbf{or} \quad & 2x+5 &= -13\\ 2x &= 8 & 2x &= -18\\ x &= 4 & x &= -9\end{aligned}$$

Sum of solutions $= 4+(-9) = -5$

Answer: d

15. $$\begin{aligned}-2x+5 &> 9\\ -2x &> 4\\ \frac{-2x}{-2} &< \frac{4}{-2}\\ x &< -2\end{aligned}$$

Answer: c

16. $$\begin{aligned}|2x-5| &\le 9\\ -9 \le 2x-5 &\le 9\\ -9+5 \le 2x-5+5 &\le 9+5\\ -4 \le 2x &\le 14\\ -2 \le x &\le 7\end{aligned}$$

Answer: d

17. $3ax^2+6a^2x = 3ax(x+2a)$

Answer: d

18. $$\begin{aligned}x^4-16 &= \left(x^2+4\right)\left(x^2-4\right)\\ &= \left(x^2+4\right)(x+2)(x-2)\end{aligned}$$

$x^2+4+x+2+x-2 = x^2+2x+4$

Answer: b

19. $8x^2-2x-3 = (4x-3)(2x+1)$

$4x-3+2x+1 = 6x-2$

Answer: a

20. $$\begin{aligned}27a^3+8 &= (3a)^3+2^3\\ &= (3a+2)\left(9a^2-6a+4\right)\end{aligned}$$

Answer: d

21. $$\begin{aligned}6x^2-5x-6 &= 0\\ (2x-3)(3x+2) &= 0\end{aligned}$$

$$\begin{aligned}2x-3 &= 0 \quad \textbf{or} \quad & 3x+2 &= 0\\ x &= \tfrac{3}{2} & x &= -\tfrac{2}{3}\end{aligned}$$

Answer: c

22. $$\begin{aligned}\frac{x^2+5x+6}{x^2-9} &= \frac{(x+2)(x+3)}{(x+3)(x-3)}\\ &= \frac{x+2}{x-3}\end{aligned}$$

Answer: b

23. $$\frac{3x+6}{x+3}-\frac{x^2-4}{x^2+x-6}=\frac{3x+6}{x+3}-\frac{(x+2)(x-2)}{(x+3)(x-2)}=\frac{3x+6}{x+3}-\frac{x+2}{x+3}$$
$$=\frac{3x+6-x-2}{x+3}=\frac{2x+4}{x+3}$$

Answer: d

24. $$\frac{y}{x+y}+\frac{x}{x-y}=\frac{y(x-y)}{(x+y)(x-y)}+\frac{x(x+y)}{(x-y)(x+y)}=\frac{xy-y^2+x^2+xy}{(x+y)(x-y)}=\frac{x^2+2xy-y^2}{(x+y)(x-y)}$$

Answer: a

25. $$\frac{\frac{1}{x}+\frac{1}{y}}{\frac{1}{y}}=\frac{\left(\frac{1}{x}+\frac{1}{y}\right)xy}{\frac{1}{y}\cdot xy}=\frac{y+x}{x}$$

Answer: c

26. $$\frac{2}{y+1}=\frac{1}{y+1}-\frac{1}{3}$$
$$3(y+1)\cdot\frac{2}{y+1}=3(y+1)\left(\frac{1}{y+1}-\frac{1}{3}\right)$$
$$6=3-(y+1)$$
$$6=3-y-1$$
$$y=-4$$

Answer: c

27.

$$x=0 \qquad\qquad y=0$$
$$2x+3y=6 \qquad 2x+3y=6$$
$$2(0)+3y=6 \qquad 2x+3(0)=6$$
$$3y=6 \qquad 2x=6$$
$$y=2 \qquad x=3$$

Sum $=2+3=5$

Answer: c

28. $$m=\frac{y_2-y_1}{x_2-x_1}=\frac{-1-(-2)}{5-3}=\frac{1}{2}$$

Answer: d

29.

$$2x-3y=4 \qquad\qquad 3x+2y=1$$
$$-3y=-2x+4 \qquad 2y=-3x+1$$
$$y=\frac{2}{3}x-\frac{4}{3} \qquad y=-\frac{3}{2}x+\frac{1}{2}$$
$$m=\frac{2}{3} \qquad m=-\frac{3}{2}$$

perpendicular

Answer: b

30. $$m=\frac{y_2-y_1}{x_2-x_1}=\frac{7-5}{6-(-2)}=\frac{2}{8}=\frac{1}{4}$$
$$y-y_1=m(x-x_1)$$
$$y-7=\frac{1}{4}(x-6)$$
$$y-7=\frac{1}{4}x-\frac{3}{2}$$
$$y=\frac{1}{4}x+\frac{11}{2}$$

Answer: b

31. $$g(t+1)=(t+1)^2-3$$
$$=t^2+2t+1-3$$
$$=t^2+2t-2$$

Answer: d

32. $$d=kt$$
$$12=k(3)$$
$$4=k$$

Answer: b

33. $x^{a/2}x^{a/5} = x^{a/2+a/5} = x^{5a/10+2a/10} = x^{7a/10} \Rightarrow$ **Answer: b**

34.
$$\begin{aligned}\left(x^{1/2}+2\right)\left(x^{-1/2}-2\right) &= x^{1/2}x^{-1/2} - 2x^{1/2} + 2x^{-1/2} - 4 \\ &= x^0 - 2x^{1/2} + 2x^{-1/2} - 4 \\ &= 1 - 2x^{1/2} + 2x^{-1/2} - 4 = -3 - 2x^{1/2} + 2x^{-1/2}\end{aligned}$$

Answer: c

35. $\sqrt{112a^3} = \sqrt{16a^2}\sqrt{7a} = 4a\sqrt{7a} \Rightarrow$ **Answer: d**

36. $\sqrt{50} - \sqrt{98} + \sqrt{128} = \sqrt{25}\sqrt{2} - \sqrt{49}\sqrt{2} + \sqrt{64}\sqrt{2} = 5\sqrt{2} - 7\sqrt{2} + 8\sqrt{2} = 6\sqrt{2}$

Answer: b

37.
$$\frac{3}{2-\sqrt{3}} = \frac{3\left(2+\sqrt{3}\right)}{\left(2-\sqrt{3}\right)\left(2+\sqrt{3}\right)} = \frac{3\left(2+\sqrt{3}\right)}{4+2\sqrt{3}-2\sqrt{3}-3} = \frac{3\left(2+\sqrt{3}\right)}{1} = 6+3\sqrt{3}$$

Answer: b

38.
$$\begin{aligned}d &= \sqrt{\left(x_2-x_1\right)^2 + \left(y_2-y_1\right)^2} \\ &= \sqrt{\left(6-(-2)\right)^2 + \left(-8-3\right)^2} \\ &= \sqrt{8^2 + \left(-11\right)^2} \\ &= \sqrt{64+121} = \sqrt{185}\end{aligned}$$

Answer: a

39.
$$\begin{aligned}\sqrt{x+7} - 2x &= -1 \\ \sqrt{x+7} &= 2x-1 \\ \left(\sqrt{x+7}\right)^2 &= (2x-1)^2 \\ x+7 &= 4x^2 - 4x + 1 \\ 0 &= 4x^2 - 5x - 6 \\ 0 &= (x-2)(4x+3)\end{aligned}$$

$x - 2 = 0$ **or** $4x + 3 = 0$

$x = 2$ $\qquad$ $x = -\frac{3}{4}$ doesn't check

Answer: a

40. $y > 3x + 2$

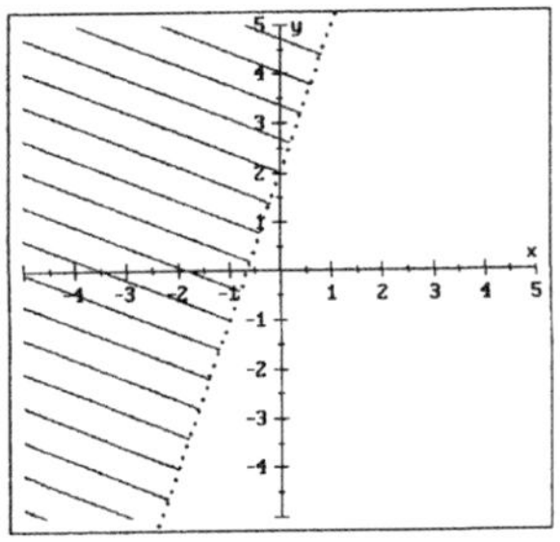

Answer: d

41. $x = \dfrac{-b \pm \sqrt{b^2 - 4ac}}{2a}$

Answer: d

42.
$$\begin{aligned}(2+3i)^2 &= (2+3i)(2+3i)\\ &= 4+6i+6i+9i^2\\ &= 4+12i+9(-1)\\ &= 4+12i-9\\ &= -5+12i\end{aligned}$$
Answer: a

43.
$$\begin{aligned}\frac{i}{3+i} &= \frac{i(3-i)}{(3+i)(3-i)}\\ &= \frac{3i-i^2}{9-3i+3i-i^2}\\ &= \frac{3i-(-1)}{9-(-1)}\\ &= \frac{1+3i}{10} = \frac{1}{10}+\frac{3}{10}i\end{aligned}$$
Answer: b

44.
$$\begin{aligned}y &= 2x^2+4x-3\\ y+3 &= 2(x^2+2x)\\ y+3+2 &= 2(x^2+2x+1)\\ y &= 2(x+1)^2-5\end{aligned}$$
Vertex: $(-1,-5)$

Answer: c

45.
$$\begin{aligned}\frac{2}{x} &< 3\\ \frac{2}{x}-3 &< 0\\ \frac{2}{x}-\frac{3x}{x} &< 0\\ \frac{2-3x}{x} &< 0\end{aligned}$$

$2-3x$ +++++++++0 − − −

x − − − 0++++++++++

⟵) —— (⟶ (at 0 and $\frac{2}{3}$)

solution set: $(-\infty,0)\cup\left(\frac{2}{3},\infty\right)$

Answer: c

46. $f(3)=2(3)^2+1=2(9)+1=19$

Answer: b

47.
$$\begin{aligned}y &= 3x+2\\ x &= 3y+2\\ x-2 &= 3y\\ \frac{x-2}{3} &= y\end{aligned}$$
Answer: c

48. $(x+2)^2+(y-4)^2=16$

Answer: a

49. $\frac{4}{x}+\frac{2}{y}=2 \Rightarrow 4m+2n=2 \Rightarrow \times 3 \quad 12m+6n=6$

$\frac{2}{x}-\frac{3}{y}=-1 \Rightarrow 2m-3n=-1 \Rightarrow \times 2 \quad 4m-6n=-2$

$$\begin{aligned} 16m &= 4 \\ m &= \tfrac{1}{4} \end{aligned}$$

Solve for n:

$$\begin{aligned} 4m+2n &= 2 \\ 4\left(\tfrac{1}{4}\right)+2n &= 2 \\ 1+2n &= 2 \\ 2n &= 1 \\ n &= \tfrac{1}{2} \end{aligned}$$

Solve for x:

$$m=\frac{1}{x} \qquad \frac{1}{4}=\frac{1}{x} \qquad 4=x$$

Solve for y:

$$n=\frac{1}{y} \qquad \frac{1}{2}=\frac{1}{y} \qquad 2=y$$

Solution: $\boxed{(4,2)}$ Sum $=4+2=6$

Answer: a

50. $y=\dfrac{\begin{vmatrix} 4 & 5 \\ 8 & 3 \end{vmatrix}}{\begin{vmatrix} 4 & 6 \\ 8 & -9 \end{vmatrix}}=\dfrac{12-40}{-36-48}=\dfrac{-28}{-84}=\dfrac{1}{3}$

Answer: b

51. $\begin{vmatrix} 2 & -3 \\ 4 & 4 \end{vmatrix}=8-(-12)=20$

Answer: b

52. $z=\dfrac{\begin{vmatrix} 1 & 1 & 4 \\ 2 & 1 & 6 \\ 3 & 1 & 8 \end{vmatrix}}{\begin{vmatrix} 1 & 1 & 1 \\ 2 & 1 & 1 \\ 3 & 1 & 2 \end{vmatrix}}=\dfrac{1\begin{vmatrix} 1 & 6 \\ 1 & 8 \end{vmatrix}-1\begin{vmatrix} 2 & 6 \\ 3 & 8 \end{vmatrix}+4\begin{vmatrix} 2 & 1 \\ 3 & 1 \end{vmatrix}}{1\begin{vmatrix} 1 & 1 \\ 1 & 2 \end{vmatrix}-1\begin{vmatrix} 2 & 1 \\ 3 & 2 \end{vmatrix}+1\begin{vmatrix} 2 & 1 \\ 3 & 1 \end{vmatrix}}=\dfrac{1(2)-1(-2)+4(-1)}{1(1)-1(1)+1(-1)}=\dfrac{0}{-1}=0$

Answer: a

53. $\log_a N=x \Rightarrow a^x=N$

Answer: a

54. $\log_2 \frac{1}{32}=x \Rightarrow 2^x=\frac{1}{32} \Rightarrow x=-5$

Answer: c

55. $\log 7+\log 5=\log(7\cdot 5)=\log 35$

Answer: d

56. $b^{\log_b x}=x$

Answer: b

57.

$$\begin{aligned} \log y+\log(y+3) &= 1 \\ \log y(y+3) &= 1 \\ 10^1 &= y^2+3y \\ 0 &= y^2+3y-10 \\ 0 &= (y+5)(y-2) \end{aligned}$$

$y+5=0$ **or** $y-2=0$

$y=-5 \qquad y=2$

Answer: b

58. In the 3rd term, the exponent on b is 2.

Variables: a^4b^2

Coef. $=\dfrac{n!}{r!(n-r)!}=\dfrac{6!}{2!4!}=15$

Term $=15a^4b^2$

Answer: c

$$= \frac{7!}{4!} = \frac{7 \cdot 6 \cdot 5 \cdot 4!}{4!}$$
$$= 7 \cdot 6 \cdot 5 = 210$$

60. $$C(7,3) = \frac{7!}{3!(7-3)!}$$
$$= \frac{7!}{3!4!}$$
$$= \frac{7 \cdot 6 \cdot 5 \cdot 4!}{3 \cdot 2 \cdot 1 \cdot 4!} = \frac{7 \cdot 6 \cdot 5}{6} = 35$$

Answer: a

61. $a_1 = 2, d = 3, n = 100$
$$a_n = a_1 + (n-1)d$$
$$a_{100} = 2 + (100-1)(3)$$
$$= 2 + 99(3)$$
$$= 2 + 297 = 299$$

Answer: b

62. $a_1 = 1, r = \frac{1}{3}$
$$S_\infty = \frac{a_1}{1-r} = \frac{1}{1-\frac{1}{3}} = \frac{1}{\frac{2}{3}} = \frac{3}{2}$$

Answer: c

63. $\frac{2}{6} = \frac{1}{3}$

Answer: b